HANDBOOK OF STATISTICS
VOLUME 4

Handbook of Statistics

VOLUME 4

General Editor

P. R. Krishnaiah

NORTH-HOLLAND
AMSTERDAM · NEW YORK · OXFORD

Nonparametric Methods

Edited by

P. R. Krishnaiah

Center for Multivariate Analysis
University of Pittsburgh, Pittsburgh, PA, U.S.A.

P. K. Sen

Department of Biostatistics
University of North Carolina, Chapel Hill, NC, U.S.A.

1984

NORTH-HOLLAND
AMSTERDAM · NEW YORK · OXFORD

ISBN: 0 444 86871 2

Published by:
ELSEVIER SCIENCE PUBLISHERS B.V.
P.O. Box 1991
1000 BZ Amsterdam
The Netherlands

Sole distributors for the U.S.A. and Canada:
ELSEVIER SCIENCE PUBLISHING COMPANY, INC.
52 Vanderbilt Avenue
New York, N.Y. 10017
U.S.A.

Library of Congress Cataloging in Publication Data
Main entry under title:

Nonparametric methods.

(Handbook of statistics ; v. 4)
Includes index.
1. Nonparametric statistics. I. Krishnaiah, Paruchuri R. II. Sen, Pranab Kumar, 1937- III. Series.
QA278.8.N65 1984 519.5 84-18776
ISBN 0-444-86871-2 (U.S.)

PRINTED IN THE NETHERLANDS

Preface

The series *Handbook of Statistics* was started to disseminate knowledge on a very broad spectrum of topics in the field of statistics. The present volume is devoted to the area of nonparametric methods. In accordance with the general objectives of the series, general methodological aspects and applications of nonparametric methods have been reviewed in this volume in a logically integrated and systematic form.

The past fortyfive years have witnessed a phenomenal growth of the literature on nonparametric procedures. With a tenuous origin of 'quick and dirty methods', nearly fifty years ago, it did not take a long time for the nonparametric methods to be established on a firm theoretical basis and to have a vital role in a variety of applications. As it stands now, the theoretical developments in the area of nonparametric methods constitute one of the most active areas of research in statistics and probability. Live applications have a dominant role to play in these developments. As in most of the other branches in statistics, the need to develop the theory arose out of demand for applications in various areas of physical, engineering, and behavioral sciences. In the forties and fifties, the developments in nonparametric methods were mostly confined to a few simple models (mostly univariate) where the exact distribution-freeness was the key factor. Today nonparametric methods are by no means confined to a restricted domain. Multivariate analysis, general linear models, biological assays, time-series analysis, meterological sciences clinical trials, and sequential analysis, among others constitute the core of applicability of modern nonparametric methods. Robustness, efficiency, and other desirable properties of nonparametric procedures have received a great deal of attention. The advent of the modern computing facilities has brought nonparametrics within the reach of practicing people in different scientific disciplines. In this volume, we have attempted to cover this broad domain of nonparametric methods in a unique way where theory and applications have merged together in a meaningful fashion.

We would like to express our most sincere thanks to all the contributors to this volume. They are very knowledgeable and active workers in this field, and without their active cooperation and first rate contributions, this volume could not have been compiled and completed. We are grateful to Professors R.J. Beaver, V.P. Bhapkar, L.P. Devorye, O. Dykstra, Jr., M. Ghosh, P. Hall, T. Hettmansperger, M. Hollander, K.V. Mardia, S.G. Mohanty, W. Phillipp, and P. Révész for reviewing the chapters in this volume. Thanks are due to North-Holland Publishing Company for their excellent cooperation in bringing out this volume. Last but not least, we wish to thank Indira Krishnaiah and Gauri Sen for their constant encouragement.

P. R. Krishnaiah
P. K. Sen

Table of Contents

PART V: STOCHASTIC APPROXIMATION AND DENSITY ESTIMATION

PART VI: LIFE TESTING AND RELIABILITY

PART IX: TABLES

Contributors

J. N. Adichie, *Department of Statistics, University of Nigeria, Nsukka, Nigeria* (Ch. 11)

J. C. Aubuchon, *317 Pond Lab., Pennslyvania State University, University Park, PA 16802, USA* (Ch. 12)

A. P. Basu, *Department of Statistics, University of Missouri, Columbia, MO 65201, USA* (Ch. 25)

C. B. Bell, *Department of Mathematical Sciences, San Diego State University, San Diego, CA 92182, USA* (Ch. 1)

R. Beran, *Department of Statistics, University of California, Berkeley, CA 94720, USA* (Ch. 30)

V. P. Bhapkar, *Department of Statistics, University of Kentucky, Levington, KY 40506, USA* (Ch. 2)

P. K. Bhattacharya, *University of California, Davis Division of Statistics, Davis, CA 95616, USA* (Ch. 18)

G. K. Bhattacharyya, *Department of Statistics, University of Wisconsin-Madison, 1210 W. Dayton Street, Madison, WI 53706, USA* (Ch. 5)

R. A. Bradley, *Department of Statistics, The University of Georgia, Athens, GA 30602, USA* (Ch. 14)

S. K. Chatterjee, *Department of Statistics, Calcutta University, 35, Ballygunge Circular Rd., Calcutta-700019, India* (Ch. 15)

E. Csáki, *Mathematical Institute of the Hungarian Academy of Sciences, 1053 Budapest V, Realtanoda U, 13–15, Hungary* (Ch. 19)

M. Csörgő, *Department of Mathematics, Carleton University, Colonel By Drive, Ottawa, Ontario, Canada K1S 5B6* (Ch. 20)

K. A. Doksum, *Department of Statistics, University of California, Berkeley, CA* 94720, *USA* (Ch. 26)

V. Dupač, *Department of Statistics, Charles University, Sokolovska U1 83, Prague B, Czechoslovakia* (Ch. 23)

J. L. Folks, *Statistics Laboratory, Oklahoma State University, Stillwater, OK 74074, USA* (Ch. 6)

M. H. Gail, *Department of Health and Human Services, National Institutes of Health, Landow Building, Room 5C09 Bethesda, MD 20205, USA* (Ch. 33)

J. Galambos, *Department of Mathematics, Temple University, Philadelphia, PA 19122, USA* (Ch. 17)

M. Ghosh, *Statistics Department, Iowa State University, Ames, IA 50011, USA* (Ch. 8)

T. P. Hettmansperger, *Department of Statistics, Pennsylvania State University, 219 Pond Laboratory, University Park, PA 16802, USA* (Ch. 12)

M. Hollander, *Department of Statistics, Florida State University, Tallahassee, FL 32306, USA* (Ch. 27)

M. Hušková, *Department of Statistics, Charles University, Sokolovska U1 83, Praha 8, Czechoslovakia* (Ch. 3, Ch. 16)

S. R. Jammalamadaka, *Department of Mathematics, University of California, Santa Barbara, CA 93106, USA* (Ch. 31)

K. Joag-Dev, *Department of Mathematics, University of Illinois, Urbana, IL 61801, USA* (Ch. 4)

J. Jurečková, *Department of Statistics, Charles University, Sokolovska U1 83, Praha 8, Czechoslovakia* (Ch. 21)

J. C. Keegel, *Department of Statistics, The George Washington University Washington, DC 20052, USA* (Ch. 35)

P. R. Krishnaiah, *Center For Multivariate Analysis, University of Pittsburgh, 516 Thackeray Hall, Pittsburgh, PA 15260, USA* (Ch. 36, Ch. 37)

S. Kullback, *10143 41 Trail #175, Boynton Beach, FL 33436, USA* (Ch. 35)

P. W. Mielke, Jr., *Department of Statistics, Colorado State University, Ft. Collins, CO 80503, USA* (Ch. 34)

U. Müller-Funk, *Institut fur Mathematische Stochastik, der Albrecht-Ludwigs-Universitat Freiburg, D-7800 Freiburg, Hermann-Herder-Str. 10, West Germany* (Ch. 28)

F. Proschan, *Department of Statistics, Florida State University, Tallahassee, FL 32326, USA* (Ch. 27)

D. Quade, *Biostatistics Department, School of Public Health, University of North Carolina, Chapel Hill, NC 27514, USA* (Ch. 10)

P. Révész, *Mathematical Institute of the Hungarian Academy of Sciences, 1053 Budapest V, Realtanoda U, 13–15, Hungary* (Ch. 24)

A. K. M. E. Saleh, *Department of Mathematics, Carleton University, Ottawa, Canada K1S 5B6* (Ch. 13)

P. K. Sen, *Department of Biostatistics, University of North Carolina, Chapel Hill, NC 27514, USA* (Ch. 1, Ch. 13, Ch. 22, Ch. 29, Ch. 37)

K. Singh, *Department of Statistics, Rutgers University, New Brunswick, NJ, USA* (Ch. 9)

L. Takács, *Department of Statistics, Case-Western Reserve University, Cleveland, OH 44106, USA* (Ch. 7)

H. S. Wieand, *Department of Mathematics and Statistics, University of Pittsburgh, Pittsburgh, PA 15260, USA* (Ch. 32)

B. S. Yandell, *Biostatistics Department, University of California – Berkeley, Berkeley, CA 94720, USA* (Ch. 26)

PART I

CLASSICAL DEVELOPMENTS

P. R. Krishnaiah and P. K. Sen, eds., *Handbook of Statistics, Vol. 4*
© Elsevier Science Publishers (1984) 1–29

1

Randomization Procedures

C. B. Bell and P. K. Sen

1. Introduction

Randomization procedures are the precursors of the nonparametric ones, and, during the past fifty years, they have played a fundamental role in the evolution of distribution-free methods. Randomization procedures rest on less stringent regularity conditions on the sampling scheme or on the underlying probability laws, encompass a broad class of statistics (which, otherwise, may not be distribution-free), are easy to interpret and apply (do not require standard or new statistical tables) and they have a broader scope of applicability (relative to other nonparametric competitors).

Traditional developments on randomization procedures (mostly, in the thirties) were spotty and piecemeal. The so called *permutation tests* were developed for the one-sample location (symmetry) problem, one-way analysis of variance (ANOVA) models relating to the two or several sample problems, bivariate independence problem and ANOVA models relating to simple multi-factor designs (viz. Fisher (1934), Pitman (1937a,b, 1938), Welch (1937), and others). In all these problems, for the randomization procedures, it is not necessary to assume the form(s) of the underlying distribution(s) and, for the determination of the critical values of the test statistics, no statistical tables may be necessary; these critical values are obtained directly from the sample(s) by performing suitable permutations on the observations (and this feature is inherent in the terminology permutation tests). A more unified picture emerged in the forties (viz., Scheffé, 1943, and Lehmann and Stein, 1949, among others), where a characterization of randomization tests as *similar size* α ($0 < \alpha < 1$) tests having the *$S(\alpha)$-structure* enhanced the scope to a wider class of problems. In the fifties, incorporation of the concept of boundedly complete sufficient statistics (viz. Lehmann and Scheffé, 1950, 1955, and Watson, 1957, among others) further clarified the structure of randomization tests and paved the way for *tests with the Neyman-structure.* Characterization of various nonparametric hypotheses in terms of *invariance* (of the (joint) distribution of the sample point) under certain (finite) *groups of transformations* (which map the sample space onto itself) led to the constitution of *maximal invariants* which play the key role in the construction of randomization tests. In the sixties, randomization principles provided the basic structure for various multivariate nonparametric procedures. In these multivariate problems, procedures based on (coordinate-wise) ranks are, in general, not genuinely distribution-free. Nevertheless, they can be rendered permutationally (conditionally) distribution-free by appeal to the conventional randomization principles (viz. Chatterjee and Sen, 1964; Chatterjee, 1966; Sen and Puri, 1967; Sen, 1968; Quade, 1967; Bell and Haller, 1969; Bell and Smith, 1969; Bell and Donoghue, 1969; among others). Randomization procedures can be adapted for the conventional parametric statistics without making explicit assumptions on the underlying distribution(s) and they may also be applied on the nonparametric statistics based on ranks, empirical distributions and other characteristics. Randomization procedures allow easy adjustments for ties or some other irregularities which may be encountered in practical applications. A variant form of randomization procedures due to Bell and Doksum (1965) deserves mention in this context. Randomization procedures have also been developed for drawing statistical inference from some stochastic processes and some non-standard problems too (viz. M. N. Ghosh, 1954; Bell, Woodroofe and Avadhani, 1970, and Bell, 1975; among others). The current chapter attempts to provide a basic survey of the randomization procedures in the wide spectrum of applications mentioned above.

2. The structure of randomization tests

To motivate the general theory, we consider first a simple example. Let (X_i, Y_i), $i = 1, \ldots, n$, be n independent and identically distributed (i.i.d.) random vectors (r.v.) with a distribution function (d.f.) $F(x, y)$, defined on the Euclidean plane $\mathbb{R}^2$. F need not be a bivariate normal d.f. or may not even be continuous almost everywhere (a.e.). Suppose that we want to test for the null hypothesis H_0 that X_i and Y_i are independent, i.e., $F(x, y) = F(x, \infty)F(\infty, y)$, for every $(x, y) \in \mathbb{R}^2$. Denote the *sample point* E_n (a $2 \times n$ matrix) by

$$E_n = \begin{pmatrix} X_1, \ldots, X_n \\ Y_1, \ldots, Y_n \end{pmatrix}. \tag{2.1}$$

Now let $\mathcal{P}_n = \{r_1, \ldots, r_n\}$ be the set of all possible $n!$ permutations $\boldsymbol{r} = (r_1, \ldots, r_n)$ of the first n natural integers $1, \ldots, n$. For each $\boldsymbol{r}$ in $\mathcal{P}_n$, define

$$E_n(\boldsymbol{r}) = \begin{pmatrix} X_1, \ldots, X_n \\ Y_{r_1}, \ldots, Y_{r_n} \end{pmatrix}. \tag{2.2}$$

The sample point E_n and the set $\mathcal{P}_n$ (of $n!$ permutations) give rise to a set

$$\mathcal{E}_n = \{E_n(\boldsymbol{r}) \colon \boldsymbol{r} \in \mathcal{P}_n\}. \tag{2.3}$$

The cardinality of this set is equal to $n!$. Now, under the null hypothesis, X_i and Y_i are independent, for each i $(=1,\ldots,n)$, while the n vectors are also i.i.d. Thus, the (joint) distribution of E_n is the $2n$-fold product of the distributions of its $2n$ arguments, and, for each $\boldsymbol{r} \in \mathcal{P}_n$, $E_n(\boldsymbol{r})$ has the same (joint) distribution as E_n. Also, we may write

$$E_n(\boldsymbol{r}) = g_n \cdot E_n, \quad g_n \in \mathcal{G}_n\,, \tag{2.4}$$

where $\mathcal{G}_n$ is the *group* of all $n!$ permutations (transformations) $\{g_n\}$ which map the sample space $(\mathbb{R}^{2n})$ onto itself. Thus, $\mathcal{E}_n = \{g_n \cdot E_n \colon g_n \in \mathcal{G}_n\}$ is the *orbit*, and under H_0, the distribution of E_n remains invariant under the permutation group $\mathcal{G}_n$ (though this distribution may depend on the marginals $F(x, \infty)$ and $F(\infty, y)$). Under H_0, the conditional distribution of E_n on the orbit $\mathcal{E}_n$ is therefore given by

$$P\{E_n = E_n(\boldsymbol{r}) \mid \mathcal{E}_n, H_0\} = (n!)^{-1}, \quad \boldsymbol{r} \in \mathcal{P}_n\,, \tag{2.5}$$

(and 0 mass points elsewhere). Thus, for a given significance level α $(0<\alpha<1)$, if we consider a test function $\phi_\alpha(E_n)$ (where $0 \le \phi_\alpha(\cdot) \le 1$), such that

$$\sum_{\boldsymbol{r} \in \mathcal{P}_n} \phi_\alpha(E_n(\boldsymbol{r})) = (n!)\alpha\,, \tag{2.6}$$

then, by (2.5) and (2.6), we have

$$E_{H_0}\{\phi_\alpha(E_n) \mid \mathcal{E}_n\} = \alpha \quad (\text{a.e. } \mathcal{E}_n)\,, \tag{2.7}$$

so that

$$E_{H_0}\phi_\alpha(E_n) = E\{E_{H_0}(\phi_\alpha(E_n) \mid \mathcal{E}_n)\} = \alpha\,. \tag{2.8}$$

Thus, $\phi_\alpha(E_n)$ is a test function of *size* α $(0<\alpha<1)$. In this setup, it is not necessary to assume that the marginal d.f.'s of F are of given forms, and hence, the test is nonparametric in character. Operationally, to make use of (2.5)–(2.6), the first step is to characterize the orbit through the appropriate group of transformations. Next, one may consider a real valued statistic $T_n = T(E_n)$ and enumerate the set of realizations: $\{T(E_n(\boldsymbol{r}))\colon \boldsymbol{r} \in \mathcal{P}_n\}$. In this context, one may choose for T_n the classical product moment correlation coefficient, the Spearman rank correlation or the Kendall tau statistic, among other possibilities. If we denote the ordered values of these $n!$ realizations by $T_n^{(1)} \le \cdots \le T_n^{(n!)}$ (note that all of these may not be distinct), then, for a given n (and $\mathcal{E}_n$) and α, we may find two positive integers M_1 and M_2, such that

$$M_1 < (n!)(1-\alpha) \le M_2\,, \tag{2.9}$$

$$T_n^{(M_1-1)} < T_n^{(M_1)} = \cdots = T_n^{(M_2-1)} < T_n^{(M_2)}\,, \tag{2.10}$$

and consider a (randomized) test function

$$\phi_\alpha(T_n) = \begin{cases} 1, & T_n \geq T_n^{(M_2)}, \\ \delta_n = (M_2 - (1-\alpha)(n!))/(M_2 - M_1), & T_n = T_n^{(M_1)}, \\ 0, & T_n \leq T_n^{(M_1-1)}, \end{cases} \tag{2.11}$$

then, (2.6) holds, and hence, (2.7)–(2.8) hold. A similar case may be worked out for the two-sided (randomized) test.

It may be noted that the permutation group $\mathcal{G}_n$ in (2.4) is basically related to the hypothesis of *matching invariance.* If both the marginal d.f.'s are continuous (so that the ties among the observations may be neglected, in probability), and if we consider the coordinatewise transformations:

$$X_i \to h_1(X_i) \quad \text{and} \quad Y_i \to h_2(Y_i), \qquad i = 1, \ldots, n, \tag{2.12}$$

where both $h_1(\cdot)$ and $h_2(\cdot)$ are monotone, then, we may like to have a test for H_0 which remains invariant under all such (h_1, h_2). Let $X_{(1)} < \cdots < X_{(n)}$ and $Y_{(1)} < \cdots < Y_{(n)}$ be the two sets of order statistics (for the two rows in (2.2)), and define $\boldsymbol{Q} = (Q_1, \ldots, Q_n)$ by letting

$$X_{(i)} = X_{S_i} \quad \text{and} \quad Y_{S_i} = Y_{(Q_i)} \quad \text{for } i = 1, \ldots, n. \tag{2.13}$$

Note that the *induced-rank vector* $\boldsymbol{Q}$ takes on (under H_0) each permutation of $(1, \ldots, n)$ with the common probability $(n!)^{-1}$, while the two order statistics vectors are (jointly) sufficient for F, when H_0 holds. Thus, under H_0, $\boldsymbol{Q}$ is a maximal invariant (with respect to the class of transformations in (2.12)). A test statistic (say, ϕ_α^*) which depends on E_n through $\boldsymbol{Q}$ only is an invariant test statistic, and under H_0, this will be genuinely distribution-free. On the other hand, a randomization test function, such as the one in (2.11), may not be solely dependent on $\boldsymbol{Q}$, and under H_0, its unconditional distribution may depend on the unknown marginal d.f.'s of F. Thus, it need not be genuinely distribution-free. As in (2.11), the critical values of a randomization test statistic are generally r.v. (assuming constant values on each orbit) tacitly so chosen that (2.6) holds. This explains why a randomization test (if not based solely on the maximal invariant) may fail to be truely distribution-free. Nevertheless, it is, at least, conditionally (permutationally) distribution-free. We have also noticed that the group $\mathcal{G}_n$ has $n!$ elements. If in (2.1), the elements in the same row are not all distinct, the effective number of distinct $E_n(\boldsymbol{r})$ in (2.2) will be equal to M_n where $(n!)/M_n$ is a positive integer. In such a case, an effective group $\mathcal{G}_n^0$ of M_n transformations $\{g_n^0\}$ may be defined in a similar manner, and, in (2.5)–(2.6), $n!$ has to be replaced by M_n. The structure of the randomization test in (2.7)–(2.8) remains the same. In (2.13), the elements of $\boldsymbol{Q}$ may not be all distinct and adjustments for ties are to be made. The unconditional null distribution of such adjusted Q may therefore depend on the marginal d.f.'s (which are not necessarily continuous a.e.), so that genuinely distribution-free tests based on the adjusted Q may not exist. However, the conditional (permutational) distribution of the ties-adjusted Q over M_n equally likely realizations remains intact, so that randomization tests can be constructed as in (2.11). Since (2.5), in this extended form, insures that for some (possibly vector valued) U_n assuming constant values on the different orbits, the condition of similarity of the test $\phi(E_n)$ is satisfied and U_n behaves as a sufficient statistic (vector), we may conclude that the randomization tests have *Neyman Structure.* We may also remark that as regards the permutational distribution, the i.i.d. character of the r.v.'s may be replaced by exchangeability, and this may well arise in sampling from a finite population or in some other sampling schemes where the independence may not hold. With these observations, we may summarize the basic structure of randomization tests in the following unified manner:

Let n be a positive integer and let X_n be a random element with the sample space $\mathscr{X}_n$. The probability distribution $P(A_n) = P\{X_n \in A_n\}$ is defined on an additive class $\mathscr{A}_n$ of subsets A_n of $\mathscr{X}_n$. Consider a partition Π_n of $\mathscr{X}_n$, which is a class of mutually exclusive subsets S of $\mathscr{X}_n$, such that every point x_n of $\mathscr{X}_n$ belongs to one of the subsets S. The set $O(x_n)$ of all points in $\mathscr{X}_n$ belonging to the same subset containing x_n is termed the *orbit* of x_n ($\in \mathscr{X}_n$), and the number of points belonging to $O(x_n)$ is denoted by $M_n(x_n)$. Typically, the partitioning Π_n (and the orbit $O(x_n)$) are characterized by the invariance of $P(A_n)$ under appropriate *groups of transformations* which map the sample space ($\mathscr{X}_n$) onto itself. We assume that for every (finite) n and $x_n \in \mathscr{X}_n$, $M_n(x_n)$ is finite. Let S_{M_n} be the set theoretic union of all those sets $S \in \Pi_n$ containing exactly M_n elements. Then, we assume that there exist mutually exclusive subsets $S^{(i)}_{M_n}$, $i = 1, \ldots, M_n$, which are measurable, and, the cardinality of $S \cap S^{(i)}_{M_n}$ is equal to 1 for each i ($=1, \ldots, M_n$). The null hypothesis (H_0) relates to the invariance of $P(A_n)$ under Π_n. If $\phi_\alpha(x_n)$ be a test function ($0 \leq \phi_\alpha(\cdot) \leq 1$), for which

$$\sum_{x_n^* \in O(x_n)} \phi_\alpha(x_n^*) = \alpha M_n(x_n) \quad (\text{a.e. } \mathscr{X}_n)\,, \tag{2.14}$$

then, we have

$$E\{\phi_\alpha(X_n) \mid H_0\} = E_{H_0}\{E(\phi_\alpha(X_n) \mid O(x_n))\} = \alpha\,, \tag{2.15}$$

so that $\phi_\alpha(X_n)$ is a similar size α ($0 < \alpha < 1$) test function.

Ideally, if a maximal invariant T_n (possibly vector valued) exists, then $\phi_\alpha(X_n)$ may be based solely on T_n (i.e., $\phi_\alpha(X_n) = \phi_\alpha(T_n)$). Otherwise, the unconditional null hypothesis distribution of $\phi_\alpha(X_n)$ may depend on the underlying probability law, though, (2.14) insures the conditional distribution-freeness. In the literature, the maximal invariants have also been referred to as the maximal statistical noise (MSN). We now present the basic structure of randomization tests in terms of MSS's (minimal sufficient statistics), orbits and the MSN. Randomization tests in current use are almost exclusively employed in testing nonparametric (NP) hypotheses. Hence, one needs to introduce several entities related to NP statistics.

Definition 1. A statistic $T(X_n)$ is NPDF (nonparametric distribution-free) with respect to a family Ω of distributions (on some sample space $\mathscr{X}_n$) if there exists a distribution $G(\cdot)$ such that for all t, and for all $J(\cdot)$ in Ω,

$$P\{T(X_n) \leq t \mid J(\cdot)\} = G(t).$$

A set C^* is similar with respect to Ω if there exists an α such that $P\{C^* \mid J(\cdot)\} = \alpha$, for all $J(\cdot)$ in Ω. A collection $\{A_\nu\}$ of sets is called a MESP (maximal essential similar partition) with respect to Ω if

(a) each A_ν is similar with respect to Ω,
(b) each $P(A_\nu) > 0$,
(c) $D^* = \mathscr{X}_n - \cup A_\nu$ is similar with $P(D^*) = 0$,
(d) the $\{A_\nu\}$ are disjoint, and
(e) no A_ν has a nontrivial similar subset.

This definition leads us to the following

Lemma 2.1. (i) *C^* is similar with respect to Ω if and only if its indicator function is NPDF.* (ii) *T is NPDF with respect to Ω if and only if each set $\{T \leq t\}$ is similar with respect to Ω.* (iii) *Let $S(X_n)$ be a sufficient statistic for Ω. Then, any statistic T independent of $S(X_n)$ is NPDF with respect to Ω.*

In the construction of randomization tests one uses the above lemma in conjunction with the concept of orbits. Recall that if $S(X_n)$ be a MSS for Ω, then each set of the form $\{S(x_n) = s^*\}$ is called an S-orbit. The structure of similar sets, and, hence, NPDF statistics then follows from the Neyman-Structure Theorem as discussed earlier in (2.14)–(2.15). This leads us to the following

Theorem 2.2. *Let $S(X_n)$ be a complete sufficient statistic for Ω. Then, a set C^* is similar with respect to Ω with $P(C^*) = \alpha$ if and only if for almost all s^* and all $J(\cdot)$ in Ω, $P\{C^* \mid S(X_n) = s^*\} = \alpha$.*

Roughly, this says that a similar set must contain a 'proportion' α of the a.e. orbit. Randomization statistics arise as a method of choosing these points on the orbits. Generally, for nonparametric hypotheses, the orbits are finite, and one employs permutation functions in selecting the points of the orbits.

Definition 2. Let S be a finite set of permutations with respect to which each element (i.e., likelihood functions of Ω) is invariant, and let k^* be the number of elements of S. (i) $h(\cdot)$ is called a B-Pitman function with respect to Ω and S if, for all $J(\cdot)$ in Ω and all γ in S, $P\{h(X_n) = h(\gamma(X_n))\} = 0$ unless $X_n = \gamma(X_n)$. (ii) $R(h(X_n))$ is called a permutation (randomization) statistic generated by $h(\cdot)$ if $R(h(X_n)) = \Sigma\, u(h(X_n) - h((X_n))$, where $u(t)$ is 1 or 0 according as t

is $\geqslant$ or. <0, and the summation extends over all γ in S. (iii) $\{[R(h(X_n)) = r]: r = 1, \ldots, k^*\} = \mathscr{P}(h(\cdot))$ is called the partition generated by $h(\cdot)$.

One sees immediately that the B-Pitman function assigns different values to the points on a.e. given orbit. Hence, it can be employed to select subsets of points from a.e. orbit. (The 'a.e.' is used here because in the more common nonparametric cases, ties may cause some difficulties as to the orbit size etc., and are excluded.) It is also clear that there are many possible permutation statistics and associated partitions. The results of several authors (including Bell and Doksum, 1967; Bell and Donoghue, 1969, and Lehmann and Stein, 1949) allow one to prove the following

THEOREM 2.3. *Let Ω be a family of distributions admitting a finite set S of permutations with respect to which each set of Ω is invariant.*

(i) *If $h(\cdot)$ is a B-Pitman function with respect to Ω and S then $R(h(X_n))$ is NPDF with respect to Ω and*

(ii) *$R(h(X_n))$ has a discrete uniform distribution over $\{1, \ldots, k^*\}$. If $S(X_n)$ is a complete sufficient statistic, then*

(iii) *a statistic $T(X_n)$ is NPDF with respect to Ω if and only if there exists a B-Pitman function $h(\cdot)$ and a function W^* such that $T(X_n) \equiv W^*[h(X_n)]$.*

(iv) *$P(h(\cdot))$ is a MESP, and*

(v) *$R(h(X_n))$ and $S(X_n)$ are independent.*

The examples given in the next Section illustrate this theorem. The following results relate to further generalizations of the structure results presented above.

Let Ω admit a MSS, $S(X_n)$ and let $\delta(X_n) = [S(X_n), N(X_n)]$.

DEFINITION 3. (i) $\delta(\cdot)$ is called a BDT (basic data transformation) and (ii) $N(X_n)$ is called a version of MSN if (a) $\delta(\cdot)$ is one-to-one a.e., and (b) $S(X_n)$ and $N(X_n)$ are independent.

Here, the MSN is an ancillary statistic which is complementary to the MSS, and $\delta(\cdot)$ gives the new 'coordinate' system. The relation between MSN and NPDF statistics is given by the following

THEOREM 2.4. *Let Ω be a family of distributions admitting a MSS $S(X_n)$, a BDT $\delta(\cdot)$ and MSN $N(X_n)$. Then,* (i) *each statistic of the form $W^*[N[X_n)]$ is NPDF with respect to Ω;* (ii) *if $S(X_n)$ is complete, and Ω and $S(X_n)$ are sufficiently regular (see e.g. Bell and Smith,* 1972a), *then $T(X_n)$ is NPDF with respect to Ω if and only if there exists (a version of) MSN $N(X_n)$ and $W^*(\cdot)$ such that $T(X_n) \equiv W^*[N(X_n)]$.*

The relation between permutation statistics and BDT's is given by

THEOREM 2.5. *Let Ω be a family of distributions each of which is invariant with respect to a finite set of permutations S. Let $S(X_n)$ be a MSS. Then, for each B-Pitman function $h(\cdot)$,* (i) *$R(h(X_n))$ is (a version of) MSN,* (ii) *$\delta_h(\cdot)$ is a BDT if*

$\delta_h(X_n) = [S(X_n), R_h(X_n)]$, *and* (iii) *if* $S(X_n)$ *is complete, then, for any version* $N(X_n)$ *of MSN there exists a B-Pitman function* $h(\cdot)$, *such that* $N(X_n) \equiv R(h(X_n))$.

This means that under appropriate circumstances the family of all permutation (randomization) statistics for a given NP hypothesis can be parameterized by the family of all B-Pitman functions or by the family of all versions of MSN. Further, noninvariant B-Pitman functions are associated with MSN versions which are not (maximal) invariants. (See Section 7.) For the MSS there is essentially only one version. Some modifications of these theorems are needed when $S(X_n)$ is not complete.

3. Randomization tests for various hypotheses of invariance

For most NP hypotheses, there are a vast number of permutation (randomization) statistics available. The choice of which statistics to employ depends on a variety of circumstances. This section presents some common NP hypotheses and the basic entities of the structure of the associated randomization tests. One may generally view each NP hypothesis considered as a generalization of some classical parametric hypothesis. Hence, the randomization statistics given will bear some simple relation to the corresponding classical parametric statistics. Side by side, invariant randomization tests are also presented.

3.1. Hypothesis of sign-invariance

Let $X_1, \ldots, X_n$ be n i.i.d.r.v. with a d.f. F, defined on the real line $\mathbb{R}$ $(=(-\infty, \infty))$. Consider the null hypothesis $H_0^{(1)}$ of the symmetry of F around 0, i.e.,

$$H_0^{(1)}: \quad F(x) + F(-x-) = 1 \ \ \forall x \in \mathbb{R} . \tag{3.1}$$

For the sample point $E_n = (X_1, \ldots, X_n) \in \mathbb{R}^n$, consider the group $\mathcal{G}_n$ of transformations $\{g_n\}$, where, typically,

$$g_n \cdot E_n = ((-1)^{j_1} X_{i_1}, \ldots, (-1)^{j_n} X_{i_n}) , \tag{3.2}$$

with each $j_i = 0, 1$ $(1 \leq i \leq n)$ and $(i_1, \ldots, i_n)$ a permutation of $(1, \ldots, n)$. Thus, $\mathcal{G}_n$ has $(n!)2^n$ possible elements, and under $H_0^{(1)}$, for each $g_n \in \mathcal{G}_n$, $g_n \cdot E_n$ has the same (joint) distribution as E_n. For any $e_n = (x_1, \ldots, x_n) \in \mathbb{R}^n$, the *orbit* $O(e_n)$ is defined by

$$O(e_n) = \{e_n^* = ((-1)^{j_1}|x_{i_1}|, \ldots (-1)^{j_n}|x_{i_n}|) : i_k = 0, 1;\ k = 1, \ldots, n ;$$
$$(i_1, \ldots, i_n) \text{ permutations of } (1, \ldots, n)\} . \tag{3.3}$$

Fisher (1934) exploited this in the proposal of the permutation test based on

$$t_n = n^{1/2}\bar{X}_n / \Big(\sum_{i=1}^{n} X_i^2\Big)^{1/2} \quad \text{where } \bar{X}_n = n^{-1}\sum_{i=1}^{n} X_i\,. \tag{3.4}$$

Note that $\Sigma_{i=1}^n X_i^2$ remains invariant under $\mathcal{G}_n$, while, $\bar{X}_n$ is a symmetric function of $X_i, \ldots, X_n$ (and hence, remains invariant under any permutation of the subscripts). Thus, conditionally on $|X_1|, \ldots, |X_n|$ given, t_n in (3.4) has 2^n equally likely realizations over the 2^n equally likely sign-inversions. Thus, we may proceed as in (2.9) through (2.11), replace $n!$ by 2^n and construct an exact size α test based on this permutation distribution of t_n generated by the 2^n equally likely sign inversions. Note that the test function (or the critical region) for such a randomization test, generally, depends on $|X_1|, \ldots, |X_n|$, and has to be determined from the given sample. In this setup, standard statistical tables are therefore of not much use. Further, for this test, we have not used the permutational-invariance, and hence, it is not necessary to assume that the X_i are i.i.d.r.v. It suffices to reframe the null hypothesis $H_0^{(1)}$ as that of the symmetry of the d.f. of each X_i without imposing the identity of these d.f.'s. Thus, in this permutation procedure, we not only eliminate the normality assumption needed with the parametric (Student) t-test, but also allow for possible nonhomogeneity of the d.f.'s This explains the broad scope of this randomization test.

Suppose now that the d.f. F in (3.1) is continuous. Let R_i^+ be the rank of $|X_i|$ among $|X_1|, \ldots, |X_n|$, for $i = 1, \ldots, n$, and let $Z_1 < \cdots < Z_n$ be the ordered r.v.'s corresponding to $|X_1|, \ldots, |X_n|$. We denote by

$$\begin{aligned} &\boldsymbol{S}_n = (\operatorname{sgn} X_1, \ldots, \operatorname{sgn} X_n), \\ &\boldsymbol{R}_n^+ = (R_1^+, \ldots, R_n^+) \quad \text{and} \quad \boldsymbol{Z}_n = (Z_1, \ldots, Z_n)\,. \end{aligned} \tag{3.5}$$

Then, under $H_0^{(1)}$, $\boldsymbol{S}_n$, $\boldsymbol{R}_n^+$ and $\boldsymbol{Z}_n$ are mutually independent. $\boldsymbol{Z}_n$ is a sufficient statistic for F^* (where $F^*(x) = F(x) - F(-x) = 2F(x) - 1$, $x \geq 0$), while $(\boldsymbol{S}_n, \boldsymbol{R}_n^+)$ is a maximal invariant. $\boldsymbol{S}_n$ has 2^n equally likely realizations (generated by the 2^n possible sign inversions) and $\boldsymbol{R}_n^+$ has $n!$ equally likely realizations (generated by the $n!$ permutations of $(1, \ldots, n)$). A test function $\phi_\alpha(\cdot)$ depending on E_n through $(\boldsymbol{S}_n, \boldsymbol{R}_n^+)$ only, though a randomization test function, will be genuinely distribution-free under $H_0^{(1)}$. Actually, the test function $\phi_\alpha(\boldsymbol{S}_n, \boldsymbol{R}_n^+)$ remains invariant under a monotone transformation $h: X \to h(X)$ where $h(-x) = -h(x)$, $x \in R$ and h is otherwise quite arbitrary. On the other hand, t_n in (3.4) is scale-invariant i.e., for $h(x) = cx$, $c \neq 0$, but may not be invariant under nonlinear $h(\cdot)$. This distinction led to formulation of *component randomization* and *rank randomization tests* (viz. Wilks, 1962, p. 462), where the former is only conditionally (permutationally) distribution-free, while the latter is genuinely distribution-free. However, we may note that tests based solely on $(\boldsymbol{S}_n, \boldsymbol{R}_n^+)$ need not be genuinely distribution-free if we drop the assumption of continuity of F or proceed on to the general multivariate case. Towards this, we consider the following.

Let $\boldsymbol{X}_1, \ldots, \boldsymbol{X}_n$ be n i.i.d.r.v. having a p-variate d.f. F, defined on $\mathbb{R}^p$, for some $p \geq 1$. The d.f. F is said to be diagonally symmetric about $\mathbf{0}$, if both $\boldsymbol{X}_1$ and

$(-1)\boldsymbol{X}_1$ have the same d.f. F. For $p=1$, this reduces to (3.1), while for $p>1$, this is a natural generalization of the notion of symmetry of F and is less restrictive than the *total symmetry* of F. If we define the sample point E_n by $(\boldsymbol{X}_1,\ldots,\boldsymbol{X}_n)$ (so that $\mathscr{X}_n=\mathbb{R}^{pn}$) and consider the group $\mathscr{G}_n^0$ of transformations $\{g_n^0\}$, where

$$g_n^0\cdot E_n=((-1)^{j_1}\boldsymbol{X}_1,\ldots,(-1)^{j_n}\boldsymbol{X}_n),\quad j_i=0,1 \text{ for } i=1,\ldots,n, \tag{3.6}$$

then the joint distribution of E_n remains invariant under $\mathscr{G}_n^0$. Note that E_n is a $p\times n$ matrix, so that for each row, we may consider a set of order statistics for the absolute values and define the absolute ranks and signs in the same manner as in the univariate case. This leads us to three matrices of (i) absolute order statistics, (ii) absolute ranks and (iii) signs of the observations. Unfortunately, for $p\geqslant 2$, these three matrices are not generally mutually independent. Thus, the maximal invariance character of the matrices of signs and absolute ranks may not be true, in general and hence, tests, based on these matrices, are not generally genuinely distribution-free. Nevertheless, the group $\mathscr{G}_n^0$ relates to the invariance under 2^n conditionally equally likely sign inversions (under $H_0^{(1)}$), and hence, a randomization test may be constructed as in (2.14)–(2.15) with $M_n=2^n$. For the multivariate sign-test, this intelligent observation is due to Chatterjee (1966), while, Sen and Puri (1967) adapted this for general multivariate signed-rank tests. We may refer to Chapter 3 (by Hušková) for more details.

3.2. *Hypothesis of randomness*

Let $X_1,\ldots,X_n$ be n independent r.v.'s with d.f.'s $F_1,\ldots,F_n$, all defined on the real line R. Consider the null hypothesis

$$H_0^{(2)}:\quad F_1=\cdots=F_n=F \ \text{(unknown)}, \tag{3.7}$$

against the alternatives that they are not all the same. For the sample point $E_n=(X_1,\ldots,X_n)$ $(\in\mathbb{R}^n)$, consider the group $\mathscr{G}_n$ of transformations $\{g_n\}$, where

$$g_n\cdot E_n=(X_{r_1},\ldots,X_{r_n}),\quad \boldsymbol{r}=(r_1,\ldots,r_n)\in\mathscr{P}_n, \tag{3.8}$$

and $\mathscr{P}_n=\{r_1,\ldots,r_n\}$ is the set of all possible $n!$ permutations of $(1,\ldots,n)$. Thus, for every $e_n=(x_1,\ldots,x_n)\,(\in\mathbb{R}^n)$, the orbit $O(e_n)$ is defined by

$$O(e_n)=\{e_n^*\colon e_n^*=(x_{r_1},\ldots,x_{r_n}) \text{ and } \boldsymbol{r}\in\mathscr{P}_n\}. \tag{3.9}$$

Under $H_0^{(2)}$, the conditional distribution of E_n over $O(e_n)$ is a (discrete) uniform one, with the probability mass $(n!)^{-1}$ attached to the $n!$ elements of $O(e_n)$, and 0 mass elsewhere. This provides the basis for a randomization (permutation) test for $H_0^{(2)}$. The choice of the test function depends on the alternative hypotheses.

First, consider the two-sample problem which can be treated as a special case of (3.7). Let $X_1, \ldots, X_{n_1}$ be i.i.d.r.v. with the d.f. F and $X_{n_1+1}, \ldots, X_n$ be i.i.d.r.v. with the d.f. G, where $n_1 \geq 1, n_2 = n - n_1 \geq 1$ and F and G are unknown d.f.'s, defined on R. Here (3.7) reduces to H_0: $F = G$ against one or two sided alternatives. When both F and G are normal with the common (unknown) variance σ^2 ($<\infty$) and possibly different means, an optimal parametric test is based on the Student t-statistic

$$t_n = (\bar{X}^{(1)} - \bar{X}^{(2)})(n_1 n_2/n)^{1/2}/s\,, \tag{3.10}$$

where

$$\bar{X}^{(1)} = n_1^{-1} \textstyle\int_{i=1}^{n_1} X_i, \qquad \bar{X}^{(2)} = n_2^{-1} \sum_{i=n_1+1}^{n} X_i$$

and

$$s^2 = (n-2)^{-1}\{\textstyle\sum_{i=1}^{n_1}(X_i - \bar{X}^{(1)})^2 + \sum_{i=n_1+1}^{n}(X_i - \bar{X}^{(2)})^2\}\,.$$

If we let $\bar{\bar{X}} = n^{-1}\sum_{i=1}^{n} X_i$, $T_n^2 = \sum_{i=1}^{n}(X_i - \bar{\bar{X}})^2$ and $Z_n = (n_1 n_2/n)^{1/2} \cdot (\bar{X}^{(1)} - \bar{X}^{(2)})$, then, we may write $t_n = (Z_n/T_n)(1 - Z_n^2/T_n^2)^{-1/2}$, so that t_n is a monotone function of $t_n^* = Z_n/T_n$. Now, T_n^2 remains invariant under $\mathcal{G}_n$, while $\mathcal{G}_n$ leads to a set $\{g_n \cdot Z_n\colon g_n \in \mathcal{G}_n\}$ of $n!$ realizations for Z_n on $O(E_n)$. This generates the permutation distribution of t_n^* and (2.14)–(2.15) hold with $M_n = n!$. (Actually, Z_n remains invariant under any permutation of the n_1 arguments $X_1, \ldots, X_{n_1}$ among themselves, as well as the n_2 arguments $X_{n_1+1}, \ldots, X_n$ among themselves, so that M_n may be reduced to $\binom{n}{n_1}$.) This permutation test is due to Pitman (1937a). In passing, we may remark that the critical values of t_n^* (and hence, t_n), determined from this permutation distribution, generally, depend on $O(E_n)$, and hence, are r.v. This eliminates the use of standard statistical tables and requires direct enumeration from the observed data. The situation is somewhat different with the rank statistics. If we assume that the d.f. are continuous, then, the ties among the observations may be neglected, with probability one. Let $Z_1 < \cdots < Z_n$ be the ordered r.v. for the combined sample $X_1, \ldots, X_n$, and, we write $X_i = Z_{R_i}$, $i = 1, \ldots, n$, so that R_i is the rank of X_i among the n observations, for $i = 1, \ldots, n$. Under $H_0^{(2)}$, $\boldsymbol{Z}_n = (Z_1, \ldots, Z_n)$ is a complete, sufficient statistic, and $\boldsymbol{R}_n = (R_1, \ldots, R_n)$ is independent of Z_n and it assumes each permutation of $(1, \ldots, n)$ with the common probability $(n!) - 1$. Thus, $\boldsymbol{R}_n$ is a maximal invariant. Thus, if a test function $\phi_\alpha(E_n)$ depends on E_n through $\boldsymbol{R}_n$ only, then, the test is genuinely distribution-free. Note that such a test function $\phi_\alpha(\boldsymbol{R}_n)$ remains invariant under any monotone transformation on the observations, whereas (3.10) is only scale-invariant. Tests based on the Kolmorogov-Smirnov statistics, linear rank statistics and some other (generalized) U-statistics are of this invariant type when the d.f. F is continuous. However, if we drop the assumption of continuity of F (so that ties among the X_i may occur with a positive probability) or proceed on to the general multivariate case (where the rank collection matrix is not necessarily the maximal invariant), then rank based tests, though conditionally (permutationally) distribution-free, are generally not genuinely distribution-free.

Let us next consider the several sample problems. Let X_{ij}, $j = 1, \ldots, n_i$ be n_i i.i.d.r.v.'s with a d.f. F_i, $i = 1, \ldots, c$ (≥ 2), $n = n_1 + \cdots + n_c$, and, referred to (3.7), we want to test for H_0: $F_1 = \cdots = F_c = F$ (unknown). For the normal model, assuming that all the d.f.'s have a common variance σ^2 ($<\infty$), one usually considers the classical one-way ANOVA test statistic

$$\mathscr{F}_n = \left\{\sum_{i=1}^{c} n_i(\bar{X}_i - \bar{\bar{X}})^2/(c-1)\right\}/s^2, \tag{3.11}$$

where $\bar{X}_i = (\Sigma_{j=1}^{n_i} X_{ij})/n_i$, $i = 1, \ldots, c$, $\bar{\bar{X}} = (\Sigma_{i=1}^{c} \Sigma_{j=1}^{n_i} X_{ij})/n$ and $s^2 = (n-c)^{-1} \Sigma_{i=1}^{c} \Sigma_{j=1}^{n_i} (X_{ij} - \bar{X}_i)^2$. Again $\mathscr{F}_n$ is a monotone function of Z_n^2/T_n^2 where $Z_n^2 = \Sigma_{i=1}^{c} n_i(\bar{X}_i - \bar{\bar{X}})^2$ and $T_n^2 = \Sigma_{i=1}^{c} \Sigma_{j=1}^{n_i} (X_{ij} - \bar{\bar{X}})^2$. Note that T_n^2 is $\mathscr{G}_n$-invariant. Thus, a randomization test based on Z_n^2/T_n^2 can be constructed by reference to (2.14)–(2.15), where M_n is reducible to $n!/(n_1! \ldots n_c!)$. This is the classical permutational ANOVA test, due to Pitman (1938). Here also, if the d.f.'s are all continuous, one may base a test on the vector $\boldsymbol{R}_n$ (of the ranks of the n combined sample observations), where $\boldsymbol{R}_n$ is a maximal invariant, and hence, the test will be genuinely distribution-free. The Kruskal–Wallis statistic and various other multi-sample rank statistics, considered in Chapter 2 (by Bhapkar) are all usable in this context, and, for such rank tests, standard tables for the critical values are available.

We consider now the multivariate case. Here, $\boldsymbol{X}_1, \ldots, \boldsymbol{X}_n$ have d.f.'s $F_1, \ldots, F_n$, all defined on the p (≥ 1)-dimensional space $\mathbb{R}^p$. In the two-sample (location) problem, one may consider the Hotelling T^2 statistic (a natural generalization of the Student t-statistic) and base a permutation test on this statistic; this was done by Wald and Wolfowitz (1944). For the case of multivariate multi-sample location problem (MANOVA), one may similarly consider permutation procedures based on the Lawley–Hotelling trace statistics or other parametric forms. In this case also, (3.9) provides the orbit, where we have to keep in mind that the x_{r_i} are p-vectors. However, the structure of the permutation distribution remains the same, and hence, we may use (2.14) and (2.15) to obtain some permutationally (conditionally) distribution-free tests. The critical values for these randomization tests are themselves r.v. As in the multivariate one-sample problem, here also, one may replace the observation vectors by the rank vectors, where separate ranking is made for each of the p variates (over all the sample observations), and consider suitable rank order statistics on which the tests may be based. We may refer to Chapter 2 (by Bhapkar) for some details. Here also, for $p \geq 2$, in general, the rank collection matrix, as obtained above, is not a maximal invariant, so that the multivariate rank tests are not generally genuinely distribution-free. However, the uniform (conditional) distribution over the orbit provides a conditionally (permutationally) distribution-free test. By reference to Chatterjee and Sen (1964), these may therefore be termed rank randomization tests. Here also, the critical values of the test statistics are themselves r.v., and standard statistical tables may not provide their exact values.

Tests for randomness in (3.7) may also arise in the context of *trend alternatives* where, one conceives of the model

$$H^{>}: \quad F_1 \geq \cdots \geq F_n \text{ (a.e.)} \tag{3.12}$$

with at least one strict inequality on a set of measure nonzero; the case of $H^{<}$

may be formulated with the opposite inequalities. Regression alternatives are also sometimes considered by letting

$$F_i(x) = F(x - \beta_0 - \beta c_i), \quad i = 1, \dots, n, \; x \in \mathbb{R}, \tag{3.13}$$

where F is unspecified, β_0 and β are unknown parameters and the c_i are known constants. The null hypothesis relates to $\beta = 0$ against one or two sided alternatives. One may even consider a more general form where βc_i may be replaced by $\boldsymbol{\beta}'\boldsymbol{c}_i$ with the $\boldsymbol{c}_i$ known q-vectors for some $q \geq 1$ and $\boldsymbol{\beta}' = (\beta_1, \dots, \beta_q)$ an unknown vector of parameters. The null hypothesis relates to $\boldsymbol{\beta} = \boldsymbol{0}$. The two sample location model is a special case of (3.13) (where the c_i can assume only the values 0 and 1), while the multi-sample location model is a special case of this general model where the $\boldsymbol{c}_i$ may have only q different realizations. When F is assumed to be normal, classical tests for these hypotheses are based on the least squares estimators of the parameters and the residual mean squares due to error. Randomization tests can be based on the same statistics, and again, (2.14)–(2.15) provide the desired exact size α property of these tests. For continuous F, rank based tests for these hypotheses may also be based on linear rank statistics [viz., Chapters 11 (by Adichie) and 12 (Aubuchon and Hettmansperger)] and these will be genuinely distribution-free. For the multivariate case, again, such rank tests are generally only conditionally (permutationally) distribution-free.

For testing randomness against *serial dependence* one may also use randomization procedures. Against H_0 in (3.7), one may be interested in the stochastic dependence of successive observations. One possible simple model is the following:

$$X_i = \rho X_{i-1} + e_i, \quad i = 1, \dots, n, \tag{3.14}$$

where the e_i are i.i.d.r.v. with d.f. F, defined on R. Under H_0, $\rho = 0$ and we may want to test this against $\rho > 0$ or $\neq 0$. Normal theory tests are based on a suitable version (circular or not) of the serial statistic $\Sigma\,(X_{i-1}X_i)$. Here also, under H_0, the joint distribution of $(X_1, \dots, X_n)$ remains invariant under any permutation of the arguments, so that a randomization test procedure may be based on the statistic using (2.14)–(2.15) for $M_n = n!$. More general serial statistics were considered by M. N. Ghosh (1954) and the theory of randomization tests was developed. Randomization tests based on pure or mixed rank statistics can also be constructed by the same permutation principle; we may refer to Chapter 5 (by G. K. Bhattacharyya) for some of these. Tests based on the pure rank statistics are genuinely distribution-free (when F is continuous), while the ones based on mixed rank statistics are only permutationally (conditionally) distribution-free. The model (3.14) can also be generalized to the multivariate case by replacing the X_i, X_{i-1} and e_i by appropriate p-vectors and ρ by $\boldsymbol{P}$, a $p \times p$ matrix, where again the null hypothesis relates to $\boldsymbol{P} = \boldsymbol{0}$. For $p \geq 2$, the pure rank based tests are generally only permutationally distribution-free, while the others are so even for $p = 1$. Note that the structure of the orbit remains the same for univariate as well as multivariate situations.

3.3. Hypothesis of multivariate independence

Since the bivariate case has been treated in Section 2, we consider here the more general case of multivariate d.f.'s. Let $\boldsymbol{X}_i' = (\boldsymbol{X}_i^{(1)\prime}, \boldsymbol{X}_i^{(2)\prime}) = (X_{1i}, \ldots, X_{pi}, X_{(p+1)i}, \ldots, X_{(p+q)i})$, $i = 1, \ldots, n$, be n i.i.d.r.v. having a $(p+q)$-variate d.f. F, defined on $\mathbb{R}^{p+q}$, where $p \geq 1$ and $q \geq 1$. Also, let F_1 (and F_2) be the (joint) d.f. of $\boldsymbol{X}_i^{(1)}$ (and $\boldsymbol{X}_i^{(2)}$), and, we consider the null hypothesis

$$H_0^{(3)}:\quad F(\boldsymbol{x}^{(1)}, \boldsymbol{x}^{(2)}) = F_1(\boldsymbol{x}^{(1)})F_2(\boldsymbol{x}^{(2)}), \quad (\boldsymbol{x}^{(1)}, \boldsymbol{x}^{(2)}) \in \mathbb{R}^{p+q}, \tag{3.15}$$

that is, $\boldsymbol{X}_i^{(1)}$ and $\boldsymbol{X}_i^{(2)}$ are stochastically independent (though the variates within the same vector need not be independent). As in (2.1)–(2.2), we have here

$$E_n = \begin{pmatrix} X_1^{(1)} \cdots X_n^{(1)} \\ X_1^{(2)} \cdots X_n^{(2)} \end{pmatrix} \quad \text{and} \quad E_n(\boldsymbol{r}) = \begin{pmatrix} X_1^{(1)} \cdots X_n^{(1)} \\ X_{r_1}^{(2)} \cdots X_{r_n}^{(2)} \end{pmatrix}, \quad \boldsymbol{r} \in \mathcal{P}_n, \tag{3.16}$$

where $\mathcal{P}_n$ is the set of $n!$ permutations of $(1, \ldots, n)$. Then, the definition of the orbit $\mathcal{E}_n$ in (2.3) and the group of transformations $\mathcal{G}_n$ in (2.4) remain intact, and hence, (2.5) through (2.8) hold.

For the normal theory model (i.e. F multinormal), canonical correlations are employed for testing $H_0^{(3)}$. If $\boldsymbol{S}_n = (n-1)^{-1}\Sigma_{i=1}^n (\boldsymbol{X}_i - \bar{\boldsymbol{X}}_n)(\boldsymbol{X}_i - \bar{\boldsymbol{X}}_n)'$ (with $\bar{\boldsymbol{X}}_n = (\Sigma_{i=1}^n \boldsymbol{X}_i)/n$) be the sample covariance matrix and we partition it as

$$\boldsymbol{S}_n = \begin{pmatrix} \boldsymbol{S}_{n11} & \boldsymbol{S}_{n12} \\ \boldsymbol{S}_{n21} & \boldsymbol{S}_{n22} \end{pmatrix} \tag{3.17}$$

where $\boldsymbol{S}_{n11}$ is a $p \times p$ matrix, $\boldsymbol{S}_{n22}$ $q \times q$ and $\boldsymbol{S}_{n12} = \boldsymbol{S}_{n21}'$ is a $p \times q$ matrix, then we may consider the matrix (of canonical correlations)

$$\boldsymbol{S}_{n11}^{-1}\boldsymbol{S}_{n12}\boldsymbol{S}_{n22}^{-1}\boldsymbol{S}_{n21} = \boldsymbol{S}_n^* \text{ (say) of order } p \times p. \tag{3.18}$$

The largest characteristic root (or the trace) of $\boldsymbol{S}_n^*$ is taken as the test statistic for testing the null hypothesis. Note that both $\boldsymbol{S}_{n11}$ and $\boldsymbol{S}_{n22}$ remain invariant under $\mathcal{G}_n$ while $\boldsymbol{S}_{n12}$ has $n!$ possible (conditionally, equally likely) realizations over the orbit. Hence, for a real valued function $T_n = T(\boldsymbol{S}_n^*)$ of $\boldsymbol{S}_n^*$, we may generate the permutational (conditional) distribution of T_n by using (2.5), and with this, we may proceed as in (2.9)–(2.11) to get an exact size α (randomization) test. For $p = q = 1$ (i.e., the bivariate case), this reduces to the two-sided version of (2.11). If at least one of p or q is greater than 1, as in Section 3.1 or 3.2, we may conclude that the randomization tests are only permutationally distribution-free, while for $p = q = 1$, we may proceed as in after (2.12) and obtain some genuinely distribution-free rank tests. Puri, Sen and Gokhale (1970) have considered multivariate rank tests for this independence problem and their procedures are generally permutationally distribution-free. In their formulation, the sample covariance matrix has been replaced by some (generalized) grade (rank) correlation matrix, while the structure of the orbit and the permutation principle remain the same. These authors have also considered the case of testing for independence of more than two subsets of variates. In particular, if all the coordinates are stochastically independent, so that we have a *total independence* model, then, one may consider an extended group $\mathcal{G}_n^0$ of transformations $\{g_n^0\}$ for which genuinely distribution-free rank tests exist and are obtainable by the permutation procedure described above.

3.4. *Hypothesis of exchangeability*

Let $\boldsymbol{X}_i = (X_{1i}, \ldots, X_{pi})'$, $i = 1, \ldots, n$ be n i.i.d.r.v. with a p (≥ 2)-variate d.f. F, defined on $\mathbb{R}^p$. The null hypothesis of interest is that the coordinates of $\boldsymbol{X}_i$ are interchangeable or exchangeable r.v.'s. This may be framed as

$$H_0^{(4)}:\quad F(x_{r_1}, \ldots, x_{r_p}) = F(x_1, \ldots, x_p), \quad \boldsymbol{r} \in \mathcal{P},\ \boldsymbol{x} \in \mathbb{R}^p, \tag{3.19}$$

where $\mathcal{P}$ is the set of all possible ($p!$) permutations of $(1, \ldots, p)$. In this case, the sample point $E_n = (\boldsymbol{X}_1, \ldots, \boldsymbol{X}_n)$ is a $p \times n$ matrix. Let $\boldsymbol{r}_i = (r_{1i}, \ldots, r_{pi})'$, $i = 1, \ldots, n$, be n independent vectors, where each $\boldsymbol{r}_i$ takes on the permutations of $(1, \ldots, p)$ with the common probability $(p!)^{-1}$. Also, let $\mathcal{P}_n = \{i_1, \ldots, i_n\}$ be the set of all possible $n!$ permutations of $(1, \ldots, n)$. Consider then the group $\mathcal{G}_n$ of $(n!)(p!)^n$ transformations $\{g_n\}$, where, typically,

$$g_n \cdot E_n = \begin{pmatrix} X_{i_1 r_{11}} & \cdots & X_{i_n r_{1n}} \\ \vdots & & \vdots \\ X_{i_1 r_{p1}} & \cdots & X_{i_n r_{pn}} \end{pmatrix}, \quad (i_1, \ldots, i_n) \in \mathcal{P}_n,\ \boldsymbol{r}_i \in \mathcal{P},\ i = 1, \ldots, n. \tag{3.20}$$

The group $\mathcal{G}_n$ maps the sample space onto itself, and, under $H_0^{(4)}$ in (3.19), $g_n \cdot E_n$ has the same distribution as E_n for every $g_n \in \mathcal{G}_n$. Thus, we may appeal to (2.14)–(2.15) with $M_n = (n!)(p!)^n$. Here also, if the test function $\phi_\alpha(E_n)$ is symmetric in $X_1, \ldots, X_n$, then, we may reduce M_n to $(p!)^n$. For the normal theory model, (3.19) reduces to the equality of the means, of the variances of the p coordinate variates and the equality of the covariances for any pair of them; in the literature, this is known as the H_{MVC} (viz., Wilks 1946). The same test statistic may be used in the randomization procedure wherein the assumed normality of F will no longer be needed; this test will only be conditionally distribution-free. Alternatively, as in Sen (1967), we may consider a rank statistic where the X_{ij} in (3.20) are replaced by their ranks (with respect to all the $np = N$ observations). Since the coordinates in each column of E_n are not necessarily independent, such rank statistics may not be genuinely distribution-free (under $H_0^{(4)}$), although they are conditionally so. This hypothesis of multivariate interchangeability also arises in a very natural manner in the analysis of randomized block designs, which we present below.

Suppose that there are n (≥ 2) blocks of p (≥ 2) plots each where p different treatments are applied, and let X_{ij} stand for the response of the plot in the ith block receiving the jth treatment, for $i = 1, \ldots, n$, $j = 1, \ldots, p$. We denote the d.f. of X_{ij} by F_{ij}, defined on $\mathbb{R}$, and frame the null hypothesis

$$H_0^{(5)}: \quad F_{i1} = F_{i2} = \cdots = F_{ip} = F_i \text{ (unknown)}, \quad i = 1, \ldots, n, \tag{3.21}$$

where the F_i are arbitrary. Again, under $H_0^{(5)}$, the vectors $\boldsymbol{r}_i$, $i = 1, \ldots, n$, defined as in before (3.20), are independent, each having $p!$ possible equally likely realizations (when the F_i are assumed to be continuous). Hence, we have a group $\mathcal{G}_n$ of $(p!)^n$ transformations $\{g_n\}$, where, typically,

$$g_n \cdot E_n = \begin{pmatrix} X_{1r_{11}} \cdots X_{nr_{1n}} \\ X_{1r_{p1}} \cdots X_{nr_{pn}} \end{pmatrix}, \quad \boldsymbol{r}_i \in \mathcal{P}, \ i = 1, \ldots, n, \tag{3.22}$$

and, for every $g_n \in \mathcal{G}_n$, $g_n \cdot E_n$ has the same distribution as E_n. Thus, we may again appeal to (2.14)–(2.15) with $M_n = (p!)^n$, and in this setup, it is not necessary to assume that the F_i are all the same. It is also possible to replace the assumption that for each i, $X_{i1}, \ldots, X_{ip}$ are independent, by their exchangeability, i.e., the joint d.f. of $(X_{i1}, \ldots, X_{ip})$ is symmetric in its arguments, a case that may arise in 'mixed models' where the block effects may be random and the treatment effects are nonstochastic. Thus, in a randomization procedure, the additivity of the block or treatment effects and the independence of the errors may be eliminated along with the normality of the F_i. Randomization tests based on the classical ANOVA test statistics, dating back to Fisher (1934), are conditionally distribution-free, while the procedures solely based on the vectors $\boldsymbol{r}_i$, $i = 1, \ldots, n$, (such as the tests due to Friedman, 1937; Brown and Mood, 1951, and others), are genuinely distribution-free when the F_i are all continuous. One of the characteristics of the 'within block rankings' is that they do not incorporate the inter-block comparisons. Incorporation of this inter-block information is possible through suitable alignment processes: Tests are based on 'ranking after alignments' and rest on a somewhat different group-invariance structure. If we conceive of additive block effects (i.e., the F_i differ in shifts only), then, it seems intuitive to eliminate the block-effects through substitution of their estimates and adapting an overall ranking on the residuals (or aligned observations). Based on $X_{i1}, \ldots, X_{ip}$, let $\tilde{X}_i$ be some translation-invariant estimator of the block average, and let $Y_{ij} = X_{ij} - \tilde{X}_i$, for $j = 1, \ldots, p$, $i = 1, \ldots, n$. If $\tilde{X}_i$ is symmetric in $(X_{i1}, \ldots, X_{ip})$, then, under $H_0^{(5)}$, $Y_{i1}, \ldots, Y_{ip}$ are exchangeable r.v., for each i and these aligned vectors are independent for different i. Let $Y_{i;1} < \cdots < Y_{i;p}$ be the ordered r.v. corresponding to the Y_{ij}, $1 \leq j \leq p$, for $i = 1, \ldots, n$, and let $\boldsymbol{Y}^*$ be the matrix of these order statistics. Then, under $H_0^{(5)}$, conditional on $\boldsymbol{Y}^*$, one obtains a set of $(p!)^n$ possible realizations of the aligned observations, corresponding to the intra-block permutations of the order statistics, and these are all conditionally equally likely. Hence, we may again appeal to (2.14)–(2.15) with $M_n = (p!)^n$, and rank tests based on these aligned observations are conditionally (permutationally) distribution-free, while the intra-block rankings (for continuous d.f.) lead to genuinely distribution-free tests. For more details, we may refer to Chapter 10 (by Quade). The hypothesis of interchangeability or exchangeability has also been extended to the multivariate case (and adapted for the multivariate analysis of variance models). Suppose that each $\boldsymbol{X}_i$ is a $p \times k$ matrix, and the hypothesis of interchangeability relates to the exchangeability of the p rows in the sense that the (joint) d.f. of $\boldsymbol{X}_i$ remains invariant under any permutation of the rows of the $p \times k$ matrix argument. This permutational invariance argument is the same as in the univariate case, and this has been exploited by Gerig (1969) and Sen (1969) (among others) in considering permutationally distribution-free rank tests for the MANOVA models. Note that for $k \geq 2$, these rank tests (like the ones in earlier sub-sections) are only conditionally distribution-free.

The hypothesis of *spherically exchangeability* is also of some interest. Let $\boldsymbol{X} = (X_1, \ldots, X_p)'$ be a r.v. having a p ($\geq$2)-variate d.f. F, defined on $\mathbb{R}^p$, and, for every $\boldsymbol{t} \in \mathbb{R}^p$, define the characteristic function of $\boldsymbol{X}$ by $\psi(\boldsymbol{t}) = E_F\{\exp(i\boldsymbol{t}'\boldsymbol{X})\}$. If $\psi(\boldsymbol{t})$ is expressible as $\psi_0(\boldsymbol{t}'\boldsymbol{t})$, $\forall \boldsymbol{t} \in R^p$, where $\psi_0(a)$ depends only on a ($\geq$0), then $\boldsymbol{X}$ is termed a spherically exchangeable r.v. For the particular case of $p = 2$, if one lets $X_1 = R \cos \theta$ and $X_2 = R \sin \theta$, then for such a model, the r.v. θ has a uniform d.f. on $[0, 2\pi]$, so that for a sample $\boldsymbol{X}_1, \ldots, \boldsymbol{X}_n$ of size n (from F), the problem reduces to that of testing for the goodness of fit of the uniform distribution for the corresponding angles $\theta_1, \ldots, \theta_n$. For $p > 2$, one may similarly consider the transformation $X \to (R, \boldsymbol{\theta})$ where $R = (\boldsymbol{X}'\boldsymbol{X})^{1/2}$ and $\boldsymbol{\theta}$ is a $(p-1)$-vector of angles, and under the spherically exchangeability model, we have a uniform distribution on the surface of the unit sphere. Tests for uniformity on the surface of the unit sphere have also been developed (see Chapter 31 by J.R. Jammalamadaka) and these may be called on for the present purpose. Alternatively, the surface of the unit sphere may be divided into r ($\geq$2) exclusive and exhaustive regions and a multinomial distribution (for the frequencies) may be used for the goodness of fit problem. Bell and Smith (1972a) considered some generalized permutation statistics for this problem. The set of permutations here is the set of rotations of a sphere. This relates to an infinite set of permutations, and the basic structure for such a case is somewhat different, and will not be considered here.

3.5. *Randomization tests for NP hypotheses for stochastic processes*

The literature on NP tests for stochastic processes is relatively limited. However, some results and references can be found, e.g., in Bell, Avadhani and Woodroofe (1970), Basawa, Rao and Prakasa (1980), Bell (1982) and Bell and Ahmad (1982). The types of processes treated are as follows.

(a) *Processes with stationary independent increments.* If $\{X_t; t \geq 0\}$ is a stochastic process with stationary independent increments, then, for every h (>0), $Y_1 = X_h - X_0$, $Y_2 = X_{2h} - X_h, \ldots, Y_n = X_{nh} - X_{(n-1)h}$ are i.i.d.r.v.'s. The common distribution function F_0 is uniquely determined by the law $\mathscr{L}_0$ of the process and the time interval $(0, h]$, and conversely F_0 and h determine $\mathscr{L}_0$ if $X_0 = 0$, with probability 1. With the Y_j defined as in above, the theory developed in Section 3.2 applies, and hence, randomization procedures based on the Y_j workout for the problems related to such processes with stationary independent increments.

(b) *Processes with stationary symmetric increments.* Define the Y_j as in (a). Here, additionally, one has the information that F_0 is symmetric. As such, we may appeal to the theory developed in Section 3.1, and the randomization tests developed there are applicable for problems related to such processes.

(c) *Spherically exchangeable processes.* The problems here lead to permutation tests analogous to those in Section 3.4 (relating to tests for sphericity).

(d) *Exchangeable processes.* For such processes, the Y_i, defined as in (a), are not necessarily independent, but are exchangeable r.v. As such, the theory developed in Section 3.4 remains applicable here.

It should be mentioned here that throughout this section the tests discussed are not 'omnibus' tests, but are rather 'directional' in the sense of being directed towards certain parametric alternatives. In fact, in his original papers, Pitman (1937a,b; 1938) was concerned with the classical normal alternatives, and was seeking NPDF tests which would yield high power against these alternatives. These early tests were based on the classical statistics for the problem at hand, and hence, the associated B-Pitman functions were closely related to the classical parametric statistics. In Section 5, we shall discuss the problem of choosing randomization tests statistics for a specific NP problem, and consider some optimality criteria too.

4. Randomized rank procedures

Bell and Doksum (1965) have considered a modification of randomization tests where a resampling scheme provides a means of obtaining genuinely distribution-free tests for many hypotheses of common interest. This avoids the problem of computing the critical values of the test statistics from the sample (which tends to be very cumbrous as the sample sizes increase), and enables one to use some standard statistical tables. We illustrate their procedure with the same example as in Section 2.

Let (X_i, Y_i), $i = 1, \ldots, n$, be n i.i.d.r.v. with an unknown bivariate d.f. F, and we want to test for the null hypothesis H_0: $F(x, y) = F(x, \infty)F(\infty, y) = F_1(x)F_2(y)$, $\forall(x, y) \in \mathbb{R}^2$ (i.e., the X and Y are independent). We assume that the marginal d.f.'s F_1 and F_2 are continuous (a.e.) and define $\boldsymbol{Q}$, the vector of induced ranks as in (2.13). The Bell–Doksum procedure is then based on $\boldsymbol{Q}$ and the following resampling scheme. Let $X_1^0, \ldots, X_n^0$ (and $Y_1^0, \ldots, Y_n^0$) be independent samples from a standard normal distribution, and let $X_{n:1}^0 < \cdots < X_{n:n}^0$ (and $Y_{n:1}^0 < \cdots < Y_{n:n}^0$) be the corresponding ordered r.v. Consider then the usual sample correlation statistic

$$r_n(\boldsymbol{Q}) = \left\{\sum_{i=1}^{n} (X^0_{n:i} - \bar{X}^0_n)(Y^0_{n:Q_i} - \bar{Y}^0_n)\right\} \Big/ \left\{\left(\sum_{i=1}^{n} (X^0_i - \bar{X}^0_n)^2\right)\left(\sum_{i=1}^{n} (Y^0_i - \bar{Y}^0_n)^2\right)\right\}^{1/2}, \tag{4.1}$$

where $\bar{X}^0_n = n^{-1}\Sigma^n_{i=1} X^0_i$ and $\bar{Y}^0_n = n^{-1}\Sigma^n_{i=1} Y^0_i$. Note that $r_n(\boldsymbol{Q}) \in [-1, 1]$ for all $\boldsymbol{Q}$ and the permutation distribution of $r_n(\boldsymbol{Q})$ is given by

$$G_n(r) = (n!)^{-1}\{\#\boldsymbol{Q} \in \mathcal{P}_n : r_n(\boldsymbol{Q}) \leq r\}, \quad -1 \leq r \leq 1 . \tag{4.2}$$

Further, the unconditional null distribution of $r_n(\boldsymbol{Q})$ is given by

$$G^*_n(r) = P\{r_n(\boldsymbol{Q}) \leq r \mid H_0\} = EG_n(r), \quad r \in [-1, 1] . \tag{4.3}$$

On the other hand, under H_0, r_n has the same distribution as r^0_n, where

$$r^0_n = \left\{\sum_{i=1}^{n} (X^0_i - \bar{X}^0_n)(Y^0_i - \bar{Y}^0_n)\right\} \Big/ \left\{\sum_{i=1}^{n} (X^0_i - \bar{X}^0_n)^2 \sum_{i=1}^{n} (Y^0_i - \bar{Y}^0_n)^2\right\}^{1/2} . \tag{4.4}$$

But, by definition, the d.f. of r^0_n has the probability density function (d.d.f.)

$$\left\{\frac{(n-1)}{2} \Big/ \frac{1}{2}\,\frac{(n-2)}{2}\right\}^{1/2} (1 - r^2)^{(n-4)/2}, \quad r \in (-1, 1) , \tag{4.5}$$

i.e., $(n-2)^{1/2} r^0_n (1 - r^{02}_n)^{-1/2}$ has the Student t-distribution with $n - 2$ degrees of freedom. Thus, the normal theory critical values can be adapted to the distribution-free test based on $r_n(\boldsymbol{Q})$. This simplification is achieved through the resampling scheme along with the use of the maximal invariant $\boldsymbol{Q}$.

In the two or several sample problems, one may similarly use the maximal invariant $\boldsymbol{R}_n$ (the vector of ranks), draw a random sample of size n for a standard normal (or some other standard) distribution and incorporating $\boldsymbol{R}_n$, one can construct a test statistic which has the same null distribution as the standard parametric theory test statistic, and hence, standard statistical tables are usable for obtaining the critical values of their modified randomization test statistics. For the one sample problem, one may similarly use the maximal invariants (signs and ranks of the absolute values) and draw a random sample from a chi distribution with 1 degree of freedom and obtain a test statistic having the Student t-distribution with $n - 1$ degrees of freedom.

The tactful use of the maximal invariants along with the prescribed resamping scheme form the basis of the Bell–Doksum procedures. However, the resampling scheme introduces some arbitrariness into the testing procedure. Two or more statisticians using independent resampling data may end up with possibly different test statistics values leading to different conclusions, though in either case, the same maximal invariants are used and the same formula for the test statistic is adapted. It has also shown by Jogdeo (1966) that the

Bell–Doksum procedure may not be consistent against distant-alternatives (in the sense that the power of their procedure may not converge to 1 even if the hypothesis moves away from the null one). For small sample sizes, the local optimality of the Bell–Doksum procedure may not follow readily. Nevertheless, their procedure provides an alternative simple manner of obtaining distribution-free tests avoiding the problem of computing the critical values from the sample, and, for large sample sizes, at least, they possess good properties.

5. Optimal randomization tests

Note that invariance, similarity, Neyman-structure and maximal invariants provide the basis of permutation of randomization tests. However, in this way, one may end up with a class of competing tests. The choice of a desirable one (within this class) may depend on the particular type of alternatives one may have in mind, and, no single test may remain uniformly optimal (or desirable or good) for all possible alternatives. In this context, it is not uncommon to choose a particular null hypothesis of invariance (inducing a nonparametric structure) against alternatives belonging to some specific parametric families. For example, for the two-sample problem, we may be interested in the null hypothesis of equality of the two d.f.'s against a shift alternative, where under the alternative, the d.f.'s are assumed to be normal. In this way, the test would remain valid for a broad class of distributions, and, at the same time, it would have some desirable (or optimal) properties when the specific alternatives hold. For such a problem, often, the randomization tests may be so constructed that the above objectives are met. The following result due to Lehmann and Stein (1949) deserves special mention:

Referred to (2.14)–(2.15), let H_0 be the null hypothesis of invariance (under the partition Π_n), and consider an alternative hypothesis for which the sample point X_n has a probability density function q_n (not in H_0). Then, for the orbit $O(x_n)$ of M_n points, we define an ordering of these points by

$$q_n(x_n^{(1)}) \leq \cdots \leq q_n(x_n^{(M_n)}), \tag{5.1}$$

and with this, in (2.9)–(2.11), we replace the T_n by $q_n(x_n)$. Then, for testing H_0 against q_n, the test function $\phi_\alpha(\cdot)$ in (2.11) leads to a most powerful size α randomization test. If $Q_n = \{q_n\}$ be the family of all probability densities (for the X_n) for which the same ordering in (5.1) holds, then, (2.11) provides a uniformly most powerful size α (randomization) test for H_0 against the family Q_n.

It is known (Theorem 2.5) that rank statistics must be based on permutation (B-Pitman) statistics. Hence, when one chooses an optimal rank statistic, one is in some sense choosing a permutation statistic on which it is based. A variety of authors (e.g., Hoeffding, 1951; Terry, 1952; Lehmann, 1959; Capon, 1961, and Hájek and Šidák, 1967; among others) have developed most powerful (MP) and locally most powerful (LMP) rank tests for a variety of situations. However, the alternatives against which these tests are so optimal are, more often than not, parametric families. It turns out that for appropriate parametric families of alternatives, the MP NPDF test is always a permutation test. This optimal test can only be a rank test if the B-Pitman function is constant on rank sets. In this context, we have the following result due to Bell and Donoghue (1969).

THEOREM 5.1. *Let Ω and S be as in Theorem 2.5. Then against a simple alternative H_1 with likelihood function L_1 the MP level α NPDF test is of the form*

$$\psi(X_n)=\begin{cases}1 & \text{if } R(h(X_n))>K(\alpha,n),\\ \eta & \text{if } R(h(X_n))=K(\alpha,n),\\ 0 & \text{if } R(h(X_n))<K(\alpha,n),\end{cases}$$

where $h(\cdot)$ is a B-Pitman function such that $L_1(X_n')<L_1(X_n'')$ implies $h(X_h')<h(X_n'')$ for all X_n' and X_n'' on the same orbit, and for almost all orbits.

We illustrate this theorem with the following example from Bell and Donoghue (1969). For the hypothesis H_0: $F_1=\cdots=F_n=F$ (unknown) (see Section 3.2), consider the parametric alternative such that $X_1,\ldots,X_n$ are independent with $X_r\sim N(\mu_r,\sigma^2)$, where $\mu_r=\mu_r(\theta)=(-1)^r\cos(r\pi/\theta)$, $r\geq 1$. Here,

$$\log L_1(\boldsymbol{X}_n,\theta)=-\frac{n}{2}\log(2\pi)-\tfrac{1}{2}\sum_{r=1}^{n}(X_r-\mu_r)^2=K^*+\sum_{r=1}^{n}\mu_rX_r$$

(and $\boldsymbol{X}_n=(X_1,\ldots,X_n)$). This implies that the MP NPDF test should be based on the permutation (B-Pitman) statistic with $h(\boldsymbol{X}_n)=\Sigma_{r=1}^{n}(-1)^rX_r\cos(r\pi/\theta)$. One may wish to avoid the dependence on θ. For θ close to 1, one has

$$h(\boldsymbol{X}_n)=1-\left\{\pi^2\sum_{r=1}^{n}r^2X_r\right\}(\theta-1)^2+o((\theta-1)^2),$$

so that a LMP NPDF test should be based on $h(\boldsymbol{X}_n)=\Sigma_{r=1}^{n}r^2X_r$.

At this point, it is clear that the recommendation is that the randomization (permutation) test be chosen on the basis of a parametric family of alternatives. The resulting test would, in general, not be an 'omnibus' test, but would be optimal against those alternatives as well as some nearby ones. In fact, Bell and Donoghue (1969) proved the following result for the randomness hypothesis (in Section 3.2). It can be readily extended to other NP hypotheses as well, but only the original form is presented here.

THEOREM 5.2. *Let $h(\boldsymbol{X}_n)=\Sigma_{i=1}^{n}A_i(X_i)$ be a B-Pitman function. Then*

(i) *$R(h(\boldsymbol{X}_n))$ is the statistic of the MP NPDF statistic against the family*

$$\{\Pi F_i\colon F_i'(x,\theta)=\exp\{a(\theta)A_i(x)+k(i,\theta)+t(x,\theta)\},\text{ with } a(\theta)>0\}.$$

(ii) *If, further, $a(\theta)$ is continuous in θ, $a(\theta_0) = 0 = k(i, \theta_0)$, and $Q(x, \theta, i) = o(a(\theta))$, for all $i \geq 1$ and x, then $R(h(\mathbf{X}_n))$ is the statistic of the LMP NPDF test for $\theta = \theta_0$ (randomness) vs. the family*

$$\{\Pi F_i\colon F_i'(x, \theta) = \exp\{a(\theta)A_i(x) + k(i, \theta) + t(x, \theta) + Q(x, \theta, i)\}, \theta > \theta_0\}.$$

As is generally the case, for a composite alternative hypothesis, the density $q_n(x_n)$ (as well as the ordering in (5.1)) depends on the unknown (nuisance) parameters; a very similar case arises with the $h(X_n)$ in Theorem 5.1 (or 5.2). Thus, the (generalized) Neyman–Pearson lemma on which the above optimality property is based may not hold for the entire set of alternatives. The situation is comparable to the classical Neyman–Pearson testing theory, where a (uniformly) most powerful similar region may not generally exist, and, one may have to choose some restricted optimal one (such as the locally most powerful, asymptotically most powerful, most powerful unbiased, $c(\alpha)$ test etc.,). Such a restricted optimal parametric test statistic may then be employed as in Section 3 in the construction of randomization tests whenever under the null hypothesis the orbit and the conditional probability law on the orbit can be defined unambiguously. Such a randomization test will have the similar (restricted) optimality property within the class of randomization tests. For example, for the two-sample problem with the shift alternative in mind, under normality on the d.f. (when the alternative holds), the one-sided uniformly most powerful randomization test is based on the classical Student t-statistic with the rule in (2.9)–(2.11). For some other d.f., for shift alternatives, we may similarly use the locally most powerful test statistic and the use of that in the randomization procedures in Sections 2 and 3 would lead to locally most powerful randomization tests against such specific alternatives. If we use invariant tests (such as the rank tests), then, we would get locally most powerful invariant tests in this manner. The recent papers of Sen (1981a) and Basu, Ghosh and Sen (1983) cast some light on locally optimal tests of this type. Use of Neyman's $c(\alpha)$-test statistic in a randomization procedure (permitting the availability of the orbit) may be generally recommended on the ground of local asymptotic optimality, where the sample size is taken large and the alternative hypotheses are chosen in the neighborhood of the null hypothesis, so that the asymptotic power function does not converge to the degenerate limit 1.

We have discussed the optimality or asymptotic optimality of randomization or permutation tests for certain parametric alternatives. One very practical question is: How good are these tests against wider (NP) families of alternatives? No attempt is made to answer this question in this article. A more practical question concerns the amount of computation involved in actually performing such tests. Some comments on this aspect are made in the next section.

6. Approximations to randomization tests

One should note that in testing various of the NP hypotheses of Section 3, when $n = 10$, the actual number of permutations called for could be (a)

$n! = 3\,628\,000$, (b) $2^n = 1024$ or (c) $(2^n)(n!) = 3\,715\,891\,200$, depending on the hypothesis at hand. This being the case, one immediately seeks some modification of the original form of the randomization tests when n is not very small. Several such modifications have been treated in the literature. Brief outlines of only three such modifications will be presented here. They are (a) random permutations, (b) matching moments and (c) central limit theorems on orbits. (A rigorous discussion of much of this material is in Puri and Sen (1971).)

6.1. *Random permutations*

Several authors (e.g., Dwass, 1957) have suggested utilizing a random sample $\gamma_1, \ldots, \gamma_m$ of permutations from the relevant set S. In this case, the randomized permutation statistic becomes

$$R^*(h(X_n)) = \sum_{r=1}^{m} u(h(X_n) - h(\gamma_r(X_n)))\,. \tag{6.1}$$

For several practical reasons, one may wish to exclude the identity permutation, e^*, from the random sample. That is, one chooses $\{\gamma_1, \ldots, \gamma_m\}$ to be a random sample (with replacement) from $S - \{e^*\}$. In this case, some of the analysis becomes more tractable. For (6.1) adapted to this sampling scheme, one has the following

THEOREM 6.1. *Under* H_0, (i) $R^*(h(X_n))$ *is distributed as* $(k^*)^{-1}\sum_{r=1}^{k^*} Z_r$ *where* k^* *is the cardinality of the set* S *and* Z_r *has the binomial law with parameters* (m, p_r) *with* $p_r = (r-1)/(k^*-1)$, (ii) $E[R^*(h(X_n))] = m/2 = \mu_m$, (iii) $V[R^*(h(X_n))] = (m/12)[6 - 3m + 2(m-1)(2k-1)/(k-1)] = \sigma_m^2$, *and* (iv) $[R^*(h(X_n)) - \mu_m]/\sigma_m$ *has approximately a normal distribution with* 0 *mean and unit variance* (*when* m *is large*).

This means that in performing the randomized permutation test with the statistic in (6.1), the critical region of the form: $\{R^*(h(X_n)) > C(\alpha, k^*, m)\}$, for large m, one may take $C(\alpha, k^*, m) \simeq \mu_m + \sigma_m \tau_\alpha$, where τ_α is the upper $100\alpha\%$ point of the standard normal d.f. There are other approaches too, but they will not be treated here.

6.2. *Matching moments*

Several authors (e.g., Hoeffding, 1952) have found that for several of the NP hypotheses (treated in Section 3), the distribution of the permutation statistic is asymptotically the same as some corresponding classical statistics. Actually, this asymptotic equivalence has not only been studied under the null hypothesis, but also under alternatives, and the latter leads one to the asymptotic optimality of such randomization tests. Consider the following example (cited in Kendall and Stuart, 1979, p. 501). For the bivariate independence problem, treated in Section 2, $h_1(\cdot) = \sum_{i=1}^{n} X_i Y_i$ is the B-Pitman function of interest. This function is equivalent to the sample correlation coefficient $h_2(\cdot) = r_n$. The permutational (conditional) moments of $h_2(\cdot)$ are (i) $E(r_n) = 0$, (ii) $V(r_n) = (n-1)^{-1}$, (iii) $E(r_n^3) = O(n^{-2})$ and (iv) $E(r_n^4) = 3(n^2-1)^{-1}[1 + O(n^{-1})]$. For large n, these moments correspond to those in the normal theory case when the population correlation coefficient is equal to 0. Thus, a permutation test based on $h_1(\cdot)$ or $h_2(\cdot)$ is approximately equal to the (parametric) test based on $T = r_n\{(n-2)/(1-r_n^2)\}^{1/2}$. A more comprehensive picture is given in Hoeffding (1952).

6.3. Central limit theorems on orbits

In view of the preceding developments, notably the construction of the randomization (permutation) tests statistics and the Neyman Structure Theorem (Theorem 2.4), one sees that the permutation tests are conditional tests. This means that all relevent calculations involve solely the S-orbit of the original data. Further, if $h(\cdot)$ is a B-Pitman function, one can relate the $h(\cdot)$ values on the orbit to 'Scores' such as ranks. In this process, as the critical region is constructed directly from the set of points in the orbit (excepting in the case of some genuinely distribution-free tests), standard statistical tables are of not much use in providing these exact critical values. Further, the cardinality of the orbit is generally a rapidly increasing function of n (such as 2^n, $n!$, $(p!)^n$). Thus, the amount of labor involved in the exact enumeration of the permutational (conditional) distributions needed to derive the permutational critical values may increase prohibitively with the increase in the sample size(s), and hence, for large sample sizes, one is forced to use some suitable approximations for these. Fortunately, during the last forty years, the permutational limit theorems have been studied very systematically by a host of researchers, and these provide satisfactory solutions to the majority of the cases. For various types of permutational statistics, large sample distribution theory, under the title of 'permutational central limit theorem' have been studied by Wald and Wolfowitz (1944), Noether (1949), Hoeffding (1951, 1952), Motoo (1957), Hájek (1961), among many others. We may refer to Chapter (by M. Ghosh) where these are discussed. Basically, asymptotic normality (or chi square distributional) results are available under quite general regularity conditions, and these provide simple asymptotic expressions for the critical values of the randomization test statistics. These also provide some asymptotic equivalence results on these randomization tests and some asymptotically distribution-free tests based on the asymptotic results (without using the permutational invariance structure). In this context, the pioneering work of Hoeffding (1952) deserves special mention. He studied not only the asymptotic behaviour of the randomization tests under the hypotheses of invariance, but also exhibited the asymptotic power-equivalence of a randomization test based on some parametric form of test statistic and the corresponding parametric test. Thus, if we have some optimal parametric test statistic (viz., locally most powerful, asymptotically most powerful etc.) and if we employ the same in the construction of a randomization test, then, the resulting randomization test will be asymptotically optimal in the same sense. For invariant tests, the asymptotic equivalence-results are one-step further generalization in the sense that the optimal invariant test statistics may be the projections of the optimal unconditional test statistics into the space of the maximal invariants, and hence, the asymptotic equivalence results demand that the proportional loss in the conditional arguments converges to 0 as the sample sizes increase. For rank tests, this theory has been developed in Terry (1952), Hájek and Šidák (1967, pp. 63–71), and some further developments are sketched in Sen (1981a).

In the final section, we briefly discuss the issue of invariance of permutation tests.

7. Invariance and permutation tests

An empirical study of the NP literature (both applied and theoretical) would lead one to conclude that rank tests are studied significantly more than randomization tests. One reason is, of course, the problem of computing permutation tests. Another reason is related to the fact that for many NP problems, the rank statistics are invariant with respect to the relevant natural groups of transformations, while the permutation statistics are, in general, not invariant with respect to these groups of transformations. Bell and Kurotschka (1971) and Junge (1976) have investigated this question in some detail. Before presenting their approach, we consider the following

EXAMPLE 7.1. Consider H_0: $F_1 = F_2 = F_3 = F$ (unknown), where F is strongly increasing continuous. Let $g(\cdot)$ be a strictly increasing, continuous transformation of the real line $\mathbb{R}$ onto itself, and let $G'' = \{g'': g''(z) = [g(x_1), g(x_2), g(x_3)]\}$, where $z = (x_1, x_2, x_3)$. Let $A_1 = \{x_1 < x_2 < x_3\}$, $A_2 = \{x_2 < x_1 < x_3\}, \ldots, A_6 = \{x_3 < x_2 < x_1\}$; and $h^*(z) = r$ if $z \in A_r$, $=0$ if $z \in R_3 - \bigcup_1^6 A_r = D^*$. It is clear that each A_r is invariant wrt G''; that $\{A_r\}$ is a maximal (essential) similar partition of R_3; and, hence, that this partition is also invariant wrt G''. Further, it is clear that $h^*(\cdot)$ is a B-Pitman function wrt G'' and $G'' = G_3$. In this case, the permutation statistic $R(h^*(z))$ is a rank statistic, and, is necessarily invariant wrt G''.

If one considers only statistics invariant wrt G'', then one will consider only rank statistics. This restriction may be quite acceptable for many purposes, but one is certain to never achieve, say, the MP NPDF test vs. normal regression alternatives if one considers only rank tests.

The questions then arises: What is the 'natural' group of transformations for a statistics problem? What are its essential properties?

DEFINITION 7.1. Let Ω'' be a family of distributions (on some sample space χ). A group G'' of 1–1 transformations (of χ onto itself) is called an Ω''-generating group, if $\{J(g''(\cdot)): g'' \in G''\} = \Omega''$ for each $J(\cdot)$ in Ω''.

One notes immediately, that in Example 7.1, G'' is an Ω''-generating group.

If a family Ω'' admits a group G'' as in Definition 7.1 above and a finite set S'' with the usual properties, then one can establish some useful results, as in Bell and Kurotschka (1971), Bell (1964) and Wijsman (1957).

THEOREM 7.1. *Let Ω'' admit both an Ω''-generating group G'' of transformations, and a finite permutation set S'' of k^* elements with $S''(z)$ being a sufficient statistic. Then*

(i) *each set A invariant wrt G'' is similar wrt Ω'',*

(ii) *the set-valued random variable $G''(\mathbf{Z})$ is NPDF wrt Ω'', and assumes k^* distinct values each with probability $[k^*]^{-1}$;*

(iii) *for a.a. z, $\{G''(y)\colon y \in S''(z)\}$ is a maximal (essential) similar partition (of the sample space χ);*

(iv) $\tau(z) = G''(z) \cap S''(z)$ *is* 1–1 *a.e.,*

(v) $\delta(z) = [S''(z), G''(z)]$ *is* 1–1 *a.e.,*

(vi) *$G''(z)$ and $S''(z)$ are independent, and*

(vii) *$G''(z)$ is M–S–N.*

On the basis of the above result, one might wish to define a 'natural' group of transformations as one satisfying Theorem 7.1.

Under the conditions of that theorem, one wishes to construct new groups of transformations as follows.

Let $h(\cdot)$ be a B-Pitman function wrt Ω'' and S'', and let $\{[R(h(z)) = s], s = 1, 2, \ldots, k^*\}$ be the associated maximal (essential) similar partition.

DEFINITION 7.2. (i) for $z \in \{R(h) = s\}$, let $g_h(z) = S''(g''(z)) \cap \{R(h) = s\}$; and for $z \in D^*(h) = \chi - \bigcup_1^{k^*} \{R(h) - s\}$, define $g_h(z)$ arbitrarily;

(ii) $G''_h = \{g_h(\cdot)\colon g'' \in G''\}$.

About this new set of transformations, one readily establishes.

THEOREM 7.2. *For each B-Pitman function $h(\cdot)$,*

(i) *G_h is a group of* 1–1 *transformation of $\chi - D^*(h)$ onto itself.*

(ii) *G_h is an Ω''-generating group; and*

(iii) *each statistic $W[R(h(z))]$ is invariant wrt G_h.*

Consequently,

(iv) *for each statistic, $T(z)$, NPDF wrt Ω'', there exists an Ω''-generating group wrt which $T(z)$ is essentially invariant.*

In this sense, then, similarity and invariance are equivalent properties.

References

[1] Basawa, I. V. and Rao, B. L. S. Prakasa (1980). *Statistical Inference for Stochastic Processes.* Academic Press, New York.

[2] Basu, A. P., Ghosh, J. K. and Sen, P. K. (1983). A unified way of deriving LMP rank tests from censored data. *J. Roy. Statist. Soc. Ser. B* **45**, 384–390.

[3] Basu, D. (1955). On statistics independent of a complete sufficient statistics. *Sankhyā* **15**, 377–380.

[4] Basu, D. (1958). On statistics independent of sufficient statistics. *Sankhyā* **18**, 223–226.

[5] Bell, C. B. (1960). On the structure of distribution-free statistics. *Ann. Math. Statist.* **31**, 703–709.

[6] Bell, C. B. (1964a). Some basic theorems of distribution-free statistics. *Ann. Math. Statist.* **35**, 150–156.

[7] Bell, C. B. (1964b). A characterization of multisample distribution-free statistics. *Ann. Math. Statist.* **35**, 735–738.

[8] Bell, C. B. (1965). Enige problemen in de verdelingsvrije statistiek. Rapport S340 Mathematisch Centrum, Amsterdam.

[9] Bell, C. B. (1970). Less methods non-parametriques et a distribution libre. Tech. Rep. Ctr. Res. Math. Univ. Montreal.

[10] Bell, C. B. (1982). Signal detection for spherically exchangeable stochastic processes. Tech. Rep. 4-82, Ser. Statist. Biostatist., San Diego State Univ.

[11] Bell, C. B. and Ahmad, R. (1982). Signal detection for stochastic processes with stationary independent symmetric increments. Tech. Rep. 2-82, Ser. Statist. Biostatist. San Diego State Univ.

[12] Bell, C. B. and Bellot, F. (1965). Nota sobre los estadisticos no parametricos de Pitman. *Trabajos de Estadis.* **16**, 25–39.

[13] Bell, C. B., Blackwell, D. and Breiman, L. (1960). On the completeness of order statistics. *Ann. Math. Statist.* **31**, 794–797.

[14] Bell, C. B. and Doksum, K. A. (1965). Some new distribution-free statistics. *Ann. Math. Statist.* **36**, 203–214.

[15] Bell, C. B. and Doksum, K. A. (1967). Distribution-free tests for independence. *Ann. Math. Statist.* **38**, 619–628.

[16] Bell, C. B. and Donoghue, J. (1969). Distribution-free tests of randomness. *Sankhya Ser. A* **31**, 157–176.

[17] Bell, C. B. and Haller, H. S. (1969). Bivariate symmetry tests: Parametric and nonparametric. *Ann. Math. Statist.* **40**, 259–269.

[18] Bell, C. B. and Kurotschka, V. (1971). Einige prinzipien zur behandlung nichtparameterischer hypothesen. In: *Studi di Probabilita, Statistica, e Ricera Operative in onore di Giuseppe Pompilj*, Oderrisi-Buggio, pp. 164–186.

[19] Bell, C. B. and Smith, P. J. (1969). Some nonparametric tests for the multivariate goodness of fit, multisample, independence and symmetry problems. In: P. R. Krishnaiah, ed., *Multivariate Analysis II.* Academic Press, New York, pp. 3–23.

[20] Bell, C. B. and Smith, P. J. (1972a). Some aspects of the concept of symmetry in nonparametric statistics. In: D. S. Tracy, ed., *Symmetric Functions in Statistics* (Symp. in honor of Paul Dwer). Univ. Windsor, pp. 143–181.

[21] Bell, C. B. and Smith, P. J. (1972b). Completeness theorems for characterizing distribution-free statistics. *Ann. Inst. Statist. Math.* **24**, 435–453.

[22] Bell, C. B., Woodroofe, M. and Avadhani, T. V. (1970). Some nonparametric tests for stochastic processes. In: M. L. Puri, ed., *Nonparametric Techniques in Statistical Inference.* Cambridge Univ. Press, New York, pp. 215–258.

[23] Berk, R. H. and Bickel, P. J. (1968). On invariance and almost invariance. *Ann. Math. Statist.* **39**, 1573–1576.

[24] Brown, G. W. and Mood, A. M. (1951). On median tests for linear models. In: *Proc. Second Berkeley Symp. Math. Statist. Prob.*, pp. 159–166.

[25] Capon, J. (1961). Asymptotic efficiency of certain locally most powerful rank tests. *Ann. Math. Statist.* **32**, 88–100.

[26] Chatterjee, S. K. (1966). A bivariate sign test for location. *Ann. Math. Statist.* **37**, 1771–1781.

[27] Chatterjee, S. K. and Sen, P. K. (1964). Nonparametric tests for the bivariate two-sample location problem. *Calcutta Statist. Assoc. Bull.* **13**, 18–58.

[28] Dwass, M. (1957). Modified randomization tests for nonparametric hypotheses. *Ann. Math. Statist.* **28**, 181–187.

[29] Edgington, E. S. (1980). *Randomization Tests*. Marcel Dekker, New York.

[30] Fisher, R. A. (1934). *Statistical Methods for Research Workers*. Oliver and Boyd, Edinburgh.

[31] Fraser, D. A. S. (1957). *Nonparametric Methods in Statistics*. Wiley, New York.

[32] Friedman, M. (1937). The use of ranks to avoid the assumption of normality implicit in the analysis of variance. *J. Amer. Statist. Assoc.* **32**, 675–701.

[33] Gerig, T. M. (1969). A multivariate extension of the Friedman's χ_r^2-test. *J. Amer. Statist. Assoc.* **64**, 1595–1608.

[34] Ghosh, M. N. (1954). Asymptotic distribution of serial statistics and applications to nonparametric tests of hypotheses. *Ann. Math. Statist.* **25**, 218–251.

[35] Hájek, J. (1961). Some extensions of the Wald–Wolfowitz–Noether Theorem. *Ann. Math. Statist.* **32**, 506–523.

[36] Hájek, J. and Šidák, Z. (1967). *Theory of Rank Tests*. Academic Press, New York.

[37] Hoeffding, W. (1951a). A combinatorial central limit theorem. *Ann. Math. Statist.* **22**, 558–566.

[38] Hoeffding, W. (1951b). Optimum nonparametric tests. *Proc. Second Berkeley Symp. Math. Statist. Prob.*, pp. 83–92.

[39] Hoeffding, W. (1952). The large sample power of tests based on permutations of observations. *Ann. Math. Statist.* **23**, 169–192.

[40] Jogdeo, K. (1966). On randomized rank score procedure of Bell and Doksum. *Ann. Math. Statist.* **37**, 1697–1702.

[41] Junge, K. (1976). Charakterisierunggassatze Virteilungsfreier Statistiken und Ahnlicker Mengen. *Diplomarbert, Univ. Gottingen.*

[42] Kendall, M. G. and Stuart, A. (1979). *The Advanced Theory of Statistics, Vol 2: Inference and Relationship*. Hafner Publ. New York.

[43] Lehmann, E. L. (1959). *Testing of Statistical Hypotheses*. Wiley, New York.

[44] Lehmann, E. L. and Scheffé, H. (1950). Completeness, similar regions and unbiased estimation, Part I. *Sankhyā* **10**, 305–340.

[45] Lehmann, E. L. and Scheffé, H. (1955). Completeness, similar regions and unbiased estimation, II. *Sankhyā* **15**, 219–236.

[46] Lehmann, E. L. and Stein, C. (1949). On the theory of some nonparametric hypotheses. *Ann. Math. Statist.* **20**, 28–45.

[47] Mielke, P. W., Berry, K. J., Brockwell, P. J. and Williams, J. S. (1981). A class of nonparametric tests based on multiresponse permutation procedures. *Biometrika* **68**, 720–724.

[48] Motoo, M. (1957). On the Hoeffding combinatorial central limit theorem. *Ann. Inst. Statist. Math.* **8**, 145–154.

[49] Noether, G. E. (1949). On a theorem by Wald and Wolfowitz. *Ann. Math. Statist.* **20**, 455–458.

[50] Pitman, E. J. G. (1937a). Significance tests which may be applied to samples from any population. *Suppl. JRSS* **IV**, 119–130.

[51] Pitman, E. J. G. (1937b). Significance tests which may be applied to samples from any population II. *Suppl. JRSS* **IV**, 225–232.

[52] Pitman, E. J. G., (1938). Significance tests which may be applied to samples from any population III. The Analysis of variance test. *Biometrika* **29**, 322.

[53] Puri, M. L. and Sen, P. K. (1971). *Nonparametric Methods in Multivariate Analysis*. Wiley, New York.

[54] Puri, M. L., Sen, P. K. and Gokhale, D. V. (1970). On a class of rank order tests for independence in multivariate distributions. *Sankhya Ser. A.* **32**, 271–298.

[55] Quade, D. (1967). Rank analysis of covariance. *J. Amer. Statist. Assoc.* **62**, 1187–1200.

[56] Robinson, J. (1980). An asymptotic expansion for permutation tests with several samples, *Ann. Statist.* **8** (4), pp. 851–864.

[57] Savage, I. R. (1962). *Bibliography of Nonparametric Statistics*. Harvard University Press, Cambridge, Massachusetts.

[58] Savage, I. R. (1969). Nonparametric Statistics: A personal review, *Sankhya Series A* **31**, 107–144.

[59] Scheffé, H. (1943). Statistical inference in the nonparametric case. *Ann. Math. Statist.* **14**, 305–332.

[60] Sen, P. K. (1967). Some nonparametric generalizations of Wilks' tests for H_{M}, H_{VC} and H_{MVC}, I. *Ann. Inst. Statist. Math.* **19**, 451–467.

[61] Sen, P. K. (1968). On a class of aligned rank order tests in two-way layouts. *Ann. Math. Statist.* **39**, 1115–1124.

[62] Sen, P. K. (1969). Nonparametric tests for multivariate interchangeability. Part II. The problem of MANOVA in two-way layouts. *Sankhya Ser. A* **31**, 145–156.

[63] Sen, P. K. (1981a). On invariance principles for LMP conditional test statistics. *Calcutta Statist. Assoc. Bull.* **30**, 41–56.

[64] Sen, P. K. (1981b). *Sequential Nonparametrics: Invariance Principles and Statistical Inference.* Wiley, New York.

[65] Sen, P. K. (1982). The UI-principle and LMP rank tests. In: B. V. Gnedenko et al., eds., *Colloquia Mathematica Societatis Janos Bolyai 32: Nonparametric Statistical Inference.* North-Holland, Amsterdam, pp. 843–858.

[66] Sen, P. K. (1983). On permutational central limit theorems for general multivariate linear rank statistics. *Sankhya Ser. A* **45**, 141–149.

[67] Sen, P. K. and Puri, M. L. (1967). On the theory of rank order tests for location in the multivariate one-sample problem. *Ann. Math. Statist.* **38**, 1216–1228.

[68] Smith, P. J. (1969). Structure of nonparametric tests of some multivariate hypotheses. Ph.D. Thesis, Case Western Reserve Univ.

[69] Terry, M. E. (1952). Some rank order tests which are most powerful against specific parametric alternative. *Ann. Math. Statist* **23**, 346–366.

[70] Wald, A. and Wolfowitz, J. (1944). Statistical tests based on the permutations of the observations. *Ann. Math. Statist.* **15**, 358–372.

[71] Warner, J. E. (1969). Asymptotic properties of multivariate permutation tests with applications to signal detection. Ph.D. Thesis, Case Western Reserve Univ.

[72] Watson, G. S. (1957). Sufficient statistics, similar regions and distribution-free tests. *J. Roy. Statist. Soc. Ser. B* **19**, 262–267.

[73] Welch, B. L. (1937). On the z-test in randomized blocks and Latin squares. *Biometrika* **29**, 21–52.

[74] Wijsman, R. A. (1967). Cross-sections of orbits and their applications to densities of maximal invariants. *Proc. Fifth Berkeley Symp. Mata. Statist Probab.*, **1**, pp. 389–400.

[75] Wilks, S. S. (1946). Sample criteria for testing equality of means, equality of variances and equality of covariances in a normal multivariate distribution. *Ann. Math. Statist.* **17**, 257–281.

[76] Wilks, S. S. (1962). *Mathematical Statistics.* Wiley, New York.

P. R. Krishnaiah and P. K. Sen, eds., *Handbook of Statistics, Vol. 4*
© Elsevier Science Publishers (1984) 31–62 2

Univariate and Multivariate Multisample Location and Scale Tests

Vasant P. Bhapkar

1. Introduction

In the analysis of several samples from a number of populations statistical inference is generally based on the assumption of normality of underlying distributions. Thus, in the *analysis of variance* (ANOVA) of several independent univariate samples, the classical F-tests require normality, along with some other assumptions, for their validity. Similarly, in *multivariate analysis of variance* (MANOVA) of several multivariate samples, joint normality of the variables involved is usually a standard assumption.

However, in many applications, normality assumption might not be reasonable. If the form of underlying distributions is at least approximately known, then it is reasonable to base statistical inference procedures on the corresponding known parametric models. Such nonnormal parametric models include exponential and Weibull distribution models, for example, along with Poisson and binomial or multinomial models for grouped data or data in the form of counts. Such an approach is feasible only when a natural parametric model is available and it is reasonably tractable. Even in such situations, it is only for relatively simple problems that *exact* procedures happen to be available and in most realistic problems one has to be content with procedures based on large sample normality approximations (see e.g. Bhapkar, 1980).

Another option that is available is to study the performance of standard ANOVA and MANOVA procedures based on normality and other assumptions, when some of these assumptions are violated in their use. Such robustness studies have been extensively carried out especially for ANOVA problems. However, there is still a need for an alternative approach in the absence of appropriate robustness investigation or especially if such studies have pointed out lack of robustness property.

The third option is to adopt the general nonparametric approach. Here no specific distributional model is assumed for the populations to be studied, and some general broad assumptions are made as needed. While these broad assumptions are similar in nature to those made under the classical normal models, or under other parametric models, these are usually less stringent than the overall assumptions generally made under the parametric models. As a result, the procedures developed in this approach are more generally applicable, so far as validity is concerned, than those developed under the parametric models.

In this chapter we confine to the discussion of the third viz. the nonparametric approach under which the statistical procedures are either strictly distribution-free or at least asymptotically distribution-free. It turns out that this approach has the same feature as the first general parametric approach, viz. that usually one has to be content with large sample normality approximation based procedures, that are feasible for most realistic problems and it is only for relatively simple problems, especially in the univariate case, that *exact* distribution-free procedures happen to be available.

2. Location and scale models

Denote by $\boldsymbol{X}$ the column-vector of p variables $X^{(\alpha)}$, $\alpha = 1, \ldots, p$. Let $\boldsymbol{X}_i$ denote the random vector $[X_i^{(1)}, \ldots, X_i^{(p)}]'$ from the i-th population with distribution function (c.d.f.) F_i, $i = 1, \ldots, k$, k being the number of populations. Denote by $\boldsymbol{F} = [F_1, \ldots, F_k]$, the c.d.f.'s of the k populations.

Suppose now that $\boldsymbol{X}_{it}$, $t = 1, \ldots, n_i$, are independent and identically distributed (i.i.d.) random vectors with c.d.f. F_i, and denote by $\boldsymbol{X}_{(in_i)}$ the random sample $\{\boldsymbol{X}_{it}, t = 1, \ldots, n_i\}$ of size n_i from the i-th population, $i = 1, \ldots, k$. Assume that these k random samples are independent. Denote by $\boldsymbol{X}_{(n)}$ this collection of k independent random samples $\boldsymbol{X}_{(in_i)}$ from distributions $\boldsymbol{F} = [F_1, \ldots, F_k]$ respectively.

We shall say that the distributions $\boldsymbol{F}$ belong to the *location family* $\mathscr{F}_{\mathrm{L}}$ if

$$F_i(\boldsymbol{x}) = F(\boldsymbol{x} - \boldsymbol{\mu}_i), \quad i = 1, \ldots, k, \tag{2.1}$$

for some p-variate c.d.f. F; $\boldsymbol{\mu}_i = (\mu_i^{(1)}, \ldots, \mu_i^{(p)})'$ is then the vector of *location parameters* for F_i. The distributions $\boldsymbol{F}$ are said to belong to the *scale family* $\mathscr{F}_{\mathrm{S}}$ if

$$F_i(\boldsymbol{x}) = F(\boldsymbol{\theta}_i^{-1}\boldsymbol{x}), \quad i = 1, \ldots, k; \tag{2.2}$$

here $\boldsymbol{\theta}_i$ is a diagonal matrix with positive scale parameters $\boldsymbol{\theta}_i^{(\alpha)}$, $\alpha = 1, \ldots, p$, along the diagonal. The distributions are said to belong to the *location and scale family* $\mathscr{F}_{\mathrm{LS}}$ if

$$F_i(\boldsymbol{x}) = F[\boldsymbol{\theta}_i^{-1}(\boldsymbol{x} - \boldsymbol{\mu}_i)], \quad i = 1, \ldots, k. \tag{2.3}$$

For the univariate case ($p = 1$) we shall, of course, suppress the superscript α; then the general location and scale family of distributions is

$$F_i(x) = F\left(\frac{x - \mu_i}{\theta_i}\right), \quad i = 1, \ldots, k. \tag{2.4}$$

Thus, without loss of generality, we may assume $\theta_i = 1$ in (2.4) for the univariate location problem and $\boldsymbol{\theta}_i = \boldsymbol{I}$ in (2.3) for the multivariate location problem. Similarly, for the multivariate scale family, we may assume without loss of generality $\boldsymbol{\mu}_i = \boldsymbol{0}$ in (2.3).

3. Univariate several-sample rank statistics

Suppose that the random observations X_{it}, $t = 1, \ldots, n_i$, $i = 1, \ldots, k$ are arranged in increasing order. If the underlying distributions F_i are continuous, then the probability of any tie is zero. Let R_{it} be the *rank* of X_{it} in this increasing sequence of N random observations in the combined sample of size $N = \Sigma_i n_i$. Thus

$$R_{it} = \sum_{j=1}^{k} \sum_{u=1}^{n_j} \Psi(X_{it} - X_{ju}), \tag{3.1}$$

where Ψ is a function of real variable y such that $\Psi(y) = 1$ for $y \geq 0$, and $=0$ for $y < 0$.

Let $\boldsymbol{R}_i$ denote the column-vector of ranks R_{it} of the i-th random sample, and denote by

$$\boldsymbol{R} = [\boldsymbol{R}_1', \boldsymbol{R}_2', \ldots, \boldsymbol{R}_k']' \tag{3.2}$$

the column-vector of ranks of the k random samples. $\boldsymbol{R}$ will be referred to as the *rank order vector.* A (measurable) function of $\boldsymbol{R}$ will be referred to as a *rank statistic.*

If the k populations are homogeneous, i.e. $F_1 = F_2 = \cdots = F_k = F$ say, then $P[\boldsymbol{R} = \boldsymbol{r}] = 1/N!$ for every permutation vector $\boldsymbol{r}$ of the first N natural numbers, for all continuous univariate c.d.f. F. Thus the distribution of any rank statistic is strictly distribution-free in the homogeneous case.

Two types of rank statistics have been mainly proposed as test criteria for hypotheses concerning the underlying populations especially with reference to the location and scale parameters. The first type of rank statistic is based on *linear rank statistics,* while the second is based on *generalized U-statistics.*

In either case, we begin with a rank statistic, S_i, that compares the i-th sample with respect to the remaining samples. This comparison is usually aimed at either the location parameters of the corresponding populations, or their scale parameters. The fact that these statistics S_i, $i = 1, 2, \ldots, k$, are comparing k samples among themselves is characterized by the property that these k S_i's are subject to one linear constraint leaving only $k - 1$ S's functionally independent.

One then considers a suitable nonnegative over-all function of S_i's, or rather their deviations from the expected values, under the null hypothesis being tested. Although usually the over-all statistic happens to be strictly distribution-free under the homogeneity hypothesis, the form of the over-all statistic is chosen in such a way that in practice with large samples (and hopefully even for moderate size samples) its null-distribution is closely approximated by one of the 'standard' probability distributions for which extensive tabulations are available.

In order to develop the relevant asymptotic theory, we need some regularity assumptions $\mathcal{A}$; we now introduce the subscript $\boldsymbol{n}$ to denote the vector of sample sizes.

$\mathcal{A}$:

(i) The sample sizes $n_i \to \infty$ in such a way that $n_i/N \to p_i$, $0 < p_i < 1$, for all $i = 1, \ldots, k$.

(ii) The $S_{i\boldsymbol{n}}$'s are subject to the linear constraint

$$\sum_{i=1}^{k} a_{in} S_{in} = \eta_{\boldsymbol{n}}, \tag{3.3}$$

where $\Sigma_i a_{in} \equiv 1$ and $\eta_{\boldsymbol{n}}$ is a suitable constant.

Assume that $a_{in} \to a_i$, $i = 1, \ldots, k$ and $\eta_{\boldsymbol{n}} \to \eta$ under condition (i).

(iii) Let $\boldsymbol{S}' = [S_1, S_2, \ldots, S_k]$, $\mathcal{F}$ be a class of continuous distributions $\boldsymbol{F} = [F_1, F_2, \ldots, F_k]$, and that

$$N^{1/2}[\boldsymbol{S}_{\boldsymbol{n}} - \boldsymbol{\eta}_{\boldsymbol{n}}(\boldsymbol{F})] \xrightarrow{\text{L}} \mathcal{N}(\boldsymbol{0}, \boldsymbol{T}(\boldsymbol{F})) \tag{3.4}$$

for suitable functionals $\boldsymbol{\eta}_{\boldsymbol{n}}$ and $\boldsymbol{T}$ of $\boldsymbol{F}$, as $n_i \to \infty$ under (i). Here L denotes convergence in distribution, $\mathcal{N}$ denotes a normal vector and $\boldsymbol{T}(\boldsymbol{F})$ is a positive semi-definite matrix of rank $k-1$. Assume that $\boldsymbol{\eta}_{\boldsymbol{n}}(\boldsymbol{F}) \to \boldsymbol{\eta}(\boldsymbol{F})$ under condition (i).

(iv) For $\boldsymbol{F}$ in $\mathcal{F}_0 = \{F \in \mathcal{F} \mid F_1 = F_2 = \cdots = F_k = F, \text{ say}\}$,

$$\boldsymbol{\eta}_{\boldsymbol{n}}(\boldsymbol{F}) \equiv \boldsymbol{\eta}_{\boldsymbol{n}}(F) = \eta_{\boldsymbol{n}} \boldsymbol{j}, \qquad \boldsymbol{T}(\boldsymbol{F}) \equiv \boldsymbol{T}(F) = \boldsymbol{T}, \tag{3.5}$$

where $\boldsymbol{j} = [1, 1, \ldots, 1]'$ and $\eta_{\boldsymbol{n}}$ is defined in (ii).

Remarks on $\mathcal{A}$. (a) Since $\Sigma_i a_{in} \equiv 1$, we have $\Sigma_i a_i = 1$. Letting $\boldsymbol{a} = [a_1, a_2, \ldots, a_k]'$, in view of (3.3) and (3.4) we have, for every $\boldsymbol{F}$ in $\mathcal{F}$,

$$\sum_{i=1}^{k} a_i \eta_i(F) = \eta, \qquad \boldsymbol{T}(\boldsymbol{F})\boldsymbol{a} = \boldsymbol{0}, \tag{3.6}$$

where $\boldsymbol{\eta}(\boldsymbol{F}) = [\eta_1(F), \ldots, \eta_k(\boldsymbol{F})]'$. In particular, for $\boldsymbol{F}$ in $\mathcal{F}_0$, $\boldsymbol{Ta} = \boldsymbol{0}$, in view of (3.5), while $\Sigma_i a_i \eta_i = \eta$ is automatically satisfied for $\eta_i = \eta$.

(b) Assumptions (iii) and (iv) only require that the 'asymptotic mean' $\boldsymbol{\eta}_n(\boldsymbol{F})$ of $\boldsymbol{S}$, and the 'asymptotic covariance matrix' $\boldsymbol{T}(\boldsymbol{F})$ of $N^{1/2}\,\boldsymbol{S}$ to be distribution-free in $\mathscr{F}_0$. This asymptotic distribution-free characteristic of $\boldsymbol{S}$ will be guaranteed automatically if $\boldsymbol{S}$ is, in fact, strictly distribution-free in $\mathscr{F}_0$.

(c) $\boldsymbol{\eta}_n(\boldsymbol{F}) \equiv \boldsymbol{\eta}(\boldsymbol{F})$ and $\eta_{\boldsymbol{n}} \equiv \eta$ for all $\boldsymbol{n}$ for many rank statistics. For instance, this is the case with those based on generalized U-statistics (see Section 6). This is also true for statistics based on linear rank statistics (see Section 5) for location parameters with scores defined in a *skew-symmetric* manner.

4. Univariate rank statistics for homogeneity

The rank statistics that have been proposed in the literature (see, e.g., Kruskal and Wallis, 1952, 1953; Bhapkar, 1961; Puri, 1964; Bhapkar and Deshpande, 1968; Puri and Sen, 1971) for homogeneity of several populations have one of the general forms

$$T_0 = N(\boldsymbol{S} - \eta_{\boldsymbol{n}}\boldsymbol{j})'\boldsymbol{T}^-(\boldsymbol{S} - \eta_{\boldsymbol{n}}\boldsymbol{j})\,, \tag{4.1}$$

or

$$T_0^* = N(\boldsymbol{S} - \eta\boldsymbol{j})'\boldsymbol{T}^-(\boldsymbol{S} - \eta\boldsymbol{j})\,. \tag{4.2}$$

Of course, these forms coincide whenever $\eta_{\boldsymbol{n}} \equiv \eta$.

Here $\boldsymbol{T}^-$ is any g-inverse of $\boldsymbol{T}$, i.e. a matrix that satisfies the property that $\boldsymbol{T}\boldsymbol{T}^-\boldsymbol{T} = \boldsymbol{T}$. $\boldsymbol{T}$ is determined by the functions used for $\boldsymbol{S}$, and this determination is in terms of the limits p_i of the sampling fractions n_i/N. In practice, with finite size samples, p_i is substituted by the actual fraction n_i/N; similarly, the limit a_i is replaced by the actual constant a_{in} so far as relations (3.6) are concerned.

Thus, we have $(\boldsymbol{S} - \eta_{\boldsymbol{n}}\boldsymbol{j})'\boldsymbol{a} = 0$, in view of (3.3). Thus $\boldsymbol{S} - \eta_{\boldsymbol{n}}\boldsymbol{j}$ belongs to the column-space of $\boldsymbol{T}$ and, hence, the quadratic form (4.1) is invariant under any choice of g-inverse of $\boldsymbol{T}$. If $\eta_{\boldsymbol{n}} \neq \eta$, then the form (4.2) is not necessarily invariant in the strict sense for every g-inverse, although the invariance would hold in the asymptotic sense. T_0^* could be defined, in any case, for a particular g-inverse (see e.g. (4.3)).

The statistic T_0 is used as a test criterion to reject the hypothesis of homogeneity

$$\mathscr{H}_0\colon\quad F_1 = F_2 = \cdots = F_k = F \ \text{ say}\,,$$

for sufficiently large values of T_0, say for $T_0 > T_{0,\alpha}$ at level of significance α. The distribution-free property of the rank statistic T_0 for $\boldsymbol{F} \in \mathscr{F}_0$ guarantees that $T_{0,\alpha}$ is a constant independent of F. Theoretically, $T_{0\alpha}$ can be determined by considering the $N!$ equally likely values of T_0 corresponding to distinct permutation vectors $\boldsymbol{r}$ of $(1, 2, \ldots, N)'$. In practice one need consider only the $N!/n_1! \ldots n_k!$ distinct combinations for samples of size n_i.

For large number of combinations, one uses the approximation provided by

$\chi^2_{k-1,\alpha}$, the upper 100α % critical value of chi-squared distribution with $k-1$ degrees of freedom (df). This approximation is justified in view of the next theorem.

THEOREM 4.1. *Assume the regularity conditions $\mathcal{A}$ for univariate random samples $\boldsymbol{X}_{(n)}$. If $\boldsymbol{F} \in \mathcal{F}_0$, then $T_{0n} \xrightarrow{L} \chi^2_{k-1}$. Furthermore, in case $\eta_n \neq \eta$, if $\eta_n - \eta = o(N^{-1/2})$, then $T^*_{0n} \xrightarrow{L} X^2_{k-1}$ when $\boldsymbol{F} \in \mathcal{F}_0$.*

PROOF. The proof for T_{0n} follows immediately from (3.4), (3.5) and (4.1) since $\boldsymbol{T}$ has rank $k-1$.

In case $\eta_n \neq \eta$, $N^{1/2}(\eta_n - \eta) \to 0$ and, thus, $N^{1/2}(\boldsymbol{S}_n - \eta \boldsymbol{j})$ has the same limiting distribution as $N^{1/2}(\boldsymbol{S}_n - \eta_n \boldsymbol{j})$, which is $\mathcal{N}(\boldsymbol{0}, \boldsymbol{T})$ when $\boldsymbol{F} \in \mathcal{F}_0$. Then the limiting distribution of T^*_{0n} is χ^2_{k-1}. □

The T_0-test (or, equivalently, the T^*_0-test) is seen to be consistent against all alternatives $\boldsymbol{F}$ for which $\boldsymbol{\eta}(\boldsymbol{F}) \neq \eta \boldsymbol{j}$, in case $\boldsymbol{\eta}_n(\boldsymbol{F}) \equiv \boldsymbol{\eta}(\boldsymbol{F})$ for all $\boldsymbol{n}$. This assertion depends on the following lemma:

LEMMA 4.1. *Suppose $N^{1/2}(\boldsymbol{Y}_N - \boldsymbol{\xi}_N) \xrightarrow{L} \mathcal{N}(\boldsymbol{0}, \boldsymbol{\psi})$ as $N \to \infty$ and $Q_N = N(\boldsymbol{Y}_n - \boldsymbol{\delta}_N)'\boldsymbol{A}(\boldsymbol{Y}_N - \boldsymbol{\delta}_N)$, where $\boldsymbol{A}$ is positive definite. Assume that the constants $\boldsymbol{\xi}_N$, $\boldsymbol{\delta}_N$, $\boldsymbol{\xi}$ and $\boldsymbol{\delta}$ satisfy the conditions $N^{1/2}(\boldsymbol{\xi}_N - \boldsymbol{\xi}) = \boldsymbol{0}(1)$, $N^{1/2}(\boldsymbol{\delta}_N - \boldsymbol{\delta}) = \boldsymbol{0}(1)$ as $N \to \infty$. Then $Q_n \xrightarrow{p} \infty$, in the sense that $P[Q_N > c] \to 1$, as $N \to \infty$, for every fixed c, if and only if $\boldsymbol{\xi} \neq \boldsymbol{\delta}$.*

In order to establish consistency of T_0, note that $\boldsymbol{T} \equiv [\tau_{ij}]$ is positive semi-definite of rank $k-1$. Then without loss of generality we may assume that $\boldsymbol{T}_{11}$ is positive definite, where $\boldsymbol{T}_{11}$ is cofactor of τ_{11} in $\boldsymbol{T}$. In view of invariance of T_0 under any choice of g-inverse of $\boldsymbol{T}$, we have

$$T_0 = N(\boldsymbol{S}^* - \eta \boldsymbol{j})'\boldsymbol{T}_{11}^{-1}(\boldsymbol{S}^* - \eta \boldsymbol{j}), \tag{4.3}$$

where $S' = (S_1, S^{*\prime})$. Letting $\boldsymbol{\eta}'(\boldsymbol{F}) = [\boldsymbol{\eta}_1(\boldsymbol{F}), \boldsymbol{\eta}^{*\prime}(\boldsymbol{F})]$, we note from (4.3) that $T_{0n} \xrightarrow{p} \infty$ whenever $\boldsymbol{\eta}^*(\boldsymbol{F}) \neq \eta j$, in view of Lemma 4.1. However, $\boldsymbol{\eta}^*(\boldsymbol{F}) = \eta \boldsymbol{j}$ if and only if $\boldsymbol{\eta}(\boldsymbol{F}) = \boldsymbol{j}$ as a consequence of relation (3.6). Thus it follows that $T_{0n} \xrightarrow{p} \infty$ whenever $\boldsymbol{\eta}(F) \neq \boldsymbol{nj}$. Thus we have established

THEOREM 4.2. *Under the regularity assumptions $\mathcal{A}$ for univariate random samples $\boldsymbol{X}_{(n)}$, if $\boldsymbol{\eta}_n(\boldsymbol{F}) \equiv \boldsymbol{\eta}(\boldsymbol{F})$, then the T_0-test is consistent against all alternatives $\boldsymbol{F}$ in $\mathcal{F}$ for which $\boldsymbol{\eta}(\boldsymbol{F}) \neq \eta \boldsymbol{j}$.*

In order to establish consistency of T_0 and T^*_0 tests for the case where $\boldsymbol{\eta}_n(\boldsymbol{F}) \neq \boldsymbol{\eta}(\boldsymbol{F})$, we assume the condition that

$$\boldsymbol{\eta}_n(\boldsymbol{F}) - \boldsymbol{\eta}(\boldsymbol{F}) = \mathbf{o}(N^{-1/2}) \tag{4.4}$$

for all $\boldsymbol{F}$ in $\mathscr{F}$. For $\boldsymbol{F}$ in $\mathscr{F}_0$, the condition (4.4) ensures $\eta_n - \eta = o(N^{-1/2})$, which was used as a condition for T_0 and T_0^* to have the χ^2_{k-1} limiting distribution under $\mathscr{H}_0$: $F \in \mathscr{F}_0$, in Theorem 4.1.

Under the condition (4.4), $N^{1/2}[\boldsymbol{\eta}_n(\boldsymbol{F}) - \boldsymbol{\eta}(\boldsymbol{F})] \to \boldsymbol{0}$ and, then, $N^{1/2}[\boldsymbol{S}_n - \boldsymbol{\eta}(\boldsymbol{F})]$ has the limiting distribution $\mathcal{N}(\boldsymbol{0}, \boldsymbol{T}(\boldsymbol{F}))$ in view of (3.4). Arguing as in the proof of Theorem 4.2, the consistency of T_0-test follows for $\boldsymbol{F}$ for which $\boldsymbol{\eta}(\boldsymbol{F}) \neq \eta \boldsymbol{j}$. Furthermore, the same is true for the T_0^*-test, since T_{0n} and T_{0n}^* have the same limiting distributions. Thus we have established

THEOREM 4.3. *Under the regularity assumptions $\mathscr{A}$ for univariate random samples $\boldsymbol{X}_{(n)}$, assume condition* (4.4). *Then the T_0 and T_0^*-tests are consistent against alternatives $\boldsymbol{F}$ for which $\boldsymbol{\eta}(\boldsymbol{F}) \neq \eta \boldsymbol{j}$.*

For the investigation of asymptotic power of T_0, as a test statistic for $\mathscr{H}_0$, one needs to consider a sequence of alternatives, say $\boldsymbol{F}_{(N)}$ in $\mathscr{F}_{(N)}$, converging to some $\boldsymbol{F}$ in $\mathscr{F}_0$ at a suitable rate. Thus, we have

THEOREM 4.4. *In addition to regularity assumptions $\mathscr{A}$ for univariate random samples $\boldsymbol{X}_{(n)}$, assume further the existence of sequence $\{\boldsymbol{F}_{(N)}\}$ in $\mathscr{F}_{(N)} \subset \mathscr{F}$, such that*

$$\boldsymbol{\eta}_n(\boldsymbol{F}_{(N)}) = \eta_n \boldsymbol{j} + N^{-1/2}\boldsymbol{\delta} + \boldsymbol{o}(N^{-1/2}) \tag{4.5}$$

and

$$\boldsymbol{T}(\boldsymbol{F}_{(N)}) = \boldsymbol{T} + \boldsymbol{o}(1)\,. \tag{4.6}$$

Then, under the sequence $\{\boldsymbol{F}_{(N)}\}$ of distributions

$$T_{0n} \xrightarrow{L} \chi^2(k-1, \xi)\,,$$

where the noncertrality parameter $\xi = \boldsymbol{\delta}' \boldsymbol{T}^- \boldsymbol{\delta}$. Furthermore if $\eta_n - \eta = o(N^{-1/2})$ then $T_{0n}^* \xrightarrow{L} \chi^2(k-1, \xi)$.

The proof is straightforward in view of relations (3.4), (4.5) and (4.6) and, hence, the details are omitted. Note here in view of the first relation in (3.6) that we have

$$\sum_{i=1}^{k} a_i \delta_i = 0\,. \tag{4.7}$$

Thus $\boldsymbol{\delta}$ belongs to the column-space of $\boldsymbol{T}$ and, hence, ξ is invariant under any choice of g-inverse of $\boldsymbol{T}$.

5. Univariate tests based on linear rank statistics

One important class of rank statistics T_0 (or T_0^*) is based on linear rank statistics. Here we take

$$S_i = \frac{1}{n_i}\sum_{t=1}^{n_i} s_{R_{it}} = \frac{1}{n_i}\sum_{u=1}^{N} s_u Z_{iu}\,, \quad i = 1, \ldots, k\,, \tag{5.1}$$

where R_{it} is the rank of X_{it} defined in (3.1), $\{s_u, u = 1, 2, \ldots, N\}$ is a system of *scores* and $Z_{iu} = 1$ or 0 according as the u-th smallest among the N $\{X_{jt}$, all $t, j\}$ is, or is not from the i-th sample.

The condition (3.3) is obviously satisfied for $a_{in} = n_i/N$; then $a_i = p_i$, $i = 1, \ldots, k$ and $\eta_n = \Sigma_{u=1}^N s_u/N = \bar{s}$, say. Let the function J_N be defined by

$$J_N\left(\frac{u}{N+1}\right) = s_u\,, \quad u = 1, \ldots, N\,, \tag{5.2}$$

with J_N constant valued over intervals $(u/(N+1), (u+1)/(N+1)]$, and suppose

$$\begin{aligned} &J(v) = \lim_{N\to\infty} J_N(v)\,, \quad 0 < v < 1\,, \qquad \eta = \int_0^1 J(v)\,\mathrm{d}v\,, \\ &H(x) = \sum_{i=1}^{k} p_i F_i(x)\,. \end{aligned} \tag{5.3}$$

Then it can be shown that the conditions $\mathscr{A}$(ii)–(iv) are satisfied, provided $\eta_n - \eta = \mathrm{o}(N^{-1/2})$ and some regularity conditions are satisfied for scores s_u (or functions J_N, J; see the chapter by Ghosh in this volume). Then, we have

$$\begin{aligned} &\eta_{ni}(\boldsymbol{F}) = \frac{1}{n_i}\sum_{u=1}^{N} s_u \xi_{iu}(\boldsymbol{F})\,, \qquad \eta_i(\boldsymbol{F}) = \int_{-\infty}^{\infty} J(H(x))\,\mathrm{d}F_i(x)\,, \\ &\boldsymbol{T} = \lambda[\boldsymbol{\Delta}_p^{-1} - \boldsymbol{J}]\,, \end{aligned} \tag{5.4}$$

where $\boldsymbol{\Delta}_p$ is a diagonal matrix with elements $p_1, \ldots, p_k$ along the diagonal, $\boldsymbol{J} = [1]$,

$$\xi_{iu}(\boldsymbol{F}) = P_{\boldsymbol{F}}[Z_{iu} = 1] \quad \text{and} \quad \lambda = \int_0^1 J^2(v)\,\mathrm{d}v - \eta^2\,. \tag{5.5}$$

After replacing p_i by n_i/N, the test-criterion T_0^*, given by (4.2), simplifies to

$$T_0^* = \frac{1}{\lambda}\sum_{i=1}^{k} n_i(S_i - \eta)^2\,; \tag{5.6}$$

in case $\eta_n \neq \eta$, one could consider alternatively T_0 on replacing η in (5.6) by $\eta_n = \bar{s}$.

Hajek and Sidak (1967) have considered a modified version of T_0, say T_0', as a test criterion for $\mathscr{H}_0$; here

$$T_0' = \frac{1}{\lambda_N} \sum_{i=1}^{k} n_i (S_i - \bar{s})^2 ,$$

where

$$\lambda_N = \frac{1}{N-1} \sum_{u=1}^{N} (s_u - \bar{s})^2 . \tag{5.7}$$

In view of (5.2), (5.3) and (5.5) we note that $\lambda_N \to \lambda$ and the limiting distributional properties of T_{0n}' are the same as those of T_{0n}.

It can be shown (Puri and Sen, 1971) that the conditions (4.5) and (4.6) hold for sequences of location and scale alternatives, under regularity assumptions.

Locally most powerful rank tests

Consider the class of criteria T_0 (or T_0^*, T_0') where the scores s_u are defined by the formula

$$s_u = E\left[\frac{\partial}{\partial\theta} \ln g_\theta(V_u)\right]_{\theta=\theta_0}, \quad u = 1, \ldots, N, \tag{5.8}$$

where g_θ is the probability density function (p.d.f.) of a distribution with c.d.f. G_θ, θ being a real-valued parameter, and $V_1 < V_2 < \cdots < V_N$ are the order statistics in a random sample of size N from the distribution with c.d.f. $G \equiv G_{\theta_0}$ for some specified θ_0.

For the case $k = 2$,

$$T_0 = \frac{Nn_1}{n_2\lambda} (S_1 - \eta_n)^2 = Z^2 , \text{ say} ; \tag{5.9}$$

thus the large-sample χ^2 test with 1 df, using criterion T_0, is equivalent to the two-sided standard normal test using $Z = (S_1 - \eta_n)(Nn_1/n_2\lambda)^{1/2}$. It is known (see, e.g., Capon, 1961; Hajek and Sidak, 1967) that the right tailed standard normal test using Z is the asymptotic version of the *locally most powerful rank* (l.m.p.r.) test for the hypothesis $\mathscr{H}_0$: $F_1 = F_2$ against *one-sided* alternatives

$$\mathscr{H}: \quad F_1(x) = G_{\theta_1}(x) , \quad F_2(x) = G_{\theta_2}(x) , \quad \theta_1 > \theta_2 . \tag{5.10}$$

For the location parameter θ, we have

$$G_\theta(x) = G(x - \theta) , \qquad g_\theta(x) = g(x - \theta)$$

with $\theta_0 = 0$. Then the l.m.p.r. test ($k = 2$) has scores

$$s_u = E\left[-\frac{g'(V_u)}{g(V_u)}\right], \quad u = 1, \ldots, N ; \tag{5.11}$$

here the expectation is with respect to distribution G and $g'(x) = [-\partial g(x-\theta)/\partial\theta]_{\theta=0} = [dg(x)/dx]$.

For the scale parameter θ, we have

$$G_\theta(x) = G\left(\frac{x}{\theta}\right), \qquad g_\theta(x) = \frac{1}{\theta} g\left(\frac{x}{\theta}\right)$$

with $\theta_0 = 1$. Then the l.m.p.r. test ($k = 2$) has scores essentially given by

$$s_u = E\left[-V_u \frac{g'(V_u)}{g(V_u)}\right], \qquad u = 1, \ldots, N. \tag{5.12}$$

Location tests based on linear rank statistics

Here we consider some specific T_0^* (or T_0') criteria of the type (5.6) or (5.7) based on linear rank statistics (5.1) for the hypothesis $\mathscr{H}_0$: $F_1 = F_2 = \cdots = F_k$ against *location alternatives* $\mathscr{H}_1$: $F_i(x) = F(x - \mu_i)$, $i = 1, \ldots, k$ for some F, with not all μ_i equal.

Consider the scores $s_u = E(Y_u)$, $u = 1, \ldots, N$ where $Y_1 < Y_2 < \cdots < Y_N$ are order statistics in a random sample of size N from a distribution which is symmetric around zero; then $\eta_n \equiv \eta = 0$. In this case, as pointed out in Remark 3 on the conditions $\mathscr{A}$ in Section 3, the criterion T_0^* coincides with T_0. Two such criteria are presented below.

Kruskal–Wallis H-statistic. Here $s_u = u/(N+1) - \frac{1}{2}$ for $u = 1, \ldots, N$, and thus $J_N(v) \equiv J(v) = v - \frac{1}{2}$ for $0 < v < 1$. Then $\eta_n = \bar{s} = 0 = \eta$ for all $\boldsymbol{n}$ and $\lambda = \frac{1}{12}$. Also $S_i = \Sigma_{t=1}^{n_i} R_{it}/n_i(N+1) - \frac{1}{2} = \{\bar{R}_i - (N+1)/2\}/(N+1)$, $\bar{R}_i$ being the average rank for the i-th sample, and we have

$$T_0 = T_0^* = \frac{12}{(N+1)^2} \sum_{i=1}^{k} n_i \left(\bar{R}_i - \frac{N+1}{2}\right)^2.$$

Using λ_N, instead of λ, from (5.7) we get

$$T_0' = \frac{12}{N(N+1)} \sum_{i=1}^{k} n_i \left(\bar{R}_i - \frac{N+1}{2}\right)^2 = H \quad \text{say}. \tag{5.13}$$

It is in this form, viz. (5.13), that the Kruskal–Wallis H-statistic is usually presented.

Observe that s_u here may be represented as $E(Y_u)$, where Y_u's are order statistics in a sample of size N from uniform distribution on $[-\frac{1}{2}, \frac{1}{2}]$, which is symmetrical around zero, leading to $\eta_n \equiv \eta = 0$.

For the case $k = 2$, the Z-version of the H-statistic is in terms of S_1 or, equivalently, in terms of the average rank $\bar{R}_1$ of the first sample. In this form, the test is referred to as the *Wilcoxon test* for two samples. This right-tailed test can be seen to be l.m.p.r. test for $\mathscr{H}_0$: $F_1 = F_2$ against one-sided *location alternatives* $\mathscr{H}_1$ with $\mu_1 - \mu_2 > 0$ for $G(x) = e^x/(1+e^x)$, $-\infty < x < \infty$, which is the c.d.f. of logistic distribution.

Normal scores statistic. Here the scores are given by (5.11), where $V_1 < V_2 < \cdots < V_N$ are the order statistics in a random sample of size N from the standard normal distribution with c.d.f. Φ. Then $s_u = E(V_u)$, $u = 1, \ldots, N$. Since the standard normal distribution is symmetric around zero, $\eta_n \equiv \eta = 0$. Note that here

$$s_u = E(V_u) = E[\Phi^{-1}(Y_u)] ;$$

for this system of scores, we have $J(v) = \Phi^{-1}(v)$, $0 < v < 1$, in view of (5.2) and (5.3). Hence $\lambda = 1$, and thus

$$T_0 = T_0^* = \sum_{i=1}^{k} n_i S_i^2 . \tag{5.14}$$

This test statistic or, rather, the T_0' version of the statistic is originally due to Fisher and Yates (1938). An asymptotically equivalent form due to Van der Waerden (1953) uses *approximate* scores $s_u = \Phi^{-1}(u/(N+1))$.

For the case $k = 2$, the right-tailed test in terms of $Z = S_1(Nn_1/n_2)^{1/2}$ is l.m.p.r. test for $\mathscr{H}_0$: $F_1 = F_2$ against one-sided *location alternatives* $\mathscr{H}_1$ with $\mu_1 - \mu_2 > 0$ for $G(x) = \Phi(x)$.

Scale tests based on linear rank statistics

Now we consider the problem of testing $\mathscr{H}_0$ for the scale family $\mathscr{F}_S$, given by (2.4) with $\mu_i = 0$. Thus, test criteria T_0 (or T_0^*, T_0') of type (5.6) or (5.7) can be constructed for testing the hypothesis $\mathscr{H}_0$ against *scale alternatives* $\mathscr{H}_2$: $F_i(x) = F(x/\theta_i)$, $i = 1, \ldots, k$, for some F, with not all θ_i equal. The following are some of the systems of scores used in literature:

$$\begin{aligned} &\text{(i)} \quad s_u = \left| \frac{u}{N+1} - \tfrac{1}{2} \right| , \\ &\text{(ii)} \quad s_u = \left(\frac{u}{N+1} - \tfrac{1}{2} \right)^2 , \\ &\text{(iii)} \quad s_u = \left[\Phi^{-1} \left(\frac{u}{N+1} \right) \right]^2 , \end{aligned} \tag{5.15}$$

for $u = 1, 2, \ldots, N$. In view of (5.3) the corresponding J-functions are $J(v) = |v - \frac{1}{2}|$, $(v - \frac{1}{2})^2$ and $[\Phi^{-1}(v)]^2$ for $0 < v < 1$. Several authors have essentially used system (i) for the case $k = 2$, (ii) corresponds to the statistic due to Mood (1954), while (iii) produces the normal scores test for scale parameters. A test that is asymptotically equivalent to the one based on the choice (iii) is produced by taking $s_u = E(V_u^2)$, where $V_1 < V_2 < \cdots < V_N$ are the order statistics in a random sample of size N from the standard normal distribution.

For the case $k=2$, the right-tailed Z-test, given by (5.9) with the + sign for the square root, happens to be a l.m.p.r. test for $\mathscr{H}_0$: $F_1 = F_2$ against one-sided *scale alternatives* $\mathscr{H}_2$ for which $\theta_1/\theta_2 > 1$ and $G = \Phi$, for the normal scores $s_u = E(V_u^2)$ or its approximate version (iii).

6. Univariate tests based on U-statistics

Another important class of rank statistics T_0 is based on U-statistics. Here we take

$$S_i = \frac{1}{n_1 n_2 \cdots n_k} \sum_{t_1=1}^{n_1} \sum_{t_2=1}^{n_2} \cdots \sum_{t_k=1}^{n_k} \phi_i(X_{1t_1}, X_{2t_2}, \ldots, X_{kt_k}), \tag{6.1}$$

where the *kernel* function ϕ_i is defined as

$$\phi_i(x_1, x_2, \ldots, x_k) = \phi(r_i), \tag{6.2}$$

r_i being the rank of x_i in the k-tuplet $\{x_1, \ldots, x_k\}$. The condition (3.3) is satisfied for $a_{in} \equiv a_i = 1/k$, and $\eta_n \equiv \eta = \Sigma_{j=1}^k \phi(j)/k$.

It can be shown under suitable regularity assumptions (see Chapter 8 by Ghosh in this volume; Bhapkar, 1961, 1966) that conditions $\mathscr{A}$ (iii)–(iv) are satisfied for

$$\begin{aligned} \eta_{in}(F) \equiv \eta_i(\boldsymbol{F}) &= \sum_{j=1}^{k} \phi(j)\nu_{ij}(\boldsymbol{F}), \\ \boldsymbol{T} &= \frac{\lambda}{(k-1)^2}[k^2\boldsymbol{\Delta}_p^{-1} - k\boldsymbol{q}\boldsymbol{j}' - k\boldsymbol{j}\boldsymbol{q}' + q\boldsymbol{J}]. \end{aligned} \tag{6.3}$$

Here $\boldsymbol{\Delta}_p$ is as defined in Section 5, $\boldsymbol{q} = [1/p_1, \ldots, 1/p_k]'$, $q = \Sigma_{i=1}^k 1/p_i$ and

$$\nu_{ij}(\boldsymbol{F}) = P_{\boldsymbol{F}}[R_i = j]; \tag{6.4}$$

in (6.4), R_i is the rank of X_i in the k-tuplet $(X_1, \ldots, X_k)$, X_i's being independently distributed random variables with c.d.f. F_i respectively. Also, in (6.3)

$$\begin{aligned} \lambda &= E_F[\phi_1(X_1, \ldots, X_k)\phi_1(X_{k+1}, \ldots, X_{2k})] - \eta^2, \\ &= \sum_{j=1}^{k}\sum_{l=1}^{k} \phi(j)\phi(l)\binom{k-1}{j-1}\binom{k-1}{l-1}B(j+l-1, 2k-j-l+1) - \eta^2; \end{aligned} \tag{6.5}$$

here the expectation is for i.i.d.r.v.'s X_i, except that $X_1 = X_{k+1}$. After replacing p_i by n_i/N, the criterion T_0 (or T_0^*) simplifies to (see, e.g., Bhapkar, 1961, 1966)

$$T_0 = \frac{1}{\lambda}\frac{(k-1)^2}{k^2}\left[\sum_{i=1}^{k} n_i(S_i-\eta)^2 - \frac{1}{N}\left\{\sum_{i=1}^{k} n_i(S_i-\eta)\right\}^2\right]$$

$$= \frac{1}{\lambda}\frac{(k-1)^2}{k^2}\sum_{i=1}^{k} n_i(S_i-\bar{S})^2\,, \tag{6.6}$$

where $\bar{S} = \Sigma\, n_i S_i / N$. It can be shown (see, e.g., Ghosh, 1982) that the conditions (4.5) and (4.6) hold for sequences of location and scale alternatives.

For *location* tests, Bhapkar and Deshpande (1968) have suggested T_0-tests, denoted by letters V, B, L and W corresponding to the following choices:

$$\begin{aligned} &\phi_V(1) = 1\,, \quad \phi_V(j) = 0 \quad \text{for } j = 2,\ldots,k\,,\\ &\phi_B(k) = 1\,, \quad \phi_B(j) = 0 \quad \text{for } j = 1,\ldots,k-1\,,\\ &\phi_L(1) = -1\,, \quad \phi_L(k) = 1\,, \quad \phi_L(j) = 0 \quad \text{for } j = 2,\ldots,k-1\,,\\ &\phi_W(j) = j\,. \end{aligned} \tag{6.7}$$

For the *scale* tests with common (possibly unknown) location μ in the family (2.4), the functions suggested corresponding to tests D and C are

$$\begin{aligned} &\phi_D(1) = \phi_D(k) = 1\,, \qquad \phi_D(j) = 0\,, \quad j = 2,\ldots,k-1\,,\\ &\phi_C(j) = \left|j - \frac{k+1}{2}\right|\,. \end{aligned} \tag{6.8}$$

The location tests V or B can also be used as criteria for testing the equality of scale parameters for asymmetrical distributions, e.g., those encountered in life studies or survival analysis. Reference may also be made to Chatterjee (1966) for scale test based on U-statistics.

Deshpande (1970) has considered the problem of determining the relative 'weights', say $\phi(j)/\Sigma_i\,\varphi(i)$, so as to maximize the efficiency (see Section 11) of the T_0^2-test for testing the hypothesis $\mathcal{H}_0$ against location (or scale) alternatives for the normal family.

For the special case $k = 2$, the T_0^2-statistic (6.5) (except those for which $\phi(j)$ is constant valued, e.g. for the D test) reduces essentially to the two-sided Mann–Whitney (1947) criterion. The Mann–Whitney U is defined as the number of pairs (X_{1t_1}, X_{2t_2}) for which $X_{2t_2} > X_{1t_1}$, and it can be shown that the Mann–Whitney test is equivalent to the Wilcoxon rank sum test for two samples.

7. Multivariate several-sample rank statistics

In the notation of Section 2, suppose $\boldsymbol{X}_{(\boldsymbol{n})}$ denotes independent samples $\{\boldsymbol{X}_{it}, t = 1,\ldots,n_i\}$ of p-vectors with distribution F_i, $i = 1,\ldots,k$, and $\boldsymbol{F} = (F_1,\ldots,F_k)$.

Suppose now $S_i^{(\alpha)}$ denotes a rank statistic comparing the i-th sample to the remaining samples with respect to variable $\alpha = 1, 2, \ldots, p$, for each $i = 1, 2, \ldots, k$. We assume that these statistics, then, satisfy constraints of the type (3.3), viz.

$$\sum_{i=1}^{k} a_{in} S_{in}^{(\alpha)} = \eta_n^{(\alpha)}, \quad \alpha = 1, \ldots, p .$$

Letting $\boldsymbol{S}_{in} = [S_{in}^{(1)}, \ldots, S_{in}^{(p)}]'$, $\boldsymbol{\eta}_n = [\eta_n^{(1)}, \ldots, \eta_n^{(p)}]'$, we assume the regularity conditions $\mathscr{C}$ which are multivariate analogs of conditions $\mathscr{A}$ in Section 3.
$\mathscr{C}$:

(i) Same as $\mathscr{A}$(i).

(ii) The $\boldsymbol{S}_{in}$'s are subject to the linear constraint

$$\sum_{i=1}^{k} a_{in} \boldsymbol{S}_{in} = \boldsymbol{\eta}_n , \tag{7.1}$$

where $\Sigma_i a_{in} \equiv 1$ and $\boldsymbol{\eta}_n$ is a constant vector. It is assumed that $a_{in} \to a_i$ and $\boldsymbol{\eta}_n \to \boldsymbol{\eta}$.

(iii) Let $\boldsymbol{S}_n' = (\boldsymbol{S}_{1n}', \boldsymbol{S}_{2n}', \ldots, \boldsymbol{S}_{kn}')$, $\mathscr{F}$ be a class of *nonsingular* continuous c.d.f.'s $\boldsymbol{F}$, and

$$N^{1/2}[\boldsymbol{S}_n - \boldsymbol{\eta}_n(\boldsymbol{F})] \xrightarrow{\mathrm{L}} \mathscr{N}(\boldsymbol{0}, \boldsymbol{T}(\boldsymbol{F})) , \tag{7.2}$$

where $\boldsymbol{\eta}_n(\boldsymbol{F}) \to \boldsymbol{\eta}(\boldsymbol{F})$ under condition (i).

(iv) For $\boldsymbol{F}$ in $\mathscr{F}_0 = \{\boldsymbol{F} \in \mathscr{F}: F_1 = F_2 = \cdots = F_k = F \text{ say}\}$,

$$\boldsymbol{\eta}_n(\boldsymbol{F}) \equiv \boldsymbol{\eta}_n(F) \equiv \boldsymbol{j} \otimes \boldsymbol{\eta}_n , \qquad \boldsymbol{T}(\boldsymbol{F}) = \boldsymbol{\Sigma} \otimes \boldsymbol{\Lambda}(F) , \tag{7.3}$$

where $\boldsymbol{A} \otimes \boldsymbol{B}$ is the Kronecker product matrix $[a_{ij}\boldsymbol{B}]$ for $\boldsymbol{A} = [a_{ij}]$, $\boldsymbol{\Sigma}$ is a $k \times k$ positive semi-definite (p.s.d.) matrix of rank $k-1$ such that $\boldsymbol{\Sigma a} = \boldsymbol{0}$, and $\boldsymbol{\Lambda}(F)$ is positive definite of order p.

Remarks on $\mathscr{C}$. (1) As in (3.6) we have

$$\sum_{i=1}^{k} a_i \boldsymbol{\eta}_i(\boldsymbol{F}) = \boldsymbol{\eta} \tag{7.4}$$

in view of $\mathscr{C}$(ii) and $\mathscr{C}$(iii).

(2) In condition (ii) we have required the same type of linear constraint, i.e. the same coefficients a_{in}, on statistics $S_i^{(\alpha)}$ for different variables $\alpha = 1, \ldots, p$. In particular, this assumes that the same type of statistic (e.g. linear rank statistic, U-statistic) is used for different variables. However, these comparisons between samples could be based on different systems for variables $\alpha = 1, \ldots, p$. Thus, with linear rank statistics, different scores systems $s_u^{(\alpha)}$ may be

used and, similarly, with U-statistics different kernels $\phi^{(\alpha)}$ may be used. In a specific application (see 'profile analysis' in Section 9) it will be convenient to use the same comparisons for different variables α (i.e. $s_u^{(\alpha)} \equiv s_u$ or $\phi^{(\alpha)} \equiv \phi$) leading to $\boldsymbol{\eta}_n = \eta_n \boldsymbol{j}$.

Rank statistics for homogeneity

The rank statistics for testing $\mathscr{H}_0$: $\boldsymbol{F} \in \mathscr{F}_0$ that have been proposed in the literature (see e.g., Sugiura, 1965; Bhapkar, 1966; Tamura, 1966; Puri and Sen, 1971) are of the forms (dropping $\boldsymbol{n}$ for convenience)

$$T_0 = N(\boldsymbol{S} - \boldsymbol{j} \otimes \boldsymbol{\eta}_n)'(\boldsymbol{\Sigma}^- \otimes \boldsymbol{L}^{-1})(\boldsymbol{S} - \boldsymbol{j} \otimes \boldsymbol{\eta}_n), \tag{7.5}$$

or T_0^* with $\boldsymbol{\eta}$ replacing $\boldsymbol{\eta}_n$ in (7.5). Here $\boldsymbol{L}$ is some consistent estimator of $\boldsymbol{\Lambda}(F)$ under $\mathscr{H}_0$. As in section (4), $T_0^* = T_0$ whenever $\boldsymbol{\eta}_n \equiv \boldsymbol{\eta}$; furthermore, T_0 is invariant under any choice of g-inverse $\boldsymbol{\Sigma}^-$, while T_0^* is invariant only asymptotically, in case $\boldsymbol{\eta}_n \neq \boldsymbol{\eta}$.

The use of T_0 (or T_0^*) as a large sample χ^2 criterion with $(k-1)p$ d.f. for testing $\mathscr{H}_0$ is justified in view of

THEOREM 7.1. *Assume conditions $\mathscr{C}$ and suppose $\boldsymbol{F} \in \mathscr{F}_0$. If $\boldsymbol{L} \xrightarrow{\text{P}} \boldsymbol{\Lambda}(F)$, then $T_{0n} \xrightarrow{\text{L}} \chi^2((k-1)p)$. If $\boldsymbol{\eta}_n \neq \boldsymbol{\eta}$ and $\|\eta_n - \boldsymbol{\eta}\| = \text{o}(N^{-1/2})$, then $T_{0n}^* \xrightarrow{\text{L}} \chi^2((k-1)p)$.*

The proof is straightforward and, hence, is deleted. The consistency property can be established as in the univariate case in proving the following combined multivariate version of Theorems 4.2 and 4.3:

THEOREM 7.2. *Assume conditions $\mathscr{C}$ and suppose $\|\boldsymbol{\eta}_n - \boldsymbol{\eta}\| = \text{o}(N^{-1/2})$. Then the T_0 (or T_0^*) test is consistent against all alternatives for which $\boldsymbol{\eta}(\boldsymbol{F}) \neq \boldsymbol{j} \otimes \boldsymbol{\eta}$.*

The asymptotic power is obtained next under a suitable sequence of alternatives:

THEOREM 7.3. *In addition to conditions $\mathscr{C}$, suppose that there exists sequence $\{\boldsymbol{F}_{(N)}\}$ in $\mathscr{F}_N \subset \mathscr{F}$, and F such that*

$$\boldsymbol{\eta}_n(\boldsymbol{F}_{(N)}) = \boldsymbol{j} \otimes \boldsymbol{\eta}_n + N^{-1/2}\boldsymbol{\delta}(F) + \mathbf{o}(N^{-1/2}) \tag{7.6}$$

and

$$\boldsymbol{T}(\boldsymbol{F}_{(N)}) = \boldsymbol{\Sigma} \otimes \boldsymbol{\Lambda}(F) + \mathbf{o}(1).$$

Then under the sequence of distributions $\{F_{(N)}\}$, $T_{0n} \xrightarrow{\text{L}} \chi^2((k-1)p, \xi)$, where

$$\xi = \boldsymbol{\delta}'(F)(\boldsymbol{\Sigma}^- \otimes \boldsymbol{\Lambda}^{-1}(F))\boldsymbol{\delta}(F). \tag{7.7}$$

If $\|\boldsymbol{\eta}_n - \boldsymbol{\eta}\| = \text{o}(N^{-1/2})$, then T_{0n}^* has the same limiting distribution.

8. Some specific multivariate T_0 (or T_0^*) criteria

An important class of statistics is based on linear rank statistics (see (5.1))

$$S_i^{(\alpha)} = \frac{1}{n_i} \sum_{t=1}^{n_i} s_{R_{it}(\alpha)}^{(\alpha)} = \frac{1}{n_i} \sum_{u=1}^{N} s_u^{(\alpha)} Z_{iu}^{(\alpha)}, \tag{8.1}$$

$i = 1, \ldots, k$; here $R_{it}^{(\alpha)}$ is the rank of $X_{it}^{(\alpha)}$ among $\{X_{jt}^{(\alpha)},\ t = 1, \ldots, n_j,\ j = 1, \ldots, k\}$ and $Z_{iu}^{(\alpha)} = 1$ or 0 according as the u-th smallest among $\{X_{jt}^{(\alpha)}$, all $t, j\}$ is, or is not, from the i-th sample.

Then $J_N^{(\alpha)}$, $J^{(\alpha)}$, $\eta^{(\alpha)}$, $H(\boldsymbol{x})$ are defined as in (5.2) and (5.3). Also $\eta_{\boldsymbol{\eta}}^{(\alpha)} = \bar{s}^{(\alpha)} = \Sigma\, s_u^{(\alpha)}/N$,

$$\eta_{ni}^{(\alpha)}(\boldsymbol{F}) = \frac{1}{n_i} \sum_{u=1}^{N} s_u^{(\alpha)} \xi_{iu}^{(\alpha)}(\boldsymbol{F}), \qquad \eta_i^{(\alpha)}(\boldsymbol{F}) = \int_{-\infty}^{\infty} J^{(\alpha)}(H^{(\alpha)}(y))\, \mathrm{d}F_i^{(\alpha)}(y),$$

$$\boldsymbol{\Sigma} = \boldsymbol{\Delta}_p^{-1} - \boldsymbol{J}, \tag{8.2}$$

$$\lambda_{\alpha\beta}(F) = \int_{-\infty}^{\infty} \int_{-\infty}^{\infty} J^{(\alpha)}[F^{(\alpha)}(y)] J^{(\beta)}[F^{(\beta)}(z)]\, \mathrm{d}F_{(y,z)}^{(\alpha,\beta)} - \eta^{(\alpha)}\eta^{(\beta)}.$$

In (8.2), $H^{(\alpha)}$, $F^{(\alpha,\beta)}$ refer to marginal c.d.f.'s for variables α, (α, β) respectively for distributions H, F etc., $\boldsymbol{\Lambda}(f) = [\lambda_{\alpha\beta}(F)]$ and

$$\xi_{iu}^{(\alpha)}(\boldsymbol{F}) = P_{\boldsymbol{F}}[Z_{iu}^{(\alpha)} = 1], \quad u = 1, \ldots, N. \tag{8.3}$$

It can be shown (see Puri and Sen, 1971) that, under suitable regularity assumptions, $\mathscr{C}$(iii) and (iv) are satisfied for $\boldsymbol{S_n}$ defined by linear rank statistics (8.1); furthermore, the conditions (7.6) are satisfied for suitable sequences of location and scale alternatives.

The estimates $\boldsymbol{L}$ of $\boldsymbol{\Lambda}(F)$ can be obtained on replacing $F^{(\alpha)}$, $F^{(\alpha,\beta)}$ in (8.2) by the corresponding sample distribution functions.

For the special case where $s_u^{(\alpha)} \equiv s_u$, we have $J^{(\alpha)} = J$, $\eta_{\boldsymbol{n}}^{(\alpha)} = \eta_{\boldsymbol{n}}$ and

$$\lambda_{\alpha\alpha}(F) = \int_{-\infty}^{\infty} J^2(F^{(\alpha)}(y))\, \mathrm{d}F^{(\alpha)}(y) - \eta^{2(\alpha)} = \int_0^1 J^2(v)\, \mathrm{d}v - \eta^2 = \lambda, \tag{8.4}$$

with λ defined as before in (5.5). Then $\boldsymbol{\Lambda}(F) = \lambda \boldsymbol{P}(F)$, where $\boldsymbol{P}(F)$ is some correlation matrix, and the T_0 criterion reduces to

$$T_0 = \frac{1}{\lambda} \sum_{i=1}^{k} n_i (\boldsymbol{S}_i - \eta_n \boldsymbol{j})' \boldsymbol{E}^{-1} (\boldsymbol{S}_i - \eta_n \boldsymbol{j}), \tag{8.5}$$

where $\boldsymbol{E}$ is a consistent estimate of the correlation matrix $\boldsymbol{P}(F)$, with diagonal elements of $\boldsymbol{E}$ equal to one. The T_0' analog replaces λ by λ_N as in (5.7), while T_0^* uses η instead of $\eta_{\boldsymbol{n}}$ in (8.5).

As in Section 5, the T_0^* and T_0 criteria coincide when $\boldsymbol{\eta}_n \equiv \boldsymbol{\eta}$, which is often the case with linear rank statistics for the location problem. In particular, for the multivariate analog of Kruskal–Wallis statistic (5.13) we have

$$T_0 = T_0^* = \frac{12}{(N+1)^2} \sum_{i=1}^{k} n_i \Big(\bar{\boldsymbol{R}}_i - \frac{N+1}{2}\boldsymbol{j}\Big)' \boldsymbol{E}^{-1} \Big(\bar{\boldsymbol{R}}_i - \frac{N+1}{2} j\Big); \tag{8.6}$$

here $s_u^{(\alpha)} \equiv s_u = u/(N+1) - \frac{1}{2}$, $\bar{\boldsymbol{R}}_i$ is the vector of average ranks for the i-th sample, and the T_0' analog is the multivariate version

$$H = \frac{12}{N(N+1)} \sum_{i=1}^{k} n_i \Big(\bar{\boldsymbol{R}}_i - \frac{N+1}{2}\boldsymbol{j}\Big)' \boldsymbol{E}^{-1} \Big(\bar{\boldsymbol{R}}_i - \frac{N+1}{2}\boldsymbol{j}\Big), \tag{8.7}$$

of the Kruskal–Wallis statistic (5.13).

Another class of rank statistics is based on U-statistics (see 6.1)

$$S_i^{(\alpha)} = \frac{1}{n_1 n_2 \cdots n_k} \sum_{t_1=1}^{n_1} \cdots \sum_{t_k=1}^{n_k} \phi_i^{(\alpha)}(\boldsymbol{X}_{1t_1}, \ldots, \boldsymbol{X}_{kt_k}), \tag{8.8}$$

where the kernel function $\phi_i^{(\alpha)}$ is defined as

$$\phi_i^{(\alpha)}(\boldsymbol{x}_1, \boldsymbol{x}_2, \ldots, \boldsymbol{x}_k) = \phi^{(\alpha)}(r_i^{(\alpha)}). \tag{8.9}$$

Here $r_i^{(\alpha)}$ is the rank of $x_i^{(\alpha)}$ in the k-tuplet $\{x_1^{(\alpha)}, x_2^{(\alpha)}, \ldots, x_k^{(\alpha)}\}$. Then we have, from (6.3) and (6.4),

$$\eta_{\boldsymbol{n}}^{(\alpha)} \equiv \eta^{(\alpha)} = \sum_{j=1}^{k} \frac{\phi^{(\alpha)}(j)}{k}, \qquad \eta_{i\boldsymbol{n}}^{(\alpha)}(\boldsymbol{F}) \equiv \eta_i^{(\alpha)}(\boldsymbol{F}) = \sum_{j=1}^{k} \phi^{(\alpha)}(j)\nu_{ij}^{(\alpha)}(\boldsymbol{F}),$$

$$\boldsymbol{\Sigma} = \frac{1}{(k-1)^2}[k^2\boldsymbol{\Delta}_p^{-1} - k\boldsymbol{qj}' - k\boldsymbol{jq}' + q\boldsymbol{J}], \tag{8.10}$$

in the notation of Section 6. Also

$$\nu_{ij}^{(\alpha)}(\boldsymbol{F}) = P_{\boldsymbol{F}}[R_i^{(\alpha)} = j], \tag{8.11}$$

with $R_i^{(\alpha)}$ the rank of $X_i^{(\alpha)}$ in the k-tuplet $\{X_1^{(\alpha)}, \ldots, X_k^{(\alpha)}\}$, $\boldsymbol{X}_i$'s being independent random vectors with c.d.f.'s F_i, respectively. Furthermore,

$$\lambda_{\alpha\beta}(F) = E[\phi_1^{(\alpha)}(\boldsymbol{X}_1, \ldots, \boldsymbol{X}_k)\phi_1^{(\beta)}(\boldsymbol{X}_{k+1}, \ldots, \boldsymbol{X}_{2k})] - \eta^{(\alpha)}\eta^{(\beta)} \tag{8.12}$$

where $\boldsymbol{X}_1, \ldots, \boldsymbol{X}_{2k}$ are i.i.d. vectors with c.d.f. F except that $\boldsymbol{X}_1 = \boldsymbol{X}_{k+1}$.

If the same kernel function ϕ is used for all variables α, we have $\eta^{(\alpha)} = \eta$ and

$$\lambda_{\alpha\alpha}(F) = E[\phi^2(R_1^{(\alpha)})] - \eta^2 = \lambda,$$

with λ defined as in Section 6. Then $\boldsymbol{\Lambda}(F) = \lambda \boldsymbol{P}(F)$, where $\boldsymbol{P}(F)$ is some correlation matrix; also it has been shown (see Bhapkar, 1966) that conditions $\mathscr{C}$(iii) and (iv) are satisfied for $\boldsymbol{S}_n$ defined by (8.8). The T_0-criterion then reduces to

$$T_0 = \frac{(k-1)^2}{k^2\lambda} \sum_{i=1}^{k} n_i(\boldsymbol{S}_i - \bar{\boldsymbol{S}})' \boldsymbol{E}^{-1}(\boldsymbol{S}_i - \bar{\boldsymbol{S}}), \tag{8.13}$$

where $\boldsymbol{E}$ is a consistent estimator of the correlation matrix $\boldsymbol{P}(\boldsymbol{F})$, with diagonal elements of $\boldsymbol{E}$ equal to one, and $\bar{\boldsymbol{S}} = \Sigma_i n_i \boldsymbol{S}_i / N$.

Unbiased and consistent estimator $[l_{\alpha\beta}] = \boldsymbol{L} = \lambda \boldsymbol{E}$ of $\boldsymbol{\Lambda}(F) = \lambda \boldsymbol{P}(F)$ has been suggested by Bhapkar (1966) as

$$l_{\alpha\beta} = \frac{\Sigma_{i=1}^{k} \Sigma_P \phi_i^{(\alpha)}(\boldsymbol{X}_{it_1}, \ldots, \boldsymbol{X}_{it_k}) \phi_i^{(\beta)}(\boldsymbol{X}_{it_{k+1}}, \ldots, \boldsymbol{X}_{it_{2k}})}{\Sigma_{i=1}^{k} n_i(n_i - 1) \cdots (n_i - 2k + 2)} - \eta^2,$$

where P denotes the sum over all permutations of $2k-1$ integers $(t_1, \ldots, t_{2k})$ chosen from $(1, 2, \ldots, n_i)$ with $t_i = t_{k+i}$; it is, of course, assumed that the first sum is over those i for which $n_i \geq 2k$. Also it is assumed that we have at least one i for which $n_i \geq 2k$.

It can be shown (see Puri and Sen, 1971) that conditions (7.6) are satisfied for sequences of location or scale alternatives.

9. Rank statistics for profile analysis

In MANOVA with k normal populations with unknown means $\boldsymbol{\mu}_i$, $i = 1, \ldots, k$, respectively, and common unknown nonsingular covariance matrix, the *population profiles* are said to be *parallel* if

$$\mu_i^{(1)} - \mu_1^{(1)} = \mu_i^{(2)} - \mu_1^{(2)} = \cdots = \mu_i^{(p)} - \mu_1^{(p)}, \tag{9.1}$$

for $i = 2, \ldots, k$. Test criteria are available (see, e.g., Morrison, 1976), based on the characteristic roots of a suitable determinantal equation, for testing such a hypothesis of parallelism of profiles. This hypothesis is of interest with *commensurable* variables, e.g. repeated measurements at different time points.

Discarding the distributional assumption of normality, Bhapkar and Patterson (1977) formulated a nonparametric analog of the hypothesis of parallelism, say $\mathscr{H}_1$, in the form

$$\mathscr{H}_1: \quad \nu_{ij}^{(1)}(\boldsymbol{F}) = \nu_{ij}^{(2)}(\boldsymbol{F}) = \cdots = \nu_{ij}^{(p)}(\boldsymbol{F}), \quad i, j = 1, \ldots, k, \tag{9.2}$$

for $\nu_{ij}^{(\alpha)}(\boldsymbol{F})$ defined by (8.11). For the location family $\mathscr{F}_{\mathrm{L}}$ (2.1), if the variables α

have the same marginal distribution, $\alpha = 1, \ldots, p$, then the condition (9.1) implies (9.2). For the scale family $\mathscr{F}_S$ (2.2) with the same marginal distribution for all variables α, Bhapkar (1979) has shown that the condition

$$\frac{\theta_i^{(1)}}{\theta_1^{(1)}} = \frac{\theta_i^{(2)}}{\theta_1^{(2)}} = \cdots = \frac{\theta_i^{(p)}}{\theta_1^{(p)}}, \quad i = 2, \ldots, k, \tag{9.3}$$

again implies (9.2). (9.3) may be interpreted as the condition of parallelism for parametric scalar profiles.

The formulation (9.2) is convenient for criteria based on U-statistics. Bhapkar and Patterson (1977) have proposed the criterion

$$T_1 = \frac{(k-1)^2}{k^2\lambda} \sum_{i=1}^{k} n_i(\boldsymbol{S}_i - \bar{\boldsymbol{S}})'[\boldsymbol{E}^{-1} - g\boldsymbol{E}^{-1}\boldsymbol{J}\boldsymbol{E}^{-1}](\boldsymbol{S}_i - \bar{\boldsymbol{S}}) = T_0 - T_2, \quad \text{say}, \tag{9.4}$$

in the notation of Section 8, for statistics $\boldsymbol{S}$ defined by (8.8) with the same kernel $\phi^{(\alpha)} = \phi$ for all variables α. Here $g = 1/\boldsymbol{j}'\boldsymbol{E}^{-1}\boldsymbol{j}$ and T_0 is the statistic (8.13) to be used as a $\chi^2((k-1)p)$ criterion for $\mathscr{H}_0$, while T_1 is proposed as a $\chi^2((k-1)(p-1))$ criterion for testing parallelism hypothesis formulated as $\mathscr{H}_1$. T_1 may be interpreted as a statistic testing *interaction* between populations and variables; if this interaction is absent and, thus, the profiles may be considered to be parallel, then T_2 may be used as a *conditional* criterion for testing that the profiles coincide, given that they are parallel.

For criteria based on linear rank statistics, Bhapkar (1982) has proposed the corresponding formulation of nonparametric analog of parallelism hypothesis, say

$$\mathscr{H}_1': \quad \xi_{iu}^{(1)}(\boldsymbol{F}) = \xi_{iu}^{(2)}(\boldsymbol{F}) = \cdots = \xi_{iu}^{(p)}(\boldsymbol{F}), \quad i = 1, \ldots, k, \quad u = 1, \ldots, N, \tag{9.5}$$

for $\xi_{iu}^{(\alpha)}(\boldsymbol{F})$ defined by (8.3). The corresponding T_1 criterion offered by Bhapkar is

$$T_1 = \frac{1}{\lambda} \sum_{i=1}^{k} (\boldsymbol{S}_i - \eta_n\boldsymbol{j})'[\boldsymbol{E}^{-1} - g\boldsymbol{E}^{-1}\boldsymbol{J}\boldsymbol{E}^{-1}](\boldsymbol{S}_i - \eta_n\boldsymbol{j}) = T_0 - T_2, \quad \text{say}, \tag{9.6}$$

where S's are linear rank statistics defined by (8.1), T_0 is (8.5) and $g = 1/\boldsymbol{j}'\boldsymbol{E}^{-1}\boldsymbol{j}$. As with U-statistics, T_1, T_2 are to be used as $\chi^2((k-1)(p-1))$ and $\chi^2((k-1))$ criteria for testing $\mathscr{H}_1'$ and $\mathscr{H}_0$, conditional on $\mathscr{H}_1'$ respectively. Alternatively, we could consider T_1^* (or T_1') on replacing η_n in (9.6) by η (or λ by λ_N in (5.7)).

The use of T_1 criteria (9.4) or (9.6) for testing $\mathscr{H}_1$ or $\mathscr{H}_1'$ respectively, is justified in view of Theorems 10.1 and 10.2 in the next section.

10. Rank statistics for subhypotheses

When the preliminary test of significance based on T_0 criterion (8.5) or (8.6) (or its versions T_0^*, T_0') indicates significant deviation from the hypothesis $\mathcal{H}_0$ of homogeneity, we are frequently interested in testing subhypotheses concerning differences either on a subgroup of variables or among specified subgroups of k populations. Similarly, if the T_1 criterion (9.4) or (9.6) (or its versions T_1^*, T_1') indicate significant deviations from parallelism hypotheses $\mathcal{H}_1$, given by (9.2), or $\mathcal{H}_1'$, given by (9.5), respectively, we would like to know whether such statements may reasonably be assumed to hold for specified subgroups of variables, possibly in conjunction with choice of specified groups of populations.

Such investigations can be undertaken by selecting appropriate $m \times k$ matrix $\boldsymbol{M}$ of rank $m < k$, and $p \times q$ matrix $\boldsymbol{Q}$ of rank $q \leqslant p$; $\boldsymbol{M}$ corresponds to the specified comparisons among k (or, possibly, only m) populations, while $\boldsymbol{Q}$ could correspond to the specified subgroup of q variables.

If the conditions $\mathscr{C}$ are met by statistics S, we have from (7.2)

$$N^{1/2}[\boldsymbol{M} \otimes \boldsymbol{Q}][\boldsymbol{S}_n - \boldsymbol{\eta}_n(\boldsymbol{F})] \xrightarrow{\text{L}} \mathcal{N}(\boldsymbol{0}, [\boldsymbol{M} \otimes \boldsymbol{Q}]\boldsymbol{T}(\boldsymbol{F})[\boldsymbol{M} \otimes \boldsymbol{Q}]'). \tag{10.1}$$

This leads to the statistic (see (7.5))

$$T_{\boldsymbol{M},\boldsymbol{Q}} = \frac{N}{\lambda}[\boldsymbol{S}_n - \eta_n\boldsymbol{j}]'[\boldsymbol{M} \otimes \boldsymbol{Q}]'[(\boldsymbol{M\Sigma M}')^{-1} \otimes (\boldsymbol{QEQ}')^{-1}]$$
$$\times [\boldsymbol{M} \otimes \boldsymbol{Q}][\boldsymbol{S}_n - \eta_n\boldsymbol{j}], \tag{10.2}$$

assuming that the same comparisons are made across all the variables, as in remark (2) in Section 7. It is assumed that $\boldsymbol{M\Sigma M}'$ is of full rank $m < k$. In (10.2) it is the understanding that for the case $m = k$, e.g. with $\boldsymbol{M} = \boldsymbol{I}_k$, the inverse of $\boldsymbol{M\Sigma M}'$ is to be understood as a g-inverse.

The statistic $T_{\boldsymbol{M},\boldsymbol{Q}}$ has been suggested (Bhapkar, 1982) as a large-sample $\chi^2(mq)$ criterion for $m < k$, and $\chi^2((k-1)q)$ criterion for $m = k$, for $q \leqslant p$.

Observe that $T_{\boldsymbol{I}_k,\boldsymbol{I}_p} = \boldsymbol{T}_0$, the criterion given by (7.5) with $\boldsymbol{j} \otimes \boldsymbol{\eta}_n$ replaced by $\eta_n\boldsymbol{j}$ for this special case of the same comparisons for all the variables. We also note that $T_{\boldsymbol{I}_k,\boldsymbol{P}} = T_1$, the criterion given by (9.4) or (9.6) for parallelism of profiles, when $\boldsymbol{P}$ is a $(p-1) \times p$ matrix of rank $(p-1)$ such that $\boldsymbol{Pj} = \boldsymbol{0}$. This assertion holds in view of the identity

$$\boldsymbol{P}'(\boldsymbol{PEP}')^{-1}\boldsymbol{P} = \boldsymbol{E}^{-1} - g\boldsymbol{E}^{-1}\boldsymbol{JE}^{-1}.$$

The validity of criteria like $T_{\boldsymbol{M},\boldsymbol{Q}}$ for testing appropriate subhypotheses follows in view of the following theorems concerning asymptotic properties of the statistic $T_{\boldsymbol{M},\boldsymbol{Q},n}$.

THEOREM 10.1. *Assume conditions $\mathscr{C}$ for $\{\boldsymbol{S}_n\}$ and suppose that the sequence of distributions $\{\boldsymbol{F}_{(N)}\}$ satisfies the conditions* (7.6). *Then, under $\{\boldsymbol{F}_{(N)}\}$, for the sequence of statistics defined by* (10.2)

$$T_{M,Q,N} \xrightarrow{\mathrm{L}} \begin{cases} \chi^2(mq, \xi) & \text{for } m > k\,, \\ \chi^2((k-1)q, \xi) & \text{for } m = k\,, \end{cases}$$

for $q \leq p$, *where the noncentrality parameter*

$$\xi = \boldsymbol{\delta}'(F)[(\boldsymbol{M\Sigma M}')^{-1} \otimes (\boldsymbol{Q\Lambda}(F)\boldsymbol{Q}')^{-1}]\boldsymbol{\delta}(F)\,;$$

for $m = k$, *the inverse is replaced by g-inverse for* $\boldsymbol{M\Sigma M}'$.

COROLLARY 10.1. *Under conditions of Theorem* 10.1, *the limiting distributions are central* χ^2 *if*

$$[\boldsymbol{M} \otimes \boldsymbol{Q}]\boldsymbol{\delta}(F) = \boldsymbol{0}\,.$$

The proof of Theorem 10.1 is straightforward for the case $m < k$, and follows for the case $m = k$ in view of the condition $\Sigma_i\, a_i \delta_i^{(\alpha)}(F) = 0$, which follows from the condition (7.4). These details are omitted.

THEOREM 10.2. *Under conditions* $\mathscr{C}$ *with the same comparisons for all the variables,* $T_{M,Q,n} \xrightarrow{\mathrm{P}} \infty$ *for all* $\boldsymbol{F}$ *in* $\mathscr{F}$ *for which*

$$[\boldsymbol{M} \otimes \boldsymbol{Q}][\boldsymbol{\eta}(\boldsymbol{F}) - \eta \boldsymbol{j}] \neq \boldsymbol{0}\,.$$

The proof is straightforward for the case $m < k$ in view of (10.1) and (4.4). For the case $m = k$, it follows in view of the relation (7.4). These details are omitted.

These techniques are illustrated in Section 15 with a numerical example.

There are a couple of points that should be noted regarding the statistics of type $T_{M,Q}$, given by (10.2), which include as special cases statistics T_1, given by (9.4) or (9.6). Here in Theorem 10.1, the sequence of distributions $\{\boldsymbol{F}_{(N)}\}$ is explicitly assumed to satisfy condition (7.6); then $\boldsymbol{E}$ is some consistent estimator of the correlation matrix $\boldsymbol{P}(F)$, where $\boldsymbol{\Lambda}(F) = \lambda \boldsymbol{P}(F)$, as in Section 8. It is, then, desirable to use some weighted 'within samples' type estimator $\boldsymbol{E}$, as at the end of Section 8, in view of (7.3).

The second point to be noted is the increase in the over-all significance level if such statistics $T_{M,Q}$ are to be used simultaneously for several subhypotheses with different $\boldsymbol{M}$ and/or $\boldsymbol{Q}$, each at the same nominal significance level. To guard against the possibility that the over-all significance level increases beyond the nominal level, it is desirable to restrict such subhypotheses tests only to those planned in advance. Furthermore, the significance level, say α_j, for the j-th subhypothesis being tested, should satisfy the restriction

$$\sum_{j=1}^{r} \alpha_j = \alpha\,,$$

where r is the number of such subhypotheses, and α is the over-all nominal level. With such a choice, the over-all significance level $\leqslant\alpha$. The drawback of such a choice is that each subhypothesis test becomes too conservative, and hence relatively insensitive, if r is moderately large and we use, for example, $\alpha_j = \alpha/r$ for each j. The alternative that is available in such cases is to use some collection of 'simultaneous comparisons' procedures, which guarantee the nominal level regardless of the number of subhypotheses to be tested. However, these too have a drawback that, generally, each individual subhypothesis test tends to be overly conservative. See Chinchilli and Sen (1982) for the discussion and details regarding such simultaneous tests.

11. Efficiency results

When two or more statistics are available for testing a given hypothesis, one statistic is considered more efficient if it is more powerful than other statistics, using the same level of signification, at the same fixed alternative.

Such a comparison of powers for two statistics based on the same data is usually dependent on α, the level of significance, N, the sample size (or some measure of sample sizes with several samples) and the fixed alternative at which the powers are compared. In order to define a suitable measure of efficiency an alternative approach is adopted comparing the corresponding sample sizes necessary to attain equal power, say β, at the same alternative for two tests using the same level α. A limit argument is usually needed for this measure to be independent of particular values α, β; furthermore, one needs to use then a sequence of alternatives converging to the null hypothesis at a suitable rate in order to come up with a meaningful definition.

Accordingly, the *asymptotic relative efficiency* (ARE) measure $e(S_1, S_2)$ of test S_1 relative to test S_2, developed by Pitman (1948), is defined as

$$e(S_1, S_2) = \lim \frac{N_2}{N_1},$$

where $N_1, N_2 \to \infty$ is such a way that the tests S_i at level α, based on samples of size N_i, have the same limiting power β for a suitable sequence of alternatives. It is understood that the limit $e(S_1, S_2)$ is independent of α and β; furthermore, it should also be independent of the sequence of alternatives if possible.

Consider the simple case of null hypothesis $\mathcal{H}_0$: $\theta = \theta_0$, for a real parameter θ, and tests S_j rejecting $\mathcal{H}_0$ if the corresponding test statistic $S_j \geqslant c_{j\alpha}$, $j = 1, 2$. Under some regularity conditions it has been shown (see, e.g., Noether, 1955) that

$$e(S_1, S_2) = \left(\frac{\eta_1'}{\eta_2'}\right) \frac{\zeta_2}{\zeta_1}, \tag{11.1}$$

where $N^{1/2}(S_{jN} - \eta_j(\theta)) \xrightarrow{L} \mathcal{N}(0, \zeta_j(\theta))$, when θ is the parameter, $\zeta_j = \zeta_j(\theta_0)$ and $\eta_j' = \mathrm{d}\eta_j(\theta)/\mathrm{d}\theta$ at θ_0, $j = 1, 2$.

More generally, consider the case where $\boldsymbol{S}_j = (S_{j1}, \ldots, S_{jk})'$ and the univariate statistics $\{S_{ij},\ i = 1, \ldots, k\}$ satisfy the conditions $\mathcal{A}$ in Section 3 for $j = 1, 2$. In particular, from (3.4) we have

$$N^{1/2}(\boldsymbol{S}_{jn} - \boldsymbol{\eta}_{jn}(\boldsymbol{F})) \xrightarrow{L} \mathcal{N}(\boldsymbol{0}, \boldsymbol{T}(\boldsymbol{F})),$$

where $\boldsymbol{\eta}_{jn}(\boldsymbol{F}) \rightarrow \boldsymbol{\eta}_j(\boldsymbol{F})$. Let T_{0j} be the test rejecting $\mathcal{H}_0$ if $T_{0j} \geqslant c_{j\alpha}$, where the statistic T_{0j} is given by the expression (4.1) in terms of $\boldsymbol{S}_j$, $j = 1, 2$. Assuming conditions (4.5) and (4.6), from Theorem 4.4 we have under the sequence $\{\boldsymbol{F}_{(N)}\}$

$$T_{0jn} \xrightarrow{L} \chi^2(k-1, \xi_j), \quad j = 1, 2$$

where the noncentrality parameter ξ_j is $\boldsymbol{\delta}_j' \boldsymbol{T}^- \boldsymbol{\delta}_j$. It can be shown (see, e.g., Andrews, 1954) that the ARE of test T_{01} relative to T_{02} is given by

$$e(T_{01}, T_{02}) = \frac{\xi_1}{\xi_2}. \tag{11.2}$$

In particular, consider the sequence of location alternatives $\boldsymbol{F}_{(N)} = (F_{1(n)}, \ldots, F_{k(N)})$,

$$F_{i(N)}(x) = F(x - N^{-1/2}\mu_i) \tag{11.3}$$

where not all μ_i are equal with $\Sigma_i\, \mu_i = 0$. For the test criteria T_0 in Section 5, based on linear rank statistics, it can be shown (e.g. Andrews, 1954; Puri, 1964) that $\boldsymbol{\delta}$ in (4.5) is then equal to $\gamma_{\mathrm{L}}(F)\boldsymbol{\mu}$, so that $\xi = (\gamma_{\mathrm{L}}^2/\lambda)\, \Sigma_i\, p_i(\mu_i - \bar{\mu})^2$, where $\bar{\mu} = \Sigma_i\, p_i\mu_i$, λ is given by (5.5) and

$$\gamma_{\mathrm{L}}(F) = \int_{-\infty}^{\infty} \left\{\frac{\mathrm{d}}{\mathrm{d}x} J[F(x)]\right\} \mathrm{d}F(x). \tag{11.4}$$

For the tests using statistics T_0, given by (6.6) based on U-statistics, $\boldsymbol{\delta}$ in (4.5) can be shown to be again equal to $\gamma_{\mathrm{L}}'(F)\boldsymbol{\mu}$, so that $\xi = \{[\gamma_{\mathrm{L}}'(k-1)]^2/\lambda k^2\}\, \Sigma_i\, p_i(\mu_i - \bar{\mu})^2$, where λ is given by (6.5) and

$$\gamma_{\mathrm{L}}'(F) = k \sum_{j=2}^{k} \{\phi(j) - \phi(j-1)\} \binom{k-2}{j-2} a(j-2, k-j, F), \tag{11.5}$$

with $a(b, c, F) = \int_{-\infty}^{\infty} F^b(1-F)^c f\, \mathrm{d}F$ for $b \geqslant 0$, $c \geqslant 0$.

The ANOVA F-test for comparing the means of k normal populations is asymptotically valid as a χ_{k-1}^2 test against location alternatives, and its noncentrality parameter ξ can be shown (see Andrews, 1954) to be equal to $\{\Sigma_i p_i(\mu_i - \bar{\mu})^2\}/\sigma^2$, where $\sigma^2(F)$ is the variance of the distribution F in (11.3).

Thus, for any test T_0 in Section 5, based on linear rank statistics, we have the A.R.E. with respect to the ANOVA F-test (say test T)

$$e(T_0, T; F) = \frac{\sigma^2(F)\gamma_L^2(F)}{\lambda}. \tag{11.6}$$

Note that this expression is the same for all k, the number of samples. In particular, for the Kruskal–Wallis H-test, given by (5.13), we have

$$e(H, T; F) = 12\sigma^2(F)\gamma_L^2(F) \tag{11.7}$$

where $\gamma_L(F) = \int_{-\infty}^{\infty} f(x)\,dF(x)$, f being the density function of F. It is known (Hodges and Lehmann, 1956) that the expression (11.7) has no finite upper bound, but it is bounded below by 0.86.

For statistics T_0 in Section 6, based on U-statistics, the A.R.E. with respect to the ANOVA test is

$$e(T_0, T; F) = \frac{(k-1)^2}{k^2}\frac{\sigma^2(F)\gamma_L'^2}{\lambda} \tag{11.8}$$

for λ given by (6.5) and γ_L' by (11.5). For any two tests, say T_{01} and T_{02}, the efficiency can be obtained as $e(T_{01}, T_{02}) = e(T_{01}, T)/e(T_{02}, T)$.

For the univariate scalar problem, consider the sequence of scale alternatives

$$F_{i(N)}(x) = F\left(\frac{x}{1 + N^{-1/2}\Delta_i}\right) \tag{11.9}$$

where not all Δ_i are equal and $\Sigma_i \Delta_i = 0$. Then for statistics in Section 5, $\boldsymbol{\delta} = \gamma_S(F)\boldsymbol{\Delta}$, where

$$\gamma_S(F) = -\int_{-\infty}^{\infty} x\left\{\frac{d}{dx} J[F(x)]\right\} dF(x), \tag{11.10}$$

and $\xi = (\gamma_S^2/\lambda)\Sigma_i p_i(\Delta_i - \bar{\Delta})^2$ with $\bar{\Delta} = \Sigma_i p_i \Delta_i$. For statistics in Section 6, $\boldsymbol{\delta} = \gamma_S'(F)\boldsymbol{\Delta}$, where

$$\gamma_S'(F) = -k\sum_{j=2}^{k}\{\phi(j) - \phi(j-1)\}\binom{k-2}{j-2} b(j-2, k-j, F), \tag{11.11}$$

with $b(a, c, F) = \int_{-\infty}^{\infty} y\{F(y)\}^a\{1 - F(y)\}^c f(y)\,dF(y)$ for $a \geq 0$, $c \geq 0$; then $\xi = \{[\gamma_S'(k-1)]^2/\lambda k^2\}\Sigma_i p_i(\Delta_i - \bar{\Delta})^2$. The A.R.E.'s can then be obtained as the ratios of corresponding noncentrality parameters ξ, as in the location case.

In the multivariate case, however, the efficiency comparisons are not as simple and clear-cut as in the univariate case. The A.R.E. of test T_1 with respect to T_2 for testing some hypothesis $\mathscr{H}$ is still given by the formula (11.2), provided T_1 and T_2 have the same (viz. χ^2 distribution with the same d.f.) distribution under $\mathscr{H}$ and noncentral $-\chi^2$ distributions, under a suitable sequence of alternatives, with noncentrality parameters ξ_1 and ξ_2, respectively. However, in the multivariate case the ratio ξ_1/ξ_2 depends not only on F, the family of distributions under conditions (7.6), as in the univariate case, but also on the 'directions' in which the sequence of alternative distributions $\{\boldsymbol{F}_{(N)}\}$ converges to the null case F.

For instance, consider the location model with the sequence of location alternatives $\boldsymbol{F}_{(N)} = [F_{1(N)}, \ldots, F_{k(N)}]$, where

$$F_{i(N)}(\boldsymbol{x}) = F(\boldsymbol{x} - N^{-1/2}\boldsymbol{\mu}_i), \tag{11.12}$$

with $\boldsymbol{\mu}_i$ not all equal and $\Sigma_i \boldsymbol{\mu}_i = \boldsymbol{0}$. Let T_0 be a rank statistic in Section 7 for testing the homogeneity hypothesis $\mathscr{H}_0$. Then from Theorem 7.3 we have the noncentrality parameter ξ, given by (7.7) for the limiting distribution of T_0, with $\boldsymbol{\delta}_i = \boldsymbol{\Gamma}(F)\boldsymbol{\mu}_i$, where $\boldsymbol{\Gamma}(F)$ is a diagonal matrix with elements $\gamma^{(\alpha)}(F)$, $\alpha = 1, \ldots, p$; here $\gamma^{(\alpha)}$ is defined as in (11.4) or (11.5), depending upon the nature of the rank statistic, for variable α. The noncentrality parameter then can be shown to reduce to

$$\xi = \sum_{i=1}^{k} p_i(\boldsymbol{\mu}_i - \bar{\boldsymbol{\mu}})'\boldsymbol{\Gamma}'(F)\boldsymbol{\Lambda}^{-1}(F)\boldsymbol{\Gamma}(F)(\boldsymbol{\mu}_i - \bar{\boldsymbol{\mu}}), \tag{11.13}$$

or

$$\xi = \frac{(k-1)^2}{k^2} \sum_{i=1}^{k} p_i(\boldsymbol{\mu}_i - \bar{\boldsymbol{\mu}})'\boldsymbol{\Gamma}'(F)\boldsymbol{\Lambda}^{-1}(F)\boldsymbol{\Gamma}(F)(\boldsymbol{\mu}_i - \bar{\boldsymbol{\mu}}), \tag{11.14}$$

for linear rank statistics and U-statistics cases respectively, with $\boldsymbol{\mu} = \Sigma_i p_i\boldsymbol{\mu}_i$ and $\boldsymbol{\Lambda}(F)$ defined as in Section 8. We note here that the expressions for $\boldsymbol{\Lambda}(F)$, as for $\boldsymbol{\Gamma}(F)$, differ from one statistic to another.

In the special case with the same comparison function (or system of scores) for all variables α, the expressions (11.13) and (11.14) reduce respectively to

$$\xi = \frac{\gamma^2}{\lambda} \sum_{i=1}^{k} p_i(\boldsymbol{\mu}_i - \bar{\boldsymbol{\mu}})'\boldsymbol{P}^{-1}(F)(\boldsymbol{\mu}_i - \bar{\boldsymbol{\mu}}), \tag{11.15}$$

and

$$\xi = \frac{\gamma^2(k-1)^2}{\lambda k^2} \sum_{i=1}^{k} p_i(\boldsymbol{\mu}_i - \bar{\boldsymbol{\mu}})'\boldsymbol{P}^{-1}(F)(\boldsymbol{\mu}_i - \bar{\boldsymbol{\mu}}). \tag{11.16}$$

These expressions, thus, depend upon a certain (correlation) matrix $\boldsymbol{P}(F)$, which differs from one statistic to another.

Then it turns out that the ARE of one rank test relative to another is not free of $\boldsymbol{\mu}$ terms and, thus, depends on the particular sequence of alternatives. In the multivariate case, therefore, it is usually not possible to come up with a single number as the Pitman ARE of one rank test relative to another. However, under suitable regularity conditions, it can be shown that the test based on linear rank statistics, using 'optimal' score functions (see, e.g., (5.11) and (5.12)), is asymptotically equivalent to, and hence as efficient as, the likelihood-ratio χ^2-test. Thus, for instance, the statistic T_0, given by (8.5), using the 'normal' scores in Section 5, is asymptotically as efficient as the likelihood-ratio χ^2 statistic for testing equality of mean vectors of several normal populations with the same covariance matrix. See Puri and Sen (1969) for the discussion of such asymptotically 'optimal' rank tests, and some other details concerning efficiency comparisons in more general problems.

12. Some other tests

In this chapter we have described location and scale tests for several samples based on linear rank statistics and U-statistics. Some other types of statistics have been considered, mostly in the univariate case, in this problem of comparing several samples.

One such class of statistics is an extension of Kolmogorov–Smirnoff statistics and Cramér–von Mises statistic based on the *empirical* distribution functions (see, e.g., Kiefer, 1959; Birnbaum and Hall, 1960). The distribution theory of such statistics based on empirical distribution functions uses techniques different from the ones used here; also the limiting distributions, even in the null case, turn out to be somewhat *nonstandard* compared to the standard chi-squared (or, in simpler cases, normal) distributions encountered in this chapter. For this reason, tests based on empirical distribution functions have not been discussed in this chapter.

Another class of tests uses statistics which are based on counts of observations from various samples that happen to fall between the specified order-statistics of a *particular* sample (see, e.g., Sen and Govindarajulu, 1966). Some other tests use counts of observations from the sample that contains the largest observation (see, e.g., Mosteller, 1948). Again, most of these tests deal only with the univariate case; furthermore, the techniques and the limiting distributions happen to be different. For these reasons, and also in view of possible arbitrary choice of *particular* samples, such procedures have not been discussed here in this chapter.

Also, in this chapter we have considered rank tests based on U-statistics of only the *simplest* type, i.e., which have kernel function ϕ defined for k-tuplets with just one observation selected from each of the k samples. In theory, U-statistics can be based on more general type of kernel functions. Apart from the notational complexity and minor modifications needed to accommodate such more general kernel functions, the essentials are covered by our discussion of the simplest case in Sections 6 and 8. As illustrations of univariate rank tests based on U-statistics with more general kernel functions for the case of two samples, see for example Sukhatme (1958) and Tamura (1960).

13. Tables of significance points

As pointed out in Section 4, the univariate nonparametric tests for homogeneity of several populations are distribution free under the null hypothesis $\mathcal{H}_0$. Thus, it is possible, at least in theory, to construct tables of exact critical points $t_\alpha(\boldsymbol{n})$, at level of significance α, for a test statistic T_0 (or T_0^*) given by (4.1) (or (4.2)) such that

$$P_{\boldsymbol{F}}[T_0 \geq t_\alpha(\boldsymbol{n})] = \alpha\,, \quad \boldsymbol{F} \in \mathcal{F}_0 \tag{13.1}$$

The discrete nature of the distribution of T_0 makes such a determination of t_α possible only for a certain set of values of α depending on $\boldsymbol{n}$. It is then desirable to construct tables of values of $t_\alpha(\boldsymbol{n})$ for all possible values of α, or at least for values of α close to commonly used nominal values of α, e.g., 0.05, 0.01, etc., for small values of k and sample sizes.

Fairly extensive tables of this type are available for the (univariate) Kruskal–Wallis H-statistic (see, e.g., Iman, Quade and Alexander, 1975) for $k = 3, 4, 5$ and sample sizes $n_i \leq 6, 4, 3$, respectively. Similar less extensive tables are available for Bhapkar V-statistic (see Bhapkar and Schwartz, 1979). Some other k-sample tables are available, see, e.g., Kiefer (1959), Birnbaum and Hall (1960).

A much wider choice is available for the special case $k = 2$, e.g. for the Wilcoxon–Mann–Whitney procedure (Verdooren, 1963; Milton, 1964).

14. Studentization

The rank statistics discussed so far in the earlier sections compare the k samples with respect to the feature of interest assuming that the population distributions have identical other features, e.g., the shape or functional form. Thus, in the location problem, we postulate the location family $\mathcal{F}_{\mathrm{L}}$ defined by (2.1) with common c.d.f. F. This does presuppose equal scale parameter $\theta_i^{(\alpha)}$, $i = 1, \ldots, k$, for each variable α if a more general location and scale family $\mathcal{F}_{\mathrm{LS}}$ defined by (2.3) is taken into consideration.

Similarly, in a scalar problem we postulate the scale family $\mathcal{F}_{\mathrm{S}}$ defined by (2.2); in the more general $\mathcal{F}_{\mathrm{LS}}$ family it does then, in effect, presuppose equal locations or known locations, if they are unequal.

The rank statistic discussed earlier need some modifications if these are to be used for the more general location and scale family $\mathcal{F}_{\mathrm{LS}}$ with unknown, and possibly unequal, location or scale parameters, while comparing the scale or location parameters, respectively, of the k populations. Such a process of modification is usually termed studentization of the corresponding statistics. Furthermore, additional regularity assumptions are usually needed in order that the modified tests remain at least asymptotically distribution-free.

For instance, consider the scalar problem with k univariate samples. Exact distribution-free rank procedures are available (see Sections 5 and 6) assuming either equal or known location parameters μ_i in the family $\mathscr{F}_{LS}$ given by (2.4). However, if μ_i are unequal and unknown, then a studentized form of a rank statistic would involve $\hat{\mu}_i$, which is some consistent estimator of μ_i, $i = 1, \ldots, k$. Such a modification would usually disturb the strict distribution-free property, under the null hypothesis, of the initial procedure for the more restricted scalar family $\mathscr{F}_S$. Also, additional regularity conditions are needed to assert that the studentized procedure is asymptotically distribution-free. As illustrations of such studentizations in the univariate scale problem with two samples, see Raghavachari (1965) and Sukhatme (1958).

Similarly, in the location problem with several samples, rank tests (in Sections 5 and 6) which are strictly distribution-free under the null hypothesis in the more restrictive location family $\mathscr{F}_L$ can be modified in order to remain valid under the more general model of $\mathscr{F}_{LS}$. See as an illustration the modified form of the Wilcoxon–Mann–Whitney statistic due to Fligner and Policello (1981). The modified form is still a rank statistic and thus, it remains strictly distribution-free in the null case under the more restricted location model, although this null distribution is no longer the same as for the initial statistic. Furthermore, the modified statistic is asymptotically distribution-free in the more general model $\mathscr{F}_{LS}$. An extension of this work leading to the studentized form of the Kruskal–Wallis H-statistic (5.13), which is valid to test equality of location parameters under more general models (including models $\mathscr{F}_{LS}$) has been discussed by Rust and Fligner (1982).

15. Numerical illustrations

We consider here a data set consisting of $p = 4$ measurements on $k = 6$ groups of ten year old children. The children were classified according to sex: G for girls, B for boys, and father's occupation code: 0 for professional, 4 for sales and 7 for laborer. This data set is a subset of data that may be found in Hodges, Krech and Crutchfield (1975). The illustrations are reported by Patterson (1975).

The sample sizes for the 6 groups are: 139, 20, 53 for the groups (G, 0), (G, 4), (G, 7), and 146, 21, 66 for the groups (B, 0), (B, 4), (B, 7) respectively. The four variables measured for each child are: 1, height in inches; 2, weight in pounds; 3, score on the Peabody picture vocabulary test; and 4, score on the Raven progressive matrices test.

Since variables 1 and 2 are so different conceptually from variables 3 and 4, the 4 variables cannot be considered commensurable. Thus, the parametric methods should not be reasonably applied to these data, at least for their profile analysis. Hence we apply the techniques using rank statistics, discussed in Sections 8, 9 and 10. By no means we are attempting a full analysis of the data; rather these data are used only for the purpose of illustration of these techniques.

For reference purposes we first give the sample means and standard deviations of these variables for different groups. See Table 1.

Table 1

Sample/Variables	1	2	3	4
$G, 0$	53.3 ± 2.9	70.0 ± 14.2	79.3 ± 9.2	34.6 ± 9.6
$G, 4$	52.5 ± 2.5	68.7 ± 12.7	77.5 ± 9.3	30.7 ± 7.8
$G, 7$	53.0 ± 3.4	72.1 ± 15.3	70.9 ± 9.6	25.0 ± 8.1
$B, 0$	53.7 ± 2.3	70.3 ± 11.2	84.5 ± 10.1	33.5 ± 8.9
$B, 4$	53.6 ± 2.6	68.3 ± 11.7	81.3 ± 9.5	30.4 ± 8.8
$B, 7$	54.2 ± 2.8	73.3 ± 15.3	74.1 ± 9.8	24.2 ± 8.1

The parametric profiles may be plotted and these could be suggestive of population patterns. However, since the distance on variable 3 score, for example, is not at all comparable to that on height variable 1, we shall discuss only nonparametric techniques.

For this discussion we shall use the rank statistic $S_i^{(\alpha)}$, which are U-statistics defined by (8.8). Furthermore, we shall mainly use the kernel function given by ϕ_V in (6.7); the corresponding statistics will therefore be denoted by letter V.

The values of $S_i^{(\alpha)}$ for the data are shown in Table 2.

Table 2

Sample/Variables	1	2	3	4
$G, 0$	0.21	0.20	0.10	0.09
$G, 4$	0.23	0.20	0.16	0.10
$G, 7$	0.19	0.18	0.37	0.30
$B, 0$	0.13	0.13	0.04	0.07
$B, 4$	0.12	0.16	0.08	0.10
$B, 7$	0.12	0.13	0.24	0.34

Note here that large values of sample mean generally correspond to small values of $S_i^{(\alpha)}$ in view of the definition of function ϕ_V in (6.7). Thus, the sample profiles plotted in terms of $S_i^{(\alpha)}$ show general patterns opposite to those shown by sample means; however, the differences among S-values are commensurable in contrast to those among sample means.

The value of the over-all statistic, given by (8.13), for testing the hypothesis $\mathcal{H}_0$ of homogeneity is

$$V_0 = 114, \quad \text{d.f.} = 20,$$

which is highly significant. The corresponding values of univariate statistics based on variables 1, 2, 3 and 4, separately are

$$8.6, \quad 5.8, \quad 60 \quad \text{and} \quad 56,$$

each on 5 d.f. The first two are not significant, while the latter two are highly significant.

For testing the parallelism hypothesis $\mathcal{H}_1$, as formulated by (9.2), the value of statistic (9.4) is

$$V_1 = 80\,, \quad \text{d.f.} = 15\,.$$

This also is highly significant; indicating that the population profiles are not parallel.

Suppose we look at the sets of variables {1, 2} and {3, 4} separately, on the basis of the nature of these variables. In the notation of Section 10, we are then considering the two choices for matrix Q as

$$\begin{bmatrix} 1 & 0 & 0 & 0 \\ 0 & 1 & 0 & 0 \end{bmatrix} \quad \text{and} \quad \begin{bmatrix} 0 & 0 & 1 & 0 \\ 0 & 0 & 0 & 1 \end{bmatrix}$$

respectively. For these sets, the statistics are

{1, 2}	{3, 4}	
$V_0 = 9$	$V_0 = 85$	d.f. = 10
$V_1 = 1$	$V_1 = 12$	d.f. = 5 ;

again the results for {1, 2} are not significant, while they are highly significant for the set {3, 4}.

Next, we wish to consider specific comparisons among the 6 samples. First we wish to compare Girls with Boys, i.e. the first three populations with the latter three. This can be accomplished by the technique in Section 13 by letting $\boldsymbol{M} = [1\ \ 1\ \ 1\ \ -1\ \ -1\ \ -1]$. Taking $\boldsymbol{Q} = \boldsymbol{I}_4$ and $\boldsymbol{P}$ in (10.2) with $\boldsymbol{P}$ defined as in Section 3, we have

$$V_{\boldsymbol{M},\boldsymbol{I}} = 12\,, \quad \text{d.f.} = 4\,; \qquad V_{\boldsymbol{M},\boldsymbol{P}} = 9\,, \quad \text{d.f.} = 3\,,$$

respectively. These values indicate that the sex effect is significant on the 4 variables, and so is the sex-variables interaction.

For testing the occupation effects, we take

$$\boldsymbol{M} = \begin{bmatrix} 1 & -1 & 0 & 1 & -1 & 0 \\ 1 & 0 & -1 & 1 & 0 & -1 \end{bmatrix};$$

again taking $\boldsymbol{Q} = \boldsymbol{I}$ and $\boldsymbol{P}$, respectively, in (10.2) we have

$$V_{\boldsymbol{M},\boldsymbol{I}} = 93\,, \quad \text{d.f.} = 8\,; \qquad V_{\boldsymbol{M},\boldsymbol{P}} = 65\,, \quad \text{d.f.} = 6\,.$$

The occupation effects on variables, and similarly the occupation-variables interactions are highly significant.

A similar analysis produces insignificant values for the subset $\{1, 2\}$ of variables, but significant values for the subset $\{3, 4\}$.

Similar findings are reached on the basis of statistics B, L or W corresponding to kernel functions ϕ_B, ϕ_L or ϕ_W in (6.7).

References

Andrews, F. C. (1954). Asymptotic behavior of rank tests for analysis of variance. *Ann. Math. Stat.* **25**, 724–736.

Bhapkar, V. P. (1961). A nonparametric test for the problem of several samples. *Ann. Math. Stat.* **32**, 1108–1117.

Bhapkar, V. P. (1966). Some nonparametric tests for the multivariate several sample location problem. In: *Proc. 1st Internat. Symp. Mult. Analysis*, pp. 29–42.

Bhapkar, V. P. (1979). Nonparametric tests for scalar profile analysis of several multivariate samples. *Annals Inst. Stat. Math.* **31**, 9–20.

Bhapkar, V. P. (1980). ANOVA and MANOVA: Models for categorical data. *Handbook of Statistics, Vol. 1.* North-Holland, Amsterdam, pp. 343–387.

Bhapkar, V. P. (1982). On nonparametric profile analysis of several multivariate samples. Dept. of Stat., Univ. of Kentucky, Technical Report No. 194.

Bhapkar, V. P. and Deshpande, J. V. (1968). Some nonparametric tests for multisample problems. *Technometrics* **10**, 578–585.

Bhapkar, V. P. and Patterson, K. W. (1977). On some nonparametric tests for profile analysis of several multivariate samples. *J. Multiv. Anal.* **7**, 265–277.

Bhapkar, V. P. and Schwartz, J. H. (1979). Efficient computation of Bhapkar's V Statistic and significance points for its use in small samples. *J. Stat. Comp. Simul.* **10**, 1–14.

Birnbaum, Z. W. and Hall, R. A. (1960). Small sample distributions for multisample statistics of the Smirnov type. *Ann. Math. Stat.* **31**, 710–720.

Capon, J. (1961). Asymptotic efficiency of certain locally most powerful rank tests. *Ann. Math. Stat.* **32**, 88–100.

Chatterjee, S. K (1966). A multisample nonparametric scale test based on U-Statistics. *Bull. Calcutta Stat. Assoc.* **15**, 109–119.

Chinchilli, V. M. and Sen, P. K. (1982). Multivariate linear rank statistics for profile analysis. *J. Multiv. Anal.* **12**, 152–171.

Deshpande, J. V. (1970). A class of multisample distribution-free tests. *Ann. Math. Stat.* **41**, 227–236.

Fisher, R. A. and Yates, F. (1938). *Statistical Tables for Biological, Agricultural and Medical Research.* Oliver and Boyd, Edinburgh.

Fligner, M. A. and Policello, G. E. (1981). Robust rank procedures for the Behrens-Fisher problem. *J. Amer. Statist. Assoc.* **76**, 162–168.

Ghosh, M. (1982). Rank statistics and limit theorems. *Handbook of Statistics, Vol. 4.* North-Holland, Amsterdam.

Hajek, J. and Sidak, Z. (1967). *Theory of Rank Tests.* Academic Press, New York.

Hodges, J. L., Jr., Kretch, D. and Crutchfield. R. S. (1975). *Statlab.* McGraw-Hill, New York.

Hodges, J. L., Jr. and Lehmann, E. L. (1956). The efficiency of some nonparametric competitors of the t-test. *Ann. Math. Stat.* **27**, 324–335.

Iman, R. L., Quade, D. and Alexander, D. A. (1975). Exact probability levels for the Kruskal-Wallis test. *Selected Tables in Mathematical Statistics* **3**, 329–384.

Kiefer, J. (1959). K-sample analogues of the Kolmogorov–Smirnov and Cramer–von Mises tests. *Ann. Math. Stat.* **30**, 420–447.

Kruskal, W. H. (1952). A nonparametric test for the several sample problem. *Ann. Math. Stat.* **23**, 525–540.

Kruskal, W. H. and Wallis, W. A. (1952). Use of ranks in one-criterion variance analysis. *J. Amer. Statist. Assoc.* **47**, 583–621.

Mann, H. B. and Whitney, D. R. (1947). On a test of whether one of two random variables is stochastically larger than the other. *Ann. Math. Stat.* **18**, 50–60.

Milton, R. C. (1964). An extended table of critical values for the Mann-Whitney (Wilcoxon) two-sample statistic. *J. Amer. Statist. Assoc.* **59**, 925–934.

Mood, A. M. (1954). On the asymptotic efficiency of certain nonparametric two-sample tests. *Ann. Math. Stat.* **25**, 514–522.

Morrison, D. F. (1976). *Multivariate Statistical Methods.* McGraw-Hill, New York, 2nd ed.

Mosteller, F. (1948). A k-sample slippage test for an extreme population. *Ann. Math. Stat.* **19**, 58–65.

Noether, G. E. (1955). On a theorem of Pitman. *Ann. Math. Stat.* **26**, 64–68.

Patterson, K. W. (1975). A study of multivariate U statistic techniques with applications to profile analysis. Unpublished Ph.D. Dissertation, Univ. of Kentucky.

Pitman, E. J. G. (1948). Notes on nonparametric statistical inference. Unpublished notes, Columbia University, New York.

Puri, M. L. (1965). Asymptotic efficiency of a class of c-sample tests. *Ann. Math. Stat.* **35**, 102–121.

Puri, M. L. and Sen, P. K. (1969). A class of rank order tests for a general linear hypothesis. *Ann. Math. Stat.* **40**, 1325–1343.

Puri, M. L. and Sen, P. K. (1971). *Nonparametric Methods in Multivariate Analysis.* Wiley, New York.

Raghavachari, M. (1965). The two sample scale problem when the locations are unknown. *Ann. Math. Stat.* **36**, 1236–1242.

Rust, S. W. and Fligner, M. A. (1982). A modification of the Kruskal-Wallis statistic for the generalized Behrens–Fisher problem. Dept. of Stat., University of Kentucky, Technical Report No. 192.

Sen, P. K. and Govindarajulu, Z. (1966). On a class of c sample weighted rank-sum tests for location and scale. *Ann. Inst. Stat. Math.* **18**, 87–105.

Sugiura, N. (1965). Multisample and multivariate nonparametric tests based on U-statistics and their asymptotic efficiencies. *Oskaka J. Math.* **2**, 385–426.

Sukhatme, B. V. (1958). Testing the hypothesis that two populations differ only in location. *Ann. Math. Stat.* **29**, 60–78.

Tamura, R. (1960). On the nonparametric tests based on certain U-statistics. *Bull. Math. Stat.* (Japan) **9**, 61–68.

Tamura, R. (1966). Multivariate nonparametric several sample tests. *Annals Math. Stat.* **37**, 611–618.

Van der Waerden, B. L. (1953). Ein neuer test fur das Problem der zwei Stichproben. *Math. Annalen.* **126**, 93–107.

Verdooren, L. R. (1963). Extended tables of critical values of Wilcoxon's test statistic. *Biometrika* **50**, 177–186.

P. R. Krishnaiah and P. K. Sen, eds., *Handbook of Statistics, Vol. 4*
© Elsevier Science Publishers (1984) 63–78 3

Hypothesis of Symmetry

Marie Hušková

1. Introduction

The hypothesis of symmetry is the hypothesis corresponding to a nonparametric variant of the classical testing problem on paired samples which can be formulated as follows:

$(\boldsymbol{Z}_j, \boldsymbol{Y}_j)$, $j = 1, \ldots, n$, are pairs of independent, identically distributed, p-dimensional random vectors; the cumulative distribution function $H(z, y)$ of $(\boldsymbol{Z}_j, \boldsymbol{Y}_j)$ fulfills some symmetry property.

Here $\boldsymbol{Z}_j = (Z_{j1}, \ldots, Z_{jp})'$, $\boldsymbol{Y}_j = (Y_{j1}, \ldots, Y_{jp})'$; z' denotes the transpose of z. The symmetry property is usually introduced in the following way:

$$H(z, y) = H(y, z) \quad \text{for all } z, y\,. \tag{1.1}$$

Some discussion of a nonparametric version of the paired sample testing problem can be found, e.g., in Sen and David (1968), Davidson and Bradley (1969, 1970), Shane and Puri (1969), Puri and Shane (1970), Sen (1967). Sen (1971) considered the problem of paired samples as a problem of an incomplete block design.

Usually, one treats the differences $\boldsymbol{X}_j = \boldsymbol{Z}_j - \boldsymbol{Y}_j$, $j = 1, \ldots, n$, and the symmetry property (1.1) implies that $\boldsymbol{X}_j$ and $-\boldsymbol{X}_j$ have the same distribution. This kind of symmetry is called diagonal symmetry. The corresponding cumulative distribution function is called diagonally symmetric. Note that the diagonal symmetry implies that the median is zero.

In the classical approach, the distribution of $\boldsymbol{X}_j$ is assumed to be normal with zero expectation, $j = 1, \ldots, p$ (diagonal symmetry is fulfilled), and the test for symmetry is based on either Hotelling's T^2-statistic in the multivariate case or on one-sample Student test in the univariate case.

The following definition of the *hypothesis of symmetry* is adopted:

$H_{1,p}$: $\boldsymbol{X}_j = (X_{j1}, \ldots, X_{jp})'$, $j = 1, \ldots, n$, are independent, identically distributed random vectors, $\boldsymbol{X}_j$ are distributed according to an arbitrary cumulative distribution function $F \in \mathscr{F}_p$, where $\mathscr{F}_p$ is the family of p-variate continuous diagonally symmetric distribution functions.

The hypothesis $H_{1,p}$ is invariant under the groups of transformations G_{n1} and G_{n2}. The group G_{n1} consists of all transformations

$$H_n : (x_1, \ldots, x_p) \to (h_1(x_1), \ldots, h(x_p)),$$

where the h_i's are continuous, odd, and strictly increasing. The group G_{n2} consists of all mappings

$$g_n : (\boldsymbol{x}_1, \ldots, \boldsymbol{x}_n) \to ((-1)^{j_1}\boldsymbol{x}_1, \ldots, (-1)^{j_n}\boldsymbol{x}_n), \quad j = 0, 1; \; i = 1, \ldots, n,$$

i.e. G_{n2} consists of 2^n distinct elements. The maximal invariant under G_{n1} and G_{n2} is the set

$$\{R_{ji}^+, \operatorname{sign}(x_{ji}); j = 1, \ldots, n, i = 1, \ldots, p\},$$

where R_{ji}^+ is the rank of $|X_{ji}|$ in the sequence $|X_{1i}|, \ldots, |X_{ni}|$ and $\operatorname{sign}(x) = -1$ if $x < 0$, $\operatorname{sign}(x) = 0$ if $x = 0$, $\operatorname{sign}(x) = 1$ if $x > 0$ (see Lehman, 1959). Thus the test for $\{H_{1,p}, K_p(\Theta^+)\}$ invariant under G_{n1} and G_{n2} will depend on $\boldsymbol{X}_1, \ldots, \boldsymbol{X}_n$ through $\{R_{ji}^+, \operatorname{sign}(X_{ji}); j = 1, \ldots, n, \; i = 1, \ldots, p\}$. The statistics generated by this maximal invariant are called *rank statistics for hypothesis of symmetry* and the test generated by these statistics are called the *rank statistics for $H_{1,p}$*.

The most frequent alternative hypothesis is hypothesis of shift in location, i.e.:

$K_p(\Theta)$: $\boldsymbol{X}_1, \ldots, \boldsymbol{X}_n$ are independent, identically distributed, p-dimensional random vectors, $\boldsymbol{X}_j$ is distributed according to the distribution function $F(\boldsymbol{x} - \boldsymbol{\theta})$, where $F \in \mathscr{F}_p$, $\boldsymbol{\theta} = (\theta_1, \ldots, \theta_p)'$ is a vector parameter, $\boldsymbol{\theta} \in \Theta, \emptyset \neq \Theta \otimes R_p$.

The most frequent alternatives correspond to

$$\Theta^+ = \{\boldsymbol{\theta} = (\theta_1, \ldots, \theta_p)'; \theta_i > 0, i = 1, \ldots, p\},$$

$$\Theta^- = \{\boldsymbol{\theta}_1 = (\theta_1, \ldots, \theta_p)'; \theta_i < 0, i = 1, \ldots, p\} \quad \text{and}$$

$$\Theta^* = \{\boldsymbol{\theta} = (\theta_1, \ldots, \theta_p)' \neq \boldsymbol{0}\}.$$

REMARK 1. Bell and Haller (1969) and Rüschendorf (1976) among others were interested in different concept of multivariate symmetry hypothesis. They considered the hypotheses generated by some group G of transformations for the class $\mathscr{F}_G$ of distributions such that $P \in \mathscr{F}_G \Leftrightarrow gP \in \mathscr{F}_G$ for all $g \in G$.

REMARK 2. Other alternatives than $K_p(\Theta)$ were also considered; e.g. (see Hušková 1971):

$\boldsymbol{X}_1, \ldots, \boldsymbol{X}_n$ are independent random vectors, $\boldsymbol{X}_j$ has the continuous distribution function $F(\boldsymbol{x}; \boldsymbol{\theta})$, $\boldsymbol{\theta} \in \Theta$; $F(\boldsymbol{x}, \boldsymbol{0})$ is diagonally symmetric. (1.2)

$\boldsymbol{X}_1, \ldots, \boldsymbol{X}_n$ are independent random vectors, $\boldsymbol{X}_j$ has the continuous distribution function $F(\boldsymbol{x} - c_j\boldsymbol{\theta})$, $\boldsymbol{\theta} \in \Theta$, where $c_1, \ldots, c_n$ are known constants and $F(\boldsymbol{x}) \in \mathscr{F}_p$. (1.3)

More general alternatives were studied by several authors; e.g., Hollander (1971), Yanagimoto and Sibuya (1972, 1976), Snijders (1981).

REMARK 3. It should be noted that most statistics for $H_{1,p}$ arise as analogs of statistics for the hypothesis of randomness. Many properties (not all) of statistics for the hypothesis of randomness can be easily transferred to statistics for $H_{1,p}$. Consequently, not all results for statistics for $\tilde{H}_{1,p}$ have been published.

In the next two sections tests for the univariate and the multivariate hypotheses of symmetry are surveyed separately.

2. Univariate case

We shall write here H_1, $K(\Theta)$, and X_j instead of $H_{1,1}$, $K_1(\Theta)$ and X_{j1}, respectively.

Usually, the test for H_1 against $K(\Theta)$ is based on the signed rank statistic

$$S_n^+ = \sum_{i=1}^{n} \operatorname{sign}(X_i) a_n(R_i^+) , \tag{2.1}$$

where $a_n(1), \ldots, a_n(n)$ are known scores. These scores are often related to some square-integrable function φ, defined on (0, 1), in the following way:

$$\lim_{n\to\infty} \int_0^1 (a_n([un]+1) - \varphi(u))^2 \,\mathrm{d}u = 0 , \tag{2.2}$$

where $[b]$ denotes the largest integer not exceeding b. Then we shall write $S_n^+(\varphi)$ instead of S_n^+. Typically, one chooses

$$a_n(j) = E\varphi(U_{(j)}) , \quad j = 1, \ldots, n , \tag{2.3}$$

or

$$a_n(j) = \varphi\Big(\frac{j}{n+1}\Big), \quad j = 1, \ldots, n , \tag{2.4}$$

where $U_{(1)} < \cdots < U_{(n)}$ is the ordered sample corresponding to a sample of size

n from the uniform-(0, 1) distribution. It is known that the scores (2.3) satisfy (2.2), and if φ is expressible as a difference of monotone (square-integrable) functions (Hájek and Šidák, 1967, p. 164), the scores (2.4) fulfill (2.2).

The statistic S_n^+ can be rewritten in a very useful way:

$$S_n^+ = \sum_{j=1}^{n} a_n(j)T_j - \sum_{j=1}^{n} a_n(j)/2\,, \tag{2.5}$$

$T_j = 1$ if the jth smallest observation in absolute value is positive,
$ = 0$ otherwise.

Under H_1 the vectors $(\mathrm{sign}(X_1), \ldots, \mathrm{sign}(X_n))'$ and $(R_1^+, \ldots, R_n^+)'$ are independent Hájek and Šidák, 1967, p. 40);

$$P_{H_1}(\mathrm{sign}(X_1) = s_1, \ldots, \mathrm{sign}(X_n) = s_n) = 2^{-n}\,,$$
$$P_{H_1}(R_1^+ = r_1, \ldots, R_n^+ = r_n) = (n!)^{-1}\,,$$

for all $s_j = 0$ or $1, j = 1, \ldots, n$, and all permutations $(r_1, \ldots, r_n)$ of $(1, \ldots, n)$. Further, under H_1 the distribution of S_n^+ is symmetric about 0, it does not depend on the distribution of X_j, and

$$E_{H_1}S_n^+ = 0, \qquad \mathrm{var}_{H_1}S_n^+ = \sum_{i=1}^{n} a_n^2(i)\,.$$

The critical regions corresponding to the test of H_1 against $K(\Theta^*)$, $K(\Theta^+)$, $K(\Theta^-)$ are as follows:

$$|S_n^+| > v_{\alpha/2}, \qquad S_n^+ > v_\alpha, \qquad S_n^+ < -v_\alpha$$

where v_α is the $100\alpha\%$ upper critical value, i.e. $P_{H_1}(S_n^+ > v_\alpha) = \alpha$.

To formulate asymptotic properties we shall often restrict F to the subfamily $\mathcal{F}_1^* \subset \mathcal{F}_1$, where $\mathcal{F}_1^*$ is the family of continuous distribution functions with absolutely continuous density f and finite Fisher's information,

$$0 < \int_{-\infty}^{+\infty} (f'(x))^2(f(x))^{-1}\,\mathrm{d}x < +\infty\,.$$

The corresponding hypotheses are denoted by H_1^* and $K^*(\Theta)$.

The main results on the asymptotic distribution of S_n^+ can be summarized as follows.

Theorem 2.1. (a) *If*

$$\lim_{n\to\infty} \max_{1\leq j\leq n} a_n^2(j)\Big(\sum_{j=1}^{n} a_n^2(j)\Big)^{-1} = 0 \tag{2.6}$$

then

$$\left(S_n^+ \Big(\sum_{j=1}^{n} a_n^2(j) \Big)^{-1/2} \,\Big|\, H_1 \right) \xrightarrow{w} N(0, 1) \quad as\ n \to \infty\,.$$

If $a_n(j)$, $j = 1, \ldots, n$ *satisfy* (2.2) *with square-integrable* φ *then* $\Sigma_{j=1}^{n} a_n^2(j)$ *can be replaced by* $n \int_0^1 \varphi^2(u)\,\mathrm{d}u$.

(b) *If* $a_n(j)$, $j = 1, \ldots, n$, *satisfy* (2.2) *with square-integrable* φ *then*

$$\mathscr{L}\left(\Big(S_n^+ - \Delta \int_0^1 \varphi(u)\varphi_f(u)\,\mathrm{d}u \Big) \Big(\int_0^1 \varphi^2(u)\,\mathrm{d}u \Big)^{-1/2} \,\Big|\, K^*(\Delta n^{-1/2}) \right) \xrightarrow{w} N(0, 1)$$

as $n \to \infty$

for arbitrary fixed $\Delta \neq 0$, $\varphi_f(u) = f'(F^{-1}(u))/f(F^{-1}(u))$.

For the proof see e.g. Hájek and Šidák (1967).

The alternative hypothesis $K^*(\Delta n^{-1/2})$ is usually called the contiguous alternatives hypothesis. This notion was introduced by LeCam (1960). The asymptotic distribution of S_n^+ under general alternatives was studied e.g. by Hušková (1971) and Koul and Staudte (1972). Albers, Bickel and van Zwet (1976) derived Edgeworth expansion for S_n^+ under H_1 and $K(\Delta n^{-1/2}, \Delta \neq 0)$. Prášková (1976) obtained an Edgeworth expansion and a local limit theorem for the one-sample Wilcoxon statistic. Now, we turn the problem of the optimal choice of scores $a_n(j)$. Define

$$\varphi_f(u) = -\frac{f'(F^{-1}(u))}{f(F^{-1}(u))}, \quad u \in (0, 1)\,,$$

and

$$\varphi_f^+(u) = \varphi_f((u+1)/2), \quad u \in (0, 1)\,,$$

where

$$F^{-1}(u) = \inf\{x\,;\, F(x) \geqslant u\}\,.$$

f is the density (with respect to Lebesgue measure) corresponding to the distribution function F and f' denotes the derivative of f.

The test with critical region

$$S_{nf}^+ = \sum_{j=1}^{n} \operatorname{sign}(X_j) a_n(R_j^+, f) > u\,, \tag{2.7}$$

where

$$a_n(j, f) = E\varphi_f(U_{(j)})\,, \tag{2.8}$$

is locally optimal (more precisely, locally most powerful test for $\{H_1, K^*(\Theta)^+\}$; Hájek and Šidák, 1967, p. 174). The test with critical region

$$S_{nf}^+ > \Phi^{-1}(1-\alpha) \Big(\int_0^1 \varphi_f^2(u)\,\mathrm{d}u \Big)^{1/2}, \tag{2.9}$$

where Φ^{-1} is the quantile function of $N(0, 1)$, α is asymptotic level, is asymptotically optimal (i.e. it reaches asymptotically the maximum power for the problem $\{H_1^*, K^*(n^{-1/2}\Delta; \Delta > 0\}$; Hájek and Šidák, 1967, p. 251).

For more general testing problem of H_1 against the alternative hypothesis (1.2) the rank statistic

$$S_c^+ = \sum_{i=1}^{n} c_i \operatorname{sign}(X_i) a_n(R_i^+),$$

where $c_1, \ldots, c_n$ are known regression constants, is used. The properties of S_c^+ are analogous to those of S_n^+.

$$E_{H_1} S_c^+ = 0, \qquad \operatorname{var}_{H_1} S_c^+ = \sum_{i=1}^{n} a_n^2(i) \sum_{j=1}^{n} c_j^2 .$$

If $\lim_{n\to\infty} \max_{1\le j\le n} c_j^2 (\sum_{j=1}^n c_j^2)^{-1} = 0$ and (2.6) are satisfied then

$$\left(S_c^+ \left(\sum_{i=1}^{n} a_n^2(i) \sum_{j=1}^{n} c_j^2 \right)^{-1/2} \,\middle|\, H_1 \right) \xrightarrow{w} N(0, 1) \quad \text{as } n \to \infty .$$

Similarly, as in the previous case, the rank statistic

$$\sum_{j=1}^{n} c_j \operatorname{sign}(X_j) a_n(R_j^+, f),$$

where $a_N(i)$ is given by (2.8), provides both the locally and the asymptotically optimal tests for H_1^* against (1.4) with $\Theta = \{\Delta; \Delta > 0\}$ and $\Theta_n = \{\Delta n^{-1/2}; \Delta > 0\}$, respectively. These results can be derived in a manner analogous to the case $\{H_1, K(\Theta)\}$ or $\{H_1^*, K^*(\Theta)\}$. For other results the reader is referred to, e.g. Hušková (1971).

There is also another class of statistics for H_1 based on the empirical distribution function and introduced by Smirnov (1947). They are called Kolmogorov–Smirnov type statistics for H_1. They can be used for a rather broad class of alternatives; for example testing $\{H_1, K(\Theta)\}$, Θ-arbitrary nonempty or for testing H_1 versus the general alternative:

> $X_1, \ldots, X_n$ are independent identically distributed random variables distributed according to some continuous distribution function F for which there exists an x such that $F(x) + F(-x) \neq 1$. (2.10)

Smirnov (1947) proposed the statistic

$$U_n = (n/2)^{1/2} \sup_x |F_n(x) + F_n(-x) - 1| ,$$

where $F_n(x)$ is the empirical distribution function corresponding to $X_1, \ldots, X_n$. Butter (1969) and Koul and Staudte (1972b) treated its asymptotic properties. Later, several authors suggested different statistic for H_1 of this kind (Orlov, 1972; Rothmann and Woodroofe, 1972; Hell and Rao, 1977; Koziol, 1979, 1980; Sen, 1979; Srinivan and Godio, 1974). For instance Orlov (1972) investigated the Cramér–von Mises statistic for H_1:

$$V_n = 2^{-1} \int n(F_n(x) + F_n(-x) - 1)^2 \, dF_n(x) .$$

If some discontinuity of the distribution function F of X_j in H_1 is admitted then symmetry is defined as follows:

$$F(x) + F(-x_+) = 1 \quad \text{for all } x ,$$

where F is right continuous and $F(-x_+) = \lim_{y \to -x_+} F(y)$. Information on rank statistics used in this case and their properties can be found in Hemelrijk (1952), Putter (1955), Vorlíčková (1972), Conover (1973), Shirahata (1974).

In the rest of this section, the most usual rank tests for H_1 are reviewed.

The sign test. Putting $a_n(i) = 1$, $i = 1, \ldots, n$, in (2.1) we get the rank statistic

$$S_n^+ = \sum_{j=1}^{n} \operatorname{sign}(X_j) ,$$

called the sign statistic. S_n^+ equals the difference of the number of positive and negative X_j. The test based on S_n^+ is called the sign test. Under H_1,

$$E_{H_1} S_n^+ = 0, \qquad \operatorname{var}_{H_1} S_n^+ = n$$

$$P_{H_1}(S_n^+ = k) = 2^{-n} \binom{n}{n-k}, \quad k = -n, -n+2, \ldots, n-2, n .$$

If the underlying density is double exponential ($f(x) = 2^{-1}\lambda \exp\{-\lambda|x|\}$, $\lambda > 0$, $x \in R_1$), the test with critical region $\{S_n^+ > \Phi^{-1}(1-\alpha)n^{1/2}\}$ is asymptotically optimal for $\{H_1^*, K^*\{\Delta n^{-1/2}, \Delta > 0\}\}$ and S_n^+ leads to the locally optimal test. Thus the test is preferable to others if the density is double exponential.

This test is often recommended to test whether the medians of X_j's are zero versus at least one of the medians is positive without additional assumptions on the form of the distribution of X_j, $j = 1, \ldots, n$.

Further discussion of the sign test can be found in Dixon and Mood (1946), van der Waerden and Nievergelt (1956), Walsh (1951), Klotz (1963) and Conover, Wehmanen and Ramsey (1978). For tables see Dixon–Mood (1946), van der Waerden and Nievergelt (1956) and Mac Kinnon (1964). The normal approximation is good for $n = 12$.

The Wilcoxon one-sample test. The one-sample Wilcoxon statistic is defined as follows:

$$W_n^+ = \sum_{j=1}^{n} \operatorname{sign}(X_j) R_j^+ .$$

The corresponding test is called the one-sample Wilcoxon test. Under H_1,

$$E_{H_1} W_n^+ = 0, \qquad \operatorname{var}_{H_1} W_n^+ = n(n+1)(2n+1)/6 .$$

W_n^+ leads to both the locally and asymptotically optimal test if the underlying distribution is logistic. The test was introduced by Wilcoxon (1945). For tables see Wilcoxon (1947), Hájek (1955), Owen (1962), Selected tables in mathematical statistics, Vol. 1. Normal approximation is good for $n \geq 15$.

The Fraser test. This test employs the statistic

$$S_n^+ = \sum_{j=1}^{n} \operatorname{sign}(X_j) a_n(R_j^+) ,$$

where $a_n(i) = E|V|_{(i)}$, with $|V|_{(1)} \leq \cdots \leq |V|_{(n)}$ being the ordered absolute values corresponding to a sample from $N(0, 1)$ of size n. Under H_1,

$$E_{H_1} S_n^+ = 0, \qquad \operatorname{var}_{H_1} S_n^+ = \sum_{j=1}^{n} (E|V|_{(j)})^2 .$$

If the underlying distribution is normal S_n^+ leads to the locally and asymptotically optimal test. The test was derived by Fraser (1957) and later studied and tabulated by Klotz (1963).

The one-sample van der Waerden test. This test is also asymptotically optimal for the normal distribution and is determined by the statistic

$$S_n^+ = \sum_{j=1}^{n} \operatorname{sign}(X_j) \Phi^{-1}((R_j^+(n+1)^{-1} + 1)/2)$$

Under H_1,

$$E_{H_1} S_n^+ = 0, \qquad \operatorname{var}_{H_1} = \sum_{j=1}^{n} (\Phi^{-1}((i(n+1)^{-1} + 1)/2))^2 .$$

This test was introduced by van Eeden (1963). For tables see Owen (1962).

3. Multivariate case

The tests for $H_{1,p}$ in the multivariate case are based on the vector of univariate rank statistics, i.e. on

$$S_n^+ = (S_{n1}^+, \ldots, S_{np}^+)' , \tag{3.1}$$

$$S_{ni}^{+} = \sum_{j=1}^{n} \operatorname{sign}(X_{ji}) a_{ni}(R_{ji}^{+}), \quad i = 1, \ldots, p, \tag{3.2}$$

where $(a_{ni}(1), \ldots, a_{ni}(n))$, $i = 1, \ldots, p$, are scores usually related to some square-integrable functions φ_i as follows:

$$\int_0^1 (a_{ni}([un] + 1) - \varphi_i(u))^2 \, du \to 0 \quad \text{as } n \to \infty, \ i = 1, \ldots, p. \tag{3.3}$$

Denote

$$\boldsymbol{\Sigma}_{p,n} = (\sigma_{p,ik}) \tag{3.4}$$

$$\sigma_{p,ik} = \sum_{j=1}^{n} \operatorname{sign}(X_{ji}) \operatorname{sign}(X_{jk}) a_{ni}(R_{ji}^{+}) a_{nk}(R_{jk}^{+}). \tag{3.5}$$

For testing $H_{1,p}$ against $K_p(\Theta)$ the *quadratic rank statistic*

$$Q_n^{+} = \boldsymbol{S}_n^{+\prime} \boldsymbol{\Sigma}_{p,n}^{-} \boldsymbol{S}_n^{+}, \tag{3.6}$$

where $\boldsymbol{\Sigma}_{p,n}^{-}$ is the generalized inverse matrix of $\boldsymbol{\Sigma}_{p,n}$, is usually used. Denote by $\mathscr{A}_n$ the minimal σ-field generated by

$$\{|X_{ij}|, \operatorname{sign}(X_{ji})(\operatorname{sign}(X_{1j}))^{-1}, j = 1, \ldots, n; i = 1, \ldots, p\}.$$

The distribution of Q_n^{+} under $H_{1,p}$ generally depends on the underlying distribution, but the conditional distribution of Q_n under $H_{1,p}$ given $\mathscr{A}_n$ does not. Consequently, the conditional test for $H_{1,p}$ is established. Namely, for testing $H_{1,p}$ against $K_p(\Theta^{+})$ the conditional test with the crticial region (α-level)

$$Q_n^{+} > q_{n,\alpha}(\boldsymbol{X}_1, \ldots, \boldsymbol{X}_n), \tag{3.7}$$

where

$$P_{H_{1,p}}(Q_n^{+} > q_{n,\alpha}(\boldsymbol{X}_1, \ldots, \boldsymbol{X}_n) \mid \mathscr{A}_n) \leq \alpha$$
$$< P_{H_{1,p}}(Q_n^{+} \geq q_{n,\alpha}(\boldsymbol{X}_1, \ldots, \boldsymbol{X}_n) \mid \mathscr{A}_n)) \tag{3.8}$$

is performed. To apply this test in practice, we need to know $q_{n,\alpha}(\boldsymbol{X}_1, \ldots, \boldsymbol{X}_n)$. While for small sample size n it is necessary to use the exact values, for large sample size the distribution of Q_n^{+} can be approximated by the χ^2-distribution (see Theorem 3.1). Consequently, for n large the test with critical region

$$Q_n^{+} > \chi_\alpha^2(p), \tag{3.9}$$

where $\chi_\alpha^2(p)$ is the upper critical value of the χ^2-distribution with p degrees of freedom, can be used for $\{H_{1,p}, K_p(\Theta)\}$. In contrast to the univariate case, the vectors

$$(\operatorname{sign}(X_{11}), \ldots, \operatorname{sign}(X_{n1}), \ldots, \operatorname{sign}(X_{1p}), \ldots, \operatorname{sign}(X_{np}))'$$

and

$$(R_{11}^+, \ldots, R_{n1}^+, \ldots, R_{1p}^+, \ldots, R_{np}^+)', \quad p \geq 2,$$

are not generally independent under $H_{1,p}$. Under $H_{1,p}$

$$E_{H_{1,p}}(\boldsymbol{S}_n^+ \mid \mathcal{A}_n) = E_{H_{1,p}}(\boldsymbol{S}_n^+) = \boldsymbol{0},$$

$$\operatorname{var}_{H_{1,p}}(\boldsymbol{S}_n^+ \mid \mathcal{A}_n) = \boldsymbol{\Sigma}_{p,n}, \qquad \operatorname{var}_{H_{1,p}} \boldsymbol{S}_n^+ = E_{H_{1,p}} \boldsymbol{\Sigma}_{p,n},$$

$$n^{-1}(\boldsymbol{\Sigma}_{p,n} - \operatorname{var}_{H_{1,p}} \boldsymbol{S}_n^+) \to \boldsymbol{0} \quad \text{in probability as } n \to \infty$$

holds. Denote by F_j and F_{jk} the distribution functions of X_j and (X_j, X_k), $j, k = 1, \ldots, p$, resp.; and by F_j^* and F_{jk}^* the distribution functions of $|X_j|$ and $(|X_j|, |X_k|)$, $j, k = 1, \ldots, p$, resp. Put

$$\boldsymbol{\Sigma} = (\sigma_{jk})_{j,k=1,\ldots,p}, \quad \boldsymbol{\mu} = (\mu_1, \ldots, \mu_p)',$$

$$\sigma_{jk} = \int_0^{+\infty} \int_0^{+\infty} \varphi_j(F_j^*(x)) \varphi_k(F_k^*(y)) \, \mathrm{d}F_{jk}^*(x, y), \tag{3.10}$$

$$\mu_j = \Delta_j \int \operatorname{sign}(x) \varphi_j(F_j^*(|x|)) f_j(x) \, \mathrm{d}x.$$

In the following two theorems the main results on the asymptotic distributions of $\boldsymbol{S}_n^+$ and Q_n^+ under $H_{1,p}$ and under local alternatives are summarized.

THEOREM 3.1. *Let* (3.3) *be satisfied, and let the matrix* $\boldsymbol{\Sigma}$ *defined by* (3.10) *be positive definite. Then, under* $H_{1,p}$,

$$n^{-1} \operatorname{var}_{H_{1,p}} \boldsymbol{S}_n^+ \to \boldsymbol{\Sigma} \quad \text{as } n \to \infty,$$

$$\mathcal{L}_{\mathcal{A}_n}(\boldsymbol{S}_n^+ \boldsymbol{\Sigma}_{p,n}^{-1/2} \mid H_{1,p}) \xrightarrow{w} N_p(\boldsymbol{0}, \boldsymbol{I}_p) \quad \text{as } n \to \infty,$$

$$\mathcal{L}_{\mathcal{A}_n}(Q_n^+ \mid H_{1,p}) \xrightarrow{w} \chi^2(p) \quad \text{as } n \to \infty,$$

$$\mathcal{L}(\boldsymbol{S}_n^+ \mid H_{1,p}) \xrightarrow{w} N_p(\boldsymbol{0}, \boldsymbol{\Sigma}) \quad \text{as } n \to \infty,$$

and

$$\mathcal{L}(Q_n^+ \mid H_{1,p}) \xrightarrow{w} \chi^2(p) \quad \text{as } n \to \infty,$$

where $\mathcal{L}_{\mathcal{A}_n}(\ldots \mid H_{1,p})$ *denotes the conditional distribution, given* $\mathcal{A}_n$, *under* $H_{1,p}$, $\boldsymbol{I}_p$ *is the* $p \times p$ *identity matrix and* $\chi^2(p)$ *is the* χ^2*-distribution with* p *degrees of freedom.*

THEOREM 3.2. *Let* $\boldsymbol{X}_1, \ldots, \boldsymbol{X}_n$ *be independent identically distributed, p-dimensional random vectors such that the* $\boldsymbol{X}_j$ is distributed according to a continuous *distribution function* $F(\boldsymbol{x} - \boldsymbol{\Delta} n^{-1/2})$, $\boldsymbol{\Delta} = (\Delta_1, \ldots, \Delta_p)' \neq \boldsymbol{0}$, *where F is diagonally symmetric and the marginal distribution function* F_i *has absolutely continuous density* f_i *with finite Fisher's information,* $i = 1, \ldots, p$. *Let* (3.3) *be satisfied and the matrix* $\boldsymbol{\Sigma}$ *given by* (3.10) *be regular. Then*

$$n^{-1}\boldsymbol{\Sigma}_{p,n} \to \boldsymbol{\Sigma} \quad \textit{in probability as } n \to \infty\,,$$

$$n^{-1/2}(E\boldsymbol{S}_n^+ - \boldsymbol{\mu}) \to 0 \quad \textit{as } n \to \infty\,,$$

$$\mathscr{L}(n^{-1/2}\boldsymbol{S}_n^+) \xrightarrow{w} N_p(\boldsymbol{\mu}, \boldsymbol{\Sigma}) \quad \textit{as } n \to \infty\,,$$

and

$$\mathscr{L}(Q_n^+) \xrightarrow{w} \chi^2(p, \boldsymbol{\mu}'\boldsymbol{\Sigma}^{-1}\boldsymbol{\mu})\,,$$

where $\chi^2(.\,,.)$ *denotes the noncentral* χ^2*-distribution.*

The simplest test for $\{H_{1,p}, K_p(\Theta)\}$ are the multivariate extensions of the univariate sign test which are due to Hodges (1955), Blumen (1958), and Bennet (1962) among others. Bickel (1965) considered a quadratic rank statistic based on the coordinatewise one-sample Wilcoxon statistics. Puri and Sen (1967) introduced and studied the general class of quadratic rank statistics Q_n^+. They derived the asymptotic distribution of $\boldsymbol{S}_n^+$ and Q_n^+ under general alternatives. Hušková (1971) treated more general rank statistics (for testing $H_{1,p}$ against the alternative (1.4)). Namely, she considered

$$Q_c^+ = \boldsymbol{S}_c^{+\prime}\boldsymbol{\Sigma}_{p,c}^{-}\boldsymbol{S}_c^+ \quad \text{and} \quad \boldsymbol{S}_c^+ = (S_{1c}^+, \ldots, S_{pc}^+)'\,,$$

where

$$S_{ic}^+ = \sum_{j=1}^{n} c_{ji}\,\mathrm{sign}(X_{ji})a_{ni}(R_{ji}^+), \quad i = 1, \ldots, p\,,$$

and

$$\boldsymbol{\Sigma}_{p,c} = \mathrm{var}_{H_{1,p}}(\boldsymbol{S}_c \mid \mathscr{A}_n)\,,$$

and obtained the asymptotic distribution of $\boldsymbol{S}_c^+$ and Q_c^+ under the hypothesis $H_{1,p}$ under local alternatives, and under general alternatives.

Most of the known results on rank tests for $H_{1,p}$ can be found in the book by Puri and Sen (1971).

The multivariate sign test. This test is based on Q_n^+ given by (3.6) with

$$S_{ni}^+ = \sum_{j=1}^{n} \mathrm{sign}(X_{ji}), \qquad \sigma_{p,ik} = \sum_{j=1}^{n} \mathrm{sign}(X_{ji})\,\mathrm{sign}(X_{jk}), \quad k, i = 1, \ldots, p\,. \tag{3.11}$$

By direct computations one obtains the elements of the variance matrix of $\boldsymbol{S}_n^+$ under $H_{1,p}$:

$$(4F_{ik}(0,0) - 1)n, \quad i, k = 1, \ldots, p\,.$$

The multivariate sign test is often used for testing the hypothesis that the $\boldsymbol{X}_j$ have zero median vectors against the alternative hypothesis (the distribution functions of the $\mathbf{X}_j$ need not be either identical or diagonally symmetric about some point).

In the literature special attention is paid to the bivariate case ($p = 2$). Particularly, when $\mathbf{X}_j$, $j = 1, \ldots, n$, are identically distributed Bennet (1964) proposed the statistic Q_n^+ with S_{ni} and $\sigma_{p,ik}$ given by (3.11) for testing the hypothesis that the median vector is a zero vector. Chatterjee (1966) generalized the test based on Q_n^+ with (3.11) to the case when $\mathbf{X}_j$'s are not necessarily identically distributed and studied its properties. Hodges (1955) and Blumen (1958) considered different sign tests. Hodges's test was studied further by Klotz (1964), Joffe and Klotz (1962), Dietz (1982) and tabulated for $n \leqslant 30$ (Joffe and Klotz, 1962).

The multivariate Wilcoxon one-sample test is the test based on Q_n^+ with

$$S_{ni}^+ = (n+1)^{-1} \sum_{i=1}^{n} R_{ji}^+ \operatorname{sign}(X_{ji}), \quad i = 1, \ldots, p, \tag{3.12}$$

and

$$\sigma_{p,ik} = (n+1)^{-2} \sum_{j=1}^{n} R_{ji}^+ R_{jk}^+ \operatorname{sign}(X_{ji}) \operatorname{sign}(X_{jk}), \quad i, k = 1, \ldots, p. \tag{3.13}$$

The elements σ_{ik} of the matrix $\boldsymbol{\Sigma}$ can be written in the form:

$$\sigma_{ik} = \int_0^\infty \int_0^\infty F_i^*(x) F_k^*(y) \, \mathrm{d}F_{ik}^*(x, y), \quad i, k = 1, \ldots, p.$$

This test was introduced by Bickel (1965). Putting

$$a_{ni}(j) = \Phi^{-1}\left(\left(\frac{j}{n+1} + 1\right)\Big/ 2\right), \quad j = 1, \ldots, n; \; i = 1, \ldots, p,$$

in (3.6), we get the *multivariate normal scores test*. Then, particularly,

$$S_{ni}^+ = \sum_{j=1}^{n} \operatorname{sign}(X_{ji}) \Phi^{-1}\left(\left(\frac{R_{ji}^+}{n+1} + 1\right)\Big/ 2\right), \quad i = 1, \ldots, p, \tag{3.14}$$

and

$$\sigma_{p,ik} = \sum_{j=1}^{n} \operatorname{sign}(X_{ji}) \operatorname{sign}(X_{jk}) \Phi^{-1}\left(\left(\frac{R_{ji}^+}{n+1} + 1\right)\Big/ 2\right)$$

$$\times \Phi^{-1}\left(\left(\frac{R_{jk}^+}{n+1} + 1\right)\Big/ 2\right), \quad k, i = 1, \ldots, p. \tag{3.15}$$

The elements of the asymptotic variance matrix $\boldsymbol{\Sigma}$ have the form:

$$\sigma_{ik} = \int_0^\infty \int_0^\infty \Phi^{-1}\left(\frac{F_i^*(x) + 1}{2}\right) \Phi^{-1}\left(\frac{F_k^*(y) + 1}{2}\right) \mathrm{d}F_{ik}^*(x, y),$$

$$i, k = 1, \ldots, p.$$

If the distribution function of $\boldsymbol{X}_j$ is multivariate normal with zero mean vector and nonsingular variance matrix $\boldsymbol{W}$, $j = 1, \ldots, n$, then $\boldsymbol{\Sigma} = \boldsymbol{W}$.

We close with several remarks on Hotteling's T_n^2-statistic and asymptotic efficiency.

If the distribution of $\boldsymbol{X}_j$ is normal with nonsingular variance matrix $\boldsymbol{W}$, $j = 1, \ldots, n$, then the most powerful invariant test for $H_{1,p}$ against $K_p(\Theta^+)$ is based on Hotelling's T_n^2 defined as follows:

$$T_n^2 = n\bar{\boldsymbol{X}}_n' \boldsymbol{W}_n^{-1} \bar{\boldsymbol{X}}_n ,$$

where

$$\bar{\boldsymbol{X}}_n = \frac{1}{n} \sum_{j=1}^{n} \boldsymbol{X}_j \quad \text{and} \quad \boldsymbol{W}_n = \frac{1}{n-1} \sum_{j=1}^{n} (\boldsymbol{X}_j - \bar{\boldsymbol{X}}_n)(\boldsymbol{X}_j - \bar{\boldsymbol{X}}_n)' .$$

The basic asymptotic properties are:

$$\mathcal{L}(T_n^2 \mid H_{1,p}) \xrightarrow{w} \chi^2(p; 0) \quad \text{as } n \to \infty$$

and

$$\mathcal{L}(T_n^2 \mid K_p(\sqrt{n}\,\theta)) \xrightarrow{w} \chi^2(p; \boldsymbol{\theta}' \boldsymbol{W}^{-1} \boldsymbol{\theta}) \quad \text{as } n \to \infty .$$

Thus T_n^2 and Q_n^+ have, asymptotically, χ^2 distributions with the same degrees of freedom, differing possibly only in their noncentrality parameters.

The asymptotic efficiency $e_{Q_n^+, T_n^2}$ of the test based on Q_n^+ with respect to Hotelling's T_n^2 is defined as the ratio of the noncentrality parameters corresponding to Q_n^+ and T_n^2, i.e.

$$e_{Q_n^+, T_n^2} = \frac{\boldsymbol{\mu}' \boldsymbol{\Sigma}^{-1} \boldsymbol{\mu}}{\boldsymbol{\theta}' \boldsymbol{W}^{-1} \boldsymbol{\theta}} ,$$

where $\boldsymbol{\mu}$ and $\boldsymbol{\Sigma}$ are defined by (3.10) and $\boldsymbol{W}$ is the variance matrix of $\boldsymbol{X}_j$.

If the distribution of $\boldsymbol{X}_j$ is multivariate normal with nonsingular variance matrix $\boldsymbol{W}$ then

$$\lim_{n \to \infty} e_{Q_n^+, T_n^2} = 1 ,$$

for Q_n^+ with S_{ni}^+ and $\sigma_{p,ik}$ given by (3.14) and (3.15) resp.

Some inequalities for further cases can be found in Puri and Sen (1971), Bickel (1965), Bhattacharya (1967).

References

Adichie, J. N. (1967). Asymptotic efficiency of a class of nonparametric test for regression parameters. *Ann. Math. Statist.* **38**, 884–893.

Albers, W. (1979). Asymptotic expansions for the power of one-sample against contiguous nonparametric alternatives. In: *Proceedings of the 2nd Prague Symposium on Asymptotic Statistics*. North-Holland, Amsterdam, 105–117.

Albers, W., Bickel, P. and van Zwet, W. R. (1976). Asymptotic expansion for the power of distribution free tests in the one-sample problem. *Ann. Statist.* **4**, 108–156.

Anderson, T. W. (1966). Some nonparametric multivariate procedures based on statistically equivalent blocks. In P. R. Krishnaiah, ed., *Proc. 1st Internat. Symp. Mult. Analysis*, pp. 5–27.

Bell, C. B. and Haller, H. S. (1969). Bivariate symmetry tests: parametric and nonparametric. *Ann. Math. Statist.* **40**, 259–269.

Bennet, B. M. (1964). On bivariate sign test, *JRSS B* **26**, 457–461.

Bennet, B. M. (1962). On multivariate sign tests. *JRSS B* **24**, 159–161.

Bhattacharya, G. K. (1967). Asymptotic efficiency of multivariate normal score test. *Ann. Math. Statist.* **38**, 1753–1758.

Bickel, P. (1965). On some asymptotically nonparametric competitors of Hotelling's T^2. *Ann. Math. Statist.* **36**, 160–173.

Blumen, J. (1958). A new bivariate sign test. *JASA* **53**, 448–456.

Butler, C. C. (1969). A test for symmetry using the sample distribution function. *Ann. Math. Statist.* **40**, 2209–2210.

Chatterjee, S. K. (1966). A bivariate sign test for location. *Ann. Math. Statist.* **37**, 1771–1782.

Conover, W. J. (1973). Rank test one sample, two samples and k samples without the assumption of a continuous distribution function. *Ann. Statist.* **1**, 823–831.

Conover, W. J., Wehnmanen and Ramsey, J. B. (1978). A note on the small-sample power functions for nonparametric tests of location in the double exponential family. *Journ. Amer. Statist. Assoc.* **73**, 188–190.

Davidson, R. R. and Bradley, R. A. (1969). Multivariate paired comparisons; the extension of a univariate model and associated estimation and test procedures. *Biometrika* **56**, 81–95.

Davidson, R. R. and Bradley, R. A. (1970). Multivariate paired comparisons: Some large sample results on estimation and tests of equality of preference. In: M. L. Puri, ed., *Nonparametric Tech. in Statist. Inf.* Cambridge Univ. Press, pp. 111–125.

Dietz, E. J. (1982). Bivariate nonparametric tests for the one-sample location problem. *J. Amer. Statist. Assoc.* **77**, 163–169.

Dixon, W. J. and Mood, A. M. (1946). The statistical sign test. *J. Amer. Statist. Assoc.* **41**, 557–566.

Doksum, K. and Thompson, R. (1971). Power bounds and asymptotic minimax results for one sample rank tests. *Ann. Math. Statistis.* **42**, 12–34.

Fellingham, S. A. and Stoker, D. J. (1964). An approximation for the exact distribution of the Wilcoxon test for symmetry. *J. Amer. Statist. Assoc.* **59**, 899–905.

Finch, S. J. (1977). Robust univariate test of symmetry. *J. Amer. Statist. Assoc.* **72**, 387–392.

Fraser, D. A. S. (1957). Most powerful rank-type tests. *Ann. Math. Statist.* **28**, 1040–1043.

Gupta, M. K. (1967). An asymptotically nonparametric test of symmetry. *Ann. Math. Statist.* **38**, 849–866.

Hájek, J. (1955). Some rank distribution and their use (Czech.) *Časopis pro pěst. matematiky* **80**, 17–31.

Hájek, J. (1969). *Nonparametric Statistics*. Holden-Day, San Francisco.

Hájek, J. and Šidák, Zb. (1967). *Theory of rank tests*. Academia, Prague.

Hemelrijk, J. (1952). A theorem on the sign test when ties are present. *Indagationes Math.* **14**, 322–326.

Hill, D. L. and Rao, P. V. (1977). Tests for symmetry based on Cramér-von Mises Statistics. *Biometrika* **64**, 489–494.

Hodges, J. L. Jr. (1955). A bivariate sign test. *Ann. Math. Statist.* **26**, 523–527.

Hollander, M. (1971). A nonparametric test for bivariate symmetry. *Biometrika* **58**, 203–212.

Hollander, M. and Wolfe, D. A. (1973). Nonparametric statistical methods. Wiley, New York.

Hušková, M. (1970). Asymptotic distribution of linear rank statistics used for testing symmetry. *Z. Wahrsch. Verw. Gebiete* **14**, 308–322.

Hušková, M. (1971). Asymptotic distribution of rank statistics used for multivariate testing symmetry. *J. of Multivariate Analysis* **1**, 461–484.

Irle, A. and Klösener, K. (1980). Note on the sign test in the presence of ties. *Ann. Statist.* **5**, 1168–1170.

Joffe, A. and Klotz, J. (1962). Null distribution and Bahadur efficiency of the Hodges bivariate sign test. *Ann. Math. Statist.* **33**, 803–807.

Klotz, J. (1963). Small sample power and efficiency for the one sample Wilcoxon and normal scores test. *Ann. Math. Statist.* **34**, 624–632.

Klotz, J. (1964). Small sample power of the bivariate sign tests of Blumen and Hodges. *Ann. Math. Statist.* **35**, 1576–1582.

Klotz, J. (1965). Alternative efficiences for signed rank tests. *Ann. Math. Statist.* **36**, 1759–1766.

Koul, H. L. and Staudte, R. G. Jr. (1972). Asymptotic normality of signed rank statistics. *Zeitschrift Wahrsch. Verw. Gebiete* **22**, 293–300.

Koul, H. L. and Staudte, R. G. Jr. (1972b). Power bounds for Smirnov statistics in testing the hypothesis of symmetry. Michigan State University.

Koziol, J. (1979). A test for bivariate symmetry based on the empirical distribution function. *Comm. Statist. A* **8**, 207–221.

Koziol, J. (1980). On a Cramer–von Mises type statistic for symmetry. *J. Amer. Statist. Assoc.* **75**, 161–167.

LeCam, L. (1960). Locally asymptotically normal families of distributions. *Univ. Calif. Publ. in Stat.* **3**, 37–98.

Lehman, E. L. (1959). *Testing Statistical Hypothesis.* Wiley, New York.

Mac Kinnon, W. J. (1964). Table for both the sign test and distribution-free confidence intervals of the median for sample sizes to 1000. *J. Amer. Statist. Assoc.* **59**, 935–956.

Orlov, A. J. (1972). On testing symmetry of distribution. *Th. Probab. Applic.* **17**, 357–361.

Owen, B. (1962). *Handbook of Statistical Tables.* Addison-Wesley, Reading, MA.

Prášková, Z. (1976). Asymptotic expansion and local limit theorem for the signed Wilcoxon statistic. *CMUC* **17**, 335–344.

Pratt, J. W. (1959). Remark on zeros and ties in some nonparametric tests. *Ann. Math. Statist.* **26**, 368–386.

Puri, M. L. and Sen, P. K. (1969). On the asymptotic normality of one sample rank order statistics. *Teoria Verojatnost. i Primenen.* **1**, 167–172.

Puri, M. L. and Sen, P. K. (1971). *Nonparametric Methods in Multivariate Analysis.* Wiley, New York.

Puri, M. L. and Shane, H. D. (1970). Statistical inference in incomplete block designs. In: M. L. Puri, ed., *Nonparametric Techniques in Statistical Inference.* Cambridge Univ. Press, pp. 131–153.

Putter, J. (1955). The treatment of ties in some nonparametric tests. *Ann. Math. Statist.* **26**, 368–386.

Rothaman, E. D. and Woodroofe, M. (1972). A Cramér–von Mises type statistic for testing symmetry. *Ann. Math. Statist.* **34**, 2035–2038.

Rüschendorf, L. (1976). Hypothesis generating groups for testing multivariate symmetry. *Annals of Statist.* **4**, 791–795.

Russel, C. T. and Puri, M. L. (1974). Joint asymptotic multinormality for a class of rank order statistics in multivariate paired comparisons. *J. Multivariate Analysis* **4**, 88–105.

Savage, J. R. (1959). Contribution to the theory of rank order statistics; one-sample case. *Ann. Math. Statist.* **30**, 1018–1023.

Selected Tables in Mathematical Statistics Vol. 1. Edited by the Institute of Mathematical Statistics, Vol. 2, pp. 171–260.

Sen, P. K. (1967). Nonparametric tests for multivariate interchangeability. Part I. *Sankhya Ser. A* **29**, 351–372.

Sen, P. K. (1970). On the distribution of the one-sample rank order statistics. In: M. L. Puri, ed., *Nonparametric Tech. Statist. Inf.* Cambridge Univ. Press, pp. 53–72.

Sen, P. K. (1971). On a class of aligned rank order tests for multiresponse experiments in some incomplete block design. *Annals of Math. Statist.* **42**, 1104–1112.

Sen, P. K. (1979). On some distribution-free tests for affine symmetry. *Calcutta Statist. Assoc. Bull.* **27**, 105–108, 59–79.

Sen, P. K. and David, H. A. (1968). Paired comparison for paired characteristics. *Ann. Math. Statist.* **39**, 200–208.

Sen, P. K. and Puri, M. L. (1967). On the theory of rank order tests for location in the multivariate one sample problem. *Ann. Math. Statist.* **38**, 1216–1228.

Shane, H. D. and Puri, M. L. (1969). Rank order tests for multivariate paired comparisons, *Ann. Math. Statist.* **40**, 2101–2117.

Shirahata, S. (1974). On tests of symmetry for distrete populations. *Austral. J. Statist.* **6**, 83–90.

Smirnov, N. V. (1947). On criteria for the symmetry of distribution laws of random variables. *Doklady Akademii Nauk SSSR*, 11–14.

Snijders, T. (1981). Rank tests for bivariate symmetry. *Ann. Statist.* **9**, 1087–1095.

Srinivan, R. and Godio, L. B. (1974). A Cramér–von Mises type statistic for testing symmetry. *Biometrika* **61**, 196–198.

van Eeden, C. and Benard, A. (1957). A general class of distribution-free tests for symmetry containing the tests of Wilcoxon and Fisher I, II, III. *Indagationes Math.* **19** (Proc. Kon. Nederl. Akad. Wet. 60), 381–391, 392–400, 401–408.

van Eeden. C. (1963). The relation between Pitman's asymptotic relative efficiency of two tests and the correlation coefficient between their test statistics. *Ann. Math. Statist.* **34**, 1442–1451.

van der Waerden, B. L. and Nievergelt, E. (1956). Tafeln zum Vergleich zweier Stichproben mittels χ-Tests und Zeichentest, Springer, Berlin.

Vorličková, D. (1972). Asymptotic properties of rank tests of symmetry under discrete distributions. *Ann. Math. Statist.* **43**, 2013–2018.

Wilcoxon, F. (1945). Individual comparisons by ranking methods. *Biometrics Bull. I*, 80–83.

Wilcoxon, F. (1947). Probability tables for individual comparisons by ranking methods. *Biometrics* **3**, 119–122.

Wilcoxon, F. (1949). *Some rapid approximation procedures.* American Cyanamid Company, Standord.

Yanagimoto, T. and Sibuya, M. (1972). Test of symmetry of a one-dimensional distribution against positive biasedness. *Ann. Inst. Statist. Math.* **24**, 423–434.

Yanagimoto, T. and Sibuya, M. (1976). Test on symmetry of a bivariate distribution. *Sankhya Ser. A* **38**, 105–115.

Walsh, J. E. (1951). Some bounded signification level properties of the equal-tail sign test. *Ann. Math. Statist.* **22**, 408–417.

P. R. Krishnaiah and P. K. Sen, eds., *Handbook of Statistics, Vol. 4*
© Elsevier Science Publishers (1984) 79–88

4

Measures of Dependence

Kumar Joag-Dev

0. Introduction

Stochastic dependence between random measurements is one of the important aspects of many statistical investigations. The quantification of this concept for the bivariate distribution is attempted with two entirely different, and in fact, opposing viewpoints. The viewpoint which we will consider first, is concerned with concordance of the pairs. To explain this more precisely, suppose (X_i, Y_i), $i = 1, \ldots, n$ constitutes a random sample from a bivariate population with the joint distribution function $F(X, Y)$ and marginal distribution functions F_1 and F_2. Further let $X_{(1)} \leqslant \cdots \leqslant X_{(n)}$ be the order statistic of X observations. Then, the first type of motivation for constructing a measure of dependence is to ascertain the degree of conformity of this ascending order followed by Y observations. Clearly such measures are (deliberately) insensitive to monotone transformations of X and Y observations. In some sense, such measures are well suited for the families exhibiting positive or negative dependence. The above invariance also makes these measures 'nonparametric'. In fact, most of these are used as test statistics for testing the hypothesis of independence.

The other perspective is related to the utilization of dependence for the purpose predicting one variable from the other. It is hoped that if X is closer to a function of Y, say $g(Y)$, then the measure should be higher. One of the basic measures, namely, the product moment correlation, is based on

$$\operatorname{cov}(X, Y) = EXY - EX \cdot EY. \tag{0.1}$$

In order to make this independent of the units with which X and Y are measured, the covariance is divided by the product of standard deviations, making it a coefficient. As an added bonus, this normalization makes its range to be -1 to $+1$.

In this article the discussion is limited only to the bivariate distributions. Also, the measures of association which are mainly used in the categorical data analysis will not be dealt here. In order to strip the discussion of some

Research supported by AFOSR-81-0038.

(non-essential) technical terms, throughout it will be assumed that the moments, derivatives etc. in the assertions will exist. All the functions will be measureable, and unless specified otherwise all the distributions will be absolutely continuous.

1. Measures of concordance

With the same notation as in the Introduction, the general idea of constructing a measure is to develop a notion of deviation between F and the product measure $F_1 \cdot F_2$. Thus for the fixed marginals if F deviates more, the measure would be higher. Usually for every such a concept, there is a population version and a sample version of a measure. In most cases the population version is either the mean or the asymptotic mean of the sample version. We begin with the

(a) *Spearman's rank correlation coefficient.* This is one of the oldest measure based on ranks. Spearman proposed this in 1904 and the idea came from Francis Galton's use of the 'grades' (or the percentiles) of a distribution. Let R_i and S_i denote the ranks of X_i and Y_i among X and Y observations respectively and write

$$d_i = R_i - S_i .$$

The initial proposal of Spearman was to use $\Sigma\,|d_i|$, now known as Spearman's 'foot rule'. However, due to mathematical convenience, the product moment correlation between R_i and S_i, namely

$$\rho_s = \frac{12}{n(n^2-1)} \sum_{i=1}^{n} \left(R_i - \frac{n+1}{2}\right)\left(S_i - \frac{n+1}{2}\right) = 1 - \frac{6\Sigma_{i=1}^{n} d_i^2}{n(n^2-1)}, \tag{1.1}$$

gained much more popularity. An excellent historical account of the measure is contained in Hotelling and Pabst (1936). This paper is remarkable for other reasons as well. The authors not only list justifications for using the 'nonparametric' approach, well before this very term came in the vogue, but even found the Pitman efficiency relative to the product moment correlation under the assumption of normality, (now well known to be $9/\pi^2$) even before the concept of asymptotic efficiency was well defined. To find the population analog, observe that the ranks are unchanged if X_i and Y_i are replaced by $F_1(X_i)$ and $F_2(Y_i)$ respectively, each of which have uniform distribution on $(0, 1)$. It can be shown that the asymptotic mean of ρ_s is the same as

$$\rho_{S,P} = 3 \int\int \{2F_1(x) - 1\}\{2F_2(y_1) - 1\}\, \mathrm{d}F(x, y)\,, \tag{1.2}$$

where $2F_1(x) - 1$ is chosen instead of $F_1(x)$, to make the uniform distribution symmetric about 0. However, it is the product moment correlation between $F_1(X)$ and $F_2(Y)$. The coefficient in (1.2) is known as the 'grade correlation'. Of particular interest is the value of this grade correlation when the distribution is bivariate normal,

$$\rho_{S,P}(N) = 12 \int\int \Phi(x)\Phi(y)\psi(x, y; \sigma_{12})\, \mathrm{d}x\, \mathrm{d}y - \tfrac{3}{4}, \tag{1.3}$$

where Φ is the distribution function of the standard normal and ψ is the bivariate normal density with mean vector 0 and unit standard deviations and correlation σ_{12}. To evaluate the integral, Hotelling and Pabst (1936) used an ingeneous trick. First they observe

$$\frac{\partial}{\partial \sigma_{12}} \psi(x, y : \sigma_{12}) = \frac{\partial^2}{\partial x\, \partial y} \psi(x, y : \sigma_{12}), \tag{1.4}$$

an identity usually attributed to Plackett (1954). Thus from (1.3) it follows, after integration by parts (the same method employed by Slepian, 1962), that

$$\frac{\partial}{\partial \sigma_{12}} \rho_{S,P}(N) = \frac{6}{\pi\sqrt{4 - \sigma_{12}^2}},$$

and hence

$$\rho_{S,P}(N) = \frac{6}{\pi} \arcsin(\sigma_{12}/2). \tag{1.5}$$

Clearly, $\rho_{S,P}$ varies from -1 to 1 as σ_{12} does.

Finally, to express the rank correlation in terms of the indicators, define the signum function as

$$s(x) = \begin{cases} 1 & \text{if } x > 0, \\ \frac{1}{2} & \text{if } x = 0, \\ -1 & \text{if } x < 0. \end{cases} \tag{1.6}$$

Then

$$\left(R_i - \frac{n+1}{2}\right) = \tfrac{1}{2}\left\{\sum_{j=1}^{n} s(x_i - x_j) - \tfrac{1}{2}\right\} \tag{1.7}$$

and hence

$$\rho_S = \frac{3}{n(n^2 - 1)} \sum_{i=1}^{n} \sum_{j=1}^{n} \sum_{k=1}^{n} s(X_i - X_j) s(Y_i - Y_k), \tag{1.8}$$

which is in the form of the well-known U-statistic.

(b) *Kendall's tau.* This measure is defined as the product moment correlation of 'signs of concordance',

$$\tau = \frac{1}{n(n-1)} \sum\sum_{i \neq j} s(X_i - X_j)s(Y_i - Y_j)\,. \tag{1.9}$$

In this form, it looks like a simplified version of (1.8), however, the motivation behind it, as explained by Kendall (1938), was to see how well the two sequences follow the monotone order. For example, suppose in a psychological experiment each child is asked to order four glass bottles according to the amount of water contained in them. If a child orders them 3, 1, 2, 4 instead of the correct order 1, 2, 3, 4, the measure τ can be computed as follows. Compare the first number 3 in the sequence with the remaining. For each reverse order, such as (3, 1) and (3, 2) the child scores -1, while for the correct order, such as (3, 4), the score is $+1$. So the comparing 3, one obtains a subtotal -1. Comparing each following number with all its successors, the total score obtained from the six comparisons is $+2$. Thus τ equals $\frac{2}{6}$. Kendall (1938) justifies the idea of such comparisons by saying that ".... τ has a natural significance. An observer who is given a set of objects (such as coloured discs) to rank, appears to follow a process something like this: First of all he searches for the beginning of the series say the disc of the lightest shade. Having selected a disc, he compares it with each of the remainder to verify the propriety of his choice. The coefficient τ gives him one mark for each comparison he has done correctly and subtracts a mark for each error".

It should be noted that the same measure was also considered by Esscher (1924) and Lindeberg (1925 and 1929).

Although, the measure was primarily defined for the sample, the corresponding population analog can be seen to be

$$\tau_p = 2P[(X_1 - X_2)(Y_1 - Y_2) > 0] - 1\,,$$

which is same as the correlation coefficient of the indicators of $X_1 - X_2 > 0$ and $Y_1 - Y_2 > 0$. Again, for the bivariate normal density, this measure equals

$$\tau_P(\sigma_{12}) = 4\int^{\infty}\int^{\infty} \psi(x, y; \sigma_{12})\, dx\, dy - 1\,.$$

Using (1.4), it follows that

$$\frac{\partial}{\partial \sigma_{12}} \tau_P(\sigma_{12}) = 4\psi(0, 0, \sigma_{12}) = \frac{2}{\pi\sqrt{1-\sigma_{12}^2}}\,. \tag{1.10}$$

Hence

$$\tau_P(\sigma_{12}) = \frac{2}{\pi} \arcsin \sigma_{12}\,. \tag{1.11}$$

A systematic study of some of the above measures as metrics on the set of permutations of the n positive integers was made by Diaconis and Graham (1977). Besides Spearman's 'footrule', Spearman's rank correlation coefficient and Kendall's τ, they also studied a metric $I(\pi, \sigma)$ between two permutations π and σ, obtained by considering the minimum number pairwise transpositions required to achieve certain alignment. They obtained some interesting inequalities for these metrices.

Some multivariate analogs of the above measures have been studied in the literature. Ehrenberg (1952) did some numerical comparisons of the following two measures of concordance between m judges who were asked to rank n objects. The first measure is obtained by taking an average of the $\binom{m}{2}$ rank correlations while the second is obtained by using Kendall's τ instead of the rank correlation. Hays (1960) studied sampling distributions of the above measures and made a numerical comparison with some χ^2 approximations. More recently Alvo, Cabilio and Feign (1982) made the asymptotic study of these measures and showed that the average τ is superior (asymptotically) to average rank correlation coefficient.

(c) *Blomquist's q*. Blomquist (1950) proposed a measure which is similar in the spirit as 'tau', where symmetric differences $X_i - X_j$ and $Y_i - Y_j$ are replaced by the deviations from the sample medians $\tilde{X}_n$ and $\tilde{Y}_n$ respectively. Thus

$$q = \frac{\Sigma\, s(X_i - \tilde{X}_n)s(Y_i - \tilde{Y}_n)}{n}. \tag{1.12}$$

Another interpretation to this measure can be given in terms of the number of pairs residing in the four quadrants of the plane created at the point $(\tilde{X}_n, \tilde{Y}_n)$. If n_1 is the sum of the numbers in $(+,+)$ and $(-,-)$ and n_2 in the remaining two quadrants then

$$q = \frac{n_1 - n_2}{n_1 + n_2}. \tag{1.13}$$

Such a measure was also considered by Mosteller (1946), who called it as one of the 'useful inefficient statistics'. The measure can be construed as the bivariate version of the so called 'median test' for testing the hypothesis whether the two random samples came from the same population.

The population analog of q is clearly

$$q_P = 2P[(X - m_1)(Y - m_2) > 0] - 1\,.$$

The evaluation of this for the case of the bivariate normal distribution is the same as given in (1.10) and hence $2/(\pi \arcsin \sigma_{12})$.

(d) *Hoeffding's Δ*. The following measure, proposed by Hoeffding (1948b), is similar to the well known notion of the 'distance' between two distribution functions suggested by Kolmogorov, Smirnov, Cramèr and von Mises.

$$\Delta(F) = \int\!\!\int \{F(x, y) - F_1(x)F_2(y)\}^2\, \mathrm{d}F(x, y)\,. \tag{1.14}$$

This is an appropriate measure only when F is absolutely continuous. For this case $\Delta(F) = 0$ implies the independence of the components X and Y. For example, if $P[X = 0, Y = 1] = P[X = 1, Y = 0] = \frac{1}{2}$, then it is easy to see that $\Delta = 0$. However, X and Y completely determine each other and thus totally dependent. A more detailed study of this measure and its relation to equiprobable rankings and independence was done by Yanagimoto (1970). The sample measure is defined after a succession of definitions of the following functions. Let

$$c(u) = \begin{cases} 1 & \text{if } u \geq 0\,, \\ 0 & \text{otherwise}\,. \end{cases}$$

$$\psi(x_1, x_2, x_3) = c(x_1 - x_2) - c(x_1 - x_3)\,,$$

$$\phi(x_1, y_1; x_2, y_2; \ldots; x_5, y_5)$$
$$= \tfrac{1}{4}\psi(x_1, x_2, x_3)\psi(x_1, x_4, x_5)\psi(y_1, y_2, y_3)\psi(y_1, y_4, y_5)\,.$$

Then

$$D_n = \frac{1}{nP_5} \sum \phi(X_{\alpha_1}, Y_{\alpha_1}; X_{\alpha_2}, Y_{\alpha_2}; \ldots; X_{\alpha_5}, Y_{\alpha_5})\,, \tag{1.15}$$

where the sum is taken over all 5 tuples with $\alpha_i = 1, \ldots, n$, $\alpha_i \neq \alpha_j$; for $i \neq j$ and nP_k is same as $n!/(n-k)!$. Although, the expression in (1.15) has an advantage of being recognized as a U-statistic, Hoeffding (1948b) showed that an alternative expression in terms of ranks can be given as follows. Let

$$A = \sum_1^2 (R_i - 1)(R_i - 2)(S_i - 1)(S_i - 2), \quad B = \sum_1^n (R_i - 2)(S_i - 2)T_i,$$

$$C = \sum_1^n T_i(T_i - 1)\,,$$

where, as before, R_i and S_i are the ranks of X_i and Y_i and

$$T_i = \{j \colon X_j < X_i \text{ and } Y_j < Y_i\}$$

or the number of pairs in the lower quadrant with the vertex (X_i, Y_i). Then

$$D_n = \frac{A - 2(n-2)B + (n-2)(n-3)C}{nP_5}\,.$$

Blum, Kiefer and Rosenblatt (1961), while studying the asymptotic behaviour, proposed a statistic asymptotically equivalent to D_n,

$$B_n = \frac{1}{n^4} \sum_{i=1}^n [N_1(i)N_3(i) - N_2(i)N_4(i)]^2\,,$$

where N_1, N_2, N_3, N_4 are the number of pairs residing in the four quadrants of the plane formed at (X_i, Y_i).

It should be noted that unlike the measures defined in (a), (b), (c), the measures (1.14) or (1.15) due to its similarity with the notion of distance, do not take negative values.

Other distance measures can be easily constructed by replacing $\mathrm{d}F(x, y)$ in (1.14) by $\mathrm{d}F(x)\,\mathrm{d}F(y)$. Other notions such as Kolmogorov–Smirnov distance could also be used for measuring the deviation.

2. Properties of concordant measures

As commented before, the measures defined in Section 1 are designed for detecting positive or negative dependence. A notion of positive quadrant dependence (PQD) studied by Lehmann (1966), is defined by the condition,

$$F(x, y) \geqslant F_1(x)F_2(y)\,, \tag{2.1}$$

for every (x, y). From an elegant formula of Hoeffding (1940),

$$\operatorname{cov}(X, Y) = \int_{-\infty}^{\infty}\!\!\int \{F(x, y) - F_1(x)F_2(y)\}\,\mathrm{d}x\,\mathrm{d}y \tag{2.2}$$

it is immediate that (2.1) implies

$$\operatorname{cov}(X, Y) \geqslant 0\,,$$

equality holding if and only if X, Y are independent. Further, if f and g is a pair of nondecreasing functions, clearly $f(X)$ and $g(Y)$ inherit PQD property from (X, Y) and hence (2.1) becomes equivalent to

$$\operatorname{cov}[f(X), g(Y)] \geqslant 0\,, \tag{2.3}$$

for every pair of nondecreasing functions f, g. Yanagimoto and Okamoto (1969) introduced the natural partial ordering, namely larger PQD, for distributions with fixed marginals. Thus if F and F^* have the same marginals and

$$F^*(X, Y) \geqslant F(X, Y) \tag{2.4}$$

then F^* is said to possess larger PQD than F. It was observed by Hoeffding (1940) and Fréchet (1951) that the class of distributions, with fixed marginals F_1, F_2 has maximal and minimal elements, in the sense,

$$\begin{aligned} H^-(x, y) = \max\{F_1(x) + F_2(y) - 1, 0\} &\leqslant F(x, y) \\ &\leqslant \min\{F_1(x), F_2(y)\} = H^+(x, y)\,, \end{aligned}$$

for every (x, y).

It is clear that larger PQD makes cov(f, g) larger and clearly H^+ and H^- provide upper and lower bounds. Yanagimoto and Okamoto (1969) show that the product moment correlation, Kendall's τ, Blomquist's q and such measures are ordered along with the ordering (2.4), and attain upper and lower bounds when the underlying distributions are H^- and H^+. When the sample measures are used for testing hypothesis of independence, this result shows the montonicity of the power function against the alternatives with PQD ordering.

Most of the authors who proposed the measures of dependence, have also studied their asymptotic distributions. A very general method, now well known as U-statistics, was developed by Hoeffding (1948a). Blum, Kiefer and Rosenblatt (1961) used a method involving stochastic processes to study the asymptotic behaviour. In (1.1) if the ranks are replaced by rank scores, such as normal scores, one could easily obtain a new measure. The asymptotic normality of such measures which may not be U-statistics, was established by Bhuchongkul (1964).

3. Dependence measures for functional relationship

The most commonly used yardstick for measuring the dependence between two variables is the product moment correlation coefficient. It is well known that this measures *linear dependence*. If the goal is to measure functional dependence, such as $Y = g(X)$, or in general, whether $h(X, Y)$ equals constant, then correlation coefficient is certainly unsuitable. The standard example where the mass is uniformly distributed on $x^2 + y^2 = 1$ shows that the correlation coefficient can be 0 while a perfect functional dependence exists.

There have been several articles, where one starts with certain desirable conditions on the measure of dependence and then examining how some of the measures commonly used live up to those conditions. By modifying some of these conditions some authors have proposed new measures. Hall (1970) has given an excellent survey of the developments in this direction and has proposed a new measure having different desirable properties. We give a brief outline of these results.

The following is a list of the conditions which one would like a measure of dependence $\xi(X, Y)$ to satisfy. The systematic search of a measure via such conditions was first made by Renyi (1959).

(a) $\xi(X, Y) = \xi(Y, X)$.

(b) $0 \leq \xi \leq 1$.

(c) $\xi = 0$ if and only if X and Y are independent.

(d) $\xi = 1$ if either $X = g(Y)$, almost surely, or $Y = g(X)$ almost surely, for some measurable function.

(e) $\xi(f(X), g(Y)) \leq \xi(X, Y)$.

(f) If (X, Y) has a bivariate normal distribution then ξ agrees with the absolute value of the correlation coefficient.

In many instances the condition (a) may be undesirable. This is clear if one wants to predict Y by utilizing the knowledge about X. It is possible that Y may be a function of X, but the other way around may not be true.

The concept of 'maximal correlation' has been studied by Gebelein (1941), Saramonov (1958a,b) and Renyi (1959). This measure is the least upper bound for the correlation coefficient between $f(X)$ and $g(Y)$ where f and g are chosen from the class of all measurable functions. Although such a measure satisfies several conditions on the list, it has tendency to be large when X and Y are almost independent. Hall (1970) gives the following example. Let $1-\varepsilon$ mass be uniformly distributed on the unit square and the remaining ε over the upper quadrant formed at (1, 1). Choosing f and g to be indicators of $X>1$ and $Y>1$ respectively, it is seen that the 'maximal correlation' is 1, while with ε small, X and Y are almost independent. Assuming the existence of the probability density, a measure of dependence based on information theoretic concepts was proposed by Linfoot (1957). It is given by

$$r_0(X, Y) = E\left\{\ln\left[\frac{p(X, Y)}{p_1(X)p_2(Y)}\right]\right\},$$

where p_1 and p_2 are the marginal densities of P.

A nonsymmetric measure which has been used in applications is the correlation ratio η of X on Y. This measure is defined as the square of the correlation coefficient between $M(Y)=E[X\mid Y]$ and Y. The measure η does satisfy many conditions on the list although, it may not exist. Also, $M(Y)$ could be 0 and hence $\eta=0$, while X, Y could be dependent. For example, consider the uniform distribution on a disc centered at 0. Hall (1970) considers some indices or 'dependence characteristics' which are correlation ratios of e^{itX} on Y for certain values of t. It turns out that these possess several desirable properties.

Kimeldorf and Sampson (1978) define a notion of 'monotone dependence' which avoids some drawbacks of the 'maximal correlation'. This is achieved by restricting to f, g which are monotone.

Although, the effort of constructing the measures in this section is mathematically interesting, one may wonder whether one number can tell anything about the complex dependence which might exist in a bivariate distribution. Without some further assumptions, two entirely different conditions of dependence might produce the same index and will hardly carry any information. The situation is similar to guessing similarity between two distributions having the same mean. Finally, the detailed knowledge required to arrive at an index may not be available. Also, usually there do not exist sample analogs of the proposed indices from which one can estimate their population values.

References

[1] Alvo, M., Cabilio, P. and Feign, P. D. (1982). Asymptotic theory for measures of concordance with special reference to average Kendall tau. *Ann. Statist.* **10**, 1269–1276.

[2] Bhuchongkul, S. (1964). A class of nonparametric tests for independence in bivariate populations. *Ann. Math. Stat.* **35**, 138–149.

[3] Blomquist, N. (1950). On a measure of dependence between two random variables. *Ann. Math. Stat.* **21**, 593–600.

[4] Blum, J. R., Kiefer, J. and Rosenblatt, M. (1961). Distribution free tests of independence based on the sample distribution function. *Ann. Math. Stat.* **32**, 485–498.

[5] Diaconis, P. and Graham, R. L. (1977). Spearman's footrule as a measure of disarray. *J. Roy. Statist. Soc. Ser. B.* **39**, 262–268.

[6] Ehrenberg, A. S. C. (1952). On sampling from a population of rankers. *Biometrika* **39**, 82–87.

[7] Esscher, F. (1924). On a method of determining correlation from the ranks of the variates. *Aktur. tids.* **7**, 201–219.

[8] Fréchet, M. (1951). Sur les tableux de corrélation don les marges sont données. *Ann. Univ. Lyon Sect. Ser. 3.* **14**, 53–77.

[9] Gebelein, H. (1941). Das Statistiche Problem der Korrelation als Variations und Eigenwertproblem und sein Zusaminenhang mit der Ausgleichungsrechnung. *Zeit. für Agnew. Math. und Mech.* **21**, 364–379.

[10] Hall, W. J. (1970). On characterizing dependence in joint distributions. In: *Essays in Probability and Statistics*, Univ. of N.C. Press. 339–376.

[11] Hays, W. L. (1960). A note on average tau as a measure of concordance. *J. Amer. Statist. Assoc.* **55**, 331–341.

[12] Hoeffding, W. (1940). Masstabinvariante Korrelationstheorie. *Schriften des Math. Inst. des Inst. für Ange. Math. der Univ. Berlin* **5**, 179–233.

[13] Hoeffding, W. (1948a). A class of statistics with asymptotically normal distribution. *Ann. Math. Stat.* **19**, 293–325.

[14] Hoeffding, W. (1948b). A nonparametric test of independence. *Ann. Math. Stat.* **19**, 546–557.

[15] Hotelling and Pabst (1936). Rank correlation and tests of significance involving no assumptions of normality. *Ann. Math. Stat.* **7**, 29–43.

[16] Kendall, M. (1938). A new measure of rank correlation. *Biometrika.* **30**, 81–93.

[17] Kimeldorff, G. and Sampson, A. R. (1978). Monotone Dependence. *Ann. Stat.* **6**, 895–903.

[18] Lehmann, E. L. (1966). Some concepts of dependence. *Ann. Math. Stat.* **37**, 1137–53.

[19] Lindeberg (1925). Uber die Korrelation. *Skand. Mathematiker kongres i Kobenhaun.* **1**, 437–446.

[20] Lindeberg (1929). Some remarks on the mean error of the percentage of correlation. *Nordic Stat. J.* **1**, 137–41.

[21] Linfoot, E. H. (1957). *Inf. Control.* **1**, 85–89.

[22] Mosteller, F. (1946). On some useful 'inefficient' statistics. *Ann. Math. Stat.* **17**, 377.

[23] Plackett, R. L. (1954). A reduction formula for normal multivariate integrals. *Biometrika* **41**, 351–360.

[24] Renyi, A. (1959). On measures of dependence. *Acta-Math. Acad. Sc. Hung.* 441–51.

[25] Saramonov, O. V. (1958a). Maximum correlation coefficient (symmetric case). *Dokl. Akad. Na. SSSR* **120**, 715–718.

[26] Saramonov, O. V. (1958b). Maximum correlation coefficient (nonsymmetric case). *Dokl. Akad. Na. SSSR* **121**, 52–55.

[27] Slepian, D. (1962). The one sided barrier problem for Gaussian noise. *Bell. Syst. Tech. J.* **41**, 463–501.

[28] Yanagimoto, T. and Okamoto, M. (1969). Partial orderings of permutations and monotonicity of a rank correlation statistic. *Ann. Inst. Stat. Math.* **21**, 489–506.

P. R. Krishnaiah and P. K. Sen, eds., *Handbook of Statistics, Vol. 4*

5

Tests of Randomness against Trend or Serial Correlations

Gouri K. Bhattacharyya

1. Introduction

In this chapter we are concerned with the situations where observations $X_1, \ldots, X_N$ of some response (X) are collected sequentially over time or, more generally, according to the level of presence of some other factor. The problem of interest is to test the null hypothesis that the time order or the intensity of the concomitant factor has no effect on the response. The null hypothesis of *randomness of the series*, as it is generally called, is then formalized by the statement that $X_1, \ldots, X_N$ are independent and identically distributed (iid) random variables with an unknown distribution function. This amounts to the assertion that any ordering of their magnitudes is as likely to occur as any other or that the arrangement of the members in the sequence is random. The appropriate nonparametric procedure for testing randomness would largely depend on the nature of deviation from randomness that the investigator seeks to detect from the data.

A general description of the main types of the alternative hypotheses along with some brief motivating examples will help delineate the scope of this chapter. For brevity of discussion we will consider time as the covariate in which case the observations $X_1, X_2, \ldots$ constitute a time series. The ideas readily extend to such other concomitant factors as spatial proximity, the order of birth in a sibling, or the intensity of an agent in a chemical treatment.

Perhaps the most natural aspect of a possible time effect on the response (X) is whether or not the overall level of the response exhibits a pattern of change over time. The alternative hypothesis of an upward (downward) *trend* arises when one suspects that the successive responses $X_1, \ldots, X_N$ constitute a stochastically increasing (decreasing) sequence. Denoting by $F_i(x)$ the distribution function (df) of X_i, a stochastically increasing sequence is defined by the requirement that $F_i(x)$ is non-increasing in i for every fixed x. The monotone location model $F_i(x) = F(x - \theta_i)$, $\theta_1 \leq \theta_2 \leq \cdots \leq \theta_N$ is an important special case. Studies of severity of the winter, ozone concentration in the atmosphere, pollution in a lake, social or economic indicators over time are some examples where detection of a trend is of paramount importance to the investigator. Tests of randomness against trend alternatives are discussed in Section 2.

A somewhat different case of the trend alternatives arises when instead of a steady upward (downward) movement of the series one suspects that an unforeseen one-time disturbance may have occurred at an unknown time point m causing the distribution of X to remain the same at $i = 1, \ldots, m-1$, shift up (or down) at m and remain stable thereafter. An alternative hypothesis of this type is designated as the hypothesis of *jump at an unknown time point*, and the relevant tests are discussed in Section 3. As an example, the performance of a production process may be affected by an abrupt internal disorder which would cause the quality level of the product to slide down at one point of time and then remain stable. This is in contrast to the situation of a progressive wear where the general trend model is more appropriate.

One other important mode of deviation from the null hypothesis of randomness is conceived in terms of the alternatives of *serial dependence*. Here the level of the series, say the location of the distribution of X_i, is not suspected to change. However, one is concerned that the null hypothesis of independence of the X_i's may have been violated in the direction that the neighboring members are more strongly dependent than those which are farther apart. The autoregressive and moving average models of time series are stimulated by the concept of serial dependence, and their applications are wide ranging. In Section 4 we survey the important nonparametric tests for detection of serial dependence.

Finally, a departure from the null hypothesis of randomness may arise in a less clearly defined manner, for instance by a mixture of the several types described above. Lacking any a priori knowledge of a specific mode of departure from randomness, the investigator may only wish to examine whether the observations $X_1, \ldots, X_N$ contradict the model of randomness. The alternatives to randomness are left unspecified except for a vague description in such general terms as a tendency of the series to cluster. A broad and unstructured class of alternatives is referred to as the *omnibus alternatives* and some simple and heuristic tests are discussed in Section 5.

2. Tests for detecting a trend

Let the observations $X_1, \ldots, X_N$ at N successive time points be independent random variables with unknown continuous df's $F_1, \ldots, F_N$, respectively. We consider the problem of testing the null hypothesis H_0: $F_1 = \cdots = F_N$ against the alternative of an upward trend in the series. A general model of an upward trend can be expressed by the requirement that the X_i's are stochastically increasing with i, that is, $F_1(x) \geqslant \cdots \geqslant F_N(x)$ for all x, and not all F_i's are identical. While this general formulation suffices for motivating the form of the nonparametric tests, considerations of power and asymptotic relative efficiency would require a more specific structure. An important class of alternatives, called a *trend in location*, is described by $F_i(x) = F(x - \theta_i)$, $i = 1, \ldots, N$, where

F is a fixed df and $\theta_1 \leqslant \cdots \leqslant \theta_N$ are unknown location parameters. A *linear regression* structure $\theta_i = \alpha + \beta b_i$ is a further special case where the b_i's are known and increasing constants, and α and β are unknown parameters. The null hypothesis corresponds to $\beta = 0$, and the alternative is $\beta > 0$.

A fairly large number of nonparametric tests for trend have been proposed in the literature. However, they can be grouped into two broad classes, albeit with some overlap. In the first, the tests are motivated from heuristic reasoning where simplicity of the procedure and ease of computation are the prime considerations. These tests are based on the signs of the differences of some or all pairs (X_i, X_j). Letting $U_{ij} = 1$ (0) according as $X_i < (>) X_j$, the test statistics have the general form

$$\sum_{1 \leqslant i < j \leqslant N} c_{ij} U_{ij} \tag{2.1}$$

where the nonnegative constants c_{ij} are viewed as the weights attached to the signs of $(X_j - X_i)$. Some prominent members of this class are

$$\begin{aligned} M &= \sum_{1 \leqslant i < j \leqslant N} U_{ij}, \qquad D = \sum_{1 \leqslant i < j \leqslant N} (j - i) U_{ij}, \\ CS_1 &= \sum_{i=1}^{N/2} (N - 2i + 1) U_{i,N-i+1}, \qquad CS_2 = \sum_{i=1}^{N/3} U_{i,2N/3+i}. \end{aligned} \tag{2.2}$$

The first two are due to Mann (1945) and Daniels (1950), respectively. The last two are among a subclass of tests studied by Cox and Stuart (1955) where the test statistic is of the form (2.1) with the added requirement that only the disjoint U_{ij}'s be involved, that is, $c_{ij}c_{i'j'} = 0$ whenever i, j, i' and j' have any element repeated.

The null distributions of these statistics derive from the probability structure that all $N!$ orderings of $X_1, \ldots, X_N$ are equally likely. Moreover, being linear functions of the indicator variables U_{ij}, their exact means and variances are easy to calculate, and they are asymptotically normal under mild conditions on the c_{ij}'s. In particular,

$$\begin{aligned} E(M) &= N(N-1)/4, \qquad \operatorname{Var}(M) = N(N-1)(2N+5)/72, \\ E(D) &= N(N^2-1)/12, \qquad \operatorname{Var}(D) = N^2(N+1)^2(N-1)/144. \end{aligned}$$

Moreover, M and D are directly related to two important nonparametric statistics for testing independence in a bivariate sample. Considering the pairs (i, X_i), $i = 1, \ldots, N$, as the bivariate data, these latter statistics are:

$$\text{Kendall's tau:} \quad \tau = (2M - N) \Big/ \binom{N}{2},$$

$$\text{Spearman's rank correlation:} \quad r_S = [12/(N^3 - N)] \sum_{i=1}^{N} \left(i - \frac{N+1}{2}\right) R_{Ni},$$

where $R_{N1}, \ldots, R_{NN}$ denote the ranks of $X_1, \ldots, X_N$, respectively. The linear relation (hence equivalence) between M and τ is obvious. With some algebraic manipulation, it can be seen that

$$D = [(N^3 - N)/12](1 + r_S)\,.$$

Tables of critical values of τ and r_S are available, and these can be used to perform the M and D tests for trend. Cox and Stuart (1955) motivated the tests CS_1 and CS_2 by the features that these are 'simplified versions' of D and M, respectively, and they do not incur a substantial loss of asymptotic efficiency. In fact, their Pitman asymptotic relative efficiencies (ARE) with respect to D are 0.87 and 0.84, respectively, for the normal linear trend model.

The tests in the other class are based on the linear rank statistics

$$T_N = \sum_{i=1}^{N} (c_{Ni} - \bar{c}_N) a_N(R_{Ni}) \tag{2.3}$$

where c_{Ni}, $i = 1, \ldots, N$ are constants, $\bar{c}_N = N^{-1}\sum_{i=1}^{N} c_{Ni}$, and $a_N(\cdot)$ is a rank-score function. To motivate these we consider the family of location trend alternatives $F_i(x) = F(x - \beta b_i)$ with $\beta \geq 0$ and known constants $b_1 \leq \cdots \leq b_N$. Assume that $F(x)$ has an absolutely continuous probability density function (pdf) $f(x)$ and define the scores

$$a_N(i, f) = E\phi_f(U_N^{(i)}), \qquad \phi_f(u) = -f'(F^{-1}(u))/f(F^{-1}(u)), \quad 0 < u < 1\,,$$

where $U_N^{(1)} < \cdots < U_N^{(N)}$ denote the order statistics of a random sample of size N from the uniform $(0, 1)$ distribution, and $f'(x) = \mathrm{d}f(x)/\mathrm{d}x$. Following Theorem II.4.8 of Hájek and Šidák (1967), the locally most powerful (LMP) rank test of H_0: $\beta = 0$ rejects the null hypothesis for large values of the statistic

$$T_N^* = \sum_{i=1}^{N} (b_i - \bar{b}_N) a_N(R_{Ni}, f) \tag{2.4}$$

which obviously belongs to the class of linear rank statistics defined in (2.3). In particular, if the trend is linear in time ($b_i = i$), the LMP rank tests for the logistic and normal parent distributions are respectively given by

$$\sum_{i=1}^{N} \left(i - \frac{N+1}{2}\right) R_{Ni}, \qquad \sum_{i=1}^{N} \left(i - \frac{N+1}{2}\right) E_\Phi(W^{(R_{Ni})}) \tag{2.5}$$

where $W^{(\cdot)}$ are the order statistics from the standard normal df Φ. The first expression in (2.5) is a linear function of r_S and hence of D as well. Consequently, Daniel's test is the LMP rank test for a linear trend in location when the underlying distribution is logistic. The second statistic in (2.5) gives the normal scores test for trend.

In regard to the asymptotic efficiencies of the tests for trend, some scattered results were only available in the early literature confined especially to the normal distribution (cf. Cox and Stuart, 1955, and Stuart, 1956). Evidently, a rank test of the form (2.3) involves a choice of the regression constants c_{Ni} and the score function $a_N(\cdot)$. In the case of a location-trend model, its power properties would depend upon the underlying distribution F as well as the rapidity of the growth of the trend. Aiyar et al. (1979) provide a comprehensive study of the effects of these factors upon the Pitman ARE's of various trend tests including the linear rank test (2.3) and some relatively important members of the class (2.1). Considering the local alternatives $F_i(x) = F(x - \beta_N b_i)$ such that for some $\Delta > 0$, $0 < \lim_{N\to\infty} N^\Delta \beta_N < \infty$, and assuming some regularity conditions on F, b_i and c_{Ni}, the ARE of the test T_N relative to the asymptotically most powerful test T_N^* is given by

$$e(T:T^*) = [\rho_{bc}^2 \cdot \rho^2(\phi, \phi_f)]^{1/2\Delta} \tag{2.6}$$

where

$$\rho_{bc} = \lim \frac{\Sigma_{i=1}^N (b_i - \bar{b}_N)(c_{Ni} - \bar{c}_N)}{\{\Sigma_{i=1}^N (b_i - \bar{b}_N)^2 \Sigma_{i=1}^N (c_{Ni} - \bar{c}_N)^2\}^{1/2}},$$

$$\rho(\phi, \phi_f) = \frac{\int_0^1 \phi(u)\phi_f(u)\,du}{\int_0^1 \phi_f(u)\,du)(\int_0^1 \phi^2(u)\,du - \bar{\phi}^2)\}^{1/2}}$$

and $\bar{\phi} = \int_0^1 \phi(u)\,du$. Thus, the ARE (2.6) is a product of two factors; one relates to the correlation between the regression constants c_{Ni} and the constants b_i which govern the growth of the trend; the other concerns the correlation between the optimum score function $\phi_f(u)$ and the score function $\phi(u)$ actually involved in the test. Aiyar et al. (1979) provide an extensive study of the ARE behaviors of the rank and other tests for trend under several growth rates for the location trend. For the rank tests, they show that the choice of the regression constants c_{Ni} may be crucial in the situations of a slow trend (for instance, $b_i = i^\alpha$, $|\alpha| < \frac{1}{2}$ and $b_i = \log i$) whereas the choice is usually less serious in the case of a rapidly increasing trend such as $b_i = i^\alpha$, $\alpha \geqslant \frac{1}{2}$.

Estimation of the slope of a linear trend

For the dual advantage of simplicity and high ARE, the tests due to Mann and Daniels remain by far the most popular nonparametric tests for the detection of trend. Together they can be viewed as the counterpart of the Mann–Whitney–Wilcoxon test for the two-sample problem. An additional advantage of these tests draws from their easy invertibility towards generating some nonparametric estimators of the slope of a linear trend. To discuss this aspect, we consider estimation of β for the linear trend model $F_i(x) = F(x - \beta i)$, $i = 1, \ldots, N$. When the time points $t_1, \ldots, t_N$ are equispaced, an apparently more general formulation $F_i(x) = F(x - \mu - \beta t_i)$ is equivalent to this simple form because F is unspecified and the order relations of $X_1, \ldots, X_N$ are invariant under a common location and scale change.

Note that Daniels' statistic D can be written in the two equivalent forms

$$L_N = \frac{12}{N(N^2-1)}\sum_{i=1}^{N}\left(i-\frac{N+1}{2}\right)R_{Ni}$$

$$= \frac{6}{N(N^2-1)}\sum_{1\le i<j\le N}(j-i)\,\mathrm{sgn}(X_j-X_i) \tag{2.7}$$

where the sign function is defined as $\mathrm{sgn}(u)=1, 0, -1$ according as $u>, =, <0$. Let $\boldsymbol{\nu}=(1,2,\ldots,N)'$ and denote by $L_N(\boldsymbol{x})$ the value of L_N computed from a specific realization $\boldsymbol{x}$ of $\boldsymbol{X}=(X_1,\ldots,X_N)'$. The following properties can be easily verified:

(i) $L_N(\boldsymbol{x}+b\boldsymbol{\nu})$ is nondecreasing in b for every fixed $\boldsymbol{x}$,

(ii) When $\beta=0$, the distribution of $L_N(\boldsymbol{X})$ is symmetric around 0 irrespective of F.

Because of these properties, the idea of Hodges and Lehmann (1963) can be employed to construct a point estimator $\hat{\beta}_L$ of β based on the test statistic L_N. Letting

$$\beta_* = \sup\{b\colon L_N(\boldsymbol{x}-b\boldsymbol{\nu})>0\}, \qquad \beta^* = \inf\{b\colon L_N(\boldsymbol{x}-b\boldsymbol{\nu})<0\},$$

the Hodges–Lehmann type estimator of β is

$$\hat{\beta}_L = (\beta_* + \beta^*)/2\,.$$

To provide an explicit form of this estimator, we denote by $Y_1\le\cdots\le Y_k$ the ordered values of the $k=\binom{N}{2}$ ratios $(X_j-X_i)/(j-i)$, $1\le i<j\le N$, and let $\omega_1,\ldots,\omega_k$ denote the corresponding values of $(j-i)$. Letting

$$\gamma_i = \sum_{j=1}^{i}\omega_j\Big/\sum_{j=1}^{k}\omega_j,\quad i=1,\ldots,k\,,$$

define the integer m by the requirement that $\gamma_{m-1}<\frac{1}{2}\le\gamma_m$. Then

$$\begin{aligned}\hat{\beta}_L &= Y_m &&\text{if } \gamma_m>\tfrac{1}{2}\,,\\ &= (Y_m+Y_{m+1})/2 &&\text{if } \gamma_m=\tfrac{1}{2}\,.\end{aligned} \tag{2.8}$$

Thus $\hat{\beta}_L$ is the weighted median obtained from the ratios $(X_j-X_i)/(j-i)$ with the associated frequency counts $(j-i)$. It has an interesting similarity with the usual least squares estimator which is the weighted mean of the ratios $(X_j-X_i)/(j-i)$ with the corresponding weights $(j-i)$.

In a similar manner, one can also derive the Hodges–Lehmann estimator of β from the test statistic M. The resulting estimator $\hat{\beta}_M$ turns out to be the (unweighted) median of the ratios $(X_j-X_i)/(j-i)$. For the more general situation where $F_i(x)=F(x-\mu-\beta t_i)$ and the t_i's are distinct but not necessarily equispaced, the estimator $\hat{\beta}_M$ takes the form

$$\hat{\beta}_M = \text{median}\{(X_j - X_i)/(t_j - t_i), 1 \leq i < j \leq N\}. \tag{2.9}$$

Such median estimators of the slope of a line were originally proposed by Theil (1950) on intuitive ground. Sen (1968) and Bhattacharyya (1968) highlight their relation to the test statistics M and L, and derive some distributional properties. The ARE of these estimators relative to the least squares estimator is the same as the ARE of the M (or L) test relative to the t test. In particular, when $F_i(x) = F(x - \beta i)$, the ARE has the lower bound 0.953 for all continuous F, and its value is 0.985 when F is normal.

Tests for multivariate trend

Generalizing from a sequence of one-dimensional random variables, we consider here a p-variate time series observed at N consecutive time points so the observations constitute a $p \times N$ random matrix $\boldsymbol{X} = (\boldsymbol{X}_1, \ldots, \boldsymbol{X}_N)$ where $\boldsymbol{X}_i = (X_{1i}, \ldots, X_{pi})'$. The columns of $\boldsymbol{X}$ correspond to the different time points and the rows to the different variables observed. We assume that $\boldsymbol{X}_i$ has a continuous p-variate df F_i and denote its α-th marginal df by $F_{\alpha i}$, $\alpha = 1, \ldots, p$, $i = 1, \ldots, N$. The problem of interest is to test the null hypothesis H_0: $F_1 = \cdots = F_N$ against the alternative (H_1) that an upward or downward trend exists in at least one coordinate variable, that is, for at least one α $(1 \leq \alpha \leq p)$, the marginal series $(X_{\alpha 1}, \ldots, X_{\alpha N})$ is stochastically increasing or decreasing.

A formulation of rank tests for this multivariate problem is motivated from the structure of the likelihood ratio test under the multivariate normal linear trend model: $F_i = \mathrm{N}_p(\boldsymbol{\mu} + i\boldsymbol{\beta}, \boldsymbol{\Sigma})$. The likelihood ratio test of H_0: $\boldsymbol{\beta} = \boldsymbol{0}$ rejects for large values of the quadratic test statistic $W_N = \boldsymbol{b}'\boldsymbol{S}^{-1}\boldsymbol{b}(N - p - 1)/p$ where

$$\boldsymbol{b} = [12/(N^3 - N)] \sum_{i=1}^{N} \left(i - \frac{N+1}{2}\right)\boldsymbol{X}_i$$

and S corresponds to the estimated covariance matrix of $\boldsymbol{b}$. The exact null distribution of W_N is a central F with $(p, N - p - 1)$ degrees of freedom, and the test based on W_N is the uniformly most powerful invariant test under the normal model.

To construct a rank test we denote by $\boldsymbol{R}_N = (R_{\alpha i})$, the $p \times N$ rank matrix obtained by ranking each row of $\boldsymbol{X}$ separately. The basic ingredients for the construction of a generalized Daniels–Spearman statistic are

$$L_{\alpha N} = [12/(N^3 - N)] \sum_{i=1}^{N} \left(i - \frac{N+1}{2}\right) R_{\alpha i},$$

$$Q_{\alpha\beta} = [12/(N^3 - N)] \sum_{i=1}^{N} \left(R_{\alpha i} - \frac{N+1}{2}\right)\left(R_{\beta i} - \frac{N+1}{2}\right), \quad 1 \leq \alpha, \beta \leq p, \tag{2.10}$$

$$L'_N = (L_{1N}, \ldots, L_{pN}), \qquad \boldsymbol{Q}_N = (Q_{\alpha\beta}).$$

Note that $L_{\alpha N}$ and $Q_{\alpha\beta}$ are the Spearman rank correlations for the pairs $(i, X_{\alpha i})$, $i = 1, \ldots, N$ and $(X_{\alpha i}, X_{\beta i})$, $i = 1, \ldots, N$, respectively. Under the null hypothesis H_0: $F_1 = \cdots = F_N$ and conditionally given the collection of the column vectors of $\boldsymbol{R}_N$, all $N!$ permutations of the columns are equally likely. From this conditional or permutation probability measure, one can verify that the mean vector and the covariance matrix of $\boldsymbol{L}_N$ are $\mathbf{0}$ and $(N-1)\boldsymbol{Q}_N$, respectively. In analogy with the normal theory test statistic W_N, the p-variate Daniels–Spearman statistic is then defined as the quadratic form

$$L_N^{(p)} = (N-1)\boldsymbol{L}_N'\boldsymbol{Q}_N^{-1}\boldsymbol{L}_N \tag{2.11}$$

with its large values leading to the rejection of H_0. Unlike the univariate situation, the rank statistic $L_N^{(p)}$ is not unconditionally distribution-free under H_0. However, a strictly distribution-free test is obtained by performing the test conditionally on the observed configuration of $\boldsymbol{R}_N$, that is by referring to the permutation distribution of $L_N^{(p)}$. Note that the column permutations of $\boldsymbol{R}_N$ only affect $\boldsymbol{L}_N$ while the matrix $\boldsymbol{Q}_N$ remains invariant. If $\boldsymbol{Q}_N$ is singular, a generalized inverse is to be used in (2.12) in place of the regular inverse. The choice is of little consequence in large samples because the probability of $\boldsymbol{Q}_N$ being singular tends to zero as $N \to \infty$ whenever the parent distribution $F(x)$ does not degenerate to a dimension less than p.

As $N \to \infty$, the permutation distribution of $L_N^{(p)}$ converges weakly to χ_p^2, the central chi-squared distribution with p degrees of freedom. So, in large samples, an approximate level α test is given by the rejection region $L_N^{(p)} \geq c$ where c denotes the upper α-point of χ_p^2. The permutation test is asymptotically power-equivalent to this unconditional test, and for this reason, both will be referred to as the $L_N^{(p)}$ test.

The $L_N^{(p)}$ test remains consistent for a wide class of general trend alternatives where the trend in at least one variable does not damp out too fast. Specifically, the property of consistency holds for all sequences $\{F_N\}$ such that, for some $\eta < \frac{1}{2}$ and some $1 \leq \alpha \leq p$,

$$\liminf N^{\eta} \left| \int_{-\infty}^{\infty} [2F_{\alpha N} - 1] \, \mathrm{d}F_{\alpha, N+1} \right| > 0$$

where $F_{\alpha N}$ denotes the α-th marginal of F_N.

Under a sequence of local linear trend alternatives defined by the probability measures $P_N = \Pi_{i=1}^N F(\boldsymbol{x} - \boldsymbol{\mu} - N^{-3/2} i \boldsymbol{\delta})$, the Pitman ARE of $L_N^{(p)}$ relative to W_N is given by

$$e(\boldsymbol{\delta}, F) = [(\boldsymbol{\delta}'\boldsymbol{B}^{-1}\boldsymbol{\delta})/(\boldsymbol{\delta}'\boldsymbol{\Sigma}^{-1}\boldsymbol{\delta})]^{1/3} \tag{2.12}$$

where $\boldsymbol{\Sigma}$ is the covariance matrix of F,

$$\boldsymbol{B} = (b_{\alpha\beta}), \qquad b_{\alpha\beta} = \lambda_{\alpha\beta} \left[12 \int_{-\infty}^{\infty} f_\alpha^2(x) \, \mathrm{d}x \int_{-\infty}^{\infty} f_\beta^2(y) \, \mathrm{d}y \right]^{-1},$$

$$\lambda_{\alpha\beta} = 12 \int_{-\infty}^{\infty} \int_{-\infty}^{\infty} F_\alpha(x) F_\beta(y) \, \mathrm{d}F_{\alpha\beta}(x, y) - 3 .$$

The preceding development of a multivariate trend test is due to Bhattacharyya and Klotz (1966) who also provide an extension of Mann's test to the multivariate case and give applications to the analysis of the freezing and thawing dates for a midwestern lake. Bhattacharyya (1968) considers estimation of the slope using the tests of Mann and Daniels, and discusses the ARE properties of those estimators in a multivariate setting. Dietz and Killeen (1981) also give a multivariate extension of Mann's test and an application to the analysis of blood chemistry measurements of patients. Their test statistic uses a slightly different consistent estimator for the covariance matrix and is asymptotically equivalent to the statistic proposed by Bhattacharyya and Klotz (1966).

3. Tests for change at an unknown time point

Detection of a possible abrupt change in the distribution of sequentially observed random variables is of interest in many studies, especially in industrial quality control and economic or social phenomena observed over time. Assuming that the successive observations $X_1, \ldots, X_N$ are independent univariate random variables with continuous df's $F_1, \ldots, F_N$, respectively, we are again concerned with testing the null hypothesis H_0: $F_1 = \cdots = F_N$. However, our alternative hypothesis, in the present context, specifies that the sequence $X_1, \ldots, X_N$ segments at an unknown time point m so the first $(m-1)$ members are iid and are stochastically smaller than the other members which are also iid. To discuss the main ideas, we will confine attention to the model of location shift:

$$F_1(x) = \cdots = F_{m-1}(x) = F_m(x+\Delta) = \cdots = F_N(x+\Delta)\,, \qquad \Delta \geqslant 0, 2 \leqslant m \leqslant N\,, \tag{3.1}$$

for which the null hypothesis is H_0: $\Delta = 0$ and m is a discrete nuisance parameter. Moreover, the common df F under H_0 is unknown.

A general formulation of stochastically ordered alternatives and consideration of invariance lead to the ranks $\boldsymbol{R}_N = (R_{N1}, \ldots, R_{NN})$ of $(X_1, \ldots, X_N)$ in the usual way. Under the shift model, the distribution of $\boldsymbol{R}_N$ and hence the power of a rank test depends on F, Δ and m. As a general notation, let $\beta(\Delta \mid m)$ represent the power function of a rank test whose dependence on F is understood.

A class of rank tests is derived by Bhattacharyya and Johnson (1968) from the criterion of maximizing the average local power $\bar{\beta}(\Delta) = \Sigma_{i=1}^N q_i\beta(\Delta \mid i)$ with respect to an arbitrary set of weights q_i which satisfy $q_1 = 0$, $q_i \geqslant 0$, $i = 2, \ldots, N$, and $\Sigma\, q_i = 1$. From a Bayesian view, q_i may be regarded as the prior probability of a change to occur at the time i. To formalize the optimal rank test we assume that F has an absolutely continuous pdf $f(x)$ and denote

$$\phi_f(u) = -f'(F^{-1}(u))/f(F^{-1}(u)), \quad 0 < u < 1,$$

$$Q_i = \sum_{j=1}^{i} q_j, \qquad \bar{Q}_N = \frac{1}{N}\sum_{i=1}^{N} Q_i,$$

$$a_N(i, f) = E\phi_f(U_N^{(i)})$$

where $U_N^{(i)}$ are order statistics from the uniform (0, 1) distribution. Then the rank test, that maximizes the average power $\bar{\beta}(\Delta)$ uniformly for Δ in a positive neighborhood of 0, rejects for large values of the statistic

$$T_N^* = \sum_{i=1}^{N} (Q_i - \bar{Q}_N) a_N(R_{Ni}, f). \tag{3.2}$$

Comparing (3.2) with (2.4) we observe that the LMP rank test for the trend problem has a similar structure as that of the optimal rank test for the change problem of this section. However, the regression constants b_i in the former relate to the assumed trend model. In the latter, the analogue of the regression constants are the terms Q_i which relate to the weights used to average the power. For the special case of uniform weights $q_i = 1/(N-1)$, $i = 2, \ldots, N$, the test statistics

$$T_{(1)} = \sum_{i=1}^{N} (i-1)R_{Ni}, \qquad T_{(2)} = \sum_{i=1}^{N} (i-1)E_{\Phi}(W^{(R_{Ni})})$$

correspond to the choices of the logistic and normal scores, respectively. Note that $T_{(1)}$ is equivalent to the Daniels–Spearman test. Moreover, for the degenerate weight function $q_m = 1$, $q_i = 0$, $i \neq m$, $T_{(1)}$ and $T_{(2)}$ reduce to the two-sample Wilcoxon and normal scores tests, respectively.

Motivated by (3.2) one can consider a general rank statistic of the form

$$T_N = \sum_{i=1}^{N} (Q_i - \bar{Q}_N) a_N(R_{Ni}) \tag{3.3}$$

where $a_N(\cdot)$ corresponds to a square integrable score function $\phi(u)$. To present the asymptotic results for these tests we consider the sequence k_N of local change alternatives defined by $F_i(x) = F(x)$, $i = 1, \ldots, m-1$ and $F_i(x) = F(x - \theta N^{-1/2})$, $i = m, \ldots, N$, where $m/N \to \lambda$ as $N \to \infty$, $0 < \lambda < 1$. If the sequence of weights satisfies

$$\lim_{N\to\infty} \sum_{i=m}^{N} (Q_i - \bar{Q}_N)/N = a < \infty,$$

$$\lim_{N\to\infty} \sum_{i=1}^{N} (Q_i - \bar{Q}_N)^2/N = c^2, \quad 0 < c^2 < \infty,$$

then the limiting distribution of $N^{-1/2}T_N$ (under k_N) is $\mathrm{N}(\mu, c^2d^2)$ where

$$\mu = \theta a \int_0^1 \phi(u)\phi_f(u)\,\mathrm{d}u, \qquad d^2 = \int_0^1 \phi^2(u)\,\mathrm{d}u - \left(\int_0^1 \phi(u)\,\mathrm{d}u\right)^2 .$$

In the setting of a normal distribution with known σ, Chernoff and Zacks (1964) propose the test $Z_N = \Sigma\,(i-1)(X_i - \bar{X})$ from the likelihood ratio criterion. Confining to the uniform weights, and using the above results and asymptotic normal distribution of Z_N, it follows that the ARE of T_N relative to Z_N is the same as the ARE of the corresponding two-sample rank test relative to the t test. In particular,

$$\mathrm{ARE}(T_{(1)}:Z) = 12\sigma^2\left(\int f^2(x)\,\mathrm{d}x\right)^2$$

and

$$\mathrm{ARE}(T_{(2)}:Z) = \sigma^2 \int f^2(x)[\varphi(\Phi^{-1}F(x))]^{-1}\,\mathrm{d}x$$

where φ denotes the standard normal pdf.

In contrast with the linear rank tests discussed before, some interesting nonlinear rank tests have been proposed by Sen and Srivastava (1975) and Pettitt (1979). To describe these, let $T_{m,N}$ denote a two-sample rank statistic (for instance, the Mann–Whitney statistic) based on the sample sizes m and $n = (N-m)$, and denote by $\mu_{m,N}$ and $\sigma^2_{m,N}$ its null mean and variance. For the problem of detecting a change at an unknown time point, Sen and Srivastava (1975) propose the statistics of the form

$$A_N = \max\{(T_{m,N} - \mu_{m,N})/\sigma_{m,N},\ m = 1, \ldots, N-1\}, \tag{3.4}$$

while Pettitt (1979) suggests the maximum of the nonstandardized Mann–Whitney statistic:

$$B_N = \max\left\{\sum_{i=1}^{m} \sum_{j=m+1}^{N} U_{ij},\ m = 1, \ldots, N-1\right\} \tag{3.5}$$

where $U_{ij} = 1\ (0)$ according as $X_i < (>)\, X_j$. Using a heuristic argument, Pettitt states that $N^{-1}\{3/(N+1)\}^{1/2}B_N$ is asymptotically distributed as the Kolmogorov goodness of fit statistic, and recommends the use of the Kolmogorov–Smirnov table. On the other hand, Sen and Srivastava (1975) only provide some estimated critical values based on Monte Carlo Simulation. Asymptotic properties of these tests remain to be investigated.

We digress for a moment to remark that a somewhat different testing problem arises when the initial level of the process is known. Let X_0 denote the process at the initial time $t = 0$, and assume that the df F_0 of X_0 is symmetric with the known center μ_0. Based on the later observations $X_1, \ldots, X_N$, we wish to test if the center has shifted above μ_0 at some time point m $(1 \leq m \leq N)$. For this problem, Page (1955) proposes a nonparametric test based on the statistic

$$\max_{0 \leq m \leq N} \{S_m - \min_{0 \leq i \leq m} S_i\} \tag{3.6}$$

where $S_0 = 0$ and $S_m = \Sigma_{i=1}^m \operatorname{sgn}(X_i - \mu_0)$, and provides a table of 5% and 1% critical values for some selected sample sizes. Also, a class of locally best signed rank test is derived by Bhattacharyya and Johnson (1968) using the criterion of average power.

Use of sequential ranks

We now return to the problem of detecting a change in the series $X_1, X_2, \ldots$ when the initial process level is unknown. A different type of nonparametric procedure is formulated by Bhattacharya and Frierson (1981) through the use of *sequential ranks* in contrast with the single or one-time ranking involved in the tests described before.

Let $X_i = X_{iN}$, $i = 1, \ldots, N$, $N = 1, 2, \ldots$, denote a sequence of independent random variables whose first $[N\theta]$ members have a common df F and the last $N - [N\theta]$ members are distributed as G_N. Here $0 < \theta < 1$, $[a]$ denotes the integer part of a, and F and G_N are unknown but assumed continuous. While observing the X_i's sequentially, the object is to stop soon after the time point $[N\theta] + 1$ where the process shifts from F to G_N.

The basic elements of a nonparametric detection scheme are the sequential ranks $R_1, \ldots, R_N$ where R_i denotes the (ordinary) rank of X_i when X_i is ranked among the subset $\{X_1, \ldots, X_i\}$. Alternatively, we have

$$R_i = 1 + \sum_{j=1}^{i-1} U_{ji}$$

where $U_{ji} = 1\ (0)$ according as $X_j < (>)\, X_i$. Although the same symbol R is again used for these new ranks, we stress that the distributional properties of the sequential ranks are altogether different from those of the ordinary ranks. In particular, the sequential ranks of iid random variables are themselves independent random variables with R_i uniformly distributed on the integers $\{1, \ldots, i\}$ so $E(R_i) = (i + 1)/2$ and $\operatorname{Var}(R_i) = (i^2 - 1)/12$. Let

$$\begin{gathered} Z_i = i^{-1}\left(R_i - \frac{i+1}{2}\right), \qquad V_k = \sum_{i=1}^{k} Z_i\,, \\ V_N(t) = (12/N)^{1/2}\{V_{[Nt]} + (Nt - [Nt])V_{[Nt]+1}\}, \quad 0 \leq t \leq 1\,. \end{gathered} \tag{3.7}$$

In essence, the continuous time stochastic process $\{V_N(t), 0 \leq t \leq 1\}$ is constructed from the normalized sequential ranks by linear interpolation.

A sequential scheme for detecting a change in the distribution can then be formalized by means of a *nonparametric control chart* which graphs $V_N(t)$ versus t, $0 \leq t \leq 1$. With the arrival of a new observation, the graph is updated and this process is continued until when the latest plot is found to have crossed a horizontal line c, called the nonparametric control limit. If this happens, observation is terminated and a significant change in the process level is inferred.

It is desired to set the control limit c so the probability of a false alarm (type I error) is controlled at a specified value α. To obtain an approximate determination of c and to study the performance of the scheme, we consider the limiting behavior of the $V_N(t)$ process. Under a sequence of alternatives (F, G_N) such that G_N tends to be close to F, the process $V_N(t)$ converges weakly to

$$X(t) = B(t) + (12)^{1/2}\delta\theta \log(t/\theta) I(\theta, t),$$

where $B(t)$ is the standard Brownian motion process, $\delta = \lim N^{1/2} \int (F - \frac{1}{2})\, \mathrm{d}G_N$ and $I(\theta, t) = 1$ (0) according as $\theta \leq (>)\, t$. Since $\delta = 0$ under the null hypothesis of no change, and since

$$P[\sup_{t\in[0,1]} B(t) \geq c] = 2P[B(1) \geq c],$$

the control limit is given by $c = \Phi^{-1}((1-\alpha)/2)$ where Φ is the standard normal df. The nonparametric detection scheme based on the sequential ranks is then asymptotically equivalent to the

Logarithmic stopping rule: Stop at the smallest t for which $B(t) \geq \psi(t)$ where

$$\psi(t) = c - (12)^{1/2}\delta\theta \log(t/\theta) I(\theta, t). \tag{3.8}$$

(The reason for this name is that the drift of the process $X(t)$ is proportional to $\log t$).

The parametric counterpart of the above control chart is one that is based on the cumulative sums $N^{-1/2} \sum_{i=1}^{k} (X_i - \mu_F)/\sigma_F$ where μ_F and σ_F respectively denote the (known) mean and standard deviation of F. The corresponding standardized process $V_N^*(t)$ converges weakly to

$$X^*(t) = B(t) + \Delta(t-\theta) I(\theta, t)$$

where $\Delta = \lim N^{1/2}(\mu_{G_N} - \mu_F)/\sigma_F$. Asymptotic performance of this control chart is then shared by the

Linear stopping rule: Stop at the smallest t for which $B(t) \geq \psi^*(t)$, where

$$\psi^*(t) = c - \Delta(t-\theta) I(\theta, t). \tag{3.9}$$

Bhattacharya and Frierson (1981) investigate the performance of their nonparametric detection scheme relative to the control chart based on the cumulative sums. The asymptotic comparison is effected by studying the functions $\psi(t)$ and $\psi^*(t)$, defined in (3.8) and (3.9), for some important models for change such as a location shift, scale change, or a contamination of two distributions. The behavior of the function $h(\theta) = [\mathrm{d}\psi(\theta)/\mathrm{d}\theta][\mathrm{d}\psi^*(\theta)/\mathrm{d}\theta]^{-1}$ turns out to be crucial in determining whether or not one method is superior to the other. Interestingly, in the case of a location shift, $h^2(\theta)$ is the same as the Pitman ARE of the Wilcoxon test relative to the t test for the two-sample problem. It is found that for distributions with heavy tails, the nonparametric scheme is superior to the scheme based on the cumulative sums both in the sense of asymptotic power and the expected stopping time.

4. Tests for detecting serial dependence

In Sections 2 and 3 we discussed the appropriate nonparametric procedures for testing the null hypothesis of randomness when one contemplates that a departure from the iid property of $X_1, \ldots, X_N$ may occur in the form of a steady trend or an abrupt jump in the distributions of the X_i's. The assumption of independence was not questioned. A different situation arises when the stationarity of the process is not questioned or is not the target of an investigation. Instead, one is concerned that the model of randomness may be violated due to dependence of the observations occurring at adjacent points of time. In this section we review the important nonparametric tests of randomness which are apt to detect a serial dependence of the successive observations $X_1, \ldots, X_N$.

Some relatively simple formulations of serial dependence are described in the framework of a stationary autoregressive or moving average process. For instance, a first order autoregressive process is given by the structure $X_i = \alpha X_{i-1} + E_i$, $i = 1, 2, \ldots, X_0 = 0$, where $E_1, E_2, \ldots$ are iid random variables with a continuous df F, mean 0 and variance σ^2. The null hypothesis of randomness then corresponds to H_0: $\alpha = 0$ where σ is a nuisance parameter, and a positive serial dependence corresponds to $\alpha > 0$. If F is assumed to be normal, then the appropriate parametric test is based on the sample first order serial correlation

$$r_1 = \sum_{i=1}^{N-1} (X_i - \bar{X})(X_{i+1} - \bar{X})/S^2 \tag{4.1}$$

where

$$\bar{X} = \sum_{i=1}^{N} X_i/N \quad \text{and} \quad S^2 = \sum_{i=1}^{N} (X_i - \bar{X})^2 .$$

One general construct of a distribution-free test is to employ a parametric test statistic in the nonparametric mode by using the idea of the permutation (or randomization) distribution. To describe this method, let $\boldsymbol{a} = (a_1, \ldots, a_N)$ denote the observed values of the order statistics $\boldsymbol{X}^{(\cdot)}$ of $X_1, \ldots, X_N$ and define

$\Gamma = \Gamma(\boldsymbol{a})$ to be the set of $n!$ vectors which are obtained by permuting the coordinates of $\boldsymbol{a}$. Under the null hypothesis $X_1, \ldots, X_N$ are exchangeable, which implies that, conditionally given $\boldsymbol{X}^{(\cdot)} = \boldsymbol{a}$, the elements of Γ are equally likely. The assignment of equal probabilities $1/N!$ to the elements of Γ generates what is called the permutation probability measure. Now, for $z \in \Gamma$, let $r_1(z)$ denote the value of r_1 obtained from (4.1) by using $X_i = z_i$, $i = 1, \ldots, N$. A level α permutation test based on r_1 is then given by the test function $\phi(z) = 1$, $\gamma(\boldsymbol{a})$ or 0 according as $r_1(z) >, =$ or $< c(\boldsymbol{a})$ where $c(\boldsymbol{a})$ and $\gamma(\boldsymbol{a})$ are determined from the requirement that $E[\phi(\boldsymbol{Z}) \mid \boldsymbol{a}] = \alpha$. The conditional level α for each realization of $X^{(\cdot)}$ guarantees that the test ϕ has the level α unconditionally, so it is indeed a distribution-free test.

The computation involved in a permutation test rapidly increases with the sample size, and an approximation is desirable for large N. In regard to the statistic r_1 we first note that the quantities $\bar{X}$ and S^2 are invariant under permutations and hence they are constants over the set Γ. Therefore, the test can be based on the simpler statistic

$$T_1 = \sum_{i=1}^{N-1} X_i X_{i+1} \,. \tag{4.2}$$

Under the permutation distribution given $\boldsymbol{X}^{(\cdot)} = \boldsymbol{a}$, the mean and variance of T_1 are respectively given by

$$\begin{aligned} \mu_1 &= N^{-1}(A_1^2 - A_2)\,, \\ \sigma_1^2 &= N^{-1}(A_2^2 - A_4) + [N(N-1)]^{-1} 2(A_1^2 A_2 - A_2^2 - 2A_1A_2 + 2A_4) \\ &\quad + [N(N-1)]^{-1}(A_1^4 - 6A_1^2A_2 + 8A_1A_3 + 3A_2^2 - 6A_4) \\ &\quad - N^{-2}(A_1^2 - A_2)^2 \end{aligned} \tag{4.3}$$

where $A_t = \Sigma_{j=1}^N a_j^t$, $t = 1, 2, \ldots$. Noether (1950) provides expressions for the mean and variance of a serial correlation statistic of lag h, and establishes its asymptotic normality. In particular, the limiting permutation distribution of $(T_1 - \mu_1)/\sigma_1$ is N(0, 1) provided the constants $(a_1, a_2, \ldots)$ satisfy

$$\left[\sum_{i=1}^N (a_i - \bar{a}_N)^t\right]\left[\sum_{i=1}^N (a_i - \bar{a}_N)^2\right]^{-t/2} = o(N^{(2-t)/4}) \quad \text{for } t = 3, 4, \ldots. \tag{4.4}$$

Also, if F has a finite moment of the order $4 + \delta$ for some $\delta > 0$, then the unconditional limit distribution of $(T_1 - \mu_1)/\sigma_1$ is N(0, 1).

Remark. Permutation tests based on the serial correlations of lag $h \geq 1$ were proposed by Wald and Wolfowitz (1943) for the problem of detecting a trend or cyclic movement in a series. However, these tests have extremely low asymptotic efficiency for the alternatives of location trend. Indeed, they are more sensible tests to detect a serial dependence than trend, and for this reason they are included in the present section.

In the definition of the serial correlation r_1 given in (4.1), if we replace the observations $X_1, \ldots, X_N$ by their ranks $R_1, \ldots, R_N$, we obtain the *rank serial correlation*

$$12(N^3-N)^{-1}\sum_{i=1}^{N-1}\left(R_i-\frac{N+1}{2}\right)\left(R_{i+1}-\frac{N+1}{2}\right).$$

This motivates the rank statistic

$$T_2=\sum_{i=1}^{N-1} R_iR_{i+1} \tag{4.5}$$

as a natural candidate for testing randomness against serial dependence. It was also proposed by Wald and Wolfowitz (1943) in the inappropriate context of testing for trend. The condition (4.4) holds in the case of ranks, and consequently, T_2 is asymptotically normal with the mean and variance given by

$$E(T_2)=(N^2-1)(3N+2)/12,$$
$$\operatorname{Var}(T_2)=N^2(N+1)(N-3)(5N+6)/720\,.$$

More general serial rank statistics can also be constructed by using some scores in place of the ranks. For instance, a van der Waerden type normal scores statistic for serial correlation is

$$T_3=\sum_{i=1}^{N-1}\Phi^{-1}\left(\frac{R_i}{N+1}\right)\Phi^{-1}\left(\frac{R_{i+1}}{N+1}\right). \tag{4.6}$$

The aforementioned tests draw from intuitive reasoning in which the normal theory test is used as the starting point, and then the distribution-free property is attained by means of either a permutation argument or use of the ranks. Gupta and Govindarajulu (1981) derive a locally most powerful (LMP) rank test for the autoregressive process

$$X_j=E_j+h_1(\rho)X_{j-1}+\cdots+h_k(\rho)X_{j-k},\quad |h_j(\rho)|<1\,,$$

where E_j's are iid $\mathrm{N}(0,\sigma^2)$, $h_j(\rho)$ are nondecreasing and $h_j(0)=0$. For the simple special case $k=1$, the LMP rank test is based on the statistic

$$T_4=\sum_{i=1}^{N-1}a_N(R_i,R_{i+1}) \tag{4.7}$$

where $a_N(\alpha,\beta)=E(W_N^{(\alpha)}W_N^{(\beta)})$ and $W_N^{(\cdot)}$ denotes the order statistics from the standard normal. Its mean is 0 and the variance is given by

$$\operatorname{Var}(T_4) = \frac{2}{N-1} + \frac{1}{N}\left\{\sum_{\alpha=1}^{N}\sum_{\beta=1}^{N} a_N^2(\alpha,\beta) - \frac{N+1}{N-1}\sum_{\alpha=1}^{N} a_N^2(\alpha,\alpha)\right\}.$$

The statistic

$$T_4^* = \sum_{i=1}^{N-1} a_N(R_i)a_N(R_{i+1}),$$

with $a_N(\alpha) = E(W_N^{(\alpha)})$, provides an approximation to T_4 in the sense of mean squared error, and is asymptotically equivalent to T_3. Gupta and Govindarajulu (1981) tabulate some selected upper percentage points of T_4 for $4 \leqslant N \leqslant 10$ and study its normal approximation and power by means of simulation.

For the first order autoregressive model $X_i = \alpha X_{i-1} + E_i$, Aiyar (1981) studies the Pitman ARE of the rank tests T_2 and T_3 relative to the best normal theory test based on the sample serial correlation r_1. The common df F of E_i is assumed to have a pdf f, mean 0 and variance σ^2. Under some regularity conditions on f, the ARE's of T_2 and T_3 relative to r_1 are respectively given by

$$\begin{aligned} e_2(F) &= 144\left[\int_{-\infty}^{\infty} xF(x)\,\mathrm{d}F(x)\right]^2\left[\int_{-\infty}^{\infty} f^2(x)\,\mathrm{d}x\right]^2, \\ e_3(F) &= \left[\int_{-\infty}^{\infty} x\Phi^{-1}(F(x))\,\mathrm{d}F(x)\right]^2\left[\int_{-\infty}^{\infty} f^2(x)/g(x)\,\mathrm{d}x\right]^2 \end{aligned} \tag{4.8}$$

where $g(x) = \varphi[\Phi^{-1}(F(x))]$ and $\varphi(x)$ is the pdf of N(0, 1). For all F satisfying the regularity conditions, we have $e_3(F) \geqslant 1$, and the lower bound 1 is attained by $F = \Phi$. Also, $e_2(F) \geqslant 9\pi^4/1024 \approx 0.8561$.

Knoke (1977) reports an extensive Monte Carlo study to examine the adequacy of the normal approximation under the null hypothesis and to compare the power of these tests for the first and second order models under the normal as well as some non-normal parent distributions. A few other tests which are not strictly distribution-free are also included in this study. Overall, the r_1 test is found to be fairly robust and generally more powerful than T_2 except for extreme departures from normality.

All the preceding rank tests of this section are applicable when the null hypothesis of randomness only specifies that the random variables $X_1, \ldots, X_N$ are iid with a continuous df F. It is often the case that the investigator has knowledge about the level of the process so the median of F can be assumed known. If, in addition, it is reasonable to assume that F is symmetric, then an altogether different class of nonparametric tests holds considerable promise for reason of its certain elegant features.

To describe these tests we now assume that the random variables $X_1, \ldots, X_N$ have symmetric and continuous (possibly non-identical) df's $F_1, \ldots, F_N$, respectively, with a common and known median which can be taken to be 0 without loss of generality. The null hypothesis of independence of these variables then entails that $\mathrm{med}(X_iX_{i+1}) = 0$, $i = 1, \ldots, N-1$, where $\mathrm{med}(X)$ stands for the median of the distribution of X. On the other hand, the alternative (H_1) of a positive serial dependence can be modeled by the criterion that $\mathrm{med}(X_iX_{i+1}) > 0$. In fact, a strictly positive quadrant dependence of the pair (X_i, X_{i+1}) implies that $\mathrm{med}(X_iX_{i+1}) > 0$ and so does a positive autocorrelation of lag 1.

Motivated by these considerations, Dufour (1981) proposes a class of linear rank statistics based on the products

$$Z_i = X_iX_{i+1},\ i = 1, \ldots, n = N-1. \tag{4.9}$$

Under H_0, the distribution of Z_i is symmetric around 0 whereas under H_1, it is more slanted to the positive values. Consequently, the testing problem is analogous to the one-sample location problem though a fundamental difference arises in that the Z_i's are not independent. The proposed linear rank statistics have the form

$$T = \sum_{i=1}^{n} u(Z_i)a_n^+(R_{ni}^+) = \sum_{i=1}^{n} V_i a_N^+(i) \tag{4.10}$$

where, in the first expression of (4.10), $u(z) = 1\ (0)$ if $z > (\leq)\, 0$, R_{ni}^+ is the rank of $|Z_i|$ among $|Z_1|, \ldots, |Z_n|$, and $a_n^+(\cdot)$ denotes a score function. In the second expression, $V_i = 1\ (0)$ according as the absolute rank i is associated with a positive (negative) Z. Under the null hypothesis the distribution of a one-sample linear rank statistic holds in this case even though the Z_i's are dependent. In fact, the Z_i's are symmetrically distributed around 0, and $(u(Z_1), \ldots, u(Z_n))$ and R_n^+ are independent. Consequently,

$$E(T) = \sum_{i=1}^{n} a_n^+(i)/2, \qquad \mathrm{Var}(T) = \sum_{i=1}^{n} a_n^{+2}(i)/4.$$

Moreover, the distribution of T is symmetric, and is asymptotically normal whenever

$$\max_{1\leq i\leq n} a_n^{+2}(i) \Big/ \Big[\sum_{i=1}^{n} a_n^{+2}(i)\Big] \to 0.$$

A simple and interesting special case of (4.10) arises when we choose $a_n(i) = 1$. The resulting T is the sign test statistic based on $Z_1, \ldots, Z_n$, and its null distribution is binomial $(n, \frac{1}{2})$. Alternatively, T is the number of pairs (X_i, X_{i+1}) having the same signs so $(N-T)$ is the total number of runs in the sequence of signs of $X_1, \ldots, X_N$ (assuming that there are no zeros), and $(n-T)$ is binomial $(n, \frac{1}{2})$. Other interesting special cases of (4.10) are the Wilcoxon signed rank test and the normal scores tests based on $Z_1, \ldots, Z_n$.

The principal advantages of Dufour's tests are their simplicity, rapid convergence to normality, and the availability of extensive tables for some special cases. In addition, the validity of these tests does not require that $X_1, \ldots, X_N$ be identically distributed. On the other hand, the assumption of symmetric distributions with a common known median is a serious limitation.

5. Other miscellaneous tests – Omnibus alternatives

Although the null hypothesis of randomness is clearly conceived in the framework of an iid sequence of random variables, the manner of a possible violation of randomness is often quite elusive. While the alternatives discussed in the previous sections have transparent physical interpretation, together they by no means exhaust all possible departures from the hypothesis of randomness. Moreover, in many practical situations, it is not reasonable to narrow the choice to any of these special alternatives because the investigator has only a vague perception of the possible lack of randomness as a tendency of the series to cluster. A clustering of the observations may, in general terms, be described by the feature that high (or low) values tend to occur together or that the pattern of the mix of high and low values deviate from the pattern (more precisely, the lack of a pattern) inherent in a random mix. Such ill-structured alternatives are designated as the *omnibus alternatives* to randomness.

By leaving the alternatives so broad and unstructured, it is futile to search for rank tests with such optimality criteria as the maximum local power or asymptotic efficiency. One would instead settle for some quick-and-dirty procedures based on some sensible aspects such as the runs of like elements, sign changes, local peaks and troughs, etc. Numerous heuristic tests have been proposed along these lines with the primary objective of achieving simplicity rather than considerations of the sensitivity of the tests to detect any specific form of nonrandomness.

To set the ideas, it will be convenient to begin with the case where the observations $X_1, \ldots, X_N$ are dichotomous, that is, each X_i has only two possible values 0 and 1. Here the null hypothesis of randomness states that $X_1, \ldots, X_n$ constitute a sequence of N Bernoulli trials with an unknown success probability $p = P[X = 1]$. Our attention is focused on the independence of the X_i's rather than the magnitude of p which is essentially a nuisance parameter.

Two somewhat related features suggest themselves as reasonable detectors of clustering; these are the number of *runs* and the *run-lengths*. A run is defined as an uninterrupted sequence of identical elements not preceded or succeeded by an element of the same kind. The number of elements in a run is called the run-length. For example, the sequence 0100001101 has a total of 6 runs whose lengths are 1, 1, 4, 2, 1 and 1, respectively. A significant clustering or violation of randomness of a series would be indicated by fewer runs and hence larger run lengths.

In its simplest form, a *run test* is based on the total number of runs (R) in the sequence $X_1, \ldots, X_N$. Denote by m and n the observed numbers of 0's and 1's, respectively, so $m + n = N$. The null distribution of R involves the nuisance parameter p but its conditional distribution, given m, is free of p and has the form

$$P[R = 2k \mid m] = 2\binom{m-1}{k-1}\binom{n-1}{k-1}\Big/\binom{N}{m},$$

$$P[R = 2k+1 \mid m] = \left[\binom{m-1}{k-1}\binom{n-1}{k} + \binom{m-1}{k}\binom{n-1}{k-1}\right]\Big/\binom{N}{m}. \tag{5.1}$$

An exact level α test can be based on the conditional distribution (5.1) with the rejection region set on the lower tail.

Lehmann (1959) demonstrates a strong optimality property of this run test in the context of a stationary Markov chain which is defined by the structure

$$P[X_i = 1 \mid x_1, \ldots, x_{i-1}] = P[X_i = 1 \mid x_{i-1}],$$

$$P[X_i = 1 \mid X_{i-1} = 1] = p_1, \qquad q_1 = 1 - p_1,$$

$$P[X_i = 1 \mid X_{i-1} = 0] = p_0, \qquad q_0 = 1 - p_0,$$

$$P[X_i = 1] = p_0/(p_0 + q_1)$$

where p_0 and p_1 are the same for all i. With this model, the probability of any realization $(x_1, \ldots, x_N)$ is given by

$$(p_0 + q_1)^{-1} p_0^v p_1^{n-v} q_1^u q_0^{m-u}$$

where u and v are the numbers of runs of 0's and 1's, respectively. The null hypothesis of independence is equivalent to H_0: $p_0 = p_1$. Because of an exponential-family structure of the likelihood, a uniformly most powerful unbiased test exists for testing H_0 versus H_1: $p_0 < p_1$, and the test is based on the statistic $(U + V) = R$, conditional on m. This is precisely the run test described above. Under the null hypothesis, R is asymptotically normal with mean $2N\lambda(1-\lambda)$ and standard deviation $2\lambda(1-\lambda)N^{1/2}$ where $\lambda = m/N$. So a normal approximation can be used to perform the run test in large samples.

We now return to the problem of testing randomness in a series $X_1, \ldots, X_N$ where the X_i's, instead of being dichotomous, are continuous variables. The run test can be adapted to this situation once we devise a reasonable mode of converting the continuous measurements to a dichotomy. One natural criterion is the sample median $\tilde{X}$, and we write 1(0) for an X_i according as $X_i > (<) \tilde{X}$. Then the run statistic R counts the number of runs above and below the median of the original sequence. Its null distribution is given by

$$P[R = 2k] = 2\binom{[N/2]-1}{k-1}^2 \Big/ \binom{N}{[N/2]},$$

$$P[R = 2k+1] = 2\binom{[N/2]-1}{k-1}\binom{[N/2]-1}{k} \Big/ \binom{N}{[N/2]}$$

where $[N/2]$ denotes the integer part of $N/2$. For large N, R is approximately normal with mean $= N/2$ and standard deviation $= N^{1/2}/2$.

By introducing a dichotomy with reference to the median of $X_1, \ldots, X_N$ one is prone to forfeit useful information regarding the pattern of progression of the series. On the other hand, such information is captured more by the differences $(X_{i+1} - X_i)$ of the successive members. Another class of run statistics arises when we consider the signs of these differences. This species of runs is quite different from the runs above and below the median, especially in regard to the distribution theory.

By assuming continuous distributions we ignore the possibility of ties among $X_1, \ldots, X_N$, and define the vector of signs of the successive differences

$$\boldsymbol{Y} = (Y_1, \ldots, Y_{N-1}) \tag{5.2}$$

by letting $Y_i = +$ or $-$ according as $X_{i+1} >$ or $< X_i$, respectively. As candidates for tests of randomness, various statistics based on the difference-sign vector $\boldsymbol{Y}$ have been explored by Moore and Wallis (1943), Wolfowitz (1944), Levene and Wolfowitz (1944), and Levene (1952). The basic ingredients are the *runs up and down*. A sequence of h consecutive $+$ signs not immediately preceded or followed by a $+$ sign is called a *run up* of length h (similarly a *run down* is defined). Some relevant run statistics of this type are: # runs up (down) of length h, # of runs up (down) of length $\geq h$, # runs up (down), $r =$ total # runs up or down, $K = \# +$ signs in $\boldsymbol{Y}$.

Two particular test statistics deserve special mention for the reasons of their simplicity and popularity. One of them is r which counts the total number of runs up or down. It is essentially the same as the number of turning points or the local peaks and troughs in the original series $X_1, \ldots, X_N$, and is widely known as the *turning point* test of Moore and Wallis (1943). Under H_0, r is asymptotically normal with mean $(2N-1)/3$ and variance $(16N-29)/90$. The other simple test is based on K and is known as the *difference-sign test*. A normal approximation also holds for K with mean $= (N-1)/2$ and variance $= (N+1)/12$.

As a general designation for the run statistics illustrated in (5.3), Levene (1952) uses the name 'u-run statistics'. Their properties, including the exact means, variances and covariances, asymptotic joint normality under H_0 as well as under a number of different parametric alternatives, and asymptotic power, are studied by Levene and Wolfowitz (1944) and Levene (1952). The use of a function of several u-run statistics is also suggested with the intent of making the test sensitive to different departures from randomness. For instance, K and

r can be used together by forming the mixed statistic

$$T_{K,r} = Z_K^2 + Z_r^2$$

where Z_K and Z_r are the standardized forms of K and r, respectively. Since K and r are asymptotically independent, the limiting null distribution of $T_{K,r}$ is χ_2^2.

Historically, nonparametric tests were often constructed from intuitive ground with the goal of detecting a trend in a time series, but the prime consideration was given to the simplicity of the statistics and easy applicability of the test procedures. It turns out that although these tests possess the property of consistency for a wide class of trend alternatives, their asymptotic efficiency is often miserably low when compared to the nonparametric tests which are especially designed for the alternatives of trend. For example, the difference-sign test K was originally designated as a test for trend but relative to Mann's test, its ARE is 0 under the location trend alternatives. The rank serial correlation test (T_2) of Section 4 was also called a test for trend but its ARE relative to Mann's test is again 0. The turning point test r is often used as a test for serial dependence. Knoke (1977) shows that for the normal first order autoregressive model, the ARE of the turning point test relative to the sample autocorrelation is $15/(8\pi^2) \approx 0.19$.

References

Aiyar, R. J. (1981). Asymptotic efficiency of rank tests of randomness against autocorrelation. *Annals of the Institute of Statistical Mathematics* (Tokyo) **33** (2), 255–262.

Aiyar, R. J., Guillier, C. L. and Albers, W. (1979). Asymptotic relative efficiencies of rank tests for trend alternatives. *Journal of the American Statistical Association* **74**, 226–231.

Bhattacharya, P. K. and Frierson, D. Jr. (1981). A nonparametric control chart for detecting small disorders. *Annals of Statistics* **9** (3), 544–554.

Bhattacharyya, G. K. (1968). Robust estimates of linear trend in multivariate time series. *Annals of the Institute of Statistical Mathematics* (Tokyo) **20**, 299–310.

Bhattacharyya, G. K. and Johnson, R. A. (1968). Nonparametric tests for shift at unknown time point. *Annals of Mathematical Statistics* **39**, 1731–1743.

Bhattacharyya, G. K. and Klotz, J. H. (1966). The bivariate trend of Lake Mendota. Technical Report No. 98, Dept. of Statistics, University of Wisconsin.

Chernoff, H. and Zacks, S. (1964). Estimating the current mean of a normal distribution which is subject to change in time. *Annals of Mathematical Statistics* **35**, 999–1018.

Cox, D. R. and Stuart, A. (1955). Some quick sign tests for trend in location and dispersion. *Biometrika* **42**, 80–95.

Daniels, H. E. (1950). Rank correlation and population models. *Journal of the Royal Statistical Society Ser. B* **12**, 171–181.

Dietz, E. J. and Killeen, T. J. (1981). A nonparametric multivariate test for monotone trend with pharmaceutical applications. *Journal of the American Statistical Association* **76**, 169–174.

Dufour, J.-M. (1978). Rank tests for serial correlation. In: *Proc. Sec. of Bus. and Econ.* American Statistical Association, pp. 748–753.

Gupta, G. D. and Govindarajulu, Z. (1980). Nonparametric tests of randomness against autocorrelated normal alternatives. *Biometrika* **67**, 375–379.

Hájek, J. and Šidák, Z. (1967). *Theory of Rank Tests.* Academic Press, New York, and Academia, Prague.

Hodges, J. L., Jr. and Lehmann, E. L. (1963). Estimates of location based on rank tests. *Annals of Mathematical Statistics* **34**, 598–611.

Kander, Z. and Zacks, S. (1966). Test procedures for possible changes in parameters of statistical distributions occurring at unknown time points. *Annals of Mathematical Statistics* **37**, 1196–1210.

Knoke, J. D. (1977). Testing for randomness against autocorrelation: alternative tests. *Biometrika* **64**, 523–529.

Lehmann, E. L. (1959). *Testing Statistical Hypotheses.* Wiley, New York.

Levene, H. (1952). On the power function of tests of randomness based on runs up and down. *Annals of Mathematical Statistics* **23**, 34–56.

Levene, H. and Wolfowitz, J. (1944). The covariance matrix of runs up and down. *Annals of Mathematical Statistics* **15**, 58–69.

Mann, H. B. (1945). Nonparametric tests against trend. *Econometrica* **13**, 245–259.

Moore, G. H. and Wallis, W. A. (1943). Time series significance tests based on signs of differences. *Journal of the American Statistical Association* **38**, 153–164.

Noether, G. E. (1950). Asymptotic properties of the Wald-Wolfowitz test of randomness. *Annals of Mathematical Statistics* **21**, 231–246.

Page, E. S. (1955). A test for a change in a parameter occurring at an unknown point. *Biometrika* **42**, 523–526.

Pettitt, A. N. (1979). A nonparametric approach to the change-point problem. *Applied Statistics* **28** (2), 126–135.

Sen, P. K. (1968). Estimates of the regression coefficient based on Kendall's tau. *Journal of American Statistical Association* **63**, 1379–1389.

Sen, A. and Srivastava, M. S. (1975). On tests for detecting change in mean. *Annals of Statistics* **3**, 90–108.

Stuart, A. (1956). The efficiencies of tests of randomness against normal regression. *Journal of the American Statistical Association* **51**, 285–287.

Theil, H. (1950). A rank-invariant method of linear and polynomial regression analysis. *Proc. Kon. Ned. Akad. v. Wetensch. A.* **53**, 386–392, 521–525, 1397–1412.

Wald, A. and Wolfowitz, J. (1943). An exact test for randomness in the nonparametric case based on serial correlation. *Annals of Mathematical Statistics* **14**, 378–388.

Wolfowitz, J. (1944). Asymptotic distribution of runs up and down. *Annals of Mathematical Statistics* **15**, 163–172.

P. R. Krishnaiah and P. K. Sen, eds., *Handbook of Statistics, Vol. 4*

6

Combination of Independent Tests

J. Leroy Folks

1. The setting

We have obtained k independent test statistics T_i for the null hypotheses H_{0i}, $i = 1, 2, \ldots, k$. For each test the significance level L_i (also called the P-value or observed significance level) is given by an upper tail probability. That is,

$$L_i(t_i) = P(T_i > t_i \mid H_{0i} \text{ is true}) = 1 - F_{T_i}(T_i \mid H_{0i})$$

where F_{T_i} is the cdf of T_i.

How should we combine the L's to test the hypothesis H_0: All of H_{01}, $H_{02}, \ldots, H_{0k}$ are true? Sometimes the problem has been how best to combine the T's or even the raw data sets from which the T's were obtained. The combination of statistics T_i and of raw data sets will not be discussed in this article. In some applications all of the separate null hypotheses H_{0i} will be the same; in other applications the separate null hypotheses will be different. The independent test statistics may have come from independent data sets or from the same data set.

We now cite several examples for which this combination problem could arise:

(1) t tests for the equality of two treatment effects versus the one-sided alternative that one treatment is better than another. One test may be from a two-group experiment, another from a randomized block experiment, another from a Latin square experiment, etc.

(2) t tests for the equality of two treatment effects versus the two-sided alternative that treatments are unequal. This fits the established format because the two-tailed probability is the upper tail probability for the absolute value of the t statistic.

(3) F tests for equality of two variances versus one sided alternatives. The two-sided alternative is not simply accommodated as in the case of the t test.

(4) F tests for equality of several treatment means. The F tests need not all involve the same treatments.

(5) Chi squared tests for variances versus one-sided alternatives.

(6) Chi squared tests for independence in contingency tables

In all of these examples, we usually assume that the test statistic is continuous (approximately so for the contingency table chi square). In cases like these, the significance levels L_i are independent uniform (0, 1) random variables when the H_{0i} are true. With two-sided chi square and F tests, for example, this will not be true.

2. Combination methods

2.1. The probability integral transformation

It is well known that a right (or left) tail area probability of a continuous random variable is a uniform random variable on [0, 1]. Therefore under H_0, the L's are independent uniform [0, 1] random variables. Most of the combination methods make use of this fact. One of the well-known ways of generating a random variable with distribution function G is to generate a uniform (0, 1) variable U and then to calculate $X = G^{-1}(U)$. Then X has the distribution specified by G.

2.2. Tippett–Wilkinson method

The literature on combining independent tests apparently began with Tippett (1931) who proposed using the largest $F_i(T_i)$, or equivalently, the smallest L_i. Twenty years later, Wilkinson (1951) proposed using the m-th smallest significance level. Wilkinson motivated his suggested statistic by considering the probability of obtaining m or more significant statistics by chance in a group of k. Let $L_{(m)}$ be the smallest L_i among $L_1, L_2, \ldots, L_k$. It can be shown that $L_{(m)}$ is a beta variable. That is, $L_{(m)} \sim \beta(m, k-m+1)$. Then the combined significance level, L_i is given by

$$L = P(X \leqslant L_{(m)}) \quad \text{where } X \sim \beta(m, k-m+1)\,.$$

Because we can transform beta variables to F variables, it is given equivalently by

$$L = P\left(X \geqslant \frac{m}{k-m+1}\,\frac{1-L_{(m)}}{L_{(m)}}\right)$$

where $X \sim F(2(k-m+1), 2m)$.

2.3. The Fisher–Pearson method

Fisher (1932) presented and illustrated with a numerical example the use of $-2\,\Sigma_{i=1}^{k} \log L_i$. Because under H_0, L_i is a uniform random variable, $-2 \log L_i$ is a chi squared random variable with two degrees of freedom. Because of the additivity property of independent chi squares, Fisher's method leads to the combined significance level

$$L = P\left(\chi^2_{2k} \geqslant -2\sum_{i=1}^{k} \log L_i\right).$$

Fisher made it clear that the test was based upon the product $\Pi_1^k L_i$ of significance levels. Pearson (1933) proposed what appears to be the same test but suggested that L_i should be the minimum of the right and left tail areas. Once this defect was corrected (David, 1934), Fisher's test and Pearson's test were the same. Even today some writers refer to Fisher's test as Pearson's P_λ test.

2.4. Lancaster method

We noted in Section 2.1 that L_i can be transformed into any desired distribution by using the inverse of the distribution function. Fisher's method transforms to chi squares and uses the chi square additivity property. Lancaster (1961) noted that the additivity property would still hold if L_i were transformed to gamma $(\alpha_i, \frac{1}{2})$ variables. That is, let

$$S_i = \Gamma^{-1}_{\alpha_i}(1 - L_i)$$

where $\Gamma^{-1}_{\alpha_i}$ is the inverse of the gamma distribution function with parameters α_i and $\frac{1}{2}$. Then $S = \Sigma_{i=1}^k S_i$ has the gamma distribution with parameters $(\Sigma\alpha_i, \frac{1}{2})$. Then the combined significance level is given by

$$L = P\left(S \geqslant \sum_{i=1}^{k} \Gamma^{-1}_{\alpha_i}(1 - L_i)\right)$$

where $S \sim \Gamma(\Sigma\alpha_i, \frac{1}{2})$.

The choice of α's is entirely arbitrary and can be done to weight the L's differently. If α_i is integer valued then $S_i \sim \chi^2_{2\alpha_i}$. Even if all the α_i are not integer valued but Σ_{α_i} is, then $S \sim \chi^2_{2\Sigma\alpha_i}$ and the combined significance level is given by

$$L = P\left(\chi^2_{2\Sigma\alpha_i} \geqslant \sum_{i=1}^{k} \Gamma^{-1}_{\alpha_i}(1 - L_i)\right).$$

2.5. Liptak–Stouffer method

Given that the L_i can be transformed to any variable we wish, it is strange that one of the most obvious choices was for so long ignored. The L_i can be transformed to standard normal variables by $\Phi^{-1}(L_i)$ where Φ is the standard normal distribution function. Then they can be combined in many ways. Liptak (1958) suggested

$$\sum_{i=1}^{k} \alpha_i \Phi^{-1}(1 - L_i)/\sqrt{\Sigma\, \alpha_1^2}\,.$$

Then the combined level is given by

$$L = P\left(Z \geqslant \sum_{i=1}^{k} \alpha_i \Phi^{-1}(1 - L_i)/\sqrt{\Sigma\, \alpha_i^2}\right).$$

This gives another way of weighting the L_i unequally. Previously Stouffer et al. (1949) had suggested using the unweighted sum

$$(\alpha_i = 1,\ i = 1, 2, \ldots, k)\,.$$

2.6. Good–Zelen method

A rather different way of weighting the individual L_i was suggested by Good (1955). He suggested using $\Pi_1^k L_i^{\alpha_i}$, or equivalently the linear combination $W = -2\sum_{i=1}^{k} \alpha_i \log L_i$. In general, the distribution of a linear combination of chi squares is unknown but, in this case, each of the chi-squares has only two degrees of freedom and it is possible to get a closed expression for the distribution function of W. Good showed that the combined significance level is given by

$$L = \sum_{i=1}^{k} \Lambda_i\, \mathrm{e}^{-W/2\alpha_i}$$

where

$$\Lambda_i = \alpha_i^{k-1} \Big/ \prod_{\substack{j=1 \\ j \neq i}}^{k} (\alpha_i - \alpha_j)\,.$$

Zelen (1957) was interested in combining the inter and intrablock F tests for equality of treatments in an incomplete block design and proposed weights 1 and α_2. Zelen and Joel (1959) determined that the type II error is minimized over a wide range of parameters when α_2 is proportional to the ratio of intrablock and interblock variances.

2.7. George–Mudholkar method

George and Mudholkar suggest transforming the L_i to logistic variables by using the transformation $Y_i = -\log[L_i/(1 - L_i)]$. Then each Y_i is a logistic variable and a probability statement can be made concerning the sum ΣY_i. They propose the approximation

$$L = P(t_{5k+4}) > \left(-\sum_i \log \frac{L_i}{1 - L_i}\right) \Big/ (\Pi^2 k/3)^{1/2}\Big)\,.$$

2.8. Edgington method

One method of simulating a standard normal variable is to standardize the mean of independent uniform (0, 1) variables. This forms the basis for the method proposed by Edgington (1972). Because the mean of a uniform (0, 1) variable is $\frac{1}{2}$ and the variance is $\frac{1}{12}$, he proposed to approximate the combined significance level L by

$$L = P\left(Z \leqslant \frac{\bar{L} - 0.50}{\sqrt{1(12k)}}\right)$$

where Z is a standard normal variable.

3. Example

We now illustrate the combination of L's for the methods just discussed with the following example:

Test	Level	Weight
1	0.07	2
2	0.20	1
3	0.04	3

The weights will be used for several of the methods.

Tippett–Wilkinson method. Using the smallest $L_i = 0.04$,

$$L = P\left(X \geqslant \frac{1}{3-1+1}\,\frac{1-0.04}{0.04}\right) \quad \text{where } X \sim F(2(3-1+1), 2)$$

$$= P(X \geqslant 8) \quad \text{where } X \sim F(6, 2) = 0.115264.$$

Fisher–Pearson method

$$L = P\left(\chi_6^2 \geqslant -2 \sum_{i=1}^{3} \log L_i\right) = P(\chi_6^2 \geqslant 14.9751) = 0.020451\,.$$

Lancaster method

L_i	$1 - L_i$	α_i	$F^{-1}_{\chi^2, \alpha_i}(1 - L_i)$
0.07	0.93	2	8.66
0.20	0.80	1	3.22
0.04	0.96	3	13.20
		Sum	25.08

$$L = P(\chi_{12}^2 \geqslant 25.08) = 0.014448\,.$$

Liptak–Stouffer method

L_i	$1-L_i$	α_i	$\Phi^{-1}(1-L_i)$	$\alpha_i\Phi^{-1}(1-L_i)$	α_i^2
0.07	0.93	2	1.4758	2.9516	4
0.20	0.80	1	0.8416	0.8416	1
0.04	0.96	3	1.7507	5.2521	9
			Sum	9.0453	14

$$L = P(Z > 9.0453/\sqrt{14}) = 0.007815 .$$

Good–Zelen method

$$\Lambda_1 = \alpha_1^2/(\alpha_1-\alpha_2)(\alpha_1-\alpha_3) = 4/(1)(-1) = -4 ,$$
$$\Lambda_2 = \alpha_2^2/(\alpha_2-\alpha_1)(\alpha_2-\alpha_3) = 1/(-1)(-2) = \tfrac{1}{2} ,$$
$$\Lambda_3 = \alpha_3^2/(\alpha_3-\alpha_1)(\alpha_3-\alpha_2) = 9/(1)(2) = \tfrac{9}{2} ,$$
$$L = \sum_{i=1}^{3} \Lambda_i \, e^{-W/2} = 0.016877 .$$

George–Mudholkar method

$$\sum_i \log \frac{L_i}{1-L_i} = 7.151037535 , \qquad \left(\frac{\pi^2 k}{3}\right)^{1/2} = \pi .$$
$$L = P\left(t_{5k+4)} > -\pi \sum_i \log \frac{L_i}{1-L_i}\right) = 0.017297 .$$

Edgington method

$$\frac{\bar{L}-0.50}{\sqrt{1/(12k)}} = 6\left[\frac{0.31}{3} - 0.50\right] ,$$
$$L = P(Z < -2.38) = 0.008656 .$$

4. Comparison of methods

Comparisons of different methods of combining independent tests have been made by many authors. Birnbaum (1954) showed that if the method is monotonic increasing in each of the L_i's, then the procedure is admissible; that is, there exists a problem for which it is most powerful against some parameter value. When each T_i has an exponential family density Birnbaum showed that Tippett's method was admissible and that Pearson's original P_λ and Wilkinson's methods were inadmissible. He also strongly suggested that Fisher's method was admissible.

Extensive studies of power have been made when the distribution of the T_i's is assumed. Bhattacharya (1961) and Koziol and Perlman (1978) assumed that the T_i's were independent χ^2 variables. They found that Fisher's test performed well over a wide range of parameter values and that the Liptak–Stouffer inverse-normal method performed very poorly. Marden (1982) obtained theoretical results proving that under the chi-square assumption, Fisher's method and Tippett's method are admissible. The Liptak–Stouffer method, George–Mudholkar method and Edgington method were shown to be inadmissible. It was conjectured that Lancaster's method was admissible.

Zelen and Joel (1959) and Pape (1972) suggest using the Good–Zelen method for combination of tests when the T_i's are F statistics. Monti and Sen (1976) present combinations of F statistics with locally optimum properties. Marden (1982) showed that Fisher's method is admissible if and only if all degrees of freedom are greater than one and that Tippett's method is always admissible.

Oosterhoff (1969) found a complete class of tests in a problem involving the combinations of nonindependent noncentral t-tests.

Littell and Folks (1971) showed that Fisher's test was optimal in the sense of Bahadur efficiency. Essentially this means that as the sample sizes for the T_i's go to infinity the overall level, L, converges to zero under the alternative as fast or faster than any other test. Berk and Cohen (1979) showed that there are many methods which are Bahadur efficient, e.g. Lancaster's method and the George–Mudholkar method. On the other hand the Good–Zelen method is not Bahadur efficient.

Cohen, Marden and Singh (1982) study second-order optimality properties of classes of statistics which are Bahadur efficient. If both T_1 and T_2 are Bahadur efficient, consider $D = \log L_1 - \log L_2$. Then T_2 is said to be more efficient (second order) than T_1 if under the alternative hypothesis the limit of D as $n \to \infty$ is positive with probability one. If the limit in probability is greater than zero, T_2 is said to be weakly superior to T_1. Under some weak conditions, it is shown that the statistic based on normal transforms has the greatest second order efficiency.

5. Other methods

Several authors have assumed that not only the levels but the statistics, T_i, and their distributions are available. Van Zwet and Oosterhoff (1967) consider the case where the T_i's are asymptotically normal and use the corresponding asymptotically optimal methods for the small-sample case.

Littell and Louv (1981) consider inversion of combined tests as a way of generating confidence intervals.

Several authors consider methods, particularly Bayesian methods, other than the ones described in this paper.

Finally, we should mention that Lancaster (1949) and E. S. Pearson (1950) studied the effect of discreteness upon the combination of tests.

6. Summary

Fisher's method is strongly supported by the literature. It has good power for a large set of alternative parameter values. Tippett's method has good power against alternatives for which a few of the null hypotheses $H_{01}, H_{02}, \ldots, H_{0k}$ are false but not for which many are false.

For almost all problems studied, Fisher's method and Tippett's method are admissible.

References

Berk, R. H. and Cohen, A. (1978). Asymptotically optimal methods of combining tests. *J. Amer. Stat. Assoc.* **74**, 812–814.

Bhattacharya, N. (1961). Sampling experiments on the combination of independent χ^2-tests. *Sankhya, Ser. A* **23**, 191–196.

Birnbaum, A. (1954). Combining independent tests of significance. *J. Amer. Stat. Assoc.* **49**, 559–574.

Cohen, A., Marden, J. I. and Singh, Kesar (1982). Second order asymptotic and non-asymptotic optimality properties of combined tests. *J. Statist. Plann. Inference* **6**, 253–256.

David, F. N. (1934). On the P_{λ_n} test for randomness: remarks, further illustration, and table of P_{λ_n} for given values of $-\log_{10}\lambda_n$. *Biometrika* **26**, 1–11.

Edgington, E. S. (1972). A normal curve method for combining probability values from independent experiments. *J. Psych.* **82**, 85–89.

Fisher, R. A. (1932). *Statistical Methods for Research Workers.* Oliver and Boyd, Edinburgh, 4th Edition.

George, E. O. and Mudholkar, G. S. (1977). The logit method for combining independent tests. *Inst. Math. Stat. Bull.* **6**, 212.

Good, I. J. (1955). On the weighted combination of significance tests. *J. Roy. Stat. Soc. Ser. B* **17**, 264–265.

Koziol, J. A. and Perlman, M. D. (1978). Combining independent chi-squared tests. *J. Amer. Stat. Assoc.* **73**, 753–763.

Lancaster, H. O. (1949). The combination of probabilities arising from data in discrete distributions. *Biometrika* **36**, 370–382.

Lancaster, H. O. (1961). The combination of probabilities: An application of orthonormal functions. *Austral. J. Statist.* **3**, 20–33.

Littell, R. C. and Folks, J. L. (1981). Confidence regions based on methods of combining test statistics. *J. Amer. Statist. Assoc.* **76**, 125–130.

Littell, R. C. and Louv, W. C. (1981). Confidence regions based on methods of combining test statistics. *J. Amer. Statist. Assoc.* **76**, 125–130.

Marden, J. I. (1982). Combining independent noncentral chi squared or F tests. *Ann. Statist.* **10**, 266–277.

Monti, K. L. and Sen, P. K. (1976). The locally optimum combination of independent test statistics. *J. Amer. Statist. Assoc.* **71**, 903–911.

Oosterhoff, J. (1969). *Combination of One-Sided Statistical Tests*, Mathematical Centre Tracts, 28, Amsterdam.

Pape, E. S. (1972). A combination of F-statistics. *Technometrics* **14**, 89–99.

Pearson, E. S. (1950). On questions raised by the combination of tests based on discontinuous distributions. *Biometrika* **37**, 383–398.

Pearson, K. (1933). On a method of determining whether a sample of size n supposed to have been drawn from a parent population having a known probability integral has probably been drawn at random. *Biometrika* **25**, 379–410.

Stouffer, S. A., Suchman, E. A., DeVinney, L. C., Star, S. A. and Williams, R. M. (1949). *The American Soldier: Adjustment During Army Life (Vol. 1)*. Princeton Univ. Press, NJ.
Tippett, L. H. C. (1931). *The Methods of Statistics*. Williams and Norgate, London, 1st ed.
Van Zwet, W. R. and Oosterhoff, J. (1966). On the combination of independent test statistics. *Ann. Math. Stat.* **38**, 659–680.
Wilkinson, B. (1951). A statistical consideration in psychological research. *Psych. Bull.* **48**, 156–158.
Zelen, M. (1957). The analysis of incomplete block designs. *J. Amer. Statist. Assoc.* **52**, 204–216.
Zelen, M. and Joel, L. S. (1959). The weighted compounding of two independent significance tests. *Ann. Math. Statist.* **30**, 885–895.

P. R. Krishnaiah and P. K. Sen, eds., *Handbook of Statistics, Vol. 4*
© Elsevier Science Publishers (1984) 123–173 7

Combinatorics

Lajos Takács

The term *combinatorics* was introduced by G. W. Leibniz [28] in 1666 and he gave a systematic study of the subject. The basic combinatorial problems are concerned with the enumeration of the possible arrangements of several objects under various conditions.

Permutations. The number of ordered arrangements of n distinct objects, marked $1, 2, \ldots, n$, without repetition is

$$n! = 1 \cdot 2 \cdots n .$$

The $n!$ arrangements are called *permutations without repetition.* By convention, $0! = 1$.

If we have k_1 objects marked 1, k_2 objects marked $2, \ldots, k_n$ objects marked n, the number of ordered arrangements of the $k_1 + k_2 + \cdots + k_n$ objects is

$$\frac{(k_1 + k_2 + \cdots + k_n)!}{k_1! k_2! \cdots k_n!} .$$

These arrangements are called *permutations with repetition.*

If n is large, $n!$ can be calculated approximately by the following formula

$$n! \sim \sqrt{2\pi n}(n/\mathrm{e})^n$$

where $\pi = 3.1415926535\ldots$ and $\mathrm{e} = 2.7182818284\ldots$. Here the left side is asymptotically equal to the right side, that is, the ratio of the two sides tends to 1 as $n \to \infty$. This formula is the consequence of the collaboration of A. De Moivre and J. Stirling. See De Moivre [13]. More precisely, we have the inequalities

$$\sqrt{2\pi n}(n/\mathrm{e})^n < n! < \sqrt{2\pi n}(n/\mathrm{e})^n \mathrm{e}^{1/12n}$$

for all $n \geq 1$.

Combinations and variations. The number of unordered arrangements of k

different objects chosen among n distinct objects, marked $1, 2, \ldots, n$, is

$$\binom{n}{k} = \frac{n!}{k!(n-k)!} = \frac{n(n-1)\cdots(n-k+1)}{k!}$$

for $1 \leqslant k \leqslant n$; these arrangements are called *combinations without repetition.*

The number of ordered arrangements of k different objects chosen among n distinct objects is

$$\frac{n!}{(n-k)!} = n(n-1)\cdots(n-k+1)$$

for $1 \leqslant k \leqslant n$; these arrangements are called *variations without repetition.*

The number of unordered arrangements of k objects chosen in such a way that each object may be any of n distinct objects is

$$\binom{n+k-1}{k} = \frac{(n+k-1)}{k!(n-1)!}$$

for $k = 1, 2, \ldots$ and $n \geqslant 1$; these arrangements are called *combinations with repetition.*

The number of ordered arrangements of k objects chosen in such a way that each object may be any of n distinct objects is simply

$$n^k$$

for $k = 1, 2, \ldots$ and $n \geqslant 1$; these arrangements are called *variations with repetition.*

See Table 1 for a display of the above formulas. For more details see Netto [38], Riordan [42], and Ryser [45].

Binomial coefficients. The oldest combinatorial problems are connected with the notion of binomial coefficients. For any x the k-th binomial coefficient

Table 1

	No repetition ($k \leqslant n$)	Repetition ($k \geqslant 1$)
Unordered arrangements (Combinations)	$\binom{n}{k}$	$\binom{n+k-1}{k}$
Ordered arrangements (Variations)	$\frac{n!}{(n-k)!}$	n^k

$(k = 1, 2, \ldots)$ is defined as

$$\binom{x}{k} = \frac{x(x-1)\cdots(x-k+1)}{k!}$$

and $\binom{x}{0} = 1$.

In the early age of mathematics binomial coefficients appeared in three different disguises.

The first appearance is connected with the solution of the problem of finding the number of ways in which k objects can be chosen among n distinct objects without regard to order. As we have already seen, the solution is $\binom{n}{k}$ for $k = 1, 2, \ldots, n$. See Table 2.

Table 2
$\binom{n}{k}$

n \ k	0	1	2	3	4
0	1	0	0	0	0
1	1	1	0	0	0
2	1	2	1	0	0
3	1	3	3	1	0
4	1	4	6	4	1

The second appearance is connected with the notion of figurate numbers F_n^k $(n \geq 0,\ k \geq 1)$. The numbers F_0^k, $F_1^k, \ldots$ can be obtained from the sequence $F_0^1 = 1$, $F_1^1 = 1, \ldots$ by repeated summations, namely

$$F_n^{k+1} = F_0^k + F_1^k + \cdots + F_n^k$$

for $k \geq 1$ and $n \geq 0$, and $F_n^1 = 1$ for $n \geq 0$. See Table 3.

Table 3
F_n^k

k \ n	0	1	2	3	4	5
1	1	1	1	1	1	1
2	1	2	3	4	5	6
3	1	3	6	10	15	21
4	1	4	10	20	35	56
5	1	5	15	35	70	126

We have

$$F_n^k = \binom{n+k-1}{k-1}$$

for $n \geq 0$ and $k \geq 1$. F_n^k can be interpreted as the number of different ways in which n pearls (or other indistinguishable objects) can be distributed in k boxes. F_n^k is also the number of combinations of size n with repetition of k distinct objects.

The third appearance is connected with the problem of finding the n-th power ($n = 1, 2, \ldots$) of the binomial $a + b$. We have

$$(a+b)^n = \sum_{k=0}^{n} C_n^k a^k b^{n-k}$$

and C_n^k, the coefficient of $a^k b^{n-k}$, can be interpreted as the number of ways in which k letters a and $n-k$ letters b can be arranged in a row. As we have seen, this number is

$$C_n^k = \frac{n!}{k!(n-k)!} = \binom{n}{k}$$

for $k = 0, 1, \ldots, n$, where the convention $0! = 1$ is used. The numbers $C_n^k = \binom{n}{k}$ $(0 \leq k \leq n)$ are usually arranged in the form of a triangle, which is called the arithmetic triangle. See Table 4.

Table 4
Arithmetic triangle

				1				
			1		1			
		1		2		1		
	1		3		3		1	
1		4		6		4		1
.	.	.	.	.	.	.	.	.

The numbers $\binom{n}{k}$, $0 \leq k \leq n$, gained importance when it was recognized that they appear as coefficients in the expansion of $(a+b)^n$. Apparently, this was known to Omar Khayyam, a Persian poet and mathematician, who lived in the eleventh century. See Woepcke [63]. In 1303 Shih-chieh Chu [8] refers to the numbers $\binom{n}{k}$, $0 \leq k \leq n$, as an old invention. (See Y. Mikami [32, p. 89], and J. Needham [37, p. 133].) The numbers $\binom{n}{k}$, $0 \leq k \leq n$, arranged in the form of a triangular array, appeared in 1303 at the front of the book of Shih-chieh Chu. Unfortunately, the original book was lost, but it was restored in the nineteenth century. The triangular array first appeared in print in 1527 on the title page of the book of P. Apianus (see Smith [47, p. 509]). In 1544, M. Stiefel showed that in the binomial expansion of $(a+b)^n$ the coefficients C_n^k $(0 \leq k \leq n)$ can be calculated by the recurrence equations

$$C_{n+1}^{k+1} = C_n^k + C_n^{k+1}$$

where $C_n^0 = C_n^n = 1$ for $n \geq 0$. He arranged the coefficients C_n^k, $0 \leq k \leq n$, in a triangular array. In 1556 N. Tartaglia claimed the triangular array as his own invention.

The numbers C_n^k, $0 \leq k \leq n$, appeared in the seventeenth century in connection with combinations. In 1634 Hérigone knew that

$$C_n^k = n(n-1)\cdots(n-k+1)/k!\,.$$

The same formula appeared also in 1665 in a treatise by Pascal [39].

Here are a few useful formulas for binomial coefficients. If a, b and n are positive integers, we have

$$\sum_{j=0}^{n} \binom{a}{j}\binom{b}{n-j} = \binom{a+b}{n} \quad \text{and} \quad \sum_{j=a}^{n} \binom{j}{a}\binom{n-j}{b} = \binom{n+1}{a+b+1}.$$

Furthermore, if a, b, c and n are positive integers we have

$$\sum_{j=0}^{n} \binom{a+bj}{j} \frac{a}{(a+bj)} \binom{c+b(n-j)}{n-j} \frac{c}{(c+b(n-j))} = \binom{a+c+bn}{n} \frac{a+c}{(a+c+bn)}.$$

The method of inclusion and exclusion. One of the most powerful combinatorial methods in probability theory and in mathematical statistics is the method of inclusion and exclusion. To formulate the main result, let us suppose that Ω is a finite set and $A_1, A_2, \ldots, A_n$ are subsets of Ω. Denote by H_k $(0 \leq k \leq n)$ the set of all those elements of Ω which belong to exactly k sets among $A_1, A_2, \ldots, A_n$. The goal is to find $N(H_k)$, the number of elements of H_k, provided that we know $N(\Omega)$, the number of elements of Ω, and $N(A_{i_1}A_{i_2}\cdots A_{i_j})$, the number of elements in the intersection of the sets $A_{i_1}, A_{i_2}, \ldots, A_{i_j}$ for all $1 \leq i_1 < i_2 < \cdots < i_j \leq n$. We have

$$N(H_k) = \sum_{j=k}^{n} (-1)^{j-k} \binom{j}{k} S_j \tag{1}$$

where

$$S_0 = N(\Omega)\,,$$

$$S_1 = N(A_1) + N(A_2) + \cdots + N(A_n)\,,$$

$$S_2 = N(A_1A_2) + N(A_1A_3) + \cdots + N(A_{n-1}A_n)\,,$$

$$\vdots$$

$$S_n = N(A_1A_2\cdots A_n)\,,$$

or, in general,

$$S_j = \sum N(A_{i_1}A_{i_2}\cdots A_{i_j}), \quad 1 \le i_1 < i_2 < \cdots < i_j \le n,$$

for $1 \le j \le n$.

Formula (1) is a consequence of the following equation:

$$S_j = \sum_{k=j}^{n} \binom{k}{j} N(H_k) \tag{2}$$

which holds for all $j = 0, 1, \ldots, n$. The proof of this equation is easy. If $x \in \Omega$ belongs to exactly r sets among $A_1, A_2, \ldots, A_n$, then its contribution to the left side of (2) is 0 if $r < j$ and $\binom{r}{j}$ if $r \ge j$. The contribution of x to the right side of (2) is also 0 if $r < j$ and $\binom{r}{j}$ if $r \ge j$. If we multiply (2) by $(-1)^{j-k}\binom{j}{k}$ and if we sum the product for $j = k, k+1, \ldots, n$, then we obtain $N(H_k)$ and this proves (1).

Example (see Montmort [34, 35] and Takács [58]). Denote by Ω the set of all the $n!$ permutations of $1, 2, \ldots, n$. Let A_i $(i = 1, 2, \ldots, n)$ be the set of all those permutations of $1, 2, \ldots, n$ in which the i-th element is i, or briefly, A_i is the set of permutations of $1, 2, \ldots, n$ in which there is a match at the i-th place. Then $N(\Omega) = n!$ and

$$N(A_{i_1}A_{i_2}\cdots A_{i_j}) = (n-j)!$$

for $1 \le i_1 < i_2 < \cdots < i_j \le n$. Thus $S_j = n!/j!$ for $j = 0, 1, \ldots, n$. The number of permutations of $1, 2, \ldots, n$ in which there are exactly k matches is

$$N(H_k) = \frac{n!}{k!} \sum_{j=k}^{n} \frac{(-1)^{j-k}}{(j-k)!}$$

for $k = 0, 1, \ldots, n$.

Let us suppose that a box contains n cards marked $1, 2, \ldots, n$ and that all the n cards are drawn one by one without replacement. All the $n!$ possible outcomes are supposed to be equally probable. Denote by ν_n the number of matches in the permutation of $1, 2, \ldots, n$ obtained by the n drawings. Then $\mathbf{P}\{\nu_n = k\} = N(H_k)/N(\Omega)$, that is,

$$\mathbf{P}\{\nu_n = k\} = \frac{1}{k!} \sum_{j=k}^{n} \frac{(-1)^{j-k}}{(j-k)!}$$

for $k = 0, 1, \ldots, n$, and the j-th binomial moment of ν_n is

$$B_j(n) = \mathbf{E}\left\{\binom{\nu_n}{j}\right\} = \frac{S_j}{N(\Omega)} = \frac{1}{j!}$$

for $j = 0, 1, \ldots, n$.

The theorem of general probability. The previous example leads to a more general theorem which was first formulated in a general form by C. Jordan [23]. See also K. Jordan [25].

Let $A_1, A_2, \ldots, A_n$ be random events and denote by ν_n the number of events occurring among $A_1, A_2, \ldots, A_n$. Then

$$\boldsymbol{P}\{\nu_n = k\} = \sum_{j=k}^{n} (-1)^{j-k}\binom{j}{k}B_j \quad \text{for } k = 0, 1, \ldots, n,$$

and

$$\boldsymbol{P}\{\nu_n \geqslant k\} = \sum_{j=k}^{n} (-1)^{j-k}\binom{j-1}{k-1}B_j \quad \text{for } k = 1, 2, \ldots, n$$

where $B_0 = 1$ and

$$B_j = \sum \boldsymbol{P}\{A_{i_1}A_{i_2}\cdots A_{i_j}\}, \quad 1 \leqslant i_1 < i_2 < \cdots < i_j \leqslant n,$$

for $j = 1, 2, \ldots, n$. We can interpret B_j as the j-th binomial moment of ν_n, that is,

$$B_j = \boldsymbol{E}\left\{\binom{\nu_n}{j}\right\}$$

for $j = 0, 1, \ldots, n$.

If in the sums for $\boldsymbol{P}\{\nu_n = k\}$ and $\boldsymbol{P}\{\nu_n \geqslant k\}$ we add less than $n - k + 1$ terms, the error is of the same sign, and has smaller absolute value than the first term neglected.

Stirling numbers. In 1730 Stirling [49, 50] introduced some remarkable numbers which we call today Stirling's numbers of the first kind and Stirling's numbers of the second kind. See Jordan [24], Riordan [42] and Pólya and Szegö [40].

Stirling's numbers of the first kind will be denoted by $S(n, k)$ for $n \geqslant 0$ and $k \geqslant 0$. See Table 5.

Table 5
$S(n, k)$

n \ k	0	1	2	3	4	5	6
0	1	0	0	0	0	0	0
1	0	1	0	0	0	0	0
2	0	1	1	0	0	0	0
3	0	2	3	1	0	0	0
4	0	6	11	6	1	0	0
5	0	24	50	35	10	1	0
6	0	120	274	225	85	15	1

The numbers $S(n, k)$ can be calculated by the recurrence equation

$$S(n+1, k) = S(n, k-1) + nS(n, k),$$

where $n \geq 0$ and $k \geq 1$. The initial conditions are $S(n, 0) = 0$ for $n \geq 1$, $S(0, k) = 0$ for $k \geq 1$ and $S(0, 0) = 1$.

The numbers $S(n, k)$ $(1 \leq k \leq n)$ have a simple combinatorial interpretation, namely, $S(n, k)$ is the number of those permutations of $1, 2, \ldots, n$ which decompose into k disjoint cycles.

We have the following explicit formula

$$S(n, k) = \sum \frac{n!}{k_1! k_2! \cdots k_n! 1^{k_1} 2^{k_2} \cdots n^{k_n}}, \quad \begin{array}{l} k_1 + 2k_2 + \cdots + nk_n = n, \\ k_1 + k_2 + \cdots + k_n = k, \end{array}$$

for $1 \leq k \leq n$ where the summation extends over all nonnegative integers $k_1, k_2, \ldots, k_n$ which satisfy the stated conditions.

We have, for any x and $n \geq 1$,

$$\sum_{k=1}^{n} S(n, k) x^k = x(x+1) \cdots (x+n-1),$$

and, for $|x| < 1$ and $k \geq 1$,

$$\sum_{n=k}^{\infty} \frac{S(n, k) x^n}{n!} = \frac{1}{k!} \left(\log \frac{1}{(1-x)} \right)^k.$$

In 1954 Sparre Andersen [2] found an interesting result concerning the numbers $S(n, k)$ $(1 \leq k \leq n)$. Let $\zeta_0 = 0$ and $\zeta_r = \xi_1 + \xi_2 + \cdots + \xi_r$ for $r = 1, 2, \ldots, n$, where $\xi_1, \xi_2, \ldots, \xi_n$ are interchangeable real random variables. Define ν_n as the number of sides of the smallest convex majorant of the set of points $\{(r, \zeta_r): 0 \leq r \leq n\}$, i.e., the upper part of the boundary of the convex hull of the set of points $\{(r, \zeta_r): 0 \leq r \leq n\}$. If

$$\mathbf{P}\left\{ \frac{\zeta_i}{i} = \frac{\zeta_j}{j} \right\} = 0$$

for $1 \leq i < j \leq n$, then

$$\mathbf{P}\{\nu_n = k\} = \frac{S(n, k)}{n!}$$

for $1 \leq k \leq n$. In particular, $S(n, 1) = (n-1)!$ and

$$\mathbf{P}\{\nu_n = 1\} = \mathbf{P}\left\{ \frac{\zeta_i}{i} < \frac{\zeta_n}{n} \text{ for } 1 \leq i < n \right\} = \frac{1}{n}.$$

Stirling's numbers of the second kind will be denoted by $\mathfrak{S}(n, k)$ for $n \geqslant 0$ and $k \geqslant 0$. See Table 6.

Table 6
$\mathfrak{S}(n, k)$

n \ k	0	1	2	3	4	5	6
0	1	0	0	0	0	0	0
1	0	1	0	0	0	0	0
2	0	1	1	0	0	0	0
3	0	1	3	1	0	0	0
4	0	1	7	6	1	0	0
5	0	1	15	25	10	1	0
6	0	1	31	90	65	15	1

The numbers $\mathfrak{S}(n, k)$ can be calculated by the recurrence equation

$$\mathfrak{S}(n+1, k) = \mathfrak{S}(n, k-1) + k\mathfrak{S}(n, k),$$

where $n \geqslant 0$ and $k \geqslant 1$. The initial conditions are $\mathfrak{S}(n, 0) = 0$ for $n \geqslant 1$, $\mathfrak{S}(0, k) = 0$ for $k \geqslant 1$ and $\mathfrak{S}(0, 0) = 1$.

The numbers $\mathfrak{S}(n, k)$ $(1 \leqslant k \leqslant n)$ have a simple combinatorial interpretation, namely, $\mathfrak{S}(n, k)$ is the number of partitions of the set $(1, 2, \ldots, n)$ into exactly k nonempty subsets.

We have the following explicit formula

$$\mathfrak{S}(n, k) = \frac{(-1)^k}{k!} \sum_{i=0}^{k} (-1)^i \binom{k}{i} i^n$$

for $0 \leqslant k \leqslant n$. Also

$$\mathfrak{S}(n, k) = \frac{1}{k!} \sum \frac{n!}{j_1! j_2! \cdots j_k!}, \quad j_1 + \cdots + j_k = n$$

where $j_1, j_2, \ldots, j_k$ are positive integers and $1 \leqslant k \leqslant n$.

For any x and $n \geqslant 1$, we have

$$\sum_{k=1}^{n} \mathfrak{S}(n, k) k! \binom{x}{k} = x^n$$

and, for $|x| < 1/k$ and $k \geqslant 1$,

$$\sum_{n=k}^{\infty} \mathfrak{S}(n, k) x^n = \frac{x^k}{(1-x)(1-2x)\cdots(1-kx)}.$$

The numbers $\mathfrak{S}(n, k)$ $(0 \leqslant k \leqslant n)$ frequently appear in the solutions of probability problems. For example, let us suppose that a box contains m cards numbered $1, 2, \ldots, m$. We draw n times with replacement. What is the probability that in the n drawings every number out of $1, 2, \ldots, m$ will be drawn at least once? If every outcome has the same probability, then the answer is $m!\mathfrak{S}(n, m)/m^n$.

The numbers

$$T_n = \sum_{k=0}^{n} \mathfrak{S}(n, k),$$

$n = 0, 1, \ldots,$ are called Bell numbers. We can interpret T_n as the number of different partitions of a set of n elements, say $(1, 2, \ldots, n)$, into any number of disjoint subsets. We have $T_0 = T_1 = 1$, $T_2 = 2$, $T_3 = 5$, $T_4 = 15$, $T_5 = 52$, $T_6 = 203, \ldots$.

The following recurrence formula can be used in calculating T_n for $n \geqslant 1$:

$$T_{n+1} = \sum_{j=0}^{n} \binom{n}{j} T_j .$$

For $n \geqslant 1$, we have

$$T_n = \frac{1}{e} \sum_{j=1}^{\infty} \frac{j^n}{j!} .$$

See Binet and Szekeres [6] and Rota [44].

Eulerian numbers. In 1736 L. Euler [15] derived a new summation formula and introduced the so-called Eulerian numbers which we denote by $A(n, k)$ $(n \geqslant 1,\ k \geqslant 0)$. See Table 7.

In 1883 J. Worpitzky [64] demonstrated that the Eulerian numbers satisfy the recurrence formula

$$A(n+1, k) = (k+1)A(n, k) + (n+1-k)A(n, k-1)$$

for $n \geqslant 1$ and $k \geqslant 1$ where $A(n, 0) = 1$ for $n \geqslant 1$ and $A(1, k) = 0$ for $k \geqslant 1$.

Table 7
$A(n, k)$

n \ k	0	1	2	3	4	5
1	1	0	0	0	0	0
2	1	1	0	0	0	0
3	1	4	1	0	0	0
4	1	11	11	1	0	0
5	1	26	66	26	1	0
6	1	57	302	302	57	1

The numbers $A(n, k)$ $(0 \leq k \leq n)$ have a simple combinatorial interpretation. (See Takács [57].) Let us consider a triangular board of size n in which the i-th row contains i cells $(i = 1, 2, \ldots, n)$. See Figure 1. Let us put a mark in a cell in each row of the triangle. We can interpret $A(n, k)$ as the number of possible arrangements in which exactly k columns are empty.

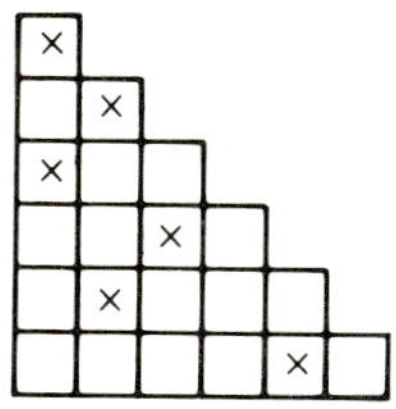

Fig. 1.

We have the following explicit formula:

$$A(n, k) = A(n, n-1-k) = \sum_{i=0}^{k+1} (-1)^i \binom{n+1}{i} (k+1-i)^n$$

for $0 \leq k \leq n-1$, and $A(n, k) = 0$ for $k \geq n \geq 1$.

If $n \geq 1$, then for any x we have

$$\sum_{k=0}^{n-1} A(n, k) \binom{x+k}{n} = x^n .$$

In 1908 P. A. MacMahon [29], by solving a problem of Simon Newcomb concerning a card game, proved that a deck of n cards numbered $1, 2, \ldots, n$ may be dealt into $k+1$ piles in $A(n, k)$ ways if cards are placed in one pack as long as they are in decreasing order of magnitude.

Compositions. The number of ways in which a positive integer n can be expressed as a sum of k positive integers is

$$\binom{n-1}{k-1}$$

for $1 \leq k \leq n$, that is, $\binom{n-1}{k-1}$ is the number of solutions of the equation

$$a_1 + a_2 + \cdots + a_k = n \tag{3}$$

in positive integers $a_1, a_2, \ldots, a_k$. Obviously $(a_1, a_2, \ldots, a_k)$ is a solution if and only if $(a_1, a_1 + a_2, \ldots, a_1 + \cdots + a_{k-1})$ is a combination of size $k-1$ of the elements $1, 2, \ldots, n-1$ without repetition.

If we impose the restriction that each $a_i \leq m$ where m is a positive integer, then the number of solutions of (3) is

$$\sum_{j=0}^{[(n-k)/m]} (-1)^j \binom{k}{j} \binom{n-1-jm}{k-1}$$

for $k \leqslant n \leqslant km$. (See Montmort [35] and De Moivre [12].)

The number of ways in which a positive integer n can be expressed as a sum of any number of positive integers is evidently

$$\sum_{k=1}^{n} \binom{n-1}{k-1} = 2^{n-1}.$$

Partitions. The above formulas are concerned with ordered arrangements. It is more difficult to find the number of unordered decompositions of a positive integer n as a sum k (or any number) of positive integers. Denote by $p(n, k)$ the number of solutions of (3) in positive integers $a_1, a_2, \ldots, a_k$ satisfying the requirements $a_1 \leqslant a_2 \leqslant \cdots \leqslant a_k$. Obviously $p(k, k) = 1$ for $k \geqslant 1$, and $p(n, 1) = 1$ for $n \geqslant 1$, and we can determine $p(n, k)$ for all $1 \leqslant k \leqslant n$ by the recurrence equation

$$p(n, k) = \sum_{i=1}^{k} p(n-k, i).$$

If $k \geqslant n > 1$, then $p(n, k) = 0$. See Table 8.

Table 8
$p(n, k)$

k \ n	1	2	3	4	5	6	7
1	1	1	1	1	1	1	1
2	0	1	1	2	2	3	3
3	0	0	1	1	2	3	4
4	0	0	0	1	1	2	3
5	0	0	0	0	1	1	1
6	0	0	0	0	0	1	1

The sum

$$p(n) = \sum_{k=1}^{n} p(n, k)$$

is the number of decompositions of n as a sum of any number ot positive integers without regard to order. We define $p(0) = 1$.

L. Euler [16, p. 318] knew that

$$\sum_{n=0}^{\infty} p(n) x^n = \prod_{n=1}^{\infty} (1 - x^n)^{-1}$$

for $|x|<1$, and that

$$\left[\sum_{n=0}^{\infty} p(n)x^n\right]\left[\sum_{k=-\infty}^{\infty} (-1)^k x^{(3k^2+k)/2}\right] = 1$$

for $|x|<1$.

The last equation yields the following recurrence formula for the determination of $p(n)$:

$$p(n) = \sum_{1\leq(3k^2\pm k)/2\leq n} (-1)^{k-1} p\left(n - \frac{3k^2\pm k}{2}\right).$$

An asymptotic series for $p(n)$ was found by G. H. Hardy and S. Ramanujan [21] and an explicit form by H. Rademacher [41]. We have

$$p(n) = \frac{2}{(24n-1)^{3/2}} \sum_{k=1}^{\infty} \left(\sum_{j\in S_k} (-1)^j \cos\frac{(6j+1)\pi}{6k}\right) \cdot$$
$$\cdot\left[e^{(\pi/6k)\sqrt{(24n-1)}}\left(\sqrt{24n-1} - \frac{6k}{\pi}\right) + e^{-(\pi/6k)\sqrt{(24n-1)}}\left(\sqrt{24n-1} + \frac{6k}{\pi}\right)\right]$$

for $n = 1, 2, \ldots$ where

$$S_k = \left\{j: \frac{j(3j+1)}{2} + n \equiv 0 \text{ (mod. } k) \text{ and } 1\leq j\leq 2k\right\}.$$

If $n \geq 576$ and if we replace the infinite sum $\Sigma_{k=1}^{\infty}$ by a finite sum $\Sigma_{k=1}^{N}$ where $N = [\frac{2}{3}\sqrt{n}]$, then the error is less than $\frac{1}{2}$, and the integer $p(n)$ can be determined precisely. If n is large, the following asymptotic formula can be used for the approximation of $p(n)$:

$$p(n) \sim \frac{1}{4n\sqrt{3}} e^{\pi\sqrt{2n/3}}$$

as $n\to\infty$.

Success runs. Let us consider a sequence of independent and identical trials in which each trial has two possible outcomes, namely, success, which has probability p, and failure, which has probability q, where $p>0$, $q>0$ and $p+q=1$. In this case we speak about a sequence of Bernoulli trials with probability p for success. Denote by ν_n the number of successes in the first n trials. Then

$$\mathbf{P}\{\nu_n = j\} = \binom{n}{j} p^j q^{n-j}$$

for $j = 0, 1, \ldots, n$.

An uninterrupted series of successes is called a success run. The probability that in n Bernoulli trials we have j successes and these form k success runs is given by

$$\binom{j-1}{j-k}\binom{n-j+1}{k}p^j q^{n-j}.$$

The probability that in n Bernoulli trials we have j successes and these form k success runs each of length $\leq m$ is given by

$$\binom{n-j+1}{k}\left\{\sum_{i=0}^{[(j-k)/m]}(-1)^i\binom{k}{i}\binom{j-im-1}{k-1}\right\}p^j q^{m-j}.$$

If we form the sum of the above probability for $k = 1, 2, \ldots,$ then we obtain the probability that in n Bernoulli trials we have j successes and no success run is of length $>m$. This probability is

$$\left\{\sum_{i=0}^{[j/m]}(-1)^i\binom{n-j+1}{i}\binom{n-im-i}{n-j}\right\}p^j q^{n-j}.$$

See Jordan [26, p. 496].

Random walk. Let us suppose that a particle performs a random walk on the x-axis. It starts at $x = 0$ and in each step independently of the others it moves either a unit distance to the right with probability p or a unit distance to the left with probability q where $p > 0$, $q > 0$ and $p + q = 1$. Denote by η_n the position of the particle at the end of the n-th step. We have

$$\boldsymbol{P}\{\eta_n = 2j - n\} = \binom{n}{j}p^j q^{n-j}$$

for $j = 0, 1, \ldots, n$. Denote by $\rho(a)$ $(a = 1, 2, \ldots)$ the smallest $n = 1, 2, \ldots$ for which $\eta_n = a$. If there is no such n, then $\rho(a) = \infty$. Accordingly, $\rho(a)$ is the first passage time through a $(a = 1, 2, \ldots)$. We have

$$\boldsymbol{P}\{\rho(a) = a + 2k\} = \frac{a}{a+2k}\binom{a+2k}{k}p^{a+k}q^k$$

for $k = 0, 1, 2, \ldots,$ and

$$\boldsymbol{P}\{\rho(a) < \infty\} = \begin{cases}(p/q)^a & \text{if } p < q,\\ 1 & \text{if } p \geq q.\end{cases}$$

Now let us consider the above random walk with the modification that there is an absorbing barrier at the point $x = a$ where $a > 0$. If the particle reaches the point $x = a$, it remains forever at this place. Denote now by η_n^* the position of the particle at the end of the n-th step. Then we have

$$\boldsymbol{P}\{\eta_n^* = 2j - n\} = \left[\binom{n}{j} - \binom{n}{j-a}\right] p^j q^{n-j}$$

if $2j < n + a$, and

$$\boldsymbol{P}\{\eta_n^* = a\} = \boldsymbol{P}\{\rho(a) \leq n\} = \boldsymbol{P}\{\eta_n \geq a\} + \left(\frac{p}{q}\right)^a \boldsymbol{P}\{\eta_n < -a\}.$$

If there are two absorbing barriers, namely at $x = a$ where $a = 1, 2, \ldots$ and at $x = -b$ where $b = 1, 2, \ldots$, and η_n^{**} denotes the position of the particle at the end of the n-th step, then

$$\boldsymbol{P}\{\eta_n^{**} = 2j - n\} = \sum_k \left[\binom{n}{j - k(a+b)} - \binom{n}{j - a - k(a+b)}\right] p^j q^{n-j}$$

for $-b < 2j - n < a$. Here the sum is formed for all $k = 0, \pm 1, \pm 2, \ldots$ for which the binomial coefficients are not 0. We note that $\binom{n}{j} = n!/j!(n-j)!$ if $j = 0, 1, \ldots, n$ and $\binom{n}{j} = 0$ otherwise.

For more details see Feller [18].

De Moivre numbers. In 1708 A. De Moivre [9] solved the following problem of games of chance: Two players, A and B, agree to play a series of games. In each game, independently of the others, either A wins a coin from B with probability p or B wins a coin from A with probability q, where $p > 0$, $q > 0$ and $p + q = 1$. Let us suppose that A has an unlimited number of coins, B has only k coins, and the series ends when B is ruined, i.e., when B loses his last coin. Denote by $\rho(k)$ the duration of the games, i.e., the number of games played until B is ruined. A. De Moivre discovered that

$$\boldsymbol{P}\{\rho(k) = k + 2j\} = \frac{k}{k+2j}\binom{k+2j}{j} p^{k+j} q^j \tag{4}$$

for $k \geq 1$ and $j \geq 0$. De Moivre stated (4) without proof. Formula (4) was proved only in 1773 by P. S. Laplace and in 1776 by J. L. Lagrange. In 1802 A. M. Ampère expressed his view that formula (4) is remarkable for its simplicity and elegance.

It is convenient to write

$$L(j,k) = \frac{k}{k+2j}\binom{k+2j}{j}$$

for $j \geq 0$ and $k \geq 1$, $L(0,0) = 1$ and $L(j,0) = 0$ for $j \geq 1$. The numbers $L(j,k)$ might appropriately be called De Moivre numbers. They can also be expressed as

$$L(j,k)=\binom{k+2j-1}{j}-\binom{k+2j-1}{j-1}$$

for $j\geq 1$, $k\geq 0$ and $L(0,k)=1$ for $k\geq 0$. See Table 9.

Table 9
$L(j,k)$

k \ j	0	1	2	3	4	5
0	1	0	0	0	0	0
1	1	1	2	5	14	42
2	1	2	5	14	42	132
3	1	3	9	28	90	297
4	1	4	14	48	165	572
5	1	5	20	75	275	1001

The numbers $L(j,k)$ $(0\leq j\leq k)$ can also be determined by the recurrence equation

$$L(j,k)=L(j-1,k+1)+L(j,k-1)$$

where $L(0,k)=1$ for $k\geq 0$ and $L(j,0)=0$ for $j\geq 1$.

By De Moivre's result, $L(j,k)$ can be interpreted in the following way: One can arrange $k+j$ letters A and j letters B in $L(j,k)$ ways so that for every $r=1,2,\ldots,k+2j$ among the first r letters there are more A than B.

The numbers $L(j,k)$ have appeared in various forms in diverse fields of mathematics.

In 1751 L. Euler found that the number of different ways of dissecting a convex polygon of n sides into $n-2$ triangles by $n-3$ nonintersecting diagonals is $L(n-2,1)$.

In 1838 E. Catalan [7] proved that the number of ways a product of n factors can be calculated by pairs is $L(n-1,1)$. The numbers $L(n-1,1)$, $n=1,2,\ldots,$ are usually called Catalan numbers.

In 1859 A. Cayley encountered the numbers $L(n-1,1)$ in the theory of graphs.

In 1879 W. A. Whitworth [62] demonstrated that one can arrange $j+k$ letters A and j letters B in

$$M(j,k)=\binom{2j+k}{j}\frac{k+1}{j+k+1}$$

ways such that for every $r=1,2,\ldots,2j+k$ among the first r letters there are at least as many A as B. Obviously, $M(j,k)=L(j,k+1)/(j+k+1)$.

Ballot theorems. In 1887 J. Bertrand [5] discovered the following ballot theorem:

If in a ballot candidate A scores a votes and candidate B scores b votes where $a \geq b$, then the probability that throughout the counting the number of votes registered for A is always greater than the number of votes registered for B is given by

$$P(a, b) = (a - b)/(a + b)$$

provided that all the possible voting records are equally probable.

$P(a, b)$ can be expressed as $N(a, b)/\binom{a+b}{a}$, where $N(a, b)$ is the number of favorable voting records, and $\binom{a+b}{a}$ is the number of possible voting records. Bertrand's formula follows from the obvious observation that $N(a, b) = L(b, a - b)$.

In 1960 L. Takács [51, 52] generalized Bertrand's ballot theorem in the following way:

Let us suppose that a box contains n cards marked $a_1, a_2, \ldots, a_n$ where $a_1, a_2, \ldots, a_n$ are nonnegative integers with sum $a_1 + a_2 + \cdots + a_n = k$, where $k \leq n$. We draw all the n cards without replacement. Let us assume that every outcome has the same probability. The probability that the sum of the first r numbers drawn is less than r for every $r = 1, 2, \ldots, n$ is given by

$$P(n, k) = (n - k)/n .$$

To obtain Bertrand's ballot theorem from the above general theorem, let us suppose that a box contains a cards each marked '0' and b cards each marked '2'. Let us draw all the $a + b$ cards from the box without replacement and suppose that a '0' corresponds to a vote for A, and a '2' corresponds to a vote for B. Then A leads throughout the counting if and only if for every $r = 1, 2, \ldots, a + b$ the sum of the first r numbers drawn is less than r. Since now $n = a + b$ and $k = 2b$, by the aforementioned formula, the probability in question is

$$P(a + b, 2b) = (a - b)/(a + b) .$$

The following theorem makes it possible to find the probability that in counting a ballot a candidate is in leading position for any given number of times.

Let $k_1, k_2, \ldots, k_n$ be integers with sum $k_1 + k_2 + \cdots + k_n = 1$. Among the n cyclic permutations of $(k_1, k_2, \ldots, k_n)$ there is exactly one for which exactly j $(j = 1, 2, \ldots, n)$ of its successive partial sums are positive. (See Takács [53].)

The ballot theorems formulated above have many applications in various fields of mathematics, and, in particular, in probability theory and in mathematical statistics. The next section provides a few examples in order statistics.

Order statistics. Most of the problems in order statistics are connected with the comparison of a theoretical and an empirical distribution function or with the comparison of two empirical distribution functions.

To consider the first case, let $\xi_1, \xi_2, \ldots, \xi_n$ be mutually independent random variables each having the same distribution function $F(x)$. Denote by $F_n(x)$ the empirical distribution function of the sample $(\xi_1, \xi_2, \ldots, \xi_n)$. Define

$$\delta_n^+ = \sup_{-\infty<x<\infty} [F_n(x) - F(x)] .$$

If $F(x)$ is a continuous distribution function, then the distribution of δ_n^+ does not depend on $F(x)$ and by the generalization of Bertrand's ballot theorem we obtain that

$$\boldsymbol{P}\left\{\delta_n^+ \leqslant \frac{k}{n}\right\} = 1 - \sum_{j=1}^{n-k} \frac{k}{(n-j)} \binom{n}{j+k} \left(\frac{j}{n}\right)^{j+k} \left(1 - \frac{j}{n}\right)^{n-j-k}$$

for $k = 1, 2, \ldots, n$. The probability $\boldsymbol{P}\{\delta_n^+ \leqslant x\}$ was found by N. V. Smirnov [46] in 1944.

To consider the second case, let $\xi_1, \xi_2, \ldots, \xi_m, \eta_1, \eta_2, \ldots, \eta_n$ be mutually independent random variables each having the same distribution function $F(x)$. Denote by $F_m(x)$ and $G_n(x)$ the empirical distribution functions of the samples $(\xi_1, \xi_2, \ldots, \xi_m)$ and $(\eta_1, \eta_2, \ldots, \eta_n)$ respectively. If $F(x)$ is a continuous distribution function, the distribution of

$$\delta^+(m, n) = \sup_{-\infty<x<\infty} [F_m(x) - G_n(x)]$$

does not depend on $F(x)$. The probability $\boldsymbol{P}\{\delta^+(m, n) \leqslant x\}$ was found by B. V. Gnedenko and V. S. Koroljuk [19] for $m = n$, and V. S. Koroljuk [27] for $n = mp$ where p is a positive integer. If $n = mp$, where p is a positive integer, and if k is a nonnegative integer, then by the generalization of Bertrand's ballot theorem we obtain that

$$\boldsymbol{P}\left\{\delta^+(m, n) \leqslant \frac{k}{n}\right\} = 1 - \frac{1}{\binom{m+n}{m}}$$

$$\times \sum_{(k+1)/p \leqslant s \leqslant m} \frac{(k+1)}{(n+k+1-sp)} \binom{sp+s-k-1}{s} \binom{m+n+k-sp-s}{m-s} .$$

See also Takács [54] for other related results.

Fluctuation theory. Many problems in probability theory and in mathematical statistics are connected with the fluctuations of sums of interchangeable random variables, or, in particular, sums of independent and identically distributed random variables. The following two theorems are very useful in such studies.

In 1953 E. Sparre Anderson [1] discovered the following combinatorial theorem. See also W. Feller [17].

Let $c_1, c_2, \ldots, c_n$ be real numbers and consider the $n!$ permutations of $(c_1, c_2, \ldots, c_n)$. There are as many permutations in which precisely k ($k = 0, 1, \ldots, n$) among its n successive partial sums are strictly positive (non-negative) as there are permutations in which the first (last) maximal element in the sequence of 0 and the n successive partial sums occur at the k-th place ($k = 0, 1, \ldots, n$).

In 1956 F. Spitzer [48] discovered another combinatorial theorem which too has many applications in fluctuation theory.

Let $c_1, c_2, \ldots, c_n$ be real numbers such that $c_1 + c_2 + \cdots + c_n = 0$ but $c_{i_1} + c_{i_2} + \cdots + c_{i_j} \neq 0$ for $1 \leq i_1 < i_2 < \cdots < i_j \leq n$ if $j < n$. For each $k = 0, 1, \ldots, n-1$ there is exactly one cyclic permutation of $(c_1, c_2, \ldots, c_n)$ such that exactly k of its partial sums are positive.

References

[1] Andersen, E. S. (1953). On sums of symmetrically dependent random variables. *Skandinavisk Aktuarietidskrift* **36**, 123–138.

[2] Andersen, E. S. (1954). On the fluctuations of sums of random variables. *Mathematica Scandinavica* **2**, 195–223.

[3] Andrews, G. E. (1976). The theory of partitions. In: *Encyclopedia of Mathematics and its Applications, Vol. 2.* Addison-Wesley, Reading, MA.

[4] Barton, D. E. and Mallows, C. L. (1965). Some aspects of the random sequence. *Annals of Mathematical Statistics* **36**, 236–260.

[5] Bertrand, J. (1887). Solution d'un problème. *Comptes Rendus Acad. Sci. Paris* **105**, 369.

[6] Binet, F. E. and Szekeres, G. (1957). On Borel fields over finite sets. *Annals of Mathematical Statistics* **29**, 494–498.

[7] Catalan, E. (1838). Note sur une équation aux différences finies. *Journal de Mathématiques Pures et Appliquées* **3**, 508–516.

[8] Shih-chieh Chu (1303). *Ssu Yuan Yü Chien* (Precious Mirror of the Four Elements).

[9] De Moivre, A. (1711). De mensura sortis, seu, de probabilitate eventuum in ludis a casu fortuito pendentibus. *Philosophical Transactions* **27**, 213–264.

[10] De Moivre, A. (1718). *The Doctrine of Chances: or, A Method of Calculating Probability of Events in Play.* London.

[11] De Moivre, A. (1738). *The Doctrine of Chances: or, A Method of Calculating the Probabilities of Events in Play, second edition.* London. [Reprinted by Frank Cass and Co., London, 1967.]

[12] De Moivre, A. (1756). *The Doctrine of Chances: or, A Method of Calculating the Probabilities of Events in Play, third edition.* London. [Reprinted by Chelsea, New York, 1967.]

[13] De Moivre, A. (1730). *Miscellanea Analytica de Seriebus et Quadraturis.* London.

[14] Erdös, P. and Spencer, J. (1974). *Probabilistic Methods in Combinatorics.* Akadémiai Kiadó, Budapest.

[15] Euler, L. (1741). Methodus universalis series summandi ulterius promota. *Commentarii Academiae Scientiarum Petropolitanae* **8**, 147–158. [Leonhardi Euleri Opera Omnia. Ser. I. vol. 14, B. G. Teubner, Leipzig and Berlin, 1925, pp. 124–137.]

[16] Euler, L. (1748). *Introductio in Analysin Infinitorum. Tom. I.* Lausanne. [Leonhardi Euleri Opera Omnia. Ser. I. Vol. 8. B. G. Teubner, Leipzig and Berlin, 1922, pp. 1–390.]

[17] Feller, W. (1959). On combinatorial methods in fluctuation theory. In: Ulf Grenander, ed., *Probability and Statistics.* The Harald Cramér Volume. Almqvist and Wiksell, Stockholm and Wiley, New York, pp. 75–91.

[18] Feller, W. (1968). *An Introduction to Probability Theory and its Applications, Volume 1, third edition.* Wiley, New York.

[19] Gnedenko, B. V. and Koroljuk, V. S. (1951). On the maximum discrepancy between two empirical distribution functions. (In Russian) *Dokl. Akad. Nauk SSSR* **80**, 525–528. [English translation: Selected Translations in Mathematical Statistics and Probability, IMS and AMS, **1** (1961) 13–16.]

[20] Hall, M. Jr. (1967). *Combinatorial Theory.* Blaisdell Publishing Co., Waltham, MA.

[21] Hardy, G. H. and Ramanujan, S. (1918). Asymptotic formulae in combinatory analysis. *Proceedings of the London Mathematical Society* **17**, 75–115. [Collected Papers of G. H. Hardy. Vol. 1. Oxford University Press, 1966, pp. 306–339.]

[22] Johnson, N. L. and Kotz, S. (1977). *Urn Models and Their Application. An Approach to Modern Probability Theory.* Wiley, New York.

[23] Jordan, C. (1867). De quelques formules de probabilité. *Comptes Rendus Acad. Sci. Paris* **65**, 993–994.

[24] Jordan, Ch. (1939). *Calculus of Finite Differences.* Budapest. [Reprinted by Chelsea, New York, 1947.]

[25] Jordan, K. (1927). A valószinüségszámitás alapfogalmai (Les fondements du calcul des probabilitiés). *Mathematikai és Physikai Lapok* **34**, 109–136.

[26] Jordan, K. (1972). *Chapters on the Classical Calculus of Probability.* (English translation of the Hungarian original published in 1956.) Akadémiai Kiadó, Budapest.

[27] Koroljuk, V. S. (1955). On the discrepancy of empirical distribution functions for the case of two independent samples. (In Russian) *Izv. Akad. Nauk SSSR Ser. Math.* **19**, 81–96. [English translation: Selected Translations in Mathematical Statistics and Probability, IMS and AMS, **4** (1963) 105–121.]

[28] Leibniz, G. W. (1666) *Dissertatio de Arte Combinatoria.*

[29] MacMahon, P. A. (1908). Second memoir on the compositions of numbers. *Philosphical Transactions of the Royal Society of London. Ser. A* **207**, 65–134.

[30] MacMahon, P. A. (1915–1916). *Combinatory Analysis. I–II.* Cambridge University Press. [Reprinted by Chelsea, New York, 1960.]

[31] MacMahon, P. A. (1978). *Collected Papers, Volume I: Combinatorics.* The MIT Press, Cambridge, Mass.

[32] Mikami, Y. (1913). *The Development of Mathematics in China and Japan.* Teubner, Leipzig. [Reprinted by Chelsea, New York, 1961.]

[33] Mohanty, S. G. (1979). *Lattice Path Counting and Applications.* Academic Press, New York.

[34] Montmort, P. R. (1708). *Essay d'Analyse sur les Jeux de Hazard.* Paris.

[35] Montmort, P. R. (1713). *Essay d'Analyse sur les Jeux de Hazard, second edition.* Paris. [Reprinted by Chelsea, New York, 1980.]

[36] Narayana, T. V. (1979). *Lattice Path Combinatorics with Statistical Applications.* University of Toronto Press, Toronto.

[37] Needham, J. (1959). *Science and Civilisation in China, Vol. 3: Mathematics and the Sciences of the Heavens and Earth.* Cambridge University Press.

[38] Netto, E. (1901). *Lehrbuch der Combinatorik.* B. G. Teubner, Leipzig. Second edition, 1927. [Reprinted by Chelsea, New York, 1964]

[39] Pascal, B. (1665). *Traité du Triangle Arithmétique.* Paris.

[40] Pólya, G. and Szegö, G. (1972 and 1976). *Problems and Theorems in Analysis, Vol. I–II.* Springer, New York.

[41] Rademacher, H. (1937). On the partition function $p(n)$. *Proceedings of the London Mathematical Society (2)* **43**, 241–254. [Collected Papers of Hans Rademacher. Vol. II. Edited by E. Grosswald. The MIT Press, Cambridge, Mass., 1974, pp. 108–122.]

[42] Riordan, J. (1958). *An Introduction to Combinatorial Analysis.* Wiley, New York.

[43] Riordan, J. (1968). *Combinatorial Identities.* Wiley, New York.

[44] Rota, G.-C. (1964). The number of partitions of a set. *American Mathematical Monthly* **71**, 498–504. [Reprinted in Finite Operator Calculus by Gian-Carlo Rota. Academic Press, New York, 1975, pp. 1–6.]

[45] Ryser, H. J. (1963). Combinatorial Mathematics. *The Carus Mathematical Monographs No. 14.* The Mathematical Association of America.

[46] Smirnov, N. V. (1944). Approximate laws of distribution of random variables from empirical data. (In Russian) *Uspechi Mat. Nauk* **10**, 179–206.

[47] Smith, D. E. (1925). *History of Mathematics, Vol. II.* [Reprinted by Dover, New York, 1958.]

[48] Spitzer, F. (1956). A combinatorial lemma and its application to probability theory. *Transactions of the American Mathematical Society* **82**, 323–339.

[49] Stirling, J. (1730). *Methodus Differentialis, sive Tractatus de Summatione et Interpolatione Serierum Infintarum*. London.

[50] Stirling, J. (1749). *The Differential Method: or, A Treatise Concerning Summation and Interpolation of Infinite Series.* London.

[51] Takács, L. (1961). The probability law of the busy period for two types of queuing processes. *Operations Research* **9**, 402–407.

[52] Takács, L. (1962). A generalization of the ballot problem and its application in the theory of queues. *Journal of the American Statistical Association* **57**, 327–337.

[53] Takács, L. (1962). Ballot problems. *Zeitschrift für Wahrscheinlichkeitstheorie und verwandte Gebiete* **1**, 154–158.

[54] Takács, L. (1967). *Combinatorial Methods in the Theory of Stochastic Processes.* Wiley, New York.

[55] Takács, L. (1967). On the method of inclusion and exclusion. *Journal of the American Statistical Association* **62**, 102–113.

[56] Takács, L. (1969). On the classical ruin problem. *Journal of the American Statistical Association* **64**, 889–906.

[57] Takács, L. (1979). A generalization of the Eulerian numbers. *Publicationes Mathematicae (Debrecen)* **26**, 173–181.

[58] Takács, L. (1980). The problem of coincidences. *Archive for History of Exact Sciences* **21**, 229–244.

[59] Takács, L. (1982). Ballot problems. In: S. Kotz and N. L. Johnson, eds., *Encyclopedia of Statistical Sciences, Vol. 1.* Wiley, New York, pp. 183–188.

[60] Takács, L. (1982). Combinatorics. In: S. Kotz and N. L. Johnson, eds., *Encyclopedia of Statistical Sciences, Vol. 2.* Wiley, New York, pp. 53–60.

[61] Wallis, J. (1693). *De Algebra Tractatus.* Oxford. [John Wallis: Opera Mathematica. II. Georg Olms Verlag, Hildesheim, 1972.]

[62] Whitworth, W. A. (1879). Arrangements of m things of one sort and n things of another sort, under certain conditions of priority. *Messenger of Mathematics* **8**, 105–114.

[63] Woepcke, F. (1851). *L'Algèbre d'Omar Alkâyyami.* French translation. Duprat, Paris.

[64] Worpitzky, J. (1883). Studien über die Bernoullischen and Eulerschen Zahlen. *Journal für die Reine und Angewandte Mathematik* **94**, 203–232.

P. R. Krishnaiah and P.K. Sen, eds., *Handbook of Statistics, Vol. 4*
© Elsevier Science Publishers (1984) 145–171 8

Rank Statistics and Limit Theorems*

Malay Ghosh

1. Introduction

Methods based on ranks have proved to be a very useful alternative to the classical parametric theory in many situations. A systematic development of the rank-based methods seems to have been sparked by the two pioneering papers of Wilcoxon (1945), and Mann and Whitney (1947), although rank tests for certain individual problems can be traced back even earlier. Wilcoxon (1945) proposed the famous rank sum test as a test for location in the two-sample problem. Let $X_1, \ldots, X_m$ and $Y_1, \ldots, Y_n$ denote independent random samples from continuous distribution functions $F(x)$ and $G(x) = F(x - \Delta)$, where the shift parameter Δ (real) is unknown. We want to test H_0: $\Delta = 0$ against the one-sided alternatives H_1: $\Delta > 0$ (or $\Delta < 0$), or the two-sided alternatives H: $\Delta \neq 0$. Writing $Z_i = X_i$ $(i = 1, \ldots, m)$, $Z_{m+j} = Y_j$ $(j = 1, \ldots, n)$, let R_l denote the rank of Z_l in the combined sample of size $m + n$. The Wilcoxon rank sum statistic is given by $W = \Sigma_{j=1}^{n} R_{m+j}$. For the one-sided alternatives H_1: $\Delta > 0$ $(\Delta < 0)$ we reject for large (small) values of W, because that indicates that the Y's have a distribution stochastically larger (smaller) than the distribution of the X's. For the two-sided alternatives H, we reject for very large or very small values of W.

Linear rank statistics (sometimes also called simple linear rank statistics) are generalized versions of W. A linear rank statistic based on a sample of size N is defined by

$$S_N = \sum_{i=1}^{N} c(i)a(R_i), \tag{1.1}$$

where the $c(i)$'s are referred to as *regression constants*, and the $a(i)$'s are referred to as scores. Such statistics arise quite naturally in obtaining locally most powerful rank tests against certain regression alternatives (see e.g. Hájek and Sidák, 1967). The special case when $c(1) = \cdots = c(m) = 0$ and $c(m+1) =$

*Research supported by the Army Research Office Durham Grant Number DAAG29-82-K-0136.

$\cdots = c(N) = 1$ and $a(1) \leqslant \cdots \leqslant a(N)$ is a generalized version of the Wilcoxon rank sum statistic in the two-sample problem. Two important special choices for the $a(i)$'s are given by (i) $a(i) = i$ where the score function a is referred to as the *Wilcoxon score*, and (ii) $a(i) = EZ_{(i)}$, where $Z_{(1)} \leqslant \cdots \leqslant Z_{(N)}$, are the order statistics in a random sample of size N from the $N(0, 1)$ distribution. In this case, the score function a is referred to as the *normal score*.

Next consider the one-sample situation when $Z_1, \ldots, Z_N$ are iid with a common continuous distribution function $F(z - \Delta)$, where the shift parameter Δ (real) is unknown, and $F(z) + F(-z) = 1$. In this case, one wants to test H_0: $\Delta = 0$ against the one-sided alternatives H_1: $\Delta > 0$ (or $\Delta < 0$) or the two-sided alternatives H: $\Delta \neq 0$. For such a testing problem, Wilcoxon (1945) proposed the signed rank test statistic $W_N^+ = \Sigma_{i=1}^N \operatorname{sgn} Z_i R_i^+$, where R_i^+ is the rank of $|Z_i|$ among $|Z_1|, \ldots, |Z_N|$, while $\operatorname{sgn} u = 1, 0$ or -1 according as $u >, =$ or <0. Generalized versions of W^+ are given by

$$S_N^+ = \sum_{i=1}^N c(i) \operatorname{sgn} Z_i a(R_i^+), \tag{1.2}$$

and such statistics are referred to as *signed rank regression statistics*. When $c(1) = \cdots = c(N) = 1$, the statistics are used for one-sample location tests. In this case the important special choices for the score function a are (i) $a(1) = \cdots = a(N) = 1$ which corresponds to the *sign test statistic*, and (iii) $a(i) = E|Z|_{(i)}$ where $|Z|_{(1)} < \cdots < |Z|_{(N)}$ denote the order statistics corresponding to the absolute values in a random sample of size N from the $N(0, 1)$ distribution. The resulting test statistic is referred to as the *one sample normal scores test statistic*.

We shall review the literature on limit theorems for linear rank statistics and signed linear rank statistics. Some of these statistics are closely related to the so-called U-statistics. For example, the Wilcoxon signed rank statistic can also be expressed as

$$\sum_{1 \leqslant i \leqslant j \leqslant n} \sum \operatorname{sgn}(Z_i + Z_j)$$

(see Tukey, 1949), and we shall see later that $\binom{n+1}{2}^{-1}$ times the above statistic is expressible as a weighted average of two one-sample U-statistics. The Wilcoxon rank sum statistic W is expressible as $W = U + \binom{n+1}{2}$, where $U = \Sigma_{i=1}^m \Sigma_{j=1}^n I_{[Y_j \geqslant X_i]}$, where U is the celebrated Mann–Whitney U-statistic. In the above, and in what follows $I_A = 1$ if the event A occurs, and $I_A = 0$, otherwise. The Mann–Whitney U-statistic is an example of a two-sample U-statistic.

Section 2 presents the central limit theorem, rates of convergence and laws of large numbers for one-sample U-statistics. Multisample extension of these results is given in Section 3. Section 4 deals with jackknifed and bootstrapped U-statistics, while Section 5 provides functional central limit theorems for U-statistics. Some miscellaneous remarks related to U-statistics are made in Section 6.

Section 7 involves a discussion of central limit theorems for linear rank statistics under the null and alternative hypotheses. Strong laws of large numbers are given for linear rank statistics in Section 8. Finally, in Section 9, functional central limit theorems are given for linear rank statistics.

2. One sample U-statistics: Central limit theorems, rates of convergence and laws of large numbers

We start with n iid random variables $X_1, \ldots, X_n$ each having a distribution function (df) $F(x)$. It is assumed that $F \in \mathcal{F}$, a class of df's in $\mathbb{R}^k$, the k-dimensional Euclidean space. Let $\theta(F)$ be a functional with domain space $\mathcal{F}$ and range space $\mathbb{R}^s$. For simplicity, we confine ourselves to $s = 1$. $\theta(F)$ is called a *regular functional* over $\mathcal{F}$ if for all $F \in \mathcal{F}$, $\theta(F)$ admits an unbiased estimator, say $\phi(X_1, \ldots, X_n)$, that is,

$$\int \cdots \int \phi(x_1, \ldots, x_n) F(\mathrm{d}x_1) \cdots F(\mathrm{d}x_n) = \theta(F) \quad \text{for all } F \in \mathcal{F}. \tag{2.1}$$

If (2.1) holds, we say that $\theta(F)$ is *estimable.* If $\theta(F)$ is estimable, the smallest sample size, say m for which (2.1) holds, is called the *degree* of $\theta(F)$, and $\phi(X_1, \ldots, X_m)$ is called the *kernel* of $\theta(F)$. Without any loss of generality, we may assume that ϕ is symmetric in its arguments, as otherwise, we can define $\phi_0(X_1, \ldots, X_m) = (m!)^{-1} \sum \phi(X_{\alpha_1}, \ldots, X_{\alpha_m})$, where the summation extends over all possible permutations $(\alpha_1, \ldots, \alpha_m)$ of the first m positive integers.

Corresponding to a symmetric kernel ϕ of $\theta(F)$, we define a one-sample U-statistic (see Hoeffding, 1948) by

$$U_n = \binom{n}{m}^{-1} \sum_{1 \leq \alpha_1 < \cdots < \alpha_m \leq n} \cdots \sum \phi(X_{\alpha_1}, \ldots, X_{\alpha_m}). \tag{2.2}$$

Such statistics have proved to be quite useful in the theory of point and confidence estimation, and in hypothesis testing. For a very detailed account of U-statistics, the reader is referred to Puri and Sen (1971, Chapter 3), Randles and Wolfe (1979, Chapter 3), and Serfling (1980, Chapter 5).

Examples of one-sample U-statistics include the sample mean

$$\bar{X}_n = n^{-1} \sum_{i=1}^{n} X_i,$$

the sample variance

$$S_n^2 = (n-1)^{-1} \sum_{i=1}^{n} (X_i - \bar{X}_n)^2 = \binom{n}{2}^{-1} \sum_{1 \leq i < j \leq n} \sum (X_i - X_j)^2/2,$$

Kendall's tau statistic

$$T_n = \binom{n}{2}^{-1} \sum\sum_{1 \le i < j \le n} (X_{i1} - X_{j1})(X_{i2} - X_{j2}),$$

where $X_i = (X_{i1}, X_{i2})$, $i = 1, \ldots, n$, along with many other important statistics.

The Tukey signed rank statistic

$$\binom{n+1}{2}^{-1} \sum\sum_{1 \le i \le j \le n} \operatorname{sgn}(X_i + X_j)$$

can be expressed as

$$(n+1)^{-1}(2U_{n1} + (n-1)U_{n2})$$

where

$$U_{n1} = n^{-1} \sum_{i=1}^{n} \operatorname{sgn} X_i$$

is the sign statistic and

$$U_{n2} = \binom{n}{2}^{-1} \sum\sum_{1 \le i < j \le n} \operatorname{sgn}(X_i + X_j).$$

Both U_{n1} and U_{n2} are U-statistics. This fact was mentioned in Section 1.

Hoeffding (1948) obtained a very useful central limit theorem for suitably normalized U-statistics. Later, Hoeffding (1961) introduced a decomposition for U-statistics which facilitated considerably the development of the asymptotics for U-statistics. Similar decomposition for more general statistics (not necessarily symmetric) is available in the later work of Hájek (1968), Efron and Stein (1981), and Karlin and Rinott (1982).

To arrive at Hoeffding's decomposition, we first introduce the following notations. For any $c = 1, \ldots, m$, let

$$\phi_c(x_1, \ldots, x_c) = E[\phi(X_1, \ldots, X_m) \mid X_1 = x_1, \ldots, X_c = x_c]. \tag{2.3}$$

Define

$$\psi_1(x_1) = \phi_1(x_1) - \theta(F); \tag{2.4}$$

$$\psi_2(x_1, x_2) = [\phi_2(x_1, x_2) - \theta(F)] - \psi_1(x_1) - \psi_1(x_2); \tag{2.5}$$

$$\psi_3(x_1, x_2, x_3) = [\phi_3(x_1, x_2, x_3) - \theta(F)] - \sum_{j=1}^{3} \psi_1(x_j) - \sum\sum_{i \le j < j' \le 3} \psi_2(x_j, x_{j'});$$

$$\ldots \tag{2.6}$$

$$\psi_m(x_1, x_2, \ldots, x_m) = [\phi(x_1, \ldots, x_m) - \theta(F)] - \sum_{j=1}^{m} \psi_1(x_j)$$

$$- \sum\sum \psi_2(x_j, x_{j'}) - \cdots - \sum\cdots\sum \psi_{m-1}(x_{j_1}, \ldots, x_{j_{m-1}}). \tag{2.7}$$

Then U_n has the decomposition

$$U_n = \theta(F) + \frac{m}{n}\sum_{j=1}^{n}\psi_1(X_j) + \frac{\binom{m}{2}}{\binom{n}{2}}\sum\sum_{1\leq j<j'\leq n^2}\psi_2(X_j, X_j') + \cdots + \frac{\binom{m}{n}}{\binom{n}{m}}$$
$$\times \sum \cdots \sum_{1\leq j_1<\cdots<j_m\leq n}\psi_m(X_{j_1}, \ldots, X_{j_m}). \tag{2.8}$$

The above decomposition has been recently referred to in the literature as a ANOVA type decomposition (see, for example, Efron and Stein, 1981; or Karlin and Rinott, 1982). If $E\phi^2(X_1, \ldots, X_m) < \infty$, then using the conditional version of Jensen's inequality, it follows that $E\psi_j^2(X_1, \ldots, X_j) < \infty$ for all $j = 1, \ldots, m-1$. Also, it can be shown that the individual terms involved in (2.8) have zero means, and are mutually orthogonal. Accordingly,

$$E(U_n) = \theta(F); \tag{2.9}$$

$$V(U_n) = \frac{m^2}{n}V[\psi_1(X_1)] + \frac{\binom{m}{2}^2}{\binom{n}{2}}V[\psi_2(X_1, X_2)] + \cdots$$
$$+ \frac{\binom{m}{m}^2}{\binom{n}{m}}V[\psi_m(X_1, \ldots, X_m)]. \tag{2.10}$$

The first main result of this section is as follows:

THEOREM 1. *Assume that* $E[\phi^2(X_1, \ldots, X_m)] < \infty$ *and* $\zeta_1 = V[\psi_1(X_1)] > 0$. *Then*

$$\sqrt{n}(U_n - \theta(F))/(m^2\zeta_1)^{1/2} \xrightarrow{L} N(0, 1).$$

where $\xrightarrow{L}$ *means convergence in distribution.*

The above theorem, first obtained by Hoeffding (1948), follows easily from (2.8) by applying the classical central limit theorem on the mean like term $mn^{-1}\sum_{j=1}^{n}\psi_1(X_j)$ and showing that $E(R_n^2) \to 0$, where R_n is the remainder term in (2.8) in the decomposition of $U_n - \theta(F)$. This proof involves the 'projection' idea of breaking up a statistic as a sample mean plus a remainder term converging to zero at a rate faster than the centered sample mean.

As an application of Theorem 1, consider the example when

$$U_n = \binom{n}{2}^{-1}\sum\sum_{1\leq i<j\leq n}(X_i - X_j)^2/2,$$

the sample variance. In this case,

$$\phi_1(x) = E[\tfrac{1}{2}(X_1 - X_2)^2 \mid X_1 = X] = \tfrac{1}{2}[(x_1 - \mu)^2 + \sigma^2]\,,$$

where $\mu = E(X_1)$ and $\sigma^2 = V(X_1)$. Hence,

$$\zeta_1 = V[\phi_1(X_1)] = V[\psi_1(X_1)] = \tfrac{1}{4}(\mu_4 - \sigma^4)\,,$$

where $\mu_4 = E(X_1 - \mu)^4$. Accordingly, if $\sigma^4 < \mu_4 < \infty$, applying Theorem 1, one gets

$$\sqrt{n}(U_n - \sigma^2)/(\mu_4 - \sigma^4)^{1/2} \xrightarrow{L} N(0, 1)\,.$$

One may make a note that (2.10) is not the usual way the variance of a U-statistic is expressed in the literature (see, for example, Puri and Sen, 1971; Randles and Wolfe, 1979; or Serfling, 1980). Writing $\zeta_c = V[\phi(X_1, \ldots, X_c)]$, $V(U_n)$ is usually expressed as $V(U_n) = \binom{n}{m}^{-1} \sum_{c=1}^{m} \binom{m}{c}\binom{n-m}{m-c}\zeta_c$.

Situations could arise though when $\zeta_1 = 0$. For instance, in the sample variance example, one could have $\mu_4 = \sigma^4$. In such a situation, Theorem 1 is not applicable.

Hoeffding (1948) obtained the inequalities $\zeta_c/c \leq \zeta_d/d$ for $1 \leq c \leq d \leq m$. If $\zeta_1 = \cdots = \zeta_c = 0 < \zeta_{c+1}$, one can still obtain asymptotic distribution of suitably normalized U_n.

The case $\zeta_1 = 0 < \zeta_2$ has been addressed by Gregory (1977) (see also Serfling, 1980). If $E\phi^2(X_1, \ldots, X_m) < \infty$ and $\zeta_1 = 0 < \zeta_2$, then

$$n(U_n - \theta(F)) \xrightarrow{L} \binom{m}{2} Y\,,$$

where Y is a weighted average of possibly infinitely many chi-squared random variables each with one degree of freedom. As an application, consider once again the sample variance example where the X_i's are Bernoulli random variables with probability of success equal to $\frac{1}{2}$.

Recently, interest has been focussed on obtaining rates of convergence to normality for U-statistics. The problem has been addressed by Grams and Serfling (1973), Bickel (1974), Chan and Wierman (1977), and Callaert and Janssen (1978), the latter obtaining the sharpest rate as follows.

THEOREM 2. *If $E_F|\phi(X_1, \ldots, X_m)|^3 < \infty$ and $\zeta_1 > 0$ then*

$$\sup_{-\infty < x < \infty} \left| P_F\left(\frac{\sqrt{n}(U_n - \theta(F))}{m\zeta_1^{1/2}} \leq x\right) - \Phi(x)\right| \leq C\nu(m^2\zeta_1)^{-3/2} n^{-1/2}\,, \tag{2.11}$$

where C is an absolute constant not depending on n or the moments of the distribution of the X_i's, $\nu = E_F|\psi_1(x_1)|^3$, and $\Phi(x)$ is the distribution function of a $N(0, 1)$ random variable.

The moment assumptions of Callaert and Janssen (1978) have been weakened further by Helmers and Van Zwet (1982). They obtained the same error rate as given in the right hand side of (2.11) under the assumptions that

(i) $E_F[\phi^2(X_1, \ldots, X_m)] < \infty$, and (ii) $\nu < \infty$. Note that the condition $E_F|\phi(X_1, \ldots, X_m)|^3 < \infty$ implies both conditions (i) and (ii), but not vice versa. An example to this effect appears in a paper of Borovskikh (1979).

Central limit theorems for U-statistics with random indices were obtained by Sproule (1974), and rates of convergence to normality for such random U-statistics were obtained by Ghosh and Dasgupta (1982). Callaert, Janssen, and Veraverbeke (1980) obtained certain asymptotic Edgeworth type expansions for U-statistics.

Next, in this section, we mention briefly the laws of large numbers for one sample U-statistics. The classical strong law of large numbers generalized to U-statistics is as follows.

THEOREM 3. *If $E_F|\phi(X_1, \ldots, X_m)| < \infty$ then $U_n \to \theta(F)$ almost surely as $n \to \infty$.*

The strong convergence result was first obtained by Sen (1960) under the stronger moment condition $E|\phi(X_1, \ldots, X_m)|^{2-m^{-1}} < \infty$. Later, Hoeffding (1961) obtained the result under the assumption of finiteness of the first moment of ϕ using the decomposition given in (2.8). Berk (1966) showed that U-statistics are backward martingales, and proved Theorem 3 using the strong convergence results for backward martingales.

For most practical purposes, it is useful to obtain an estimate of the variance of U_n or at least of $m^2\zeta_1$, since $nV(U_n)/m^2\zeta_1 \to 1$ as $n\infty$. The following Jackknifed estimator of $V(U_n)$ was proposed by Sen (1960). To introduce Sen's estimator, first let

$$U_n^{(1)}(X_i) = \binom{n-1}{m-1}^{-1} \sum \cdots \sum \phi(X_i, X_{\alpha_2}, \ldots, X_{\alpha_m}),$$

where the summation extends over all possible $1 \leq \alpha_2 < \cdots < \alpha_m \leq n$ with $\alpha_j \neq i$ for any $j = 2, \ldots, m$. Define

$$S_{1n}^2 = (n-1)^{-1} \sum_{i=1}^{n} [U_n^{(1)}(X_i) - U_n]^2 .$$

Sen (1960) (and later Sproule, 1969) showed that if $E\phi^2(X_1, \ldots, X_m) < \infty$, then

$$S_{1n}^2 \xrightarrow{P} \zeta_1 \quad \text{as } n \to \infty .$$

Accordingly,

$$\sqrt{n}(U_n - \theta(F))/(mS_{1n}) \xrightarrow{L} N(0, 1) .$$

Very recently, rate of convergence to normality for such studentized U-statistics has been obtained by Callaert and Veraverbeke (1981). They show that if $E|\phi(X_1, \ldots, X_m)|^{4.5} < \infty$, then,

$$\sup_{-\infty<x<\infty} \left| P_F\left(\frac{\sqrt{n}(U_n - \theta(F))}{mS_{1n}} \leq x\right) - \Phi(x)\right| = O(n^{-1/2}) .$$

For a generalization of Theorem 1 to vector-valued U-statistics, consider a vector $(U_{n1}, \ldots, U_{nk})$, where U_{ni} is a one sample U-statistic with kernel ϕ_i and degree m_i, i.e.

$$U_{ni} = \binom{n}{m_i}^{-1} \sum_{1 \le j_1 < \cdots < j_{m_i} \le n} \phi_i(X_{j_1}, \ldots, X_{j_{m_i}}),$$

where $n \geq \max(m_1, \ldots, m_k)$. Also, let $\theta_i = E_F\phi_i(X_1, \ldots, X_{m_i})$, $\phi_{1i}(x) = E_F[\phi_i(X_1, \ldots, X_{m_i}) \mid X_1 = x]$, and $\sigma_{ij} = \mathrm{Cov}_F[\phi_{1i}(X_1), \phi_{1j}(X_1)]$, $1 \le i, j \le k$. It is assumed that σ_{ii} is positive and finite for all $i = 1, \ldots, k$. Then, $\sqrt{n}(U_{n1} - \theta_1, \ldots, U_{nk} - \theta_k)$ converges in distribution to $N(\mathbf{0}, \boldsymbol{\Sigma})$, where $\boldsymbol{\Sigma} = ((\sigma_{ij}))$. To prove this result, one uses a decomposition similar to (2.3)–(2.8) for each U_{ni}, show that the remainder term converges to zero in mean square for each component, and apply the multidimensional central limit theorem to the vector of principal terms $(n^{-1}\sum_{j=1}^n \phi_{11}(X_j), \ldots, n^{-1}\sum_{j=1}^n \phi_{1k}(X_j))$.

Hoeffding (1948) also obtained a central limit theorem for U-statistics in the case when X_i's are independent, but not iid. This can be proved by using a decomposition similar to (but more complicated than) (2.8). In fact, such a decomposition for more general statistics is now available in the literature (see Hájek, 1968; Efron and Stein, 1981; and Karlin and Rinott, 1982).

We do not exhibit this decomposition in detail, but introduce a few notations needed to obtain the variances of U-statistics in the non-iid case, and in stating a corresponding central limit theorem. Let

$$\begin{aligned} &\phi_{c(\alpha_1, \ldots, \alpha_c)\beta_1, \ldots, \beta_{m-c}}(X_{\alpha_1}, \ldots, X_{\alpha_c}) \\ &\quad = E[\phi(X_{\alpha_1}, \ldots, X_{\alpha_c}, X_{\beta_1}, \ldots, X_{\beta_{m-c}}) \mid X_{\alpha_1} = x_{\alpha_1}, \ldots, X_{\alpha_c} = x_{\alpha_c}], \\ &\zeta_{c(\alpha_1, \ldots, \alpha_c)\beta_1, \ldots, \beta_{m-c}; \gamma_1, \ldots, \gamma_{m-c}} = [E\phi_{c(\alpha_1, \ldots, \alpha_c)\beta_1, \ldots, \beta_{m-c}}(X_{\alpha_1}, \ldots, X_{\alpha_c}) \\ &\qquad \cdot \phi_{c(\alpha_1, \ldots, \alpha_c)\gamma_1, \ldots, \gamma_{m-c}}(X_{\alpha_1}, \ldots, X_{\alpha_c})], \end{aligned}$$

$$\zeta_{c,n} = \left\{ \binom{n}{2m-c} \binom{2m-c}{c} \binom{2m-2c}{m-c} \right\}^{-1} \sum \zeta_{c(\alpha_1, \ldots, \alpha_c)\beta_1, \ldots, \beta_{m-c}; \gamma_1, \ldots, \gamma_{m-c}}$$

Then, one can write

$$V(U_n) = \binom{n}{m}^{-1} \sum_{c=1}^{m} \binom{m}{c} \binom{n-m}{m-c} \zeta_{c,n}.$$

Also, let $\bar{\phi}_{1(i)}(X_i) = \binom{n-1}{m-1}^{-1} \sum \phi_{1(i)\alpha_2, \ldots, \alpha_m}(X_i)$, where the summation extends over all $1 \le \alpha_2 < \cdots < \alpha_m \le n$ with each $\alpha_j \neq i$.

Let $\bar{\psi}_{1(i)}(X_i) = \bar{\phi}_{1(i)}(X_i) - E\bar{\phi}_{1(i)}(X_i)$, $i = 1, \ldots, n$. Then we have the following theorem.

THEOREM 4. *Suppose*

(i) $\sup_{1 \le \alpha_1 < \cdots < \alpha_m \le n} E\phi^2(X_{\alpha_1}, \ldots, X_{\alpha_m}) < \infty$;

(ii) $E|\bar{\psi}_{1(i)}(X_i)|^{2+\delta} < \infty$ *for all* $i = 1, \ldots, n$, *and*

(iii) $\lim_{n\to\infty} \Sigma_{i=1}^{n} E|\bar{\psi}_{1(i)}(X_i)|^{2+\delta}/\Sigma_{i=1}^{n}\{E\bar{\psi}_{2(i)}^{2}(X_i)\}^{1+1/2\delta} = 0.$ *Then,*

$$(U_n - EU_n)/V^{1/2}(U_n) \xrightarrow{\text{L}} N(0, 1).$$

Rates of convergence to normality for U-statistics in the non-iid case are available in Ghosh and Dasgupta (1982), and also in the Ph.D. thesis of Janssen (1978).

3. Multisample U-statistics

U-statistics can be usefully extended to the multisample case. Lehmann (1951) considered the two-sample extension of U-statistics. The general c-sample extension of such statistics is mentioned in Lehmann (1963).

In the two-sample case, let $X_1, \ldots, X_m$ and $Y_1, \ldots, Y_n$ be independent random samples from distributions with distribution functions $F(x)$ and $G(y)$ respectively. A parameter $\theta(F, G)$ is said to be *estimable of degree* (m_1, m_2) for distributions (F, G) in a family $\mathcal{F}$ if m_1 and m_2 are the smallest sample sizes for which there exists an estimator of $\theta(F, G)$ that is unbiased for every $(F, G) \in \mathcal{F}$, that is, there exists a function ϕ such that

$$E_{F,G}\phi(X_1, \ldots, X_m, Y_1, \ldots, Y_n) = \theta(F, G)\,. \tag{3.1}$$

Once again, without any loss of generality, ϕ can be assumed to be symmetric in its components, and separately in the Y_j components.

For an estimable parameter θ of degree (m_1, m_2), and with a symmetric kernel ϕ, a two-sample U-statistic is defined as

$$U_{n_1,n_2} = \binom{n_1}{m_1}^{-1}\binom{n_2}{m_2}^{-1} \sum \phi(X_{\alpha_1}, \ldots, X_{\alpha_{m_1}}, Y_{\beta_1}, \ldots, Y_{\beta_{m_2}})\,, \tag{3.2}$$

where the summation extends over all possible $1 \leq \alpha_1 < \cdots < \alpha_{m_1} \leq n_1$, and $1 \leq \beta_1 < \cdots < \beta_{m_2} \leq n_2$.

Important examples of two-sample U-statistics are $\bar{Y}_{n_2} - \bar{X}_{n_1}$ and the Mann–Whitney U-statistic

$$U_{n_1,n_2} = (n_1 n_2)^{-1} \sum_{i=1}^{n_1} \sum_{j=1}^{n_2} I_{[Y_j \geq X_i]}\,.$$

Using the notation

$$\begin{aligned} &\phi_{c,d}(x_1, \ldots, x_c, y_1, \ldots, y_d) \\ &\quad = E[\phi(X_1, \ldots, X_{m_1}, Y_1, \ldots, Y_{m_2}) \mid X_1 = x_1, \ldots, X_c = x_c\,, \\ &\qquad Y_1 = y_1, \ldots, Y_d = y_d] \end{aligned} \tag{3.3}$$

and $\zeta_{c,d} = V_{F,G}[\phi_{c,d}(X_1, \ldots, X_c, Y_1, \ldots, Y_d)]$, one gets an expression for

$V[U_{n_1,n_2}]$ as

$$V[U_{n_1,n_2}] = \binom{n_1}{m_1}^{-1}\binom{n_2}{m_2}^{-1}\sum_{c=0}^{m_1}\sum_{d=0}^{m_2}\binom{m_1}{c}\binom{n_1-m_1}{m_1-c}\binom{m_2}{d}\binom{n_2-m_2}{m_2-d}\zeta_{c,d}\,, \tag{3.4}$$

where $\zeta_{0,0}$ is defined as 0. The central limit theorem for two-sample U-statistics is now stated as follows.

THEOREM 5. *If*

(i) $E\phi^2(X_1, \ldots, X_{n_1}, Y_1, \ldots, Y_{n_2}) < \infty$,

(ii) $0 < \lambda = \lim_{N\to\infty} n_1/N < 1$ (where $N = n_1 + n_2$) *and*

(iii) $\sigma^2 = m_1^2\zeta_{10}/\lambda + m_2^2\zeta_{01}/(1-\lambda) > 0$,

then, $\sqrt{N}[U_{n_1,n_2} - \theta(F, G)]/\sigma \xrightarrow{L} N(0, 1)$ *as* $N \to \infty$.

As an application of this theorem, consider the case of Mann–Whitney U-statistics. In this case, $\phi_{1,0}(x) = 1 - G(x-)$, $\phi_{0,1}(y) = F(y)$ so that $\zeta_{1,0} = V_F[1 - G(X-)] = V_F[G(X-)]$ and $\zeta_{0,1} = V_G[F(Y)]$. Condition (i) of Theorem 5 is trivially satisfied, since ϕ is bounded. Hence, if (ii) and (iii) of Theorem 5 hold, then $\sqrt{N}(U_{n_1,n_2} - \int F(x)\,dG(x))/\sigma \xrightarrow{L} N(0, 1)$.

A Hoeffding type decomposition for two-sample U-statistics is possible, and using such a decomposition, one can prove Theorem 5. Rates of convergence to normality for two-sample U-statistics is given in the Ph.D. thesis of Jannsen (1978). A strong law of large numbers for two-sample U-statistics extending the one-sample backward martingale argument is given in Arvesen (1969). For c-sample U-statistics, Sen (1977b) obtained the strong convergence of U-statistics under the condition $E\{|\phi| \log^+ |\phi|^{c-1}\} < \infty$.

To extend the above ideas to c-sample U-statistics, let X_{ij} ($j = 1, \ldots, n_i$, $i = 1, \ldots, c$) be c independent samples, the ith sample having distribution function $F_i(x)$ ($i = 1, \ldots, c$). A parameter $\theta(F_1, \ldots, F_c)$ is said to be estimable of degree $(m_1, \ldots, m_c)$ for distributions $(F_1, \ldots, F_c)$ in a family $\mathscr{F}$ if $m_1, \ldots, m_c$ are the smallest sample sizes for which there exists an estimator of $\theta(F_1, \ldots, F_c)$ that is unbiased for every $(F_1, \ldots, F_c) \in \mathscr{F}$, i.e., there exists a function ϕ such that

$$E_{F_1}, \ldots, {}_{F_c}\phi(X_{11}, \ldots, X_{1m_1}, \ldots, X_{c1}, \ldots, X_{cm_c}) = \theta(F_1, \ldots, F_c)\,. \tag{3.5}$$

Also, ϕ is assumed to be separately symmetric in the arguments $X_{i1}, \ldots, X_{in_i}$ for each $i = 1, \ldots, c$.

The c-sample U-statistic with kernel ϕ and degree $(m_1, \ldots, m_c)$ is defined as

$$U_{n_1,\ldots,n_c} = \binom{n_1}{m_1}^{-1} \cdots \binom{n_c}{m_c}^{-1}$$
$$\times \sum_{\alpha_1 \in A_1} \cdots \sum_{\alpha_c \in A_c} \phi(X_{1\alpha_{11}}, \ldots, X_{1\alpha_{1m_1}}, \ldots, X_{c\alpha_{11}}, \ldots, X_{c\alpha_{cm_c}})\,, \tag{3.6}$$

where $\boldsymbol{\alpha}_i = (\alpha_{i1}, \ldots, \alpha_{im_i})$, and A_i is the collection of all subsets of m_i integers chosen without replacement from the integers $\{1, \ldots, n_i\}$, $i = 1, \ldots, c$.

To obtain the asymptotic normality of a properly standardized c-sample U-statistic, we proceed as follows. Let i be a fixed integer satisfying $1 \leq i \leq c$. Define

$$\phi_{i1} = \phi(X_{1\alpha_{11}}, \ldots, X_{1\alpha_{1m_1}}, \ldots, X_{c\alpha_{c1}}, \ldots, X_{c\alpha_{cm_c}})$$

and

$$\phi_{i2} = \phi(X_{1\beta_{11}}, \ldots, X_{1\beta_{1m_1}}, \ldots, X_{c\beta_{c1}}, \ldots, X_{c\beta_{cm_c}})$$

where the two sets $(\alpha_{j1}, \ldots, \alpha_{jm_j})$ and $(\beta_{j1}, \ldots, \beta_{jm_j})$ have no integers in common whenever $j \neq i$ and exactly one integer in comnon when $j = i$. Let

$$\delta_{0,\ldots,0,1,0,\ldots,0} = \mathrm{Cov}_{F_1,\ldots,F_c}(\phi_{i1}, \phi_{i2}),$$

where the only 1 in the subscript of $\delta_{0,\ldots,0,1,0,\ldots,0}$ is in the ith position. Also, let $N = \Sigma_{i=1}^{k} n_i$. The following c-sample U-statistic theorem is due to Lehmann (1963).

Theorem 5(a). *Let $U_N \equiv U(X_{11}, \ldots, X_{1n_1}, \ldots, X_{c1}, \ldots, X_{cn_c})$ be a c-sample U-statistic with kernel ϕ and degree $(m_1, \ldots, m_c)$ for the parameter $\theta \equiv \theta(F_1, \ldots, F_c)$. If*

$$\lim_{N\to\infty} (n_i/N) = \lambda_i (0 < \lambda_i < 1) \quad \text{for } i = 1, \ldots, c$$

and

$$E_{F_1,\ldots,F_c}\phi^2(X_{11}, \ldots, X_{1n_1}, \ldots, X_{c1}, \ldots, X_{cn_c}) < \infty,$$

then $\sqrt{N}(U_N - \theta) \xrightarrow{\mathrm{L}} N(0, \sigma^2)$, *where* $\sigma^2 = \Sigma_{i=1}^{c} m_i^2 \lambda_i^{-1} \zeta_{0,\ldots,0,1,\ldots,0}$.

4. Jackknifed and bootstrapped U-statistics

Jackknife, originally proposed by Quenouille (1956), and later by Tukey (1957) has proved to be a very useful tool in nonparametric bias and variance estimation. To introduce the idea of jackknife, let η be an unknown parameter of interest, and let $X_1, \ldots, X_N$ be N $(=nk)$ iid random variables with common distribution function F_η. The N observations are divided into n groups of k observations each. Let $\hat{\eta}_n$ be an estimator of η based on all the N observations, and let $\hat{\eta}_{n-1}^{i}$ denote the estimator obtained after deletion of the ith group of observations. Let

$$\hat{\eta}_i = n\hat{\eta}_n - (n-1)\hat{\eta}_{n-1}^{i}, \quad i = 1, \ldots, n. \tag{4.1}$$

The $\hat{\eta}_i$'s are called pseudo-values by Tukey. Then the jackknife estimator of η

is given by

$$\bar{\eta}_n = n^{-1} \sum_{i=1}^{n} \hat{\eta}_i \,. \tag{4.2}$$

Motivated from the study of robust procedures in Model II ANOVA problems, Arvesen (1969) studied very extensively the limit laws for jackknifed U-statistics and functions of them. We consider U-statistics with kernel ϕ and degree m. For simplicity, consider the case $k = 1$. Let

$$U^i_{n-1} = \binom{n-1}{m}^{-1} \sum_{C^i_{n-1}} \phi(X_{\beta^i_1}, \ldots, X_{\beta^i_m}) \,,$$

where C^i_{n-1} denotes that the summation is over all combinations $(\beta^i_1, \ldots, \beta^i_m)$ of m integers chosen from $(1, \ldots, i-1, i+1, \ldots, n)$. Let g be a real valued function, and let

$$\hat{\eta}_n = g(U_n), \quad \eta = g(\theta(F)), \quad \hat{\eta}^i_{n-1} = g(U^i_{n-1}),$$

$$\hat{\eta}_i = n\hat{\eta}_n - (n-1)\hat{\eta}^i_{n-1}\,, \quad \bar{\eta}_n = n^{-1} \sum_{i=1}^{n} \hat{\eta}_i$$

and

$$S^2_g = (n-1)^{-1} \sum_{i=1}^{n} (\eta_i - \bar{\eta}_n)^2 \,.$$

The following two theorems are proved by Arvesen (1969). The first one provides a central limit theorem for suitable normalized $\bar{\eta}_n$. The second theorem provides a consistent estimator of the variance of the limiting distribution of $\bar{\eta}_n$.

THEOREM 6. *Suppose $E\phi^2(X_1, \ldots, X_m) < \infty$. Let g be a real valued function defined in a neighbourhood of θ which has a bounded second derivative. Then*

$$\sqrt{N}(\bar{\eta}_n - \eta) \xrightarrow{\mathrm{L}} N(0, m^2 \zeta_1 (g'(\theta))^2) \quad \text{as } n \to \infty.$$

THEOREM 7. *Suppose $E\phi^2(X_1, \ldots, X_m) < \infty$. Let g be a real valued function defined in a neighbourhood of θ which has a continuous first derivative. Then,*

$$S^2_g \xrightarrow{\mathrm{P}} m^2 \zeta_1 (g'(\theta))^2.$$

A further generalization of jackknife was introduced by Efron (1979) under the rubric bootstrap. Bootstrapping is a resampling scheme in which the statistician learns the sampling properties of a statistic by recomputing its value on the basis of a new sample realized from the original one in a certain way. This is made more precise as follows.

Consider the one-sample situation in which a random sample $\boldsymbol{X} = (X_1, \ldots, X_n)$ is observed from a completely unspecified distribution F. The observed realization of $\boldsymbol{X}$ is denoted by $\boldsymbol{x} = (x_1, \ldots, x_n)$. The bootstrap method of realizing a sample can now be described as follows.

A. Construct the empirical distribution function F_n based on the observed realization $\boldsymbol{x}$ of $\boldsymbol{X}$.

B. With F_n fixed, draw a random sample of size N from F_n, say $\boldsymbol{X}^* = (X_1^*, \ldots, X_N^*)$. Then, conditional on $\boldsymbol{X} = \boldsymbol{x}$ fixed, the X_i^*'s are iid with distribution function F_n. The bootstrapped sample is now denoted by $\boldsymbol{X}^*$.

C. The above procedure can be repeated any number of times. A U-statistic U_N^* with kernel ϕ and degree m $(N \geq m)$ based on the bootstrapped sample $\boldsymbol{X}^*$ is given by

$$U_N^* = \binom{N}{m}^{-1} \sum_{1 \leq j_1 < \cdots < j_m \leq N} \cdots \sum \phi(X_{j_1}^*, \ldots, X_{j_m}^*), \quad N \geq m . \tag{4.3}$$

Bickel and Freedman (1981) have developed a very extensive central limit theory for many bootstrapped statistics including U-statistics. Under the assumption $E\phi^2(X_1, \ldots, X_m) < \infty$, they showed that

$$\sqrt{N}(U_N^* - U_n) \xrightarrow{\mathrm{L}} N(0, m^2\zeta_1)$$

as $\min(n, N) \to \infty$. Following Sen (1960), Athreya, Ghosh, Low and Sen (1984) proposed the estimator

$$S_N^2 = (N-1)^{-1} \sum_{i=1}^{N} \left[\binom{N-1}{m-1}^{-1} \sum_{\substack{1 \leq j_2 < \cdots < j_m \leq N \\ j_k \neq i}} \cdots \sum \phi(X_i^*, X_{j_2}^*, \ldots, X_{j_m}^*) - U_N \right]^2 \tag{4.4}$$

for ζ_1. This estimator multiplied by m^2 can be easily recognized as the jackknifed estimator of the variance of the asymptotic distribution of bootstrapped U-statistics. Athreya, Ghosh, Low and Sen (1983) showed that under the assumption that $E\phi^2(X_1, \ldots, X_m) < \infty$,

$$S_N^2 \xrightarrow{\mathrm{P}} m^2\zeta_1 \quad \text{as } \min(n, N) \to \infty .$$

Accordingly, the bootstrapped pivot

$$\sqrt{N}(U_N^* - U_n)/(mS_N) \xrightarrow{\mathrm{L}} N(0, 1) \quad \text{as } \min(n, N) \to \infty .$$

Also, a strong law of large numbers for bootstrapped U-statistics was proved by Athreya et al. (1983) under the assumption that $E_F|\phi(X_1, \ldots, X_m)|^{1+\delta} < \infty$ for some $\delta > 0$.

5. Functional central limit theorems for U-statistics

Central limit theorems for U-statistics developed earlier in Sections 2 and 3 can be generalized in a *stochastic process* setting. The need for such a generalization is discussed in Billingsley (1968), and in Serfling (1980).

Consider once again U-statistics with kernel ϕ and degree m with $E\phi^2(X_1, \ldots, X_m) < \infty$ and $\zeta_1 > 0$. With the sequence $\{U_n; n \geq m\}$ of U-statistics, consider two associated sequence of stochastic process on the unit interval $[0, 1]$.

The process pertaining to the past was introduced by Miller and Sen (1972). Define

$$Y_n(t) = 0, \quad 0 \leq t \leq \frac{m-1}{n}; \qquad Y_n(k/n) = \frac{k(U_k - \theta)}{(m^2\zeta_1)^{1/2} n^{1/2}}, \qquad k = m, m+1, \ldots, n; \tag{5.1}$$

$Y_n(t)$ $(0 \leq t \leq 1)$ is defined elsewhere by linear interpolation.

The process pertaining to the future was introduced by Loynes (1970). Define

$$\begin{aligned} &Z_n(0) = 0; \\ &Z_n(t_{nk}) = (U_k - \theta)/V^{1/2}(U_n), \; k \geq n, \; t_{nk} = V(U_k)/V(U_n); \\ &Z_n(t) = Z_n(t_{nk}), \; t_{n,k+1} < t < t_{nk}. \end{aligned} \tag{5.2}$$

For each n, $t_{nn}, t_{n,n+1}, \ldots,$ form a sequence tending to 0 and $Z_n(\cdot)$ is a step function continuous from the left.

Before stating the limit laws for the processes $Y_n(\cdot)$ and $Z_n(\cdot)$, we need certain preliminaries. Consider a collection of random variables $T_1, T_2, \ldots,$ and T taking values in an arbitrary metric space S, and having respective probability measures $P_1, P_2, \ldots,$ and P defined on the Borel sets in S (i.e. on the σ-field generated by the open sets with respect to the metric associated with S). We say that $P_n \to P$ if $\lim_{n\to\infty} P_n(A) = P(A)$ for every Borel set A is S satisfying $P(\partial A) = 0$, where ∂A = boundary of A = closure of A − interior of A. In particular if S is a metrizable *function* space, then $P_n \to P$ denotes 'convergence in distribution' of a sequence of stochastic processes to a limit stochastic process.

In the Miller and Sen (1972) situation, we take $S = C[0, 1]$, the collection of all continuous functions on the unit interval $[0, 1]$. The space $C[0, 1]$ is metrized by

$$\rho(x, y) = \sup_{0 \leq t \leq 1} |x(t) - y(t)| \tag{5.3}$$

for $x = x(\cdot)$ and $y = y(\cdot)$ in $C[0, 1]$. Denote by $\mathscr{B}$ the class of Boret sets in $C[0, 1]$ relative to ρ. Denote by Q_n the probability distribution of $Y_n(\cdot)$ in $C[0, 1]$, i.e. the probability measure on $(C, \mathscr{B})$ induced by the measure P through the relation

$$Q_n(B) = P(\{w\colon Y_n(\cdot, w) \in B\}), \quad B \in \mathscr{B}. \tag{5.4}$$

In order that the sequence of processes $\{Y_n(\cdot)\}$ converges to a limit process

$Y(\cdot)$ in the sense of convergence in distribution, we seek a measure Q on $(C, \mathscr{B})$ such that $Q_n \to Q$.

The measure Q in the Miller and Sen (1972) situation turns out to be the Wiener measure, or the probability measure Q defined by the properties

(a) $Q(\{x(\cdot): x(0) = 0\}) = 1$;

(b) $Q(\{x(\cdot): x(t) \leq a\}) = (2\pi t)^{-1/2} \int_{-\infty}^{a} \exp(-x^2/2t)\,dx$ for all $0 < t \leq 1$, $-\infty < a < \infty$;

(c) for $0 \leq t_0 \leq \cdots \leq t_k \leq 1$ and $-\infty < a_1, \ldots, a_k < \infty$,

$$Q\left(\bigcap_{i=1}^{k} \{x(\cdot): x(t_i) - x(t_{i-1}) \leq a_i\}\right) = \prod_{i=1}^{k} Q(\{x(\cdot): x(t_i) - x(t_{i-1}) \leq a_i\}).$$

A random element of $C[0, 1]$ having the distribution Q is called a *Wiener process*, and is denoted by $\{W(t), 0 \leq t \leq 1\}$ or simply by W. From (a), (b) and (c), it follows that

(a)′ $W(0) = 0$ with probability 1;

(b)′ $W(t)$ is $N(0, t)$ for each $t \in (0, 1)$;

(c)′ for $0 \leq t_0 \leq \cdots \leq t_k \leq 1$, the increments $W(t_1) - W(t_0), \ldots, W(t_k) - W(t_{k-1})$ are mutually independent. We are now in a position to state the main theorem of Miller and Sen (1972).

THEOREM 8. *Assume that* $E\phi^2(X_1, \ldots, X_m) < \infty$ *and* $\zeta_1 > 0$. *Then* $Y_n(\cdot) \xrightarrow{d} W(\cdot)$ *in* $C[0, 1]$.

For Loynes (1970), $S = D[0, 1]$, the class of all functions which are right continuous and for which left hand limit exists. The $D[0, 1]$ space is endowed with the following topology.

Let Λ denote the class of all strictly increasing continuous mappings of $[0, 1]$ onto itself. The Skorohod distance between x and y (both belonging to the $D[0, 1]$ space) is defined by

$$d(x, y) = \inf\{\varepsilon > 0;\ \text{there exists a } \lambda \text{ in } \Lambda \text{ such that} \sup_t |\lambda(t) - t| < \varepsilon \text{ and } \sup_t |x(t) - y(\lambda(t))| < \varepsilon\}. \tag{5.5}$$

Note that d is a metric (called the Skorohod metric) which generates a topology on $D[0, 1]$. This is the so-called Skorohod topology which we refer to as the J_1-topology. The main result of Loynes (1970) is as follows.

THEOREM 9. *Assume the conditions of Theorem 8. Then,* $Z_n(\cdot) \xrightarrow{d} W(\cdot)$ *in the* J_1*-topology on* $D[0, 1]$.

REMARK. Theorems 8 and 9 are useful in sequential analysis. Loynes (1970) cited Theorem 9, but did not check the validity of the needed regularity conditions. These conditions were verified later by Sen and Bhattacharyya (1977), and Loynes (1978).

For c-sample U-statistics, functional CLT's are due to Sen (1974) in the nondegenerate case, and Neuhaus (1977) in the degenerate case. Hall (1979) obtained a single result from which invariance principles in both the degenerate and nondegenerate cases followed as immediate corollaries.

6. Miscellaneous remarks

Closely related to the U-statistics are the von Mises V-statistics defined by

$$V_n = n^{-m} \sum_{\alpha_1=1}^{n} \cdots \sum_{\alpha_m=1}^{n} \phi(X_{\alpha_1}, \ldots, X_{\alpha_m}) . \tag{6.1}$$

It can be shown that (see, for example, Ghosh and Sen, 1970) if $E[\phi^2(X_{\alpha_1}, \ldots, X_{\alpha_m})] < \infty$ for all $1 \leq \alpha_1, \ldots, \alpha_m \leq n$, then $E(U_n - V_n)^2 = O(n^{-2})$. This implies that,

$$n^r(U_n - V_n) \xrightarrow{P} 0 \quad \text{for any } r < 1 .$$

Consequently, $n^{1/2}(V_n - \theta)$ has the same limiting distribution as $n^{1/2}(U_n - \theta)$. Miller and Sen (1972) obtained functional central limit theorems (of the type discussed in Section 5) for von Mises V-statistics. This requires a more subtle analysis than just showing $E(U_n - V_n)^2 = O(n^{-2})$.

Central limit laws for U-statistics in the m-dependent case were obtained by Sen (1963, 1965) and for sampling without replacement from a finite population by Nandi and Sen (1963). A functional central limit theorem for jackknifed U-statistics was obtained by Sen (1977a).

7. Linear rank statistics: Central limit theorems and rates of convergence

There are many linear rank statistics which are not U-statistics, and the development of limit distributions for such statistics requires a different analysis. Notable among such statistics are the so-called normal scores statistics introduced in Section 1. In this section, we present certain results concerning null and nonnull distributions of linear rank statistics.

Recall from Section 1 that a linear rank statistic is expressible in the form

$$S_N = \sum_{i=1}^{N} c_N(i) a_N(R_{Ni}) \tag{7.1}$$

where R_{Ni} is the rank of X_i among $X_1, \ldots, X_N$. Both the regression constants and the scores are indexed by N, i.e. they may change as N changes.

Suppose that X_i has a continuous distribution function F_i, $i = 1, \ldots, N$. First consider the null situation, i.e. where $F_1 \equiv \cdots \equiv F_N$. In this case the vector $(R_{N1}, \ldots, R_{NN})$ is equiprobable over the $N!$ permutations of the first N positive integers. In this situation (see, for example, Hájek and Sidák, 1967; Randles and Wolfe, 1979; or Sen, 1981)

$$E(S_N) \equiv \mu_N = N\bar{c}_N\bar{a}_N;$$
$$V(S_N) \equiv \sigma_N^2 = (N-1)^{-1}\sum_{i=1}^{N}\{c_N(i)-\bar{c}_N\}^2\sum_{i=1}^{N}\{a_N(i)-\bar{a}\}^2 \tag{7.2}$$

where $\bar{c}_N = N^{-1}\sum_{i=1}^{N} c_N(i)$, $\bar{a}_N = N^{-1}\sum_{i=1}^{N} a_N(i)$.

Central limit theorems for suitably normalized S_N were first obtained by Wald and Wolfowitz (1944), and subsequently by Noether (1949), Hoeffding (1951), Dwass (1956), and Hájek (1961). Hájek (1961) obtained the result under minimal regularity conditions. His theorem is as follows.

THEOREM 10. *Let S_N be defined as in* (7.1) *with regression constants $c_N(i)$ satisfying Noether's uniform asymptotic negligibility condition, namely,*

$$\max_{1\le i\le N}(c_N(i)-\bar{c}_N)^2 \Big/ \sum_{i=1}^{N}(c_N(i)-\bar{c}_N)^2 \to 0 \quad \text{as } N\to\infty. \tag{7.3}$$

Assume also that the scores $a_N(i)$ are generated by $a_N(i) = \phi(i/(N+1))$, $i = 1,\dots,N$, where ϕ is a squared integrable score function on the unit interval [0,1]. *Then,*

$$(S_N-\mu_N)/\sigma_N \xrightarrow{L} N(0,1).$$

REMARK. In the above theorem, one could also take $a_N(i) = E\phi(U_{Ni})$, $i = 1,\dots,N$, where $U_{N1} \le \cdots \le U_{NN}$ are the order statistics in a random sample of size N from the uniform [0, 1] distribution.

In the case $a_N(i) = E\phi(U_{Ni})$ where $\int_0^1 |\phi(u)|\,du < \infty$, there is an alternate way of proving (7.3). Let $\mathcal{F}_N$ denote the σ-algebra generated by $R_{N1},\dots,R_{NN}$. Then, Sen and Ghosh (1972) have shown that $\{(S_N, \mathcal{F}_N); N \ge 1\}$ is a martingale sequence. The result can now be proved by appealing to some suitable martingale central limit theorem (see, for example, Mcleish, 1974). The martingale idea has been fruitfully exploited in obtaining functional central limit theorems for linear rank statistics (see Section 9).

In the two-sample case, Chernoff and Savage (1958) obtained in their pioneering paper the asymptotic nonnull distribution of S_N under nonlocal alternatives. For local alternatives, Hájek (1962) employed the idea of 'contiguity' of probability measures, and obtained some useful limit theorems. Finally, Hájek (1968) obtained the limit distribution of S_N under nonlocal alternatives. His main results are given below.

THEOREM 11. *Consider the statistic S_N given in* (7.1) *where the score function $a_N(i)$ is given either by $\phi(i/(N+1))$ or $E\phi(U_{Ni})$ where ϕ has a bounded second derivative. Assume that*

$$\max_{1\le i\le N}(c_N(i)-\bar{c}_N)^2/V(S_N) \to 0 \quad \text{as } N\to\infty. \tag{7.4}$$

Then

$$(S_N - ES_N)/V^{1/2}(S_N) \xrightarrow{L} N(0,1). \tag{7.5}$$

The assertion remains true, if in (7.4) *and* (7.5), *we replace* $V(S_N)$ *by* $\sigma_N^2 = \sum_{i=1}^N V[l_{Ni}(X_i)]$, *where*

$$l_{Ni}(x) = N^{-1}\sum_{j=1}^{N}(c_N(j) - c_N(i))\int [I_{[y\geq x]} - F_i(y)]\phi'[\bar{F}_N(y)]\, dF_j(y),$$
$$i = 1, \ldots, N,$$

where $\bar{F}_N(y) = N^{-1}\sum_{j=1}^N F_j(y)$.

The next theorem of Hájek (1968) does not require the assumption of boundedness of the second derivative of ϕ.

THEOREM 12. *Let* $\phi(u) = \phi_1(u) - \phi_2(u)$, $0 < u < 1$, *where the* $\phi_i(u)$*'s are nondecreasing square integrable and absolutely continuous inside* (0, 1), $i = 1, 2$. *Assume that*

$$N \max_{1\leq i\leq N} (c_N(i) - \bar{c}_N)^2/V(S_N) \to 0 \quad \text{as } N\to\infty. \tag{7.6}$$

Then (7.5) *holds. Also, in* (7.5) *and* (7.6), $V(S_N)$ *can be replaced by* σ_N^2 *as defined in Theorem* 11.

The proofs of these theorems employ the 'projection method' introduced earlier in Section 2 in connection with U-statistics. Hájek's (1968) projection lemma does not require the statistics to be symmetric in their arguments. The other major tool developed by Hájek (1968) is a powerful 'variance inequality' majorizing $V(S_N)$ when the distributions $F_1, \ldots, F_N$ are not necessarily equal.

An important question left open in Hájek's (1968) paper is whether the centering constant $E(S_N)$ can be replaced by the simpler $\mu_N = \sum_{i=1}^N c_N(i)\int_{-\infty}^{\infty}\phi(\bar{F}_n(x))\, dF_i(x)$. This question has been answered in the affirmative by Hoeffding (1973) if the square integrability condition of ϕ_1 and ϕ_2 is slightly strengthened. Hoeffding's (1973) main theorem is as follows.

THEOREM 13. *Let* $\phi(u) = \phi_1(u) - \phi_2(u)$ *where each* ϕ_i *is nondecreasing, absolutely continuous in* (0, 1), *and satisfies*

$$\int_0^1 u^{1/2}(1-u)^{1/2}\, d\phi_i(u) < \infty, \quad i = 1, 2. \tag{7.7}$$

Then, if $a_N(i) = E\phi(U_{Ni})$, $1\leq i\leq N$, *the conclusion of Theorem* 12 *remains valid with* $E(S_N)$ *replaced by* μ_N. *If* $a_N(i) = \phi(i/(N+1))$, $1\leq i\leq N$, *then the conclusion of Theorem* 12 *remains valid with* $E(S_N)$ *replaced by*

$$\mu'_N = \mu_N + \bar{c}_N \left\{ \sum_{i=1}^{N} \phi(i/(N+1)) - N \int_0^1 \phi(u)\, du \right\}.$$

If, in addition, $|\bar{c}_N| / \max_{i \leq i \leq N} |c_N(i) - \bar{c}_N|$ *is bounded,* $E(S_N)$ *can be replaced by* μ_N *even when* $a_N(i) = \phi(i/(N+1))$.

For one sample signed rank statistics S_N^+, analogues of Theorems 11 and 12 have been obtained by Husková (1970). Pyke and Shorack (1968) used an alternative approach in obtaining the limit distributions of linear rank statistics in the two sample case.

Next, in this section, we consider rates of convergence to normality for the simple linear rank statistics S_N. Results in this direction were obtained by Jurecková and Puri (1975), and later by Husková (1977) and Bergstrom and Puri (1977). The method of proof consists in approximating the simple linear rank statistic by a sum of independent random variables, and establishing for arbitrary r, a suitable bound on the rth moment of the error of approximation. The following assumptions are made.

I. The scores are generated by a function $\phi(u)$, $0 \leq u \leq 1$, in either of the following two ways:

$$a_N(i) = \phi(i/(N+1)) \quad \text{or} \quad E\phi(U_{Ni}), \qquad 1 \leq i \leq N.$$

II. ϕ has a bounded second derivative.

III. The regression constants satisfy

$$\max_{1 \leq i \leq N} (c_N(i) - \bar{c}_N)^2 \Big/ \sum_{i=1}^{N} (c_N(i) - \bar{c}_N)^2 = O(N^{-1} \log N). \tag{7.8}$$

IV. $\liminf_{N \to \infty} V(S_N) > 0$.

Then the following theorem is true (see Serfling, 1980).

THEOREM 14. *Assume* I–IV. *Then, for every* $\zeta > 0$,

$$\sup_x \left| P\left(\frac{S_N - ES_N}{V^{1/2}(S_N)} \leq x \right) - \Phi(x) \right| = O(N^{-1/2+\zeta}). \tag{7.9}$$

The assertion (7.9) *remains true if* $V(S_N)$ *is replaced by* σ_N^2, *where* σ_N^2 *is defined in Theorem* 12. *Both the assertions remain true with* $E(S_N)$ *replaced by* μ_N, *where* μ_N *is defined in Theorem* 13.

REMARK. Edgeworth expansions for linear rank statistics have been obtained by Albers, Bickel and van Zwet (1976), and Bickel and van Zwet (1978).

8. Linear rank statistics: Strong laws

Sen (1970) obtained strong law of large numbers for statistics of the form S_N^+ (introduced in (1.2)) when the c_i's are all equal to 1 and $a_N(i) = \phi(i/(N+1))$, $1 \leq i \leq N$. The ideas of Sen can be extended to more general regression rank statistics of the type S_N^+ provided the c_i's satisfy certain uniform asymptotic negligibility conditions. Sen and Ghosh (1972) have a result to this effect.

To state the result of Sen and Ghosh (1972), let

$$H_i(x) = F_i(x) - F_i(-x) \quad \text{and} \quad \bar{H}_N(x) = N^{-1} \sum_{i=1}^{N} H_i(x).$$

Define

$$C_{n0}^2 = \sum_{i=1}^{n} c^2(i) \quad \text{and} \quad A_{n0}^2 = n^{-1} \sum_{i=1}^{n} \{\phi^*(i/(n+1))\}^2,$$

where $\phi^*(u) = \phi(\tfrac{1}{2} + 1/2u)$. It is also assumed that $\phi(u) + \phi(1-u) = 0$ for all $0 < u < 1$. Define

$$\tilde{S}_N = N^{-1/2} A_{N0}^{-1} \sum_{i=1}^{N} \tilde{c}_{Ni} \operatorname{sgn} X_i \phi^*(R_{Ni}^+/(N+1)),$$

where $\tilde{c}_{Ni} = c_i/(\Sigma_{i=1}^N c_i^2)^{1/2}$, $1 \leq i \leq N$. Also let,

$$\tilde{\tau}_N = N^{-1/2} A_{N0}^{-1} \sum_{i=1}^{N} \tilde{c}_{ni} \int_{-\infty}^{\infty} \operatorname{sgn}(x) \phi^*(\bar{H}_N(|x|)) \, \mathrm{d}F_i(x),$$

where $\bar{H}_N(|x|) = N^{-1} \Sigma_{i=1}^N H_i(|x|)$. Then, we have the following theorem.

THEOREM 15. *Suppose that* $\int_0^1 |\phi(u)|^r \, \mathrm{d}u < \infty$ *for some* $r > 2$. *Assume that* $\max_{i \leq i \leq N} |\tilde{c}_{Ni}| = O(N^{-1/2})$. *Then,* $\tilde{S}_N - \tilde{\tau}_N \to 0$ *a.s. as* $N \to \infty$.

For the statistics S_N, a corresponding strong convergence result as proved in Sen and Ghosh (1972) is as follows.

THEOREM 16. *Suppose that the score function* $a_N(i)$ *is the same as in Theorem* 15 *with* $\int_0^\infty |\phi(u)|^r \, \mathrm{d}u < \infty$ *for some* $r > 2$. *Assume that* $\max_{1 \leq i \leq N} |c_{Ni}^*| = O(N^{-1/2})$, *where* $c_{Ni}^* = (c_i - \bar{c}_N)/\{\Sigma_{i=1}^N (c_i - \bar{c}_N)^2\}^{1/2}$. *Define*

$$S_N^* = N^{-1/2} A_N^{-1} C_N^{-1} S_N \quad \textit{and} \quad \tau_N^* = \int_{-\infty}^{\infty} J(\bar{F}_N(x)) \, \mathrm{d}F_N^*(x),$$

where $F_N^*(x) = N^{-1/2} A_N^{-1} \Sigma_{i=1}^N c_{Ni}^* F_i(x)$ *and* $\bar{F}_N(x) = N^{-1} \Sigma_{i=1}^N F_i(x)$. *Then,* $S_N^* - \tau_N^* \to 0$ *a.s. as* $N \to \infty$.

REMARK. Hájek (1974) proved a result similar to Theorem 16.

9. Linear rank statistics: Functional central limit theorems

Consider once again the statistics $S_N = \Sigma_{i=1}^N c(i)E[\phi(U_{NR_{Ni}})]$ where $U_{N1} \leq \cdots \leq U_{NN}$ are the order statistics in a random sample of size N from the uniform (0, 1) distribution. Assume that the score function ϕ is square integrable, and define $\bar{\phi} = \int_0^1 \phi(u)\,du$ and $A^2 = \int_0^1 \phi^2(u)\,du - \bar{\phi}^2$. Defining $A_N^2 = \bar{N}^1 \Sigma_{i=1}^N [a_N(i) - \bar{a}_N]^2$, where $a_N(i) = E\phi(U_{Ni})$, $1 \leq i \leq N$, one gets the inequality $A_N^2 \leq A^2$. In fact (see Sen, 1981, pp. 92–93), using the dominated convergence theorem one can show that $A_N^2 \to A^2$ as $N \to \infty$. Write $C_N^2 = \Sigma_{i=1}^N (c_N(i) - \bar{c}_N)^2$. For every $N \geq 1$, consider the stochastic process

$$Y_N(t) = S_{\tau_0(t)}/(C_N A_N), \quad \tau_0(t) = \max\{k: C_k^2 \leq tC_N^2\}, \qquad t \in (0.1). \tag{9.1}$$

(Conventionally, let $S_0 = S_1 = 0$ so that $Y_N(0) = 0$). Then Y_N belongs to $D[0, 1]$ for every $N \geq 1$. The following theorem, proved in Sen (1981) is an improved version of a corresponding result of Sen and Ghosh (1972).

THEOREM 17. *Suppose $\int_0^1 \phi^2(u)\,du < \infty$. Then, under $F_1 \equiv \cdots \equiv F_N$ and under (7.3),*

$$Y_N \xrightarrow{\text{d}} W \quad \text{as } N \to \infty$$

in the J_1-topology on $D[0, 1]$.

The sequence $\{Y_N(\cdot), N \geq 1\}$ describes a process pertaining to the *past.* Similarly, one can define a process pertaining to the *future.* With this end, define $S_k^* = S_k/C_k^2$, $k \geq N \geq 2$; $S_1^* = 0$. Now, define the stochastic process

$$Z_N(t) = (C_N/A_N)S^*_{\tau_0(t)}, \quad \tau_0(t) = \min\{k: C_N^2/C_k^2 \leq t\}, \qquad t \in (0, 1). \tag{9.2}$$

Then Z_N belongs to the $D[0, 1]$ space for every N. The following theorem is proved in Sen (1981).

THEOREM 18. *Under the conditions of Theorem 17, as $N \to \infty$, $Z_N \to W$ in the J_1-topology on $D[0, 1]$.*

Theorems 17 and 18 relate to the null situation, namely, when $F_1 \equiv \cdots \equiv F_N$. For Contiguous alternatives, functional central limit theorems of the above type were obtained by Sen and are reported very extensively in his recent book (see Sen, 1981, pp. 98–104). Functional central limit theorems for linear rank statistics under nonlocal alternatives is an important open question.

Next, consider the one sample case where $Z_1, \ldots, Z_N$ are iid with a common continuous distribution function F. Consider signed rank statistics of the form (1.2) with $c(i) = 1$ for all i, $a_N(i) = E[\phi(U_{Ni})]$, $1 \leq i \leq N$. Consider the hypothesis of sign invariance, namely, F is symmetric about zero, i.e. $F(x) + F(-x) = 1$ for all $x > 0$. Let $R_{Ni}^+ =$ rank of $|X_1|, \ldots, |X_N|$, $1 \leq i \leq N$. Let $\mathcal{B}_N$ denote the σ-algebra generated by $(\operatorname{sgn} X_1, \ldots, \operatorname{sgn} X_N)$ and $(R_{N1}^+, \ldots, R_{NN}^+)$. Sen and

Ghosh (1971) have shown that if $\int_0^1 |\phi(u)|\,du < \infty$, then $\{(S_N^+, \mathscr{B}_N), N \geq 1\}$ is a martingale sequence. Using this martingale structure, weak and strong invariance principles (functional central limit theorems) for one sample signed rank statistics can be established. The results are discussed in detail in Sen (1981). In what follows, we present a few of the important results.

Once again, let $A_N^2 = N^{-1}\sum_{i=1}^N [a_N(i) - \bar{a}_N]^2$. Define $\tau_0(t) = \max\{k: k/N \leq t\}$, $0 < t < 1$. Let $Y_N^+(t) = S_{N(t)}^+/(N^{1/2}A_N)$, $Y_N^+(0) = 0$. The following theorem is proved in Sen and Ghosh (1973).

THEOREM 19. *If $\int_0^1 \phi^2(u)\,du < \infty$ and $F(x) + F(-x) = 1$ for all x, then $Y_N \to W$ as $N \to \infty$ in the J_1-topology on $D[0, 1]$.*

An analogous result for the tail sequence $Z_N = \{Z_N(t), 0 < t < 1\}$ is proved in Sen (1981). Define $\tau_0(t) \equiv \min\{k: N/k \leq t\}$, and $Z_N(t) = \sqrt{N} S_{\tau_0(\varepsilon)}^+/(A_N \tau_0(t))$. Then the following theorem is true.

THEOREM 20. *Under the same assumptions as in Theorem 19, $Z_N \xrightarrow{d} W$ as $N \to \infty$ in the J_1-topology on $D[0, 1]$.*

The above theorem provides a functional central limit theorem for signed rank statistics under the null situation. Results similar to Theorems 19 and 20 are proved in Sen (1981) under contiguous alternatives.

10. Multivariate linear rank statistics: Permutational central limit theorems

Let $\boldsymbol{X}_i = (X_{i1}, \ldots, X_{ip})'$, $i = 1, \ldots, n$, be n iid random variables with a common continuous distribution function F defined on $\mathbb{R}^p$, p being some positive integer. Let $\boldsymbol{c}_i = (c_{i1}, \ldots, c_{iq})'$, $i = 1, \ldots, n$, $q \geq 1$. For each j $(= 1, \ldots, p)$, let R_{ij} denote the rank of X_{ij} among $X_{1j}, \ldots, X_{nj}$ $(i = 1, \ldots, n)$; F being assumed to be continuous, ties among the observations can be neglected in probability. Also, let $a_{nj}(1), \ldots, a_{nj}(n)$ $(j = 1, \ldots, p)$ be a set of scores. Then, a set of multivariate linear rank statistics may be defined by

$$L_{njk} = \sum_{i=1}^{n} (c_{ik} - \bar{c}_{nk}^{(0)}) a_{nj}(R_{ij}), \tag{10.1}$$

where $\bar{c}_{nk}^{(0)} = n^{-1}\sum_{i=1}^n c_{ik}$.

The statistics L_{njk} defined in (10.1) appear in Puri and Sen (1969) in connection with hypothesis testing in general linear model. Special cases have been considered earlier by Chatterjee and Sen (1964) for the bivariate two sample location problem when $p = 2$, $q = 1$, and the c_{i1}'s are 1's or zeroes. Puri and Sen (1966) considered the multivariate multisample location problem, where p and q are general, but c_{ik}'s are still 1's or zeroes.

It should be noted that unconditionally $\boldsymbol{L}_n = ((L_{njk}))$ is not distribution-free.

However, it is permutationally (conditionally) distribution-free under the following rank permutation model due to Chatterjee and Sen (1964). Consider the *rank-collection matrix* $\mathcal{R}_n$ (of order $p \times n$) defined by

$$\mathcal{R}_n = (\boldsymbol{R}_1, \ldots, \boldsymbol{R}_n) = \begin{pmatrix} R_{11} \cdots R_{n1} \\ R_{1p} \cdots R_{np} \end{pmatrix}, \tag{10.2}$$

where $\boldsymbol{R}_i' = (R_{i1}, \ldots, R_{ip})$, $i = 1, \ldots, n$. Consider now a permutation of the columns of $\mathcal{R}_n$ such that the top row is in the natural order (i.e. $1, \ldots, n$), and denote the resulting matrix (called *reduced rank-collection matrix*) by $\mathcal{R}_n^*$. The totality of $(n!)^p$ rank-collection matrices may thus be partitioned into $(n!)^{p-1}$ subsets, where each subset corresponds to a particular reduced rank-collection matrix $\mathcal{R}_n^*$, and the subset $S(\mathcal{R}_n^*)$ has cardinality $n!$. The conditional distribution of $\mathcal{R}_n$ over the $S(\mathcal{R}_n^*)$ is uniform irrespective of what F is. This conditional (permutational) probability measure is denoted by $\mathcal{P}_n$. Puri and Sen (1969) have shown that

$$E_{P_n}(\boldsymbol{L}_n) = \boldsymbol{0} \quad \text{and} \quad V_{P_n}(\boldsymbol{L}_n) = \boldsymbol{V}_n \times \boldsymbol{C}_n, \tag{10.3}$$

where $\boldsymbol{V}_n = ((v_{njj'}))$ has the elements

$$v_{njj'} = (n-1)^{-1} \sum_{i=1}^{n} (a_n(R_{ij}) - \bar{a}_{nj})(a_n(R_{ij'}) - \bar{a}_{nj'}); \tag{10.4}$$

$$\bar{a}_{nj} = n^{-1} \sum_{i=1}^{n} a_{nj}(i), \quad i = 1, \ldots, p. \tag{10.5}$$

Also

$$\boldsymbol{C}_n = \sum_{i=1}^{n} (\boldsymbol{c}_i - \bar{\boldsymbol{c}}_n)(\boldsymbol{c}_i - \bar{\boldsymbol{c}}_n)', \qquad \bar{\boldsymbol{c}}_n = (\bar{c}_{n1}^{(0)} \cdots \bar{c}_{np}^{(0)})'. \tag{10.6}$$

Recently, Sen (1983) has established the asymptotic multinormality of $\boldsymbol{L}_n$ under the permutational model $\mathcal{P}_n$ assuming minimal regularity conditions. We call this result the multivariate permutational central limit theorem. Earlier Puri and Sen (1969) established the permutational central limit theorem under more stringent regularity conditions.

To state Sen's (1983) result, assume that $\boldsymbol{C}_n$ has full rank (i.e. q) for large n (say $n \geq n_0$). Define

$$\xi_n = \max_{1 \leq i \leq n} (\boldsymbol{c}_i - \bar{\boldsymbol{c}}_n)' \boldsymbol{C}_n^{-1} (\boldsymbol{c}_i - \bar{\boldsymbol{c}}_n)', \tag{10.7}$$

for $n \geq n_0$. Also, for each j $(=1, \ldots, p)$, let,

$$a_{nj}^0(u) = a_{nj}(i), \qquad (i-1)/n < u \leq i/n, \quad i = 1, \ldots, n. \tag{10.8}$$

The following assumptions are made.

I. There exist score functions $\phi_j(u)$ $(0<u<1)$, $j=1,\ldots,p$ such that $\phi_j(u)=\phi_{j1}(u)-\phi_{j2}(u)$, where $\phi_{jk}(u)$ is nondecreasing, absolutely continuous and square integrable inside $(0, 1)$, $k = 1, 2$.

II. For the $a^0_{nj}(u)$ defined in (10.8), $\max_{1\le j\le p}\int_0^1\{a^0_{nj}(u)-\phi_j(u)\}^2\,du\to 0$ as $n\to\infty$.

III. $\xi_n \equiv \max_{1\le i\le n}(\boldsymbol{c}_i-\bar{\boldsymbol{c}}_n)'\boldsymbol{C}_n^{-1}(\boldsymbol{c}_i-\bar{\boldsymbol{c}}_n)\to 0$ as $n\to\infty$.

The main theorem of Sen (1983) is as follows.

THEOREM 21. *Assume* (I)–(III). *Then, under* $\mathcal{P}_n$, $\boldsymbol{L}_n$ *is asymptotically* (*in probability*) *normal with mean vector* $\boldsymbol{0}$, *and variance–covariance matrix* $\boldsymbol{V}_n\otimes\boldsymbol{C}_n$.

REMARK. Condition (III) can be viewed as an extended Noether condition (see (7.3)).

11. Bilinear rank statistics: The independence problem

Let $(X_1, Y_1),\ldots,(X_n, Y_n)$ be a random sample from a continuous bivariate distribution with distribution function $F(x, y)$. The problem is to construct nonparametric distribution-free tests of H_0: $F(x, y)=F_1(x)F_2(y)$, for every pair (x, y) against H_1: $F(x, y)\neq F_1(x)F_2(y)$ for at least one pair (x, y), where F_1 and F_2 denote the marginal distribution functions of the X_i's and the Y_j's respectively. Thus, the null hypothesis of interest is that the X and Y variables are independent.

Let R_i denote the rank of X_i among $X_1,\ldots,X_n$, and let Q_i denote the rank of Y_i among $Y_1,\ldots,Y_n$. In addition, let $a(1)\le\cdots\le a(n)$ with $a(1)\neq a(n)$, and $c(1)\le\cdots\le c(n)$ with $c(1)\neq c(n)$ be two sets of scores. Let

$$T_n=\sum_{i=1}^n(c(R_i)-\bar{c})(a(Q_i)-\bar{a})\Big/\left\{\sum_{i=1}^n(c(R_i)-\bar{c})^2\right\}^{1/2}\left\{\sum_{i=1}^n(a(Q_i)-\bar{a})^2\right\}^{1/2}, \tag{11.1}$$

with $\bar{c}=n^{-1}\sum_{i=1}^n c_i$ and $\bar{a}=n^{-1}\sum_{i=1}^n a_i$ denote the correlation coefficient for the group of scored rank pairs $(c(R_1), a(Q_1)),\ldots,(c(R_n), a(Q_n))$. The special case $a(i)=i$, $i=1,\ldots,n$ corresponds to the Spearman rank correlation coefficient.

Note that T_n has the alternate representation

$$T_n=\left[\sum_{i=1}^n c(R_i)a(Q_i)-n\bar{a}\bar{c}\right]\Big/\left\{\sum_{i=1}^n(c_i-\bar{c})^2\right\}^{1/2}\left\{\sum_{i=1}^n(a_i-\bar{a})^2\right\}^{1/2}. \tag{11.2}$$

Write $S_n=\sum_{i=1}^n c(R_i)a(Q_i)$. When H_0 is true, the rank vectors $\boldsymbol{R}=(R_1,\ldots,R_n)'$ and $\boldsymbol{Q}=(Q_1,\ldots,Q_n)'$ are both uniformly distributed over the set of $n!$ possible permutations of the first n positive integers. Moreover, under

H_0, $\boldsymbol{Q}$ and $\boldsymbol{R}$ are independently distributed. In this case $E_{H_0}(S_n) = n\bar{a}\bar{c}$ and

$$V_{H_0}(S_n) = (n-1)^{-1} \sum_{i=1}^{n} (c(i) - \bar{c})^2 \sum_{i=1}^{n} (a(i) - \bar{a})^2 .$$

Also, Hoeffding (1952) showed that if

$$\left\{ \max_{1 \leq i \leq n} (c(i) - \bar{c})^2 \Big/ \sum_{i=1}^{n} (c(i) - \bar{c})^2 \right\} \left\{ \max_{1 \leq i \leq n} (a(i) - \bar{a})^2 \Big/ \sum_{i=1}^{n} (a(i) - \bar{a})^2 \right\} = \mathrm{O}(n^{-1}), \tag{11.3}$$

then, under H_0, $\sqrt{n}\, T_n$ is asymptotically $N(0, 1)$. The asymptotic nonnull (when H_0 is not true) distribution of T_n has been derived by Bhuchongkul (1964). A class of rank order tests for independence in multivariate distributions was proposed by Puri, Sen and Gokhale (1970), and the asymptotic distributions of the relevant test statistics in the null and nonnull cases were derived.

References

Albers, W., Bickel, P. J. and van Zwet, W. R. (1976). Asymptotic expansions for the power of distribution free tests in the one-sample problem. *Ann. Statist.* **4**, 108–156.

Arvesen, J. N. (1969). Jackknifing U-statistics. *Ann. Math. Statist.* **40**, 2076–2100.

Athreya, K. B., Ghosh, M., Low, L. and Sen, P. K. (1984). Laws of large numbers for bootstrapped means and U-statistics. *J. Stat. Planning Inference* **9**, 185–194.

Berk, R. H. (1966). Limiting behavior of posterior distributions when the model is incorrect. *Ann. Math. Statist.* **37**, 51–58.

Bergström, H. and Puri, M. L. (1977). Convergence and remainder terms in linear rank statistics. *Ann. Statist.* **5**, 671–680.

Bickel, P. J. (1974). Edgeworth expansions in nonparametric statistics. *Ann. Statist.* **2**, 1–20.

Bickel, P. J. and Freedman, D. A. (1981). Some asymptotic theory for the bootstrap. *Ann. Statist.* **9**, 1196–1217.

Bickel, P. J. and van Zwet, W. R. (1978). Asymptotic expansions for the power of distribution free tests in the two sample problem. *Ann. Statist.* **6**, 937–1004.

Billingsley, P. (1968). *Convergence of Probability Measures.* Wiley, New York.

Bhuchongkul, S. (1964). A class of nonparametric tests for independence in bivariate populations. *Ann. Math. Statist.* **35**, 138–149.

Borovskikh, Yu. V. (1979). Approximation of U-statistics distribution (in Ukrainian). *Proc. Ukrainian Acad. Sci. A* **9**, 695–698.

Callaert, H. and Janssen, P. (1978). The Berry–Esseen theorem for U-statistics. *Ann. Statist.* **6**, 417–421.

Callaert, H., Janssen, P. and Veraverbeke, N. (1980). An Edgeworth expansion for U-statistics. *Ann. Statist.* **8**, 299–312.

Callaert, H. and Veraverbeke, N. (1981). The order of the normal approximation for a studentized U-statistic. *Ann. Statist.* **9**, 194–200.

Chan, Y. K. and Wierman, J. (1977). On the Berry–Esseen Theorem for U-statistics. *Ann. Prob.* **5**, 136–139.

Chatterjee, S. K. and Sen, P. K. (1964). Nonparametric tests for bivariate two sample location problem. *Cal. Statist. Assoc. Bull.* **13**, 18–58.

Chernoff, H. and Savage, I. R. (1958). Asymptotic normality and efficiency of certain nonparametric test statistics. *Ann. Math. Statist.* **29**, 972–994.

Dwass, M. (1956). The large sample power of rank tests in two sample problems. *Ann. Math. Statist.* **27**, 352–374.

Efron, B. (1979). Bootstrap methods: another look at the jackknife. *Ann. Statist.* **7**, 1–26.

Efron, B. and Stein, C. (1981) The jackknife estimate of variance. *Ann. Statist.* **9**, 586–596.

Ghosh, M. and Dasgupta, R. (1982). Berry–Esseen theorems for U-statistics in the non iid case. In: *Proc. of the Conf. on Nonparametric Inference*, organized by the Janos Bolyai Math. Soc., held in Budapest, Hungary, pp. 293–313.

Ghosh, M. and Sen, P. K. (1970). On the almost sure convergence of von Mises' differentiable statistical functions. *Cal. Statist. Assoc. Bull.* **19**, 41–44.

Grams, W. F. and Serfling, R. J. (1973). Convergence rates for U-statistics and related statistics. *Ann. Statist.* **1**, 153–160.

Gregory, G. G. (1977). Large sample theory for U-statistics and tests of fit. *Ann. Statist.* **5**, 110–123.

Hájek, J. (1961). Some extensions of the Wald–Wolfowitz–Noether Theorem. *Ann. Math. Statist.* **32**, 506–523.

Hájek, J. (1962). Asymptotically most powerful rank tests. *Ann. Math. Statist.* **33**, 1124–1147.

Hájek, J. (1974). Asymptotic sufficiency of the vector of ranks in the Bahadur sense. *Ann. Statist.* **2**, 75–83.

Hájek, J. (1968). Asymptotic normality of simple linear rank statistics under alternatives. *Ann. Math. Statist.* **39**, 325–346.

Hájek, J. and Sidak, Z. (1967). *Theory of Rank Tests.* Academic Press, New York.

Hall, P. (1979). On the invariance principle for U-statistics. *Stoch. Proc. and Appl.* **9**, 163–174.

Hoeffding, W. (1948). A class of statistics with asymptotically normal distribution. *Ann. Math. Statist.* **19**, 293–325.

Hoeffding, W. (1951). A combinatorial central limit theorem. *Ann. Math. Statist.* **22**, 558–566.

Hoeffding, W. (1952). The large sample power of tests based on permutations of observations. *Ann. Math. Statist.* **23**, 169–192.

Hoeffding, W. (1961). The strong law of large numbers for U-statistics. Inst. Stat. Univer. North Carolina Mimeo Series No. 302.

Hoeffding, W. (1973). On the centering of a simple linear rank statistic. *Ann. Statist.* **1**, 54–66.

Husková, M. (1970). Asymptotic distribution of simple linear rank statistic for testing symmetry. *Z. Wahrsch. Verw. Geb.* **12**, 308–322.

Husková, M. (1977). The rate of convergence of simple linear rank statistics under hypothesis and alternatives. *Ann. Statist.* **5**, 658–670.

Janssen, P. (1978). De Berry–Esseen stellingen: Een asymptotische ontwikkeling voor U-statistieken. Unpublished Ph.D. dissertation, Belgium.

Jurecková, J. and Puri, M. L. (1975). Order of normal approximation for rank test statistic distribution. *Ann. Probab.* **3**, 526–533.

Karlin, S. and Rinott, Y. (1982). Applications of ANOVA type decompositions for comparisons of conditional variance statistics including jackknife estimates. *Ann. Statist.* **10**, 485–501.

Lehmann, E. L. (1951). Consistency and unbiasedness of certain nonparametric tests. *Ann. Statist.* **22**, 165–179.

Lehmann, E. L. (1963). Robust estimation in analysis of variance. *Ann. Math. Statist.* **34**, 957–966.

Loynes, R. M. (1970). An invariance principle for reversed martingales. *Proc. Amer. Math. Soc.* **25**, 56–64.

Loynes, R. M. (1978). On the weak convergence of U-statistics processes and of the empirical processes. *Proc. Camb. Phil. Soc.* (*Math.*) **83**, 269–272.

Mann, H. B. and Whitney, D. R. (1947). On a test of whether one of two random variables is stochastically larger than the other. *Ann. Math. Statist.* **18**, 50–60.

McLeish, D. (1974). Dependent central limit theorems and invariance principles. *Ann. Probab.* **2**, 620–628.

Miller, R. G., Jr. and Sen, P. K. (1972). Weak convergence of U-statistics and von Mises' differentiable statistical functions. *Ann. Math. Statist.* **43**, 31–41.

Nandi, H. K. and Sen, P. K. (1963). On the properties of U-statistics when the observations are not independent. Part II: Unbiased estimation of the parameters of a finite population. *Cal. Statist. Assoc. Bull.* **12**, 125–143.

Neuhaus, G. (1977). Functional limit theorems for U-statistics in the degenerate case. *J. Mult. Anal.* **7**, 424–439.

Noether, G. E. (1949). On a theorem of Wald and Wolfowitz. *Ann. Math. Statist.* **20**, 455–458.

Puri, M. L. and Sen, P. K. (1966). On a class of multivariate multisample rank order tests. *Sankhyā Ser. A* **28**, 353–376.

Puri, M. L. and Sen, P. K. (1969). A class of rank order tests for a general linear hypothesis. *Ann. Math. Statist.* **40**, 1325–1343.

Puri, M. L. and Sen, P. K. (1971). *Nonparametric Methods in Multivariate Analysis.* Wiley, New York.

Puri, M. L., Sen, P. K. and Gokhale, D. V. (1970). On a class of rank order tests for independence in multivariate distributions. *Sankhyā Ser. A* **32**, 271–297.

Pyke, R. and Shorack, G. R. (1968). Weak convergence of a two-sample empirical process, and a new approach to Chernoff–Savage theorems. *Ann. Math. Statist.* **39**, 755–771.

Quenouille, M. H. (1956). Note on bias in estimation. *Biometrika* **43**, 353–360.

Randles, R. H. and Wolfe, D. A. (1979). *Introduction to the Theory of Nonparametric Statistics.* Wiley, New York.

Sen, P. K. (1960). On some convergence properties of U-statistics. *Cal. Statist. Assoc. Bull.* **10**, 1–18.

Sen, P. K. (1963). On the properties of U-statistics when the observations are not independent. Part I: Estimation of the non-serial parameters of a stationary properties. *Cal. Statist. Assoc. Bull.* **12**, 69–92.

Sen, P. K. (1965). Some nonparametric tests for m-dependent time series. *J. Amer. Statist. Assoc.* **60**, 134–147.

Sen, P. K. (1970). On some convergence properties of one-sample rank order statistics. *Ann. Math. Statist.* **41**, 2206–2209.

Sen, P. K. (1974). Weak convergence of generalized U-statistics. *Ann. Probab.* **2**, 90–102.

Sen, P. K. (1977a). Some invariance principles relating to jackknifing, and their role in sequential analysis. *Ann. Statist.* **5**, 315–329.

Sen, P. K. (1977b). Almost sure convergence of generalized U-statistics. *Ann. Prob.* **5**, 287–290.

Sen, P. K. (1981). *Sequential Nonparametrics: Invariance Principles and Statistical Inference.* Wiley, New York.

Sen, P. K. (1983). On permutational central limit theorems for general multivariate linear rank statistics. *Sankhyā Ser. A* **45**, 141–149.

Sen, P. K. and Bhattacharyya, B. (1977). Weak convergence of the Rao–Blackwell estimator of a distribution function. *Ann. Probab.* **5**, 500–510.

Sen, P. K. and Ghosh, M. (1971). On bounded length sequential confidence intervals based on one-sample rank order statistics. *Ann. Math. Statist.* **42**, 189–203.

Sen, P. K. and Ghosh, M. (1972). On strong convergence of regression rank statistics. *Sankhyā Ser. A* **34**, 335–348.

Sen, P. K. and Ghosh, M. (1973). A law of iterated logarithm for one-sample rank order statistics and some applications. *Ann. Statist.* **1**, 568–576.

Serfling, R. J. (1980). *Approximation Theorems of Mathematical Statistics.* Wiley, New York.

Sproule, R. N. (1969). A sequential fixed width confidence interval for the mean of a U-statistic. Unpublished Ph.D. dissertation. UNC, Chapel Hill.

Sproule, R. N. (1974). Asymptotic properties of U-statistics. *Trans. Amer. Math. Soc.* **199**, 55–64.

Tukey, J. W. (1949). The simplest signed rank tests. *Princeton Univ. Stat. Res. Group Memo. Report No. 17.*

Tukey, J. W. (1957). Variances of variance components II: The unbalanced single classification. *Ann. Math. Statist.* **28**, 43–56.

Wald, A. and Wolfowitz, J. (1944). Statistical tests based on the permutations of the observations. *Ann. Math. Statist.* **15**, 358–372.

Wilcoxon, F. (1945). Individual comparisons by ranking methods. *Biometrics* **1**, 80–83.

P. R. Krishnaiah and P.K. Sen, eds., *Handbook of Statistics, Vol. 4*
© Elsevier Science Publishers (1984) 173–184

9

Asymptotic Comparison of Tests – A Review*

Kesar Singh

1. Local and nonlocal efficiencies

Asymptotic measures of relative efficiencies can be broadly classified in the above two categories. A measure of performance that requires the alternative to approach the null is a local efficiency and a measure that lets the alternative stay fixed as $n \to \infty$ is a nonlocal efficiency. It seems fair to state that the most popular local efficiency is Pitman efficiency and the most popular nonlocal efficiency is Bahadur efficiency (ratio of exact Bahadur slopes). The concept of Pitman efficiency is based on Neyman–Pearsonian view point, i.e. fixing α and making comparison in terms of powers. It turns out that for most of the reasonable tests the limiting power is 1 for any fixed alternative and α. So, in order to make a comparison of limiting powers possible one needs some way to prevent the power from approaching to 1. It was Pitman's idea (see Pitman, 1949) to allow the alternative to converge to null at such a pace that the power converges to something <1 and then compare the limiting powers. The measure is now popularly known as Pitman efficiency (P.E.). In the next section we look at some of the newer developments on P.E. where we give an up to date definition of it. Another approach to local comparison, which generally gives the same answer as P.E., is to compare limiting derivatives of power functions at the null.

The concept of Bahadur efficiency is entirely based upon a single concept. A test statistic T_1 is superior to a test statistic T_2 at a nonnull θ if the attained level of T_1 (defined later in this article) is smaller than that of T_2 under θ. Typically, the attained level dwindles exponentially fast and the decay rate is taken as a limiting measure of its smallness. There are some variations of Bahadur efficiency which will be talked about later. Bahadur (1971) is an excellent source for efficiencies based on level attained. Cochran (1952) suggested that one might compare two test statistics by looking at the rates at which their α's go down to zero when their powers are held fixed against the same *fixed* alternative. But it turns out that Cochran's and Bahadur's ap-

*Research supported by NSF Grant MCS-81-02341.

proaches are one and the same. A third approach to a nonlocal comparison, which is almost always much harder to work with, is again based on Neyman–Pearsonian view point (see Hodges and Lehmann (1956) for example). Here one keeps the alternative θ fixed and compares the convergence rates of powers to 1. Yet another notion of nonlocal efficiency is due to Chernoff (1952) which looks at the exponential decay rate of the minimum value of a linear combination of the two types of errors.

One of the common criticisms of P.E. is that to define it one has to work with a specific kind of sequence of alternatives converging to the null. Indeed, the requirement that the limiting power is not 1 dictates the choice of alternative and this is a pure mathematical fact. However, as we will see in the next section the current definitions of P.E. do not make use of any specific alternative sequence. A significant advantage in working with P.E. is that almost always it is computationally more elementary as one usually requires some weak limit to compute it. It seems worthwhile to make a comment here on nonlocal comparisons. Typically, the performance of a test when the alternative is far away from the null approaches to the highest limit very rapidly as the sample size gets big. As a result, a large sample comparison of two tests for a distant alternative seems like a comparison of two practically perfect objects. However, this comment does not discredit the nonlocal efficiencies very much since a nonlocal efficiency is defined for every point in the nonnull space including the ones which are close to the null. It probably does suggest that while using a nonlocal efficiency one should attach more importance to the alternatives which are closer to the null.

2. Pitman efficiency

We present in this section a modern definition of P.E. and some very general tools which can help one investigate its existence and evaluate it in a given instance. The development is mainly due to Rothe (1981) though the definition in the present form appears in Wiend (1976) too.

Assume that $X_1, X_2, \ldots$ are i.i.d. observations from a probability space (Ω, S, P_θ) where θ is an unknown parameter taking values in a metric space Λ. Consider the testing problem $\theta = \theta_0$ vs. $\theta \neq \theta_0$ where θ_0 is an accumulation point in the metric space. Let $\{T_{1n}\}$ and $\{T_{2n}\}$ be two competing test statistics for this problem. Let us fix the sizes of all the tests at a level $\alpha \in (0, 1)$. We define, for a $\theta \neq \theta_0$ and $\beta \in (0, 1)$,

$$N_1(\theta, \alpha, \beta) = \min\{m : \text{Power of } T_{1n} \text{ at } \theta \text{ is} \geq \beta \text{ for all } n \geq m\}.$$

In other words $N_1(\theta, \alpha, \beta)$ is the first sample size to guarantee a power β or more for all sample sizes bigger than or equal to it. Similarly, let us define $N_2(\theta, \alpha, \beta)$.

DEFINITION. P.E. of $\{T_{1n}\}$ relative to $\{T_{2n}\}$ is given as

$$e_{12}(\alpha,\beta)=\lim_{\theta\to\theta_0}\frac{N_2(\theta,\alpha,\beta)}{N_1(\theta,\alpha,\beta)}$$

if the limit exists.

If the limit does not exist, one can look at lower and upper efficiencies given as follows:

$$e_{12}^{+(-)}(\alpha,\beta)=\limsup_{\theta\to\theta_0}\ (\liminf_{\theta\to\theta_0})\frac{N_2(\theta,\alpha,\beta)}{N_1(\theta,\alpha,\beta)}.$$

Rothe (1981) has considered more general lower and upper efficiencies. Given below is a set of conditions due to Rothe which imply the existence of P.E. and yield an expression for it too.

Let us say that a sequence T_n is standard if the following hold. There are functions $g:\Lambda\to[0,\infty)$ and $H:[0,\infty]\to[\alpha,1]$ such that

(i) g is continuous and $g(\theta)=0$ if and only if $\theta=\theta_0$.

(ii) H is increasing and one to one.

(iii) For θ_n satisfying $g(\theta_n)n\to\eta\geqslant 0$ as $n\to\infty$, $\beta(\theta_n,T_n)\to H(\eta)$ where $\beta(\theta,T_n)$ is the power of T_n at the alternative θ.

THEOREM (Rothe). *Suppose that T_{1n} and T_{2n} are both standard. If $\lim_{\theta\to\theta_0} g_1(\theta)/g_2(\theta)$ exists, then $e_{12}(\alpha,\beta)$ exists and is given by*

$$\frac{H_2^{-1}(\beta)}{H_1^{-1}(\beta)}\lim_{\theta\to\theta_0}\frac{g_1(\theta)}{g_2(\theta)}.$$

More generally,

$$e_{12}^{+(-)}(\alpha,\beta)=\frac{H_2^{-1}(\beta)}{H_1^{-1}(\beta)}\limsup_{\theta\to\theta_0}\ (\liminf_{n\to\infty})\frac{g_1(\theta)}{g_2(\theta)}.$$

(g_i, H_i are the functions g and H for T_{in}, $i=1,2$).

For certain statistics, with a normal or a χ^2 (central or noncentral) distribution as the weak limit, Theorem 4 of Rothe (1981) provides an explicit form of the function H.

There are two points regarding the above P.E. which deserve mention: (1) $e(\alpha,\beta)$ may depend upon α and β. A natural question is: What kind of (α,β) should be regarded more relevant? (2) The set up H_0: $\theta=\theta_0$ is certainly too restrictive. About the first point, there does not seem to exist an absolutely conclusive answer. However, this author feels that (α,β) closer to $(0,1)$ should be regarded more relevant, since these are the values of (α,β) which require higher sample sizes for moderately close θ's and thus are more suitable for a large sample efficiency. Also, one may reasonably argue that an asymptotic efficiency shouldn't be based on a mediocre performance. Regarding the second point, it seems easy to extend the definition in the case of a nonsimple null by defining appropriate lower and upper efficiencies. But the question remains: When do they coincide? Apparently there hasn't been a satisfactory study on P.E. in nonnull situations.

Some of the earlier contributions on P.E. often referred to are Pitman (1949), Noether (1955), Fraser (1957) and Olsen (1967). The reader may see Noether (1955) for a generalized version of the original definition due to Pitman.

There are numerous interesting results on various computations and bounds on P.E. For instance, Hodges and Lehman (1956) showed, among other things, that the P.E. of signed rank test relative to t-test in the one-sample location problem is ≥ 0.864 within the class of all symmetric continuous populations. A similar bound for sign test is $\frac{1}{3}$. Details on P.E. for various nonparametric statistics can be found in almost any text on nonparametric inference.

3. Bahadur's slopes

The level attained $L(T_n)$ of a test statistic T_n, whose large values are significant, is defined as $\bar{F}_n(T_n)$ where

$$\bar{F}_n(x) = \sup\{P_\theta(T_n > x) \colon \theta \in \Lambda_0\},$$

Λ_0 being the null space. The level attained is also popularly known as the p-value. A p-value is stochastically at least as large as a $U[0, 1]$ random variable under the null. Thus the size of the test which reject iff $L(T_n) \leq \alpha$ is bounded by α. If T_n has a continuous null distribution, as is the case in many practical examples, the distribution of $L(T_n)$ is exactly $U[0, 1]$. It is desirable to have a low $L(T_n)$ if a nonnull θ prevails. Since the exact distribution of $L(T_n)$, under a nonnull θ is often intractable, though there are a couple of papers having to do with the exact distribution, one has to resort to an asymptotic measure of its smallness. Under a nonnull θ, $L(T_n)$ can be typically expressed as follows:

$$L(T_n) = \exp(-nc(\theta)[1 + \mathrm{o}(1)]) \quad \text{a.s.}[\theta].$$

In such a case $c(\theta)$, known as Bahadur's exact slope, is taken as a measure of asymptotic efficiency (According to Bahadur's original definition, $2c(\theta)$ is the slope). In the definition of $L(T_n)$, if the exact distribution of T_n is replaced by its weak limit then $c(\theta)$ is called the approximate slope of T_n. See Bahadur (1960, 1967 and 1971) for the basic details on the slopes.

Relative Bahadur efficiency is defined as the ratio of exact Bahadur slopes.

For two test statistics T_{1n} and T_{2n} with slopes $c_1(\theta)$ and $c_2(\theta)$, if

$$N(T_{in}, \varepsilon) = \min\{m : L(T_{in}) < \varepsilon \text{ for all } n \geq m\} \quad \text{for} \quad i = 1, 2$$

then

$$\lim_{\varepsilon \to 0} \frac{N(T_{1n}, \varepsilon)}{N(T_{2n}, \varepsilon)} = \frac{c_2(\theta)}{c_1(\theta)} \quad \text{a.s.}[\theta]$$

(see Theorem 7.2 of Bahadur, 1971). This relationship provides an interpretation of the exact slopes in terms of the sample sizes required in order to make T_n significant at a given (small) level, when a nonnull θ obtains.

The standard method for evaluating Bahadur's slope involves a strong convergence result and a large deviation result. Suppose that $T_n \to b(\theta)$ a.s. for all $\theta \in \Lambda_0$ and $\bar{F}(x) = \exp[-nf(x)(1 + o(1))]$ where f is a continuous function defined on an open interval containing $b(\theta)$ for all nonnull θ's. Then the slope of T_n is $f(b(\theta))$. Bahadur's monograph contains a number of worked out examples all based on this procedure. Almost always, the strong convergence part is trivial and the procedure boils down to establishing the existence of the large deviation index and showing that it is continuous in its domain. It seems fair to comment here that this key link between Bahadur's slope and the large deviation has been a strong motivation for the development of the large deviation theory.

There exists an upper bound for Bahadur's exact slopes whose validity requires practically no regularity condition. This bound furnishes a basis for the notion of (first order) Bahadur optimality. We can describe it as follows:

$$\liminf_{n \to \infty} n^{-1} \log L(T_n) \geq -\inf\{K(\theta, \theta_0) : \theta \in \Lambda_0\} \quad \text{a.s. } [\theta]$$

where

$$K(\theta, \theta_0) = \int [\log r(x)] \, dP_\theta ; \quad r(\theta) = \frac{dP_\theta}{dP_{\theta_0}}$$

is assumed to exist for all $\theta \in \Lambda_0$. In its present form, the result is due to Raghawachari (1970), though various weaker versions of it appeared in the earlier papers of Bahadur. Under conditions, likelihood ratio tests are known to be Bahadur optimal (see Bahadur, 1965; Bahadur and Raghawachari, 1970; Bahadur, 1971). In many cases it is much simpler to check directly that the slope of a certain statistic equals the minimum Kullback Leibler distance, than going through the regularity conditions of a general theorem. In particular this comment applies to the t-statistic and the χ^2 statistic in the $N(\theta, \sigma^2)$ situation. Another theorem of Raghawachari (1970) states that $n^{-1} \log L(T_n) \to -c(\theta)$ in probability under a nonnull θ iff $n^{-1} \log \alpha_n \to -c(\theta)$ where α_n is the size of a test based on T_n having a fixed power in (0, 1). This theorem unifies Bahadur's and Cochran's efficiencies. A weaker version of this result is contained in Bahadur (1967). Kalenberg (1980) has extended Raghawachari's equivalence result to a 'second order' level.

A huge number of papers have been devoted to Bahadur efficiency for various specific statistics and we do not find it necessary to prepare a list here.

4. Relationship between Bahadur efficiency and Pitman efficiency

The phenomenon that the nonlocal relative efficiencies converge to the local relative efficiencies as θ tends to a null value is well known. However, the available mathematical results to this effect do not seem very satisfactory. Appendix 2 in Bahadur (1960) contains a relationship between Pitman efficiency and the ratio of approximate slopes. It is shown in the appendix that if the two competing test statistics are asymptotically normal and satisfy a few standard conditions then the ratio of approximate slopes tends to P.E. as θ tends to the null value. Bahadur's result was considerably generalized in Wiend (1976). Wiend's formulation admits some nonnormal limit situations too. In general P.E., if it exists, depends upon α and β and hence the above mentioned relationship is not expected to hold unless some further limit on P.E. is taken. Under conditions, Wiend shows that the limiting ratio of approximate slopes as θ tends to the null value, coincides with the limiting P.E. as $\alpha \to 0$. The approximate slope itself generally provides an approximation to the exact slope for a θ close to the null. It is usually not so for a fixed θ away from the null. We formally state below a variation of Wiend's theorem.

Assume that the null is H_0: $\theta = \theta_0$. Let us name the following conditions on a test statistic T_n as (A):

(1) Under null, $\sqrt{n}T_n$ converges weakly to a continuous distribution F. Assume that F admits the tail estimate

$$1 - F(x) = \exp\left[-\frac{ax^2}{2}(1 + \mathrm{o}(1))\right] \quad \text{as } x \to \infty .$$

(2) There is a continuous function $b(\theta)$ such that, under θ, the sequence $\sqrt{n}[T_n - b(\theta)]$ is tight uniformly in θ belonging to a neighborhood of θ_0, i.e. for a given $\varepsilon > 0$, there exist positive K and N_0 such that

$$P_\theta(\sqrt{n}\,|T_n - b(\theta)| > K) < \varepsilon \text{ for all } \theta \in (\theta_0 - \varepsilon, \theta_0 + \varepsilon) \text{ and } N \geq N_0 .$$

THEOREM (Wiend). *If T_{1n} and T_{2n} both satisfy condition A and P.E. $e_{12}(\alpha, \beta)$ exists then*

$$\lim_{\theta \to \theta_0} E_{12}(\theta) = \lim_{\alpha \to 0} e_{12}(\alpha, \beta)$$

where $E_{12}(\theta)$ is the relative (approx.) Bahadur efficiency which exists under A.

Wiend used this relationship for computing some previously unknown P.E.

There do not seem to be any further results available on this line. A more general study on this phenomenon is felt desirable.

5. Higher order comparisons

In a situation where an asymptotic relative efficiency is one, a finer asymptotic comparison is required to make a large sample distinction among the competing procedures. This is an area of the subject which is currently attracting the attention of many mathematical statisticians and furthermore, many of the significant contributions are very recent. The available literature in my knowledge is either higher order Pitman type comparison or a higher order Bahadur–Cochran type comparison. We first discuss below Pitman type comparisons.

Hodges and Lehmann (1970) (HL) introduced a notion known as deficiency, which is suitable for discrimination between tests with P.E. = 1. Consider two tests based on T_{1n} and T_{2n} for the same null hypothesis H_0: $\theta = \theta_0$ at the same fixed level α. Let $p_{1n}(\theta_n)$ and $p_{2n}(\theta_n)$ be the powers of the two tests against the same contiguous alternative $\theta_n = \theta_0 + K/\sqrt{n}$. Let p_{1x} and p_{2x} be the continuous extensions of the two power sequences using the linear interpolation (see HL for a stochastic interpolation). A number d_n satisfying $p_{1n} = p_{2(n+d_n)}$ is known as the deficiency of T_{2n} as compared to T_{1n}. The limiting behaviour of d_n as $n \to \infty$ can be used for a distinction in a case where P.E. = 1. In HL, the concept of deficiency has been introduced through the comparison of the Z-test with the t-test. The last section of HL contained suggestions for deficiency level comparison of some first order efficient nonparametric tests with their parametric counterparts. These questions were later picked up by Albers, Bickel and van Zwet (1976) (ABZ) in the one sample cases and Bickel and van Zwet (1978) in the two sample cases. The major mathematical task in the asymptotic evaluation of d_n is the Edgeworth expansion of the statistics under the null and under a contiguous alternative. Some of the specific findings of ABZ are as follows: If the null distribution is normal, then the deficiency of the permutation test based on ΣX_i relative to the t-test tends to zero for a contiguous location alternative. For the normal null and the normal contiguous alternative, the deficiency of both Fraser's normal score test and van der Waeder's test relative to the $\bar{X}$-test is of the order $\frac{1}{2}\log\log n$; thus it is practically bounded ($\frac{1}{2}\log\log n \simeq 1.3$ for the sample size a million!) Some numerical investigations of these deficiency related results have been carried out in Albers (1974).

There also exists in this context the phenomenon 'first order efficiency implies second order efficiency'. A result to this effect was formally established by Pfanzagl (1979), though in some special cases the phenomenon had been noticed in several earlier papers. A recent article, Bickel, Chibichov and van Zwet (1982), aimed at providing insight into this result, makes a nice reading on this topic. We have some further comments on HL deficiency in the next paragraph.

HL deficiency as evaluated in ABZ seems to be crucially dependent on contiguous sequences though the definition extends to any sequence. It seems to me that in general one would get different HL deficiency for various different convergence rates of $(\theta_n - \theta_0)$. (Typically it is not so for the first order local efficiencies.) A local deficiency which appears more appealing to me is $N_2(\theta, \alpha, \beta) - N_1(\theta, \alpha, \beta)$ as θ approaches θ_0. In the special case of the median-test vs the sign test for the double exponential family, the expansion of this difference in terms of $(\theta - \theta_0)$ has been carried out in Proposition 2 of Gronenboom and Oosterhoff (1982). It is plausible that the expansion of $N_2(\theta, \alpha, \beta) - N_1(\theta, \alpha, \beta)$ is obtainable in general from the Edgeworth expansions similar to those in ABZ. A detailed study on this latter deficiency and its possible connections with HL deficiency would be worthwhile.

Turning to the studies devoted to the discrimination between tests with the same Bahadur slope, we begin with a notion of deficiency introduced in Chandra and Ghosh (1978), named Bahadur–Cochran deficiency (BCD). BCD is based on the difference between the minimum sample sizes required to bring α below a pre-specified value δ when the power is held fixed at a level β against a fixed alternative θ. The limit is taken as $\delta \to 0$. Another equivalent formulation of BCD is given in terms of the weak convergence of the p-values. The article contains several worked out examples. Chandra and Ghosh (1980) studied some multi-parameter testing problems in the same spirit. Kalenberg (1978) and (1981) also study the same higher order comparison problem. Kalenberg (1978) defines Bahadur deficiency differently. This deficiency is defined as the limiting value of $N_2(\theta, \alpha, \beta) - N_1(\theta, \alpha, \beta)$ as α tends to zero. Kalenberg shows that the deficiency of LR test in the exponential families as compared to any other test for the same hypotheses is bounded by $c \log n$ (in special cases, including the t-test it is $0(1)$. The reader would find discussions on another related concept 'shortcoming' in Kalenberg (1978). Recently some papers appeared with interesting notions of higher order comparison entirely based upon limiting behaviour of p-values. Lambert and Hall (1982) argues that if two tests have equal slopes then the asymptotic variances of the normalized log p-values can be used to distinguish between them. In terms of smallness of the p-values, a smaller asymptotic variance does mean that there would be less underestimation of the p-values by the slope. Lambert and Hall show that the p-values with smaller asymptotic variances require fewer observations to attain a level α, provided the power is fixed at a level $\beta > \frac{1}{2}$, which surely is reasonable for large samples. Later, Bahadur, Chandra and Lambert (1982) discovered a phenomenon to the effect that the first order optimality in terms of slopes implies the second order optimality in the sense of smaller asymptotic variance of normalized log p-values. There is yet another interesting paper on this line by Berk. This paper reports a rather unexpected lower bound on log p-values. Often, the successive terms of log p-values are of the orders: $-cn$, $O_p(\sqrt{n})$, $O(\log n)$, $O_p(1)$. Berk's lower bound holds up to $O_p(1)$ and this furnishes a notion of third order optimality. This bound is attained by the Neyman–Pearsonian tests in the simple vs. simple cases. It is also attained by many (though not all) LR tests in the nonsimple null cases including the t-test for μ and the χ^2-test for σ of the $N(\mu, \sigma^2)$ population.

Thus we already have quite many notions of deficiency. It is felt desirable to unify various apparently different notions of deficiencies and examine numerically how closely the limiting deficiencies approximate the finite sample exact values.

6. Bahadur efficiency of combined tests

Let $T_{1n_1}, T_{2n_2}, \ldots, T_{kn_k}$ be k test statistics available for the same testing problem, based on k independent samples of sizes $n_1, n_2, \ldots, n_k$. Assume that large values are significant for each of the statistics. Let $g = g(T_{1n_1}, \ldots, T_{kn_k})$ be a combined statistic, where g is a function from $\mathbb{R}^k$ to $\mathbb{R}$. Let g be nondecreasing in each of its coordinates. If $L(g), L(T_{1n_1}), \ldots, L(T_{kn_k})$ denote the p-values of g, $T_{1n_1}, \ldots, T_{kn_k}$, then it is easy to see that

$$L(g) \geqslant \prod_{i=1}^{k} L_i(T_{in_i})$$

and this implies that the slope of g is $\leqslant \Sigma_1^k \lambda_i c_i(\theta)$ where $\lambda_i = \lim_{n\to\infty} n_i / \Sigma_1^k n_i$ is assumed to exist; $c_i(\theta)$ is the slope of T_{in_i} at θ. The above simple but important facts were noticed first by Littell and Folks (1973). The bound on the slope of g gives rise to a notion of first order optimality (f.o.o.) of combined tests. In particular, Fisher's combination

$$T(F) = -2 \sum_{1}^{k} \log L(T_{in_i})$$

is f.o.o. Later, it was pointed out by Berk and Cohen (1979) that in fact

$$T_G(\beta, \alpha_1, \alpha_2, \ldots, \alpha_k) = \sum_{1}^{k} \Gamma_{\beta,\alpha_i}^{-1}(1 - L(T_{in_i}))$$

for any $\alpha_1, \alpha_2, \ldots, \alpha_k > 0$ is f.o.o. where $\Gamma_{\beta,\alpha}$ denotes the d.f. of the gamma distribution with parameters β, α. A higher order asymptotic study on combining independent tests was carried out in Cohen, Marden and Singh (1982) (CMS). It turned out that the log p-values of $T_G(\beta, \alpha_1, \alpha_2, \ldots, \alpha_k)$ agree up to three leading asymptotic terms and the comparable terms, which are $O_p(1)$, depends upon α's as well as $c_i(\theta)$'s. Surprisingly enough, it was found that the normal score combination

$$T_N = \sum_{1}^{k} (n_i/n)^{1/2} \Phi^{-1}(1 - L(T_{in_i})), \quad n = \sum_{1}^{k} n_i,$$

beats any member of T_G at the third asymptotic term if the slopes c_i's are equal. The combination T_N is not even f.o.o. if the slopes are not all equal. Thus an asymptotic discrimination between the members of T_G is still an open problem. CMS also contains an unresolved conjecture regarding lower bounds for $L(g)$.

7. The asymptotic efficiencies as approximations to the finite sample efficiencies

How well do Pitman and Bahadur efficiencies approximate the actual relative efficiency given as $N_2(\theta, \alpha, \beta)/N_1(\theta, \alpha, \beta)$? Presently, there does not seem to exist adequate literature on this question. Recently, there has been an interesting study on this issue by Gronenboom and Oosterhoff (1981) (GO), which reported some numerical findings demanding general mathematical explanations. It seems worthwhile to include here a few highlights of this paper.

This paper looked at Bahadur efficiency as $\lim_{\alpha\to\infty} N_2(\theta, \alpha, \beta)/N_1(\theta, \alpha, \beta)$ (the same definition of Bahadur efficiency is given in Kalenberg (1978) too). The definition of $N_i(\theta, \alpha, \beta)$ in GO is slightly different from the one given in this review, though most of the numerical results reported in GO remain the same under the two definitions. The paper considers few specific testing problems and investigates numerically how well the asymptotic efficiencies and deficiencies (in the sense of Pitman and Bahadur) approximate the corresponding exact quantities $N_2(\theta, \alpha, \beta)/N_1(\theta, \alpha, \beta)$ and $N_2(\theta, \alpha, \beta) - N_1(\theta, \alpha, \beta)$. The first example is about the normal mean θ when the dispersion matrix is known to be I. The LR test for the hypothesis $\theta \leqslant 0$ vs. $\theta < 0$ has been compared with the most powerful test for a fixed alternative. In this comparison, Bahadur efficiency equals 1. The exact relative efficiency, which turns out to be the same as Pitman efficiency in this example, was found to be much smaller than 1 (something close to $\frac{1}{2}$) for many reasonable (θ, α, β), showing the inadequacy of Bahadur efficiency in this case. The deficiency expansions, as $\alpha \downarrow 0$, provided a fair approximation to the actual deficiencies. The second example studied is the comparison of the Z-test with the t-test for the univariate normal populations. Here Pitman efficiency is 1 whereas Bahadur efficiency is $\theta^{-2}\log(1+\theta^2)$, θ being the alternative. According to Table 3 of GO, Pitman efficiency approximates the actual relative efficiencies very well when $\theta = 0.25$ (the null θ equals zero), though it badly overestimates when $\theta = 0.5$ and 1.0. Bahadur efficiency is in good agreement with reality when $\beta = 0.5$, though it is not so when $\beta = 0.9$. One term Pitman deficiency expansion does an excellent job of estimating the actual (moderate sample) deficiencies. The other two examples studied in GO are (i) comparison of the sign test and the median test for the location parameter of the double exponential family; (ii) comparison of the LR test, Hotelling's test, Roy's test, and Pillai's test for testing multivariate linear hypotheses on the mean of a multivariate normal population. In general, Bahadur efficiency seems to display a somewhat haphazard behavior in terms of approximating moderate sample realities. On the other hand, Pitman efficiency often seems to be close to the actual moderate sample efficiencies and it appears that its behavior is more predictable. Presently the reasons behind these phenomena are not well understood and it is felt that some more general theoretical studies connecting the asymptotic relative efficiencies and the finite sample efficiencies would be of great help.

References

Albers, W. (1974). *Asymptotic expansions and the deficiency concept of statistics.* Mathematical Centre Tracts, 58, Amsterdam.

Albers, W., Bickel, P. J. and van Zwet, W. R. (1976). Asymptotic expansions for the power of distribution free tests in the one sample problem. *Ann. Statist.* **4**, 108–156.

Bahadur, R. R. (1960). Stochastic comparison of tests. *Ann. Math. Stat.* **31**, 276–295.

Bahadur, R. R. (1965). An optimal property of the likelihood ratio statistic. In: *Proc. Fifth Berkeley Symp.*

Bahadur, R. R. (1967). Rates of convergence of estimates and test statistics. *Ann. Math. Stat.* **38**, 303–324.

Bahadur, R. R. (1971). *Some Limit Theorems in Statistics.* SIAM, Philadelphia.

Bahadur, R. R. and Raghawachari, M. (1970). Some asymptotic properties of likelihood ratios on general sample spaces. In: *Proc. Sixth Berkeley Symp.*

Bahadur, R. R., Chandra, T. K. and Lambert, D. (1982). Some further properties of likelihood ratios on general sample spaces. Preprint.

Berk, R. H. (1982). Stochastic bounds for attained levels. Preprint.

Berk, R. H. and Cohen, A. (1979). Asymptotically optimal methods of combining tests. *J. Amer. Statist. Assoc.* **74**, 812–814.

Bickel, P. J., Chibisov, D. M. and van Zwet, W. R. (1981). On efficiency of first and second order. *Intern. Statist, Rev.* **49**, 169–175.

Bickel, P. J. and van Zwet, W. R. (1978). Asymptotic expansions for the power of distribution free tests in the two sample problem. *Ann. Stat.* **6**, 937–1004.

Brown, L. D. (1971). Non-local asymptotic optimality of appropriate likelihood ratio tests. *Ann. Math. Stat.* **42**, 1206–1240.

Chandra, T. K. and Ghosh, J. K. (1978). Comparison of tests with same Bahadur efficiency. *Sankhya Ser. A* **40**, 253–277.

Chandra, T. K. and Ghosh, J. K. (1980). Deficiency for multiparameter testing problem. Sixtieth Birthday Volume of C. R. Rao.

Chernoff, H. (1952). A measure of asymptotic efficiency for tests of a hypothesis based on the sum of observations. *Ann. Math. Stat.* **23**, 493–507.

Cochran, W. G. (1952). The χ^2-goodness of fit. *Ann. Math. Stat.* **23**, 493–507.

Cohen, A., Marden, J., and Singh, K. (1982). Second order asymptotic and nonasymptotic optimality properties of combined tests. *J. Statist. Planning Infer.* **6**, 253–276.

Fraser, D. A. S. (1957). *Nonparametric Methods in Statistics.* Wiley, New York.

Gronenboom, P. and Oosterhoff, J. (1981). Bahadur efficiency and small sample efficiency. *Intern. Statist. Rev.* **49**, 127–141.

Hodges, J. L., Jr. and Lehmann, E. L. (1956). On efficiency of some nonparametric competitors of the *t*-test. *Ann. Math. Statist.*

Hodges, J. L., Jr. and Lehmann, E. L. (1970). Deficiency. *Ann. Math. Stat.*, **41**, 783–801.

Hoeffding, W. (1965). Asymptotically optimal tests for multinomial distributions (with discussions). *Ann. Math. Stat.* **36**, 369–408.

Kalenberg, W. C. M. (1978). *Asymptotic Optimality of Likelihood Ratio Tests in Exponential Families.* Mathematical Centre Tracts **77**, Amsterdam.

Kalenberg, W. C. M. (1981). Bahadur deficiency of likelihood ratio tests in exponential families. *Jour. Multi. Analy.* **11**, 506–531.

Kalenberg, W. C. M. (1981). Relations between Bahadur deficiencies and attained levels. Preprint.

Lambert, D. and Hall, W. J. (1982). Asymptotic log normality of p-values. *Ann. Statist.* **10**, 44–64.

Littell, R. C. and Folks, J. L. (1973). Asymptotic optimality of Fisher's method of combining tests II. *J. Amer. Statist. Assoc.* **68**, 193–194.

Noether, G. E. (1955). On a theorem by Pitman. *Ann. Math. Stat.* **26**, 64–68.

Olshen, R. A. (1967). Sign and Wilcoxon tests for linearity. *Ann. Math. Stat.* **38**, 1763–1769.

Pfanzagl, J. (1979). First order efficiency implies second order efficiency. In: J. Jureckova, ed., *Contributions to Statist.* [J. Hajek Memorial Volume] Academia, Prague, pp. 167–196.

Pitman, E. J. G. (1949). Lecture notes on nonparametric statistical inference. Columbia University Publ.

Raghawachari, M. (1970). On a theorem of Bahadur on the rate of convergence of test statistics. *Ann. Math. Statist.* **41**, 1695–1699.

Rothe, G. (1981). Some properties of the asymptotic relative Pitman efficiency. *Ann. Statist.* **9**, 663–669.

Rubin, H. and Sethuraman, J. (1965). Bayes risk efficiency. *Sankhya Ser. A* **27**, 325–346.

van Eeden, C. (1963). The relation between Pitman's asymptotic relative efficiency of two tests and the correlation coefficient between their test statistics. *Ann. Math. Stat.* **34**, 1442–1451.

Wiend, H. S. (1976). A condition under which the Pitman and Bahadur approaches to efficiency coincide. *Ann. Statist.* **4**, 1003–1011.

PART II

LINEAR MODELS

P. R. Krishnaiah and P. K. Sen, eds., *Handbook of Statistics, Vol. 4*
© Elsevier Science Publishers (1984) 185–228

10

Nonparametric Methods in Two-Way Layouts

Dana Quade

1. Introduction

This Chapter is concerned with the *two-way layout*, in which there are $n \geq 2$ *blocks* of random observations, and each observation is given one of $m \geq 2$ *treatments*. Let Y_{ijk} denote the k-th observation given the j-th treatment in the i-th block, where $k = 1, \ldots, l_{ij}$; write $l_i = \Sigma_j l_{ij}$ for the total number of observations in the i-th block. All the analyses which we consider involve the fundamental.

Assumption I. Blocks are mutually independent.

The null hypothesis of interest, stated somewhat vaguely, is that the treatments make no difference, or that each observation would have had exactly the same value if given any other treatment. A more mathematical statement is

H_0: Observations within a block are *interchangeable* (an equivalent term is *exchangeable*), meaning that for each i the $l_i!$ permutations of observations within the i-th block are all equally likely.

This may be tested by straightforward application of Fisher's randomization principle. Let T be any test criterion chosen (perhaps intuitively) so that large values should lead to rejection. Generate each of the $N = \Pi(l_i!)$ hypothetical datasets which can be obtained by permuting observations within blocks, calculate T from each such dataset, and count how many of them (say, N^*) yield values of T as large as that actually observed. Then the exact level of significance corresponding to T is the *P-value* $P = N^*/N$. It is obvious that a *randomization test* such as just described is not generally feasible unless N is very small, since the N values of T must be calculated anew for every observed dataset. If an approximate P-value is sufficient, then an alternative approach may be used: generate a random sample of M of the N datasets, count how many of them (say, M^*) yield values of T as large as that actually observed, and set $\hat{P} = M^*/M$. Nevertheless, extensive computations are still required to obtain reasonable precision. A great simplification can be obtained, however, by restricting attention to criteria which depend only on the *within-block ranks*, defined by

$$R_{ijk} = \frac{l_i + 1}{2} + \tfrac{1}{2} \sum_{j'=1}^{m} \sum_{k'=1}^{l_{ij}} \operatorname{sgn}(Y_{ijk} - Y_{ij'k'})$$

for all i, j and k. The ranks within the i-th block are of course some permutation of the fixed set of integers $1, \ldots, l_i$ (if there are no ties – see discussion later). Thus there is only one family of hypothetical datasets corresponding to a given design, and the possibility exists of tabulating the criterion once and for all. The test is then called a *rank randomization test*, based on the *method of n rankings*. Randomization tests based on the raw data will be mentioned only briefly in this Chapter.

Another simplification involves restricting attention to the *complete blocks design* in which each block has exactly m observations, of which exactly one is given each treatment. Then each $l_{ij} = 1$, and the third subscript on Y and R may be suppressed. Discussion of more complicated designs will be limited to a few remarks in Section 9.

For any ordered subset $(j_1, j_2, \ldots, j_l)$ of the integers $1, \ldots, m$, let $P_i(j_1, j_2, \ldots, j_l)$ be the probability that within the i-th block the observation given treatment j_1 has a smaller value than the observation given treatment j_2, which has a smaller value than ... the observation given treatment j_l. Then H_0 implies that $P_i(j_1, j_2, \ldots, j_l) = 1/l!$ for all subsets of size l, $l = 2, \ldots, m$. This situation is called *random ranking*. Many of the tests presented in this Chapter require no more than Assumption I for validity in testing H_0, i.e., for keeping the Type I error probability under control. However, discussion of their behavior under alternatives will be restricted to those which satisfy

ASSUMPTION IIa. $P_i(j_1, \ldots, j_l) \equiv P(j_1, \ldots, j_l)$, i.e., that all blocks have the same distribution of ranks.

(The 'a' distinguishes this assumption from more restrictive versions IIb and IIc which will be introduced later.)

Given a complete blocks design subject to Assumptions I and IIa, the most general alternative which can be detected using ranks is that $P(j_1, \ldots, j_l) \neq 1/l!$ for at least one ordered subset $(j_1, \ldots, j_l)$. Let $\boldsymbol{R}_i = (R_{i1}, \ldots, R_{im})'$ be the ranking of treatments within the i-th block, for $i = 1, \ldots, n$, and let n_k be the observed number of times that $\boldsymbol{R}_i = \boldsymbol{r}_k$, where $k = 1, \ldots, m!$ indexes the permutations of $\{1, \ldots, m\}$. Then a suitable test of H_0 against this general alternative may be based on a simple chi-squared, such as the Pearsonian

$$X_P^2 = \frac{m!}{n} \sum_{k=1}^{m!} \left(n_k - \frac{n}{m!}\right)^2 .$$

Asymptotically (for large n) X_P^2 has the χ^2 distribution with $(m! - 1)$ degrees of freedom. But this of course requires an enormous number of blocks unless m is very small.

The simplest special case is that of *matched pairs* or blocks of size $m = 2$. The only possible rankings within a pair are $\boldsymbol{r}_1 = (1, 2)$ and $\boldsymbol{r}_2 = (2, 1)$, and each pair may be classified according to the sign of the difference between its two observations. Three standard test criteria produce the two one-sided versions and the two-sided version of the well-known *sign test* for H_0: these are

$$S_{12} = \operatorname{sgn}(Y_{i1} - Y_{i2}) \qquad \text{(against } H_{12}\text{: } P(1, 2) > P(2, 1)\text{)},$$

$$S_{21} = \operatorname{sgn}(Y_{i2} - Y_{i1}) = -S_{12} \quad \text{(against } H_{21}\text{: } P(1, 2) < P(2, 1)\text{)},$$

and

$$S = |S_{12}| = |S_{21}| \qquad \text{(against } H_1\text{: } P(1, 2) \neq P(2, 1)\text{)}.$$

(It is easily verified that $S^2 = nX^2$, where X^2 is the general chi-squared of the preceding paragraph.) Under H_0, each difference $(Y_{i1} - Y_{i2})$ is symmetrically distributed about 0, whence $(S_{12} + n)/2$ has the binomial distribution with parameters $(n, \frac{1}{2})$, and asymptotically (for large n) $S_{12}/\sqrt{n}$ has the standard normal distribution; S_{21} has of course the same distribution as S_{12}.

To illustrate, consider the following Example A ($n = 5$):

		Treatments 1	2
Blocks	1	5	4
	2	7	5
	3	9	6
	4	6	8
	5	14	10

Then the three test criteria are $S_{12} = 3$, with $P = 0.1875$ (6/32); $S_{21} = -3$, with $P = 0.96875$ (31/32); and $S = 3$, with $P = 0.3750$ (12/32).

There are several ways of generalizing the sign test to more than two treatments. The earliest was proposed by Wormleighton (1959). Let the column vector $\boldsymbol{B}$ contain the $\binom{m}{2}$ sign test statistics

$$S_{jj'} = \sum \operatorname{sgn}(Y_{ij} - Y_{ij'}) \quad \text{for } 1 \leq j, j' \leq m,$$

and let W be the corresponding variance matrix, with a complicated correlation structure which Wormleighton derives. Then his statistic is defined as $X^2_W = \boldsymbol{B}'W^{-1}\boldsymbol{B}$; under H_0, $E[X^2_W] = m(m-1)/2$, and $V[X^2_W] = m(m-1)(n+1)/n$. The statistics $(S_{jj'} + n)/2$ are each individually binomial $(n, \frac{1}{2})$ under H_0, of course, and asymptotically they are jointly normal. Thence it can be shown that the asymptotic distribution of X_W^2 is χ^2 with $m(m-1)/2$ degrees of freedom. There is a particularly simple form if $m = 3$; then

$$X_W^2 = \frac{3}{4n}\{(S_{12}^2 + S_{23}^2 + S_{31}^2) + (S_{12} + S_{23} + S_{31})^2\}$$

is asymptotically $\chi^2(3)$. Wormleighton tabulates the exact distribution in this special case, for $n = 2(1)6$.

The method may be illustrated by the following Example B, with $m = 3$ and $n = 5$:

		Raw Data			Ranks		
Treatments		1	2	3	1	2	3
Blocks	1	35	29	20	3	2	1
	2	37	34	19	3	2	1
	3	26	28	21	2	3	1
	4	38	25	36	3	1	2
	5	41	32	30	3	2	1
Totals		177	147	126	14	10	6

For these data $S_{12} = 3$, $S_{23} = 3$ and $S_{31} = -5$. Hence $X_W^2 = 6.60$, whence by Wormleighton's table $P = 0.1088$ (846/7776) exactly; or, referring to $\chi^2(3)$, $P = 0.0858$ approximately.

Other generalizations of the sign test are presented in Section 4. These tests are consistent against any alternative for which $P(j, j') \neq \frac{1}{2}$ for some $j \neq j'$. However, such alternatives may still be too broad to be interesting. The next two Sections discuss tests against somewhat more narrowly focused alternatives.

The preceding discussion has ignored the difficulties caused by the occurrence of tied values within blocks. The definition given earlier then produces *average ranks*, but these are no longer a permutation of $\{1, \ldots, m\}$, so standard tables of the test statistic will not be valid. There is considerable controversy as to what must be done in such a situation. If we suppose that the ties have resulted from imprecise measurement, so that there exists some correct way to break them, although it is unknown to us, then it seems reasonable to report the smallest and especially the largest P-values attainable by breaking them; in particular, if all ways of breaking the ties lead to the same P-value, then that P-value must be correct. Alternatively, one may break the ties at random, thus producing a *randomized test*, but then the results may be ambiguous, since different investigators might get different ranks. Some argue, however, that the *only* proper basis for inference concerning H_0 is the principle of randomization, which may be applied in particular to the average ranks, even though it yields little computational advantage over using the raw data. Another approach is presented in Section 7, in the context of a model for the alternative to H_0. In large samples, of course, where asymptotic approximations are used, one may as well adopt whatever method is simplest, since there is generally no practical difference in the results. Further discussion of ties may be found in such texts as Bradley (1968), Hájek (1969), or Pratt and Gibbons (1981); for the most part the problem is ignored in the remainder of this Chapter.

A final point worth noting is that the two-way layout may arise in several ways. If observations are assigned to treatments within each block entirely at random, producing by definition a *randomized blocks* design, then interchangeability is automatically assured when the treatments have no effect. But blocks may represent subjects, and treatments times, in a *repeated measurements design*; or blocks and treatments together may constitute a *factorial experiment*. Analyses suitable for the latter two situations are considered briefly in Section 9.

2. Average external rank correlation

It seems intuitively reasonable that the treatments might greatly increase the likelihood of some particular within-block ranking, with lesser increases for similar rankings, and corresponding decreases in likelihood for the opposite rankings. Let $\boldsymbol{P} = (P_1, \ldots, P_m)'$ be the supposedly favored, or *predicted*, ranking; then similarity to it may be measured by the *average external rank correlation*

$$\bar{C} = \frac{1}{n} \sum_{i=1}^{n} C_i ,$$

where $C_i = \gamma(\boldsymbol{R}_i, \boldsymbol{P})$ and γ is any measure of rank correlation. Under H_0 the C_i's are independent and identically distributed, so the exact distribution of $n\bar{C}$ is a simple convolution; and if $E[C_i] = \gamma_0$ (zero for typical rank correlation measures) with $\text{var}[C_i] = \sigma_0^2$, then asymptotically (for large n) $\bar{C}$ is normal with mean γ_0 and variance σ_0^2/n.

Consider using Spearman's (1904) rank correlation coefficient

$$S_i = \left\{12 \sum_j R_{ij}P_j - 3m(m+1)^2\right\} \Big/ (m^3 - m) = 1 - 6 \sum_j (R_{ij} - P_j)^2/(m^3 - m) ;$$

then the average correlation is

$$\bar{S} = \sum S_i/n = \left\{12 \sum R_jP_j - nm(m+1)^2\right\} \Big/ n(m^2 - m)$$

where

$$R_j = \sum_i R_{ij}$$

is the rank sum of the j-th treatment. This statistic was suggested by Lyerly (1952). Later Page (1963) proposed the integer-valued statistic

$$L = \sum_j R_j P_j - n(m^3 - m)\bar{S}/12 + nm(m+1)^2/4$$

and tabulated its critical values ($\alpha = 0.001,\ 0.01,\ 0.05$) for $m = 3(1)10$ with $n = 2(1)50$; more detailed exact tables for $m = 3(1)8$ and $n = 2(1)10$ were given by Odeh (1977a). Under H_0, $E[S_i] = 0$ and $V[S_i] = 1/(m-1)$, whence $E[\bar{S}] = 0$ and $V[\bar{S}] = 1/n(m-1)$, or $E[L] = nm(m+1)^2/4$ and $V[L] = nm^2(m+1)^2(m-1)/144$; and either statistic is asymptotically normal for large n and/or m. In fact, those parts of Page's tables for which $m > 8$ or $n > 12$ are based on this approximation. A continuity correction may be effected by subtracting $\frac{1}{2}$ from L. Thus the test is more convenient for computation in terms of L, although the interpretation as average correlation may be useful.

Jonckheere (1954) suggested a similar criterion using Kendall's rank correlation coefficient

$$K_i = \sum_{j<j'} \operatorname{sgn}(P_j - P_{j'}) \operatorname{sgn}(R_{ij} - R_{ij'})/\tbinom{m}{2}\,.$$

Then the average correlation is

$$\bar{K} = \sum K_i/n = (C - D)/n\tbinom{m}{2}$$

where C (D) is the number of ways to choose 2 treatments and 1 block and find them concordant (discordant) with the external ranking $\boldsymbol{P}$. Tables of critical values ($\alpha = 0.0005,\ 0.001,\ 0.005,\ 0.001,\ 0.025,\ 0.05$) were provided by Ludwig (1962) for $m = 3$ with $n = 2(1)8$, $m = 4$ with $n = 2(1)6$, $m = 5$ with $n = 2(1)5$, $m = 6$ with $n = 2$ or 3, $m = 7(1)10$ with $n = 2$. Also, tables of parts of the null hypothesis distribution may be found in Skillings (1980) for $m = 2(1)6$ with $n = 2(1)5$. The asymptotic distribution of $(C - D)$ is normal with mean 0 and variance $nm(m-1)(2m+5)/18$; a continuity correction is effected by subtracting 1 from $(C - D)$.

In Example B, let the predicted ranking be $\boldsymbol{P} = (3, 2, 1)'$; then the Spearman rank correlations with $\boldsymbol{P}$ are $S_i = 1, 1, \frac{1}{2}, \frac{1}{2}, 1$ and thus $\bar{S} = 0.80$; alternatively, the treatment rank sums are $R_j = (14, 10, 6)$, whence $L = 68$. From Odeh's tables $P = 0.0066$ (51/7776) exactly; to use the normal approximation, calculate $E[L] = 60$ and $V[L] = 10$, whence $Z = (68 - \frac{1}{2} - 60)/\sqrt{10} = 2.37$, corresponding to $P = 0.0089$ approximately. The Kendall correlations with P are $K_i = 1, 1, \frac{1}{3}, \frac{1}{3}, 1$ so $\bar{K} = 0.73$, and from Skilling's tables $P = 0.0078$ (61/7776) exactly. For the normal approximation, note that $C = 13$ and $D = 2$, and $V[C - D] = 55/3$; so $Z = (13 - 2 - 1)/\sqrt{55/3} = 2.34$, corresponding to $P = 0.0097$ approximately.

Numerous additional test criteria could of course be obtained by substituting other varieties of rank correlation coefficients. However, a somewhat different idea is due to Pirie and Hollander (1972), whose statistic is

$$\mathrm{PH} = \sum_i \sum_j P_j E_N(R_{ij})$$

where $E_N(j)$ is the expected value of the j-th order statistic of a random sample of size m from a standard normal distribution. They provide critical values ($\alpha = 0.01$, 0.05, 0.10) with exact probabilities, for $m = 4$ with $n = 1(1)10$ and $m = 5$ with $n = 1(1)5$. (For $m = 3$ the test is equivalent to using Spearman correlation.) For large n, PH is asymptotically normal with mean 0 and variance $\{nm(m+1)/12\} \Sigma E_N^2(j)$.

The Pirie–Hollander test might be considered as equivalent to an unsymmetrical – note P_j rather than $E_N(P_j)$ – average rank correlation. However, alternatively PH (or $\bar{S}$) may be thought of as a special case of the general form $\Sigma_i \Sigma_j P_j r_{R_{ij}}$, where $(r_1, \ldots, r_m)$ are general scores such that $r_1 \leqslant \cdots \leqslant r_m$; without loss of generality, we may take $\Sigma r_j = 0$. This idea is developed by Berenson (1982a).

Finally, for situations in which it is anticipated that the observed rankings may tend to agree either with the specified ranking or with its opposite, Hutchinson (1976) proposed taking the absolute values of the correlations before averaging: let $\bar{H} = \Sigma |S_i|/n$. For $m = 3$ the quantity $2n(1 - \bar{H})$ is binomial $(n, \frac{2}{3})$ under H_0. For $m > 3$ Hutchinson suggested the integral-valued test criterion $(m^3 - m)n(1 - \bar{H})/6$, and tabulated its exact critical values ($\alpha = 0.001$, 0.01, 0.05, 0.10) for $m + n \leqslant 11$. The asymptotic distribution is normal with mean $nm(m+1)\{m - 1 - \sqrt{2(m-1)/\pi}\}/6$ and variance $nm^2(m+1)^2(\pi - 2)/36\pi$. In Example B, $\bar{H} = \bar{S} = 0.8$; since $m = 3$, we calculate $2n(1 - \bar{H}) = 2$, and the exact $P = 0.2058$ (1600/7776).

3. Average internal rank correlation

If the ranking supposedly favored under the alternative to random ranking is unspecified, an appropriate form of test criterion is the *average internal rank correlation*

$$\bar{\bar{C}} = \sum_{i<i'} c_{ii'} / \tbinom{n}{2},$$

where $C_{ii'} = \gamma(\boldsymbol{R}_i, \boldsymbol{R}_{i'})$. Such a test criterion seems intuitively reasonable, since if the treatments tend to produce rankings similar to some $\boldsymbol{P}$, then these rankings will *ipso facto* be similar to each other. Once the measure γ of rank correlation has been chosen, the null distribution of $\bar{\bar{C}}$ is determined, and it can be tabulated at least for small m and n. The test is equivalent to the simple rank correlation γ if $n = 2$, of course, and to the sign test if $m = 2$.

Note that $\bar{\bar{C}}$ is a U-statistic of degree 2. Given that H_0 is true, let $\gamma_0 = E[\gamma(\boldsymbol{R}_k)]$, $\eta_0 = V[\gamma(\boldsymbol{R}_k)]$, and $\zeta_0 = \text{cov}[\gamma(\boldsymbol{R}_k), \gamma(\boldsymbol{R}_k')]$. Then if $\zeta_0 > 0$, $E[\bar{\bar{C}}] = \gamma_0$, $V[\bar{\bar{C}}] = [2(n-2)\zeta_0 + \eta_0]/\binom{n}{2}$, and (see Hoeffding, 1948) $(\bar{\bar{C}} - \gamma_0)/\sqrt{V[\bar{\bar{C}}]}$ is asymptotically $N(0, 1)$. For most measures of rank correlation, however, $\gamma_0 = \zeta_0 = 0$. Then (see Quade, 1972a; Alvo et al., 1982) $(n-1)\bar{\bar{C}}$ is asymptotically distributed as $\sum \lambda_i(Z_i^2 - 1)$, where the Z_i are independent $N(0, 1)$ variates and the λ_i are the characteristic roots of the matrix whose elements are $\gamma(\boldsymbol{r}_k, \boldsymbol{r}_{k'})$ for $1 \leq k, k' \leq m!$. A reasonable approximation to the null distribution of $\bar{\bar{C}}$ for untabulated combinations (m, n) is afforded (see Quade, 1972a) by taking $X^2 = \alpha + \beta\bar{\bar{C}}$ as $\chi^2(\alpha)$, where $\alpha = n(n-1)\eta^3/(n-2)^2\phi^2$, $\beta = n(n-1)\eta/(n-2)\phi$, and $\phi = E[\gamma(\boldsymbol{R}_k, \boldsymbol{R}_{k'})\gamma(\boldsymbol{R}_{k'}, \boldsymbol{R}_{k''})\gamma(\boldsymbol{R}_{k''}, \boldsymbol{R}_k)]$. This fits three moments exactly for all values of n. However, the degrees of freedom, α, are in general not an integer.

Average internal rank correlation based on Spearman's rho was proposed by Kelley (1923), who showed that it could be computed easily from the treatment rank sums R_j: if

$$R = \sum \{R_j - n(m+1)/2\}^2,$$

then the corresponding average internal rank correlation is

$$\bar{\bar{S}} = \{12R/n(m^3 - m) - 1\}/(n-1).$$

Thus clearly the minimum value is $-1/(n-1)$, which occurs if the rank sums are all equal, and corresponds to minimal agreement among the rankings. The maximum is of course 1, when all $\boldsymbol{R}_i$ are equal. A rescaling of $\bar{\bar{S}}$ to lie between 0 and 1 is the *coefficient of concordance*

$$W = [1 + (n-1)\bar{\bar{S}}]/n = 12R/n^2(m^3 - m)$$

which was introduced by Kendall and Smith (1939). Another variation on $\bar{\bar{S}}$ is Friedman's (1937) statistic

$$X_F^2 = (m-1)[1 + (n-1)\bar{\bar{S}}] = 12R/nm(m+1);$$

under H_0 this is asymptotically distributed as $\chi^2(m-1)$.

Let

$$E_j = E[R_{ij}] = \frac{m+1}{2} + \tfrac{1}{2}\sum_{j'=1}^{m} [P(j', j) - P(j, j')].$$

Note that $E_j \equiv (m+1)/2$ under H_0, so that

$$E[\bar{\bar{S}}] = \frac{12}{m^3 - m}\sum_{j=1}^{m}\left(E_j - \frac{m+1}{2}\right)^2.$$

Thus clearly the test based on $\bar{\bar{S}}$ (or its variants) is consistent against any alternative for which the expected ranks are not all equal.

The following summary indicates the most notable original tabulations of the null distribution for $m, n > 2$; many of these have been variously reprinted, but apparently no single source covers the full range. Each gives upper tail probabilities (i.e., P-values), in terms of X_F^2 as the argument (except that Kendall and Smith use R). The notation 'xDP' indicates 'at least x decimal places' and 'xSF' indicates 'at least x significant figures'.

Reference[a]	Precision	n for: $m = 3$	$m = 4$	$m = 5$	$m = 6$
Friedman (1937)	3*DP*/2*SF*	3(1)9	3, 4	–	–
Kendall and Smith (1939)	3*DP*/2*SF*	3(1)10	3(1)6	3	–
Michaelis (1971)	6*DP*	–	–	4	3
Quade (1972a)	5*DP*/2*SF*	3(1)15	3(1)8	–	–
Hollander and Wolfe (1973)	3*DP*	3(1)13	3(1)8	3(1)5	–
Odeh (1977b)	5*DP*	–	–	6(1)8	3(1)6

[a]Owen (1962) provides tables for $m = 3$ with $n = 3(1)15$ and $m = 4$ with $n = 3(1)8$ which unfortunately are erroneous.

The three-moment chi-squared approximation is obtained using $\eta = 1/(m-1)$ and $\phi = 1/(m-1)^2$. Another class of approximations arises from performing an ordinary two-way analysis of variance on the ranks. Let VR be the variance ratio for treatments from this analysis; then

$$VR = \frac{1+(n-1)\bar{\bar{S}}}{1-\bar{\bar{S}}} = \frac{(n-1)R}{n^2(m^3-m)/12-R} = \frac{(n-1)X_F^2}{n(m-1)-X_F^2}$$

has approximately the same distribution, F with $(m-1)$ and $(n-1)(m-1)$ degrees of freedom, as in ordinary analysis of variance; Kendall and Smith (1939) refined this by changing $(m-1)$ to $(m-1-2/n)$. These and other approximations are discussed by Quade (1972a); see also Iman and Davenport (1980). In any of these approximations, a (partial) correction for continuity is effected by subtracting 1 from R.

While the preceding, based on Spearman's rho, is the standard test for random ranking, many others have appeared in the literature. Several are included in the following generalization. Given any fixed real numbers $r_1, \ldots, r_m$, not all equal, the corresponding *scores* are $r_{R_{ij}}$ for $i = 1, \ldots, n$ and $j = 1, \ldots, m$, and the *score correlation* $O_{ii'}$ between blocks i and i' is the product-moment correlation of the scores. It is easily shown that the *average internal score correlation* is

$$\bar{\bar{O}} = \frac{\sum_{i<i'} O_{ii'}}{\binom{n}{2}} = \frac{\sum (R_j - n\bar{r})^2}{n(n-1)\sum (r_j - \bar{r})^2} - \frac{1}{n-1}$$

where $\bar{r} = \Sigma\, r_j/m$ and $R_j = \Sigma\, r_{R_{ij}}$. Let $\rho_j = E[r_{R_{ij}}]$; then

$$E[\bar{\bar{O}}] = \sum_j (\rho_j - \bar{r})^2 \Big/ \sum (r_j - \bar{r})^2 ,$$

and a test which rejects for large values of $\bar{\bar{O}}$ is consistent against alternatives for which the ρ_j are not all equal. Finally, under H_0

$$X^2 = (m-1)[1 + (n-1)\bar{\bar{O}}] = (m-1) \sum (R_j - n\bar{r})^2 \Big/ n \sum (r_j - \bar{r})^2$$

is asymptotically $\chi^2(m-1)$; and, for the three-moment approximation, $\eta = 1/(m-1)$ and $\phi = 1/(m-1)^2$.

As an example, suppose $r_j \equiv j$; then the scores are the ranks and the score correlation is Spearman correlation. Another example is the test proposed by Brown and Mood (1948, 1951; see also Mood, 1950). This is based on scores $r_j = 1$ if $j > (m+1)/2$ and 0 otherwise, and corresponds to the correlation measure of Blomqvist (1950). Thus the Brown–Mood test criterion is

$$X^2_{BM} = \begin{cases} 4(m-1) \sum \left(R_j - \dfrac{n}{2}\right)^2 \Big/ nm & \text{if } m \text{ is even}, \\[2ex] 4m \sum \left(R_j - \dfrac{n(m-1)}{2m}\right)^2 \Big/ n(m+1) & \text{if } m \text{ is odd}, \end{cases}$$

where in either case $R_j = \Sigma\, r_{R_{ij}}$ is the number of times that an observation on the j-th treatment exceeds the median for its block. A slight variation is to take $r_j = \operatorname{sgn}(2j - m - 1)$, with $\bar{r} = 0$ for all m; this leads to the same criterion if m is even, and avoids a certain arbitrariness if m is odd. (For $m = 3$ it brings us back to the Friedman case.) Blomqvist (1951) tabulated the exact distribution for $m = 4$ with $n = 3(1)8$; $m = 6$ with $n = 3, 4, 5$; $m = 8, 10$ with $n = 3, 4$; and $m = 12, 14, 16$ with $n = 3$.

Ehrenberg (1952) suggested basing an average internal rank correlation on Kendall's tau: i.e.,

$$\bar{\bar{K}} = \sum_{i<i'} \sum_{j<j'} \operatorname{sgn}(R_{ij} - R_{ij'}) \operatorname{sgn}(R_{i'j} - R_{i'j'}) \Big/ \binom{m}{2}\binom{n}{2} .$$

Hays (1960) shows that

$$E[\bar{\bar{K}}] = \sum_j \sum_{j'} [P(j, j') - P(j', j)]^2 \Big/ m(m-1) ,$$

so that a test of H_0 based on $\bar{\bar{K}}$ is consistent against the same alternatives as the sign tests, namely those such that $P(j, j') \neq \frac{1}{2}$ for some $j \neq j'$; these include as a proper subset those against which the standard test based on $\bar{\bar{S}}$ is consistent. Ehrenberg provided exact tables for $m = 3$ with $n = 4$ and 5 and for $m = 5$ with $n = 3$; these were extended by Quade (1972a) to $m = 3$ with $n = 3(1)10$, $m = 4$ with $n = 3(1)6$, $m = 5$ with $n = 3$ and 4, and $m = 6$ with $n = 3$. Van Elteren (1957) showed that for large n the quantity $3\binom{m}{2}[1 + (n-1)\bar{\bar{K}}]$ has asymptotically the same distribution as $(m+1)X_1 + X_2$, where X_1 and X_2 are independent chi-squareds with $(m-1)$ and $\binom{m-1}{2}$ degrees of freedom, respectively. Ehrenberg had already derived the three-moment chi-squared approximation; it is obtained using $\eta = 2(2m+5)/9m(m-1)$ and $\phi = 4(2m^2 + 6m + 7)/27m^2(m-1)^2$. Correction for continuity is effected by subtracting 1 from the numerator of $\bar{\bar{K}}$ if n is even, or 2 if n is odd.

A final test criterion based on average internal rank correlation was suggested by Anderson (1959; see also Schach, 1979). His statistic is related to the correlation measure

$$A(\boldsymbol{R}_i, \boldsymbol{R}_{i'}) = \left[\sum A_j(i, i') - 1\right] \Big/ (m-1)$$

where $A_j(i, i') = 1$ if treatment j has the same rank in blocks i and i', and otherwise $= 0$. Let A_{jk} be the number of blocks in which treatment j has rank k (note that $\Sigma_j A_{jk} = \Sigma_k A_{jk} = m$); then

$$\bar{\bar{A}} = \frac{\Sigma \Sigma A_{jk}^2 - n(m+n-1)}{n(n-1)(m-1)}.$$

Write P_{jk} for the probability that in any block the observation which is given treatment j will have rank k (note that $\Sigma_j P_{jk} = \Sigma_k P_{jk} = 1$); then

$$E[\bar{\bar{A}}] = \left(\sum \sum P_{jk}^2 - 1\right) \Big/ (m-1).$$

Thus a test of H_0 based on $\bar{\bar{A}}$ is consistent against all alternatives such that $P_{jk} \neq 1/m$ for some $j, k = 1, \ldots, m$. These alternatives include as a proper subset all those against which Friedman's test is consistent. However, the alternatives against which $\bar{\bar{A}}$ is consistent neither include nor are included by those against which $\bar{\bar{K}}$ is consistent. For example, with $m = 3$, if the rankings (1, 2, 3) and (3, 2, 1) each have probability $\frac{1}{2}$, then $E[\bar{\bar{A}}] = \frac{1}{2}$ while $E[\bar{\bar{K}}] = 0$; but if the rankings (1, 2, 3), (2, 3, 1) and (3, 1, 2) each have probability $\frac{1}{3}$, then $E[\bar{\bar{A}}] = 0$ while $E[\bar{\bar{K}}] = \frac{1}{9}$. It appears that no table of the exact distribution of Anderson's statistic has yet been published. However, under H_0,

$$X_A^2 = (m-1)^2[1 + (n-1)\bar{\bar{A}}] = \frac{m-1}{n}\left[\sum \sum A_{jk}^2 - n^2\right]$$

is asymptotically distributed for large n as a chi-squared with $(m-1)^2$ degrees of freedom, and the three-moment approximation can be obtained using $\eta = 1/(m-1)^2$, $\phi = 1/(m-1)^4$. To correct (partially) for continuity, subtract 1 from $\Sigma\Sigma A_{kj}^2$.

Example B may be used again to illustrate these measures and provide a standard against which readers may check their calculations. The correlations among the blocks are as follows:

Blocks	Spearman	Brown and Mood	Kendall	Anderson
1, 2	1	1	1	1
1, 3	$\frac{1}{2}$	$-\frac{1}{2}$	$\frac{1}{3}$	0
1, 4	$\frac{1}{2}$	1	$\frac{1}{3}$	0
1, 5	1	1	1	1
2, 3	$\frac{1}{2}$	$-\frac{1}{2}$	$\frac{1}{3}$	0
2, 4	$\frac{1}{2}$	1	$\frac{1}{3}$	0
2, 5	1	1	1	1
3, 4	$-\frac{1}{2}$	$-\frac{1}{2}$	$-\frac{1}{3}$	$-\frac{1}{2}$
3, 5	$\frac{1}{2}$	$-\frac{1}{2}$	$\frac{1}{3}$	0
4, 5	$\frac{1}{2}$	1	$\frac{1}{3}$	0
$\bar{\bar{C}}$	0.55	0.40	0.47	0.25

Using the Spearman correlations, $\bar{\bar{S}} = 0.55$ as shown above; or $R = 32$, from which Friedman's $X_F^2 = 6.40$ and Kendall's $W = 0.64$, and the exact P-value is then 0.0394 (306/7776). Friedman's chi-squared, with 2 degrees of freedom, gives $P = 0.0408$ approximately; the three-moment fit, with continuity correction, gives $X^2 = 11.44$ with 4.44 degrees of freedom, and thence $P = 0.0302$ approximately. Using the Brown–Mood approach, the average internal rank correlation is 0.40 and $X_{BM}^2 = 5.20$; it is not difficult to calculate that $P = 0.1358$ (1056/7776) exactly. Referring this to $\chi^2(2)$ gives $P = 0.0743$ approximately; the three-moment fit, with continuity correction, gives $X^2 = 8.78$ with 4.44 degrees of freedom, and hence $P = 0.0876$ approximately. Using Kendall correlations, $\bar{\bar{K}} = 0.47$ and $P = 0.0471$ (366/7776) exactly; Ehrenberg's three-moment chi-squared, with continuity correction, is $X_E^2 = 10.44$ with 4.80 degrees of freedom, corresponding to $P = 0.0466$ approximately. Finally, Anderson's statistic is $X_A^2 = 8.00$, corresponding to $P = 0.0895$ (696/7776) exactly. Taking this as $\chi^2(4)$ gives $P = 0.0916$ approximately; the three-moment fit, with continuity correction, gives $X^2 = 15.56$ with 8.89 degrees of freedom and hence $P = 0.0899$ approximately. For such a small experiment, of course, the occurrence of really close asymptotic approximations should be regarded as fortuitous.

4. Multiple comparisons

It is rarely sufficient to reject the null hypothesis of interchangeability of the treatments within blocks; one wants to know which pairs of treatments differ from each other. For this purpose Nemenyi (1963) proposed using a *simultaneous sign statistic* equivalent to

$$S^* = \max_{j<j'} |S_{jj'}|$$

where the $S_{jj'}$ are the sign test statistics defined in Section 1. One may then declare any two treatments j and j' significantly different if $|S_{jj'}|$ exceeds the critical value for S^*. Nemenyi provided exact critical values ($\alpha = 0.01, 0.05$) for $m = 3$ with $n = 7(1)16$. Rhyne and Steel (1967) tabulated the exact distribution of S^* – actually $(n - S^*)/2$ – for $m = 3$ with $n = 2(1)24$; they also tabulated a one-sided variant equivalent to

$$S^+ = \max_{j<j'} S_{jj'} \,,$$

which is suitable for use against the ordered alternative that $P(j, j') \leq \frac{1}{2}$ for $j < j'$ (with at least one strict inequality). Miller (1966), using an approximation to the asymptotic distribution, provided a table of critical values ($\alpha = 0.01, 0.05$) for $m = 2(1)10$ with $n = 5(1)20(5)50, 100$. His table turns out to be slightly anticonservative, however, when checked against the exact tabulations for $m = 3$.

Suppose one treatment – without loss of generality let it be the last – is actually a 'control', and the only comparisons of interest are between it and the others. Then suitable test criteria are the *many–one sign statistics*

$$S_m^* = \max_{j<m} |S_{jm}| \quad \text{and} \quad S_m^+ = \max_{j<m} S_{jm} \,,$$

to be used against two- and one-sided alternatives, respectively. These statistics – actually $(n - S_m^*)/2$ and $(n - S_m^+)/2$ – were introduced by Steel (1959). At that time he tabulated the distribution of S_m^+ for $m = 3$ with $n = 4(1)10$ and for $m = 4$ with $n = 4(1)7$; also approximate critical values ($\alpha = 0.01, 0.05$) of both S_m^* and S_m^+ for $m = 3(1)10$ with $n = 3(1)20$. Later Rhyne and Steel (1965) tabulated approximate critical values ($\alpha = 0.05, 0.10, 0.15$) of both statistics for $m = 3(1)10$ with $n = 4(1)25(5)50$; these are exact, with corresponding probabilities given, for $m = 3$ with $n = 4(1)24$, $m = 4$ with $n = 5(1)15$, and $m = 5$ with $n = 5(1)7$. Any treatment j can then be declared significantly different from the control if $|S_{jm}|$ exceeds the critical value for S_m^*, or significantly better (if smaller values of Y are better) if S_{jm} exceeds the critical value for S_m^+.

A different approach is based on the rank sums R_j; McDonald and Thompson (1967) credit this idea to unpublished work of Wilcoxon in 1956. The *simultaneous rank-sum statistic* (see also Nemenyi, 1963) is

$$R^* = \max_{j<j'} |R_j - R_{j'}| \,;$$

any two treatments j and j' are considered significantly different if $|R_j - R_{j'}|$ exceeds the critical value for R^*. For large n the asymptotic distribution of

$\sqrt{12n}R^*/\sqrt{m(m+1)}$ is the same as that of the range of m independent $N(0, 1)$ random variables. Based on this asymptotic approximation, Wilcoxon and Wilcox (1964) tabulated critical values ($\alpha = 0.01, 0.05, 0.10$) for $m = 3(1)10$ with $n = 1(1)25$. McDonald and Thompson (1967) provide critical values ($\alpha = 0.01, 0.03, 0.05$), with corresponding probabilities, some exact but most based on various bounds, for $m, n = 3(1)15$.

The rank sum approach can also be applied to the comparison of treatments with a control. Define the *many-one rank-sum statistics*

$$R_m^+ = \max_{j<m} (R_j - R_{jm}) \quad \text{and} \quad R_m^* = \max_{j<m} |R_j - R_m| ;$$

then declare treatment j significantly better than (or different from) the control if $(R_j - R_{jm})$ (or $|R_j - R_{jm}|$) exceeds the critical value for R_m^+ (or R_m^*). These procedures were first proposed by Nemenyi (1963). Wilcoxon and Wilcox (1964) provide approximate critical values ($\alpha = 0.01, 0.05$) of both R_m^+ and R_m^*, for $m = 3(1)10$ with $n = 1(1)25$, based on the asymptotic distribution: for large n, $\sqrt{6}R_m^+/\sqrt{nm(m+1)}$ is distributed as the maximum, and $\sqrt{6}R_m^*/\sqrt{nm(m+1)}$ as the maximum of the absolute values, of $(m-1)$ $N(0, 1)$ random variables with common correlation $\frac{1}{2}$. Hollander and Wolfe (1973) tabulate the exact distributions of R_m^+ and R_m^* for $m = 3$ with $n = 2(1)18$ and $m = 4$ with $n = 2(1)5$.

Youden (1963) suggested declaring a treatment significantly different from the others if its rank sum comes too close to the minimum possible value (m) or to the maximum possible value (mn). Accordingly, Thompson and Willke (1963) tabulate approximate critical values of the *extreme rank-sum statistic*

$$R_E = \min_j \min(R_j - n, mn - R_j)$$

at $\alpha = 0.01$, 0.03 and 0.05 for $m, n = 3(1)15$. For large n, $\sqrt{12/n(m^2-1)}$ $[n(m-1)/2 - R_E]$ is distributed as the maximum of m $N(0, 1)$ variables with common correlation $-1/(m-1)$.

As pointed out by Miller (1966), the rank sum approach to multiple comparisons involves a conceptual disadvantage with respect to the sign tests, in that it makes the outcome of the comparison between two treatments depend on the values of the observations on the other treatments. There may, however, be some advantage in power, particularly since the sign statistics take on relatively few distinct values.

Applying the procedures of this Section to Example B gives results as follows (with $\alpha = 0.05$ experimentwise). For the simultaneous sign statistics S^* and S^+ it is not possible to achieve significance for $m = 3$ with $n = 5$; the observed values $S^* = 5$ and $S^+ = 5$ correspond to $P = 0.1628$ (1266/7776) and $P = 0.0855$ (665/7776) respectively. With the many-one sign statistic it is also not possible to achieve significance; the observed $S_m^* = 5$ and $S_m^+ = 5$ correspond to $P = 0.1165$ (906/7776) and $P = 0.0584$ (454/7776). For the simultaneous rank-sum statistic R^* the critical value is 8, and treatments 1 and 3 are declared significantly different since $|R_1 - R_3| = 8$, with $P = 0.0394$ (306/7776). For the two-sided many-one rank-sum statistic R_m^* the critical value is again 8, so treatment 3 (now considered a control) and treatment 1 are declared significantly different, but this time with exact $P = 0.0262$ (204/7776). For the one-sided R_m^+ the critical value is 7, and treatment 1 is declared to rank below treatment 3 (control) with exact $P = 0.0131$ (102/7776). Finally, the critical value for the extreme rank-sum statistic R_E is 0, but the observed $R_E = 1$, so no treatment is declared significantly different from the others; the exact $P =$ 0.1242 (966/7776).

5. Additive block effects

In the classical or parametric analysis of variance a more restrictive assumption than IIa is imposed: that of (at most) *additive block effects*. This may be expressed as

ASSUMPTION IIb. There exist quantities $\beta_1, \ldots, \beta_n$ (block effects) such that the random vectors $(Y_{i1} - \beta_i, \ldots, Y_{im} - \beta_i)'$ are all identically distributed.

Under this assumption, meaningful information can be obtained from comparisons of observations between blocks as well as within; but this *interblock information* is wasted by the procedures discussed so far which use only within-block comparisons.

In the special case where $m = 2$, the interblock information can be recovered by the well-known signed-rank test of Wilcoxon (1945). This is based on the statistic

$$W_{12} = \sum Q_i \operatorname{sgn}(Y_{i1} - Y_{i2}),$$

where $Q_1, \ldots, Q_n$ are the ranks of the within-block ranges or absolute differences $|Y_{i1} - Y_{i2}|$. Under Assumptions I and IIB these ranges are independent and identically distributed, so that their ranks are a random permutation of $\{1, \ldots, n\}$. Under H_0 these between-block ranks are independent of the within-block rankings, or signs of $(Y_{i1} - Y_{i2})$, so the $n! \times 2^n$ possible realizations of ranks and signs considered jointly are all equally likely. Thus W_{12} is distribution-free. Extensive tabulations exist, though usually for $W_{12}^+ = W_{12}/2 + n(n+1)/4$, which is always nonnegative; see for example Wilcoxon et al. (1970). Clearly $E[W_{12}] = 0$, and it is easily shown that $V[W_{12}] = n(n+1)(2n+1)/6$; the asymptotic distribution is normal, and since consecutive values of the statistic differ by 2, a continuity correction of 1 is appropriate.

In Example A, the within-block differences are (1, 2, 3, −2, 4), with corresponding signed ranks (1, 2.5, 4, −2.5, 5), if average ranks are used for the tie. This corresponds to $W_{12} = 10$, between the tabulated values $P = 0.01875$

(6/32) at $W_{12} = 11$ and $P = 0.3125$ (10/32) at $W_{12} = 9$ for a two-sided alternative: the randomization level $P = 0.2500$ (8/32) is exactly halfway between these. Using the normal approximation, $Z = (10-1)/\sqrt{55} = 1.21$, so that $P = 0.2249$ (not adjusting the variance for the tie).

For $m > 2$ the situation is less satisfactory. Pitman (1983) proposed a randomization test based on the classical two-way analysis of variance statistic, but this has the usual difficulties associated with randomizing the raw data. A different extension to $m > 2$ proceeds as follows. First estimate each block effect β_i by a suitable statistic $\hat{\beta}_i$, for example the block mean $\Sigma\, Y_{ij}/m$; then align each block by subtracting $\hat{\beta}_i$ from every observation in it; and finally rank the mn quantities $(Y_{ij} - \hat{\beta}_i)$ without regard to block, thus obtaining values r_{ij} for analysis. This idea of *ranking after alignment* was originally suggested by Hodges and Lehmann (1962), although their development was limited to the case $m = 2$ (but $l_{ij} \geqslant 1$). Mehra and Sarangi (1967) proposed the test statistic

$$X^2_{MS} = \frac{(m-1)\,\Sigma_j\, \{\Sigma_i\, r_{ij} - n(mn+1)/2\}^2}{\Sigma_i\, \Sigma_j\, (r_{ij} - \bar{r}_i)^2}$$

where $\bar{r}_i = \Sigma\, r_{ij}/m$, and showed that this statistic has a $\chi^2(m-1)$ distribution asymptotically under H_0; equivalently, one may perform an ordinary two-way analysis of variance using the aligned ranks. A generalization due to Sen (1968b) replaces each aligned rank r_{ij} by $w_{ij} = J_N(r_{ij}/N)$, where the sequence of functions $J_N(u)$ converges to a suitably regular function $J(u)$ on $(0, 1)$.

Doksum (1967) proposed a procedure which consists essentially of performing signed rank tests on all pairs of treatments simultaneously. Let $Q_{ijj'}$ be the rank of $|Y_{ij} - Y_{ij'}|$ among the n within-block absolute differences between observations given treatments j and j', and let

$$D_{jj'} = \sum (Q_{ijj'} - 1)\, \mathrm{sgn}(Y_{ij} - Y_{ij'}) = W_{jj'} - S_{jj'}\,.$$

Note that $W_{jj'}$ is the signed-rank statistic for comparing treatments j and j', and $S_{jj'}$ the sign statistic. ($D_{jj'}$, which might be called the *diminished signed-rank statistic*, is of interest in its own right for the case $m = 2$. It can alternatively be obtained by calculating $W_{jj'}$ except ranking from 0 to $n-1$ instead of from 1 to n; and its null hypothesis distribution is the same as that of $W_{jj'}$ but for sample size $n-1$.) Then under H_0

$$D = \frac{6\, \Sigma_j\, (\Sigma_{j'}\, D_{jj'})^2/n(n-1)(m-1)}{2n[1 + (m-2)(12\lambda - 3)] + [(m-2)(13 - 48\lambda) - 1]}$$

is asymptotically distributed as $\chi^2(m-1)$; but the quantity λ in this expression is an unknown parameter. An asymptotically distribution-free test can be obtained by substituting a consistent estimator $\hat{\lambda}$ for λ. Lehmann (1964), who had proposed a test similar to Doksum's but much more complicated, suggested how to obtain such an estimate. Alternatively, since Lehmann also showed that λ never exceeds $\frac{7}{24}$, one may obtain a conservative test by substituting this upper bound. Mann and Pirie (1982) showed that λ never falls below $\frac{5}{18}$, and indicate that the conservatism may be minor. Koch and Sen (1968) proposed a test (W^*) based on the undiminished signed-rank statistics, which will be considered in Section 9.

Although Mehra and Sarangi (1967) remarked that "on account of possibly unequal (and unknown) block effects, no worthwhile information would be contained in the ranks based on joint-ranking before alignment", Conover and Iman (1980, 1981) have proposed a *rank transform* method which utilizes exactly such joint ranks in an ordinary two-way analysis of variance. (Note that this method also applies when $m = 2$, and does not reduce to any earlier test in that case.) They have so far developed no theoretical underpinnings for this approach, and the question must arise as to whether it is even valid in the null case. Their extensive Monte Carlo results suggest, however, that the intuitive approximation may be satisfactory for practical purposes, at least for $n = 10$ or more blocks. A similar question of validity arises for the other tests presented so far which utilize interblock information in comparing 3 or more treatments, since none of them is distribution-free, except asymptotically, or insofar as it may be evaluated by randomization. Gilbert (1972) simulated X^2_{MS} and the Koch–Sen W^* for $m = 3$ and $n = 3(2)9$; he concluded that 10 or more blocks may be required before their asymptotic distributions are sufficiently accurate for testing purposes. His results with respect to X^2_{MS} were confirmed by Silva (1977).

These drawbacks do not apply to the procedures recently proposed by Kepner and Robinson (1982). Consider first the case where $m = 3$. Define

$$W_{123} = \sum Q_{i123} \operatorname{sgn}(Y_{i1} + Y_{i2} - 2Y_{i3})$$

where Q_{i123} is the rank of $|Y_{i1} + Y_{i2} - 2Y_{i3}|$ among its values in the n different blocks. Kepner and Robinson show that under H_0 the special signed-rank statistic W_{123} is independent of the ordinary signed-rank statistic W_{12}, which ensures that their test statistic

$$X^2_{123} = 6(W^2_{12} + W^2_{123})/n(n+1)(2n+1)$$

is strictly distribution-free. Thus an exact small-sample tabulation could be produced; the asymptotic distribution is $\chi^2(2)$. Note, however, that there are actually three distinct test statistics, obtainable by permuting the treatments, and the choice among them is arbitrary. For the case where $m = 4$, define the special signed-rank statistic

$$W_{1234} = \sum Q_{i1234} \operatorname{sgn}(Y_{i1} + Y_{i2} - Y_{i3} - Y_{i4}),$$

where Q_{i1234} is the rank of $|Y_{i1} + Y_{i2} - Y_{i3} - Y_{i4}|$ among the n blocks. Then under H_0 the test statistic

$$X^2_{12/34} = 6(W^2_{12} + W^2_{34} + W^2_{1234})/n(n+1)(2n+1)$$

is also distribution-free, and is asymptotically distributed as $\chi^2(3)$. Again three distinct statistics can be obtained by permuting the treatments. A fourth test statistic is

$$X^2_{1234} = 6(W^2_{1234} + W^2_{1324} + W^2_{1423})/n(n+1)(2n+1)\,,$$

which has the same distribution; this is easily seen to be invariant under permutation of the treatments. Unfortunately, it does not seem possible to extend the Kepner–Robinson approach to $m > 4$; but (as will be seen in Section 7) this is perhaps not too important since there may be little interblock information to recover given larger blocks.

For Example B, an ordinary two-way analysis of variance produces $VR = 4.47$, corresponding to $P = 0.0498$ using the F-distribution, but $P = 0.0586$ (456/7776) by randomization. Subtracting the block means (28, 30, 25, 33, 34) produces aligned observations and corresponding ranks as follows (again using average ranks for ties):

	7	1	−8		14	7.5	2.5
	7	4	−11		14	11	1
(aligned observations)	1	3	−4	(aligned ranks)	7.5	9.5	4.5
	5	−8	3		12	2.5	9.5
	7	−3	−4		14	6	4.5

Thence $X^2_{MS} = 5.86$, corresponding to $P = 0.0534$, or alternatively, analysis of variance produces the variance ratio 5.66, corresponding to $P = 0.0294$; the exact level is $P = 0.0455$ (354/7776) by randomization. The rank transform is as follows:

	11	7	2
	13	10	1
(rank transform)	5	6	3
	14	4	12
	15	9	8

This yields $VR = 4.25$, corresponding to $P = 0.0552$, or $P = 0.0818$ (636/7776) by randomization. To apply Doksum's method, we calculate

	samples (j, j') 1, 2	1, 3	2, 3
$W_{jj'}$	13	15	7
$S_{jj'}$	3	5	3
$D_{jj'}$	10	10	4

Thence $\Sigma_j (\Sigma_{j'} D_{jj'})^2 = 632$ and $D = 7.24$ if the upper bound is used for λ, with $P = 0.0261$ as the conservative approximate level. Finally, for the Kepner–Robinson tests, we have:

$$\begin{array}{lll} W_{123} = W_{213} = 13, & X^2_{123} = X^2_{213} = 6.15, & P = 0.0463 \\ W_{132} = W_{312} = 0, & X^2_{132} = X^2_{312} = 4.09, & P = 0.1293 \\ W_{231} = W_{321} = -15, & X^2_{231} = X^2_{321} = 4.98, & P = 0.0828 \end{array}$$

(In all three cases there were ties, for which average ranks were used with no adjustment for the variance.)

There are also several procedures available for making use of interblock information in testing against an ordered alternative – the predicted ranking $(P_1, \ldots, P_m)'$, say. One approach, based on simultaneous signed-rank statistics, was taken in two consecutive papers in the *Annals* by Hollander (1967) and Doksum (1967). Their criteria, in standardized form, are

$$H^+ = \frac{3\Sigma\Sigma \operatorname{sgn}(P_j - P_{j'})W_{jj'}}{\{n(n+1)(2n+1)m(m-1)[3+2(m-2)\rho_n]\}^{1/2}}$$

and

$$D^+ = \left\{\frac{72/m^2(m^2-1)n(n-1)}{2n[1+(m-2)(12\lambda-3)]+[(m-2)(13-48\lambda)-1]}\right\}^{1/2} \sum_j P_j \sum_{j'} D_{jj'},$$

where λ and the ρ_n are unknown parameters, λ the same as in Doksum's test against the unordered alternative. Both H^+ and D^+ are asymptotically $N(0, 1)$ under H_0. Thus asymptotically distribution-free tests are obtained if the unknown parameters are suitably estimated, or conservative tests if upper bounds are substituted for them. Hollander shows that

$$\rho_n \leq \frac{n^2 + 2n(\sqrt{2}-1) + (3-2\sqrt{2})}{(n+1)(2n+1)},$$

and $\rho_n \to (12\lambda - 3)$ as $n \to \infty$. Puri and Sen (1968) generalized this approach by defining statistics

$$PS_{jj'} = \sum_i q_{Q_{ijj'}} \operatorname{sgn}(Y_{ij} - Y_{ij'}),$$

where the q_i are block scores; then their criteria are based on $\Sigma\Sigma \operatorname{sgn}(P_j - P_{j'})PS_{jj'}$ (generalizing Hollander's test) and $\Sigma\Sigma P_j PS_{jj'}$ (generalizing Doksum's test). In order to treat the asymptotic situation as $n \to \infty$, Puri and Sen let q_i be the expected value of the i-th order statistic of a sample of size n from a distribution function $\Psi^*(x) = \Psi(x) - \Psi(-x)$, where $\Psi(x)$ is symmetric about 0 and satisfies certain Chernoff–Savage regularity conditions. Then their criteria are asymptotically normal under H_0, each with mean 0 and standard deviation estimable from the data. Consider Example B, with (3, 2, 1) as the predicted ranking. We have $\Sigma\Sigma\,\mathrm{sgn}(P_j - P_{j'})W_{jj'} = 70$, and the upper bound on ρ_5 is $(9+4\sqrt{2})/33 = 0.444$, whence $H^+ = 2.39$, corresponding to $P = 0.0083$ conservatively. Or, $\Sigma P_j \Sigma D_{jj'} = 34$, and letting $\lambda = 7/24$ produces $D^+ = 2.11$ or $P = 0.0175$ conservatively.

In a different approach, Sen (1968b) uses the aligned ranks r_{ij} and functions J_N as in his test for the unordered alternative. Write $w_{ij} = J_N(r_{ij}/N)$ and $\bar{w}_i = \Sigma\, w_{ij}/m$; then under H_0

$$RA^+ = \frac{\Sigma_{j=1}^{m}[(P_j - (m+1)/2)\Sigma_{i=1}^{n} w_{ij}]}{\{[m(m+1)/12]\Sigma\Sigma(w_{ij} - \bar{w}_i)^2\}^{1/2}}$$

is asymptotically $N(0,1)$. The simplest special case occurs when J_N is constant, so that the w_{ij} are equivalent to the r_{ij}; in Example B this produces $RA^+ = 2.39$ and $P = 0.0084$ approximately. De (1976) and Boyd and Sen (1984) extend these ideas further by employing the union–intersection principle. Their tests are also related to a procedure of Shorack (1967) which does not utilize interblock information: he suggested forming 'amalgamated' means $\bar{R}_{(1)}, \ldots, \bar{R}_{(k)}$ from the treatment rank means $\bar{R}_j = R_j/n$ according to the process derived by Bartholomew (1959) for testing against an ordered alternative in one-way analysis of variance. Then under H_0 the distribution of

$$\bar{X}_S^2 = \frac{12n}{m(m+1)} \sum_j m_j \left[\bar{R}_{(j)} - \frac{m+1}{2}\right]^2$$

is asymptotically the same as Bartholomew's, where m_j is the number of treatments included in $\bar{R}_{(j)}$; in particular, for $m = 3$ the approximate P-value given $\bar{X}_S^2 = x > 0$ is $P = \Phi(-\sqrt{x}) + e^{-x/2}/6$, where Φ is the standard normal distribution function. Note that $\bar{X}_S^2$ reduces to Friedman's X_F^2 if the $\bar{R}_j$ are exactly in the predicted ordering. This is the case in Example B, whence $\bar{X}_S^2 = X_F^2 = 6.40$, and $P = 0.0125$ approximately.

In concluding this Section, we may mention that Nemenyi (1963) suggested multiple comparison procedures based on signed ranks, both for comparing all pairs of treatments and for comparing treatments with a control. These were further developed by Miller (1966) and Hollander (1966). Other multiple comparison methods which incorporate interblock information are due to Sen (1969) and Wei (1982).

6. Weighted rankings

Given Assumption IIb, under which the blocks are comparable, suppose the observations on different treatments are more distinct in some blocks than in the others; then it seems intuitively reasonable that the ordering of the treatments which these blocks suggest is more likely to reflect any underlying true ordering. These same blocks might more or less equivalently be described as having greater observed variability, although the word *observed* is to be emphasized because actually blocks are identically distributed except for additive block effects. Thus, these blocks, which will be referred to as more *credible* with respect to treatment ordering, may be given greater weight in the analysis.

This idea seems to have been expressed first in a rarely-cited paper of Tukey (1957), where he proposed the following procedure for the case $m = 3$: Assign *block ranks* $Q_1, \ldots, Q_n$ according to the least difference among the three responses, and let the test statistic be $T = \max(T_k)$, where T_k is the sum of the ranks assigned to those blocks which exhibit the k-th of the 6 possible orderings. He remarked, however, that the technique "is not, for the present, recommended for use". Given a two-sided ordered alternative, rank according to the least differences between responses to adjacent treatments, and relabel as necessary so that the favored ordering and its opposite are the first two; then the test statistic is $T_0 = \max(T_1, T_2)$. Tukey tabulated 5% and 1% critical values of both T and T_0 for $n = 3(1)10$. In Example B, for the unordered alternative, the block ranks are (5, 4, 2.5, 2.5, 1), so $T = 10$; for the ordered alternative, the block ranks are (4, 3, 2, 5, 1), so $T_0 = 8$. Neither statistic is significant at 5%.

Quade (1972b) independently discovered the idea of ranking the blocks, and proposed a method of *weighted rankings*, as follows. Let $0 \leq q_1 \leq \cdots \leq q_n$ be fixed *block scores* (or *block weights*), and let $r_1, \ldots, r_m$ be fixed *treatment scores* as in Section 3. Then the general weighted-rankings statistic is

$$X_Q^2 = \frac{(m-1)\sum_j \{\sum_i q_{Q_i}(r_{R_{ij}} - \bar{r})\}^2}{\sum q_i^2 \sum (r_j - \bar{r})^2}\,.$$

Quade shows that the asymptotic null-hypothesis distribution of X_Q^2 is $\chi^2(m-1)$, provided the block weights are so chosen that

$$\sum_i (q_i - \bar{q})^k \Big/ \Big[\sum (q_i - \bar{q})^2\Big]^{k/2} = O(n^{1-k/2}) \quad \text{for } k = 3, 4, \ldots;$$

he also gives formulas for a three-moment chi-squared approximation. An equivalent formulation (Quade, 1979) begins with the score correlations $O_{ii'}$, from which one may define the *weighted average internal score correlation*

$$\bar{\bar{O}}_w = \frac{\sum_{i<i'} q_{Q_i} q_{Q_{i'}} O_{ii'}}{\sum_{i<i'} q_{Q_i} q_{Q_{i'}}} = \frac{2\sum_{i<i'} q_{Q_i} q_{Q_{i'}} O_{ii'}}{(\sum q_i)^2 - \sum q_i^2}\,;$$

then

$$X_Q^2 = (m-1)\Big[1 + \frac{(\sum q_i)^2 - \sum q_i^2}{\sum q_i^2}\,\bar{\bar{O}}_w\Big].$$

The class of weighted-rankings statistics involves several dimensions of choice, including the block weights q_i, the treatment scores r_j, and the measure of credibility. A simple and intuitive special case arises from choosing *linear* block weights $q_i = i$, with treatment scores $r_j = j$. This yields a linearly weighted average internal Spearman correlation

$$\bar{\bar{S}}_L = \frac{2(2n+1)}{(n-1)(3n+2)}\left[\frac{72L}{(m^3-m)n(n+1)(2n+1)} - 1\right],$$

where

$$L = \sum_{j=1}^{m}\left\{\sum_{i=1}^{r} Q_i\left(R_{ij} - \frac{m+1}{2}\right)\right\}^2,$$

and hence

$$X_L^2 = \frac{72L}{m(m+1)n(n+1)(2n+1)}.$$

The statistic X_L^2 was suggested by Quade (1972b), and later independently proposed by Lawler (1978). Quade tabulates its exact null-hypothesis distribution in this special case for the following combinations of m and n: (3, 3), (3, 4), (3, 5), (3, 6), (3, 7), (4, 3), (4, 4) and (5, 3). The three-moment χ^2 approximation treats $\{[X_L^2 - (m-1)]\gamma_1/\gamma_2 + \alpha\}$ as $\chi^2(\alpha)$, where

$$\alpha = (m-1)\gamma_1^3/\gamma_2^2,$$

$$\gamma_1 = 1 - 6(3n^2+3n-1)/5n(n+1)(2n+1),$$

$$\gamma_2 = 3\gamma_1 - 2 + 72(3n^4+6n^3-3n+1)/7n^2(n+1)^2(2n+1)^2.$$

Because the successive values of L differ by 2 (or multiples of 2), a continuity correction of 1 to L may be employed in applying the asymptotic χ^2 approximations. The measure of credibility is still open to choice; reasonable possibilities include the block range and variance, in addition to the measure proposed by Tukey.

For testing against the ordered alternative of a predicted ranking $\boldsymbol{P} = (P_1, \ldots, P_m)'$, Salama and Quade (1981) generalized the average external rank correlation of Section 2 in the obvious manner to a *weighted average external rank correlation*

$$\bar{C}_w = \sum q_{Q_i} C_i \Big/ \sum q_i.$$

Clearly any such statistic is distribution-free under H_0, and its sampling distribution could be tabulated for small m and n. For larger experiments one may use a normal approximation: If the block weights $q_1, \ldots, q_n$ are so chosen that $\max \sum q_i^2/q_i^2 \to 0$ as $n \to \infty$, then $\bar{C}_w$ is asymptotically normally distributed under H_0, with mean equal to $E[C]$ and variance $V[C]\sum q_i^2/(\sum q_i)^2$, where $E[C]$ and $V[C]$ are the null-hypothesis mean and variance of the correlation statistic C.

As particular special cases Salama and Quade proposed using the Spearman correlations S_i or the Kendall correlations K_i, with the linear weights $q_i = i$. This produces

$$\bar{S}_L = \frac{2}{n(n+1)} \sum_{i=1}^{n} Q_i S_i = \frac{12 I_S}{n(n+1)(m^3 - m)}$$

where

$$I_S = 2 \sum_{i=1}^{n} Q_i \sum_{j=1}^{m} P_j R_{ij} - m(m+1)^2 n(n+1)/4\,,$$

and

$$\bar{K}_L = \frac{2}{n(n+1)} \sum_{i=1}^{n} Q_i K_i = \frac{4 I_K}{n(n+1)m(m-1)}$$

where

$$I_K = \sum_{i=1} Q_i \sum_{j<j'} \operatorname{sgn}(P_j - P_{j'}) \operatorname{sgn}(R_{ij} - R_{ij'})\,.$$

The quantities I_S and I_K are convenient integers; Salama and Quade tabulated exact one-sided critical values in terms of them at $\alpha = 0.1000$, 0.0500, 0.0250, 0.0100, 0.0050, 0.0025, 0.0010 and 0.0005 for $m = 3$ with $n = 2(1)15$, $m = 4$ with $n = 2(1)10$, $m = 5$ with $n = 2(1)9$, $m = 6$ with $n = 2(1)7$, and $m = 7$ with $n = 2(1)5$. Both $\bar{S}_L$ and $\bar{K}_L$ have mean 0 under H_0;

$$V[\bar{S}_L] = \frac{2(2n+1)}{3n(n+1)(m-1)} \quad \text{and} \quad V[\bar{K}_L] = \frac{4(2n+1)(2m+5)}{27n(n+1)m(m-1)}\,.$$

For large n the asymptotic normality may be used to approximate the distribution; a continuity correction of 1 to I_S or to I_K may be incorporated. Some other special cases are considered by Salama and Quade (1984).

In Example B, using either range or variance as the measure of credibility, the block ranks are (4, 5, 1, 3, 2). Considering first the unordered alternative, we have $L = 344$, where $\bar{\bar{S}}_L = 0.688$ and $X_L^2 = 6.25$ and $P = 0.0438$. Using the three-moment approximation, including continuity correction, we calculate $X^2 = 18.53$ with 8.14 degrees of freedom, corresponding to $P = 0.0190$; from the table in Quade (1972b) we find $P = 0.0193$ (150/7776) exactly. The weighted average external Spearman correlation is $\bar{S}_L = 0.867$ ($I_S = 52$), with $P = 0.0026$ (20/7776) exactly; the normal deviate (with continuity correction) is $Z = 2.43$, with $P = 0.0075$ approximately. Similarly, the weighted average external Kendall correlation is $\bar{K}_L = 0.822$ ($I_K = 37$), with $P = 0.0033$ (26/7776) exactly; the normal deviate is $Z = 2.54$ with $P = 0.0056$ approximately.

7. Estimation of treatment effects

To this point attention has been focused on testing the null hypothesis of random ranking, with the alternative expressed only in terms of the distribution of the ranks. To progress further we shall require a suitable model for the alternative, expressed in terms of the original observations. For example, we may suppose that

$$Y_{ij} = t(X_{ij}; \tau_j)$$

where the X_{ij} are interchangeable within blocks, the τ_j are fixed but unknown parameters called *treatment effects*, and t is a known *treatment transformation function* such that $t(x; 0) = x$. Note that this otherwise general model does not allow for any *interaction* between blocks and treatments, in that the same transformation applies to every block. However, it is totally nonrestrictive with respect to the nature of block-to-block differences under H_0.

Given such a model, one may test the hypothesis that the effects have any specified values $\boldsymbol{\tau} = (\tau_1, \ldots, \tau_m)'$; this hypothesis may be denoted $H(\boldsymbol{\tau})$, and the null hypothesis is then $H(\mathbf{0})$. The test is accomplished by inverting the treatment transformation function to obtain values $X_{ij} = t^{-1}(Y_{ij}; \tau_j)$ and then testing the interchangeability of the X_{ij} using any of the procedures presented earlier. In addition, of course, the confidence region for the parameter $\boldsymbol{\tau}$ consists of those values $\boldsymbol{\tau}$ which, if hypothesized, would be accepted by the test, where if the test is at level (at most) α, then the confidence coefficient is (at least) $1-\alpha$. Note how such an approach avoids the difficulties caused by ties: Since the values of $\boldsymbol{\tau}$ which lead to ties generally form only a lower-dimensional subset of the parameter space, we may declare such values acceptable if and only if they are limit points of unambiguously acceptable values. This principle amounts to taking all confidence regions as closed sets, which makes them and the corresponding tests conservative. Note also that letting the confidence coefficient decrease toward zero, and taking the limit of the correspondingly shrinking confidence region, may produce a reasonable point estimate of $\boldsymbol{\tau}$. This idea was developed by Hodges and Lehmann (1963).

From now on we specialize the model of the previous paragraph to *additive treatment effects*, as follows.

Assumption IIIa. There exist quantities $\tau_1, \ldots, \tau_m$ (treatment effects) such that for each $i = 1, \ldots, n$ the variables $\{Y_{i1} - \tau_1, \ldots, Y_{im} - \tau_m\}$ are interchangeable.

Since this assumption does not completely determine the τ_j, we shall impose the restriction that $\tau_m = 0$, without loss of generality. (Other restrictions are equally possible, of course; indeed, to make $\Sigma\, \tau_j = 0$ is more common.)

Consider first the special case of matched pairs ($m = 2$). Here we are assuming for each i that $X_{i1} = Y_{i1} - \tau_1$ and $X_{i2} = Y_{i2} - \tau_2 = Y_{i2}$ are interchangeable. We may test the hypothesis $H(\delta)$ that the difference between the treatment effects is $\tau_1 - \tau_2 = \tau_1 = \delta$, against the alternative that $\tau_1 \neq \delta$, by rejecting for large values of the sign statistic

$$S(\delta) = \left| \sum \operatorname{sgn}(Y_{i1} - Y_{i2} - \delta) \right| .$$

The corresponding confidence set, with confidence coefficient $(1-\alpha)$, is the set of values δ for which $H(\delta)$ is accepted at level α; this is an interval extended from the k-th smallest to the k-th largest of the n differences $d_i = Y_{i1} - y_{i2}$, where $(n+2-2k)$ is the critical value for $S(\delta)$. As the confidence coefficient decreases, k increases until $k = n/2$ for n even or $k = (n-1)/2$ for n odd; thus a reasonable point estimate of τ_1 is the median of the differences. Confidence intervals may also be one-sided, of course, based on the one-sided test criteria S_{12} and S_{21}.

If we now impose Assumption IIb, which was not required for point and interval estimates based on the sign test, then we may base estimates on the signed-rank tests also. Consider the $n(n+1)/2$ averages of two (not necessarily distinct) within-block differences, i.e.,

$$d_{ii'} = \frac{(Y_{i1} - Y_{i2}) + (Y_{i'1} - Y_{i'2})}{2} \quad \text{for } 1 \leqslant i \leqslant i' \leqslant n\,;$$

Tukey (1949) discovered that

$$W_{12} = \sum_{i \leqslant i'} \sum \operatorname{sgn}(d_{ii'})$$

is an alternate representation of Wilcoxon's statistic, from which it can be shown that a confidence interval for τ_1 extends from the k-th smallest to the k-th largest $d_{ii'}$, where $n(n+1)/2+2-2k$ is the critical value for $W = |W_{12}|$. Hodges and Lehmann (1963) discuss the median of the $d_{ii'}$ as a point estimator of τ_2. (If the diminished signed ranks were used, the corresponding point estimate would be the median of the $d_{ii'}$ where $i < i'$).

For Example A the within-block differences are as in the first column below, and the averages of two differences are as in all three columns:

$i = i'$	$d_{ii'} = d_i$	i, i'	$d_{ii'}$	i, i'	$d_{ii'}$
1	1.0	1, 2	1.5	2, 4	0.0
2	2.0	1, 3	2.0	2, 5	3.0
3	3.0	1, 4	−0.5	3, 4	0.5
4	−2.0	1, 5	2.5	3, 5	3.5
5	4.0	2, 3	2.5	4, 5	1.0

At $\alpha = 0.50$ (say), the critical value for $S(\delta)$ is 3, whence $k = 2$, so that the 50% confidence interval for τ_1 extends from the 2nd smallest to the 2nd largest of the d_i, i.e., [1.0, 3.0]. The corresponding point estimate is 2.0, the median of the d_i. For W the critical value at $\alpha = 0.50$ is 7, whence $k = 5$, so the 50% confidence interval extends from the 5th smallest to the 5th largest of the $d_{ii'}$, or from 1.0 to 2.5; the point estimate is 2.0 again.

A confidence procedure for the case $m = 3$ was suggested by Kraft and Van

Eeden (1968). Let $T(\tau_1, \tau_2)$ be the value of some test criterion T calculated from the within-block ranks of the values $X_{ij} = Y_{ij} - \tau_j$, for $i = 1, \ldots, n$, $j = 1, 2, 3$. Then a *confidence region* for (τ_1, τ_2) may be constructed by taking all points in the (τ_1, τ_2) plane such that $T(\tau_1, \tau_2)$ falls short of the critical value. But consider how $T(\tau_1, \tau_2)$ may vary as the point (τ_1, τ_2) moves along some continuous path. A small change in (τ_1, τ_2) may not change the within-block ranks of the X_{ij} at all, and hence may leave T unaffected. The within-block ranks can change only at a point (τ_1, τ_2) where $Y_{i1} - \tau_1 = Y_{i3}$, or $Y_{i2} - \tau_2 = Y_{i3}$, or $Y_{i1} - \tau_1 = Y_{i2} - \tau_2$, for some i. But the set of all such points form $3n$ lines in the (τ_1, τ_2) plane: one passing vertically, one horizontally, and one at a 45° angle, through each of the points $(Y_{i1} - Y_{i3}, Y_{i2} - Y_{i3})$. Within any one of the subregions bounded by these lines, the ranks remain constant, and hence the test criterion also; in different subregions the ranks are different, and the test criterion may (but may not) be different also. Thus to construct the confidence region it is necessary at worst to calculate the value of $T(\tau_1, \tau_2)$ for one point in each subregion; acceptance or rejection at that point determines acceptance or rejection of the whole subregion.

All this is quite feasible if n is small, and the shapes of the confidence regions produced may lend insight to the data, or to the nature of the test. The reader is invited to draw the three lines through each of the five points (15, 9), (18, 15), (5, 7), (2, −11) and (11, 1) determined by Example B, say, and thence construct the confidence region for (τ_1, τ_2) corresponding to Friedman's chi-squared or any other test based on the within-block ranks. (Kraft and Van Eeden applied their procedure only to Friedman's chi-squared, i.e., average internal rho; the confidence regions are easier to construct, however, using average internal or external tau.)

Extension of the Kraft and Van Eeden procedure to tests which make use of interblock information is possible in principle, but seems not generally feasible in practice. For example, to construct a confidence region based on ranking after alignment would require drawing $3n(3n-1)/2$ lines. A slightly simpler situation is presented by the weighted-ranking methods, at least if block variance is used as the measure of credibility; then, since ties in the block ranking can also be shown to correspond to lines in the (τ_1, τ_2) plane, there are in all $n(n+5)/2$ lines to be drawn. Of course, for all these tests, and for any n, an approximate confidence region may be conducted by dividing the plane into small subregions arbitrarily, calculating the test statistic at one point in each subregion, and declaring the whole subregion to lie inside the confidence region if and only if the one point chosen to represent it does so. The confidence boundaries may then be determined as closely as desired by making the subregions sufficiently small. Finally, extension to $m > 3$ obviously corresponds to plotting in higher-dimensional spaces.

Let us now consider the point estimation of arbitrary contrasts in the treatment effects. Let $\omega = \Sigma\, c_j\tau_j$, where $\Sigma\, c_i = 0$; then it seems reasonable to estimate ω by $\tilde{\omega}$, the median of the n estimates $w_i = \Sigma\, c_j Y_{ij}$; or, if interblock information is available, by the Hodges–Lehmann estimate $\hat{\omega}$, the median of the $n(n+1)/2$ averages $w_{ii'} = \Sigma\, c_j(Y_{ij} + Y_{i'j})/2$ for $1 \leqslant i \leqslant i' \leqslant n$. But a drawback of this approach is that the estimates of different contrasts are *incompatible*, in that, for example, $\omega \equiv \omega_1 + \omega_2$ does not imply $\tilde{\omega} = \tilde{\omega}_1 + \tilde{\omega}_2$ or $\hat{\omega} = \hat{\omega}_1 + \hat{\omega}_2$. To avoid this problem, Lehmann (1964) showed how to obtain a set of compatible estimates to replace the $\hat{\omega}$; Doksum (1967) adopted the same idea for the $\tilde{\omega}$, and Puri and Sen (1967) generalized the approach to estimates based on general scores. For each pair of treatments $1 \leqslant j, j' \leqslant m$, write

$$\tilde{t}(j, j') = \operatorname*{median}_{1 \leqslant i \leqslant n} (Y_{ij} - Y_{ij'})$$

and

$$\hat{t}(j, j') = \operatorname*{median}_{1 \leqslant i \leqslant i' \leqslant n} \frac{(Y_{ij} - Y_{ij'}) + (Y_{i'j} - Y_{i'j'})}{2}$$

for the median and Hodges–Lehmann estimates, respectively, of the difference $(\tau_j - \tau_{j'})$ between their effects. Then take

$$\tilde{\tau}_j = \frac{1}{m} \sum_{j'} [\tilde{t}(j, j') - \tilde{t}(m, j')] \quad \text{and} \quad \hat{t}_j = \frac{1}{m} \sum_{j'} [\hat{t}(j, j') - \hat{t}(m, j')]$$

as the compatible estimates of τ_j (note $\tilde{\tau}_m \equiv \hat{\tau}_m \equiv 0$), and from compatible estimates of the general contrast ω as $\Sigma\, c_j \tilde{\tau}_j$ or $\Sigma\, c_j \hat{\tau}_j$. It may be noted that although these new estimates resolve the problem of incompatibility, they still suffer from a possible disadvantage in that the estimate of every contrast involves data obtained on all treatments, even if the contrast depends on only some of them.

These estimators may be illustrated using the data of Example B. Note first that $(\tau_1 - \tau_2) + (\tau_2 - \tau_3) + (\tau_3 - \tau_1) \equiv 0$ but the direct estimates of these three differences are

$$\begin{array}{rllll} \text{median:} & \tilde{t}(1,2) = 6, & \tilde{t}(2,3) = 7, & \tilde{t}(3,1) = -11 & (\text{sum} = 2), \\ \text{Hodges–Lehmann:} & \hat{t}(1,2) = 6, & \hat{t}(2,3) = 5, & \hat{t}(3,1) = -10 & (\text{sum} = 1). \end{array}$$

The compatible estimates of these differences, which do sum to 0, are

$$\begin{array}{rlll} \text{median:} & 5\tfrac{1}{3}, & 6\tfrac{1}{3}, & -11\tfrac{2}{3}, \\ \text{Hodges–Lehmann:} & 5\tfrac{2}{3}, & 4\tfrac{2}{3}, & -10\tfrac{1}{3}. \end{array}$$

The estimates of treatment effects on which these are based are

$$\begin{array}{rlll} \text{median:} & \tilde{\tau}_1 = 11\tfrac{2}{3}, & \tilde{\tau}_2 = 6\tfrac{1}{3}, & \tilde{\tau}_3 = 0, \\ \text{Hodges–Lehmann:} & \hat{\tau}_1 = 10\tfrac{1}{3}, & \hat{\tau}_2 = 4\tfrac{2}{3}, & \hat{\tau}_3 = 0. \end{array}$$

These may be compared with the classical estimates by

$$\text{least-squares:} \quad t_1 = 10.2, \quad t_2 = 4.2, \quad t_3 = 0.$$

8. Efficiency

In this Section we present results concerning the *relative efficiency* of one test of H_0 with respect to another. Let n_A be the number of blocks required by some test A to detect a specified alternative, and n_B the number required by test B. Then the relative efficiency of A with respect to B is n_B/n_A, and the *asymptotic relative efficiency* (ARE) in the sense of Pitman (1948) may be interpreted as the limit of n_B/n_A as the alternative approaches the null hypothesis (in which case n_A and n_B both increase to infinity). Although defined in terms of large samples, Pitman ARE has been found a generally reliable guide for comparing tests even when the samples are quite small.

The basic situation which we consider is that Assumptions I, IIb and IIIb all hold, where Assumption IIIb is the strengthening of Assumption IIIa in which 'interchangeable' becomes 'independent and identically distributed'. We then have the linear model

$$Y_{ij} = \beta_i + \tau_j + E_{ij} \quad \text{for } i = 1, \ldots, n \text{ and } j = 1, \ldots, m\,,$$

where $\tau_m = 0$ and the E_{ij} are independently and identically distributed according to some distribution (say) F. Set $\tau_j = \theta_j/\sqrt{n}$, and let n tend to infinity; then in the limit many of the test statistics considered in this Chapter are distributed as noncentral chi-squared with $(m-1)$ degrees of freedom, and the Pitman ARE is the ratio of the noncentrality parameters (Hannan, 1956). We shall not give details of the assumptions required in each case; they can be found in the references.

The noncentrality parameter for the ordinary two-way analysis of variance statistic (VR) is well known to be $\Sigma\,(\theta_j - \bar{\theta})^2/\sigma^2(F)$ where $\bar{\theta} = \Sigma\,\theta_j/m$ (including $\theta_m \equiv 0$), and

$$\sigma^2(F) = \int x^2\,\mathrm{d}F(x)$$

is the error variance. This test will be used as the standard against which to compare the others in the unordered case. Sen (1968a) showed that any test based on an average internal score correlation $\bar{\bar{O}}$ has corresponding

$$\mathrm{ARE}(\bar{\bar{O}}, \mathrm{VR}) = \frac{m^2(m-1)\sigma^2(F)}{\Sigma\,(r_j - \bar{r})^2}\left[\sum_{k=1}^{m-1} \binom{m-2}{k-1}(r_{k+1} - r_k)\xi_k\right]^2,$$

where

$$\xi_k = \int [F(x)]^{k-1}[1 - F(x)]^{m-1-k}F'(x)\,\mathrm{d}F(x)$$

for $k = 1, \ldots, m-1$. The most important special case, that of Friedman's chi-squared (based on Spearman correlation), had previously been studied by Van Elteren and Noether (1959); for it the result simplifies to

$$\text{ARE}(\bar{\bar{S}}, \text{VR}) = \frac{m}{m+1}\sigma^2(F)\Psi^2(F)$$

where for any distribution F it is convenient to define the functional

$$\Psi^2(F) = 12\left[\int F'(x)\,\mathrm{d}F(x)\right]^2,$$

provided the density F' exists and the indicated integral converges. Bhapkar (1963) had considered the Brown–Mood test similarly. Note that for $m = 2$ all these tests reduce to the sign test, with ARE equal to $(2/3)\sigma^2(F)\Psi^2(F)$.

For ranking after alignment (RA) on block means, using scores based on the general function J, Sen (1968b) found

$$\text{ARE}(\text{RA}, \text{VR}) = \frac{\sigma^2(F)\{\int (\mathrm{d}/\mathrm{d}x)J[G_1(x)]\,\mathrm{d}G_1(x)\}^2}{\int J^2(u)\,\mathrm{d}u - \int\int J[G_1(x_1)]J[G_1(x_2)]\,\mathrm{d}G_2(x_1, x_2)},$$

where G_2 is the joint distribution under H_0 of two aligned observations in the same block, and G_1 is the corresponding marginal distribution. Mehra and Sarangi (1967) had previously treated the special case $J(u) = u$ which yields their test statistic X^2_{MS}; actual evaluation of the ARE is generally difficult, although if F is normal then

$$\text{ARE}(X^2_{\text{MS}}, \text{VR}) = \frac{3m}{4(m-1)\{\pi - 3\tan^{-1}\sqrt{(2m-3)/(2m-1)}}.$$

Doksum (1967) showed that the ARE for his test statistic D (with consistent estimator $\hat{\lambda}$ rather than upper bound $\frac{7}{24}$) is

$$\text{ARE}(D, \text{VR}) = \frac{(m/2)\sigma^2(F_2)\psi^2(F_2)}{1 + (m-2)(12\lambda - 3)},$$

where F_2 is the distribution of $(E_{i1} - E_{i2})$; the parameter $\lambda = \int F_3^2(x)\,\mathrm{d}F(x)$, where F_3 is the distribution of $(E_{i1} + E_{i2} - E_{i3})$. Kepner and Robinson (1982) derived efficiency formulas for their special signed-rank statistics; we present here only that for the statistic X^2_{1234} which is invariant under permutation of the ($m = 4$) treatments, viz., $\sigma^2(F_4)\Psi^2(F_4)$, where F_4 is the distribution function of $(E_{i1} + E_{i2} - E_{i3} - E_{i4})$. (With normal errors, the efficiency reduces to $3/\pi$ for all their statistics, both for $m = 3$ and $m = 4$.) Finally, although Silva (1977) was able to show that weighted rankings statistics also have asymptotic noncentral chi-squared distributions under Pitman alternatives, to date no explicit expressions for their efficiencies have been derived. For $m = 2$ all these tests (except for RA with general scores) reduce to Wilcoxon's signed-rank test, with ARE equal to $\sigma^2(F_2)\Psi^2(F_2)$.

Table 1 shows actual numerical values of ARE with respect to ordinary

Table 1
ARE with respect to analysis of variance of various nonparametric tests of interchangeability

Test	Uniform Errors				Normal Errors				Laplace Errors			
m:	2	3	4	∞	2	3	4	∞	2	3	4	∞
Average Rho (Friedman)	0.667	0.750	0.800	1.000	0.637	0.716	0.764	0.955	1.000	1.125	1.200	1.500
Ranking after Alignment (Mehra and Sarangi)	0.889	0.877 ±0.064	0.871 ±0.043	?	0.955	0.966	0.965	0.955	1.172	1.136 ±0.038	1.172 ±0.052	?
D (Doksum)	0.889	0.895	0.898	0.906	0.955	0.966	0.972	0.989	1.172	1.193	1.204	1.238
Special Signed-Rank Statistics (Kepner and Robinson)	0.889	{0.882 0.889	{0.889 0.919*	N.A.	0.955	0.955	0.955	N.A.	1.172	{1.150 1.172	{1.053* 1.172	N.A.
Linearly Weighted Rankings using Ranges (Quade)	0.889	1.021 ±0.068	1.151 ±0.049	?	0.955	0.858 ±0.032	0.871 ±0.028	?	1.172	1.027 ±0.035	1.021 ±0.048	?

N.A. = not applicable, ? = unknown.

analysis of variance of the nonparametric tests listed above, for three error distributions: uniform, normal and Laplace (double-exponential). The values of m included are $m = 2$ (in which case Friedman's χ^2 is equivalent to the two-sided sign test, and the others to the signed-rank test), $m = 3$, $m = 4$ and '$m = \infty$' (i.e., the limit as $m \to \infty$). The AREs shown for weighted rankings are estimates ($\pm$ standard errors) obtained by Silva and Quade (1983) for the statistic X_L^2 with linear block weights, using the range as the measure of credibility; their estimates are also shown for X_{MS}^2 in the uniform and Laplace cases, for which the exact expressions have not been evaluated. (It might be noted that the Silva–Quade method produced an underestimate of the ARE in 17 of 18 cases where an exact value was available for comparison.) For Doksum's test the AREs were calculated using $\lambda = 0.2909$ for the uniform distribution and $\lambda = 0.2902$ for the normal, as given by Lehmann (1964); for the Laplace distribution, $\lambda = 0.2894$ (3001/10368). The dual AREs shown for the Kepner–Robinson tests represent the minimum and maximum obtainable by permutation of the treatments; the starred values at $m = 4$ apply to the invariant statistic X_{1234}^2.

Turning now to the efficiencies of tests against ordered alternatives, suppose that for some constant τ we have $\tau_j = (P_j - P_m)\tau$ in the linear model $Y_{ij} = \beta_i + \tau_j + E_{ij}$, so that the hypothesis H_0: $\tau = 0$ may be tested against the alternative H_1: $\tau > 0$. Given normal errors, the maximum likelihood estimate of τ is

$$\hat{\tau} = \frac{12}{n(m^3 - m)} \sum_j \left(P_j - \frac{m+1}{2}\right) \sum_i Y_{ij},$$

and the likelihood ratio test, as given by Hollander (1967), rejects for large values of

$$T^+ = \hat{\tau}\left\{\frac{n(m-1) - 1}{12/n(m^3 - m)\sum\sum(Y_{ij} - \bar{Y}_i)^2 - \hat{\tau}^2}\right\}^{1/2}.$$

Since this test criterion is asymptotically $N(0, 1)$ under H_0, as are all the others we have presented which are aimed at the ordered alternative (except Shorack's $\bar{X}_S^2$), we are able to compare them using Pitman ARE.

Hollander (1967) found the efficiencies of Jonckheere's test (average external Kendall correlation) and Page's test (average external Spearman correlation) to be

$$\mathrm{ARE}(\bar{K}, T^+) = 2(m+1)\sigma^2(F)\Psi^2(F)/(2m+5)$$

and

$$\mathrm{ARE}(\bar{S}, T^+) = m\sigma^2(F)\Psi^2(F)/(m+1),$$

where as before F is the distribution function of any error E_{ij}. It may be noted that $\mathrm{ARE}(\bar{S}, T^+)$ is identical with $\mathrm{ARE}(\bar{\bar{S}}, \mathrm{VR})$ as given earlier in this Section. This relationship holds for the whole family of tests based on scores of which S

and PH are special cases. Thus ARE(PH, T^+), as given (implicitly) by Pirie and Hollander (1972), can be obtained by substituting $E_N(j)$ for r_j in the expression given for ARE($\bar{\bar{O}}$, VR). For $m = 2$, of course, all these tests reduce to the one-sided sign test.

Hollander (1967) found the efficiency of the test he proposed to be

$$\text{ARE}(H^+, T^+) = \frac{(m+1)\sigma^2(F_2)\Psi^2(F_2)}{3+2(m-2)(12\lambda-3)},$$

where F_2 and λ are as defined earlier in this Section, and Doksum (1967) found ARE(D^+, T^+) ≡ ARE(D, VR). For $m = 2$, of course, these results reduce to those for a one-sided signed-rank test. Puri and Sen (1968) extended these efficiency results to their generalizations of these tests, and Sen (1968b) showed that ARE(RA$^+$, T^+) = ARE(RA, VR).

Similar efficiency results pertain to the contrast estimators presented at the end of Section 7; note that the parameter τ we have been considering is itself a contrast. In particular, Doksum (1967) showed that the ARE of his compatible median estimator with respect to the least squares estimator of any contrast is the same as ARE($\bar{\bar{S}}$, VR) = ARE($\bar{S}$, T^+); and Lehmann (1964) has shown that the ARE of his compatible estimator is

$$\text{ARE}(\text{HL}, \text{LS}) = \frac{(m/2)\sigma^2(F^*)\Psi^2(F^*)}{1+(m-2)(12\lambda-3)}$$

where F^* is the distribution of the contrast estimator from any single block. The multiplier of $\sigma^2(F^*)\Psi^2(F^*)$ in this expression represents the ARE of the compatible estimators with respect to the incompatible; it equals 1 at $m = 2$, when the two estimators are identical, and increases with m to $1/2(12\lambda - 3)$, between 1 and 1.5, as $m \to \infty$. All these results were extended to general scores by Puri and Sen (1967).

The preceding presentation using Pitman ARE had to omit from consideration those statistics which are not asymptotically $\chi^2(m-1)$ (when considering unordered alternatives), or asymptotically normal (ordered). However, there have of course been numerous empirical studies to compare the various tests for small values of m and n. These studies, which began as early as the original paper of Friedman (1937), are often published only as theses or technical reports; or if more formally, they tend to be outside the mainstream of statistical literature. We cite here only a few of them, emphasizing discussions of techniques for which the Pitman ARE has not been tractable. Gilbert (1972) compared the powers of the classical variance ratio (VR), Friedman's X_F^2, the ranking-after-alignment statistic X_{MS}^2, and the Koch–Sen statistic W^*, for $m = 3$ with $n = 3(2)9$, given normal errors. Silva (1977) and Silva and Quade (1980) compared the expected significance levels of VR, X_F^2, X_{MS}^2 and 10 weighted-ranking statistics including X_L^2 (using 5 measures of credibility times 2 block scoring schemes), for $m = 3, 4, 5$ with $n = 3, 4, 5, 6$, given normal, uniform and Laplace errors. Conover and Iman (1980) compared the powers of VR, X_F^2, X_L^2, and the rank transform, for $m = 2, 3, 4, 5, 10$ with $n = 10, 20, 30$ (and 40, 50 if $m = 2$; recall that X_F^2 is equivalent to the sign test, and X_L^2 to the signed-rank test, in that case), given normal, uniform, Laplace, Cauchy and lognormal errors. Kepner and Robinson (1982) compared the powers of their statistics with X_F^2, for $m = 3$ with $n = 5$ and 10, and for $m = 4$ with $n = 4$ and 8, given normal, uniform and Laplace errors. Some studies which consider the ordered alternative are as follows: Salama and Quade (1981) compared the expected significance levels of $\bar{S}$, $\bar{K}$, $\bar{S}_L$, $\bar{K}_L$ and D^+ (using a bound on λ rather than estimating it), for $m = 3, 4, 5$ with $n = 3(1)7$, given normal, uniform and Laplace errors. Berenson (1982b) compared the powers of T_1^*, AT, $\bar{S}$, PH, H^+, RA^+ and $\bar{X}_S^2$ for $m = 4, 5, 6$ with $n = 5$ and 10, for 11 different error distributions; here T_1^* is a variant of T^*, AT is the parametric 'maximin contrast' of Abelson and Tukey (1963), and H^+ was calculated using the upper bound on ρ_n. Boyd and Sen (1984) compared the powers of $\bar{S}$ and four statistics derived from the union–intersection principle (one based on within-block ranks, two based on weighted ranks, and one on aligned ranks), for $m = 3, 4, 5$ with $n = 10(5)25$, given normal errors.

Let us now attempt to use these various efficiency results, both theoretical and empirical, to choose among nonparametric tests of the hypothesis of no treatment effects within the context of a linear model for the complete blocks design (Assumptions I, IIb and IIIb). Such tests will be considered in three stages: first, comparisons among those which ignore interblock information by relying strictly on within-block ranks; second, comparisons among procedures which utilize the interblock information; and finally, comparisons between these two major classes of tests.

For $m = 2$, all nonparametric procedures using only the within-block ranks reduce to the sign test. For $m > 2$, average Spearman correlation – Friedman's X_F^2 (or $\bar{\bar{S}}$) for the unordered alternative and Page's L (or $\bar{S}$) for the ordered – are close to being standard. With respect to the classical procedures based on assumptions of normality, they have

$$\text{ARE} = \frac{m}{m+1}\sigma^2(F)\Psi^2(F)$$

in the Pitman sense, where F is the error distribution. Hodges and Lehmann (1956) showed that $\sigma^2(F)\Psi^2(F)$ cannot fall below 0.864 if F is symmetric. Sen (1968a) showed how to choose the best test against the unordered alternative from among those based on score correlations. The choice depends on F, but Spearman correlation is optimal for $m = 3$ if F is symmetric, and for all m if F is logistic. Schach (1979) and Alvo et al. (1982) indicate that Anderson's test (based on $\bar{\bar{A}}$) and Ehrenberg's ($\bar{\bar{K}}$), respectively – recall that these are not score correlation tests – are superior to Friedman's on the basis of approximate Bahadur (1960, 1967) slope. For the ordered alternative, however, a direct comparison of Spearman with Kendall correlation (Hollander, 1967) shows

$$\text{ARE}(\bar{S}, \bar{K}) = m(2m+5)/2(m+1)^2 .$$

This increases with m from the value 1 at $m = 2$ to a maximum of 1.042 (25/24) at $m = 5$, and then decreases to 1 again as $m \to \infty$. Thus Page's test based on $\bar{S}$, in addition to being a bit simpler computationally, is asymptotically at least as efficient as Jonckheere's test based on K, for all m and all error distributions, though the difference is always small. Pirie and Hollander (1972) suggested that their test PH 'should be preferred to Page's test' because 'it exhibits better efficiency properties' in that ARE(PH, $\bar{S}$) > 1 for $m > 3$ given normal, uniform or exponential errors; but ARE(PH, $\bar{S}$) < 1 given (for example) logistic errors; furthermore, PH is more complicated to use and less well tabulated. Berenson (1982b) found Shorack's $\bar{X}^2_S$ distinctly inferior to all the other tests he studied, for the sort of ordered alternative ('trend') which has been considered here, although it did quite well for another sort of contrast ('end gap'), especially for skewed error distributions. In summary, within the class of tests not utilizing interblock information, unless one has rather specialized knowledge of the situation to suggest some other choice, Friedman's and Page's tests are to be recommended.

For $m = 2$, the standard nonparametric procedure among those which pay attention to the interblock information is Wilcoxon's matched-pairs signed-rank test. With respect to the classical procedures based on assumptions of normality, this has ARE $= \sigma^2(F_2)\Psi^2(F_2)$, which cannot fall below 0.864 since F_2 is symmetric. For $m > 3$ no one of the available tests has established itself as standard. The methods based on ranking after alignment, or simultaneous signed-rank tests (except for the Koch–Sen W^*, which actually tests a broader hypothesis than considered here), or on the rank transform all appear to recover the interblock information satisfactorily for the unordered situation. For the ordered alternative, one comparison between two of these tests is easy: we have

$$\text{ARE}(D^+, H^+) = \frac{m}{m+1} \times \frac{3 + 2(m-2)(12\lambda - 3)}{2 + 2(m-2)(12\lambda - 3)},$$

which increases with m from the value 1 at $m = 2$ to a maximum of at most 1.042 (25/24) at $m = 5$, and then decreases to 1 again as $m \to \infty$. Thus Doksum's test, in addition to being a bit simpler computationally, is asymptotically at least as efficient as Hollander's, for all m and all error distributions, though the difference is always small. The weighted rankings idea is certainly interesting, particularly since it allows unconditionally distribution-free tests, but of the large class of possibilities the only ones which have been evaluated at all extensively (X^2_L for the unordered situation, $\bar{S}_L$ and $\bar{K}_L$ for the ordered) are all limited to linear block weights; these have given mixed results, and thus cannot be recommended for general use. The Kepner–Robinson tests are also unconditionally distribution-free, and exact tables will undoubtedly appear soon; the one based on X^2_{1234} seems particularly attractive for the case $m = 4$. Otherwise, for general $m > 2$, Doksum's tests for both ordered and unordered alternatives may be slightly preferable to the others.

Consider now the comparison between those tests which do and those which do not utilize interblock information. For $m = 2$ this means comparing the Wilcoxon signed-ranks test to the sign test, yielding

$$\mathrm{ARE}(W, S) = 3\psi^2(F_2)/\psi^2(F) = 3\left\{\int F_2'(x)\,\mathrm{d}F_2(x)\Big/F_2'(0)\right\}^2,$$

which cannot exceed 3 if F_2 is unimodal: see Pratt and Gibbons (1981) for an excellent discussion of bounds on this and similar ARE expressions. Given some specific error distributions F we have

Distribution:	Normal	Uniform	Laplace	Exponential or Cauchy
ARE(W, S):	1.500	1.333	1.172	0.750

For $m > 2$ we may compare Doksum's tests, whose overall efficiency properties seem as good as those of any tests which utilize interblock information, to Friedman's test or Page's. This amounts to multiplying ARE(W, S) by the factor

$$\nu(m, \lambda) = (m + 1)/3[1 + (m - 2)(12\lambda - 3)],$$

which decreases from the value 1 at $m = 2$ to $1/3(12\lambda - 3)$ as $m \to \infty$; for all the distributions listed above, ν is close to 0.9 at $m = 3$ and to 0.7 for $m \to \infty$. These results suggest that no one test which utilizes interblock information can be expected to produce a general improvement in efficiency, unless perhaps only for very small m – no more than 4, say. In addition, such tests are generally not truly distribution-free for $m > 3$. Thus Pirie (1974) argued for using tests based on within-block rankings 'for most applications' against the ordered alternative; and his recommendation appears to be equally defensible for the unordered alternative also.

Recall that Assumption IIb (additive block effects) is not required for validity of the tests which do not attempt to utilize interblock information; it can also be weakened to some extent under the classical model. Following Sen (1968a), let the error distribution in the i-th block be F_i. Then in the noncentrality parameter for the classical VR test one need only replace $\sigma^2(F)$ by

$$\overline{\sigma^2} = \lim_{n\to\infty} \sum \sigma^2(F_i)\Big/n,$$

if this limit exists. Similarly, to obtain ARE($\bar{\bar{O}}$, VR), replace ξ_k by $\bar{\xi}_k = \Sigma\, \xi_{ki}/n$, where ξ_{ki} has the same definition as ξ_k but with F_i replacing F throughout. For the special case of Friedman's test, Sen (1967) had already found

$$\mathrm{ARE}(\bar{\bar{S}}, \mathrm{VR}) = \frac{12m\overline{\sigma^2}}{m+1}\left\{\frac{1}{N}\sum \int F_i'(x)\,\mathrm{d}F_i(x)\right\}^2.$$

In particular, suppose the distribution functions F_i differ only by scale factors, i.e., $F_i(x) = F(x/\sigma_i)$. Then

$$\mathrm{ARE}(\bar{\bar{S}}, \mathrm{VR}) = \frac{12m}{m+1}\left[\frac{1}{n}\sum \sigma_i^2\right]\left[\frac{1}{n}\sum \frac{1}{\sigma_i}\right]^2.$$

This is minimized if $\sigma_1 = \cdots = \sigma_n$, and can easily exceed 1 for only moderately heteroscedastic errors.

Furthermore, in conformity with most authors, Assumption IIIa (requiring interchangeability of errors within blocks) was strengthened in the foregoing discussion to Assumption IIIb (requiring that the errors be mutually independent). However, the tests are all valid under the weaker assumption, and the relative efficiency results apply with at most slight modifications: see Sen (1968b, 1968c, 1972) for details.

9. Miscellaneous extensions

In this final Section we consider briefly some miscellaneous extensions of the material presented in Sections 2 through 8.

For one such extension, suppose each of the n blocks contains observations on only k of the m treatments, in accordance with a balanced incomplete blocks design. Then to test the hypothesis of interchangeability within blocks Durbin (1951) proposed the criterion

$$X_{\mathrm{D}}^2 = \frac{12(m-1)}{n(k^3-k)}\sum_{j=1}^{m}\left(R_j - \frac{nk(k+1)}{2m}\right)^2,$$

where R_j is the sum of the within-block ranks of all observations which receive the j-th treatment. (Note that if $k = m$, then X_{D}^2 is Friedman's statistic X_{F}^2.) Van Der Laan and Prakken (1972) tabulated the exact null-hypothesis distribution of X_{D}^2 for 15 small designs, and discussed asymptotic approximations: the simplest is that, for large n, $X_{\mathrm{D}}^2 \dot{\sim} \chi^2(m-1)$. Skillings and Mack (1981) tabulated critical values obtained by simulation, at α nearest 0.10, 0.05 and 0.01, for 21 further designs. Noether (1967) found that the ARE with respect to the classical VR is the same as for Friedman's test, but with k replacing m. This direct use of the within-block ranks has been extended to the Brown–Mood scores by Bhapkar (1961), to general scores by Lemmer et al. (1968) and to weighted rankings by Silva (1977).

Benard and Van Elteren (1953) extended Friedman's test to the general block design described in Section 1, in which $l_{ij} \geqslant 0$ observations within the i-th block receive the j-th treatment, for $i = 1, \ldots, n$ and $j = 1, \ldots, m$. Following the simplified computational scheme laid out by Brunden and Mohberg (1976), define the vector

$$\boldsymbol{R} = (R_1, \ldots, R_m),$$

where

$$R_j = \sum_i \sum_k [R_{ijk} - (l_i + 1)/2]$$

is the corrected rank sum corresponding to the j-th treatment. Define also the $m \times m$ matrix V whose (j, j') element is

$$V_{jj'} = \sum_i l_{ij}(l_i\delta_{jj'} - l_{ij'})\Gamma_i/l_i(l_i - 1),$$

where $\delta_{jj'}$ is Kronecker's delta, and

$$\Gamma_i = \sum_j \sum_k [R_{ijk} - (l_i + 1)/2]^2 .$$

The quantity Γ_i incorporates an adjustment for ties; it equals $(l_i^3 - l_i)/12$ if there are no ties in the i-th block. Then the Benard–Van Elteren criterion for testing the hypothesis of interchangeability within blocks against the unordered alternative is

$$X^2_{\text{BVE}} = \boldsymbol{R}' V^- \boldsymbol{R}$$

where V^- is any generalized inverse of V. This reduces to Friedman's X^2_{F} if $l_{ij} = 1$ for all i and j. The approximate distribution of X^2_{BVE} is $\chi^2(r)$ where r is the rank of V, equal to $(m - 1)$ at most, and exactly $(m - 1)$ for almost any design actually used in practice. Hettmansperger (1975) produced a similar extension of Page's test against the ordered alternative. Lemmer et al. (1968) showed how general scores can be substituted for ranks. Skillings and Mack (1981) presented a variant of the Benard–Van Elteren method which is somewhat simpler computationally.

For the case $m = 2$, the Bernard–Van Elteren statistic is equivalent to the sum over all blocks of the Wilcoxon rank–sum statistics comparing the two treatments within each block. Van Elteren (1960) showed how to improve efficiency by differentially weighting the blocks if they are of unequal size or if they have error distributions which differ in a known manner. These ideas have been extended to $m > 3$ by Prentice (1979) for the unordered alternative and by Skillings and Wolfe (1977) for the ordered alternative.

Consider now the situation in which the response variable Y is dichotomous, taking the values (say) 0 and 1 only. For $m = 2$ the situation is equivalent to that considered by McNemar (1947). For $m > 2$, Cochran (1950) proposed the test criterion

$$Q_0 = \frac{m(m-1)\Sigma(T_j - \bar{T})^2}{\Sigma B_i(m - B_i)},$$

where $T_j = \Sigma_i Y_{ij}$ and $B_i = \Sigma_j Y_{ij}$. Note that blocks for which $B_i = 0$ or $B_i = m$ are irrelevant and may as well be discarded (with the 'effective' n being reduced accordingly). The test is valid for the hypothesis of interchangeability, which in this context is equivalent to

$$H_0: \quad P\{\boldsymbol{Y}_i = \boldsymbol{y}\} \text{ depends only on } \sum y_j,$$

where $\boldsymbol{Y}_i = (Y_{i1}, \ldots, Y_{im})'$ and $\boldsymbol{y} = (y_1, \ldots, y_m)'$ is any vector of 0s and 1s. The same test was discovered independently by Van Elteren (1963), who showed that it is equivalent to an average Spearman or Kendall correlation among the $\boldsymbol{Y}_i$. Van Elteren also tabulated the exact null hypothesis distribution of Q_0 for 34 cases where m and n are very small and $B_i = B$ is constant for all $i = 1, \ldots, n$. Patil (1975) presented a more convenient algorithm for calculating the exact distribution, and tabulated critical values at $\alpha = 0.10, 0.05, 0.01$ for $m = 3$ with $n = 4(1)20$. As $n \to \infty$, Q_0 is asymptotically distributed as $\chi^2(m-1)$. Madansky (1963) extended the test to nominal responses with more than 2 categories.

Madansky also suggested a test for a hypothesis of *homogeneity*, specifically

$$H_*: \quad P\{Y_{i1} = y\} = \cdots = P\{Y_{im} = y\} \text{ for } y = 0, 1,$$

it being assumed that $P\{(Y_{i1}, \ldots, Y_{im})' = \boldsymbol{y}\}$ is the same for all $i = 1, \ldots, n$. Bennett (1967) and Bhapkar (1970) proposed asymptotically equivalent tests of H_*; Bhapkar's simple general criterion is

$$Q_* = \sum \sum T_j T_{j'} c_{jj'} - \left(\sum \sum T_j c_{jj'}\right)^2 \Big/ \sum \sum c_{jj'}$$

where $c_{jj'}$ is in (j, j') element of the inverse of the matrix whose (j, j') element is $(\Sigma_i Y_{ij} Y_{ij'} - T_j T_{j'}/n)$. Note that H_* is implied by the narrower hypothesis H_0, but does not imply it. It can be shown that Q_0 is consistent for testing H_0 against H_1: (not H_*), but it is not valid for testing H_*. However, Q_* is asymptotically distributed as $\chi^2(m-1)$ under H_* (under the added assumption of identical blocks, which Q_0 does not require) and is also consistent against H_1. Bhapkar and Somes (1976) and Wackerly and Dietrich (1976) developed multiple comparisons procedures for the probabilities involved in H_*.

Let us now return from the special case of a dichotomy to the more general response variable. As noted at the end of Section 1, interchangeability of $Y_{i1}, \ldots, Y_{im}$ is the natural hypothesis expressing absence of treatment effects in a true randomized-blocks situation; but there may also be interest in a broader hypothesis, such as,

$$H_*: \quad E[R_{ij}] \equiv (m+1)/2$$

for $i = 1, \ldots, n$ and for $j = 1, \ldots, m$. This generalizes the hypothesis of homogeneity for a dichotomy. Stuart (1951), Linhart (1960) and Quade (1972a) have proposed tests for H_*. Quade's method actually applies to hypotheses of the form

$$H_C: \qquad E[\bar{\bar{C}}] = 0$$

for general correlation measures C; H_* is a special case since it is equivalent to $E[\bar{\bar{S}}] = 0$.

Alternatively, suppose the data did not arise from a randomized blocks design, but from (for example) a repeated measures design, in which the rows (blocks) represent different subjects and the columns (treatments) represent different times at which measurements are taken. Following Koch and Sen (1968), we set up a *mixed model* (their 'Case III') for this situation, in which

$$Y_{ij} = \mu + \tau_j + E_{ij} \quad \text{for } i = 1, \dots, n \text{ and } j = 1, \dots, m$$

where (without loss of generality) $\tau_m = 0$ as before and the E_{ij} have median 0. We retain Assumption I (independence of blocks), strength Assumption II to

Assumption IIIc. The random vectors $(Y_{i1}, \dots, Y_{im})'$ for $i = 1, \dots, n$ are identically distributed.

And weaken Assumption III to

Assumption IIIc. The joint distribution of any linearly independent set of contrasts among the observations in any particular row is diagonally symmetric.

This last assumption implies, in particular, that any linear combination of the E_{ij} is distributed symmetrically about 0.

Given this model, the natural hypothesis is

$$H_x: \quad \tau_1 = \cdots = \tau_m = 0 \,,$$

and the standard test of it is based on Hotelling's criterion

$$T^2 = \frac{n-1}{n} \underline{1}'U[U'(I - \underline{1}\underline{1}'/n)U]^{-1}U'\underline{1} \,,$$

where U is the $n \times (m-1)$ matrix whose (i, j) element is $(Y_{ij} - Y_{im})$. If H_x is true, then the distribution of $(n - m + 1)T^2/(n-1)(m-1)$ is F with $(m - 1, n - m + 1)$ degrees of freedom under the additional assumption that the errors in each block are jointly normal, and, as $n \to \infty$, T^2 is asymptotically $\chi^2(m-1)$ under weak assumptions. Koch and Sen (1968) proposed to test H_x using the criterion

$$W^* = \underline{1}'V[V'(I - \underline{1}\underline{1}'/n)V]^{-1}V'\underline{1},$$

where V is the $n \times (m-1)$ matrix whose (i, j) element is

$$V_{ij} = \sum_{k=1}^{m} \{Q_{ijk} \operatorname{sgn}(Y_{ij} - Y_{ik}) - Q_{imk} \operatorname{sgn}(Y_{im} - Y_{ik})\},$$

and the Qs are as defined in Section 5. Assumption IIIc implies that under H_* the two vectors $\mathbf{V}_i = (V_{i1}, \ldots, V_{im})'$ and $-\mathbf{V}_i$ are equally likely a priori, for $i = 1, \ldots, n$, so that W^* has 2^n (conditionally) equally likely realizations, and an exact P-value can be calculated; as $n \to \infty$, W^* is asymptotically distributed as $\chi^2(m-1)$. Koch and Sen suggested another test criterion (W) for H_x, which is obtained if V_{ij} is replaced by $(R_{ij} - R_{im})$ when calculating W^* as explained above; this is appropriate without requiring any version of Assumption II, and the same remarks about its exact and asymptotic distributions apply. They derived expressions for the noncentrality parameters of both W^* and W under Pitman alternatives, but explicit evaluation would be complicated and they did not carry it out. Gilbert (1972) simulated the distribution of W^* under H_x for $m = 3$, given normal errors with several different variance matrices, and found that the asymptotic approximation provides a conservative test, but with reasonable accuracy for n as small as 9. Simulations under shift alternatives, however, suggested that X_F^2, RA and VR may be farily robust under the diagonal symmetry assumption, and more powerful that W^*. T^2 and W were not considered in his study. Note, by the way, that if $m = 2$ then T^2, W^* and W reduce to the t, signed-ranks and sign tests for matched pairs, respectively.

In Section 1 we mentioned one further source of two-way layouts: the true factorial, in which the observations are completely randomized over the treatment combinations, or there is an independent sample from each. Many of the procedures presented above can be applied to such data, but we shall not provide any explicit discussion of factorials.

References

Abelson, R. P. and Tukey, J. W. (1963). Efficient utilization of non-numerical information in quantitative analysis: General theory and the case of simple order. *Annals of Mathematical Statistics* **34**, 1347–1369.

Alvo, M., Cabilio, P. and Feigin, P. D. (1982). Asymptotic theory for measures of concordance with special reference to average Kendall tau. *Annals of Statistics* **10**, 1269–1276.

Anderson, R. L. (1959). Use of contingency tables in the analysis of consumer preference studies. *Biometrics* **15**, 582–590.

Bahadur, R. R. (1960). Stochastic comparison of tests. *Annals of Mathematical Statistics* **31**, 276–295.

Bahadur, R. R. (1967). Rates of convergence of estimates and test statistics. *Annals of Mathematical Statistics* **38**, 303–324.

Bartholomew, D. J. (1959). A test of homogeneity for ordered alternatives. *Biometrika* **46**, 36–48.

Bennett, B. M. (1967). Tests of hypotheses concerning matched samples. *Journal of the Royal Statistical Society B* **29**, 468–474.

Bernard, A. and Van Elteren, P. (1953). A generalization of the method of *m* rankings. *Indagationes Mathematicae* **15**, 358–369.

Berenson, M. L. (1982a). Some useful nonparametric tests for ordered alternatives in randomized block experiments. *Communications in Statistics – Theory and Methods* **11**, 1681–1693.

Berenson, M. L. (1982b). A study of several useful tests for ordered alternatives in the randomized block design. *Communications in Statistics – Simulation and Computation* **11**, 563–581.

Bhapkar, V. P. (1961). Some nonparametric median procedures. *Annals of Mathematical Statistics* **32**, 846–863.

Bhapkar, V. P. (1963). The asymptotic power and efficiency of Mood's test for two-way classification. *Journal of the Indian Statistical Association* **1**, 24–31.

Bhapkar, V. P. (1970). On Cochran's *Q*-test and its modification. In: G. P. Patil, ed., *Random Counts in Scientific Work Vol. 2.* Penn. State., University Park and London.

Bhapkar, V. P. and Somes, G. W. (1976). Multiple comparisons of matched proportions. *Communications in Statistics – Theory and Methods* **5**, 17–25.

Blomqvist, N. (1950). On a measure of dependence between two random variables. *Annals of Mathematical Statistics* **21**, 593–600.

Blomqvist, N. (1951). Some tests based on dichotomization. *Annals of Mathematical Statistics* **22**, 362–371.

Boyd, M. N. and Sen, P. K. (1984). Union-intersection rank tests for ordered alternatives in a complete block design. To appear in *Communications in Statistics.*

Bradley, J. V. (1968). *Distribution-free Statistical Tests.* Prentice-Hall, Englewood Cliffs, N.J.

Brown, G. W. and Mood, A. M. (1948). Homogeneity of several samples. *The American Statistician* **2** (3) 22.

Brown, G. W. and Mood, A. M. (1951). On median tests for linear hypotheses. In: J. Neyman, ed., *Proceedings of the Second Berkeley Symposium on Mathematical Statistics and Probability.* University of California Press, Berkeley, pp. 159–166.

Brunden, M. N. and Mohberg, N. R. (1976). The Bernard–Van Elteren statistic and nonparametric computation. *Communications in Statistics – Simulation and Computation* **5**, 155–162.

Cochran, W. G. (1950). The comparison of percentages in matched samples. *Biometrika* **37**, 256–266.

Conover, W. J. and Iman, R. L. (1980). A comparison of distribution-free procedures for the analysis of complete blocks. Unpublished manuscript presented at the annual meeting of the American Institute of Decision Sciences, Las Vegas.

Conover, W. J. and Iman, R. L. (1981). Rank transformations as a bridge between parametric and nonparametric statistics. *The American Statistician* **35**, 124–133.

De, N. (1976). Rank tests for randomized blocks against ordered alternatives. *Calcutta Statistical Association Bulletin* **25**, 1–27.

Doksum, K. A. (1967). Robust procedures for some linear models with one observation per cell. *Annals of Mathematical Statistics* **38**, 878–883.

Durbin, J. (1951). Incomplete blocks in ranking experiments. *British Journal of Psychology (Statistical Section)* **4**, 85–90.

Ehrenberg, A. S. C. (1952). On sampling from a population of rankers. *Biometrika* **39**, 82–87.

Friedman, M. (1937). The use of ranks to avoid the assumptions of normality implicit in the analysis of variance. *Journal of the American Statistical Association* **32**, 675–701.

Gilbert, R. O. (1972). A Monte-Carlo study of analysis of variance and competing rank tests for Scheffe's mixed model. *Journal of the American Statistical Association* **67**, 71–75.

Hajek, J. (1969). *A Course in Nonparametric Statistics.* Holden-Day, San Francisco.

Hannan, E. J. (1956). The asymptotic powers of certain tests based on multiple correlations. *Journal of the Royal Statistical Society B* **18**, 227–233.

Hays, W. L. (1960). A note on average tau as a measure of concordance. *Journal of the American Statistical Association* **55**, 331–341.

Hettmansperger, T. P. (1975). Non-parametric inference for ordered alternatives in a randomized block design. *Psychometrika* **40**, 53–62.

Hodges, T. L. Jr. and Lehmann, E. L. (1956). The efficiency of some nonparametric competitors of the *t*-test. *Annals of Mathematical Statistics* **27**, 324–335.

Hodges, J. L., Jr. and Lehmann, E. L. (1962). Rank methods for combination of independent experiments in analysis of variance. *Annals of Mathematical Statistics* **33**, 482–497.

Hodges, J. L., Jr. and Lehmann, E. L. (1963). Estimates of location based on rank tests. *Annals of Mathematical Statistics* **34**, 598–611.

Hoeffding, W. (1948). A class of statistics with asymptotically normal distribution. *Annals of Mathematical Statistics* **19**, 293–325.

Hollander, M. (1966). An asymptotically distribution-free multiple comparison procedure – treatments versus control. *Annals of Mathematical Statistics* **37**, 735–738.

Hollander, M. (1967). Rank tests for randomized blocks when the alternatives have an a priori ordering. *Annals of Mathematical Statistics* **38**, 867–877.

Hollander, M. and Wolfe, D. A. (1973). *Nonparametric Statistical Methods.* Wiley, New York.

Hutchinson, T. P. (1976). Combining two-tailed rank-correlation statistics. *Applied Statistics* **25**, 21–25.

Iman, R. L. and Davenport, J. M. (1980). Approximation of the critical region of the Friedman statistic. *Communications in Statistics – Theory and Methods* **9**, 571–595.

Jonckheere, A. R. (1954). A test of significance for the relation between *m* rankings and *k* ranked categories. *British Journal of Statistical Psychology* **7**, 93–100.

Kelley, T. L. (1923). *Statistical Method.* Macmillan, New York.

Kendall, M. G. (1948). *Rank Correlation Methods*, 1st ed. Griffin, London.

Kendall, M. G. and Smith, B. B. (1939). The problem of *m* rankings. *Annals of Mathematical Statistics* **10**, 275–287.

Kepner, J. L. and Robinson, D. H. (1982). A distribution-free test for treatment effects in randomized complete block designs and repeated measures designs. Department of Statistics, University of Florida, Technical Report No. 171.

Koch, G. G. and Sen, P. K. (1968). Some aspects of the statistical analysis of the 'mixed model'. *Biometrics* **24**, 27–48.

Kraft, C. H. and Van Eeden, C. (1968). *A Nonparametric Introduction to Statistics.* Macmillan, New York.

Lawler, K. R. (1978). A distribution-free generalization of the Wilcoxon signed-ranks test for the two-way layout. Unpublished Master's thesis, Texas Tech University.

Lehmann, E. L. (1964). Asymptotically nonparametric inference in some linear models with one observation per cell. *Annals of Mathematical Statistics* **35**, 726–734.

Lemmer, H. H., Stoker, D. J. and Reinach, S. G. (1968). A distribution-free analysis of variance technique for block designs. *South African Statistical Journal* **2**, 9–32.

Linhart, H. (1960). Approximate test for *m* rankings. *Biometrika* **47**, 476–480.

Ludwig, O. (1962). Ueber Kombination von Rangkorrelationskoeffizienten aus unabhaendigen Messreihen. *Biometrische Zeitschrift* **4**, 40–50.

Lyerly, S. B. (1952). The average Spearman rank correlation coefficient. *Psychometrika* **17**, 421–428.

Madansky, A. (1963). Tests of homogeneity for correlated samples. *Journal of the American Statistical Association* **58**, 97–119.

Mann, B. L. and Pirie, W. R. (1982). Tighter bounds and simplified estimation for moments of some rank statistics. *Communications in Statistics – Theory and Methods* **11**, 1107–1117.

McDonald, B. J. and Thompson, W. A., Jr. (1967). Rank sum multiple comparisons in one- and two-way classifications. *Biometrika* **54**, 487–497.

McNemar, Q. (1947). Note on the sampling error of the difference between correlated proportions or percentages. *Psychometrika* **12**, 153–157.

Mehra, K. L. and Sarangi, J. (1967). Asymptotic efficiency of certain rank tests for comparative experiments. *Annals of Mathematical Statistics* **38**, 90–107.

Michaelis, J. (1977). Schwellenwerte des Friedman-tests. *Biometrische Zeitschrift* **13**, 118–129.

Miller, R. G. (1966). *Simultaneous Statistical Inference.* McGraw-Hill, New York.

Mood, A. M. (1950). *Introduction to the Theory of Statistics.* McGraw-Hill, New York.

Nemenyi, P. (1963). Distribution-free multiple comparisons. Unpublished doctoral thesis, Princeton University, Princeton, N.J.

Noether. G. E. (1967). *Elements of Nonparametric Statistics.* Wiley, New York.

Oden, R. E. (1977a). Extended tables of the distribution of Friendman's S-statistic in the two-way layout. *Communications in Statistics – Simulation and Computation*, **6**, 29–48.

Odeh, R. E. (1977b). The exact distribution of Page's L-statistic in the two-way layout. *Communications in Statistics – Simulation and Computation* **6**, 49–61.

Owen, D. B. (1962). *Handbook of Statistical Tables.* Addison-Wesley, Reading, Massachusetts.

Page, E. B. (1963). Ordered hypotheses for multiple treatments: a significance test for linear ranks. *Journal of the American Statistical Association* **58**, 216–230.

Patil, K. D. (1975). Cochran's Q test: exact distribution. *Journal of the American Statistical Association* **70**, 186–189.

Pirie, W. R. (1974). Comparing rank tests for ordered alternatives in randomized blocks. *Annals of Statistics* **2**, 374–382.

Pirie, W. R. and Hollander, M. (1972). A distribution-free normal scores test for ordered alternatives in the randomized block design. *Journal of the American Statistical Association* **67**, 855–857.

Pitman, E. J. G. (1938). Significance tests which may be applied to samples from any population, III: The analysis of variance test. *Biometrika* **29**, 322–335.

Pitman, E. J. G. (1948). *Nonparametric Statistical Inference.* Lecture notes, Institute of Statistics, University of North Carolina.

Pratt, J. W. and Gibbons, J. D. (1981). *Concepts of Nonparametric Theory.* Springer, New York.

Prentice, M. J. (1979). On the problem of m incomplete rankings. *Biometrika* **66**, 167–170.

Puri, M. L. and Sen, P. K. (1967). On some optimum nonparametric procedures in two-way layouts. *Journal of the American Statistical Association* **62**, 1214–1224.

Puri, M. L. and Sen, P. K. (1968). On Chernoff–Savage tests for ordered alternatives in randomized blocks. *Annals of Mathematical Statistics* **39**, 967–972.

Quade, D. (1972a). Average internal rank correlation. Report SW 16/72, Mathematical Center, Amsterdam.

Quade, D. (1972b). Analyzing randomized blocks by weighted rankings. Report SW 18/72, Mathematical Center, Amsterdam.

Quade, D. (1979). Using weighted rankings in the analysis of complete blocks with additive block effects. *Journal of the American Statistical Association* **74**, 680–683.

Rhyne, A. L. and Steel, R. G. D. (1965). Tables for a treatments versus control multiple comparisons sign test. *Technometrics* **7**, 293–306.

Rhyne, A. L. and Steel, R. G. D. (1967). A multiple comparisons sign test: all pairs of treatments. *Biometrics* **23**, 539–549.

Salama, I. A. and Quade, D. (1981). Using weighted rankings to test against alternatives in randomized blocks. *Communications in Statistics – Theory and Methods* **10**, 385–399.

Salama, I. A. and Quade, D. (1984). Some simple statistics to test against ordered alternatives in complete block designs. *Biometrical Journal*, to appear.

Schach, S. (1979). An alternative to the Friedman test with certain optimality properties. *Annals of Statistics* **7**, 537–550.

Sen, P. K. (1967). A note on the asymptotic efficiency of Friedman's χ_r^2-test. *Biometrika* **54**, 677–679.

Sen, P. K. (1968a). Asymptotically efficient tests by the method of n rankings. *Journal of the Royal Statistical Society B* **30**, 312–317.

Sen, P. K. (1968b). On a class of aligned rank order tests in two-way layouts. *Annals of Mathematical Statistics* **39**, 1115–1124.

Sen, P. K. (1968c). Robustness of some nonparametric procedures in linear models. *Annals of Mathematical Statistics* **39**, 1913–1922.

Sen, P. K. (1969). On nonparametric T-method of multiple comparisons for randomized blocks. *Annals of the Institute of Statistical Mathematics* **21**, 329–333.

Sen, P. K. (1972). A further note on the asymptotic efficiency of Friedman's χ_r^2-test. *Metrika* **18**, 234–237.

Shorack, G. L. (1967). Testing against ordered alternatives in Model I analysis of variance; normal theory and nonparametric. *Annals of Mathematical Statistics* **38**, 1740–1752.

Silva, C. (1977). Analysis of randomized blocks designs based on weighted rankings. North Carolina Institute of Statistics Mimeo Series No. 1137.

Silva, C. and Quade, D. (1980). Evaluation of weighted rankings using expected significance level. *Communications in Statistics – Theory and Methods* **9**, 1087–1096.

Silva, C. and Quade, D. (1983). Estimating the asymptotic relative efficiency of weighted rankings. *Communications in Statistics – Simulation and Computation* **12**, 511–521.

Skillings, J. H. (1980). On the null distribution of Jonckheere's statistic used in two-way models for ordered alternatives. *Technometrics* **22**, 431–436.

Skillings, J. H. and Mack, G. A. (1981). On the use of a Friedman-type statistic in balanced and unbalanced designs. *Technometrics* **23**, 171–177.

Skillings, J. H. and Wolfe, D. A. (1977). Testing ordered alternatives by combining independent distribution-free block statistics. *Communications in Statistics – Theory and Methods* **6**, 1453–1463.

Spearman, C. (1904). The proof and measurement of association between two things. *American Journal of Psychology* **15**, 72–101.

Steel, R. G. D. (1959). A multiple comparison sign test: treatments versus control. *Journal of the American Statistical Association* **54**, 767–775.

Stuart, A. (1951). An application of the distribution of the ranking concordance coefficient. *Biometrika* **38**, 33–42.

Thompson, W. A., Jr. and Willke, T. A. (1963). On an extreme rank sum test for outliers. *Biometrika* **50**, 375–383.

Tukey, J. W. (1949). The simplest signed-rank tests. Mem. Report 17, Statistical Research Group, Princeton University.

Tukey, J. W. (1957). Sums of random partitions of ranks. *Annals of Mathematical Statistics* **28**, 987–992.

Van Der Laan, P. and Prakken, J. (1972). Exact distribution of Durbin's distribution-free test statistic for balanced incomplete block designs, and comparison with the chi-square and F approximation. *Statistica Neerlandica* **26**, 155–164.

Van Elteren, P. (1957). The asymptotic distribution for large m of Terpstra's statistic for the problem of m rankings. *Proceedings Koningklijke Nederlandse Akademie van Wetenschappen* **60**, 522–534.

Van Elteren, P. (1960). On the combination of independent two-sample tests of Wilcoxon. *Bulletin of the International Statistical Institute* **37**, 351–361.

Van Elteren, P. (1963). Een permutatietoets voor alternatief verdeelde grootheden. *Statistica Neerlandica* **17**, 487–505.

Van Elteren, P. and Noether, G. E. (1959). The asymptotic efficiency of the χ_r^2-test for a balanced incomplete block design. *Biometrika* **46**, 475–477.

Wackerly, D. D. and Dietrich, F. H. (1976). Pairwise comparison of matched proportions. *Communications in Statistics – Theory and Methods* **5**, 1455–1467.

Wei, L. J. (1982). Asymptotically distribution-free simultaneous confidence region of treatment differences in a randomized complete block design. *Journal of the Royal Statistical Society B* **44**, 201–208.

Wilcoxon, F. (1945). Individual comparisons by ranking methods. *Biometrics* **1**, 80–83.

Wilcoxon, F., Katti, S. K. and Wilcox, R. A. (1964). Critical values and probability levels for the Wilcoxon rank sum test and the Wilcoxon signed rank test. In: H. L. Harter and D. B. Owen, eds., *Selected Tables in Mathematical Statistics Vol. 1.* Markham, Chicago.

Wilcoxon, F. and Wilcox, R. A. (1964). *Some Rapid Approximate Statistical Procedures.* Lederle Laboratories, Pearl River, N.Y.

Wormleighton, R. (1959). Some tests of permutation symmetry. *Annals of Mathematical Statistics* **30**, 1005–1017.

Youden, W. J. (1963). Ranking laboratories by round-robin tests. *Materials Research and Standards* **3**, 9–13.

P. R. Krishnaiah and P. K. Sen, eds., *Handbook of Statistics, Vol. 4*
© Elsevier Science Publishers (1984) 229–257

11

Rank Tests in Linear Models

J. N. Adichie

1. Introduction

In the general study of science, it is usual to start by postulating a mathematical model that would best describe the phenomena of interest. In statistics the phenomena often manifest themselves in the relationships among various characteristics; for example, a result Y of an experiment may be associated with known constants $x = (x_1, \ldots, x_q)$ in such a way that for different values of x, Y takes correspondingly different values. In that case, the expectation EY may be written as $EY = g(x_1, \ldots, x_q)$, where g is some function. The statistical problem would then be to determine the function g. It turns out that very good and useful results have been obtained by taking g as a linear function. Y can be univariate or multivariate. We shall start with univariate Y, and later in section 5 treat the multivariate model. With this convention, we have the general univariate linear regression model where the observations $Y = (Y_1, \ldots, Y_n)'$ can be written as

$$Y = \boldsymbol{\alpha} + X_n'\boldsymbol{\beta} + \boldsymbol{\varepsilon} \tag{1.1}$$

where $\boldsymbol{\alpha}$ is a vector of n equal components, X_n is a $q \times n$ matrix of known constants, $\boldsymbol{\beta}' = (\beta_1, \ldots, \beta_q)$ are regression parameters of interest, while $\boldsymbol{\varepsilon}$ represent the random part, usually with $E(\boldsymbol{\varepsilon}) = 0$.

The classical method of testing hypotheses about β assumes that $\boldsymbol{\varepsilon}$ is normally distributed. Under this assumption, the likelihood ratio criterion provides the most powerful test. Quite often however, there is good reason to doubt the normality of $\boldsymbol{\varepsilon}$. When this is the case, the likelihood ratio test statistic calculated on the wrong assumption of normality fails to perform. Its exact distribution is not even known! In such a situation, a rank method may provide an answer to the testing problem. This is so because in many rank methods, it is enough to assume that the distribution function F of the observations, is continuous. For rank tests there is no need even to assume that the variance $\sigma^2(F)$ is finite. The use of rank tests as alternative to classical tests, especially when classical assumptions can not be upheld, has been discussed by many authors, see e.g. Pitman (1948), Hodges and Lehmann (1956), Chernoff and Savage (1958).

In this chapter, we discuss the rank tests for the various hypotheses that are usually tested with respect to the parameters of a linear model. Section 2 deals with tests of hypotheses involving all $\boldsymbol{\beta}$ in (2.1), while in section 3, we discuss tests of sub-hypotheses. Section 4 treats tests involving comparisons of several regression lines while Section 5 discusses tests for the multivariate linear model. In all the discussions, we shall restrict attention to cases where the design matrix is of full rank.

2. Tests of regression of full rank

Consider independent observations $Y_1, \ldots, Y_n$ taken at $(\boldsymbol{x}_1, \ldots, \boldsymbol{x}_n)$, such that the distribution function of Y_i is

$$F_i(y) = F(y - \alpha - \boldsymbol{\beta}'\boldsymbol{x}_i), \quad i = 1, \ldots, n, \tag{2.1}$$

where α and $\boldsymbol{\beta} = (\beta_1, \ldots, \beta_q)'$ $(q \geqslant 1)$ are unknown parameters of interest, and $\boldsymbol{x}_i = (x_{1i}, \ldots, x_{qi})'$ are vectors of known constants constituting $q \times n$ design matrix $X_n = (X_1, \ldots, X_q)' = ((x_{ji}))$, $i = 1, \ldots, n$, $j = 1, \ldots, q$. The distribution function F is continuous but its functional form need not be further known. This formulation includes all distribution functions that may look like but are quite different from the normal. The commonest hypothesis tested in this set up is

$$H_0: \quad \boldsymbol{\beta} = \boldsymbol{0} \tag{2.2}$$

against the alternative $\boldsymbol{\beta} \neq \boldsymbol{0}$.

2.1. Rank test statistics

Following from the fundamental work of Hajek (1962), Hajek and Sidak (1967) and their various extensions and generalisations, a good rank test for (2.2) is a quadratic in rank order statistics. Under (2.2), Y_i's are independent identically distributed random variables to which the ranking method can easily be applied. Put

$$Z_n = (Z_1, \ldots, Z_q)' = ((z_{ji})), \quad j = 1, \ldots, q, \; i = 1, \ldots, n, \tag{2.3}$$

where $z_{ji} = (x_{ji} - \bar{x}_j)$, with $\bar{x}_j = n^{-1} \sum_i x_{ji}$. This has the effect of reparametrizing (2.1) to $F_0(y) = F(y - \alpha - \boldsymbol{\beta}'\bar{x} - \beta' \boldsymbol{z}_i)$.

To obtain the required test statistic, define

$$S_{nj} = \sum_i z_{ji} \psi_n(R_i), \quad j = 1, \ldots, q, \tag{2.4}$$

where R_i is the rank of Y_i, while $\psi_n(i) = \psi(i/(n+1))$ are scores generated by a given function $\psi(u)$, $0 < u < 1$. Writing $\boldsymbol{S}_n$ for $(S_{n1}, \ldots, S_{nq})$, the rank order statistic for testing (2.2) is given by

$$M_n = \boldsymbol{S}_n'(Z_nZ_n')^{-1}\boldsymbol{S}_n/A^2(\psi) \tag{2.5}$$

where $A^2(\psi) = \int \psi^2(u)\,\mathrm{d}u - \bar{\psi}^2$, with $\bar{\psi} = \int \psi(u)\,\mathrm{d}u$.

In practice only two types of scores are generally used: the Wilcoxon scores $\psi_n(i) = i/(n+1)$ generated by $\psi(u) = u$, which corresponds to using the ordinary rank test, (see Wilcoxon, 1945). The other are the Normal scores $\psi_n(i) = \Phi^{-1}(i/(n+1))$ where Φ is the distribution function of the standard normal random variable. (See van der Waerden, 1952, 1953). Values of $\Phi^{-1}(i|(n+1))$ for $16 \leq n \leq 40$ have been tabulated in Hajek (1969).

The M_n test in (2.5) rejects H_0 in (2.2) if M_n is large, where the cut off point is obtained from the distribution of M_n. Although it is possible to obtain the exact distribution of M_n under (2.2) for any sample size n, nowhere has this distribution been tabulated, mainly because it depends on the given z_{ji}. However, under certain regularity conditions, which we give below, M_n has been shown to have asymptotically as n becomes large, a chi-square distribution with q degrees of freedom when (2.2) holds. The conditions required, as given in Adichie (1967a, 1978) are as follows:

Condition A. The distribution function F in (2.1) has a density f which is absolutely continuous such that the Fisher Information $I(F) = \int (f'(y)/f(y))^2 f(y)\,\mathrm{d}y$ is finite

Condition B. The regression constants satisfy
 (i) $(\max_i z_{ji}^2/\sum_i z_{ji}^2) \to 0$ for each $j = 1, \ldots, q$,
 (ii) $\mathrm{rank}(Z_nZ_n') = q$ for all n,
 (iii) $n^{-1}(Z_nZ_n')$ tends to a positive definite and finite matrix as $n \to \infty$.

Condition C. $\psi(u)$ is absolutely continuous, nondecreasing and square integrable over (0, 1).

For large values of n therefore, an approximate level α test for (2.2) may be obtained if H_0 is rejected for $M_n > \chi_q^2(1-\alpha)$, the $(1-\alpha)$ fractile of the chi-square distribution with q degrees of freedom.

Remark. Observe that Condition B(ii) may not be satisfied for all cases for which the original design matrix X_n is of rank q. A particular class of X_n for which B(ii) holds, is any orthogonal design matrix with $\bar{x}_1 = \cdots = \bar{x}_q$.

Example 1. Consider the following artificial example of model (2.1) with $q = 2$, where the observations are

Y:	14	20	10	16	33	27
X_1:	1	5	3	0	8	7
X_2:	21	15	13	4	5	8

We want to test H_0: $\beta_1 = \beta_2 = 0$. Rewriting x in terms of z as in (2.3), we obtain

Y:	14	20	10	16	33	27
z_1:	-3	1	-1	-4	4	3
z_2:	10	4	2	-7	-6	-3

Using Wilcoxon scores, and noting that rank(Y_i) are 2, 4, 1, 3, 6, 5, we have

$$S_1 = \{-3(2) + 1(4) + (-1)(1) + (-4)(3) + 4(6) + 3(5)\}/7 = 3.43\,.$$

Similarly $S_2 = -4.86$. Also $(Z_nZ_n') = \left(\begin{smallmatrix} 52 & -33 \\ -33 & 214 \end{smallmatrix}\right)$ which satisfies Condition B(ii) giving $M_n = 3.156$.

2.1.1. Rank test statistics in the presence of ties

The discussion above assumes that there are no ties in the observations. In practice however, ties occur in many situations. When such is the case, we either use randomized ranks, (i.e. breaking the ties at random) or we use average scores $\bar{\psi}_n(i)$ in the definition of M_n. For practical purposes, average scores are preferred to randomized ranks because the use of the latter introduces a process that bears no relevance to the experiment of interest. For an account of treatment of ties in rank procedures see e.g. Chapter VII of Hajek (1969).

Let us write the M_n statistic calculated on average scores as

$$M_n(\bar{\psi}) = \boldsymbol{S}_n'(\bar{\psi})(Z_nZ_n')^{-1}\boldsymbol{S}_n(\bar{\psi})|A_n^2(\bar{\psi}) \tag{2.6}$$

where $A_n^2(\bar{\psi}) = n^{-1}(\Sigma_i\, \bar{\psi}_n(i) - n^{-1}\, \Sigma_i\, \bar{\psi}_n(i))^2$.

It has been shown (see e.g. Theorem 29C of Hajek, 1969) that under regularity conditions similar to those in section 2.1, $M(\bar{\psi})$ has asymptotically a chi square distribution with q degrees of freedom, under the hypothesis of no regression. It is to be observed however, that this asymptotic distribution is conditional on $\tau = (\tau_1, \ldots, \tau_g)$, the given number and configuration of ties in the observations. For example, unlike A^2 in (2.5) which depends only on $\psi(u)$, we do not know from the set of observed values, the value of $A^2(\bar{\psi}) = \lim A_n^2(\bar{\psi})$ given in (2.6), because the set of $\bar{\psi}_n(i)$ depends on τ. From the discussion above, it follows that even when ties occur in the observations, an approximate level α-test for no regression may still be obtained if the hypothesis is rejected for $M_n(\bar{\psi}) > \chi_q^2(1-\alpha)$.

EXAMPLE 2. Consider another artificial example of the model (2.1) with $q = 2$, where the observations are given as

Y:	1	5	0	4	4	+1
x_1:	1	2	1	3	3	3
x_2:	1	1	2	1	2	3

We note that the raw ranks of Y_i are 2, 6, 1, 4, 4, 2, and the average scores are

$$\begin{aligned}\bar{\psi}_n(i) &= \tfrac{1}{2}(\psi_n(2) + \psi_n(3)), && i = 1, \text{ and } 6\,,\\ &= \psi_n(6), \psi_n(1), && i = 2, 3\,,\\ &= \tfrac{1}{2}(\psi_n(4) + \psi_n(5)), && i = 4 \text{ and } 5\,.\end{aligned}$$

For the Wilcoxon scores, the average scores are the same as the midranks 2.5, 6, 1, 4.5, 4.5, 2.5. Rewriting X in terms of Z as in (2.3) we obtain $S_1 = 0.286$, $S_2 = -3.5$ with $Z_nZ_n' = \left(\begin{smallmatrix}4.83 & 1.33\\ 1.33 & 3.33\end{smallmatrix}\right)$ giving $M_n(\bar{\psi}) = 4.338$. For the Normal scores, the values of $\bar{\psi}_n(i)$ are -0.37, 1.07, -1.07, 0.37, 0.37, -0.37, giving $S_1 = 0.918$, $S_2 - 1.44$ and $M_n(\bar{\psi}) = 3.7321$.

EXAMPLE 3. Consider a sample of size 30 of real life data in Table 1 taken from page 282 of Steel and Torrie (1960). We assume the model, in (2.1) with $q = 3$, where Y is the log of leaf burn in seconds, X_1 is the percentage of nitrogen, X_2 the percentage of chlorine and X_3 the percentage of potassium. Our interest is to test H_0: $\boldsymbol{\beta} = 0$. The ranks of Y_i are as follows:

10.5	3.5	13	15	5.5	1	2	3.5	30	17
25	22	18.5	28	23	24	29	14	5.5	10.5
12	18.5	20	21	27	26	7.5	9	16	7.5

and from the given X, we easily find

$$(Z_nZ_n') = \begin{pmatrix} 9.8453 & 2.1413 & 1.6705 \\ 2.1413 & 10.6209 & 7.6367 \\ 1.6705 & 7.6367 & 33.0829 \end{pmatrix}.$$

Using Wilcoxon (average) scores, we obtain $\boldsymbol{S}_n'(\bar{\psi})(Z_nZ_n')^{-1}\boldsymbol{S}_n(\bar{\psi}) = 1.99646$, and because $A_n^2(\bar{\psi})$ in this case yields 0.0778702, we find that $M_n(\bar{\psi}) = 25.6383314$. This suggests that H_0 should be rejected, since in fact $\chi_3^2(0.99) = 11.3$. It is to be observed that the least squares estimates of $\boldsymbol{\beta}$ in this problem are: $\hat{\beta}_1 = -0.531459$, $\hat{\beta}_2 = -0.439641$ and $\hat{\beta}_3 = 0.208979$. Furthermore the classical variance-ratio criterion $Q = (Q_1/3) \div (Q_0/26)$ applied to the data gives $Q = 40.24$ which is greater than $F_{3,26}(0.99) = 4.60$. This again suggests that the hypothesis $\beta_1 = \beta_2 = \beta_3 = 0$ should not be accepted.

2.1.2. Asymptotic nature of M_n-test

From the foregoing discussions on M_n test, it would appear that the test can be performed only asymptotically. It is however emphasized that the exact

Table 1
Percentages of nitrogen X_1, chlorine X_2, potassium X_3, and log of leaf burn in seconds Y, in samples of tobacco from farmers' fields

Sample no.	Nitrogen % X_1	Chlorine % X_2	Potassium % X_3	Log of leaf burn Y, sec
1	3.05	1.45	5.67	0.34
2	4.22	1.35	4.86	0.11
3	3.34	0.26	4.19	0.38
4	3.77	0.23	4.42	0.68
5	3.52	1.10	3.17	0.18
6	3.54	0.76	2.76	0.00
7	3.74	1.59	3.81	0.08
8	3.78	0.39	3.23	0.11
9	2.92	0.39	5.44	1.53
10	3.10	0.64	6.16	0.77
11	2.86	0.82	5.48	1.17
12	2.78	0.64	4.62	1.01
13	2.22	0.85	4.49	0.89
14	2.67	0.90	5.59	1.40
15	3.12	0.92	5.86	1.05
16	3.03	0.97	6.60	1.15
17	2.45	0.18	4.51	1.49
18	4.12	0.62	5.31	0.51
19	4.61	0.51	5.16	0.18
20	3.94	0.45	4.45	0.34
21	4.12	1.79	6.17	0.36
22	2.93	0.25	3.38	0.89
23	2.66	0.31	3.51	0.91
24	3.17	0.20	3.08	0.92
25	2.79	0.24	3.98	1.35
26	2.61	0.20	3.64	1.33
27	3.74	2.27	6.50	0.23
28	3.13	1.48	4.28	0.26
29	3.49	0.25	4.71	0.73
30	2.94	2.22	4.58	0.23

distribution of M_n test statistic can be calculated using permutation technique although no ready-made tables are available.

Of all the conditions necessary for the asymptotic distribution of M_n, Conditions B(i) and (iii) are the most difficult to check for validity. This is because they involve limiting processes that can not easily be checked from available observations. As an illustration, observe that in Example 1 of Section 2.1 $(\max_i z_{ji}^2/\Sigma_i z_{ji}^2) = 4/13,\ 50/107$ for $j = 1, 2$ and $Z_n Z_n' = \begin{pmatrix} 52 & -33 \\ -33 & 214 \end{pmatrix}$. Whether these ratios will eventually tend to zero and the matrix $n^{-1}(Z_n Z_n')$ tend to a positive definite and finite matrix, will depend on subsequent observations that are not yet available.

It is perhaps necessary to state here that one can not make a categorical pronouncement on the value of n that is required for the approximation any given situation. Surprisingly, good approximations have been obtained for n as small as 10, while there have been cases where the approximation is not good enough for n as large as 50. As a general rule, it would require many more observations for a good approximation in the case of $M_n(\bar{\psi})$ than it would in the case of M_n.

2.1.3. *Asymptotic efficiency of M_n tests*

It is necessary to assess the performance of any testing procedure. The usual measure used is the power of the test against alternative hypothesis. In most nonparametric procedures, especially those that are asymptotic, it may not be easy to specify the alternative hypotheses. Even when this difficulty can be overcome, the power calculation may be intractable. A way out of these difficulties is to use the concept of asymptotic efficiency as defined either by Pitman (see Noether, 1954) or by Bahadur (1967). We shall use the Pitman version in this work. It turns out that for near alternatives tending to the hypothesis at a suitable rate, the asymptotic power of most rank procedures can be calculated, and it is related to the asymptotic efficiency. If two test statistics, under the same sequence of alternatives have noncentral chi-square limit distributions with the same degrees of freedom, it has been shown by Andrews (1954) that their relative asymptotic efficiency is given by the ratio of their noncentrality parameters.

For our M_n test, it has been shown that under a sequence of near alternatives,

$$H_1: \qquad \boldsymbol{\beta} = n^{-1/2}\boldsymbol{b}, \quad \|\boldsymbol{b}\| < c \tag{2.7}$$

and subject to conditions A–C of Section 2.1, M_n has an asymptotic noncentral chi square distribution with q degrees of freedom and noncentrality parameter

$$\Delta_M = \{\boldsymbol{b}'(ZZ')\boldsymbol{b}\}B^2(F)|A^2 \tag{2.8}$$

with

$$B^2(F) = \int \psi'(F(y))\,\mathrm{d}F(y)$$

where ψ' denotes derivative with respect to y. We note that for testing H_0 in (2.2) after reparametrization as in (2.3) the test statistic Q based on the likelihood ratio has the form $(n-q)Q_1/qQ_0$, where $Q_1 = Y_n'Z_n'(Z_nZ_n')^{-1}Z_nY_n$ and as $n \to \infty$, $Q_0/(n-q)$ tends to $\sigma^2 = \sigma^2(F)$, the variance of Y. Under (2.7) it turns out that qQ has an asymptotic noncentral chi-square distribution with q degrees of freedom and noncentraliy parameter

$$\Delta_Q = \{\boldsymbol{b}'(ZZ')\boldsymbol{b}\}/\sigma^2(F) \tag{2.9}$$

The asymptotic efficiency of M_n test relative to the usual Q (variance-ratio) test is therefore

$$e_{M,Q} = \sigma^2(F)B^2(F)/A^2 \tag{2.10}$$

which is the standard efficiency expression of rank test procedures relative to the usual normal theory ones. Expression (2.10) has been studied in detail by Hodges and Lehmann (1956).

2.2. Signed rank test statistics

If in addition to the conditions A–C of Section 2.1, we also assume that the distribution function F of Y is symmetric, then a test statistic for testing H_0 in (2.2) may be obtained as quadratic in signed rank statistics. Let

$$S_{nj}^+ = \sum_i x_{ji}\phi_n(R_i^+)\,\mathrm{sgn}(Y_i), \quad j = 1, \ldots, q, \tag{2.11}$$

where R_i^+ is the rank of $|Y_i|$ among $|Y_1|, \ldots, |Y_n|$, $\mathrm{sgn}(y) = 1$ or -1 for $y > 0$ or $y < 0$, while $\phi_n(i) = \phi((n+1+i)/2(n+1))$ are scores generated by a given function $\phi((1+u)/2)$, $0 < u < 1$. Writing $\boldsymbol{S}_n^+$ for $(S_{n1}^+, \ldots, S_{nq}^+)'$ the signed rank test statistic for testing (2.2) is

$$M_n^+ = \boldsymbol{S}_n^{+\prime}(X_nX_n')^{-1}\boldsymbol{S}_n^+/A^2 \tag{2.12}$$

where $A^2 = \int \phi^2(u)\,\mathrm{d}u$.

It has been shown by many authors e.g. Adichie (1967a, 1978), Huskova (1970) and Srivastava (1970) among others, that under regularity conditions similar to those in A–C of Section 2.1 and F being symmetric, M_n^+ has asymptotically, as n becomes large, a chi-square distribution with q degrees of freedom when (2.2) holds. For large n therefore, an approximate level α test for (2.2) may be obtained if H_0 is rejected for $M_n^+ > \chi_q^2(1-\alpha)$.

If ties occur, we use average scores $\bar{\phi}_n(i)$ as discussed in Section 2.1.1, and the resulting statistic is

$$M_n^+(\bar{\phi}) = \boldsymbol{S}_n^{+\prime}(\bar{\phi})(X_nX_n')^{-1}\boldsymbol{S}_n^+(\bar{\phi})/A_n^2(\bar{\phi}) \tag{2.13}$$

where $A_n^2(\bar{\phi}) = n^{-1}\Sigma^1\,\bar{\phi}_n^2(i)$.

For a discussion of methods of handling ties and zeros in signed rank test statistics, see Pratt (1959) and Connover (1973). It has been shown (see e.g. Vorlickova, 1972) that under conditions similar to those required for M_n^+ given in (2.12), the statistic $M_n^+(\bar{\phi})$ defined in (2.13) also has asymptotically, as n becomes large, a chi-square distribution with q degrees of freedom when (2.2) holds.

3. Testing sub hypotheses

In the general univariate linear model (1.1) it is sometimes necessary to test hypotheses about some of the β's while regarding others as nuisance parameters. In this section we consider rank methods of testing sub hypotheses about the β's in model (1.1). It is more convenient to write (1.1) in the form

$$\boldsymbol{Y} = \boldsymbol{\alpha} + X'_{n1}\boldsymbol{\beta}_1 + X'_{n2}\boldsymbol{\beta}_2 + \boldsymbol{\varepsilon} \tag{3.1}$$

where $\boldsymbol{Y}$ and $\boldsymbol{\alpha}$ are as defined in (1.1); X_{n1}, $(q \times n)$ and X_{n2}, $(q_2 \times n)$ form a partition of X_n, $(q_1 + q_2 = q)$ while $\boldsymbol{\beta}_1$ and $\boldsymbol{\beta}_2$ are corresponding subvectors of $\boldsymbol{\beta}$. The interest is to test

$$H_0: \quad \boldsymbol{\beta}_1 = 0, \quad \boldsymbol{\beta}_2 \text{ unspecified} \tag{3.2}$$

against the alternative that $\boldsymbol{\beta}_1 \neq 0$.

3.1. Rank test statistic

Although Koul (1970), Puri and Sen (1973) and Adichie (1978) among others have suggested rank test statistics for (3.2), their work is nevertheless based on the extra assumption that F, the distribution function of Y is symmetric. Since we do not need that assumption, we first reparametrize (3.1) as follows:

$$\boldsymbol{Y} = \boldsymbol{\alpha}_n + Z'_{n1}\boldsymbol{\beta}_1 + Z'_{n2}\boldsymbol{\beta}_2 + \boldsymbol{\varepsilon} \tag{3.3}$$

where

$$\boldsymbol{\alpha}_n = \boldsymbol{\alpha} + \bar{X}'_n\boldsymbol{\beta}_1 + \bar{X}'_n\beta_2; \qquad Z_n = (Z_{n1}, Z_{n2}) = (X_{n1} - \bar{X}_n, X_{n2} - \bar{X}_n)$$

corresponds to $X_n = (X_{n1}, X_{n2})$ of (3.1), while $\bar{X}_n$ is as defined in (2.3).

The effect of the unknown β_2 is removed by substituting its suitable estimate under H_0 in (3.2). It has been shown that both the least squares and the 'rank' estimates of β_2 are suitable for the problem. Rank methods of estimating regression parameters are discussed in detail in a separate chapter in this volume, see also Adichie (1967b), Jureckova (1971), Puri and Sen (1973) among others.

Let $\hat{\boldsymbol{\beta}}_2$ be any estimate of $\boldsymbol{\beta}_2$, that satisfies condition D below. Consider the aligned vector of observations, $Y - Z'_{n2}\hat{\boldsymbol{\beta}}_2$, and let

$$Y_i(\hat{\beta}_2) = \hat{Y}_i = (Y - Z'_{n2}\hat{\boldsymbol{\beta}}_2)i \tag{3.4}$$

denote the i-th aligned observation.

Define

$$\hat{\boldsymbol{T}}_{n1} = Z_{n1}(I_n - C_{n,2})\Psi_n(\hat{R}) = (\hat{T}_{n1}, \ldots, \hat{T}_{nq_1})' \tag{3.5}$$

where

$$C_{n,2} = Z'_{n2}(Z_{n2}Z'_{n2})^{-1}Z_{n2}$$

is an $n \times n$ indempotent matrix. I_n is an identity matrix of order n, while

$$\Psi_n(\hat{R}) = (\psi_n(\hat{R}_1), \ldots, \psi_n(\hat{R}_n))',$$

and $\hat{R}_i$ is the rank of $\hat{Y}_i$ given in (3.4) and the scores are as defined in (2.4).

The proposed test statistic for testing H_0 in (3.2) would be a quadratic form in the q_1 elements of $\hat{T}_{n1}$, with the discriminant being the inverse of the covariance matrix. It has been shown in Adichie (1978) that subject to conditions A–D given below, the null distribution of $(n^{-1/2}\hat{\boldsymbol{T}}_{n1})$ tends to a q_1-variate normal with zero mean and covariance matrix $A^2(\psi)C^*$, where $C^* = \lim n^{-1}C_n^*$, with

$$C_n^* = Z_{n1}(I_n - C_{n,2})Z'_{n1} \tag{3.6}$$

and $A^2(\psi)$ is as defined in (2.5). It follows that for testing H_0 in (3.2), we may use

$$\hat{M}_n = (\hat{\boldsymbol{T}}'_{n1}C_n^{*-1}\hat{T}_{n1})/A^2(\psi) \tag{3.7}$$

which under (3.2) has asymptotically a chi-square distribution with q_1 degrees of freedom. For large values of n, an approximate level α test of H_0 in (3.2) may be obtained if H_0 is rejected for $\hat{M}_n > \chi^2_{q1}(1-\alpha)$. Another variant of (3.7) was proposed in Sen and Puri (1977). If the estimate $\hat{\boldsymbol{\beta}}_2$ used in the aligned observations (3.4) is a 'rank estimate', then $\hat{\boldsymbol{T}}_{n1}$ in (3.5) can be simplified to

$$\hat{\boldsymbol{S}}_{n1} = Z_{n1}\Psi_n(\hat{R}) = (\hat{S}_{n1}, \ldots, \hat{S}_{nq_1})' . \tag{3.8}$$

It can be shown as in Sen and Puri (1977) that subject to the conditions A–D of this section, the null distribution of $(n^{-1/2}\hat{\boldsymbol{S}}_{n1})$ tends to a q_1 variate normal with mean zero and the same covariance matrix $A^2(\psi)C^*$ given in (3.6). It follows that

$$\hat{M}_n^* = (\hat{\boldsymbol{S}}'_{n1}C_n^{*-1}\hat{\boldsymbol{S}}_{n1})/A^2(\psi) \tag{3.9}$$

may also be used for testing H_0 in (3.2) particularly if the alignment (3.4) is done using 'rank' estimate. It is also clear that $\hat{M}_n$ in (3.7) and $\hat{M}_n^*$ in (3.9) have the same limiting distribution.

The conditions required for the limiting distribution of $\hat{M}_n(\hat{M}_n^*)$ are as follows:

Condition A. The distribution function in (2.1) has a density f which is absolutely continuous such that the Fisher Information $I(F) = \int (f'(y)/f(y))^2 f(y)\,dy$ is finite.

Condition B. The regression constants satisfy

(i) $\max_i(z_{ji}^2/\Sigma_i z_{ji}^2) \to 0$ for each $j = 1, \ldots, q$

(ii) $\operatorname{rank}(Z_nZ'_n) = q$ for all n.

(iii) $n^{-1}(Z_nZ'_n)$ tends to a positive definite and finite matrix (ZZ') as $n \to \infty$.

(iv) For each $i = 1, \ldots, n$, $(z_{ji} - \bar{z}_j) = (z_{ji}^{(1)} - \bar{z}_j^{(1)}) - (z_{ji}^{(2)} - \bar{z}_j^{(2)})$ where for $j = 1, \ldots, q$, each is nondecreasing.

CONDITION C. The score function satisfies $\psi(u) = \psi_1(u) - \psi_2(u)$, where each $\psi_s(u)$, $s = 1, 2$, is absolutely continuous nondecreasing square integrable on $(0, 1)$.

CONDITION D. The estimate $\hat{\boldsymbol{\beta}}_2$ of $\boldsymbol{\beta}_2$ is

(i) translation invariant i.e. for all β_2, $\hat{\beta}_2(\boldsymbol{Y} - Z'_{n2}\beta_2) = \hat{\boldsymbol{\beta}}_2(\boldsymbol{Y}) - \boldsymbol{\beta}_2$, where $\hat{\boldsymbol{\beta}}_2(Y)$ denotes estimates computed from Y.

(ii) Consistent, i.e. the difference $n^{1/2}|\hat{\boldsymbol{\beta}}_2 - \boldsymbol{\beta}_2|$ is $O_p(1)$ as $n \to \infty$, where p refers to probability under (3.2).

3.1.1. Computation of $\hat{M}_n$ ($\hat{M}^*_n$)

The computation of either $\hat{M}_n$ in (3.7) or $\hat{M}^*_n$ in (3.9) presents no special difficulty. However the $\hat{M}_n$ statistic in (3.7) has some advantage over $\hat{M}^*_n$ in (3.9) with respect to the easy in computation. We have pointed out that $\hat{M}^*_n$ requires the use of rank estimate of $\hat{\beta}_2$ for its definition (Sen and Puri, 1977). Rank estimates of regression parameters are usually not very easy to calculate because of the iterative processes involved. The definition and study of $\hat{M}_n$ on the other hand admits any estimate $\hat{\boldsymbol{\beta}}_2$ that satisfies condition D of Section 3.1; and many reasonable estimates including the usual least square estimate, satisfy the condition. Furthermore, it is easy to see that $\hat{M}_n$ in (3.7) can be written as

$$\hat{M}_n = (\Psi'_n(\hat{R}) W_n \Psi_n(\hat{R}))/A^2(\psi) \tag{3.10}$$

where

$$W_n = Z'_n(Z_n Z'_n)^{-1} Z_n - Z'_{n2}(Z_{n2} Z'_{n2})^{-1} Z_{n2} = C_n - C_{n,2}$$

is a symmetric idempotent matrix of order $n \times n$. For $q_1 > 2$, it is generally easier to calculate W_n than to obtain the inverse C^{*-1}_n of $Z_{n1}(I_n - C_{n,2})Z'_{n1}$ required for the calculation of $\hat{M}^*_n$ in (3.9). The representation of $\hat{M}_n$ in (3.10) brings out very clearly its similarity in form to the variance-ratio statistic Q used in normal theory. This can be written as $Q = (Q_1/q_1) \div (Q_0/(n-1))$, where the divisor tends to $\sigma^2 = \mathrm{Var}(Y)$ as n tends to infinity, and

$$Q_1 = \boldsymbol{Y}' W_n Y .$$

EXAMPLE 4. Consider the following artificial example adapted from page 142 of Graybill (1961), where the model is $Y = \alpha + \beta_1 x_1 + \beta_2 x_2 + \varepsilon$. We want to test H_0: $\beta_1 = 0$, β_2 unspecified. Assume that after reparametrization, as in (3.3), the adapted observations become:

Y:	6	13	13	29	33	23	46	117
z_1:	−5	−5	−3	−4	−1	0	3	15
z_2:	−5	−4	−3	−2	−1	0	2	13

Under H_0, we use the LSE of β_2 i.e. $(\Sigma_i z_{2i} Y_i)/\Sigma_i z^2_{2i}$ which gives $\hat{\beta}_2 = 10.1$. Using Wilcoxon scores we find that, by (3.8), $\hat{S}_n = (\hat{S}_{n1}, \hat{S}_{n2})' = (1/9)(59, 59)'$ and, by (3.5), $\hat{T}_{n1} = 1/9(59 - (1.158)49) = 0.24$ with

$$Z_n Z_n' = \begin{pmatrix} 310 & 264 \\ 264 & 228 \end{pmatrix}$$

and $C_n^* = 310 - (264)^2/228 \cong 4.32$, giving $\hat{M}_n = 0.16$. It is observed that for this problem the variance ratio criterion gives $Q = 7Q_1/Q_0 = 0.05$, implying as did the $\hat{M}_n$-test, that it is an obvious case of nonrejection.

EXAMPLE 5. Let us consider again the data used in example 3 with the model $Y = \beta_1 Z_1 + \beta_2 Z_2 + \beta_3 Z_3 + \boldsymbol{\varepsilon}$ (observations are given in x's in the table), where we now want to test H_0: $\beta_3 = 0$, β_1 and β_2 unspecified. Using the least squares estimate of β_1 and β_2 under H_0, we get $\hat{\beta}_1 = -0.592$, $\hat{\beta}_2 = -0.290$. Ranking the aligned observations $\hat{Y}_i = (Y_i - \hat{\beta}_1 Z_{1i} - \hat{\beta}_2 Z_{2i})$, we obtain the following:

7	17	3	16	6	1	11	2	29	30
24	15	4	28	25	26	20	22	18	12
27	10	8	14	23	19	21	5	13	9

With $(Z_n Z_n')$ already given in Example 3, we find, using simply the Wilcoxon scores that

$$\hat{\boldsymbol{S}}_n (Z_n Z_n')^{-1} \hat{\boldsymbol{S}}_n = 7.45245, \qquad \hat{\boldsymbol{S}}_n' (Z_{n2} Z_{n2}')^{-1} \hat{\boldsymbol{S}}_n = 0.00305$$

thus $\hat{M}_n$ (3.6) becomes 89.3928. This suggests that the hypothesis be rejected.

3.1.2. Asymptotic efficiency of $\hat{M}_n^*$ ($\hat{M}_n$)

It has been shown in Adichie (1978), that under

$$H_n: \quad \boldsymbol{\beta}_1 = n^{-1/2} \boldsymbol{b}_1, \qquad \|\boldsymbol{b}_1\| < k \tag{3.11}$$

and subject to Conditions A–D of Section 3.1, $n^{-1/2} \hat{T}_{n1}$ defined in (3.5) has asymptotically, as $n \to \infty$, a q_1 variate normal distribution with mean $\boldsymbol{\mu}$ and covariance matrix C^* given in (3.6), where

$$\boldsymbol{\mu} = \lim n^{-1/2} (Z_{n1}(I_n - C_{n,2}) Z_{n1}' \boldsymbol{b}_1) B(F) = (C^* \boldsymbol{b}_1) B(F), \tag{3.12}$$

with $B(F)$ as defined in (2.8).

It follows from (3.12) that under H_n in (3.11), the statistic $\hat{M}_n$ in (3.7) has asymptotically a noncentral chi square distribution with q_1 degrees of freedom and noncentrality parameter given by

$$\Delta_{\hat{M}} = (\boldsymbol{b}_1' C^* \boldsymbol{b}_1) B^2(F) / A^2(\psi) \tag{3.13}$$

That the same result holds for $\hat{M}_n^*$ in (3.9) follows from Sen and Puri (1977).

The normal theory test statistic for H_0 in (3.2) can be written as $Q_n = (Q_1/q_1) \div (Q_0/(n-q))$. When F is not normal, but has a finite variance $\sigma^2(F)$, then it can be shown that under H_n in (3.11) and Condition B of Section 3.1, Q_n has asymptotically as $n \to \infty$ a noncentral chi-square distribution with q_1 degrees of freedom and noncentrality parameter,

$$\Delta_Q = (\boldsymbol{b}_1' C^* \boldsymbol{b}_1)/\sigma^2(F) \tag{3.14}$$

From (3.13) and (3.14) it is seen that the asymptotic efficiency of $\hat{M}_n^*$ ($\hat{M}_n$) relative to Q_n is

$$\hat{e}_{\hat{M},Q} = \sigma^2(F)B^2(F)/A^2(\psi) \tag{3.15}$$

which is the same as the one given in (2.10).

4. Comparison of several regression lines

Problems concerning the comparison of many linear regression models are frequently encountered in practice. Assume we have k independent samples where each Y_{ij} the i-th observation in the j-th sample, is taken at the level x_{ij}. More precisely, let

$$Y_{ij} = \alpha_j + \beta_j x_{ij} - \varepsilon_{ij}, \quad i = 1, \dots, n_j, \; j = 1, \dots, k, \tag{4.1}$$

where for each j, ε_{ij} has the same continuous distribution function, $F(\cdot)$ whose functional form is not necessarily known. Statistical testing problems connected with (4.1) are of two types: first, testing the parallelism (i.e. $\beta_j = \beta$) and secondly, testing the coincidence ($\alpha_j = \alpha$, $\beta_j = \beta$) of the regression lines.

4.1. Testing parallelism of regression lines

A number of authors have suggested rank methods for testing parallelism of several regression lines. Notable among these are Sen (1969) and Adichie (1974). In the model (4.1), the hypothesis of interest is

$$H_0: \qquad \beta_j = \beta, \quad \alpha_j \text{ unspecified}. \tag{4.2}$$

The first step in constructing a rank test statistic for (4.2) is to align each of the k samples on β. But since the common value of β is not usually known, the alignment is on a suitable estimate of β. Sen (1969) used a rank estimate, but the least squares estimate

$$\bar{\beta} = \sum_i \sum_j (x_{ij} - \bar{x}_j) Y_{ij} \Big/ \sum_i \sum_j (x_{ij} - \bar{x}_j)^2, \quad \bar{x}_j = n_j^{-1} \sum_i x_{ij}$$

would also be suitable

4.1.1. Test based on separate rankings

Consider the aligned observations

$$\hat{Y}_{ij} = (Y_{ij} - \bar{\beta}x_{ij}), \quad j = 1, \ldots, k, \; i = 1, \ldots, n_j, \tag{4.3}$$

and rank each of the k samples separately. Let $\hat{r}_{ij}$ denote the rank of $\hat{Y}_{ij}$ in the ranking of the j-th sample.

For each $j = 1, \ldots, k$, let

$$C_j^2 = \sum_i (x_{ij} - \bar{x}_j)^2, \qquad \tau_j = C_j^2 \Big/ \sum_j C_j^2 \tag{4.4}$$

where $\bar{x}_j = n_j^{-1} \Sigma_i x_{ij}$ and $N = \Sigma_j n_j$. Now define

$$\hat{T}_{1j} = \sum_i (x_{ij} - \bar{x}_j)\phi_n(\hat{r}_{1j}). \tag{4.5}$$

The proposed test statistic for testing (4.2) is

$$\hat{L}_1 = \sum_j (\hat{T}_{1j}/A(\psi)C_j)^2 \tag{4.6}$$

where $A(\psi)$ is as defined in (2.5).

Sen (1969) showed that $\hat{L}_1$ has asymptotically a chi square distribution with $(k-1)$ degrees of freedom when (4.2) is true. This implies that for large N, an approximate level α test may be obtained if the hypothesis (4.2) is rejected for $\hat{L}_1 > \chi^2_{k-1}(1-\alpha)$.

One feature of the $\hat{L}_1$ test is that it involves ranking k different samples separately. A procedure that would take the simultaneous ranking of all the observations in the k samples would certainly be preferred. This later method has been suggested by Adichie (1974) but only for the special case of model (4.1) where $\alpha_i = \alpha$ (unknown).

4.1.2. Tests based on simultaneous ranking

Although Adichie (1974) considered aligned observations $(Y_{ij} - \hat{\beta}x_{ij})$, where $\hat{\beta}$ is rank estimate of β, it is has been shown that his method is valid also for the aligned observations given in (4.3). Let $\hat{R}_{ij}$ be the rank of $\hat{Y}_{ij}$ in the simultaneous ranking of all N $(=\Sigma_j n_j)$ observations. To simplify the notation, let

$$\begin{aligned} c_{si}^{(j)} &= \gamma_j(x_{sj} - \bar{x}_s), \qquad && s = 1, \ldots, j-1, j+1, \ldots, k, \\ &= (\gamma_j - 1)(x_{si} - \bar{x}_s), && s = j, \end{aligned} \tag{4.7}$$

so that

$$\bar{c}^{(j)} = N^{-1} \sum_s \sum_i c_{si}^{(j)} = 0, \quad j = 1, \ldots, k, \tag{4.8}$$

and

$$\sum_s \sum_i (c_{si}^{(j)} - \bar{c}^{(j)})^2 = \gamma_j(1-\gamma_j) \sum_s C_s^2 .$$

Now define

$$\hat{T}_{2j} = \sum_s \sum_i c_{si}^{(j)} \phi_n(\hat{R}_{si})$$

where for the summations, i goes from 1 to n_j; j (or s) goes from 1 to k. The proposed test statistic is

$$\hat{L}_2 = \sum_j (\hat{T}_{2j} | A(\psi) C_j)^2 . \tag{4.9}$$

Adichie (1974) gives conditions under which $\hat{L}_2$ has asymptotically a chi square distribution under (4.2) with $\alpha_j = \alpha$. For large N, an approximate level α test is obtained if the hypothesis is rejected for $\hat{L}_2 > \chi^2_{k-1}(1-\alpha)$.

4.1.3. Asymptotic efficiency of $\hat{L}$-tests

It is proved in Sen (1969) and Adichie (1974) that under a sequence of near alternatives

$$H_1: \quad \beta_j = \beta + \left(\theta_j \Big/ \sum C_s^2\right), \quad |\theta_j| \leq M, \quad i = 1, \ldots, k . \tag{4.10}$$

$\hat{L}_1$ and $\hat{L}_2$ respectively have asymptotically a noncentral chi-square distribution with $k-1$ degrees of freedom and noncentrality parameter given by

$$\Delta_L = \sum_j \gamma_j (\theta_j - \bar{\theta})^2 B^2(F)' A^2(\psi) \tag{4.11}$$

where

$$\bar{\theta} = \sum_s \gamma_s \theta_s \quad \text{and} \quad B^2(F) \text{ is as in (2.8)} .$$

The usual test statistic for (4.2) based on the least squares estimates has the form

$$Q = \sum_j C_j^2 (\bar{\beta}_j - \bar{\bar{\beta}})^2 / (k-1) s_e^2 \tag{4.12}$$

where

$$\bar{\beta}_j = \sum_i (x_{ij} - \bar{x}_j) Y_{ij} / C_j^2, \qquad \bar{\bar{\beta}} = \sum_j \gamma_j \bar{\beta}_j ,$$

and s_e^2 is the mean square due to error. It is shown that under (4.10), $(k-1)Q$ has asymptotically a noncentral chi square distribution with $(k-1)$ degrees of freedom and noncentrality parameter given by

$$\Delta_Q = \sum_j \gamma_j(\theta_j - \bar{\theta})/\sigma^2(F). \tag{4.13}$$

The asymptotic efficiency of $\hat{L}$-test relative to the Q-test is therefore

$$e(\Delta_L|\Delta_Q) = \sigma^2 B^2(F)|A^2(\psi) \tag{4.14}$$

which is same as (2.10) and (3.15).

4.5. *Testing coincidence of regression lines*

In Section 4.1 we discussed the problem of testing for parallelism of regression lines; in this section we take up the question of testing whether several regression lines are identical. We refer again to model (4.1) and want to test

$$H_0: \quad \alpha_j = \alpha, \quad \beta_j = \beta, \quad \alpha \text{ and } \beta \text{ unknown}. \tag{4.15}$$

An early attempt at providing a rank test statistic for (4.15) was made by Puri and Sen (1969) who considered the case of only two regression lines,

$$Y_{ij} = \alpha_j + \beta_j x_{ij} + \varepsilon_{ij}, \quad j = 1, 2, \ i = 1, \ldots, n_j.$$

In their approach, they introduced a 'balanced symmetric design' by assuming $n_1 = n_2$ and $x_{i1} = x_{i2} = x_i$. Their approach then is to reduce the problem to a single regression case by considering the difference

$$Y_{i1} - Y_{i2} = (\alpha_1 - \alpha_2) + (\beta_1 - \beta_2)x_i + \varepsilon_{il} - \varepsilon_{i2}, \quad i = 1, \ldots, n,$$

and applying the procedure outlined in Section 2, for $p = 2$.

A different approach to the testing problem was suggested by Adichie (1975). His method also requires some symmetric balancing of the design matrix. In considering model (4.1), he assumed that the x_{ij}'s have been centred about their group means $\bar{x}_j$. More precisely, let the model be

$$Y_{ij} = \alpha_j + \beta_j(x_{ij} - \bar{x}_j) + \varepsilon_{ij}, \quad j = 1, \ldots, k, \ i = 1, \ldots, n_j, \tag{4.16}$$

where interest is on testing the composite hypothesis given in (4.15) against the set of alternatives that violate (4.15). Consider the aligned observations $\hat{Y}_{ij} = Y_{ij} - \bar{\beta}(x_{ij} - \bar{x}_j)$, similar to those given in (4.3). Observe that because rank $(\hat{Y}_{ij} - \alpha)$ is the same as rank $\hat{Y}_{ij}$, it is not necessary to align the observations also on a suitable estimate $\hat{\alpha}$ of α. Let $\hat{R}_{ij}$ be the rank of $\hat{Y}_{ij}$ in the simultaneous ranking of all the N $(= \Sigma_j n_j)$ aligned observations. Write

$$\lambda_j = (n_j|N) \tag{4.17}$$

and let

$$d_{si} = 0, \quad s \neq j, \\ \quad = 1, \quad s = j, \tag{4.18}$$

so that

$$d^{(j)} = N^{-1} \sum_s \sum_i d_{si}^{(j)} = \lambda_j .$$

For each $j = 1, \ldots, k$, define

$$\hat{T}_{\alpha j} = \sum_s \sum_i (d_{si}^{(j)} - \bar{d}^{(j)}) \phi_n(\hat{R}_{si}) \tag{4.19}$$

and

$$\hat{T}_{\beta j} = \sum_s \sum_i c_{si}^{(j)} \phi_n(\hat{R}_{si}) \tag{4.20}$$

where $c_{si}^{(j)}$ is as defined in (4.7), and put

$$\hat{V}_j = N^{-1/2}(\hat{T}_{\alpha j}|A(\psi)), \qquad \hat{U}_j = (\hat{T}_{\beta j}|A(\psi)C_j) \tag{4.21}$$

where C_j is as defined in (4.4). The proposed test statistic is

$$M = \sum_j (\hat{V}_j^2 + \hat{U}_j^2) . \tag{4.22}$$

Adichie (1975) gave conditions under which M asymptotically has a chi square distribution with $2(k-1)$ degrees of freedom under (4.15). For large N therefore an approximate level α test is obtained if the hypothesis (4.15) is rejected for $M > \chi^2_{2(k-1)}(1-\alpha)$.

Example 6. Consider model (4.1),

$$Y_{ij} = \alpha_j + \beta_j x_{ij} + \varepsilon_{ij}, \quad j = 1, \ldots, 4, \ i = 1, \ldots, 14 ,$$

and we want to test $\beta_1 = \beta_2 = \beta_3 = \beta_4$. The observations (adapted from the data on page 395 of Brownlee's *Statistical Theory and Methodology in Science and Engineering*) are shown in Table 2.

Since we do not assume $\alpha_j = \alpha$, we shall use $\hat{L}_1$ of (4.6). With Wilcoxon scores, $\hat{T}_{ij} = (1/15) \sum_i \hat{r}_{ij} x_{ij} - 7\bar{x}_j$. The following quantities are calculated from the given data: Least squares estimate $\bar{\beta} = 0.56$; $\bar{x}_1 = 12.5$, $\bar{x}_2 = 15$, $\bar{x}_3 = 11.64$, $\bar{x}_4 = 16.21$; $C_j^2 = 1461.5$, 858, 1770.15, 5322.24. The ranks $\hat{r}_{ij}$ of the aligned observations $(Y_{ij} - \bar{\beta} x_{ij})$ are shown on Table 3. With these, $\hat{T}_{1j} = 26.43, 9.47, 27.72, 26.67$ for $j = 1, \ldots, 4$, giving $\hat{L}_1 = 13.80$, which suggests that the hypothesis of equality of the regression parameter is rejected.

Table 2

Sample No.															
1	Y_1	24	30	39	41	27	46	56	25	18	25	31	52	38	45
	x_1	4	8	18	22	4	19	25	−1	−1	2	14	32	7	22
2	Y_2	18	32	33	35	19	28	41	37	33	39	38	41	38	36
	x_2	3	13	19	16	−2	8	23	17	10	29	17	20	16	21
3	Y_3	35	31	55	43	44	28	33	13	39	58	18	17	41	25
	x_3	7	7	24	18	12	12	4	−7	19	32	−3	0	29	29
4	Y_4	28	40	34	27	47	30	24	28	21	28	39	26	46	42
	x_4	7	24	19	7	39	16	−1	14	2	8	19	9	35	29

Table 3

Sample No.														
1	2	6	9	8	5	13	14	7	1	4	3	11	12	10
2	1	7	3	8	2	5	11	10	9	4	12	14	13	6
3	10	7	14	11	12	5	9	2	8	13	4	3	6	1
4	8	10	6	5	11	2	9	3	1	7	14	4	13	12

4.2.1. Asymptotic efficiency of M test

Under a sequence of near alternatives

$$H_1:\quad \alpha_j = \alpha + \left(\xi_j \Big/ \sum_s C_s\right), \quad \beta_j = \beta + \left(\theta_j \Big/ \sum_s C_s\right), \tag{4.23}$$

where $|\xi_j| < K_1$, $|\theta_j| < K_2$, it is also proved that M test has asymptotically as $n_j \to \infty$, a noncentral chi square distribution with $2(k-1)$ degrees of freedom and noncentrality parameter

$$\Delta_M = \lim \sum_j \{\rho_j(\xi_j - \bar{\xi})^2 + \gamma_j(\theta_j - \bar{\theta})^2\} B^2(F)/A^2(\psi) \tag{4.24}$$

where $\bar{\xi} = \Sigma_j \lambda_j \xi_j$, $\bar{\theta}$ is as defined in (4.11) and $\rho_j = (n_j/\Sigma_s C_s)$.

On the other hand, the classical test statistic Q for the hypothesis (4.15) is the variance ratio criterion

$$Q = \sum_j \{n_j(\bar{Y} - \bar{\bar{Y}})^2 + C_j^2(\bar{\beta}_j - \bar{\bar{\beta}})^2\}/2(k-1)s_e^2 \tag{4.25}$$

where $\bar{Y}_j = n_j^{-1} \Sigma_i Y_{ij}$, while $\bar{\beta}_j$, $\bar{\bar{\beta}}$ and s_e^2 are as defined in (4.12). Under the assumption of normality of the underlying distribution function F, Q has the variance ratio criterion with $(2k-2,\ N-2k)$ df. When the normality assumption is dropped, the exact distribution of Q is not known. It has however been shown that for any $F(y)$ for which $\sigma^2(F)<\infty$, $2(k-1)Q$ under (4.15) has asymptotically a chi square distribution with $2(k-1)$ degree of freedom, while under (4.23), it has asymptotically a noncentral chi square with $2(k-1)$ degrees of freedom and noncentrality parameter

$$\Delta_Q = \lim \sum_j \{\rho_j(\xi_j - \bar{\xi})^2 + \gamma_j(\theta_j - \bar{\theta})^2\}/\sigma^2(F). \tag{4.26}$$

It follows from (4.25) and (4.26) that the asymptotic efficiency of M test relative to Q test is

$$e_{M/Q} = \sigma^2(F)B^2(F)/A^2(\psi) \tag{4.27}$$

which is the same familiar efficiency expression obtained in (2.10), (3.15) and (4.14) in different testing problems.

5. Rank tests in multivariate linear models

So far we have been discussing rank tests for linear hypothesis in the univariate set up. In this section, we present briefly the extensions to multivariate linear models.

Consider independent observations (random vectors) $\boldsymbol{Y}_i = (Y_{1i}, \ldots, Y_{pi})'$, $i = 1, \ldots, n$, taken at $\boldsymbol{x}_i$, such that the distribution function of $\boldsymbol{Y}_i$ is

$$P(\boldsymbol{Y}_i \leqslant y) = F_i(y) = F(y - \boldsymbol{\alpha} - \boldsymbol{\beta}\boldsymbol{x}_i), \quad i = 1, \ldots, n, \ \ y \in \mathbb{R}^p, p \geqslant 1, \tag{5.1}$$

where $\boldsymbol{\alpha} = (\alpha_1, \ldots, \alpha_p)'$, $\boldsymbol{\beta} = ((\beta_{kj}))$, $k = 1, \ldots, p$, $j = 1, \ldots, q$ is a $p \times q$ matrix of unknown parameters, and $\boldsymbol{x}_i = (x_{1i}, \ldots, x_{qi})'$ are vectors of known regression constants. The functional form of the distribution function F need not be known. Observe that (5.1) is the p-variate version of the univariate $(p = 1)$ model given in (2.1).

For the model (5.1), we shall discuss two types of test of hypothesis concerning $\boldsymbol{\beta}$, namely testing that all the pq elements of $\boldsymbol{\beta}$ are zero and secondly testing the sub-hypothesis, i.e. that some parts of $\boldsymbol{\beta}$ are zero, while the rest are nuisance parameters. We start with the first one.

5.1. *Rank test of no regression*

In model (5.1), we want to test the hypothesis

$$H_0:\quad \beta = 0\,. \tag{5.2}$$

The procedure described here is due to Puri and Sen (1969). Under (5.2), $\boldsymbol{Y}_i$, $i = 1, \ldots, n$, are independent identically distributed p-variate vectors. In multivariate cases, we rank the marginals of the observations $(\boldsymbol{Y}_1, \ldots, \boldsymbol{Y}_n)$. So let R_{ki} be the rank of Y_{nk} in the ranking of $(Y_{k1}, \ldots, Y_{kn})$ $k = 1, \ldots, p$. Let

$$\psi_{nk}(i) = \psi_k(i/(n+1)), \quad k = 1, \ldots, p\,, \tag{5.3}$$

be a set of scores generated by functions $\psi_k(u)$, $0 < u < 1$. To obtain the test statistic, define a pq row vector

$$\boldsymbol{S}'_n = (\boldsymbol{S}'_{n1}, \ldots, \boldsymbol{S}'_{np}) = (S_{n,kj};\, k = 1, \ldots, p;\, j = 1, \ldots, q) \tag{5.4}$$

where

$$\boldsymbol{S}_{nk} = Z_n \Psi_{nk}(R_k), \quad k = 1, \ldots, p\,, \tag{5.5}$$

are q-vectors, and $\Psi_{nk}(R_k) = (\psi_{nk}(R_{k1}), \ldots, \psi_{nk}(R_{kn}))'$ with the scores as defined in (5.3), and Z_n is obtained from X_n as in (2.3), after reparametrization. As in the univariate case, the statistics required would be a quadratic form in the pq elements of $\boldsymbol{S}_n$ where the discriminant of the quadratic form is the inverse of the asymptotic covariance matrix G say of $\boldsymbol{S}_n$. Unlike the situation in the univariate case, G, even under (5.2) depends on the marginal distributions $F_{(k)}$ and $F_{(k,k')}$ defined in (5.7) below. To obtain the expression for G, first define

$$A(F) = ((a_{k,k'}(F))), \quad k, k' = 1, \ldots, p\,, \tag{5.6}$$

where

$$a_{k,k'}(F) = \int\int \psi_k(F_{(k)}(y)\psi_{k'}(F_{(k')}(y))\, \mathrm{d}F_{(k,k')}(y, z) - \bar{\psi}_k\bar{\psi}_{k'} \tag{5.7}$$

with $F_{(k)}(y)$ and $F_{(k,k')}(y, z)$ being the marginal distribution functions of the respective k-th and (k, k')th component of $\boldsymbol{Y}$, and

$$\bar{\psi}_k = \int \psi_k(u)\, \mathrm{d}u, \quad k = 1, \ldots, p\,, \tag{5.8}$$

The matrix $A(F)$ in (5.6) is the multivariate analogue of $A^2(\psi)$ in (2.5). Now define a $pq \times pq$ matrix by

$$G = A(F) \otimes ZZ' = ((a_{kk'}(F)z'_{jj})) \tag{5.9}$$

where $\otimes$ denotes the Kronecker product, and $ZZ' = \lim n^{-1}(Z_n Z'_n)$. Because we do not know the functional form of F, $A(F)$ is not available for use in constructing the test statistic. Consider a sample measure of $A(F)$ given by

$$\hat{A}_n = ((\hat{a}_{n,kk'})), \quad k, k' = 1, \ldots, p\,, \tag{5.10}$$

where

$$\hat{a}_{n,kk'} = (n-1)^{-1} \sum_i \{\psi_{nk}(R_{ki})\psi_{nk'}(R_{k'i}) - \bar{\psi}_{nk}\bar{\psi}_{nk'}\} \tag{5.11}$$

and

$$\bar{\psi}_{nk} = n^{-1} \sum_i \psi_{nk}(i), \quad k = 1, \ldots, p. \tag{5.12}$$

Let

$$\hat{G}_n = \hat{A}_n \otimes Z_n Z_n' = ((\hat{a}_{n,kk'} z_{n,jj'})) \tag{5.13}$$

be a sample measure of G in (5.9). Note that $\hat{G}_n$ is also a $pq \times pq$ matrix. The test statistic for testing H_0 in (5.2) as proposed by Puri and Sen (1969a) can be written as

$$\begin{aligned}\mathcal{L}_n &= S_n' \hat{G}_n^{-1} S_n = (S_{n1}', \ldots, S_{np}') \hat{G}_n^{-1} (S_{n1}', \ldots, S_{np}')' \\ &= \sum_k^p \sum_{k'}^p \sum_j^q \sum_{j'}^q S_{n,kj}, S_{n,k'j'}, \hat{a}_n^{kk'}, z_n^{jj'};\end{aligned} \tag{5.14}$$

where

$$((\hat{a}_n^{kk'})) = \hat{A}_n^{-1} \quad \text{and} \quad ((z_n^{jj'})) = (Z_n Z_n')^{-1}. \tag{5.15}$$

The $\mathcal{L}_n$ test in (5.14) rejects H_0 in (5.2) if $\mathcal{L}_n$ is large, where the cut off point is obtained from the distribution of $\mathcal{L}_n$. It turns out that in the multivariate case, the exact distribution of the ranks of the observations, even under H_0 in (5.2), depends on $F(y)$. Since the functional form of F is not assumed to be known, it is not possible to obtain the exact distribution of $\mathcal{L}_n$. However, a conditional distribution is obtainable (see Puri and Sen (1966) for details), but even at that, the enumeration of this permutation distribution of the ranks, and hence that of $\mathcal{L}_n$, can be tedious for a given n.

A way out is to resort to limiting distribution if one is available. It has been shown in Puri and Sen (1969a) that under conditions A–C given below, $(n^{-1/2} S_n)$ has asymptotically a pq variate normal distribution with zero mean and covariance matrix G when (5.2) holds. Hence $\mathcal{L}_n$ in (5.14) has asymptotically, as n becomes large, a chi-square distribution with pq degrees of freedom when H_0 in (5.2) holds.

It follows that for large n, and under conditions A–C below, an approximate level α test for H_0 in (5.2) may be obtained if H_0 is rejected for $\mathcal{L}_n > \chi^2_{pq}(1-\alpha)$.

The conditions required for the limiting distribution of $\mathcal{L}_n$ are as follows:

Condition A. (i) Each marginal distribution function $F_{(k)}$, $k = 1, \ldots, p$, has a density $f_{(k)}$ which is absolutely continuous, with finite Fisher Information $I(f_{(k)} = \int (f'_{(k)}(y)/f_{(k)}(y))^2 f_{(k)}(y)\, dy$.

(ii) $A(F)$ defined in (5.6) is positive definite and finite.

Condition B. The regression constants are such that

(i) For $\varepsilon > 0$, there exists an integer $n_0 = n_0(\varepsilon)$ such that for $n \geqslant n_0$,

$$n^{-1}\sum_i z_{ji}^2 > \varepsilon(\max_i z_{ji}^2), \quad j = 1, \ldots, q,$$

(ii) rank $(Z_n Z_n') = q$ for all n.
(iii) $n^{-1}(Z_n Z_n')$ tends to a positive definite and finite matrix ZZ' as $n \to \infty$.

CONDITION C. For each $k = 1, \ldots, p$,
(i) $\psi_k(u) = \psi_{k1}(u) - \psi_{k2}(u)$, where $\psi_{ks}(u)$, $s = 1, 2$ is absolutely continuous, nondecreasing and square integrable over (0, 1).
(ii) $\int_0^1 |\psi_{ks}(u)|(u(1-u))^{-1/2}\, du < \infty$.

REMARKS. 1. For generating functions $\psi_k(u)$ like the Wilcoxon and the Normal Scores condition B(i) can be replaced by the milder Noether Condition, i.e. $\max_i(z_{ji}^2/\Sigma_i z_{ji}^2) \to 0, j = 1, \ldots, q$.
2. As remarked in the univariate case, Condition B(ii) may not be satisfied in all cases where the original matrix X_n of regression constants is of rank q.

5.1.1. Asymptotic efficiency of $\mathcal{L}_n$ test

It has also been proved in Puri and Sen (1969a) that under a sequence of alternatives

$$H_n: \qquad \beta = n^{-1/2}b, \quad b = ((b_{kj})), \quad k = 1, \ldots, p, \; j = 1, \ldots, q, \tag{5.16}$$

and, subject to Conditions A–C of Section 5.1, $(n^{-1/2}S_n)$ has asymptotically a pq variate normal distribution with covariance matrix G, and mean

$$\boldsymbol{\mu}' = (\boldsymbol{b}_1'(ZZ')B_1(F), \ldots, \boldsymbol{b}_p'(ZZ')B_p(F)) \tag{5.17}$$

where b in (5.16) is also written as

$$b' = (\boldsymbol{b}_1, \ldots, \boldsymbol{b}_p) \quad \text{and} \quad B_k(F) = \int \psi_k'(F_{(k)}(y)dF_{(k)}(y), \; k = 1, \ldots, p. \tag{5.18}$$

It follows from (5.17) that under (5.16) and subject to the Conditions of Section 5.1, $\mathcal{L}_n$ has asymptotically a noncentral chi square distribution with pq degrees of freedom and noncentrality parameter given by

$$\Delta_L = \mu' G^{-1}\mu = \sum_k^p \sum_{k'}^q \sum_j^q \sum_{j'}^q b_{kj} b_{k'j'}, z_{jj'}, \tau^{kk'}(F) \tag{5.19}$$

where

$$((\tau^{kk'}(F))) = ((\tau_{kk'}(F)))^{-1} \tag{5.20}$$

with

$$\tau_{kk'}(F) = a_{kk'}(F)|B_k(F)B_{k'}(F), \quad k = 1, \ldots, p.$$

Observe that for $p = 1$, $a_{kk'}(F) = \int \psi^2(u) - \bar{\psi}^2$ given in (2.5) so that (5.19) reduces to $\{\boldsymbol{b}'(ZZ')\boldsymbol{b}\}B^2(F)|A^2(\psi)$ as given in (2.8).

In the normal theory, there are at least three well known methods that can be used in testing (5.2), namely the likelihood ratio test, the maximum (minimum) characteristic criterion and the trace criteria (see e.g. Chapter 8 of Anderson, 1958). In discussing the performance of the $\mathcal{L}_n$ test, we shall compare it with the test based on the likelihood ratio or which is the same thing, on the test based on the least squares estimates (LSE) of β. The likelihood ratio test statistic can be written as $-2 \log \lambda_n$, where

$$\lambda_n = (\|n\hat{\Sigma}_n - \bar{\beta}_n Z_n Z_n' \bar{\beta}_n\| / \|n\hat{\Sigma}_n\|)^{(1/2)n} \tag{5.21}$$

with

$$\hat{\Sigma}_n = n^{-1}(Y_i - \bar{Y})(Y_i - \bar{Y})', \quad \bar{Y} = n^{-1} \sum_i Y_i,$$

and $\bar{\beta}_n$ is the LSE of β, while $\|A\|$ denotes the determinant of A. By expanding (5.21) and writing the pq elements of $\bar{\beta}_n$ as a row vector

$$\bar{\beta}_n' = (\bar{\boldsymbol{\beta}}_{n1}', \ldots, \bar{\boldsymbol{\beta}}_{np}') = (\bar{\beta}_{n,kj};\, k = 1, \ldots, p;\, j = 1, \ldots, q)$$

we can write

$$-2 \log \lambda_n \sim (\bar{\boldsymbol{\beta}}_{n1}', \ldots, \bar{\boldsymbol{\beta}}_{np}')(\hat{\Sigma}_n^{-1} \otimes Z_n Z_n')(\bar{\boldsymbol{\beta}}_{n1}', \ldots, \bar{\boldsymbol{\beta}}_{np}')'$$

$$= \sum_k^p \sum_{k'}^p \sum_j^q \sum_{j'}^q \bar{\beta}_{n,kj} \bar{\beta}_{n,k'j'}\, z_{n,jj'} \hat{\sigma}_n^{kk'} \tag{5.22}$$

where

$$((\hat{\sigma}_n^{kk'})) = \hat{\Sigma}_n^{-1}.$$

If F, the distribution function of $\boldsymbol{Y}$ has a finite and positive definite covariance matrix Σ, it can be shown that under (5.1) and subject to Condition B of Section 5.1, $\hat{\Sigma}_n \to \Sigma$ in probability. Also since $\bar{\beta}_n$ is a linear combination of independent random vectors, it is easy to establish that $(n^{-1/2}\bar{\beta}_n)$ has asymptotically a pq-variate normal distribution with mean β and covariance matrix $\Sigma \otimes (ZZ')^{-1}$. It follows then that under Condition B of Section 5.1, and provided Σ is finite and positive definite, the log likelihood ratio statistic in (5.22) has under H_0 in (5.2), asymptotically a chi-square distribution with pq degrees of freedom, and under H_n in (5.16), a noncentral chi square distribution, with pq degrees of freedom and noncentrality parameter given by

$$\Delta_\lambda = (\boldsymbol{b}_1', \ldots, \boldsymbol{b}_p')(\Sigma^{-1} \otimes ZZ')(\boldsymbol{b}_1', \ldots, \boldsymbol{b}_p')'$$

$$= \sum_k^p \sum_{k'}^p \sum_j^q \sum_{j'}^q b_{kj} b_{k'j'}\, z_{jj'} \sigma^{kk'}(F) \tag{5.23}$$

Observe that for $p = 1$ (univariate), (5.23) reduces to Δ_Q given in (2.9). The asymptotic efficiency of $\mathcal{L}_n$-test relative to the likelihood ratio test $-2 \log \lambda_n$ is therefore

$$e_{\mathcal{L},\lambda} = \Delta_{\mathcal{L}}/\Delta_{\lambda} . \tag{5.24}$$

Observe that simplification of (5.24) is not as easy as in the univariate case, because both (5.19) and (5.23) depend on b, Σ, $A(F)$ of (5.6), $B_k(F)$ of (5.18) and ZZ'. However, it is shown that for an arbitrary $F(y)$, the efficiency in (5.24) is bounded as follows:

$$\mathrm{ch}_p(\Sigma\tau^{-1}(F)) \leqslant e_{\mathcal{L},\lambda} \leqslant \mathrm{ch}_1(\Sigma\tau^{-1}(F)) \tag{5.25}$$

where $\mathrm{ch}_k(A)$ denotes the k-th largest characteristic root of A, and $\tau(F)$ is as given in (5.20).

The efficiency in (5.24) including the bounds in (5.25) has been studied in Sen and Puri (1967).

5.2. Rank tests of subhypotheses

In model (5.1), let us partition

$$\beta = (\beta_{(1)}, \beta_{(2)}) \tag{5.26}$$

where $\beta_{(s)} = ((\beta_{k,j}))$, $k = 1, \ldots, p$, $j = 1, \ldots, q_s$, $s = 1, 2$, $q = q_1 + q_2$, and consider the null hypothesis

$$H_0: \qquad \beta_{(1)} = 0, \quad \beta_{(2)} \text{ unspecified} \tag{5.27}$$

against the alternative that $\beta_{(1)} \neq 0$. Now partition

$$Z_n = \begin{pmatrix} Z_{n1} \\ Z_{n2} \end{pmatrix}, \quad Z_{ns} \text{ is of order } q_s \times n .$$

As in Section 3.1, let us reparametrize model (5.1) as follows

$$P(Y_i \leqslant y) = F_i(y) = F(y - \boldsymbol{\alpha}_n - \beta_{(1)}z_{1i} - \beta_{(2)}z_{2i}),\ i \leqslant i \leqslant n,\ y \in \mathbb{R}^p \tag{5.28}$$

where

$$\boldsymbol{\alpha}_n = \boldsymbol{\alpha} + \beta_{(1)}\bar{x}_{1i} + \beta_{(2)}\bar{\boldsymbol{x}}_{2i} \quad \text{with } z_{ji} = (x_{ji} - \bar{x}_j) .$$

Under (5.27), the distribution function given in (5.28) reduces to

$$F_i(y) = F(y - \alpha_n - \beta_{(2)}z_{2i}) . \tag{5.29}$$

To test H_0 in (5.27) in the model (5.28), we would first remove the effect of $\beta(2)$, by substituting a suitable estimate $\hat{\beta}_{(2),n}$ for $\beta_{(2)}$ in (5.29). Many traditional methods of estimating $\beta_{(2)}$ are well known. There is also the 'rank' method. We shall not discuss estimation in this chapter, but readers interested in the rank method of estimating regression parameters may consult a separate chapter in this volume where it is discussed in detail. It can be shown that both the traditional and the rank methods of estimating $\beta_{(2)}$ are suitable for this problem as long as the resultant estimate $\hat{\beta}_{(2),n}$ satisfies Condition D of Section 3.1. We shall first present and discuss an aligned rank test statistic that is valid for any suitable estimate $\hat{\beta}_{(2),n}$ and later give a test statistic that is valid specially for 'rank' estimates of $\beta_{(2)}$.

Let $\hat{\beta}_{(2),n}$ be any estimate of $\beta_{(2)}$ that satisfies Condition D of section 3.1, and let $\boldsymbol{Y}_k = (Y_{k1}, \ldots, Y_{kn})$, $k = 1, \ldots, p$, be the k-th component vector of the observations. Let $\hat{R}_{ki}$ be the rank of $\hat{Y}_{ki}$ in the ranking of $(\hat{Y}_{k1}, \ldots, \hat{Y}_{kn})$ where

$$\hat{Y}_{ki} = Y_{ki}(\hat{\beta}_{(2),n}) = (Y - \hat{\beta}_{(2),n}Z_{n2})_{ki} \tag{5.30}$$

is the k-th component of the i-th aligned observation. Define a pq_1 row vector by

$$\hat{T}'_{(1),n} = (\hat{\boldsymbol{T}}'_{n1}, \ldots, \hat{\boldsymbol{T}}'_{np}) = (\hat{T}_{n,kj}; k = 1, \ldots, p; j = 1, \ldots, q_1) \tag{5.31}$$

where

$$\hat{\boldsymbol{T}}_{nk} = Z_{n1}(I_n - C_{n,2})\Psi_{nk}(\hat{R}_k), \quad k = 1, \ldots, p, \tag{5.32}$$

and q_1 vectors $\Psi_{nk}(\hat{R}_k) = (\psi_{nk}(\hat{R}_{k1}), \ldots, \psi_{nk}(\hat{R}_{kn}))'$ while I_n denotes the identity matrix of order n and $C_{n,2}$ is as given in (3.5). As in Section 5.1, the statistics required would be a quadratic form in the pq_1 elements of $\hat{T}_{(1),n}$ with the discriminant being the inverse of the asymptotic covariance matrix G^* say of the distribution of $\hat{T}_{(1),n}$. G^* is a $pq_1 \times pq_1$ matrix given by

$$G^* = A(F) \otimes C^*$$

where

$$C^* = \lim n^{-1}C^*_n \quad \text{with } C^*_n = Z_{n1}(I_n - C_{n,2})Z'_{n1},$$

and $A(F)$ is as defined in (5.6). Since $A(F)$ is not assumed to be known, we substitute it with $\hat{A}_n$ defined in (5.10). If we then write a $pq_1 \times pq_1$ matrix

$$\hat{G}^*_n = \hat{A}_n \otimes C^*_n = ((\hat{a}_{nkk'}C^*_{nj,j'})), \quad k, k' = 1, \ldots, p, \; j, j' = 1, \ldots, q_1, \tag{5.33}$$

the proposed test statistic for testing H_0 in (5.27) can be written as

$$\begin{aligned}\hat{\mathscr{L}}_n &= \hat{T}'_{(1),n}\hat{G}^{*-1}_n\hat{T}_{(1),n} = (\hat{\boldsymbol{T}}'_{n1}, \ldots, \hat{\boldsymbol{T}}'_{np})\hat{G}^{*-1}_n(\hat{\boldsymbol{T}}'_{n1}, \ldots, \hat{\boldsymbol{T}}'_{np})' \\ &= \sum_k^p \sum_{k'}^p \sum_j^{q_1} \sum_{j'}^{q_1} \hat{T}_{n,kj}, \hat{T}_{n,k'j'}\hat{a}^{kk'}_n c^{*jj'}_n .\end{aligned} \tag{5.34}$$

It can be shown in the manner of Sen and Puri (1977) that under H_0 in (5.27) and subject to the Conditions A–C of Section 5.1 and Condition D of Section 3.1, $(n^{-1/2}\hat{T}_{(1),n})$ has asymptotically a pq_1 variate normal distribution with mean zero and covariance matrix G^* given in (5.33). It follows then that under H_0 in (5.27) and subject to the Conditions mentioned above, $\hat{\mathscr{L}}_n$ given in (5.34) has asymptotically a chi-square distribution with pq_1 degrees of freedom. Hence for large values of n, an approximate level α test of H_0 in (5.27) may be obtained if H_0 is rejected for $\hat{\mathscr{L}}_n > \chi^2_{pq1}(1-\alpha)$.

In the case where a 'rank' estimate $\hat{\beta}_{(2)n}$ is used to align the observations (5.30), Sen and Puri (1977) have proposed a rank order statistic $\mathscr{L}^*_n$ for testing H_0 in (5.27). $\mathscr{L}^*_n$ is a simplified version of $\hat{\mathscr{L}}_n$ in (5.34). To obtain the test statistic, define a pq_1 row vector by

$$\hat{\boldsymbol{S}}'_{(1)n} = (\hat{\boldsymbol{S}}'_{n1}, \ldots, \hat{\boldsymbol{S}}'_{np}) = (\hat{S}_{n,kj}\,;\, k = 1, \ldots, p\,; j = 1, \ldots, q_1) \tag{5.35}$$

where

$$\hat{\boldsymbol{S}}_{nk} = Z_{n1}\Psi_{nk}(\hat{R}_k), \quad k = 1, \ldots, p, \tag{5.36}$$

are q_1 vectors. The statistic proposed by Sen and Puri can be written as

$$\begin{aligned}\mathscr{L}^*_n &= \hat{\boldsymbol{S}}'_{(1),n}\hat{G}^{*-1}_n\hat{\boldsymbol{S}}_{(1)n} = (\hat{\boldsymbol{S}}'_{n1}, \ldots, \hat{\boldsymbol{S}}'_{np})\hat{G}^{*-1}_n(\hat{\boldsymbol{S}}'_{n1}, \ldots, \hat{\boldsymbol{S}}'_{np})' \\ &= \sum_{k}^{p}\sum_{k'}^{p}\sum_{j}^{q_1}\sum_{j'}^{q_1} \hat{S}_{n,kj}\hat{S}_{n,k'j'},\ \hat{a}^{kk'}_n c^{*jj'}_n\end{aligned} \tag{5.37}$$

where $\hat{G}^*_n$ is as defined in (5.34).

It is proved in Sen and Puri (1977) that, subject to Conditions of Section 5.1, $(n^{-1/2}\hat{\boldsymbol{S}}_{(1)n})$ has asymptotically a pq_1 variate normal distribution with mean zero and covariance matrix G^* when (5.27) holds. It is to be remarked that the 'rank' estimate $\hat{\beta}_{(2)n}$ also satisfies Condition D of Section 3.1. It follows that $\mathscr{L}^*_n$ in (5.37) has asymptotically a chi-square distribution with pq_1 degrees of freedom. An approximate level α test of H_0 in (5.27) may be obtained if H_0 is rejected for $\mathscr{L}^*_n > \chi^2_{pq_1}(1-\alpha)$.

5.2.1. Asymptotic efficiency of $\hat{\mathscr{L}}_n(\mathscr{L}^*_n)$ tests

It has also been shown in Sen and Puri (1977) that under a sequence of near alternatives

$$H_n: \quad \beta_{(1)} = n^{-1/2}b_{(1)}, \quad b_{(1)} = ((b_{kj})), \quad k = 1, \ldots, p,\ j = 1, \ldots, q, \tag{5.38}$$

and subject to the conditions of Section 5.1 and condition D of section 3.1, $n^{-1/2}(\hat{\boldsymbol{S}}_{n(1)})$ is asymptotically distributed as a qp_1 variate normal with mean μ and covariance, matrix G^*, where

$$\mu' = (\boldsymbol{b}'_1C^*B_1(F), \ldots, \boldsymbol{b}'_pC^*B_p(F)), \tag{5.39}$$

$B_k(F)$ is as given in (5.18), and $b_{(1)}$ in (5.38) has been written as $b'_{(1)} = (\boldsymbol{b}_1, \ldots, \boldsymbol{b}_p)$. It follows from (5.39) that under conditions of Section 5.1 $\mathscr{L}^*_n$ has asymptotically a noncentral chi square distribution with pq_1 degrees of freedom and noncentrality parameter.

$$\Delta_{\mathscr{L}^*} = \mu' G^{*-1}\mu = \sum_k \sum_{k'} \sum_j \sum_{j'} b_{kj} b_{k'j'} c^*_{jj'} \tau^{kk'}(F) k,$$

$$k = 1, \ldots, p, \quad j, j' = 1, \ldots, q_1, \tag{5.40}$$

where $\tau^{kk'}$ is as given in (5.20), while $c^*_{jj'}$ is defined in (5.33). As for $\hat{\mathscr{L}}_n$ of (3.4) it is straightforward to show that, under H_n of (5.38) and subject to the conditions of Section 5.1 and Condition D of Section 3.1, $n^{-1/2}(\hat{T}_{n(1)})$ is asymptotically distributed as a pq_1 variate normal with the same mean μ given in (5.39). It follows that the noncentrality parameter of $\hat{\mathscr{L}}_n$ is the same as that given in (5.40). Observe that if $p = 1$ (univariate), G^{*-1} reduces to $A^{-1}(\psi)C^{*-1}$, while (5.39) becomes $\boldsymbol{b}'C^*B(F)$ so that (5.40) reduces to

$$(\boldsymbol{b}_1'C^*\boldsymbol{b}_1)B^2(F)/A^2(\psi)$$

given in (3.13).

The usual test statistic for the hypothesis (5.27) is based on the likelihood ratio criterion, (see e.g. Chapter 8 of Anderson, 1958). The test statistic can be written as $-2\log\lambda$, where

$$\lambda_n = (\|n\hat{\Sigma}_n - \bar{\beta}_{n(1)} C^*_n \bar{\beta}'_{n(1)}\| / \|n\hat{\Sigma}_n\|)^{(1/2)n} \tag{5.41}$$

By expanding (5.41), we can write

$$-2\log\lambda_n \sim (\bar{\boldsymbol{\beta}}^{(1)\prime}_{n,1}, \ldots, \bar{\boldsymbol{\beta}}^{(1)\prime}_{n,p})(\hat{\Sigma}_n^{-1} \otimes C^*_n)(\boldsymbol{\beta}^{(1)\prime}_{n,1}, \ldots, \bar{\boldsymbol{\beta}}^{(1)\prime}_{n,p})' \tag{5.42}$$

$$= \sum_k^p \sum_{k'}^p \sum_j^{q_1} \sum_{j'}^{q_1} \bar{\beta}^{(1)}_{n,kj} \bar{\beta}^{(1)}_{n,k'j'} c_{n,jj'} \hat{\sigma}^{kk'}_n, \quad k, k' = 1, \ldots, p,$$

$$j, j' = 1, \ldots, q_1,$$

where $((\hat{\sigma}^{kk'})) = \hat{\Sigma}_n^{-1}$.

It is well known that under Condition B of Section 5.1 and provided the distribution function F of Y has a finite and positive definite covariance matrix $\Sigma(F)$, the statistic in (5.42) has under H_0 in (5.27) asymptotically a chi square distribution with pq_1 degrees of freedom. Also under H_n in (5.38), $-2\log\lambda_n$ has asymptotically a noncentral chi square distribution with qp_1 degrees of freedom and noncentrality parameter

$$\Delta_\lambda = \sum_k^p \sum_{k'}^p \sum_j^{q_1} \sum_{j'}^{q_1} b_{kj} b_{k'j'} c^*_{jj'} \sigma^{kk'}(F) \tag{5.43}$$

when $p = 1$, Δ_λ in (5.43) reduces to Δ_q given in (3.14).

From (5.40) and (5.43), it follows that the asymptotic efficiency of $\hat{\mathscr{L}}_n(\mathscr{L}^*_n)$ tests relative to the likelihood ratio test is

$$e_{L,\lambda} = \Delta_{\mathscr{L}}/\Delta_\lambda \tag{5.44}$$

which is the same as (5.24), the efficiency of rank tests relative to the likelihood ratio tests in testing for regression in the multivariate case. If follows that the remarks made about $e_{\mathcal{L},\lambda}$ in Section 5.1.1 including the bounds in (5.25) remain valid for (5.44).

References

Adichie, J. N. (1967a). Asymptotic efficiency of a class of nonparametrics tests for regression parameters. *Ann. Math. Statist.* **38**, 884–893.

Adichie, J. N. (1967b). Estimation of regression based on rank test. *Ann. Math. Statist.* **38**, 894–904.

Adichie, J. N. (1974). Rank score comparison of several regression parameters. *Ann. Statist.* **2**, 396–402.

Adichie, J. N. (1975). On the use of ranks for testing the coincidence of several regression lines *Ann. Statist.* **3**, 521–527.

Adichie, J. N. (1978). Rank tests of sub-hypotheses in the general linear regression. *Ann. Statist.* **6**, 1012–1026.

Anderson, T. W. (1958). *Introduction to Multivariate Statistical Analysis*, Wiley, New York.

Andrews, F. C. (1954). Asymptotic behaviour of some rank tests for analysis of variance. *Ann. Math. Statist.* **25**, 724–735.

Bahadur, R. R. (1967). Rates of covergence of estimates and test statistics. *Ann. Math. Statist.* **38**, 303–324.

Chernoff, H. and Savage, I. R. (1958). Asymptotic normality and efficiency of certain non-parametric test statistics. *Ann. Math. Statist.* **29**, 972–994.

Conover, W. J. (1973). On methods of handling ties in the Wilcoxon Signed rank test. *J. Amer. Statist. Assoc.* **69**, 255–258.

Graybill, F. A. (1961). *An introduction to linear Statistical Models, Vol. I.* McGraw-Hill, New York.

Hajek, J. (1962). Asymptotically most powerful rank order tests. *Ann. Math. Statist.* **33**, 1124–1147.

Hajek, J. (1969). *A course in Nonparametric Statistics.* Holden-Day, San Francisco.

Hajek, J. and Sidak, Z. (1967). *Theory of rank tests.* Academic Press, New York.

Hodges, J. L., Jr. and Lehmann, E. L. (1966). Efficiency of some nonparametric competitors of the *t*-test. *Ann. Math. Statist.* **27**, 324–335.

Huskova, M. (1970). Asymptotic distribution of simple linear rank statistics for testing symmetry. *Z. Wahrsch.* **14**, 308–322.

Jureckova, J. (1971). Nonparametric estimates of regression coefficients. *Ann. Math. Statist.* **42**, 1328–1338.

Koul, H. L. (1970). A class of ADF tests of subhypothesis in multiple linear regression. *Ann. Math. Statist.* **41**, 1273–1281.

Kraft, C. H. and van Eeden, C. (1972). Linearized estimates and signed-rank estimates for the general linear hypothesis. *Ann. Math. Statist.* **43**, 42–57.

Lehmann, E. L. (1975). *Nonparametrics: Statistical Methods based on ranks.* Holden-Day, San Francisco.

Noether, G. E. (1954). On a theorem of Pitman. *Ann. Math. Statist.* **25**, 514–522.

Pitman, E. J. G. (1948). Lecture Notes on Nonparametric Statistics. Columbia University, New York.

Pratt, John W. (1959). Remarks on Zeros and Ties in the Wilcoxon Signed rank Procedures. *J. Amer. Statist. Assoc.* **54**, 655–667.

Puri, M. L. and Sen, P. K. (1966). On a class of multivariate multisample rank order tests. *Sankyha Ser. A* **28**, 353–376.

Puri, M. L. and Sen, P. K. (1969a). A class of rank order tests for general linear hypothesis. *Ann. Math. Statist.* **40**, 1325–1343.

Puri, M. L. and Sen, P. K. (1969b). On a class of rank order tests for the identity of two multiple regression surfaces. *Z. Wahrsch. Verw. Geb.* **12**, 1–8.

Puri, M. L. and Sen, P. K. (1973). A note on asymptotic distribution-free tests for subhypotheses in multiple linear regression. *Ann. Statist.* **1**, 553–556.

Sen, P. K. (1969). On a class of rank order tests for parallelism of several regression lines. *Ann. Math. Statist.* **40**, 1668–1683.

Sen, P. K. (1972). On a class of aligned rank order tests for the identity of intercepts of several regression lines. *Ann. Math. Statist.* **43**, 2004–2012.

Sen, P. K. and Puri, M. L. (1967). On the theory of rank order tests for location in the multivariate one sample problem. *Ann. Math. Statist.* **38**, 1216–1228.

Sen, P. K. and Puri, M. L. (1977). Asymptotically Distribution-Free aligned rank order tests for composite hypotheses for general multivariate linear models. *Zeit. Wahr. verw. Geb.* **39**, 175–186.

Srivastava, M. S. (1970). On a class of nonparametric tests for regression parameters. *J. Statist. Res.* **4**, 117–132.

Steel, Robert G.D. and Torrie, J. H. (1960). *Principles and Procedures of Statistics*. McGraw-Hill, New York.

van der Waerden, B. L. (1952, 1953). Order Tests for the Two-sample Problem and Their Power. *Indag. Math.* **14**, 453–458, **15**, 303–316.

Vorlickova, Dana (1972). Asymptotic Properties of Rank Tests of Symmetry under Discrete Distributions. *Ann. Math. Statist.* **43**, 2013–2018.

Wilcoxon, Frank (1945). Individual Comparisons by Ranking Methods. *Biometrics* **1**, 80–83.

P. R. Krishnaiah and P. K. Sen, eds., *Handbook of Statistics*, Vol. 4

12

On the Use of Rank Tests and Estimates in the Linear Model

*James C. Aubuchon and Thomas P. Hettmansperger**

1. Introduction

The purpose of this paper is to review the current practical status of statistical methods based on ranks in the linear model. In addition to describing various approaches to the construction of tests and estimates in Section 2, we provide an extensive discussion of the computational issues involved in implementing the procedures in Section 3. All statistical methods are illustrated on a data set in Section 4 and, for the case of aligned rank tests, the SAS and Minitab programs needed to carry out the analysis are provided.

Twenty years ago Hodges and Lehmann (1962) proposed aligned rank tests in a two-way layout. Their approach was to remove the effect of the nuisance parameter and then construct a test of hypothesis on the residuals. Aligned rank tests have received much attention in the literature since 1962. Koul (1970) discussed aligned rank tests for regression with two independent variables, and improvements were suggested by Puri and Sen (1973). Recent papers by Adichie (1978) and Sen and Puri (1977) describe aligned rank tests for the univariate and multivariate linear models, respectively.

Surprisingly, for all the attention accorded to aligned rank tests in the past twenty years, they are not widely available to the data analyst. For example, they are not currently contained in any of the statistical packages, so some degree of special programming is required to implement them.

There are alternative approaches to the aligned rank tests. By substituting for least squares a measure of dispersion based on ranks, due to Jaeckel (1972), McKean and Hettmansperger (1976) proposed a test statistic based on the reduction in dispersion due to fitting the reduced and full models. This method uses rank estimates (R-estimates) of the regression parameters. Rank estimates in regression were first proposed by Adichie (1967) as extensions of the Hodges and Lehmann (1963) rank estimates of location. Another approach, not discussed in the literature of nonparametric tests, is to construct a quadratic form

*Research partially supported by ONR contract N00014-80-C-0741.

based on an R-estimate of the parameters in the linear model. This is sometimes referred to as a Wald test statistic.

In considering tests based on ranks we are interested in various practical aspects. We will take as given that the procedures are asymptotically equivalent in the sense of Pitman efficiency and that they have the same asymptotic null distribution. The practical questions are: (1) Does a nominal size-α test maintain its size for small to moderate sample sizes? The answer, based on simulations, usually is not quite, and some tuning of the test statistic generally results. (2) What methods are available for computing the test? Can the computations be carried out using existing packages, or are special programs required? If iterative methods are used in computing the tests what can be said about convergence? and (3) Are there small sample differences in the powers of the various tests? We will describe three methods of constructing tests based on ranks, and we will summarize the state of each method in the light of the three questions above.

It is disappointing to note that little is known about the behavior of rank tests and estimates for small samples. Further, they are not generally easy to compute. They require special programs for their implementation with the exception of an aligned rank test in Section 2i. They are not currently available in any statistical computing packages; however, in 1984 a rank-regression command will be available in the Minitab statistical computing system. The output from this command will contain the rank estimate described in Section 2ii and the rank tests described in Sections 2ii and 2iii.

In his 1981 monograph, Huber discusses another approach to robust estimation based on so-called M-estimates. In his Section 7.10, he briefly discusses tests of hypotheses in linear model. Schrader and Hettmansperger (1980) discuss two tests based on M-estimates: a test based on the reduction in dispersion due to fitting the full and reduced models and a Wald test. Sen (1982) develops an aligned test based on M-estimates.

2. Rank tests and estimates

Problems in analysis of variance, analysis of covariance and regression can often be treated in a unified manner by casting them in terms of a general linear model. The recent books by Draper and Smith (1981) and Neter and Wasserman (1974) contain many examples.

We begin with a linear model for the $n \times 1$ vector y of observations, specified by

$$\boldsymbol{y} = \mathbf{1}\alpha + \boldsymbol{X\beta} + \boldsymbol{e} = \mathbf{1}\alpha^* + \boldsymbol{X}_c\boldsymbol{\beta} + \boldsymbol{e}\,, \tag{1}$$

where α is the scalar intercept, β is a $p \times 1$ vector of regression parameters, $\boldsymbol{X}$ is an $n \times p$ design matrix and $\boldsymbol{e}$ is an $n \times 1$ vector of random errors. In the second equation, we have the centered design matrix $\boldsymbol{X}_c = \boldsymbol{X} - \mathbf{1}\bar{\boldsymbol{X}}'$ where $\bar{\boldsymbol{X}}'$ is a $1 \times p$ vector of column means from $\boldsymbol{X}$ and $\alpha^* = \alpha + \bar{\boldsymbol{X}}'\boldsymbol{\beta}$. The details of an example, with data, are given in Section 4.

Working assumptions will be listed as they are needed in the discussion. The reader should consult the primary references for the regularity conditions needed for the asymptotic theory.

ASSUMPTION A1. We suppose the n errors are independent, identically distributed according to a continuous distribution which has arbitrary shape and median 0.

ASSUMPTION A2. Suppose $\boldsymbol{X}_c$ is of full rank p. We partition $\boldsymbol{\beta}$ into two parts: $\boldsymbol{\beta}_1$ is $(p-q)\times 1$ and $\boldsymbol{\beta}_2$ is $q \times 1$. Hence the model (1) can be written

$$\boldsymbol{y} = \mathbf{1}\alpha^* + \boldsymbol{X}_{1c}\boldsymbol{\beta}_1 + \boldsymbol{X}_{2c}\boldsymbol{\beta} + \boldsymbol{e}\,. \tag{2}$$

We will consider tests of

$$H_0\colon \quad \boldsymbol{\beta}_2 = 0\,,$$

so that $\boldsymbol{\beta}_1$ is a $(p-q)\times 1$ vector of nuisance parameters.

Before turning to the rank tests we will present the F statistic in a form that will motivate the introduction of the aligned rank test.

Consider, first, the F statistic for $H_0\colon \boldsymbol{\beta} = 0$. There are no nuisance parameters, and

$$F = \frac{\boldsymbol{y}'\boldsymbol{X}_c(\boldsymbol{X}_c'\boldsymbol{X}_c)^{-1}\boldsymbol{X}_c'\boldsymbol{y}}{p\hat{\sigma}^2}\,, \tag{3}$$

where $\hat{\sigma}^2$ is the usual unbiased estimate of the error variance σ^2, assumed to be finite.

The general F statistic for $H_0\colon \boldsymbol{\beta}_2 = 0$ can be derived from (3) by removing the effects of the nuisance parameters $\boldsymbol{\beta}_1$ from both $\boldsymbol{y}$ and $\boldsymbol{X}_{2c}$ before applying (3). Hence in (3) replace $\boldsymbol{y}$ by $\boldsymbol{y} - \boldsymbol{X}_{1c}\hat{\hat{\boldsymbol{\beta}}}_1$, where $\hat{\hat{\boldsymbol{\beta}}}_1 = (\boldsymbol{X}_{1c}'\boldsymbol{X}_{1c})^{-1}\boldsymbol{X}_{1c}'\boldsymbol{y}$, the reduced model least squares estimate, and replace $\boldsymbol{X}_c$ by $\boldsymbol{Z} = \boldsymbol{X}_{2c} - \boldsymbol{X}_{1c}(\boldsymbol{X}_{1c}'\boldsymbol{X}_{ic})^{-1}\boldsymbol{X}_{1c}'\boldsymbol{X}_{2c}$. Now p is replaced by q, the dimension of $\boldsymbol{\beta}_2$ and (3) becomes the usual F statistic, written in unusual form,

$$F = \frac{(\boldsymbol{y} - \boldsymbol{X}_{1c}\hat{\hat{\boldsymbol{\beta}}}_1)'\boldsymbol{Z}(\boldsymbol{Z}'\boldsymbol{Z})^{-1}\boldsymbol{Z}'(\boldsymbol{y} - \boldsymbol{X}_{1c}\hat{\hat{\boldsymbol{\beta}}}_1)}{p\hat{\sigma}^2}\,. \tag{4}$$

Further, a bit of matrix algebra shows that

$$\boldsymbol{Z}(\boldsymbol{Z}'\boldsymbol{Z})^{-1}\boldsymbol{Z}' = \boldsymbol{X}_c(\boldsymbol{X}_c'\boldsymbol{X}_c)^{-1}\boldsymbol{X}_c' - \boldsymbol{X}_{1c}(\boldsymbol{X}_{1c}'\boldsymbol{X}_{1c})^{-1}\boldsymbol{X}_{1c}'\,. \tag{5}$$

2i. *Aligned rank tests*

These tests are easiest to implement when we use the Wilcoxon score function; other score functions are discussed in the references. Let

$$\phi(u) = (12)^{1/2}(u - 1/2)\,, \tag{6}$$

and define $a(i) = \phi(i/(n+1))$. Then $a(1) \leqslant \cdots \leqslant a(n)$ are called the Wilcoxon scores. Note that $\int_0^1 \phi(u)\,du = 0$, $\int_0^1 \phi^2(u)\,du = 1$, and $\Sigma_1^n a(i) = 0$.

We will let $a(R(\boldsymbol{\beta}))$ denote the $n \times 1$ vector whose i-th component is $a(R(y_i - \boldsymbol{x}'_{ic}\boldsymbol{\beta}))$, where $R(y_i - \boldsymbol{x}'_{ic}\boldsymbol{\beta})$ is the rank of $y_i - \boldsymbol{x}'_{ic}\boldsymbol{\beta}$ among the n uncentered residuals.

Assumption A3. Suppose $\hat{\hat{\boldsymbol{\beta}}}_1$ is a reduced model estimate such that $\hat{\hat{\boldsymbol{\beta}}}_1(\boldsymbol{y} + b) = \hat{\hat{\boldsymbol{\beta}}}_1(\boldsymbol{y}) + b$ and $n^{1/2}(\hat{\hat{\boldsymbol{\beta}}}_1 - \boldsymbol{\beta}_1) = \boldsymbol{0}_p(1)$.

If the error distribution has finite variance and if the maximum leverage tends to 0, then the results of Huber (1981, p. 157) show that the least-squares estimate satisfies A3. For testing H_0: $\boldsymbol{\beta}_2 = 0$ we first align the observations and construct $a(R(\hat{\hat{\boldsymbol{\beta}}}_1))$, the vector of rank-scored reduced-model residuals. The aligned rank test statistic is constructed from the numerator of (4) by replacing the residuals with $a(R(\hat{\hat{\boldsymbol{\beta}}}_1))$:

$$A = a'(R(\hat{\hat{\boldsymbol{\beta}}}_1))\boldsymbol{Z}(\boldsymbol{Z}'\boldsymbol{Z})^{-1}\boldsymbol{Z}'a(R(\hat{\hat{\boldsymbol{\beta}}}_1))\,. \tag{7}$$

Under regularity conditions, Adichie (1978) showed that A is asymptotically chi-squared with q degrees of freedom. Hence, a large sample test of H_0: $\boldsymbol{\beta}_2 = 0$ rejects H_0 if $A \geqslant \chi^2_\alpha(q)$. Note that A is very easy to compute with a computer package that contains both least squares and ranking capabilities. Computation is carried out in the following way:

C1. Find the reduced-model least-squares residuals,

C2. Find the ranks of these residuals and calculate

$$(12)^{1/2}((n+1)^{-1}R(y_i - \boldsymbol{x}'_{ic}\hat{\hat{\boldsymbol{\beta}}}_1) - 1/2), \quad i = 1, \ldots n\,,$$

C3. Compute the F statistic on the values in C2; then A is the numerator sum of squares.

Note that for each hypothesis of the form H_0: $\boldsymbol{\beta}_2 = 0$ which is to be tested a new set of reduced model residuals must be computed. Minitab and SAS programs are provided in Section 3.

Summary: We consider the three questions raised in the Introduction. (1) There are no published studies of the small sample properties of A. We do not know if the test, which is not nonparametric for finite sample size, maintains its level near the nominal level. We do not know how large the sample size should be before it is reasonable to use A. There are many possibilities for choice of

$\hat{\hat{\boldsymbol{\beta}}}_1$. The least-squares estimate or a rank estimate discussed in the next subsection are possible candidates. We do not know how the choice of $\hat{\hat{\boldsymbol{\beta}}}_1$ effects the level of the test for small samples; asymptotically it makes no difference. In a small, unpublished study, Hettmansperger and McKean (1981) find that in testing for parallelism, the simulated levels of A were more erratic than the levels of the other tests discussed below. Further, there was some indication that the presence of good design points with moderate leverage can make A extremely conservative. (2) As described above, if the least-squares reduced model estimate is used, A is easy to compute. It would seem more natural to align the observations using the rank estimate in (ii); see Sen and Puri (1977). Rank estimates present computational difficulties that are discussed in Section 3. (3) There are no published studies on small-sample power of A. In the unpublished report by Hettmansperger and McKean (1981), A had very little power when there was a moderate leverage point in the parallelism design.

2ii. *Wald test statistic*

This test is constructed from a quadratic form in the full model rank estimate of $\boldsymbol{\beta}$; see Rao (1973) for a general discussion of the Wald test statistic. The computation of the rank estimate requires specific programs that are not generally available in statistical packages. In Section 3 we discuss the computational problems that must be overcome to be able to implement the Wald test. By 1984 there will be a rank-regression command in the Minitab statistical computing system which will provide all of the necessary computations. Until that time special programs are required to implement the test.

We begin with Jaeckel's (1972) measure of dispersion of the residuals. Using (6), define

$$D(\boldsymbol{\beta}) = \boldsymbol{a}'(R(\boldsymbol{\beta}))(\boldsymbol{y} - \boldsymbol{X}_c\boldsymbol{\beta}) = \sum_{i=1}^{n} a(R(y_i - \boldsymbol{x}'_{ic}\boldsymbol{\beta}))(y_i - \boldsymbol{x}'_{ic}\boldsymbol{\beta}) \tag{8}$$

and

$$\boldsymbol{S}(\boldsymbol{\beta}) = \boldsymbol{X}'_c\boldsymbol{a}(R(\boldsymbol{\beta}))\,. \tag{9}$$

The j-th component of the $p \times 1$ vector $\boldsymbol{S}(\boldsymbol{\beta})$ is given by

$$S_j(\boldsymbol{\beta}) = \sum_{i=1}^{n} (x_{ij} - \bar{x}_j)a(R(y_i - \boldsymbol{x}'_{ic}\boldsymbol{\beta}))\,,$$

the rank test statistic corresponding to the j-th component of $\boldsymbol{\beta}$. Jaeckel points out that $-\boldsymbol{S}(\boldsymbol{\beta})$ is essentially the gradient of $D(\boldsymbol{\beta})$, so setting (9) to zero yields a set of nonlinear normal equations derived from $D(\boldsymbol{\beta})$.

A rank estimate $\hat{\boldsymbol{\beta}}$ minimizes $D(\boldsymbol{\beta})$ or solves $\boldsymbol{S}(\boldsymbol{\beta}) \doteq 0$. Jureckova (1971) suggests the equivalent method of minimizing $\Sigma\,|S_j(\boldsymbol{\beta})|$.

ASSUMPTION A4. Suppose $n^{-1}(\boldsymbol{X}'_c\boldsymbol{X}_c)$ converges to a positive definite matrix as $n \to \infty$.

Then the limiting distribution of $n^{1/2}(\hat{\boldsymbol{\beta}} - \boldsymbol{\beta})$ is MVN$(0, \tau^2 n(\boldsymbol{X}_c'\boldsymbol{X}_c)^{-1})$, where

$$\tau^{-2} = 12\left\{\int_{-\infty}^{\infty} f^2(x)\,\mathrm{d}x\right\}^2 \tag{10}$$

and $f(x)$ is the density of the error distribution.

Let $\boldsymbol{H} = [0, \boldsymbol{I}]$ where $\boldsymbol{I}$ is the $q \times q$ identity matrix; then H_0: $\boldsymbol{\beta}_2 = 0$ can be written as H_0: $\boldsymbol{H\beta} = \boldsymbol{0}$. Now, $n^{1/2}(\boldsymbol{H}\hat{\boldsymbol{\beta}} - \boldsymbol{H\beta})$ has a limiting covariance matrix $\tau^2 n\boldsymbol{H}(\boldsymbol{X}_c'\boldsymbol{X}_c)^{-1}\boldsymbol{H}'$, and the Wald statistic is

$$W = \frac{(\boldsymbol{H}\hat{\boldsymbol{\beta}})'[\boldsymbol{H}(\boldsymbol{X}_c'\boldsymbol{X}_c)^{-1}\boldsymbol{H}']^{-1}(\boldsymbol{H}\hat{\boldsymbol{\beta}})}{\hat{\tau}^2}. \tag{11}$$

The statistic W is analogous to the corresponding form of the F statistic where $\hat{\boldsymbol{\beta}}$ is the least-squares estimate and τ^2 is replaced by σ^2. Graybill (1979, p. 184) discusses this form of F in detail. Under regularity conditions and when $\hat{\tau}$ is a consistent estimate of τ, W has a limiting chi-squared distribution with q degrees of freedom.

To carry out the test we need a consistent estimate of τ^2. Let $w(x) = 1$ for $|x| \leqslant \frac{1}{2}$ and 0 otherwise; then the following rectangular window estimate of $\gamma = \int f^2(x)\,\mathrm{d}x$ is consistent:

$$\hat{\gamma} = \frac{1}{n^{3/2}h_n} + \frac{1}{n(n-1)h_n}\sum_{i\neq j}\sum w\left(\frac{r_i - r_j}{h_n}\right), \tag{12}$$

where $r_i = y_i - \boldsymbol{x}_{ic}'\hat{\boldsymbol{\beta}}$ and

$$h_n = (4.1)n^{-1/2}(r_{(0.75n)} - r_{(0.25n)})\,. \tag{13}$$

The window width h_n incorporates a resistant estimate of scale, the interquartile range of the residuals, and a normalizing factor. The factor 4.1, used in (13), corresponds to a normal error distribution. The estimate (12) is a modification of a window estimate in the independent, identically distributed case proposed by Schuster (1974) and independently by Schweder (1975). The extension to the linear model is discussed by Aubuchon (1982).

We now take $\hat{\tau}^2 = 1/(12\hat{\gamma}^2)$ in (10). Then the test rejects H_0: $\boldsymbol{\beta}_2 = 0$ if $W \geqslant \chi^2_\alpha(q)$. This test is illustrated in Section 4.

An alternative estimate of τ is available with the additional assumption of symmetry of the error distribution.

ASSUMPTION A5. Suppose the underlying error distribution is symmetric about 0. Let $W_{(1)} \leqslant \cdots \leqslant W_{(N)}$, $N = n(n+1)/2$ be the ordered set of all pairwise averages of the residuals, including the residuals. These averages are referred to as Walsh averages. Let c by the lower critical point of a two-sided size-α Wilcoxon signed-rank test. Then under assumption A5, McKean and Hettmansperger (1976) show that

$$\tau^* = n^{1/2}(W_{(N-c)} - W_{(c+1)})/(2Z_{\alpha/2}) \tag{14}$$

where $2Z_{\alpha/2}$ is the $1-\alpha$ interpercentile range of the standard normal distribution, is a consistent estimate of τ in (10). This extends the results of Lehmann (1963) in the independent and identically distributed case.

We complete this subsection with a short discussion of k-step rank estimates $\hat{\boldsymbol{\beta}}^{(k)}$, proposed by McKean and Hettmansperger (1978). Let $\boldsymbol{\beta}^{(0)}$ denote the least squares estimate of $\boldsymbol{\beta}$ and compute $S(\hat{\boldsymbol{\beta}}^{(0)})$ from (9) and $\hat{\tau}^{(0)}$ from (12) (or (14)), then, using a linear approximation to $\boldsymbol{S}(\boldsymbol{\beta})$ discussed by Jureckova (1971), form

$$\hat{\boldsymbol{\beta}}^{(1)} = \hat{\boldsymbol{\beta}}^{(0)} + \hat{\tau}^{(0)}(\boldsymbol{X}_c'\boldsymbol{X}_c)^{-1}\boldsymbol{S}(\hat{\boldsymbol{\beta}}^{(0)})\,. \tag{15}$$

The estimate $\hat{\boldsymbol{\beta}}^{(k)}$ found by iterating (15) has the same asymptotic distribution for any $k = 1, 2, \ldots,$ as $\hat{\boldsymbol{\beta}}$, the rank estimate that minimizes $D(\boldsymbol{\beta})$ in (8). Hence $\hat{\boldsymbol{\beta}}^{(k)}$ could be used in W to construct a test. Generally, $\hat{\boldsymbol{\beta}}^{(k)}$ does not converge to $\hat{\boldsymbol{\beta}}$ as k increases. It is probably best to take around 4 or 5 steps and then construct W.

Summary: (1) There are no published studies that show how the level of W behaves for small samples. There is no indication of how large the sample size should be before the asymptotic distribution provides a good approximation. (2) Computation of W requires special programs and cannot be carried out using existing statistical packages. In 1984 the Minitab statistical computing system will contain a command that will produce in its output the rank estimate $\hat{\beta}$ and the test statistic W. There is further discussion of computation in Section 3. (3) With the exception of the small unpublished simulation by Hettmansperger and McKean (1981) no study of the small sample power of W is available. In the simulation just mentioned the estimate (14) was used after some small sample adjustments. For example τ^* is multiplied by the bias correction $(n/(n-p))^{1/2}$. It was found in the parallelism designed that W was often liberal and needed further correction to reduce the probability of a type I error. Its small sample power was comparable to the power of the F, A and D^* (in the next subsection) tests.

2iii. *Test based on reduction due to fitting the full and reduced models*

This method in the rank case is analogous to the F statistic which can be written as the reduction in sum of squares due to fitting the full and reduced models. The aligned rank test and the Wald test are not directly based on the comparison of a reduced and full model. The aligned test is constructed from reduced-model residuals and the Wald test is a quadratic form in the full-model estimates. It might seem most natural to combine estimation with fitting in a robust fashion in order to have a set of strategies parallel to least squares. Then data analytic methods such as plotting have direct counterparts based on ranks.

The test is based on Jaeckel's measure of dispersion (8). Let

$$D^* = \frac{D(\hat{\hat{\boldsymbol{\beta}}}_1) - D(\hat{\boldsymbol{\beta}})}{\hat{\tau}/2}, \tag{16}$$

where $\hat{\hat{\boldsymbol{\beta}}}_1$ and $\hat{\boldsymbol{\beta}}$ are the reduced and full model rank estimates, respectively, and $\hat{\tau}$ is a consistent estimate of τ. Under regularity conditions, McKean and Hettmansperger (1976) show that D^* has a limiting chi-square distribution with q degrees of freedom. Hence, the test based on (16) rejects H_0: $\boldsymbol{\beta}_2 = 0$ if $D^* \geqslant \chi^2_\alpha(q)$. Computation of D^* requires special programs for $\hat{\hat{\boldsymbol{\beta}}}_1$, $\hat{\boldsymbol{\beta}}$ and $\hat{\tau}$. The forthcoming Minitab rank-regression command will incorporate this test as part of its output. See Section 3 for further aspects of the computational problems. The test is illustrated on data in Section 4.

Summary: (1) Hettmansperger and McKean (1977) provide a small simulation which indicates that D^* along with τ^*, tuned for small samples, has a significance level close to the nominal level. McKean and Hettmansperger (1978) provide simulation results for the k-step estimate, D^* and τ^*. Again, the test seems to have a stable level. There are no simulation studies of D^* with $\hat{\tau}$. There is no indication of how large the sample size should be before the asymptotic distribution of D^* with $\hat{\tau}$ provides a good approximation. (2) Because of the computational problems involved in computing $\hat{\boldsymbol{\beta}}$ and $\hat{\tau}$ or τ^* (see Section 3) special programs are required to compute D^*. In 1984 the Minitab statistical computing system will produce $\hat{\boldsymbol{\beta}}$, D^* and $\hat{\tau}$ or τ^* in the output of a rank-regression command. (3) In an unpublished simulation study by Hettmansperger and McKean (1981) the test based on D^* with τ^* had power comparable to the F test and the test based on W.

Finally, it should be emphasized that the use of τ^* requires the assumption of symmetry of the error distribution. The estimate $\hat{\tau}$ does not require symmetry. It is not yet known how well $\hat{\tau}$ will work in the asymmetric case and it is not known if $\hat{\tau}$ will be a viable substitute for τ^* in the symmetric case. Consistency of $\hat{\tau}$ has only been established for the Wilcoxon scores; see Aubuchon (1982).

3. Computations

As was mentioned in Section 2, special programs are necessary to calculate any of the test statistics other than Adichie's (using least squares to fit the reduced model). First of all, a program which minimizes the dispersion, (8), is needed to obtain rank estimates and to evaluate the dispersion for these estimates. Then, in order to use either the Wald quadratic-form test or the drop-in-dispersion test, we need a program to compute an estimate of the scaling functional τ appearing in the denominator. The discussion in this section pertains to procedures generated by general score functions. The Wilcoxon score function in (6) can be replaced by any $\phi(u)$, nonconstant, nondecreasing, square integrable such that $\int \phi(u)\,\mathrm{d}u = 0$. See the comments following (6).

An algorithm suggested by J. W. McKean (personal communication) for minimizing the dispersion is perhaps best thought of as a member of the class of iterative schemes known as gradient methods. The increment to the estimate at the K-th step is given by a positive step size $t^{(K)}$ times some symmetric, positive definite matrix C times the negative of the gradient:

$$\hat{\boldsymbol{\beta}}^{(K+1)} - \hat{\boldsymbol{\beta}}^{(K)} = t^{(K)} \boldsymbol{C}\boldsymbol{S}(\hat{\boldsymbol{\beta}}^{(K)}) . \tag{17}$$

Recall that $-\boldsymbol{S}(\boldsymbol{\beta})$ is the gradient of the dispersion, (9). Two considerations led us to set $\boldsymbol{C} = (\boldsymbol{X}_c'\boldsymbol{X}_c)^{-1}$ in (17). First of all, since the asymptotic variance-covariance structure of $\hat{\boldsymbol{\beta}}$ is given by a constant times $(\boldsymbol{X}_c'\boldsymbol{X}_c)^{-1}$, a natural norm for $\boldsymbol{\beta}$ is $\|\boldsymbol{\beta}\| = (\boldsymbol{\beta}'\boldsymbol{X}_c'\boldsymbol{X}_c\boldsymbol{\beta})^{1/2}$. Results of Ortega and Rheinboldt (1970) show that the direction of steepest descent with respect to this norm is precisely $(\boldsymbol{X}_c'\boldsymbol{X}_c)^{-1}\boldsymbol{S}(\boldsymbol{\beta}^{(K)})$. On the other hand, Jaeckel (1972) shows that the dispersion function may be approximated asymptotically by a quadratic:

$$D(\boldsymbol{\beta}) \doteq D(\boldsymbol{\beta}_0) - (\boldsymbol{\beta} - \boldsymbol{\beta}_0)'\boldsymbol{S}(\boldsymbol{\beta}_0) + (2\tau)^{-1}(\boldsymbol{\beta} - \boldsymbol{\beta}_0)'\boldsymbol{X}_c'\boldsymbol{X}_c(\boldsymbol{\beta} - \boldsymbol{\beta}_0) , \tag{18}$$

where $\boldsymbol{\beta}_0$ is the vector of true regression parameters. The minimum of this quadratic is attained for

$$\boldsymbol{\beta} = \boldsymbol{\beta}_0 + \tau(\boldsymbol{X}_c'\boldsymbol{X}_c)^{-1}\boldsymbol{S}(\boldsymbol{\beta}_0) . \tag{19}$$

Thus, if we substitute $\hat{\boldsymbol{\beta}}^{(K)}$, our current estimate, for $\boldsymbol{\beta}_0$ in (19) we are again led to take a step in the direction $(\boldsymbol{X}_c'\boldsymbol{X}_c)^{-1}\boldsymbol{S}(\hat{\boldsymbol{\beta}}^{(K)})$.

It remains to choose the step size, $t^{(K)}$. We might search for the minimum of $D(\hat{\boldsymbol{\beta}}^{(K+1)})$ as a function of $t^{(K)}$ using any good linear search method – the golden section search or one of the other methods described in Kennedy and Gentle (1980), for example. McKean suggests that this search might be conducted by making use of the asymptotic linearity of the derivative of $D[\hat{\boldsymbol{\beta}}^{(K)} + t(\boldsymbol{X}_c'\boldsymbol{X}_c)^{-1}\boldsymbol{S}(\hat{\boldsymbol{\beta}}^{(K)})]$ with respect to t, given below in (20). (Compare Hettmansperger and McKean (1975).) Specifically, he suggests application of the Illinois version of false position, as discussed by Dowell and Jarratt (1971), to find the approximate root of this derivative:

$$S^*(t) = \boldsymbol{a}'(\hat{\boldsymbol{\beta}}^{(K)})\boldsymbol{X}_c(\boldsymbol{X}_c'\boldsymbol{X}_c)^{-1}\boldsymbol{X}_c'\boldsymbol{a}[\hat{\boldsymbol{\beta}}^{(K)} + t(\boldsymbol{X}_c'\boldsymbol{X}_c)^{-1}\boldsymbol{S}(\hat{\boldsymbol{\beta}}^{(K)})] , \tag{20}$$

which is a nondecreasing step function. Whatever linear search method is employed, this approach is equivalent to transforming the linear model by obtaining an orthogonal design matrix and then using the method of steepest descent.

As with any iterative method, it is necessary to specify starting values and convergence criteria. One possibility for $\hat{\boldsymbol{\beta}}^{(0)}$ is the usual least-squares estimate, which is easy to compute and which we would most likely desire for comparative purposes in any case. Another choice would be some more resistant estimate, such as the L_1 estimate, which is, however, more expensive to compute. It is not clear what the trade-offs in computational efficiency would be in making such choices. For a convergence criterion it may be best to focus on relative change in the dispersion, since the value of $\boldsymbol{\beta}$ which minimizes the dispersion is not generally unique. Criteria which check whether the gradient is (approximately) zero will not be useful, since the gradient is a step function and may step across zero.

If, in (17), we let $\boldsymbol{C} = (\boldsymbol{X}_c'\boldsymbol{X}_c)^{-1}$ as suggested and set $t^{(K)} = \hat{\tau}^{(K)}$, an estimate of τ computed on the residuals at the K-th step, we essentially have an iterative scheme based on the K-step estimates discussed in Section 2ii. While such estimates may be of interest in their own right, early experience of McKean and others indicates that, taken as an algorithm for minimizing the dispersion, this scheme can behave rather poorly for some data sets, failing to converge to, and in fact moving away from, a minimizing point.

We should also mention that Osborne (1981) and others have developed algorithms for minimizing the dispersion using methods of convex analysis.

Although iterative methods are not needed to compute the window estimate of τ, a naive approach will not be very efficient. Schweder (1975) suggests an interesting scheme for computing $\Sigma_i \Sigma_j I\{|r_i - r_j| < h_n/2\}$ but does not give details. A time- and space-efficient algorithm based on Schweder's suggestion may be found in Aubuchon (1982).

With the assumption that the error distribution is symmetric, McKean and Hettmansperger (1976) show that a consistent estimate of τ may be obtained by applying a one-sample rank procedure to the uncentered residuals, $r_i = y_i - \boldsymbol{x}_{ic}'\hat{\boldsymbol{\beta}}$, using the one-sample score function corresponding to ϕ: $\phi^+(u) = \phi((u+1)/2)$. If $(\hat{\alpha}_L, \hat{\alpha}_U)$ is the $100(1-\alpha)\%$ confidence interval obtained for the center of symmetry in this fashion, then $\hat{\tau} = \sqrt{n}(\hat{\alpha}_U - \hat{\alpha}_L)/(2Z_{\alpha/2})$ is a consistent estimate of τ. This approach is an extension of the work of Sen (1966) to the linear model.

If Wilcoxon scores are used, there are at least three approaches to obtaining $\hat{\alpha}_L$ and $\hat{\alpha}_U$. If storage space and efficiency are not of critical importance, the $n(n+1)/2$ pairwise (Walsh) averages may be computed. Then $\hat{\alpha}_L$ and $\hat{\alpha}_U$ are the $(c+1)$st and $(n(n+1)/2 - c)$th order statistics from this set, where c is the lower critical point of a two-sided, size-α Wilcoxon signed-rank test. This critical point may be obtained from tables or from a normal approximation. Any fast algorithm for selecting order statistics might then be used to find $\hat{\alpha}_L$ and $\hat{\alpha}_U$; see, for example, Knuth (1973). An approach which is faster and which requires much less storage is based on Johnson and Mizoguchi (1978), with improvements discussed by Johnson and Ryan (1978). These papers actually present the algorithm for the two-sample problem; but simple modifications make it applicable to the present case as well. One advantage of this method is that it still selects exact order statistics from the set of Walsh averages, without computing and storing all of them. A third method, relying on the asymptotic linearity of signed-rank statistics, does not guarantee exact results but is quite fast and space-efficient. The Illinois version of false position is used to find approximate solutions to the equations (21) defining $\hat{\alpha}_L$ and $\hat{\alpha}_U$ in terms of a signed-rank statistic:

$$\sqrt{n}\,\bar{V}(\hat{\alpha}_L) = Z_{\alpha/2}, \qquad \sqrt{n}\,\bar{V}(\hat{\alpha}_U) = -Z_{\alpha/2}\,, \tag{21}$$

where $\bar{V}(\alpha) = n^{-1}\Sigma_{i=1}^{n}\phi^{+}(R_i^{+}/(n+1))\,\mathrm{sign}(r_i - \alpha)$, R_i^{+} is the rank of $|r_i - \alpha|$ among $|r_i - \alpha|, \ldots, |r_n - \alpha|$ and $Z_{\alpha/2}$ is the upper $\alpha/2$ point of the standard normal distribution. See McKean and Ryan (1977) for the use of this algorithm in the corresponding two-sample problem.

For certain other score functions, for example the scores suggested by Policello and Hettmansperger (1976), $\hat{\alpha}_L$ and $\hat{\alpha}_U$ are order statistics from a well-defined subset of the Walsh averages. In this case, the first two methods discussed above are still applicable. In general, $\hat{\alpha}_L$ and $\hat{\alpha}_U$ are weighted order statistics from the Walsh averages, with weight $a^{+}(j-i+1) - a^{+}(j-1)$ given to $(r_{(i)} + r_{(j)})/2$, where $a^{+}(i) = a^{+}(i/(n+1))$; see Bauer (1972). Thus, if a program for selecting weighted order statistics is available, the first method still works. Otherwise, the third method may be used with any score function.

4. Example

In order to illustrate the procedures discussed in this paper, we have applied them to data from an experiment described by Shirley (1981). In this section computations are based on the Wilcoxon scores in (6). Two censored observations are recorded at the censoring point for the purposes of this example. The data are displayed in Table 1. Thirty rats received a treatment intended to

Table 1
Times taken for rats to enter cages

Group 1		Group 2		Group 3	
Before treatment	After treatment	Before treatment	After treatment	Before theatment	After treatment
1.8	79.1	1.6	10.2	1.3	14.8
1.3	47.6	0.9	3.4	2.3	30.7
1.8	64.4	1.5	9.9	0.9	7.7
1.1	68.7	1.6	3.7	1.9	63.9
2.5	180.0[a]	2.6	39.3	1.2	3.5
1.0	27.3	1.4	34.0	1.3	10.0
1.1	56.4	2.0	40.7	1.2	6.9
2.3	163.3	0.9	10.5	2.4	22.5
2.4	180.0[a]	1.6	0.8	1.4	11.4
2.8	132.4	1.2	4.9	0.8	3.3

[a]Censored observations.

delay entry into a chamber. The rats were divided into three groups of ten, a control group and two experimental groups. The experimental groups each received some antidote to the treatment, while the control group received none. The time taken by each rat to enter the chamber was recorded before the treatment and again after the treatment and antidote – if any.

We consider the measurement before treatment as a covariate and test for interaction between the grouping factor and the covariate; i.e., we test for unequal slopes. The observations are strongly skewed; we applied the natural log transformation to gain some degree of symmetry so that the estimate τ^* in (14) may be applied to the data.

Computations for the aligned rank test, using least squares to fit the reduced model, can be carried out in the SAS statistical computing system (see Helwig and Council, 1979) using the following program:

```
DATA;
   INPUT BEFORE AFTER ANTIDOTE;
   LOG_AFT = LOG (AFTER);
   CARDS;
{data goes here}
PROC GLM;
   CLASS ANTIDOTE;
   MODEL LOG_AFT = ANTIDOTE BEFORE;
   OUTPUT OUT = RESID RESID = RESID;
PROC SORT DATA = RESID;
   BY RESID;
DATA RSCORE;
   SET RESID;
   RSCORE = SQRT(12) * (_N_/31 - .5);
PROC GLM DATA = RSCORE;
   CLASS ANTIDOTE;
   MODEL RSCORE = ANTIDOTE BEFORE ANTIDOTE *
   BEFORE;
```

The desired test statistic will be the Type 4 sum of squares for ANTIDOTE * BEFORE in the second GLM output.

The same calculation can be made in the Minitab statistical computing system (see Ryan, Joiner and Ryan, 1981). Some manipulation is necessary to create the design matrix so that the REGRESS command can be used. Indicator variables for the first two groups are put in 'A1' and 'A2'; then two interaction columns, 'INTER1' and 'INTER2', are produced by multiplying each of these by the covariate.

```
NAME C1 = 'BEFORE' C2 = 'AFTER' C3 = 'ANTIDOTE'
NAME C4 = 'LOG.AFT' C5 = 'A1' C6 = 'A2'
NAME C7 = 'A3' C8 = 'INTER1' C9 = 'INTER2'
NAME C10 = 'STD.RES.' C11 = 'FITS' C12 = 'RANKS'
NAME C13 = 'RSCORES' C14 = 'RESIDS'

READ 'BEFORE' 'AFTER' 'ANTIDOTE'

{Data goes here}

LET 'LOG.AFT.' = LOGE('AFTER')
INDICATORS FOR 'ANTIDOTE' IN 'A1' 'A2' 'A3'
LET 'INTER1' = 'A1' * 'BEFORE'
LET 'INTER2' = 'A2' * 'BEFORE'
REGRESS 'LOG.AFT' 3 'A1' 'A2' 'BEFORE' 'STD.RES.' &
            'FITS'
LET 'RESIDS' = 'LOG.AFT.' - 'FITS'
RANKS OF 'RESIDS' IN 'RANKS'
LET 'RSCORES' = SQRT(12) * ('RANKS'/31 - .5)
REGRESS 'RSCORES' 5 'A1' 'A2' 'BEFORE'
  'INTER1' & 'INTER2'
```

The test statistic is then the sum of the last two sums of squares in the table labeled 'SS explained by each variable when entered in the order given'. In general, the columns to be tested should be given last in the REGRESS command used to fit the full model to the rank scores.

The divisor 31 used to calculate the rank scores in the two programs corresponds to the quantity $(n+1)$. Using either program, we have the value 0.42 for the test statistic. When this is compared to a χ^2 critical point with two degrees of freedom, we fail to reject the hypothesis of equal slopes at any reasonable level. We could now proceed to perform similar tests for the group effect and for the covariate.

A program implementing the algorithms described in Section 3 in Fortran was used to perform the Wald test and the drop-in-dispersion test for the equal slopes hypothesis. Both of the estimates for τ discussed in Section 2ii were employed. The results are presented in detail for comparison with other programs for minimizing the dispersion. In Table 2, the fitting of full and reduced models is summarized. The values in parentheses correspond to the least-squares estimates, which were used as starting values.

Seven steps were required to attain convergence of the full-model estimates, while three steps were used for the reduced-model estimates. We report the two estimates of τ, given by $\hat{\tau} = (12\hat{\gamma}^2)^{-1}$ for $\hat{\gamma}$ in (12) and by τ^* in (14). The least-squares estimate of σ is also shown.

In Table 3 we present four test statistics, corresponding to use of either the Wald test statistic in (11) or the drop test statistic in (16) combined with either $\hat{\tau}$ or τ^* as an estimate of τ in the denominator. For comparison, we also list the aligned rank test statistic computed above, as well as twice the usual least-

Table 2
Fitting full and reduced models[a]

	Full		Reduced	
Dispersion	17.942	(18.126)	18.306	(18.372)
$\hat{\alpha}$ = Intercept	0.664	(0.465)	0.836	(0.874)
$\hat{\beta}_1$ = Antidote 1 − Antidote 3	2.20	(2.43)	1.60	(1.61)
$\hat{\beta}_2$ = Antidote 2 − Antidote 3	−0.410	(−0.020)	−0.235	(−0.342)
$\hat{\beta}_3$ = Before	1.20	(1.35)	1.10	(1.08)
$\hat{\beta}_4$ = Slope for Antidote 1 − Slope for Antidote 3	−0.323	(−0.502)	—	
$\hat{\beta}_5$ = Slope for Antidote 2 − Slope for Antidote 3	0.101	(−0.221)	—	

[a]Tabled numbers correspond to procedures based on Wilcoxon; numbers in parentheses correspond to least-squares.

Note: $\hat{\tau} = 0.5309$, $\tau^* = 0.5215$, $\hat{\sigma} = 0.7884$.

Table 3
Tests for equal slopes

	Test statistic	
Estimate of τ	W	D^*
$\hat{\tau}$	1.12	1.38
τ^*	1.15	1.40

Note: $A = 0.42$, $2F = 0.67$.

squares F statistic. All of these statistics may be compared to the upper α-point of the chi-square distribution with two degrees of freedom. In practice, we would most likely apply some small-sample tuning to τ^* or $\hat{\tau}$. Further, we would divide all of the test statistics except, perhaps, Adichie's by the numerator degrees of freedom q ($q = 2$ for this example) and compare the results to the upper α-point of the F distribution with q and $n-p-1$ degrees of freedom.

References

Adichie, J. N. (1978). Rank tests of sub-hypotheses in the general linear regression. *The Annals of Statistics* **5**, 1012–1026.

Aubuchon, J. C. (1982). Rank Tests in the Linear Model: Asymmetric Errors. Unpublished Ph.D. Thesis. The Pennsylvania State University.

Bauer, D. F. (1972). Constructing confidence sets using rank statistics. *Journal of the American Statistical Association* **38**, 894–904.

Dowell, M. and Jarratt, P. (1971). A modified Regula Falsi method for computing the root of an equation. *BIT* **11**, 168–174.

Draper, N. R. and Smith, H. (1981). *Applied Regression Analysis*. Wiley, New York, 2nd ed.

Graybill, F. A. (1976). *Theory and Applications of the Linear Model.* Duxbury Press, North Scituate, MA.

Helwig, J. T. and Council, K. A. (1979). *SAS User's Guide.* SAS Institute, 1979 edition.

Hettmansperger, T. P. and McKean, J. W. (1977). A robust alternative based on ranks to least squares in analyzing linear methods. *Technometrics* **19**, 275–284.

Hettmansperger, T. P. and McKean, J. W. (1981). A geometric interpretation of inferences based on ranks in the linear model. Technical Report 36, Department of Statistics, The Pennsylvania State University.

Hodges, J. L., Jr. and Lehmann, E. L. (1962). Rank methods for combination of independent experiments in analysis of variance. *The Annals of Mathematical Statistics* **33**, 482–497.

Hodges, J. L., Jr. and Lehmann, E. L. (1963). Estimation of location based on rank tests. *The Annals of Mathematical Statistics.* **34**, 598–611.

Huber, P. S. (1981). *Robust Statistics.* Wiley, New York.

Jaeckel, L. A. (1972). Estimating regression coefficients by minimizing the dispersion of the residuals. *The Annals of Mathematical Statistics* **43**, 1449–1458.

Johnson, D. B. and Mizoguchi, T. (1978). Selecting the Kth element in $X + Y$ and $X_1 + X_2 + \cdots + X_m$. *SIAM Journal of Computing* **7**, 147–153.

Johnson, D. B. and Ryan, T. A., Jr. (1978). Fast computation of the Hodges Lehmann estimator – theory and practice. in: *Proceedings of the American Statistical Association Statistical Computing Section.*

Jureckova, J. (1971a). Nonparametric estimate of regression coefficients. *The Annals of Mathematical Statistics* **42**, 1328–1338.

Jureckova, J. (1971b). Asymptotic independence of rank test statistic for testing symmetry on regression. *Sankhya Ser. A* **33**, 1–18.

Kennedy, W. J., Jr. and Gentle, J. E. (1980). *Statistical Computing.* Marcel Dekker, Inc., New York.

Knuth, D. E. (1973). *The Art of Computer Programming, Sorting and Searching,* Vol. 2. Addison-Wesley, Reading, MA.

Koul, H. L. (1970). A class of ADF tests for subhypothesis in the multiple linear regression. *The Annals of Mathematical Statistics* **41**, 1273–1281.

Kraft, C. H. and van Eeden, C. (1972). Linearized rank estimates and signed-rank estimates for the general linear model. *The Annals of Mathematical Statistics* **43**, 42–57.

Lehmann, E. L. (1963). Nonparametric confidence intervals for a shift parameter. *The Annals of Mathematical Statistics* **34**, 1507–1512.

McKean, J. W. and Hettmansperger, T. P. (1976). Tests of hypotheses based on ranks in the general linear model. *Communications in Statistics – Theory and Methods* **A5** (8), 693–709.

McKean, J. W. and Hettmansperger, T. P. (1978). A robust analysis of the general linear model based on one step R-estimates. *Biometrika* **65**, 571–579.

McKean, J. W. and Ryan, T. A., Jr. (1977). An algorithm for obtaining confidence intervals and point estimates based on ranks in the two-sample location problem. *Association for Computing Machinery Transactions on Mathematical Software* **3**, 183–185.

McKean, J. W. and Schrader, R. M. (1980). The geometry of robust procedures in linear models. *Journal of the Royal Statistical Society Ser. B* **42**, 366–371.

Neter, J. and Wasserman, W. (1974). *Applied Linear Statistical Models.* Richard D. Irwin, Inc., Homewood, IL.

Ortega, J. M. and Rheinboldt, W. C. (1970). *Iterative Solution of Nonlinear Equations in Several Variables.* Academic Press, New York.

Osborne, N. R. (1981). A Finite Algorithm for the Rank Regression Problem. Unpublished manuscript, Australian National University.

Policello, G. E., II and Hettmansperger, T. P. (1976). Adaptive robust procedures for the one sample location problem. *Journal of the American Statistical Association* **71**, 624–633.

Puri, M. L. and Sen, P. K. (1973). A note on asymptotically distribution free tests for subhypotheses in multiple linear regression. *Annals of Statistics* **1**, 553–556.

Rao, C. R. (1973). *Linear Statistical Inference and its Applications.* Wiley, New York, 2nd ed.

Ryan, T. A., Jr., Joiner, B. L. and Ryan, B. F. (1981). *Minitab Reference Manual.* Minitab Project, Pennsylvania State University, University Park, PA.

Schrader, R. M. and Hettmansperger, T. P. (1980). Robust analysis of variance based upon a likelihood ratio criterion. *Biometrika* **67**, 93–101.

Schuster, E. (1974). On the rate of convergence of an estimate of a functional of a probability density. *Scandinavian Acturial Journal* **1**, 103–107.

Schweder, T. (1975). Window estimation of the asymptotic variance of rank estimators of location. *Scandinavian Journal of Statistics* **2**, 113–126.

Sen, P. K. (1966). On a distribution-free method of estimating asymptotic efficiency of a class of nonparametric tests. *The Annals of Mathematical Statistics* **37**, 1759–1770.

Sen, P. K. (1980). On M tests in linear models. *Biometrika* **69**, 245–248.

Sen, P. K. and Puri, M. L. (1977). Asymptotically distribution-free aligned rank order tests for composite hypotheses for general multivariate linear models. *Zeitschrift fuer Wahrscheinlichkeitstheorie und Verwandte Gebiete* **39**, 175–186.

Shirley, E. A. (1981). A distribution-free method for analysis of covariance based on ranked data. *Applied Statistics* **30**, 158–162.

P. R. Krishnaiah and P. K. Sen, eds., *Handbook of Statistics, Vol. 4*

13

Nonparametric Preliminary Test Inference

A. K. Md. Ehsanes Saleh and Pranab Kumar Sen

1. Introduction

In problems of estimation and testing of hypotheses, *prior information* on (some of) the parameters, when available, generally leads to improved inference procedures (for the other parameters of interest). Usually, this prior information is incorporated in the *model* in the form of *parametric constraints.* For procedures based on such constrained models, to be termed here-after, the *constrained procedures*, the *performance characteristics* are generally better than the *unconstrained* ones (which do not take into account the constraints) when these constraints hold. On the other hand, the validity and efficiency of the constrained procedures may be restricted to the domain of validity of the constraints, while the unconstrained procedures retain their validity over a wider class of models (though may not be fully efficient for the constrained models). Thus, the lack of *validity and efficiency robustness of the constrained procedures* (against departures from the assumed constraints) has to be weighed against possible *loss of efficiency of the unconstrained ones* (when the constraints hold) in deciding on a choice between the two (particularly, when a full confidence may not be allotted to the prior information on the constraints).

It is not uncommon to encounter problems of inference with *uncertain prior information in the model*: From some extraneous considerations, some prior information may lend itself into the model, but, there may not be sufficient evidence on their validity so as to advocate an unrestricted use of the constrained procedures. In such a case, it may be more natural to perform a *preliminary test* on the validity of the parametric constraints, provided by the uncertain prior information, and then to choose between the constrained or the unconstrainted ones depending on the outcome of the preliminary test. This is termed *preliminary test inference.* The primary objective of this preliminary test inference procedure is to guard against the lack of validity robustness of the constrained procedure without much compromise on its optimality and/or desirability. On the other hand, in view of the preliminary test, the distribution theory of the final estimator or the test statistic may be quite involved, and, in many cases, exact results may not be available in closed, simple forms.

Though in the parametric case, preliminary test inference procedures have been developed in increasing generality during the past 40 years, the nonparametric counterparts are, mostly, of only recent origins. For the parametric theory, mostly, based on finite sample sizes from normal or multinormal distributions, an extensive bibliography and an useful survey are due to Bancroft and Han (1977, 1980). The main objective of the current study is to focus on some recent developments in the area of nonparametric preliminary test inference (NPTI) based on a broad class of *rank order statistics* and *derived estimators*. We intend to cover *general* (*multivariate*) *linear models* for which suitable rank procedures have already been discussed in other chapters (see Adichie, Chapter 11, and Aubuchon and Hettmansperger, Chapter 12). In a general framework, we may present the NPTI procedures as follows.

Let $\boldsymbol{Y}$ be a (possibly vector or matrix valued) random variable (r.v.) with a distribution function (d.f.) F involving some unknown parameter $\boldsymbol{\theta}$. We partition $\boldsymbol{\theta} = (\boldsymbol{\theta}_1, \boldsymbol{\theta}_2)$, and our basic problem is to draw inference on $\boldsymbol{\theta}_1$ when $\boldsymbol{\theta}_2$ is suspected to be close to some specified $\boldsymbol{\theta}_2^0$, which, for simplicity, we may take as $\mathbf{0}$.

(i) *The preliminary test estimation* (*PTE*) *problem*. For testing H_0: $\boldsymbol{\theta}_2 = \mathbf{0}$ against $\boldsymbol{\theta}_2 \neq \mathbf{0}$, let $\xi(\boldsymbol{T}_2)$ be a *test function*, based on the test statistic $\boldsymbol{T}_2$, such that $0 \leq \xi(\boldsymbol{T}_2) \leq 1$ and $E_0\xi(\boldsymbol{T}_2) \leq \alpha_2$, the level of significance of the test. We let $\xi(\boldsymbol{T}_2)$ assume the values 0 and 1, according as H_0 is accepted or not. (A randomized test function may also be considered. However, for the sake of simplicity, we confine ourselves to nonrandomized tests only.) Let $\hat{\boldsymbol{\theta}}_1$ be the constrained estimator of $\boldsymbol{\theta}$ (when $\boldsymbol{\theta}_2$ is taken as $\mathbf{0}$) and $\tilde{\boldsymbol{\theta}}_1$ is the component of the unconstrained estimator $\tilde{\boldsymbol{\theta}} = (\tilde{\boldsymbol{\theta}}_1, \tilde{\boldsymbol{\theta}}_2)$ of $\boldsymbol{\theta}$ when $\boldsymbol{\theta}_2$ is not specified. Then $\boldsymbol{\theta}_1^*$, the PTE of $\boldsymbol{\theta}_1$, is defined by

$$\boldsymbol{\theta}_1^* = \{1 - \xi(\boldsymbol{T}_2)\}\hat{\boldsymbol{\theta}}_1 + \xi(\boldsymbol{T}_2)\tilde{\boldsymbol{\theta}}_1 . \tag{1.1}$$

Though $\hat{\boldsymbol{\theta}}_1$ may be an (unbiased and) efficient estimator of $\boldsymbol{\theta}_1$ when $\boldsymbol{\theta}_2 = \mathbf{0}$ and $\tilde{\boldsymbol{\theta}}_1$ may be a similar (unbiased and) efficient estimator of $\boldsymbol{\theta}_1$ when $\boldsymbol{\theta}_2$ is unspecified, $\boldsymbol{\theta}_1^*$ may neither be unbiased nor efficient, when $\boldsymbol{\theta}_2 = 0$ or $\boldsymbol{\theta}_2$ is unspecified. Nevertheless, the PT $\xi(\boldsymbol{T}_2)$ ties $\boldsymbol{\theta}_1^*$ to $\hat{\boldsymbol{\theta}}_1$ or $\tilde{\boldsymbol{\theta}}_1$, depending on the outcome, and guards against major *bias* (due to $\hat{\boldsymbol{\theta}}_1$ when $\boldsymbol{\theta}_2 \neq 0$) or *loss of efficiency* (due to $\tilde{\boldsymbol{\theta}}_1$ when $\boldsymbol{\theta}_2 = \mathbf{0}$) when $\boldsymbol{\theta}_2$ may not be equal to $\mathbf{0}$, but is suspected to be so. The basic problem is therefore to study the distributional properties of $\boldsymbol{\theta}_1^*$ with a view to providing useful ideas about the *bias*, *mean square error* and the *robustness* of $\boldsymbol{\theta}_1^*$, as compared to $\hat{\boldsymbol{\theta}}_1$ and $\tilde{\boldsymbol{\theta}}_1$.

(ii) *The preliminary test testing* (*PTT*) *problem*. Suppose now that we want $H_0^{(1)}$: $\boldsymbol{\theta}_1 = \boldsymbol{\theta}_1^0$ against $\boldsymbol{\theta}_1 \neq \boldsymbol{\theta}_1^0$, when $\boldsymbol{\theta}_2^0$ is suspected to be close to $\mathbf{0}$. Consider a preliminary test function $\xi(\boldsymbol{T}_2) = \xi_2$ for testing $H_0^{(2)}$: $\boldsymbol{\theta}_2 = \mathbf{0}$ against $\boldsymbol{\theta}_2 \neq 0$. Further, let $H_0^{(12)}$: $\boldsymbol{\theta}_1 = \boldsymbol{\theta}_1^0$, $\boldsymbol{\theta}_2 = \mathbf{0}$ and $H_0^{(1\cdot)}$: $\boldsymbol{\theta}_1 = \boldsymbol{\theta}_1^0$, $\boldsymbol{\theta}_2$ nuisance parameter. Let then ξ_{12} and ξ_{10} be appropriate test functions for testing $H_0^{(12)}$ against $\boldsymbol{\theta}_1 \neq \boldsymbol{\theta}_1^0$ or $\boldsymbol{\theta}_2 \neq \mathbf{0}$ and $H_0^{(1\cdot)}$ against $\boldsymbol{\theta}_1 = \boldsymbol{\theta}_1^0$. Then, the PTT for $H_0^{(1)}$ has the test function

$$\xi^* = (1 - \xi_2)\xi_{12} + \xi_2\xi_1 . \tag{1.2}$$

Again, ξ_{12} may be the unbiased and optimal (in some sense) for testing $H_0^{(12)}$, while $\xi_{1\cdot}$ may be so for testing $H_0^{(1\cdot)}$, ξ^* may not be unbiased or optimal. On the top of that if α_{12}, $\alpha_{1\cdot}$ and α_2 be the size of the test ξ_{12}, $\xi_{1\cdot}$ and ξ_2, then the size of ξ^* need not be equal to $(1-\alpha_2)\alpha_{12}+\alpha_2\alpha_{1\cdot}$; indeed, it may depend on the joint distribution of the test statistics in a more involved way and also on whether $\boldsymbol{\theta}_2$ is equal to $\mathbf{0}$ or not. Nevertheless, ξ^* is likely to be more robust than ξ_{12} against $\boldsymbol{\theta}_2$ away from $\mathbf{0}$ and more efficient than $\xi_{1\cdot}$ when $\boldsymbol{\theta}_2$ is close to $\mathbf{0}$. The effect of preliminary test on the *size, power* and *robustness* of the ultimate test is the main item of study.

In the subsequent sections, we specialize this model to various specific models (where nonparametric estimators and tests work out well) and formulate NPTI procedures, and present various available results on their performance characteristics. Some general discussions are made in the concluding section.

2. NPTI for the univariate simple regression model

Let $Y_1, \ldots, Y_n$ be independent r.v. with continuous d.f.

$$F_i(x) = P\{Y_i \leq x\} = F(x-\theta-\beta c_1), \quad 1 \leq i \leq n, \ x \in E, \tag{2.1}$$

where F is of unspecified form, θ, β are unknown parameters and

$$\boldsymbol{c}_n = (c_1, \ldots, c_n)' \text{ is a vector of known constants}. \tag{2.2}$$

The two-sample location model is a special case where the c_i may only be 0 or 1. We are primarily concerned with inference on θ, when it is suspected but not evident that $\beta = 0$ (or some other specified value). As has been discussed in (1.1)–(1.2), we incorporate a preliminary test on the hypothesis $H_0^{(2)}$: $\beta = 0$ (against $\beta \neq 0$ or $\beta > 0$): If $H_0^{(2)}$ is tenable, (2.1) reduces to the classical one-sample model, so that the usual signed-rank statistics (on $Y_1, \ldots, Y_n$) may be employed to draw inference on θ, while if $H_0^{(2)}$ is not tenable, based on some rank estimator ($\hat{\beta}_n$) of β, we define the residuals $\hat{Y}_i = Y_i - \hat{\beta}_n c_i$, $1 \leq i \leq n$ and use (aligned) signed-rank statistics on these residuals for drawing inference on θ. Our primary concern is to study the effect of the preliminary test on β on the performance characteristics of the NPTI procedures for θ. For this, we introduce, first, the preliminary notions and basic regularity conditions:

(i) $F \in \mathscr{F}$, the class of all absolutely continuous, symmetric (about 0) d.f. with (almost everywhere) absolutely continuous probability density function (pdf) having finite Fisher information

$$I(f) = \int_{-\infty}^{\infty} \{f'(x)/f(x)\}^2 \, \mathrm{d}F(x) < \infty, \tag{2.3}$$

where $f'(x) = (\mathrm{d}/\mathrm{d}x)f(x) = (\mathrm{d}^2/\mathrm{d}x^2)F(x)$. Also let

$$\bar{c}_n = n^{-1}\sum_{i=1}^{n} c_i \quad \text{and} \quad Q_n = \sum_{i=1}^{n}(c_i - \bar{c}_n)^2 . \tag{2.4}$$

(ii) There exists Q^* $(0 < Q^* < \infty)$ and $\bar{c}$ $(|\bar{c}| < \infty)$, such that

$$\lim_{n\to\infty} \bar{c}_n = \bar{c} \text{ and } \lim_{n\to\infty} n^{-1}Q_n = Q^* \text{ both exist}, \tag{2.5}$$

and, further, the c_i are all bounded, so that by (2.4)–(2.5)

$$\lim_{n\to\infty}\{Q_n^{-1} \max_{1\le i\le n}(c_i - \bar{c}_n)^2\} = 0 . \tag{2.6}$$

(iii) Let $\varphi = \{\varphi(u), 0 < u < 1\}$ be a nondecreasing, skew-symmetric (i.e., $\varphi(u) + \varphi(1-u) = 0$, $0 < u < 1$) and square-integrable *score function*, $\varphi^* = \{\varphi^*(u) = \varphi((1+u)/2), 0 < u < 1\}$, and for

$$a_n(i) = E\varphi(U_{ni}) \text{ or } \left(\frac{i}{n+1}\right)$$

and

$$a_n^*(i) = E\varphi^*(U_{ni}) \text{ or } \varphi^*\left(\frac{i}{n+1}\right), \tag{2.7}$$

for $i = 1, \ldots, n$, where $0 < U_{n1} < \cdots < U_{nn} < 1$ are the ordered r.v.'s of a sample of size n from the uniform (0, 1) d.f. Note that by definition, $\bar{a}_n = n^{-1}\sum_{i=1}^{n} a_n(i) = \int_0^1 \varphi(u)\,du = 0$, $\forall n \ge 1$. Let then

$$A_n^2 = (n-1)^{-1}\sum_{i=1}^{n}[a_n(i) - \bar{a}_n]^2, \qquad A_\varphi^2 = \int_0^1 \varphi^2(u)\,du , \tag{2.8}$$

$$A_n^{*2} = n^{-1}\sum_{i=1}^{n}(a_n^*(i))^2 \quad \text{and} \quad A_\varphi^{*2} = \int_0^1 (\varphi^*(u))^2\,du . \tag{2.9}$$

Finally, let $\boldsymbol{Y}_n = (Y_1, \ldots, Y_n)'$, $\boldsymbol{1}_n = (1, \ldots, 1)'$ and for every real (a, b), let $\boldsymbol{Y}_n(a, b) = \boldsymbol{Y}_n - a\boldsymbol{1}_n - b\boldsymbol{c}_n$. Let $R_{ni}(a, b)$ $(= R_{ni}(b))$ (or $R_{ni}^+(a, b)$) be the rank of $Y_i - a - bc_i$ (or $|Y_i - a - bc_i|$) among $Y_1 - a - bc_1, \ldots, Y_n - a - bc_n$ (or $|Y_2 - a - bc_1|, \ldots, |Y_n - a - bc_n|$), for $i = 1, \ldots, n$; note that $R_{ni}(a, b)$ does not depend on a. Consider then the statistics

$$S_n(a, b) = n^{-1}\sum_{i=1}^{n} \operatorname{Sgn}(Y_i - a - bc_i)a_n^*(R_{ni}^+(a, b)) , \tag{2.10}$$

$$L_n(b) = n^{-1}\sum_{i=1}^{n}(c_i - \bar{c}_n)a_n(R_{ni}(b)) . \tag{2.11}$$

Note that given $\boldsymbol{Y}_n$, $L_n(b)$ is $\searrow$ in b and, given b and $\boldsymbol{Y}_n$, $S_n(a, b)$ is $\searrow$ in a (see

Theorems 4.5.1 and 5.5.1 of Sen, 1981). Also, if, in (2.1), $\theta = 0 = \beta$, then both $S_n(0,0)$ and $L_n(0)$ are distributed symmetrically about 0. As in Adichie (1967), we consider the following estimators of θ and β:

$$\hat{\theta}_n = \tfrac{1}{2}(\sup\{a: S_n(a,0) > 0\} + \inf\{a: S_n(a,0) < 0\}), \tag{2.12}$$

$$\tilde{\beta}_n = \tfrac{1}{2}(\sup\{b: L_n(b) > 0\} + \inf\{b: L_n(b) < 0\}), \tag{2.13}$$

$$\tilde{\theta}_n = \tfrac{1}{2}(\sup\{a: S_n(a, \tilde{\beta}_n) > 0\} + \inf\{a: S_n(a, \tilde{\beta}_n) < 0\}). \tag{2.14}$$

For various properties of these estimators, see Adichie, Chapter 11; $\hat{\theta}_n$ is prescribed for the constrained model ($\beta = 0$), while $\tilde{\theta}_n$, $\tilde{\beta}_n$ for the general case. For the preliminary test on β, we use $L_n = L_n(0)$. Thus, for the one-sided case ($\beta = 0$) vs. $\beta > 0$), the test function $\xi(L_n)$ is of the form

$$\xi(L_n) = \begin{cases} 1, & (nL_n)/(A_n Q_n^{1/2}) \geq l_{n,\alpha_2}, \\ 0, & \text{otherwise}, \end{cases} \tag{2.15}$$

where α_2 $(0 < \alpha_2 < 1)$ is the level of significance and l_{n,α_2} can be obtained (for small n) by enumeration of the exact null-distribution of L_n (over the $n!$ equally likely permutations of the ranks); for large n, $l_{n,\alpha_2} \to \tau_{\alpha_2}$, where τ_ε is the upper 100ε % point of the standard normal distribution (G), i.e., $G(\tau_\varepsilon) = 1 - \varepsilon$, $0 < \varepsilon < 1$. For the two-sided case, in (2.15), we replace L_n by $|L_n|$ and l_{n,α_2} by l^*_{n,α_2}, where $l^*_{n,\alpha_2} \to \tau_{(1/2)\alpha_2}$ as $n \to \infty$.

The PTE θ^*_n of θ, as in (1.1), is then defined by

$$\theta^*_n = \xi(L_n)\tilde{\theta}_n + \{1 - \xi(L_n)\}\hat{\theta}_n. \tag{2.16}$$

Note that though $\hat{\theta}_n$ is median-unbiased for θ when $\beta = 0$, while $\sqrt{n}(\tilde{\theta}_n - \theta)$ has median converging to 0 as $n \to \infty$, for all β, θ^*_n may not be median-unbiased for θ, even when $\beta = 0$. Naturally, the effect of the preliminary test (on β) on the bias and mean square error of θ^*_n has to be explored.

For the PTT problem, suppose that one wants to test $H_0^{(1)}$: $\theta = 0$ against $\theta >$ (or $\neq$) 0, when β is suspected to be equal to 0. Let $S_n = S_n(0,0)$ and $\hat{S}_n = S_n(0, \tilde{\beta}_n)$. For testing the hypothesis H_0^*: $\theta = 0 = \beta$, the test function $\xi_{12} = \xi(S_n)$ assumes the values 1 and 0 according as $n^{1/2}S_n/A^*_n$ (or $n^{1/2}|S_n|/A^*_n$) is $\geq$ or $> \tau^{(1)}_{n,\alpha_1}$ (or $\tau^{(1)}_{n,(1/2)\alpha_1}$), where the critical value can be obtained by enumerating the null distribution of S_n over the 2^n equally likely sign-inversions, and, for large n, $\tau^{(1)}_{n,\varepsilon} \to \tau_\varepsilon$, $0 < \varepsilon < 1$. On the other hand, when β is treated as a nuisance parameter, the test for $H_0^{(1)}$: $\theta = 0$, is based on the aligned rank statistic $\hat{S}_n$ and is only *asymptotically distribution-free* (ADF); the corresponding test function $\xi_{1\cdot} = \xi(\hat{S}_n)$ takes on the values 1 or 0 according as $n^{1/2}\hat{S}_n/A^*_n(1 + \bar{c}_n^2/Q_n)^{1/2}$ (or its absolute value) is $\geq$ or $< \tau_{\alpha_3}$ (or $\tau_{(1/2)\alpha_3}$). With the aid of the preliminary test function $\xi(L_n)$ in (2.15), we may now define the PTT test function $\xi^*_n = \xi(L_n, S_n, \hat{S}_n)$ as follows:

$$\xi_n^* = \xi(L_n)\xi(\hat{S}_n) + \{1 - \xi(L_n)\}\xi(S_n)\,. \tag{2.17}$$

It is clear from (2.15) and (2.17) that the size and the power of the PTT will not only be dependent of α_1, α_2, α_3 but also on whether $\beta = 0$ or not. The test function $\xi(S_n)$ may not perform that well when $\beta = 0$, while $\xi(\hat{S}_n)$ looses efficiency when $\beta = 0$. Thus, the robustness of ξ_n^* against fluctuations of β around 0 remains to be explored.

We intend to throw light on the performance characteristics of θ_n^* and ξ_n^*, mostly, in the asymptotic theory framework, developed by the current authors (1978, 1982a). Note that $\xi(L_n)$ in (2.15) is a consistent test-function, so that for a fixed β, different from 0, (positive, for the one-sided test), $\xi(L_n)$ will be asymptotically equal to 1, with probability 1, so that θ_n^* will be asymptotically equivalent to θ_n and ξ_n^* to $\xi(\hat{S}_n)$. For $\tilde{\theta}_n$ and $\xi(\hat{S}_n)$, detailed asymptotic theory have been developed (see Adichie, Chapter 11), and hence, the same would be applicable to θ_n^* and ξ_n^*, respectively. The picture is, however, different when $\beta = 0$ (or is very close to 0), so that $\xi(L_n)$ does not converge to 0 or 1, in probability, and θ_n^* (or ξ_n^*) does not share the common properties of either $\hat{\theta}_n$ or $\tilde{\theta}_n$ (or $\xi(\hat{S}_n)$ or $\xi(S_n)$). For this reason and the basic fact that the PTI is appropriate only when β is suspected to be close to 0, we conceive of the following sequence $\{K_n\}$ of local alternatives

$$K_n:\quad \beta = \beta_{(n)} = n^{-1/2}\lambda,\ \ \lambda\,(\text{real})\text{ fixed}\,, \tag{2.18}$$

and study the asymptotic properties of $\{\theta_n^*\}$ along with that of $\{\hat{\theta}_n\}$ and $\{\tilde{\theta}_n\}$ when $\{K_n\}$ holds. Similarly, for the PTT problem, we conceive of $\{K_n^*\}$ where

$$K_n^*:\quad (\theta, \beta) = (\theta_{(n)}, \beta_{(n)}) = n^{-1/2}(\lambda_0, \lambda)\,, \tag{2.19}$$

(λ_0, λ fixed) and study the asymptotic properties of $\{\xi_n^*\}$ along with those of $\{\xi(S_n)\}$ and $\{\xi(\hat{S}_n)\}$, when $\{K_n^*\}$ holds. For this purpose, we define

$$A_\psi^2 = \int_0^1 \psi^2(u)\,du, \qquad \psi(u) = -f'(F^{-1}(u))/f(F^{-1}(u)),\quad 0<u<1\,, \tag{2.20}$$

$$\gamma(\psi, \varphi) = \int_0^1 \varphi(u)\psi(u)\,du\,. \tag{2.21}$$

Note that $A_\psi^2 = I(f) < \infty$. Then, as in Saleh and Sen (1978), we have

$$\mathcal{L}(n^{1/2}(\tilde{\theta}_n - \theta, \tilde{\beta}_n - \beta)) \to \mathcal{N}_2\left(\mathbf{0};\frac{A_\varphi^2}{\gamma^2(\varphi,\psi)}\begin{pmatrix} 1+\bar{c}^2/Q^* & -\bar{c}/Q^* \\ -\bar{c}/Q^* & 1/Q^* \end{pmatrix}\right) \tag{2.22}$$

and when $\beta = 0$,

$$\mathcal{L}(n^{1/2}(\hat{\theta}_n - \theta)) \to \mathcal{N}_1\left(0, \frac{A_\varphi^2}{\gamma^2(\varphi,\psi)}\right). \tag{2.23}$$

Further, under $\{K_n\}$,

$$\mathscr{L}(n^{1/2}((\tilde{\theta}_n - \theta), L_n)) \to \mathcal{N}_2(0, \lambda Q^* \gamma(\varphi, \psi); \boldsymbol{\Sigma}_1), \tag{2.24}$$

$$\mathscr{L}(n^{1/2}((\hat{\theta}_n - \theta), L_n)) \to \mathcal{N}_2(\lambda \bar{c}, \lambda Q^* \gamma(\varphi, \psi); \boldsymbol{\Sigma}_2) \tag{2.25}$$

where

$$\boldsymbol{\Sigma}_1 = A_\varphi^2 \begin{pmatrix} (1 + \bar{c}^2/Q^*)/\gamma^2(\varphi, \psi) & -\bar{c}/\gamma(\varphi, \psi) \\ -\bar{c}/\gamma(\varphi, \psi) & Q^* \end{pmatrix}, \tag{2.26}$$

$$\boldsymbol{\Sigma}_2 = A_\varphi^2 \begin{pmatrix} 1/\gamma^2(\varphi, \psi) & 0 \\ 0 & Q^* \end{pmatrix}. \tag{2.27}$$

From (2.15), (2.16) and (2.24)–(2.27), we have under $\{K_n\}$,

$$P\{n^{1/2}(\theta_n^* - \theta)\gamma(\varphi, \psi)/A_\varphi \leq x \mid K_n\} \to G_1^*(x), \quad x \in E, \tag{2.28}$$

where

$$G_1^*(x) = G(x - \lambda\nu_1)G(\tau_{\alpha_2} - \lambda\nu_2) + \int_{\tau_{\alpha_2} - \lambda\nu_2}^{\infty} G(x + w\nu_1/\nu_2)\,\mathrm{d}G(w), \tag{2.29}$$

with

$$\nu_1 = \bar{c}\gamma(\varphi, \psi)/A_\varphi, \qquad \nu_2 = \sqrt{Q^*}\gamma(\varphi, \psi)/A_\varphi, \qquad \nu_1/\nu_2 = \bar{c}/\sqrt{Q^*} \tag{2.30}$$

and $G(x)$ is the standard normal d.f. For the two-sided preliminary test, the corresponding expression is given by

$$\begin{aligned} G_2^*(x) = {} & G(x - \lambda\nu_1)\{G(\tau_{(1/2)\alpha_2} - \lambda\nu_2) - G(-\tau_{(1/2)\alpha_2} - \lambda\nu_2)\} \\ & + \int_{-\infty}^{-\tau_{(1/2)\alpha_2} - \lambda\nu_2} + \int_{\tau_{(1/2)\alpha_2} - \lambda\nu_2}^{\infty} G(x + w\nu_1/\nu_2)\,\mathrm{d}G(w). \end{aligned} \tag{2.31}$$

Note that both G_1^* and G_2^* depend on α_2, $\bar{c}$, Q^*, λ and $\gamma(\varphi, \psi)$. The first and the second moments computed from G_1^* (or G_2^*) are defined to be the *asymptotic bias* and *asymptotic mean square error* of $\sqrt{n}(\theta_n^* - \theta)$ for the one-sided (or two-sided) PT case. Thus, the asymptotic bias in the one and two-sided test case are respectively

$$\mu_1^* = \bar{c}\{\lambda G(\tau_{\alpha_2} - \lambda\nu_2) - \nu_2^{-1} g(\tau_{\alpha_2} - \lambda\nu_2)\}, \tag{2.32}$$

$$\begin{aligned} \mu_2^* = \bar{c}\{ & \lambda[G(\tau_{(1/2)\alpha_2} - \lambda\nu_2) - G(-\tau_{(1/2)\alpha_2} - \lambda\nu_2)] \\ & - \nu_2^{-1}[g(\tau_{(1/2)\alpha_2} - \lambda\nu_2) - g(-\tau_{(1/2)\alpha_2} - \lambda\nu_2)]\} \end{aligned} \tag{2.33}$$

where $g(y)$ stands for the standard normal pdf. Similarly, the asymptotic mean square errors, in the two cases, are respectively,

$$\begin{aligned} \sigma_1^{*2} = {} & [A_\varphi/\gamma(\varphi, \psi)]^2(1 + \bar{c}^2/Q^*) \\ & + \bar{c}^2\{(\lambda^2 - \nu_2^{-2})G(\tau_{\alpha_2} - \lambda\nu_2) + \nu_2^{-2}(\tau_{\alpha_2} - \lambda\nu_2)g(\tau_{\alpha_2} - \lambda\nu_2)\}, \end{aligned} \tag{2.34}$$

$$\sigma_2^{*2} = [A_\varphi/\gamma(\varphi, \psi)]^2(1 + \bar{c}^2/Q^*)$$
$$+ \bar{c}^2\{(\lambda^2 - \nu_2^{-2})[G(\tau_{(1/2)\alpha_2} - \lambda\nu_2) - G(-\tau_{(1/2)\alpha_2} - \lambda\nu_2)]$$
$$+ \nu_2^{-2}[(\tau_{(1/2)\alpha_2} - \lambda\nu_2)g(\tau_{(1/2)\alpha_2} - \lambda\nu_2) + (\tau_{(1/2)\alpha_2} + \lambda\nu_2)g(\tau_{(1/2)\alpha_2})]\} \tag{2.35}$$

Further, note that by (2.24)–(2.25), the asymptotic bias of $\sqrt{n}(\tilde{\theta}_n - \theta)$, under $\{K_n\}$, is equal to 0, while for $\sqrt{n}(\hat{\theta}_n - \theta)$, it is equal to $\lambda\bar{c}$. We denote these by μ_0 and μ_c, respectively. Similarly, the asymptotic mean square errors for the unconstrained and the constrained estimators are $\sigma_\cdot^2$ and σ_c^2, respectively, where

$$\sigma_\cdot^2 = [A_\varphi/\gamma(\varphi, \psi)]^2(1 + \bar{c}^2/Q^*)\,, \tag{2.36}$$

$$\sigma_c^2 = [A_\varphi/\gamma(\varphi, \psi)]^2 + \lambda^2\bar{c}^2\,. \tag{2.37}$$

Thus, from (2.32) through (2.37), we obtain that under $\{K_n\}$,

$$\bar{c} = 0 \Rightarrow \begin{cases} \mu_1^* = \mu_2^* = \mu_\cdot = \mu_c = 0\,, \\ \sigma_1^* = \sigma_2^* = \sigma_\cdot = \sigma_c = A_\varphi/\gamma(\varphi, \psi)\,, \end{cases} \tag{2.38}$$

so that all the three estimators behave similarly and the preliminary test does not entail any difference in their (asymptotic) properties. Secondly, if $\bar{c} \neq 0$, but $\lambda = 0$ (i.e., H_0: $\beta = 0$ holds), then $\mu_2^* = \mu_\cdot = \mu_c = 0$, but $\mu_1^* = -\bar{c}\nu_2^{-1}g(\tau_{\alpha_2})$ and is $<$ or >0 according as $\bar{c}$ is >0 or not. Thus, for the null hypothesis case, the two-sided PT has an advantage over the one-sided test so far as the asymptotic bias is concerned. As regards their asymptotic mean squares errors, we have the following result due to Saleh and Sen (1978): Under H_0: $\beta = 0$ and for $\bar{c} \neq 0$,

$$0 < \sigma_c < \sigma_1^* < \sigma_2^* < \sigma_\cdot < \infty\,, \tag{2.39}$$

so that θ_n^* performs better than $\tilde{\theta}_n$, but, $\hat{\theta}_n$ is better than θ_n^*; one-sided PTE is better than the two-sided case. The picture is different when $\lambda \neq 0$. For $\bar{c} > 0$, μ_1^* is negative at $\lambda = 0$, is $\nearrow$ in $\lambda \in (-\infty, \lambda_0)$ where $\lambda_0 > 0$, at $\lambda = \lambda_0$, μ_1^* is positive but $<\bar{c}\lambda_0$ and for $\lambda > \lambda_0$, it is $\searrow$ in λ and goes to 0 as $\lambda \to \infty$. μ_2^* is a symmetric and nonnegative function of λ which vanishes at $\lambda = 0$ and as $\lambda \to \pm\infty$, and on $[0, \infty)$ (or $(-\infty, 0]$) it is bell shaped. Thus, the bias $\lambda\bar{c}$ for the constrained estimator dominates the scene. A similar case holds when $\bar{c} < 0$ (see Saleh and Sen, 1978). With respect to the asymptotic mean square errors, there exists an interval J containing 0 as an inner point, such that $\sigma_c < \sigma_1^* < \sigma_2^* < \sigma_\cdot$ for all $\lambda \in J$, while outside this interval, θ_n^* performs better than the others. The picture, of course, depends on α_2, and some discussion on the appropriate choice of α_2 will be made later on.

We proceed on to study the asymptotic properties of the PTT ξ_n^* in (2.17) under $\{K_n^*\}$ in (2.19). First, we consider the case when $\xi(L_n)$, $\xi(S_n)$ and $\xi(\hat{S}_n)$ are all one-sided test functions. Then, by (2.15), and the discussion following (2.16)–(2.17), we conclude that α_n^* = size of the PTT $= E\{\xi_n^* \mid H_0: \theta = 0\}$ is given by

$$\alpha_n^* = P\left\{\frac{nL_n}{A_n\sqrt{a_n}} < l_{n,\alpha_2}, \frac{n^{1/2}S_n}{A_n^*} \geq \tau_{n,\alpha_1}^{(1)} \mid \theta = 0\right\}$$
$$+ P\left\{\frac{nL_n}{A_n\sqrt{a_n}} \geq l_{n,\alpha_2}, \frac{n^{1/2}S_n}{A_n^*} \geq \tau_{n,\alpha_3}^{(1)}\left(1 + \frac{\bar{c}_n^2}{Q_n}\right)^{1/2} \mid \theta = 0\right\} \tag{2.40}$$

which, in general, depends on α_1, α_2, α_3 as well as on β. We would like to study (2.40) when $\{K_n\}$ in (2.18) holds, and similarly, replacing $\{\theta = 0\}$ by $\{\theta = n^{-1/2}\lambda_0\}$ in (2.40), the corresponding probability will give us the power of the PTT. We denote the asymptotic size and power of the PTT by $\alpha^*(\lambda)$ and $\pi^*(\lambda_0, \lambda)$, respectively. Also, for every real (a, b) and ρ: $-1 < \rho < 1$, let $h(a, b; \rho)$ be the quadrant probability $P\{z_1 \geq a, z_2 \geq b\}$, where

$$\boldsymbol{Z} = (z_1, z_2) \sim \mathcal{N}\left(\boldsymbol{0}, \begin{pmatrix} 1 & \rho \\ \rho & 1 \end{pmatrix}\right).$$

Then (cf. Saleh and Sen, 1982a), under $\{K_n^*\}$ in (2.19),

$$\pi^*(\lambda_0, \lambda) = G(\tau_{\alpha_2} - \lambda\nu_1)[1 - G(\tau_{\alpha_1} - (\lambda_0 - \bar{c}\lambda)\gamma(\varphi, \psi)/A_\varphi)]$$
$$+ h(\tau_{\alpha_2} - \lambda\nu_2, \tau_{\alpha_3} - \lambda_0\gamma(\varphi, \psi)/A_\varphi(1 + \bar{c}^2/Q^*)^{1/2}; -\bar{c}/(Q^* + \bar{c}^2)^{1/2}). \tag{2.41}$$

The asymptotic powers $\pi_c(\lambda_0, \lambda)$ and $\pi_.(\lambda_0, \lambda)$ of the constrained and the unconstrained tests are respectively

$$\pi_c(\lambda_0, \lambda) = 1 - G(\tau_{\alpha_1} - (\lambda_0 + \bar{c}\lambda)\gamma(\varphi, \psi)/A_\varphi), \tag{2.42}$$

$$\pi_.(\lambda_0, \lambda) = 1 - G(\tau_{\alpha_3} - \lambda_0\gamma(\varphi, \psi)/A_\varphi(1 + \bar{c}^2/Q^*)^{1/2}). \tag{2.43}$$

Note that $\pi_.(0, \lambda) = 1 - G(\tau_{\alpha_3}) = \alpha_3$, $\forall\lambda$, while $\pi^*(0, \lambda)$ and $\pi_c(0, \lambda)$ may both depend on λ. In particular, if we let $\alpha_1 = \alpha_3 = \alpha$ $(0 < \alpha < 1)$, then for $\bar{c} = 0$,

$$\pi^*(0, \lambda) = \pi_c(0, \lambda) = \pi_.(0, \lambda) = \alpha$$

and

$$\pi^*(\lambda_0, \lambda) = \pi_c(\lambda_0, \lambda) = \pi_.(\lambda_0, \lambda) = 1 - G(\tau_\alpha - \lambda\gamma(\varphi, \psi)/A_\varphi),$$

so that all the three tests have the asymptotic size α and are power equivalent under $\{K_n^*\}$, whatever be α_2 $(\in (0, 1))$. The picture is different when $\bar{c} \neq 0$. First, $\pi_c(0, \lambda)$ is $>$ or $<\alpha_1$ according as $\lambda\bar{c}$ is $>$ or <0, while $\pi_.(0, \lambda) = \alpha_3$ $(=\alpha_1)$, $\forall\lambda$. Secondly, for $\lambda\bar{c} \gtrless 0$, $\pi_c(0, \lambda) \gtrless \pi^*(0, \lambda)$, where $\pi^*(0, \lambda)$ is closer to α. Thus, as regards the asymptotic size of the tests is concerned, ξ_n^* is more robust than $\xi(S_n)$, though $\xi(\hat{S}_n)$ remains a better choice. On the other hand, for $\lambda_0 \neq 0$, the asymptotic power of the constrained test $\xi(S_n)$ is greater than that of the unconstrained test $\xi(\hat{S}_n)$ whenever

$$(\lambda_0 + \bar{c}\lambda) \geq (1 + \bar{c}^2/Q^*)^{-1/2}\lambda_0\,, \tag{2.44}$$

where for $\bar{c} \neq 0$, the right hand side is smaller than λ_0 whenever $\lambda_0 > 0$, so that unless $\bar{c}\lambda$ is largely negative, (2.44) holds. A similar case holds for the two-sided tests. In any event, the asymptotic power of the PTT lies in between that of the constrained and the unconstrained tests, where the latter does not depend on λ while the former is grossly affected by λ. The PTT is not so sensitive to extreme fluctuations of λ. This illustrates the efficiency-robustness of the PTT. The choice of α_2 (for the PT) in the context of PTI will be discussed later on.

3. NPTI for the multivariate simple regression model

We consider here the multivariate generalization of the simple regression model in (2.1); here the Y_i $\{=(Y_{i1}, \ldots, Y_{ip})'\}$ are all p-vectors for some $p \geq 1$, F is a p-variate d.f., $\boldsymbol{\theta} = (\theta_1, \ldots, \theta_p)'$ and $\boldsymbol{\beta} = (\beta_1, \ldots, \beta_p)'$ are unknown parameters (vectors) and the c_i are known scalar quantities. Again, the multivariate two-sample model is a particular case where the c_i assume the values 0 and 1 only. We are mainly concerned with the location parameter $\boldsymbol{\theta}$ (i.e., estimating $\boldsymbol{\theta}$ or testing that $\boldsymbol{\theta} = \mathbf{0}$, or some specified vector), when it is suspected but not evident that $\boldsymbol{\beta} = 0$ (or some other specified value). As regards the estimation of $\boldsymbol{\theta}$ or $\boldsymbol{\beta}$ is concerned, one may virtually repeat the procedures in Section 2 for each of the p coordinates separately. However, the PT for $\boldsymbol{\beta} = \mathbf{0}$ will be somewhat different. We summarize these as follows.

(i) The basic regularity conditions on the c_i are the same as in (2.4) through (2.6).

(ii) Here F is a p-variate d.f., and hence, we replace (2.3) by that of a finite and positive definite (p.d.) Fisher-information matrix

$$\mathcal{J} = \mathcal{J}(f) = \int_{E^p} \cdots \int [(\mathrm{d}/\mathrm{d}\boldsymbol{x})f(\boldsymbol{x})][(\mathrm{d}/\mathrm{d}\boldsymbol{x})f(\boldsymbol{x})]'\,\mathrm{d}F(\boldsymbol{x})\,, \tag{3.1}$$

where $f(\boldsymbol{x}) = (\mathrm{d}/\mathrm{d}\boldsymbol{x})F(\boldsymbol{x})$.

(iii) For each coordinate j $(=1, \ldots, p)$, we define the score functions $\varphi_j = \{\varphi_j(u), 0 < u < 1\}$ and $\varphi_j^* = \{\varphi_j^*(u), 0 < u < 1\}$ as in before (2.7) and the scores $a_{nj}(i)$, $a_{nj}^*(2)$, $1 \leq i \leq n$ as in (2.7). The φ_j for different j need not be the same.

(iv) For each j $(=1, \ldots, p)$, considering the Y_{ij}, $1 \leq i \leq n$, the ranks $R_{nij}(b)$ and $R_{nij}^+(a, b)$, $1 \leq i \leq n$ are defined as in after (2.9). Also we let $R_{nij}(0) = R_{nij}$, $1 \leq i \leq n$, for $j = 1, \ldots, p$, and denote the $p \times n$ matrix $((R_{nij}))$ by $\boldsymbol{R}_n$, the *rank-collection matrix.*

(v) Finally, for each j $(=1, \ldots, p)$, replacing the Y_i by Y_{ij}, a_n (or a_n^*) by a_{nj} (or a_{nj}^*) and the $R_{ni}(b)$ (or $R_{ni}^+(a, B)$) by $R_{nij}(b)$ or $R_{nij}^+(a, b)$), as in (2.10)–(2.11), we define $S_{nj}(a, b)$ and $L_{nj}(b)$. Let then $L_{nj}(0) = L_{nj}$, $1 \leq j \leq p$ and $\boldsymbol{L}_n = (L_{n1}, \ldots, L_{np})'$.

To formulate the PT for H_0: $\boldsymbol{\beta} = \boldsymbol{0}$ against H_1: $\boldsymbol{\beta} \neq \boldsymbol{0}$, we define $\boldsymbol{M}_n = ((m_{jl}^{(n)}))$ by letting for every $1 \leq j$, $l \leq p$,

$$m_{jl}^{(n)} = n^{-1} \sum_{i=1}^{n} \{a_{nj}(R_{nij}) - \bar{a}_{nj}\}\{a_{nl}(R_{nil}) - \bar{a}_{nl}\} \tag{3.2}$$

where $\bar{a}_{nj} = n^{-1} \sum_{i=1}^{n} a_{nj}(i)$, $1 \leq j \leq p$ and let $\boldsymbol{M}_n^-$ be a reflexible generalized inverse of $\boldsymbol{M}_n$. The following Puri and Sen (1969), we use the following test statistic

$$\mathscr{L}_n = nQ_n^{-1}\{\boldsymbol{L}_n'\boldsymbol{M}_n^-\boldsymbol{L}_n\}, \tag{3.3}$$

and the test-function $\xi(\mathscr{L}_n)$ is of the form

$$\xi(\mathscr{L}_n) = \begin{cases} 1 & \text{if } \mathscr{L}_n \geq \mathscr{L}_{n,\alpha_2}, \\ 0 & \text{otherwise}, \end{cases} \tag{3.4}$$

where α_2 $(0 < \alpha_2 < 1)$ is the level of significance of the PT. For small n, the conditional (null) distribution of $\mathscr{L}_n$ can be derived from the $n!$ equally likely column permutations of $\boldsymbol{R}_n$, so that $\mathscr{L}_{n,\alpha_2}$ can be obtained by direct enumeration, while for large n,

$$\mathscr{L}_{n,\alpha_2} \to \chi_{p,\alpha_2}^2 \quad \text{where } P\{\chi_p^2 \geq \chi_{p,\alpha_2}^2\} = \alpha_2 \tag{3.5}$$

and χ_p^2 has the central chi square d.f. with p degrees of freedom (D.F.).

For each j $(=1, \ldots, p)$, we proceed as in (2.12)–(2.14) and denote the derived rank estimates by $\hat{\theta}_{jn}$, $\tilde{\beta}_{jn}$ and $\tilde{\theta}_{jn}$, respectively. Let $\hat{\boldsymbol{\theta}}_n = (\hat{\theta}_{1n}, \ldots, \hat{\theta}_{pn})'$, $\tilde{\boldsymbol{\beta}}_n = (\tilde{\boldsymbol{\beta}}_{1n}, \ldots, \tilde{\boldsymbol{\beta}}_{pn})'$, $\tilde{\boldsymbol{\theta}}_n = (\tilde{\boldsymbol{\theta}}_{1n}, \ldots, \tilde{\boldsymbol{\theta}}_{pn})'$. Then parallel to (2.16), the PTE $\boldsymbol{\theta}_n^*$ of $\boldsymbol{\theta}$ is defined by

$$\boldsymbol{\theta}_n^* = (\theta_{1n}^*, \ldots, \theta_{pn}^*)' = \xi(\mathscr{L}_n)\hat{\boldsymbol{\theta}} + \{1 - \xi(\mathscr{L}_n)\}\tilde{\boldsymbol{\theta}}_n. \tag{3.6}$$

When F is diagonally symmetric about $\boldsymbol{0}$, for $\boldsymbol{\beta} = \boldsymbol{0}$, $\hat{\boldsymbol{\theta}}_n$ has also a diagonally symmetric d.f. around θ, while for $\boldsymbol{\beta}$ not necessarily $\boldsymbol{0}$, $n^{1/2}(\tilde{\boldsymbol{\theta}}_n - \boldsymbol{\theta})$ has coordinate wise medians all converging to 0 as $n \to \infty$. However, as in the univariate case, $\boldsymbol{\theta}_n^*$ may not have this (asymptotic) median unbiasedness, even when $\boldsymbol{\beta} = \boldsymbol{0}$. Hence, the effect of the PT on the asymptotic d.f. of $n^{1/2}(\boldsymbol{\theta}_n^* - \boldsymbol{\theta})$ remains to be studied.

Side by side, we consider the PTT problem. We are primarily concerned with tests for $\boldsymbol{\theta}$ when $\boldsymbol{\beta}$ is suspected to be close to $\boldsymbol{0}$ (or some specified vector). If one assumes that $\boldsymbol{\beta} = \boldsymbol{0}$, then a nonparametric test for $H_0^{(1)}$: $\boldsymbol{\theta} = \boldsymbol{0}$ against $\boldsymbol{\theta} \neq \boldsymbol{0}$, is based on the statistic $\boldsymbol{S}_n = (S_{n1}(\boldsymbol{0}), \ldots, S_{np}(\boldsymbol{0}))'$. We define the $R^+_{nij}(a, b)$ as in after (3.1), and let $R^+_{nij} = R^+_{nij}(\boldsymbol{0})$, for $1 \leq j \leq p$, $1 \leq i \leq n$. Consider the matrix $\boldsymbol{M}^*_n = ((m^{*(n)}_{jl}))$ with elements given by

$$m^{*(n)}_{jl} = \frac{1}{n}\sum_{i=1}^{n} a^*_{nj}(R^+_{nij})a^*_{nl}(R^+_{nil})\,\mathrm{sgn}(Y_{ij})\mathrm{sgn}(y_{il})\,, \tag{3.7}$$

for $j, l = 1, \ldots, p$. Then, as in Sen and Puri (1967), the test statistic may be formulated as

$$\bar{\mathscr{L}}_n = n\boldsymbol{S}'_n(\boldsymbol{M}^*_n)^- \boldsymbol{S}_n\,. \tag{3.8}$$

On the other hand, when $\boldsymbol{\beta}$ is treated as a nuisance parameter, we estimate $\tilde{\beta}_n$ as in before (3.1), let $\hat{\boldsymbol{S}}_n = (S_{n1}(0, \tilde{\beta}_{1n}), \ldots, S_{np}(0, \tilde{\beta}_{pn}))'$, $\hat{R}^+_{nij} = R^+_{nij}(0, \tilde{\beta}_{jn})$, $1 \leq j \leq p$, $1 \leq i \leq n$, and in (3.7), replacing the R^+_{nij} by $\hat{R}^+_{nij}$, the resulting quantity is defined as $\hat{m}^{*(n)}_{jl}$, for $j, l = 1, \ldots, p$; $\hat{\boldsymbol{M}}^*_n = ((\hat{m}^{*(n)}_{jl}))$. The allied test statistic is

$$\hat{\mathscr{L}}_n = n\hat{\boldsymbol{S}}'_n(\hat{\boldsymbol{M}}^*_n)^- \hat{\boldsymbol{S}}_n\,. \tag{3.9}$$

Finally, the test for H_0: $\boldsymbol{\beta} = \boldsymbol{0}$ vs. $\boldsymbol{\beta} \neq \boldsymbol{0}$ is based on $\mathscr{L}_n$ in (3.3).

Now, under $H^{(1)}_0$, $\bar{\mathscr{L}}_n$ is conditionally (given the reduced rank collection matrix) distribution-free, whose null distribution is generated by the 2^n (conditionally) equally likely sign-inversions of the vectors of the signs of $\boldsymbol{Y}_i$. For large n, $\bar{\mathscr{L}}_n$ has asymptotically chi square distribution with p D.F. On the other hand, if $\boldsymbol{\beta} \neq \boldsymbol{0}$, $\bar{\mathscr{L}}_n$ is neither conditionally distribution-free, nor it has asymptotically chi square distribution with p D.F. when $\boldsymbol{\theta} = \boldsymbol{0}$. $\hat{\mathscr{L}}_n$ is asymptotically distribution-free (when $\boldsymbol{\theta} = \boldsymbol{0}$), irrespective of $\boldsymbol{\beta} = \boldsymbol{0}$ or not. We define $\xi(\mathscr{L}_n)$ as in (3.4). Also, let

$$\xi(\bar{\mathscr{L}}_n) = \begin{cases} 1 & \text{if } \bar{\mathscr{L}}_n \geq \bar{\mathscr{L}}_{n,\alpha_1}\,, \\ 0 & \text{otherwise}\,, \end{cases} \tag{3.10}$$

and

$$\xi(\hat{\mathscr{L}}_n) = \begin{cases} 1 & \text{if } \hat{\mathscr{L}}_n \geq \hat{\mathscr{L}}_{n,\alpha_3}\,, \\ 0 & \text{otherwise}\,, \end{cases} \tag{3.11}$$

where, as in (3.5),

$$\bar{\mathscr{L}}_{n,\alpha_1} \to \chi^2_{p,\alpha_1} \quad \text{and} \quad \hat{\mathscr{L}}_{n,\alpha_3} \to \chi^2_{p,\alpha_3}\,. \tag{3.12}$$

We are now in a position to formulate the PTT-test function $\xi^*_n = \xi(\mathscr{L}_n, \bar{\mathscr{L}}_n, \hat{\mathscr{L}}_n)$ as follows:

$$\xi^*_n = \xi(\mathscr{L}_n)\xi(\hat{\mathscr{L}}_n) + \{1 - \xi(\mathscr{L}_n)\}\xi(\bar{\mathscr{L}}_n)\,. \tag{3.13}$$

It is clear from (3.4), (3.5) and (3.10) through (3.13) that the size and the power of the PTT will not only depend on α_1, α_2, α_3 but also on whether $\boldsymbol{\beta} = \mathbf{0}$ or not. Thus, the robustness of ξ_n^* (relative to $\xi(\bar{\mathscr{L}}_n)$ and $\xi(\hat{\mathscr{L}}_n)$) remains to be explored.

As in Section 2, we intend to study the performance characteristics of the PTE $\boldsymbol{\theta}_n^*$ and the PTT ξ_n^* in the asymptotic theory framework, developed by the authors (1979, 1983a). For the same reason, as in Section 2, we consider some local alternatives $\{K_n\}$ and $\{K_n^*\}$ (as in (2.18) and (2.19)), where we replace the scalar λ by $\boldsymbol{\lambda} = (\lambda_1, \ldots, \lambda_p)'$ and λ_0 by $\boldsymbol{\lambda}_0 = (\lambda_{10}, \ldots, \lambda_{p0})'$. Corresponding to the p variate d.f. F, the univariate and bivariate marginal d.f.'s are denoted by $F_{[j]}$ and $F_{[jl]}$, $1 \leq j \neq l \leq p$, and let φ_j, $1 \leq j \leq p$ be defined as in after (3.1). Also, in (2.20)–(2.21), we replace f and F by $f_{[j]}$ $(=F'_{[j]})$ and $F_{[j]}$, respectively, and denote the resulting quantities by ψ_j and $\gamma_j = \gamma(\varphi_j, \psi_j)$, respectively. Let $\boldsymbol{T} = ((\tau_{jl}))$ be defined by

$$\tau_{jl} = \left\{\int_{-\infty}^{\infty}\int_{-\infty}^{\infty} \varphi_j(F_{[j]}(x))\varphi_l(F_{[l]}(x, y))\right\} \Big/ \{\gamma_j\gamma_l\}\,, \tag{3.14}$$

for $j, l = 1, \ldots, p$. Then, under $\{K_n\}$ and the assumed regularity conditions, as $n \to \infty$

$$n^{1/2}(\hat{\boldsymbol{\theta}}_n - \boldsymbol{\theta}) \xrightarrow{\mathscr{D}} \mathscr{N}_p(\bar{c}\boldsymbol{\lambda}, \boldsymbol{T})\,, \tag{3.15}$$

$$n^{1/2}(\tilde{\boldsymbol{\theta}}_n - \boldsymbol{\theta}) \xrightarrow{\mathscr{D}} \mathscr{N}_p(\mathbf{0}, (1 + \bar{c}^2/Q^*)\boldsymbol{T})\,, \tag{3.16}$$

and

$$\begin{aligned} G_p^*(\boldsymbol{x}; \boldsymbol{\lambda}) &= \lim_{n\to\infty} P_{K_n}\{n^{1/2}(\boldsymbol{\theta}_n^* - \boldsymbol{\theta}) \leq \boldsymbol{x}\} \\ &= G_p(\boldsymbol{x} - \bar{c}\boldsymbol{\lambda}; \mathbf{0}, \boldsymbol{T})H_p(\chi^2_{p,\alpha_2}; \Delta) \\ &\quad + \int_{E(\boldsymbol{\lambda})} G_p(\boldsymbol{x} + (\bar{t}/\sqrt{Q^*})\boldsymbol{Z}; \mathbf{0}, \boldsymbol{T})\, \mathrm{d}G_p(\boldsymbol{Z}; \mathbf{0}, \boldsymbol{T}) \quad \forall \boldsymbol{x} \in E^p\,, \end{aligned} \tag{3.17}$$

where $G_p(\boldsymbol{x}; \boldsymbol{\mu}, \boldsymbol{\Sigma})$ is the p variate normal d.f. with mean vector $\boldsymbol{\mu}$ and dispersion matrix $\boldsymbol{\Sigma}$, $H_p(x; \delta)$ is the noncentral chi square d.f. with p D.F. and noncentrality parameter δ,

$$E(\boldsymbol{\lambda}) = \{\boldsymbol{y} \in E^p : (\boldsymbol{y} + \sqrt{Q^*}\boldsymbol{\lambda})'\boldsymbol{T}^{-1}(\boldsymbol{y} + \sqrt{Q^*}\boldsymbol{\lambda}) > \chi^2_{p,\alpha_2}\}$$

and

$$\Delta = Q^*(\boldsymbol{\lambda}'\boldsymbol{T}^{-1}\boldsymbol{\lambda})\,.$$

For later use, the p.d.f. corresponding to $G_p(\cdot)$ and $G_p^*(\cdot)$ are denoted by $g_p(\cdot)$ and $g_p^*(\cdot)$, respectively. Then, the first moment (vectors) of the d.f.'s in (3.15)–(3.17) are employed to study the asymptotic bias of these estimators. Thus, $\xi_1(\boldsymbol{\lambda})$, the asymptotic bias of $n^{1/2}(\hat{\boldsymbol{\theta}}_n - \boldsymbol{\theta})$ (under $\{K_n\}$) is $\bar{c}\lambda$, $\xi_2(\boldsymbol{\lambda})$, that of $n^{1/2}(\tilde{\boldsymbol{\theta}}_n - \boldsymbol{\theta})$ is $\mathbf{0}$ and $\xi^*(\boldsymbol{\lambda})$, the asymptotic bias of $n^{1/2}(\boldsymbol{\theta}_n^* - \boldsymbol{\theta})$ is given by $H_{p+2}(\chi^2_{p,\alpha_2}; \Delta)\bar{c}\boldsymbol{\lambda}$. Note that $H_{p+2}(\chi^2_{p,\alpha_2}; \Delta)$ is bounded by 1 and it converges to 0 as $\Delta \to \infty$. Hence, compared to $\hat{\boldsymbol{\theta}}_n$, $\boldsymbol{\theta}_n^*$ is more robust as regards the asymptotic bias is concerned, while $\tilde{\theta}_n$ performs better in this respect. The asymptotic dispersion matrix of $n^{1/2}(\boldsymbol{\theta}_n^* - \boldsymbol{\theta})$ is equal to

$$T\left\{1+\frac{\bar{c}^2}{Q^*}(1-H_{p+2}(\chi^2_{p,\alpha};\Delta))+\bar{c}^2T^{-1}\lambda\lambda'(2H_{p+2}(\chi^2_{p,\alpha};\Delta)-H_{p+4}(\chi^2_{p,\alpha};\Delta))\right\}. \tag{3.18}$$

If we denote the dispersion matrices in (3.15), (3.16) and (3.18) by $\boldsymbol{\Gamma}^*(\boldsymbol{\lambda})$, $\boldsymbol{\Gamma}_2(\boldsymbol{\lambda})$ and $\boldsymbol{\Gamma}^*(\boldsymbol{\lambda})$, respectively, then we have for $\boldsymbol{\lambda}=0$, when $\bar{c}\neq 0$, $\boldsymbol{\Gamma}_2(\mathbf{0})-\boldsymbol{\Gamma}_1(\mathbf{0})$, $\boldsymbol{\Gamma}_2(\mathbf{0})-\boldsymbol{\Gamma}^*(\mathbf{0})$ and $\boldsymbol{\Gamma}^*(\mathbf{0})-\boldsymbol{\Gamma}_1(\mathbf{0})$ are positive definite, when $\boldsymbol{T}$ is so. From this point of view, employing the 'generalized variance' as the inverse of the efficiency of the estimator, we conclude that under H_0: $\boldsymbol{\lambda}=\mathbf{0}$, $\boldsymbol{\theta}_n^*$ performs better than $\tilde{\boldsymbol{\theta}}_n$, while $\hat{\boldsymbol{\theta}}_n$ performs better than $\boldsymbol{\theta}_n^*$. The picture is thus opposite to that in the case of asymptotic bias. For $\boldsymbol{\lambda}\neq\mathbf{0}$, $\boldsymbol{\Gamma}_1(\boldsymbol{\lambda})=\boldsymbol{T}+\bar{c}^2\boldsymbol{\lambda}\boldsymbol{\lambda}'$, where $\boldsymbol{\lambda}\boldsymbol{\lambda}'$ is the rank of 1, so that Trace$(\boldsymbol{T}^{-1}\boldsymbol{\lambda}\boldsymbol{\lambda}')=$ largest root of $(\boldsymbol{T}^{-1}\boldsymbol{\lambda}\boldsymbol{\lambda}')=\Delta=\boldsymbol{\lambda}'\boldsymbol{T}^{-1}\boldsymbol{\lambda}>0$. Hence $\hat{\boldsymbol{\theta}}_n$ performs (asymptotically) better or worse than $\tilde{\boldsymbol{\theta}}_n$ according as $\boldsymbol{\lambda}'\boldsymbol{T}^{-1}\boldsymbol{\lambda}$ is $>$ or $<[(1+\bar{c}^2/Q^*)^p-1]/(\bar{c}^2/Q^*)$, so that for $\boldsymbol{\lambda}$ away from $\mathbf{0}$, $\hat{\boldsymbol{\theta}}_n$ may not perform at all well. On the other hand, for $\boldsymbol{\lambda}$ in the neighbourhood of the origin, $\boldsymbol{\theta}_n^*$ performs better than $\tilde{\boldsymbol{\theta}}_n$ and it performs even better than $\hat{\boldsymbol{\theta}}_n$ as $\boldsymbol{\lambda}$ moves away from the origin. Thus, for $\boldsymbol{\lambda}$ away from $\mathbf{0}$, $\boldsymbol{\theta}_n^*$ is preferred to both $\tilde{\boldsymbol{\theta}}_n$ and $\hat{\boldsymbol{\theta}}_n$, while for $\boldsymbol{\lambda}$ close to $\mathbf{0}$, it is a good compromise. This explains the robustness property of the PTE. For further details, we may refer to Sen and Saleh (1979).

Let us now consider the case of the PTT in (3.13); this has been studied in detail in Saleh and Sen (1983a). Like the univariate case, we consider here a sequence $\{K_n^*\}$ of local alternatives, where under K_n^*, $(\boldsymbol{\theta},\boldsymbol{\beta})=n^{-1/2}(\boldsymbol{\lambda}_0,\boldsymbol{\lambda})$. Then, by (3.9) and (3.11), $E\{\xi(\mathscr{L}_n)\mid\boldsymbol{\lambda}_0=\mathbf{0}\}\to\alpha_3$, irrespective of $\boldsymbol{\lambda}$, while $E\{\xi(\bar{\mathscr{L}}_n)\mid\boldsymbol{\lambda}_0=\mathbf{0}\}\to P\{\chi_p^2(\bar{\Delta})>\chi^2_{p,\alpha_1}\}$ $(\geqslant\alpha_1)$, where $\bar{\Delta}=(\boldsymbol{\lambda}_0+\bar{c}\boldsymbol{\lambda})'\boldsymbol{T}^{-1}(\boldsymbol{\lambda}_0+\bar{c}\boldsymbol{\lambda})$, so that for $\boldsymbol{\lambda}_0=\mathbf{0}$, $\bar{\Delta}=\bar{c}^2\boldsymbol{\lambda}'\boldsymbol{T}^{-1}\boldsymbol{\lambda}=\bar{c}^2\Delta$ $(\geqslant 0)$. Thus, the asymptotic size of $\xi(\bar{\mathscr{L}}_n)$ is $\geqslant\alpha_1$, where the equality sign holds when $\boldsymbol{\lambda}=\mathbf{0}$, or, in other words, $\xi(\bar{\mathscr{L}}_n)$ is not robust against $\boldsymbol{\lambda}\neq\mathbf{0}$, and hence, the actual significance level of $\xi(\bar{\mathscr{L}}_n)$ may be higher than α_1 when $\boldsymbol{\lambda}\neq\mathbf{0}$. In fact, this may be quite different from α_1, depending on Δ. Further, $E\{\xi_n^*\mid\boldsymbol{\lambda}_0=\boldsymbol{\lambda}=\mathbf{0}\}\to\alpha_1(1-\alpha_2)+P\{\boldsymbol{Z}\geqslant(\chi^2_{p,\alpha_3},\chi^2_{p,\alpha_2})\}$, where $\boldsymbol{Z}$ has the same distribution as of two correlated chi-square variables with the same D.F. p. Details of this are given in Saleh and Sen (1983a). On the other hand, if $\boldsymbol{\lambda}\neq 0$, we have

$$E\{\xi_n^*\mid\boldsymbol{\lambda}_0=\mathbf{0}\}\to H_p(\chi^2_{p,\alpha_2};\Delta)[1-H_p(\chi^2_{p,\alpha_1};\bar{\Delta})]+P\{\boldsymbol{Z}(\Delta)\geqslant(\chi^2_{p,\alpha_3},\chi^2_{p,\alpha_2})\},$$

where $\boldsymbol{Z}(\Delta)$ has a correlated noncentral chi square (bivariate) distribution. It follows that for $\boldsymbol{\lambda}_0=\boldsymbol{\lambda}=\mathbf{0}$, the asymptotic size of ξ_n^* is bounded from above by $\alpha_1(1-\alpha_2)+\alpha_3\wedge\alpha_2=\alpha_1+\alpha_3\wedge\alpha_2-\alpha_1\alpha_2$, while for $\boldsymbol{\lambda}\neq 0$, the asymptotic size depends on $\boldsymbol{\lambda}$, but, unlike the case of $\xi(\bar{\mathscr{L}}_n)$, it is bounded by

$$H_p(\chi^2_{p,\alpha_2};\Delta)[1-H_p(\chi^2_{p,\alpha_1};\bar{\Delta})]+\alpha_3\wedge[1-H_p(\chi^2_{p,\alpha_2};\Delta)]$$

and does not go to 1 as $\Delta \to \infty$ or $\bar{\Delta} \to \infty$. Or, in other words, it is more robust against $\boldsymbol{\lambda} \neq \mathbf{0}$ than $\xi(\bar{\mathscr{L}}_n)$. Hence, from consideration of validity-robustness, $\xi(\mathscr{L}_n^*)$ may be preferred to $\xi(\bar{\mathscr{L}}_n)$. As for the asymptotic power, for $\xi(\hat{\mathscr{L}}_n)$, we have $1 - H_p(\chi^2_{p,\alpha_3}; \Delta^0)$, where $\Delta^0 = (1 + \bar{c}^2/Q^*)^{-1}\boldsymbol{\lambda}_0' \boldsymbol{T}^{-1}\boldsymbol{\lambda}_0$, and this does not depend on $\boldsymbol{\lambda}$. On the other hand, for $\bar{\mathscr{L}}_n$, we have the asymptotic power $1 - H_p(\chi^2_{p,\alpha_1}; \bar{\Delta})$ where $\bar{\Delta} = (\boldsymbol{\lambda}_0 + \bar{c}\boldsymbol{\lambda})' \boldsymbol{T}^{-1}(\boldsymbol{\lambda}_0 + \bar{c}\boldsymbol{\lambda})$. Thus, $\xi(\bar{\mathscr{L}}_n)$ may be more powerful than $\xi(\hat{\mathscr{L}}_n)$ (where we let $\alpha_1 = \alpha_3 = \alpha$) when $\bar{\Delta} \geq \Delta^0$, and this depends on $\boldsymbol{\lambda}_0$ and $\boldsymbol{\lambda}$ both (as well as $\bar{c}$). Thus, $\xi(\bar{\mathscr{L}}_n)$ may not be efficiency-robust. Thus, $\xi(\mathscr{L}_n^*)$ emerges as an over all robust competitor of the others.

4. NPTI for general multivariate linear models

Consider now the general multivariate linear model where $\boldsymbol{Y} = (\boldsymbol{Y}_1, \ldots, \boldsymbol{Y}_n)$ follows the model

$$\boldsymbol{Y} = \boldsymbol{\beta}\boldsymbol{C}_n + \boldsymbol{e}_n; \quad \boldsymbol{e} = (\boldsymbol{e}_1, \ldots, \boldsymbol{e}_n), \tag{4.1}$$

$\boldsymbol{\beta}$ is a $p \times q$ matrix of unknown parameters, $\boldsymbol{C}_n$ is a $q \times n$ matrix of known constants and the columns of $\boldsymbol{e}$ are distributed independently according to a common d.f. F. Consider the partitioning of $\boldsymbol{\beta}$ as $\boldsymbol{\beta} = (\boldsymbol{\beta}_1, \boldsymbol{\beta}_2)$, where $\boldsymbol{\beta}_j$ is $p \times q_j$, $j = 1, 2$, $q_1 + q_2 = q$, $q_j \geq 0$, $j = 1, 2$. We are primarily interested in estimating $\boldsymbol{\beta}_1$ or in testing a linear hypothesis on $\boldsymbol{\beta}_1$, when it is suspected that $\boldsymbol{\beta}_2$ may be 'close to' $\mathbf{0}$. This is a canonical reduction of the nonparametric preliminary test inference problem in the multivariate case.

We write $\boldsymbol{C}_n = (\boldsymbol{c}_1, \ldots, \boldsymbol{c}_n)$, $\boldsymbol{D}_n = \boldsymbol{C}_n\boldsymbol{C}_n'$ and partition the $\boldsymbol{c}_i$ and $\boldsymbol{D}_n$ in accordance with that of $\boldsymbol{\beta}$, i.e., $\boldsymbol{c}_2' = (\boldsymbol{c}_i^{(1)\prime}, \boldsymbol{c}_i^{(2)\prime})'$, $i \geq 1$ and

$$\boldsymbol{D}_n = ((\boldsymbol{D}_{nij})) = \begin{pmatrix} \boldsymbol{C}_n^{(1)} \\ \boldsymbol{C}_n^{(2)} \end{pmatrix} (\boldsymbol{C}_n^{(1)\prime}, \boldsymbol{C}_n^{(2)\prime}).$$

We assume that

(i)

$$\max_{1 \leq i \leq n} \boldsymbol{c}_i' \boldsymbol{D}_n^{-1} \boldsymbol{c}_i \to 0, \text{ as } n \to \infty \tag{4.2}$$

(ii) $\boldsymbol{D}_n$ is of full rank for every $n \geq n_0$ $(\geq q)$ and

$$\lim_{n \to \infty} n^{-1}\boldsymbol{D}_n = \boldsymbol{\Delta} = ((\boldsymbol{\Delta}_{ij}))_{i,j=1,2} \text{ exists}. \tag{4.3}$$

The d.f. F is assumed to satisfy all the regularity conditions in Section 3. Also, for every j $(1 \leq j \leq p)$, we define the score functions φ_j, φ_j^* and the scores $a_{nj}(i)$, $a_{nj}^*(i)$, $1 \leq i \leq n$, $j = 1, \ldots, p$ as in Section 3. Further, replacing $\boldsymbol{b}\boldsymbol{c}_i$ by $\boldsymbol{B}\boldsymbol{c}_i$, for each j, we define the $\boldsymbol{L}_{nj}(B)$ as in Section 3. Note that since the $\boldsymbol{c}_i$ are q-vectors, the $\boldsymbol{L}_{nj}(\boldsymbol{B})$ are q-vectors and are functions of the $p \times q$ matrix B. For each j,

$R_{nij}(\boldsymbol{B})$ is defined as in Section 3 with $\boldsymbol{Y}_i(\boldsymbol{B}) = \boldsymbol{Y}_i - \boldsymbol{B}\boldsymbol{c}_i$, $1 \leq i \leq n$. To estimate $\boldsymbol{\beta}$, we proceed as in Jurečková (1971) and Sen and Puri (1977), and define

$$A_n = \left\{\boldsymbol{B}: \sum_{j=1}^{p}\sum_{k=1}^{q} |L_{njk}(\boldsymbol{B})| = \text{minimum}\right\}. \tag{4.4}$$

Then the estimator of $\boldsymbol{\beta}$ is given by the center of gravity of A_n, we denote this by $\tilde{\boldsymbol{\beta}}_n$. Similarly, under $H_0^{(2)}$: $\boldsymbol{\beta}_2 = \boldsymbol{0}$, we have $Y = \boldsymbol{\beta}_1\boldsymbol{C}_n^{(1)} + \boldsymbol{e}$, so that working with $\boldsymbol{B}^* = (\boldsymbol{B}_1, \boldsymbol{0})$, we define $\boldsymbol{Y}_i(B^*) = \boldsymbol{Y}_i - \boldsymbol{B}^*\boldsymbol{C}_n = Y_i - \boldsymbol{B}_1\boldsymbol{C}_n^{(1)}$, $1 \leq i \leq n$ and the $R_{nij}(\boldsymbol{B}^*)$ and $L_{njk}(\boldsymbol{B}^*)$ are defined in a similar manner. Let then

$$A_n^* = \left\{\boldsymbol{B}^*: \sum_{j=1}^{p}\sum_{k=1}^{q_1} |L_{njk}(\boldsymbol{B}^*)| = \text{minimum}\right\}. \tag{4.5}$$

Then the constrained estimator $\hat{\boldsymbol{\beta}}_{n1}$ of $\boldsymbol{\beta}_1$ (when $\boldsymbol{\beta}_2$ is assumed to be $\boldsymbol{0}$) is given by the center of gravity of A_n^*. We also define $\hat{R}_{nij} = R_{nij}(\hat{\boldsymbol{\beta}}_{n1})$, $1 \leq i \leq n$, $1 \leq j \leq p$ and define $\hat{\boldsymbol{M}}_n$ as in (3.2), where the R_{nij} are replaced by $\hat{R}_{nij}$. Let then

$$\boldsymbol{D}_{n22\cdot 1} = \boldsymbol{D}_{n22} - \boldsymbol{D}_{n21}\boldsymbol{D}_{n11}^{-1}\boldsymbol{D}_{n12}\,, \tag{4.6}$$

$$\hat{\boldsymbol{G}}_n = \hat{\boldsymbol{M}}_n \otimes \boldsymbol{D}_{n22\cdot 1}\,, \tag{4.7}$$

$$\hat{\boldsymbol{H}}_n = ((\hat{L}_{njk}\hat{L}_{nj'k'}))_{j,j'=1,\ldots,p;\,k,k'=q_1+1,\ldots,q} \tag{4.8}$$

Then, for testing $H_0^{(2)}$: $\boldsymbol{\beta}_2 = \boldsymbol{0}$ against $\boldsymbol{\beta}_2 \neq \boldsymbol{0}$, we proceed as in Sen and Puri (1977) and Saleh and Sen (1982b) and consider the test statistic

$$\mathcal{L}_n^{(2)} = \text{Trace}(\hat{\boldsymbol{H}}_n\hat{\boldsymbol{G}}_n^{-1})\,. \tag{4.9}$$

Under $H_0^{(2)}$, $\mathcal{L}_n^{(2)}$ has asymptotically chi square distribution with pq_2 D.F., so that its critical value $l_{n,\alpha_2}^{(2)}$ converges to $\chi^2_{pq_2,\alpha_2}$ as $n \to \infty$. Then, the preliminary test rank order estimator (PTROE) of $\boldsymbol{\beta}_1$ may be defined as

$$\boldsymbol{\beta}_{n1}^* = \xi(\mathcal{L}_n^{(2)})\tilde{\boldsymbol{\beta}}_{n1} + \{1 - \xi(\mathcal{L}_n^{(2)})\}\hat{\boldsymbol{\beta}}_{n1}\,, \tag{4.10}$$

where

$$\xi(\mathcal{L}_n^{(2)}) \text{ is 1 or 0 according as } \mathcal{L}_n^{(2)} \text{ is} \geq \text{or} < l_{n,\alpha_2}^{(2)}\,. \tag{4.11}$$

Finally, we consider the preliminary test testing problem. We are primarily interested in testing $H_0^{(1)}$: $\boldsymbol{\beta}_1 = \boldsymbol{0}$ when it is suspected but not evident that $\boldsymbol{\beta}_2$ is 'close to' $\boldsymbol{0}$. For this purpose, in addition to $\tilde{\boldsymbol{\beta}}_n$ and $\hat{\boldsymbol{\beta}}_{n1}$, defined in (4.4) and (4.5), we denote by $\hat{\boldsymbol{\beta}}_{n2}$ the estimator of $\boldsymbol{\beta}_2$ when we take $\boldsymbol{\beta}_1 = \boldsymbol{0}$, that is, letting $\boldsymbol{B}^{**} = (\boldsymbol{0}, \boldsymbol{B}_2)$, we have $\hat{\boldsymbol{\beta}}_{n2}$ as the center of gravity of

$$A_n^{**} = \left\{\boldsymbol{B}^{**}: \sum_{j=1}^{p}\sum_{k=q_1+1}^{q} |L_{njk}(B^*)| = \text{minimum}\right\}. \tag{4.12}$$

We denote by

$$\hat{R}^{(1)}_{nij} = R_{nij}(\mathbf{0}, \hat{\beta}_{n2}),\ \hat{R}^{(2)}_{nij} = R_{nij}(\boldsymbol{\beta}_{n1}, 0),\ \hat{R}_{nij} = R_{nij}(\tilde{\beta}_n)\,. \tag{4.13}$$

Then, we let

$$\hat{\boldsymbol{L}}^{(1)}_n = ((\hat{L}^{(1)}_{njk})) = ((L_{njk}(\hat{R}^{(1)}_{nij})))_{1\leq j\leq p, q_1+1\leq k\leq q}, \tag{4.14}$$

$$\hat{L}^{(2)}_n = ((\hat{L}^{(2)}_{njk})) = ((L_{njk}(\hat{R}^{(2)}_{nijk})))_{1\leq j\leq p, 1\leq k\leq q_1}, \tag{4.15}$$

and let

$$\boldsymbol{D}_{n11\cdot 2} = \boldsymbol{D}_{n11} - \boldsymbol{D}_{n12}\boldsymbol{D}^{-1}_{n22}\boldsymbol{D}_{n21}\,, \tag{4.16}$$

$$\hat{\boldsymbol{G}}^{(1)}_n = \hat{\boldsymbol{M}}^{(1)}_n \otimes \boldsymbol{D}_{n11\cdot 2}, \qquad \hat{\boldsymbol{G}}^{(2)}_n = \hat{\boldsymbol{G}}_n \text{ in (4.7)}\,, \tag{4.17}$$

where $\hat{\boldsymbol{M}}^{(1)}_n$ is defined as in (3.2) with R_{nij} replaced by $\hat{R}^{(1)}_{nij}$, for $1\leq i\leq n$, $j\geq 1$. Let then

$$\hat{\boldsymbol{H}}^{(1)}_n = ((\hat{L}^{(1)}_{njk}\hat{L}^{(1)}_{nj'k'}))_{1\leq j,j\leq p; 1\leq k,k'\leq q_1}, \tag{4.18}$$

$$\hat{\boldsymbol{H}}^{(2)}_n = ((\hat{L}^{(2)}_{njk}\hat{L}^{(2)}_{nj'k'}))_{1\leq j,j'\leq p; q_1+1\leq k,k'\leq q}. \tag{4.19}$$

Finally, let $L_{njk} = L_{njk}(\mathbf{0})$, $1\leq j\leq p$, $1\leq k$, $k'\leq q_1$ and

$$\bar{\boldsymbol{H}}_n = ((L_{njk}L_{nj'k'}))_{1\leq j\leq p'\leq p; 1\leq k,k'\leq q_1}, \tag{4.20}$$

$$\bar{\boldsymbol{G}}_n = \boldsymbol{M}_n \otimes \boldsymbol{D}_{n11}\,. \tag{4.21}$$

Then, the test statistics are $\mathscr{L}^{(2)}_n$, defined by (4.9) and

$$\mathscr{L}^{(0)}_n = \mathrm{Tr}(\hat{\boldsymbol{H}}^{(1)}_n(\hat{\boldsymbol{G}}^{(1)}_n)^{-1}), \qquad \bar{\mathscr{L}}_n = \mathrm{Tr}(\bar{\boldsymbol{H}}_n\bar{\boldsymbol{G}}^{-1}_n)\,. \tag{4.22}$$

Note that $\bar{\mathscr{L}}_n$ is the test statistic for testing $H^{(1)}_0$: $\boldsymbol{\beta}_1 = \mathbf{0}$ when $\boldsymbol{\beta}_2$ is given as $\mathbf{0}$, while $\mathscr{L}^{(0)}_n$ is the test statistic when $\boldsymbol{\beta}_2$ is treated as the nuisance parameter. Finally, the preliminary test function for testing $H^{(1)}_0\boldsymbol{\beta}_2 = \mathbf{0}$ is based on

$$\xi(\mathscr{L}^*_n) = \begin{cases} 1 & \text{if } \mathscr{L}^{(2)}_n < l^{(2)}_{n,\alpha_2},\ \bar{\mathscr{L}}_n \geq \bar{l}_{n,\bar{\alpha}}\,, \\ & \text{or } \mathscr{L}^{(2)}_n \geq l^{(2)}_{n,\alpha_2},\ \bar{\mathscr{L}}^{(0)}_n \geq l^{(0)}_{n,\alpha^0}\,, \\ 0 & \text{otherwise}\,. \end{cases} \tag{4.23}$$

As in Sections 2 and 3, the main interest centers around the robustness study of the NPTE and NPTT when $H_0^{(2)}$: $\boldsymbol{\beta}_2 = \mathbf{0}$ may or may not be tenable, and, as in earlier sections, we confine ourselves to some sequence of local alternatives, for which meaningful asymptotic results are derivable. Against $H_0^{(2)}$, we consider a sequence $\{K_n\}$ of alternative hypotheses, where

$$K_n:\quad \boldsymbol{\beta}_2 = n^{-1/2}\boldsymbol{\lambda}_2,\ \boldsymbol{\lambda}_2 \in E^{pq_2},\ \text{fixed}\,, \tag{4.24}$$

while in the context of the PTT, we consider the sequence $\{K_n^*\}$ of alternative hypotheses, where

$$K_n^*:\quad \boldsymbol{\beta} = (\boldsymbol{\beta}_1, \boldsymbol{\beta}_2) = n^{-1/2}\boldsymbol{\lambda} = n^{-1/2}(\boldsymbol{\lambda}_1, \boldsymbol{\lambda}_2)\,, \tag{4.25}$$

with $\boldsymbol{\lambda}_1$ and $\boldsymbol{\lambda}_2$ fixed.

Note that for the unconstrained estimator $\tilde{\boldsymbol{\beta}}_n = (\tilde{\boldsymbol{\beta}}_{n1}, \tilde{\boldsymbol{\beta}}_{n2})$, we have under $\{K_n\}$ and the assumed regularity conditions:

$$n^{1/2}(\tilde{\boldsymbol{\beta}}_{n1} - \boldsymbol{\beta}_1) \xrightarrow{\mathcal{D}} \mathcal{N}_{pq_1}(\mathbf{0}, \boldsymbol{T}\otimes\boldsymbol{\Delta}_{11}^*)\,, \tag{4.26}$$

where

$$\boldsymbol{\Delta}^* = \begin{pmatrix} \boldsymbol{\Delta}_{11}^* & \boldsymbol{\Delta}_{12}^* \\ \boldsymbol{\Delta}_{21}^* & \boldsymbol{\Delta}_{22}^* \end{pmatrix} = \boldsymbol{\Delta}^{-1} = \begin{pmatrix} \boldsymbol{\Delta}_{11} & \boldsymbol{\Delta}_{12} \\ \boldsymbol{\Delta}_{21} & \boldsymbol{\Delta}_{22} \end{pmatrix}^{-1}. \tag{4.27}$$

Also, under $\{K_n\}$,

$$n^{1/2}(\hat{\boldsymbol{\beta}}_{n1} - \boldsymbol{\beta}_1) \xrightarrow{\mathcal{D}} \mathcal{N}_{pq_1}(-\boldsymbol{\lambda}_2\boldsymbol{\Delta}_{21}\boldsymbol{\Delta}_{11}^{-1}, \boldsymbol{T}\otimes\boldsymbol{\Delta}_{11}^{-1})\,. \tag{4.28}$$

Finally, for the PTE $\boldsymbol{\beta}_{n1}^*$ in (4.10), we have, as $n\to\infty$,

$$\begin{aligned} &P\{n^{1/2}(\boldsymbol{\beta}_{n1}^* - \boldsymbol{\beta}_1) \leq \boldsymbol{x} \mid K_n\} \\ &\quad\to H_{pq_2}(\chi^2_{pq_2,\alpha_2}; \delta_2)G_{pq_1}(\boldsymbol{x} - \boldsymbol{\lambda}_2\boldsymbol{\Delta}_{22}^{-1}\boldsymbol{\Delta}_{21}; \mathbf{0}, \boldsymbol{T}\otimes\boldsymbol{\Delta}_{11\cdot 2}^*) \\ &\qquad + \int_{E(\boldsymbol{\lambda}_2)} G_{pq_1}(\boldsymbol{x} - \boldsymbol{z}\boldsymbol{\Delta}_{22}^{*-1}\boldsymbol{\Delta}_{21}^*; \mathbf{0}, \boldsymbol{T}\otimes\boldsymbol{\Delta}_{11\cdot 2}^*)\,\mathrm{d}G_{pq_2}(\boldsymbol{z}; \mathbf{0}, \boldsymbol{T}\otimes\boldsymbol{\Delta}_{22}^*) \end{aligned} \tag{4.29}$$

where

$$\delta_2 = \operatorname{Tr}((\boldsymbol{\lambda}_2^0\boldsymbol{\lambda}_2^{0\prime})(\boldsymbol{T}\otimes\boldsymbol{\Delta}_{22}^*)^{-1})\,, \tag{4.30}$$

$\boldsymbol{\lambda}_2^0$ is the rolledout form of $\boldsymbol{\lambda}_2$, $\boldsymbol{\Delta}_{11\cdot 2}^* = \boldsymbol{\Delta}_{11}^* - \boldsymbol{\Delta}_{12}^*\boldsymbol{\Delta}_{22}^{*-1}\boldsymbol{\Delta}_{21}^*$ and

$$E(\boldsymbol{\lambda}_2) = \{\boldsymbol{y} \in E^{pq_2}\colon \operatorname{Tr}(T\otimes\Delta_{22}^*)^{-1}(\boldsymbol{y} + \boldsymbol{\lambda}_2^0)(\boldsymbol{y} + \boldsymbol{\lambda}_2^0)' \geq \chi^2_{pq_2,\alpha}\}\,. \tag{4.31}$$

Thus, under $\{K_n\}$, the asymptotic bias of $n^{1/2}(\tilde{\boldsymbol{\beta}}_{n1} - \boldsymbol{\beta}_1)$, $n^{1/2}(\boldsymbol{\beta}_{n1} - \boldsymbol{\beta}_1)$ and $n^{1/2}(\boldsymbol{\beta}_{n1}^* - \boldsymbol{\beta}_1)$ are, respectively, $\mathbf{0}$, $-\boldsymbol{\lambda}_2\boldsymbol{\Delta}_{22}^{*-1}\boldsymbol{\Delta}_{21}^*$ and $-\boldsymbol{\lambda}_2\boldsymbol{\Delta}_{22}^{*-1}\boldsymbol{\Delta}_{21}^*\times H_{pq_2+2}(\chi^2_{pq_2+2}; \delta_2)$, so that conclusions similar to those in Section 3 also hold here. If we rollout $\boldsymbol{\lambda}_2 \boldsymbol{\Delta}_{22}^{*-1} \Delta_{21}^*$ into a pq_1-vector $\boldsymbol{\lambda}_0^*$, then the asymptotic dispersion matrices for the three estimators are

$$\boldsymbol{T} \otimes \boldsymbol{\Delta}_{11}^*, \qquad \boldsymbol{T} \otimes \boldsymbol{\Delta}_{11\cdot 2}^* + \boldsymbol{\lambda}_0^* \boldsymbol{\lambda}_0^*$$

and

$$\boldsymbol{T} \otimes \Delta_{11}^* - \boldsymbol{\lambda}_0^* \boldsymbol{\lambda}_0^{*\prime}[2H_{pq_2+2}(\chi^2_{pq_2,\alpha_2};\, \delta_2) - 2H_{pq_2+4}(\chi^2_{pq_2,\alpha_2};\, \delta_2)] \\ - \boldsymbol{T} \otimes \boldsymbol{\Delta}_{12}^* \boldsymbol{\Delta}_{22}^{*-1} \boldsymbol{\Delta}_{21}^* H_{pq_2+2}(\chi^2_{pq_2,\alpha_2};\, \delta_2)\,,$$

so that the robustness of the PTE follows as in Section 3. Note that $\boldsymbol{\Delta}_{11}^* = \boldsymbol{\Delta}_{11\cdot 2}^* + \boldsymbol{\Delta}_{12}^* \boldsymbol{\Delta}_{22}^{*-1} \boldsymbol{\Delta}_{21}^*$, so that depending on the departure of $\boldsymbol{\lambda}_0^*$ from $\mathbf{0}$, the unconstrained estimator may or may not be better than the constrained estimator.

Let us now consider the PTT problem. For the test statistic $\mathscr{L}_n^{(0)}$, the asymptotic size is equal to α^0, irrespective of $\boldsymbol{\lambda}_2$, while under $\{K_n^*\}$ (where K_n^*: $\boldsymbol{\beta} = n^{-1/2}\boldsymbol{\lambda}$, $\boldsymbol{\lambda} = (\boldsymbol{\lambda}_1, \boldsymbol{\lambda}_2)$, $\mathscr{L}_n^{(0)}$ has asymptotically noncentral chi square distribution with pq_1 D.F. and noncentrality parameter

$$\delta^0 = \operatorname{Trace}\{\boldsymbol{T}^{-1}(\boldsymbol{\lambda}_1' \boldsymbol{\Delta}_{11\cdot 2}^* \boldsymbol{\lambda}_1)\}\,.$$

For $\bar{\mathscr{L}}_n$, the asymptotic size of the test is equal to $\bar{\alpha}$ when $\boldsymbol{\beta}_1 = \mathbf{0}$, $\boldsymbol{\beta}_2 = \mathbf{0}$, while for $\boldsymbol{\beta}_2 \neq \mathbf{0}$, $\boldsymbol{\beta}_1 = \mathbf{0}$, the asymptotic size may be larger than $\bar{\alpha}$. Under $\{K_n\}$, the size of the test based on $\bar{\mathscr{L}}_n$ converges to $P\{\chi^2_{pq_1}(\bar{\delta}_0) \geqslant \chi^2_{pq_1,\bar{\alpha}}\}$ where

$$\bar{\delta}_0 = \operatorname{Trace}\{\boldsymbol{T}^{-1}(\boldsymbol{\lambda}_2 \boldsymbol{\Delta}_{22}^{*-1} \boldsymbol{\Delta}_{21}^*) \Delta_{11}^* (\boldsymbol{\lambda}_2 \boldsymbol{\Delta}_{22}^{*-1} \boldsymbol{\Delta}_{21}^*)'\}\,, \tag{4.32}$$

which exceeds $\bar{\alpha}$ whenever $\boldsymbol{\lambda}_2 \boldsymbol{\Delta}_{22}^{*-1} \boldsymbol{\Delta}_{21}^*$ is nonnull. This explains the lack of robustness (in size) of the constrained test based on $\bar{\mathscr{L}}_n$. In fact $\bar{\delta}_0 \to \infty$ as $\boldsymbol{\lambda}_2$ moves away from $\mathbf{0}$, and hence, the asymptotic size may be quite high (and even close to 1) depending on $\boldsymbol{\lambda}_2$. Under $\{K_n^*\}$, $\bar{\mathscr{L}}_n$ has asymptotically noncentral chi square distribution with pq_1 D.F. and noncentrality parameter

$$\bar{\delta} = \operatorname{Trace}\{\boldsymbol{T}^{-1}(\boldsymbol{\lambda}_1 - \boldsymbol{\lambda}_2 \boldsymbol{\Delta}_{22}^{*-1} \boldsymbol{\Delta}_{21}^*) \Delta_{11}^* (\boldsymbol{\lambda}_1 - \boldsymbol{\lambda}_2 \boldsymbol{\Delta}_{22}^{*-1} \boldsymbol{\Delta}_{21}^*)'\}$$

where $\bar{\delta}$ may or may not be greater than δ^0 in (4.31) depending on $\boldsymbol{\lambda}_1$, $\boldsymbol{\lambda}_2$ and Δ^*. As for the PTT $\mathscr{L}_n^*$, the asymptotic size is bounded from above by $\bar{\alpha}(1-\alpha_2) + \alpha_2 \wedge \alpha^0$, when $\boldsymbol{\lambda}_2 = \mathbf{0}$, where for $\bar{\alpha}$ and α^0 when chosen close to each other (as it should be the case), this upper bound provides a close approximation. For $\boldsymbol{\lambda}_2 \neq 0$, like the multivariate simple regression model case in Section 3, the size of the PTT is asymptotically expressible in terms of d.f. of correlated chi square variables, and through it will depend on $\boldsymbol{\lambda}_2$, it behaves robustly in the sense that it does not go to $+1$ as $\boldsymbol{\lambda}_2$ moves away from $\mathbf{0}$. The picture with the asymptotic power is also very similar to that in these. For details, we may refer to Saleh and Sen (1983b). This explains the robustness properties of the PTT procedures.

5. Some general remarks

Parametric procedures for PTE or PTT for the models in Sections 2, 3, and 4, when F is treated as a normal d.f., are based on the classical least squares estimators for the constrained and unconstrained models. For example, in (4.1), the unconstrained estimator of $\boldsymbol{\beta}$ is

$$(\boldsymbol{Y}\boldsymbol{C}_n')(\boldsymbol{C}_n\boldsymbol{C}_n')^{-1} = \boldsymbol{Y}\boldsymbol{C}_n'\boldsymbol{D}_n^{-1} = \tilde{\boldsymbol{\beta}}_{n(L)}, \quad \text{say}, \tag{5.1}$$

while the constrained estimator of $\boldsymbol{\beta}_1$ (when $\boldsymbol{\beta}_2$ is taken as $\mathbf{0}$) is given by

$$(\boldsymbol{Y}\boldsymbol{C}_{n1}')(\boldsymbol{C}_{n1}\boldsymbol{C}_{n1}') = \boldsymbol{Y}\boldsymbol{C}_{n1}'\boldsymbol{D}_{n11}^{-1} = \hat{\boldsymbol{\beta}}_{n1(L)}, \quad \text{say}. \tag{5.2}$$

The normal theory likelihood ratio test may be used to test for $H_0^{(2)}$: $\boldsymbol{\beta}_2 = \mathbf{0}$ and with this the PTE $\boldsymbol{\beta}_{n1(L)}^*$ may be defined as in (4.10). The case of the PTT is also similar to that in (4.23), where for the constrained and the unconstrained likelihood ratio test, the chi square approximation works out well. For the PTE based on these least squares estimators, (4.26)–(4.30) holds with the notable change that $\boldsymbol{T}$ has to be replaced by $\boldsymbol{\Sigma}$, the covariance matrix of $\boldsymbol{Y}$ (assumed to be finite), and also the same change is needed in (4.31). As such, the results run parallel to the nonparametric setup and the same asymptotic properties hold for the PTE and PTT procedure based on the least squares estimators. However, here we need $\boldsymbol{\Sigma}$ to exist, while in the nonparametric case, this is not needed.

Secondly, in the nonparametric case, one may use a quadratic form in the rank based estimates for the testing purpose too. This will give some asymptotically equivalent tests. However, in that case, one would need to use some consistent estimators of γ_j $(1 \leq j \leq p)$ in (3.14), which is not needed in our case; such estimates are available in the literature (cf. Jurečková, 1971; Sen and Puri, 1977, though are usually computationally quite tedious. Hence, we prefer not to advocate this alternative procedure.

Finally, we should like to comment on the role and choice of the significance level of preliminary test in PTI.

6. Choice of significance level of the preliminary test

For the PTE and PTT procedure, a natural question arises as to the choice of optimal level of significance of the preliminary test. For the PTE problem, this can be achieved by drawing tables and graphs of the ARE's of $\boldsymbol{\theta}_1^*$ relative to $\hat{\boldsymbol{\theta}}_1$ and $\tilde{\boldsymbol{\theta}}_1$ as a function of unknown parameters, which will dictate the choice of α_2 for which $\boldsymbol{\theta}_1^*$ will be preferred to either $\hat{\boldsymbol{\theta}}_1$ or $\tilde{\boldsymbol{\theta}}_1$ at least for certain region of values of the parameters involved. In the PTT problem, one constructs tables and graphs for the power function of ξ^*, ξ_{12} and ξ_1 as a function of unknown parameters, which will dictate the optimal choice of α_2 for which ξ^* will be preferred over ξ_{12} or ξ_1 at least for certain region of values of the parameters involved. In such studies one sets $\alpha_{12} = \alpha_1 = \alpha$ to determine optimal α_2 and it will be necessary to make detailed study of each type PTT problem to ensure proper choice of α_2.

A demonstration of analytical approach to the PTE problem of a appropriate choice of the level of significance in the univariate simple linear model discussed in Section 2 is given here. Consider the two-sided preliminary test for the estimation of the location parameter. The ARE of θ^* relative to $\tilde{\theta}$ as given in Saleh and Sen (1978) is

$$e(\theta^*: \tilde{\theta}) = (1 + \bar{c}^2/Q^*)/[1 + (\bar{c}^2/Q^*)q(\tau_{\alpha_2/2}, \delta)] \tag{6.1}$$

where $\tau_{\alpha_2/2} > 0$ for every $\alpha_2 \in (0, 1)$ and for every $\tau_{\alpha_2/2} \in [0, \infty)$ and $\delta \in (-\infty, \infty)$ and $\delta = \lambda\nu_2$,

$$\begin{aligned} q(x, y) = y^2\{G(x-y) - G(-x-y)\} + 1 - G(x-y) + G(-x-y) \\ + (x-y)g(x-y) + (x+y)g(x+y) \end{aligned} \tag{6.2}$$

The ARE $e(\theta^*: \tilde{\theta})$ is a function of α_2 and δ, say $E(\alpha_2, \delta)$ which is symmetric function of δ for fixed α_2. Restricting attention to $\delta \in [0, \infty)$. $E(\alpha_2, \delta)$ as a function of δ has a maximum at $\delta = 0$ for $\alpha_2 > 0$ and decreases crossing the line $E(\alpha_2, \delta) = 1$ as δ increases, then decreases to a minimum and then increases toward 1 as δ tends to ∞. Consider $\delta = 0$ (null hypothesis case) and α_2 varies, then one has $\max_{\alpha_2} E(\alpha_2, 0) = E(0, 0)$. On the other hand, let $\alpha_2 = 0$ and let δ very in $E(0, \delta)$, then the curves $E(0, \delta)$ and $E(1, \delta) = 1$ intersect at $\delta = 1$. Thus, for general $\alpha_2 \in (0, 1)$, $E(\alpha_2, \delta)$ and $E(1, \delta)$ intersect in the interval $[0, 1]$. Further, the values of δ decreases as α increases. Therefore, for two different values of α_2 say, α_{21} and $\alpha_{22} E(\alpha_{21}, \delta)$ and $E(\alpha_{22}, \delta)$ will never intersect above $E = 1$. Thus, one can formulate the following criteria to make a suitable choice for α_2: When the statistician wants to choose an estimator with smallest asymptotic mean square error, i.e. the estimator will choose an estimator with largest $e(\theta^*: \tilde{\theta})$. If it is known that $\delta \in (0, 1]$, the estimator $\hat{\theta}$ is used since $E(0, \delta)$ is maximum in this interval for all $\delta \in [0, 1]$. However, if $\delta > 1$, there is no way of choosing uniformly best estimator if size of δ is not known. In this case, the statistician wants an estimator with ARE equal to E^0, then among the estimator with $\alpha_2 \in A$ where $A = \{\alpha_2: E(\alpha_2, \delta) > E^0 \,\forall \delta\}$, the estimator chosen maximizes $E(\alpha_2, \delta)$ for all (α_2, δ). Thus, one solves for α_2 from the condition $\max_{\alpha_2} \min_\delta E(\alpha_2, \delta) = E^0$. The solution α^* is the appropriate level of significance of the preliminary test which will guarantee an ARE E^0. In practice, a table may be prepared for appropriate values of α^* and corresponding δ^* such that $E(\alpha^*, \delta^*) = E^0$ and the maximin ARE-value E^* always equals $E(\alpha^*, 0)$. Similar procedures may be developed for multivariate problems discussed earlier. For the PTT problem same analysis with power function as a function of two-parameters (univariate and multivariate case) and size of the preliminary test is conceivable and detailed study is necessary to determine the appropriate size.

Acknowledgement

This work has been supported by the NSERC grant No. A3088 (Canada) and by the U.S. National Heart, Lung and Blood Institute. Contract N1H-NHLB1-71-2243-L from National Health Institutes of Health.

References

[1] Adichie, J. N. (1967). Estimates of regression parameters based on rank tests. *Ann. Math. Statist.* **38**, 894–904.

[2] Ahsanullah, M. and Saleh, A. K. Md. Ehsanes (1972). Estimation of intercept in a linear regression model with one dependent variable after a preliminary test of significance. *Rev. Inst. Internat. Statist.* **40**, 139–145.

[3] Bancroft, T. A. (1944). On biases in estimation due to use of preliminary test of significance. *Ann. Math. Statist.* **15**, 190–204.

[4] Bancroft, T. A. (1953). Certain approximate formulas for the power and size of a general linear hypothesis incorporating preliminary test of significance. Unpublished prelim. report. Stat-Lab. Iowa State Univ. Ames.

[5] Bancroft, T. A. (1964). Analysis and inference for incompletely specified models involving use of preliminary tests of significance. *Biometrics* **20**, 427–442.

[6] Bancroft, T. A. (1972). Some recent advances in inference procedures using preliminary tests of significance. In: *Statististical Papers in Honour of George W. Senedecor.* Ames Iowa State Univ. Press, Chapter 2, pp. 19–30.

[7] Bancroft, T. A. and Han, C. P. (1976). On pooling of means in multivariate normal distributions. In: Ikeda et al., eds., *Essays in Probability and Statistics*: *Ogawa Volume.* Tokyo, pp. 353–366.

[8] Bancroft, T. A. and Han, C. P. (1977). Inference based on conditional specification: A Note and Bibliography. *Int. Inst. Statist. Review* **45**, 117–127.

[9] Bancroft, T. A. and Han, C. P. (1980). Inference based on conditionally specified ANOVA models incorporating preliminary testing. In: P. R. Krishnaiah, ed., *Handbook of Statistics, Vol. 1.* North-Holland, Amsterdam.

[10] Hájek, J. and Šidák, Z. (1967). *Theory of Rank Tests.* Academic Press, N.Y.

[11] Han, C. P. and Bancroft, T. A. (1968). On pooling of means when the variance is unknown. *J. Amer. Stat. Assoc.* **62**, 1333–1342.

[12] Jureckova, J. (1969). Asymptotic linearity of a rank statistic in regression parameter. *Ann. Math. Statist.* **40**, 1889–1900.

[13] Jurečková, J. (1971). Asymptotic independence of a rank statistic for testing symmetry on regression. *Sankhyá Ser. A* **33**, 1–18.

[14] Jurečková, J. (1971). Non parametric estimates of regression coefficients. *Ann. Math. Statist.* **42**, 1328–1338.

[15] Kraft, C. H. and Van Eeden, C. (1972). Linearized rank estimates and signed rank estimates for the general linear hypothesis. *Ann. Math. Statist.* **43**, 42–57.

[16] Mosteller, F. (1948). On pooling data. *JASA* **43**, 231–242.

[17] Puri, M. L. and Sen, P. K. (1969). A class of rank order tests for general linear hypotheses. *Ann. Math. Statist.* **40**, 1325–1343.

[18] Puri, M. L. and Sen, P. K. (1971). *Nonparametric methods in Multivariate Analysis.* Wiley, New York.

[19] Saleh, A. K. Md. Ehsanes (1973). Pretest estimators of means of multivariate normal distribution. Carleton Math. Ser. 94.

[20] Saleh, A. K. Md. Ehsanes (1977). Pretest estimators of component means of multivariate normal distribution. In: Proceedings of ISI meeting at Delhi, India, pp. 454–457.

[21] Saleh, A. K. Md. Ehsanes and Sen, P. K. (1978). Non parametric estimation of location parameter after a preliminary test on regression. *Ann. Statist.* **6**, 154–168.
[22] Saleh, A. K. Md. Ehsanes and Sen, P. K. (1982a). Nonparametric tests for location after preliminary test on regression. *Comm. Statist.* **11**(6), 639–652.
[23] Saleh, A. K. Md. Ehsanes and Sen, P. K. (1982b). Nonparametric estimation following a preliminary test on regression (a collection of four papers). Carleton Mathematical Lecture Notes No. 41.
[24] Saleh, A. K. Md. Ehsanes and Sen, P. K. (1983a). Nonparametric tests for location after a preliminary test on regression in the multivariate case. *Common. Statist.* **12**(16), 1855–1872.
[25] Saleh, A. K. Md. Ehsanes and Sen, P. K. (1983b). Asymptotic properties of tests of hypothesis following a preliminary test. *Statistics and Decisions* **1**, 455–477.
[26] Sen, P. K. (1982). Asymptotic properties of likelihood ratio tests based on conditional specification. *Statistics and Decisions* **1**, 81–106.
[27] Sen, P. K. and Puri, M. L. (1969). On robust nonparametric estimation in some multivariate linear models. In: P. R. Krishnaiah, ed., *Multivariate Analysis II*. Academic Press, New York, pp. 3–52.
[28] Sen, P. K. and Puri, M. L. (1977). Asymptotically distribution free aligned rank order tests for composite hypotheses for general multivariate linear models. *Z. Wahrsch. Verw. Gebiete* **39**, 175–186.
[29] Sen, P. K. and Saleh, A. K. Md. Ehsanes (1979). Nonparametric estimation of location parameter after a preliminary test on regression in multivariate case. *J. Multivariate Analysis* **9**, 322–331.

P. R. Krishnaiah and P. K. Sen, eds., *Handbook of Statistics, Vol. 4*
© Elsevier Science Publishers (1984) 299–326

14

Paired Comparisons: Some Basic Procedures and Examples

*Ralph A. Bradley**

1. Introduction

Interest in paired comparisons in statistics and psychometrics has developed in the contexts of the design of experiments, nonparametric statistics, and scaling, including multidimensional scaling. Applications have arisen in many areas, but most notably in food technology, marketing research, and sports competition. An extensive bibliography on paired comparisons by Davidson and Farquhar (1976) contains some 400 references.

Paired comparisons have been considered in design of experiments as incomplete block designs with block size two by Clatworthy (1955) and others. Scheffé (1952) developed an analysis of variance for paired comparisons with consideration for possible order effects for the two treatments within blocks. When the usual parametric models of analysis of variance are imposed, the analysis of such designs follows standard methods and will not be discussed here.

The emphasis in this chapter will be on paired comparisons as a means of designing comparative experiments when no natural measuring scale is available. The author's interest in paired comparisons arose in consideration of statistical methods in sensory difference testing. When responses of individuals to items under comparison are subjective, and particularly when sensory responses to taste, odor, color or sound are involved, evaluation is easier when the number of items or samples to be considered at one time is small and the effects of sensory fatigue are minimized. Probabilistic models for paired comparisons may be devised to represent the experimental situation and permit appropriate data analysis. The models provide probabilities of possible choices of items or treatments from pairs of items and hence depend on orderings. The statistical methods devised are thus ranking methods and, while they are not literally nonparametric methods, they are often so classified.

*The work of the author was supported in part by the Office of Naval Research under Contract N00014-80-C-0093 with the Florida State University. Reproduction in whole or in part is permitted for any purpose of the United States Government.

The basic paired comparisons experiment has t treatments, $T_1, \ldots, T_t$, and $n_{ij} \geq 0$ comparisons of T_i with T_j, $n_{ji} = n_{ij}$, $i \neq j$, $i, j = 1, \ldots, t$. For each comparison, preference or order is designated by $a_{ij\alpha}$, $a_{ij\alpha} = 1$ if T_i is 'preferred' to T_j in the α-th comparison of T_i and T_j, $a_{ij\alpha} = 0$ otherwise, $a_{ij\alpha} + a_{ji\alpha} = 1$. In further definition of notation, let $a_{ij} = \Sigma_{\alpha=1}^{n_{ij}} a_{ij\alpha}$ and $a_i = \Sigma_{j,j\neq i} a_{ij}$, the total number of preferences for T_i. In sensory evaluations, responses may be preferences or attribute order judgments on such characteristics as sweetness, smoothness, whiteness, etc. We shall loosely refer to preference judgments.

Dykstra (1960) provides typical data on a paired comparisons preference taste test involving four variations of the same product. The data are summarized in Table 1. Note that the experiment is not balanced: $n_{12} = 140$, $n_{13} = 54$, $n_{14} = 57$, $n_{23} = 63$, $n_{24} = 58$, $n_{34} = 0$; treatments T_3 and T_4 were not compared. Unbalanced experiments are permissible as long as the design is connected: it is not possible to select a subset of the treatments such that no treatment in the subset is compared directly with a treatment in the complementary subset. Balanced experiments are more efficient when there is equal interest in all treatments and treatment comparisons.

Table 1
Summary of results of a taste test

	T_1	T_2	T_3	T_4	a_i
T_1	—	28	15	23	66
T_2	112	—	46	47	205
T_3	39	17	—	—	56
T_4	34	11	—	—	45

We shall return to analysis of the data of Table 1, which gives values of a_{ij}, after discussion of models for paired comparisons and establishment of basic procedures.

This chapter is organized in such a way as to give initial attention to the analysis of basic paired comparisons data like those of Table 1. Then extensions of the method are developed for factorial treatment combinations and for multivariate responses, responses on several attributes for each paired comparison. The emphasis is on the methodology and applications, although properties of procedures are noted and references given. We conclude with comments on additional methods of analysis.

2. Models for paired comparisons

When $t = 2$, a paired comparisons experiment with treatments T_1 and T_2 might be modelled as $n_{12} > 0$ independent Bernoulli trials with probabilities of choices for T_1 and T_2 being π_1 and π_2, $\pi_i \geq 0$, $i = 1, 2$, $\pi_1 + \pi_2 = 1$. Then in some sense π_1 and π_2 are measures of 'worth' of T_1 and T_2. Binomial theory applies and the sign test may be used to test the hypothesis H_0: $\pi_1 = \pi_2$.

Bradley and Terry (1952a) proposed a basic model for paired comparisons, extended by Dykstra (1960) to include unequal values of the n_{ij}. The approach was a heuristic extension of the special binomial when $t = 2$. Treatment parameters, $\pi_1, \ldots, \pi_t$, $\pi_i \geqslant 0$, $i = 1, \ldots, t$, are associated with the t treatments, $T_1, \ldots, T_t$. It was postulated that these parameters represent relative selection probabilities for the treatments so that the probability of selection of T_i when compared with T_j is

$$P(T_i \rightarrow T_j) = \pi_i/(\pi_i + \pi_j), \quad i \neq j, \; i, j = 1, \ldots, t. \tag{2.1}$$

Since the right-hand member of (2.1) is invariant under change of scale, specificity was obtained by the requirement that

$$\sum_{i=1}^{t} \pi_i = 1 . \tag{2.2}$$

The model proposed imposes structure in that the most general model might postulate binomial parameters π_{ij} and $\pi_{ji} = 1 - \pi_{ij}$ for comparisons of T_i and T_j so that the totality of functionally independent parameters is $\binom{t}{2}$ rather than $(t-1)$ as specified in (2.1) and (2.2).

The basic model (2.1) for paired comparisons has been discovered and rediscovered by various authors. Zermelo (1929) seems to have proposed it first in consideration of chess competition. Ford (1957) proposed the model independently. Both Zermelo and Ford concentrated on solution of normal equations for parameter estimation and Ford proved convergence of the iterative procedure for solution.

The model arises as one of the special simple realizations of more general models developed from distributional or psychophysical approaches. Bradley (1976) has reviewed various model formulations and discussed them under such categories as linear models, the Lehmann model, psychophysical models, and models of choice and worth.

David (1963, Section 1.3) supposes that T_i has 'merit' V_i, $i = 1, \ldots, t$, when judged on some characteristic, and that these merits may be represented on a merit scale. He defined 'linear' models to be such that

$$P(T_i \rightarrow T_j) = H(V_i - V_j) , \tag{2.3}$$

where H is a distribution function for a symmetric distribution, $H(-x) = 1 - H(x)$. Model (2.1) is a linear model since it may be written in the form,

$$P(T_i \rightarrow T_j) = \tfrac{1}{4}\int_{-(\log \pi_i - \log \pi_j)}^{\infty} \operatorname{sech}^2 y/2 \, dy = \pi_i/(\pi_i + \pi_j) , \tag{2.4}$$

as described by Bradley (1953) using the logistic density function.

Thurstone (1927) proposed a model for paired comparisons, that is also a linear model, through the concept of a subjective continuum, an inherent sensation scale on which order, but not physical measurement, could be discerned. Mosteller (1951) provides a detailed formulation and an analysis of Thurstone's important Case *V*. With suitable scaling, each treatment has a location point on the continuum, say μ_i for T_i, $i = 1, \ldots, t$. An individual is assumed to receive a sensation X_i in response to T_i, with responses X_i normally distributed about μ_i. When an individual compares T_i and T_j, he in effect is assumed to report the order of sensations X_i and X_j which may be correlated; $X_i > X_j$ may be associated with $T_i \to T_j$. Case *V* takes all such correlations equal and the variances of all X_i equal. The probability of selection may be written

$$P(T_i \to T_j) = P(X_i > X_j) = \frac{1}{\sqrt{2\pi}} \int_{-(\mu_i - \mu_j)}^{\infty} e^{-y^2/2}\, dy\,. \tag{2.5}$$

It is apparent from (2.4) and (2.5) that the two models are very similar. The choice between the models is much like the choice between logits and probits in biological assay. The use of $\log \pi_i$ as a measure of location for T_i in the first model is suggested.

Models (2.4) and (2.5) give very similar results in applications. Comparisons are made by Fleckenstein, Freund and Jackson (1958) with test data on comparisons of typewriter carbon papers. In general, more extensions of model (2.4) exist and we shall use that model in this chapter.

3. Basic procedures

The general approach to analysis of paired comparisons based on the model (2.1) is through likelihood methods. On the assumption of independent responses for the n_{ij} comparisons of T_i and T_j, the binomial component of the likelihood function for this pair of treatments is

$$\left(\frac{\pi_i}{\pi_i + \pi_j}\right)^{a_{ij}} \left(\frac{\pi_j}{\pi_i + \pi_j}\right)^{a_{ji}} = \pi_i^{a_{ij}} \pi_j^{a_{ji}} / (\pi_i + \pi_j)^{n_{ij}}\,,$$

ties or no preference judgments not being permitted. The complete likelihood function, on the assumption of independence of judgments between pairs of treatments, is

$$L = \prod_i \pi_i^{a_i} \Big/ \prod_{i<j} (\pi_i + \pi_j)^{n_{ij}}\,. \tag{3.1}$$

It is seen that $a_1, \ldots, a_t$ constitute a set of sufficient statistics for the estimation of $\pi_1, \ldots, \pi_t$ and that a_i is the total number of preferences or selections of T_i, $i = 1, \ldots, t$, for the entire experiment.

3.1. Likelihood estimation

ML estimators, p_i for π_i, $i = 1, \ldots, t$, are obtained through maximization of $\log L$ in (3.1) subject to the constraint (2.2). After minor simplifications, the resulting likelihood equations are

$$\frac{a_i}{p_i} - \sum_{\substack{j \\ j \neq i}} \frac{n_{ij}}{p_i + p_j} = 0, \quad i = 1, \ldots, t, \tag{3.2}$$

and

$$\sum_i p_i = 1 . \tag{3.3}$$

Solution of equations (3.2) and (3.3) is done iteratively. If $p_i^{(k)}$ is the k-th approximation to p_i,

$$p_i^{(k)} = p_i^{*(k)} \Big/ \sum_i p_i^{*(k)},$$

where

$$p_i^{*(k)} = a_i \Big/ \sum_{\substack{j \\ j \neq i}} [n_{ij}/(p_i^{(k-1)} + p_j^{(k-1)})], \quad k = 1, 2, \ldots .$$

The iteration is started with initial specification of the $p_i^{(0)}$; one may take $p_i^{(0)} = 1/t$, $i = 1, \ldots, t$, and this is adequate, although Dykstra (1956, 1960) has suggested better initial values.

We return to the example of Table 1. Values of a_i are given in the table and values of n_{ij} precede the table. Solution of equations (3.2) and (3.3) was begun with $p_i^{(0)} = \frac{1}{4}$, $i = 1, \ldots, 4$. Results for initial iterations are summarized in Table 2 along with final values for p_i; typically approximately 10 iterations are sufficient for four-decimal accuracy in the final values. It is this iterative procedure that Ford (1957) has shown to converge. The procedure is easy to program on computers because of the symmetry of the equations to be solved. Bradley and Terry (1952a) and Bradley (1954a) have provided tables giving values of the p_i for equal values of the $n_{ij} = n$, $t = 3$, $n = 1, \ldots, 10$; $t = 4$, $n = 1, \ldots, 8$; $t = 5$, $n = 1, \ldots, 5$.

In small experiments, small values of the n_{ij}, perhaps with poorly selected

Table 2
Values of the estimators in the iterative solution

T_i	$p_i^{(0)}$	$p_i^{(1)}$	$p_i^{(2)}$	$p_i^{(3)}$	$p_i^{(4)}$	$p_i^{(5)}$	p_i
1	0.25	0.1371	0.1188	0.1137	0.1112	0.1101	0.1082
2	0.25	0.4094	0.4656	0.4918	0.5049	0.5131	0.5193
3	0.25	0.2495	0.2413	0.2357	0.2327	0.2290	0.2294
4	0.25	0.2040	0.1743	0.1588	0.1512	0.1478	0.1431

treatments, the estimates p_i may define a point on a boundary of the parameter space. These situations may be recognized from tables like Table 1 and require special consideration. As an example, refer to Table 1 and suppose that T_2 and T_3 are always preferred to T_1 and T_4 and Table 1 is unchanged otherwise. Then $a_1 = 23$, $a_2 = 244$, $a_3 = 71$ and $a_4 = 34$. Treatments T_2 and T_3 dominate T_1 and T_4 and information on the relative values of T_2 and T_3 comes only from the direct comparisons of T_2 and T_3. It follows that $p_1 = 0$, $p_2 = \frac{46}{63} = 0.7302$, $p_3 = \frac{17}{63} = 0.2698$, and $p_4 = 0$. But there is also information on the relative values of π_1 and π_4. We find $p_1/p_4 = 23/34 = 0.4035/0.5965$ and can write $p_1 = 0.4035\delta$ and $p_4 = 0.5965\delta$, δ infinitesimal. A formal analysis may be conducted through maximization of $\log L$ with respect to π_1^*, π_2, π_3, π_4^*, $\pi_2 + \pi_3 = 1$, $\pi_1^* + \pi_4^* = 1$, where $\pi_1 = \delta\pi_1^*$, $\pi_4 = \delta\pi_4^*$ and δ is small. Indeed, the maximum value of $\log L$ may be found in this way and it is needed in the computation of likelihood ratios as discussed below. Bradley (1954a) provides additional discussion of these special boundary problems, problems not usually encountered in applications.

3.2. *Tests of hypotheses*

(i) The major test proposed by Bradley and Terry (1952a) was that of treatment preference or selection equality. The null hypothesis is

$$H_0: \quad \pi_1 = \pi_2 = \cdots = \pi_t = 1/t \tag{3.4}$$

and the general alternative hypothesis is

$$H_a: \quad \pi_i \neq \pi_j \text{ for some } i, j, i \neq j, i, j = 1, \ldots, t. \tag{3.5}$$

If we designate the likelihood ratio as λ_1, it is easy to show that

$$-2\log\lambda_1 = 2N\log 2 - 2B_1, \quad N = \sum_{i<j} n_{ij},$$
$$B_1 = \sum_{i<j} n_{ij}\log(p_i + p_j) - \sum_i a_i \log p_i. \tag{3.6}$$

For large n_{ij}, $-2\log\lambda_1$ has the central chi-square distribution with $(t-1)$ degrees of freedom under H_0. Values of B_1, together with exact significance levels, were provided with the cited tables[1] of estimators p_i. Comparison of significance levels for the large-sample test with small-sample exact significance levels in the tables suggest that the former may be used for modest values of the n_{ij}, a situation perhaps comparable to use of the normal approximation to the binomial.

[1]Common logarithms were used to compute B_1 in these tables. In this paper, natural logarithms are used throughout.

For the values of the a_i of Table 1, the noted values of the n_{ij} above that table, $N = 372$, and the values of the p_i in Table 2, we have $B_1 = 206.3214$ and $-2 \log \lambda_1 = 103.06$ with 3 degrees of freedom. There is a clear indication that the π_i are not equal and that treatment preferences differ.

(ii) It is always encumbent on statisticians to check the validity of models used in statistical analyses when possible. We have noted above that a general 'multi-binomial' model with $\binom{t}{2}$ functionally independent parameters π_{ij} may be posed that ignores the structure of paired comparisons in the sense that the same treatment is compared with more than one other treatment. The multi-binomial model fits the data of tables like Table 1 perfectly. This permits a test of the more restrictive model of (2.1).

The following likelihood ratio test was proposed by Bradley (1954b) and extended by Dykstra (1960). Consider the null hypothesis,

$$H_0: \quad \pi_{ij} = \pi_i/(\pi_i + \pi_j), \; i \neq j, \; i, j = 1, \ldots, t, \tag{3.7}$$

and the alternative hypothesis,

$$H_a: \quad \pi_{ij} \neq \pi_i/(\pi_i + \pi_j) \text{ for some } i, j, i \neq j. \tag{3.8}$$

Under H_a, the likelihood estimator of π_{ij} is $p_{ij} = a_{ij}/n_{ij}$ when $n_{ij} > 0$ and the estimator is not needed when $n_{ij} = 0$. Under H_0, p_i is the estimator of π_i from equations (3.2) and (3.3). Designating λ_2 as the likelihood ratio statistic, we have

$$-2 \log \lambda_2 = 2\Big(\sum_{i \neq j} a_{ij} \log a_{ij} - \sum_{i<j} n_{ij} \log n_{ij} + B_1\Big). \tag{3.9}$$

For large n_{ij}, $-2 \log \lambda_2$ is taken to have the chi-square distribution with $\binom{t}{2} - (t-1) = \frac{1}{2}(t-1)(t-2)$ degrees of freedom under H_0. An alternative statistic, asymptotically equivalent to that of (3.9), is

$$\chi^2 = \sum_{i \neq j} (a_{ij} - a'_{ij})^2/a'_{ij}, \tag{3.10}$$

where $a'_{ij} = n_{ij}p_i/(p_i + p_j)$. This alternate form may be rewritten,

$$\chi^2 = \sum_{i \neq j} n_{ij}\{p_{ij} - [p_i/(p_i + p_j)]\}^2/[p_i/(p_i + p_j)]. \tag{3.11}$$

Dykstra has noted that the test statistics may be distorted when some n_{ij} are small. Since there is no basis for pooling terms in this case, he suggested omitting terms in (3.11) with very small values of n_{ij} (and hence n_{ji}) and deleting one degree of freedom for each pair of terms so deleted.

For the data of Table 1, $n_{34} = 0$ and the tests for the fit of the model have

$\frac{1}{2}(3)(2)-1=2$ degrees of freedom. From (3.9), $-2\log\lambda_2=2.02$ and there seems to be no reason to doubt the appropriateness of the model (2.1). The statistic in (3.10) is evaluated also for illustrative purposes. Values of the a'_{ij} are given in Table 3 and they may be compared directly with the values of a_{ij} in Table 1. Computation yields $\chi^2=2.00$; the close agreement of the two computations is typical.

Table 3
Estimated frequencies for the data of Table 1

	T_1	T_2	T_3	T_4	Row Sums
T_1	—	24.14	17.31	24.54	65.99
T_2	115.86	—	43.70	45.47	205.03
T_3	36.69	19.30	—	—	55.99
T_4	32.46	12.53	—	—	44.99

In the author's fairly extensive experience in fitting model (2.1) to data in food technology and consumer testing, the model is usually found to fit well. When the model does not fit, one or more treatments are often found to possess a characteristic not found in the others, possibly leading to preference judgments influenced by this attribute when such treatments are in a comparison.

(iii) In some uses of paired comparisons, responses may be obtained for several demographic groups, under different evaluation conditions, or other criterion for grouping responses. The possibility of group by treatment interaction or preference disagreement arises and this may be tested.

Let $u=1,\dots,g$ index groups of responses in paired comparisons, let π_i^u be the treatment parameter for T_i in group u, and suppose that sufficient comparisons are made within each group to obtain p_i^u, the estimator of π_i^u, $i=1,\dots,t$. Interest is in the hypotheses,

$$H_0:\quad \pi_i^u=\pi_i,\ i=1,\dots,t,\ u=1,\dots,g, \tag{3.12}$$

and

$$H_a:\quad \pi_i^u\neq\pi_i \text{ for some } i \text{ and } u. \tag{3.13}$$

The likelihood ratio test depends on

$$-2\log\lambda_3=2\Big(B_1-\sum_{u=1}^{g}B_{1u}\Big),$$

where B_{1u} is computed from (3.6) for the data within group u and B_1 is computed similarly for the pooled data from all of the groups. For large values of the n_{iju}, the number of comparisons of T_i and T_j in group u, $-2\log\lambda_3$ has the central chi-square distribution with $(g-1)(t-1)$ degrees of freedom under H_0 of (3.12).

An omnibus test of treatment equality may be described:

$$H_0: \quad \pi_i^u = 1/t, \quad i = 1, \ldots, t, \quad u = 1, \ldots, g,$$

$$H_a: \quad \pi_i^u \neq 1/t \text{ for some } i \text{ and } u,$$

$$-2 \log \lambda_4 = 2N \log 2 - 2 \sum_{u=1}^{g} B_{1u}, \quad N = \sum_{u=1}^{g} N_u = \sum_u \sum_{i<j} n_{iju}.$$

The test statistic is taken to have the chi-square distribution with $g(t-1)$ degrees of freedom under H_0. An analysis of chi-square table may be formed: $-2 \log \lambda_4 = -2 \log \lambda_3 - 2 \log \lambda_1$, where $-2 \log \lambda_1$ is the test statistic of (3.6) based on the pooled data.

Bradley and Terry (1952a) gave a small example for two tasters evaluating pork roasts from hogs with differing diets, $t = 3$, $g = 2$, $n_{iju} = 5$ for all $i, j, u, i \neq j$. The data are summarized in Table 4, and Table 5 is the analysis of chi-square table. The large total treatment effect is seen to be due to disagreement of the two judges on preferences.

Table 4
Roast pork preference data for two judges

Diet	Judge 1		Judge 2		Pooled Data	
T_i	$a_i^{(1)}$	$p_i^{(1)}$	$a_i^{(2)}$	$p_i^{(2)}$	a_i	p_i
1	1	0.0526	7	0.5324	8	0.2479
2	7	0.4737	5	0.2993	12	0.4268
3	7	0.4737	3	0.1683	10	0.3253
	$B_{11} = 6.7166$		$B_{12} = 9.2895$		$B_1 = 20.2565$	

Table 5
Analysis of chi-square, roast pork data

Test	Statistic	d.f.	χ^2
Treatments, given agreement	$-2 \log \lambda_1$	2	1.07
Judge by Treatment Interaction	$-2 \log \lambda_3$	2	8.50
Treatments	$-2 \log \lambda_4$	4	9.58

(iv) Tests for specified treatment contrasts, contrasts on the log π_i, may be made by the method of Section 5.

Bradley and Terry (1952a) proposed one additional test. It was assumed that the treatments fell into two groups, say $T_1, \ldots, T_s$ and $T_{s+1}, \ldots, T_t$, with $\pi_1 = \cdots = \pi_s = \pi$ and $\pi_{s+1} = \cdots = \pi_t = (1 - s\pi)/(t - s)$. The test is of the equality of π and $(1 - s\pi)/(t - s)$, or equivalently of $\pi_i = 1/t$, $i = 1, \ldots, t$, against the two-group alternative of the assumption. The reader is referred to the reference for details.

3.3. Confidence regions

Large-sample theory may be used to obtain variances and covariances for the estimators $p_1, \ldots, p_t$ or their logarithms in paired comparisons. Bradley (1955) considered this theory with each $n_{ij} = n$ and Davidson and Bradley (1970), considering the multivariate model discussed in Section 6, obtained results for general n_{ij} as a special case.

Let $\mu_{ij} = n_{ij}/N$, $N = \Sigma_{i<j}\, n_{ij}$. Then $\sqrt{N}(p_1 - \pi_1), \ldots, \sqrt{N}(p_t - \pi_t)$ have the singular multivariate normal distribution of dimensionality $(t-1)$ in a space of t dimensions with zero mean vector and dispersion matrix $\boldsymbol{\Sigma} = [\sigma_{ij}]$ such that

$$\sigma_{ij} = \text{cofactor of } \lambda_{ij} \text{ in } \begin{bmatrix} \boldsymbol{\Lambda} & \mathbf{1} \\ \mathbf{1}' & 0 \end{bmatrix} \Big/ \begin{vmatrix} \boldsymbol{\Lambda} & \mathbf{1} \\ \mathbf{1}' & 0 \end{vmatrix}, \tag{3.14}$$

where $\boldsymbol{\Lambda} = [\lambda_{ij}]$, $\mathbf{1}'$ is the t-dimensional unit row vector, and

$$\lambda_{ii} = \frac{1}{\pi_i} \sum_{\substack{j \\ j \neq i}} \mu_{ij}\pi_j/(\pi_i + \pi_j)^2, \quad i = 1, \ldots, t,$$

and

$$\lambda_{ij} = -\mu_{ij}/(\pi_i + \pi_j)^2, \quad i \neq j,\ i, j = 1, \ldots, t. \tag{3.15}$$

In order to use these results in applications, σ_{ij} must be estimated; this is done through substitution of p_i for π_i in (3.15) to obtain the $\hat{\lambda}_{ij}$, and subsequent substitution in (3.14) yields the $\hat{\sigma}_{ij}$'s.

For the data of Table 1, values of $p_1, \ldots, p_4$ in Table 2 are used to obtain

$$\hat{\boldsymbol{\Lambda}} = \begin{bmatrix} 10.4963 & -0.9558 & -1.2740 & -20.4259 \\ -0.9558 & 0.4304 & -0.3022 & -0.3553 \\ -1.2740 & -0.3022 & 0.7441 & 0 \\ -2.4259 & -0.3553 & 0 & 3.1237 \end{bmatrix}$$

from whence

$$\hat{\boldsymbol{\Sigma}} = \begin{bmatrix} 0.0800 & -0.0695 & -0.0314 & 0.0208 \\ -0.0695 & 0.6644 & -0.4689 & -0.1260 \\ -0.0314 & -0.4689 & 0.6784 & -0.1781 \\ 0.0208 & -0.1260 & -0.1781 & 0.2833 \end{bmatrix}. \tag{3.16}$$

Note that $\hat{\boldsymbol{\Sigma}}$ is singular, the row and column sums being zero.

Approximate confidence regions may be obtained. The confidence interval on π_i is developed from the fact that $\sqrt{N}(p_i - \pi_i)/\sqrt{\hat{\sigma}_{ii}}$ is standard normal for large N. In the example, the 0.95-confidence interval for π_1 is (0.0795, 0.1369). Let $\boldsymbol{\pi}^*$ be a vector containing any subset of t^* distinct parameters of the set, $t^* < t$. The $(1-\alpha)$-confidence region for these t^* parameters is that ellipsoidal region of the parameter subspace for which

$$N(\boldsymbol{\pi}^* - \boldsymbol{p}^*)'\hat{\boldsymbol{\Sigma}}^{*-1}(\boldsymbol{\pi}^* - \boldsymbol{p}^*) \leq \chi^2_{\alpha,t^*} . \tag{3.17}$$

In (3.17), $\boldsymbol{p}^*$ is the vector of estimates corresponding to $\boldsymbol{\pi}^*$, $\hat{\boldsymbol{\Sigma}}^*$ is the dispersion matrix for $\sqrt{N}(\boldsymbol{p}^* - \boldsymbol{\pi}^*)$ obtainable from (3.16), and χ^2_{α,t^*} is the $(1-\alpha)$-percentage point of the central chi-square distribution with t^* degrees of freedom. As an example, let $\boldsymbol{\pi}^* = (\pi_1, \pi_2)'$ and then $\boldsymbol{p}^* = (0.1082, 0.5193)'$,

$$\hat{\boldsymbol{\Sigma}}^* = \begin{bmatrix} 0.0800 & -0.0695 \\ -0.0695 & 0.6644 \end{bmatrix} \quad \text{and} \quad \hat{\boldsymbol{\Sigma}}^{*-1} = \begin{bmatrix} 13.7441 & 1.4372 \\ 1.4372 & 1.6553 \end{bmatrix} .$$

With $\alpha = 0.01$, $t^* = 2$, $\chi^2_{0.01,2} = 9.210$, it may be verified that (3.17) yields the 0.99-confidence region,

$$13.7441(\pi_1 - 0.1082)^2 + 1.6553(\pi_2 - 0.5193)^2$$
$$+ 2.8744(\pi_1 - 0.1082)(\pi_2 - 0.5193) \leq 0.0248 .$$

Since it may be appropriate to regard $\log \pi_i$ as the location parameter for T_i, $i = 1, \ldots, t$, in view of (2.4) and (2.5), confidence intervals or regions on the $\log \pi_i$ may be desired. It follows that $\sqrt{N}(\log p_1 - \log \pi_1), \ldots, \sqrt{N}(\log p_t - \log \pi_t)$ have the singular multivariate normal distribution with zero mean vector and dispersion matrix $\boldsymbol{D\Sigma D}$ where $\boldsymbol{D}$ is the diagonal matrix with typical element $1/\pi_i$. Estimated variances and covariances are as follows: est.var.$(\sqrt{N} \log p_i) = \hat{\sigma}_{ii}/p_i^2$, est.covar.$(\sqrt{N} \log p_i,\ \sqrt{N} \log p_j) = \hat{\sigma}_{ij}/p_ip_j$, $i \neq j$. Confidence intervals or regions on the $\log \pi_i$ may be obtained analogously to those shown above for the π_i. If a method of multiple comparisons is to be used, the necessary variances and covariances may be obtained from the information given.

In the very special case when each $n_{ij} = n$, approximate variances and covariances may be obtained if the treatments are not too disparate. Then, on the assumption that $\pi_i = 1/t$, $i = 1, \ldots, t$, $\sigma_{ii} = 2(t-1)^2/t^3$ and $\sigma_{ij} = -2(t-1)/t^3$, $i \neq j$, while $N = n\binom{t}{2}$. Like the binomial with its stable variance for its parameter in a middle range, so are the variances and covariances stable in paired comparisons when the π_i are near $1/t$ and the $n_{ij} = n$. This can reduce computational effort for balanced experiments.

3.4. *Asymptotic relative efficiency*

It is well known that the asymptotic relative efficiency of the sign test to the Student test is $2/\pi$ when assumptions for the latter apply and appropriate data could be obtained. Bradley (1955) showed that, under similar conditions, the asymptotic relative efficiency of paired comparisons relative to a randomized complete block design with the same number of treatment replications is $t/\pi(t-1)$, when each $n_{ij} = n$. This result may be adjusted to show that the relative efficiency of paired comparisons relative to the analysis of variance for the similar balanced incomplete block design is $2/\pi$ by the methods of Raghavarao (1971, Sections 4.3 and 4.5).

While the asymptotic relative efficiency factor of $2/\pi$ suggests loss of efficiency through use of the ranking or preference designations of paired comparisons, the method is usually used because measurement scales are not available for sensory or judgment evaluations.

4. Extensions of the basic model

4.1. Adjustments for ties

The basic paired comparisons experiment forces decision on the part of the respondent and data like those of Table 1 result. Nevertheless, ties or 'non-selection' judgments often arise, for example, in consumer testing.

The treatment of ties in the sign test has received considerable attention. Hemelrijk (1952) demonstrated that the most powerful test of significance was obtained by omission of ties and use of a conditional binomial test on the sample results so reduced. But the treatment ties must depend on experimental objectives, see Gridgeman (1959), and estimation of potential share of a consumer market surely must require other considerations. Decisions for paired comparisons must be similar to those for the sign test. Two formal methods for the treatment of ties in paired comparisons are available.

Rao and Kupper (1967) introduced a parameter $\theta \geqslant 1$ and adjusted probabilities associated with the comparison of T_i and T_j to obtain

$$P(T_i \rightarrow T_j) = \pi_i/(\pi_i + \theta\pi_j) = \tfrac{1}{4}\int_{-(\log \pi_i - \log \pi_j)+\eta}^{\infty} \operatorname{sech}^2 y/2 \, \mathrm{d}y ,$$

and

$$P(T_i = T_j) = (\theta^2 - 1)\pi_i\pi_j/(\pi_i + \theta\pi_j)(\theta\pi_i + \pi_j)$$
$$= \int_{-(\log \pi_i - \log \pi_j)-\eta}^{-(\log \pi_i - \log \pi_j)+\eta} \operatorname{sech}^2 y/2 \, \mathrm{d}y, \quad i \neq j, \tag{4.1}$$

where $\eta = \log \theta$. It is seen that the model extends the linear model of (2.4) and that $\log \theta$ is, in a sense, a threshold parameter associated with discriminatory ability.

Rao and Kupper extended the theory in parallel with that given above. Unfortunately, they assumed that $n_{ij} = n$, but the work is easily extended. We summarize only the results leading to the test of treatment equality, although they provide other asymptotic results including variances and covariances for their estimators. We use our notation. Let $N = \Sigma_{i<j}\, n_{ij}$ and b_{ij} be the sum of the number of ties and the number of preferences for T_i in the n_{ij} comparisons of T_i and T_j. Let $b_i = \Sigma_{j, j\neq i}\, b_{ij}$ and let b_0 be the total number of ties in the experiment. The likelihood equations are:

$$\begin{gathered}\frac{b_i}{p_i} - \sum_{\substack{j\\ j\neq i}} \frac{b_{ij}}{p_i + \hat{\theta}p_j} - \sum_{\substack{j\\ j\neq i}} \frac{b_{ij}}{\hat{\theta}p_i + p_j} = 0, \quad i = 1, \ldots, t, \\ \sum_i p_i = 1, \qquad \frac{b_0\hat{\theta}}{\hat{\theta}^2 - 1} - \tfrac{1}{2}\sum_{i\neq j} \frac{b_{ij}p_j}{p_i + \hat{\theta}p_j} = 0\,,\end{gathered} \tag{4.2}$$

where p_i is the estimator of π_i and $\hat{\theta}$ of θ. The likelihood ratio test of H_0: $\pi_i = 1/t$, $i = 1, \ldots, t$, versus H_a: $\pi_i \neq 1/t$ for some i, leads to the statistic,

$$-2 \log \lambda_1^* = 2N \log 2N - 2b_0 \log 2b_0 - 2(N - b_0)\log(N - b_0) - 2B_1^*\,, \tag{4.3}$$

where

$$B_1^* = \sum_{i\neq j} b_{ij} \log(p_i + \hat{\theta}p_j) - \sum_i b_i \log p_i - b_0 \log(\hat{\theta}^2 - 1)\,. \tag{4.4}$$

Again, for large N and under H_0, $-2\log\lambda_1^*$ has the central chi-square distribution with $(t-1)$ degrees of freedom. An iterative solution of equations (4.2) is suggested by Rao and Kupper. They provided also a test of the hypothesis, $\theta = \theta_0$, against the alternative, $\theta \neq \theta_0$.

Davidson (1970) proposed probabilities corresponding to those of (4.1) as

$$\begin{gathered}P(T_i \to T_j) = \pi_i/(\pi_i + \pi_j + \nu\sqrt{\pi_i\pi_j}) \\ \text{and} \\ P(T_i = T_j) = \nu\sqrt{\pi_i\pi_j}/(\pi_i + \pi_j + \nu\sqrt{\pi_i\pi_j})\,,\end{gathered} \tag{4.5}$$

$\nu \geq 0$. This model preserves the odds ratio, $P(T_i \to T_j)/P(T_j \to T_i) = \pi_i/\pi_j$, consistent with the Luce (1959) choice axiom. In addition, the probability of a tie is a maximum when $\pi_i = \pi_j$ and diminishes as π_i and π_j differ, an intuitively desirable effect.

Let b_{ij}^* be the sum of the number of ties and twice the number of preferences for T_i in the n_{ij} comparisons of T_i and T_j and let $b_i^* = \Sigma_{j,j\neq i}\, b_{ij}$. Davidson's likelihood equations are

$$\begin{gathered}\frac{b_i^*}{p_i} - \sum_{\substack{j\\ j\neq i}} n_{ij}(2 + \hat{\nu}\sqrt{p_j/p_i})/(p_i + p_j + \hat{\nu}\sqrt{p_ip_j}) = 0, \quad i = 1, \ldots, t, \\ \sum_i p_i = 1, \qquad \frac{b_0}{\hat{\nu}} - \sum_{i<j} n_{ij}\sqrt{p_ip_j}/(p_i + p_j + \hat{\nu}\sqrt{p_ip_j}) = 0\,,\end{gathered} \tag{4.6}$$

where p_i is the estimator of π_i and $\hat{\nu}$ of ν. The likelihood ratio statistic corresponding to (4.3) is of the same form with B_1^* replaced by

$$B_1^{**} = \sum_{i<j} n_{ij} \log(p_i + p_j + \hat{\nu}\sqrt{p_ip_j}) - \tfrac{1}{2}\sum_i b_i^* \log p_i - b_0 \log \hat{\nu}\,. \tag{4.7}$$

Davidson also proposed an iterative solution for the equations (4.6) and examined large-sample theory. He showed that the Rao–Kupper test and the Davidson test for treatment equality are asymptotically equivalent.

The choice between the two methods for extending the basic paired comparisons model to a model allowing for ties seems to be a matter of intuitive appeal. Both give very similar results in applications.

4.2. Adjustments for order

In paired comparisons, there is often concern for the effects of order of presentation of the two items in a pair. Experiments are conducted so that, for each pair of treatments, each order of presentation is used equally frequently in an effort to 'balance out' the effects of order. Scheffé (1952) addressed this problem in the analysis of variance. Beaver and Gokhale (1975) extended our basic model to allow for order effects.

Davidson and Beaver (1977) describe the Beaver–Gokhale model as having additive order effects and discuss also a model with multiplicative order effects suggested by Beaver (1976). For the ordered pair (T_i, T_j), Beaver and Gokhale defined

$$P_{ij}(T_i \to T_j) = \frac{\pi_i + \delta_{ij}}{\pi_i + \pi_j}, \qquad P_{ij}(T_j \to T_i) = \frac{\pi_j - \delta_{ij}}{\pi_i + \pi_j} \tag{4.8}$$

and, for the ordered pair (T_j, T_i),

$$P_{ji}(T_i \to T_j) = \frac{\pi_i - \delta_{ij}}{\pi_i + \pi_j}, \qquad P_{ji}(T_j \to T_i) = \frac{\pi_j + \delta_{ij}}{\pi_i + \pi_j}. \tag{4.9}$$

The corresponding probabilities for the model with multiplicative order effects are

$$\begin{aligned} P_{ij}(T_i \to T_j) &= \frac{\theta_{ij}\pi_i}{\theta_{ij}\pi_i + \pi_j}, \qquad & P_{ij}(T_j \to T_i) &= \frac{\pi_j}{\theta_{ij}\pi_i + \pi_j}, \\ P_{ji}(T_i \to T_j) &= \frac{\pi_i}{\pi_i + \theta_{ij}\pi_j}, \qquad & P_{ji}(T_j \to T_i) &= \frac{\theta_{ij}\pi_j}{\pi_i + \theta_{ij}\pi_j}. \end{aligned} \tag{4.10}$$

The model given by (4.8) and (4.9) requires that $|\delta_{ij}| \leq \min(\pi_i, \pi_j)$, an awkward feature, while the model (4.10) only requires that $\theta_{ij} > 0$. Advantages of the multiplicative model (4.10) are:

(i) Preference probabilities depend on the worth parameters π_i and π_j only through the ratio π_i/π_j.

(ii) Model (4.10) admits a sufficient statistic whose dimension is that of the parameter space.

(iii) Model (4.10) is a linear model and, for example,

$$P_{ij}(T_i \to T_j) = \tfrac{1}{4} \int_{-(\log \pi_i - \log \pi_j) - \log \theta_{ij}}^{\infty} \operatorname{sech}^2 y/2 \, dy\,.$$

For these reasons, we limit further discussion to (4.10).

Explicit methodology for model (4.10) does not appear in the statistical literature. Various likelihood ratio tests and associated estimation procedures can be developed easily when needed. We consider only the special case when $\theta_{ij} = \theta$ for all $i \neq j$. Then the likelihood equations are

$$\begin{aligned} &\frac{a_i}{p_i^*} - \sum_{\substack{j \\ j\neq i}} \frac{n_{ij}\hat{\theta}}{(\hat{\theta}p_i^* + p_j^*)} - \sum_{\substack{j \\ j\neq i}} \frac{n_{ij}}{(p_i^* + \hat{\theta}p_j^*)} = 0, \quad i = 1, \ldots, t, \\ &\sum_i p_i^* = 1, \qquad \frac{f}{\hat{\theta}} - \sum_{i\neq j} \frac{n_{ij}p_i^*}{(\hat{\theta}p_i^* + p_j^*)} = 0\,, \end{aligned} \tag{4.11}$$

where f is the total number of preferences for the first presented item of a pair, p_i^* is the estimator of π_i and $\hat{\theta}$ of θ, while n_{ij} is the number of judgments on the ordered pair (T_i, T_j) and n_{ji} is the number of judgments on the ordered pair (T_j, T_i). The likelihood ratio statistic for H_0: $\pi_i = 1/t$, $i = 1, \ldots, t$, versus H_a: $\pi_i \neq 1/t$ for some i in the presence of an order effect is

$$-2 \log \lambda_1^\dagger = 2N \log N - 2f \log f - 2(N-f) \log(N-f) - 2B_1^\dagger\,, \tag{4.12}$$

where

$$B_1^\dagger = \sum_{i\neq j} n_{ij} \log(\hat{\theta}p_i^* + p_j^*) - \sum_i a_i \log p_i^* - f \log \hat{\theta}\,. \tag{4.13}$$

Again, under H_0, $-2 \log \lambda_1^\dagger$ has the central chi-square distribution with $(t-1)$ degrees of freedom. A test for the presence of a common order effect, H_0: $\theta = 1$ versus H_a: $\theta \neq 1$, follows immediately. For this test,

$$-2 \log \lambda = 2(B_1 - B_1^\dagger) \tag{4.14}$$

has the central chi-square distribution with 1 degree of freedom when $\theta = 1$. In (4.14), B_1 is taken from (3.6).

Other tests could be developed. One of interest is the test for a common order effect: H_0: $\theta_{ij} = \theta$ for all $i \neq j$, H_a: $\theta_{ij} \neq \theta$ for some $i, j, i \neq j$. Such a test could be described as a test of order by treatment pair interaction.

Note that neither model for order effects suggests that an effort to balance out the effects of order is exactly right. Note also that both order effects and ties could be important and this is the situation addressed by Davidson and Beaver (1977), who note that the results in (4.11) to (4.14) can be obtained with minor adaptation.

4.3. A Bayesian approach

Davidson and Solomon (1973) considered a Bayesian approach to the estimation of the worth parameters $\pi_1, \ldots, \pi_t$ of paired comparisons. Let $\boldsymbol{a}^0 = [a^0_{ij}]$ and $\boldsymbol{n}^0 = [n^0_{ij}]$, $n^0_{ii} = a^0_{ii} = 0$, $n^0_{ij} = n^0_{ji}$. They formulated a conjugate prior distribution for the parameters,

$$\begin{aligned}\phi(\boldsymbol{\pi}) &= A(\boldsymbol{a}^0, \boldsymbol{n}^0) \prod_{i<j} \pi_i^{a^0_{ij}} \pi_j^{a^0_{ji}} / (\pi_i + \pi_j)^{n^0_{ij}}, \quad \boldsymbol{\pi} \in \Omega, \\ &= A(\boldsymbol{a}^0, \boldsymbol{n}^0) \prod_i \pi_i^{a^0_i} \Big/ \prod_{i<j} (\pi_i + \pi_j)^{n^0_{ij}},\end{aligned} \tag{4.15}$$

where $\Omega = \{\boldsymbol{\pi}: \pi_i \geq 0, i = 1, \ldots, t, \Sigma_i \pi_i = 1\}$. They restricted attention to densities (4.15) for which $a^0_{ij} \geq 0$ and $a^0_{ij} + a^0_{ji} = n^0_{ij}$. They noted that, even with these restrictions, each $(\boldsymbol{a}^0, \boldsymbol{n}^0)$ determines a distinct prior distribution and that the family of priors can represent a wide spectrum of prior beliefs. Davidson and Solomon suggested that the experimenter think of his prior beliefs in terms of a conceptual experiment with n^0_{ij} responses to the pair (T_i, T_j) with a^0_{ij} of them being preferences for T_i. Choice of n^0_{ij} is to be made as a measure of the strength of the experimenter's beliefs on the pair (T_i, T_j).

It is noted that the selection of an estimator for the vector of worth parameters $\boldsymbol{\pi}$ is of central interest. This is to be done on the basis of the prior distribution (4.15) and the results of experimentation summarized in the likelihood function conditioned on $\boldsymbol{\pi}$,

$$l(\boldsymbol{a} \mid \boldsymbol{\pi}) = \prod_i \pi_i^{a_i} \prod_{i<j} \binom{n_{ij}}{a_{ij}} (\pi_i + \pi_j)^{-n_{ij}}. \tag{4.16}$$

The estimator of $\boldsymbol{\pi}$ can be used to estimate pairwise preference probabilities or to provide a ranking of the items or treatments in the experiment.

One estimator of $\boldsymbol{\pi}$ is the mode $\boldsymbol{p}^*$ of the posterior distribution of $\boldsymbol{\pi}$. This mode is shown to be the solution of the set of equations,

$$\frac{a'_i}{p^*_i} - \sum_{\substack{j \\ j\neq i}} \frac{n'_{ij}}{(p^*_i + p^*_j)} = 0, \quad i = 1, \ldots, t, \qquad \sum_i p^*_i = 1, \tag{4.17}$$

where $n'_{ij} = n^0_{ij} + n_{ij}$ and $a'_i = a^0_i + a_i$, $i < j$, $i, j = 1, \ldots, t$. It is seen that the choice of prior distribution led to a natural combination of prior and experimental information as seen from the definitions of n'_{ij} and a'_j. Further, equations (4.17) have the form of equations (3.2) and (3.3).

Davidson and Solomon considered also the Bayes estimator of $\boldsymbol{\pi}$ under a quadratic loss function, namely $\bar{\boldsymbol{p}}$, the mean of the posterior distribution of $\boldsymbol{\pi}$. While they did not obtain a closed expression for $\bar{\boldsymbol{p}}$, they did show that, if $n'_{ij} = n'$ for all $i < j$, the rankings determined by $\boldsymbol{p}^*$ and $\bar{\boldsymbol{p}}$ are identical with the Bayes ranking determined by the posterior score $\boldsymbol{a}'$.

4.4. Triple comparisons

The basic model for paired comparisons can be extended to triple comparisons in at least two ways. Bradley and Terry (1952b) proposed the model,

$$P(T_i \to T_j \to T_k) = \pi_i \pi_j / (\pi_i + \pi_j + \pi_k)(\pi_j + \pi_k) \tag{4.18}$$

for comparison of T_i, T_j and T_k in a triplet, $i \neq j \neq k$, $i, j, k = 1, \ldots, t$. Pendergrass and Bradley (1960) proposed the model,

$$P(T_i \to T_j \to T_k) = \pi_i^2 \pi_j / [\pi_i^2(\pi_j + \pi_k) + \pi_j^2(\pi_i + \pi_k) + \pi_k^2(\pi_i + \pi_j)] . \tag{4.19}$$

In both models, the π's may again be regarded as worth parameters with $\Sigma_i \pi_i = 1$. Both models have some desirable properties as discussed in the second reference. Model (4.18) is consistent with the Luce choice axiom and can be written as a Lehmann model (see Bradley (1976)). Model (4.19) has the property that the set of treatment rank sums constitutes a set of sufficient statistics for the estimation of $\pi_1, \ldots, \pi_t$. Basic methodology for the second model is well developed including estimation procedures, tests of hypotheses including goodness of fit, and asymptotic theory.

We show only the estimating equations and the basic test for model (4.19). If $p_1, \ldots, p_t$ are the estimators of $\pi_1, \ldots, \pi_t$, they result from solution of the equations,

$$\frac{a_i}{p_i} - \sum_{\substack{j<k \\ j,k \neq i}} \frac{n_{ijk}[2p_i(p_j + p_k) + p_j^2 + p_k^2]}{D_{ijk}(\boldsymbol{p})} = 0 , \quad i = 1, \ldots, t, \qquad \sum_i p_i = 1 , \tag{4.20}$$

where

$$D_{ijk}(p) = p_i^2(p_j + p_k) + p_j^2(p_i + p_k) + p_k^2(p_i + p_j) \tag{4.21}$$

and n_{ijk} is the number of repetitions or rankings on the triplet (T_i, T_j, T_k), $i < j < k$. The quantity a_i in (4.20) is such that

$$a_i = 3 \sum_{\substack{j<k \\ j,k \neq i}} n_{ijk} - R_i ,$$

where R_i is the total sum of ranks for T_i in the experiment. Pendergrass and Bradley suggest iterative means of solution of the equations (4.20) although they held each $n_{ijk} = n$ for all $i < j < k$.

The likelihood ratio test of H_0: $\pi_i = 1/t$, $i = 1, \ldots, t$, versus H_a: $\pi_i \neq 1/t$ for some i, is based on

$$-2 \log \lambda_5 = 2N \log 6 + 2 \sum_i a_i \log p_i - 2 \sum_{i<j<k} n_{ijk} \log D_{ijk}(\boldsymbol{p}) , \tag{4.22}$$

where $N = \Sigma_{i<j<k} n_{ijk}$. Under H_0, $-2 \log \lambda_5$ has the central chi-square distribution with $(t - 1)$ degrees of freedom for large N.

Park (1961) applied the Pendergrass–Bradley procedures to experimental data and compared the results with those from companion experiments using paired comparisons. He found good model fits and estimator agreement.

5. Treatment contrasts and factorials

It became apparent very early in applications of paired comparisons to sensory experimentation that there was need for special analyses when the treatments represent factorial treatment combinations. Abelson and Bradley (1954) attempted to address this need with very limited success and it remained an open problem until solved by Bradley and El-Helbawy (1976). They considered factorial treatment combinations in the more general framework of specified treatment contrasts. This simplified both notation and theory.

In Table 6, we show paired comparisons data for treatments representing a 2^3 factorial set of treatment combinations. The data are taken from Bradley and El-Helbawy (1976) and arise from a consumer preference taste test on coffees, where the factors are brew strength, roast color and coffee brand, each at two levels. Twenty-six preference judgments were obtained on each of the 28 distinct treatment comparisons. Note that it is convenient to replace the typical treatment T_i by $T_{\alpha_1\alpha_2\alpha_3}$, $\alpha_i = 1$ or 0, $i = 1, 2, 3$, so that the subscripts indicate the chosen levels of the factors. We shall return to these data to illustrate use of the general method explained below with factorials.

A general set of orthonormal treatment contrasts is considered in terms of the basic model (2.1) and parameters, $\log \pi_1, \ldots, \log \pi_t$. The scale-determining constraint (2.2) is replaced for convenience by a new one,

$$\sum_{i=1}^{t} \log \pi_i = 0 . \tag{5.1}$$

The change does not affect the estimation of probabilities of pairwise pref-

Table 6
Preference data in coffee testing

		Treatment not preferred, T_β β								
		000	001	010	011	100	101	110	111	a_α
Treatment preferred, T_α	000	—	15	15	16	19	14	19	16	114
	001	11	—	10	15	15	14	15	12	92
	010	11	16	—	15	15	14	18	15	104
	011	10	11	11	—	14	11	15	13	85
	α 100	7	11	11	12	—	9	14	13	77
	101	12	12	12	15	17	—	16	18	102
	110	7	11	8	11	12	10	—	12	71
	111	10	14	11	13	13	8	14	—	83

erences and the two sets of parameters are related: Given $\pi_1, \ldots, \pi_t$ subject to (5.1) and $\pi_1^*, \ldots, \pi_t^*$ with $\Sigma_i \pi_i^* = 1$, $\pi_i^* = \pi_1/\Sigma_i \pi_i$ and $\pi_i = \pi_i^*/(\Pi_i \pi_i^*)^{1/t}$. Let $\boldsymbol{B}_m$ be an $m \times t$ matrix consisting of m zero-sum orthonormal rows, $0 \leq m \leq (t-1)$. A set of m orthonormal treatment contrasts is defined by

$$\boldsymbol{B}_m \log \boldsymbol{\pi}, \tag{5.2}$$

where $\log \boldsymbol{\pi}$ is the t-element column vector with typical element $\log \pi_i$. Clearly a set of contrast representing factor effects or their interactions can be formulated by choice of the elements of $\boldsymbol{B}_m$ exactly as done for usual analysis of variance for factorials.

Only one estimation problem need be considered, the estimation of the elements of $\boldsymbol{\pi}$ given constraints,

$$\begin{bmatrix} \mathbf{1}_t' \\ \boldsymbol{B}_m \end{bmatrix} \log \boldsymbol{\pi} = \boldsymbol{O}_{m+1}, \tag{5.3}$$

where $\mathbf{1}_t'$ is the t-element row vector of unit elements, $\boldsymbol{\pi}' = (\pi_1, \ldots, \pi_t)$, and $\boldsymbol{O}_{m+1}$ is the $(m+1)$-element column vector of zeros. Estimation is by the method of maximum likelihood and the resulting reduced estimating equations are

$$\begin{aligned} &a_i/p_i - \phi_i(\boldsymbol{p}) = 0, \quad i = 1, \ldots, t, \\ &\sum_i \log p_i = 0, \qquad \boldsymbol{B}_m \log \boldsymbol{p} = \boldsymbol{O}_m, \end{aligned} \tag{5.4}$$

where $\boldsymbol{p} = (p_1, \ldots, p_t)$,

$$\phi_i(\boldsymbol{p}) = \sum_{\substack{j \\ j \neq i}} \frac{n_{ij}}{p_i + p_j} - \frac{1}{p_i} \sum_{\substack{j \\ j \neq 1}} E_j(\boldsymbol{p}) \frac{D_{ij}}{D_{ii}}, \tag{5.5}$$

$$E_i(\boldsymbol{p}) = a_i - \sum_{\substack{j \\ j \neq 1}} n_{ij} p_i/(p_i + p_j), \tag{5.6}$$

$i = 1, \ldots, t$, and D_{ij} is the typical element of

$$\boldsymbol{D} = \boldsymbol{I}_t - \boldsymbol{B}_m' \boldsymbol{B}_m, \tag{5.7}$$

$D_{ii} > 0$, $\boldsymbol{I}_t$, the t-square identity matrix. Note the similar forms in (5.6) and (3.2). If $m = 0$, the estimation process involves solution of (3.2) with (3.3) replaced by the second equation of (5.4).

Iterative solution of equations (5.4) is discussed briefly by Bradley and El-Helbawy (1976) and in detail by El-Helbawy and Bradley (1977). In the latter reference, it is shown that the proposed iterative procedure converges and yields a maximum of the likelihood function over the parameter space $\{\boldsymbol{\pi}: \pi_i > 0, i = 1, \ldots, t, \Sigma_i \log \pi_i = 0, \boldsymbol{B}_m \log \boldsymbol{\pi} = \boldsymbol{O}_m\}$.

A class of likelihood ratio tests may be developed. Let $\boldsymbol{B}_{m_a}$, $\boldsymbol{B}_{m_1}$, and

$$\boldsymbol{B}_{m_0} = \begin{bmatrix} \boldsymbol{B}_{m_a} \\ \boldsymbol{B}_{m_1} \end{bmatrix}$$

be matrices like $\boldsymbol{B}_m$, $0 \leq m_a$, $m_1 \leq m_0 \leq (t-1)$, $m_0 = m_a + m_1$. With the condition that $\Sigma_i \log \pi_i = 0$, we test

$$H_0: \quad \boldsymbol{B}_{m_0} \log \boldsymbol{\pi} = \boldsymbol{O}_{m_0} \tag{5.8}$$

against

$$H_a: \quad \boldsymbol{B}_{m_a} \log \boldsymbol{\pi} = \boldsymbol{O}_{m_a} . \tag{5.9}$$

The test statistic is

$$-2 \log \lambda_{m_0, m_a} = 2[B_1(\boldsymbol{p}_0) - B_1(\boldsymbol{p}_a)] , \tag{5.10}$$

where B_1 is defined in (3.6), and, for large $N = \Sigma_{i<j} n_{ij}$ and under H_0 in (5.8), the statistic has the central chi-square distribution with m_1 degrees of freedom. In (5.10), $\boldsymbol{p}_0$ is the solution of (5.4) where $\boldsymbol{B}_m = \boldsymbol{B}_{m_0}$ and $\boldsymbol{p}_a$, the solution when $\boldsymbol{B}_m = \boldsymbol{B}_{m_a}$. Basically, the test involves the assumption that

$$\begin{bmatrix} \boldsymbol{1}'_t \\ \boldsymbol{B}_{m_a} \end{bmatrix} \log \boldsymbol{\pi} = \boldsymbol{O}_{m_a+1}$$

and a test of the additional constraints,

$$\boldsymbol{B}_{m_1} \log \boldsymbol{\pi} = \boldsymbol{O}_{m_1} ,$$

$\boldsymbol{B}_{m_1}$ consisting of m_1 orthonormal rows orthogonal to those of $\boldsymbol{B}_{m_a}$.

The test procedure is illustrated with the data of Table 6. Treatments T_i have subscripts in the lexicographic order of $T_{\boldsymbol{\alpha}}$ in the table. Suppose that we wish to test the hypothesis that there are no two-factor interactions on the assumption that there is no three-factor interaction. Then $t = 8$, $m_a = 1$, $m_1 = 3$, $m_0 = 4$ with

$$\boldsymbol{B}_{m_a} = \frac{1}{\sqrt{8}}(1, -1, -1, 1, -1, 1, 1, -1)$$

and

$$\boldsymbol{B}_{m_1} = \frac{1}{\sqrt{8}} \begin{bmatrix} 1 & 1 & -1 & -1 & -1 & -1 & 1 & 1 \\ 1 & -1 & 1 & -1 & -1 & 1 & -1 & 1 \\ 1 & -1 & -1 & 1 & 1 & -1 & -1 & 1 \end{bmatrix} .$$

Necessary calculations yield:

$$p_0 = (1.300,\ 1.275,\ 1.060,\ 1.040,\ 0.962,\ 0.944,\ 0.784,\ 0.769)\,,$$

$$p_a = (1.515,\ 1.060,\ 1.342,\ 0.855,\ 0.790,\ 1.193,\ 0.647,\ 0.890)\,,$$

$$B_1(p_0) = 497.81, \quad B_1(p_a) = 490.14,$$

$$-2 \log \lambda_{m_0, m_a} = 2(497.81 - 490.14) = 15.34\,.$$

The statistic $-2 \log \lambda_{m_0,m_a}$ has the central chi-square distribution with 3 degrees of freedom and is large. It is possible also to partition this chi-square into three chi-squares, each with 1 degree of freedom, as is done in Table 7.

The general test procedure for hypothesis (5.8) versus (5.9) based on the statistic (5.10) may be used repeatedly to produce an analysis of chi square table. Two such analyses are given in Tables 7 and 8 for the data of Table 6. Rows in these tables correspond to rows of the usual analysis of variance table for a 2^3 factorial and similar descriptive terms have been used. In order to preserve orthogonality of the various chi-squares, they must be sequenced properly; each row requires that certain conditions be assumed, equivalent to the specification of $\boldsymbol{B}_{m_a}$. Both Tables 7 and 8 are shown to illustrate two different sequencings of the rows and to suggest that the choice of sequencing does not have substantial effects on the inferences that may be made. Additional details on computations for Tables 7 and 8 are given by Bradley and El-Helbawy (1976).

The analyses below were done through recognition of the factorial structure of the treatments. Factorial parameters may be introduced formally, although it is not necessary to do so. We illustrate with the 2^3 factorial. Let $\pi_{\boldsymbol{\alpha}}$ replace π_i for the treatment $T_{\boldsymbol{\alpha}} \equiv T_i$, where $\boldsymbol{\alpha} = (\alpha_1, \alpha, \alpha_3)$, $\alpha_r = 0$ or 1, $r = 1, 2, 3$. We reparameterize by writing

$$\pi_\alpha = \prod_{r=1}^{3} \pi^{(r)}_{\alpha_r} \cdot \prod_{r<s} \pi^{(rs)}_{\alpha_r \alpha_s} \cdot \pi^{(123)}_{\alpha_1 \alpha_1 \alpha_3}\,. \tag{5.11}$$

The parameters on the right-hand side of (5.11) are new factorial parameters. The transformation is linear if logarithms are taken; the logarithms of the new factorial parameters are subject to the usual linear constraints for factorial parameters in the analysis of variance in order to make the transformation one-to-one. Estimators of the factorial parameters are functions of the estimators $p_{\boldsymbol{\alpha}}$. A full explanation of these procedures is given by El-Helbawy and Bradley (1976).

Special treatment contrasts may be of interest in paired comparisons. Suppose that, in a coffee taste test experiment with $t = 4$, T_4 represents an experimental coffee produced by a new process while the other treatments came from a standard process. One may wish to compare T_4 with the other three treatments. Two approaches are possible. The first assumes nothing, $m_a = 0$, and takes

Table 7
An analysis of chi-square for the coffee data

Hypothesis tested[a]	Conditions assumed	Degrees of freedom	Chi-square
No F_1 effect	No F_2, F_3, no interactions	1	9.28
No F_2 effect	No F_3, no interactions	1	4.29
No F_3 effect	No interactions	1	0.04
No F_1F_2, F_1F_3, F_2F_3 interactions	No $F_1F_2F_3$ interaction	3	15.34
No F_2F_3 interaction	No $F_1F_2F_3$ interaction	1	0.22
No F_1F_3 interaction	No F_2F_3, $F_1F_2F_3$ interactions	1	14.96
No F_1F_2 interaction	No F_1F_3, F_2F_3, $F_1F_2F_3$ interactions	1	0.15
No $F_1F_2F_3$ interaction	None	1	0.63
No treatment effects	None	7	29.58

[a] F_1 is brew strength, F_2 is roast colour, F_3 is brand.

Table 8
An alternative analysis of chi-square for the coffee data

Hypothesis tested[a]	Conditions assumed	Degrees of freedom	Chi-square
No F_1 effect	None	1	9.47
No F_2 effect	No F_1 effect	1	4.33
No F_3 effect	No F_1, F_2 effects	1	0.04
No F_1F_2, F_1F_3, F_2F_3 interactions	No main effects	3	15.12
No F_1F_2 interaction	No main effects	1	0.16
No F_1F_3 interaction	No main effects, no F_1F_2 interaction	1	14.73
No F_2F_3 interaction	No main effects, no F_1F_2, F_1F_3 interactions	1	0.24
No $F_1F_2F_3$ interaction	No main effects, no two-factor interactions	1	0.62
No treatment effects	None	7	29.58

[a] F_1 is brew strength, F_2 is roast colour, F_3 is brand.

$$B_{m_1} = \frac{1}{\sqrt{12}}(1, 1, 1, -3).$$

The second approach assumes that $\pi_1 = \pi_2 = \pi_3$, $m_a = 2$,

$$B_{m_a} = \begin{bmatrix} 1/\sqrt{2} & -1/\sqrt{2} & 0 & 0 \\ 1/\sqrt{6} & 1/\sqrt{6} & -2/\sqrt{6} & 0 \end{bmatrix}$$

and retains the same B_{m_1}. With these matrices defined, the general test procedure of this section is used.

We have presented a method for the examination of specified treatment contrasts and the analysis of factorial paired comparison experiments together with examples. These methods provide much new flexibility.

6. Multivariate paired comparisons

Multivariate responses to paired comparisons are often obtained. For example, this happens in consumer testing where, on paired samples, preferences on a number of characteristics are solicited.

Davidson and Bradley (1969) extended the paired comparisons model to the multivariate case. Let $s = (s_1, \ldots, s_p)$, $s_\alpha = i$ or j, be the response vector on attributes $\alpha = 1, \ldots, p$ for the treatment pair (T_i, T_j), $s_\alpha = i$ indicating preference for T_i on attribute α. The probability of response s on (T_i, T_j) is

$$P(s \mid i, j) = p^{(1)}(s \mid i, j) h(s \mid i, j), \tag{6.1}$$

where

$$p^{(1)}(s \mid i, j) = \prod_{\alpha=1}^{p} \pi_{\alpha s_\alpha} / (\pi_{\alpha i} + \pi_{\alpha j}) \tag{6.2}$$

and

$$h(s \mid i, j) = 1 + \sum_{\alpha<\beta} \delta(s_\alpha, s_\beta) \rho_{\alpha\beta} (\pi_{\alpha i}/\pi_{\alpha j})^{-\delta(i, s_\alpha)/2} (\pi_{\beta i}/\pi_{\beta j})^{-\delta(i, s_\beta)/2}, \tag{6.3}$$

for all s, $i<j$, $i, j = 1, \ldots, t$. Notation is as follows: $\pi_{\alpha i}$ is the worth parameter for T_i on attribute α, $\Sigma_i \pi_{\alpha i} = 1$, $\rho_{\alpha\beta}$ is a 'correlation' parameter for attributes α and β assumed constant for all treatment pairs, and $\delta(s_\alpha, s_\beta) = 1$ or -1 as the two arguments of the indicator function agree or disagree. Note $\boldsymbol{\rho} = \mathbf{0}$ implies independence of responses on attributes $\boldsymbol{\rho}$ has typical element $\rho_{\alpha\beta}$. It is necessary to restrict the parameter space so that $\pi_{\alpha i} \geq 0$, $\alpha = 1, \ldots, p$, $i = 1, \ldots, t$, and $h(s \mid i, j) \geq 0$ for each of the 2^p cells associated with each of the $\binom{t}{2}$ treatment pairs.

Let

$$B(\boldsymbol{\pi}) = -\sum_{\alpha=1}^{p} B_1(\boldsymbol{\pi}_\alpha) \tag{6.4}$$

and

$$C(\boldsymbol{\pi}, \boldsymbol{\rho}) = \sum_{i<j} \sum_{s} f(s \mid i, j) \log h(s \mid i, j), \tag{6.5}$$

where $\boldsymbol{\pi}$ has typical element $\pi_{\alpha i}$ and $\boldsymbol{\pi}_\alpha$ is the α-th row of $\boldsymbol{\pi}$. The quantity $B_1(\boldsymbol{\pi}_\alpha)$ is the function B_1 of (3.6) with p_i there replaced by $\pi_{\alpha i}$ and a_i replaced by $a_{\alpha i}$, the total number of preferences for T_i on attribute α. In addition, $f(s \mid i, j)$ is the number of times the preference vector s occurs among the n_{ij} responses to the pair (T_i, T_j). We may express the logarithm of the likelihood function as

$$\log L = C(\boldsymbol{\pi}, \boldsymbol{\rho}) + B(\boldsymbol{\pi}). \tag{6.6}$$

Consider first a test for independence: H_0: $\boldsymbol{\rho} = \mathbf{0}$ versus H_a: $\rho_{\alpha\beta} \neq 0$ for some $\alpha < \beta$, $\alpha, \beta = 1, \ldots, p$. Under H_0, the likelihood equations reduce to equations (3.2) and (3.3) for each $\alpha = 1, \ldots, p$. If $\boldsymbol{p}^0_\alpha$ is the solution for the α-th set of equations and becomes the α-th row of $\boldsymbol{p}^0$, $\boldsymbol{p}^0$ estimates $\boldsymbol{\pi}$ under H_0. Under H_a, the equations to be solved are:

$$\begin{aligned} &\sum_{i<j} f(s \mid i, j) h^{-1}(s \mid i, j) \delta(s_\alpha, s_\beta)(\pi_{\alpha i}/\pi_{\alpha j})^{-\delta(i, s_\alpha)/2}(\pi_{\beta i}/\pi_{\beta j})^{-\delta(i, s_\beta)/2}\Big|_{\substack{\boldsymbol{\pi}=\boldsymbol{p}\\ \boldsymbol{\rho}=\hat{\boldsymbol{\rho}}}} = 0, \\ &\qquad \alpha < \beta,\ \alpha, \beta = 1, \ldots, p, \\ &\frac{a_{\alpha i} + R_{\alpha i}}{p_{\alpha i}} - \sum_{\substack{j\\ j\neq i}} \frac{n_{ij}}{p_{\alpha i} + p_{\alpha j}} = 0, \quad i = 1, \ldots, t,\ \alpha = 1, \ldots, p, \\ &\sum_i p_{\alpha i} = 1, \quad \alpha = 1, \ldots, p, \end{aligned} \tag{6.7}$$

where

$$\begin{aligned} R_{\alpha i} = {} & -\tfrac{1}{2} \sum_{\substack{j\\ j\neq i}} \sum_{s} f(s \mid i, j) h^{-1}(s \mid i, j)(\pi_{\alpha i}/\pi_{\alpha j})^{-\delta(i, s_\alpha)/2} \\ & \times \sum_{\substack{\beta\\ \beta\neq\alpha}} \delta(i, s_\beta)\rho_{\alpha\beta}(\pi_{\beta i}/\pi_{\beta j})^{-\delta(i, s_\beta)/2}. \end{aligned} \tag{6.8}$$

Solution of equations (6.7) is discussed by Davidson and Bradley (1969). If we let $\boldsymbol{p}$ and $\hat{\boldsymbol{\rho}}$ be the estimators of $\boldsymbol{\pi}$ and $\boldsymbol{\rho}$ from equations (6.7), the likelihood ratio test statistic is

$$-2 \log \lambda_6 = 2\{B(\boldsymbol{p}) - B(\boldsymbol{p}^0) + C(\boldsymbol{p}, \hat{\boldsymbol{\rho}})\} \tag{6.9}$$

and, under H_0, it has the central chi-square distribution with $\frac{1}{2}p(p-1)$ degrees of freedom.

If it is assumed that $\boldsymbol{\rho} = \mathbf{0}$, tests on the parameters $\boldsymbol{\pi}_\alpha$ may be made separately as in the univariate case for each $\alpha = 1, \ldots, p$.

An overall test of no treatment preferences may be made in the presence of correlations. Then we have H_0: $\boldsymbol{\pi} = [1/t]$ and H_a: $\pi_{\alpha i} \neq 1/t$ for some α and i. Under H_a, the estimators from equations (6.7) are again $\boldsymbol{p}$ and $\hat{\boldsymbol{\rho}}$. Under H_0, the estimators of $\boldsymbol{\pi}$ and $\boldsymbol{\rho}$ are $[1/t]$ and $\hat{\boldsymbol{\rho}}_0$, the latter obtained from solution of (6.7) with $\boldsymbol{p} = [1/t]$. The test statistic is

$$-2 \log \lambda_7 = 2\{B(\boldsymbol{p}) + C(\boldsymbol{p}, \hat{\boldsymbol{\rho}}) + pN \log 2 - C(1/t, \hat{\boldsymbol{\rho}}_0)\} \tag{6.10}$$

with the central chi-square distribution with $p(t-1)$ degrees of freedom under H_0.

A likelihood ratio test of the fit of the model (6.1) is given by Davidson and Bradley. An alternative test may be based on

$$\chi^2 = \sum_{i<j} \sum_{s} \{f(s \mid i, j) - \hat{f}(s \mid i, j)\}^2 / \hat{f}(s \mid i, j) \tag{6.11}$$

and, under the model, has the central chi-square distribution for large N with $\{(2^p - 1)\binom{t}{2} - p(t-1) - \binom{p}{2}\}$ degrees of freedom. The estimators $\boldsymbol{p}$ and $\hat{\boldsymbol{\rho}}$ are substituted in (6.1) to obtain expected cell frequencies $\hat{f}(s \mid i, j) = n_{ij}\hat{P}(s \mid i, j)$.

Davidson and Bradley (1970) examine large-sample properties of procedures discussed above. Davidson and Bradley (1971) examine regression relationships among the characteristics in the multivariate problem.

We conclude this section with one of the examples given by Davidson and Bradley (1969). Table 9 shows the observed and expected cell frequencies, the latter in parentheses, for a chocolate pudding test with $t = 3$, $p = 3$, the treatments being brands, and the attributes being taste, color and texture.

Details on calculations are not given. However, as a possible check on computer programming, the solution of (6.7) is as follows:

$$\boldsymbol{p} = \begin{bmatrix} 0.312 & 0.360 & 0.328 \\ 0.307 & 0.321 & 0.372 \\ 0.338 & 0.288 & 0.374 \end{bmatrix}, \quad \hat{\rho}_{12} = 0.675, \ \hat{\rho}_{13} = 0.654, \ \hat{\rho}_{23} = 0.588\,.$$

Table 9
Observed and expected cell frequencies for a chocolate pudding test

Treatment pair	Cell frequencies $f(s \mid i, j)$								Frequency
i, j	Cells s								n_{ij}
	(iii)	(jii)	(iji)	(jji)	(iij)	(jij)	(ijj)	(jjj)	
1, 2	8	1	1	1	0	2	0	9	22
	(7.93)	(1.09)	(1.15)	(1.69)	(0.76)	(0.97)	(0.37)	(8.03)	
1, 3	6	0	1	1	1	0	1	9	19
	(6.25)	(0.60)	(1.24)	(0.92)	(1.12)	(0.62)	(0.64)	(7.61)	
2, 3	7	1	1	1	3	1	1	6	21
	(6.92)	(0.37)	(1.26)	(0.60)	(1.70)	(0.75)	(1.10)	(8.31)	

Table 10
Test statistics for hypotheses for the chocolate pudding data

Test	Statistic	Ref. No.	Value	d.f.
Test of Independence	$-2 \log \lambda_6$	(6.9)	62.665	3
Test of Equal Preferences	$-2 \log \lambda_7$	(6.10)	2.362	6
Test of Model Fit	χ^2	(6.11)	7.557	12

Tests are summarized in Table 10. It is seen that the major effects are the high correlations among responses on attributes.

As a final comment on the example, cell frequencies are small and asymptotic theory must be regarded only as approximate. The tests do, however, seem to work well and be adequately indicative.

7. Other methods of paired comparisons

Our efforts in this chapter have concentrated on one method of paired comparisons and extensions. This was done because it has been most fully developed and has been found to work well in applications. Even so, it has been necessary to be brief and applications require computer programs that are easily developed after review of pertinent references for additional detail.

We have seen that the Thurstone model is very similar to the one used here. It has had less attention. However, three papers do extend the Thurstone model: Harris (1957) generalized the model to allow for possible order effects, Glenn and David (1960) allowed for ties, and Sadasivan (1982) permitted unequal numbers of judgments on pairs.

Other approaches to the analysis of paired comparisons exist. Kendall and Babington Smith (1940) considered the count of circular triads as a measure of consistency of judgments and also developed a coefficient of concordance as a measure of agreement of judgments by several judges. Guttman (1946) developed a method of scaling treatments in paired comparisons, the objective of Zermello. Saaty (1977) proposed a consensus method through evaluation by group discussion to provide treatment or item scores on a ratio scale. Bliss, Greenwood and White (1956) used 'rankits' in the analysis of paired comparisons. Mehra (1964) and Puri and Sen (1969) extended the idea of signed ranks to paired comparisons. Fienberg and Larntz (1976) have proposed a log-linear analysis of paired comparisons and Grizzle, Starmer and Koch (1969) have used a weighted least squares approach. Wei (1952) and Kendall (1955) have proposed an iterative scoring system that takes into account not only direct comparisons but also roundabout comparisons involving other items.

No attention has been given here to the design of tournaments. There is an extensive litetature on this subject included in the Davidson–Farquhar bibliography.

References

Abelson, R. M. and Bradley, R. A. (1954). A 2×2 factorial with paired comparisons. *Biometrics* **10**, 487–502.

Beaver, R. J. (1976). Discussion: Science, statistics and paired comparisons. *Biometrics* **32**, 233–235.

Beaver, R. J. and Gokhale, D. V. (1975). A model to incorporate within-pair order effects in paired comparisons. *Comm. Statist.-Theor. Meth.* **4**, 923–939.

Bliss, C. I., Greenwood, M. L. and White, E. S. (1956). A rankit analysis of paired comparisons for measuring the effects of sprays on flavor. *Biometrics* **12**, 381–403.

Bradley, R. A. (1953). Some statistical methods in taste testing and quality evaluation. *Biometrics* **9**, 22–38.

Bradley, R. A. (1954a). The rank analysis of incomplete block designs. II. Additional tables for the method of paired comparisons. *Biometrika* **41**, 502–537.

Bradley, R. A. (1954b). Incomplete block rank analysis: On the appropriateness of the model for a method of paired comparisons. *Biometrics* **10**, 375–390.

Bradley, R. A. (1955). Rank analysis of incomplete block designs. III. Some large-sample results on estimation and power for a method of paired comparisons. *Biometrika* **42**, 450–470.

Bradley, R. A. (1976). Science, statistics and paired comparisons. *Biometrics* **32**, 213–232.

Bradley, R. A. and El-Helbawy, A. T. (1976). Treatment contrasts in paired comparisons: Basic procedures with application to factorials. *Biometrika* **63**, 255–262.

Bradley, R. A. and Terry, M. E. (1952a). The rank analysis of incomplete block designs. I. The method of paired comparisons. *Biometrika* **39**, 324–345.

Bradley, R. A. and Terry, M. E. (1952b). *Statistical Methods for Sensory Difference Tests of Food Quality, Appendix A.* Biannual Rpt. No. 4, Virginia Agric. Exp. Sta., Blacksburg, Va.

Clatworthy, W. H. (1955). Partially balanced incomplete block designs with two associate classes and two treatments per block. *J. Res. Nat. Bur. Stand.* **54**, 177–190.

David, H. A. (1963). *The Method of Paired Comparisons.* Griffin, London.

Davidson, R. R. (1970). On extending the Bradley-Terry model to accommodate ties in paired comparison experiments. *J. Amer. Statist. Assoc.* **65**, 317–328.

Davidson, R. R. and Beaver, R. J. (1977). On extending the Bradley-Terry model to incorporate within-pair order effects. *Biometrics* **33**, 245–254.

Davidson, R. R. and Bradley, R. A. (1969). Multivariate paired comparisons: The extension of a univariate model and associated estimation and test procedures. *Biometrika* **56**, 81–95.

Davidson, R. R. and Bradley, R. A. (1970). Multivariate paired comparisons: Some large sample results on estimation and tests of equality of preference. In: M. L. Puri, ed., *Nonparametric Techniques in Statistical Inference.* Cambridge University Press, 111–125.

Davidson, R. R. and Bradley, R. A. (1971). A regression relationship for multivariate paired comparisons. *Biometrika* **58**, 555–560.

Davidson, R. R. and Farquhar, P. H. (1976). A bibliography on the method of paired comparisons. *Biometrics* **32**, 241–252.

Davidson, R. R. and Solomon, D. L. (1973). A Bayesian approach to paired comparison experimentation. *Biometrika* **60**, 477–487.

Dykstra, O. (1956). A note on the rank analysis of incomplete block designs – applications beyond the scope of existing tables. *Biometrics* **12**, 301–306.

Dykstra, O. (1960). Rank analysis of incomplete block designs: A method of paired comparisons employing unequal repetitions on pairs. *Biometrics* **16**, 176–188.

El-Helbawy, A. T. and Bradley, R. A. (1976). Factorial treatment combinations in paired comparisons. In: *Proc. Int. Conf. on Statistics, Computer Science and Social Research* **1**, 121.1-144.1, Cairo University Press, Cairo.

El-Helbawy, A. T. and Bradley, R. A. (1977). Treatment contrasts in paired comparisons: Convergence of a basic iterative scheme for estimation. *Commun. Statistic. – Theor. Meth.* **6**, 197–207.

El-Helbawy, A. T. and Bradley, R. A. (1978). Treatment contrasts in paired comparisons: Large-sample results, applications and some optimal designs. *J. Amer. Statist. Assoc.* **73**, 831–839.

Fienberg, S. E. and Larntz, K. (1976). Log linear representation for paired and multiple comparisons models. *Biometrika* **63**, 245–262.

Fleckenstein, M., Freund, R. A. and Jackson, J. E. (1958). A paired comparison test of typewriter carbon papers. *Tappi* **41**, 128–130.

Ford, L. R. Jr. (1957). Solution of a ranking problem from binary comparisons. *Amer. Math. Monthly* **64**(8), 28–33.

Glenn, W. A. and David, H. A. (1960). Ties in paired-comparison experiments using a modified Thurstone-Mosteller method. *Biometrics* **16**, 86–109.

Gridgeman, N. T. (1959). Pair comparison, with and without ties. *Biometrics* **15**, 382–388.

Grizzle, J. E., Starmer, C. F. and Koch, G. G. (1969). Analysis of categorical data by linear models. *Biometrics* **25**, 489–504.

Guttman, L. (1946). An approach for quantifying paired comparisons and rank order. *Ann. Math. Statist.* **17**, 144–163.

Harris, W. P. (1957). A revised law of comparative judgment. *Psychometrika* **22**, 189–198.

Hemelrijk, J. (1952). A theorem on the sign test when ties are present. *Indag. Math.* **14**, 322–326.

Kendall, M. G. (1955). Further contributions to the theory of paired comparisons. *Biometrics* **11**, 43–62.

Kendall, M. G. and Babington Smith, B. (1940). On the method of paired comparisons. *Biometrika* **31**, 324–345.

Luce, R. D. (1959). *Individual Choice Behavior.* Wiley, New York.

Mehra, K. L. (1964). Rank tests for paired-comparison experiments involving several treatments. *Ann. Math. Statist.* **35**, 122–137.

Mosteller, F. (1951). Remarks on the method of paired comparisons: I. The least squares solution assuming equal standard deviations and equal correlations. *Psychometrika* **16**, 3–9.

Park, G. T. (1961). Sensory testing by triple comparisons. *Biometrics* **17**, 251–260.

Pendergrass, R. N. and Bradley, R. A. (1960). Ranking in triple comparisons. In: I. Olkin et al., eds., *Contributions to Probability and Statistics.* Stanford Univ. Press, 331–351.

Puri, M. L. and Sen, P. K. (1969). On the asymptotic theory of rank order tests for experiments involving paired comparisons. *Ann. Inst. Statist. Math.* **21**, 163–173.

Raghavarao, D. (1971). *Construction and Combinatorial Problems in Design of Experiments.* Wiley, New York.

Rao, P. V. and Kupper, L. L. (1967). Ties in paired-comparison experiments: A generalization of the Bradley-Terry model. *J. Amer. Statist. Assoc.* **62**, 194–204, Corrigenda **63**, 1550.

Saaty, T. L. (1977). A scaling method for priorities in hierarchical structures. *J. Math. Psych.* **15**, 234–280.

Sadasivan, G. (1982). A Thurstone-type model for paired comparisons with unequal numbers of repetitions. *Comm. Statist. – Theor. Meth.* **11**, 821–833.

Scheffé, H. (1952). An analysis of variance for paired comparisons. *J. Amer. Statist. Assoc.* **47**, 381–400.

Thurstone, L. L. (1927). Psychophysical analysis. *Amer. J. Psych.* **38**, 368–389.

Wei, T. H. (1952). The Algebraic Foundations of Ranking Theory. Unpublished thesis, Cambridge University.

Zermelo, E. (1929). Die Berechnung der Turnier-Ergebnisse als ein Maximumproblem der Wahrscheinlichkeitsrechnung. *Math. Zeit.* **29**, 436–460.

P. R. Krishnaiah and P. K. Sen, eds., *Handbook of Statistics, Vol. 4*
© Elsevier Science Publishers (1984) 327–345

15

Restricted Alternatives

Shoutir Kishore Chatterjee

1. Introduction

In many testing problems under both the parametric and the nonparametric set-up the aspect of the underlying model which is of direct interest to the experimenter may be described by a k-dimensional ($k \geq 1$) real vector $(\theta_1, \theta_2, \ldots, \theta_k)' = \boldsymbol{\theta}$ and the null hypothesis is H_0: $\boldsymbol{\theta} = \mathbf{0}$. In such situations, so far as the choice of the structure of the test is concerned, the most important consideration should be which alternative values of $\boldsymbol{\theta}$ we are especially interested in detecting.

Traditional tests for H_0 have been developed keeping in view all possible alternative values of $\boldsymbol{\theta}$. These are tests against unrestricted alternatives or *unrestricted tests* (of Chapters 1, 5). However, in practice often experimental conditions are such that $\boldsymbol{\theta}$, if it deviates from $\mathbf{0}$, can do so only in some but not all directions. For example, in an agricultural trial $\theta_1, \theta_2, \ldots, \theta_k$ may represent the effects (reparametrized so that $\Sigma\, \theta_i = 0$) of increasing doses of a manure, and if the manure is effective at all, it may be expected that $\theta_1, \theta_2, \ldots, \theta_k$ would have progressively increasing values in that order. Similarly, $\theta_1, \theta_2, \ldots, \theta_k$ may represent the changes in the levels of some physiological character produced by a drug, and it may be that, if the drug is effective, one or more of the θ_i's would be positive. While it would be perfectly legitimate to use the unrestricted tests in such situations, it is obvious that it would be more paying if the test, instead of being sensitive to deviations in all directions, concentrates on the relevant directions only. In other words, in such situations, specific tests against restricted alternatives or *restricted tests* are called for.

2. A brief review

When we have a real parameter θ ($k = 1$) and want to test H_0: $\theta = 0$ against a onesided restricted alternative like $\theta > 0$, under both parametric and non-parametric set-ups the standard practice is to use a one tailed test based on some appropriate statistic. However, difficulty in formulating the test arises when we have $k > 1$. In the parametric context solutions to such problems have been derived mostly by the likelihood ratio technique (Bartholomew, 1959, 1961; Chacko, 1963; Kudo, 1963; Nueseh, 1966; Shorack, 1967; Perlman, 1969, Pincus, 1975; a detailed account of many of these results is given by Barlow et al., 1972). Krishnaiah (1965) derives what he calls finite intersection tests for certain multiparameter linear hypotheses against restricted alternatives of special types under the multinormal set-up. These are union-intersection tests (Roy, 1957) based on a finite family of statistics corresponding to a representation of the alternative in terms of a finite number of sub-hypotheses.

Under the nonparametric set-up the tool of likelihood is no longer available. There is also difficulty in extending the union-intersection technique in a straightforward manner since, generally, no obvious choice for the optimal tests against the sub-hypotheses is available; nor is the distribution problem under the nonparametric set-up easily resolvable in terms of independent sets of statistics as in Krishnaiah (1965). From intuitive considerations nonparametric restricted tests for a number of testing problems have been proposed and studied by several workers. Particular attention has been received by the problem of testing homogeneity against ordered alternatives. For this problem rank solutions have been considered in the case of one-way layout by Jonekheere (1954), Chacko (1963) and Shorack (1967), and in the case of two-day layout, by Jonckheere (1954), Page (1963), Hollander (1967), Doksum (1967), Shorack (1967) and Sen (1968). All these solutions except those of Chacko (1963) and Shorack (1967) are based on some intuitively suggested function of rankscores which tends to be large when a trend alternative of the appropriate type obtains. The solutions of Chacko and Shorack are derived by using ranks in place of observations in the corresponding parametric likelihood ratio tests. Among other nonparametric restricted testing problems considered, we may mention that of comparing the locations of two bivariate populations against directional alternatives. David and Fix (1961) studied a test for this based on marginal Wilcoxon statistics whereas Bhattacharya and Johnson (1970) proposed a test based on 'layer ranks'. Studies of performance of these nonparametric tests indicate that against the restricted class of alternatives most of them are fairly sensitive in certain directions and relatively insensitive in others. This is obviously a consequence of the element of arbitrariness present in their formulation. To devise a method of derivation of comprehensive restricted tests for nonparametric problems, an adaptation of the union–intersection technique was considered by Chatterjee and De (1972, 1974), De (1976, 1977) and Chinchilli and Sen (1981a, 1981b, 1982). In the next section we describe this approach in some detail.

3. An adaptation of the union–intersection technique

In the classical union–intersection approach (Roy, 1957) to the problem of testing a null hypothesis H_0 against a composite alternative H_1, it is customary to represent H_1 as $\bigcup_{\lambda\in\Lambda_1} H_{1\lambda}$, where λ is a labelling index with Λ_1 for its domain. (More generally, H_0 also is represented as $\bigcap_{\lambda\in\Lambda_1} H_{0\lambda}$, but we need not consider this here). Here the $H_{1\lambda}$'s are so defined that, for each λ, the component problem of testing H_0 against $H_{1\lambda}$ is relatively simple and, whatever ε $(0<\varepsilon<1)$ a 'good' size-ε test for testing H_0 against $H_{1\lambda}$ is available. To test H_0 against H_1 we test it against each $H_{1\lambda}$ by an appropriate size-ε test. H_0 is rejected when it is rejected in favour of at least one of the $H_{1\lambda}$'s; otherwise, it is accepted. Here ε is so adjusted that the overall test has the requisite size.

In our context, however, the classical union-intersection approach will have to be geared to the asymptotic theory of rank tests. Further, this will involve the definition of a 'good' test in a specialised sense.

Let us suppose we are concerned with a sequence of observable random (real or vector) variables $X_1, X_2, \ldots,$ such that for each $N \geq 1$, the joint distribution of $X_1, X_2, \ldots, X_N$ depends on the parameter vector $\boldsymbol{\theta} \in \mathbb{R}^k$. Our problem is to test H_0: $\boldsymbol{\theta} = \mathbf{0}$ against H_1: $\boldsymbol{\theta} \in \Theta_1$ where Θ_1 is a suitable subset of $\mathbb{R}^k - \{\mathbf{0}\}$. Suppose that utilizing the known symmetries of the joint distribution of $X_1, \ldots, X_N$, it is possible to construct for each N a vector statistic $\boldsymbol{T}_N = (T_{1N}, \ldots, T_{qN})'$, $q \geq 1$, such that, as $N \to \infty$, (i) under H_0, irrespective of the form of the parent distribution, $\boldsymbol{T}_N$ is asymptotically distributed as $N_q(\mathbf{0}, I_q)$, (ii) the sequence of right tailed tests for H_0 based on linear compounds of the form $\boldsymbol{\lambda}'\boldsymbol{T}_N$, where $\boldsymbol{\lambda}$ is any fixed nonnull q-vector, is consistent against some $\boldsymbol{\theta} \neq \mathbf{0}$. In our applications q would be either k or $k-1$ and $\boldsymbol{T}_N$ would be canonical transforms of suitable linear rank statistics. Consider the class of component tests with critical region $\boldsymbol{\lambda}'\boldsymbol{T}_N > c$, where c is a fixed number and $\boldsymbol{\lambda}$ lies on the q-dimensional unit sphere $\Lambda = \{\boldsymbol{\lambda} : \lambda \in \mathbb{R}^q, \boldsymbol{\lambda}'\boldsymbol{\lambda} = 1\}$. Assume that for each $\boldsymbol{\lambda}$ it is possible to demarcate in $\mathbb{R}^q - \{\mathbf{0}\}$ a subset $\Theta_{(\lambda)}$, such that against $\boldsymbol{\theta} \in \Theta_{(\lambda)}$, in some reasonable sense, the test $\boldsymbol{\lambda}'\boldsymbol{T}_N > c$ performs better than any other test $\boldsymbol{\lambda}^{*\prime}\boldsymbol{T}_N > c$ of the same class. We can then say that within the class of component tests considered, the test $\boldsymbol{\lambda}'\boldsymbol{T}_N > c$ performs best against the subclass of alternatives $\Theta_{(\lambda)}$. We then determine a subset $\Lambda_1 \subset \Lambda$ such that

$$\bigcup_{\boldsymbol{\lambda}\in\Lambda_1} \boldsymbol{\Theta}_{(\lambda)} = \boldsymbol{\Theta}_1 \,. \tag{3.1}$$

In conformity with the spirit of the union–intersection technique we would then take

$$\bigcup_{\boldsymbol{\lambda}\in\Lambda_1} \{\boldsymbol{\lambda}'\boldsymbol{T}_N > c\} = \{Q_N > c\}\,, \tag{3.2}$$

where

$$Q_N = \sup_{\boldsymbol{\lambda}\in\Lambda_1} \boldsymbol{\lambda}'\boldsymbol{T}_N\,, \tag{3.3}$$

as our critical region for testing H_0 against H_1. The cut-off point c would have to be adjusted with reference to the asymptotic distribution of Q_N so that the asymptotic size of the overall test has the stipulated value. In many problems it would happen that $\Theta_{(\lambda)}$ and hence, Λ_1 would depend on some unknown parameters of the parent distribution. Naturally then, Q_N given by (3.3), as also its limiting distribution under H_0, would involve the same. In such cases it would generally be possible to find consistent estimates of the unknown parameters, and substituting these in Q_N we would get a statistic $\hat{Q}_N$ determined solely by the sample. Generally, continuity considerations will show that Q_N and $\hat{Q}_N$ would be asymptotically equivalent. A similar substitution may have to be made in the limiting distribution for determining the cut-off point up to an asymptotic approximation.

The approach outlined above, may seem rather too narrow, the component tests being chosen as right-tailed tests based on linear compounds of the form $\boldsymbol{\lambda}'\boldsymbol{T}_N$. Nevertheless, for all the restricted testing problems that we have in mind we get reasonable solutions even confining ourselves to such a special class. In contrast with the classical union–intersection approach, the proposed adaptation has one distinctive feature. In the classical approach we first identify the subhypotheses $H_{1\lambda}$ comprising H_1 and choose a good test against each $H_{1\lambda}$, whereas here we start from a class of component tests and demarcate the regions $\Theta_{(\lambda)}$ against which the component tests perform best. This reversal of priorities helps us to utilize known results on the asymptotics of standard statistics and simplifies the problem.

One crucial technical question in carrying through the above programme is how to demarcate the region $\Theta_{(\lambda)}$ against which the right-tailed tests $\boldsymbol{\lambda}'\boldsymbol{T}_N > c$ perform best. In Chatterjee and De (1972) and De (1976,77) the performance of a sequence of tests against any fixed alternative $\boldsymbol{\theta} \neq \mathbf{0}$ was measured in terms of the corresponding Bahaudur slope and $\Theta_{(\lambda)}$ was defined as consisting of those points $\boldsymbol{\theta}$ for which this slope is maximum for the $\boldsymbol{\lambda}$ considered. One difficulty in this approach is that in it one often requires to have $\boldsymbol{\theta}$-free estimates of certain correlation coefficients. As this creates difficulty in the more complex restricted testing problems, in this article, we follow Chinchilli and Sen (1981a) to judge performance of a sequence of tests in an entire direction such as $\{\boldsymbol{\theta}: \boldsymbol{\theta} = M\boldsymbol{\tau}, M > 0\}$ (M is a variable scalar; $\boldsymbol{\tau} \neq \mathbf{0}$ determines the direction) by the asymptotic power against contiguous alternatives approaching H_0 along that direction. $\Theta_{(\lambda)}$ consists of the direction in which the tests $\boldsymbol{\lambda}'\boldsymbol{T}_N > c$ maximize the asymptotic power (The assumption of consistency of the tests $\boldsymbol{\lambda}'\boldsymbol{T}_N > c$ against some $\boldsymbol{\theta}$ ensures that for each $\boldsymbol{\lambda}$ the asymptotic power can be meaningfully defined in some directions). Here the region Θ_1 is assumed to be *positively homogeneous* in the sense that it consists of a collection of complete directions and the union–intersection technique is applied on that basis. However, for two major types of restricted testing problems namely, those involving orthant alternatives and order alternatives (See Section 4), in single-sample and multi-sample situations, the Bahadur slope approach and the asymptotic power approach lead to identical or almost identical tests.

4. Application of the union–intersection technique

We now proceed to apply the technique outlined in the preceding section to various restricted testing problems.

4.1. p-Variate p-parameter problems – orthant alternative

We first consider the p-variate *two-sample location* problem. Let

$$\boldsymbol{X}_\alpha, \quad \alpha = 1, \ldots, n_1; \qquad \boldsymbol{X}_\alpha, \quad \alpha = n_1 + 1, \ldots, N(= n_1 + n_2);$$
$$\boldsymbol{X}_\alpha = (X_{1\alpha}, \ldots, X_{p\alpha})' \tag{4.1}$$

be independent random samples from two populations with density functions $f(\boldsymbol{x})$ and $f(\boldsymbol{x} - \boldsymbol{\theta})$, $\boldsymbol{x} = (x_1, \ldots, x_p)'$, $\boldsymbol{\theta} = (\theta_1, \ldots, \theta_p)'$, respectively, the functional form of f being unspecified. Our problem is to test H_0: $\boldsymbol{\theta} = \boldsymbol{0}$ against the orthant alternative H_1: $\boldsymbol{\theta} \geqslant \boldsymbol{0}$, $\boldsymbol{\theta} \neq \boldsymbol{0}$ (i.e. $\theta_i \geqslant 0$, $i = 1, \ldots, p$, with at least one inequality strict). We first convert the observations into variate-wise ranks. Let $R_{i\alpha}$ be the rank of $X_{i\alpha}$ among all the N observations on the i-th variable. We choose for each i a set of N scores

$$a_{iN}(\alpha), \quad \alpha = 1, \ldots, N, \tag{4.2}$$

and form the linear rank statistics

$$S_{iN} = \sum_{\alpha = n_1+1}^{N} a_{iN}(R_{i\alpha}), \quad i = 1, \ldots, p. \tag{4.3}$$

For the location problem the scores (4.2) are taken so that these form a (nonconstant) nondecreasing sequence. We assume that for each i there is a function $\varphi_i(u)$, $u \in (0, 1)$ satisfying (a) $\varphi_i(u)$ is (nonconstant) nondecreasing, (b) $\varphi_i(u)$ has atmost a finite number of discontinuities in (0, 1), (c) $\int_0^1 \varphi_i^2(u)\,du < \infty$, such that, as $N \to \infty$

$$\int_0^1 \{a_{iN}(1 + [uN]) - \varphi_i(u)\}^2\,du \to 0. \tag{4.4}$$

Standard location scores like the Wilcoxon scores ($a_{iN}(\alpha) = \alpha/(N+1)$), median scores ($a_{iN}(\alpha) = 0$ or 1 according as $\alpha <$ or $\geqslant \frac{1}{2}(N+1)$) and normal scores ($a_{iN}(\alpha) = EV_{N:\alpha}$, $V_{N:\alpha} = \alpha$-th orderstatistic for N observations from $N(0, 1)$) are known to satisfy the above requirement.

We suppose there is a number ν $(0 < \nu < 1)$ such that, as $N \to \infty$,

$$n_2/N \to \nu. \tag{4.5}$$

So that

$$h_N = n_1 n_2/N^2 \to h = \nu(1-\nu)\,. \tag{4.6}$$

We write

$$\bar{a}_{iN} = \frac{1}{N}\sum_{\alpha=1}^{N} a_{iN}(\alpha), \qquad v_{ijN} = \frac{1}{N}\sum_{\alpha=1}^{N}\{a_{iN}(R_{i\alpha}) - \bar{a}_{i\cdot N}\}\{a_{jN}(R_{j\alpha}) - \bar{a}_{j\cdot N}\}$$

$$g_{ij\cdot N} = v_{ij\cdot N}/(v_{ii\cdot N}v_{jjN})^{1/2}, \quad i, j = 1, \ldots, p, \qquad G_N = (g_{ij\cdot N})\,, \tag{4.7}$$

$$\bar{\varphi}_i = \int_0^1 \varphi_i(u)\,\mathrm{d}u, \qquad v_{ij} = \int \{\varphi_i(F_{[i]}(x_i)) - \bar{\varphi}_i\}\{\varphi_j(F_{[j]}(x_j)) - \bar{\varphi}_j\} f(\boldsymbol{x})\,\mathrm{d}\boldsymbol{x}\,,$$

$$g_{ij} = v_{ij}/(v_{ii}v_{jj})^{1/2}, \quad i, j = 1, \ldots, p, \qquad G = (g_{ij})\,, \tag{4.8}$$

where $F_{[i]}(x)$ denotes the cdf of the i-th variable for the distribution $f(\boldsymbol{x})$, $\mathrm{d}x$ stands for $\Pi\,\mathrm{d}x_i$ and the integral in v_{ij} is taken over $\mathbb{R}^p$.

Let

$$D_{i\cdot N} = (Nv_{ii\cdot N})^{-1/2}(S_{i\cdot N} - n_2\bar{a}_{i\cdot N}), \quad i = 1, \ldots, p,$$
$$\boldsymbol{D}_N = (D_{1\cdot N}, \ldots, D_{p\cdot N})'\,. \tag{4.9}$$

It can be shown that under the assumptions made, under H_0, $N \to \infty$,

$$\boldsymbol{D}_N \xrightarrow{\mathscr{L}} N_p(\boldsymbol{0}, hG), \qquad G_N \xrightarrow{\mathrm{P}} G \tag{4.10}$$

(see Hájek and Sidák, 1965, Chapter V; and Puri and Sen, 1971, Chapter V).

If the variables in the parent distribution are not a.s. functionally related, G is p.d., and hence, G_N is p.d. in probability, so that from (4.10) we get

$$\boldsymbol{T}_N = h_N^{-1/2} G_N^{-1/2} \boldsymbol{D}_N \xrightarrow{\mathscr{L}} N_p(\boldsymbol{0}, I_P)\,. \tag{4.11}$$

When an alternative $\boldsymbol{\theta} \neq \boldsymbol{0}$ is true G_N converges in probability to a p.d. matrix (different from G) and $N^{-1/2}D_{i\cdot N}$ to a number with sign same as that of θ_i. Hence the asymptotically distribution-free tests $\boldsymbol{\lambda}'\boldsymbol{T}_N > c$ for each $\boldsymbol{\lambda} \neq \boldsymbol{0}$ is consistent against some $\boldsymbol{\theta} \neq \boldsymbol{0}$. $\boldsymbol{T}_N$ thus meets the requirements formulated in Section 3.

Consider next the sequence of alternatives

$$\boldsymbol{\theta}^{(N)} = N^{-1/2}\boldsymbol{\tau}, \quad \boldsymbol{\tau} \neq \boldsymbol{0}\,. \tag{4.12}$$

Here the fixed vector $\boldsymbol{\tau}$ represents the direction along which $\boldsymbol{\theta}^{(N)}$ approaches $\boldsymbol{0}$. We now assume that the density $f(\boldsymbol{x})$ has continuous first partial derivatives and the Fisher information matrix for $f(x)$ exists. Then (4.5) implies that the sequence of alternatives (4.12) is contiguous (see Chinchilli and Sen, 1981a, Theorem 3.1) so that $G_N \xrightarrow{P} G$ holds under $\boldsymbol{\theta}^{(N)}$ as well. Further, writing

$$
\begin{aligned}
& f_{[i]}(x) = \frac{\mathrm{d}}{\mathrm{d}x} F_{[i]}(x), \ \dot{f}_{[i]}(x) = \frac{\mathrm{d}}{\mathrm{d}x} f_{[i]}(x) \\
& \psi_i(u) = -\dot{f}_{[i]}(F_{[i]}^{-1}(u))/f_{[i]}(F_{[i]}^{-1}(u)) \qquad (4.13) \\
& \gamma_i = v_{ii}^{-1/2} \int_0^1 \varphi_i(u)\psi_i(u)\,\mathrm{d}u, \qquad \Gamma = \mathrm{diag}(\gamma_1, \ldots, \gamma_p),
\end{aligned}
$$

by the application of standard techniques, from contiguity we get that, under $\boldsymbol{\theta}^{(N)}$,

$$\boldsymbol{D}_N \xrightarrow{\mathcal{L}} N_p(h\Gamma\boldsymbol{\tau}, hG), \ \boldsymbol{T}_N \xrightarrow{\mathcal{L}} N_P(h^{1/2}G^{-1/2}\Gamma\boldsymbol{\tau}, I_p) \qquad (4.14)$$

and hence, for any $\boldsymbol{\lambda}$, $\boldsymbol{\lambda}'\boldsymbol{\lambda} = 1$,

$$\boldsymbol{\lambda}'\boldsymbol{T}_N \xrightarrow{\mathcal{L}} N(h^{1/2}\boldsymbol{\lambda}'G^{-1/2}\Gamma\boldsymbol{\tau}, 1).$$

Therefore using $\Phi(\cdot)$ to denote the standard normal cdf, the asymptotic power of the tests $\boldsymbol{\lambda}'\boldsymbol{T}_N > c$ against $\boldsymbol{\theta}^{(N)}$ comes out as

$$1 - \Phi(c - h^{1/2}\boldsymbol{\lambda}'G^{-1/2}\Gamma\boldsymbol{\tau}). \qquad (4.15)$$

Since $\boldsymbol{\lambda}'\boldsymbol{\lambda} = 1$, by Schwarż inequality we get

$$\boldsymbol{\lambda}'G^{-1/2}\Gamma\boldsymbol{\tau} \leqslant (\boldsymbol{\tau}'\Gamma G^{-1}\Gamma\boldsymbol{\tau})^{1/2},$$

with equality holding iff

$$\boldsymbol{\lambda} = \text{const } G^{-1/2}\Gamma\boldsymbol{\tau} \quad \text{i.e. } \boldsymbol{\tau} = \text{const } \Gamma^{-1}G^{1/2}\boldsymbol{\lambda}. \qquad (4.16)$$

Thus, given $\boldsymbol{\tau}$, the asymptotic power would be maximum for $\boldsymbol{\lambda}$ given by (4.16). Reciprocally, given $\boldsymbol{\lambda}$, we can say the tests $\boldsymbol{\lambda}'\boldsymbol{T}_N > c$ perform best in the direction $\boldsymbol{\tau}$ given by (4.16) (in the sense that for that direction no other $\boldsymbol{\lambda}^*$ gives a higher asymptotic power). Therefore, we can define

$$\Theta_{(\lambda)} = \{\boldsymbol{\theta}\colon \boldsymbol{\theta} = M\Gamma^{-1}G^{1/2}\boldsymbol{\lambda}, M > 0\} \qquad (4.17)$$

as the subclass of alternatives against which the sequence of tests $\boldsymbol{\lambda}'\boldsymbol{T}_N > c$ has best performance.

We now assume that $\gamma_i > 0$, $i = 1, \ldots, p$. From (4.13) it may be shown that this holds if, for each i, $\psi_i(u)$ is increasing and/or $\lim_{x\to\pm\infty} \varphi_i(F_{[i]}(x))f_{[i]}(x) = 0$.

Under this assumption (3.1) holds for $\Theta_{(\Lambda)}$ given by (4.17) and $\Theta_1 = \{\boldsymbol{\theta}: \boldsymbol{\theta} \geq \mathbf{0}, \boldsymbol{\theta} \neq \mathbf{0}\}$ is we take

$$\Lambda_1 = \{\boldsymbol{\lambda}: G^{1/2}\boldsymbol{\lambda} \geq \mathbf{0}, \boldsymbol{\lambda}'\boldsymbol{\lambda} = 1\}. \tag{4.18}$$

In accordance with (3.3), then the union–intersection statistic would be

$$Q_N = \max_{\substack{G^{1/2}\boldsymbol{\lambda} \geq 0 \\ \boldsymbol{\lambda}'\boldsymbol{\lambda}=1}} \boldsymbol{\lambda}'\boldsymbol{T}_N = \max_{\substack{u \geq 0 \\ u'G^{-1}u=1}} uG^{-1/2}T_N,$$

where we write $G^{1/2}\boldsymbol{\lambda} = \boldsymbol{u}$. (We replace 'sup' by 'max' since the function is continuous and the domain compact.) As noted in the paragraph following (3.3), in practice, G would have to be replaced by its estimate G_N (cf. (4.10). Then, by (4.11), the approximate union–intersection statistic so obtained would be

$$\hat{Q}_N = h_N^{-1/2} \max_{\substack{\boldsymbol{u} \geq \mathbf{0} \\ \boldsymbol{u}'G_N^{-1}\boldsymbol{u}=1}} \boldsymbol{u}'G_N^{-1}\boldsymbol{D}_N. \tag{4.19}$$

The maximum in (4.19) would be attained at an interior point $\boldsymbol{u} > \mathbf{0}$ (i.e. $u_i > 0$, $i = 1, \ldots, p$) only if the Lagrange conditions for maximum subject to $uG_N^{-1}u = 1$ hold at u. This requires $\boldsymbol{D}_N > \mathbf{0}$ and the maximal point comes out as $\boldsymbol{u} = \boldsymbol{D}_N/(\boldsymbol{D}_N'G_N^{-1}\boldsymbol{D}_N)^{1/2}$. On the other hand, if $\boldsymbol{D}_N \not> \mathbf{0}$ (i.e. if '$D_{iN} > 0$ for all i' does not hold), the maximum would be attained at a boundary point where one or more of the u_i's are zero. The explicit expression comes out in a particularly simple form when $p = 2$. There,

$$\begin{aligned} \hat{Q}_N &= h_N^{-1/2}(\boldsymbol{D}_N'G_N^{-1}\boldsymbol{D}_N)^{1/2} \quad \text{if } \boldsymbol{D}_N > \mathbf{0}, \\ &= h_N^{-1/2} \max\{D_{pN} - g_{12\cdot N}D_{2N}, D_{2N} - g_{12\cdot N}D_{1N}\}/(1 - g_{12\cdot N}^2)^{1/2} \\ &\qquad\qquad \text{if } \boldsymbol{D}_N \not> \mathbf{0}, \end{aligned} \tag{4.20}$$

whence by (4.10), for $c \geq 0$,

$$\lim_{N\to\infty} P_{H_0}\{\hat{Q}_N > c\} = \text{Prob}\{\chi_2^2 > c^2\}(\tfrac{1}{2} - \tfrac{1}{2}\cos^{-1}g_{12}) + 1 - \Phi(c), \tag{4.21}$$

χ_ν^2 denoting a chi-square r.v. with d.f. ν. As noted earlier, in practice, we have to use $g_{12\cdot N}$ in place of g_{12} while determining the cut off point c from (4.21).

For general p, the boundary of the domain of $\boldsymbol{u}$ in (4.19) is more complex and as such the determination of $\hat{Q}_N$ is rather complicated. Viewing the problem as one of nonlinear programming, we can apply the Kuhn–Tucker theory (see Hadley, 1964) and observe the following.

(i) The maximum in (4.19) would be attained at a point

$$\boldsymbol{u}: \quad u_1 > 0, \ldots, u_k > 0, u_{k+1} = \cdots = u_p = 0 \quad (1 \leq k \leq p) \tag{4.22}$$

if and only if for the partition

$$G_N = \begin{pmatrix} \overset{k\times k}{G_{11\cdot N}} & \overset{k\times(p-k)}{G_{12\cdot N}} \\ G_{21\cdot N} & \underset{G_{22\cdot N}}{(p-k)\times(p-k)} \end{pmatrix}, \quad \boldsymbol{D}_N = \begin{pmatrix} \overset{k\times 1}{\boldsymbol{D}_{1\cdot N}} \\ (p-k)\times 1 \\ \boldsymbol{D}_{2\cdot N} \end{pmatrix}, \tag{4.23}$$

$$\boldsymbol{D}_{1\cdot N} - G_{12\cdot N} G_{22\cdot N}^{-1} \boldsymbol{D}_{2\cdot N} > \boldsymbol{0}, \qquad G_{22\cdot N}^{-1} \boldsymbol{D}_{2\cdot N} \leq \boldsymbol{0}\,. \tag{4.24}$$

In this case,

$$\hat{Q}_N = h_N^{-1/2} [D'_{1\cdot 2\cdot N} G_{11\cdot 2\cdot N}^{-1} \boldsymbol{D}_{1\cdot 2\cdot N}]^{1/2}\,, \tag{4.25}$$

where

$$\boldsymbol{D}_{1\cdot 2\cdot N} = \boldsymbol{D}_{1\cdot N} - G_{12\cdot N} G_{22\cdot N}^{-1} \boldsymbol{D}_{2\cdot N},\ G_{11\cdot 2\cdot N} = G_{11\cdot N} - G_{12\cdot N} G_{22\cdot N}^{-1} G_{21\cdot N}\,. \tag{4.26}$$

(ii) If for some $1 \leq i_1 < \cdots < i_k \leq p$, the maximum is attained at a point

$$\boldsymbol{u}: \qquad u_{i_1} > 0, \ldots, u_{i_k} > 0 \quad u_i = 0 \text{ for } i \neq i_1, \ldots, i_k \ (1 \leq k \leq p)\,, \tag{4.27}$$

we can permute the elements of $\boldsymbol{u}$ so that $u_{i_1}, \ldots, u_{i_k}$ occur in the first k positions. If we now make the same permutation of the rows and columns of G_N and the elements of $\boldsymbol{D}_N$ (so that the maximization problem remains same) conditions (4.24) would hold for the new G_N and $\boldsymbol{D}_N$ and $\hat{Q}_N$ would have the expression (4.24) in terms of these.

(iii) If $G_N^{-1}\boldsymbol{D}_N \nleq \boldsymbol{0}$ conditions (4.24) will hold for some k and some permutation of the elements. If $G_N^{-1}\boldsymbol{D}_N \leq \boldsymbol{0}$, this is no longer true, and there, writing $G_N^{-1} = (g^{ij\cdot N})$

$$\hat{Q}_N = h_N^{-1/2} \max\left\{ \sum_{j=1}^{p} g^{ij\cdot N} D_{j\cdot N} \Big/ (g^{ii\cdot N})^{1/2},\ i = 1, \ldots, p \right\}. \tag{4.28}$$

Verification of the condition (4.24) for different choices of $i_1, \ldots, i_k$ would admittedly be troublesome unless p is small. Workable computer programs would come handy here.

To express the tail probability of the asymptotic null distribution of $\hat{Q}_N$ for general p, we use the notation $\pi_k(\boldsymbol{\Sigma})$ to denote the probability of the positive orthant of the distribution $\boldsymbol{N}_k(\boldsymbol{0}, \boldsymbol{\Sigma})$. Let

$$W(k \mid G) = \sum_{1 \leq i_1 \cdots < i_k \leq p} \pi_k(G_{11\cdot 2}) \pi_{p-k}(G_{22}^{-1}), \quad 1 \leq k \leq p\,, \tag{4.29}$$

where it is to be understood that for each choice of $i_1, \ldots, i_k$, rows and columns of G are to be permuted so that i_1-th, ..., i_k-th rows (columns) occur in the first k positions and from the transformed matrix $G_{11\cdot 2}$ and G_{22} are to be determined as in (4.23) and (4.26). By standard techniques we can then deduce that, for $c \geqslant 0$,

$$\lim_{n\to\infty} P_{H_0}\{\hat{Q}_N > c\} = \sum_{k=1}^{p} \text{Prob}\{\chi_k^2 > c^2\} W(k|G). \tag{4.30}$$

The details we have worked out above for the location case hold with minor modifications for other p-parameter problems against orthant alternative. For the p-variate *two-sample scale* problem, the samples (4.1) are supposed to be from two populations with density functions $f(x_1 - \mu_1, \ldots, x_p - \mu_p)$ and $e^{-\Sigma\theta_i} f(e^{-\theta_1}(x_1 - \mu_1), \ldots, e^{-\theta_p}(x_p - \mu_p))$, $\mu_1, \ldots, \mu_p$ being nuisance parameters. The hypotheses, with $\boldsymbol{\theta}$ as defined here, are same as before. The scores (4.2) and $\varphi_i(u)$ in (4.4) are no longer nondecreasing as in the location case. One should use here some system of scores appropriate for the scale problem, such as $a_i(\alpha) = |(\alpha/N+1) - \frac{1}{2}|$, $((\alpha/N+1) - \frac{1}{2})^2$ or $EV_{N:\alpha}^2$ (see e.g. Hájek and Sidák, 1965, Chapter III). The functions $\varphi_i(u)$ would now be expressible as the difference of two functions each satisfying the conditions (a)–(c). The rest of the development remains same except that the functions $\psi_i(u)$ in (4.13) have to be replaced by their counterparts for the scale problem (Hájek and Sidák, 1965, Chapter VI). The expressions for the test statistic and the tail probability remain same.

For the p-variate regression problem we have N independent observations $\boldsymbol{X}_\alpha$, $\alpha = 1, \ldots, N$, $\boldsymbol{X}_\alpha$ following the distribution $f(\boldsymbol{x} - c_{\alpha N}\boldsymbol{\theta})$ where $c_{\alpha N}$, $\alpha = 1, \ldots, N$, are known numbers. We want to test H_0: $\boldsymbol{\theta} = \boldsymbol{0}$ against H_1: $\boldsymbol{\theta} \geqslant \boldsymbol{0}$, $\boldsymbol{\theta} \neq \boldsymbol{0}$ as before. Converting the observations into rank scores just as in the location case, we now set up

$$S_{iN} = \sum_{\alpha=1}^{N} (c_{\alpha N} - \bar{c}_N) a_{iN}(R_{i\alpha}), \quad \bar{c}_N = \frac{1}{N}\sum_{\alpha=1}^{N} c_{\alpha N},\ i = 1, \ldots, p. \tag{4.31}$$

We now assume that, as $N \to \infty$,

$$\max_\alpha \frac{1}{N}(c_{\alpha N} - \bar{c}_N)^2 \to 0, \qquad h_N = \frac{1}{N}\sum_{\alpha=1}^{N}(c_{\alpha N} - \bar{c}_N)^2 \to h, \tag{4.32}$$

where h is some positive number. This is the counterpart of (4.5)–(4.6). Now, using the notations (4.7)–(4.8) and defining $\boldsymbol{D}_N$ and $\boldsymbol{T}_N$ as in (4.9) and (4.11), we get that, under H_0, (4.10) and (4.11) hold. Contiguity of the alternatives $\boldsymbol{\theta}^{(N)}$ given by (4.12) follows from (4.32). Hence Γ being as in (4.13), (4.14) holds. Then, proceeding as in (4.15)–(4.30) we get $\hat{Q}_N$ and its tail probability in the same forms as before.

For the p-variate *single sample location* problem the observations $\boldsymbol{X}_\alpha$, $\alpha =$

$1, \ldots, N$, are supposed to be a random sample from $f(\boldsymbol{x} - \boldsymbol{\theta})$ where $f(x)$ has each univariate marginal symmetric about 0. To test H_0: $\boldsymbol{\theta} = \boldsymbol{0}$ against the positive orthant alternative H_1, we find for each α the rank $R^+_{i\alpha}$ of $|X_{i\alpha}|$ among $|X_{i1}|, \ldots, |X_{iN}|$. For each i a set of N scores (4.2) forming a nondecreasing sequence of positive numbers is chosen and the statistics

$$S_{iN} = \sum_{\alpha=1}^{N} \operatorname{sign}(X_{i\alpha}) a_{iN}(R^+_{i\alpha}), \quad i = 1, \ldots, p, \tag{4.33}$$

are set up. The scores satisfy (4.4) with respect to functions $\varphi_i(u)$ which are positive nondecreasing and satisfy the conditions (b)–(c) stated earlier. (Some possible choices of the scores are $a_{iN}(\alpha) = 1$, $\alpha/(N+1)$, or $EV^+_{N:\alpha}$, where $V^+_{N:\alpha}$ is the α-th smallest absolute value for N observations from $N(0, 1)$). Writing

$$v_{ij\cdot N} = \frac{1}{N} \sum_{\alpha=1}^{N} a_{iN}(R^+_{i\alpha}) a_{jN}(R^+_{j\alpha}) \operatorname{sign}(X_{i\alpha}) \operatorname{sign}(X_{j\alpha}),$$

$$v_{ij} = \int \varphi_i(2F_{[i]}(|x_i|) - 1)\varphi_j(2F_{[j]}(|x_j|) - 1) \operatorname{sign}(x_i) \operatorname{sign}(x_j) f(\boldsymbol{x})\, \mathrm{d}\boldsymbol{x},$$

$$i, j = 1, \ldots, p,$$

$$D_{i\cdot N} = (Nv_{ii\cdot N})^{-1/2} S_{i\cdot N}, \; D_N = (D_{1\cdot N}, \ldots, D_{p\cdot N})'$$

and defining G_N and G as in (4.7) and (4.8), we see that (4.10)–(4.11) hold with $h_N = h = 1$. After modifying the definition of the functions $\psi_i(u)$ appropriately (cf. Hájek and Sidák, 1965, Chapter VI), from contiguity (4.14) follows. The subsequent development leading to the expressions for $\hat{Q}_N$ and its tail probability carries through as before.

In all the applications considered above we have assumed that under H_1, $\boldsymbol{\theta}$ lies in the positive orthant. If H_1 is represented by some other orthant the definition of Λ_1 given by (4.18) would have to be appropriately modified. This would entail obvious modifications in the form of $\hat{Q}_N$ and its tail probability. For the two-sample and single sample location problems and the regression problem, an alternative approach would be to make appropriate initial changes in the signs of the variables so that the problem is transformed into a positive orthant problem. The earlier expressions can then be used for the transformed set-up.

4.2. p-Variate p-parameter problems – Other alternatives

As noted at the end of Section 3 the union–intersection approach chalked out by us can be followed whenever the alternative subset Θ_1 is positively homogeneous. The form of the test-statistic, however, would depend very much on the structure of Θ_1. In the case of a full orthant alternative Λ_1 (cf. (4.18)) turns out to be free from Γ. In general, Λ_1 would involve Γ which would have to be replaced by an estimate. In certain cases, however, the problem can be reduced to an orthant alternative problem by an initial transformation of variables. We describe both the approaches with reference to the two-sample location problem with

$$\Theta_1 = \{\boldsymbol{\theta}: C\theta \geqslant \mathbf{0}, \boldsymbol{\theta} \neq \mathbf{0}\} \tag{4.34}$$

where C is a given nonsingular matrix.

To consider the approach based on initial transformation first, let us transform to

$$C^{-1}\boldsymbol{X}_\alpha = \boldsymbol{X}^*_\alpha, \quad \alpha = 1, \ldots, N.$$

Then writing

$$f^*(\boldsymbol{x}^*) = |C| f(C\boldsymbol{x}^*), \quad C^{-1}\boldsymbol{\theta} = \boldsymbol{\theta}^*,$$

we see that $\boldsymbol{X}^*_\alpha$, $\alpha = 1, \ldots, n_1$ is a random sample from $f^*(x^*)$ and $\boldsymbol{X}_\alpha$, $\alpha = n_1 + 1, \ldots, N$ is a random sample from $f^*(\boldsymbol{x}^* - \boldsymbol{\theta}^*)$. The problem is that of testing H_0: $\boldsymbol{\theta}^* = \mathbf{0}$ against H_1: $\boldsymbol{\theta}^* \geqslant \mathbf{0}$, $\boldsymbol{\theta}^* \neq \mathbf{0}$. This can be handled as a positive orthant problem by ranking the transformed observations $X^*_{i\alpha}$ for each i.

In the straight-forward approach in view of (4.17) and (4.34), to realise (3.1) we should take

$$\Lambda_1 = \{\boldsymbol{\lambda}: C\Gamma^{-1}G^{1/2}\boldsymbol{\lambda} \geqslant \mathbf{0}, \boldsymbol{\lambda}'\boldsymbol{\lambda} = 1\}.$$

If we then write $\boldsymbol{u} = C\Gamma^{-1}G^{1/2}\boldsymbol{\lambda}$, we get

$$Q_N = \max_{\boldsymbol{u} \geqslant \mathbf{0}, \boldsymbol{u}'C'^{-1}\Gamma G^{-1}\Gamma C^{-1}\boldsymbol{u} = 1} \boldsymbol{u}'C'^{-1}\Gamma C^{-1/2}\boldsymbol{T}_N \tag{4.35}$$

In (4.35) we have to replace G by G_N and Γ by an appropriate estimate. If the scores (4.2) are derived from the function $\varphi_i(u)$ as

$$a_{iN}(\alpha) = \varphi_i\left(\frac{\alpha}{N+1}\right) \quad \text{or} \quad a_{iN}(\alpha) = E\varphi_i(U_{N:\alpha})$$

($U_{N:\alpha}$ is the α-th smallest of N observations from the distribution rect. (0, 1)) an estimate of Γ can be derived from the results of Jurečková (1969, 1971) (see Chinchilli and Sen, 1981a). Thus in the two-sample location case

$$\gamma_{i \cdot N} = (Nv_{ii \cdot N})^{-1/2}(S_{i \cdot N} - S^{(1)}_{i \cdot N}) \tag{4.36}$$

where, writing $S_{i \cdot N}$ in (4.3) as $S_{iN}(X_{i1}, \ldots, X_{in_1}; X_{in_1+1}, \ldots, X_{iN})$,

$$S^{(1)}_{iN} = S_{iN}(X_{i1}, \ldots, X_{in_1}; X_{in_1+1} - 1, \ldots, X_{iN} - 1),$$

represents a consistent estimate of γ_i. From (4.36) we get an estimate Γ_N of Γ. Using Γ_N and G_N in place of Γ and G in (4.35) and arguing as before we can derive $\hat{Q}_N$. The expression for the tail probability remains same as before with G replaced by $C\Gamma^{-1}G\Gamma^{-1}C'$.

Both the approaches described above for the two-sample location problem can be adapted to the regression and the single sample location problem. The techniques can be directly extended to the two-sample scale problem (see paragraph following (4.30)) only if $\mu_i = 0$, $i = 1, \ldots, p$.

Among other forms of Θ_1 that have been considered we may mention $\Theta_1 = \{\boldsymbol{\theta}: \boldsymbol{\theta}_1 \geq \mathbf{0}, \boldsymbol{\theta} \neq \mathbf{0}\}$ where $\boldsymbol{\theta}_1$ is a specified subvector of $\boldsymbol{\theta}$. One approach to this based on an appropriate partitioning of $\Gamma_N^{-1}\boldsymbol{D}_N$ is described in Chinchilli and Sen (1981b).

4.3. Homogeneity against order alternative

Testing homogeneity of several treatments against a specified order of their effects is a commonly occurring form of restricted testing problem. Such a problem may come up in a one-way or multi-way layout. We discuss this with reference to a randomized block experiment (De, 1976).

Let X_{ij} denote the yield of the i-th treatment in the j-th block of a RBD with p treatments and n blocks. We assume the model

$$X_{ij} = \mu + \theta_i + \alpha_j + e_{ij}, \quad i = 1, \ldots, p,\ j = 1, \ldots, n,\ \sum \theta_i = 0\,, \tag{4.37}$$

where θ_i's are treatment effects, α_j's are block effects, and e_{ij}'s are errors. The vectors $\boldsymbol{e}_j = (e_{1j}, \ldots, e_{pj})'$, $i = 1, \ldots, n$, are i.i.d. with a permutation-invariant common distribution. The problem is to test H: $\theta = 0$ against a simple order alternative H_1: $\theta_1 \leq \theta_2 \leq \cdots \leq \theta_p$, $\theta_1 < \theta_p$. We consider here an approach to the problem based on ranking after alignment (cf. Puri and Sen, 1971, Chapter VII). The approach is quite flexible and can be extended to less simple designs.

The first step is to align the observations to eliminate the block effects. For simplicity, we consider only alignment based on block means. If $X_{0j} = \Sigma_{i=1}^{p} X_{ij}/p$, we take $Y_{ij} = X_{ij} - X_{0j}$, $i = 1, \ldots, p$, $j = 1, \ldots, n$, and rank together all the $N = pn$ Y_{ij}'s. If R_{ij} is the rank of Y_{ij} so obtained, taking a set of N nondecreasing scores $a_N(\alpha)$, $\alpha = 1, \ldots, N$, satisfying (4.4) with respect to some $\varphi(u)$, we form

$$S_{in} = \sum_{j=1}^{n} a_N(R_{ij}), \quad i = 1, \ldots, p\,. \tag{4.38}$$

Writing

$$v_n = \frac{1}{n(p-1)} \sum_{j=1}^{n} \sum_{i=1}^{p} \{a_N(R_{ij}) - \bar{a}_{j\cdot N}\}^2, \quad \bar{a}_{j\cdot N} = \frac{1}{p} \sum_{i=1}^{p} a_N(R_{ij})\,, \tag{4.39}$$

we set up

$$D_{i\cdot n} = (nv_n)^{-1/2}(S_{i+1\cdot n} - S_{i\cdot n}), \quad i = 1, \ldots, p-1,$$
$$D_n = (D_{1\cdot n}, \ldots, D_{p-1\cdot n})' . \tag{4.40}$$

Then defining

$$G = (g_{ij}),\ g_{ii} = 2,\ g_{i\cdot i+1} = -1,\ g_{ij} = 0 \text{ for } |i-j| > 1,\ i, j = 1, \ldots, p-1, \tag{4.41}$$

and making appropriate assumptions about $\varphi(u)$ and the parent distribution (see Puri and Sen, 1971, Section 7.3), it can be shown that under H_0, $\boldsymbol{D}_n \xrightarrow{\mathcal{L}} N_{p-1}(\boldsymbol{0}, G)$ and under a sequence of alternatives of the form $\boldsymbol{\theta}^{(n)} = n^{-1/2}\boldsymbol{\tau}$, $\Sigma\, \tau_i = 0$, $\tau_{i+1} = \tau_i + \delta_i$, $i = 1, \ldots, p-1$, $\boldsymbol{\delta} = (\delta_1, \ldots, \delta_{p-1})' \geq \boldsymbol{0}$, $\boldsymbol{\delta} \neq \boldsymbol{0}$, $D_n \xrightarrow{\mathcal{L}} N_{p-1}(\gamma\boldsymbol{\delta}, G)$, $\gamma > 0$, being determined by φ and the parent distribution. Note that unlike in Subsection 4.1, G here is a known p.d. matrix. Taking now $\boldsymbol{T}_n = G^{-1/2}\boldsymbol{D}_n$, and proceeding as in Subsection 4.1 we get the union–intersection statistic in the form given by (4.20) and (4.22)–(4.28) with p, N, h_N, G_N replaced by $p-1$, n, 1, G respectively.

For the present problem since G is explicitly given by (4.41), it is possible to give a computational algorithm for the statistic based on the S_{in}'s of (4.38). The algorithm is same as the 'amalgamation process' well known in the context of parametric likelihood ratio tests against order alternatives and consists in first splitting up the sequence $S_{1\cdot n}, S_{2\cdot n}, \ldots, S_{p\cdot n}$ into a number of subsets of consecutive terms such that certain inequalities hold among the members of the subsets and their means. Another sequence $S^*_{1\cdot n}, S^*_{2\cdot n}, \ldots, S^*_{p\cdot n}$ is then formed by replacing all the members of some of the subsets by corresponding subset means. The union–intersection statistic is then computed as $(nv_n)^{-1/2}[\Sigma_1^p(S^*_{i\cdot n} - n\bar{a}_N)^2]^{1/2}$ $(\bar{a}_N = \Sigma_1^N a_N(\alpha)/N)$. Details are similar to those in Chacko (1963).

The expression for the tail probability of the asymptotic null distribution of the statistic is same as that given by (4.29)–(4.30) with p replaced by $p-1$ and G as in (4.41). However, since G is completely known it is possible to evaluate the weights $W(k \mid G)$, $k = 1, \ldots, p-1$, explicitly and it turns out that $W(k \mid G) =$ coeff. of z^{k+1} in $z(z+1)\cdots(z+p-1)/p!$ (see Barlow et al., 1972, p. 143).

In the above we have discussed the problem of testing H_0 against a complete ordering of the θ_i's. If the alternative specifies an incomplete ordering, the solution can be developed along the same lines.

4.4. Other problems

The union–intersection approach can be followed in tackling many restricted testing problems apart from those considered in the preceding subsections. Chinchilli and Sen (1981b) consider the problem of testing against an orthant alternative for the p-variate regression set up with more than one predictor. If there are r predictors the observation vector $\boldsymbol{X}_\alpha$ follows a distribution having density $f(x - c_{(1)\alpha N}\boldsymbol{\theta}_{(1)} - \cdots - c_{(r)\alpha N}\boldsymbol{\theta}_{(r)})$, $\alpha = 1, \ldots, N$ where $\boldsymbol{\theta}_{(1)}, \ldots, \boldsymbol{\theta}_{(r)}$ are the r regression vectors. Writing $\boldsymbol{\theta}$ for the pr-vector $(\boldsymbol{\theta}'_{(1)}, \ldots, \boldsymbol{\theta}'_{(r)})'$ and partitioning $\boldsymbol{\theta} = (\boldsymbol{\theta}^{*\prime}, \boldsymbol{\theta}^{**\prime})'$ where θ^* represents a subset of a parameters $(1 \leq a \leq pr)$, the problem considered is that of testing H_0: $\boldsymbol{\theta} = \mathbf{0}$ against H_1: $\boldsymbol{\theta}^* \geq \mathbf{0}$, $\boldsymbol{\theta} \neq \mathbf{0}$. Starting with r sets of p statistics of the form (4.31) corresponding to the r predictors we can proceed as earlier invoking the Kuhn–Tucker theory to determine the test criterion.

The union–intersection technique can of course be applied conveniently when the alternative puts constraints on the parameters in the form of linear or nonlinear equations and not inequalities. Some such applications in the context of profile analysis have been made by Chinichilli and Sen (1982).

So far, in all the applications that we have considered, the number of basic statistics from which the test is developed has been equal to the number of parameters whose values are specified by H_0. While the tests constructed in this way are quite comprehensive, one practical difficulty is that the determination of the test criterion and evaluation of the tail probability become complicated when the number of inequality restrictions imposed by H_1 is large. One way of resolving this may be to take a small number of suitable linear compounds of the basic statistic and to develop the union–intersection test on their basis. Thus, for the homogeneity problem of Subsection 4.3, we may take two linear compounds, say

$$\sum_1^p \left(i - \frac{p+1}{2}\right) S_{in} \quad \text{and} \quad \sum_{i=1}^{p-1} (S_{i+1\cdot n} - S_{i\cdot n})\,.$$

Simple expressions similar to (4.20) and (4.21) would hold for a test derived from two linear compounds. Some studies made by De (1977) indicate that such compromise tests would generally have satisfactory performance and that they would represent definite improvements over earlier tests based on single linear compounds (see references given in Section 2).

5. Tables and their uses

To apply the restricted tests described in the preceding section one would require tables for determining the critical value c corresponding to different probability levels, or equivalently, the tail areas corresponding to different values of the statistic. As has been seen, for the problems considered in Subsections 4.1–4.3, the tail area can be expressed as the weighted sum of the tail areas of certain χ^2-distributions. The tail probability of likelihood ratio criteria for certain parametric restricted testing problems (Barlow et al., 1972, Chapters III–IV) comes out in the same form. In both situations the problem of determination of the tail probability would be solved if one can give workable procedures and tables for computing the weights, which are in reality orthant probabilities of a multinormal distribution.

In the case of bivariate two-parameter problems the weights in the expression (4.21) for the tail probability are simple trigonometric functions of the single correlation coefficient g_{12} and hence, can be readily determined. A table for c based on these is reproduced in Barlow et al. (1972, Table A.1) for negative values of g_{12} (such values only are possible in the context of the problem of homogeneity of three ordered means in connexion with which, the table was prepared. Note, however, that the values tabulated in Barlow et al. (1972) correspond to $\hat{Q}_N^2$ in our notation). An unpublished table covering both negative and positive values of g_{12} is contained in De (1977).

Real difficulty in tabulation, however, arises in the case of problems involving more than two parameters. The weights in the expression (4.30) depend on $\frac{1}{2}p(p-1)$ correlation coefficients represented by the $p \times p$ matrix G and it would be impracticable to attempt a table covering the ranges of so many parameters. One needs here workable expressions for these weights in terms of G. Useful expressions are available for $p = 3$. For $4 \leq p \leq 6$ one can follow a procedure for computer evaluation of multinormal orthant probabilities described by Milton (1972). In the general case a satisfactory solution to the problem valid for all G remains to be evolved. An account of the available results and further references are given in Barlow et al. (1972, Chapter III).

For the problem of testing homogeneity against ordered alternatives in one-way or multi-way lay-outs, tabulation is relatively simple if all the treatments are equally replicated. (This condition naturally obtains in the case of a RBD.) The expression for the tail probability here depends only on p, the number of treatments. In this case a table of critical values covering $p \leq 12$ is reproduced by Barlow et al. (1972, Table A.3).

6. Gain in sensitivity

Is a restricted test specifically designed for detecting some subclass of alternatives more sensitive against these than the corresponding unrestricted test? If so, what is the extent of the gain in sensitivity? – These questions naturally arise in the context of our development. In the present state of our knowledge we cannot give complete answers to these questions in the general setting. (Incidentally, the position with regard to restricted likelihood ratio tests in the parametric case is same.) We have definite knowledge about the superiority of the restricted test over the unrestricted test in certain special cases only.

Chatterjee and De (1974) showed that in the case of problems essentially involving quadrant restriction of two parameters, the asymptotic power of the restricted union–intersection test in the relevant quadrant is always greater than that of the unrestricted test. To describe the essence of their approach, let us consider the bivariate two-sample location problem against quadrant alternative described in Section 4. The unrestricted test statistic here is always given by the first expression under (4.20). Asymptotically, it has a χ_2-distribution under H_0 and the distribution of $\chi'_{2,\delta}$ (noncentral chi with d.f. 2 and noncentrality parameter δ) where $\delta = h\boldsymbol{\tau}'\Gamma G^{-1}\Gamma\boldsymbol{\tau}$. Thus the asymptotic power of the size-α unrestricted test is given by

$$P_{UR}(\delta) = \text{Prob}\{\chi'^2_{2,\delta} > d_1^2\}, \qquad \text{Prob}\{\chi_2^2 > d_1^2\} = \alpha\,. \tag{6.1}$$

Asymptotic power, of the size-α restricted test based on $\hat{Q}_N$, derived from (4.14) and (4.20), depends not only on δ but also on g_{12} and the ratio $\tau_1\gamma_1/\tau_2\gamma_2$. Using the properties of a MLR family of distributions, for $\tau > 0$ a lower bound to this can be derived in the form

$$\begin{aligned} &LB_R(\delta) = \tfrac{1}{2}\text{Prob}\{\chi'^2_{2,\delta} > d_2^2\} + \tfrac{1}{2}\text{Prob}\{\chi'^2_{1,\delta} > d_2^2\},\\ &\tfrac{1}{2}\text{Prob}\{\chi_2^2 > d_2^2\} + \tfrac{1}{2}\text{Prob}\{\chi_1^2 > d_2^2\} = \alpha\,. \end{aligned} \tag{6.2}$$

Now, using the notation $f_\delta(x)$ to denote the density function of $\chi'^2_{1,\delta}$ and Z to denote a r.v. following $N(0, 1)$, from (6.1) and (6.2),

$$P_{UR}(\delta) = \text{Prob}\{Z^2 + \chi'^2_{1,\delta} > d_1^2\} = \int_0^\infty \varphi_1^*(x) f_\delta(x)\,\mathrm{d}x \quad \text{(say)}\,, \tag{6.3}$$

$$\begin{aligned} LB_R(\delta) &= \tfrac{1}{2}\text{Prob}\{Z^2 + \chi'^2_{1,\delta} > d_2^2\} + \tfrac{1}{2}\text{Prob}\{\chi'^2_{1,\delta} > d_2^2\}\\ &= \int_0^\infty \varphi_2^*(x) f_\delta(x)\,\mathrm{d}x \quad \text{(say)}\,. \end{aligned} \tag{6.4}$$

It is readily seen that $\varphi_i^*(x)$, $i = 1, 2$, in (6.3)–(6.4) are functions such that

$$0 \leq \varphi_i^*(x) \leq 1, \qquad \int_0^\infty \varphi_i^*(x) f_0(x)\,\mathrm{d}x = \alpha, \quad i = 1, 2\,. \tag{6.5}$$

Thus $\varphi_i^*(\chi)$, $i = 1, 2$ can be regarded as the critical functions of two randomized size-α tests for testing $\delta = 0$ on the basis of a single observation from $f_\delta(x)$. By a careful examination of the forms of $\varphi_i^*(x)$, $i = 1, 2$, it can be shown that there exists a number x_0 $(0 < x_0 < \infty)$ such that $\varphi_1^*(x) > \varphi_2^*(x)$ for $0 < x < x_0$ and $\varphi_1^*(x) \leq \varphi_2^*(x)$ for $x_0 < x < \infty$. Since $f_\delta(x)$, $\delta > 0$ forms a MLR family from this it can be shown that $P_{UR}(\delta) < LB_R(\delta)$ for all $\delta > 0$.

What we have described above with reference to the bivariate two-sample location problem applies to other bivariate two-parameter problems (and indeed, to the homogeneity problem against order alternative with three treatments). If, instead of a quadrant alternative we have an alternative represented by the angle between two lines meeting at the origin, the same approach can be adopted. Adaptation to a particular form of the multiparameter orthant restriction problem is described in Chinchilli and Sen

Table 1
Maximum and minimum asymptotic power of the restricted test against quadrant alternative versus the asymptotic power of the unrestricted test for bivariate two-parameter problems

Level of significance = 0.05		Noncentrality parameter δ			
		1.00		4.00	
	g_{12}	Maximum	Minimum	Maximum	Minimum
Restricted test	0.8	0.194	0.159	0.514	0.469
	0.5	0.210	0.173	0.543	0.494
	0.2	0.219	0.185	0.555	0.513
	−0.2	0.233	0.197	0.579	0.560
	−0.5	0.244	0.211	0.605	0.569
	−0.8	0.250	0.218	0.618	0.599
Unrestricted test		0.130		0.402	

(1981b). The extension of the result to the general case, however, is still an open problem.

For bivariate two-parameter problems with quadrant alternative, the maximum and the minimum asymptotic power of the restricted test depend on the noncentrality parameter δ and the correlation coefficient g_{12}. Table 1 reproduced from Chatterjee and De (1972) compares their values with the asymptotic power of the unrestricted test at 5 per cent level of significance for two values for δ. The figures show a considerable gain for the restricted test the improvement being more marked with g_{12} becoming negatively large.

Numerical studies performed in the trivariate case (see Chinchilli and Sen, 1918b) and in the case of the homogeneity problem against order alternative with multiple treatments (Barlow et al., 1972, Chapter III) seem to indicate even more remarkable gains for the restricted tests.

In the above we have compared the performances of tests in terms of asymptotic power. Since the power functions of restricted and unrestricted tests have different forms no single value of ARE can be given here. One can think of comparison in terms of some other measures of test performance. It can be shown that, for the type of problem we have considered here, the unrestricted and restricted tests have identical Bahadur slopes (Bahadur, 1960) so that their relative Bahadur efficiency is 1. Chandra and Ghosh (1978) discuss the comparison of tests with relative Bahadur efficiency 1 in terms of what they call Bahadur Cochran deficiency. This approach may throw some light on the problem considered by us. The details, however, would require considerable working out.

References

Bahadur, R. R. (1960). Stochastic comparison of tests. *Ann. Math. Statist.* **31**, 276–295.

Barlow, R. E., Bartholomew, D. J., Bremner, J. M. and Brunk, H. D. (1972). *Statistical Inference under Order Restrictions*, Wiley, New York.

Bartholomew, D. J. (1959). A test of homogeneity for ordered alternatives. *Biometrika* **46**, 36–48.

Bartholomew, D. J. (1961). A test of homogeneity of means under restricted alternatives, *J. Roy. Statist. Soc. Ser. B* **23**, 239–281.

Bhattacharya, G. K. and Johnson, R. A. (1970). A layer rank test for ordered bivariate alternatives. *Ann. Math. Statist.* **41**, 1296–1309.

Chacko, V. J. (1963). Testing homogeneity against ordered alternatives. *Ann. Math. Statist.* **34**, 945–956.

Chandra, T. K. and Ghosh, J. K. (1978). Comparison of tests with same Bahadur efficiency. *Sankhya Ser. A* **40**, 253–277.

Chatterjee, S. K. and De, N. K. (1972). Bivariate nonparametric tests against restricted alternatives. *Calcutta Statist. Assoc. Bull.* **21**, 1–20.

Chatterjee, S. K. and De, N. K. (1974). On the power superiority of certain bivariate location tests against restricted alternatives. *Calcutta Statist. Assoc. Bull.* **23**, 73–84.

Chinchilli, V. M. and Sen, P. K. (1981a). Multivariate linear rank statistics and the union-intersection principle for hypothesis testing under restricted alternatives. *Sankhya Ser. B* **43**, 135–151.

Chinchilli, V. M. and Sen, P. K. (1981b). Multivariate linear rank statistics and the union–intersection principle for the orthant restriction problem. *Sankhya Ser. B* **43**, 152–171.

Chinchilli, V. M. and Sen, P. K. (1982). Multivariate linear rank statistics for profile analysis. *J. Multivariate Anal.* **12**, 219–229.

David, F. N. and Fix, E. (1961). Rank correlation and regression in non-normal surface, in: *Proc. Fifth Berkeley Symposium, Vol. 1*, 177–197.

De, N. K. (1976). Rank tests for randomized blocks against ordered alternatives. *Calcutta Statist. Assoc. Bull.* **25**, 1–27.

De, N. K. (1977). Multivariate nonparametric tests against restricted alternatives. Ph.D. thesis. Calcutta University.

Doksum, K. (1967). Robust procedures for some linear models with one observation per cell. *Ann. Math. Statist.* **38**, 878–883.

Hadley, G. (1964). *Non linear and Dynamic Programming*, Addison-Wesley, Reading, MA.

Hájek, J. and Sidák, Z. (1967). *Theory of Rank Tests.* Academic Press, New York.

Hollander, M. (1967). Rank tests for randomized blocks when the alternatives have a priori ordering. *Ann. Math. Statist.* **38**, 867–877.

Jonckheere, A. R. (1954). A distribution-free k-sample test against ordered alternatives. *Biometrika* **41**, 135–145.

Krishnaiah, P. R. (1965). Multiple comparison tests in multiresponse experiments. *Sankhyā Ser. A* **27**, 65–72.

Kudo, A. (1963). A multivariate analogue of the onesided test. *Biometrika* **50**, 403–418.

Milton, R. C. (1972). Computer evaluation of the multivariate normal integral. *Technometrics* **14**, 881–889.

Nüesch, P. E. (1966). On the problem of testing location in multivariate populations for restricted alternatives. *Ann. Math. Statist.* **37**, 113–119.

Page, E. B. (1963). Ordered hypothesis for multiple treatments, a significance test for linear ranks. *J. Amer. Statist. Assoc.* **58**, 216–230.

Perlman, M. D. (1969). One-sided testing problems in multivariate analysis. *Ann. Math. Statist.* **40**, 549–567.

Pincus, R. (1975). Testing linear hypotheses under restricted alternatives. *Math. Operationsforsch. Statist.* **6**, 733–751.

Puri, M. L. and Sen, P. K. (1971). *Nonparametric Methods in Multivariate Analysis.* Wiley, New York.

Roy, S. N. (1957). *Some Aspects of Multivariate Analysis.* Wiley, New York.

Sen, P. K. (1968). On a class of aligned rank order tests in two-way lay-outs. *Ann. Math. Statist.* **39**, 1115–1124.

Shorack, S. R. (1967). Testing against ordered alternatives in model I analysis of variance; Normal theory and nonparametric. *Ann. Math. Statist.* **38**, 1740–1753.

P. R. Krishnaiah and P. K. Sen, eds., *Handbook of Statistics, Vol. 4*
© Elsevier Science Publishers (1984) 347–358 **16**

Adaptive Methods

M. Hušková

1. Introduction

The aim of this chapter is to present the main ideas of adaptive procedures, to summarize their basic structure, to state their properties and to review the typical ones. In this context, the main emphasis is laid on the one-sample and two-sample models, which are presented below.

One-sample model: Let $X_1, \ldots, X_n$ be a sample from a distribution with the probability density function (p.d.f.) $f(x-\theta)$, $x \in \mathbb{R}$ (the real line), where f is symmetric about 0, absolutely continuous and has a finite Fisher information

$$0 < I(f) = \int_{-\infty}^{\infty} \{f'(x)/f(x)\}^2 f(x)\, \mathrm{d}x < \infty . \tag{1.0}$$

Two-sample model: Let $X_1, \ldots, X_n$ and $Y_1, \ldots, Y_m$ be two independent samples from distributions with p.d.f.'s $f(x)$ and $f(x-\theta)$, respectively, where f (need not be symmetric) is absolutely continuous with a finite Fisher information.

For both the models, θ is an unknown parameter (location or shift), and one is interested in drawing inference on θ. For testing the null hypothesis H_0: $\theta = 0$ against A: $\theta > 0$, the general form of rank test statistic is the following:

$$S_N(\phi) = \sum_{i=1}^{n} \phi(R_i/(N+1)) \quad \text{(two-sample case)}, \tag{1.1}$$

$$S_n^+(\phi) = \sum_{i=1}^{n} \operatorname{sign} X_i \phi(R_i^+/(n+1)) \quad \text{(one-sample case)}, \tag{1.2}$$

where $N = n + m$, R_i is the rank of X_i among $X_1, \ldots, X_n$, $Y_1, \ldots, Y_m$, for $i = 1, \ldots, n$, R_i^+ is the rank of $|X_i|$ among $|X_1|, \ldots, |X_n|$, for $i = 1, \ldots, n$, and ϕ is some square integrable function on the unit interval (0, 1). For the estimation of θ, in the one-sample model, there are three broad types of estimators: Maximum likelihood type (M-)estimators, rank based (R-)estimators and linear ordered (L-)estimators (see Chapter 21 (by Jurečková)). An M-estimator $\hat{\theta}_M(\psi)$ of θ is defined as a solution of the following

$$\sum_{i=1}^{n} \psi(X_i - t) \simeq 0\,, \tag{1.3}$$

where ψ is a suitable score function (defined on $\mathbb{R}$) and $\simeq$ means approximate equality in some sense. The R-estimator $\hat{\theta}_R(\phi)$ is defined as a solution of

$$\sum_{i=1}^{n} \operatorname{sign}(X_i - t)\phi(R_i^+(t)/(n+1)) \simeq 0\,, \tag{1.4}$$

where $R_i^+(t)$ is the rank of $|X_i - t|$ among $|X_1 - t|, \ldots, |X_n - t|$, for $i = 1, \ldots, n$, and ϕ is a monotone, square integrable score function (defined on (0, 1)). The L-estimator has the form

$$\hat{\theta}_L(J) = \sum_{i=1}^{n} J(i/(n+1))X_{(i)} \tag{1.5}$$

where $X_{(1)} \leqslant \cdots \leqslant X_{(n)}$ stand for the ordered random variables (r.v.) corresponding to $X_1, \ldots, X_n$. Sometimes, linearized versions of M- and R-estimators are used; these may be defined as follows:

$$\hat{\theta}_M^L(\psi) = \tilde{\theta} - n^{-1}\Big(\int \psi(x)f'(x)\,\mathrm{d}x\Big)^{-1} \sum_{i=1}^{n} \psi(X_i - \tilde{\theta})\,, \tag{1.6}$$

$$\hat{\theta}_R^L(\phi) = \tilde{\theta} - n^{-1}\Big(\int_0^1 \phi^2(u)\,\mathrm{d}u\Big)^{-1} \sum_{i=1}^{n} \operatorname{sign}(X_i - \tilde{\theta})(R_i^+(\tilde{\theta})/(n+1))\,, \tag{1.7}$$

where $\tilde{\theta}$ is a preliminary (equivariant) estimator of θ, such that

$$n^{1/2}(\tilde{\theta} - \theta) = \mathrm{O}_p(1) \quad \text{as } n \to \infty\,. \tag{1.8}$$

For the two-sample model, these estimators are defined in an analogous way (see Chapter 21 (by Jurečková)).

If the pdf f is known and satisfies some regularity conditions, then the test for H_0 against A, based on $S_n^+(\phi_f^+)$ and $S_N(\phi_f)$ leads to asymptotically most powerful tests for contiguous alternatives (see Chapter 3 (by Hušková) for the one-sample model); here,

$$\phi_f(u) = -f'(F^{-1}(u))/f(F^{-1}(u)), \quad u \in (0, 1)\,, \tag{1.9}$$

$$\phi_f^+(u) = \phi_f((1+u)/2), \qquad F^{-1}(u) = \inf\{x\colon F(x) \geqslant u\}, \quad 0 < u < 1\,. \tag{1.10}$$

In the estimation problems [see Chapter 21 (by Jurečková)], the functions

$$\begin{aligned}\psi_f(x) &= -f'(x)/f(x), \quad x \in R,\\ J_f(u) &= \phi_f(u) f(F^{-1}(u)) I^{-1}(f), \quad 0 < u < 1\,,\end{aligned} \tag{1.11}$$

and ϕ_f^+, given by (1.10), generate the asymptotically optimal estimators, denoted by $\hat{\theta}_M(\psi_f)$, $\hat{\theta}_L(J_f)$ and $\hat{\theta}_R(\phi_f)$, respectively, where asymptotic optimality relates to the property:

$$(nI(f))^{1/2}(\hat{\theta} - \theta) \xrightarrow{\mathscr{D}} \mathcal{N}(0, 1) \quad \text{as } n \to \infty\,.$$

Among the L-estimators, the α-trimmed mean, defined below, is of special interest. Here we take, for some $\alpha \in (0, \frac{1}{2})$,

$$\begin{aligned} J(t) &= (1 - 2\alpha)^{-1} && \text{for } \alpha \leq t \leq 1 - \alpha\,,\\ &= 0 && \text{otherwise}\,. \end{aligned}$$

In practice, the form of the density f is not known. Nevertheless, we are interested in having the asymptotically optimal (or at least reasonable) tests or estimators. The procedures leading to such desirable solutions are termed the *adaptive procedures*; these are adapted according to the data. These procedures are generally of two kinds: *restrictive and nonrestrictive procedures.*

The basic structure of the restrictive procedures is the following:

(1) Choose a reasonable family $\mathscr{F}$ of distributions and a type (M-, R- or L-) of statistics to be used for the decision rule.

(2) Choose according to the decision rule a density $f_0 \in \mathscr{F}$.

(3) Provide the test (or estimator) of the pre-chosen type, which is optimal for the density f_0.

The nonrestrictive adaptive procedure consists in

(1) prechoosing a type of the test statistic (or the estimator),

(2) estimating ϕ_f, ψ_f, J_f or the density f from the data, and

(3) using the decision rule with these estimated scores or density.

Adaptive procedures of both kinds have been considered and their properties were studied by various workers. Generally, the restrictive procedures are simpler, but they are optimal or desirable only when the true density belongs to the chosen family $\mathscr{F}$. Small sample behavior of these procedures has mostly been studied by monte carlo methods. On the other hand, the nonrestrictive decision rules are generally asymptotically optimal for a broad class of densities. But, these may involve very tedious computations and the convergence rates for the asymptotic results may be very slow. For some good review of these adaptive procedures, we may also refer to Hogg (1974, 1976).

We review the developments in the area of restricted adaptive procedures in Section 2, and the parallel results for the nonrestricted procedures are then presented in Section 3.

2. Restrictive procedures

The attention here is concentrated to the review of possible families $\mathscr{F}$ of distributions and decision rules for selection of the distribution in the one-sample model. In the following, we shall often work with the subfamily $\mathscr{F}(f)$ of densities generated by a density f as follows:

$$\mathscr{F}(f)=\{g;\, g(x)=\lambda f(\lambda x-u),\, -\infty<u<+\infty\,,\, \lambda>0\}\,.$$

First, we mention the procedures with decision rules motivated by the behavior of the tails of distribution (e.g. Hájek, 1970; Hogg and Randles, 1973; Hogg, Fisher and Randles, 1975; Jones, 1979). In such a case the family $\mathscr{F}$ contains densities ranging from the light-tailed (like uniform) to heavy-tailed (like Cauchy).

Randles and Hogg (1973) considered the family $\mathscr{F}$ consisting of three type densities $\mathscr{F}(f_1)$, $\mathscr{F}(f_2)$, $\mathscr{F}(f_3)$, where f_1 is double exponential density (heavy-tailed), f_2 is logistic one (medium-tailed), f_3 is uniform one (light-tailed). The decision rule is the following:

$$\begin{aligned}&\text{choose } \mathscr{F}(f_1) \quad \text{if } Q>2.96-5.5/n\\ &\text{choose } \mathscr{F}(f_2) \quad \text{if } 2.96-5.5/n\geqslant Q\geqslant 2.08-2/n\\ &\text{choose } \mathscr{F}(f_3) \quad \text{if } 2.08-2/n>Q\,,\end{aligned}$$

where

$$Q=(X_{(n)}-X_{(1)})n\left\{\sum_{i=1}^{n}\left|\ X_{(i)}-\text{median of } X_i\text{'s}\right|\ 2\right\}^{-1} \quad \text{for } n\leqslant 20\,, \tag{2.1}$$

$$Q=10(\bar{U}_{0.05}-\bar{L}_{0.05})(\bar{U}_{0.5}-\bar{L}_{0.5})^{-1} \quad \text{for } n>20\,, \tag{2.2}$$

with $\bar{U}_\alpha$ and $\bar{L}_\alpha$ being the average of 100α % the largest and smallest order statistics, resp. The motivation for this decision rule, if $n\leqslant 20$, comes from the fact that the optimal translation and scale invariant test that the considered sample is from the uniform distribution versus double exponential one can be (approximately) based on Q given by (2.1). As for $n>20$, notice that

$$\left.\begin{aligned}Q&\to 3.3\\ Q&\to 2.6\\ Q&\to 1.96\end{aligned}\right\} \text{ in probability as } n\to\infty \text{ for } \left\{\begin{aligned}&f\in\mathscr{F}(f_1)\,,\\ &f\in\mathscr{F}(f_2)\,,\\ &f\in\mathscr{F}(f_3)\,.\end{aligned}\right.$$

Then the test statistics are chosen according to the general rule except $\mathscr{F}(f_3)$ – the authors recommend to use some modified Wilcoxon one-sample test.

Some modifications of this decision rule were used to obtain other adaptive procedures (also for two-sample case), e.g. Hogg, Fisher and Randles (1975) considered the decision rule based on Q and a measure asymmetry

$$Q^* = \frac{\bar{U}_{0.05} - \bar{M}_{0.5}}{\bar{M}_{0.5} - \bar{L}_{0.05}},$$

where $\bar{M}_\alpha$ is the average of 100α % middle order statistics. Harter, Moore and Curry (1979) suggested adaptive estimators of location and scale parameter with three kinds of decision rules (the sample kurtosis, Q and the sample likelihood). Another version of the decision rule based on Q was developed by Moberg, Ramberg and Randles (1978, 1980) to obtain adaptive M-type estimators with application to a selection problem and in a general regression model.

Another procedure based on tail behavior was suggested by Hájek (1970) for the family $\mathcal{F} = \{\mathcal{F}(f_1), \ldots, \mathcal{F}(f_k)\}$, where f_i are distinct symmetric densities. The decision rule consists, in fact, in choosing $\mathcal{F}(f_i)$ for which the quantile function corresponding to f_i is close to the sample quantile function. The procedure is very quick but no properties were studied.

Jones (1979) introduced the family $\mathcal{F} = \{f_\lambda, \lambda \in \mathbb{R}_1\}$, where f_λ satisfies

$$F_\lambda^{-1}(u) = (u^\lambda - (1-u)^\lambda)/\lambda$$

(i.e. $\varphi(u, f_\lambda) = (\lambda - 1)(u^{\lambda-2} - (1-u)^{\lambda-2})(u^{\lambda-1} + (1-u)^{\lambda-1})^{-2}$). This family contains densities ranging from light-tailed ones ($\lambda > 0$) to heavy-tailed ones ($\lambda < 0$). Particularly, for $\lambda = 1$ and $\lambda = 2$, f_λ is uniform; for $\lambda = 0.135$, f_λ is approximately normal; for $\lambda = 0$, f_λ is logistic. The author proposed to estimate λ through the ordered sample as follows:

$$\hat{\lambda} = (\log 2)^{-1} \log\{[|X|_{(n-2M+1)} - |X|_{(n-4M+1)}] \cdot [|X|_{(n-M+1)} - |X|_{(n-2M+1)}]^{-1}\}$$

where M is chosen in some proper way reflecting the behavior of the tail. As the resulting φ-function is taken $\varphi(u, f_\lambda)$.

As examples of procedures not motivated by the behavior of tails, we shall sketch two procedures published by Hájek (1970) for a general family $\mathcal{F} = \{\mathcal{F}(f_1), \ldots, \mathcal{F}(f_k)\}$, where $f_1, \ldots, f_k$ are distinct densities and the procedure by Albers (1979). In the first procedure, the decision rule is the Bayesian translation and scale invariant rule and the second one is based on the asymptotic linearity of rank statistics, the third one utilizes the estimate of the kurtosis. In order to have the decision rule dependent only on the ordered sample $|X|_{(1)} = \cdots = |X|_{(n)}$ (corresponding to $|X_1|, \ldots, |X_n|$) we define new random variables $X_i^* = V_i |X|_{(Q_i)}$, $i = 1, \ldots, n$, where $(Q_1, \ldots, Q_n)$ is a random permutation of $(1, \ldots, n)$ and $(V_1, \ldots, V_n)$ are i.i.d. with $P(V_i = 1) = P(V_i = -1) = \frac{1}{2}$ independent of $X_1, \ldots, X_n$. Then under H the random vector $(X_1^*, \ldots, X_n^*)$ is independent of $(R_1^+, \ldots, R_n^+)$ and $(\operatorname{sign} X_1, \ldots, \operatorname{sign} X_n)$ and distributed as $(X_1, \ldots, X_n)$.

The Baysian translations and scale invariant rule (if all types are a priori equiprobable) yields the following: choose $\mathcal{F}(f_1)$ if $\max_{1 \leq j \leq k} p_{jn}(X_1^*, \ldots, X_n^*) =$

$p_{in}(X_1^*, \ldots, X_n^*)$ where

$$p_{jn}(X_1^*, \ldots, X_n^*) = \int_0^{+\infty} \int_{-\infty}^{+\infty} \prod_{i=1}^{n} f(\lambda X_i^* - u)\lambda^{n-2}\, du\, d\lambda, \quad j = 1, \ldots, k.$$

Uthoff (1970) derived $p_n(X_1, \ldots, X_n)$ for some well-known distributions (e.g. normal, uniform, exponential). Sometimes there are computational problems with evaluating $p_n(X_1, \ldots, X_n)$. For such cases Hogg et al. (1972) recommended to use

$$p_{jn}^*(X_1^*, \ldots, X_n^*) = \prod_{i=1}^{n} \hat{\sigma}_{jn}^{-1} f((X_i^* - \hat{\mu}_{jn})\hat{\sigma}_{jn}^{-1}),$$

where $\hat{\mu}_{jn}$ and $\hat{\sigma}_{jn}$ are maximum likelihood estimators of location and scale for j-th distribution, instead of $p_{jn}(X_1^*, \ldots, X_n^*)$.

The decision rule based on the asymptotic linearity of rank statistics ($f_1, \ldots, f_k$ absolutely continuous and $I(f_j) < +\infty$, $j = 1, \ldots, k$) is defined as follows:

$$\text{choose } \mathcal{F}(f_i) \quad \text{if } \max_{1 \le j \le n} L_{jn} = L_{in},$$

where

$$L_{jn} = [S_n^*(j, n^{-1/2}) - S_n^*(j, 0)] \cdot [\text{var}_H S_n^*(j, 0)]^{-1/2}$$

$$S_n^*(j, t) = \sum_{i=1}^{n} \text{sign}(X_i^* + t)\varphi_{f_j}^+(R_i^+(t)(n+1)^{-1})$$

with $R_i^+(t)$ being the rank of $|X_i^* + t|$ in the sequence $|X_1^* + t|, \ldots, |X_n^* + t|$. Notice that by the asymptotic linearity (see van Eeden, 1972)

$$L_{jn} \to \int_0^1 \varphi_f^+(u)\varphi_{f_j}^+(u)\, du \left(\int_0^1 \varphi_{f_j}^{+2}(u)\, du\right)^{-1/2},$$

in probability as $n \to \infty$, $j = 1, \ldots, k$, if the true density is f. By Schwarz inequality the right-hand side is smaller or equal $(\int_0^1 \varphi_f^{+2}(u)\, du)^{1/2}$ and the maximum is achieved if $f \in \mathcal{F}(f_i)$.

Albers (1979) considered instead of the family of distributions $\mathcal{F}$ the family (say J) of functions φ generating the test statistics $S_N^+(\varphi)$ given by (1.2), where $J = \{\varphi_r; \varphi_r = \varphi_0 + rh, -D_1 \le r \le D_2\}$, φ_0 and h are smooth functions on $(0, 1)$, $D_1 > 0$, $D_2 > 0$. He recommended to choose such $\varphi_{\tilde{r}}$, where $\tilde{r}$ is minimizing

$$\left| \frac{n^{-1} \sum_{i=1}^{n} (|X|_{(i)})^q}{(n^{-1} \sum_{i=1}^{n} (|X|_{(i)})^p)^{q/p}} - \frac{\int |x|^q\, dF_r(x)}{(\int |x|^p\, dF_r(x))^{q/p}} \right|,$$

with F_r being the distribution function corresponding to φ_r, $0 < p < q$. It is also indicated how p and q should be chosen for given J.

The procedures with the decision rules based on Q, the linearity of rank statistic and the Bayesian rule can be easily modified to the two-sample case (see Randles and Hogg, 1973; Hájek, 1970).

At the end of this section let us mention some adaptive α-trimmed means, more exactly, possible choice of α according to the data. Hogg (1967) recommended to choose α according to the kurtosis, later, in 1974, he developed the rule based on Q. Prescott (1978) and de Wet and van Wyk (1979) studied some modifications of these rules. Another procedure was suggested by Jaeckel (1971).

3. Nonrestrictive procedures

Here will be presented estimators of the functions φ_f, ψ_f, J_f, some of their asymptotic properties and also the resulting test statistics and the estimators.

Hájek (1962) proposed the following estimator of φ_f based on ordered sample $X_{(1)} \leqslant \cdots \leqslant X_{(n)}$:

$$\tilde{\varphi}_n(u) = 2\frac{q_n t_n}{n+1}\{(X_{(h_{nj}+q_n)} - X_{(h_{nj}-q_n)})^{-1} - (X_{(h_{n,j+1}+q_n)} - X_{(h_{n,j+1}-q_n)})^{-1}\}$$

$$\text{for } h_{nj}n^{-1} \leqslant u \leqslant h_{n,j+1},\ 1 \leqslant j \leqslant t_n, \tag{3.1}$$

where $q_n = [n^{3/4}\varepsilon_n^{-2}]$, $t_n = [n^{1/4}\varepsilon_n^3]$, $\varepsilon_n \to 0$, $n^{1/4}\varepsilon_n^3 \to \infty$, as $n \to \infty$, $h_{nj} = [jn/(t_n+1)]$, $1 \leqslant j \leqslant t_n$, with $[a]$ denoting the largest integer part of a.

The motivation of this estimator comes from the following two facts:

(1) $X_{(nu)} \to F^{-1}(u)$ in probability as $n \to \infty$, $u \in (0, 1)$,

(2) $$\lim_{r \downarrow 0, s \downarrow 0} 2rs\left\{\frac{1}{F^{-1}(u+r) - F^{-1}(u-r)} - \frac{1}{F^{-1}(u+s+r) - F^{-1}(u+s-r)}\right\} = \varphi_f(u),\ u \in (0, 1).$$

Beran (1974) suggested another estimator (say $\hat{\varphi}_n$) of φ_f through the estimators $\hat{c}_k$ of the Fourier coefficient c_k belonging to $\varphi_f(u)$. Namely,

$$\hat{\varphi}_n(u) = \sum_{|k|=1}^{M_n} \hat{c}_k \exp\{2\pi i k u\}, \qquad \hat{c}_k = T_n(\boldsymbol{X}, \exp\{-2\pi i k u\}) \tag{3.2}$$

where $\boldsymbol{X} = (X_1, \ldots, X_n)'$, $T_n(\boldsymbol{X}, g)$ is the functional defined on $L_2(0, 1)$ given by

$$T_n(\boldsymbol{X}, g) = (2n\theta_n)^{-1} \sum_{\nu=1}^{n} \Big\{ g\Big((n-1)^{-1} \sum_{\substack{j=1 \\ j\neq\nu}}^{n} u(X_\nu - X_j + \theta_n)\Big) - g\Big((n-1)^{-1} \sum_{\substack{j=1 \\ j\neq\nu}}^{n} u(X_\nu - X_j - \theta_n)\Big)\Big\},$$

$u(x) = 1$ if $x \geqslant 0$, $u(x) = 0$ if $x < 0$, $\theta_n = bn^{-1/2}$ for some $b \neq 0$. In fact, $T_n(\boldsymbol{X}, g)$ is an estimator of the functional

$$T(g) = \int_0^1 \varphi_f(u) g(u)\, \mathrm{d}u = \int \frac{\mathrm{d}g(F(x))}{\mathrm{d}x} \mathrm{d}F(x),$$

obtained by replacing the theoretical distribution by the empirical one and replacing the derivative by difference.

The estimator $\tilde{\varphi}_n(u)$ is consistent in the sense that

$$\int_0^1 (\tilde{\varphi}_n(u) - \varphi_f(u))^2\, \mathrm{d}u \to 0 \quad \text{in probability as } n \to \infty \tag{3.3}$$

and

$$\int_0^1 \tilde{\varphi}_n^2(u)\, \mathrm{d}u \to I(f) \quad \text{in probability as } n \to \infty. \tag{3.4}$$

If, moreover, $M_n \to \infty$ and $M_n^{7/2} n^{-1} \to 0$ as $n \to \infty$ then in both relations $\tilde{\varphi}_n(u)$ can be replaced by $\hat{\varphi}_n(u)$.

The respective adaptive test statistics in the two-sample model are $S_N(\tilde{\varphi}_N^*)$ and $S_N(\hat{\varphi}_N^*)$, where $S_N(\varphi)$ is given by (1.1) and

$$\tilde{\varphi}_N^*(u) = N^{-1}(n\tilde{\varphi}_n(u, \boldsymbol{X}) + m\tilde{\varphi}_m(u, \boldsymbol{Y})), \quad u \in (0, 1), \tag{3.5}$$

writing $\tilde{\varphi}_n(u, \boldsymbol{X})$ and $\tilde{\varphi}_m(u, \boldsymbol{Y})$ for $\tilde{\varphi}_n$ obtained from the sample $\boldsymbol{X} = (X_1, \ldots, X_n)'$ and $\boldsymbol{Y} = (Y_1, \ldots, Y_m)'$, resp. $\hat{\varphi}_N^*$ is defined similarly. Under H

$$(S_N(\tilde{\varphi}_N^*) - \hat{E}_N)\tilde{D}_N^{-1} \xrightarrow{\mathscr{D}} N(0, 1) \quad \text{as } \min(n, m) \to \infty, \tag{3.6}$$

where $\tilde{E}_N = N^{-1} \Sigma_{i=1}^N \tilde{\varphi}_N^*(i(N+1)^{-1})$, $\tilde{D}_N^2 = (N-1) \Sigma_{i=1}^N (\tilde{\varphi}_N^*(i(N+1)^{-1}) - \tilde{E}_N)^2$. If, moreover, $M_N \to \infty$, $M_N^4(\min(n, m))^{-1} \to 0$, then $\tilde{\varphi}_N^*$ can be replaced by $\hat{\varphi}_N^*$ in (3.6). The tests based on either $S_N(\tilde{\varphi}_N^*)$ or $S_N(\hat{\varphi}_N^*)$ provide the asymptotically most powerful tests for H versus contiguous alternatives.

As for estimators, Beran (1974) proved asymptotic optimality of $\hat{\theta}_R^L(\hat{\varphi}_N^*)$ given by (1.4). Van Eeden (1970) and Kraft and van Eeden (1970) got the same properties (under somewhat stronger conditions) of $\hat{\theta}_R^L(\tilde{\varphi}_N^{**})$ and $\hat{\theta}_R(\tilde{\varphi}_N^{**})$, resp., where $\tilde{\varphi}_N^{**}$ is a modified form of the estimator $\tilde{\varphi}_N^*$ obtained from vanishingly small fraction of the data. Both the test statistics and the estimators can be easily modified for one-sample case.

Rieder (1980) using the estimator $\tilde{\varphi}_N$ suggested and studied the adaptive correlation coefficient.

Behnen and Neuhaus (1981) and Behnen, Neuhaus and Ruymgaart (1982) considering general two-sample model: $(X_1, \ldots, X_m)$ and $(Y_1, \ldots, Y_n)$ are two independent samples from the populations with the cumulative distribution functions F and G (both unknown), respectively, developed several adaptive procedures based on ranks for the testing problem $H: F = G$ against $A: F \leqslant$

G, $F \neq G$ (i.e. F is stochastically larger than G). If F and G are known the score-generating function

$$\psi_N(u) = \bar{f}_N(u) - \bar{g}_N(u), \quad u \in (0, 1),$$

where $\bar{f}_N$ and $\bar{g}_N$ are the densities of $F(H_N^{-1}(X_i))$ and $G(H_N^{-1}(Y_i))$, respectively, $N = n + m$,

$$H_N = \frac{m}{N} F + \frac{n}{N} G,$$

is an appropriate one. The authors constructed an estimator $\hat{\psi}_N$ of ψ_N based on ranks for the case F and G unknown and recommended as a test statistic $S_N(\hat{\psi}_N)$. Asymptotic results and simulation studies were done.

Sacks (1975) proposed an adaptive L-type estimator of θ in the one-sample model. Using the same ideas as Hájek, he estimated $f(F^{-1}(t))$ by

$$\tilde{h}_n(t) = 2^{-1}\gamma_n(X_{((t+\gamma_n/2)n)} - X_{((t-\gamma_n/2)n)})^{-1} + (X_{((1-t+\gamma_n/2)n)} - X_{((1-t-\gamma_n/2)n)})^{-1}.$$

Let $0 \leqslant t_0 < t_1 < \cdots < t_{k+1} \leqslant 1$, $t_i + t_{k-i+1} = 1$, $t_1 > \gamma_n$, $t_{i+1} - t_i = d_n$; $(t_i \pm \gamma_n/2)n$ are integers. The adaptive estimator is defined as follows:

$$\hat{\theta}_L = n^{-1} \sum_{i=I}^{k-I} (h_n(t_{i+1}) - h_n(t_i))(h(t_{i+1})X_{(nt_{i+1})} - h(t_i)X_{(nt_i)}) \sum_{j=I}^{k-I} (h(t_{j+1}) - h(t_j))^2,$$

where $i = \min\{i; i \leqslant j \leqslant k - i + 1, h(t_j) \geqslant \lambda\} + 1$. Under very mild conditions $\hat{\theta}_L$ is asymptotically optimal. Other adaptive L-type estimators were presented by Takeuchi (1971) and Johns (1974).

Stone (1975) developed an adaptive estimator of maximum likelihood type.

Write g for the density of X (i.e. $g(x) = f(x - \theta)$). The following estimator of the function ψ is used:

$$\hat{\psi}_n(y) = \int (\hat{d}_n(x + y; r_n)z(x + y)/c_n)\beta(x; r_n)\,\mathrm{d}x,$$

$$\hat{d}_n(x; r_n) = \hat{g}'_n(x; r_n)(\hat{g}_n(x; r_n))^{-1},$$

$$\hat{g}_n(x; r_n) = (2n)^{-1} \sum_{j=1}^{n} (\beta(x + \bar{\theta} - X_j; r_n) + \beta(-x + \bar{\theta} - X_j; r_n)),$$

where $z: \langle -1, 1\rangle \to \langle 0, 1\rangle$ is a function twice continuously differentiable, symmetric, $z(0) = 1$, $\beta(x, r)$ is the density of $N(0, r^2)$, c_n and r_n are some random variables, $\bar{\theta}$ is a preliminary estimator. The function g_n is the kernel type density estimator.

If $c_n n^{\varepsilon-1} r_n^{-6} = O_p(1)$ for $\varepsilon > 0$ then the estimator $\hat{\theta}_M(\hat{\psi}_n)$ is asymptotically optimal, translation and scale equivariant. Another adaptive estimator of maximum likelihood type was considered by Samanta (1974).

There exist some other kinds of estimators that can be classified as adaptive ones, e.g. Parr and Schucany (1980), Beran (1977, 1978), Rao, Schuster and Littel (1975). Let us mention briefly two of them. Rao, Schuster and Littel (1975) and Beran (1978) suggested estimators of θ in the one-sample model motivated by the symmetry of f. Writing again G and g for the distribution function and the density, respectively, of X, we have

$$g(x) = g(-x + 2\theta), \quad G(x) = 1 - G(-x + 2\theta) \qquad \text{for all } x .$$

Replacing in these relations the theoretical distribution function and the theoretical density by their suitable estimators $\hat{G}_n$ and $\hat{g}_n$ respectively, we expect that the relations will hold approximately. This leads to the following definitions of estimators $\hat{\theta}_{1n}$, $\hat{\theta}_{2n}$. $\hat{\theta}_{1n}$ is defined as a value t minimizing the Kolmogorov distance of $\hat{G}_n(x)$ and $1 - \hat{G}_n(x + 2t)$, i.e. minimizing

$$\sup_x |1 - \hat{G}_n((-x + 2t)_+) - \hat{G}_n(x)| .$$

Using the Hellinger distance instead of the Kolmogorov one, Beran (1978) arrived at the estimator $\hat{\theta}_{2n}$; $\hat{\theta}_{2n}$ is that value of t minimizing

$$\int (\sqrt{\hat{g}_n(x)} - \sqrt{\hat{g}_n(-x + 2t)})^2 \, \mathrm{d}x$$

where

$$\hat{g}_n(x) = \sum_{i=1}^{n} w(c_n^{-1}(x - X_i)) z_n^2(x - \bar{\theta}) \left\{ \sum_{j=1}^{n} \int w(c_n^{-1}(x - X_j)) z_n^2(x - \bar{\theta}) \, \mathrm{d}x \right\}^{-1}$$

with $\bar{\theta}$ being a preliminary estimator of θ, w being a nonvanishing density, symmetric about zero, absolutely continuous and $w'(x)/w(x)$ is bounded over R_1 and

$$\begin{aligned} z_n(x) &= 1, && |x| \leq b_n , \\ &= z(x - b_n \operatorname{sign} x), && b_n \leq |x| \leq b_n + 1 , \\ &= 0, && \text{otherwise} , \end{aligned}$$

z is the function considered in Stone's estimator, $\{b_n\}_{n=1}^{\infty}$, $\{c_n\}_{n=1}^{\infty}$ are constants suitably chosen.

As for the properties of these estimators, $\hat{\theta}_{2n}$ is asymptotically optimal (for $c_n \to 0$, $b_n \to \infty$, $(nc_n^4)^{-1} b_n^2 \to 0$ as $n \to \infty$), $\hat{\theta}_{1n}$ is not asymptotically optimal only $(\hat{\theta}_{1n} - \theta)\sqrt{n} = O_p(1)$ holds. The assertion on $\hat{\theta}_{2n}$ remains true under small departures from the symmetry in the one-sample model.

Recently, Bickel (1982) published a (largely survey) paper on general problem of adaptive estimation.

References

Albers, W. (1979). Asymptotic deficiencies of one-sample rank tests under restricted adaptation. *Ann. Statist.* **7**, 944–954.

Behnen, K. and Neuhaus, G. (1981). Two-sample rank tests with estimated scores and the Galton test. Preprint 81-10, Hamburg Universität.

Behnen, K., Neuhaus, G. and Ruymgaart, F. (1982). Chernoff–Savage theorem for rank statistics with estimated scores and rank estimators of score functions. Preprint 82-2, Hamburg Universität.

Beran, R. (1974). Asymptotically efficient adaptive rank estimates in location model. *Ann. Statist.* **2**, 63–74.

Beran, R. (1977). Minimum Hellinger distance estimates for parametric models. *Ann. Statist.* **5**, 445–463.

Beran, R. (1978). An efficient and adaptive estimator of location. *Ann. Statist.* **6**, 292–313.

Bickel, P. J. (1982). On adaptive estimation. *Ann. Statist.* **10**, 647–671.

van Eeden, C. (1970). Efficiency-robust estimation of location. *Ann. Math. Statist.* **41**, 172–181.

van Eeden, C. (1972). An analogue for signed rank statistics of Jurečková's asymptotic linearity theorem for rank statistics. *Ann. Math. Statist.* **43**, 791–802.

Hájek, J. (1962). Asymptotically most powerful rank order tests. *Ann. Math. Statist.* **33**, 1124–1147.

Hájek, J. (1969). *A Course in Nonparametric Statistics.* Holden-Day, San Francisco.

Hájek, J. (1970). Miscellaneous problems of rank test theory. In: *Nonparameteric Techniques in Statistic Inference.* London-Cambridge University Press, pp. 3–17.

Harter, H. L., Moore, A. H. and Curry, T. F. (1979). Adaptive robust estimation of location and scale parameter. *Comm. Statist., A* **8**, 1473–1491.

Hogg, R. V. (1967). Some observations on robust estimation. *J. Amer. Statist. Assoc.* **62**, 1179–1186.

Hogg, R. V. (1972). More light on the kurtosis and related statistics. *J. Amer. Statist. Assoc.* **67**, 422–424.

Hogg, R. V. (1974). Adaptive robust procedures: partial review and some suggestions for future applications and theory. *J. Amer. Statist. Assoc.* **69**, 909–923.

Hogg, R. V. (1976). A new dimension to nonparametric tests. *Comm. Statist. Th. Methods A* **5**, 1313–1325.

Hogg. R. V., Fisher, D. M. and Randles, R. G. (1975). A two-sample adaptive distribution-free test. *J. Amer. Statist. Assoc.* **70**, 1020–1034.

Hogg, R. V. et al. (1972). On the selection of the underlying distribution and adaptive estimation. *J. Amer. Statist. Assoc.* **67**, 579–600.

Jaeckel, L. A. (1971). Some flexible estimates of location. *Ann. Math. Statist.* **42**, 1540–1552.

Jones, D. H. (1977). A one-sample adaptive distribution-free test with a stable power function. *Commun. Statist. A* **5**, 251–260.

Jones, D. H. (1979). An efficient adaptive distribution-free test for location. *J. Amer. Statist. Assoc.* **74**, 822–828.

Johns, M. V. (1974). Nonparametric estimation of location. *J. Amer. Statist. Assoc.* **69**, 453–460.

Jurečková, J. (1969). Asymptotic linearity of a rank statistic. *Ann. Math. Statist.* **40**, 1889–1900.

Kraft, C. and van Eeden, C. (1970). Efficient linearized estimates based on ranks, in M. L. Puri (ed.), *Nonparametric Techniques in Statistical Inference.* Cambridge University Press, London.

Moberg, T. F., Ramberg, J. S. and Randles, R. H. (1978). An adaptive M-estimator and its applications to a selection problem. *Technometrics* **20**, 255–263.

Moberg, T. F., Ramberg, J. S. and Randles, R. H. (1980). An adaptive multiple regression procedure based on M-estimators. *Technometrics* **22**, 213–224.

Parr, W. C. and Schucany, W. R. (1980). Minimum distance and robust estimation. *J. Amer. Statist. Assoc.* **75**, 616–624.

Policello, G. E. and Hettmansperger, T. P. (1976). Adaptive robust procedures for the one-sample location problem. *J. Amer. Statist. Assoc.* **71**, 624–633.

Prescott, P. (1978). Selection of trimming proportions for robust adaptive trimmed means. *J. Amer. Statist. Assoc.* **73**, 133–140.

Rao, P. V., Schuster, E. F. and Littel, R. C. (1975). Estimation of shift and center of symmetry based on Kolmogorov–Smirnov statistics. *Ann. Statist.* **3**, 862–873.

Randles, R. H. and Hogg, R. V. (1973). Adaptive distribution-free tests. *Comm. Statist.* **2**, 337–356.

Randles, R. H., Ramberg, J. S. and Hogg, R. V. (1973). An adaptive procedure for selecting the population with largest location parameters. *Technometrics* **15**, 769–778.

Rieder, H. (1980). Locally robust correlation coefficient. *Comm. Statist. A* **9**, 803–819.

Sacks, J. (1975). An asymptotically efficient sequence of estimators of a location parameter. *Ann. Statist.* **4**, 285–298.

Samanta, M. (1974). Efficient nonparametric estimation of a shift parameter. *Sankhya Ser. A* **36**, 273–292.

Stone, C. J. (1975). Adaptive maximum likelihood estimators of a location parameter. *Ann. Statist.* **3**, 267–284.

Takeuchi, K. (1971). A uniformly asymptotically efficient estimator of a location parameter. *J. Amer. Statist. Assoc.* **66**, 292–301.

Uthoff, V. A. (1970). An optimum test property of two-well-known statistics. *J. Amer. Statist. Assoc.* **65**, 1597–1600.

de Wet, T. and van Wyk, J. W. J. (1979). Efficiency and robustness of Hogg's adaptive trimmed means. *Comm. Statist. A* **8**, 117–128.

PART III

ORDER STATISTICS AND EMPIRICAL DISTRIBUTION

P. R. Krishnaiah and P. K. Sen, eds., *Handbook of Statistics, Vol. 4*
© Elsevier Science Publishers (1984) 359–382

17

Order Statistics

Janos Galambos

1. Introduction

When the values of a sequence $X_1, X_2, \ldots, X_n$ of random variables are rearranged in an increasing order $X_{1:n} \leqslant X_{2:n} \leqslant \cdots \leqslant X_{n:n}$ of magnitude, then the r-th member $X_{r:n}$ of this new sequence is called the r-th order statistic of the X_j, $1 \leqslant j \leqslant n$. The two terms $X_{1:n} = \min(X_1, X_2, \ldots, X_n)$ and $X_{n:n} = \max(X_1, X_2, \ldots, X_n)$ are called extremes. If, for some r, $X_{r:n} = X_{r+1:n}$, and thus, for some $j \neq t$, $X_j = X_t$, we do not apply any rule for determining whether $r = j$ or $r = t$. Such an ambiguity will not influence the theory discussed in the present chapter.

Order statistics play an important role both in model building and in statistical inference. For example, if $X_1, X_2, \ldots, X_{365}$ are the water levels of a river on consecutive days, then properties of $X_{1:365}$ are used for modeling drought, while $X_{365:365}$ is relevant in a model for flood. As another example, consider n light-bulbs with useful life $X_1, X_2, \ldots, X_n$. If these bulbs are switched on in a test room concurrently, then, as the bulbs burn out, we actually record the order statistics $X_{1:n} \leqslant X_{2:n} \leqslant \cdots \leqslant X_{n:n}$. In statistical inference, the range $X_{n:n} - X_{1:n}$ is widely used to estimate the standard deviation of the population distribution, when the X_j, $1 \leqslant j \leqslant n$, constitute a random sample. Additional examples, when the order statistics play a central role are discussed in Sections 5 and 6.

In the present chapter we will discuss the distribution theory of order statistics. Although a considerable portion will be devoted to the case when the X_j are independent and identically distributed (i.i.d.), we shall also include important results when the X_j are dependent or not identically distributed. These latter cases are not just mathematical curiosities: when the X_j and their order statistics are part of the model which we investigate, such as flood, drought, strength of material, failure of systems of components and others, then we usually have no control over their dependence structure. Hence, for

This work was supported by Grant MCS-7912139 of the National Science Foundation to Temple University.

understanding such models, we should investigate the order statistics of dependent random variables.

The reader who would like to study in more detail one or several aspects of the theory discussed here, can consult the following books: David (1981) (statistical methods, basic theory, extensive references), Galambos (1978) (extreme value theory in the univariate and multivariate case, including dependent models, extensive references), Gumbel (1958) (statistics of extremes in the i.i.d. case) and the four volume set by Johnson and Kotz (1968–1972) (the order statistics of random samples for specific distributions). Additional references are given within the text. The development of the theory here, however, is self-contained and it is substantially different from any one of the books listed above.

2. The distribution of order statistics in the i.i.d. case

In the present section, we assume that $X_1, X_2, \ldots, X_n$ are independent and identically distributed. The common distribution function is denoted by $F(x) = P(X_j < x)$.

Let $m_n(x)$ be the number of those X_j, $1 \leqslant j \leqslant n$, which satisfy $\{X_j < x\}$. Evidently

$$\{X_{r:n} < x\} = \{m_n(x) \geqslant r\}. \tag{2.1}$$

Since the distribution of $m_n(x)$ is binomial with parameters n and $F(x)$, (2.1) yields

$$\begin{aligned} F_{r:n}(x) = P(X_{r:n} < x) &= \sum_{k=r}^{n} \binom{n}{k} F^k(x)[1 - F(x)]^{n-k} \\ &= r\binom{n}{r} \int_0^{F(x)} t^{r-1}(1-t)^{n-r}\, dt. \end{aligned} \tag{2.2}$$

In particular,

$$F_{n:n}(x) = P(X_{n:n} < x) = F^n(x) \tag{2.3}$$

and

$$F_{1:n}(x) = P(X_{1:n} < x) = 1 - [1 - F(x)]^n. \tag{2.4}$$

If $F(x)$ has a density function $F'(x) = f(x)$, then we get from (2.2) through (2.4)

$$F'_{r:n}(x) = r\binom{n}{r} F^{r-1}(x)[1 - F(x)]^{n-r} f(x), \tag{2.5a}$$

$$F'_{n:n}(x) = nF^{n-1}(x) f(x) \tag{2.5b}$$

and

$$F'_{1:n}(x) = n[1 - F(x)]^{n-1} f(x). \tag{2.5c}$$

We also deduce a formula for the joint density function $f_{r_1, r_2, \ldots, r_k:n} \times (x_1, x_2, \ldots, x_k)$ of the order statistics $X_{r_j:n}$, $1 \le j \le k$, $1 \le r_1 < r_2 < \cdots < r_k \le n$, when the population density function $f(x)$ exists. First notice that

$$f_{r_1, r_2, \ldots, r_k:n}(x_1, x_2, \ldots, x_k) = 0$$

if the inequalities $x_1 < x_2 < \cdots < x_k$ fail to hold. Now, for $x_1 < x_2 < \cdots < x_k$,

$$P(x_j \le X_{r_j:n} < x_j + \Delta x_j, 1 \le j \le k) \sim f_{r_1, r_2, \ldots, r_k:n}(x_1, x_2, \ldots, x_k)\, \Delta x_1\, \Delta x_2 \cdots \Delta x_k .$$

On the other hand, the event $\{x_j \le X_{r_j:n} < x_j + \Delta x_j, 1 \le j \le k\}$, for small Δx_j, $1 \le j \le k$, means that, for each j, one of $X_1, X_2, \ldots, X_n$ is approximately x_j, and the other X_t are spread between the values $x_0 = -\infty < x_1 < \cdots < x_k < x_{k+1} = +\infty$ with the following rule: there are exactly $r_j - r_{j-1} - 1$ among the X_t which satisfy $x_{j-1} < X_t < x_j$, $1 \le j \le k+1$, where $r_0 = 0$ and $r_{k+1} = n$. Hence, an easy combinatorial calculation yields that, for $x_1 < x_2 < \cdots < x_k$,

$$f_{r_1, r_2, \ldots, r_k:n}(x_1, x_2, \ldots, x_k)$$
$$= n! f(x_1) f(x_2) \cdots f(x_k) \prod_{j=1}^{k} \frac{[F(x_j) - F(x_{j-1})]^{r_j - r_{j-1} - 1}}{(r_j - r_{j-1} - 1)!} . \tag{2.6}$$

This last formula can be used to evaluate the joint distribution of order statistics as well as the distribution of functions of order statistics. Out of these possibilities, we record the following three results.

First, with the choice of $k = 2$, $r_1 = 1$ and $r_2 = n$, we get the following formula for the density of the range $R = X_{n:n} - X_{1:n}$:

$$f_R(z) = n(n-1) \int_{-\infty}^{+\infty} f(x) f(x+z) [F(x+z) - F(x)]^{n-2}\, dx ,$$

from which the distribution function of R is

$$F_R(z) = n \int_{-\infty}^{+\infty} f(x) [F(x+z) - F(x)]^{n-1}\, dx . \tag{2.7}$$

The other two consequences of (2.6) are formulated as theorems.

THEOREM 2.1. *For the uniform distribution $F(x) = x$, $0 < x < 1$, the differences $X_{r+1:n} - X_{r:n}$ are identically distributed with common density function*

$$f^*_{r,r+1:n}(x) = n(1-x)^{n-1}, \quad 0 < x < 1 .$$

The joint distribution of the order statistics $X_{r:n}$, $1 \le r \le n$ is the same as the distribution of the vector $(U_1, U_2, \ldots, U_n)$, where $U_j = S_j / S_n$, $S_j = Y_1 + Y_2 + \cdots + Y_j$, and the Y's are independent with $P(Y_t < x) = 1 - e^{-x}$, $x > 0$.

THEOREM 2.2. *For the exponential distribution* $F(x)=1-e^{-\lambda x}$, $x>0$, *the differences* $X_{r+1:n}-X_{r:n}$ *are independent exponential variables with*

$$P(X_{r+1:n}-X_{r:n}<x)=1-e^{-\lambda(n-r)x},$$

with $x>0$, $0\leq r\leq n-1$, *and* $X_{0:n}=0$. *Consequently,*

$$X_{r+1:n}=n^{-1}Y_1+(n-1)^{-1}Y_2+\cdots+(n-r)^{-1}Y_{r+1}, \tag{2.8}$$

where $Y_1, Y_2, \ldots$ *are i.i.d. with common distribution function* $F(x)$.

Both Theorems 2.1 and 2.2 follow from (2.6) by an appropriate substitution, and thus their proof is left as an exercise.

In spite of their simplicity, Theorems 2.1 and 2.2 can serve as the foundation of a large variety of investigations (such as life tests, initiated by Epstein and Sobel (1953), asymptotic distributions for order statistics as demonstrated by Rényi (1953), goodness of fit tests, for which see Csörgö et al. (1975) and several nonparametric tests). The strength of Theorems 2.1 and 2.2 lies in the fact that they can be applied to an arbitrary continuous population distribution. Namely, if X has the continuous distribution function $F(x)$, then $X^*=F(X)$ is uniform on $(0, 1)$, and $X^{**}=-\log F(X)$ is exponential.

3. Asymptotic results in the i.i.d. case

We again assume that $X_1, X_2, \ldots, X_n$ are independent with a common distribution function $F(x)$. Hence, the formulas of the previous section are applicable.

In the present section we concentrate on the asymptotic behavior of

$$X_{r:n}(a_n, b_n)=(X_{r:n}-a_n)/b_n,$$

where a_n and $b_n>0$ are two sequences of constants. Our particular interest is whether, with a proper choice of a_n and b_n, $X_{r:n}(a_n, b_n)$ has a limiting distribution, where $r=r(n)$ is of a given form.

We distinguish three cases for $r=r(n)$: (i) we call $X_{r:n}$ a lower order statistic and $X_{n-r:n}$ an upper order statistic if r does not depend on n; (ii) we speak of moderately low $(X_{r:n})$ and moderately upper order statistics $(X_{n-r:n})$ when $r=r(n)\to+\infty$ and $r/n\to 0$, as $n\to+\infty$; and, finally, (iii) $X_{r:n}$ is a central order statistic, if $r/n\to\lambda$, $0<\lambda<1$, as $n\to+\infty$.

Let us remark that every statement on a lower order statistic, or a moderately low order statistic becomes a statement on an upper order statistic or moderately upper order statistic if we change each X_j to $(-X_j)$, and vice versa. This observation will be exploited throughout the present chapter.

We first deal with the upper and lower order statistics.

THEOREM 3.1. *If, with some a_n and $b_n > 0$, the normalized maximum $(X_{n:n} - a_n)/b_n$ has a limiting distribution $H(z)$, then, with the same constants, each normalized upper order statistic $X_{n-r:n}(a_n, b_n)$ has a limiting distribution $H_r(z)$. The relation*

$$H_r(z) = -H(z) \sum_{t=0}^{r} \frac{1}{t!} \{\log H(z)\}^t \tag{3.1}$$

holds, whenever $H(z) > 0$.

The reader is advised to restate Theorem 3.1 for lower order statistics.

PROOF. By the assumption, and in view of (2.3),

$$F^n(a_n + b_n z) \to H(z)$$

(at each continuity point z of $H(z)$), as $n \to +\infty$. It is therefore evident that $F(a_n + b_n z) \to 1$, and thus, for large n, and for those z for which $H(z) > 0$,

$$\begin{aligned} \log H(z) \sim n \log F(a_n + b_n z) &= n \log\{1 - [1 - F(a_n + b_n z)]\} \\ &= -n[1 - F(a_n + b_n z)](1 + o(1)), \end{aligned} \tag{3.2}$$

where we used the Taylor expansion $\log(1-u) = -u(1 + o(1))$ as $u \to 0$. Now, if r does not depend on n, then, writing (2.2) as

$$P(X_{n-r:n} < a_n + b_n z) = \sum_{k=0}^{r} \binom{n}{r-k} F^{n-r+k}(a_n + b_n z)[1 - F(a_n + b_n z)]^{r-k},$$

and observing that

$$F^{n-r+k}(a_n + b_n z) \sim F^n(a_n + b_n z)$$

and

$$\binom{n}{r-k}[1 - F(a_n + b_n z)]^{r-k} \sim \frac{\{n[1 - F(a_n + b_n z)]\}^{r-k}}{(r-k)!},$$

the asymptotic equations at (3.2) lead to (3.1).

The case of the upper and lower order statistics is thus reduced to the behaviour of the extremes $X_{1:n}$ and $X_{n:n}$. Since this latter case is adequately discussed in Chapter 2 of Galambos (1978), we restrict ourselves here to the following statements.

THEOREM 3.2. *Assume that with some constants a_n and $b_n > 0$, the distribution function of $(X_{n:n} - a_n)/b_n$ converges to a distribution function $H(z)$ (at each of its continuity points). Then, with suitable numbers A and $B > 0$, $H(A + Bz)$ is one of the following three functions:*

$$H_{3,0}(z) = \exp(-e^{-z}), \quad -\infty < z < +\infty,$$

$$H_{1,\gamma}(z) = \begin{cases} \exp(-z^{-\gamma}) & \text{if } z > 0, \\ 0 & \text{otherwise} \end{cases},$$

and

$$H_{2,\gamma}(z) = \begin{cases} 1 & \text{if } z \geq 0, \\ \exp(-(-z)^{\gamma}) & \text{if } z < 0, \end{cases}$$

where $\gamma > 0$ is arbitrary.

Notice that the three functions above can be combined into a single parametric family. Namely, the family

$$H_c(z) = \exp\{-(1+cz)^{-1/c}\}, \quad 1 + cz > 0,$$

where $H_0(z) = \lim_{c\to 0} H_c(z)$, reduces to the above distribution functions according as $c = 0$, $c > 0$, or $c < 0$.

THEOREM 3.3. (i) *If $F(x) < 1$ for all x, and if, for all $x > 0$,*

$$\lim_{t=+\infty} \frac{1 - F(tx)}{1 - F(t)} = x^{-\gamma}, \tag{3.3}$$

with some $\gamma > 0$, then there are constants a_n and $b_n > 0$ such that the distribution of $(X_{n:n} - a_n)/b_n$ converges to $H_{1,\gamma}(z)$. One can choose $a_n = 0$ and b_n as the smallest x such that $1 - F(x) \leq 1/n$.

(ii) *If there is a finite number $w(F)$ such that $F(x) = 1$ for all $x > w(F)$ and $F(x) < 1$ for $x < w(F)$ and if $F^*(x) = F[w(F) - 1/x]$, $x > 0$, satisfies* (3.3), *then there are constants a_n and $b_n > 0$ such that the distribution function of $(X_{n:n} - a_n)/b_n$ converges to $H_{2,\gamma}(z)$. One can choose $a_n = w(F)$ and $b_n = w(F) - c_n$, where c_n is the smallest x such that $1 - F(x) \leq 1/n$.*

(iii) *If, for some finite a such that $F(a) < 1$,*

$$\int_a^{+\infty} [1 - F(y)]\, dy < +\infty,$$

and if

$$\lim_{t=w(F)} \frac{1 - F(t + xR(t))}{1 - F(t)} = e^{-x}, \tag{3.4}$$

where $w(F)$ is the largest x such that $F(x) < 1$, and

$$R(t) = \frac{1}{1 - F(t)} \int_t^{w(F)} [1 - F(y)]\, dy,$$

then there are constants a_n and $b_n > 0$ such that the distribution function of

$(X_{n:n}-a_n)/b_n$ converges to $H_{3,0}(z)$. One can choose a_n as c_n in part (ii) *and $b_n=R(a_n)$.*

If, to $F(x)$, none of (i), (ii) *and* (iii) *applies, then there are no constants a_n and $b_n>0$ such that the distribution of $(X_{n:n}-a_n)/b_n$ would converge.*

An easy check of the conditions of the preceding theorem immediately yields the following negative result.

COROLLARY. *If X_1 is discrete which takes the nonnegative integers only, and if the limit*

$$\lim_{n=+\infty}\frac{P(X_1\geqslant n)}{P(X_1\geqslant n+1)}=1 \tag{3.5}$$

fails, then, whatever be the constants a_n and $b_n>0$, $(X_{n:n}-a_n)/b_n$ does not have a limiting distribution. The limit (3.5) *fails for the geometric and Poisson distributions.*

Since several familiar distributions of applied statistics belong to part (iii) of Theorem 3.3 and since (3.4) is difficult to check even for the most popular ones, we give one classical result covering these distributions. The meaning of $w(F)$ below is as at (3.4).

THEOREM 3.4. *If $F(x)$ is such that, for some $x_1<w(F)$, $f(x)=F'(x)$ and $F''(x)$ exist for all $x_1\leqslant x<w(F)$, $f(x)\neq 0$, and*

$$\lim_{x=w(F)}\frac{\mathrm{d}}{\mathrm{d}x}\left\{\frac{1-F(x)}{f(x)}\right\}=0\,,$$

then there are constants a_n and $b_n>0$ such that the distribution function of $(X_{n:n}-a_n)/b_n$ converges to $H_{3,0}(z)$.

We would delay the discussion of specific distributions $F(x)$ until the case of other order statistics, not just the extremes and the upper and lower order statistics, have been covered. We would, however, pause here for some comments on the literature.

Theorem 3.3, with part (iii) in a weaker form, is due to Gnedenko (1943). The present form of part (iii) was obtained by de Haan (1970), who also gives a number of variants of the sufficiency part of Theorem 3.3. Chapter 2 of Galambos (1978) provides additional rules for choosing a_n and b_n. Further detailed discussion of the literature concerning the material covered so far can be found on pages 117–120 of Galambos (1978), which we do not repeat here; we would rather supplement the extensive list appearing in the discussion of the mentioned place. It has long been suspected, and it is supported by empirical data, that the distribution of extremes is very slowly approaching its limiting form. This is certainly true when the population distribution is normal, but not for all population distributions. The fact that the population distribution is a major factor in the speed of convergence in Theorems 3.1 and 3.2 is clearly seen from the estimates given by Galambos (1978, p. 113). It should be noted that these estimates of Galambos are applicable to any population distribution $F(x)$, and for any values of a_n, b_n and n. For the special cases when $F(x)$ is normal, P. Hall (1979), and when $F(x)$ is exponential, Hall and Wellner (1979), obtained uniform bounds for the difference

$$P(X_{n:n} - a_n < b_n z) - \exp(-e^{-z}) .$$

In the case of the normal distribution, the bound is $c/\log n$, while for the exponential distribution, it is of the magnitude of $1/n$. Nair (1981) extended Theorem 3.2 to an asymptotic expansion of several terms for the case when the population distribution is normal, while Reiss (1981) gives an asymptotic expansion, together with a uniform estimate of his remainder term, to the distribution of $X_{n-k:n}(a_n, b_n)$ for all bounded k, and for a large family of population distributions.

Let us return to the investigation of $X_{r:n}(a_n, b_n)$, where now neither r nor $n-r$ is bounded. We have two options to proceed. One is the relation at (2.1), in which the fact that the distribution of $m_n(x)$ is binomial with parameters n and $F(x)$ provides an easy tool for approximating the distribution of $X_{r:n}(a_n, b_n)$ by the normal distribution. On the other hand, if $F(x)$ is continuous, then, with the transformation $X_j^{**} = -\log F(X_j)$, (2.8) leads to another form of normal approximation to the distribution of $X_{r:n}$. We develop both of these approximations below.

We start with the latter case, i.e., we assume that the population distribution function $F(x)$ is continuous. If we transform the data X_j into $X_j^{**} = -\log F(X_j)$, then

$$X^{**}_{n-r+1:n} = -\log F(X_{r:n}) ,$$

and since the distribution function of X_j^{**} is $F^{**}(x) = 1 - e^{-x}$, $x > 0$, we can apply the representation (2.8) with $\lambda = 1$ to $-\log F(X_{r:n})$. We thus have

$$-\log F(X_{r:n}) = n^{-1}Y_1 + (n-1)^{-1}Y_2 + \cdots + r^{-1}Y_{n-r+1} , \tag{3.6}$$

where the Y_j, $j \geq 1$, are independent random variables with common distribution function $1 - e^{-x}$, $x > 0$. Now, since both r and $n-r$ are unbounded as $n \to +\infty$, the Ljapunov form of the central limit theorem (Loéve, 1963, p. 277) provides a normal approximation to the distribution of a properly normalized form of $\{-\log F(X_{r:n})\}$, in which approximation the error term can also be estimated by the Berry–Esseen bound (Loéve, 1963, p. 288). As a matter of fact, since $E(Y_j) = 1$, $V(Y_j) = 1$ and $E(|Y_j - 1|^3) < 3$, the quoted normal approximation theorem and (3.6) yield that uniformly for all x and with constant

c occurring in the Berry–Esseen bound,

$$|P(X_{r:n} > F^{-1}[\exp(-a_{n,r} - b_{n,r}x)]) - \Phi(x)| < 3cd_{n,r}b_{n,r}^{-3}, \tag{3.7}$$

where $\Phi(x)$ is the standard normal distribution function, and

$$a_{n,r} = \sum_{k=r}^{n} \frac{1}{k}, \qquad b_{n,r}^2 = \sum_{k=r}^{n} \frac{1}{k^2}, \qquad d_{n,r} = \sum_{k=r}^{n} \frac{1}{k^3}. \tag{3.8}$$

Since the estimate is uniform in x, one can, of course, substitute $y = F^{-1}[\exp(-a_{n,r} - b_{n,r}x)]$, from which $x = -\{\log F(y) + a_{n,r}\}/b_{n,r}$, thus obtaining the same estimate as above on

$$|P(X_{r:n} > y) - \Phi(x)|.$$

Although Loéve gives an actual value for c, his bound can be improved considerably (we do not carry out such a computation here, since in our second approach we give a numerical value in the error term, which value was computed by Englund (1980)). However, the present form is sufficient for asymptotic results and Loéve's c also suffices for large n and 'large' r. In particular, the given estimate implies that if $r = r(n)$ is such that

$$n^{1/2}|r(n)/n - p| \to 0 \tag{3.9}$$

as $n \to +\infty$, and $0 < p < 1$, and if, with some $a < b$, $F(a) < p < F(b)$, and $F'(x) = f(x)$ exists, continuous and positive for $a \leq x \leq b$, then $X_{r:n}$ is asymptotically normal with expectation $F^{-1}(p)$ and standard deviation $\sqrt{p(1-p)}/f(F^{-1}(p))\sqrt{n}$. As a matter of fact, when $r(n)$ satisfies (3.9), then simple asymptotic expressions for the values at (3.8) yield

$$a_{n,r} = \log\frac{1}{p} + o\left(\frac{1}{\sqrt{n}}\right), \qquad b_{n,r}^2 = \frac{1-p}{np} + O(n^{-3/2}), \qquad d_{n,r} = O(n^{-2}).$$

Hence, the error term $d_{n,r}/b_{n,r}^3 \to 0$ as $n \to +\infty$. We also get from these relations that

$$\begin{aligned} &\lim_{n=+\infty} P(-\log F(X_{r:n}) < a_{n,r} + b_{n,r}x) \\ &= \lim_{n=+\infty} P\left(-\log F(X_{r:n}) < \log\frac{1}{p} + \sqrt{\frac{1-p}{np}}\,x\right) \\ &= \lim_{n=+\infty} P\left(X_{r:n} > F^{-1}\left[\exp\left(\log p - \sqrt{\frac{1-p}{np}}\,x\right)\right]\right). \end{aligned} \tag{3.10}$$

Now, by the differentiability assumptions on $F(x)$,

$$F^{-1}\left[\exp\left(\log p - \sqrt{\frac{1-p}{np}}\,x\right)\right] = F^{-1}(p) + \frac{p\exp(-\sqrt{(1-p)/npx}) - p}{f(\vartheta_n F^{-1}(p))}\,,$$

where $\vartheta_n - 1$ as $n \to +\infty$. Furthermore,

$$p\exp\left(-\sqrt{\frac{1-p}{np}}\,x\right) - p = -\sqrt{\frac{p(1-p)}{n}}\,x + \mathrm{O}\left(\frac{1}{n}\right).$$

Since $f(x)$ is continuous at $F^{-1}(p)$ by assumption, (3.7) and (3.10) imply

$$\begin{aligned}\Phi(x) &= \lim P(X_{r:n} > F^{-1}[\exp(-a_{n,r} - b_{n,r}x)]) \\ &= \lim_{n=+\infty} P\left(X_{r:n} - F^{-1}(p) < \sqrt{\frac{p(1-p)}{n}}\,x\right),\end{aligned}$$

which was to be proved.

The advantage of this approach is that, without any change in the calculations, except that instead of a one-dimensional central limit theorem one needs a multidimensional one, one immediately gets a multivariate extension of this limit theorem, which we formulate below.

THEOREM 3.5. *Let the population distribution $F(x)$ satisfy the following properties: for some $a < b$, $F'(x) = f(x)$ exists and is continuous for all $a \leq x \leq b$. Let $F(a) < p_1 < p_2 < \cdots < p_k < F(b)$, and assume that $f(F^{-1}(p_j)) \neq 0$, $1 \leq j \leq k$. If $r_j = r_j(n)$ satisfies* (3.9) *with p being replaced by p_j, then the vector*

$$\sqrt{n}(X_{r_j:n} - F^{-1}(p_j)),\quad 1 \leq j \leq k\,,$$

is asymptotically a k-dimensional normal vector with zero expectation and with covariance matrix

$$\frac{p_j(1-p_t)}{f(F^{-1}(p_j))f(F^{-1}(p_t))}\,,\quad j \leq t\,.$$

For a review of the literature concerning Theorem 3.5, see pp. 257–258 of David (1981).

For getting another form of approximation to the distribution of $X_{r:n}$, we turn to (2.1). Rewriting (2.1) as

$$\{X_{r:n} < x\} = \left\{\frac{m_n(x) - nF(x)}{(nF(x)[1-F(x)])^{1/2}} \geq \frac{r - nF(x)}{(nF(x)[1-F(x)])^{1/2}}\right\},$$

the classical central limit theorem, together with the Berry–Esseen bound, yields

$$\left| P(x_{r:n} < x) - \Phi\left\{\frac{nF(x) - r}{(nF(x)[1 - F(x)])^{1/2}}\right\}\right| \leq c\,\frac{E_3}{\sqrt{n}\{F(x)[1 - F(x)]\}^{3/2}},$$

where c is a suitable constant, and

$$E_3 = E(|I_F - F(x)|^3),$$

where $I_F = 1$ or 0 according as $X_1 < x$ or $X_1 \geq x$. In the above estimate one can use

$$E_3 = F(x)[1 - F(x)]^3 + F^3(x)[1 - F(x)] \leq F(x)[1 - F(x)].$$

Englund (1980) computes c and shows that the above estimate can be modified to replacing $F(x)[1 - F(x)]$ by $\{(r/n)[1 - (r/n)]\}$, thus achieving an estimate which is uniform in x. In fact, he gets that for an arbitrary population distribution $F(x)$, and for any values of n, r and x,

$$\left| P(X_{r:n} < x) - \Phi\left\{\frac{nF(x) - r}{[r(1 - r/n)]^{1/2}}\right\}\right| \leq \frac{3}{[r(1 - r/n)]^{1/2}}. \tag{3.11}$$

This inequality is suitable for obtaining asymptotic results as well as for numerical computation. In particular, the one dimensional case of Theorem 3.5 follows from (3.11), for which an error estimate is also obtained from it (in this regard, see also the results of Reiss, 1976). Additional limit theorems can be deduced from (3.11) by choosing r different from those at (3.9). While such limit theorems are interesting from a theoretical point of view, their practical usefulness is limited.

For example, (3.11) implies that if $F(x) = 1 - e^{-x}$, $x > 0$, and $n - r \sim \log n$, then

$$P\left(X_{r:n} < \log n - \log\log n + \frac{x}{\sqrt{\log n}}\right) \to \Phi(x). \tag{3.12}$$

Indeed, putting $a_n = \log n - \log\log n$ and $b_n = (\log n)^{-1/2}$, we have

$$\frac{nF(a_n + b_n x) - r}{\{r(1 - r/n)\}^{1/2}} \sim \frac{n\{(\log n)/n - \exp(-a_n - b_n x)\}}{(\log n)^{1/2}}$$

$$= (\log n)^{1/2}\left\{1 - \exp\left(-\frac{x}{(\log n)^{1/2}}\right)\right\} \sim x.$$

We also have from (3.11) that the error term in (3.12) is of the order of magnitude of $1/(\log n)^{1/2}$, uniformly in x. But the real meaning of (3.12) for the applied scientist is rather confusing, because $\log n$ is so slowly increasing that for all practical value of n, $X_{r:n}$ with $r = n - \log n$ would be viewed as an upper order statistic, to which (3.1) applies, rather than a moderately upper order statistic in which $n-r$ depends on n (note that $\log 3\,000 = 8$).

However, as remarked earlier, the theoretical results are quite interesting and quite thorough. For example, Chibisov (1964) obtained that, for moderately upper and lower order statistics, besides the normal distribution, $X_{r:n}(a_n, b_n)$ may have as its asymptotic distribution the log normal distribution. On the other hand, Smirnov (1949) has shown that when the limit in (3.9) is positive then the asymptotic distribution of $X_{r:n}(a_n, b_n)$ may be that of a certain power of a normally distributed random variable (but again, e.g., in $X_{459:900}$, $r = 459$ can be viewed as $r = 0.5n + 0.3\sqrt{n}$, clearly violating (3.9), but satisfying Smirnov's assumption, or one can view $r = 459$ as $r = 0.51n$, clearly satisfying (3.9); hence only very good error terms make such asymptotic results practical). A more thorough analysis for the case when (3.9) is violated is given by Balkema and de Haan (1978).

Let us look at the results of the present section through some specific distributions.

1. *The exponential distribution.* Let the population distribution be

$$F(x) = 1 - \exp\{-(x-a)/b\}, \quad x \geq a .$$

If the distribution function of X is above $F(x)$, then the distribution function of $(X-a)/b$ is the same $F(x)$ with $a = 0$ and $b = 1$. We can therefore restrict our attention to this case ($a = 0, b = 1$), which we call the unit exponential distribution.

Let now $X_1, X_2, \ldots, X_n$ be a random sample from a unit exponential distribution. Then (Theorem 3.3) the asymptotic distribution of $(X_{n:n} - \log n)$ is $H_{3,0}(x) = \exp(-e^{-x})$. Hence, by Theorem 3.1, for any r not depending on n,

$$\lim_{n=+\infty} P(X_{n-r:n} - \log n < x) = H_{3,0}(x) \sum_{t=0}^{r} \frac{1}{t!} e^{-tx} . \tag{3.13}$$

By applying the same theorems as above to $(-X_j)$, $1 \leq j \leq n$, (or directly from Theorem 2.2), the (asymptotic) distribution of $nX_{1:n}$ is $F(x) = 1 - e^{-x}$ itself, and, for any r, not dependent on n,

$$\lim_{n=+\infty} P(X_{r:n} < nx) = 1 - e^{-x}\left(\sum_{t=0}^{r-1} \frac{x^t}{t!}\right) . \tag{3.14}$$

Now, if r_j, $1 \leq j \leq k$, satisfy (3.9) with some $0 < p_j < 1$, $1 \leq j \leq k$, then (Theorem 3.5) the vector $\sqrt{n}(X_{r_j:n} + \log(1-p_j))$, $1 \leq j \leq k$, is asymptotically a k-dimensional normal vector with zero expectation and with covariance matrix $p_j/(1-p_j)$, $j \leq t$ (the value itself does not depend on t).

Finally, if $r \to +\infty$ with n and $r/n \to 0$ or if $n - r \to +\infty$ and $r/n \to 1$, then (3.11) can be applied to see that $X_{r:n}(a_n, b_n)$ is asymptotically normal (although see our previous discussion on its practical meaning when r or $n-r$ increases very slowly to infinity). In fact, if $r \to +\infty$ with n and if the order of magnitude of r is smaller than $n^{2/3}$, then

$$\lim_{n=+\infty} P\left(X_{r:n} - \frac{r}{n} < \frac{\sqrt{r}}{n} x\right) = \Phi(x) \tag{3.15}$$

(if r is of larger order of magnitude than mentioned then further normalizing terms are required in the form of powers of (r/n)), and if $n - r \to +\infty$ with n and $r/n \to 1$, then

$$\lim_{n=+\infty} P\left(X_{r:n} - \log \frac{n}{n-r} < \frac{x}{\sqrt{n-r}}\right) = \Phi(x). \tag{3.16}$$

We have pointed out that (3.15) would rarely be utilized when r goes to infinity very slowly (such as $\log n$), in which case (3.14) would be used. On the other hand, when r is a power of n, but r/n is still small, then (3.15) is applicable. In boundary cases, results computed in two different ways should be equal within the error of approximation. As an example, let us take $n = 900$ and $r = 36$. Then, if $r = 36$ is viewed as $r = 0.04n$, that is a central term, the result is that $30(X_{36:900} - \log(1 - 0.04))$ is asymptotically normal with zero expectation and with variance $= 0.04/0.96$, i.e., $147(X_{36:900} - 0.04)$ is asymptotically standard normal. On the other hand, if $r = 36$ is viewed as $r = 1.2\sqrt{n}$, then (3.15) yields that $150(X_{36:900} - 0.04)$ is asymptotically standard normal. The difference is reasonably small.

2. *The logistic distribution.* When brought into standard form, the distribution of X is logistic if

$$F(x) = \frac{1}{1 + e^{-x}}, \quad -\infty < x < +\infty.$$

If $X_1, X_2, \ldots, X_n$ are observations on X, then $X_{n:n}$ and $X_{n-r:n}$, with r not depending on n, have the same asymptotic properties as for an exponential sample. Indeed, since in Theorem 3.3 the only important factor is the behaviour of $F(x)$ as $x \to \infty$, the fact that

$$\frac{1}{1 + e^{-x}} \sim 1 - e^{-x} \quad \text{as } x \to +\infty. \tag{3.17}$$

implies that (3.13) is again applicable. On the other hand, since $F(x)$ is symmetric with respect to zero, we get a formula similar to (3.13) for $X_{r:n}$, if r is fixed. We get

$$\lim_{n=+\infty} P(X_{r:n} + \log n < x) = 1 - e^{-e^{x}} \left(\sum_{t=0}^{r-1} \frac{e^{tx}}{t!} \right).$$

For the central terms r_j, $1 \leq j \leq k$, satisfying (3.9) with $0 < p_j < 1$, the normalizing term $F^{-1}(p_j) = \log(p_j/(1 - p_j))$, and thus the vector

$$\sqrt{n}(X_{r_j:n} - \log(p_j/(1-p_j))), \quad 1 \leq j \leq k,$$

is normal with zero expectation and covariance matrix $1/(1-p_j)p_t$, $j \leq t$.

In view of (3.17), (3.16) also applies, again with the limitation that the order of magnitude of $n-r$ be smaller than $n^{2/3}$. By the symmetry of $F(x)$ about zero, the normalizing constants for $X_{r:n}$, $r \to +\infty$, $r/n \to 0$, are similar to those at (3.16) except that $\log\{n/(n-r)\}$ is to be replaced by $\log\{(n-r)/n\}$.

3. *The standard normal distribution.* We now turn to the case when the population distribution function

$$F(x) = \frac{1}{\sqrt{2\pi}} \int_{-\infty}^{x} e^{-(1/2)t^2} \, dt.$$

By Theorems 3.4, 3.3 (iii) and 3.1, we get that for each fixed r,

$$\lim_{n=+\infty} P(X_{n-r:n} - a_n < b_n x) = \exp(-e^{-x}) \sum_{t=0}^{r} \frac{1}{t!} e^{-tx},$$

where

$$a_n = (2 \log n)^{1/2} - \frac{\frac{1}{2}(\log \log n + \log 4\pi)}{(2 \log n)^{1/2}} \quad \text{and} \quad b_n = (2 \log n)^{-1/2}.$$

Although, as was mentioned earlier, the uniform bound on the speed of convergence in the above limit relation is of the order of magnitude of $1/\log n$, it should not discourage us from applying this approximation. The fit is remarkably good for $n \geq 100$ (see Gumbel, 1958, p. 222). (This is not a contradiction; in a uniform estimate of errors 'large' values of x can distort the errors of approximation which actually rarely occur for most data.)

Because $F(x)$ is symmetric about zero, $X_{r:n}$ has the same asymptotic properties as $-X_{n-r+1:n}$.

If $n - r \to +\infty$ with n and $r/n \to 1$, then one can deduce from (3.11) that, with suitable constants A_n and $B_n > 0$, $X_{r:n}(A_n, B_n)$ is asymptotically normal. However, since $F(x)$ is well tabulated, a direct application of (3.11), by possibly ignoring the error estimate in it, provides a better approximation than what could be obtained through long expressions for A_n and B_n.

4. *The Cauchy distribution.* Let the population distribution

$$F(x) = \frac{1}{2} + \frac{1}{\pi} \arctan x.$$

Then, by L'Hospital's rule, (3.3) applies with $\gamma = 1$. Hence, we get from Theorems 3.3 and 3.1 that, for r not depending on n,

$$\lim_{n=+\infty} P\left(X_{n-r:n} < x \tan \pi\left(\frac{1}{2} - \frac{1}{n}\right)\right) = \exp(-1/x) \sum_{t=0}^{r} \frac{1}{x^t t!}.$$

For the lower order statistics, a similar formula applies in view of the symmetry of $F(x)$ about zero.

Turning to the central terms, since $F^{-1}(y) = \tan \pi(y - \frac{1}{2})$, Theorem 3.5 implies that, for r satisfying (3.9),

$$\frac{\sqrt{n}\cos \pi(p-\frac{1}{2})(X_{r:n} - \tan \pi(p-\frac{1}{2}))}{\pi\{p(1-p)\}^{1/2}}$$

is asymptotically standard normal.

We leave it to the reader to analyze the asymptotic behaviour of $X_{r:n}$ for other values of r.

Notice that for each of the exponential, logistic and normal distributions, the growth of $X_{n-r:n}$ with r constant is so small that $(1/n)(X_1 + X_2 + \cdots + X_n)$ is not affected if some of these upper order statistics are dropped (censoring). However, since for large n, $\tan \pi(\frac{1}{2} - (1/n)) \sim (n/\pi)$, $X_{n:n}$ alone is of the same order of magnitude as $X_1 + X_2 + \cdots + X_n$. Hence, censoring can lead to considerable effect on the mean. For a detailed study of the effect of extremes on the mean, see Mori (1981). A somewhat related question is whether the extremes are independent of the mean. Very general results on this line can be found in the works of Tiago de Oliveira (1961) and Rossberg (1965). For additional references, see Section 2.11 of Galambos (1978).

Due to lack of space, multivariate extensions are not discussed in the present paper. See Chapter 5 of Galambos (1978), and the recent contributions by Deheuvels (1978) and (1980), Tiago de Oliveira (1980) and Pickands (1981).

Let us conclude the present section with a few short remarks on some problems of statistical inference where order statistics proved to be useful tools. For a detailed discussion of this topic, see David (1981).

It is a textbook example to show that the maximum likelihood estimator of Θ, if the population distribution is uniform on $(0, \Theta)$, is the extreme order statistic $X_{n:n}$. The normalized extreme $(n+1)X_{n:n}/n$ is the unique uniformly minimum variance unbiased estimator of Θ. By Theorem 3.3, $X_{n:n}$, when properly normalized, is asymptotically exponential.

If the parameters of the population distribution are location and scale only, then linear functions $L = a_1X_{1:n} + a_2X_{2:n} + \cdots + a_nX_{n:n}$ of order statistics (so called L-estimators or L-statistics) can be found as unbiased estimators of both location and scale (with suitable coefficients a_i) which have minimum variance in the class of linear unbiased estimators (see Sarhan and Greenberg, 1962; David, 1981). Their small sample theory is simple, and with mild conditions on their coefficients, L-statistics are known to be asymptotically normal (see Serfling, 1980). The special L-statistic obtained by choosing $a_i = 0$ if $i \leq k$ or $i \geq n-k$ and $a_i = 1$ otherwise, is known as the trimmed mean, which is used in place of the mean as estimator in order to avoid outliers. Its theory is part of that of L-estimators.

4. General case: Inequalities

With no assumption on the sequence $X_1, X_2, \ldots, X_n$ of random variables, the order statistic $X_{r:n}$ satisfies the relation at (2.1). That is, if $m_n(x)$ is the number of those X_j, $1 \leq j \leq n$, for which $\{X_j < x\}$, then

$$\{X_{r:n} < x\} = \{m_n(x) \geq r\}. \tag{4.1}$$

We put $y_r = P(m_n(x) \geq r)$, $S_0 = 1$ and

$$S_k = S_{k,n}(x) = E\left[\binom{m_n(x)}{k}\right] = \sum P(X_{i_1} < x, X_{i_2} < x, \ldots, X_{i_k} < x), \tag{4.2}$$

where the summation is over all $1 \leq i_1 < i_2 < \cdots < i_k \leq n$. Evidently, $1 \geq y_1 \geq y_2 \geq \cdots \geq y_n \geq 0$ and, for $k \geq 1$,

$$S_k = \sum_{j=k}^{n} \binom{j-1}{k-1} y_j. \tag{4.3}$$

If we multiply (4.3) by $(-1)^{k-r}\binom{k-1}{r-1}$ and sum the terms with respect to k, we get the classical Bonferroni–Jordan inequalities: for $r \geq 1$, and for any nonnegative integer u,

$$\sum_{k=r}^{r+2u-1} (-1)^{k-r}\binom{k-1}{r-1} S_k \leq P(m_n(x) \geq r) \leq \sum_{k=r}^{r+2u} (-1)^{k-r}\binom{k-1}{r-1} S_k. \tag{4.4}$$

Notice that the two bounds coincide if $m + 2u - 1 \geq n$, and thus an identity is obtained for $P(m_n(x) \geq r) = y_r = P(X_{r:n} < x)$. It was shown by Galambos (1977), that the above inequalities can be improved as follows: for $r \geq 1$, and for any nonnegative integer u such that $2 \leq 2u \leq n - r - 1$,

$$\sum_{k=r}^{r+2u-1} (-1)^{k-r}\binom{k-1}{r-1} S_k + \frac{2u}{n-r}\binom{r+2u-1}{r-1} S_{r+2u} \leq P(m_n(x) \geq r)$$

$$\leq \sum_{k=r}^{r+2u} (-1)^{k-r}\binom{k-1}{r-1} S_k - \frac{2u+1}{n-r}\binom{r+2u}{r-1} S_{r+2u+1}. \tag{4.5}$$

Without restating these inequalities, we remark that if we replace the binomial coefficients $\binom{k-1}{r-1}$ by $\binom{k}{r}$ and by appropriately changing the last terms in the bounds of (4.5), inequalities similar to those at (4.4) and (4.5) are obtained for $P(m_n(x) = r)$. See Galambos (1978, pp. 19–20) and Walker (1981), who gave new proofs for these inequalities.

The inequalities are very useful for finding limit theorems for $y_r = P(m_n(x) \geq r) = P(X_{r:n} < x)$ in the non-i.i.d. case, while (4.5) gives better bounds in actual numerical calculations. The case $r = 1$ received considerable attention, in which case all binomial coefficients in (4.5) become one. Additional reduction of (4.5) is usually necessary since S_k defined at (4.2) assumes that the k-dimensional distribution of the X_j is known. In particular, when only bivariate distributions are available then (4.4) and (4.5) reduce to

$$S_1 - S_2 \leq P(m_n(x) \geq 1) \leq S_1 - \frac{1}{n-1} S_2 . \tag{4.6}$$

No better upper bound is known if only S_1 and S_2 are available. However, the best lower bound in the form of $aS_1 + bS_2$ was determined in Kwerel (1975a) and Galambos (1977); namely if $k_0 = 1 + [2S_2/S_1]$, where $[y]$ signifies the integer part of y, then

$$\frac{2S_1}{k_0+1} - \frac{2S_2}{k_0(k_0+1)} \leq P(m_n(x) \geq 1) . \tag{4.7}$$

Using the method of Galambos (1977), Sathe et al. (1980) extended (4.7) by showing that if

$$2S_2 < (n+r-2)S_1 - n(r-1) \tag{4.8}$$

then

$$P(m_n(x) \geq r) \geq 2 \frac{(k_r-1)(S_1-r+1) - S_2 + \frac{1}{2}(r-1)(r-2)}{(k_r-r)(k_r-r+1)} , \tag{4.9}$$

where $k_r + r - 3$ is the integer part of $\{2S_2 - (r-1)(r-2)\}/(S_1 - r + 1)$.

Sathe et al. (1980) give upper bounds, too, under some restrictions on r, n, S_1 and S_2. One can expect further improvements of some of these inequalities in the light of the discovery of new methods of proof by Galambos and Mucci (1980) and Walker (1981). In particular, further sharpening of the bounds in (4.5) can be expected when more than two binomial moments S_k are utilized. For lower bounds on $P(m_n(x) \geq 1)$, Kwerel (1975b,c) developed a method of finding the best estimation in terms of $S_1, S_2, \ldots, S_k$ (which is conveniently applicable for $k \leq 3$). No optimal bounds are known for $P(m_n(x) \geq r)$, $r \neq 1$, with any set of binomial moments. In fact, the only known alternative to (4.5) is the following inequality, due to Galambos (1969):

$$P(m_n(x) \geq r) \geq \frac{(S_1 - r + 1)S_r}{(r+1)S_{r+1} + rS_r} .$$

For a historical account of early results in this direction, see Takács (1958).

5. General case: Asymptotic results, with applications to model building

The asymptotic results of Section 3 are not applicable when the random variables $X_1, X_2, \ldots, X_n$ are not i.i.d. In fact, normalized central terms may have asymptotic distributions different from normal; the distribution of extremes, when normalized, may converge to an arbitrary prescribed distribution function, and extremes may possess limiting distribution and at the same time, no other upper or lower order statistics could be normalized so that an asymptotic distribution would exist. All these facts point to caution in applications, since the i.i.d. case is mainly an abstraction and thus rare in practice. On the other hand, the fact that a departure from the i.i.d. case can lead to a larger family of possible limiting distributions is reassuring in some fields such as reliability theory. In the present section we describe some asymptotic results which appear useful in applications. Others are included in this list for supporting the claims of the present paragraph.

1. *Central terms.* Introduce the indicator variables $I_j(x)=1$ if $X_j<x$, and zero otherwise, $1\leq j\leq n$. Then the number $m_n(x)$ of those X_j which satisfy $X_j<x$ can be written as

$$m_n(x)=I_1(x)+I_2(x)+\cdots+I_n(x)\,.$$

Now, since just as at (2.1), $\{X_{r:n}<x\}=\{m_n(x)\geq r\}$, the behaviour of the central terms ($r\sim np$, $0<p<1$) is related to the behaviour of the arithmetical mean $(1/n)\sum_{j=1}^{n} I_j(x)$. This latter is known to be asymptotically normal for a large variety of dependence structures (see e.g. Sen, 1968 and 1972) but it is not always normal. For example, if the events $\{X_j<x\}$ are exchangeable, which means that for every choice $1\leq i_1<i_2<\cdots<i_k$ of the subscripts

$$P(X_{i_1}<x, X_{i_2}<x,\ldots,X_{i_k}<x)=P(X_1<x, X_2<x,\ldots,X_k<x)\,,$$

then the asymptotic distribution of normalized central terms is usually not normal. See the book Chow and Teicher (1978) for the most general results.

2. *Strength of bundles of threads.* Consider a bundle of n parallel threads of equal length. Let $X_1, X_2,\ldots,X_n$ denote the strength of the individual threads. We assume that the X_j are identically distributed. Let us also assume that a free load on the bundle is distributed equally on the individual threads. Then the bundle will not break under a load S if there are at least k threads in the bundle each of which can withstand a load S/k. In other words, if $X_{1:n}\leq X_{2:n}\leq\cdots\leq X_{n:n}$ are the ordered strengths of the individual threads, then the strength S_n of the bundle can be represented as

$$S_n=\max\{(n-k+1)X_{k:n}\colon 1\leq k\leq n\}\,.$$

With our usual meaning for $m_n(x)$, the above expression for S_n can also be written as

$$\frac{S_n}{n}=\max_{1\leq j\leq n}\left\{X_j\left[1-\frac{m_n(X_j)}{n}\right]\right\}=\sup\left\{x\left(1-\frac{m_n(x)}{n}\right)\colon -\infty<x<+\infty\right\},$$

from which, under quite general conditions on the dependence of the X_j and on their common distribution function $F(x)$ one can deduce that, with a suitable constant A, $\sqrt{n}((S_n/n)-A)$ is asymptotically normal. See Suh et al. (1970), Sen et al. (1973), Sen (1973a,b) and Phoenix and Taylor (1973).

3. *Strength of a sheet of metal.* The random strength S of a sheet of metal can be represented as an extreme of some dependent random variables. Indeed, if the sheet is subdivided (hypothetically) into n smaller pieces of equal size, and if the strength of the j-th piece is X_j, then, by the weakest link principle,

$$S = \min(X_1, X_2, \ldots, X_n) = X_{1:n}. \tag{5.1}$$

Evidently, the X_j are identically distributed and their common distribution is similar to S (when normalized for size). Notice the exact equation at (5.1). Now, if we can develop an asymptotic distribution for $X_{1:n}$, then it is the distribution of S. Under very general conditions it is shown in Galambos (1978) and (1981) that the distribution of S is necessarily Weibull:

$$P(S < x) = 1 - \exp\{-A(x-B)^{\gamma}\}, \quad x \geq B,$$

where $A > 0$, $\gamma > 0$ and B are parameters. See also the relevant references in Harter (1978).

4. *Reliability applications.* There are two special systems of components whose life distribution is directly related to extreme value theory. Let a system consist of n components with useful random lives $X_1, X_2, \ldots, X_n$. The system is called parallel if it functions until at least one component functions, while it is called series if it fails as soon as one component fails. If the life of the system is denoted by L, then

$$L = \max(X_1, X_2, \ldots, X_n) = X_{n:n} \quad \text{(parallel)}$$

and

$$L = \min(X_1, X_2, \ldots, X_n) = X_{1:n} \quad \text{(series)}.$$

A basic result of reliability theory is that every coherent system can be reduced to a parallel or to a series system (see the first two chapters in Barlow and Proschan, 1975). Therefore, a good approximation to the distribution of L is an asymptotic distribution of the extremes in a dependent model. It should be recognized that in the decomposition into a parallel or series system, one has no control over the dependence of the 'new' components, hence an assumption of independence is unrealistic. A number of dependent models are discussed in Chapter 3 of Galambos (1978). One of these models concludes that the set of possible limiting distributions of the extremes is all distributions with monotonic hazard rates, which is in agreement with the experience of engineers. It also covers such general models as stationary sequences (for which see also Leadbetter, 1974), exchangeable sequences (see also Berman, 1962; Chernick 1980) and Gaussian sequences. References to the original works are also provided. For additional references on reliability applications see David (1981, pp. 156–158).

An excellent newer review of univariate extreme value theory and applications is given by Deheuvels (1981).

6. Model building through characterizations

A very instructive classroom example for model building is to ask the following question: if a new insurance company issues 10 000 accident insurance policies to a group where the expected time period until an accident for each individual in the group is 10 years, when can the company expect the first claim to arrive?

In the question there is hardly any recognizable mathematical assumption and yet, there is a unique answer to the question. Namely, if X_j, $1 \leq j \leq 10\,000$, is the length of time until a claim by the j-th individual, then, the X_j can be assumed to be i.i.d. with $E(X_j) = 10$ (years). Furthermore, insurance is against accident only, hence, for any $s > 0$ and $t > 0$,

$$P(X_j \geq s + t \mid X_j \geq t) = P(X_j \geq s)\,, \tag{6.1}$$

which is the exact form of emphasizing that not age was the factor in a claim. The question is then $E(X_{1:10\,000})$.

The equation at (6.1) is known as the lack of memory property and it is well known to have a unique solution among distribution functions, namely, $F(x) = 1 - e^{-\lambda x}$, $x > 0$, with some $\lambda > 0$ (see Galambos and Kotz, 1978, p. 8). Since $E(X_j) = 10$, $\lambda = 1/10$, and by Theorem 2.2, $X_{1:n}$ is also exponential with $E(X_{1:n}) = (1/n)E(X_j)$ (hence, the answer is that the first claim is expected in 1/1000 year, which is about 1/3 day, i.e., in the first few hours after the completion of the deal).

When in a practical question there is only one underlying distribution for the random quantity we investigate, we speak of a characterization theorem. There are a large number of characterization results which are based on properties of order statistics. Since the works Galambos (1975) and Galambos and Kotz (1978) give a good collection of such results we present here a few typical ones only. For these characterizations, we return to the assumption of independence and identical distributions.

The simplest but useful characterization is due to Huang (1974a). He remarked that, in view of the last expression at (2.2), the distribution of a single order statistic uniquely determines the population distribution.

Another characterization of general nature is in terms of moments. If we put

$$F^{-1}(u) = \inf\{x : F(x) \geq u\}\,,$$

then, in view of (2.2),

$$E(X_{r:n}) = r\binom{n}{r}\int_0^1 F^{-1}(u)u^{r-1}(1-u)^{n-r}\,du\,, \tag{6.2}$$

whenever the left hand side is finite. Now, since

$$(n-r)\binom{n}{r}(1-u)+(r+1)\binom{n}{r+1}u = n\binom{n-1}{r},$$

we get from (6.2) that for $0<r<n$, $n\geqslant 2$,

$$(n-r)E(X_{r:n}) + rE(X_{r+1:n}) = nE(X_{r:n-1})\,.$$

(For these, and other recurrence formulas, see Section 3.4 in David, 1981). Hence, if $E(X_{r(n):n}) = E(Y_{r(n):n})$ for some $r(n)$ and all $n\geqslant 1$, then, in fact, $E(X_{r:n}) = E(Y_{r:n})$ for all r and all n. But then (6.2) is the same value for all r and n whether $F^{-1}(u)$ is the inverse of the distribution of X or of Y, from which (by the classical moment problem) it follows that $F_x^{-1}(u) = F_y^{-1}(u)$, i.e., X and Y have the same distribution. As an example we mention that it thus follows that if $E(X_{1:n}) = c/n$, $n\geqslant 1$, with some c, then the population distribution is exponential, while if $E(X_{1:n}) = c/(n+1)$, $n\geqslant 1$, then it is uniform. The shocking fact in this theorem is that, although there is hardly any difference between $E(X_{1:n}) = c/n$ or $E(X_{1:n}) = c/(n+1)$ for large values of n, the requirement 'for all $n\geqslant 1$' cannot be dropped entirely; in fact, 'for infinitely many n' is insufficient (see Huang, 1974b and 1975; Mejzler, 1965; Galambos and Kotz, 1978, Section 3.4).

A weaker form of characterization than the one mentioned in the previous paragraph is the equivalence of the following two statements: (i) the population distribution is exponential and (ii) $nX_{1:n}$, $n\geqslant 1$ is distributed as X_1. Here, 'all $n\geqslant 1$' can, however, be reduced to two values of n, n_1 and n_2, say, if n_1 and n_2 do not satisfy $n_1^k = n_2^t$ with some positive integers k and t. Interestingly, further reduction is possible by requiring (ii) to hold for a single random value of n, which takes at least two values with the just mentioned property (see Shimizu and Davies, 1979). If we modify (ii) to require that, with some function $g(x)$, $g(n)X_{1:n}$ is distributed as X_1, where n is a random variable, then the just mentioned characterization easily extends to a characterization of the Weibull distribution through $g(x) = x^\alpha$, $\alpha > 0$. However, there is no general solution for any other $g(x)$. When $g(x)$ is constant, then within a limited family of distributions for n, Baringhaus (1980) shows that only for a logistic population distribution and geometric n can $X_1 + c$ and $X_{1:n}$ be identically distributed.

A detailed account of characterizations based on order statistics can be found in Chapter 3 of Galambos and Kotz (1978), which is supplemented in the review of Galambos (1982). The following papers contain interesting characterizations of the geometric distribution: Arnold (1980), Arnold and Ghosh (1976) and El-Neweihi and Govidarajulu (1979). See also the work of Gerber (1980), which is only indirectly related to order statistics, and the survey of Kotz (1974).

References

Arnold, B. C. (1980). Two characterizations of the geometric distribution. *J. Appl. Probab.* **17**, 570–573.

Arnold, B. C. and Ghosh, M. (1976). A characterization of geometric distributions by distributional properties of order statistics. *Scand. Actuar. J.* **4**, 232–234.

Balkema, A. A. and de Haan, L. (1978). Limit distributions for order statistics. *Theor. Veorjatnost. i Primen.* I:**23**, 80–96; II:**23**, 358–375.

Baringhaus, L. (1980). Eine simultane Characterisierung der geometrischen Verteilung und der logistischen Verteilung. *Metrika* **27**, 237–242.

Barlow, R. E. and Proschan, F. (1975). *Statistical Theory of Reliability and Life Testing: Probability Models.* Holt, Rinehart and Winston, New York.

Berman, S. M. (1962). Limiting distribution of the maximum term in a sequence of dependent random variables. *Ann. Math. Statist.* **33**, 894–908.

Chernick, M. R. (1980). A limit theorem for the maximum term in a particular ERMA(1,1) sequence. *J. Appl. Probab.* **17**, 869–873.

Chibisov, D. M. (1964). On limit distributions for members of a variational series. *Teor. Verojatnost. i Primen.* **9**, 150–165.

Chow, Y. S. and Teicher, H. (1978). *Probability Theory.* Springer, New York.

Csörgő, M., Seshadri, V. and Yalovsky, M. (1975). Applications of characterizations in the area of goodness of fit. In: G. P. Patil et al., eds., *Statistical Distributions in Scientific Work, Vol. 2.* Reidel, Dordrecht, pp. 79–90.

David, H. A. (1981). *Order Statistics.* Wiley, New York, 2nd ed.

Deheuvels, P. (1978). Characterisation complede des lois extremes mutivariees et de la convergence des types extremes. *Publ. Inst. Statist. Univ. Paris* **23**, 1–36.

Deheuvels, P. (1980). The decomposition of infinite order and extreme multivariate distributions. In: *Asymptotic Theory of Statistical Tests and Estimation* (Proc. Adv. Internat. Symp. Univ. North Carolina, Chapel Hill, NC 1979). Academic Press, New York, pp. 259–286.

Deheuvels, P. (1981). Univariate extreme values-theory and applications. Bull. ISI (Proceedings of the 43rd Session), pp. 837–858.

El-Neweihi, E. and Govindarajulu, Z. (1979). Characterizations of geometric distribution and discrete IFR (DRF) distributions using order statistics. *J. Statist. Plann. Inference* **3**, 85–90.

Englund, G. (1980). Remainder term estimates for the asymptotic normality of order statistics. *Scand. J. Statist.* **7**, 197–202.

Epstein, B. and Sobel, M. (1953). Life testing. *J. Amer. Statist. Assoc.* **48**, 486–502.

Galambos, J. (1969). Quadratic inequalities among probabilities. *Ann. Univ. Sci. Budapest Sectio Math.* **12**, 11–16.

Galambos, J. (1975). Characterizations of probability distributions by properties of order statistics. I–II. In: G. P. Patil et al., eds., *Statistical Distributions in Scientific Work, Vol. 3.* Reidel, Dordrecht, pp. 71–88 and 89–101.

Galambos, J. (1977). Bonferroni inequalities. *Ann. Probab.* **5**, 577–581.

Galambos, J. (1978). *The Asymptotic Theory of Extreme Order Statistics.* Wiley, New York.

Galambos, J. (1981). Extreme value theory in applied probability. *Math. Sci.* **6**, 13–26.

Galambos, J. (1982). The role of functional equations in stochastic model building. *Aequationes Math.* **25**, 21–41.

Galambos, J. and Mucci, R. (1980). Inequalities for linear combination of binomial moments. *Publ. Math. Debrecen* 27, 263–268.

Galambos, J. and Kotz, S. (1978). *Characterizations of Probability Distributions.* Lecture Notes in Mathematics 675, Springer, Berlin.

Gerber, H. U. (1980). A characterization of certain families of distributions via Esscher transforms and independence. *J. Amer. Statist. Assoc.* **75**, 1015–1018.

Gnedenko, B. V. (1943). Sur la distribution limite du terme maximum d'une serie aleatoire. *Ann. Math.* **44**, 423–453.

Gumbel, E. J. (1958). *Statistics of Extremes.* Columbia University Press, New York.

Haan, L. de (1970). On regular variation and its application to the weak convergence of sample extremes. Math. Centre Tracts 32, Amsterdam.

Hall, P. (1979). On the rate of convergence of normal extremes. *J. Appl. Probab.* **16**, 433–439.

Hall, W. J. and Wellner, J. A. (1979). The rate of convergence in law of the maximum of exponential sample. *Statist. Neerlandica* **33**, 151–154.

Harter, H. L. (1978). A bibliography of extreme value theory. *Internat. Statist. Rev.* **46**, 279–306.

Huang, J. S. (1974a). Personal communication.

Huang, J. S. (1974b). Characterizations of the exponential distribution by order statistics. *J. Appl. Probab.* **11**, 605–608.

Huang, J. S. (1975). Characterization of distributions by the expected values of the order statistics. *Ann. Inst. Statist. Math.* **27**, 87–93.

Johnson, N. L. and Kotz, S. (1968–1972). *Distributions in Statistics*, Vol. I–IV. Wiley, New York.

Kotz, S. (1974). Characterizations of statistical distributions: A supplement to recent surveys. *Internat. Statist. Rev.* **42**, 39–65.

Kwerel, S. M. (1975a). Most stringent bounds on aggregated probabilities of partially specified dependent probability systems. *J. Amer. Statist. Assoc.* **70**, 472–479.

Kwerel, S. M. (1975b). Bounds on the probability of the union and intersection of m events. *Adv. Appl. Probability* **7**, 431–448.

Kwerel, S. M. (1975c). Most stringent bounds on the probability of the union and intersection of m events for systems partially specified by $S_1, S_2, \ldots, S_k$ $2 \leq k \leq m$. *J. Appl. Probab.* **12**, 612–619.

Leadbetter, M. R. (1974). On extreme values in stationary sequences. *Z. Wahrsch. Verw. Geb.* **28**, 289–303.

Loeve, M. (1963). *Probability Theory.* Van Nostrand, New York, 3rd ed.

Mejzler, D. (1965). On a certain class of limit distributions and their domain of attraction. *Trans. Amer. Math. Soc.* **117**, 205–236.

Mori, T. (1981). The relation of sums and extremes of random variables. Bull. ISI (Proc. of the 43rd Session), pp. 879–894.

Nair, K. A. (1981). Asymptotic distribution and moments of normal extremes. *Ann. Probab.* **9**, 150–153.

Phoenix, S. L. and Taylor, H. M. (1973). The asymptotic strength distribution of a general fiber bundle. *Adv. Appl. Probab.* **5**, 200–216.

Pickands, J. (1981). Multivariate extreme value distributions. Bull. ISI (Proc. of the 43rd Session), pp. 859–878.

Reiss, R. D. (1976). Asymptotic expansions for sample quantiles. *Ann. Probab.* **4**, 249–258.

Reiss, R. D. (1981). Uniform approximation to distributions of extreme order statistics. *Adv. Appl. Probab.* **13**, 533–547.

Rényi, A. (1953). On the theory of order statistics. *Acta Math. Acad. Sci. Hungar.* **4**, 191–231.

Rossberg, H. J. (1965). Die asymptotische Unabhangigkeit der kleinsten und grossten Werte einer Stichprobe von Stichprobenmittel. *Math. Nachr.* **28**, 305–318.

Sarhan, A. E. and Greenberg, B. G. (1962). *Contributions to Order Statistics.* Wiley, New York.

Sathe, Y. S., Pradhan, M. and Shah, S. P. (1980). Inequalities for the probability of the occurrence of at least m out of n events. *J. Appl. Probab.* **17**, 1127–1132.

Sen, P. K. (1968). Asymptotic normality of sample quantiles for m-dependent processes. *Ann. Math. Statist.* **39**, 1724–1730.

Sen, P. K. (1972). On the Bahadur representation of sample quantiles for sequences of ϕ-mixing random variables. *J. Multivariate Anal.* **2**, 77–95.

Sen, P. K. (1973a). On fixed size confidence bands for the bundle strength of filaments. *Ann. Statist.* **1**, 526–537.

Sen, P. K. (1973b). An asymptotically efficient test for the bundle strength of filaments. *J. Appl. Probab.* **10**, 586–596.

Sen, P. K., Bhattacharyya, B. B. and Suh, M. W. (1973). Limiting behavior of the extremum of certain sample functions. *Ann. Statist.* **1**, 297–311.

Serfling, R. J. (1980). *Approximation Theorems of Mathematical Statistics.* Wiley, New York.

Shimizu, R. and Davies, L. (1979). General characterization theorems for the Weibull and the stable distributions. Technical Report, Inst. Statist. Math., Tokyo.

Smirnov, N. V. (1949). Limit laws for members of a variational series. *Trudi Math. Steklov* **25**, 1–60.

Suh, M. W., Bhattacharyya, B. B. and Grandage, A. (1970). On the distribution and moments of the strength of a bundle of filaments. *J. Appl. Probab.* **7**, 712–720.

Takács, L. (1958). On a general probability theorem and its applications in the theory of stochastic processes. *Proc. Cambridge Philos. Soc.* **54**, 219–224.

Tiago de Oliveira, J. (1961). The asymptotical independence of the sample mean and the extremes. *Revista da Fac. Ciencias Univ. Lisboa A* **8**, 299–310.

Tiago de Oliveira, J. (1980). Bivariate extremes: Foundations and statistics. In: P. R. Krishnaiah, ed., *Multivariate Analysis V.* North-Holland, Amsterdam, 349–366.

Walker, A. M. (1981). On the classical Bonferroni inequalities and the corresponding Galambos inequalities. *J. Appl. Probab.* **18**, 757–763.

P. R. Krishnaiah and P. K. Sen, eds., *Handbook of Statistics, Vol. 4*
© Elsevier Science Publishers (1984) 383–403

18

Induced Order Statistics: Theory and Applications

P. K. Bhattacharya

1. Introduction

Suppose X and Y are two numerical characteristics defined for each individual in a population. In a random sample, order the X-values and consider the Y-value associated with the r-th order statistic of the X's. We call this the r-th *induced order statistic* or the *concomitant of the r-th order statistic.* Induced order statistics arise naturally in the context of selection where individuals ought to be selected by their ranks in respect of Y, but are actually selected by their ranks in a related variate X due to unavailability of Y at the time of selection. Induced order statistics are also useful in regression analysis, especially when the observations are subject to a type II censoring scheme with respect to the dependent variable, or when the regression function at a given quantile of the predictor variable is of interest.

A systematic study of the induced order statistics, their ranks, their extremes and their partial sums has been undertaken only very recently. We shall present here some of these developments and their applications.

2. Definitions and notations

Suppose $(X_1, Y_1), (X_2, Y_2), \ldots$ are independent and identically distributed (iid) as a two-dimensional random vector (X, Y). For each $\boldsymbol{x}_n = (x_1, \ldots, x_n)$ in the sample space of $(X_1, Y_1), \ldots, (X_n, Y_n)$, arrange $x_1, \ldots, x_n$ in ascending order of magnitude, treating ties in an arbitrary but well-defined manner (e.g., among ties x_i is placed before or after x_j depending on whether $i < j$ or $j < i$). Thus the coordinates of each $\boldsymbol{x}_n$ are arranged as $x_{n1} \leqslant \cdots \leqslant x_{nn}$. This defines the X-order statistics $X_{n1} \leqslant \cdots \leqslant X_{nn}$ obtained from $\boldsymbol{X}_n = (X_1, \ldots, X_n)$. We now define the r-th induced Y-order statistic Y_{nr} to be $Y_{nr} = Y_j$ if $X_{nr} = X_j$. In other words, the r-th induced Y-order statistic is the Y-value associated with the r-th X-order statistic and is, in general, not the same as the r-th smallest Y-value.

Research supported by National Science Foundation Grant MCS 8101976.

If Y has a continuous cumulative distribution function (cdf), then the Y_{nr} are distinct with probability 1, and the rank R_{nr} of Y_{nr} among $Y_1, \ldots, Y_n$ is unambiguously defined. This rank can be written as

$$R_{nr} = \sum_{i=1}^{n} 1\{Y_{nr} \geq Y_i\},$$

where $1\{S\}$ denotes the indicator function of the event S.

In what follows, $F(x)$ and $G(y)$ are the marginal cdf's of X and Y respectively, $G(y \mid x)$ the conditional cdf of Y given $X = x$, $m(x) = E(Y \mid X = x)$ the regression of Y on X and $\sigma^2(x) = \mathrm{Var}(Y \mid X = x)$ the residual variance. We shall also denote the mean and the variance of X, the mean and the variance of Y and the correlation between X and Y by μ_x, σ_x^2, μ_y, σ_y^2 and ρ respectively. If $\sigma^2(x)$ is a constant, we write $\sigma^2(x) = \sigma^2$. For the case of linear regression with constant residual variance, as in the bivariate normal distribution, $\sigma^2 = (1 - \rho^2)\sigma_y^2$. We shall also denote the standard normal cdf by Φ.

3. Dependence structure

In this section we present two results. The first one, due to Bhattacharya (1974), shows that the $\{Y_{nr}\}$ are conditionally independent given $\boldsymbol{X}_n$, while the other one, due to Sen (1976), shows that their partial sums, adjusted for the regression, form a martingale sequence.

LEMMA 3.1. *For every n, the induced order statistics $Y_{n1}, \ldots, Y_{nn}$ are conditionally independent given $X_1, \ldots, X_n$ with conditional cdf's $G(\cdot \mid X_{n1}), \ldots, G(\cdot \mid X_{nn})$ respectively.*

PROOF. For $\boldsymbol{X}_n = (X_1, \ldots, X_n)$, define $\lambda(r, \boldsymbol{X}_n)$ to be that j for which $X_{nr} = X_j$. Then the random permutation $(\lambda(1, \boldsymbol{X}_n), \ldots, \lambda(n, \boldsymbol{X}_n))$ of $(1, \ldots, n)$ is determined only by $\boldsymbol{X}_n$ and $Y_{nr} = Y_{\lambda(r, \boldsymbol{X}_n)}$. Argue conditionally given $\boldsymbol{X}_n = \boldsymbol{x}_n$. Since, for each i, Y_i is independent of $\{(X_j, Y_j), j \neq i\}$, it follows that

$$\begin{aligned}
&P[Y_{nr} \leq y_r, k = 1, \ldots, n \mid \boldsymbol{X}_n = \boldsymbol{x}_n] \\
&\quad = P[Y_{\lambda(r, \boldsymbol{x}_n)} \leq y_r, r = 1, \ldots, n \mid X_{\lambda(r, \boldsymbol{x}_n)} = x_{\lambda(r, \boldsymbol{x}_n)}, r = 1, \ldots, n] \\
&\quad = \prod_{r=1}^{n} P[Y_{\lambda(r, \boldsymbol{x}_n)} \leq y_r \mid X_{\lambda(r, \boldsymbol{x}_n)} = x_{\lambda(r, \boldsymbol{x}_n)}] \\
&\quad = \prod_{r=1}^{n} G(y_r \mid x_{\lambda(r, \boldsymbol{x}_n)}) = \prod_{r=1}^{n} G(y_r \mid x_{nr}),
\end{aligned}$$

proving the desired result.

REMARK. Although X_{nr} is the r-th X-order statistic in the sample $(X_1, Y_1), \ldots, (X_n, Y_n)$ in our discussion so far, the key fact on which the proof of Lemma 3.1 depends, is that $(X_{n1}, \ldots, X_{nn})$ is a random permutation of $(X_1, \ldots, X_n)$ determined only by the X's. Indeed, the random variable X may take values in an arbitrary space. The scope of Lemma 3.1 and many of its consequences (including Lemma 3.2 below) is, therefore, much broader than the immediate context of our main discussion. For example, the induced order statistics Y_{nr} associated with the r-th nearest neighbor X_{nr} of a fixed x among $X_1, \ldots, X_n$ in the d-dimensional Euclidean space comes within the scope of Lemma 3.1. The average

$$k_n^{-1} \sum_{r=1}^{k_n} Y_{nr}$$

of the first k_n of these induced order statistics with $k_n \to \infty$ at an appropriate rate slower than n is of interest as a nonparametric estimator of the regression $m(x)$. Another use of Lemma 3.1 where the induced order statistics are not concomitants of X-order statistics will be discussed in Section 5 in connection with a double sampling plan.

Using Lemma 3.1, we can analyze the behavior of the induced order statistics in a relatively simple manner by first conditioning with respect to the X's and then removing the condition. This will be done in most of the subsequent derivations. In particular, the following lemma is obtained like this.

LEMMA 3.2. *For arbitrary* $\{C_k, n \geq 1, 1 \leq r \leq n\}$, *let*

$$S^*_{nk} = \sum_{r=1}^{k} C_{nr}\{Y_{nr} - m(X_{nr})\}\,.$$

Then for each n, $\{S^*_{nk}, 1 \leq k \leq n\}$ *is a martingale.*

PROOF. For $1 \leq k \leq n-1$,

$$\begin{aligned} &E[S^*_{n,k+1} \mid S^*_{n1}, \ldots, S^*_{nk}] \\ &\quad = EE[S^*_{nk} + C_{n,k+1}\{Y_{n,k+1} - m(X_{n,k+1})\} \mid S^*_{n1}, \ldots, S^*_{nk}, \boldsymbol{X}_n]\,, \end{aligned}$$

equals S^*_{nk}, since, by Lemma 3.1, $Y_{n,k+1}$ is conditionally independent of $Y_{n1}, \ldots, Y_{nk}$ given $\boldsymbol{X}_n$.

In Sections 4–7 we discuss small sample and large sample properties of induced order statistics and their ranks. These results were developed in various degrees of generality, by Watterson (1964), David (1973, 1981), David and Galambos (1974), David, O'Connell and Yang (1977) and Yang (1977). In Section 5, the special case of linear regression with homogeneous residuals is discussed and a number of applications are outlined. In Section 8, extremes of induced order statistics are briefly discussed. The main result here is a theorem of Galambos

(1978) on the asymptotic distribution of Y_{nn}. The last section deals with partial sums of induced order statistics and their applications in regression analysis. This aspect of induced order statistics was developed by Bhattacharya (1974, 1976) and Sen (1976). In this section, we also briefly describe estimates of $m(x)$ due to Yang (1981), which are weighted averages of Y_{nr} using kernel weights, and some results on mixed rank statistics due to Sen (1981).

4. Small sample properties of induced order statistics and their ranks

The joint distribution of $Y_{n1}, \ldots, Y_{nn}$ is obtained by integrating their conditional joint distribution given in Lemma 3.1 with respect to the joint distribution of $X_{n1}, \ldots, X_{nn}$. In particular, if $F(x)$ and $G(y \mid x)$ have densities $f(x)$ and $g(y \mid x)$ respectively, the Y_{nr} has probability density function

$$p_{nr}(y) = r\binom{n}{r} \int_{-\infty}^{\infty} g(y \mid x) F^{r-1}(x)\{1 - F(x)\}^{n-r} f(x)\, dx .$$

More generally, we have

THEOREM 4.1. *For* $1 \leq r_1 < \cdots < r_k \leq n$,

$$P[Y_{nr_i} \leq y_i,\ 1 \leq i \leq k] = E\left[\prod_{i=1}^{k} G(y_i \mid X_{nr_i})\right].$$

Formulas for the mean and the variance of Y_{nr} and covariances between Y_{nr} and Y_{ns} and between X_{nr} and Y_{ns} are also obtained by conditioning with respect to $\boldsymbol{X}_n$ using Lemma 3.1. These formulas are given below.

THEOREM 4.2. (a) $E(Y_{nr}) = E[m(X_{nr})]$,
(b) $\operatorname{Var}(Y_{nr}) = E[\sigma^2(X_{nr})] + \operatorname{Var}[m(X_{nr})]$,
(c) $\operatorname{Cov}(Y_{nr}, Y_{ns}) = \operatorname{Cov}[m(X_{nr}), m(X_{ns})]$ *for* $r \neq s$,
(d) $\operatorname{Cov}(X_{nr}, Y_{ns}) = \operatorname{Cov}[X_{nr}, m(X_{ns})]$.

PROOF. (a) and (b) follow immediately from Lemma 3.1. For (c), observe that $\operatorname{Cov}(Y_{nr}, Y_{ns} \mid \boldsymbol{X}_n) = 0$ for $r \neq s$ by the conditional independence of Y_{nr} and Y_{ns}, and for (d), observe that $\operatorname{Cov}(X_{nr}, Y_{ns} \mid \boldsymbol{X}_n) \equiv 0$.

We next consider the distribution of the rank R_{nr} of Y_{nr} among $Y_1, \ldots, Y_n$.

THEOREM 4.3.

$$P[R_{nr} = s] = n \int_{-\infty}^{\infty} \int_{-\infty}^{\infty} \sum_{k=0}^{t} C_k \theta_1^k \theta_2^{r-1-k} \theta_3^{s-1-k} \theta_4^{n-r-s+1+k}\, dF(x)\, dG(y \mid x),$$

where $t = \min(r-1, s-1)$,

$$C_k(r, s, n) = \frac{(n-1)!}{k!(r-1-k)!(s-1-k)!(n-r-s+1+k)!},$$

$$\theta_1(x, y) = P[X \leq x, Y \leq y], \qquad \theta_2(x, y) = P[X \leq x, Y > y],$$

$$\theta_3(x, y) = P[X > x, Y \leq y], \qquad \theta_4(x, y) = P[X > x, Y > y].$$

PROOF. Decompose the event $\{R_{nr} = s\}$ into events

$$\{R_{nr} = s, X_{nr} = j\} = \{\text{rank}(X_j) = r, \text{rank}(Y_j) = s\}, \quad 1 \leq j \leq n,$$

each of which has the same probability. Hence

$$\begin{aligned} P[R_{nr} = s] &= nP[\text{rank}(X_n) = r, \text{rank}(Y_n) = s] \\ &= nP\Big[\sum_{i=1}^{n-1} 1\{X_n \geq X_i\} = r-1, \sum_{i=1}^{n-1} 1\{Y_n \geq Y_i\} = s-1\Big] \\ &= n\int_{-\infty}^{\infty}\int_{-\infty}^{\infty} P\Big[\sum_{i=1}^{n-1} 1\{X_i \leq x\} = r-1, \\ &\qquad \sum_{i=1}^{n-1} 1\{Y_i \leq y\} = s-1\Big] \mathrm{d}F(x)\,\mathrm{d}G(y \mid x). \end{aligned}$$

To complete the proof we only have to recognize the integrand in the last expression as the probability of obtaining marginal frequencies $r-1$ for $\{X_i \leq x\}$ and $s-1$ for $\{Y_i \leq y\}$ in a 2×2 contingency table in which a total frequency of $n-1$ is classified according to the occurrence of the events $\{X_i \leq x\}$ and $\{Y_i \leq y\}$ and their complements.

An interesting variation of the event $\{R_{nn} = n\}$, i.e., the same pair obtains the highest rank in both X and Y, arises when in addition to the bivariate observations $(X_1, Y_1), \ldots, (X_n, Y_n)$, we also have $m_1 + m_2$ other independent observations $X'_1, \ldots, X'_{m_1}$ from the X-distribution and $Y''_1, \ldots, Y''_{m_2}$ from the Y-distribution. Now the X-ranking involves $X_1, \ldots, X_n, X'_1, \ldots, X'_{m_1}$, the Y-ranking involves $Y_1, \ldots, Y_n, Y''_1, \ldots, Y''_{m_2}$, and one of the pairs (X_i, Y_i) obtaining the highest ranks in both X and Y is the event of interest. The probabilities of this and related events have been studied by Spruill and Gastwirth (1981) in the context of professional couples and singles completing for employment in two departments of a university.

5. Linear regression model with homogeneous residuals

We now specialize the results of Theorem 4.2. First suppose that the residual

$$Z = Y - m(X)$$

is independent of X. If in addition, the regression $m(x)$ is linear, then Y is expressed as

$$Y = \mu_y + \rho\sigma_y\left(\frac{X - \mu_x}{\sigma_x}\right) + Z, \qquad E(Z) = 0,$$
$$\mathrm{Var}(Z) = \sigma^2 = \sigma_y^2(1 - \rho^2)\,. \tag{5.1}$$

We call this the linear regression model with homogeneous residuals. For example, this holds when (X, Y) follows a bivariate normal distribution, in which case Z is $N(0, \sigma^2)$. Let

$$\alpha_{nr} = (E(X_{nr}) - \mu_x)/\sigma_x, \qquad \beta_{nr}^2 = \beta_{n,rr} = \mathrm{Var}(X_{nr})/\sigma_x^2,$$
$$\beta_{n,rs} = \mathrm{Cov}(X_{nr}, X_{ns})/\sigma_x^2\,. \tag{5.2}$$

Then the following formulas, originally due to Watterson (1959) are obtained as special cases of Theorem 4.2.

THEOREM 5.1. *Suppose* (5.1) *holds with Z independent of X. Then*
(a) $E(Y_{nr}) = \mu_y + \rho\sigma_y\alpha_{nr}$,
(b) $\mathrm{Var}(Y_{nr}) = \rho^2\sigma_y^2\beta_{nr}^2 + \sigma_y^2(1 - \rho^2)$,
(c) $\mathrm{Cov}(Y_{nr}, Y_{ns}) = \rho^2\sigma_y^2\beta_{n,rs}$ *for* $r \neq s$,
(d) $\mathrm{Cov}(X_{nr}, Y_{ns}) = \rho\sigma_x\sigma_y\beta_{n,rs}$.

We now indicate some applications of induced order statistics under this model.

(i) *A selection problem*. Consider a selection procedure in which a score Y is predicted on the basis of predictor variables $\xi_1, \ldots, \xi_p$ by means of a linear regression formula, and s out of a sample of n individuals with highest predicted scores are selected. Let

$$E(Y \mid \boldsymbol{\xi}) = \beta_0 + \beta_1\xi_1 + \cdots + \beta_p\xi_p = \beta_0 + \boldsymbol{\beta}'\boldsymbol{\xi}$$

denote the true regression formula. However, in practice the true regression formula is estimated by the method of least squares from a previous sample of size N from the same population, for which all $p + 1$ variables $\xi_1, \ldots, \xi_p, Y$ were observed. Let $\hat{\beta}_0, \hat{\beta}_1, \ldots, \hat{\beta}_p$ denote the estimated coefficients. Then the above procedure calls for selection of s out of n individuals with the highest values of $X = \hat{\beta}_1\xi_1 + \cdots + \hat{\beta}_p\xi_p = \hat{\boldsymbol{\beta}}'\boldsymbol{\xi}$. Let (X_i, Y_i), $i = 1, \ldots, n$ denote the predicted score and the true score in the sample from which the s individuals with the highest X-values are selected. The average true score of the selected individuals is then given by

$$\bar{Y}_s = s^{-1} \sum_{r=n-s+1}^{n} Y_{nr}\,,$$

where the induced Y-order statistics Y_{nr} clearly depend on $\hat{\boldsymbol{\beta}}$. The standardized form of the conditional expectation $E(\bar{Y}_s \mid \hat{\boldsymbol{\beta}})$, viz.

$$\{E(\bar{Y}_s \mid \hat{\boldsymbol{\beta}}) - \mu_y\}/\sigma_y ,$$

which is a measure of the predictive accuracy of the estimated regression formula, has been called by Gross (1973) the expected gain from selection. By Theorem 5.1,

$$\{E(\bar{Y}_s \mid \hat{\boldsymbol{\beta}}) - \mu_y\}/\sigma_y = \rho(\hat{\boldsymbol{\beta}})s^{-1} \sum_{r=n-s+1}^{n} \alpha_{nr} ,$$

where $\rho(\hat{\beta})$ is the conditional correlation coefficient between $\hat{\boldsymbol{\beta}}'\boldsymbol{\xi}$ and Y given $\hat{\beta}$, i.e.,

$$\rho(\hat{\boldsymbol{\beta}}) = \frac{\Sigma_1^p \hat{\beta} \operatorname{Cov}(\xi_1, Y)}{\sigma_y \sqrt{\Sigma_1^p \Sigma_1^p \hat{\beta}_1 \hat{\beta}_j \operatorname{Cov}(\xi_i, \xi_j)}} .$$

Comparing the expected gain from selection using $\hat{\boldsymbol{\beta}}'\boldsymbol{\xi}$ with the maximum expected gain that can be achieved by any linear prediction formula, we see that a relative measure of efficacy of the formula $\hat{\boldsymbol{\beta}}'\boldsymbol{\xi}$ is provided by the ratio

$$e = \rho(\hat{\boldsymbol{\beta}})/\rho ,$$

where ρ is the population multiple correlation coefficient of Y on $\xi_1, \ldots, \xi_p$. Note that a particular training sample may produce an estimate $\hat{\boldsymbol{\beta}}$ for which e is small, or even negative. Gross (1973) has computed the training sample sizes N necessary for $P(e > 0.90) = 0.90$ or 0.95 for $p = 3, 5, 7, 9$ under the assumption of multivariate normality with the population ρ following certain prior distributions.

(ii) *A double sampling plan.* Suppose (5.1) holds with Z independent of X, and consider the problem of estimating μ_y when we are allowed to observe X on a sample of size n and Y on a subsample of size k from these n. Such a double sampling plan may yield better estimates of μ_y for the same cost when X is relatively inexpensive to observe in comparison with Y. The simplest plan is to observe Y on k individuals selected at random from n for double sampling, and reindexing the X's we may assume $(X_1, Y_1), \ldots, (X_k, Y_k)$ and $X_{k+1}, \ldots, X_n$ to constitute the totality of all observations. For such a sampling scheme, Cochran [4] studies the regression estimator

$$\hat{Y}_k = \bar{Y}_k + b_k(\bar{X}_n - \bar{X}_k) , \tag{5.3}$$

where

$$\bar{X}_k = \sum_1^k X_i \Big/ k, \quad \bar{Y}_k = \sum_1^k Y_i \Big/ k, \quad \bar{X}_n = \sum_1^n X_i \Big/ n,$$
$$b_k = \sum_1^k (X_i - \bar{X}_k)(Y_i - \bar{Y}_k) \Big/ \sum_1^k (X_i - \bar{X}_k)^2 . \tag{5.4}$$

However, it is natural to explore the possibility of improving this sampling scheme by using the information in $X_1, \ldots, X_n$ to determine which Y-values to observe. To this end, O'Connell and David (1976) proposed to observe Y for those individuals who occupy suitable ranks $1 \leq r_1 < \cdots < r_k \leq n$. The resulting Y-observations are, therefore, the induced order statistics $Y_{nr_1}, \ldots, Y_{nr_k}$ from $(X_1, Y_1), \ldots, (X_n, Y_n)$. Let

$$\bar{X}_{(k)} = \sum_1^k X_{nr_i} \Big/ k, \quad \bar{Y}_{(k)} = \sum_1^k Y_{nr_i} \Big/ k, \qquad b_{(k)} = s_{xy,(k)}/s^2_{x,(k)}, \tag{5.5}$$

where

$$s^2_{x,(k)} = k^{-1} \sum_1^k (X_{nr_i} - \bar{X}_{(k)})^2, \qquad s_{xy,(k)} = k^{-1} \sum_1^k (X_{nr_i} - \bar{X}_{(k)})(Y_{nr_i} - \bar{Y}_{(k)}), \tag{5.6}$$

and, for future reference,

$$s^2_{y,(k)} = k^{-1} \sum_1^k (Y_{nri} - \bar{Y}_{(k)})^2, \qquad r_{(k)} = s_{xy,(k)}/[s_{x,(k)}s_{y,(k)}]. \tag{5.7}$$

Then a modified version of the regression estimator (5.3) is

$$\hat{Y}_{(k)} = \bar{Y}_{(k)} + b_{(k)}(\bar{X}_n - \bar{X}_{(k)}). \tag{5.8}$$

By Lemma 3.1, it follows that under model (5.1),

$$E(\hat{Y}_{(k)} \mid \boldsymbol{X}_n) = \mu_y + \rho \frac{\sigma_y}{\sigma_x}(\bar{X}_n - \mu_x),$$

$$\operatorname{Var}(\hat{Y}_{(k)} \mid \boldsymbol{X}_n) = k^{-1}(1-\rho^2)\sigma_y^2(1+V),$$

where

$$V = (\bar{X}_{(k)} - \bar{X}_n)^2/s^2_{x,(k)}. \tag{5.9}$$

Hence $\hat{Y}_{(k)}$ is an unbiased estimator of μ_y, with

$$\operatorname{Var}(\hat{Y}_{(k)})/\sigma_y^2 = k^{-1}(1-\rho^2)[1+E(V)] + n^{-1}\rho^2. \tag{5.10}$$

An optimum choice of $r_1, \ldots, r_k$ thus requires the minimization of $E(V)$. In the bivariate normal case, confining attention to symmetrically spaced ranks and approximating $E(V)$ by Taylor expansion, O'Connell and David (1976) have computed the optimal ranks for some n and k. They have also studied other estimators of the form

$$Y^* = \sum_1^k a_i Y_{nr_r},$$

which are unbiased if $\Sigma_1^k a_i = 1$ and $\Sigma_1^k a_i \alpha_{nr_i} = 0$ hold, and have variance

$$\operatorname{Var}(Y^*)/\sigma_y^2 = (1-\rho^2)\sum_1^k a_i^2 + \rho^2 \operatorname{Var}(X^*) \quad \text{where } X^* = \sum_1^k a_i X_{nr_i}\,.$$

In the above approach, the choice of $\{r_1, \ldots, r_k\}$ is restricted to *fixed* sets of k elements from $\{1, \ldots, n\}$. We now let $\{r_1, \ldots, r_k\}$ be a *random* set depending on $\boldsymbol{X}_n$. The induced order statistics Y_{nr_i} are no longer concomitants of X-order statistics, but Lemma 3.1 is still applicable by virtue of the remark following that lemma. Consequently, $\hat{Y}_{(k)}$ given by (5.8) is still an unbiased estimator of μ_y and $\operatorname{Var}(\hat{Y}_{(k)})$ is still given by (5.10). We now optimize by choosing the random $\{r_1, \ldots, r_k\}$ so as to minimize $V = V(r_1, \ldots, r_k; \boldsymbol{X}_n)$ given by (5.9) for each $\boldsymbol{X}_n$. This optimization procedure is entirely data-driven and it leads to a more efficient estimator of μ_y than the one considered by O'Connell and David [15] irrespective of how (X, Y) is distributed in the population. It is interesting to note that the choice of $\{r_1, \ldots, r_k\}$ to minimize V has the natural interpretation of being the optimum solution to a familiar design problem in linear regression analysis. The actual search for $r_1, \ldots, r_k$ to minimize $V(r_1, \ldots, r_k; \boldsymbol{X}_n)$ for given $\boldsymbol{X}_n$ is easy for small n, but would need an efficient algorithm even for moderately large n. We leave this as an open problem. However, it would often be possible to obtain suboptimal but fairly good solutions relatively easily by restricting the choice near the two extremes of the X-observations.

(iii) *Maximum likelihood estimation from censored bivariate normal sample.* Consider a life-testing problem in which X is the failure-time of a randomly selected unit and Y is a covariate whose effect on X is of interest. If in an experiment with a random sample of n units only k units are observed according to a type II censoring scheme, i.e., the experiment is run until the first k failures and (X, Y) observations are made only on these k units, then the observations consist of the first k order statistics $X_{n1} < \cdots < X_{nk}$ of the X-variate and the corresponding induced order statistics $Y_{n1}, \ldots, Y_{nk}$ of the covariate. In the bivariate normal case, Harrell and Sen (1979) have obtained the maximum likelihood estimators (mle) of μ_x, μ_y, σ_x, σ_y and ρ, and the likelihood ratio test for H_0: $\rho = 0$ under such a censoring scheme. Using Lemma 3.1 in conjunction with (5.1), the log likelihood function of $(X_{n1}, \ldots, X_{nk})$ and the conditional log likelihood function of $(Y_{n1}, \ldots, Y_{nk})$ given $\boldsymbol{X}_n$ are obtained as, respectively,

$$\begin{aligned} L_{n,k}^{X} &= C_1 - k \log \sigma_x - (2\sigma_x^2)^{-1} \sum_1^k (X_{nr} - \mu_x)^2 \\ &\quad + (n-k)\log[1 - \Phi((X_{nk} - \mu_x)/\sigma_x)]\,, \end{aligned}$$

$$\begin{aligned} L_{n,k}^{Y|X} &= C_2 - k \log(\sigma_y \sqrt{1-\rho^2}) \\ &\quad - \{2\sigma_y^2(1-\rho^2)\}^{-1} \sum_1^k \left\{ Y_{nr} - \mu_y - \frac{\rho\sigma_y}{\sigma_x}(X_{ni} - \mu_x) \right\}^2 . \end{aligned}$$

Adding these, we obtain the log likelihood based on the complete data, from which the mle's $\hat{\mu}_x$, $\hat{\mu}_y$, $\hat{\sigma}_x$, $\hat{\sigma}_y$ and $\hat{\rho}$ are calculated. Of these, $\hat{\mu}_x$ and $\hat{\sigma}_x$ turn out to be the same as the mle's based on X alone, and they are computed by an iterative process. In terms of $\hat{\mu}_x$ and $\hat{\sigma}_x$, we have

$$\hat{\mu}_y = \bar{Y}_{(k)} + b_{(k)}(\hat{\mu}_x - \bar{X}_{(k)}),$$

while $\hat{\sigma}_y$ and $\hat{\rho}$ are obtained by solving

$$\hat{\rho}\hat{\sigma}_y/\hat{\sigma}_x = b_{(k)} \quad \text{and} \quad \hat{\sigma}_y^2(1-\hat{\rho}^2) = s_{y,(k)}^2(1-r_{(k)}^2),$$

where $\bar{X}_{(k)}$, $\bar{Y}_{(k)}$, $s_{y,(k)}^2$, $b_{(k)}$ and $r_{(k)}$ are given by (5.5)–(5.7). Since $\hat{\mu}_x$ and $\hat{\sigma}_x$ are known to have considerable bias under heavy censoring, it may be preferable to estimate these by other suitable estimates throughout (Saw (1959, 1961)).

The likelihood ratio statistic for testing H_0 is a function of $r_{(k)}$ and conditionally, given $\boldsymbol{X}_n$, the statistic

$$T = (k-2)r_{(k)}^2/(1-r_{(k)}^2)$$

follows a noncentral F-distribution with degrees of freedom $(1, k-2)$ and noncentrality parameter $\Delta = k(\rho^2/(1-\rho^2))s_{x,(k)}^2/\sigma_x^2$.

6. Asymptotic distribution of a finite set of induced order statistics

As before, let the residual $Z = Y - m(X)$ be independent of X. Let Y_{nr}^* denote the deviation of Y_{nr} from its mean. For a fixed set of distinct positive integers $r_1, \ldots, r_k$, we have by Theorem 4.2(a),

$$Y_{nr_i}^* = Y_{nr_i} - Em(X_{nr_i}) = \{m(X_{nr_i}) - Em(X_{nr_i})\} + Z_{nr_i}, \tag{6.1}$$

where Z_{nr} is the r-th induced order statistic of the random sample $(X_1, Z_1), \ldots, (X_n, Z_n)$ from the distribution of (X, Z). Since X and Z are independent, it follows from Lemma 3.1 that $Z_{nr_1}, \ldots, Z_{nr_k}$ are independent, each being distributed as Z. Now if for large n, the random variables $m(X_{nr_i})$ are approximately the same as their mean values, then $Y_{nr_i}^* \simeq Z_{nr_i}$, and are therefore approximately iid as Z. The following theorem due to David and Galambos (1974) makes this precise.

THEOREM 6.1. *Suppose the residual $Z = Y - m(X)$ is independent of X. If, moreover, the marginal distribution of X and the regression function $m(x)$ are such that, for each r,*

$$\lim_{n\to\infty} E[\{m(X_{nr}) - Em(X_{nr})\}^2] = 0, \tag{6.2}$$

then for any fixed $1 \leq r_1 < \cdots < r_k \leq n$, *the induced order statistics* $Y_{nr_1}, \ldots, Y_{nr_k}$ *are asymptotically independent, the deviation of each from its mean being distributed as* Z.

PROOF. Let $\varepsilon_{nr} = m(X_{nr}) - Em(X_{nr})$. Then (6.1) is written as

$$(Y^*_{nr_1}, \ldots, Y^*_{nr_k}) = (Z_{nr_1}, \ldots, Z_{nr_k}) + (\varepsilon_{nr_1}, \ldots, \varepsilon_{nr_k}),$$

where $(\varepsilon_{nr1}, \ldots, \varepsilon_{nr_k})$ converges to $(0, \ldots, 0)$ in probability as $n \to \infty$ by condition (6.2). It therefore follows that the limiting distribution of $(Y^*_{nr_1}, \ldots, Y^*_{nr_k})$ is the same as that of $(Z_{nr_1}, \ldots, Z_{nr_k})$. But the Z_{nr_i} are iid as Z as a consequence of the independence of X and Z, which completes the proof.

In particular, condition (6.2) holds if (X, Y) follows a bivariate normal distribution, in which case $Z \sim N(0, \sigma^2)$, where $\sigma^2 = \sigma_y^2(1-\rho^2)$. We thus have

COROLLARY 6.2. *If* $(X_1, Y_1), \ldots, (X_n, Y_n)$ *is a random sample from a bivariate normal distribution, then for any* $1 \leq r_1 < \cdots < r_k \leq n$,

$$\lim_{n\to\infty} P[Y^*_{nr_1} \leq y_1, \ldots, Y^*_{nr_k} \leq y_k] = \prod_{i=1}^{k} \Phi(y_i/\sigma).$$

In Theorem 4.1, we gave an exact formula for the joint distribution of $(Y_{nr_1}, \ldots, Y_{nr_k})$. The next result due to Yang (1977) gives the limiting form of this result when $1 \leq r_1 < \cdots < r_k \leq n$ are sequences of integers such that $r_i/n \to \lambda_i$ as $n \to \infty$, with $0 < \lambda_i < 1$.

THEOREM 6.3. *Suppose the marginal distribution of* X *has a density* $f(x)$ *which is bounded away from* 0 *in a neighborhood of* $F^{-1}(\lambda_i)$, $i = 1, \ldots, k$, *and the conditional cdf* $G(y \mid x)$ *at* $y_1, \ldots, y_k$ *are continuous in* x. *Then for* $1 \leq r_1 < \cdots < r_k \leq n$ *such that* $r_i/n \to \lambda_i \in (0, 1)$ *as* $n \to \infty$,

$$\lim_{n\to\infty} P[Y_{nr_1} \leq y_1, \ldots, Y_{nr_k} \leq y_k] = \prod_{i=1}^{k} G(y_i \mid F^{-1}(\lambda_i)).$$

PROOF. By the hypotheses of the theorem, $\Pi_{i=1}^{k} G(y_i \mid X_{nri})$ is a bounded continuous function of $X_{nr1}, \ldots, X_{nr_k}$ and $X_{nr_r} \to F^{-1}(\lambda_i)$ in probability as $n \to \infty$. Taking limits in Theorem 4.1, the desired result follows.

7. Asymptotic distribution of ranks of induced order statistics

In Section 4, the exact distribution of R_{nr} was given. We now consider the asymptotic distribution of R_{nr}/n.

THEOREM 7.1. *Suppose (X, Y) has a joint pdf $p(x, y)$ satisfying*
(i) *$p(x, y)$ is continuous and bounded by an integrable function $q(y)$ for all x in a neighborhood of $F^{-1}(\lambda)$, and*
(ii) *the marginal density $f(x)$ of X is bounded away from 0 in a neighborhood of $F^{-1}(\lambda)$, where $0<\lambda<1$.*
Then, for $r/n \to \lambda$ as $n \to \infty$,

(a) $$\lim_{n\to\infty} E[(R_{nr}/n)^k] = \int_{-\infty}^{\infty} G^k(y)\,\mathrm{d}G(y \mid F^{-1}(\lambda)), \quad \textit{and}$$

(b) $$\lim_{n\to\infty} P[R_{nr} \leq na] = G[G^{-1}(a) \mid F^{-1}(\lambda)]\,.$$

PROOF. The derivation of (b) is based on convergence of moments. Calculations based on the leading term in the expansion of $R_{nr}^k = [\Sigma_{i=1}^n 1\{Y_{nr} \geq Y_i\}]^k$ yield

$$\lim_{n\to\infty} E[(R_{nr}/n)^k] = \int_{-\infty}^{\infty} G^k(y)\,\mathrm{d}G(y \mid F^{-1}(\lambda)) = E[G^k(Y) \mid X = F^{-1}(\lambda)]\,,$$

and (a) is proved, which in turn implies that the asymptotic distribution of R_{nr}/n is the same as the conditional distribution of $G(Y)$ given $X = F^{-1}(\lambda)$. Thus

$$\lim_{n\to\infty} P[R_{nr} \leq na] = P[G(Y) \leq a \mid X = F^{-1}(\lambda)] = G[G^{-1}(a) \mid F^{-1}(\lambda)]\,,$$

and (b) is proved.

COROLLARY 7.2. *Suppose (X, Y) is bivariate normal with $|\rho|<1$. Then, for $r/n \to \lambda \in (0, 1)$ as $n \to \infty$,*

(a) $$\lim_{n\to\infty} E[(R_{nr}/n)^k] = \int_{-\infty}^{\infty} \Phi^k(\sqrt{1-\rho^2}u + \rho\Phi^{-1}(\lambda))\,\mathrm{d}\Phi(u)\,.$$

In particular,

$$\lim_{n\to\infty} E[R_{nr}/n] = \Phi(\rho\Phi^{-1}(\lambda)/\sqrt{2-\rho^2})\,.$$

(b) $$\lim_{n\to\infty} P[R_{nr} \leq na] = \Phi\left(\frac{\Phi^{-1}(a) - \rho\Phi^{-1}(\lambda)}{\sqrt{1-\rho^2}}\right).$$

PROOF. Substitute $G(y \mid F^{-1}(\lambda)) = \Phi(u)$ and $G(y) = \Phi(\sqrt{1-\rho^2}u + \rho\Phi^{-1}(\lambda))$ with $u = \{(y-\mu_y)/\sigma_y - \rho\Phi^{-1}(\lambda)\}/\sqrt{1-\rho^2}$ in Theorem 7.1. For $k = 1$ in part (a), recognize $\int_{-\infty}^{\infty} \Phi(\sqrt{1-\rho^2}u + \rho\Phi^{-1}(\lambda))\,\mathrm{d}\Phi(u)$ as the convolution of the cdf's of $N(0, 1)$ and $N(0, (1-\rho^2)^{-1})$ evaluated at $\rho\Phi^{-1}(\lambda)/\sqrt{1-\rho^2}$.

8. Extremes of induced order statistics

We mentioned earlier the distinction between the induced Y-order statistic Y_{nr} and the r-th smallest among $Y_1, \ldots, Y_n$, which we denote by $Y_{n(r)}$. Indeed, for $r = n$ the extremes Y_{nn} and $Y_{n(n)}$ behave quite differently as n gets large. For example, in the bivariate normal case with $\mu_x = \mu_y = 0$ and $\sigma_x = \sigma_y = 1$,

$$(2 \log n)^{-1/2} Y_{n(n)} \to 1 \quad \text{but} \quad (2 \log n)^{-1/2} Y_{nn} \to \rho$$

in probability as $n \to \infty$. The same results remain true if $Y_{n(n)}$ and Y_{nn} are replaced by $Y_{n,(n-k)}$ and $Y_{n,n-k}$ for any fixed k. Thus, if X and Y represent measurements of a characteristic for a parent and an offspring, then in a large population, the offspring of the top individuals will not be in the top part of the next generation.

We state without proof the following general result due to Galambos (1978), describing the asymptotic behavior of the extreme induced order statistics.

THEOREM 8.1. *Suppose the cdf $F(x)$ and pdf $f(x)$ of X satisfy*
(i) $F(x) < 1$ *for all* x,
(ii) $F''(x)$ *exist for all large* x,
(iii) $f(x) \neq 0$, *and*
(iv) $\lim_{x \to +\infty} \dfrac{\mathrm{d}}{\mathrm{d}x}\left[\dfrac{1 - F(x)}{f(x)}\right] = 0.$

Suppose further that for sequences $a_n, b_n > 0$ and $A_n, B_n > 0$,

$$\lim_{n \to \infty} F^n(a_n + b_n z) = \exp(-e^{-z})$$

and

$$\lim_{n \to \infty} P[Y < A_n + B_n u \mid X = a_n + b_n z] = T(u, z),$$

a nondegenerate distribution function. Then

$$\lim_{n \to \infty} P[Y_{nn} \leq A_n + B_n u] = \int_{-\infty}^{\infty} T(u, z) \exp(-\mathrm{e}^{-z}) \mathrm{e}^{-z}\, \mathrm{d}z .$$

Suppose $(X_i^{(1)}, Y_i^{(1)})$, $1 \leq i \leq n_1$, and $(X_i^{(2)}, Y_i^{(2)})$, $1 \leq i \leq n_2$, are random samples from two nonsingular bivariate normal populations. Let $Y_{n_1 r}^{(1)}$, $1 \leq r \leq n_1$, and $Y_{n_2 r}^{(2)}$, $1 \leq r \leq n_2$, denote the induced order statistics in the two samples, and for a fixed k consider the probability

$$p(n_1, n_2, k) = P\left[\sum_{r=n_1-k+1}^{n_1} Y_{n,r}^{(1)} > \sum_{r=n_2-k+1}^{n_2} Y_{n_2 r}^{(2)}\right].$$

Obviously, if $\rho_1 > 0$ then for fixed k and n_2, this probability tends to 1 as $n_1 \to \infty$. Kaminsky (1978) has derived an approximate formula for n_1 for which $p(n_1, n_2, k)$ attains an arbitrary level γ when n_2 is a given large number.

9. Sums of induced order statistics and the integrated regression function

The convergence property of the sample t-quantile $X_{n,nt}$ of X to $F^{-1}(t)$ suggests that under regularity conditions, when n is large,

$$E(Y_{n,nt}) \simeq E(Y \mid X = F^{-1}(t)) = m(F^{-1}(t)) = h(t)$$

for $0 < t < 1$. Here for simplicity, we have written nt for the largest integer $[nt]$ not exceeding nt. The problem of estimating $h(t)$, the regression of Y on X evaluated at the t-quantile of X, can be approached by attempting to estimate $E(Y_{n,nt})$. In a model-free setting, the natural way to attempt this would be to take a weighted average of the Y_i, attaching predominantly large weights to the Y-values corresponding to the few X_i which are close to $X_{n,nt}$. Examples of such estimators are:

$$m_n(x) = n^{-1} \sum_1^n b(n)^{-1} K\left(\frac{r/n - F_n(x)}{b(n)}\right) Y_{nr}$$

of $m(x)$ and

$$m_n^*(t) = n^{-1} \sum_1^n b(n)^{-1} K\left(\frac{r/n - t}{b(n)}\right) Y_{nr}$$

of $h(t)$ constructed with kernel weights, where $F_n(x)$ is the empirical cdf, K is a pdf and $b(n) \downarrow 0$ as $n \to \infty$. Using asymptotic properties of linear functions of induced order statistics, Yang (1981) has shown the above estimators to be mean square consistent under fairly mild conditions.

Now consider inference about the integrated regression function

$$H(t) = \int_0^t h(s)\,\mathrm{d}s = \int_{-\infty}^{F^{-1}(t)} m(x)\,\mathrm{d}F(x), \quad 0 < t < 1 .$$

Then natural estimates of $H(t)$ are

$$H_n(t) = n^{-1} \sum_1^{nt} Y_{nr} \quad \text{or} \quad H_n^*(t) = n^{-1} \sum_{\{r \mid F(X_{nr}) \leq t\}} Y_{nr}$$

when F is known.

The functions H and h are related to one another in a manner analogous to the relation between the cdf and the pdf of a random variable. Many questions concerning the regression function can be equivalently posed in terms of the integrated regression function $H(t)$, while others are expressed directly in terms of $H(t)$. As an example in the latter category, suppose X is the income of a family and Y its consumption of a particular commodity. Then $H(t)/H(1)$ is the proportion of the total national consumption of the commodity consumed by the poorest $100t$ percent of the families. For most commodities the regression $m(x)$ is an increasing function of x, so that $H(t)/H(1)$ is a convex function connecting the points $(0, 0)$ and $(1, 1)$. The area enclosed below the 45° line and above the graph of $H(t)/H(1)$ is typically small for a necessity and large for a luxury. The function $H(t)$ is therefore an important indicator of the consumption pattern of a commodity, and can be useful in determining taxation policy.

Define $\{S_{nk}, 1 \leqslant k \leqslant n\}$ and $\{S_n(t), 0 \leqslant t \leqslant 1\}$ by

$$S_{nk} = \sum_{r=1}^{k} \{Y_{nr} - m(X_{nr})\}, \qquad S_n(t) = S_{n,nt}. \tag{9.1}$$

We shall now discuss convergence properties of the processes

$$U_n(t) = n^{-1/2} S_n(t), \qquad U_n^*(t) = n^{-1/2} \sum_{F(X_{nr}) \leqslant t} \{Y_{nr} - m(X_{nr})\}, \tag{9.2}$$

$H_n(t)$ and $H_n^*(t)$ and their applications in regression analysis.

A sequence of random functions $\{Z_n(t), 0 \leqslant t \leqslant 1\}$ is said to converge weakly to a random function $\{Z(t), 0 \leqslant t \leqslant 1\}$ if $E[g(Z_n)] \to E[g(Z)]$ as $n \to \infty$ for all bounded continuous functions g defined on the relevant space of functions. Here continuity of g, i.e., $g(z_n(\cdot)) \to g(z(\cdot))$ as $z_n(\cdot) \to z(\cdot)$ is to be understood in terms of uniform convergence of z_n to z in Theorems 9.1 and 9.2. In particular, the weak convergence $Z_n \Rightarrow Z$ implies that the asymptotic distributions of statistics such as

$$\sup_{0 \leqslant t \leqslant 1} |Z_n(t)| \quad \text{or} \quad \int_0^1 Z_n(t)\, \mathrm{d}t$$

are obtained by replacing $Z_n(t)$ by $Z(t)$ and then finding the distributions of the corresponding statistics.

We shall denote a standard Brownian motion by $B(t)$, $t \geqslant 0$, and a Brownian bridge on $[0, 1]$, i.e., a process having the same distribution as $B(t) - tB(1)$, by $B^*(t)$. We assume that the following conditions hold.

CONDITION 1. $\beta(x) = E(\{Y - m(x)\}^4 \mid X = x)$ is bounded.

CONDITION 2. $\sigma^2(x) = \mathrm{Var}(Y \mid X = x)$ is of bounded variation.

We now state Theorem 9.1 which gives the joint weak convergence of

$(U_n(t), V_n(t))$ and $(U_n^*(t), V_n(t))$, where $U_n(t)$ and $U_n^*(t)$ are as in (9.2) and $V_n(t)$ is the empirical process

$$V_n(t) = n^{1/2}\{G_n(t) - t\} \quad \text{where } G_n(t) = F_n(F^{-1}(t)) . \tag{9.3}$$

The proof of the theorem is based on a conditional application of a well-known theorem of Skorokhod (1965, page 163), by which sums of independent random variables can be represented on a Brownian path at random stopping times. For details see Bhattacharya (1974, 1976).

THEOREM 9.1. *Under Conditions 1 and 2, both* $\{(U_n(t), V_n(t))\}$ *and* $\{(U_n^*(t), V_n(t))\}$ *converge weakly to* $\{(B(\psi(t)), B^*(t))\}$ *where* $B(t)$ *is a standard Brownian motion,* $B^*(t)$ *is a Brownian bridge independent of* $B(t)$, *and*

$$\psi(t) = \int_{-\infty}^{F^{-1}(t)} \sigma^2(x)\, dF(x) .$$

From this, we now have the following.

THEOREM 9.2. *Suppose Conditions 1 and 2 hold. Then*

(a) $U_n(t) \Rightarrow B(\psi(t)), \quad 0 \leq t \leq 1$,

(b) $\sqrt{n}(H_n^*(t) - H(t)) \Rightarrow B(\psi(t)) + \int_0^t B^*(s)\, dh(s) - B^*(t)h(t), \quad 0 \leq t \leq 1$,

(c) *Moreover, if h is continuous, then, for all* $\varepsilon > 0$,

$$\sqrt{n}(H_n(t) - H(t)) \Rightarrow B(\psi(t)) + \int_0^t B^*(s)\, dh(s), \quad \varepsilon \leq t \leq 1 - \varepsilon ,$$

where $B(t)$, $B^*(t)$ *and* $\psi(t)$ *are as in Theorem* 9.1.

PROOF. Part (a) is immediate from Theorem 9.1. Now write $\sqrt{n}(H_n - H) = U_n - J_n$, where

$$J_n(t) = \sqrt{n}\left[\int_0^t h(s)\, ds - \int_0^{F(X_{n,nt})} h(s)\, dG_n(s)\right] = \int_0^t V_n(s)\, dh(s) + R_n(t)$$

where V_n is given by (9.3) and R_n converges to 0 in probability uniformly in $\varepsilon \leq t \leq 1 - \varepsilon$ by continuity of h. Hence R_n can be neglected for our purpose and (c) follows from Theorem 9.1. The proof of (b) is accomplished similarly by writing $\sqrt{n}(H_n^* - H) = U_n^* - J_n^*$, where

$$J_n^*(t) = \int_0^t V_n(s)\, dh(s) - V_n(t)h(t) .$$

We now apply the results of Theorem 9.2 to some problems in regression analysis in a model-free setting.

(i) *Testing H_0: $m = m_0$ where m_0 is a specified function.* Under the null hypothesis, $U_n(t) = n^{-1/2} S_{n,nt} \Rightarrow B(\psi(t))$, so that

$$S_{n,nt}/(n\psi(1))^{1/2} \Rightarrow B(\psi(t)/\psi(1)),$$

where S_{nk} is obtained by using $m = m_0$ in (9.1). We can use

$$\max_{1 \le k \le n} |S_{nk}|/(n\psi(1))^{1/2} \quad \text{or} \quad \left| \int_0^1 S_{n,nt}\, dt \right| \Big/ (n\psi(1))^{1/2}$$

as a measure of discrepancy between m_0 and the true regression. To construct test statistics from these, we use the uniformly consistent estimate

$$\hat{\psi}_n(t) = n^{-1} \sum_1^{nt} \{Y_{nr} - m_0(X_{nr})\}^2$$

of $\psi(t)$ under H_0. The test statistics thus obtained are

$$W_n^{(1)} = \max_{1 \le k \le n} |S_{nk}|/(n\hat{\psi}_n(1))^{1/2}$$

and

$$W_n^{(2)} = \int_0^1 S_{n,nt}\, dt \Big/ \left[n \int_0^1 \int_0^1 \hat{\psi}_n(\min(s,t))\, ds\, dt \right]^{1/2}$$
$$= \sum_{i=1}^{n-1} (n - R^*_{ni})\{Y_i - m_0(X_i)\} \Big/ \left[\sum_{i=1}^{n-1} (n^2 - R^{*2}_{ni})\{Y_i - m_0(X_i)\} \right]^{1/2},$$

where R^*_{ni} is the rank of X_i among $X_1, \ldots, X_n$. Under H_0, $W_n^{(2)}$ is asymptotically $N(0,1)$, while

$$\lim_{n\to\infty} P[W_n^{(1)} \le \lambda] = P[\sup_{0 \le t \le 1} |B(\psi(t)/\psi(1))| \le \lambda]$$
$$= P[\sup_{0 \le t \le 1} |B(t)| \le \lambda] = K(\lambda)$$
$$= \sum_{j=-\infty}^{n-1} (-1)^j \{\Phi((2j+1)\lambda) - \Phi((2j-1)\lambda)\}. \tag{9.4}$$

Hence the critical regions

$$W_n^{(1)} \ge K^{-1}(1-\alpha) \quad \text{and} \quad |W_n^{(2)}| \ge \Phi^{-1}(1-\alpha/2)$$

are asymptotically of level α.

(ii) *Testing for a constant regression.* Assume $\sigma^2(x)$ to be constant so that $\psi(t) = t\psi(1)$, and consider H_0: $m(x) = \mu_y$, which is equivalent to $H(t) - tH(1) \equiv$

0. Since $H_n(1) = \bar{Y}$ and $s^2 = n^{-1}\Sigma_1^n (Y_i - \bar{Y})^2$ is a consistent estimate of $\psi(1)$ under H_0, it follows from Theorem 9.2(c) that, under H_0,

$$\begin{aligned}\sqrt{n}(H_n(t) - t\bar{Y})/s &= \sqrt{n}[(H_n(t) - H(t)) - t(H_n(1) - H(1))]/s \\ &\Rightarrow (B(\psi(t)) - tB(\psi(1)))/(\psi(1))^{1/2} = (B(t\psi(1)) - tB(\psi(1)))/(\psi(1))^{1/2},\end{aligned}$$

which has the same distribution as that of the Brownian bridge $B^*(t)$. Hence the critical region

$$\sup_{0 \le t \le 1} \sqrt{n}|H_n(t) - t\bar{Y}|/s \ge D_\alpha^*,$$

where D_α^* is the upper 100α % point of the cdf of $\sup_{0\le t\le 1} |B^*(t)|$, is asymptotically of level α. Note that the Kolmogorov–Smirnov statistic has the same asymptotic null distribituon.

(iii) *Testing equality of two regressions.* Let (X_i, Y_i), $1 \le i \le n$, and (X'_i, Y'_i), $1 \le i \le n'$ be random samples from two bivariate populations with regression functions $m(x)$ and $m'(x)$ respectively. We want to test H_0: $m = m'$ under the assumption of common marginal cdf F of X and X' and common residual variance $\sigma^2(x)$. Suppose F is unknown, and define

$$H_n(t) = \sum_1^{nt} Y_{nr} \Big/ n, \qquad H'_{n'}(t) = \sum_1^{n't} Y'_{n'r} \Big/ n'$$

in terms of the induced order statistics Y_{nr} and $Y'_{n'r}$ from the two samples. Consider a test based on

$$T_n = (n^{-1} + n'^{-1})^{-1/2} \int_0^1 (H_n(t) - H'_{n'}(t))\, \mathrm{d}t,$$

which is asymptotically normal under H_0 with mean 0 and variance v of which a consistent estimate v_n can be easily obtained. Thus the statistic $T_n/\sqrt{v_n}$ is asymptotically $N(0, 1)$ under H_0, so that the critical region $|T_n/\sqrt{v_n}| \ge \Phi^{-1}(1 - \alpha/2)$ is asymptotically of level α. A similar test based on

$$T_n^* = (n^{-1} + n'^{-1})^{-1/2} \int_0^1 (H_n^*(t) - H_{n'}^{*\prime}(t))\, \mathrm{d}t$$

when F is known can also be constructed.

(iv) *Confidence band for H.* To obtain a confidence band for the function H around H_n on $[a, b] \subset (0, 1)$, the distribution of

$$\sup_{a \le t \le b} \left| B(\psi(t)) + \int_0^t B^*(s)\, \mathrm{d}h(s) \right|$$

is to be derived, which is unfortunately an intractable problem. However in Theorem 9.1 it is possible to show that conditionally, given $X_1, X_2, \ldots, U_n(t) \Rightarrow B(\psi(t))$ and the convergence is uniform in $X_1, X_2, \ldots$. Using this,

$$H_n(t) \pm nK^{-1}(1-\alpha)\left\{\sum_1^n (Y_j - m_n(X_j))^2\right\}^{1/2}, \quad 0 \leq t \leq 1,$$

can be seen to be a conditional confidence band for

$$\hat{H}_n(t) = n^{-1} \sum_1^{nt} m(X_{nr}), \quad 0 \leq t \leq 1,$$

with confidence coefficient $1-\alpha$, where K is the cdf of $\sup_{0 \leq t \leq 1} |B(t)|$ given in (9.4) and m_n is a regression estimate of the type considered by Nadaraya [14] and Watson [23]. Using the results of Theorem 9.2(c), it is also possible to construct a confidence interval for $H(t)$ around $H_n(t)$ for a fixed t.

Replacing $U_n(t)$ by a slightly different process and changing from uniform convergence to convergence in the Skorokhod J_1-topology (see Billingsley (1968)), Sen (1976) has established a sequential version of Theorem 9.2(a) solely under the uniform integrability condition

$$\lim_{s \to \infty} \sup_{x \in \mathbb{R}} E[(Y - m(x))^2 1\{|Y - m(x)| > s\} | X = x] = 0,$$

which is less restrictive than Conditions 1 and 2 under which the other results are proved. Sen uses Lemma 3.2 to establish the above result from which sequential analogues of the tests discussed in (i) and (ii) can be constructed.

We conclude with a brief discussion of statistics of the form

$$M_{nt} = \sum_{i=1}^{nt} [a_n(R^*_{ni}) - \bar{a}_n] Y_i, \quad 0 \leq t \leq 1,$$

R^*_{ni} being the rank of X_i among $X_1, \ldots, X_n$. Here the scores are defined in the usual manner as $a_n(i) = E\varphi(U_{ni})$ where U_{ni} is the i-th order statistic in a random sample of size n from the uniform distribution on (0, 1), $\bar{a}_n = n^{-1} \sum_1^n a_n(i)$ and φ is a square-integrable function on (0, 1). These statistics have been called *mixed rank statistics* by Ghosh and Sen (1971). The structural affinity of M_n with linear combinations of induced order statistics is identified by writing

$$M_n = \sum_{i=1}^{n} [a_n(i) - \bar{a}_n] Y_{ni}.$$

The following weak and strong convergence properties of these statistics have been obtained by Sen (1981) subject to regularity conditions on φ.

(i) *Weak convergence under independence.* If X_i and Y_i are independent, then

$$(\sqrt{n}A\sigma_y)^{-1}M_{nt} \Rightarrow B(t), \quad 0 \leq t \leq 1,$$

in the Skorokhod J_1-topology, where $A^2 = \int_0^1 \varphi^2(u)\,du - (\int_0^1 \varphi(u)\,du)^2$.

(ii) *Weak convergence under alternative distributions contiguous to independence.* If the joint distribution of (X_i, Y_i) follows a sequence of the form

$$F(x)G(y)[1 + \Omega_n(x, y)]$$

which is contiguous to the product distribution given by $F(x)G(y)$, then there exists a sequence of non-stochastic functions $\{\omega_n(t)\}$ in $D[0, 1]$ such that

$$(\sqrt{n}A\sigma_y)^{-1}M_{nt} - \omega_n(t) \Rightarrow B(t), \quad 0 \leq t \leq 1.$$

(iii) *Strong convergence under independence.* If X_i and Y_i are independent, $E(|Y_i|^{2+\delta}) < \infty$ for some $\delta > 0$ and if the underlying probability space is rich enough, then there exists a Brownian motion $\{B(t), t \geq 0\}$ on the same space such that, for some $\eta > 0$,

$$(A\sigma_y)^{-1}M_n = B(n) + O(n^{1/2-\eta}) \quad \text{a.s.}$$

as $n \to \infty$. A law of iterated logarithm for M_n, viz., that the lim sup and lim inf of $M_n/\sqrt{2nA^2\sigma_y^2 \log\log n}$ are $+1$ and -1 respectively with probability 1, follows from this.

Sen [19] has also proved weak and strong convergence results for M_n under general alternatives and for arbitrary scores $a_n^*(i)$ which are sufficiently close to $a_n(i)$ defined above and has indicated the use of these results in deriving asymptotic properties of sequential tests of the hypothesis of independence.

References

Bhattacharya, P. K. (1974). Convergence of sample paths of normalized sums of induced order statistics. *Ann. Statist.* **2**, 1034–1039.

Bhattacharya, P. K. (1976). An invariance principle in regression analysis. *Ann. Statist.* **4**, 621–624.

Billingsley, P. (1968). *Convergence of probability measures.* Wiley, New York.

Cochran, W. G. (1963). *Sampling techniques.* Wiley, New York.

David, H. A. (1973). Concomitants of order statistics. *Bull. International Stat. Inst.* **45**, 295–300.

David, H. A. (1981). *Order statistics.* Wiley, New York.

David, H. A. and Galambos, J. (1974). The asymptotic theory of concomitants of order statistics. *J. Appl. Probab.* **11**, 762–770.

David, H. A., O'Connell, M. J. and Yang, S. S. (1977). Distribution and expected value of the rank of a concomitant of an order statistic. *Ann. Statist.* **5**, 216–223.

Galambos, J. (1978). *The asymptotic theory of extreme order statistics.* Wiley, New York.

Ghosh, M. and Sen, P. K. (1971). On a class of rank order tests for regression with partially informed stochastic predictors. *Ann. Math. Statist.* **42**, 650–661.

Gross, A. L. (1973). Prediction in future samples studied in terms of gain from selection. *Psychometrika* **38**, 151–172.

Harrell, F. E. and Sen, P. K. (1979). Statistical inference for censored bivariate normal distributions based on induced order statistics. Biometrika **66**, 293–298.

Kaminsky, K. S. (1981). A note on concomitants of order statistic in two bivariate normal samples. Proc. 43rd Session of ISI, Buenos Aires, Vol. 1, Contributed Papers, pp. 161–164.

Nadaraya, E. A. (1964). On estimating regression. *Theor. Probab. Appl.* **9**, 141–142.

O'Connell, M. J. and David, H. A. (1976). Order statistics and their concomitants in some double sampling situations. In: S. Ikeda et al., eds., *Essays in Probability and Statistics, Ogawa Volume.* Shinko Tsusho, Japan.

Saw, J. G. (1959). Estimation of the normal population parameters given a singly censored sample. *Biometrika* **46**, 150–155.

Saw, J. G. (1961). The bias of the maximum likelihood estimates of the location and scale parameters given a type II censored normal sample. *Biometrika* **48**, 448–451.

Sen, P. K. (1976). A note on invariance principles for induced order statistics. *Ann. Prob.* **4**, 474–479.

Sen, P. K. (1981). Some invariance principles for mixed rank statistics and induced order statistics and some applications. *Comm. Statist.* **A10**, 1691–1718.

Skorokhod, A. V. (1965). *Studies in the theory of random processes.* Addison-Wesley, Reading, MA.

Spruill, N. L. and Gastwirth, J. L. (1981). A probability model illustrating the employment problems a professional coupld faces. Unpublished manuscript.

Watterson, G. A. (1959). Linear estimation in censored samples from multivariate normal populations. *Ann. Math. Statist.* **30**, 814–824.

Watson, G. S. (1964). Smooth regression analysis. *Sankhyā Ser. A.* **26**, 359–372.

Yang, S. S. (1977). General distribution theory of the concomitants of order statistics. *Ann. Statist.* **5**, 996–1002.

Yang, S. S. (1981). Linear functions of concomitants of order statistics with application to nonparametric estimation of a regression function. *JASA* **76**, 658–662.

P. R. Krishnaiah and P. K. Sen, eds., *Handbook of Statistics*, Vol. 4

19

Empirical Distribution Function

Endre Csáki

1. Introduction

If X is a real valued random variable, then its distribution function is defined by

$$F(x) = P(X \leqslant x) \tag{1.1}$$

for $x \in \mathbb{R}$. A statistical estimation of $F(x)$ based on a random sample $(X_1, \ldots, X_n)$ is the so called empirical or sample distribution function defined by

$$F_n(x) = \frac{1}{n} \sum_{i=1}^{n} I(X_i \leqslant x), \tag{1.2}$$

where $I(\cdot)$ denotes the indicator function of the event in the brackets. With other words, for all real x, $F_n(x)$ is the relative frequency of the event whose probability is given by the distribution function $F(x)$. Hence, for fixed x, we have

$$P\left(F_n(x) = \frac{k}{n}\right) = \binom{n}{k}(F(x))^k(1 - F(x))^{n-k}, \quad k = 0, \ldots, n, \tag{1.3}$$

$$\lim_{n\to\infty} F_n(x) = F(x) \quad \text{a.s.} \tag{1.4}$$

etc. $F_n(x)$ is an unbiased and consistent estimator of $F(x)$.

$F_n(x)$ is considered also a (random) function of x. Kolmogorov (1933) in his fundamental paper introduced the statistic

$$D_n = \sup_{x\in\mathbf{R}} |F_n(x) - F(x)| \tag{1.5}$$

and determined its limiting distribution for continuous $F(x)$. Later Smirnov (1939a,b) considered the one-sided statistics

$$D_n^+ = \sup_{x \in \mathbf{R}} (F_n(x) - F(x)) \tag{1.6}$$

and

$$D_n^- = \sup_{x \in \mathbf{R}} (F(x) - F_n(x)) \tag{1.7}$$

and also their two sample analogous. These are the so called Kolmogorov–Smirnov statistics. It is well known that they are distribution free as long as $F(x)$ is assumed to be continuous.

Empirical measure can be defined in a much more general context. In this paper however we consider the case when $X_1, X_2, \ldots, X_n, \ldots$ is a sequence of i.i.d. real valued random variables. We study the more general statistics introduced by Anderson and Darling (1952):

$$K_{n,\tau} = \sup_{x \in \mathbf{R}} (|F_n(x) - F(x)|\tau(F(x))), \tag{1.8}$$

$$K_{n,\tau}^+ = \sup_{x \in \mathbf{R}} ((F_n(x) - F(x))\tau(F(x))), \tag{1.9}$$

$$K_{n,\tau}^- = \sup_{x \in \mathbf{R}} ((F(x) - F_n(x))\tau(F(x))), \tag{1.10}$$

where $\tau(u) \geq 0$ $(0 \leq u \leq 1)$ is a weight function. $\tau(u) = 1$ $(0 \leq u \leq 1)$ corresponds to D_n, D_n^+ and D_n^-, but it sometimes is not satisfactory for the statistician, because this does not give enough weight on the tails, i.e. when $F(x)$ is near 0 or 1. There are situations (e.g. in reliability, biometry, etc.) when small or large values of $F(x)$ are more important then the middle values of $F(x)$. On the other hand, it may occur that the sample is censored or truncated so that $F_n(x)$ is not available on the whole real line. Particular weight functions were studied e.g. by Maniya (1949), Rényi (1953), etc.

In this paper we assume that the weight function $\tau(u) \geq 0$ satisfies the following condition:

There exist δ_1, δ_2 $(0 \leq \delta_1 \leq \delta_2 \leq 1)$ such that $\tau(u)$ is nonincreasing in $(0, \delta_1]$ while $\tau(u)$ is nondecreasing in $[\delta_2, 1)$ and $\tau(u)$ is bounded in (δ_1, δ_2).

The class of $\tau(u)$ satisfying this condition will be denoted by $\mathcal{T}$.

For the empirical distribution function a fundamental result is the Glivenko–Cantelli theorem (Glivenko, 1933; Cantelli, 1933) stating that

$$\lim_{n \to \infty} D_n = 0 \quad \text{a.s.} \tag{1.11}$$

This theorem has been extended in various direction (cf. the survey paper by Gaenssler and Stute, 1979).

For the general statistic $K_{n,\tau}$ defined by (1.8), Wellner (1977b) has shown that

$$\lim_{n\to\infty} K_{n,\tau} = 0 \quad \text{a.s.} \tag{1.12}$$

provided

$$\int_0^1 \tau(u)\,\mathrm{d}u < \infty\,. \tag{1.13}$$

On the other hand, $\int_0^1 \tau(u)\,\mathrm{d}u = \infty$ implies

$$\limsup_{n\to\infty} K_{n,\tau} = \infty \quad \text{a.s.} \tag{1.14}$$

The condition (1.13) fails for $\tau(u) = 1/u$, or $\tau(u) = 1/(1-u)$ $(0 < u < 1)$. For this particular weight functions we have (1.14). It can be seen furthermore that for continuous $F(x)$, we have also

$$\sup_{0<F(x)\leq 1} \frac{F(x) - F_n(x)}{F(x)} = 1 \quad \text{a.s.} \tag{1.15}$$

and by a theorem of Daniels (1945)

$$P\left(\sup_{0<F(x)\leq 1} \frac{F_n(x) - F(x)}{F(x)} \leq z\right) = \frac{z}{1+z}, \quad 0 \leq z \leq 1\,. \tag{1.16}$$

To eliminate the effect of the heavy tail caused by $\tau(u)$ it is reasonable to consider also weight functions of the form

$$\tau_n(u) = \begin{cases} \tau(u) & \text{if } a_n \leq u \leq b_n\,, \\ 0 & \text{otherwise}\,, \end{cases} \tag{1.17}$$

where $0 \leq a_n < b_n \leq 1$ and $\lim_{n\to\infty} a_n = 0$, $\lim_{n\to\infty} b_n = 1$; $\tau(\cdot) \in \mathcal{T}$.

For this type of weight function Chang (1955) has shown:

If $F(x)$ is continuous and $na_n \to \infty$ as $n \to \infty$, then

$$\sup_{a_n\leq F(x)\leq 1} \frac{|F_n(x) - F(x)|}{F(x)} \xrightarrow{\text{P}} 0 \tag{1.18}$$

where $\xrightarrow{\text{P}}$ denotes convergence in probability.

The almost sure version of the above theorem is due to Wellner (1978):

If $F(x)$ is continuous and $na_n/\log\log n \to \infty$ as $n \to \infty$, then

$$\lim_{n\to\infty} \sup_{a_n\leq F(x)\leq 1} \frac{|F_n(x) - F(x)|}{F(x)} = 0 \quad \text{a.s.} \tag{1.19}$$

In the following sections further properties of the empirical distribution function will be investigated. In Section 2 we study exact distributions of statistics based on empirical distribution function. In Section 3 limit theorems will be given, including limit distributions, strong limit theorems, etc. Here we study also some martingale and Markov properties and certain inequalities. Section 4 is devoted to the main properties of tests based on empirical distribution function.

The literature on empirical distribution function is very huge. In this paper we cannot give a complete outline on this subject. The interested reader is referred to the following survey articles: Barton and Mallows (1965), Sahler (1968), Pyke (1972), Durbin (1973), Gupta and Panchapakesan (1973), Gaenssler and Stute (1979), where further references can be found. Further important results are given also in the following books: Hájek and Šidák (1967), Takács (1967), Puri and Sen (1971), Mohanty (1979), Pitman (1979), Csörgő and Révész (1981), Korolyuk and Borovskih (1981).

2. Exact distributions

In order to apply statistical methods based on empirical distribution, such as goodness of fit or two-sample tests, confidence intervals, etc. one needs the exact or limiting distributions of statistics concerned. In this section we give explicit formulas for some exact distributions. Here we assume that $F(x)=x$ $(0 \leq x \leq 1)$, i.e. the theoretical distribution is the uniform distribution over $(0, 1)$. The general case can be reduced to this one, even when $F(x)$ has discontinuities.

The first result is due to Smirnov (1944) and Birnbaum and Tingey (1951) who determined the distribution of D_n^+ and D_n^-. Korolyuk (1955), gave the distribution of D_n. Later several methods and formulas have been developed to determine the exact distribution of the general statistics $K_{n,T}$ (or $K_{n,T}^+$ and $K_{n,T}^-$). (See e.g. Whittle (1961), Suzuki (1967), Durbin (1968), Epanechnikov (1968), Noé and Vandewiele (1968), Steck (1971), Niederhausen (1981)). The problem can be reduced to determining the probability of the form $P(G_1(x) \leq F_n(x) \leq G_2(x),\ 0 \leq x \leq 1)$. An explicit expression is given by Steck (1971):

$$P(G_1(x) \leq F_n(x) \leq G_2(x),\ 0 \leq x \leq 1) = n!\,\det\left[\frac{(v_i - u_j)_+^{j-i+1}}{(j-i+1)!}\right], \tag{2.1}$$

where G_1 is assumed to be nondecreasing and left-continuous, G_2 is nondecreasing and right-continuous. Moreover

$$v_i = G_1^{-1}\left(\frac{i-1}{n}\right), \qquad u_j = G_2^{-1}\left(\frac{j}{n}\right) \tag{2.2}$$

and

$$(v-u)_+ = \begin{cases} v-u & \text{if } v \geq u, \\ 0 & \text{otherwise}. \end{cases} \tag{2.3}$$

Pitman (1972) and Sarkadi (1973) gave further proofs of (2.1).

For computational reason however the following recurrence relation due to Epanechnikov (1968) is more convenient:

$$P_n = \sum_{i=1}^{n} (-1)^{i-1} \frac{(v_{n-i+1} - u_n)_+^i}{i!} P_{n-i}, \tag{2.4}$$

where

$$P_n = P(G_1(x) \leqslant F_n(x) \leqslant G_2(x), 0 \leqslant x \leqslant 1). \tag{2.5}$$

In the two sample case, i.e. when we have two independent samples $(X_1, \ldots, X_{n_1})$ and $(Y_1, \ldots, Y_{n_2})$ with distribution functions $F^{(1)}(x)$ and $F^{(2)}(x)$, resp., and empirical distribution functions $F_{n_1}^{(1)}(x)$ and $F_{n_2}^{(2)}(x)$, resp., we may consider the following general statistics:

$$K_{n_1,n_2,\tau} = \sup_{x \in \mathbf{R}} |F_{n_1}^{(1)}(x) - F_{n_2}^{(2)}(x)| \tau(F_N(x)), \tag{2.6}$$

$$K_{n_1,n_2,\tau}^{+} = \sup_{x \in \mathbf{R}} (F_{n_1}^{(1)}(x) - F_{n_2}^{(2)}(x)) \tau(F_N(x)), \tag{2.7}$$

$$K_{n_1,n_2,\tau}^{-} = \sup_{x \in \mathbf{R}} (F_{n_2}^{(2)}(x) - F_{n_1}^{(1)}(x)) \tau(F_N(x)), \tag{2.8}$$

where $N = n_1 + n_2$ and $F_N(x)$ denotes the empirical distribution function of the combined sample. The problem of finding the distributions of the above statistics can be reduced to the determination of the probability $P(a_i \leqslant R_i \leqslant b_i, i = 1, \ldots, n_1)$, where $R_1 \leqslant R_2 \leqslant \cdots \leqslant R_{n_1}$ are the ordered ranks of $(X_1, \ldots, X_{n_1})$ in the combined sample. Steck (1969) gives the following formula for $F^{(1)}(x) \equiv F^{(2)}(x)$:

$$P(a_i \leqslant R_i \leqslant b_i, i = 1, \ldots, n_1) = \frac{1}{\binom{n_1 + n_2}{n_1}} \det\left[\binom{b_i - a_j + 1}{j - i + 1}_+ \right], \tag{2.9}$$

where

$$\binom{A}{B}_+ = \begin{cases} \binom{A}{B} & \text{when } A \geqslant B > 0, \\ 0 & \text{when } B < 0 \text{ or } A < B, \\ 1 & \text{when } B = 0. \end{cases}$$

(2.9) can also be obtained from a lattice path counting result due to Kreweras (1965).

The recurrence relation for $P_{n_1}^{(1)} = P(a_i \leqslant R_i \leqslant b_i, i = 1, \ldots, n_1)$ reads as follows (Mohanty, 1982):

$$P_{n_1}^{(1)} = \sum_{i=1}^{n_1} (-1)^{i-1} \binom{b_{n_1-i+1} - a_{n_1} + 1}{i} + P_{n_1-i}^{(1)} \tag{2.10}$$

For computational aspects of the above probabilities we refer to Mohanty (1982).

In particular cases however more powerful methods and simpler formulas are available. In the two sample case when $n_1 = n_2 = n$, Gnedenko and Korolyuk (1951) have developed a method, based on random walk, to determine the distribution of

$$D_{n,n} = \sup_{x\in\mathbf{R}} |F_n^{(1)}(x) - F_n^{(2)}(x)| \tag{2.11}$$

and

$$D_{n,n}^{+} = \sup_{x\in\mathbf{R}} (F_n^{(1)}(x) - F_n^{(2)}(x)) . \tag{2.12}$$

Their result states

$$P\Big(D(n,n) < \frac{c}{n}\Big) = \frac{1}{\binom{2n}{n}} \sum_{k=-\infty}^{\infty} (-1)^k \binom{2n}{n+ck}, \quad c = 1, 2, \ldots, n, \tag{2.13}$$

$$P\Big(D^{+}(n,n) < \frac{c}{n}\Big) = 1 - \frac{\binom{2n}{n+c}}{\binom{2n}{n}}, \quad c = 1, 2, \ldots, n . \tag{2.14}$$

By using the random walk approach, distributions and joint distributions have been derived for random quantities, such as runs, number of intersections, number of zeros, Galton statistic, maximum, place of the maximum, etc. See Takács (1967), Mohanty (1979), Korolyuk and Borovskih (1981) and the references in these books. Vincze (1970) studied the case of discontinuous distribution. Dwass (1967) developed an alternative method to derive distributions and joint distributions of the above quantities (cf. also Mohanty (1979)).

In the one sample case Smirnov (1944) and Birnbaum and Tingey (1951) have derived the following distribution:

$$P(D_n^{+} \leqslant z) = P(D_n^{-} \leqslant z) = 1 - z \sum_{j=0}^{[n(1-z)]} \binom{n}{j} \Big(\frac{j}{n} + z\Big)^{j-1} \Big(1 - \frac{j}{n} - z\Big)^{n-j}, \quad 0 \leqslant z \leqslant 1 . \tag{2.15}$$

Later a number of authors have obtained the distributions of one sided statistics $K_{n,\tau}^{+}$ and $K_{n,\tau}^{-}$ for particular weight functions (see the references in Csáki, 1977a, where we have also tried to sort out and correct the ones which appeared to be wrong). Takács (1964, 1967) has developed a method based on an extension of the ballot theorem to find the distributions of $K_{n,\tau}^{+}$ and $K_{n,\tau}^{-}$ for particular weight functions, such as (i) $\tau(u) = 1$ $(a_1 \leqslant u \leqslant a_2)$, (ii) $\tau(u) = 1/u$

$(a_1 \leqslant u \leqslant a_2)$, (iii) $\tau(u) = 1/(1-u)$ $(a_1 \leqslant u \leqslant a_2)$, etc. The weight function given by (i) has been introduced by Maniya (1949), While Rényi (1953) introduced the weight functions (ii) and (iii). The distributions $P(K^+_{n,\tau} \leqslant z)$ and $P(K^-_{n,\tau} \leqslant z)$ for the above weight functions are determined by the following formula:

$$P(F_n(x) \geqslant cx + b,\ a_1 \leqslant x \leqslant a_2)$$

$$= \begin{cases} \displaystyle\sum_{\alpha > n(ca_1+b)} \binom{n}{\alpha} a_1^\alpha (1-a_1)^{n-\alpha} - \sum_{\alpha > n(ca_1+b)} (x_0 - a_1) \\ \displaystyle\times \sum_{j=0}^{[n(ca_2+b)]-\alpha} \frac{n!}{\alpha!\, j!\,(n-\alpha-j)!} a_1^\alpha (x_j - a_1)^{j-1} (1-x_j)^{n-\alpha-j} & \\ \qquad\qquad\qquad\qquad\qquad\qquad\qquad\qquad \text{if } ca_1 + b > 0\,, \\ \displaystyle 1 + \frac{b}{c} \sum_{j=0}^{[n(ca_2+b)]} \binom{n}{j} \left(\frac{j}{nc} - \frac{b}{c}\right)^{j-1} \left(1 - \frac{j}{nc} + \frac{b}{c}\right)^{n-j} & \text{if } ca_1 + b \leqslant 0\,, \end{cases} \tag{2.16}$$

where

$$x_j = \frac{\alpha + j}{nc} - \frac{b}{c}$$

and $0 \leqslant a_1 < a_2 \leqslant 1$, $c > 0$, $ca_2 + b \leqslant 1$. Then

$$\begin{aligned} P\Big(\sup_{a_1 \leqslant x \leqslant a_2} (x - F_n(x)) \leqslant z\Big) &= P\Big(\sup_{1-a_2 \leqslant x \leqslant 1-a_1} (F_n(x) - x) \leqslant z\Big) \\ &= P(F_n(x) \geqslant x - z,\ a_1 \leqslant x \leqslant a_2)\,, \end{aligned} \tag{2.17}$$

$$\begin{aligned} P\left(\sup_{a_1 \leqslant x \leqslant a_2} \left(\frac{x - F_n(x)}{1-x}\right) \leqslant z\right) &= P\left(\sup_{1-a_2 \leqslant x \leqslant 1-a_1} \left(\frac{F_n(x) - x}{x}\right) \leqslant z\right) \\ &= P(F_n(x) \geqslant (1+z)x - z,\ a_1 \leqslant x \leqslant a_2)\,, \end{aligned} \tag{2.18}$$

$$\begin{aligned} P\left(\sup_{a_1 \leqslant x \leqslant a_2} \left(\frac{x - F_n(x)}{x}\right) \leqslant z\right) &= P\left(\sup_{1-a_2 \leqslant x \leqslant 1-a_1} \left(\frac{F_n(x) - x}{1-x}\right) \leqslant z\right) \\ &= P(F_n(x) \geqslant (1-z)x,\ a_1 \leqslant x \leqslant a_2)\,. \end{aligned} \tag{2.19}$$

For $a_1 = 0$, $a_2 = 1$, $c = 1 + z$, $b = -z$, by using Abel's identity, (2.16) reduces to (1.16), i.e. to Daniel's result, which plays a basic role in this investigation.

In Csáki (1977a) several other distributions are also given including the case when $F(x)$ is discontinuous.

For distributions, joint distributions of other quantities as the point of the maximum, number of intersections, etc. the reader is referred to Birnbaum and Pyke (1958), Dempster (1959), Dwass (1959, 1974), Nef (1964), Takács (1965b, 1967, 1970, 1971), Csáki and Tusnády (1972), Csáki (1977a), Wellner (1977d).

3. Limit theorems

3.1. Limiting distributions

Kolmogorov (1933) in his fundamental paper has proved

$$\lim_{n\to\infty} P(n^{1/2}D_n \leqslant y) = \sum_{k=-\infty}^{\infty} (-1)^k e^{-2k^2y^2}, \quad 0 < y. \tag{3.1}$$

Smirnov (1939a,b) has determined the one-sided and the two sample analogues of (3.1):

$$\lim_{n\to\infty} P(n^{1/2}D_n^+ \leqslant y) = \lim_{n\to\infty} P(n^{1/2}D_n^- \leqslant y) = 1 - e^{-2y^2}, \quad 0 \leqslant y \tag{3.2}$$

$$\lim_{\min(n_1,n_2)\to\infty} P\left(\left(\frac{n_1n_2}{n_1+n_2}\right)^{1/2} D_{n_1,n_2} \leqslant y\right) = \sum_{k=-\infty}^{\infty} (-1)^k e^{-2k^2y^2}, \quad 0 < y, \tag{3.3}$$

$$\lim_{\min(n_1,n_2)\to\infty} P\left(\left(\frac{n_1n_2}{n_1+n_2}\right)^{1/2} D_{n_1,n_2}^+ \leqslant y\right) = 1 - e^{-2y^2}, \quad 0 \leqslant y, \tag{3.4}$$

where D_{n_1,n_2}, D_{n_1,n_2}^+ and D_{n_1,n_2}^- are $K_{n_1,n_2,\tau}$, $K_{n_1,n_2,\tau}^+$ and $K_{n_1,n_2,\tau}^-$ resp. corresponding to the weight function $\tau(u) = 1$, $0 \leqslant u \leqslant 1$.

In the sequel we assume again that the basic distribution of the sample elements X_i is uniform on (0, 1).

In the one sided case the asymptotic formula corresponding to (2.16) reads as follows:

$$\lim_{n\to\infty} P\left(F_n(x) > \left(1 + \frac{u}{n^{1/2}}\right)x - \frac{v}{n^{1/2}},\ a_1 \leqslant x \leqslant a_2\right)$$
$$= \Phi(A_1, B_1; -r) - e^{-2v(v-u)}\Phi(A_2, B_2; r), \tag{3.5}$$

where $\Phi(A, B; r)$ denotes the bivariate normal distribution function with means 0, variances 1 and correlation coefficient r, i.e.

$$\Phi(A, B; r) = \frac{1}{2\pi(1-r^2)^{1/2}} \int_{-\infty}^{A} \int_{-\infty}^{B} \exp\left\{-\frac{1}{2(1-r^2)}(z_1^2 - 2rz_1z_2 + z_2^2)\right\} dz_1\, dz_2 .$$

Furthermore

$$r = \left(\frac{a_1(1-a_2)}{a_2(1-a_1)}\right)^{1/2},$$

$$A_1 = \frac{v - ua_1}{(a_1(1-a_1))^{1/2}}, \qquad B_1 = \frac{v - ua_2}{(a_2(1-a_2))^{1/2}},$$

$$A_2 = \frac{v(1-2a_1) + ua_1}{(a_1(1-a_1))^{1/2}}, \qquad B_2 = \frac{v(2a_2-1) - ua_2}{(a_2(1-a_2))^{1/2}}.$$

(3.5) can be obtained from (2.16) by using normal approximation of binomial probabilities. From (3.5) we can get the following limiting distributions:

$$\begin{aligned}
&\lim_{n\to\infty} P(n^{1/2} \sup_{a_1\le x\le a_2} (x - F_n(x)) \le y) \\
&\quad = \lim_{n\to\infty} P(n^{1/2} \sup_{1-a_2\le x\le 1-a_1} (F_n(x) - x)) \le y) \\
&\quad = \lim_{n\to\infty} P\left(F_n(x) \ge x - \frac{y}{n^{1/2}},\ a_1 \le x \le a_2\right). \qquad (3.6)
\end{aligned}$$

This gives also the limiting distribution of the two-sample statistic

$$\left(\frac{n_1 n_2}{n_1+n_2}\right)^{1/2} \sup_{a_1\le F_N(x)\le a_2} (F^{(1)}_{n_1}(x) - F^{(2)}_{n_2}(x)).$$

Furthermore

$$\begin{aligned}
&\lim_{n\to\infty} P\left(n^{1/2} \sup_{a_1\le x\le a_2} \left(\frac{x - F_n(x)}{1-x}\right) \le y\right) \\
&\quad = \lim_{n\to\infty} P\left(n^{1/2} \sup_{1-a_2\le x\le 1-a_1} \left(\frac{F_n(x) - x}{x}\right) \le y\right) \\
&\quad = \lim_{n\to\infty} P\left(F_n(x) \ge \left(1 + \frac{y}{n^{1/2}}\right)x - \frac{y}{n^{1/2}},\ a_1 \le x \le a_2\right) \\
&\quad = \lim_{\min(n_1,n_2)\to\infty} P\left(\left(\frac{n_1 n_2}{n_1+n_2}\right)^{1/2} \sup_{a_1<F^{(1)}_{n_1}(x)<a_2} \left(\frac{F^{(1)}_{n_1}(x) - F^{(2)}_{n_2}(x)}{1 - F^{(1)}_{n_1}(x)}\right) \le y\right). \qquad (3.7)
\end{aligned}$$

$$\begin{aligned}
&\lim_{n\to\infty} P\left(n^{1/2} \sup_{a_1\le x\le a_2} \left(\frac{x - F_n(x)}{x}\right) \le y\right) \\
&\quad = \lim_{n\to\infty} P\left(n^{1/2} \sup_{1-a_2\le x\le 1-a_1} \left(\frac{F_n(x) - x}{1-x}\right) \le y\right) \\
&\quad = \lim_{n\to\infty} P\left(F_n(x) \ge \left(1 - \frac{y}{n^{1/2}}\right)x,\ a_1 \le x \le a_2\right) \\
&\quad = \lim_{\min(n_1,n_2)\to\infty} P\left(\left(\frac{n_1 n_2}{n_1+n_2}\right)^{1/2} \sup_{a_1<F^{(1)}_{n_1}(x)<a_2} \left(\frac{F^{(1)}_{n_1}(x) - F^{(2)}_{n_2}(x)}{F^{(1)}_{n_1}(x)}\right) \le y\right). \qquad (3.8)
\end{aligned}$$

References: Maniya (1949), Rényi (1953), Csörgő (1965a,c). The case of discontinuous distribution is studied in Schmid (1958), Vincze (1970). Further limiting distributions are given in Csáki (1977a).

An effective method to obtain limiting distributions is based on invariance principle (see Csörgő, 1984) stating weak and strong convergence of empirical process to the Brownian bridge. From O'Reilly (1974) it follows that the limiting distributions of the statistics $n^{1/2}K_{n,\tau}$ and $n^{1/2}K^+_{n,\tau}$, are the distributions of $\sup_{0\le x\le 1}(|B(x)|\tau(x))$ and $\sup_{0\le x\le 1}(B(x)\tau(x))$, resp., if the conditions

$$\int_0^1 t^{-1}\exp(-\varepsilon h_i^{-2}(t))\,\mathrm{d}t<\infty, \quad i=1,2, \tag{3.9}$$

hold for all $\varepsilon>0$, where $h_1(t)=t^{1/2}\tau(t)$, $h_2(t)=t^{1/2}\tau(1-t)$ and $B(x)$ is a Brownian bridge (tied down Wiener process). Anderson and Darling (1953) studied the limiting distribution of $K_{n,\tau}$ but no explicit formula is known in the general case. In particular cases, e.g. for Maniya- and Rényi-type weight functions the following limiting distributions are given. Maniya (1949) shows

$$\begin{aligned}
&\lim_{n\to\infty} P(n^{1/2}\sup_{a_1\le x\le a_2}|F_n(x)-x|\le y)\\
&\quad=\lim_{\min(n_1,n_2)\to\infty} P\left(\left(\frac{n_1n_2}{n_1+n_2}\right)^{1/2}\sup_{a_1\le F_N(x)\le a_2}|F^{(1)}_{n_1}(x)-F^{(2)}_{n_2}(x)|\le y\right)\\
&\quad=\frac{1}{2\pi(1-r^2)^{1/2}}\sum_{k=-\infty}^{\infty}(-1)^k\,\mathrm{e}^{-2k^2y^2}\\
&\qquad\times\int_{-A_k}^{A_k}\int_{-B_k}^{B_k}\exp\left\{-\frac{1}{2(1-r^2)}(z_1^2-2rz_1z_2+z_2^2)\right\}\mathrm{d}z_1\,\mathrm{d}z_2,
\end{aligned}\tag{3.10}$$

where

$$r=\left(\frac{a_1(1-a_2)}{a_2(1-a_1)}\right)^{1/2},$$

$$A_k=\frac{y-2kya_1}{\sqrt{a_1(1-a_1)}},\qquad B_k=\frac{y-2ky(1-a_2)}{\sqrt{a_2(1-a_2)}}.$$

Rényi (1953) gives

$$\begin{aligned}
&\lim_{n\to\infty} P\left(n^{1/2}\sup_{a_1\le x\le a_2}\left|\frac{F_n(x)-x}{x}\right|\le y\right)\\
&\quad=\lim_{\min(n_1,n_2)\to\infty} P\left(\left(\frac{n_1n_2}{n_1+n_2}\right)^{1/2}\sup_{a_1\le F^{(1)}_{n_1}(x)\le a_2}\frac{|F^{(1)}_{n_1}(x)-F^{(2)}_{n_2}(x)|}{F^{(1)}_{n_1}(x)}\le y\right)\\
&\quad=\frac{4}{\pi}\sum_{k=0}^{\infty}\frac{(-1)^k}{2k+1}\exp\left\{-\frac{(2k+1)^2\pi^2(1-a_1)}{8y^2a_1}\right\}E_k,\qquad 0\le y,
\end{aligned}\tag{3.11}$$

where

$$\begin{aligned}
E_k&=1-\frac{2}{(2\pi)^{1/2}}\int_{y(a_2/1-a_2)^{1/2}}^{\infty}\exp\left\{-\frac{u^2}{2}\right\}du\\
&\quad+\frac{2\exp\{-y^2a_2/2(1-a_2)\}}{y(2\pi)^{1/2}(a_1/(1-a_1))^{1/2}}\int_0^{(2k+1)(\pi/2)}\exp\left\{\frac{u^2(1-a_2)}{2a_2y^2}\right\}\sin u\,du.
\end{aligned}$$

The joint limiting distribution of $n^{1/2}D_n^+$ and $n^{1/2}D_n^-$ is given in Smirnov (1939b):

$$\lim_{n\to\infty} P(n^{1/2}D_n^+ \le y_1,\, n^{1/2}D_n^- \le y_2)$$

$$= \lim_{\min(n_1,n_2)\to\infty} P\left(\left(\frac{n_1 n_2}{n_1+n_2}\right)^{1/2} D_{n_1,n_2}^+ \le y_1,\, \left(\frac{n_1 n_2}{n_1+n_2}\right)^{1/2} D_{n_1,n_2}^- \le y_2\right)$$

$$= \sum_{k=-\infty}^{\infty} \left(\mathrm{e}^{-2k^2(y_1+y_2)^2} - \mathrm{e}^{-2((k+1)y_1+ky_2)^2}\right), \quad 0 \le y_1, 0 \le y_2\,. \tag{3.12}$$

Gnedenko (1954) for two sample case and Kuiper (1960) for one sample case determined the following limiting distribution:

$$\lim_{n\to\infty} P(n^{1/2}(D_n^+ + D_n^-) \le y) = \lim_{\min(n_1,n_2)\to\infty} P\left(\left(\frac{n_1 n_2}{n_1+n_2}\right)^{1/2} (D_{n_1,n_2}^+ + D_{n_1,n_2}^-) \le y\right)$$

$$= 1 + 2\sum_{k=1}^{\infty} (1-4k^2y^2)\,\mathrm{e}^{-2k^2y^2}, \quad y \ge 0\,. \tag{3.13}$$

The weight function $\tau(u) = (u(1-u))^{-1/2}$ $(0<u<1)$ does not satisfy (3.9). The limiting distributions in this case have been found by Jaeschke (1979):

$$\lim_{n\to\infty} P\left(A_n \sup_{0<x<1} \left(n^{1/2}\frac{F_n(x)-x}{(x(1-x))^{1/2}}\right) - B_n \le y\right) = \exp\left\{-\frac{\mathrm{e}^{-y}}{\pi^{1/2}}\right\},$$
$$-\infty < y < \infty\,, \tag{3.14}$$

and

$$\lim_{n\to\infty} P\left(A_n \sup_{0<x<1} \left(n^{1/2}\frac{|F_n(x)-x|}{(x(1-x))^{1/2}}\right) - B_n \le y\right) = \exp\left\{\frac{2\mathrm{e}^{-y}}{\pi^{1/2}}\right\},$$
$$-\infty < y < \infty\,, \tag{3.15}$$

where $A_n = (2\log\log n)^{1/2}$, $B_n = 2\log\log n + \frac{1}{2}\log\log\log n$. Some related results are given in Jaeschke (1979) and Eicker (1979).

For weight function of the form (1.17) we have (Csáki, 1977a)

$$\lim_{n\to\infty} P\left((na_n)^{1/2} \sup_{a_n\le x\le 1} \left(\frac{x-F_n(x)}{x}\right) \le y\right)$$

$$= \lim_{n\to\infty} P\left((na_n)^{1/2} \sup_{0\le x\le 1-a_n} \left(\frac{x-F_n(x)}{1-x}\right) \le y\right)$$

$$= \Phi(y) - \Phi(-y), \quad 0 \le y \tag{3.16}$$

where $\lim_{n\to\infty} a_n = 0$ and $\lim_{n\to\infty} na_n = \infty$.

Furthermore

$$\lim_{n\to\infty} P\Big(\sup_{v/n\leq x\leq 1}\Big(\frac{x-F_n(x)}{x}\Big)\leq y\Big)$$
$$=\sum_{k\geq v(1-y)}\frac{v^k}{k!}\mathrm{e}^{-v}\Big(1-\Big(\frac{k}{1-y}-v\Big)\sum_{j=0}^{\infty}\frac{((k+j)/(1-y)-v)^{j-1}}{j!}$$
$$\times\exp\Big\{-\frac{k+y}{1-y}+v\Big\}\Big),\quad 0<y, \tag{3.17}$$

and

$$\lim_{n\to\infty} P\Big(\sup_{0\leq x\leq 1-v/n}\Big(\frac{x-F_n(x)}{1-x}\Big)\leq y\Big)$$
$$=1-\frac{y}{1+y}\sum_{k\geq v(1+y)}\frac{k^k}{(1+y)^k k!}\exp\Big\{-\frac{k}{1+y}\Big\},\quad 0<y. \tag{3.18}$$

There are also a number of results concerning asymptotic expansions of the distributions discussed in this Section. See Gaenssler and Stute (1979), Korolyuk and Borovskih (1981) and references therein.

3.2. Strong limit theorems

Glivenko–Cantelli theorems (see Section 1) assert convergence to zero of statistics $K_{n,\tau}$, $K^+_{n,\tau}$ and $K^-_{n,\tau}$. The rate of this convergence is given by the law of the iterated logarithm (LIL). The first result is due to Smirnov (1944) and Chung (1949):

$$\limsup_{n\to\infty}\frac{n^{1/2}D_n}{(\log\log n)^{1/2}}=\frac{1}{2^{1/2}}\quad \text{a.s.} \tag{3.19}$$

and the same is true for D_n replaced by either D^+_n or D^-_n. For the general statistic $K_{n,\tau}$ James (1975) gives

$$\limsup_{n\to\infty}\frac{n^{1/2}K_{n,\tau}}{(\log\log n)^{1/2}}=\sup_x\,(\tau(x)x(1-x))^{1/2}\quad \text{a.s.} \tag{3.20}$$

provided that the weight function τ satisfies the condition

$$\int_0^1\frac{\tau^2(u)}{\log\log(1/u(1-u))}\,\mathrm{d}u<\infty. \tag{3.21}$$

(3.21) is also necessary for (3.20) in the sense that if (3.21) fails then

$$\limsup_{n\to\infty}\frac{n^{1/2}K_{n,\tau}}{(\log\log n)^{1/2}}=\infty\quad \text{a.s.} \tag{3.22}$$

Finkelstein (1971) established Strassen-type (functional) LIL for D_n and James (1975) extended it to $K_{n,\tau}$, with τ satisfying (3.21). It is readily seen that the weight functions $\tau(u)=(u(1-u))^{-1/2}$, $\tau(u)=1/u$, $\tau(u)=1/(1-u)$, $\tau(u)=(u(1-u))^{-1}$, $0<u<1$ do not satisfy (3.21) and hence (3.22) holds true for these weight functions. For $\tau(u)=(u(1-u))^{-1/2}$ we have shown the following result (Csáki, 1974, 1982): Let δ_n be an increasing sequence of positive constants, then

$$P\left(n^{1/2}\sup_{0<x<1}\left(\frac{|F_n(x)-x|}{(x(1-x))^{1/2}}\right)\geq \delta_n \text{ i.o.}\right)=0 \text{ or } 1 \tag{3.23}$$

according as $\Sigma_{n=1}^{\infty} n^{-1}\delta_n^{-2}$ converges or diverges.

A corollary to this result yields

$$\limsup_{n\to\infty}\frac{\log(n^{1/2}\sup_{0<x<1}(|F_n(x)-x|/(x(1-x))^{1/2}))}{\log\log n}=\tfrac{1}{2} \quad \text{a.s.} \tag{3.24}$$

For $\tau(u)=1/u$ Shorack and Wellner (1978) show for increasing δ_n,

$$P\left(\sup_{0<x<1}\frac{|F_n(x)-x|}{x}\geq \delta_n \text{ i.o.}\right)=0 \text{ or } 1 \tag{3.25}$$

according as $\Sigma\, n^{-1}\delta_n^{-1}$ converges or diverges. Hence

$$\limsup_{n\to\infty}\frac{\log(\sup_{0\leq x<1}((|F_n(x)-x|/x)))}{\log\log n}=1 \quad \text{a.s.} \tag{3.26}$$

The same is true if $\tau(u)=1/u$ is replaced by either $\tau(u)=1/(1-u)$ or $\tau(u)=1/(u(1-u))$, $0<u<1$.

Recently Mason (1981) gave a result which is a common generalization of (3.24) and (3.25):

Let δ_n be an increasing sequence and $0\leq\alpha\leq\frac{1}{2}$. Then

$$P\left(n^{\alpha}\sup_{0<x<1}\frac{|F_n(x)-x|}{(x(1-x))^{-1+\alpha}}\geq \delta_n \text{ i.o.}\right)=0 \text{ or } 1 \tag{3.27}$$

according as $\Sigma_{n=1}^{\infty} n^{-1}\delta_n^{-1/(1-\alpha)}$ converges or diverges. Hence

$$\limsup_{n\to\infty}\frac{\log(n^{\alpha}\sup_{0<x<1}(|F_n(x)-x|/(x(1-x))^{-1+\alpha}))}{\log\log n}=1-\alpha \quad \text{a.s.} \tag{3.28}$$

On the other hand, as shown by Jaeschke (1979), the weak version of the law of the iterated logarithm holds for the weight function $\tau(u)=(u(1-u))^{-1/2}$:

$$\frac{n^{1/2}\sup_{0<x<1}(|F_n(x)-x|/(x(1-x))^{1/2})}{(2\log\log n)^{1/2}}\xrightarrow{\text{P}}1\,. \tag{3.29}$$

For one sided version it is shown in Csáki (1977b) and Shorack (1980) that

$$\limsup_{n\to\infty}\left(\frac{n}{\log\log n}\right)^{1/2}\sup_{0<x<1/2}\left(\frac{x-F_n(x)}{(x(1-x))^{1/2}}\right)=2 \quad \text{a.s.} \tag{3.30}$$

We consider again weight functions of the form (1.17). First we mention the following result of Kiefer (1972b). Define $\beta_d'>1$ as the solution of the equation

$$\beta(\log\beta-1)=\frac{1-d}{d} \quad (d>0), \tag{3.31}$$

Let $\beta_d''=0$ if $0<d\leq 1$, and the other solution ($\beta_d''<1$) of the equation (3.31) if $1<d<\infty$. Put $a_n=dn^{-1}\log\log n$. Then

$$\limsup_{n\to\infty}\frac{F_n(a_n)}{a_n}=\beta_d' \quad \text{a.s.} \tag{3.32}$$

and

$$\limsup_{n\to\infty}\frac{a_n}{F_n(a_n)}=\frac{1}{\beta_d''} \quad \text{a.s.} \tag{3.33}$$

A related result of Eicker (1970a) and Kiefer (1972b) says that if $\lim_{n\to\infty} na_n(\log\log n)^{-1}=\infty$ then

$$\limsup_{n\to\infty}\left(\frac{n}{2\log\log n}\right)^{1/2}\frac{|F_n(a_n)-a_n|}{(a_n(1-a_n))^{1/2}}=1 \quad \text{a.s.} \tag{3.34}$$

Csörgő and Révész (1975) proved for $a_n=(\log n)^4/n$ that

$$\limsup_{n\to\infty}\left(\frac{n}{\log\log n}\right)^{1/2}\sup_{a_n\leq x\leq 1-a_n}\left(\frac{|F_n(x)-x|}{(x(1-x))^{1/2}}\right)=2 \quad \text{a.s.} \tag{3.35}$$

An extension of these results is given in Csáki (1977b) (see also Shorack, 1980):

If a_n is a decreasing sequence such that

$$\lim_{n\to\infty} na_n(\log\log n)^{-1}=\infty \quad \text{and} \quad \lim_{n\to\infty}(\log\log a_n^{-1})(\log\log n)^{-1}=c$$

with $0\leq c\leq 1$, then

$$\limsup_{n\to\infty}\left(\frac{n}{\log\log n}\right)^{1/2}\sup_{a_n\leq x\leq 1-a_n}\left(\frac{|F_n(x)-x|}{(x(1-x))^{1/2}}\right)=(2(c+1))^{1/2} \quad \text{a.s.} \tag{3.36}$$

Moreover, if $a_n=dn^{-1}\log\log n$, then

$$\limsup_{n\to\infty}\left(\frac{n}{\log\log n}\right)^{1/2}\sup_{a_n\le x\le 1-a_n}\left(\frac{|F_n(x)-x|}{(x(1-x))^{1/2}}\right)$$
$$=\max(2,\, d^{1/2}(\beta'_d-1))\quad \text{a.s.} \tag{3.37}$$

Wellner (1978) has shown the following results:
If $a_n\downarrow$ and $na_n(\log\log n)^{-1}\uparrow\infty$, then

$$\limsup_{n\to\infty}\left(\frac{na_n}{2\log\log n}\right)^{1/2}\sup_{a_n\le x<1}\left(\frac{|F_n(x)-x|}{x}\right)=1\quad \text{a.s.} \tag{3.38}$$

If $a_n = dn^{-1}\log\log n$, then

$$\limsup_{n\to\infty}\sup_{a_n\le x<1}\left(\frac{|F_n(x)-x|}{x}\right)=\beta'_d-1\quad \text{a.s.}, \tag{3.39}$$

$$\limsup_{n\to\infty}\sup_{a_n\le x<1}\left(\frac{F_n(x)-x}{x}\right)=\beta'_d-1\quad \text{a.s.}, \tag{3.40}$$

$$\limsup_{n\to\infty}\sup_{a_n\le x<1}\left(\frac{x-F_n(x)}{x}\right)=1-\beta''_d\quad \text{a.s.} \tag{3.41}$$

Much less is known for liminf of $K_{n,\tau}$ and related statistics. Mogulskii (1979) and Kuelbs (1979) show

$$\liminf_{n\to\infty}(D_n(n\log\log n)^{1/2})=\frac{\pi}{8^{1/2}}\quad \text{a.s.} \tag{3.42}$$

and Csáki (1982) shows

$$\liminf_{n\to\infty}\left(\frac{n}{2\log\log n}\right)^{1/2}\sup_{0<x<1}\left(\frac{|F_n(x)-x|}{(x(1-x))^{1/2}}\right)=1\quad \text{a.s.} \tag{3.43}$$

Finally we note that strong limit theorems discussed in this Section are useful in studying functions of order statistics, rank tests, etc. In this respect we refer to Wellner (1977c) and Sen (1981).

3.3. Martingale properties and inequalities

A number of results in the previous section are based on certain martingale and Markov properties of empirical distribution function and on certain inequalities.

It is obvious that for fixed x, both $n(x-F_n(x))\tau(x)$ and $n(F_n(x)-x)\tau(x)$ are martingales in n, since they are sums of i.i.d. random variables with zero expectation. It can be seen also (cf. Kiefer, 1972a) that, for fixed n, $(F_n(x)-x)/(1-x)$ is a martingale in x. Hence

$$U(n,x)=\frac{n(F_n(x)-x)}{1-x}, \qquad n=1,2,\ldots, \quad 0\leq x<1, \tag{3.44}$$

is a martingale in both n and x. It is also shown in Hmaladze (1981) that

$$F_n(x)-\int_0^x \frac{1-F_n(s)}{1-s}\,\mathrm{d}s \tag{3.45}$$

is a martingale in x.

An easy consequence of the above properties is that the sequences $nK_{n,\tau}$, $nK^+_{n,\tau}$ and $nK^-_{n,\tau}$ are all submartingales. Furthermore (cf. Sen, 1973), $K_{n,\tau}$, $K^+_{n,\tau}$ and $K^-_{n,\tau}$ are reverse submartingales.

Moreover (Hmaladze, 1981; Stute, 1982) the process $nF_n(x)$, $0\leq x\leq 1$ is a Markov process.

A basic inequality for D_n is due to Dvoretzky, Kiefer and Wolfovitz (1956):

$$P(D_n\geq y)\leq c\,\mathrm{e}^{-2ny^2}, \quad 0<y<\infty \tag{3.46}$$

with some constant c. Devroye and Wise (1979) evaluated $c\approx 611$ in the original proof. Shorack (1980) claims that (3.46) holds with $c=58$. It has been conjectured (see Birnbaum and McCarty, 1958) that (3.46) is true with $c=2$ but this conjecture has not been proved so far. Similar inequalities are true for D^+_n and D^-_n with c replaced by $c/2$.

For the general statistic $K_{n,\tau}$ however an exponential bound similar to (3.46) cannot be expected. By the use of some martingale properties and the Birnbaum–Marshall inequality it can be shown that

$$P(K_{n,\tau}\geq y)\leq \frac{c}{ny^2}\int_0^1 \tau^2(u)\,\mathrm{d}u, \quad y>0, \tag{3.47}$$

with some constant c (refer to Govindarajulu, LeCam and Raghavachari, 1967; Pyke and Shorack, 1968; Puri and Sen, 1971; Shorack, 1972; Wellner, 1977a).

For particular weight functions however better inequalities can be obtained.

Ghosh (1972) has shown that for every $\delta\geq 0$ there exist $K>0$, $\varepsilon>0$, n_0 such that for $n\geq n_0$

$$P\left(\sup_{0<x<1}\left(\frac{|F_n(x)-x|}{(x(1-x))^{1/2-\varepsilon}}\geq\frac{K\log n}{n^{1/2}}\right)\leq\frac{2}{n^{1+\delta}}. \tag{3.48}$$

Based on martingale properties and Hoeffding inequality, James (1975), Wellner (1978) and Shorack (1980) gave some inequalities for the particular weight functions $\tau(u)=1/u$ and $\tau(u)=1/(1-u)$.

$$\begin{aligned}P\left(n^{1/2}\sup_{0<x\leq b}\left(\frac{|F_n(x)-x|}{1-x}\right)>\frac{y}{1-b}\right)&=P\left(n^{1/2}\sup_{1-b\leq x<1}\frac{|F_n(x)-x|}{x}>\frac{y}{1-b}\right)\\&\leq 2\exp\{-y^2\psi(y/bn^{1/2})/(2b(1-b))\}, \quad y>0,\end{aligned} \tag{3.49}$$

where

$$\psi(u) = \frac{(u+1)\log(u+1) - u}{u^2}.$$

Mason (1981) proves an inequality equivalent to

$$P\left(n^{1/2-\alpha} \sup_{0<x\leq a_n} \left(\frac{F_n(x) - x}{x^{1/2+\alpha}}\right) \geq y\right) \leq cy^{-2/(1+2\alpha)}, \quad y>0, \tag{3.50}$$

with some constant c, where $0 \leq \alpha < \frac{1}{2}$, and

$$a_n = \frac{1}{n}\left(\frac{(1-2\alpha)y}{3-2\alpha}\right)^{2/(1-2\alpha)}$$

Further inequalities can be found in Butler and McCarty (1960), Whittle (1961), Darling and Robbins (1968).

Further inequalities concern the supremum of $K_{n,\tau}$ with respect to n. By the submartingale property of $nK_{n,\tau}$, $nK^+_{n,\tau}$ and $nK^-_{n,\tau}$ we can obtain a Chernoff–type inequality

$$P\left(\sup_{n_1<n\leq n_2} nK^0_{n,\tau} \geq y\right) \leq \inf_{t>0}\left(e^{-ty}E(e^{tK^0_{n_2,\tau}})\right) \tag{3.51}$$

where $K^0_{n,\tau}$ denotes either of $K_{n,\tau}$, $K^+_{n,\tau}$, $K^-_{n,\tau}$. This method however requires evaluation or a good estimation of the moment generating function of $K^0_{n,\tau}$. See Csáki (1968, 1977a,b), Stanley (1972) and Khan (1977).

James (1975) gives an inequality improved by Shorack (1980) as follows:

Let $\tau(u)$ be a positive function on some $(0, \delta]$ such that $\tau(u)u^{1/2}$ is increasing on $(0, \delta]$ and $\tau(u)u^{1/2} \to 0$ as $u \to 0$. Let $y>0$, $0<c<1$, $\alpha>1$ be given. Then there exists $0<b_1\leq\delta$ such that for all integers $n_1 \leq n_2$ having $n_2/n_1 \leq \alpha$ and for all $0 \leq a \leq b \leq b_1$ we have

$$\begin{aligned} &P\left(\max_{n_1\leq n\leq n_2} n^{1/2} \sup_{a\leq x\leq b} (|F_n(x) - x|\tau(x)) > y\right) \\ &\leq 2P\left(n_2^{1/2} \sup_{a\leq x\leq b} (|F_{n_2}(x) - x|\tau(x)) > cy\alpha^{-1/2}\right). \end{aligned} \tag{3.52}$$

For the applications of inequalities and martingale properties in the theory of rank tests, order statistics and quantiles refer to Shorack (1972), Wellner (1977c), Sen (1981), Csörgő and Révész (1981), etc.

4. Tests based on e.d.f.

The one sample statistics $K_{n,\tau}$, $K^+_{n,\tau}$ and $K^-_{n,\tau}$ defined by (1.8), (1.9) and (1.10), resp., are used in goodness of fit tests, i.e. to test the null hypothesis that the true distribution function $F(x)$ is identically equal to the given hypothetical distribution function $F_0(x)$. In these tests the null hypothesis is rejected when the test statistics are large, i.e. the critical regions are of the form $\{K^0_{n,\tau}, \geq k_\alpha\}$, where $K^0_{n,\tau}$ denotes one of $K_{n,\tau}$, $K^+_{n,\tau}$ or $K^-_{n,\tau}$ and the crtical value k_α is determined so that under the null hypothesis we have

$$P(K^0_{n,\tau} \geq k_\alpha) \sim \alpha \,. \tag{4.1}$$

In order to find the value of k_α we need the exact or limit distribution of the statistic used. For the most important weight functions the values of k_α are given in tables. In this Section we study some properties of these tests and their two sample analogues.

4.1. Powers

The power of the test is the probability of the critical region under the alternative. The problem of determining the power can always be reduced to computing probabilities of the form (2.1) in the one sample case or (2.9) in the two sample case.

Consider e.g. the null hypothesis $F_0(x) = x$, $0 \leq x \leq 1$ and use the test statistic $D_n^- = \sup_{0 \leq x \leq 1}(x - F_n(x))$. Reject the null hypothesis if $D_n^- \geq d_\alpha$, where d_α is determined by $P(D_n^- \geq d_\alpha / H_0) = \alpha$. Consider two alternatives:

$$G_1(x) = \begin{cases} x & \text{if } 0 \leq x \leq x_0 - \Delta \,, \\ x_0 - \Delta & \text{if } x_0 - \Delta \leq x < x_0 \,, \\ x & \text{if } x_0 \leq x \leq 1 \,; \end{cases} \tag{4.2}$$

$$G_2(x) = \begin{cases} x - \Delta & \text{if } \Delta \leq x < 1 \,, \\ 1 & \text{if } 1 \leq x \,. \end{cases} \tag{4.3}$$

Then the powers under G_1 and G_2 are given by

$$\begin{aligned} P(D_n^- \geq d_\alpha / G_1(x)) = 1 - \sum_{k \geq n(x_0 - d_\alpha)} \binom{n}{k} (x_0 - \Delta)^k (1 - x_0 + \Delta)^{n-k} \\ \times \left(1 - \frac{d_\alpha}{x_0 - \Delta} \sum_{j=0}^{[n(x_0 - \Delta - d_\alpha)]} \binom{k}{j} \left(\frac{j/n + d_\alpha}{x_0 - \Delta}\right)^{j-1} \left(1 - \frac{j/n + d_\alpha}{x_0 - \Delta}\right)^{k-j}\right) \\ \times \left(1 - \frac{k/n + d_\alpha}{1 - x_0 + \Delta} \sum_{i=0}^{[n(1-d_\alpha)]-k} \binom{n-k}{i} \right. \\ \times \left. \left(\frac{(k+i)/n + d_\alpha}{1 - x_0 + \Delta}\right)^{i-1} \left(1 - \frac{(k+i)/n + d_\alpha}{1 - x_0 + \Delta}\right)^{n-k+i}\right) \end{aligned} \tag{4.4}$$

and

$$\begin{aligned} P(D_n^- \geq d_\alpha \mid G_2(x)) = 1 - (d_\alpha - \Delta) \sum_{j=0}^{[n(1-d_\alpha)]} \binom{n}{j} \\ \times \left(d_\alpha - \Delta + \frac{j}{n}\right)^{j-1} \left(1 - d_\alpha + \Delta - \frac{j}{n}\right)^{n-j}, \quad d_\alpha > \Delta. \end{aligned} \tag{4.5}$$

$G_1(x)$ is a minimal alternative, while $G_2(x)$ is a maximal alternative (cf. Chapman, 1958) in the sense that if $G(x)$ is an arbitrary distribution function such that $G_2(x) \leq G(x) \leq G_1(x)$ for $0 \leq x \leq 1$, then

$$P(D_n^- \geq d_\alpha \mid G_1(x)) \leq P(D_n^- \geq d_\alpha \mid G(x)) \leq P(D_n^- \geq d_\alpha \mid G_2(x)) . \quad (4.6)$$

For further investigations of the power refer to Massey (1952), Birnbaum (1953), Chapman (1958), Vincze (1968), Yu (1975, 1977), Csáki (1977a). For normal alternatives see Knott (1970), and for exponential alternatives see Durbin (1971), where also asymptotic results can be found. For further asymptotic results refer to Quade (1965), Andél (1967), Klotz (1967). Lehmann alternatives are studied by Steck (1974).

4.2. Consistency

Recall that a test is consistent against an alternative if the power under this alternative tends to 1 as the sample size tends to infinity.

For the consistency of test based on $K_{n,\tau}$, $K_{n,\tau}^+$ or $K_{n,\tau}^-$ we need the (weak version) of Glivenko-Cantelli theorem, i.e. for the consistency of a test based on $K_{n,\tau}^0$ against an alternative it is sufficient that

$$K_n^0 \xrightarrow{\mathrm{P}} 0 \quad \text{as } n \to \infty \quad (4.7)$$

under the null hypothesis and

$$K_n^0 \xrightarrow{\mathrm{P}} \Delta > 0 \quad \text{as } n \to \infty \quad (4.8)$$

under the alternative. Similar results hold for two sample tests. For the validity of Glivenko–Cantelli theorem see the Introduction. Hence the test based on

$$\sup_{a_n \leq F(x) \leq 1} \left(\frac{|F_n(x) - F(x)|}{F(x)} \right)$$

is consistent against any alternative provided a_n satisfies $\lim_{n\to\infty} na_n = \infty$, while it is not consistent against any alternative if $a_n = 0$.

4.3. Efficiency

There are several definitions of efficiency in the literature. The Pitman efficiency for tests based on empirical distribution has been studied by Capon (1965), Ramachandramurty (1966), Yu (1971), Kalish and Mikulski (1971).

Chernoff (1952, 1956) and Bahadur (1960, 1967) have introduced efficiency concepts based on large deviations.

Let $\{T_n, n \geq 1\}$ be a sequence of statistics for testing H_0 against H_1. Let the critical region be defined by $\{T_n \geq t\}$.

Chernoff (1956) introduces the following efficiency: assume that there exist $\rho > 0$ and $\{t_n\}$ such that

$$\lim_{n\to\infty}(P(T_n \geq t_n \mid H_0))^{1/n} = \lim_{n\to\infty}(P(T_n < t_n \mid H_1))^{1/n} = \rho . \tag{4.9}$$

(4.9) says that both errors of first and second kind are about the same order ρ^n. Then if $\{T_n^{(1)}\}$ and $\{T_n^{(2)}\}$ are two sequences of test statistics testing H_0 against H_1, whose ρ's are ρ_1 and ρ_2, resp. then the Chernoff efficiency is defined by

$$e_c = \frac{\log \rho_1}{\log \rho_2} . \tag{4.10}$$

Bahadur (1960, 1967) introduces the following efficiency: let

$$H_n(t) = P(T_n \leq t \mid H_0) \tag{4.11}$$

and

$$L_n = 1 - H_n(T_n) . \tag{4.12}$$

Assume that

$$P(\lim_{n\to\infty} n^{-1} \log L_n = -b/2 \mid H_1) = 1 . \tag{4.13}$$

b is called the exact Bahadur slope of T_n. Now if $\{T_n^{(1)}\}$ and $\{T_n^{(2)}\}$ are two sequences of test statistics whose exact Bahadur slope are b_1 and b_2, resp., then the Bahadur efficiency is defined by

$$e_b = \frac{b_1}{b_2} . \tag{4.14}$$

Both efficiency concepts require exponential convergence of large deviations. In this respect we quote a recent result of Groeneboom and Shorack (1981):

$$\lim_{n\to\infty} \frac{1}{n} P(K_{n,\tau}^0 \geq y) = -g_\tau(y) \tag{4.15}$$

where $K_{n,\tau}^0$ is any of statistics $K_{n,\tau}$, $K_{n,\tau}^+$ or $K_{n,\tau}^-$ and the function $g_\tau(y)$ is defined as follows:

Let

$$f(a,t) = \begin{cases} (a+t)\log\dfrac{a+t}{t} + (1-a-t)\log\dfrac{1-a-t}{1-t} & \text{if } 0 \leq a \leq 1-t , \\ \infty & \text{if } a > 1-t , \end{cases} \tag{4.16}$$

$$g_\tau^+(y) = \inf_{0<t<1} f\left(\frac{y}{\tau(t)}, t\right), \tag{4.17}$$

$$g_\tau^-(y) = \inf_{0<t<1} f\left(\frac{y}{\tau(t)}, 1-t\right), \tag{4.18}$$

$$g_\tau(y) = \min(g_\tau^+(y), g_\tau^-(y)). \tag{4.19}$$

If $\tau(t)$ is such that either

$$\liminf_{t \downarrow 0} \frac{\log(1/t)}{\tau(t)} = 0 \tag{4.20}$$

or

$$\liminf_{t \downarrow 0} \frac{\log(1/t)}{\tau(1-t)} = 0 \tag{4.21}$$

then $g_\tau(y) = 0$ for all $y \geq 0$.

A consequence of this result is that if

$$K_{n,\tau}^0 \xrightarrow{\text{a.s.}} \xi \quad \text{as } n \to \infty \tag{4.22}$$

under H_1, then the exact Bahadur slope is $b = 2g_\tau(\xi)$.

The conditions (4.20) and (4.21) show that $\tau(u) = -\log(u(1-u))$ $(0 < u < 1)$ is a key weight function so that exponential convergence still holds for this weight function but does not hold for weight functions with heavier tail. This corrects a result of Abrahamson (1967).

For previous and related large deviation results see the references in Groeneboom and Shorack (1981).

References

Abrahamson, I. G. (1967). The exact Bahadur efficiencies for the Kolmogorov–Smirnov and Kuiper one- and two-sample statistics. *Ann. Math. Statist.* **38**, 1475–1490.

Anděl, J. (1967). Local asymptotic power and efficiency of tests of Kolmogorov–Smirnov type. *Ann. Math. Statist.* **38**, 1705–1725.

Anderson, T. W. and Darling, D. A. (1952). Asymptotic theory of certain 'goodness of fit' criteria based on stochastic processes. *Ann. Math. Statist.* **23**, 193–212.

Bahadur, R. R. (1960). Stochastic comparison of tests. *Ann. Math. Statist.* **31**, 276–295.

Bahadur, R. R. (1967). Rates of convergence of estimates and test statistics. *Ann. Math. Statist.* **38**, 303–324.

Barton, D. E. and Mallows, C. L. (1965). Some aspects of the random sequence. *Ann. Math. Statist.* **36**, 236–260.

Birnbaum, Z. W. (1953). On the power of a one-sided test of fit for continuous probability functions. *Ann. Math. Statist.* **24**, 484–489.

Birnbaum, Z. W. and Lientz, B. P. (1969). Exact distributions for some Rényl-type statistics. *Zastos. Mat.* **10** (Hugo Steinhaus Jubilee Vol.), 179–192.

Birnbaum, Z. W. and McCarthy, R. C. (1958). A distribution-free upper confidence bound for $\Pr\{Y<X\}$, based on independent samples of X and Y. *Ann. Math. Statist.* **29**, 558–562.

Birnbaum, Z. W. and Pyke, R. (1958). On some distributions related to the statistic D_n^+. *Ann. Math. Statist.* **29**, 179–187.

Birnbaum, Z. W. and Tingey, F. H. (1951). One-sided confidence contours for probability distribution functions. *Ann. Math. Statist.* **22**, 592–596.

Butler, J. B. and McCarthy, R. C. (1960). A lower bound for the distribution of the statistic D_n^+. *Notices Amer. Math. Soc.* **7**, 80–81. (Abstract no. 565-2.)

Cantelli, F. P. (1933). Sulla determinazione empirica delle leggi di probabilita. *Giorn. Ist. Ital. Attuari* **4**, 421–424.

Capon, J. (1965). On the asymptotic efficiency of the Kolmogorov–Smirnov test. *J. Amer. Statist. Assoc.* **60**, 843–853.

Carnal, H. (1962). Sur les théorèmes de Kolmogorov et Smirnov dans le cas d'une distribution discontinue. *Comment. Math. Helv.* **37**, 19–35.

Chang, Li-Chien (1955). On the ratio of the empirical distribution function to the theoretical distribution function. *Acta Math. Sinica* **5**, 347–368; English transl. in: *Selected Transl. Math. Statist. and Probability, Amer. Math. Soc.* **4**, 17–38. Providence, R.I., 1963.

Chapman, D.G. (1958). A comparative study of several one-sided goodness-of-fit tests. *Ann. Math. Statist.* **29**, 655–674.

Chernoff, H. (1952). A measure of asymptotic efficiency for tests of a hypothesis based on the sum of observations. *Ann. Math. Statist.* **23**, 493–507.

Chernoff, H. (1956). Large-sample theory: Parametric case. *Ann. Math. Statist.* **27**, 1–22.

Chibisov, D. M. (1964). Some theorems on the limiting behaviour of the empirical distribution function. *Trudy Mat. Inst. Steklov.* **71**, 104–112; English transl. in: *Selected Transl. Math. Statist. and Probability, Amer. Math. Soc.* **6**, 147–156. Providence, R.I., 1964.

Chung, K. L. (1949). An estimate concerning the Kolmogorov limit distribution. *Trans. Amer. Math. Soc.* **67**, 36–50.

Coberly, W. A. and Lewis, T. O. (1972). A note on a one-sided Kolmogorov–Smirnov test of fit for discrete distribution functions. *Ann. Inst. Statist. Math.* **24**, 183–187.

Conover, W. J. (1972). A Kolmogorov goodness-of-fit test for discontinuous distributions. *J. Amer. Statist. Assoc.* **67**, 591–596.

Csáki, E. (1968). An iterated logarithm law for semimartingales and its application to empirical distribution functions. *Studia Sci. Math. Hungar.* **3**, 287–292.

Csáki, E. (1975). Some notes on the law of the iterated logarithm for empirical distribution function. *Coll. Math. Soc. J. Bolyai* **11**, Limit theorems of probability theory, Keszthely (Hungary), 1974, 47–57.

Csáki, E. (1977a). Investigations concerning the empirical distribution function. *Magyar Tud. Akad. Mat. Fiz. Oszt. Közl.* **23**, 239–327. English transl. in: *Selected Transl. Math. Statist. and Probability, Amer. Math. Soc.* **15**, 229–317. Providence, R.I., 1981.

Csáki, E. (1977b). The law of the iterated logarithm for normalized empirical distribution function. *Z. Wahrsch. Verw. Gebiete* **38**, 147–167.

Csáki, E. (1982). On the standardized empirical distribution function. *Coll. Math. Soc. J. Bolyai* **32**, *Nonparametric Statistical Inference*, Budapest (Hungary), 1980, pp. 123–138.

Csáki, E. and Tusnády, G. (1972). On the number of intersections and the ballot theorem. *Period. Math. Hungar.* **2** (A. Rényi Mem. Vol.), 1–13.

Csörgő, M. (1965a). Exact and limiting probability distributions of some Smirnov type statistics. *Canad. Math. Bull.* **8**, 93–103.

Csörgő, M. (1965b). Exact probability distribution functions of some Rényi type statistics. *Proc. Amer. Math. Soc.* **16**, 1158–1167.

Csörgő, M. (1965c). Some Rényi type limit theorems for empirical distribution functions. *Ann. Math. Statist.* **36**, 322–326; correction: 1069.

Csörgő, M. (1965d). Some Smirnov type theorems of probability theory. *Ann. Math. Statist.* **36**, 1113–1119.

Csörgő, M. (1984). Empirical processes. This Volume.

Csörgő, M. and Révész, P. (1975). Some notes on the empirical distribution function and the quantile process. *Coll. Math. Soc. J. Bolyai* **11**, Limit theorems of probability theory, Keszthely (Hungary), 1974, pp. 59–71.

Csörgő, M. and Révész, P. (1981). *Strong Approximations in Probability and Statistics.* Academic Press, New York.

Daniels, H. E. (1945). The statistical theory of the strength of bundles of threads, I. *Proc. Roy. Soc. London Ser. A* **183**, 405–435.

Darling, D. A. and Robbins, H. (1968). Some nonparametric sequential tests with power one. *Proc. Nat. Acad. Sci. U.S.A.* **61**, 804–809.

Dempster, A. P. (1959). Generalized D_n^+ statistics. *Ann. Math. Statist.* **30**, 593–597.

Devroye, L. P. and Wise, G. L. (1979). On the recovery of discrete probability densitites from imperfect measurements. *J. Franklin Inst.* **307**, 1–20.

Durbin, J. (1968). The probability that the sample distribution function lies between two parallel straight lines. *Ann. Math. Statist.* **39**, 398–411.

Durbin, J. (1971). Boundary-crossing probabilities for the Brownian motion and Poisson processes and techniques for computing the power of the Kolmogorov–Smirnov test. *J. Appl. Probability* **8**, 431–453.

Durbin, J. (1973). Distribution theory for tests based on the sample distribution function. *Regional conference series in appl. math.*, No. 9, Siam, Phil.

Dvoretzky, A., Kiefer, J. and Wolfowitz, J. (1956). Asymptotic minimax character of the sample distribution function and of the classical multinomial estimator. *Ann. Math. Statist.* **27**, 642–669.

Dwass, M. (1959). The distribution of a generalized D_n^+ statistic. *Ann. Math. Statist.* **30**, 1024–1028.

Dwass, M. (1967). Simple random walk and rank order statistics. *Ann. Math. Statist.* **38**, 1042–1053.

Dwass, M. (1974). Poisson process and distribution free statistics. *Adv. Appl. Probability* **6**, 359–375.

Eicker, F. (1970a). A log log-law for double sequences of random variables. *Z. Wahrsch. Verw. Gebiete* **16**, 107–133.

Eicker, F. (1970b). On the probability that a sample distribution function lies below a line segment. *Ann. Math. Statist.* **41**, 2075–2092.

Eicker, F. (1979). The asymptotic distribution of the suprema of the standardized empirical processes. *Ann. Statist.* **7**, 116–138.

Epanechnikov, V. A. (1968). The significance level and power of the two-sided Kolmogorov test in the case of small sample sizes. *Theory Probab. Appl.* **13**, 686–690.

Feller, W. (1948). On the Kolmogorov–Smirnov limit theorems for empirical distributions. *Ann. Math. Statist.* **19**, 177–189.

Finkelstein, H. (1971). The law of the iterated logarithm for empirical distributions. *Ann. Math. Statist.* **42**, 607–615.

Gaenssler, P. and Stute, W. (1979). Empirical process: a survey of results for independent and identically distributed random variables. *Ann. Probability* **7**, 193–243.

Ghosh, M. (1972). On the representation of linear functions of order statistics. *Sankhyā Ser. A* **34**, 349–356.

Glivenko, V. (1933). Sulla determinazione empirica della leggi di probabilitá. *Giorn. Ist. Ital. Attuari* **4**, 92–99.

Gnedenko, B. V. (1954). Kriterien für die Unveränderlichkeit der Wahrscheinlichkeitsverteilung von zwei unabhängigen Stichprobenreihen. *Math. Nachrichten* **12**, 29–66.

Gnedenko, B. V. and Korolyuk, V. S. (1951). On the maximum discrepancy between two empirical distribution functions. *Dokl. Akad. Nauk SSSR* **80**, 525–528. English transl. in: *Selected Transl. Math. Statist. and Probability, Amer. Math. Soc.* **1**, 13–16. Providence, RI, 1961.

Goodman, V., Kuelbs, J. and Zinn, J. (1981). Some results on the LIL in Banach space with applications to weighted empirical processes. *Ann. Probability* **9**, 713–752.

Govindarajulu, Z., LeCam, L. and Raghavachari, M. (1967). Generalizations of theorems of Chernoff and Savage on the asymptotic normality of test statistics. *Proc. Fifth Berkeley, Sympos. Math. Statist. and Probability Vol. I.* Univ. of California Press, Berkeley, CA, pp. 609–638.

Groeneboom, P. and Shorack, G. R. (1981). Large deviations of goodness of fit statistics and linear combinations of order statistics. *Ann. Probability* **9**, 971–987.

Gupta, S. S. and Panchapakesan, S. (1973). On order statistics and some applications of combinatorial methods in statistics. In: *A Survey of Combinatorial Theory* (R. C. Bose Seventieth Birthday Volume) (Proc. Internat. Sympos. Combinatorial Math. and Appl., Fort Collins, CO, 1971). North-Holland, Amsterdam, pp. 217–250.

Hájek, J. and Šidák, Z. (1967). *Theory of Rank Tests*. Academia, Prague and Academic Press, New York.

Hmaladze, E. V. (1981). Martingale approach in the theory of goodness-of-fit tests. *Teor. Verojatnost. i Primenen.* **26**, 246–265 (Russian).

Ishii, G. (1958). Kolmogorov–Smirnov test in life test. *Ann. Inst. Statist. Math.* **10**, 37–46.

Ishii, G. (1959). On the exact probabilities of Rényi's tests. *Ann. Inst. Statist. Math.* **11**, 17–24.

Jaeschke, D. (1979). The asymptotic distribution of the supremum of the standardized empirical distribution function on subintervals. *Ann. Statist.* **7**, 108–115.

James, B. R. (1975). A functional law of the iterated logarithm for weighted empirical distributions. *Ann. Probability* **3**, 762–772.

Kalish, G. and Mikulski, P. W. (1971). The asymptotic behavior of the Smirnov test compared to standard "optimal procedures". *Ann. Math. Statist.* **42**, 1742–1747.

Khan, R. A. (1977). On strong convergence of Kolmogorov–Smirnov statistic and a sequential detection procedure. *Tamkang J. Math.* **8**, 157–165.

Kiefer, J. (1972a). Skorohod embedding of multivariate r.v.'s and the sample d.f. *Z. Wahrsch. Verw. Gebiete* **24**, 1–35.

Kiefer, J. (1972b). Iterated logarithm analogues for sample quantiles when $p_n \downarrow 0$. *Proc. Sixth Berkeley Sympos. Math. Statist. and Probability, Vol. I*. Univ. of California Press, Berkeley, CA, pp. 227–244.

Klotz, J. (1967). Asymptotic efficiency of the two-sample Kolmogorov–Smirnov test. *J. Amer. Statist. Assoc.* **62**, 932–938.

Knott, M. (1970). The small sample power of one-sided Kolmogorov tests for a shift in location of the normal distribution. *J. Amer. Statist. Assoc.* **65**, 1384–1391.

Kolmogorov, A. N. (1933). Sulla determinazione empirica di una legge di distribuzione. *Giorn. Ist. Ital. Attuari* **4**, 83–91.

Korolyuk, V. S. (1955). On the discrepancy of empiric distributions for the case of two independent samples. *Izv. Akad. Nauk SSSR Ser. Mat.* **19**, 81–96 (Russian).

Korolyuk, V. S. and Borovskih, Yu.V. (1981). *Analytical problems of asymptotics of probability distributions*. Kiev: Naukova Dumka (Russian).

Kreweras, G. (1965). Sur une classe de problemes de denombrement liès au treillis des partitions de entiers. *Cahiers du Bur. Univ. de Rech. Oper.* **6**, 5–105.

Krumbholz, W. (1976). The general form of the law of the iterated logarithm for generalized Kolmogorov–Smirnov–Rényi statistics. *J. Multivariate Anal.* **6**, 653–658.

Kuelbs, J. (1979). Rates of growth for Banach space valued independent increment processes. Lecture Notes in Mathematics, No. 709. Springer, New York, pp. 151–169.

Kuelbs, J. and Dudley, R. M. (1980). Log log laws for empirical measures. *Ann. Probability* **8**, 405–418.

Kuiper, N. H. (1960). Tests concerning random points on a circle. *Nederl. Akad. Wetensch. Proc. Ser. A* **63**, *Indag. Math.* **22**, 38–47.

Lauwerier, H. A. (1963). The asymptotic expansion of the statistical distribution of N. V. Smirnov. *Z. Wahrsch. Verw. Gebiete* **2**, 61–68.

Malmquist, S. (1954). On certain confidence contours for distribution functions. *Ann. Math. Statist.* **25**, 523–533.

Maniya, G. M. (1949). Generalization of Kolmogorov's criterion for an estimate for the law of distribution for empirical data. *Dokl. Akad. Nauk SSSR* **69**, 495–497 (Russian).

Mason, D. M. (1981). Bounds for weighted empirical distribution functions. *Ann. Probability* **9**, 881–884.

Massey, F. J. Jr. (1950). A note on the power of a non-parametric test. *Ann. Math. Statist.* **21**, 440–443; correction: **23** (1952), 637–638.

Mogulskiĭ, A. A. (1979). On the law of the iterated logarithm in Chung's form for functional spaces. *Theory Probab. Appl.* **24**, 405–413.

Mohanty, S. G. (1979). *Lattice Path Counting and Applications*. Academic Press, New York.

Mohanty, S. G. (1982). On some computational aspects of rectangular probabilities. *Coll. Math. Soc. J. Bolyai* **32**, Nonparametric Statistical Inference, Budapest (Hungary), 1980, 597–617.

Nef, W. (1964). Über die Differenz zwischen theoretischer und empirischer Verteilungsfunktion. *Z. Wahrsch. Verw. Gebiete* **3**, 154–162.

Niederhausen, H. (1981). Scheffer polynomials for computing exact Kolmogorov–Smirnov and Rényi type distributions. *Ann. Statist.* **9**, 923–944.

Noé, M. and Vandewiele, G. (1968). The calculation of distributions of Kolmogorov–Smirnov type statistics including a table of significant points for a particular case. *Ann. Math. Statist.* **39**, 233–241.

Noether, G. E. (1963). Note on the Komogorov statistic in the discrete case. *Metrika* **7**, 115–116.

Pitman, E. J. G. (1972). Simple proofs of Steck's determinantal expressions for probabilities in the Kolmogorov and Smirnov tests. *Bull. Austral. Math. Soc.* **7**, 227–232.

Pitman, E. J. G. (1972). *Some Basic Theory for Statistical Inference.* Chapman and Hall, London.

Puri, M. L. and Sen, P. K. (1971). *Nonparametric methods in multivariate analysis.* Wiley, New York.

Pyke, R. (1972). Empirical Processes. Jefferey-Williams Lecture, Can. Math. Cong., Montreal, 13–143.

Pyke, R. and Shorack, G. R. (1968). Weak convergence of a two sample empirical process and a new approach to Chernoff-Savage theorems. *Ann. Math. Statist.* **39**, 755–771.

Quade, D. (1965). On the asymptotic power of the one-sample Kolmogorov–Smirnov tests. *Ann. Math. Statist.* **36**, 1000–1018.

Raghavachari, M. (1973). Limiting distributions of Kolmogorov–Smirnov type statistics under the alternative. *Ann. Statist.* **1**, 67–73.

Ramachandramurty, P. V. (1966). On the Pitman efficiency of one-sided Kolmogorov and Smirnov tests for normal alternatives. *Ann. Math. Statist.* **37**, 940–944.

O'Reilly, N. E. (1974). On the weak convergence of empirical processes in sup-norm metrics. *Ann. Probability* **2**, 642–651.

Rényi, A. (1953). On the theory of order statistics. *Acta Math. Acad. Sci. Hungar.* **4**, 191–231.

Rényi, A. (1968). On a group of problems in the theory of ordered samples. *Magyar Tud. Akad. Mat. Fiz. Oszt. Közl.* **18**, 23–30; English transl. in: *Selected Transl. Math. Statist. and Probability, Amer. Math. Soc.* **13**, 289–298, Providence, R.I., 1973.

Sahler, W. (1968). A survey of distribution-free statistics based on distances between distribution functions. *Metrika* **13**, 149–169.

Sarkadi, K. (1973). On the exact distributions of statistics of Kolmogorov–Smirnov type. *Period Math. Hungar.* **3** (A. Rényi Mem. Vol. II), 9–12.

Schmid, P. (1958). On the Kolmogorov and Smirnov limit theorems for discontinuous distribution functions. *Ann. Math. Statist.* **29**, 1011–1027.

Sen, P. K. (1973). An almost sure invariance principle for multivariate Kolmogorov–Smirnov statistics. *Ann. Probability* **1**, 488–496.

Sen, P. K. (1981). *Sequential Nonparametrics: Invariance Principles and Statistical Inference.* Wiley, New York.

Shorack, G. R. (1972). Functions of order statistics. *Ann. Math. Statist.* **43**, 412–427.

Shorack, G. R. (1980). Some law of the iterated logarithm type results for the empirical process. *Austral J. Statist.* **22**, 50–59.

Shorack, G. R. and Wellner, J. A. (1978). Linear bounds on the empirical distribution function. *Ann. Probability* **6**, 349–353.

Smirnov, N. V. (1939a). Sur les ècarts de la courbe de distribution empirique. *Mat. Sb.* **6** (48), 3–26 (Russian; French summary).

Smirnov, N. V. (1939b). On the estimation of the discrepancy between empirical curves of distribution for two independent samples. *Bull. Math. Univ. Moscow* **2**, 3–14 (Russian).

Smirnov, N. V. (1944). Approximate laws of distribution of random variables from empirical data. *Uspehi Mat. Nauk.* **10**, 179–206 (Russian).

Smirnov, N. V. (1961). Probabilities of large values of nonparametric one-sided goodness-of-fit statistics. *Trudy Mat. Inst. Steklov.* **64**, 185–210; English transl. in: *Selected Transl. Math. Statist. and Probability, Amer. Math. Soc.* **5**, 210–239. Providence, R.I., 1965.

Stanley, R. M. (1972). Boundary crossing probabilities for the Kolmogorov–Smirnov statistics. *Ann. Math. Statist.* **43**, 664–668.

Steck, G. P. (1969). The Smirnov two-sample tests as rank tests. *Ann. Math. Statist.* **40**, 1449–1466.

Steck, G. P. (1971). Rectangular probabilities for uniform order statistics and the probability that the empirical distribution function lies between two distribution functions. *Ann. Math. Statist.* **42**, 1–11.

Steck, G. P. (1974). A new formula for $P(R_i \leq b_i, 1 \leq i \leq m/m, n, F = G^k)$. *Ann. Probability* **2**, 147–154.

Stute, W. (1982). The oscillation behavior of empirical processes. *Ann. Probability* **10**, 86–107.

Suzuki, G. (1967). On exact probabilities of some generalized Kolmogorov's D-statistics. *Ann. Inst. Statist. Math.* **19**, 373–388.

Takács, L. (1964). The use of a ballot theorem in order statistics. *J. Appl. Probability* **1**, 389–392.

Takács, L. (1965a). Applications of a ballot theorem in physics and in order statistics. *J. Roy. Statist. Soc. Ser. B* **27**, 130–137.

Takács, L. (1965b). The distributions of some statistics depending on the deviations between empirical and theoretical distribution functions. *Sankhyā Ser. A* **27**, 93–100.

Takács, L. (1967). *Combinatorial methods in the theory of stochastic processes.* Wiley, New York.

Takács, L. (1970). Combinatorial methods in the theory of order statistics. In: M. L. Puri, ed., *Nonparametric Techniques in Statistical Inference.* Cambridge Univ. Press, London, pp. 359–384.

Takács, L. (1971). On the comparison of a theoretical and an empirical distribution function. *J. Appl. Probability* **8**, 321–330.

Vincze, I. (1961). On two-sample tests based on order statistics. In: *Proc. Fourth Berkeley Sympos. Math. Statist. and Probability, Vol. I.* Univ. of California Press, Berkeley, CA, pp. 695–705.

Vincze, I. (1967). On some questions connected with two-sample tests of Smirnov type. In: *Proc. Fifth Berkeley Sympos. Math. Statist. and Probability, Vol. I.* Univ. of California Press, Berkeley, CA, pp. 654–666.

Vincze, I. (1968). On the power of the Kolmogorov–Smirnov two-sample test and related non-parametric tests. *Studies in Math. Statist.: Theory and Appl.* (Sympos. Budapest, 1964). Akad. Kiadó, Budapest, pp. 201–210.

Vincze, I. (1970). On Kolmogorov–Smirnov type distribution theorems. In: M. L. Puri, ed., *Nonparametric Techniques in Statistical Inference.* Cambridge Univ. Press, London, pp. 385–401.

Vincze, I. (1972). On some results and problems in connection with statistics of the Kolmogorov–Smirnov type. In: *Proc. Sixth Berkeley Sympos. Math. Statist. and Probability, Vol. I.* Univ. of California Press, Berkeley, CA, pp. 459–470.

Wald, A. and Wolfowitz, J. (1939). Confidence limits for continuous distribution functions. *Ann. Math. Statist.* **10**, 105–118.

Walsh, J. E. (1963). Bounded probability properties of Kolmogorov–Smirnov and similar statistics for discrete data. *Ann. Inst. Statist. Math.* **15**, 153–158.

Wellner, J. A. (1977a). A martingale inequality for the empirical process. *Ann. Probability* **5**, 303–308.

Wellner, J. A. (1977b). A Glivenko–Cantelli theorem and strong laws of large numbers for functions of order statistics. *Ann. Statist.* **5**, 473–480. Correction: **6** (1978), 1391.

Wellner, J. A. (1977c). A law of the iterated logarithm for functions of order statistics. *Ann. Statist.* **5**, 481–494.

Wellner, J. A. (1977d). Distributions related to linear bounds for the empirical distribution function. *Ann. Statist.* **5**, 1003–1016.

Wellner, J. A. (1978). Limit theorems for the ratio of the empirical distribution function to the true distribution function. *Z. Wahrsch. Verw. Gebiete* **45**, 73–88.

Whittle, P. (1961). Some exact results for one-sided distribution tests of the Kolmogorov–Smirnov type. *Ann. Math. Statist.* **32**, 499–505.

Yu, G. C. S. (1971). Pitman efficiencies of Kolmogorov–Smirnov tests. *Ann. Math. Statist.* **42**, 1595–1605.

Yu, G. C. S. (1975). Bounds on the power for some Rényi-type statistics. *J. Amer. Statist. Assoc.* **70**, 233–237.

Yu, G. C. S. (1977). Power bounds on some non-parametric test procedures for censored data. *Sankhyā Ser. B*, **39**, 279–283.

P. R. Krishnaiah and P. K. Sen, eds., *Handbook of Statistics, Vol. 4*

20

Invariance Principles for Empirical Processes

Miklós Csörgő

1. Introduction: Basic notions and definitions

Let $X_1, X_2, \ldots$ be independent identically distributed random variables (i.i.d. rv) on a probability space $(\Omega, \mathcal{A}, P)$ with values in a sample space $(R, \mathcal{B})$. Denote by μ the probability distribution of X_1 on $\mathcal{B}$, i.e.,

$$\mu(B) = P\{\omega \in \Omega: X_1(\omega) \in B\} \quad \text{for all } B \in \mathcal{B}. \tag{1.1}$$

For each $\omega \in \Omega$ and $B \in \mathcal{B}$ let $n\mu_n(B)$ be the number of those $X_1(\omega), \ldots, X_n(\omega)$ which fall into the set B. The number $\mu_n(B)$ is called the *empirical measure* of B for the random sample $X_1, \ldots, X_n$, conveniently written as

$$\mu_n(B) = n^{-1} \sum_{j=1}^{n} 1_B(X_j) \tag{1.2}$$

where

$$1_B(x) = \begin{cases} 1 & \text{if } x \in B, \\ 0 & \text{if } x \notin B. \end{cases} \tag{1.3}$$

The corresponding *empirical measure process* β_n is defined by

$$\beta_n(B) = n^{1/2}(\mu_n(B) - \mu(B)), \quad B \in \mathcal{B}. \tag{1.4}$$

Usually $X_1, X_2, \ldots$ will be random vectors in the Euclidean space $\mathbb{R}^d$ $(d \geq 1)$, i.e., $R = \mathbb{R}^d$, and $\mathcal{B}$ is the Borel subsets of $\mathbb{R}^d$. In this case let F be the right continuous distribution function of X_1, i.e., $F(x) = \mu((-\infty, x]) = P\{\omega \in \Omega: X_1(\omega) \in (-\infty, x)\}$, where $(-\infty, x]$ is a d-dimensional interval with $x \in \mathbb{R}^d$. The corresponding *empirical distribution function* $\mu_n((-\infty, x])$ of the random sample $X_1, \ldots, X_n$ will be denoted by $F_n(x)$, i.e., for each $\omega \in \Omega$ $nF_n(x)$ is the number of those $X_j(\omega) = (X_{j1}(\omega), \ldots, X_{jd}(\omega))$ $(j = 1, \ldots, n)$

Research supported by a NSERC Canada Grant at Carleton University, Ottawa.

whose components are less than or equal to the corresponding components of $x = (x_1, \ldots, x_d) \in \mathbb{R}^d$, conveniently written as

$$F_n(x) = n^{-1} \sum_{j=1}^{n} \prod_{i=1}^{d} 1_{(-\infty, x_i]}(X_{ji}) . \tag{1.5}$$

Whence on denoting $\beta_n((-\infty, x])$ by $\beta_n(x)$, the empirical measure process of (1.4) in terms of these distribution functions F, F_n is

$$\beta_n(x) = n^{1/2}(F_n(x) - F(x)), \quad x \in \mathbb{R}^d \ (d \geq 1), \tag{1.6}$$

and it will be simply called the *empirical process*. If X_1 is uniformly distributed over the unit cube $I^d = [0, 1]^d$ $(d \geq 1)$ then for F_n, F, β_n we use the symbols E_n, λ, α_n and the corresponding *uniform empirical process* then is

$$\alpha_n(y) = n^{1/2}(E_n(y) - \lambda(y)), \quad y \in I^d \ (d \geq 1), \tag{1.7}$$

where $\lambda(y) = \prod_{i=1}^{d} y_i$ with $y = (y_1, \ldots, y_d) \in I^d$.

In the context of *continuous* distribution functions F on $\mathbb{R}^d$ the uniform empirical process occurs the following way. Let $\mathcal{F}$ be *the class of continuous distribution functions* on $\mathbb{R}^d$, and let $\mathcal{F}_0$ be *the subclass consisting of every member of* $\mathcal{F}$ *which is a product of its associated one-dimensional marginal distribution functions*. Let $y_i = F_{(i)}(x_i)$ $(i = 1, \ldots, d)$ be *the i-th marginal* distribution of $F \in \mathcal{F}$ and let $F_{(i)}^{-1}(y_i) = \inf\{x_i \in \mathbb{R}^1 \colon F_{(i)}(x_i) \geq y_i\}$ be its *inverse*. Define the mapping $L^{-1} : I^d \to \mathbb{R}^d$ by

$$L^{-1}(y) = L^{-1}(y_1, \ldots, y_d) = (F_{(1)}^{-1}(y_1), \ldots, F_{(d)}^{-1}(y_d)), \quad y = (y_1, \ldots, y_d) \in I^d . \tag{1.8}$$

Then (cf. (1.6) and (1.7)), *whenever* $F \in \mathcal{F}_0$,

$$\alpha_n(y) = \beta_n(L^{-1}(y)), \quad y = (y_1, \ldots, y_d) \in I^d \ (d \geq 1), \tag{1.9}$$

i.e., if $F \in \mathcal{F}_0$, then the empirical process $\beta_n(L^{-1}(y)) = \alpha_n(y)$ is distribution free (does not depend on the distribution function F). *In statistical terminology* we say that when we are testing the *independence null hypothesis*

$$H_0\colon F \in \mathcal{F}_0 \quad \text{versus the alternative} \quad H_1\colon F \in \mathcal{F} - \mathcal{F}_0 \quad (d \geq 2), \tag{1.10}$$

then the null distribution of $\beta_n(L^{-1}(y))$ is that of $\alpha_n(y)$, i.e., the same for all $F \in \mathcal{F}_0$ and for $d = 1$ with F simply continuous. Otherwise, i.e., if H_1 obtains, the empirical process β_n is a function of F and so will be also its distribution.

Hoeffding (1948), and Blum, Kiefer and Rosenblatt (1961) suggested an *alternate empirical process* for handling H_0 of (1.10). Let F_{ni} be the *marginal empirical distribution function* of the i-th component of X_j ($j = 1, \ldots, n$), i.e.,

$$F_{ni}(x_i) = n^{-1} \sum_{j=1}^{n} 1_{(-\infty, x_i]}(X_{ji}), \quad i = 1, \ldots, d, \tag{1.11}$$

and define

$$T_n(x) = T_n(x_1, \ldots, x_d) = n^{1/2}\Big(F_n(x) - \prod_{i=1}^{d} F_{ni}(x_i)\Big), \quad d \geq 2, \tag{1.12}$$

with F_n as in (1.5). In terms of the mapping L^{-1} of (1.8) we define t_n, the uniform version of T_n, by

$$\begin{aligned} t_n(y) = T_n(L^{-1}(y)) &= n^{1/2}\Big(F_n(F_{(1)}^{-1}(y_1), \ldots, F_{(d)}^{-1}(y_d)) - \prod_{i=1}^{d} F_{ni}(F_{(i)}^{-1}(y_i))\Big) \\ &= n^{1/2}\Big(E_n(y) - \prod_{i=1}^{d} E_{ni}(y_i)\Big), \end{aligned} \tag{1.13}$$

where $E_{ni}(y_i)$ ($i = 1, \ldots, d$) is the i-th uniform empirical distribution function of the i-th component of $L(X_j) = (F_{(1)}(X_{j1}), \ldots, F_d(X_{jd}))$ ($j = 1, \ldots, n$), i.e., L is the inverse of L^{-1} of (1.8). Consequently H_0 of (1.10) is equivalent to H_0: $F(L^{-1}(y)) = \Pi_{i=1}^{d} y_i = \lambda(y)$, i.e., given H_0, $T_n(L^{-1}(y)) = t_n(y)$ is distribution free. Hence, in order to study the distribution of T_n under H_0, we may take F to be the uniform distribution on I^d ($d \geq 2$) and study the distribution of t_n instead.

If F is *a continuous distribution function* on $\mathbb{R}^1$ then its inverse function F^{-1} *will be called the quantile function* Q of F, i.e.,

$$\begin{aligned} Q(y) = F^{-1}(y) &= \inf\{x \in \mathbb{R}^1 : F(x) \geq y\} \\ &= \inf\{x \in \mathbb{R}^1 : F(x) = y\}, \quad 0 \leq y \leq 1 . \end{aligned} \tag{1.14}$$

Thus $F(Q(y)) = y \in [0, 1]$, and if we put $U_1 = F(X_1)$, then U_1 is a uniformly distributed rv on $[0, 1]$. Also $U_1 = F(X_1)$, $U_2 = F(X_2), \ldots$ are independent uniform-$[0, 1]$ rv. Let $X_{1:n} \leq X_{2:n} \leq \cdots \leq X_{n:n}$ be the order statistics of a random sample on F which, in turn, induce the uniform-$[0, 1]$ order statistics $U_{1:n} = F(X_{1:n}) \leq U_{2:n} = F(X_{2:n}) \leq \cdots \leq U_{n:n} = F(X_{n:n})$ of $U_1, U_2, \ldots, U_n$. The empirical distribution function F_n of (1.5) can be written as

$$F_n(x) = \begin{cases} 0 & \text{if } X_{1:n} > x, \\ k/n & \text{if } X_{k:n} \leq x < X_{k+1:n}, \\ 1 & \text{if } X_{n:n} \leq x, \end{cases} \tag{1.15}$$

$x \in \mathbb{R}^1$, and the uniform empirical distribution function E_n of $U_1, \ldots, U_n$ as

$$E_n(y) = F_n(Q(y)) = \begin{cases} 0 & \text{if } U_{1:n}, \\ k/n & \text{if } U_{k:n} \leq y < U_{k+1:n}, \\ 1 & \text{if } U_{n:n} \leq y, \end{cases} \tag{1.16}$$

$0 \leq y \leq 1$. Then (1.7) takes up the simple form

$$\alpha_n(y) = \beta_n(Q(y)) = n^{1/2}(E_n(y) - y), \quad 0 \leq y \leq 1. \tag{1.17}$$

In terms of the latter empirical distribution functions F_n and E_n we define now the *empirical quantile function* Q_n by

$$\begin{aligned} Q_n(y) = F_n^{-1}(y) &= \inf\{x \in \mathbb{R}^1 \colon F_n(x) \geq y\} \\ &= X_{k:n} \quad \text{if } (k-1)/n < y \leq k/n \quad (k = 1, \ldots, n) \end{aligned} \tag{1.18}$$

and the uniform empirical quantile function U_n by

$$\begin{aligned} U_n(y) = E_n^{-1}(y) &= \inf\{u \in [0, 1] \colon F_n(Q(u)) \geq y\} \\ &= U_{k:n} = F(X_{k:n}) \quad \text{if } (k-1)/n < y \leq k/n \quad (k = 1, \ldots, n) \\ &= F(Q_n(y)). \end{aligned} \tag{1.19}$$

If F is an *absolutely continuous distribution function* on $\mathbb{R}^1$, then let $f = F'$ be its density function (with respect to Lebesgue measure). We define the *quantile process* ρ_n by

$$\rho_n(y) = n^{1/2} f(Q(y))(Q_n(y) - Q(y)), \quad y \in (0, 1), \tag{1.20}$$

and the corresponding *uniform quantile process* u_n by

$$u_n(y) = n^{1/2}(U_n(y) - y). \tag{1.21}$$

A simple relationship like that of (1.17) for α_n and β_n does not exist for u_n and ρ_n. However by (1.19) we have

$$\rho_n(y) = n^{1/2} f(Q(y))(Q(F(Q_n(y))) - Q(y)) = u_n(y)(f(Q(y))/f(Q(\theta_{y,n}))), \tag{1.22}$$

where $U_n(y) \wedge y < \theta_{y,n} < U_n(y) \vee y$. Since $u_n(k/n) = -\alpha_n(U_{k:n})$ $(k = 1, \ldots, n)$, it is reasonable to expect that the asymptotic distribution theory of α_n and u_n should be the same. This, in turn, implies that via (1.22) ρ_n should also have the same kind of asymptotic theory if f is 'nice'. We are going to see in Section 4 that this actually is true under appropriate conditions of f.

Let $C(t) = \int_{\mathbb{R}^d} \exp(i\langle t, x\rangle)\,\mathrm{d}F(x)$ be the characteristic function of $F(x)$ on $\mathbb{R}^d$, where $\langle t, x\rangle = \sum_{k=1}^d t_k x_k$, the usual inner product of $t = (t_1, \ldots, t_k)$, $x = (x_1, \ldots, x_d) \in \mathbb{R}^d$. With F_n as in (1.5) the *empirical characteristic function* C_n of the sample $X_1, \ldots, X_n$ is defined by

$$C_n(t) = n^{-1} \sum_{k=1}^n \exp(i\langle t, X_k\rangle) = \int_{\mathbb{R}^d} \exp(i\langle t, x\rangle)\,\mathrm{d}F_n(x), \quad t \in \mathbb{R}^d. \tag{1.23}$$

Next we define a number of Gaussian processes which play a basic role in the asymptotic distribution of some of the empirical processes introduced so far.

Wiener process: A real valued separable Gaussian process

$$\{W(x); x \in \mathbb{R}_+^d\} = \{W(x_1, \ldots, x_d); 0 \leq x_i < \infty, i = 1, \ldots, d\}$$

with continuous sample paths is called a Wiener process if $EW(x) = 0$ and

$$EW(x)W(y) = \lambda(x \wedge y) \quad \text{with } \lambda(x \wedge y) = \prod_{i=1}^{d} (x_i \wedge y_i),$$

where $x = (x_1, \ldots, x_d)$, $y = (y_1, \ldots, y_d) \in \mathbb{R}_+^d$ $(d \geq 1)$.

Brownian bridge: $\{B(x); x \in I^d\} = \{W(x) - \lambda(x)W(1, \ldots, 1); x \in I^d\}$. Whence $EB(x) = 0$ and $EB(x)B(y) = \lambda(x \wedge y) - \lambda(x)\lambda(y)$, $x = (x_1, \ldots, x_d)$, $y = (y_1, \ldots, y_d) \in I^d$ $(d \geq 1)$, where $\lambda(x) = \lambda(x \wedge x)$.

Kiefer process: $\{K(x, t); (x, t) \in I^d \times \mathbb{R}_+^1\} = \{W(x, t) - \lambda(x)W(1, \ldots, 1, t); x \in I^d, t \geq 0\}$. Whence $EK(x, t) = 0$ and

$$EK(x, t_1)K(y, t_2) = (t_1 \wedge t_2)(\lambda(x \wedge y) - \lambda(x)\lambda(y)),$$

where $x = (x_1, \ldots, x_d)$, $y = (y_1, \ldots, y_d) \in I^d$, $t_1, t_2 \geq 0$.

For a proof of existence of the multitime parameter Wiener process $\{W(x); x \in \mathbb{R}_+^d\}$, which is sometimes called the Yeh–Wiener process or Brownian (Wiener) sheet, we refer to Yeh (1960), Čencov (1956), or Csörgő and Révész (1981a, Section 1.11). A Kiefer process at integer valued arguments $t = n$ can be also viewed as the partial sum process of a sequence of *independent* Brownian bridges $\{B_i(x); x \in I^d\}_{i=1}^{\infty}$:

$$\{K(x, n); x \in I^d, n = 1, 2, \ldots\} = \left\{\sum_{i=1}^{n} B_i(x); x \in I^d, n = 1, 2, \ldots\right\},$$

and

$$\{B_n(x); x \in I^d\} = \{K(x, n) - K(x, n-1); x \in I^d\} \quad (n = 1, 2, \ldots) \tag{1.24}$$

is a sequence of independent Brownian bridges.

For further properties of a Kiefer process we refer to M. Csörgő and P. Révész (1981a, Section 1.15 and Theorem S.1.15.1).

Strong approximations of empirical processes are going to be described in Section 2. For distribution theory of tests based on the sample distribution function on $\mathbb{R}^1$ we refer to Durbin (1973a). A direct treatment of the empirical process on $\mathbb{R}^1$ and many of its statistical applications can be seen in this volume by Csáki (1982b), and Doksum and Yandell (1982) (cf. Also Csáki, 1977a, 1977b, 1982a). For the parameters estimated empirical process on $\mathbb{R}^d$ $(d \geq 1)$ we refer to Durbin (1973b, 1976), Neuhaus (1976) and M. D. Burke, M. Csörgő, S. Csörgő and P. Révész (1979). For a theory of, and further references on strong and weak convergence of the empirical characteristic function we refer to S. Csörgő (1980, 1981a,b).

Recent work on the limiting distribution of and critical values for the multivariate Cramér–von Mises and Hoeffding–Blum–Kiefer–Rosenblatt independence criteria is reviewed in Section 3.

An up-to-date review of strong and weak approximations of the quantile process, including that of weak convergence in $\|\cdot/q\|$-metrics, is given in Section 4. For further readings, references on this subject and its statistical applications, we refer to Chapters 4, 5 in Csörgő and Révész (1981a), Csörgő and Révész (1981), M. Csörgő, (1983), and Csörgő, Csörgő, Horváth and Révész (1982). For the parameters estimated quantile process we refer to Carleton Mathematical Lecture Note No. 31 by Aly and Csörgő (1981) and references therein. For recent results on the product-limit empirical and quantile processes we refer to Carleton Mathematical Lecture Note No. 38 by Aly, Burke, Csörgő, Csörgő and Horváth (1982) and references therein and to Chapter VIII in M. Csörgő (1983).

2. Strong and weak approximations of empirical processes by Gaussian processes

There is an excellent recent survey of results concerning empirical processes and measures of i.i.d. rv by Gaenssler and Stute (1979). Chapter 4 of Csörgő and Révész (1981a) is also devoted to the same subject on $\mathbb{R}^1$. For further references, in addition to the ones mentioned in this exposition, we refer to these sources. Even with the latter references added, this review is not and, indeed, does not intend to be complete. In this section we are going to list, or mention, essentially only those results for α_n and β_n which are best possible or appear to be the best available ones so far in their respective categories.

First, on strong approximations of the *uniform empirical process* α_n of (1.7) or, equivalently, that of $\alpha_n = \beta_n(L^{-1})$ of (1.9) in terms of a *sequence* of Brownian bridges $\{B_n(x); x \in I^d\}_{n=1}^{\infty}$ we have

THEOREM 2.1. *Let $X_1, \ldots, X_n$ $(n = 1, 2, \ldots)$ be independent random d-vectors, uniformly distributed on I^d, or with distribution function $F \in \mathcal{F}_0$ on $\mathbb{R}^d$. Let α_n be as in* (1.7) *or in* (1.9). *Then one can construct a probability space for $X_1, X_2, \ldots$ with a sequence of Brownian bridges $\{B_n(y); y \in I^d\ (d \geq 1)\}_{n=1}^{\infty}$ on it so that:*

(i) *for all n and x we have* (*cf. Komlós, Major and Tusnády*, 1975a)

$$P\{\sup_{y\in I^1} |\alpha_n(y) - B_n(y)| > n^{-1/2}(C\log n + x)\} \leq Le^{-\lambda x}, \tag{2.1}$$

where C, L, λ are positive absolute constants,

(ii) *for all n and x we have* (*cf. Tusnády*, 1977a)

$$P\{\sup_{y\in I^2} |\alpha_n(y) - B_n(y)| > n^{-1/2}(C\log n + x)\log n\} \leq Le^{-\lambda x}, \tag{2.2}$$

where C, L, λ are positive absolute constants,

(iii) *for any $\lambda > 0$ there exists a constant $C > 0$ such that for each n (cf. Csörgő and Révész*, 1975a)

$$P\{\sup_{y\in I^d} |\alpha_n(y) - B_n(y)| > C(\log n)^{3/2} n^{-1/2(d+1)}\} \leq n^{-\lambda}, \quad d \geq 1 . \tag{2.3}$$

The constants of (2.1) *for example can be chosen as* $C = 100$, $L = 10$, $\lambda = 1/50$ (*cf. Tusnády*, 1977b,c).

COROLLARY 2.1. (2.1), (2.2), (2.3) *in turn imply*

$$\sup_{y\in I^1} |\alpha_n(y) - B_n(y)| \overset{\text{a.s.}}{=} O(n^{-1/2} \log n) , \tag{2.4}$$

$$\sup_{y\in I^2} |\alpha_n(y) - B_n(y)| \overset{\text{a.s.}}{=} O(n^{-1/2} \log^2 n) , \tag{2.5}$$

$$\sup_{y\in I^d} |\alpha_n(y) - B_n(y)| \overset{\text{a.s.}}{=} O(n^{-1/2(d+1)} (\log n)^{3/2}), \quad d \geq 1 . \tag{2.6}$$

Due to Bártfai (1966) and/or to the Erdős–Rényi (1970) theorem, the O(·) rate of convergence of (2.4) is best possible (cf. Komlós, Major, Tusnády, 1975a, or Theorem 4.4.2 in Csörgő and Révész, 1981a).

Next, on strong approximations of the *uniform empirical process* α_n of (1.7) or, equivalently, that of $\beta_n(L^{-1})$ of (1.9) in terms of a *single* Gaussian process, the Kiefer process $\{K(x, t); (x, t) \in I^d \times \mathbb{R}^1_+\}$, we have

THEOREM 2.2. *Let $X_1, \ldots, X_n$ $(n = 1, 2, \ldots)$ be independent random d-vectors, uniformly distributed on I^d, or with distribution function $F \in \mathcal{F}_0$ on $\mathbb{R}^d$. Let α_n be as in* (1.7) *or in* (1.9). *Then one can construct a probability space for $X_1, X_2, \ldots$ with a Kiefer process $\{K(y, t); (y, t) \in I^d \times \mathbb{R}^1_+\}$ on it so that*:

(i) *for all n and x we have (cf. Komlós, Major and Tusnády*, 1975a)

$$P\{\sup_{1\leq k\leq n} \sup_{y\in I^1} |k^{1/2}\alpha_k(y) - K(y, k)| > (C \log n + x) \log n\} < Le^{-x\lambda} , \tag{2.7}$$

where C, L, λ are positive absolute constants,

(ii) *for any $\lambda > 0$ there exists a constant $C > 0$ such that for each n (cf. Csörgő and Révész*, 1975a)

$$P\{\sup_{1\leq k\leq n} \sup_{y\in I^d} |k^{1/2}\alpha_k(y) - K(y, k)| > Cn^{(d+1)/2(d+2)} \log^2 n\} \leq n^{-\lambda}, \quad d \geq 1 . \tag{2.8}$$

COROLLARY 2.2. (2.7), (2.8) *in turn imply*

$$n^{-1/2} \sup_{1\leq k\leq n} \sup_{y\in I^1} |k^{1/2}\alpha^k(y) - K(y, k)| \overset{\text{a.s.}}{=} O(\log^2 n/n^{1/2}) , \tag{2.9}$$

$$n^{-1/2} \sup_{1\le k\le n} \sup_{y\in I^d} |k^{1/2}\alpha_k(y) - K(y,k)| \overset{\text{a.s.}}{=} O(n^{-1/(2d+4)} \log^2 n), \quad d \geq 1. \tag{2.10}$$

The first result of the type of (2.4) is due to Brillinger (1969) with the a.s. rate of convergence $O(n^{-1/4}(\log n)^{1/2}(\log\log n)^{1/4})$. Kiefer (1969) was first to call attention to the desirability of viewing the one dimensional (in y) empirical process $\alpha_n(y)$ as a two-time parameter stochastic process in y and n and that it should be a.s. approximated in terms of an appropriate two-time parameter Gaussian process. Müller (1970) introduced $\{K(y,t); (y,t) \in I^1 \times \mathbb{R}^1_+\}$ and proved a corresponding two dimensional weak convergence of $\{\alpha_n(y); y \in I^1, n = 1, 2, \ldots\}$ to the latter stochastic process. Kiefer (1972) gave the first strong approximation solution of the type (2.9) with the a.s. rate of convergence $O(n^{-1/6}(\log n)^{2/3})$.

Both Corollaries 2.1 and 2.2 imply the weak convergence of α_n to a Brownian bridge B on the Skorohod space $D[0,1]^d$. Writing $k = [ns]$ ($s \in [0,1]$), $[ns]^{1/2}\alpha_{[ns]}(y)/n^{1/2}$ is a random element in $D[0,1]^{d+1}$ for each integer n, and (2.10) implies also

COROLLARY 2.3. $[n\cdot]^{1/2}\alpha_{[n\cdot]}(\cdot)/n^{1/2} \xrightarrow{D} K(\cdot,\cdot)$ *on* $D[0,1]^{d+1}$.

For $d = 1$ the latter result is essentially the above mentioned result of Müller (1970) (cf. also discussion of the latter on page 217 in Gaenssler and Stute, 1979).

Corollary 2.2 also provides strong invariance principles, i.e. laws like the Glivenko–Cantelli theorem, LIL, etc. are inherited by $\alpha_n(y)$ from $K(y,n)$ or vice versa (cf., e.g., Section 5.1, Theorem S.5.1.1 in Csörgő and Révész, 1981a). The main difference between the said Corollaries is that such strong laws like the ones mentioned do not follow from Corollary 2.1. In the latter we have no information about the finite dimensional distributions in n of the sequence of Brownian bridges $\{B_n(y)\}_{n=1}^\infty$. On the other hand, the inequalities (2.1), (2.2) and (2.3) can be used to estimate the rates of convergence for the distributions of some functionals of α_n (to those of a Brownian bridge B), and those of the appropriate Prohorov distances of measures generated by the sequences of stochastic processes $\{\alpha_n(y); y \in I^d\}_{n=1}^\infty$ and $\{B_n(y); y \in I^d\}_{n=1}$ (cf., e.g., Komlós, Major and Tusnády, 1975b; Theorem 1.16 in M. Csörgő, 1981a; Theorem 2.3.1 in Gaenssler and Stute, 1979 and references therein; Borovkov, 1978; and review of the latter paper by M. Csörgő, 1981, MR 81j; 60044; the latter two to be interpreted in terms of $\{\alpha_n(y); y \in I^d\}_{n=1}^\infty$ and $\{B_n(y); y \in I^d\}_{n=1}^\infty$ instead of the there considered partial sum and Wiener processes).

When the distribution function F of (1.6) is not a product of its marginals for all $x \in \mathbb{R}^d$ ($d \geq 2$), then strong and weak approximations of β_n can be described in terms of the following *Gaussian processes associated with the distribution function* $F(x)$ ($x \in \mathbb{R}^d$, $d \geq 2$).

Brownian bridge B_F *associated with* F *on* $\mathbb{R}^d$ ($d \geq 2$): A separable d-

parameter real valued Gaussian process with the following properties

$$EB_F(x)=0, \qquad EB_F(x)B_F(y)=F(x\wedge y)-F(x)F(y),$$

$$\lim_{x_i\to-\infty} B_F(x_1,\ldots,x_d)=0 \quad (i=1,\ldots,d),$$

$$\lim_{(x_1,\ldots,x_d)\to(\infty,\ldots,\infty)} B_F(x_1,\ldots,x_d)=0,$$

where for $x, y\in\mathbb{R}^d$ we write $x\wedge y=(x_1\wedge y_1,\ldots,x_d\wedge y_d)$.

Kiefer process $K_F(\cdot,\cdot)$ on $\mathbb{R}^d\times[0,\infty)$ associated with the distribution function F on $\mathbb{R}^d$ ($d\geq 2$): A separable $(d+1)$-parameter real valued ($x\in\mathbb{R}^d, 0\leq t<\infty$) Gaussian process with the following properties:

$$K_F(x,0)=0,$$

$$\lim_{x_i\to-\infty} K_F(x_1,\ldots,x_d,t)=0 \quad (i=1,\ldots,d),$$

$$\lim_{(x_1,\ldots,x_d)\to(\infty,\ldots,\infty)} K_F(x_1,\ldots,x_d,t)=0,$$

$$EK_F(x,t)=0 \quad\text{and}\quad EK_F(x,t_1)K_F(y,t_2)=(t_1\wedge t_2)(F(x\wedge y)-F(x)F(y))$$

for all $x, y\in\mathbb{R}^d$ and $t_1, t_2\geq 0$.

A more tractable description of B_F and K_F can be given in terms of the mapping $L:\mathbb{R}^d\to I^d$ defined by

$$L(x)=L(x_1,\ldots,x_d)=(F_{(1)}(x_1),\ldots,F_{(d)}(x_d))=(y_1,\ldots,y_d)\in I^d,$$
$$(x_1,\ldots,x_d)\in\mathbb{R}^d, \tag{2.11}$$

the inverse map of L^{-1} of (1.8), where, just as in the latter map, $y_i=F_{(i)}(x_i)$ ($i=1,\ldots,d$) are the i-th marginals of F. It is well known that (cf. p. 293 in Wichura, 1973; or Lemma 1 in Philipp and Pinzur, 1980; or Lemma 3.2 in Moore and Spruill, 1975) there is a d-variate distribution function G on I^d with uniform marginals on [0, 1] such that

$$F(x)=G(L(x)), \qquad G(y)=F(L^{-1}(y)), \tag{2.12}$$

i.e., G on I^d has uniform marginals $y_i=F_{(i)}(x_i)$ ($i=1,\ldots,d$) on [0, 1]. (We note for example that if $F\in\mathcal{F}_0$ then $G(y)=\lambda(y)$ with $\lambda(\cdot)$ as in (1.7), i.e., *in the latter case $G(y)$ is the uniform distribution function on I^d*.)

Now consider the d-parameter Wiener process W_G associated with the distribution function G on I^d, defined as follows.

Wiener process W_G on I^d associated with the distribution function G on I^d ($d\geq 2$): A real valued d-parameter Gaussian process $\{W_G(y); y\in I^d\}$ with $EW_G(y)=0$, $EW_G(x)W_G(y)=G(x\wedge y)$, and $W_G(y_1,\ldots,y_d)=0$ whenever $y_i=0$ ($i=1,\ldots,d$).

Then the d-parameter Gaussian process

$$\{B_G(y);\, y\in I^d\}=\{W_G(y_1,\ldots,y_d)-G(y_1,\ldots,y_d)W_G(1,\ldots,1);\\ y=(y_1,\ldots,y_d)\in I^d\} \tag{2.13}$$

is a *Brownian bridge process on* I^d *associated with* G *on* I^d, and *the Brownian bridge process* B_F *associated with* F *on* $\mathbb{R}^d$ *can be represented* via (2.12) and (2.13) *as*

$$\{B_F(x);\, x\in\mathbb{R}^d\}=\{W_G(L(x))-G(L(x))W_G(1,\ldots,1);\, x\in\mathbb{R}^d\}$$

and

$$\{B_F(L^{-1}(y));\, y\in I^d\}=\{B_G(y);\, y\in I^d\}. \tag{2.14}$$

Consider also the $(d+1)$-parameter Wiener process $W_G(\cdot,\cdot)$ on $I^d\times[0,\infty)$ associated with the distribution function G on I^d, defined as follows.

Wiener process $W_G(\cdot,\cdot)$ *on* $I^d\times[0,\infty]$ *associated with the distribution function* G *on* I^d $(d\geq 2)$: A real valued $(d+1)$-parameter Gaussian process $\{W_G(y,t);\, y\in I^d,\ t\geq 0\}$ with $W_G(y_1,\ldots,y_d,t)=0$ whenever any of $y_1,\ldots,y_d$ or t is zero, $EW_G(y,t)=0$, and $EW_G(y,t_1)W_G(x,t_2)=(t_1\wedge t_2)G(x\wedge y)$.

Then the $(d+1)$-parameter Gaussian process

$$\{K_G(y,t);\, y\in I^d, t\geq 0\}=\{W_G(y_1,\ldots,y_d,t)-G(y_1,\ldots,y_d)W_G(1,\ldots,1,t);\\ y=(y_1,\ldots,y_d)\in I^d, t\geq 0\} \tag{2.15}$$

is a *Kiefer process on* $I^d\times[0,\infty]$ *associated with* G *on* I^d, and *the Kiefer process* $K_F(\cdot,\cdot)$ *associated with* F *on* $\mathbb{R}^d$ *can be represented* via (2.12) and (2.15) *as*

$$\{K_F(x,t);\, x\in\mathbb{R}^d, t\geq 0\}\\ =\{W_G(L(x),t)-G(L(x))W_G(1,\ldots,1,t);\, x\in\mathbb{R}^d,\ r\geq 0\}$$

and

$$\{K_F(L^{-1}(y),t);\, y\in I^d, t\geq 0\}=\{K_G(y,t);\, y\in I^d,\ t\geq 0\}. \tag{2.16}$$

We note that if $F\in\mathscr{F}_0$ or $d=1$, then the latter Kiefer processes $K_G(\cdot,\cdot)$ and $K_F(\cdot,\cdot)$ coincide with our originally defined Kiefer process $K(\cdot,\cdot)$ on $I^d\times\mathbb{R}^1_+$. The same is true concerning our originally defined Brownian bridge B, and the Brownian bridges B_G and B_F in the context of $F\in\mathscr{F}_0$. Note also that in general β_n of (1.16) can be written as

$$\beta_n(L^{-1}(y))=n^{1/2}(F_n(L^{-1}(y))-F(L^{-1}(y))),\\ =n^{1/2}(F_n(L^{-1}(y))-G(y)),\quad y\in I^d, \tag{2.17}$$

which, in turn, reduces to the equality of (1.9) whenever $F\in\mathscr{F}_0$ or $d=1$.

As far as we know, the best available strong approximation of the empirical process β_n of (1.6) or, equivalently, that of $\beta_n(L^{-1})$ of (2.17) is (for more recent information we refer to Borisov, 1982).

THEOREM 2.3 (Philipp and Pinzur 1980). *Let* $X_1, \ldots, X_n$ $(n = 1, 2, \ldots)$ *be independent d-vectors on* $\mathbb{R}^d$ *with distribution function F. Let* β_n *be as in* (1.6) *or, equivalently, as in* (2.17). *Then one can construct a probability space for* $X_1, X_2, \ldots$ *with a Kiefer process* $\{K_F(x, t);\ x \in \mathbb{R}^d,\ t \geq 0\}$ *associated with F on it such that*

$$
\begin{aligned}
&P\{\sup_{1 \leq k \leq n} \sup_{x \in \mathbb{R}^d} |k^{1/2}\beta_k(x) - K_F(x, k)| > C_1 n^{(1/2)-\lambda}\} \\
&\quad = P\{\sup_{1 \leq k \leq n} \sup_{y \in I^d} |k^{1/2}\beta_k(L^{-1}(y)) - K_G(y, k)| > C_1 n^{(1/2)-\lambda}\} \leq C_2 n^{-(1+1/36)}
\end{aligned}
\tag{2.18}
$$

for $\lambda = 1/(5000d^2)$, *where* C_1, C_2 *are positive constants depending only on F and d.*

COROLLARY 2.4. (2.18) *in turn implies*

$$
\begin{aligned}
&n^{-1/2} \sup_{1 \leq k \leq n} \sup_{x \in \mathbb{R}^d} |k^{1/2}\beta_k(x) - K_F(x, k)| \\
&\quad = n^{-1/2} \sup_{1 \leq k \leq n} \sup_{y \in I^d} |k^{1/2}\beta_k(L^{-1}(y)) - K_G(y, k)| \overset{\text{a.s.}}{=} O(n^{-\lambda})
\end{aligned}
\tag{2.19}
$$

with λ *as in Theorem* 2.3.

REMARK 2.1. We note that Philipp and Pinzur (1980) state only (2.19) in their Theorem 1. S. Csörgő (1981b) noted that going through their proof one can see that they had in fact proved the somewhat stronger (2.18).

REMARK 2.2. Theorem 2.3 and Corollary 2.4 are best available in the sense that in them there are no assumptions made on F. The a.s. rates of convergence of Corollary 2.2 are of course better than that (2.19), but in the former F is assumed to be uniformly distributed on I^d. While in case of $d = 1$ the latter assumption is not a restriction (for $d = 1$ and F continuous we have (1.9), and if F is arbitrary in the latter case, then (2.1) and (2.7) remain true (cf. Remark 1 in S. Csörgő, 1981a)), for $d \geq 2$ it is. Csörgő and Révész (1975b) actually proved (2.18) with the better rate $n^{(d+1)/(2d+4)} \log^2 n$ (cf. (2.8)) replacing its present rate of $n^{(1/2)-\lambda}$ $(\lambda = 1/(5000d^2))$, but only for a class of d-variate distribution functions satisfying a rather strict regularity condition.

Clearly, for all fixed $t > 0$,

$$
\begin{aligned}
&\{t^{-1/2}K_F(L^{-1}(y), t);\ y \in I^d\} = \{t^{-1/2}K_G(y, t);\ y \in I^d\} \\
&\quad \overset{\mathrm{D}}{=} \{B_F(L^{-1}(y));\ y \in I^d\} = \{B_G(y);\ y \in I^d\}\,.
\end{aligned}
\tag{2.20}
$$

Therefore Corollary 2.4 implies

COROLLARY 2.5. $\beta_n(L^{-1}(\cdot)) \xrightarrow{\mathrm{D}} B_G(\cdot)$ *on* $D[0, 1]^d$. (2.21)

Also on writing $k = [ns]$ ($s \in [0, 1]$), $[ns]^{1/2}\beta_{[ns]}(L^{-1}(y))/n^{1/2}$ is a random element in $D[0, 1]^{d+1}$ for each integer n, and (2.19) implies

COROLLARY 2.6. $[n\cdot]^{1/2}\beta_{[n\cdot]}(L^{-1}(\cdot))/n^{1/2} \xrightarrow{\mathrm{D}} K_G(\cdot, \cdot)$ *on* $D[0, 1]^{d+1}$. (2.22)

The result of (2.21) was first proved by Dudley (1966) (cf. also Neuhaus, 1971; Straf, 1971; Bickel and Wichura, 1971; Theorem 2.1.3 in Gaenssler and Stute, 1979, and discussion of the latter theorem therein). The result of (2.22) was first proved by Bickel and Wichura (1971) (cf. Theorems 2.1.4 and 2.1.5 in Gaenssler and Stute, 1979; see also Neuhaus and Sen, 1977).

A common property of the quoted results so far is that they are uniform approximations of measures over intervals only (of I^d or those of $\mathbb{R}^d$ mapped onto I^d). Concerning now the more general problem of approximating the empirical measure process β_n of (1.4) by appropriate Gaussian measure processes, the question is over how rich a class of subsets $\mathscr{C} \subset \mathscr{B}$ of $(R, \mathscr{B})$ (cf. paragraph one of Section 1) could we possibly have theorems like, for example, Theorems 2.1, 2.2 and 2.3. Let $(R, \mathscr{B}, \mu) = (I^d, \mathscr{B}, \lambda)$, λ the uniform Lebesgue measure on I^d, i.e., $X_1, \ldots, X_n$ ($n = 1, 2, \ldots$) are distributed as in Theorems 2.1 and 2.2. In this case write $\alpha_n(B) = n^{1/2}(\lambda_n(B) - \lambda(B))$, $B \in \mathscr{B}$, instead of β_n of (1.4). Then, while it is true (cf. Philipp, 1973) that

$$\limsup_{n\to\infty} \sup_{B\in\mathscr{C}} (2n \log\log n)^{1/2}\alpha_n(B) \overset{\text{a.s.}}{=} \tfrac{1}{2}, \tag{2.23}$$

where $\mathscr{C} = \mathscr{C}(2)$ is the class of convex sets of I^2, it is also known that the law of the iterated logarithm (2.23) fails for $\mathscr{C} = \mathscr{C}(d)$, the class of convex sets of I^d ($d \geq 3$). The latter negative result for $d = 3$ was only recently proved by Dudley (1982) (for a discussion of previous results see e.g. Gaenssler and Stute 1979). In spite of (2.23) dimension two ($d = 2$) is also critical, for Dudley (1982) showed also that if $\mathscr{C}$ is the collection of lower layers in I^2 (a lower layer in I^2 is a set B such that if $(x, y) \in B$, $u \leq x$ and $v \leq y$, then $(u, v) \in B$), then the law of the iterated logarithm (2.23) fails again (for previous mostly negative results and references along these lines for higher dimensions we refer to Stute, 1977; and Gaenssler and Stute, 1979). The same kind of negative results hold true concerning the problem of central limit theorem for $\alpha_n(B)$ (cf. Dudley, 1979), i.e., concerning empirical measure processes on I^d or $\mathbb{R}^d$, for the central limit theorem as well as law of the iterated logarithm the critical dimension is 2 for the lower layers and 3 for the convex sets. Hence any extension of results like those of Theorems 2.1, 2.2 and Corollaries 2.1, 2.2 in terms of uniform distances over a class of sets $\mathscr{C}$ of I^d, other than the intervals already considered, can only be true for somewhat restricted classes $\mathscr{C} \subset \mathscr{B}$. Révész (1976a,b) extended (2.6) and (2.10) over sets in I^d, defined by differentiability conditions. For example, instead of (2.6) we have (Révész, 1976a)

$$\sup_{B\in\mathscr{C}} |\alpha_n(B) - B_n(B)| \overset{\text{a.s.}}{=} O(n^{-1/19}) \tag{2.24}$$

and, instead of (2.10) we can have (Révész, 1976a)

$$n^{-1/2} \sup_{B\in\mathscr{C}} |n^{1/2}\alpha_n(B) - K(B, n)| \overset{\text{a.s.}}{=} O(n^{-1/50}), \tag{2.25}$$

where $\mathscr{C}$ is the class of those Borel sets of I^2 which have twice differentiable boundaries, and $\{B_n(B); B \in \mathscr{C}\}_{n=1}^{\infty}$ respectively $\{K(B, n); B \in \mathscr{C}, n \geq 1\}$ are Gaussian measure processes with mean zero and covariance function $EB_n(B)B_n(D) = \lambda(B \cap D) - \lambda(B)\lambda(D)$ $(n = 1, 2, \ldots)$ respectively $EK(B, n)K(D, m) = (n \wedge m)(\lambda(B \cap D) - \lambda(B)\lambda(D))$ for all $B, D \in \mathscr{C}$. Similar extensions hold true over sets in I^d $(d \geq 2)$ with differentiable boundaries (cf. Révész, 1976b; Ibero, 1979a,b).

A common feature of the results of Theorem 2.2, Corollary 2.2 and that of the first one of these types by Kiefer (1972) is not only that they improve (they imply for instance functional laws of iterated logarithm (cf., e.g., Section 5.1 in Csörgő and Révész, 1981a)) and are conceptually simpler than the original weak convergence result of Donsker (1952) on empirical distribution functions, but they also avoid the problem of measurability and topology caused by the fact that $D[0, 1]^d$ endowed with the supremum norm is not separable (cf., e.g., Billingsley, 1968, p. 153). This idea of proving a.s. or in probability invariance principles à la Kiefer (1972) also works for distribution functions on $\mathbb{R}^d$ (cf. Theorem 2.3 and its predecessors by Révész (1976a, Theorem 3), and Csörgő and Révész (1975b, Theorem 1)) and, as we have just seen in (2.24) and (2.25), for uniform distances over sets of I^d, defined by differentiability conditions (see also Révész, 1976b; Ibero, 1979a,b). Recently Dudley and Philipp (1981) used the same idea to reformulate and strengthen the results of Dudley (1978, 1981a,b), Kuelbs (1976) on empirical measure processes while removing their previously assumed measurability conditions. They do this by proving invariance principles for sums of not necessarily measurable random elements with values in a not necessarily separable Banach space and by showing that empirical measure processes fit easily into the latter setup. We refer for example to Theorems 1.5 and 7.1 in Dudley and Philipp (1981) which can be viewed as far reaching generalizations (with slower but adequate rates of convergence) of Theorems 2.2, 2.3 and their Corollaries 2.2, 2.4, and also that of (2.25), in terms of Kiefer Measures $\{K_\mu(B, n); B \in \mathscr{C}, n \geq 1\}$ associated with probability measures μ on $(R, \mathscr{B})$ over a subclass $\mathscr{C}$ (of some generality) of $\mathscr{B}$.

The strong and weak approximations of the empirical characteristic function C_n of (1.23) can be accomplished, on $\mathbb{R}^1$ in terms of Gaussian processes built on Kiefer and Brownian bridge processes (cf. S. Csörgő, 1981a) and on $\mathbb{R}^d$ $(d \geq 2)$ in terms of Gaussian processes built on Kiefer and Brownian bridge processes associated with F on $\mathbb{R}^d$ (cf. S. Csörgő, 1981b). For further references we refer to the just mentioned two papers of S. Csörgő.

3. On the limiting distribution of and critical values for the multivariate Cramér–von Mises and Hoeffding–Blum–Kiefer–Rosenblatt independence criteria

A study of empirical and quantile processes on $\mathbb{R}^1$ with the help of strong approximation methodologies is given in Chapters 4 and 5 of Csörgő and Révész (1981a). We are also going to touch upon some of these problems in the light of some recent developments in Section 4. of this exposition. An excellent *direct* theoretical and statistical study of the empirical process on $\mathbb{R}^1$ can be seen in this volume by Csáki (1982b). The latter is recommended as parallel reading to the material covered in this paper. In this section we add details to Theorems 2.1, 2.2 and Corollaries 2.1, 2.2 while studying Cramér–von Mises functionals of the empirical processes α_n (cf. (1.7) and (1.9)) and $t_n = T_n(L^{-1})$ (cf. (1.13)).

When testing the null hypothesis H_0 of (1.10), or that of F being a completely specified continuous distribution function on $\mathbb{R}^1$, one of the frequently used statistics is the Cramér–von Mises statistic $\omega^2_{n,d}$, defined by

$$\omega^2_{n,d} = \int_{R^d} \beta^2_n(x) \prod_{i=1}^{d} \mathrm{d}F_{(i)}(x_i) = \int_{I^d} \alpha^2_n(y) \prod_{i=1}^{d} \mathrm{d}y_i$$
$$= n^{-1} \sum_{k=1,j=1}^{n} \left\{ \prod_{i=1}^{d} (1-(y_{ki} \vee y_{ji})) - \prod_{i=1}^{d} 2^{-1}(1-y^2_{ki}) - \prod_{i=1}^{d} 2^{-1}(1-y^2_{ji}) + 3^{-d} \right\}, \tag{3.1}$$

$d \geq 1$, where $(y_{j1}, \ldots, y_{jd})_{j=1}^n$ with $y_{ji} = F_{(i)}(X_{ji})$ $(i = 1, \ldots, d)$ are the observed values of the random sample $X_j = (X_{j1}, \ldots, X_{jd})$, $j = 1, \ldots, n$. One rejects H_0 of (1.10), or that of F being a given continuous distribution function on $\mathbb{R}^1$, if for a given random sample $X_1, \ldots, X_n$ on F the computed value of $\omega^2_{n,d}$ is too large for a given level of significance (fixed size Type I error). Naturally, in order to be able to compute the value of $\omega^2_{n,d}$ for a sample, H_0 of (1.10), i.e. the marginals of F, or F itself on $\mathbb{R}^1$, will have to be completely specified (simple statistical hypothesis). While it is true that the distribution of $\omega^2_{n,d}$ will not depend on the specific form of these marginals (cf. (1.9)), the problem of finding and tabulating this distribution is not an easy task at all.

Let $V_{n,d}$ be the distribution function of the rv $\omega^2_{n,d}$, i.e.,

$$V_{n,d}(x) = P\{\omega^2_{n,d} \leq x\}, \quad 0 < x < \infty. \tag{3.2}$$

Csörgő and Stachó (1979) gave a recursion formula for the exact distribution function $V_{n,1}$ of the rv $\omega^2_{n,1}$. The latter in principle is applicable to tabulating $V_{n,1}$ exactly for any given n. Naturally, much work has already been done to compile tables for $V_{n,1}$. A survey and comparison of these can be found in Knott (1974), whose results prove to be the most accurate so far. All these results and tables are based on some kind of an approximation of $V_{n,1}$. As to higher dimensions $d \geq 2$, no analytic results appear to be known about the exact distribution function $V_{n,d}$. It follows by (2.6) that we have

$$\lim_{n\to\infty} V_{n,d}(x) = P\{\omega_d^2 \leq x\} := V_d(x), \quad 0 < x < \infty, \ d \geq 1, \tag{3.3}$$

where $\omega_d^2 = \int_{I^d} B^2(y)\,dy$, $\{B(y);\, y \in I^d\}$ a Brownian bridge, and $dy = \Pi_{i=1}^d dy_i$ from now on.

For the sake of describing the speed of convergence of the distribution functions $\{V_{n,d}\}_{n=1}^{\infty}$ to the distribution function V_d of ω_d^2 (cf. (3.3)) we define

$$\Delta_{n,d} = \sup_{0<x<\infty} |V_{n,d}(x) - V_d(x)|. \tag{3.4}$$

S. Csörgő (1976) showed that $\Delta_{n,1} = O(n^{-1/2} \log n)$ and, on the basis of his complete asymptotic expansion for the Laplace transform of the rv $\omega_{n,1}^2$ (cf. (3.1)), he conjectured that $\Delta_{n,1}$ is of order $1/n$. Indeed, the latter turned out to be correct (cf. Corollary 1: $\Delta_{n,1} = O(n^{-1})$, in Cotterill and M. Csörgő, 1982), and it can be deduced from the ground breaking work of Götze (1979). Actually the latter work when combined with Dugue (1969), and Bhattacharya and Ghosh (1978) implies (cf. Section 2 in Cotterill and M. Csörgő, 1982) an asymptotic expansion of arbitrary order for the distribution function $V_{n,d}$ of (3.2) and also that

$$\Delta_{n,d} = O(n^{-1}), \quad d \geq 1, \tag{3.5}$$

(cf. Corollary 3 in Cotterill and M. Csörgő, 1982).

An extensive tabulation of the distribution function V_1 (cf. (3.3)) can be found in the monograph of Martynov (1978), where the theory and applications of a wide range of univariate Cramér–von Mises types statistics are also surveyed.

There appear to be no tables available for the distribution function $V_{n,d}$ ($d \geq 2$) (cf. (3.2)). Hence, and in the light of the just quoted result of (3.5), tables for the distribution function V_d ($d \geq 2$) of (3.3) are of special interest. Durbin (1970) tabulated V_d for $d = 2$, and Krivyakova, Martynov and Tyurin (1977) for $d = 3$. Using the characteristic function of the distribution function V_d (cf. Dugue, 1969; Durbin, 1970), Cotterill and M. Csörgő (1982) obtained a recursive equation for the cumulants of the rv ω_d^2 and, using the first six of these cumulants in the Cornish–Fisher asymptotic expansion, tabulated its critical values for $d = 2, 3, \ldots, 50$ at various levels of significance. These critical values are within 3% of Durbin's values for $d = 2$ and those of Krivyakova, Martynov and Tyrin for $d = 3$. We note also that errors in the said tables for higher dimensions should be further reduced due to the fact that cumulants of ω_d^2 are $O(e^{-d})$ (cf. Corollary 7 and Remark 3.2 in D. S. Cotterill and M. Csörgő, 1982). As far as we know, for the present there exist no further tables of V_d for $d \geq 4$.

As mentioned already, for the sake of computing the value of $\omega_{n,d}^2$ for a sample, H_0 of (1.10) will have to be completely specified. An alternate route to testing for H_0 of (1.10) can be based on the empirical process $t_n = T_n(L^{-1})$ of (1.13) which will not require the specification of the marginals of F under H_0, i.e., it will work also when H_0 of (1.10) is a *composite* statistical hypothesis.

For the sake of describing the latter approach due to Hoeffding (1948), and Blum, Kiefer and Rosenblatt (1961), we define the sequence of Gaussian processes $\{T^{(n)}(y); y \in I^d\}_{n=1}$ by

$$\{T^{(n)}(y); y \in I^d\} = \Big\{B_n(y) - \sum_{i=1}^{d} B_n(1, \ldots, 1, y_i, 1, \ldots, 1) \prod_{j \neq i} y_i; \quad y = (y_1, \ldots, y_d) \in I^d (d \geq 2)\Big\} \tag{3.6}$$

where $\{B_n(y); y \in I^d (d \geq 2)\}_{n=1}^{\infty}$ is a sequence of Brownian bridges.

Define also the Gaussian process $\{T(y, t); y \in I^d, t \geq 0\}$ by

$$\{T(y, t); y \in I^d, t \geq 0\} = \Big\{K(y, t) - \sum_{i=1}^{d} K(1, \ldots, 1, y_i, 1, \ldots, 1, t) \times \prod_{j \neq i} y_j; \; y \in I^d (d \geq 2), t \geq 0\Big\}, \tag{3.7}$$

where $\{K(y, t); y \in I^d (d \geq 2), t \geq 0\}$ is a Kiefer process.

Obviously $ET^{(n)}(y) = ET(y, t) = 0$, and simple but somewhat tedious calculations yield the covariance functions

$$ET^{(n)}(x)T^{(n)}(y) = \prod_{i=1}^{d} (x_i \wedge y_i) + (d-1) \prod_{i=1}^{d} x_i y_i - \sum_{i=1}^{d} (x_i \wedge y_i) \prod_{j \neq i} x_j y_j := \rho(x, y) \quad \text{for all } n, \tag{3.8}$$

and

$$ET(x, s)T(y, t) = (s \wedge t)\rho(x, y), \tag{3.9}$$

where $x = (x_1, \ldots, x_d)$, $y = (y_1, \ldots, y_d) \in I^d$ $(d \geq 2)$ and $s, t \geq 0$.

Strong approximations of $t_n = T_n(L^{-1})$ in terms of the latter Gaussian processes follow quite directly by Theorems 2.1 and 2.2. The following results are known (cf. Theorems 3 and 4 in M. Csörgő, 1979).

THEOREM 3.1. (Csörgő, 1979). *Let $X_1, \ldots, X_n$ $(n = 1, 2, \ldots)$ be independent random d-vectors on $\mathbb{R}^d$ with distribution function $F \in \mathcal{F}_0$ and let t_n be as in* (1.13). *Then one can construct a probability space for $X_1, X_2, \ldots$ with a sequence of Gaussian processes $\{T^n(y); y \in I^d$ $(d \geq 2)\}_{n=1}^{\infty}$, defined as in* (3.6), *and a Gaussian process $\{T(y, t); y \in I^d$ $(d \geq 2)$, $t \geq 0\}$, defined as in* (3.7), *on it so that*

(i) *for all n and x we have*

$$P\{\sup_{y\in I^2}|t_n(y)-T^{(n)}(y)|>n^{-1/2}(C\log n+x)\log n\}\leqslant L\,\mathrm{e}^{-\lambda x} \tag{3.10}$$

where C, L and λ *are positive absolute constants,*

(ii) *for any* $\lambda>0$ *there exists a constant* $C>0$ *such that*

$$P\{\sup_{y\in I^d}|t_n(y)-T^{(n)}(y)|>C(\log n)^{3/2}n^{-1/2(d+1)}\}\leqslant n^{-\lambda},\quad d\geqslant 2, \tag{3.11}$$

and

$$P\{\sup_{1\leqslant k<n}\sup_{y\in I^d}|k^{1/2}t_k(y)-T(y,k)|>Cn^{(d+1)/2(d+2)}\log^2 n\}\leqslant n^{-\lambda},\quad d\geqslant 2. \tag{3.12}$$

COROLLARY 3.1. (3.10), (3.11), (3.12) *in turn imply*

$$\sup_{y\in I^2}|t_n(y)-T^{(n)}(y)|\overset{\text{a.s.}}{=}\mathrm{O}(n^{-1/2}\log^2 n), \tag{3.13}$$

$$\sup_{y\in I^d}|t_n(y)-T^{(n)}(y)|\overset{\text{a.s.}}{=}\mathrm{O}(n^{-1/2(d+1)}(\log n)^{3/2}),\quad d\geqslant 2, \tag{3.14}$$

$$n^{-1/2}\sup_{1\leqslant k\leqslant n}\sup_{y\in I^d}|k^{1/2}t_k(y)-T(y,k)|\overset{\text{a.s.}}{=}\mathrm{O}(n^{-1/(2d+4)}\log^2 n),\quad d\geqslant 2. \tag{3.15}$$

It follows from (3.6) and (3.7), or by (3.8) and (3.9), that for each n

$$\begin{aligned}\{T^{(n)}(y);\,y\in I^d\,(d\geqslant 2)\}&\overset{\mathrm{D}}{=}\{n^{-1/2}T(y,n);\,y\in I^d\ (d\geqslant 2)\}\\&\overset{\mathrm{D}}{=}\{T(y,1);\,Y\in I^d\ (d\geqslant 2)\}.\end{aligned} \tag{3.16}$$

Define the Gaussian process $\{T(y);\,y\in I^d\ (d\geqslant)\}$ by

$$\begin{aligned}\{T(y);\,y\in I^d\ (d\geqslant 2)\}&=\{T(y,t);\,y\in I^d\ (d\geqslant 2),\,t=1\}\\&\overset{\mathrm{D}}{=}\Big\{B(y)-\sum_{i=1}^{d}B(1,\ldots,1,y_i,1,\ldots,1)\prod_{j\neq i}y_j;\\&\qquad y=(y_1,\ldots,y_d)\in I^d\ (d\geqslant 2)\Big\},\end{aligned} \tag{3.17}$$

where $\{B(y);\,y\in I^d\ (d\geqslant 2)\}$ is a Brownian bridge. Thus $T(\cdot)$ has mean zero and covariance function $\rho(\cdot,\cdot)$ of (3.8), and weak convergence of t_n to the Gaussian process T of (3.17) on the Skorohod space $D[0,1]^d$ follows by (3.14) say. Also, a Corollary 2.3 type weak convergence of $[n\cdot]^{1/2}t_{[n\cdot]}(\cdot)/n^{1/2}$ to $T(\cdot,\cdot)$ of (3.7) on $D[0,1]^{d+1}$ follows by (3.15).

Blum, Kiefer and Rosenblatt (1961) proposed the following Cramér–von Mises type test statistic for H_0 of (1.10):

$$C_{n,d}=\int_{\mathbb{R}^d}T_n^2(x)\prod_{i=1}^{d}\mathrm{d}F_{(i)}(x_i)=\int_{I^d}t_n^2(y)\,\mathrm{d}y,\quad d\geqslant 2. \tag{3.18}$$

One rejects H_0 of (1.10) if for a given random sample $X_1, \ldots, X_n$ on F the computed value of $C_{n,d}$ is too large for a given level of significance.

Let $\Gamma_{n,d}$ be the distribution function of the rv $C_{n,d}$, i.e.,

$$\Gamma_{n,d}(x) = P\{C_{n,d} \leqslant x\}, \quad 0 < x < \infty, \ d \geqslant 2 . \tag{3.19}$$

Then, by (3.14) say, we have

$$\lim_{n\to\infty} \Gamma_{n,d}(x) = P\{C_d \leqslant x\} := \Gamma_d(x), \quad 0 < x < \infty, \ d \geqslant 2 , \tag{3.20}$$

where $C_d = \int_{I^d} T^2(y)\,\mathrm{d}y$ with $\{T(y);\, y \in I^d\ (d \geqslant 2)\}$ as in (3.17).

There does not seem to be anything known about the exact distribution function $\Gamma_{n,d}$ of the rv $C_{n,d}$. As to the speed of convergence in (3.20) via (3.10) and (3.11) we get (cf. Theorem 1 in Cotterill and Csörgő, 1980)

$$\nabla_{n,d} := \sup_{0<x<\infty} |\Gamma_{n,d}(x) - \Gamma_d(x)| = \begin{cases} \mathrm{O}(n^{-1/2} \log^2 n) & \text{if } d = 2 , \\ \mathrm{O}(n^{-1/(2d+2)} (\log n)^{3/2}) & \text{if } d \geqslant 3 . \end{cases} \tag{3.21}$$

As far as we know the rates of convergence in (3.21) are the only ones available so far.

Concerning tables for the distribution function Γ_d, for $d = 2$, Blum, Kiefer and Rosenblatt (1961) obtained the characteristic function of the distribution function Γ_d of the rv C_d and tabulated its distribution via numerical inversion of the said characteristic function. The statistic $C_{n,d}$ of (3.18) itself cannot be computed unless $F \in \mathscr{F}_0$ of H_0 of (1.10) is completely specified. Hoeffding (1948), and Blum, Kiefer and Rosenblatt (1961) suggested, as critical region for H_0 of (1.10) when it is viewed as a *composite* statistical hypothesis, large values of

$$\hat{C}_{n,d} = \int_{\mathrm{R}^d} T_n^2(x)\,\mathrm{d}F_n(x), \quad d \geqslant 2 , \tag{3.22}$$

or those of

$$\tilde{C}_{n,d} = \int_{\mathrm{R}^d} T_n^2(x) \prod_{i=1}^{d} \mathrm{d}F_{ni}(x_i), \quad d \geqslant 2 . \tag{3.23}$$

These two statistics are equivalent to $C_{n,d}$ in that both converge in distribution to the rv C_d. This was already noted by Blum, Kiefer and Rosenblatt (1961), and for a detailed proof of this statement we refer to Section 4 in Cotterill and Csörgő (1980). Recently DeWet (1980) studied a version of (3.23) in the case of $d = 2$ with some nonnegative weight functions multiplying the integrand T_n^2 of $\tilde{C}_{n,d}$. Koziol and Nemec (1979) studied $\hat{C}_{n,d}$ of (3.22) and its performance (power properties) in testing for independence with bivariate normal observations.

As to tables for the distribution function Γ_d for $d \geq 2$, Cotterill and Csörgő (1980, Section 4) find an expression for the characteristic function of the rv C_d, $d \geq 2$, via utilizing the representation of the stochastic process $\{T(y); y \in I^d\ (d \geq 2)\}$ of (3.17) in terms of Brownian bridges. This in turn enables them to find the first five cumulants of the rv C_d, and using these in the Cornish–Fisher asymptotic expansion, they tabulated its critical values for $d = 2, \ldots, 20$ at the 'usual' levels of significance. These tables and details as to how to calculate approximate critical values of the rv C_d for all $d \geq 2$ are given in Sections 5 and 6 of the said paper. Compared with the figures of Blum, Kiefer and Rosenblatt (1961) for $d = 2$, the Cornish–Fisher approximation seems to work quite well. For $d > 2$ we do not know of any other tables for the rv C_d.

Another approach to this problem was suggested by Deheuvels (1981), who showed that the Gaussian process $\{T(y); y \in I^d\ (d \geq 2)\}$ of (3.17) which approximates the empirical process t_n of (1.13) (cf. Theorem 3.1) can be decomposed into $2^d - d - 1$ independent Gaussian processes whose covariance functions are of the same structure for all $d \geq 2$ as that of $T(y)$ for $d = 2$. If tables for the Cramér–von Mises functionals of these $2^d - d - 1$ independent rv were available, then one could test asymptotically independently whether there are dependence relationships within each subset of the coordinates of $X \in \mathbb{R}^d$, $d \geq 2$.

4. On strong and weak approximations of the quantile process

In this section we are going to give an up-to-date summary of strong and weak invariance principles for the quantile process ρ_n of (1.20). For further readings, references on this subject and its applications to statistics we refer to Doksum (1974), Doksum and Sievers (1976), Doksum, Fenstad and Aaberge (1977), Parzen (1979, 1980), Chapters 4, 5 in Csörgő and Révész (1981a), Csörgő and Révész (1981b), M. Csörgő (1981b, 1983), and Csörgő, Csörgő, Horváth and Révész (1982). Random variables are $\mathbb{R}^1$-valued throughout this section. We start with comparing the general quantile process ρ_n of (1.20) to its corresponding uniform version, the uniform quantile process u_n of (1.21) (cf. also (1.22), and (1.14)–(1.19) for definitions used in this section).

THEOREM 4.1 (Csörgő and Révész, 1978). *Let $X_1, X_2, \ldots$ be i.i.d. rv with a continuous distribution function F and assume*

(i) *is twice differentiable on* (a, b), *where* $a = \sup\{x: F(x) = 0\}$, $b = \inf\{x: F(x) = 1\}$, $-\infty \leq a < b \leq +\infty$,

(ii) $F'(x) = f(x) > 0$ *on* (a, b),

(iii) *for some* $\gamma > 0$ *we have*

$$\sup_{0<y<1} y(1-y)\frac{|f'(Q(y))|}{f^2(Q(y))} \leq \gamma .$$

Then, with $\delta_n = 25n^{-1}\log\log n$,

$$\sup_{\delta_n \le y \le 1-\delta_n} |\rho_n(y) - u_n(y)| \overset{\text{a.s.}}{=} O(n^{-1/2}\log\log n). \tag{4.1}$$

If, in addition to (i), (ii) *and* (iii), *we also assume*
(iv) $A = \lim_{x\downarrow a} f(x) < \infty$, $B = \lim_{x\uparrow b} f(x) < \infty$,
(v) *one of* (v, α) $A \wedge B > 0$, (v, β) *if* $A = 0$ (*resp.* $B = 0$) *then* f *is nondecreasing* (*resp. nonincreasing*) *on an interval to the right of* a (*resp. to the left of* b), *then, if* (v, α) *obtains,*

$$\sup_{0\le y\le 1} |\rho_n(y) - u_n(y)| \overset{\text{a.s.}}{=} O(n^{-1/2}\log\log n), \tag{4.2}$$

and if (v, β) *obtains,*

$$\sup_{0\le y\le 1} |\rho_n(y) - u_n(y)| \overset{\text{a.s.}}{=} \begin{cases} O(n^{-1/2}\log\log n) & \text{if } \gamma < 1, \\ O(n^{-1/2}(\log\log n)^2) & \text{if } \gamma = 1, \\ O(n^{-1/2}(\log\log n)^{\gamma}(\log n)^{(1+\varepsilon)(\gamma-1)}) & \text{if } \gamma > 1, \end{cases} \tag{4.3}$$

where $\varepsilon > 0$ *is arbitrary, and* γ *is as in* (iii).

The above theorem also implies approximations of ρ_n in terms of appropriate sequences of Brownian bridges $\{B_n(y), 0 \le y \le 1\}$ (cf. Csörgő and Révész, 1978; Section 3.1 in M. Csörgő, 1983) due to

THEOREM 4.2 (Csörgő and Révész, 1975c, 1978). *For an i.i.d. sequence of rv* $X_1, X_2, \ldots$ *there exists a probability space with a sequence of Brownian bridges* $\{B_n\}$ *on it such that*

$$\sup_{0\le y\le 1} |u_n(y) - B_n(y)| \overset{\text{a.s.}}{=} O(n^{-1/2}\log n). \tag{4.4}$$

Naturally, from the above two theorems it follows that

$$\rho_n(\cdot) \overset{\text{D}}{\to} B(\cdot) \tag{4.5}$$

in Skorohod's space $D[0, 1]$.

Let $q(y) \ge 0$ be a continuous function on $[0, 1]$ which is strictly positive on $(0, 1)$, nondecreasing on $[0, \frac{1}{2}]$, and symmetric about $y = \frac{1}{2}$, and let

$$h(y) = \left(y(1-y)\log\log\frac{1}{y(1-y)}\right)^{1/2}, \quad 0 \le y \le 1. \tag{4.6}$$

Define also $g(y) = q(y)/h(y)$ so that

$$g(y) = q(y)/h(y) \to \infty \quad \text{as } y \to 0 . \tag{4.7}$$

Then $g(y) = g(1-y)$ by definition.

Shorack (1979) showed that with q and g as above, the condition (4.7) is sufficient for O'Reilly's (1974) sufficient condition on q for

$$\sup_{1/(n+1) \leq y \leq n/(n+1)} |(u_n(y) - B_n(y))/q(y)| \xrightarrow{\text{P}} 0 \tag{4.8}$$

to be true with the Brownian bridges B_n of (4.4). For an up-to-date discussion of O'Reilly's (1974) theorems in the light of strong approximations we refer to Chapter V in M. Csörgő (1983). Similar work to that of O'Reilly's was done earlier by Chibisov (1964), and Pyke and Shorack (1968). S. Csörgő (1982) showed that on assuming conditions (i), (ii), (iii), (iv) and (v) of Theorem 4.1 and the condition (4.7), we have

$$\sup_{1/(n+1) \leq y \leq n/(n+1)} |(\rho_n(y) - B_n(y))/q(y)| \xrightarrow{\text{P}} 0 \tag{4.9}$$

with B_n as in (4.4), provided that γ of condition (iii) is less than 1. Otherwise, i.e., if in (iii) $\gamma \geq 1$, growth conditions had to be introduced for the function $G(t) = \inf\{g(y): 0 \leq y \leq t\} \uparrow \infty$ as $t \downarrow 0$. M. Csörgő (1983, Theorem 5.1.1) verified (4.9) only under the conditions (i), (ii), (iii) of Theorem 4.1 and (4.7). Stute (1982) proved (4.9) with $q \equiv 1$ under (4.7) and only the conditions (iv), (v) of Theorem 4.1 as follows: we have (4.9) with $q \equiv 1$ if (v, α) obtains, or if (v, β) obtains, provided that in the latter case both

$$g(y)f(Q(y/\lambda))/f(Q(y)) \to \infty \quad \text{as } y \to 0 \tag{4.10}$$

and (symmetrically)

$$g(y)f(Q(1-y)/\lambda)/f(Q(1-y)) \to \infty \quad \text{as } y \to 0 \tag{4.11}$$

for each $\lambda \geq 1$ (note that in Stute's (1982) case $g = 1/h$ on account of $q \equiv 1$). Shorack (1982) announced the latter result with q and g as in (4.7) (for a proof we may, for example, refer to M. Csörgő, 1983, Corollary 5.3.2). All the afore-mentioned results concerning (4.9) are contained in

THEOREM 4.3 (M. Csörgő, 1983, Theorem 5.3.1). *Let a, b be as in Theorem 4.1 and assume that F has a continuous density function $F' = f$ that is positive on (a, b), the support of F. Let q be any given O'Reilly weight function with* (4.7). *Then, as $n \to \infty$, with the sequence of Brownian bridges $\{B_n\}$ of* (4.4) *we have* (4.9) *under* (4.7), *provided that with g of the latter the following assumption also holds true*:

For any given $0 < \eta < 1$ and $\varepsilon > 0$ there exist $0 < c < 1$ and n_0 such that

$$P\Big\{\sup_{1/(n+1)\le y\le c}\sup_{\theta_{y,n}}\frac{f(Q(y))}{f(Q(\theta_{y,n}))}\frac{1}{g(y)}>\varepsilon\Big\}\le\eta$$

and similarly

$$P\Big\{\sup_{1-c\le y\le n/(n+1)}\sup_{\theta_{y,n}}\frac{f(Q(y))}{f(Q(\theta_{y,n}))}\frac{1}{g(y)}>\varepsilon\Big\}\le\eta \tag{4.12}$$

for all $n\geq n_0$, *where* $U_n(y)\wedge y<\theta_{y,n}<U_n(y)\vee y$.

All the afore quoted results concerning (4.9) can be put in terms of weak convergence on $D[0,1]$, *provided* we redefine u_n (and hence also ρ_n) to be equal to zero on $[0,1/(n+1))$ and $(n/(n+1),1]$.

The sufficient conditions of Theorem 4.3 (cf. (4.12)) for the weak approximation of ρ_n on $[1/(n+1),n/(n+1)]$ are nearly necessary as well. For convenience, a weight function q will be called an O'Reilly weight function from now on if $g=q/h$ satisfies (4.7). We have

THEOREM 4.4 (Csörgő, Csörgő, Horváth and Révész, 1982). *Let* a, b *be as in Theorem* 4.1 *and assume that* F *has a continuous density function* $F'=f$ *that is positive on* (a,b), *the support of* F. *If for any given* g *of an O'Reilly weight function* q *we have that the rv*

$$\Big(f\Big(Q\Big(\frac{1}{n}\Big)\Big)\Big/f(Q(\theta_{1/n,n}))\Big)\Big/g\Big(\frac{1}{n}\Big)(\log\log n)^{1/2},$$

or $$\Big(f\Big(Q\Big(\frac{n-1}{n}\Big)\Big)\Big/f(Q(\theta_{(n-1)/n,n}))\Big)\Big/g\Big(\frac{1}{n}\Big)(\log\log n)^{1/2} \tag{4.13}$$

is, contrary to (4.12), *bounded away from zero in probability as* $n\to\infty$, *then the rv of* (4.9) *is also bounded away from zero in probability for any sequence of Brownian bridges* $\{B_n\}$.

As to the problem of weak convergence of ρ_n/q in sup-norm metric over $[0,1]$, we first observe that for a Brownian bridge B

$$\sup_{0<y\le 1/n}|B(y)/q(y)|\xrightarrow{\text{P}}0\quad\text{and}\quad\sup_{1-1/n\le y<1}|B(y)/q(y)|\xrightarrow{\text{P}}0$$

as $n\to\infty$ for any O'Reilly weight function q. Hence the only way for ρ_n/q to converge weakly to B/q over $[0,1]$ in sup-norm metric is that we have the latter two statements holding true also with ρ_n replacing B in them. In this context we have

THEOREM 4.5 (Csörgő, Csörgő, Horváth and Révész, 1982). *Let* a, b *be as in Theorem* 4.1 *and assume that* F *has a continuous density function* $F'=f$ *that is positive on* (a,b), *the support of* F. *Let* q *be any given O'Reilly weight function so that as* $n\to\infty$

$$\sup_{0\le y\le 1/n}\sup_{\theta_{y,n}}\frac{n^{-1/2}}{q(y)}\frac{f(Q(y))}{f(Q(\theta_{y,n}))}\xrightarrow{P}0\,,$$
$$\sup_{(n-1)/n\le y\le 1}\sup_{\theta_{1,n}}\frac{n^{-1/2}}{q(y)}\frac{f(Q(y))}{f(Q(\theta_{y,n}))}\xrightarrow{P}0\,, \tag{4.14}$$

where $\theta_{y,n}$ *is as in* (4.12). *Then, as* $n\to\infty$,

$$\sup_{0\le y\le 1/n}|\rho_n(y)/q(y)|\xrightarrow{P}0,\qquad \sup_{(n-1)/n\le y\le 1}|\rho_n(y)/q(y)|\xrightarrow{P}0\,. \tag{4.15}$$

If, on the other hand, for the given q *and* g *there exists a sequence of rv* $\tau_n\le 1/n$ *so that the rv*

$$n^{-1/2}(f(Q(\tau_n))/f(Q(\theta_{\tau_n,n})))/q(\tau_n),\quad or$$
$$n^{-1/2}(f(Q(1-\tau_n))/f(Q(\theta_{1-\tau_n,n})))/q(\tau_n) \tag{4.16}$$

with $\tau_n\wedge U_{1:n}\le\theta_{\tau_n,n}\le\tau_n\vee U_{1:n}$ *and* $(1-\tau_n)\wedge U_{n:n}\le\theta_{1+\tau_n,n}\le(1-\tau_n)\vee U_{n:n'}$ *is bounded away from zero in probability as* $n\to\infty$, *then so will be also the rv of* (4.15).

We now mention some *implications* (examples) of interest which follow from Theorems 4.3, 4.4 and 4.5:

(1) We have already noted that Theorem 4.3 implies the Stute (1982)–Shorack (1982) result: (4.9) holds true under the conditions (iv) and (v) of Theorem 4.1 if (v, α) obtains, or if (v, β) obtains and (4.10), (4.11) are also assumed. If (v, α) obtains then F has finite support and hence (4.9) holds immediately. If (v, β) obtains and, say, A of (iv) is zero, then for small enough c we have

$$\sup_{1/(n+1)\le y\le c}\sup_{\theta_{y,n}}\frac{f(Q(y))}{f(Q(\theta_{y,n}))}\frac{1}{g(y)}\le\sup_{1/(n+1)\le y\le c}\frac{f(Q(y))}{f(Q(y/\lambda))}\frac{1}{g(y)}$$

for some $\lambda\ge 1$ and all $n\ge 1$ with probability arbitrarily near to one by Remark 1 in Wellner (1978). Hence (4.17) implies the first condition of (4.12) and the Stute (1982)–Shorack (1982) theorem follows from Theorem 4.3.

(2) (Taken from Csörgő, Csörgő, Horváth and Révész, 1982). For any O'Reilly weight function q such that $q(y)\to 0$ we have $\sup_{0<y\le 1/n}1/q(y)=\infty$. Hence in case of the uniform quantile process u_n we can choose $\{\tau_n\}$ of (4.16) such that for any given constant $K>0$ we have $(1/n^{1/2}q(\tau_n))>K$. Consequently, for any q with $q(0)=0$ and for any sequence of Brownian bridges $\{B_n\}$ we have

$$p\{\sup_{0\le y\le 1}|(u_n(y)-B_n(y))/q(y)|=\infty\}=1\,.$$

(3) (Taken from Csörgő, Csörgő, Horváth and Révész, 1982). As to the

problem of having

$$\sup_{0\le y\le 1} |(\rho_n(y) - B_n(y))/q(y)| \xrightarrow{P} 0 \tag{4.17}$$

with B_n as in (4.4), we note that choosing the g function of q appropriately, for some specific distributions (4.17) might turn out to be true. For example in case of $F(x) = 1 - e^{-x}$, $x \geq 0$, we may choose $g(y) = Q(y)a(y)/(\log\log(1/1-y))^{1/2}$, where now $Q(y) = \log(1/(1-y))$, $a(y) \to \infty$ as $y \to 1$, and then (4.17) will hold true. Naturally a similar statement holds true symmetrically for $F(x) = 1 - e^{x}$, $x \leq 0$. We note also that with the same g function as above, (4.17) will hold true also for the Weibull distribution.

(4) As mentioned already, M. Csörgő (1983, Theorem 5.1.1) verified (4.9) only under the conditions (i), (ii), (iii) of Theorem 4.1, and noted also that Theorem 4.3 also contained the latter result. In order to see this, consider

$$\sup_{1/(n+1)\le y\le n/(n+1)} \sup_{\theta_{y,n}} \frac{f(Q(y))}{f(Q(\theta_{y,n}))} \leq \sup_{1/(n+1)\le y\le n/(n+1)} \left(\frac{U_n(y)\vee y}{U_n(y)\wedge y}\,\frac{1-(U_n(y)\wedge y)}{1-(U_n(y)\vee y)}\right)^{\gamma}, \tag{4.18}$$

where γ is as in (iii) of Theorem 4.1, and the inequality is by Lemma 1 in Csörgő and Révész (1978). Using now Lemma 2 in Wellner (1978) on the right hand side rv of the inequality of (4.18) it follows (cf. e.g., (2.8) in M. Csörgő, 1983) that

$$\lim_{\varepsilon\to\infty} \overline{\lim_{n\to\infty}}\, P\left\{\sup_{1/(n+1)\le y\le n/(n+1)} \sup_{\theta_{y,n}} \frac{f(Q(y))}{f(Q(\theta_{y,n}))} > \varepsilon\right\} = 0\,. \tag{4.19}$$

Hence condition (4.12) of Theorem 4.3 is satisfied and consequently (4.9) holds true with *any* O'Reilly weight function q under conditions (i), (ii), (iii) of Theorem 4.1.

(5) (Taken from Csörgő, Csörgő, Horváth and Révész, 1982). Given the conditions (i), (ii) of Theorem 4.1 and replacing its condition (iii) by requiring the existence of the limits (cf. (vi) of Theorem 4.7)

$$\lim_{y\downarrow 0} y\frac{f'(Q(y))}{f^2(Q(y))} = \gamma_1, \qquad \lim_{y\uparrow 1} (1-y)\frac{f'(Q(y))}{f^2(Q(y))} = \gamma_2\,,$$

where γ_1 and γ_2 are real numbers, then it can be shown (cf. Theorem 3.A in Parzen (1980), or page 7 of Seneta (1976), or Mason (1982)) that we have

$$\begin{aligned} &f(Q(y)) = y^{\gamma_1}L_1(y) \quad \text{as } y\downarrow 0,\\ &f(Q(y)) = (1-y)^{\gamma_2}L_2(y) \quad \text{as } y\uparrow 1\,, \end{aligned} \tag{4.20}$$

where L_1 and L_2 are slowly varying functions at 0 resp. at 1. If we simply *assume the forms of* (4.20) *for* $f(Q)$ *on the tails*, then these are weaker assumptions on f than that of (iii), for then we do not require the existence of f' on (a, b), the support of F. So let us *assume that* $f(Q)$ *is as in* (4.20) *on the tails*, and consider its first statement (regarding the second one, similar conclusions will hold true). It follows from Corollary on page 274 in Feller (1966) that for a slowly varying function L_1 we have: for any $\varepsilon > 0$ there exist some positive constants K_1, K_2 and $0 < y_0 < 1$ such that we have

$$K_1 y^{\varepsilon} < L_1(y) < K_2 y^{-\varepsilon} \quad \text{for all } 0 < y \leq y_0 . \tag{4.21}$$

Consider now (cf. first statement of (4.20) with $\gamma_1 \equiv \gamma$)

$$\begin{aligned}
\sup_{1/(n+1) \leq y \leq c} \sup_{\theta_{y,n}} \frac{f(Q(y))}{f(Q(\theta_{y,n}))} \frac{1}{q(y)} &= \sup_{1/(n+1) \leq y \leq c} \sup_{\theta_{y,n}} \left(\frac{y}{\theta_{y,n}}\right)^{\gamma} \frac{L_1(y)}{L_1(\theta_{y,n})} \frac{1}{g(y)} \\
&\leq \sup_{1/(n+1) \leq y \leq c} \sup_{\theta_{y,n}} \left(\frac{y}{\theta_{y,n}}\right)^{\gamma} \frac{K_2 y^{-\varepsilon}}{K_1 \theta_{y,n}^{\varepsilon}} \frac{1}{g(y)} \\
&= \sup_{1/(n+1) \leq y \leq c} \sup_{\theta_{y,n}} \left(\frac{y}{\theta_{y,n}}\right)^{\gamma+\varepsilon} \frac{K_2}{K_1} \frac{1}{y^{2\varepsilon}} \frac{1}{g(y)} .
\end{aligned} \tag{4.22}$$

It follows from (2.8) in M. Csörgő (1983) that for both $\gamma + \varepsilon > 0$ and $\gamma + \varepsilon < 0$ we have

$$\lim_{a\to\infty} \overline{\lim_{n\to\infty}} P\left\{ \sup_{1/(n+1) \leq y \leq c} \sup_{\theta_{y,n}} \left(\frac{y}{\theta_{y,n}}\right)^{\gamma+\varepsilon} > a \right\} = 0 . \tag{4.23}$$

Hence choosing $g(y) = y^{-\delta}$, $\delta > 0$ and then $\varepsilon > 0$ of (4.21) so that $\varepsilon < \delta/2$, then the first condition of (4.12) holds true by (4.22) and (4.23) combined. This means that having assumed (4.20), which is weaker than (i), (ii), (iii) of Theorem 4.1 combined, the statement of (4.9) holds only with $g(y) = y^{-\delta}$ for $\delta > 0$, arbitrary otherwise (cf. the last sentence of our example (4)).

(6) We note that the conditions discussed in (4) and (5) for the validity of (4.9) cannot, in general, insure also the validity of (4.17). Namely we have the following (Csörgő, Csörgő, Horváth and Révész, 1982)

Observation. *If* $\overline{\lim}_{y\to 0} f(Q(y))/q(y) = \infty$ *or* $\overline{\lim}_{y\to 1} f(Q(y))/q(y) = \infty$, *then*

$$P\{ \sup_{0 \leq y \leq 1} |\rho_n(y)/q(y)| = \infty\} = 1 . \tag{4.24}$$

Next we mention some *new strong approximations* of ρ_n. First we recall that conditions (i), (ii), (iii) of Theorem 4.1 imply (4.1), and given also the tail monitonicity assumptions of (iv), (v) we have also (4.2) and (4.3). We have just seen in Examples (4) and (5) that conditions (i), (ii), (iii) of Theorem 4.1 alone, or the somewhat weaker assumptions of (4.20) for the tail behaviour of $f(Q)$, imply (4.9) (in (4) without any further restrictions on q, while in (5) with $g(y)$ of the form $y^{-\delta}$ ($\delta>0$) only). As to the possibility of extending the statement of (4.1) over a wider range than $[\delta_n, 1-\delta_n]$, $\delta_n = 25n^{-1}\log\log n$, but using only the assumptions (i), (ii), (iii) of Theorem 4.1 and not those of its conditions (iv), (v) which, when combined with (i), (ii), (iii), made (4.2) and (4.3) possible, we have

THEOREM 4.6 (Csörgő, Csörgő, Horváth and Révész, 1982). *Assume the conditions* (i), (ii), (iii) *of Theorem* 4.1. *Then*

$$\sup_{1/(n+1)\leq y\leq n/(n+1)} |\rho_n(y)-u_n(y)| \overset{\text{a.s.}}{=} \begin{cases} O(n^{-1/2}(\log\log n)^{1+\gamma}) & \text{if } \gamma\leq 1, \\ O(n^{-1/2}(\log\log n)(\log n)^{(1+\varepsilon)(\gamma-1)}) & \text{if } \gamma>1, \end{cases} \tag{4.25}$$

where $\varepsilon>0$ *is arbitrary, and* γ *is as in condition* (iii).

It is clear from Theorem 4.6 that, when proving (4.2) and (4.3) the conditions (iv) and (v) of Theorem 4.1 come into play only because of the tail regions $[0, 1/(n+1))$, $(n/(n+1), 1]$. Having replaced δ_n of (1.8) by $1/(n+1)$ in (4.25), we have only paid the price of slightly weakened rates of convergence. While they render (4.2) and (4.3) true, the extra conditions (iv) and (v) of Theorem 4.1 are somewhat disjoint from that of (iii). Next we modify the latter somewhat for the sake of seeking further insight into the effect of the tail behaviour of the density-quantile function $f(Q)$ on a statment like (4.25). We are going to formulate this over the interval $[0,\frac{1}{2}]$ only and note that similar statements can be made over $[\frac{1}{2}, 1]$.

THEOREM 4.7 (Csörgő, Csörgő, Horváth and Révész, 1982). *Assume the conditions* (i), (ii), (iii) *of Theorem* 4.1 *and, instead of its conditions* (iv), (v), *we assume now that*

(vi) $$\lim_{y\downarrow 0} y\frac{f'(Q(y))}{f^2(Q(y))} = \gamma_1 .$$

Then, if $\gamma_1>0$, *we have* (4.3), *and if* $\gamma_1>0$ *we have*

$$\sup_{n^{-\alpha}\leq y\leq 1/2} |\rho_n(y)-u_n(y)| \xrightarrow{\text{a.s.}} 0 \quad \text{as } n\to\infty, \tag{4.26}$$

provided that $\alpha<1+1/(2|\gamma_1|)$. *On the other hand, when* $\gamma_1<0$, *for* $\alpha> 1+1/(2|\gamma_1|)$ *there exists positive constants* $K=K(\alpha)$ *and* $\lambda=\lambda(\alpha)$ *such that*

$$\varlimsup_{n\to\infty} n^{-\lambda} \sup_{n^{-\alpha}\leq y\leq 1/2} |\rho_n(y)|\geq K \quad \text{a.s.} \tag{4.27}$$

One of the interesting consequences of Theorem 4.1 is the following law of iterated logarithm (LIL) for ρ_n:

$$\overline{\lim_{n\to\infty}} \left(\frac{2}{\log\log n}\right)^{1/2} \sup_{0\le y\le 1} |\rho_n(y)| \stackrel{\text{a.s.}}{=} 1 . \tag{4.28}$$

An interesting consequence of (4.26) of Theorem 4.7 is that it throws new light on the latter LIL. We have

COROLLARY 4.1 (Csörgő, Csörgő, Horváth and Révész, 1982). *Assume the conditions* (i), (ii), (iii) *of Theorem* 4.1 *and condition* (vi) *of Theorem* 4.7. *Then if* $\gamma_1 > 0$ *we have* (4.28), *and if* $\gamma_1 < 0$ *we have*

$$\overline{\lim_{n\to\infty}} \left(\frac{2}{\log\log n}\right)^{1/2} \sup_{n^{-\alpha}\le y\le 1/2} |\rho_n(y)| \stackrel{\text{a.s.}}{=} \begin{cases} 1 & \text{if } \alpha < 1 + 1/(2|\gamma_1|) , \\ \infty & \text{if } \alpha > 1 + 1/(2|\gamma_1|) . \end{cases} \tag{4.29}$$

PROOF. We have $\overline{\lim}_{n\to\infty} (2/\log\log n)^{1/2} \sup_{n^{-\alpha}\le y\le 1/2} |u_n(y)| \stackrel{\text{a.s.}}{=} 1$.

Now the first statement of (4.29) follows from the latter combined with (4.26). The second statement of (4.29) is by (4.17).

We note that Corollary 4.1 implies the non existence of LIL for ρ_n over [0, 1] under (i), (ii), (iii) and (vi) if $\gamma_1 < 0$. On the other hand it follows from Theorem 3 in Mason (1982) that if we replace the weight function $f(Q)$ in $n^{-1/2}\rho_n$ by y^{ε}, $\varepsilon > 0$, then

$$\lim_{n\to\infty} \sup_{0\le y\le 1/2} y^{\varepsilon} |Q_n(y) - Q(y)| \stackrel{\text{a.s.}}{=} 0$$

for every $\varepsilon > 0$, given the conditions (i), (ii), (iii) and (vi), i.e., under the latter conditions we always have a Mason type Glivenko–Cantelli theorem for $(Q_n(y) - Q(y))$.

Summarizing the main features of the problem of strong approximation of ρ_n by u_n over [0, 1] in general, we have seen so far that under the conditions (i), (ii), (iii), (iv), (v) of Theorem 4.1 we have (4.2) and (4.3), on dropping the conditions (iv) and (v) we have (4.25), and when we replace the conditions (iv), (v) by that of (vi) we have (4.26) and (4.27). Now we give an example which will amount to saying that for results like (4.2) and (4.3) neither the conditions (iv), (v), nor the condition (vi) are necessary. This example is due to Parzen (1979, page 116). Continuing the numbering of examples of this section, we now have

(7) Parzen's example (1979) (Result of (4.32) is quoted from Csörgő, Csörgő, Horváth and Révész, 1982): Let

$$1 - F(x) = \exp(-x - C\sin x), \quad x \ge 0, \ 0.5 < C < 1.$$

Letting $x = Q(y)$ $(0 \le y \le 1)$ we get

$$-\log(1-y) = Q(y) + C \sin Q(y)$$

and

$$f(Q(y)) = (1-y)(1 + C \cos Q(y)).$$

Hence

$$Q(y) \leq |\log(1-y)| + C \tag{4.30}$$

and

$$f(Q(y)) \leq (1-y)(1+C). \tag{4.31}$$

Also

$$f'(Q(y)) = -(1-y)((1 + C \cos Q(y))^2 + C \sin Q(y)).$$

Clearly then

$$\sup_{0<x<\infty} F(x)(1-F(x)) \frac{|f'(x)|}{f^2(x)} = \sup_{0<y<1} y(1-y) \frac{|f'(Q(y))|}{f^2(Q(y))} \leq \frac{1}{1-C},$$

i.e., *conditions* (i), (ii), (iii) *of Theorem* 4.1 *are satisfied. Hence by Theorem* 4.6 *we have* (4.25) *with* $\gamma = 1/(1-C)$. *On the other hand, as* $y \to 1$, $Q(y) \to \infty$ *and* $f(Q(y))$ *oscillates. Hence conditions* (iv), (v) *of Theorem* 4.1 *are not satisfied. Also, as* $y \to 1$,

$$(1-y) \frac{f'(Q(y))}{f^2(Q(y))}$$

oscillates, i.e., the right tail version limit requirement of condition (vi) *is also not satisfied. Nevertheless in case of this example we have*

$$\sup_{0 \leq y \leq 1} |\rho_n(y) - u_n(y)| \overset{\text{a.s.}}{=} O(n^{-1/2}(\log \log n)^{1/(1-C)}(\log n)^{(1+\varepsilon)C/(1-C)}) \tag{4.32}$$

where $C \in (0.5, 1)$ *as above.*

For a discussion of Bahadur's (1966) representation of sample quantiles and extension of Kiefer's (1970) theory of deviations between the sample quantile and empirical processes in the light of Theorem 4.1, we refer to Section 5.2 in Csörgő and Révész (1981a) and Chapter VI in M. Csörgő (1983).

References

Aly, E.-E., Burke, M. D., Csörgő, M., Csörgő, S. and Horváth, L. (1982). On the product-limit empirical and quantile processes: A collection of four papers. Carleton Mathematical Lecture Note No. 38, Carleton University, Ottawa.

Aly, E.-E. and Csörgő, M. (1981). Three papers on quantiles and the parameters estimated quantile process. Carleton Mathematical Lecture Note No. 31, Carleton University, Ottawa.

Bahadur, R. R. (1966). A note on quantiles in large samples. *Ann. Math. Statist.* **37**, 577–580.

Bártfai, P. (1966). Die Bestimmung der zu einem wiederkehrenden Process gehörenden Verteilungsfunktion aus den mit Fehlern behafteten Daten einer Einzigen Relation. *Studia Sci., Math. Hung.* **1**, 161–168.

Bhattacharya, R. N. and Ghosh, J. K. (1978). On the validity of the formal Edgeworth expansion. *Ann. Statist.* **6**, 434–451.

Bickel, P. J. and Wichura, M. J. (1971). Convergence criteria for multiparameter stochastic processes and some applications. *Ann. Math. Statist.* **42**, 1656–1670.

Billingsley, P. (1968). *Convergence of Probability Measures.* Wiley, New York.

Blum, J. R., Kiefer, J. and Rosenblatt, M. (1961). Distribution free tests of independence based on the sample distribution function. *Ann. Math. Statist.* **32**, 485–498.

Borisov, I. S. (1982). An approximation of empirical fields. *Coll. Math. Soc. J. Bolyai* **32**, *Nonparametric Statistical Inference,* Budapest 1980.

Borovkov, A. A. (1978). Rate of convergence and large deviations in invariance principle. In: *Proceedings of the International Congress of Mathematicians* (Helsinki, 1978), 725–731. Acad. Sci. Fennica, Helsinki (1980).

Brillinger, D. L. (1969). An asymptotic representation of the sample distribution function. *Bull. Amer. Math. Soc.* **75**, 545–547.

Burke, M. D., Csörgő, M., Csörgő, S. and Révész, P. (1979). Approximations of the empirical process when parameters are estimated. *Ann. Probability* **7**, 790–810.

Čencov, N. N. (1956). Wiener random fields depending on several parameters. *Dokl. Akad. Nank. SSSR* **106**, 607–609.

Chibisov, D. (1964). Some theorems on the limiting behaviour of empirical distribution function. *Selected Translations Math. Statist. Probability* **6**, 147–156.

Cotterill, D. S. and Csörgő, M. (1980). On the limiting distribution of and critical values for the Hoeffding–Blum–Kiefer–Rosenblatt independence criterion. Carleton Mathematical Lecture Note No. 24, Carleton University, Ottawa.

Cotterill, D. S. and Csörgő, M. (1982). On the limiting distribution of and critical values for the multivariate Cramér–von Mises statistic. *Ann. Statist.* **10**, 233–244.

Csáki, E. (1977a). Investigations concerning the empirical distribution function. *Magyar Tud. Akad. Mat. Fiz. Oszt. Kozl.* **23**, 239–327. English transl. in *Selected Transl. Math. Statist. and Probability* **15**, 229–317. Amer. Math. Soc., Providence, R.I. 1981.

Csáki, E. (1977b). The law of the iterated logarithm for normalized empirical distribution function. *Z. Wahrsch. Verw. Gebiete* **38**, 147–167.

Csáki, E. (1982a). On the standardized empirical distribution function. *Coll. Math. Soc. J. Bolyai* **32**. *Nonparametric Statistical Inference,* Budapest 1980.

Csáki, E. (1982b). Empirical distribution function. *This Volume.*

Csörgő, M. (1979). Strong approximations of the Hoeffding, Blum, Kiefer, Rosenblatt empirical process. *J. Multivariate Anal.* **9**, 84–100.

Csörgő, M. (1981a). Gaussian processes, strong approximations: an interplay. Colloques Internationaux Du Centre National De La Recherche Scientifique No. 307, *Aspects Statistiques Et Aspects Physiques Des Processus Gaussiens,* Saint-Flour 22–29 juin 1980, 131–229, Éditions du CNRS, Paris 1981.

Csörgő, M. (1981b). On a test for goodness-of-fit based on the empirical probability measure of Foutz and testing for exponentiality. In: D. Dugue et al., Eds., *Analytical Methods in Probability Theory,* Lecture Notes in Mathematics 861, Springer, Berlin, pp. 25–34.

Csörgő, M. (1981). MR81j:60044.

Csörgő, M. (1983). *Quantile Processes with Statistical Applications.* Regional Conference Series in Applied Mathematics, Philadelphia, SIAM (To appear).

Csörgő, M., Csörgő, S., Horváth, L. and Révész, P. (1982). On weak and strong approximations of the quantile process. To appear in *Proceedings of the Seventh Conference on Probability Theory* (Brasov, Aug. 29 – Sept. 4, 1982).

Csörgő, M. and Révész, P. (1975a). A new method to prove Strassen type laws of invariance principle, II. *Z. Wahrsch. Verw. Gebiete* **31**, 261–269.

Csörgő, M. and Révész, P. (1975b). A strong approximation of the multivariate empirical process. *Studia Sci. Math. Hungar.* **10**, 427–434.

Csörgő, M. and Révész, P. (1975c). Some notes on the empirical distribution function the quantile process. *Coll. Math. Soc. J. Bolyai* **11**, *Limit Theorems of Probability Theory* (P. Révész, Ed.). North-Holland, Amsterdam–London, pp. 59–71.

Csörgő, M. and Révész, P. (1978). Strong approximations of the quantile process. *Ann. Statist.* **6**, 882–894.

Csörgő, M. and Révész, P. (1981a). *Strong Approximations in Probability and Statistics*. Academic Press, New York – Akadémiai Kiadó, Budapest.

Csörgő, M. and Révész, P. (1981b). Quantile processes and sums of weighted spacings for composite goodness-of-fit. In: M. Csörgő et al., eds., *Statistics and Related Topics*. North-Holland, Amsterdam, pp. 69–87.

Csörgő, S. (1976). On an asymptotic expansion for the von Mises ω^2 statistic. *Acta Sci. Math. (Szeged)* **38**, 45–67.

Csörgő, S. (1980). On the quantogram of Kendall and Kent. *J. Appl. Probab.* **17**, 440–447.

Csörgő, S. (1981a). Limit behaviour of the empirical characteristic function. *Ann. Probab.* **9**, 130–144.

Csörgő, S. (1981b). Multivariate empirical characteristic function. *Z. Wahrsch. Verw. Gebiete* **55**, 203–229.

Csörgő, S. (1982). On general quantile processes in weighted sup-norm metrics. *Stoch. Process. Appl.* **12**, 215–220.

Csörgő, S. and Stacho, L. (1979). A step toward an asymptotic expansion for the Cramér–von Mises statistic. *Coll. Math. Soc. J. Boyai* **21**, Analytic Function Methods in Probability Theory (B. Gyires, Ed.), 53–65, North-Holland, Amsterdam.

Deheuvels, P. (1981). An asymptotic decomposition for multivariate distribution-free tests of independence. *J. Multivariate Anal.* **11**, 102–113.

DeWet, T. (1980). Cramér–von Mises tests for independence. *J. Multivariate Anal.* **10**, 38–50.

Doksum, Kjell (1974). Empirical probability plots and statistical inference for nonlinear models in the two-sample case. *Ann. Statist.* **2**, 267–277.

Doksum, K. A., Fenstad, G. and Aaberg, R. (1977). Plots and tests for symmetry. *Biometrika* **64**, 473–487.

Doksum, K. A. and Sievers, G. L. (1976). Plotting with confidence: Graphical comparisons of two populations. *Biometrika* **63**, 421–434.

Doksum, K. A. and Yandell, B. S. (1982). Tests for exponentiality. *This volume.*

Donsker, M. (1952). Justification and extension of Doob's heuristic approach to the Kolmogorov–Smirnov theorems. *Ann. Math. Statist.* **23**, 277–283.

Dudley, R. M. (1966). Weak convergence of probabilities on nonseparable metric spaces and empirical measures on Euclidean spaces. *Ill. J. Math.* **10**, 109–126.

Dudley, R. M. (1978). Central limit theorems for empirical measures. *Ann. Probab.* **6**, 899–929; Correction (1979), ibid. **7**, 909–911.

Dudley, R. M. (1979). Lower layers in $\mathbb{R}^2$ and convex sets in $\mathbb{R}^3$ are not GB classes. *Probability in Banach Spaces II*, Lecture Notes in Mathematics 709. Springer, Berlin, pp. 97–102.

Dudley, R. M. (1981a). Donker classes of functions. In: M. Csörgő et al., eds., *Statistics and Related Topics*. North-Holland, Amsterdam, pp. 341–352.

Dudley, R. M. (1981b). Vapnik–Červonenkis Donsker classes of functions. Colloques Internationaux Du Centre National De La Recherche Scientifique No. 307, *Aspects Statistiques Et Aspects Physiques Des Processus Gaussiens*, Saint-Flour 22–29 juin 1980, 251–269, Editions du CNRS, Paris.

Dudley, R. M. (1982). Empirical and Poisson processes on classes of sets or functions too large for central limit theorems. Preprint.

Dudley, R. M. and Philipp, W. (1981). Invariance principles for sums of Banach space valued random elements and empirical processes. Preprint.

Dugue, D. (1969). Characteristic functions of random variables connected with Brownian motion and of the von Mises multidimensional ω_n^2. In: P. R. Krishnaiah, ed., *Multivariate Analysis 2*. Academic Press, New York, 289–301.

Durbin, J. (1970). Asymptotic distributions of some statistics based on the bivariate sample distribution function. In: M. L. Puri, ed., *Nonparametric Techniques in Statistical Inference*. Cambridge University Press, London, pp. 435–451.

Durbin, J. (1973a). *Distribution Theory for Tests Based on the Sample Distribution Function*. Regressional Conference Series in Applied Mathematics 9, Philadelphia, SIAM.

Durbin, J. (1973b). Weak convergence of the sample distribution function when parameters are estimated. *Ann. Statist.* **1**, 279–290.

Durbin, J. (1976). Kolmogorov–Smirnov tests when parameters are estimated. In: P. Gaenssler and P. Révész, eds., *Empirical Distributions and Processes,* Lecture Notes in Mathematics 566. Springer, Berlin, pp. 33–44.

Erdös, P. and Rényi, A. (1970). On a new law of large numbers. *J. Analyse Math.* **23**, 103–111.

Feller, W. (1966). *An Introduction to Probability Theory and Its Applications, Vol. II.* Wiley, New York.

Gaenssler, P. and Stute, W. (1979). Empirical processes: a survey of results for independent and identically distributed random variables. *Ann. Probab.* **7**, 193–243.

Götze, F. (1979). Asymptotic expansions for bivariate von Mises functionals. *Z. Wahrsch. Verw. Gebiete* **50**, 333–355.

Hoeffding, W. (1948). A nonparametric test of independence. *Ann. Math. Statist.* **19**, 546–557.

Ibero, M. (1979a). Approximation forte du processus empirique fonctionnel multidimensionnel. *Bull. Sci. Math.* **103**, 409–422.

Ibero, M. (1979b). Intégrale stochastique par rapport au processus empirique multidimensionell. *C.R. Acad. Sci. Paris Sér. A* **288**, 165–168.

Kiefer, J. (1969). Old and new methods for studying order statistics and sample quantiles. In: *Proc. First International Conf. Nonparametric Inference.* Cambridge University Press 1970, London, pp. 349–357.

Kiefer, J. (1970). Deviations between the sample quantile process and the sample D.F. In: M. L. Puri, ed., *Nonparametric Techniques in Statistical Inference.* Cambridge University Press, London, pp. 299–319.

Kiefer, J. (1972). Skorohod embedding of multivariate RV's and the sample D.F. *Z. Wahrsch. Verw. Gebiete* **24**, 1–35.

Knott, M. (1974). The distribution of the Cramér–von Mises statistic for small sample sizes. *J. Roy. Statist. Soc. Ser. B.* **36**, 430–438.

Komlós, J., Major, P. and Tusnády, G. (1975a). An approximation of partial sums of independent R.V.'s and the sample D.F. I. *Z. Wahrsch. Verw. Gebiete* **32**, 111–131.

Komlós, J., Major, P. and Tusnády, G. (1975b). Weak convergence and embedding. *Coll. Math. Soc. J. Bolyai* **11**, Limit Theorems of Probability Theory (P. Révész, Ed.), North-Holland, Amsterdam–London, pp. 149–165.

Koziol, J. A. and Nemec, A. F. (1979). On a Cramér–von Mises type statistic for testing bivariate independence. *Canad. J. Statist.* **7**, 43–52.

Krivyakova, E. N., Martynov, G. V. and Tyurin Yu. N. (1977). On the distribution of the ω^2 statistic in the multidimensional case. *Th. Probab. Appl.* **22**, 406–410.

Kuelbs, J. (1976). A strong convergence theorem for Banach space valued random variables. *Ann. Probab.* **4**, 744–771.

Martynov, G. V. (1978). *Omega-Square Criteria* (in Russian). Nauka, Moscow.

Mason, D. M. (1982). Some characterizations of almost sure bounds for weighted multidimensional empirical distributions and a Glivenko–Cantelli theorem for sample quantiles. *Z. Wahrsch. Verw. Gebiete* **59**, 505–513.

Moore, D. S. and Spruill, M. C. (1975). Unified large-sample theory of general chi-squared statistics for tests. *Ann. Statist.* **4**, 599–616.

Müller, D. W. (1970). On Glivenko–Cantelli convergence. *Z. Wahrsch. Verw. Gebiete* **16**, 195–210.

Neuhaus, G. (1976). Weak convergence under contiguous alternatives of the empirical process when parameters are estimated: The D_k approach. In: P. Gaenssler and P. Révész, eds., *Empirical Distributions and Processes,* Lecture Notes in Mathematics 566, Springer, Berlin, pp. 68–82.

Neuhaus, G. (1971). On weak convergence of stochastic processes with multidimensional time parameter. *Ann. Math. Statist.* **42**, 1285–1295.

Neuhaus, G. and Sen, P. K. (1977). Weak convergence of tail-sequence processes for sample distributions and averages. *Mitt. Math. Sem. Giessen* **123**, 25–35.

O'Reilly, N. (1974). On the weak convergence of empirical processes in sup-norm metrics. *Ann. Probability* **2**, 624–651.

Parzen, E. (1979). Nonparametric statistical data modelling. *J. Amer. Statist. Assoc.* **58**, 13–30.

Parzen, E. (1980). Quantile functions, convergence in quantile, and extreme value distribution theory. Texas A & M University Technical Report No. B-3.

Philipp, W. (1973). Empirical distribution functions and uniform distribution mod 1. In: C. F. Osgood, ed., *Diophantine Approximation and its Applications.* Academic Press, New York.

Philipp, W. and Pinzur, L. (1980). Almost sure approximation theorems for the multivariate empirical process. *Z. Wahrsch. Theorie Verw. Gebiete* **54**, 1–13.

Pyke, R. and Shorack, G. R. (1968). Weak convergence of the two-sample empirical process and a new approach to Chenoff–Savage theorems. *Ann. Math. Statist.* **39**, 755–771.

Révész, P. (1976a). On strong approximation of the multidimensional empirical process. *Ann. Probab.* **4**, 729–743.

Révész, P. (1976b). Three theorems of multivariate empirical process. In: P. Gaenssler and P. Révész, eds., *Empirical Distributions and Processes,* Lecture Notes in Mathematics 566. Springer, Berlin, pp. 106–126.

Seneta, E. (1976). Regularly Varying Functions. In: Lecture Notes in Mathematics 508, Springer, Berlin.

Shorack, G. R. (1979). Weak convergence of empirical and quantile processes in sup-norm metrics via KMT-construction. *Stoch. Process. Appl.* **9**, 95–98.

Shorack, G. R. (1982). Weak convergence of the general quantile process in $\|\cdot/q\|$-metrics (preliminary report). *IMS Bull.* **11**, Abstract 82t-2.

Straf, M. L. (1971). Weak convergence of stochastic processes with several parameters. *Proc. Sixth Berkeley Symp. Math. Statist. Prob.* **2**, University of California Press, pp. 187–221.

Stute, W. (1977). Convergence rates for the isotrope discrepancy. *Ann. Probab.* **5**, 707–723.

Stute, W. (1982). The oscillation behaviour of empirical processes. *Ann. Probab.* **10**, 86–107.

Tusnády, G. (1977a). A remark on the approximation of the sample DF in the multidimensional case. *Periodica Math. Hungar.* **8**, 53–55.

Tusnády, G. (1977b). *A Study of Statistical Hypotheses* (in Hungarian). Candidatus Dissertation, Hungarian Academy of Sciences.

Tusnády, G. (1977c). Strong invariance principles. In: J. R. Barra et al., eds., *Recent Developments in Statistics.* North-Holland, Amsterdam, pp. 289–300.

Wellner, J. A. (1978). Limit theorems for the ratio of the empirical distribution function to the true distribution function. *Z. Wahrsch. Verw. Gebiete* **45**, 73–88.

Wichura, M. (1973). Some Strassen-type laws of the iterated logarithm for multiparameter stochastic processes with independent increments. *Ann. Probab.* **1**, 272–296.

Yeh, I. (1960). Wiener measure in space of functions of two variables. *Trans. Amer. Math. Soc.* **95**, 433–450.

PART IV

ESTIMATION PROCEDURES

P. R. Krishnaiah and P. K. Sen, eds., *Handbook of Statistics, Vol. 4*
© Elsevier Science Publishers (1984) 463–485

21

M-, L- and R-estimators

Jana Jurečková

1. Introduction

Let $X_1, \ldots, X_n$ be independent observations, X_i distributed according to the distribution function $F(x - \sum_{j=1}^{p} c_{ij}\theta_j)$, $i = 1, \ldots, n$, where c_{ij}, $1 \leq i \leq n$; $1 \leq j \leq p$ are given constants. We want to construct an estimator $T_n = T_n(X_1, \ldots, X_n)$ of the parameter $\boldsymbol{\theta} = (\theta_1, \ldots, \theta_n)'$ whose definition is independent of the form of F. Such is, e.g., the classical least squares estimator (and the sample mean in the location case), but these classical estimators are highly sensitive to the outlying observations and to the long-tailed distributions and are ineffective for other than the normal distribution. It was illustrated in a variety of Monte Carlo studies (e.g., Andrews et al. (1972)), in the results on the characterization of the normal law through the admissibility of the sample mean and the least squares estimator with respect to the quadratic loss (Kagan, Linnik and Rao, 1965, 1972); in the studies of tail-behavior of location estimators (Jurečková, 1979, 1980, 1981), among others.

Among the robust alternatives of the classical estimators, which are less sensitive to the deviations from a specific distribution shape, three broad classes play the most important role: M-estimators, L-estimators and R-estimators.

The aim of the present chapter is to describe these three classes of estimators, their finite-sample properties as well as asymptotic properties, first on the location and then on the regression case. We shall also touch the computionally easier one-step versions of these estimators and the mutual relations of the estimators. This account is far from being exhaustive: various other results concerning the robust estimators may be found in the bibliography.

2. Estimation of location

Let $X_1, X_2, \ldots$ be a sequence of independent observations from the population with the distribution function (d.f.) $F(x - \theta)$. The problem is that of estimating θ after observing $X_1, \ldots, X_n$. We shall assume, unless otherwise stated, that F is absolutely continuous with the symmetric density f. If we do not impose any other special conditions on F, we cannot take the sample mean as a convenient estimator of θ. We must then look for alternative procedures which are *robust*, i.e. relatively insensitive to the special shape of distribution.

Assume that F is an unknown member of a given family $\mathcal{F}$ of distribution functions. The choice of the estimation procedure then depends on $\mathcal{F}$ which may be as large as the family of all [symmetric] absolutely continuous d.f.'s, it may be a neighborhood of a fixed distribution or a finite set of distribution shapes. Estimating θ over a large $\mathcal{F}$ corresponds to the global point of view; the estimator which is not very poor whatever is $F \in \mathcal{F}$ is then paid for by the lower efficiency. Estimating θ over a small neighborhood $\mathcal{F}$ of a given distribution corresponds to the local point of view; for convenient neighborhoods there often exists an estimator which is minimax over $\mathcal{F}$.

2.1. *M-extimators* (*estimators of maximum likelihood type*)

The class of M-estimators was suggested by Huber (1964), who then studied their properties in a series of papers; the results may be also found in Huber's recent monograph (1981).

Let $X_1, X_2, \ldots$ be a sequence of independent observations from the population with the d.f. $F(x-\theta)$ such that F is absolutely continuous and $F(x)+F(-x)=1$, $x \in \mathbb{R}^1$. The *M-estimator* $M_n = M_n(X_1, \ldots, X_n)$ is defined implicitly as a solution of the equation

$$\sum_{i=1}^{n} \psi(x_i - t) = 0 \tag{2.1}$$

with respect to t, where ψ is an appropriate function attaining positive as well as negative values. If there are more solutions of (2.1) then M_n may be defined as that the nearest to a preliminary consistent estimator T_n of θ (and larger one, if there are two solutions equally distant from T_n; we may put $M_n = 0$, if there is no solution).

The function ψ is often selected nondecreasing and skew-symmetric, i.e. $\psi(-x) = -\psi(x)$, $x \in \mathbb{R}^1$. In such a case M_n may be defined as

$$M_n = \tfrac{1}{2}(M_n^+ + M_n^-) \tag{2.2}$$

where

$$M_n^- = \sup\left\{t: \sum_{i=1}^{n} \psi(X_i - t) > 0\right\}, \qquad M_n^+ = \inf\left\{t: \sum_{i=1}^{n} \psi(X_i - t) < 0\right\} \tag{2.3}$$

or, alternatively, M_n may be defined through the randomization: M_n is equal either to M_n^- or M_n^+, both with probability $\frac{1}{2}$.

If F happens to be known and smooth, we can put

$$\psi(x) = -f'(x)/f(x), \quad x \in \mathbb{R}^1$$

and then M_n coincides with the maximum likelihood estimator (m.l.e.) of θ. Particularly, for $\psi(x) \equiv x$ is $M_n = \bar{X}_n$, which turns out to be the m.l.e. for the normal distribution. The class of M-estimators covers also the sample median (which corresponds to $\psi(x) = \text{sign } x$).

Various ψ-functions lead to various M-estimators; the question is then that of the proper choice of ψ. We intuitively feel that, if M_n, is to be resistant to the outliers and to long-tailed distributions, we should take a bounded ψ-function. The most utilized function ψ, suggested by Huber (1964), is given by

$$\psi(x) = \begin{cases} x & \text{if } |x| \leq c, \\ c \text{ sign } x & \text{if } |x| > c, \end{cases} \tag{2.4}$$

with given $c > 0$. Various alternative choices of ψ are described, e.g., in Andrews et al. (1972). If we wish to get a better performance of M_n at very long-tailed distributions, we should select a function satisfying

$$\psi(x) = 0 \quad \text{if } |x| > c \tag{2.5}$$

for some $c > 0$. The pertaining M-estimators, called *redescending*, are studied by Collins (1977), Portnoy (1977), Collins and Portnoy (1981); see also Huber (1981) and Hampel, Rousseeuw and Ronchetti (1981).

2.1.1. *Finite-sample properties of M-estimators*

Assume that $X_1, X_2, \ldots, X_n$ are i.i.d. random variables distributed according to d.f. $F(x - \theta)$ such that

$$F(x) + F(-x) = 1, \quad x \in \mathbb{R}^1 .$$

Let M_n be an M-estimator defined in (2.2) and (2.3) with a nondecreasing, nonconstant and skew-symmetric function ψ. Then

(i) $M_n(x_1 + c, \ldots, x_n + c) = M_n(x_1, \ldots, x_n) + c \quad$ for $\boldsymbol{x} \in \mathbb{R}^n$, $c \in \mathbb{R}^1$,

(ii) $P_\theta\left\{\sum_{i=1}^{n} \psi(X_i - t) < 0\right\} \leq P_\theta(M_n \leq t) \leq P_\theta\left\{\sum_{i=1}^{n} \psi(X_i - t) \leq 0\right\} \quad$ for $t, \theta \in \mathbb{R}^1$;

(iii) $\frac{1}{2} - \frac{1}{2}\varepsilon \leq P_\theta(M_n < \theta) \leq P_\theta(M_n \leq \theta) \leq \frac{1}{2} + \frac{1}{2}\varepsilon \quad$ for $\theta \in \mathbb{R}^1$,

with $\varepsilon = P_0\left(\sum_{i=1}^{n} \psi(X_i) = 0\right)$.

By (iii), M_n is median unbiased provided $P_0(\sum_{i=1}^{n} \psi(X_i) = 0) = 0$. By (i), M_n is translation-equivariant; however, M_n is generally not scale-equivariant, i.e., it generally does not satisfy

$$M_n(cx_1, \ldots, cx_n) = cM_n(x_1, \ldots, x_n), \quad c > 0 .$$

In practice it means that M-estimators of location should be supplemented by an estimator of scale.

The following theorem, due to Huber (1968), shows that M_n generated by ψ of (2.4) has an interesting minimax property over the Kolmogorov neighborhood of the normal distribution.

THEOREM 2.1. *Let $X_1, \dots, X_n$ be i.i.d. random variables distributed according to the d.f. $F(x-\theta)$ such that $F(x)+F(-x)=1$, $x \in \mathbb{R}^1$ and F is an unknown element of the family*

$$\mathcal{F} = \{F: \sup_{x \in \mathbf{R}^1} |F(x) - \Phi(x)| \leq \varepsilon\} \tag{2.6}$$

with Φ being the d.f. of the standard normal distribution, $\varepsilon > 0$. Let M_n be defined as

$$P(M_n = M_n^+) = P(M_n = M_n^-) = \tfrac{1}{2} \tag{2.7}$$

where M_n^-, M_n^+ are defined in (2.3) *and ψ is given in* (2.4) *with $c>0$ satisfying*

$$e^{-2ac}\Phi(a-c) - \Phi(-a-c) = \varepsilon(1+e^{-2ac}), \quad a > 0\,. \tag{2.8}$$

Then M_n minimizes

$$\sup_{F \in \mathcal{F}} \sup_{\theta \in \mathbf{R}^1} \max[P_\theta(T_n - \theta < -a), P_\theta(T_n - \theta > a)] \tag{2.9}$$

over the set of estimators T_n of θ.

PROOF. The theorem is proved in Huber (1968); see also Huber (1969). Further and more general finite-sample minimax results may be found in Huber and Strassen (1973), Rieder (1977, 1980); see also Huber (1981).

2.1.2. Asymptotic efficiency of M-estimators

If $T_n = T_n(X_1, \dots, X_n)$, $n = 1, 2, \dots$, is a sequence of estimators which is asymptotically normally distributed as $n \to \infty$, then the efficiency of T_n is usually measured through the variance of its asymptotic distribution. The M-estimators are asymptotically normally distributed under mild conditions on ψ and F; this was first proved in Huber (1964, 1965).

If ψ is skew-symmetric and has bounded variation in every interval, i.e., it may be written as $\psi = \psi_1 - \psi_2$ where ψ_1 and ψ_2 are nondecreasing, and if F has an absolutely continuous symmetric density $f(x)$ such that

$$I(F) = \int (f'(x)/f(x))^2 \, dF(x) < \infty \tag{2.10}$$

(finite Fisher information) and

$$\int \psi^2(x)\,\mathrm{d}F(x) < \infty \tag{2.11}$$

then the M-estimator M_n is consistent and asymptotically normally distributed in the sense that

$$\sqrt{n}(M_n - \theta) \xrightarrow{\mathscr{D}} N(0, \sigma^2(\psi, F)) \quad \text{as } n \to \infty \tag{2.12}$$

where

$$\sigma^2(\psi, F) = \int \psi^2(x)\,\mathrm{d}F(x) \cdot \left(\int f(x)\,\mathrm{d}\psi(x)\right)^{-2}. \tag{2.13}$$

The more we assume about ψ, the less we need to impose on F to achieve the asymptotic normality of M_n; for instance, if ψ is a step-function, then the derivative of F should exist only in a neighborhood of the jump-points of ψ.

We see that, for ψ bounded, $\sigma^2(\psi, F)$ is finite for a large class of distributions. The characteristic $\sup_{F\in\mathscr{F}} \sigma^2(\psi, F)$ may be considered as a measure of robustness of the M-estimator generated by ψ over the family $\mathscr{F}$. If $\mathscr{F}$ is a neighborhood of a given distribution, for instance of the normal one, there may exist an optimal ψ which minimizes $\sup_{F\in\mathscr{F}} \sigma^2(\psi, F)$. Let us illustrate one of such minimax results (established by Huber (1964)) corresponding to the case that $\mathscr{F}$ forms a special neighborhood of the normal distribution.

Theorem 2.2 (Huber, 1964). *Let $\mathscr{F}_\varepsilon$ be in the family of ε-contaminated normal distributions, i.e.,*

$$\mathscr{F}_\varepsilon = \{F : F = (1-\varepsilon)\Phi + \varepsilon H,\ H \in \mathcal{M}\} \tag{2.14}$$

where $\mathcal{M}$ is the set of all symmetric distribution functions, ε is a fixed number, $0 \leq \varepsilon < 1$, and Φ is the standard normal d.f. Denote by $\psi_0(x)$ the function defined in (2.4) with c satisfying

$$2[(f^*(c)/c) - 1 + \Phi(c)] = \varepsilon/(1-\varepsilon), \quad f^*(x) = \frac{\mathrm{d}\Phi}{\mathrm{d}x}. \tag{2.15}$$

Then

$$\sup_{F\in\mathscr{F}_\varepsilon} \sigma^2(\psi_0, F) = \inf_{\psi} \sup_{F\in\mathscr{F}_\varepsilon} \sigma^2(\psi, F) \tag{2.16}$$

and the supremum on the left-hand side of (2.16) is attained for the d.f. F_0 with the density

$$f_0(x) = \begin{cases} (1-\varepsilon) f^*(x) & \text{if } |x| \leq c, \\ (1-\varepsilon)(2\pi)^{-1/2} \exp\left\{\dfrac{c^2}{2} - c|x|\right\} & \text{if } |x| > c. \end{cases} \tag{2.17}$$

REMARK 1. The distribution (2.17) is the least informative one in $\mathcal{F}_\varepsilon$, i.e. $I(F_0) = \inf\{I(F): F \in \mathcal{F}_\varepsilon\}$; the M-estimator generated by ψ_0 is the maximum likelihood estimator for F_0.

REMARK 2. The characteristic $\sup_{F\in\mathcal{F}} \sigma^2(\psi, F)$ is also studied in Collins (1977), Portnoy (1977) and in Collins and Portnoy (1981).

2.1.3. Some further developments

Hampel (1964) introduced the influence curve of an estimator T_n as a measure of the local sensitivity of T_n to the infinitesimal deviations from the underlying distribution. It is the measure of the sensitivity of the functional counterpart $T(F)$ of T_n and is defined as

$$\mathrm{IC}(x; F, T) = \lim_{h\to 0} \frac{1}{h}[T((1-h)F + h\delta_x) - T(F)]\,, \tag{2.18}$$

where δ_x is the degenerate d.f. of the constant x, $x \in \mathbb{R}^1$. The influence curve of the M-estimator generated by ψ is

$$\mathrm{IC}(x; T, F) = \psi(x - T(F))\Big(\int \psi'(x - T(F))\,\mathrm{d}F(x)\Big)^{-1}. \tag{2.19}$$

Field and Hampel (1978) (cf. Field (1978)) developed an Edgeworth-type expansion for $(-g_n'/g_n)$ with g_n being the density of the M-estimator M_n. Their method provides very precise approximations even for small samples.

Boos and Serfling (1980) derived the law of iterated logarithm for M_n (cf. Serfling, 1980). Bahadur's type representations of M_n were established by Carroll (1978) and Jurečková (1980). Jurečková and Sen (1982) proved the moment convergence of M_n and derived the asymptotically risk-efficient sequential versions of M_n with respect to the loss $L(a, c) = a(T_n - \theta)^2 + cn$; $a, c > 0$.

2.2. R-estimators (estimators derived from the signed-rank tests)

The signed-rank test of the hypothesis $H: \theta = \theta_0$ is typically based on the statistic

$$S_n(\boldsymbol{X} - \theta_0) = \sum_{i=1}^{n} \operatorname{sign}(X_i - \theta_0)\varphi^+\Big(\frac{R_{ni}^+(\theta_0)}{n+1}\Big) \tag{2.20}$$

where $R_{ni}^+(\theta_0)$ is the rank of $|X_i - \theta_0|$ among $|X_1 - \theta_0|, \ldots, |X_n - \theta_0|$ and $\varphi^+(t) = \varphi((t+1)/2)$, $0 < t < 1$, where $\varphi(t)$ is nondecreasing and square-integrable function, $\varphi(1-t) = -\varphi(t)$, $0 < t < 1$. The statistic $S_n(\boldsymbol{X} - t)$ is then nonincreasing in t and attains positive as well as negative values with probability 1 and $E_{\theta_0} S_n(\boldsymbol{X} - \theta_0) = 0$. The R-estimator of θ is then defined as a solution of the equation $S_n(\boldsymbol{X} - t) = 0$; more precisely, it is defined as

$$R_n = \tfrac{1}{2}(R_n^- + R_n^+) \tag{2.21}$$

where

$$R_n^- = \sup\{t: S_n(\boldsymbol{X} - t) > 0\}, \qquad R_n^+ = \inf\{t: S_n(\boldsymbol{X} - t) < 0\}. \tag{2.22}$$

The R-estimators of location, which are the inversions of the signed-rank tests, were suggested by Hodges and Lehmann (1963). Only some single R-estimators could be given a simple explicit form: besides the sample median (which is the inversion of the sign test), the most well-known is the R-estimator corresponding to the Wilcoxon one-sample test (usually called Hodges–Lehmann's estimator); it can be written as

$$R_n = \text{med}\left\{\frac{X_i + X_j}{2} : 1 \le i, j \le n\right\}. \tag{2.23}$$

The trimmed version of the Hodges–Lehmann's estimator, namely,

$$R_n = \text{med}\left\{\frac{X_i + X_j}{2} : [n\alpha] + 1 \le i, j = n - [n\alpha]\right\}, \tag{2.24}$$

$0 < \alpha < \frac{1}{2}$, appears in some contexts (cf. Miura, 1981); this estimator corresponds to the trimmed Wilcoxon test.

2.2.1. Finite-sample properties of R-estimators

Let $X_1, \ldots, X_n$ be i.i.d. random variables distributed according to an absolutely continuous d.f. $F(x - \theta)$ such that $F(x) + F(-x) = 1$, $x \in \mathbb{R}^1$. Let R_n be an R-estimator defined in (2.21) and (2.22), generated by a nondecreasing score function φ such that $\varphi(1 - t) = -\varphi(t)$, $0 < t < 1$. Then

(i) $R_n(x_1 + c, \ldots, x_n + c) = R_n(x_1, \ldots, x_n) + c$ for $\boldsymbol{x} \in \mathbb{R}^n$, $c \in \mathbb{R}^1$ (translation-equivariance);

(ii) $R_n(cx_1, \ldots, cx_n) = cR_n(x_1, \ldots, x_n)$, $\boldsymbol{x} \in \mathbb{R}^n$, $c > 0$ (scale-equivariance);

(iii) $P_\theta(S_n(\boldsymbol{X} - t) < 0) \le P_\theta(R_n \le t) \le P_\theta(S_n(\boldsymbol{X} - t) \le 0)$, $t, \theta \in \mathbb{R}^1$;

(iv) $\frac{1}{2} - \frac{1}{2}\varepsilon \le P_\theta(R_n < \theta) \le P_\theta(R_n \le \theta) \le \frac{1}{2} + \frac{1}{2}\varepsilon$, $\theta \in \mathbb{R}^1$, with $\varepsilon = P_\theta(S_n(\boldsymbol{X} - \theta) = 0) = P_0(S_n(\boldsymbol{X}) = 0)$.

Remark. The properties (i), (iii) and (iv) are analogous to these of M-estimators. The value ε in (iv) is independent of F. The property (ii) is the only one which we miss in the case of M-estimators. On the other hand, R-estimators do not have the finite-sample minimax property of Huber's M-estimator (see Theorem 2.1).

2.2.2. Asymptotic efficiency of R-estimators

Hodges and Lehmann (1963) proved that the asymptotic efficiency of the R-estimator coincides with the Pitman efficiency of the corresponding signed-rank test. Thus, to establish the efficiency of an R-estimator, we need to known the asymptotic distribution of the signed-rank statistics under contiguous location alternatives; and this was studied in details in the monograph Hájek and Šidák (1967).

Assume that the score-generating function φ is skew-symmetric, square-integrable and is of bounded variation on every subinterval of (0, 1), i.e., $\varphi = \varphi_1 - \varphi_2$ where φ_1 and φ_2 are nondecreasing; then, provided F has an absolutely continuous symmetric density with finite Fisher's information (see (2.10)), then

$$\sqrt{n}(R_n - \theta) \xrightarrow{\mathcal{D}} N(0, \sigma^2(\varphi, F)) \tag{2.25}$$

with

$$\sigma^2(\varphi, F) = \int_0^1 \varphi^2(t)\, dt \cdot \left(\int \varphi(F(x)) f'(x)\, dx \right)^{-2}. \tag{2.26}$$

We see that $0 < \sigma^2(\varphi, F) < \infty$ under general conditions. If we put

$$\varphi(t) = \varphi_f(t) = -f'(F^{-1}(t))/f(F^{-1}(t)), \quad 0 < t < 1, \tag{2.27}$$

then $\sigma^2(\varphi, F) = 1/I(F)$; it means that the class of R-estimators also contains an asymptotically efficient element. Similarly as in the case of M-estimators, we are interested in the behavior of $\sup_{F \in \mathcal{F}} \sigma^2(\varphi, F)$ over some family $\mathcal{F}$ of distributions, e.g. over the family $\mathcal{F}_\varepsilon$ of contaminated normal distributions (2.14). Then (cf. Jaeckel, 1971), if we put $\varphi_0(t) = \varphi_{f_0}(t)$, $0 < t < 1$, with f_0 being the least informative distribution (2.17), i.e.,

$$\varphi_0(t) = \begin{cases} -c & \text{if } t < \alpha, \\ \Phi^{-1}((t - \frac{1}{2}\varepsilon)/(1 - \varepsilon)) & \text{if } \alpha \le t \le 1 - \alpha, \\ c & \text{if } t > 1 - \alpha, \end{cases} \tag{2.28}$$

where $\alpha = \frac{1}{2}\varepsilon + (1 - \varepsilon)\Phi(-c)$, we get an R-estimator satisfying

$$\sup_{F \in \mathcal{F}_\varepsilon} \sigma^2(\varphi_0, F) = \inf_{\varphi} \sup_{F \in \mathcal{F}_\varepsilon} \sigma^2(\varphi, F).$$

It could be shown similarly that the trimmed Hodges–Lehmann's estimator (2.24) provides the saddle-point for the family of contaminated logistic distributions (cf. Miura, 1981).

2.2.3. Some further developments

Antille (1974) established the Bahadur's type representation of Hodges–Lehmann's estimator and Hušková and Jurečková (1981) for a more general R-estimator. Van Eeden (1970) and Beran (1974) developed asymptotically uniformly efficient (adaptive) R-estimators of location. Sen (1980) proved the moment convergence and developed the asymptotically risk-efficient sequential versions of R-estimators.

2.3. L-estimators (linear combinations of order statistics)

Let $X_1, X_2, \ldots$ be a sequence of independent random variables, identically distributed according to d.f. $F(x-\theta)$, $F(x)+F(-x)=1$, $x \in \mathbb{R}^1$; let $X_{n:1} \leqslant \cdots \leqslant X_{n:n}$ be the order statistics corresponding to $X_1, \ldots, X_n$. The L-estimator of θ is defined as

$$L_n = \sum_{i=1}^{n} c_{ni} X_{n:i} \tag{2.29}$$

where the coefficients $c_{n1}, \ldots, c_{nn}$ satisfy

$$c_{ni} = c_{n,n-i+1} \geqslant 0, \; i = 1, \ldots, n; \quad \sum_{i=1}^{n} c_{ni} = 1 . \tag{2.30}$$

This class of estimators covers the sample mean as well as the sample median. The L-estimators are computionally more appealing than M- and R-estimators. If we wish to get a robust L-estimator, insensitive to the extreme observations, we must put $c_{ni} = 0$ for $i \leqslant k_n$ and $i \geqslant n - k_n + 1$ with a proper k_n. Typical examples of such estimators are the α-trimmed mean,

$$L_n = (1/(n-2)[n\alpha]) \sum_{i=[n\alpha]+1}^{n-[n\alpha]} X_{n:i} \tag{2.31}$$

and the α-Winsorized mean,

$$L_n = \frac{1}{n}\left\{[n\alpha]X_{n:[n\alpha]} + \sum_{i=[n\alpha]+1}^{n-[n\alpha]} X_{n:i} + [n\alpha]X_{n:n-[n\alpha]+1}\right\}; \tag{2.32}$$

$0 < \alpha < \frac{1}{2}$ and $[x]$ is the largest integer k satisfying $k \leqslant x$.

Most of the L-estimators may be expressed in the following way

$$L_n = \frac{1}{n}\sum_{i=1}^{n} J\left(\frac{i}{n+1}\right) X_{n:i} + \sum_{j=1}^{k} a_j X_{n:[np_j]} \tag{2.33}$$

where $J(u)$, $0<u<1$, is a proper weight function satisfying $J(u) = J(1-u) \geqslant 0$, $0<u<1$ and $p_1, \ldots, p_k$; $a_1, \ldots, a_k$ are given constants satisfying $0 < p_1 < \cdots < p_k < 1$, $p_j = 1 - p_{k-j+1}$, $a_j = a_{k-j+1} = 0$, $j = 1, \ldots, k$. L_n is then of the form (2.29) with c_{ni} equal to $n^{-1}J(i/(n+1))$ plus an additional contribution a_j if $i = [np_j]$ for some j $(1 \leqslant j \leqslant k)$. It is usually assumed that J is continuous and differentiable up to a finite number of points. Many L-estimators which we encounter in the practice coincide either with the first or with the second component of (2.33).

2.3.1. Finite-sample properties of L-estimators

Let L_n be the L-estimator defined in (2.29) and satisfying (2.30); then, provided that $X_1, X_2, \ldots$ are distributed according to the d.f. $F(x-\theta)$, $F(x)+F(-x)=1$, $x \in \mathbb{R}^1$, the following holds:

(i) $L_n(x_1+c, \ldots, x_n+c) = L_n(x_1, \ldots, x_n) + c; \quad \boldsymbol{x} \in \mathbb{R}^n,\ c \in \mathbb{R}^1.$
(ii) $L_n(cx_1, \ldots, cx_n) = cL_n(x_1, \ldots, x_n); \quad \boldsymbol{x} \in \mathbb{R}^n,\ c > 0.$
(iii) If F is absolutely continuous, then

$$P_\theta(L_n < \theta) = P_\theta(L_n \leq \theta) = \tfrac{1}{2}, \quad \theta \in \mathbb{R}^1 .$$

2.3.2. Asymptotic efficiency of L-estimators

The asymptotic normality of L-estimators was studied by many authors under various conditions on F and on c_{ni}'s. We may mention Bickel (1965, 1967), Boos (1979), Boos and Serfling (1980), Chernoff, Gastwirth and Johns (1967), Huber (1969), Shorack (1969, 1972), Stigler (1969, 1974), among others. A good review of the asymptotic results on L-estimators may be found in Serfling (1980, Chapter 8).

Let us first consider the L-estimators of the form (2.33) with vanishing second component. Then, again, the more we assume on J, the less we must assume on F. For robust L-estimators, it is more convenient to put more restrictions on J rather than on F. From the various theorems on asymptotic normality of L_n, let us describe one proved in Stigler (1974; see also Stigler (1979)).

THEOREM 2.3. *Let $X_1, X_2, \ldots$ be the sequence of independent observations from the d.f. $F(x-\theta)$ such that $F(x)+F(-x)=1$, $x \in \mathbb{R}^1$. Let $J(u)$ be a function such that $J(u)=J(1-u)$, $0<u<1$ and $\int_0^1 J(u)\,\mathrm{d}u = 1$. Then, under the assumptions*

(i) *$J(u)=0$ for $0<u<\alpha$ and $1-\alpha<u<1$, is bounded and satisfies a Lipschitz condition of order $>\frac{1}{2}$ (except possibly a finite number of points of F^{-1} measure 0);*

(ii) *$\int |F(x)(1-F(x))|^{1/2}\,\mathrm{d}x < \infty$ and*

$$\sigma^2(J, F) = \int\int J(F(x))J(F(y))[F(\min(x,y)) - F(x)F(y)]\,\mathrm{d}x\,\mathrm{d}y \tag{2.34}$$

is positive.

Then the estimator

$$L_n = \frac{1}{n}\sum_{i=1}^{n} J\left(\frac{i}{n+1}\right) X_{n:i} \tag{2.35}$$

satisfies

$$\sqrt{n}(L_n - \theta) \xrightarrow{\mathcal{D}} N(0, \sigma^2(J,F)) \quad \text{as } n \to \infty . \tag{2.36}$$

If L_n is of the form (2.33) and the second component does not vanish, then, under the assumptions of Theorem 2.3, $\sqrt{n}(L_n - \theta)$ is asymptotically normally distributed $N(0, \sigma^2(F))$ with

$$\sigma^2(F) = \mathrm{var}\Big\{-\int (I[X_1 \leq y] - F(y))J(F(y))\,\mathrm{d}y + \sum_{j=1}^{k} [a_j/f(F^{-1}(p_j))](p_j - I[X_1 \leq F^{-1}(p_j)])\Big\}, \tag{2.37}$$

provided $F^{-1}(p)$ has positive derivative at p_j, $j = 1, \ldots, k$ (cf. Boos, 1979).

Under additional assumptions on F, the asymptotic normality with the variance $\sigma^2(J, F)$ of (2.34) may be established even for J which puts positive weights on the extremes (Stigler, 1974; Shorack, 1972). On the other hand, the assumption (ii) may be weakened in many special cases.

If F has an absolutely continuous density f and finite Fisher's information $I(F)$, then the L-estimator (2.35) with

$$J(t) = J_F(t) = \psi_F'(F^{-1}(t))/I(F), \quad 0 < t < 1, \tag{2.38}$$

where $\psi_F(x) = -f'(x)/f(x)$, $x \in \mathbb{R}^1$, satisfies $\sigma^2(J_F, F) = 1/I(F)$. It means that the class of L-estimators also contains an asymptotically efficient element.

If we put $J_0(t) = J_{F_0}(t)$, $0 < t < 1$, with F_0 being the d.f. of the least informative distribution (2.17), i.e.,

$$J_0(t) = \begin{cases} 1/(1-2\alpha) & \text{if } \alpha \leq t \leq 1-\alpha, \\ 0 & \text{otherwise}, \end{cases} \tag{2.39}$$

with $\alpha = F_0(-c) = \frac{1}{2}\varepsilon + (1-\varepsilon)\Phi(-c)$, we get an L-estimator satisfying

$$\sup_{F \in \mathscr{F}_\varepsilon} \sigma^2(J_0, F) = \inf_J \sup_{F \in \mathscr{F}_\varepsilon} \sigma^2(J, F) \tag{2.40}$$

where $\mathscr{F}_\varepsilon$ is the family of ε-contaminated normal distributions (2.14). The L-estimator generated by $J_0(t)$ is the α-trimmed mean. We may conclude that the α-trimmed mean is the most recommendable estimator of the center of symmetry of the contaminated normal distribution. It is computationally simple and it is not only translation- but also scale-equivariant. Bickel and Lehmann (1975) proved another attractive property of the α-trimmed mean: its asymptotic efficiency relatively to the sample mean $\bar{X}_n$ cannot get below $(1-2\alpha)^2$ not only for symmetric F but also for every strictly increasing and continuous F.

2.3.3. *Some further developments*

Berry–Esseen bounds for L-estimators were studied, among others, by Bickel (1967), Bjerve (1977), Boos and Serfling (1979), Helmers (1977, 1980, 1981); the law of iterated logarithm and almost sure asymptotic results were established by Wellner (1977a,b) and van Zwet (1980). Invariance principles for L-estimators were proved by Sen (1977, 1978); see also Sen (1981). Moment convergence of L-estimators and their asymptotically risk-efficient versions were studied by Jurečková and Sen (1982). The tail behavior of L-estimators in the finite sample case was studied by Jurečková (1979, 1981).

3. Estimation of regression

Let $X_n = (X_{n1}, \ldots, X_{nn})'$ be the vector of independent observations satisfying

$$\boldsymbol{X}_n = \boldsymbol{C}_n\boldsymbol{\theta} + \boldsymbol{E} \tag{3.1}$$

where $\boldsymbol{\theta} = (\theta_1, \ldots, \theta_p)'$ is the vector of unknown regression parameters, $\boldsymbol{E} = (E_1, \ldots, E_n)'$ is the vector of errors and $\boldsymbol{C}_n$ is a known $(n \times p)$ design matrix of the rank p.

The problem is that of estimating $\boldsymbol{\theta}$. We shall assume throughout that E_i, $i = 1, \ldots, n$, are independent and identically distributed according to a common d.f. F which is an unknown member of a family $\mathscr{F}$ of d.f.'s. The coordinates of $\boldsymbol{X}_n$ and of $\boldsymbol{C}_n$ depend on n; we shall not indicate explicitly this dependence unless it could cause a confusion. Let us denote

$$\delta_i(t) = X_i - \sum_{j=1}^{p} c_{ij}t_j, \quad i = 1, \ldots, n\,, \tag{3.2}$$

the residuals corresponding to the vector $\boldsymbol{t} = (t_1, \ldots, t_p)'$.

What was said about the sensitivity of the sample mean to the outlying observations and to the long-tailed distributions, holds also for the least-squares estimator (l.s.e.); and the outliers are more difficult to track in the linear model. The M-, R- and L-estimators extend, in a more or less straightforward way, to the linear model.

3.1. M-estimators

The M-estimator $\boldsymbol{M}_n$ of $\boldsymbol{\theta}$ is defined as a solution of the system of equations

$$\sum_{i=1}^{n} c_{ij}\psi\Big(X_i - \sum_{k=1}^{p} c_{ik}t_k\Big) = 0, \quad j = 1, \ldots, p\,, \tag{3.3}$$

with respect to $t_1, \ldots, t_p$. If there are more solutions of (3.3), then $\boldsymbol{M}_n$ may be defined as that the nearest to some proper preliminary consistent estimator of $\boldsymbol{\theta}$. If F has an absolutely continuous density f and we put $\psi(x) = -f'(x)/f(x)$, $x \in \mathbb{R}^1$, we get the m.l.e. of $\boldsymbol{\theta}$; $\boldsymbol{M}_n$ coincides with the l.s.e. if $\psi(x) = x$, $x \in \mathbb{R}^1$.

Similarly as in the location case, $\boldsymbol{M}_n$ is translation-equivariant but generally not scale-equivariant, so that, unless the scale of F is supposed to be known, $\boldsymbol{M}_n$ should be supplemented by an appropriate estimator of scale.

The asymptotic behavior of $\boldsymbol{M}_n$ as $n \to \infty$ was studied by Relles (1968), Huber

(1972, 1973), Yohai and Maronna (1979), among others. Under the assumptions on ψ and on F analogous to those in the location case (besides the assumption of symmetry of F), it was shown that, as $n \to \infty$, $\boldsymbol{\Sigma}_n^{1/2}(\boldsymbol{M}_n - \boldsymbol{\theta})$ is asymptotically p-dimensionally normally distributed with expectation $\mathbf{0}$ and with the covariance matrix $\sigma^2(\psi, F)\boldsymbol{I}_p$ with $\sigma^2(\psi, F)$ given in (2.13) and $\boldsymbol{\Sigma}_n = \boldsymbol{C}_n'\boldsymbol{C}_n$; the matrix $\boldsymbol{\Sigma}_n$ is assumed to be positive and of the rank p for $n \geq n_0$. We see that, the sequence $\{\boldsymbol{C}_n\}$ being fixed, the efficiency properties of $\boldsymbol{M}_n$ depend only on the constant $\sigma^2(\psi, F)$ and are analogous to these in the location case. This further implies that the asymptotic minimax property of M-estimators over the family $\mathcal{F}_\varepsilon$ of ε-contaminated normal distributions (see Section 2.1.2), extends to the linear model (3.1).

Huber (1973) considered the asymptotic behavior of M_n in the case that $p \to \infty$ simultaneously with n. An extension of M-estimators to the multivariate linear model and its asymptotic behavior was studied by Maronna (1976) and Carroll (1978). M-estimators of regression parameters with random design matrix were studied by Maronna, Bustos and Yohai (1979). Bahadur type representation of M-estimators in the linear model was considered by Jurečková and Sen (1981a,b).

3.2. *R-estimators*

R-estimators of regression parameters are inversions of linear rank tests of regression. The general rank test of the hypothesis H: $\boldsymbol{\theta} = \boldsymbol{\theta}_0$ in the model (3.1) is based on the vector of statistics

$$S_{nj}(\boldsymbol{\theta}_0) = \sum_{i=1}^{n} (c_{ij} - \bar{c}_j)\varphi\left(\frac{R_{ni}(\boldsymbol{\theta}_0)}{n+1}\right), \quad j = 1, \ldots, p, \tag{3.4}$$

where $R_{ni}(\boldsymbol{\theta}_0)$ is the rank of the residual $\delta_i(\boldsymbol{\theta}_0)$ among $\delta_1(\boldsymbol{\theta}_0), \ldots, \delta_n(\boldsymbol{\theta}_0)$ and $\varphi(t)$ is a nondecreasing square-integrable score function, $0 < t < 1$. Denote $\boldsymbol{S}_n(\boldsymbol{t}) = (S_{n1}(\boldsymbol{t}), \ldots, S_{np}(\boldsymbol{t}))'$; then, under $\boldsymbol{\theta} = \boldsymbol{\theta}_0$, $E_{\boldsymbol{\theta}_0}\boldsymbol{S}_n(\boldsymbol{\theta}_0) = \mathbf{0}$ and analogously as in the location case, we may define the R-estimator of $\boldsymbol{\theta}_0$ as any solution of the system of 'equations'

$$S_{nj}(\boldsymbol{t}) \cong 0, \quad j = 1, \ldots, p, \tag{3.5}$$

with respect to $\boldsymbol{t}$.

The statistics (3.4) are invariant to the translation, so that they are not able to estimate the main additive effect (i.e., the component θ_j for which $c_{ij} = 1$, $i = 1, \ldots, n$). The main additive effect should be estimated with the aid of the signed rank statistics on the same line as the location parameter (cf. Jurečková, 1971b).

Adichie (1967), following the ideas of Hodges and Lehmann suggested an estimator of (θ_1, θ_2) in the regression model $X_i = \theta_1 + \theta_2 c_i + E_i$, $i = 1, \ldots, n$, based on the Wilcoxon tests and derived its asymptotic distribution. Jurečková (1971a), Koul (1972) and Jaeckel (1972) then extended the procedure to the

p-parameter regression and to the general linear rank tests. The three respective R-estimators are asymptotically equivalent and thus they have the same asymptotic distributions and efficiencies. The estimators differ in the way how they describe the solution of (3.5). Jurečková (1971a) suggested the estimator $\boldsymbol{R}_n$ as any solution of the minimization problem

$$\sum_{j=1}^{p}\sum_{i=1}^{n}|S_{nj}(\boldsymbol{t})| := \min \tag{3.6}$$

and proved that $\boldsymbol{\Sigma}_n^{1/2}(\boldsymbol{R}_n - \boldsymbol{\theta})$ is asymptotically normally distributed with the expectation $\boldsymbol{0}$ and with the covariance matrix $\sigma^2(\varphi, F)\boldsymbol{I}_p$ where $\sigma^2(\varphi, F)$ is given in (2.25) and $\boldsymbol{\Sigma}_n = \boldsymbol{C}_n'\boldsymbol{C}_n$. The assumption on φ and on F are similar as in the location case while the assumptions on $\boldsymbol{C}_n$ were rather restrictive in Jurečková (1971a) and some related papers (concordance–discordance condition on the columns of $\boldsymbol{C}_n$ for $n \geqslant n_0$). Later on Heiler and Willers (1979) proved that the asymptotic normality of $\boldsymbol{R}_n$ holds also without corcordance–discordance condition.

Jaeckel (1972) suggested R-estimator of $\boldsymbol{\theta}$ as a solution of the minimization problem

$$\sum_{i=1}^{n}\left[\varphi\left(\frac{R_{ni}(\boldsymbol{t})}{n+1}\right) - \bar{\varphi}_n\right]\delta_i(\boldsymbol{t}) := \min \tag{3.7}$$

with respect to $\boldsymbol{t}$; $\bar{\varphi}_n = (1/n)\sum_{i=1}^{n}\varphi(i/(n+1))$. The idea is that (3.7) could be considered as a measure of the dispersion of the residuals $\delta_i(\boldsymbol{t})$, $i = 1, \ldots, n$, instead of the proper variance of the residuals which is used in the method of least squares. Jaeckel proved the asymptotic equivalence of the solution of (3.7) and of (3.6), respectively, as $n \to \infty$.

Koul (1971) suggested the R-estimator as minimizing, instead of (3.6) and (3.7), an appropriate quadratic form in the statistics $S_{nj}(\boldsymbol{t})$, $j = 1, \ldots, p$, with respect to $\boldsymbol{t}$. All three estimators are asymptotically equivalent, as $n \to \infty$.

3.3. L-estimators

While being computionally very appealing in the location case, the L-estimators do not have any straightforward extension to the linear model. Let us mention some of the attempts which appeared in the literature.

Koenker and Bassett (1978) extended the concept of quantiles to the linear model. For a fixed α, $0 < \alpha < 1$, put

$$\psi_\alpha(x) = \alpha - I[x < 0] \tag{3.8}$$

and

$$\rho_\alpha(x) = x \cdot \psi_\alpha(x), \quad x \in \mathbb{R}^1 . \tag{3.9}$$

Koenker and Bassett defined the α-th regression quantile as the solution

$\boldsymbol{T}_n(\alpha) = (T_{n1}(\alpha), \ldots, T_{np}(\alpha))'$ of the minimization

$$\sum_{i=1}^{n} \rho_\alpha\Big(X_i - \sum_{j=1}^{p} c_{ij}t_j\Big) := \min \tag{3.10}$$

with respect to $\boldsymbol{t} = (t_1, \ldots, t_p)'$. They proved that the asymptotic behavior of the regression quantiles is similar to that of the standard sample quantiles and suggested the following α-trimmed least squares estimator:

Remove X_i from the sample if $\delta_i(\boldsymbol{T}_n(\alpha)) < 0$ (the i-th residual from $\boldsymbol{T}_n(\alpha)$ is negative) or if $\delta_i(\boldsymbol{T}_n(1-\alpha)) > 0$, $i = 1, \ldots, n$; $0 < \alpha < \frac{1}{2}$; and calculate the least-squares estimator using the remaining observations.

The resulting estimator $\boldsymbol{L}_n^*$ was later studied by Ruppert and Carroll (1980) who proved that $\boldsymbol{\Sigma}_n^{1/2}(\boldsymbol{L}_n^* - \boldsymbol{\theta})\boldsymbol{\Sigma}_n^{1/2}$ is asymptotically normally distributed with the expectation $\boldsymbol{0}$ and with the covariance matrix $\sigma^2(\alpha, F)\boldsymbol{I}_p$ where $\sigma^2(\alpha, F)$ is the asymptotic variance of the α-trimmed mean in the location case. Jurečková (1983) proved that $\boldsymbol{L}_n^*$ is asymptotically equivalent, in probability, to the Huber estimator of $\boldsymbol{\theta}$ generated by the function ψ of (2.4) with $c = F^{-1}(1-\alpha)$. The regression quantiles seem to provide a basis for the extension of various L-estimators from the location to the regression model.

Ruppert and Carroll (1980) also suggested another extension of the α-trimmed mean to the linear model. Starting with some reasonable preliminary estimator $\boldsymbol{L}_0$, one calculates the residuals $\delta_i(\boldsymbol{L}_0)$ from $\boldsymbol{L}_0$, $i = 1, \ldots, n$, and removes the observations corresponding to $[n\alpha]$ smallest and $[n\alpha]$ largest residuals. The estimator $\boldsymbol{L}_n^{**}$ is then defined as the least-squares estimator calculated from the remaining observations. The asymptotic behavior of $\boldsymbol{L}_n^{**}$ depends on $\boldsymbol{L}_0$ and generally is not similar to that of the trimmed mean; $\boldsymbol{L}_n^{**}$ is asymptotically equivalent to $\boldsymbol{L}_n^*$ provided $\boldsymbol{L}_0 = \frac{1}{2}(\boldsymbol{T}_n(\alpha) + \boldsymbol{T}_n(1-\alpha))$.

Bickel (1973) proposed a general class of one-step L-estimators of $\boldsymbol{\theta}$ depending on a preliminary estimate of $\boldsymbol{\theta}$. The estimators have the best possible efficiency properties, i.e. analogous to those of the corresponding location L-estimators but they are computationally complex and are not invariant under a reparametrization of the vector space spanned by the columns of $\boldsymbol{C}_n$.

4. Computational aspects – One step versions of the estimators

Besides the L-estimators of location and the Hodges–Lehmann estimator, the estimators considered so far are not very computionally appealing. They are generally defined in the implicit form or as a solution of a complex minimization problem.

Thus, it is often convenient to use the one-step versions of the estimators, which are characterized as follows: we start with some reasonably good consistent preliminary estimator $\hat{\boldsymbol{\theta}}$ and then apply one step of Gauss–Newton method to the equation defining the estimator. Under mild conditions, it could be shown that the result of one-step Gauss–Newton approximation behaves asymptotically as the root of the equation. This idea was applied by Kraft and van Eeden (1972a,b) to the R-estimators of location and regression, respectively. Bickel (1975) studied the one-step versions of the M-estimators in the linear model.

Let us first describe the one-step version of the M-estimator. Let M_n be the M-estimator of θ in the linear model (3.1), defined as the solution of the system of equations (3.3). Assume that the design matrix $\boldsymbol{C}_n$ satisfies the condition $n^{-1}\boldsymbol{C}_n'\boldsymbol{C}_n \to \boldsymbol{\Sigma}$ as $n\to\infty$ where $\boldsymbol{\Sigma}$ is a positive $p\times p$ matrix. Then, provided F has an absolutely continuous density f, $I(F)<\infty$ and ψ has bounded variation on any compact interval,

$$\sqrt{n}(\boldsymbol{M}_n - \boldsymbol{\theta}) = (1/\gamma\sqrt{n})\boldsymbol{\Sigma}^{-1}\boldsymbol{C}_n'v_n(\boldsymbol{\theta}) + o_p(1) \tag{4.1}$$

with

$$\boldsymbol{v}_n(\boldsymbol{\theta}) = (\psi(\delta_1(\boldsymbol{\theta})), \ldots, \psi(\delta_n(\boldsymbol{\theta})))' \tag{4.2}$$

and

$$\gamma = -\int \psi(x)f(x)\,\mathrm{d}x \tag{4.3}$$

(cf. Bickel, 1975; Jurečková, 1977). Let $\hat{\boldsymbol{\theta}}_n$ be a consistent preliminary estimator of $\boldsymbol{\theta}$ which is shift-equivariant, i.e. $\hat{\boldsymbol{\theta}}_n(\boldsymbol{x}+\boldsymbol{C}_n\boldsymbol{t}) = \hat{\boldsymbol{\theta}}_n(\boldsymbol{x})+\boldsymbol{t}$, $\boldsymbol{x}\in\mathbb{R}^n$, $\boldsymbol{t}\in\mathbb{R}^p$, and which satisfies

$$\|\hat{\boldsymbol{\theta}}_n - \boldsymbol{\theta}\| = o_p(1) \quad \text{as } n\to\infty\,. \tag{4.4}$$

The one-step version of $\boldsymbol{M}_n$ is defined as

$$\boldsymbol{M}_n^* = \hat{\boldsymbol{\theta}}_n + (1/n\hat{\gamma}_n)\boldsymbol{\Sigma}^{-1}\boldsymbol{C}_n'\boldsymbol{v}_n(\hat{\boldsymbol{\theta}}_n) \tag{4.5}$$

where $\hat{\gamma}_n$ is an appropriate consistent estimator of γ; one of the possible estimators of γ is

$$\hat{\gamma}_n = n^{-1/2}\|\boldsymbol{t}_2-\boldsymbol{t}_1\|^{-1}\|\boldsymbol{\Sigma}^{-1}\cdot\boldsymbol{C}_n'(\boldsymbol{v}_n(\hat{\boldsymbol{\theta}}_n - n^{-1/2}\boldsymbol{t}_2) - \boldsymbol{v}_n(\hat{\boldsymbol{\theta}}_n - n^{-1/2}\boldsymbol{t}_1))\| \tag{4.6}$$

where $\boldsymbol{t}_1, \boldsymbol{t}_2$ is a fixed pair of $p\times 1$ vectors, $\boldsymbol{t}_1\neq\boldsymbol{t}_2$. Then it could be proved that

$$\sqrt{n}\|\boldsymbol{M}_n - \boldsymbol{M}_n^*\| \xrightarrow{p} 0 \quad \text{as } n\to\infty\,. \tag{4.7}$$

Let us briefly describe possible one-step versions of R-estimator $\boldsymbol{R}_n$ defined in (3.6) or (3.7). Assume that

$$n^{-1}\sum_{i=1}^{n}(c_{ij}-\bar{c}_j)(c_{ik}-\bar{c}_k)\to\sigma_{jk}^* \quad \text{as } n\to\infty \tag{4.8}$$

for $1\leq j,\ k\leq p$, where $\boldsymbol{\Sigma}^* = [\sigma_{jk}^*]_{j,k=1,\ldots,p}$ is a positive matrix. Then, provided φ

is of bounded variation on any compact subinterval of $(0,1)$ and square integrable, F has an absolutely continuous density and $I(F)<\infty$, it could be proved (Jurečková, 1971a)

$$\boldsymbol{R}_n=\boldsymbol{\theta}+(1/n\gamma)\boldsymbol{\Sigma}^{*-1}\boldsymbol{S}_n(\boldsymbol{\theta})+\mathrm{o}_p(n^{-1/2}) \tag{4.9}$$

with $\boldsymbol{S}_n(\boldsymbol{\theta})=(S_{n1}(\boldsymbol{\theta}),\ldots,S_{np}(\boldsymbol{\theta}))'$ given in (3.4) and

$$\gamma=-\int\varphi(F(x))f'(x)\,\mathrm{d}x\,. \tag{4.10}$$

Let $\hat{\boldsymbol{\theta}}_n$ be the preliminary shift-equivariant estimator satisfying (4.4). Then, if we know γ, we could consider the one-step version of $\boldsymbol{R}_n$ in the form

$$\boldsymbol{R}'_n=\hat{\boldsymbol{\theta}}_n+(1/n\gamma)\boldsymbol{\Sigma}^{*-1}\boldsymbol{S}_n(\hat{\boldsymbol{\theta}}_n)\,, \tag{4.11}$$

and $\sqrt{n}\|\boldsymbol{R}_n-\boldsymbol{R}'_n\|\xrightarrow{p}0$ as $n\to\infty$. However, γ is generally unknown; Kraft and van Eeden (1972b) suggested to replace γ in (4.11) by $\int_0^1(\varphi(t)-\bar{\varphi})^2\,\mathrm{d}t$, $\bar{\varphi}=\int_0^1\varphi(t)\,\mathrm{d}t$. The resulting estimator (say, $\boldsymbol{R}''_n$) is generally not asymptotically equivalent to $\boldsymbol{R}_n$; it could be proved (cf. Humak, 1983) that $\sqrt{n}\|\boldsymbol{R}_n-\boldsymbol{R}''_n\|\xrightarrow{p}0$ as $n\to\infty$ if and only if either both $\boldsymbol{R}_n$ and $\boldsymbol{R}''_n$ are asymptotically efficient (i.e., where $\varphi(t)=-f'(F^{-1}(t))/f(F^{-1}(t))$) or if $\boldsymbol{R}_n$ (and thus also $\boldsymbol{R}''_n$) is asymptotically equivalent to the preliminary estimator $\hat{\boldsymbol{\theta}}_n$. In order to get an estimator asymptotically equivalent to $\boldsymbol{R}_n$, we should replace γ in (4.11) by an appropriate estimator $\hat{\gamma}_n$, similarly as in the case of M-estimator. One of such possible estimators is

$$\hat{\gamma}_n=n^{-1/2}\|\boldsymbol{t}_2-\boldsymbol{t}_1\|^{-1}\cdot\|\boldsymbol{\Sigma}^{*-1}(\boldsymbol{S}_n(\hat{\boldsymbol{\theta}}_n-n^{-1/2}\boldsymbol{t}_2)-\boldsymbol{S}_n(\hat{\boldsymbol{\theta}}_n-n^{-1/2}\boldsymbol{t}_1))\|\,, \tag{4.12}$$

where $\boldsymbol{t}_1,\boldsymbol{t}_2$ are fixed $p\times1$ vectors, $\boldsymbol{t}_1\neq\boldsymbol{t}_2$.

5. Asymptotic relations of M-R-L-estimators

We have seen that the three groups of estimators, though being defined in different ways, follow the same idea: to cut-off the influence of outliers and to diminish the sensitivity to the long-tailed distributions. It turns out that these three classes of estimators are even nearer than one would expect; in fact, they become asymptotically equivalent as $n\to\infty$.

The asymptotic relations of M-R-L-estimators were studied by Jaeckel (1971), Bickel and Lehmann (1975), Jurečková (1977, 1978, 1981), Hušková and Jurečková (1981), among others. Let us briefly illustrate some of the results on the location submodel.

Let $X_1,X_2,\ldots$ be the sequence of independent observations, identically distributed according to the distribution function $F(x-\theta)$ such that

$F(x)+F(-x)=1$, $x\in\mathbb{R}^1$. Let M_n be the M-estimator generated by the function $\psi(x)$, $x\in\mathbb{R}^1$ and R_n be the R-estimator generated by the function $\varphi(t)$, $0<t<1$. Then, under some regularity conditions, $\sqrt{n}(M_n-R_n)=o_p(1)$ as $n\to\infty$ if and only if

$$\psi(x)=a\varphi(F(x)),\quad a>0\,, \tag{5.1}$$

for almost all $x\in\mathbb{R}^1$. The relation (5.1) means that, given the distribution F, there exists an M-estimator to every R-estimator (and vice versa) such that both estimators are asymptotically equivalent. Being dependent on the unknown d.f. F, the relation (5.1) does not enable to determine the value of the M-estimator once we have calculated the value of R-estimator; it rather indicates which type of M-estimators belongs to a given type of R-estimators etc.

Let L_n be the L-estimator (2.34) generated by the function $J(t)$ such that $J(t)=J(1-t)\geq 0$, $0<t<1$. Then, under some smoothness conditions on J and F, $\sqrt{n}(L_n-M_n)=o_p(1)$ as $n\to\infty$ for the M-estimator M_n generated by the function

$$\psi(x)=\int_0^1 J(t)(I[F(x)\leq t]-t)\,\mathrm{d}F^{-1}(t),\quad x\in\mathbb{R}^1\,. \tag{5.2}$$

Let L_n be the α-trimmed mean (2.31); then $\sqrt{n}(L_n-M_n)=o_p(1)$ as $n\to\infty$ where M_n is Huber estimator generated by ψ given in (2.4), more precisely,

$$\psi(x)=\begin{cases} F^{-1}(\alpha) & \text{if } x<F^{-1}(\alpha)\,,\\ x & \text{if } F^{-1}(\alpha)\leq x\leq F^{-1}(1-\alpha)\,,\\ F^{-1}(1-\alpha) & \text{if } x>F^{-1}(1-\alpha)\,.\end{cases} \tag{5.3}$$

If L_n is a linear combination of single sample quantiles, $L_n=\sum_{j=1}^k a_jX_{n:[np_j]}$ (cf. 2.33), then $\sqrt{n}(L_n-M_n)=o_p(1)$ where M_n is the M-estimator generated by the function

$$\psi(x)=-\sum_{j=1}^k [a_j/f(F^{-1}(p_j))](I[F(x)\leq p_j]-p_j)\,, \tag{5.4}$$

$x\in\mathbb{R}^1$. Especially, the M-estimator counterpart of the α-Winsorized mean is generated by the function (Jurečková, 1983a)

$$\psi(x)=\begin{cases} F^{-1}(\alpha)-\dfrac{\alpha}{f(F^{-1}(\alpha))} & \text{for } x<F^{-1}(\alpha)\,,\\ x & \text{for } F^{-1}(\alpha)\leq x\leq F^{-1}(1-\alpha)\,,\\ F^{-1}(1-\alpha)+\dfrac{\alpha}{f(F^{-1}(1-\alpha))} & \text{for } x>F^{-1}(1-\alpha)\,.\end{cases} \tag{5.5}$$

The relations of R- and L-estimators could be derived by combining the relations of M- and R-estimators and these of M- and L-estimators, respectively.

References

Adichie, J. N. (1967). Estimate of regression parameters based on rank tests. *Ann. Math. Statist.* **38**, 894–904.

Andrews, D. F., Bickel, P. J., Hampel, F. R., Huber, P. J., Rogers, W. H. and Tukey, J. W. (1972). *Robust Estimates of Location. Survey and Advances.* Princeton University Press.

Antille, A. (1974). A linearized version of the Hodges–Lehmann estimator. *Ann. Statist.* **2**, 1308–1313.

Azencott, R., Birgé, L., Costa, V., Dacunha-Castelle, D., Deniau, C., Deshayes, J., Huber, C., Jolivaldt, P., Oppenheim, G., Picard, D., Trécourt, P. and Viano, C. (1977). Théorie de la robustesse et estimation d'un parametre. *Astérisque* 43–44.

Beran, R. (1974). Asymptotically efficient adaptive rank estimates in location models. *Ann. Statist.* **2**, 63–74.

Beran, R. (1977a). Robust location estimates. *Ann. Statist.* **5**, 431–444.

Beran, R. (1977b). Minimum Hellinger distance estimates for parametric models. *Ann. Statist.* **5**, 445–463.

Beran, R. (1978). An efficient and robust adaptive estimator of location. *Ann. Statist.* **6**, 292–313.

Bickel, P. J. (1965). On some robust estimates of location. *Ann. Math. Statist.* **36**, 847–858.

Bickel, P. J. (1967). Some contributions to the theory of order statistics. *Proc. 5th Berkeley Symp.* **1**, 575–591.

Bickel, P. J. (1975). One-step Huber estimates in the linear model. *J. Amer. Statist. Assoc.* **70**, 428–434.

Bickel, P. J. (1976). Another look at robustness: A review of reviews and some new developments. *Scand. Journ. of Statistics* **3**, 145–168.

Bickel, P. J. and Lehmann, E. L. (1975). Descriptive statistics for nonparametric model. I. Introduction, II. Location. *Ann. Statist.* **3**, 1038–1044 and 1045–1069.

Birnbaum, A. and Laska, E. (1967). Optimal robustness: A general method, with applications to linear estimates of location. *J. Amer. Statist. Assoc.* **62**, 1230–1240.

Bjerve, S. (1977). Error bounds for linear combinations of order statistics. *Ann. Statist.* **5**, 357–369.

Boos, D. D. (1979). A differential for L-statistics. *Ann. Statist.* **7**, 955–959.

Boos, D. and Serfling, R. J. (1979). On Berry–Essen rates for statistical functions, with application to L-estimates. Technical Rep., Dept. Statist., Florida State University.

Boos, D. and Serfling, R. J. (1980). A note on differentials and the CLT and LIL for statistical functions, with application to M-estimates. *Ann. Statist.* **8**, 618–624.

Carroll, R. J. (1978). On the asymptotic distribution of multivariate M-estimates. *Journ. Multivar. Analysis* **8**, 361–371.

Chernoff, H., Gastwith, J. L. and Johns, M. V. (1967). Asymptotic distribution of linear combinations of order statistics, with application to estimation. *Ann. Math. Statist.* **38**, 52–72.

Collins, J. R. (1976). Robust estimation of a location parameter in the presence of asymmetry. *App. Statist.* **25**, 228–237.

Collins, J. R. (1977). Upper bounds on asymptotic variance of M-estimators of location. *Ann. Statist.* **5**, 646–657.

Collins, S. L. and Portnoy, J. R. (1981). Maximizing the variance of M-estimators using the generalized method of moment spaces. *Ann. Statist.* **9**, 567–577.

Dutter, R. (1975a). Robust regression: Different approaches to numerical solution and algorithms. Res. Report No. 6/ETH Zürich.

Dutter, R. (1975b). Numerical solution of robust regression problems: Computional aspects, a comparison. Res. Report No. 7, ETH Zürich.

Dutter, R. (1977). Numerical solution of robust regression problems: Computional aspects, a comparison. *J. Statist. Comput. Simul.* **5**, 207–238.

Dutter, R. (1978). Robust regression: LINDWDR and NLWDR. In: L. C. A. Corsten, ed., *COMPSTAT 1978, Proc.* Physica-Verlag, Vienna.

Field, C. A. (1978). Summary of small size asymptotics for location estimates. *Proc. 2nd. Prague Symp. on Asymptotic Statistics.* North-Holland, pp. 173–179.

Field, C. A. and Hampel, F. R. (1982). Small sample asymptotic distributions of M-estimators of location. *Biometrika* **69**, 221–226.

Gastwirth, J. and Rubin, H. (1969). On robust linear estimators. *Ann. Math. Statist.* **40**, 24–39.

Hájek, J. and Šidák, Z. (1967). *Theory of Rank Tests.* Academia, Prague.

Hampel, F. R. (1971). A general qualitative definition of robustness. *Ann. Math. Statist.* **42**, 1887–1896.

Hampel, F. R. (1973). Robust estimation: A condensed partial survey. *Z. Wahrsch. Verw. Geb.* **27**, 87–104.

Hampel, F. R. (1974a). Small sample asymptotics. *Proc. Prague Symp. on Asympt. Statistics II*, 109–126.

Hampel, F. R. (1974b). The influence curve and its role in robust estimation. *J. Amer. Statist. Assoc.* **62**, 1179–1186.

Hampel, F. R., Rousseeuw, P. J. and Ronchetti, E. (1981). The change-of-variance curve and optimal redescending M-estimators. *J. Amer. Statist. Assoc.* **76**, 643–648.

Heiler, S. and Willers, R. (1979). Asymptotic normality of R-estimates in the linear model. Forschungsbericht 79/6, Univ. Dortmund.

Helmers, R. (1977). The order of the normal approximation for linear combinations of order statistics with smooth weight functions. *Ann. Prob.* **5**, 940–953.

Helmers, R. (1980). Edgeworth expansions for linear combinations of order statistics. *Math. Centre Tract.* 105, Amsterdam.

Helmers, R. (1981). A Berry–Essen theorem for linear combination of order statistics. *Ann. Probab.* **9**, 342–347.

Hodges, J. L. and Lehmann, E. L. (1963). Estimates of location based on rank tests. *Ann. Math. Statist.* **34**, 598–611.

Hogg, R. V. (1967). Some observations on robust estimation. *J. Amer. Statist. Assoc.* **62**, 1179–1186.

Hogg, R. V. (1974). Adaptive robust procedures. A partial review and some suggestions for future. Applications and theory. *J. Amer. Statist. Assoc.* **69**, 909–923.

Hogg, R. V. (1979). Statistical Robustness: One View of its use in applications today. *The American Statistician* **33**, 108–115.

Hollander, M. and Wolfe, D. A. (1973). *Nonparametric Statistical Methods.* Wiley, New York.

Huber, P. J. (1964). Robust estimation of a location parameter. *Ann. Math. Statist.* **35**, 73–101.

Huber, P. J. (1965a). A robust version of a probability ratio test. *Ann. Math. Statist.* **36**, 1753–1758.

Huber, P. J. (1965b). The behaviour of maximum likelihood estimates under nonstandard conditions. *Proc. 5th Berkeley Symp.* **1**, 221–233.

Huber, P. J. (1968). Robust confidence limits. *Z. Wahrsch. Verw. Geb.* **10**, 269–278.

Huber, P. J. (1969). Théorie de l'inférence statistique robuste. Presses de l'Université Montréal.

Huber, P. J. (1970). Studentizing robust estimates. In: M. L. Puri, ed., *Nonparametric Techniques in Statistical Inference.* Cambridge University Press.

Huber, P. J. (1973). Robust regression: Asymptotics, conjectures and Monte Carlo. *Ann. Statist.* **1**, 799–821.

Huber, P. J. (1977). Robust Statistical Procedures. Regional Conference Series in Applied Math. No. 27. SIAM Philadelphia.

Huber, P. J. (1981). *Robust Statistics.* Wiley, New York.

Huber, P. J. and Dutter, R. (1977). Numerical solutions of robust regression problems. In: Bruckmann, ed., *COMPSTAT 1974.* Physica-Verlag, Vienna.

Huber, P. J. and Strassen, V. (1973). Minimax tests and Neyman–Pearson lemma for capacities. *Ann. Statist.* **1**, 251–263; **2**, 223–224.

Humak, K. M. S. (1983). *Statistische Methoden der Modelbildung. Band II.* Akademie-Verlag, Berlin.

Hušková, M. and Jurečková, J. (1981). Second order asymptotic relations of M-estimators and L-estimators in two-sample location model. *Journ. Statist. Planning Infer.* **5**, 309–328.

Jaeckel, L. A. (1971a). Robust estimates of location: Symmetry and asymmetry contamination. *Ann. Math. Statist.* **42**, 1020–1034.

Jaeckel, L. A. (1971b). Some flexible estimates of location. *Ann. Math. Statist.* **42**, 1540–1552.

Jaeckel, L. A. (1972). Estimating regression coefficients by minimizing the dispersion of the residuals. *Ann. Math. Statist.* **43**, 1449–1458.

Jurečková, J. (1969). Asymptotic linearity of a rank statistic in regression parameter. *Ann. Math. Statist.* **40**, 1889–1900.

Jurečková, J. (1971a). Nonparametric estimate of regression coefficients. *Ann. Math. Statist.* **42**, 1328–1338.

Jurečková, J. (1971b). Asymptotic independence of rank test statistic for testing symmetry on regression. *Sankhya Ser. A* **33**, 1–18.

Jurečková, J. (1973a). Central limit theorem for Wilcoxon rank statistics process. *Ann. Statist.* **1**, 1046–1060.

Jurečková, J. (1973b). Asymptotic behaviour of rank and signed-rank statistics from the point of view of applications. *Proc. Prague Symp. on Asymptotic Statistics I*, 139–155.

Jurečková, J. (1977). Asymptotic relations of M-estimates and R-estimates in linear regression model. *Ann. Statist.* **5**, 464–472.

Jurečková, J. (1978). Asymptotic relations of least-squares estimate and of two robust estimates of regression parameter vector. *Trans. 7th Prague Conf. and European Meeting of Statisticians* **II**, 231–237.

Jurečková, J. (1979). Finite sample comparison of L-estimators of location. *Comment. Math. Univ. Carolinae* **70**, 507–518.

Jurečková, J. (1980). Asymptotic representation of M-estimators of location. *Math. Operationsforsch. Statistik, Ser. Statistics* **11**, 61–73.

Jurečková, J. (1981). Tail behavior of location estimators. *Ann. Statist.* **9**, 578–585.

Jurečková, J. (1983a). Robust estimators of location and their second order asymptotic relations. To appear in ISI Centennial Volume.

Jurečková, J. (1983b). Robust estimators of location and regression parameters and their second order asymptotic relations. *Trans. 9th Prague Conf. Inform. Th., Random Processes and Statist. Decision Functions* (Reidel), pp. 19–32.

Jurečková, J. and Sen, P. K. (1981a). Invariance principles for some stochastic processes related to M-estimators and their role in sequential statistical inference. *Sankhya Ser. A* **43**, 191–210.

Jurečková, J. and Sen, P. K. (1981b). Sequential procedures based on M-estimators with discontinuous score functions. *Journ. Statist. Planning Infer.* **5**, 253–266.

Jurečková, J. and Sen, P. K. (1982a). M-estimators and L-estimators of location: Uniform integrability and asymptotically risk-efficient sequential versions. *Commun. Statist. C* **1**, 27–56.

Jurečková, J. and Sen, P. K. (1982b). Simultaneous M-estimator of the common location and the scale-ratio in the two-sample problem. *Math. Operationsforsch. Statistik, Ser. Statistics* **13**, 163–169.

Kagan, A. M., Linnik, Yu. V. and Rao, C. R. (1965). On a characterization of the normal law based on a property of the sample average. *Sankhya Ser. A* **27**, 405–406.

Kagan, A. M., Linnik, Yu. V. and Rao, C. R. (1972). *Characteristic Problems of Mathematical Statistics* (in Russian). Nauka, Moscow.

Koenker, R. and Bassett, C. (1978). Regression quantiles. *Econometrica* **40**, 33–50.

Koul, H. L. (1971). Asymptotic behavior of a class of confidence regions based on ranks in regression. *Ann. Math. Statist.* **42**, 466–476.

Koul, H. L. (1977). Behavior of robust estimators in the regression model with dependent errors. *Ann. Statist.* **5**, 681–699.

Kraft, C. and van Eeden, C. (1972a). Asymptotic efficiencies of quick methods of computing efficient estimates based on ranks. *J. Amer. Statist. Assoc.* **67**, 199–202.

Kraft, C. and van Eeden, C. (1972b). Linearized rank estimates and signed rank estimates for the general linear hypothesis. *Ann. Math. Statist.* **43**, 42–57.

Krasker, W. S. and Welsch, R. E. (1982). Efficient bounded-influence regression estimation. *Journ. Amer. Statist. Assoc.* **77**, 595–604.

Launer, R. L. and Wilkinson, G. N., eds. (1979). *Robustness in Statistics.* Academic Press, New York.

Maronna, R. A. (1976). Robust M-estimates of multivariate location and scatter. *Ann. Statist.* **4**, 51–67.

Maronna, R., Bustos, O. and Yohai, V. (1979). Bias- and efficiency robustness of general M-estimators for regression with random carriers. In: T. Gasser and M. Rosenblatt, eds., Smoothing Techniques for Curve Estimation, 91–116. *Lecture Notes in Math. 757.* Springer, Berlin.

Millar, W. S. (1981). Robust estimation via minimum distance methods. *Z. Wahrsch. Verw. Geb.* **55**, 73–89.

Miura, R. (1981). Adaptive confidence intervals for a location parameter. *The Keiei Kenkyu* **XXXI**, 197–218.

Nowak, H. and Zentgraf, R., eds. (1980). Robust Verfahren. *Medizinische Informatik und Statistik No. 20.* Springer, Berlin.

Portnoy, S. L. (1977). Robust estimation in dependent situations. *Ann. Statist.* **5**, 22–43.

Raoult, J. P. ed. (1981). *Statistique Non Paramétrique Asymptotique.* Lecture Notes in Math. 821, Springer, Berlin.

Relles, D. A. (1968). Robust regression by modified least squares. Ph.D. Thesis, New York.

Relles, D. A. and Rogers, W. H. (1977). Statisticians are fairly robust estimators of location. *J. Amer. Statist. Assoc.* **72**, 107–111.

Rey, W. J. J. (1978). *Robust Statistical Methods.* Lecture Notes in Math. 690. Springer, Berlin.

Rieder, H. (1977). Least favorable pairs for special capacities. *Ann. Statist.* **5**, 909–921.

Rieder, H. (1980). Estimates derived from robust tests. *Ann. Statist.* **8**, 106–115.

Rocke, D. M. and Downs, G. W. (1981). Estimating the variances of estimators of location: Influence curve, jackknife and bootstrap. *Comm. Statist. B* **10**, 221–248.

Rocke, D. M., Downs, G. W. and Rocke, A. J. (1982). Are robust estimators really necessary? *Technometrics* **24**, 95–101.

Rousseeuw, P. J. (1981). A new infinitesimal approach to robust estimation. *Z. Wahrsch. Verw. Geb.* **56**, 127–132.

Ruppert, D. and Carrol, R. J. (1980). Trimmed least squares estimation in the linear model. *J. Amer. Statist. Assoc.* **75**, 828–838.

Sacks, J. and Ylvisaker, D. (1982). L- and R-estimation and the minimax property. *Ann. Statist.* **10**, 643–645.

Sen, P. K. (1977). On Wiener process embedding for linear combinations of order statistics. *Sankhya Ser. A* **39**, 138–143.

Sen, P. K. (1978). An invariance principle for linear combinations of order statistics. *Z. Wahrsch. Verw. Geb.* **42**, 327–340.

Sen, P. K. (1980). On nonparametric sequential point estimation of location based on general rank order statistics. *Sankhya Ser. A* **42**, 202–218.

Sen, P. K. (1981). *Sequential Nonparametrics. Invariance Principles and Statistical Inference.* Wiley, New York.

Serfling, R. J. (1980). *Approximation Theorems of Mathematical Statistics.* Wiley, New York.

Shorack, G. R. (1969). Asymptotic normality of linear combinations of functions of order statistics. *Ann. Math. Statist.* **40**, 2041–2050.

Shorack, G. R. (1972). Functions of order statistics. *Ann. Math. Statist.* **43**, 412–427.

Shulenin, V. P. (1978). On the stability of a class of estimators of Hodges–Lehmann's type (in Russian). *Proc. Conf. on Theory of Coding and Inform. Trans.* **VI**, 147–151 (Vilnius, USSR).

Shulenin, V. P. (1980). Minimax estimators and numerical characteristics of their stability (in Russian). *Math. Statistics and its Applications* 32–47, University of Tomsk.

Sievers, G. L. (1978). Estimation of location: A large deviation comparison. *Ann. Statist.* **6**, 610–618.

Stigler, S. M. (1969). Linear functions of order statistics. *Ann. Math. Statist.* **40**, 770–788.

Stigler, S. M. (1974). Linear functions of order statistics with smooth weight functions. *Ann. Statist.* **2**, 676–693.

Stigler, S. M. (1977). Do robust estimators work with real data? *Ann. Statist.* **5**, 1055–1098.

Torgenson, E. N. (1971). A counterexample on translation invariant estimators. *Ann. Math. Statist.* **42**, 1450–1451.

van Eeden, C. (1970). Efficiency-robust estimation of location. *Ann. Math. Statist.* **41**, 172–181.

van Zwet, W. R. (1980). A strong law for linear functions of order statistics. *Ann. Probab.* **8**, 986–990.

Wegman, E. J. and Carroll, R. J. (1977). A Monte Carlo study of robust estimation of location. *Comm. Statist. A* **6**, 795–812.

Wellner, J. A. (1977a). A Glivenko–Cantelli-theorem and strong laws of large numbers for functions of order statistics. *Ann. Statist.* **5**, 473–480; correction note *Ann. Statist.* **6**, 1394.

Wellner, J. A. (1977b). A law of iterated logarithm for functions of order statistics. *Ann. Statist.* **5**, 481–494.

P. R. Krishnaiah and P. K. Sen, eds., *Handbook of Statistics. Vol. 4*
© Elsevier Science Publishers (1984) 487–514

22

Nonparametric Sequential Estimation

Pranab Kumar Sen

1. Introduction

There are certain basic estimation problems for which procedures based on a prefixed sample size may not workout, and for valid solutions, some multi-stage or sequential procedures are needed. There are other situations where the observations are gathered sequentially, so that for the desired inferential purpose, a *stopping rule* needs to be adapted along with the *estimation rule*. A sequential estimation procedure is defined by a stopping rule and an estimation rule; this characterization applies to both the *point* and *interval estimation* problems. For these two problems, the associated stopping rules may or may not be isomorphic, but the estimation rules are generally different. Note that if $(\Omega, \mathscr{B}, P)$ be a probability space and $\{\mathscr{B}_n;\, n \geq 1\}$ be an increasing sequence of sub-sigma fields of $\mathscr{B}$, then a measurable function N $(=N(\omega))$ taking values in $\mathbb{N}$ $(=\{1, \ldots, \infty\})$ is called a stopping rule if $\{N = n\} \in \mathscr{B}_n$ $\forall n \geq 1$ and $P\{N = \infty\} = 0$. In a sequential scheme, sampling is curtailed at the n-th stage if $N = n$, and, based on the observations gathered up to the n-th stage, an estimation rule is then employed for the desired inference on the parameter(s) of interest. In such a sequential scheme, if the stopping and estimation rules are based on appropriate (sequence of) statistics without making explicit assumptions on the form of the underlying distribution(s) of the random variables, then the resultant procedures are termed *nonparametric sequential estimation procedures.*

We shall discuss mainly the developments on nonparametric methods in the two basic sequential estimation problems: (i) minimum risk point estimation of a parameter, and (ii) confidence interval for a parameter with a prefixed coverage probability and having a width bounded by some prespecified positive constant, where in either case, the parameter is interpreted as a functional of the underlying distribution function. If the form of the underlying distribution is not specified or if it involves some unknown algebraic constants (appearing as parameters), then the minimum sample size needed to achieve the desired goal (for either of the problems) cannot be determined in advance, so that an estimation procedure based on a prefixed sample size may not workout. In such a case, multi-stage or sequential procedures can be adapted with success, and these will be considered here. In passing, we may remark that the two or multi-stage procedures are the precursors of the sequential ones; these provide the desired solutions, but may not be fully efficient, even in some convenient asymptotic setup. We shall review these briefly as we proceed along. In the developments to follow, the *asymptotic theory* plays a vital role: It enables one to encompass a broad class of statistics in a broad class of problems, and these sequential procedures usually retain their *efficiency* in some asymptotic setup (viz., the cost of sampling or the width of the confidence interval is made to converge to 0). Though the asymptotic theory may not be strictly applicable in a nonasymptotic case, yet the asymptotic solutions provide reasonably good approximations for the 'moderate case' as well. Further, in the asymptotic case, the solutions can be derived in a unified manner, whereas in the nonasymptotic case, exact solutions may either demand more stringent regularity conditions or depend on the specific problem at hand.

In Section 2, we consider the minimum risk sequential point estimation problem. We motivate the procedure through a parametric approach and then present the nonparametric generalizations. Section 3 is devoted to the study of the bounded-width sequential confidence interval problem, where also the procedure is motivated through a parametric case. Section 4 deals with the properties of the stopping times for either of these problems. In Section 5, nonparametric and parametric procedures are compared in the light of the asymptotic efficiency results. Some general discussions are made in the concluding section. For the dual sequential testing problems, we may refer to Chapter 28 due to Müller-Funk.

2. Sequential point estimation

Let $\{X_i; i \geq 1\}$ be a sequence of independent and identically distributed random variables (i.i.d.r.v.) with the distribution function (d.f.) F, defined on E^p, the $p(\geq 1)$-dimensional Euclidean space. Let $\theta = \theta(F)$ be an *estimable parameter*, and for a sample $(X_1, \ldots, X_n)$ of size n, let $T_n = T(X_1, \ldots, X_n)$ be a suitable estimator of θ. Let $c(n)$ be the *cost* of drawing a sample of size n, and consider the *loss* due to the estimation of θ by T_n:

$$L_n = g(|T_n - \theta|) + c(n) \tag{2.1}$$

where $g(\cdot)$ is a suitable (nonnegative) function with $g(0) = 0$, and, conceivably, $g(y)$ should be a nondecreasing function of y on E^+ $(=(0, \infty))$, though $g(\cdot)$ may or may not be bounded. The expected loss or *risk* due to the estimation of θ by T_n is

$$R_n = EL_n = Eg(|T_n - \theta|) + c(n)\,. \tag{2.2}$$

Here, we assume tacitly that there exists an n_0 (≥ 1), such that

$$\gamma_n^2 = Eg(|T_n - \theta|) \quad \text{exists for every } n \geq n_0 . \tag{2.3}$$

Further, for a sequence $\{T_n\}$ of estimators, in view of the desired consistency property, it may be quite reasonable to assume that for $n \geq n_0$,

$$\gamma_n^2 \quad \text{monotonically converges to 0 as } n \to \infty . \tag{2.4}$$

On the other hand, $c(n)$ is conceived to be a monotinically nondecreasing function of n, so that for $n \geq n_0$, R_n in (2.2) is the sum of the two terms, one nonincreasing and the other nondecreasing in n. Naturally, one would like to choose n in such a way that R_n is a minimum. For a given $g(\cdot)$ and $c(n)$, let n^* be defined by

$$n^* = \min\{n \geq n_0 \colon R_n = \inf_m R_m\} . \tag{2.5}$$

Then, for the sequence $\{T_n\}$ of estimators and corresponding to the risk function in (2.2), T_n^* is the minimum risk estimator of θ. (Later on, we shall comment on the choice of $g(\cdot)$ and $c(n)$ in this context.) In actual practice, for a nontrivial $g(\cdot)$, generally γ_n^2 depends not only on n but also on some other (unknown) parameter(s) (functional) of the d.f. F. For example, when the X_i are the real valued r.v. and we take $\theta = \int x \, \mathrm{d}F(x) =$ mean (μ) of the d.f. F, and let $T_n = \bar{X}_n = n^{-1}\Sigma_{i=1}^n X_i$, $n \geq 1$ and $g(y) = ay^2$, $y \in E$, for some $a > 0$, then $\gamma_n^2 = an^{-1}\sigma^2$ depends on n as well as $\sigma^2 = \int (x-\mu)^2 \, \mathrm{d}F(x)$, which we need to assume to be finite. If $c(n) \equiv c_0 + cn$ for some c_0, c (>0), then

$$R_{n+1} - R_n \text{ is } \{\lesseqqgtr\}\ 0 \quad \text{according as} \quad n(n+1) \text{ is } \{\lesseqqgtr\}\ ac^{-1}\sigma^2 , \tag{2.6}$$

so that by (2.5) and (2.6), $n^*(n^*-1) \leq ac^{-1}\sigma^2 < n^*(n^*+1)$. Thus, n^* depends on both (c, a) and σ^2, and hence, no fixed-sample size (n) can lead to the minimum risk estimation (MRE) simultaneously for all σ (>0). A similar case arises if we let $g(y) = a|y|^b$ and $c(n) = c_0 + cn^d$ for some $b > 0$, $d > 0$. In this setup, for normal F, n^* can be explicitly obtained in terms of a, b, d, c_0, c and σ^2, while for nonnormal F, the solution will depend on F through the functional $\nu_b = \int |x|^b \, \mathrm{d}F(x)$. Hence, in any case, MRE depends on the unknown σ or some other functional of the d.f. F, and a prefixed sample size may fail to yield the MRE when γ_n^2 is not completely specified. For this reason, we take recourse to multi-stage or sequential procedures.

To motivate the sequential procedures, we go back to the normal mean problem, stated earlier, and take $g(y) = ay^2$, $a > 0$, $y \in E$. Let $S_n^2 = (n-1)^{-1}\Sigma_{i=1}^n (X_i - \bar{X}_n)^2$, $n \geq 2$, be the sequence of sample variances. For some initial sample size n_0 (≥ 2), we define a stopping variable N ($= N_c$) by letting

$$N = \max\{n_0, [(a/c)^{1/2} S_{n_0}] + 1\} , \tag{2.7}$$

and consider the *two-stage* estimator (Stein, 1945) $T_n = \bar{X}_N$. Note that for normal F, $\{\bar{X}_n, n \geq 1\}$ and $\{S_n^2, n \geq 2\}$ are independent and hence, using the definition of n^* following (2.6), we obtain that for $c(n) \equiv cn$, $c > 0$, the relative risk of the two-stage procedure with respect to the optimal procedure (if σ were known) is given by

$$\begin{aligned}
&\{aE(\bar{X}_N - \mu)^2 + cEN\}/\{aE(\bar{X}_{n^*} - \mu)^2 + cn^*\} \\
&\quad = \{a\sigma^2 E(N^{-1}) + cEN\}/\{a\sigma^2(n^*)^{-1} + cn^*\} \\
&\quad = \tfrac{1}{2}[E(N^{-1}n^*) + E(N/n^*)] , \qquad (2.8)
\end{aligned}$$

where we take, for simplicity, $n^* = (a/c)^{1/2}\sigma$. Since $\frac{1}{2}(x + 1/x) \geq 1$ $\forall x \geq 0$, (2.8) exceeds one, unless $N = n^*$, with probability 1. On the other hand, $(n_0 - 1)S_{n_0}^2/\sigma^2$ has chi-square distribution with $n_0 - 1$ degrees of freedom, and hence, (N/n^*) has a nondegenerate distribution, so that (2.8) exceeds one, for every fixed c (>0) and n_0. Thus, for any given n_0 (≥ 2) and $c > 0$, the two-stage procedure in (2.7) fails to be a MRE. It was observed by Mukhopadhyay (1980) that if we allow n_0 ($= n_0(c)$) to depend on c (>0), in such a way that as $c \downarrow 0$, $n_0(c) \to \infty$ but $cn_0^2(c) \to 0$, then writing n^* as n_c^*, we have $N_c/N_c^* \xrightarrow{\text{P}} 1$, as $c \downarrow 0$, and hence, by some standard steps, (2.8) converges to 1, as $c \downarrow 0$. Thus, the modified two-stage procedure is asymptotically ($c \downarrow 0$) MRE. For nonnormal distributions, $\bar{X}_n$ and $[N = n]$ may not be independent, and the above simple arguments may not hold.

Basically, in a two-stage procedure, one does not update S_n^2 (see (2.7)), and hence, the MRE property may not hold. Based on updated versions of $\{S_n^2\}$, sequential procedures for the normal mean problem, were considered by Robbins (1959), Starr (1966), Starr and Woodroofe (1969), and others. The first nonparametric attempt (where F is of unspecified form) is due to Ghosh and Mukhopadhyay (1979); their regularity conditions were relaxed by Chow and Yu (1981). Sen and Ghosh (1981) extended the theory of asymptotic MRE to a general class of estimable parameters based on U-statistics. Sen (1980b) has developed the theory also for the rank based estimators in the location problem, while Jurečková and Sen (1982) have considered robust nonparametric procedures based on M- and L-estimators of location and estimators of their variation. These will be systematically reviewed here.

We consider first (asymptotically) risk-efficient sequential point estimation of location of a (symmetric) distribution. Procedures based on rank statistics, M-statistics and linear combinations of order statistics (L-statistics) will be discussed here. Consider the model as in (2.1) through (2.4), where θ stands for the location parameter of a d.f. $F_\theta(x) = F(x - \theta)$, $x \in E$, where F is symmetric about 0. The form of the d.f. F is not assumed to be specified. For simplicity, in (2.1), we take $g(y) = ay^2$ and $c(n) = cn$, where $a > 0$ and $c > 0$ are given constants. Later on, we shall discuss briefly the other cases of L_n in (2.1).

For the statistic T_n, defined as in before (2.1), we assume that there exists an n_0 (≥ 1), such that for every $n \geq n_0$, $\sigma_n^2 = nE(T_n - \sigma)^2$ exists , and $\sigma_n^2 \to \sigma^2$ as $n \to \infty$, where $0 < \sigma < \infty$. In this case, the risk function R_n in (2.2) is therefore given by

$$R_n = R_n(a, c) = an^{-1}\sigma_n^2 + cn \quad \forall n \geq n_0 . \tag{2.9}$$

Since, in general, σ_n^2 is unknown, minimization of R_n poses a problem. We overcome this problem by an appeal to an asymptotic situation, where c is made to converge to 0 (keeping a fixed). We may, without any loss of generality, set $a = 1$, $R_n(a, c) = R_n(c)$. Then, noting that $\sigma_n^2 \to \sigma^2$, we observe that for small c, we may write $R_n(c) = n^{-1}\sigma^2 + cn + o(n^{-1})$, so that if σ were known, an optimal sample size can be approximated by

$$n_c^0 = [\sigma c^{-1/2}] + 1 , \tag{2.10}$$

and the corresponding risk function is

$$R_{n_c^0}(c) = 2\sigma c^{1/2} + o(c^{1/2}) \quad \text{as } c \downarrow 0 . \tag{2.11}$$

Since σ is unknown, we assume that there exists a sequence $\{\hat{\sigma}_n\}$ of consistent estimators of σ, and keeping (2.7) and (2.10) in mind, we define a *stopping number* N_c by

$$N_c = \inf\{n \geq n_0 \colon n \geq c^{-1/2}(\hat{\sigma}_n + n^{-h})\} , \tag{2.12}$$

where h (>0) is an arbitrary constant. Note that $\hat{\sigma}_n = \hat{\sigma}_n(X_1, \ldots, X_n)$ is $\mathcal{B}_n$-measurable, for $n \geq n_0$, where $\mathcal{B}_n = \mathcal{B}(X_1, \ldots, X_n)$, $n \geq 1$. Thus, whenever, $\hat{\sigma}_n$ stochastically converges to σ, as $n \to \infty$, (2.12) defines a stopping rule unambiguously, and the estimation rule relates to the point estimator T_{N_c}, where $T_{N_c} = T_n$ whenever $N_c = n$, $n \geq n_0$. The risk of this (sequential) point estimator is given by

$$R_c^* = E(T_{N_c} - \theta)^2 + cEN_c, \quad c > 0 . \tag{2.13}$$

Note that for $T_n = \bar{X}_n$, $S_n^2 = \hat{\sigma}_n^2$, $n \geq n_0 = 2$, (2.12) closely resembles (2.7), with the notable difference that the first-sample estimator S_{n_0} is replaced by S_n, $n \geq n_0$, and this updating of the estimator is expected to enhance the efficiency of the sequential procedure over the two-stage procedure. Basically, the main objective is to show that under appropriate regularity conditions, as $c \downarrow 0$,

$$R_c^*/R_{n_c^0}(c) \to 1 \quad \text{and} \quad EN_c/n_c^0 \to 1 , \tag{2.14}$$

so that the sequential procedure is asymptotically (as $c \downarrow 0$) efficient. In the literature, (2.14) is referred to as the *first-order asymptotic efficiency* of the sequential procedure. In many situations, it may also be possible to make some deeper analysis. First, we may note that by (2.11), $R_{n_c^0}(c) = 0(c^{1/2})$ as $c \downarrow 0$. As such, we may like to know whether

$$R_c^* - R_{n_c^0}(c) = 0(c) \quad \text{as } c \downarrow 0 , \tag{2.15}$$

$$EN_c - n_c^0 = 0(1) \quad \text{as } c \downarrow 0 . \tag{2.16}$$

These are referred to as the *second order asymptotic efficiency* results. Secondly, we may also like to study the behavior of N_c as $c \downarrow 0$. Specifically, we may like to show that as $c \downarrow 0$,

$$(n_c^0)^{-1/2}(N_c - n_c^0) \xrightarrow{\mathscr{D}} \mathcal{N}(0, \nu^2) , \tag{2.17}$$

for some finite ν $(0 < \nu < \infty)$. This result is classically known as the *asymptotic normality of the stopping time*. This provides useful information on the excess of N_c over n_c^0 when c is small.

With these objectives in mind, we introduce now the R-, M- and L-estimators of location; these have already been discussed in the nonsequential case by Jurečková in Chapter 21.

For every n $(\geqslant 1)$, consider a signed-rank statistic

$$S_n = \sum_{i=1}^{n} \operatorname{sign} X_i a_n^+(R_{ni}^+) \tag{2.18}$$

where R_{ni}^+ = rank of $|X_i|$ among $|X_1|, \ldots, |X_n|$, $i = 1, \ldots, n$, and for every n $(\geqslant 1)$, $a_n^+(1) \leqslant \cdots \leqslant a_n^+(n)$ are *scores* generated by a *score function* $\phi^+ = \{\phi^+(u), 0 < u < 1\}$ in the following way:

$$a_n^+(i) = E\phi^+(U_{ni}) \quad \text{or} \quad \phi^+\left(\frac{i}{n+1}\right), \qquad 1 \leqslant i \leqslant n , \tag{2.19}$$

where $U_{n1} < \cdots < U_{nn}$ are the ordered r.v. of a sample of size n from the uniform (0, 1) d.f. and $\phi^+(u) = \phi(1+u)/2$, $0 < u < 1$, ϕ being a nondecreasing, skew-symmetric function. Suppose that we write S_n in (2.18) as $S(\mathbf{X}_n)$, $\mathbf{X}_n = (X_1, \ldots, X_n)$. If we replace $\mathbf{X}_n$ by $\mathbf{X}_n - a\mathbf{1}_n$, a real, $\mathbf{1}_n = (1, \ldots, 1)$ and recompute the signed-rank statistic; we denote the same by $S_n(a)$. Then $S_n(a)$ is $\searrow$ in a, and the *rank-based* (R-) *estimator* of θ is defined by

$$\hat{\theta}_{n(R)} = \tfrac{1}{2}(\sup\{a: S_n(a) > 0\} + \inf\{a: S_n(a) < 0\}) . \tag{2.20}$$

Note that $\hat{\theta}_{n(R)}$ is a translation-invariant, robust and consistent estimator of θ. If the score function ϕ is square integrable, we denote by

$$A_\phi^2 = \int_0^1 \phi^2(u)\,\mathrm{d}u = \int_0^1 \{\phi^+(u)\}^2\,\mathrm{d}u . \tag{2.21}$$

Further, we assume that the d.f. F has an absolutely continuous probability density function f with a finite Fisher information

$$I(f) = \int_{-\infty}^{\infty} \{f'(x)/f(x)\}^2\,\mathrm{d}F(x) \ (<\infty) \tag{2.22}$$

where $f'(x) = (\mathrm{d}/\mathrm{d}x)f(x)$. Let then

$$\psi(u) = -f'(F^{-1}(u))/f(F^{-1}(u)), \quad 0 < u < 1, \tag{2.23}$$

$$\rho(\phi, \psi) = \left(\int_0^1 \phi(u)\psi(u)\,\mathrm{d}u\right) \Big/ (A_\psi I^{1/2}(f)). \tag{2.24}$$

Then, we have

$$n^{1/2}(\hat{\theta}_{n(R)} - \theta) \xrightarrow{\mathscr{D}} \mathcal{N}(0, (\rho^2(\phi, \psi)I(f))^{-1}). \tag{2.25}$$

Thus, if we let

$$\sigma^2_{(R)} = \{\rho^2(\phi, \psi)I(f)\}^{-1} = A^2_\phi \Big/ \left(\int_0^1 \phi(u)\psi(u)\,\mathrm{d}u\right)^2, \tag{2.26}$$

then, we may note that for any adopted score function ϕ, A^2_ϕ is known, while ψ as well as $\int_0^1 \phi(u)\psi(u)\,\mathrm{d}u$ remain unknown. To estimate $\sigma^2_{(R)}$, we note that for every n (≥ 2) and every $\alpha \in (0, 1)$, there exists an α_n (≤ 2) and an $S_{n,\alpha}$, such that

$$P\{-S_{n,\alpha} \leqslant S_n \leqslant S_{n,\alpha} \mid H_0\colon \theta = 0\} = 1 - \alpha_n \geqslant 1 - \alpha \tag{2.27}$$

where $\alpha_n \to \alpha$ as $n \to \infty$, and $n^{-1/2}S_{n,\alpha} \to A_\phi \tau_{\alpha/2}$; here τ_ε is the upper 100ε % point of the standard normal d.f.

Then let

$$\hat{\theta}^{(R)}_{L,n} = \sup\{a\colon S_n(a) > S_{n,\alpha}\}, \tag{2.28}$$

$$\hat{\theta}^{(R)}_{U,n} = \inf\{a\colon S_n(a) < -S_{n,\alpha}\}, \tag{2.29}$$

and let

$$D^{(R)}_n = 2S_{n,\alpha}/n(\hat{\theta}^{(R)}_{U,n} - \hat{\theta}^{(R)}_{L,n}). \tag{2.30}$$

Finally, we let

$$\hat{\sigma}^2_{n(R)} = A^2_n/(D^{(R)}_n)^2, \qquad A^2_n = \frac{1}{n}\sum_{i=1}^n \{a^+_n(i)\}^2. \tag{2.31}$$

The (strong) consistency of $\hat{\sigma}_{n(R)}$ are an estimator or $\sigma_{(R)}$ has been studied by Sen and Ghosh (1971), and others. For the rank order estimator $\hat{\theta}_{n(R)}$, we do not need to assume that F has a finite first or second moment. In fact, if for some $a > 0$,

$$\int_{-\infty}^{\infty} |x|^a \,\mathrm{d}F(x) < \infty, \tag{2.32}$$

then, it follows from Sen (1980b) that for every $k>0$, there exists a positive integer n_{0k}, such that

$$E|\hat{\theta}_{n(R)}|^k<\infty \quad \forall n \geqslant n_{0k}\,. \tag{2.33}$$

Further, if we assume that for the score function ϕ,

$$|(\mathrm{d}^r/\mathrm{d}u^r)\phi(u)| \leqslant k[u(1-u)]^{-\delta-r}, \qquad r=0,1,2, \quad 0<u<1\,, \tag{2.34}$$

holds for some $\delta<(4+2\tau)^{-1}$, $\tau>0$, where k is a generic constant, then for every $k<2(1+\tau)$,

$$\lim_{n\to\infty} E\{n^{k/2}|\hat{\theta}_{n(R)}-\theta|^k\} = \sigma_{(R)}^k E|Z|^k\,, \tag{2.35}$$

where Z has the standard normal distribution. Now, (2.35) ensures that $nE(\hat{\theta}_{n(R)}-\theta)^2=\sigma_{n(R)}^2$ exists for all $n \geqslant n_{02}$ and $\sigma_{n(R)}^2 \to \sigma_{(R)}^2$ as $n\to\infty$. As such, we may proceed as in (2.9) through (2.12), and consider the stopping number

$$N_c(R)=\inf\{n\geqslant n_0\colon n\geqslant c^{-1/2}(\hat{\sigma}_{n(R)}+n^{-h})\} \tag{2.36}$$

where n_0 is $\geqslant 2$ and h (>0) is an arbitrary number. In passing, we may remark that for the Wilcoxon signed rank statistic, $\phi(u)=2u-1$, $0<u<1$, while for the normal scores statistics $\phi(u)=\Phi^{-1}(u)$, $0<u<1$, is the inverse of the standard normal d.f., and for both these scores, (2.34) holds. Actually, (2.34) holds for all the well-known signed rank statistics. Then, we have the following result due to Sen (1980b):

If (2.34) *holds for some* $\delta<(4+2\tau)^{-1}$ *where* $\tau>1+2h$, *with* h *being defined in* (2.36), *then under* (2.22) *and* (2.32), *for the proposed sequential rank procedure, the asymptotic risk efficiency in* (2.14) *is attained.*

Let us next consider sequential procedures based on suitable M-statistics. The M-estimators, originally proposed by Huber (1964), may be introduced by considering some function ξ ($=\{\xi(x), -\infty<x<\infty\}$) and defining

$$W_n(t)=\sum_{i=1}^{n}\xi(X_i-t) \tag{2.37}$$

and the solution $W_n(t)=0$ leads to the estimator $\hat{\theta}_{n(M)}$ of θ. In particular, if we assume that ξ is nondecreasing a skew-symmetric, so that $E_\theta\xi(X_i-\theta)=0$, $\hat{\theta}_{n(M)}$ may formally be defined by

$$\hat{\theta}_{n(M)}=\tfrac{1}{2}(\sup\{t\colon W_n(t)>0\}+\inf\{t\colon W_n(t)<0\}). \tag{2.38}$$

For various properties of $\hat{\theta}_{n(M)}$, we may again refer to Chapter 21 (by Jurečková). It may be noted that if we let $\xi(u) = \text{sign}\, u$, $u \in (-\infty/\infty)$, then (2.38) relates to the sample median. Another popular ξ, due to Huber (1964), is given by

$$\xi(x) = \begin{cases} x, & |x| \leq k, \\ k \text{ sgn } x, & |x| > k, \end{cases} \tag{2.39}$$

where k (>0) is some specified constant. We assume that

$$\xi(x) = \xi_1(x) + \xi_2(x), \quad x \in (-\infty, \infty), \tag{2.40}$$

where both ξ_1 and ξ_2 are nondecreasing and skew-symmetric functions, ξ_1 is absolutely continuous on any bounded interval and ξ_2 is a step-function having finitely many jumps. Actually we restrict ourselves to ξ such that $\xi(x) = \xi(C) \text{ sgn } x$ for $|x| \geq C$, where C is some finite positive numbers. This is mainly to reduce the influence of outliers. Let then

$$\sigma^2_{(\xi)} = \int_{-\infty}^{\infty} \xi^2(x)\, \mathrm{d}F(x), \tag{2.41}$$

$$\gamma_{(\xi)} = -\int_{-\infty}^{\infty} \xi(x) f'(x)\, \mathrm{d}x, \tag{2.42}$$

where we assume that (2.22) holds. Then, parallel to (2.25), we have

$$n^{1/2}(\hat{\theta}_{n(M)} - \theta) \xrightarrow{\mathscr{D}} \mathscr{N}(0, \sigma^2_{(\xi)}/\gamma^2_{(\xi)}). \tag{2.43}$$

Thus, if we let

$$\sigma^2_{(M)} = \sigma^2_{(\xi)}/\gamma^2_{(\xi)}, \tag{2.44}$$

we have both $\sigma^2_{(\xi)}$ and $\gamma^2_{(\xi)}$ unknown, and hence, $\sigma^2_{(M)}$ is also so. To estimate $\sigma^2_{(M)}$, we define

$$S^2_{n(M)} = n^{-1} \sum_{i=1}^{n} \xi^2(X_i - \hat{\theta}_{n(M)}), \quad n \geq 1. \tag{2.45}$$

Again, as in the case of R-estimators, we let (for some prefixed $\alpha: 0 < \alpha < 1$),

$$\hat{\theta}^{(M)}_{L,n} = \sup\{t: n^{-1/2} W_n(t) > \tau_{\alpha/2} S_{n(M)}\}, \tag{2.46}$$

$$\hat{\theta}^{(M)}_{U,n} = \inf\{t: n^{-1/2} W_n(t) < -\tau_{\alpha/2} S_{n(M)}\}; \tag{2.47}$$

$$d_{n(M)} = \hat{\theta}^{(M)}_{U,n} - \hat{\theta}^{(M)}_{L,n}.$$

Then, it follows from the results of Jurečková (1977) that

$$\hat{\sigma}_{n(M)} = n^{1/2} d_{n(M)}/2\tau_{\alpha/2} \xrightarrow{p} \sigma_{(M)} = \sigma_{(\xi)}/\gamma_{(\xi)} . \tag{2.48}$$

Strong consistency of $\hat{\sigma}_{n(M)}$ has been established by Jurečková and Sen (1981a,b).

If, in addition to (2.22) and (2.32), we assume that f is unimodal, then, it follows from Jurečková and Sen (1982) that for every $k>0$, there exists a positive integer N_{0k}, such that

$$E|\hat{\theta}_{n(M)}|^k < \infty \quad \forall n \geq n_{0k} , \tag{2.49}$$

and further, as $n \to \infty$,

$$E_\theta(\sqrt{n}|\hat{\theta}_{n(M)} - \theta|)^{2k} \to \frac{(2k)!}{k!2^k} (\sigma_{(M)})^{2k} , \tag{2.50}$$

for every $k = 1, 2, \ldots$. As such, we may proceed as in (2.9) through (2.12), and consider the stopping number

$$N_c(M) = \inf\{n \geq n_0 \colon n \geq c^{-1/2}(\hat{\sigma}_{n(M)} + n^{-h})\} , \tag{2.51}$$

where h (>0) is any arbitrary positive number. Under the regularity conditions mentioned before, for the sequential M-procedure in (2.51) (with the sequential estimator $\hat{\theta}_{N_c(M)(M)}$), the asymptotic risk efficiency in (2.14) is attained.

For the location parameter θ, an L-estimator $\hat{\theta}_{n(L)}$ is typically of the form

$$\hat{\theta}_{n(L)} = \sum_{i=1}^{n} c_{ni} X_{n:i} , \tag{2.52}$$

where $X_{n:1} \leq \cdots \leq X_{n:n}$ are the order statistics corresponding to the r.v.'s $X_1, \ldots, X_n$ (ties neglected, with probability 1, by the assumed continuity of F) and the c_{ni} are suitable constants. By virtue of the assumed symmetry of F, we would have ideally

$$c_{ni} = c_{nn-i+1} \geq 0 \ \forall i \ (1 \leq i \leq n) \quad \text{and} \quad \sum_{i=1}^{n} c_{ni} = 1 . \tag{2.53}$$

We express the c_{ni} in a functional form by introducing a function $J_n = \{J_n(t), 0 \leq t \leq 1\}$ by letting

$$\frac{1}{n} J_n(t) = c_{ni}, \qquad (i-1)/n < t \leq i/n, \quad 1 \leq i \leq n , \tag{2.54}$$

where for some smooth function $J = \{J(t), 0 \leq t \leq 1\}$, we assume that

$$J_n(t) \to J(t) \text{ a.e.} \quad \text{and} \quad \int_0^1 J(t)\,\mathrm{d}t = \int_0^1 J_n(t)\,\mathrm{d}t = 1 . \tag{2.55}$$

Let then

$$\sigma^2_{(L)} = \int_{-\infty}^{\infty}\int_{-\infty}^{\infty} \{F(x \wedge y) - F(x)F(y)\} J(F(x)) J(F(y))\, \mathrm{d}x\, \mathrm{d}y\,. \tag{2.56}$$

Parallel to the case of the M-procedure, we assume here that for some $\varepsilon\colon (0 < \varepsilon < \frac{1}{2})$, $J_n(t) = 0\ \forall\ 0 \leqslant t \leqslant \varepsilon$ and $1 - \varepsilon \leqslant t \leqslant \varepsilon$, while $J(t)$ is of bounded variation on $[0, 1]$. Here (2.22) may be replaced by a weaker assumption that

$$\sup\{|f'(x)|\colon F^{-1}(\varepsilon) \leqslant x \leqslant F^{-1}(1-\varepsilon)\} < \infty\,. \tag{2.57}$$

Now, $\sigma^2_{(L)}$ in (2.56) depends on the unknown F, and hence, we need to provide a suitable estimator of it. As in Sen (1978), we let

$$\hat\sigma^2_{n(L)} = \sum_{i=1}^{n-1}\sum_{j=1}^{n-1} c_{ni}c_{nj}\{n(i \wedge j) - ij\}(X_{n:i+1} - X_{n:i})(X_{n:j+1} - X_{n:j})\,, \tag{2.58}$$

which converges a.s. to $\sigma^2_{(L)}$, under quite general regularity conditions. Further asymptotic theory of this estimator has been explored by Gardiner and Sen (1979).

It follows from Jurečková and Sen (1982) that under the assumed regularity conditions, for every $k > 0$, there exists a positive integer n_{0k}, such that

$$E|\hat\theta_{n(L)}|^k < \infty \quad \text{for every } n \geqslant n_{0k}\,, \tag{2.59}$$

and further, as $n \to \infty$, for every k $(=1, 2, \ldots)$,

$$E_\theta\{\sqrt{n}|\hat\theta_{n(L)} - \theta|\}^{2k} \to \frac{(2k)!}{2^k \cdot k!}\sigma^{2k}_{(L)}\,. \tag{2.60}$$

As such, we may proceed as in (2.9) through (2.12), and consider a stopping number

$$N_c(L) = \inf\{n \geqslant n_0\colon n \geqslant c^{-1/2}(\hat\sigma_{n(L)} + n^{-h})\} \tag{2.61}$$

where h (>0) is arbitrary, and the sequential L-estimator is then taken as $\hat\theta_{N_c(L)}(L)$. Under the regularity conditions mentioned above, for this sequential L-procedure, the asymptotic risk efficiency in (2.14) is attained.

It may be remarked that for the sequential R-, L- and M-procedures, described above, (2.32) suffices for some $a > 0$ (need not be $\geqslant 1$), so that the scope of these procedures is not confined to d.f.'s having finite first or second moment. Actually, these are robust against outliers. The Winsorized and trimmed means (viz., $(n-2k)^{-1}\sum_{i=k+1}^{n-k} X_{n:i}$ and $n^{-1}\{kX_{n:k} + kX_{n:n-k+1} + \sum_{i=k+1}^{n-k} X_{n:i}\}$, for some k $(1 \leqslant k < n/2)$) are covered by the class considered here. If, however, we are in a position to assume that (2.32) holds for some $a \geqslant 2$, then the boundedness condition on ξ (in the M-procedure) or the condition that $c_{ni} = 0$, $i \leq [n\varepsilon]$ and $i \geq n - [n\varepsilon]$ (in the L-procedure) can be omitted. For all the three procedures, in addition to (2.14), we may claim that with n_c^0 being defined by (2.10) and N_c by (2.12),

$$(n_c^0)^{1/2}(T_{N_c} - \theta)/\sigma \xrightarrow{\mathcal{D}} \mathcal{N}(0, 1) \quad \text{as } c \downarrow 0 . \tag{2.62}$$

Asymptotic normality of the stopping time (viz. (2.17)) has also been established by Jurečková and Sen (1982) for both the L- and M-procedures in (2.51) and (2.61). In this context, it may be remarked that if in (2.40), ξ_2 is absent, then in (2.17), the order $(n_c^0)^{-1/2}$ is achieved, while if ξ_2 is present, then $(n_c^0)^{-1/2}$ has to be replaced by $(n_c^0)^{-1/4}$, indicating a slower rate of convergence. Verification of (2.15) and/or (2.16) remains as open problems. Also, generalizations of these procedures to general linear models pose harder problems, and are being investigated now.

Asymptotically risk efficient sequential point estimation theory has been considered in a more general setup by Sen and Ghosh (1981). They considered the case of general estimable parameters (which are functionals of the underlying d.f.) and provided solutions based on U-statistics and their estimated variances. For these U-statistics, we refer to Chapter 8 (by Ghosh).

Corresponding to an estimable parameter $\theta(F) = \int \cdots \int \phi(x_1, \ldots, x_m)$ $\mathrm{d}F(x_1) \cdots \mathrm{d}F(x_m)$, where the *kernel* ϕ of degree m (≥ 1) is symmetric in its m arguments, we define the U-statistic U_n by

$$U_n = \binom{n}{m}^{-1} \sum_{1 \leq i_1 < \cdots < i_m \leq n} \phi(X_{i_1}, \ldots, X_{i_m}), \quad n \geq m . \tag{2.63}$$

Thus, U_n is a symmetric (in $X_1, \ldots, X_n$) and unbiased estimator of $\theta(F)$, with variance

$$E\{(U_n - \theta(F))^2\} = \binom{n}{m}^{-1} \sum_{i=1}^{m} \binom{m}{i}\binom{n-m}{m-i} \zeta_i = m^2 n^{-1} \zeta_1 + \mathrm{O}(n^{-2}) , \tag{2.64}$$

where the ζ_i are functionals of the d.f. F ($0 \leq \zeta_1 < \cdots < \zeta_m$); for their expressions, we may refer to Chapter 8. The case of $\theta = \mu = \int x \, \mathrm{d}F(x)$ is a special one, where $m = 1$, $\phi(x) = x$ and $\zeta_1 = \sigma^2 = \int x^2 \, \mathrm{d}F(x) - \mu^2$. The case of σ^2 is also a special one, where $\phi(x, y) = \frac{1}{2}(x-y)^2$, $m = 2$, and ζ_1 and ζ_2 are defined in terms of the 4th moment of X. As such, for such U-statistics, we are confronted with the same problem as in (2.1) through (2.5), where n^* in (2.5) and R_{n^*} in (2.2) depends on the unknown $\zeta_1, \ldots, \zeta_m$. If we let $\sigma_n^2 = nV(U_n - \theta(F))$, $n \geq m$, where we assume that $E\phi^2 < \infty$, then $\sigma_n^2 = m^2\zeta_1 + \mathrm{O}(n^{-1})$, and the following estimator of σ_n^2 is due to Sen (1960). Let

$$V_{ni} = \binom{n-1}{m-1}^{-1} \sum_{\substack{1 \leq i_2 < \cdots < i_m \leq n \\ i_j \neq i}} \phi(X_i, X_{i_2}, \ldots, X_{i_m}), \quad k = 1, \ldots, n , \tag{2.65}$$

and let

$$S_n^2 = m^2(n-1)(n-m)^{-2} \sum_{i=1}^{n} (V_{ni} - U_n)^2 . \tag{2.66}$$

Then S_n^2 is a consistent estimator of $m^2\zeta_1$. As such, we may proceed as in (2.9) through (2.12) and consider a stopping number

$$N_e = \inf\{n \geq n_0 : n \geq c^{-1/2}(S_n + n^{-h})\}, \quad c > 0 , \tag{2.67}$$

where h (>0) is arbitrary. The desired sequential estimator is then U_{N_c}. In this case, (2.10) simplifies to (when $\zeta_1 > 0$)

$$n_c^0 \sim mc^{-1/2}\zeta_1 \quad \text{as } c \downarrow 0 , \tag{2.68}$$

while (2.11) reduces to $2cn_2^0 \sim 2m(c\zeta_1)^{1/2}$. Unlike the case of the R-, L- and M-procedures, here, instead of (2.32), one needs to assume appropriate moment conditions on ϕ, though (2.22) may not be necessary. $\theta(F)$ is termed stationary of order 0 if $\zeta_1 > 0$. Then, we have the following results due to Sen and Ghosh (1981):

If $\theta(F)$ is stationary or order 0, for some $\delta > 0$,

$$E|\phi(X_1, \ldots, X_m)|^{2+\delta} < \infty \tag{2.69}$$

and in (2.67), $h \in (0, \delta^2/2(2+\delta))$, then for the sequential procedure based on (2.67), the asymptotic risk efficiency in (2.14) holds. Actually, without any restriction on h (>0), $EN_c/n_c^0 \to 1$ as $c \downarrow 0$, when (2.69) holds. If (2.69) be replaced by $E|\phi|^{4+\delta} < \infty$, for some $\delta > 0$ and in (2.67), h is restricted to $(\frac{1}{2}, \infty)$ then, as $c \downarrow 0$,

$$2m^2\zeta_1(N_c - n_c^0)/\{\nu^2 n_c^0\}^{1/2} \xrightarrow{\mathcal{D}} \mathcal{N}(0, 1) \tag{2.70}$$

where

$$\nu^2 = \lim_{n\to\infty} \{nE(S_n^2 - m^2\zeta_1)^2\} . \tag{2.71}$$

Thus, for the case of U-statistics, (2.14) as well as (2.17) hold under quite general conditions. There is some interest in studying (2.15)–(2.16) fot these procedures too. For the special case of $m = 1$ (where nU_n is a sum of i.i.d.r.v.'s), some of these results have recently been studied by Chow and Martinsek (1982). Their technique encounters considerable difficulties when $m > 1$, and these are being investigated now.

In the developments so far, we have restricted ourselves to the case of the squared error loss ($g(y) = y^2$) and proportional cost ($c(n) \equiv cn$). Parallel solutions for some other choices of $g(\cdot)$ and $c(n)$ can be worked out in similar ways.

Basically, one needs to assume (2.3) and (2.4), and further, to approximate γ_n^2 by an estimable function of n. For example, if $g(y) \equiv |y|$, then γ_n^2 in (2.4) may be approximated by $n^{-1/2}(2/\pi)^{1/2}\sigma$, where σ^2 is the asymptotic variance of $n^{1/2}(T_n - \theta)$. So that, an estimator of σ, as in (2.9)–(2.13), may be employed to define a stopping rule, and showing that for such a rule, (2.14) holds for the modified risk function too. In any case, the stopping rule depends on the choice of $g(\cdot)$ and $c(\cdot)$, and hence, adherence to practicality in their choice is a good criterion.

Multiparameter extensions of the theory have also been worked out by several workers. For example, if $\boldsymbol{T}_n$ is a q-vector (for some $q \geqslant 1$) estimating $\boldsymbol{\theta}$, also a q-vector, then in (2.1), one may take $g(\|\boldsymbol{T}_n - \boldsymbol{\theta}\|) = E\{(\boldsymbol{T}_n - \boldsymbol{\theta})'\boldsymbol{A}(\boldsymbol{T}_n - \boldsymbol{\theta})\}$, where $\boldsymbol{A}$ is some positive semi-definite matrix. Actually, if we let

$$\boldsymbol{\Gamma}_n = E(\boldsymbol{T}_n - \boldsymbol{\theta})(\boldsymbol{T}_n - \boldsymbol{\theta})' , \tag{2.72}$$

then we have $g(\|\boldsymbol{\Gamma}_n - \boldsymbol{\theta}\|) = \text{Trace}(\boldsymbol{A}\boldsymbol{\Gamma}_n) =$ sum of the characteristic roots of $\boldsymbol{A}\boldsymbol{\Gamma}_n$. As such, if $\boldsymbol{\Gamma}_n$ behaves like $n^{-1}\boldsymbol{\Gamma} + \mathrm{o}(n^{-1})$, where $\boldsymbol{\Gamma}$ is possibly unknown, then for (2.2) we have

$$EL_n = R_n = n^{-1}T_r(\boldsymbol{A}\boldsymbol{\Gamma}) + cn + \mathrm{o}(n^{-1}) , \tag{2.73}$$

so that, parallel to (2.10), here we have for $c \downarrow 0$,

$$n_c^0 \sim \{c^{-1}T_r(\boldsymbol{A}\boldsymbol{\Gamma})\}^{1/2} . \tag{2.74}$$

As such, if for a given $\boldsymbol{A}$, we have a sequence $\{\hat{\boldsymbol{\Gamma}}_n\}$ of estimators of $\boldsymbol{\Gamma}$, then parallel to (2.12), we may define

$$N_c = \inf\{n \geqslant n_0 \colon n \geqslant \{c^{-1}T_r(\boldsymbol{A}(\hat{\boldsymbol{\Gamma}}_n + n^{-2h}\boldsymbol{I}_q))\}^{1/2}\} \tag{2.75}$$

where h (>0) is defined as in after (2.12). Under conditions on $\{\hat{\boldsymbol{\Gamma}}_n\}$, paralleling the uniparameter case, the asymptotic risk efficiency as well as the normality of the stopping time hold for the above sequential procedure too. We may refer to Sen and Ghosh (1981) and Sen (1984) for some details. Some other related loss functions (e.g. $g(\|\boldsymbol{T}_n = \boldsymbol{\theta}\|) =$ generalized variance of $(\boldsymbol{T}_n - \boldsymbol{\theta})$ etc.) may also be adapted in this context.

For the uniparameter as well as multiparameter case, the stopping rule as well as the optimality properties are motivated and established under an asymptotic setup where $c \downarrow 0$. A natural question may arise in this context: How good are the sequential procedures where c is not very small? Some numerical studies made in this context provide quite encouraging pictures. However, the picture depends a lot on the underlying d.f. F as well as on $\{T_n\}$; the uniformity of this asymptotic solution over a class of F may not generally hold. Nor for a given F, the uniformity is expected over a class of $\{T_n\}$. Indeed, more numerical studies are needed to throw light on the behavior of the sequential procedures for moderate values of c and possibly nonnormal parent distributions.

3. Sequential confidence intervals

As in Section 2, we conceive of a sequence $\{X_i; i \geq 1\}$ of i.i.d.r.v. with a d.f. F, and let θ be a real valued parameter (a functional of the d.f. F), which we desire to estimate. Based on $\boldsymbol{X}_n = (X_1, \ldots, X_n)$, let $\hat{\theta}_{L,n}$ and $\theta_{U,n}$ be two statistics, such that $\hat{\theta}_{L,n} \leq \hat{\theta}_{U,n}$ and

$$P_\theta\{\hat{\theta}_{L,n} \leq \theta \leq \hat{\theta}_{U,n}\} \geq 1 - \alpha\,, \tag{3.1}$$

for some specified α $(0 < \alpha < 1)$. Then $[\hat{\theta}_{L,n}, \hat{\theta}_{U,n}]$ is a *confidence interval* for θ with *confidence coefficient* (coverage probability) $1-\alpha$ and width $\delta_n = \hat{\theta}_{U,n} - \hat{\theta}_{L,n}$ (≥ 0). In this setup, $\hat{\theta}_{L,n}$ and $\hat{\theta}_{U,n}$ are termed the *lower* and *upper confidence limits*, respectively. In practice, often, one wants to prescribe a confidence interval for θ, for which, in addition to (3.1), for some preassigned d (>0),

$$\delta_n \leq 2d\,, \tag{3.2}$$

i.e., the width of the confidence interval is bounded from above by a preassigned number $2d$. Since, in general, the (joint) distribution of $(\hat{\theta}_{L,n}, \hat{\theta}_{U,n})$ may depend on the underlying F, which is either of unspecified form or involves unknown parameters, it may not be possible to determine a value of n, for which (3.1) and (3.2) hold simultaneously for all such F. For this reason, one may take recourse to sequential schemes. To motivate such sequential schemes, we again consider the simple normal mean problem (as in after (2.5)). If F is normal with mean θ and variance σ^2, then if σ were known, one would, by letting

$$n_d = \inf\{k : k^{1/2}d/\sigma \geq \tau_{\alpha/2}\}\,, \tag{3.3}$$

obtain that for $n = n_d$, $\hat{\theta}_{L,n} = \bar{X}_n - d$, $\hat{\theta}_{U,n} = \bar{X}_n + d$, (3.1) and (3.2) hold. But n_d actually depends on σ, and hence, if σ were unknown, the solution n_d cannot satisfy (3.1) and (3.2), for all σ. Dantzig (1940) proved the nonexistence of fixed-sample size confidence intervals for θ for which both (3.1) and (3.2) hold, for all σ. Stein (1945) overcame this problem, by considering a two-stage procedure. Parallel to (2.7), define

$$N = \max\{n_0, [t^2_{n_0-1,\alpha} S^2_{n_0}/d^2] + 1\} \tag{3.4}$$

where $S^2_{n_0}$ is the sample variance computed from the initial sample of size n_0, and for Student's t-statistic with $n_0 - 1$ degrees of freedom, $t_{n_0-1,\alpha}$ is the upper

50α % point. If now we consider the interval $[\bar{X}_N - d, \bar{X}_N + d]$, then it is easy to verify that both (3.1) and (3.2) hold. However, the validity of (3.1) depends very crucially on the underlying d.f. F being normal. Moreover, the r.v. N may be stochastically much larger than n_d in (3.3), unless n_0 is chosen large. Nearly twenty years later on, Chow and Robbins (1965) considered a sequential procedure, which actually uses updated versions of S_n^2 in (3.4), and the validity and efficiency are thereby extended to a broader class of distributions. Generalizations of the Chow–Robbins procedure in various nonparametric setups have been considered by a host of workers; a general account of these developments is given in Sen (1981, Chapter 10). A characteristic feature of the nonparametric procedures is that the confidence intervals are obtained by inversion of suitable rank statistics, so that the stopping rules may be defined more naturally in terms of the width. Also, the various other situations, robust statistics have been employed, and these extend the Chow–Robbins theory to a much wider setup. As in Section 2, we consider first the location model, and then, we will proceed on to other problems as well.

First, we consider the sequential rank procedure. Here, we take $F = F_\theta$, where $F_\theta(x) = F(x - \theta)$, $-\infty < x < \infty$, F symmetric about 0, and θ is the location parameter (median) of F. We define the signed-rank statistic S_n as in (2.18)–(2.19), and for every n (≥ 2), we define $\hat{\theta}_{L,n}^{(R)}$ and $\hat{\theta}_{U,n}^{(R)}$ as in (2.28)–(2.29). Note that by (2.7)–(2.9), for every n,

$$P_\theta\{\hat{\theta}_{L,n}^{(R)} \leq \theta \leq \hat{\theta}_{U,n}^{(R)}\} = 1 - \alpha_n \quad (\geq 1 - \alpha)\,, \tag{3.5}$$

whatever be the form of F (assumed to be continuous, of course). Given this distribution-free confidence interval for θ, one may define the width

$$\delta_n^{(R)} = \hat{\theta}_{U,n}^{(R)} - \hat{\theta}_{L,n}^{(R)}\,, \tag{3.6}$$

so that a natural stopping variable would be

$$N_d^{(R)} = \min\{n \geq 2\colon \delta_n^{(R)} \leq 2d\}\,. \tag{3.7}$$

The sequential confidence interval for θ is then

$$[\hat{\theta}_{L,N_d^{(R)}}^{(R)}, \hat{\theta}_{U,N_d^{(R)}}^{(R)}]\,, \tag{3.8}$$

where by definition [in (3.7)], the width of the interval in (3.8) is $\leq 2d$, so that (3.2) holds. The crucial point is to verify that (3.1) holds in some sense, and to show that such a procedure is efficient in some sense too. In this context, we take shelter in an asymptotic setup, where, in (3.2), we allow $d \to 0$. Towards this, we have the following results:

Under essentially the same set of regularity conditions (on the score function and the d.f.) as in Section 2, $N_d^{(R)}$ is a nonincreasing function of d (>0), it is finite a.s., $E(N_d^{(R)}) < \infty$ for all $d > 0$, $\lim_{d \downarrow 0} N_d^{(R)} = +\infty$ a.s., and $\lim_{d \downarrow 0} E(N_d^{(R)}) =$

$+\infty$. Further

$$\lim_{d\downarrow 0}\{N_d^{(R)}/n_d\} = 1 \quad \text{a.s.}\,, \tag{3.9}$$

$$\lim_{d\downarrow 0} P_\theta\{\hat{\theta}^{(R)}_{L,N_d^{(R)}} \le \theta \le \hat{\theta}^{(R)}_{U,N_d^{(R)}}\} = 1-\alpha\,, \tag{3.10}$$

$$\lim_{d\downarrow 0} (EN_d^{(R)})/n_d = 1\,, \tag{3.11}$$

where

$$n_d = \left[A_\phi^2\tau_{\alpha/2}^2/d^2\Big(\int_0^1 \phi(u)\psi(u)\,du\Big)^2\right], \quad d>0\,, \tag{3.12}$$

and A_ϕ, ψ, ϕ and $\tau_{\alpha/2}$ are all defined in (2.21)–(2.28).

Note that (3.10) insures that as $d\downarrow 0$, the coverage probability converges to the preassigned $1-\alpha$, and is termed the *asymptotic (as $d\downarrow 0$) consistency*. Also, if $\int_0^1 \phi(u)\psi(u)\,du$ were known, then for small values of d (>0), a nonsequential procedure (based on $\hat{\theta}^{(R)}_{L,n_d}$ and $\hat{\theta}^{(R)}_{U,n_d}$) would also have the converge probability close to $1-\alpha$ and would be the desired procedure. Therefore, (3.11) provides the efficiency of the sequential procedure relative to the nonsequential one, and it is termed the *asymptotic (as $d\downarrow 0$) efficiency*. These results for the particular cases of the sign and signed rank statistics are due to Geertsema (1970), and, in this generality, these are due to Sen and Ghosh (1971). A similar procedure has also been worked out by Ghosh and Sen (1972) for the regression problem. Consider the simple regression model where $X_i = \beta_0 + \beta c_i + e_i$, $i\ge 1$, where β_0 and β are unknown parameters, the c_i are known regression constants and the e_i are i.i.d.r.v. with a continuous (unknown) d.f. F, defined on the real line E. We desire to provide a confidence interval for the regression slope β, for which (3.1) and (3.2) hold. For this purpose, for every n ($\ge$1), we consider a linear rank statistic

$$L_n = \sum_{i=1}^n (c_i - \bar{c}_n)a_n(R_{ni}), \qquad \bar{c}_n = n^{-1}\sum_{i=1}^n c_i\,, \tag{3.13}$$

where R_{ni} = rank of X_i among $X_1,\ldots,X_n$, for $i=1,\ldots,n$, and the scores $a_n^{(i)}$ ($=E\phi(U_{ni})$ or $\phi(i/(n+1))$), $i=1,\ldots,n$ are defined as in after (2.19). We write $L_n = L_n(\boldsymbol{X}_n)$, and, if in (3.13), we replace $\boldsymbol{X}_n$ by $\boldsymbol{X}_n - b\boldsymbol{c}_n$, b real, $\boldsymbol{c}_n = (c_1,\ldots,c_n)'$, then the resulting statistic is denoted by $L_n(b)$. Note that for $\phi\nearrow$, $L_n(b)$ is $\searrow$ in $b\in E$, and $L_n(\beta)$ has the same distribution as of $L_n(0)$ when H_0: $\beta=0$ holds, and the later has mean 0 and variance $C_n^2A_n^2$, where $C_n^2 = \Sigma_{i=1}^n (c_i-\bar{c}_n)^2$ and $A_n^2 = (n-1)^{-1}\ \{\Sigma_{i=1}^n a_n^2(i) - n^{-1}(\Sigma_{i=1}^n a_n(i))^2\}$. Further, under H_0, $L_n(0)$ has a distribution independent of F. As such, parallel to (2.20), the R-estimator of β may be defined by

$$\hat{\beta}_{n(R)} = \tfrac{1}{2}(\sup\{b\colon L_n(b)>0\} + \inf\{b\colon L_n(b)<0\})\,, \tag{3.14}$$

and it is possible to locate two values $L_{n,\alpha}^{(1)}$, $L_{n,\alpha}^{(2)}$, such that for every α $(0<\alpha<1)$ and n (≥ 2),

$$P\{L_{n,\alpha}^{(1)} \leq L_n(0) \leq L_{n,\alpha}^{(2)} \mid H_0\} = 1-\alpha_n \geq 1-\alpha \tag{3.15}$$

where α_n does not depend on F and $\alpha_n \to \alpha$ as $n \to \infty$, and

$$C_n^{-1} A_n^{-1} L_{n,\alpha}^{(i)} \to (-1)^i \tau_{\alpha/2} \quad \text{as } n \to \infty\,. \tag{3.16}$$

Thus, if we let

$$\hat{\beta}_{L,n}^{(R)} = \sup\{b\colon L_n(b) > L_{n,\alpha}^{(2)}\}\,, \tag{3.17}$$

$$\hat{\beta}_{U,n}^{(R)} = \inf\{b\colon L_n(b) < L_{n,\alpha}^{(1)}\}\,, \tag{3.18}$$

then by (3.15)–(3.18), $[\hat{\beta}_{L,n}^{(R)}, \hat{\beta}_{U,n}^{(R)}]$ provides a distribution-free confidence interval for β. Parallel to (3.6), we let $\delta_n^{(R)} = \hat{\beta}_{U,n}^{(R)} - \hat{\beta}_{L,n}^{(R)}$ and let the stopping time $N_d^{(R)}$ be defined as in (3.7). Then, the desired sequential confidence interval for β is

$$[\hat{\beta}_{L,N_d^{(R)}}^{(R)}, \hat{\beta}_{U,N_d^{(R)}}^{(R)}]\,. \tag{3.19}$$

If we define $Q(x) = (n+1-x)C_n^2 + (x-n)C_{n+1}^2$ for $n \leq x \leq n+1$, $n \geq 0$, $C_0^2 = C_1^2 = 0$, and assume that $Q(x)$ is $\uparrow$ in x $(\in E^+)$ and

$$\lim_{n\to\infty} Q(na_n)/Q(n) = S(a) \quad \text{exists } \forall\{a_n\} \text{ with } \lim_{n\to\infty} a_n = a\,, \tag{3.20}$$

where $s(a)$ is $\uparrow$ in a (>0) with $s(1)=1$, then the stopping number $N_d^{(R)}$ satisfies all the properties mentioned before (3.9), and further, (3.9)–(3.11) hold with

$$n_d = Q^{-1}\left(A^2\tau_{\alpha/2}^2/d^2\left(\int_0^1 \phi(u)\psi(u)\,\mathrm{d}u\right)^2\right) \tag{3.21}$$

and $Q^{-1}(y) = \inf\{x\colon Q(x) \geq y\}$. Thus, the asymptotic consistency and efficiency both hold for the sequential procedure in (3.19). For the least square estimator, parallel results are due to Gleser (1965).

Let us next consider procedures based on L-statistics. Parallel to (2.52), we consider a L-statistic of the form

$$L_n = n^{-1} \sum_{i=1}^{n} J_n\left(\frac{i}{n+1}\right) g(X_{n:i}) \tag{3.22}$$

where the $X_{n:i}$ are defined as there, $g(\cdot)$ is a suitable function and the score function $J_n(\cdot) \to J(\cdot)$ on $(0, 1)$, for some smooth J. Parallel to (3.22), the population counterpart is

$$\mu = \int_E J(F(x))g(x)\,\mathrm{d}F(x)\,. \tag{3.23}$$

Let then

$$\sigma^2_{(L)} = \int_{E^2}\int J(F(x))J(F(y))\{F(x \wedge y) - F(x)F(y)\}\,\mathrm{d}g(x)\,\mathrm{d}g(y) \tag{3.24}$$

and, in (2.58), we replace the $X_{n:i}$ by $g(X_{n:i})$ and denote the resulting statistic by $\hat{\sigma}^2_{n(L)}$. Then, under fairly general regularity conditions on $J(\cdot)$ and $g(\cdot)$, as $n \to \infty$,

$$n^{1/2}(L_n - \mu)/\hat{\sigma}_{n(L)} \xrightarrow{\mathcal{D}} \mathcal{N}(0, 1)\,, \tag{3.25}$$

so that an asymptotic confidence interval for μ may be based on (3.25). As such, in such a case, we may define a stopping number N_L $(=N_L(d))$ by

$$N_L(d) = \min\{n \geq n_0\colon \hat{\sigma}^2_{n(L)} \leq nd^2/\tau^2_{\alpha/2}\}\,, \tag{3.26}$$

where n_0 (≥ 2) is some initial sample size. Also, let

$$n_L(d) = [d^{-2}\tau^2_{\alpha/2}\sigma^2_{(L)}]\,. \tag{3.27}$$

If we let $b(u) = g(F^{-1}(u))$, $0 < u < 1$, assume that $b(u)$ is of bounded variation on $[\varepsilon, 1-\varepsilon]$ $\forall 0 < \varepsilon < \frac{1}{2}$, and that some generic constant K $(0 < k < \infty)$, for every $0 < u < 1$,

$$|b(u)| \leq K[u(1-u)]^{-\alpha}, \quad |\phi(u)| \leq K[u(1-u)]^{-\beta}, \quad |\phi'(u)| \leq K[u(1-u)]^{-1-\beta}\,, \tag{3.28}$$

where α, β are real and

$$\alpha + \beta = \tfrac{1}{2} - \delta \quad \text{for some } \delta > 0\,, \tag{3.29}$$

then, for the sequential confidence interval

$$[L_{N_L(d)} - d,\ L_{N_L(d)} + d]\,, \tag{3.30}$$

all the properties mentioned before (3.9) hold and also (3.9)–(3.11) hold with n_d, defined by (3.27). For details of the proofs, we may refer to Sen (1981, Chapter 10). Whereas, in the rank statistics case, we use distribution-free confidence intervals to define the stopping number, in this case, (3.25) provides only an asymptotically distribution-free (ADF) setup.

We consider next the sequential M-procedure. Let $\{X_i, i \geq 1\}$ be a sequence of independent r.v. with d.f.'s $\{F_i, i \geq 1\}$, where

$$F_i(x) = F(x - \Delta c_i), \qquad i \geq 1, \quad x \in E, \tag{3.31}$$

the c_i are known regression constants (not all equal to 0), Δ is an unknown parameter and the d.f. F satisfies the regularity conditions of Section 2. The location model is a special case where $c_i = 1\ \forall i \geq 1$. Analogous to (2.37), we define here

$$W_n(t) = \sum_{i=1}^{n} c_i \xi(X_i - tc_i), \quad t \in E, \tag{3.32}$$

where the score function ξ is monotone, skew-symmetric and satisfies the same regularity conditions as in after (2.39). Parallel to (2.38), we define here

$$\hat{\Delta}_{n(M)} = \tfrac{1}{2}(\sup\{t\colon W_n(t) > 0\} + \inf\{t\colon W_n(t) < 0\}), \tag{3.33}$$

while, we replace (2.45) by

$$S^2_{n(M)} = n^{-1} \sum_{i=1}^{n} \xi^2(x_i - c_i\hat{\Delta}_{n(M)}). \tag{3.34}$$

Further, we write $C_n^2 = \Sigma_{i=1}^n c_i^2$, and for some prefixed α $(0 < \alpha < 1)$, we define

$$\hat{\Delta}^{(M)}_{L,n} = \sup\{t\colon W_n(t) > C_n S_{n(M)} \tau_{\alpha/2}\}, \tag{3.35}$$

$$\hat{\Delta}^{(M)}_{U,n} = \inf\{t\colon W_n(t) < -C_n S_{n(M)} \tau_{\alpha/2}\}; \tag{3.36}$$

$$d_{n(M)} = \hat{\Delta}^{(M)}_{U,n} - \hat{\Delta}^{(M)}_{L,n}. \tag{3.37}$$

Then, following Jurečková and Sen (1981b), we may consider the stopping number

$$N_M(d) = \inf\{n \geq n_0\colon d_{n(M)} \leq 2d\}, \quad d > 0, \tag{3.38}$$

where n_0 is some initial sample size, and the (sequential) confidence interval for Δ is then $(\hat{\Delta}^{(M)}_{L,N_M(d)}, \hat{\Delta}^{(M)}_{U,N_M(d)})$.

We assume that $\sigma^2_{(\xi)}$, defined by (2.41) is finite and strictly positive and defining ξ as in (2.40) (but, without necessarily assuming that ξ is a constant outside a compact interval), we further assume that (i) $\int (\xi_1'(x))^2\, dF(x) < \infty$, (ii) as $t \downarrow 0$, $\int_E \{\xi_1'(x+t) - \xi_1'(x)\}^2\, dF(x) \to 0$, (iii) at the points of jumps of ξ_2, f' is bounded and (iv) $\max\{c_i^2/C_n^2\colon 1 \leq i \leq n\} \to 0$, as $n \to \infty$. Then, it follows from

Jurečková and Sen (1981b) that for this sequential M-procedure, the properties mentioned before (3.9) hold, and also (3.9)–(3.11) hold with n_d defined by

$$n_d = \inf\{n : C_n^2 \geq d^{-2}\tau_{\alpha/2}^2\sigma_{(\xi)}^2\} . \tag{3.39}$$

Here also, (3.35)–(3.36) have justifications on ADF grounds only.

For a general class of estimable parameters, sequential confidence intervals based on U-statistics have been considered by Sproule (1974) and others (see Section 10.2 of Sen (1981)). We consider the same notations as in (2.63) through (2.66), and define $\theta(F)$, U_n and S_n^2 as in there. The problem is to find a bounded width confidence interval for $\theta(F)$, satisfying (3.1) and (3.2). For small d (>0), here

$$n_d \sim d^{-2}m^2\zeta_1 \cdot \tau_{\alpha/2}^2 \tag{3.40}$$

where the unknown parameter ζ_1 is defined as in (2.64). We define our stopping variable by

$$N_U(d) = \inf\{n \geq m+1 : S_n^2 \leq nd^2/\tau_{\alpha/2}^2\} \tag{3.41}$$

and the (sequential) confidence interval for $\theta(F)$ is then

$$[U_{N_U(d)} - d, U_{N_U(d)} + d] \tag{3.42}$$

This procedure is a direct generalization of the Chow–Robbins (1965) procedure, which corresponds to the special case of $U_n = \bar{X}_n$, $\theta(F) = \int x \, \mathrm{d}F(x)$. If the kernel $\phi(X_1, \ldots, X_m)$ is square integrable, $S_n^2 \to \zeta_1$ a.s., as $n \to \infty$, and if $\zeta_1 \in (0, \infty)$, invariance principles hold for U_n (viz. Miller and Sen (1972)). For the particular case of $\bar{X}_n$, we may note that

$$S_n^2 = (n-1)^{-1} \sum_{i=1}^{n} (X_i - \bar{X}_n)^2 \leq (n-1)^{-1} \sum_{i=1}^{n} (X_i - \mu)^2$$

where the $(x_i - \mu)^2$ i.i.d.r.v. with mean σ^2. Hence, in this case, we may proceed as in Theorem 10.2.1 of Sen (1981) and show that $EN(d) < \infty \ \forall d > 0$ and $EN(d)/n_d \to 1$ as $d \downarrow 0$. However, for general $m \geq 1$, this may require a slightly more stringent condition that $E\{\sup_{n \geq n_0} S_n^2\} < \infty$, and for this it suffices to assume that $E\{\phi^2 \log \phi^2\} < \infty$ or $E|\phi|^r < \infty$ for some $r > 2$. Under either condition, for $N_U(d)$, the properties listed before (3.9) hold, and also (3.9)–(3.11) hold with n_d, defined by (3.40).

A natural generalization of this problem is the sequential confidence region for $\boldsymbol{\theta}$, a vector of unknown parameters, where instead of (3.1), we need to construct a closed (and possibly convex) region I_n such that $P\{\theta \in I_n\} > 1 - \alpha$, and instead of (3.2), we like to have the property that the maximum diameter of I_n is $\leq 2d$, for some $d > 0$. These are discussed in detail in Section 10.2.5 of

Sen (1981). The (joint) asymptotic normality of the estimates of $\boldsymbol{\theta}$ and the strong consistency property of their variance–covariance estimators are used in this context. Basically, in (3.41) (or in other appropriate places), S_n^2 is to be replaced by the largest characteristic root of $\boldsymbol{S}_n$, the estimated covariance matrix, and $\tau_{\alpha/2}^2$ by $\chi_{r,\alpha}^2$, the upper 100α % point of the chi square distribution with r degrees of freedom, where r is the dimension of $\boldsymbol{\theta}$. The regularity conditions are essentially the same. For some specific problems of special interest, we may refer to Ghosh and Sen (1973) and Sen and Ghosh (1973b).

4. Asymptotic properties of the stopping time

For both the sequential point and interval procedures in Sections 2 and 3, a variety of stopping numbers has been considered. In the point estimation problem, the main emphasis has been laid on (2.14), while in the interval estimation problem, the main theme was to show that $N_d/n_d \to 1$ a.s. or in 1st mean, as $d \downarrow 0$. These may be regarded as the first order asymptotic efficiency results of the sequential procedures. In (2.15)–(2.16) we have sketched the second order asymptotic efficiency results in the context of the sequential point estimation problem. One of the problems with the sequential confidence intervals is that the procedures considered may not satisfy (3.1); we have only the asymptotic equality (to $1-\alpha$) as $d \downarrow 0$. Thus, it may be quite appropriate to put this question: for any given d (>0), it is possible to have a procedure for which (3.1) holds and if so, then what is the order of magnitude of $EN_d - n_d$? For normal population, this problem has been considered by Simons (1968) and others. For the nonparametric problems, though some studies have been made on $EN_d - n_d$, a complete or satisfactory answer to this question is still unavailable. However, in the majority of the cases, (2.17) or its parallel form in the interval estimation problem has been studied under suitable regularity conditions.

For U-statistics or related von Mises' functionals, the asymptotic normality of the stopping time in (2.17), for N_c defined by (2.67), has already been considered in (2.70)–(2.71). For the sequential interval estimation problem, for $N_U(d)$ in (3.41), the same result holds whenever $E\phi^4 < \infty$.

For the sequential M-procedures, for both the point and interval estimation problems, the asymptotic normality of the stopping time has been studied by Jurečková and Sen (1981b, 1982). It has been observed that in either case, $n_d^{-a}(N_d - n_d)$ (or $n_c^{-a}(N_c - n_c)$) has asymptotically a normal distribution, under quite general regularity conditions, where referred to (2.40),

$$a = \begin{cases} \frac{1}{2} & \text{if } \xi = \xi_1, \text{ i.e. } \xi_2 = 0 \text{ a.e.}, \\ \frac{1}{4} & \text{if not } \xi_2 = 0 \text{ a.e.} \end{cases} \tag{4.1}$$

Thus, the effect of jump discontinuities on the score function ξ is to induce a smaller denominator ($n^{1/4}$ instead of $n^{1/2}$). Asymptotic normality results on the variance estimators of L-statistics, considered by Gardiner and Sen (1979), can similarly be used to show that for the sequential confidence interval problem, the asymptotic normality result holds for $N_L(d)$, where as in (2.17), we have $n_d^{-1/2}$ as the normalizing factor. For the point estimation problem, for $N_c(L)$ in (2.61), (2.17) has been established by Jurečková and Sen (1982). In both these cases, the score function $J(\cdot)$ has been assumed to be quite smooth. In the nonregular case, the rate $n_d^{-1/2}$ may not hold, and $n_d^{-1/4}$ may hold in some cases.

For the R-estimation procedure, the asymptotic normality of the stopping time rests on some deeper linearity results on rank statistics. Some of these results are studied in some special cases (viz., two-sample problem) by Hušková and Jurečková (1981) and Hušková (1982), while the general model remains to be studied.

5. Asymptotic efficiency results

Consider the sequential point estimation problem first. Let T be the set of sequences $\{T_n\}$ of estimators which are asymptotically normally distributed such that the minimum risk $R_{n_c^0}(c)$ in (2.11) exists and satisfies

$$\lim_{c \downarrow 0} \{R^2_{n_c^0}(c)/4c\} = \sigma^2(T; F) \quad \forall F \in \mathscr{F}, \tag{5.1}$$

where $\sigma^2(T, F)$ is the asymptotic variance of $\sqrt{n}(T_n - \theta)$ if F is the underlying d.f., and for which there exists a sequential point estimation procedure (based on $\{T_{N_c}\}$) with the risk R_c^*, defined in (2.13), satisfying (2.14) i.e.,

$$\lim_{c \downarrow 0} \{R_c^*/R_{n_c^0}(c)\} = 1 \quad \forall F \in \mathscr{F}. \tag{5.2}$$

Then, we may consider the limit

$$e(T; F) = \lim_{c \downarrow 0} \{\sqrt{c}/R_c^*\}, \quad F \in \mathscr{F}, \tag{5.3}$$

as a measure of *efficacy* of the sequential point estimator T_{N_c} when F is the underlying d.f., defined over the class $\mathscr{F}$ of d.f.'s. Thus, if $\{T_{N_c}\}$ and $\{T^*_{N_c^*}\}$ be two sequential point estimators (defined for $c > 0$) for which (5.1)–(5.3) hold, then the *asymptotic relative efficiency* (A.R.E.) of $\{T_{N_c}\}$ with respect to $\{T^*_{N_c^*}\}$ is defined by

$$e(T, T^*; F) = e(T; F)/e(T^*; F) = \sigma^2(T^*; F)/\sigma^2(T; F), \tag{5.4}$$

and this agrees with the conventional measure of A.R.E. in the nonsequential case. Also, note that for any asymptotically unbiased and normally distributed estimator $\{T_n\}$,

$$\sigma^2(T; F) \geqslant \{\mathscr{I}(F)\}^{-1}, \tag{5.5}$$

where $\mathscr{I}(F)$ is the *Fisher information* on θ. Thus, by (5.4) and (5.5), the sequential point estimator $\{T_N\}$ is asymptotically fully efficient when, in (5.1), $\sigma^2(T; F)$ is equal to the information limit $(\phi(F))^{-1}$.

By virtue of (5.4) and the results on the A.R.E. of (nonsequential) nonparametric estimators available in the literature, we may conclude that the nonparametric sequential procedures may be advocated over the normal theory procedures when the underlying F is not normal. In particular, the use of normal scores statistics for the location problem in Section 2 leads to a value of (5.4) (against the procedure based on the sample mean and variance) bounded from below by 1, for all F, where the lower bound is attained only when F is normal. Also, from the robustness point of view, for the (local) error-contamination model, via (5.5), the asymptotic minimax property of sequential M-, L- and R-estimators may be established as in Jurečková and Sen (1982).

Let us next consider the interval estimation problem. We have noticed in (3.11)–(3.12) and elsewhere in Section 3 that for small values of d (>0),

$$EN_d \sim \tau^2_{\alpha/2} d^{-2} \sigma^2(T; F); \quad \sigma^2(T, F) = \lim_{n\to\infty} nE(T_n - \theta)^2, \tag{5.6}$$

where F ($\in \mathscr{F}$) is the true d.f. Therefore, for two competing sequential interval procedures (corresponding to a common d (>0) and a coverage probability $1-\alpha$), equating the expected sample sizes (up to the ratio being asymptotically (as $d \downarrow 0$) equal to 1), we arrive at the same measure of A.R.E. as in (5.4). Hence, what has been discussed following (5.4) also pertains to the confidence interval problem.

6. Some general remarks

In the sequential point estimation problem, the minimum risk as well as the expected sample size depend very much on the form of the loss function (viz., $g(x)$ and $c(n)$ in (2.1)). For example, if instead of $g(x) = x^2$, we choose $g(x) = |x|$, then in (2.10)–(2.11) we would have for small c (>0),

$$n^0_c \sim (\gamma/2c)^{2/3} \quad \text{and} \quad R_{n^0_c}(c) \sim 3c^{1/3}(\gamma/2)^{2/3} \tag{6.1}$$

where $\gamma = \lim_{n\to\infty} E\{n^{1/2}|T_n - \theta|\}$. Though the choice of $g(x) = x^2$ is more conventional and somewhat justified on the ground of the 'mean square error' being a popular criterion, the case of $g(x) = |x|$ may also be advocated on parallel grounds. In fact, for the normal mean problem, Robbins (1959) considered the case of $g(x) = |x|$. In some other cases, where θ is regarded as a positive quantity, $g(x) = \theta^{-1}|x|$ or $\theta^{-2}x^2$ has also been some other workers (viz., Chow and Martinsek, 1982). It seems desirable to work out the general case with some bowl-shaped loss function.

Mukhopadhyay (1980) has shown that for the normal mean problem, a Stein-type two-stage procedure where the initial sample size n_0 ($= n_{0c}$ or n_{0d}) depends on c (or d), such that as c (or d) $\downarrow 0$, $n_{0c} \to \infty$ but $c^{1/2}n_{0c} \to 0$ (or $n_{0d} \to \infty$ but $d^2 n_{0d} \to 0$) has also the first order asymptotic efficiency in (2.14) or (3.11), though it does not have the second order efficiency (i.e., finite regret as c (or d) $\downarrow 0$). The characteristic is shared by two-stage nonparametric procedures too. This raises the question whether or not the second order efficiency can be attained by a three or multi-stage procedure. If the answer is in the affirmative, then much of the labour involved in a genuine sequential procedure may be avoided by adapting a multi-stage or group-sequential procedure.

Mostly relating to the parametric cases, some attempts have been made to obtain some asymptotic expansions for $EN_c - n_c^0$ (or $EN_d - n_d$), so that some idea of the regrets may be gathered. However, in most of the nonparametric problems, one encounters nonlinear statistics, and such expansions may be quite involved. More work is needed in this area. For both the point and interval estimation problems, in the nonparametric case, the theory has mostly been justified on an asymptotic ground where c (or d) is made to converge to 0. Though these approximations work out quite well for small values of c or d, they may depend on the statistics used and also on the underlying distributions. Therefore, there remains good scope for numerical studies on the adequacy of the asymptotic theory for moderate values of c or d.

For both the problems in Sections 2 and 3, the sequential procedures are based on some well defined stopping times. In some situations, one may face the problem of providing a *confidence sequence* for a parameter, where there may be any role for a stopping number. We may conceive of a sequence $\{X_i;\, i \geq 1\}$ of independent r.v.'s defined on a common probability space, and we desire to form a sequence $\{J_n\}$ of (confidence) intervals, such that for some parameter θ and positive integer m,

$$P\{\theta \in J_n\ \forall n \geq m\} \geq 1 - \alpha_m\,, \tag{6.2}$$

$0 < \alpha_m < 1$, α_m can be made to converge to 0 as $m \to \infty$ and the length of $J_n \to 0$ as $n \to \infty$. This may be done without much difficulty. As an example, we consider the case of the location parameter based on signed rank statistics. Suppose that in (2.27), we choose the $S_{n,\alpha}$ in such a way that for some suitable sequence $\{e_n\}$ of (slowly) increasing function of n,

$$S_{n,\alpha} \sim A_\phi n^{1/2} e_n\,, \tag{6.3}$$

where $\{e_n\}$ is so chosen that under H_0: $\theta = 0$, $|S_n| \leq A_\phi n^{1/2} e_n$ a.s., as $n \to \infty$. We may for example, let $e_n = (2 \log\log n)^{1/2}$, for which the above holds, though in (6.2), a precise order of α_m (even for large m) may be difficult to obtain.

However, for $e_n \sim c \log n$, precise order of (as well as suitable bounds for) α_m may be specified. With such a choice of $\{e_n\}$ is (6.3), we then proceed as in (2.8)–(2.29) and define the corresponding limits by $\tilde{\theta}_{L,n}^{(R)}$ and $\tilde{\theta}_{U,n}^{(R)}$, respectively. Then, we conclude that

$$P\{\tilde{\theta}_{L,n}^{(R)} \leq \theta \leq \tilde{\theta}_{U,n}^{(R)} \quad \text{for all } n \geq n_0\} = \alpha_{n_0}, \tag{6.4}$$

where $\alpha_{n_0} \to 0$ as $n_0 \to \infty$, or some suitable bounds can also be attached to α_{n_0}. In a similar manner, for the regression model in Section 3, in (3.15)–(3.16), we may choose $L_{n,\alpha}^{(j)}$ as $C_n A_n (-1)^j e_n$, $j = 1, 2$, where the $\{e_n\}$ are chosen as in after (6.3), and then proceeding as in (3.17)–(3.18), we arrive at a confidence sequence for β. Procedures based on M-statistics and L-statistics can also be worked in a similar manner. For some details, we may refer to Section 10.4 of Sen (1981).

References

Anscombe, F. J. (1952). Large sample theory of sequential estimation. *Proc. Cambridge Phil. Soc.* **48**, 600–607.

Carroll, R. J. (1977). On the asymptotic normality of stopping times based on robust estimators. *Sankhyā, Ser. A* **39**, 355–377.

Chatterjee, S. K. (1977). Sequential inference procedures of Stein's type for a class of multivariate regression problems. *Ann. Math. Statist.* **33**, 1039–1064.

Chow, Y. S. and Martinsek, A. T. (1982). Bounded regret of a sequential procedure for estimation of the mean. *Ann. Statist.* **10**, 909–914.

Chow, Y. S. and Robbins, H. (1965). On the asymptotic theory of fixed-width sequential confidence intervals for the mean. *Ann. Math. Statist.* **36**, 457–462.

Chow, Y. S. and Yu, K. F. (1981). The performance of a sequential procedure for the estimation of the mean. *Ann. Statist.* **9**, 184–188.

Dantzig, G. B. (1940). On the non-existence of tests of "Student's hypothesis" having power function independent of σ. *Ann. Math. Statist.* **11**, 186–192.

Darling, D. A. and Robbins, H. (1967). Confidence sequences for mean, variances and median. *Proc. Nat. Acad. Soc. USA* **58**, 66–68.

Gardiner, J. C. and Sen, P. K. (1979). Asymptotic normality of a variance estimator of a linear combination of a function of order statistics. *Zeit. Wahrsch. Verw. Geb.* **50**, 205–221.

Geertsema, J. C. (1970). Sequential confidence intervals based on rank tests. *Ann. Math. Statist.* **41**, 1016–1026.

Ghosh, M. and Mukhopadhyay, N. (1979). Sequential point estimation of the mean when the distribution is unspecified. *Comm. Statist: Th. Methods, Ser. A.* **8**, 637–652.

Ghosh, M. and Mukhopadhyay, N. (1981). Consistency and asymptotic efficiency of two-stage and sequential estimation procedures. *Sankhyā Ser. A.* **43**, 220–227.

Ghosh, M. and Sen, P. K. (1971). Sequential confidence intervals for the regression coefficient based on Kendall's tall. *Calcutta Statist. Assoc. Bull.* **20**, 23–36.

Ghosh, M. and Sen, P. K. (1972). On bounded length confidence intervals for the regression coefficient based on a class of rank statistics. *Sankhyā, Ser. A* **34**, 33–52.

Ghosh, M. and Sen, P. K. (1973). On some sequential simultaneous confidence intervals procedures. *Ann. Inst. Statist. Math.* **25**, 123–134.

Ghosh, M., Sinha, P. K. and Mukhopadhyay, N. (1976). Multivariate sequential point estimation. *J. Multiv. Anal.* **6**, 281–294.

Gleser, L. J. (1965). On the asymptotic theory of fixed-size sequential confidence bounds for linear regression parameters. *Ann. Math. Statist.* **36**, 463–467.

Hoeffding, W. (1948). A class of statistics with asymptotically normal distribution. *Ann. Math. Statist.* **19**, 293–325.

Huber, P. J. (1964). Robust estimation of a location parameter. *Ann. Math. Statist.* **35**, 73–101.

Huber, P. J. (1981). *Robust Statistics.* Wiley, New York.

Hušková, M. (1982). On bounded length sequential confidence interval for parameters in regression model based on ranks. *Coll. Nonparametric Infer., Janos Bolyai Math. Soc.* **32**, 435–463.

Jurečková, J. (1977). Asymptotic relations of M-estimates and R-estimates in linear regression models. *Ann. Statist.* **5**, 664–672.

Jurečková, J. and Sen, P. K. (1981a). Invariance principles for some stochastic processes related to M-estimators and their role in sequential statistical inference. *Sankhyā, Ser. A.* **43**, 190–210.

Jurečková, J. and Sen, P. K. (1981b). Sequential procedures based on M-estimators with discontinuous score functions. *J. Statist. Plann. Inference* **5**, 253–266.

Jurečková, J. and Sen, P. K. (1982). M-estimators and L-estimators of location: Uniform intergrability and asymptotically risk-efficient sequential versions. *Sequen. Anal.* **1**, 27–56.

Koul, H. L. (1969). Asymptotic behaviour of Wilcoxon type confidence regions in multiple regression. *Ann. Math. Statist.* **40**, 1950–1979.

Lai, T. L. and Siegmund, D. (1977). A nonlinear renewal theory with applications to sequential analysis. *Ann. Statist.* **5**, 946–954.

Lai, T. L. and Siegmund, D. (1979). A nonlinear renewal theory with applications to sequential analysis, II. *Ann. Statist.* **7**, 60–76.

Miller, R. G. Jr. and Sen, P. K. (1972). Weak convergence of U-statistics and von Mises' differentiable statistical functions. *Ann. Math. Statist.* **43**, 31–41.

Mukhopadhyay, N. (1980). Consistent and asymptotically efficient two-stage procedure to construct fixed-width confidence interval for the mean. *Mertika* **27**, 281–284.

Robbins, H. (1959). Sequential estimation of the mean of a normal population. In: *Probability and Statistics* (H. Crameŕ vol.). Almquist and Wicksell, Uppsala, pp. 235–245.

Sen, P. K. (1960). On some convergence properties of U-statistics. *Calcutta Statist. Assoc. Bull.* **10**, 1–18.

Sen, P. K. (1977). Some invariance principles relating to jackknifing and their role in sequential analysis. *Ann. Statist.* **5**, 319–329.

Sen, P. K. (1978). An invariance principle for linear combinations of order statistics. *Z. Wahrsch. Verw. Geb.* **42**, 327–340.

Sen, P. K. (1980a). Nonparametric Simultaneous inference for some MANOVA models. In: P. R. Krishnaiah, ed., *Handbook of Statistics, Vol. 1.* North-Holland, Amsterdam, pp. 673–702.

Sen, P. K. (1980b). On nonparametric sequential point estimation of location based on general rank statistics. *Sankhyā Ser. A* **42**, 201–220.

Sen, P. K. (1981). *Sequential Nonparametrics: Invariance principles and statistical inference.* Wiley, New York.

Sen, P. K. (1983). Sequential R-estimation of location in the general Behrens-Fisher model. *Sequen. Anal.* **2**, 311–335.

Sen, P. K. (1984). On sequential nonparametric estimation of multivariate location. *Proc. Third Prague Conf. Asymp. Meth.*, to appear.

Sen, P. K. and Ghosh, M. (1971). On bounded length sequential confidence intervals based on one-sample rank order statistics. *Ann. Math. Statist.* **42**, 189–203.

Sen, P. K. and Ghosh, M. (1973a). A law of iterated logarithm for one-sample rank order statistics and some applications. *Ann. Statist.* **1**, 568–576.

Sen, P. K. and Ghosh, M. (1973b). Asymptotic properties of some sequential nonparametric estimators in some multivariate linear models. In: P. R. Krishnaiah, ed., *Multivariate Analysis* III. Academic Press, New York, pp. 299–316.

Sen, P. K. and Ghosh, M. (1981). Sequential point estimation of estimable parameters based on U-statistics. *Sankhyā Ser. A* **43**.

Simons, G. (1978). The cost of not knowing the variance when making a fixed-width confidence interval for the mean. *Ann. Math. Statist.* **39**, 1946–1952.

Sproule, R. N. (1974). Asymptotic properties of U-statistics. *Trans. Amer. Math. Soc.* **199**, 55–64.

Starr, N. (1966). On the asymptotic efficiency of a sequential point estimation. *Ann. Math. Statist.* **37**, 1173–1185.

Starr, N. and Woodroofe, M. (1969). Remarks on sequential point estimation. *Proc. Nat. Acad. Sci. USA* **63**, 285–288.

Starr, N. and Woodroofe, M. (1972). Further remarks on sequential point estimation: the exponential case. *Ann. Math. Statist.* **43**, 1147–1154.

Stein, C. (1945). A two-sample test for a linear hypothesis whose power function is independent of σ. *Ann. Math. Statist.* **16**, 243–258.

Williams, G. W. and Sen, P. K. (1973). Asymptotically optimal sequential estimation of regular functionals of several distributions based on generalized U-statistics. *J. Multivar. Anal.* **3**, 469–482.

Williams, G. W. and Sen, P. K. (1974). On bounded maximum width sequential confidence ellipsoids based on generalized U-statistics. *J. Multivar. Anal.* **4**, 453–468.

Woodroofe, M. (1977). Second order approximations for sequential point and interval estimation. *Ann. Statist.* **5**, 984–995.

Woodroofe, M. (1982). *Nonlinear Renewal Theory in Sequential Analysis.* S.I.A.M. Publication, Philadelphia.

PART V

STOCHASTIC APPROXIMATION AND DENSITY ESTIMATION

P. R. Krishnaiah and P. K. Sen, eds., *Handbook of Statistics, Vol. 4*

23

Stochastic Approximation

Václav Dupač

1. A simple iterative procedure for solving deterministic equations

Suppose we have to find the solution θ to the equation $M(x) = 0$, where M is a real function of a real variable. One of the commonly used numerical methods of solution is Newton's method, described by the iteration

$$x_{t+1} = x_t - (M'(x_t))^{-1}M(x_t) .$$

However, this method obviously cannot be made use of in situations when we are able to calculate numerically the function values $M(x)$ only, not the values of the derivative $M'(x)$. For some of these situations, a simple iterative formula can be recommended, viz.

$$x_{t+1} = x_t + aM(x_t) , \tag{1.1}$$

where a is a constant.

Assume that the graph of the function M lies between two lines of negative slopes, passing through the point $[\theta; 0]$. More precisely, assume that

$$M(x) \gtreqqless 0 \text{ for } x \lesseqqgtr \theta \text{ and } K_1|x - \theta| \leqq |M(x)| \leqq K_2|x - \theta|$$

for all $x \in \mathbb{R}$ and some positive constants K_1, K_2. Then the approximations x_t determined by (1.1) tend to the solution θ for every initial value $x_1 \in \mathbb{R}$, provided that $0 < a < 2/K_2$. The rate of convergence is at least that of a geometrical sequence with the quotient

$$q = \max(1 - aK_1, aK_2 - 1) .$$

Of course, the same holds true if the graph of M lies between two lines of positive slopes, and if the plus in (1.1) is replaced by a minus. An analogous remark applies throughout the paper.

2. The Robbins–Monro stochastic approximation method

Consider now a situation, when the values $M(x)$ can be found up to an observational or experimental error only, i.e., a situation, when an observation or an experiment at point x results in $Y(x) = M(x) + e(x)$, where $e(x)$ is a random error with zero expectation. We can try to find θ by means of the procedure (1.1) again, with $Y_t = M(X_t) + e(t+1, X_t)$ in the place of $M(X_t)$. (Here we utilize capital letters for denoting random variables; we have also indicated that the random errors are different for different values of t.)

It is easy to see that this time a cannot be chosen as a constant, but has to be replaced by a sequence of positive constants a_t, tending to 0 and such that the sum $\Sigma_{t=1}^{\infty} a_t$ diverges to infinity. If namely a were a constant for all t, then owing to random errors, the differences $X_{t+1} - X_t = aY_t$ could no longer converge to 0 and, consequently, nor the sequence of approximations X_t could be convergent. Hence, we must have $a_t \to 0$. On the other hand, if the convergence of a_t to 0 were so fast that $\Sigma_{t=1}^{\infty} a_t = A < +\infty$, then the approximations X_t could tend to another limit than θ. As an example consider $Y(x)$ attaining values ± 1 only, for all $x \in \mathbb{R}$, and an initial approximation X_1 at a distance greater than A from θ.

If convergence of X_t to θ with probability one is to be proven, then the additional condition $\Sigma_{t=1}^{\infty} a_t^2 < +\infty$ will be required, similarly as in theorems on convergence with probability one of sums of independent random variables.

The described procedure was proposed and investigated first by Robbins and Monro (1951) and called a stochastic approximation method; the original motivations having been a sequential determination of the so-called LD50, the dose of a drug, which is lethal to 50% of experimental animals. Since 1951, hundreds of papers appeared devoted to the mathematical theory as well as to applications of stochastic approximation to diverse problems. Two monographs can also be quoted: Wasan (1969) and Nevel'son and Has'minskiĭ (1972/76). The latter will be our reference book; results stated without quotation can be found there.

3. Convergence theorems

We start with a general multivariate convergence theorem. The measurability conditions involved are more or less formal and can be assumed as fulfilled in all practical problems.

Let $M : \mathbb{R}^p \to \mathbb{R}^p$ be a $\mathcal{B}$-measurable mapping ($\mathcal{B}$ denotes the Borel sets in $\mathbb{R}^p$). Let $(\Omega, \mathcal{A}, P)$ be a probability space with a sequence of σ-fields $\mathcal{F}_1 \subset \mathcal{F}_2 \subset \cdots \subset \mathcal{A}$ and a sequence of p-vector valued random functions $(e(t, x), x \in \mathbb{R}^p)_{t=2}^{\infty}$, measurable with respect to $\mathcal{B} \times \mathcal{F}_t$, such that $e(t, x)$, $x \in \mathbb{R}^p$, is independent of $\mathcal{F}_{t-1}$ and $Ee(t, x) = 0$, for all $x \in \mathbb{R}^p$, $t \geqq 2$.

Let us call M the regression function; assume that at any time-instant t, the function value of M at an arbitrarily chosen point x can be unbiasedly estimated by an observation Y.

Choose X_1 ($\mathscr{F}_1$-measurable) arbitrarily; then define recursively

$$X_{t+1} = X_t + a_t Y_t, \quad 1 \leqq t < +\infty, \tag{3.1}$$

where $Y_t = M(X_t) + e(t+1, X_t)$ and a_t are positive constants.

In what follows, K with or without subscript denotes a positive constant, possibly different in different theorems; the symbol ∇ denotes the gradient.

THEOREM 3.1. *Let $V: \mathbb{R}^p \to \mathbb{R}$ be a twice continuously differentiable function, with bounded second order derivatives, such that, for some $\theta \in \mathbb{R}^p$,*

$$V(\theta) = 0, \qquad V(x) > 0 \quad \forall x \neq \theta, \qquad \lim_{|x| \to \infty} V(x) = +\infty, \tag{3.2}$$

$$\sup_{\varepsilon < |x-\theta| < \varepsilon^{-1}} \langle M(x), \nabla V(x) \rangle < 0 \quad \forall \varepsilon > 0; \tag{3.3}$$

further assume

$$|M(x)|^2 + E|e(t,x)|^2 \leqq K(1 + V(x) + |\langle M(x), \nabla V(x) \rangle|), \tag{3.4}$$

$$\sum_{t=1}^{\infty} a_t = +\infty, \qquad \sum_{t=1}^{\infty} a_t^2 < +\infty. \tag{3.5}$$

Then we have $\lim_{t\to\infty} X_t = \theta$ *with probability* 1.

Choosing $V(x) = \langle C(x-\theta), x-\theta \rangle$ with C a positive definite matrix, we get the following

THEOREM 3.2. *Let C be a positive definite matrix such that*

$$\sup_{\varepsilon < |x-\theta| < \varepsilon^{-1}} \langle M(x), C(x-\theta) \rangle < 0 \quad \forall \varepsilon > 0; \tag{3.6}$$

further assume

$$|M(x)|^2 + E|e(t,x)|^2 \leqq K_1(1 + |x|^2) \tag{3.7}$$

and (3.5). *Then we have* $\lim_{t\to\infty} X_t = \theta$ *with probability* 1.

The geometrical meaning of the assumption (3.3) will be clear by the following consideration. Let us neglect for a moment the random errors $e(t, x)$ and at the same time smooth the iterative scheme $X_{t+1} = X_t + a_t M(X_t)$ into a time-continuous one, $\mathrm{d}X = a_t M(X)\,\mathrm{d}t$. The existence of a function V with properties as in Theorem 3.1 (called Ljapunov function in stability theory) ensures that any trajectory X_t of the above differential equation moves continuously towards the point of the minimum of V, because of $\mathrm{d}V(X)/\mathrm{d}t = a_t \langle M(X), \nabla V(X) \rangle < 0$, $X \neq \theta$.

In the one-dimensional case, (3.6) reduces to the assumption

$$\sup_{\varepsilon<|x-\theta|<\varepsilon^{-1}} M(x)\,\mathrm{sgn}(x-\theta)<0 \quad \forall\varepsilon>0\,.$$

This is especially satisfied by any function M which is everywhere continuous and of opposite sign as $x-\theta$.

Strengthening the assumption (3.6) we get the mean square convergence, even without requiring $\Sigma_{t=1}^{\infty} a_t^2<+\infty$:

THEOREM 3.3. *Let C be a positive definite matrix such that*

$$\langle M(x), C(x-\theta)\rangle \leqq -K\langle C(x-\theta), x-\theta\rangle \quad \forall x\in\mathbb{R}^p\,, \tag{3.8}$$

further assume (3.7) *and*

$$a_t\to 0, \qquad \sum_{t=1}^{\infty} a_t=+\infty\,. \tag{3.9}$$

Then we have $\lim_{t\to\infty} E|X_t-\theta|^2=0$, *provided that* $E|X_1|^2<+\infty$.

4. Rate of convergence – The choice of a_t's

The constants a_t in the Robbins–Monro procedure are usually chosen in the form

$$a_t=at^{-\alpha}, \quad a>0,\ 0<\alpha\leqq 1\,.$$

The rate of mean square convergence can be then found, under the assumptions of Theorem 3.3:

$$E|X_t-\theta|^2=\begin{cases} \mathrm{O}(t^{-\alpha}) & \text{for } 0<\alpha<1\,,\\ \mathrm{O}(t^{-1}) & \text{for } \alpha=1 \text{ and } 2Ka>1\,,\\ \mathrm{O}(t^{-1}\log t) & \text{for } \alpha=1 \text{ and } 2Ka=1\,,\\ \mathrm{O}(t^{-2Ka}) & \text{for } \alpha=1 \text{ and } 2Ka<1\,. \end{cases}$$

Hence the choice $\alpha=1$ and $2Ka>1$ is optimal from this point of view.

The following result, due to Révész (1973), concerns the rate of decrease of large deviation probabilities: Assume

$$\langle M(x), x-\theta\rangle \leqq -K|x-\theta|^2, \qquad |M(x)|^2\leqq K_1(1+|x|^2)\,,$$
$$|e(t,x)|\leqq K_2\,\forall x\in\mathbb{R}^p, \quad t\geqq 2, \qquad a_t=at^{-1}, \quad a>0\,.$$

Then we have

$$\liminf_{t\to\infty} |t^{-1} \log P(|X_t - \theta| \geqq \varepsilon)| > 0 \quad \forall \varepsilon > 0 .$$

5. Asymptotic normality

THEOREM 5.1 (Fabian, 1968). *Let the sequence X_t, $t \geqq 1$, defined by* (3.1) *with $a_t = at^{-1}$, tend to θ with probability* 1. *Additionally assume*

$$M(x) = B(x - \theta) + \delta(x), \quad \text{where } \delta(x) = o(|x - \theta|) \text{ for } x \to \theta \tag{5.1}$$

and the matrix B is negative definite,

$$\|E(e(t, x)e^{\mathrm{T}}(t, x))\| \leqq K \tag{5.2}$$

for all $t \geqq 2$ and all x from a neighborhood of θ,

$$\lim_{t\to\infty,\, x\to\infty} E(e(t, x)e^{\mathrm{T}}(t, x)) = \Sigma , \tag{5.3}$$

$$\lim_{t\to\infty,\, x\to\theta} E(|e(t, x)|^2 I_{[|e(t, x)|^2 \geqq rt]}) = 0 \quad \forall r > 0 . \tag{5.4}$$

Let P be an orthogonal matrix such that $-P^{\mathrm{T}}BP = \Lambda$ is diagonal; let $2\lambda_* a > 1$, where $\lambda_* = \min_i \Lambda^{(ii)}$. Then the asymptotic distribution of the random vector $t^{1/2}(X_t - \theta)$ is the normal one with zero expectation and covariance matrix $aPQP^{\mathrm{T}}$, where

$$Q^{(ij)} = \frac{a(P^T \Sigma P)^{(ij)}}{a\Lambda^{(ii)} + a\Lambda^{(jj)} - 1} . \tag{5.5}$$

If the random errors $e(t, x) = e(t)$ do not depend on x and are all identically distributed, then all the assumptions (5.2), (5.3) and (5.4) are satisfied, provided that $E|e(2)|^2 < +\infty$.

Let us specialize Theorem 5.1 for the one-dimensional case:

Let the sequence X_t, $t \geqq 1$, defined by (3.1) with $p = 1$ and with $a_t = at^{-1}$ tend to θ with probability 1. Let the derivative of M at point θ exist, be negative and let $2|M'(\theta)|a > 1$. Denote $m = |M'(\theta)|$. Assume

$$Ee^2(t, x) \leqq K \tag{5.6}$$

for all $t \geqq 2$ and all x from some neighborhood of θ,

$$\lim_{t\to\infty, x\to\theta} Ee^2(t, x) = \sigma^2 < +\infty , \tag{5.7}$$

$$\lim_{t\to\infty, x\to\theta} E(e^2(t,x) I_{[e^2(t,x)\geq rt]}) = 0 \quad \forall r > 0 . \tag{5.8}$$

Then $t^{1/2}(X_t - \theta)$ is asymptotically normally distributed with parameters

$$\left(0, \frac{a^2\sigma^2}{2ma - 1}\right) . \tag{5.9}$$

Conditions ensuring the asymptotic normality of the X_t's are not very severe; they are the differentiability of the regression function at θ and a sort of boundedness and continuity of the covariance matrix of the errors, together with the usual Lindeberg-type condition.

6. An adaptive procedure

In this section, we confine ourselves to the one-dimensional case for simplicity.

From the formula (5.9) we can deduce, that the asymptotic variance of $t^{1/2}(X_t - \theta)$ is minimized by choosing $a = 1/m$; its minimal value is σ^2/m^2. Apparently, this piece of knowledge cannot be made use of, as $m = |M'(\theta)|$ is not known to us; nevertheless, m can be estimated in the course of the approximation process and the estimate inserted in the iteration scheme:

In the t-th step of the procedure, we take two observations Y'_t and Y''_t of the regression function M, at points $X_t + c_t$ and $X_t - c_t$ respectively, make the quotient $(Y'_t - Y''_t)/(2c_t)$ and estimate $M'(\theta)$ be the average of these quotients,

$$W_t = \frac{1}{t-1} \sum_{i=1}^{t-1} \frac{Y'_i - Y''_i}{2c_i} , \tag{6.1}$$

i.e., we estimate m by $|W_t|$.

We assume that there are two numbers $0 < r_1 < r_2 < +\infty$ known to us, such that $r_1 \leq m \leq r_2$. Utilizing this prior information, we modify the estimate $|W_t|$ of m, viz. we replace it by

$$[|W_t|] = [|W_t|]_{r_1}^{r_2} . \tag{6.2}$$

(We use the symbol $[y]_c^d$ for $\max(c, \min(y, d))$; $c < d$.) The modified estimate is finally inserted in the iteration scheme,

$$X_{t+1} = X_t + \frac{1}{t[|W_t|]} Y_t . \tag{6.3}$$

As we have no observation of the function M at point X_t, we replace it by the average of observations at $X_t \pm c_t$, i.e. Y_t in (6.3) means $Y_t = \frac{1}{2}(Y'_t + Y''_t)$.

The described adaptive procedure was proposed by Venter (1967); he has also proven

THEOREM 6.1. *Let $M(x) \gtreqless 0$ for $x \gtreqless \theta$,*

$$M^2(x) + Ee^2(t, x) \leqq K_1(1 + x^2) \quad \forall t \geqq 2, \; x \in \mathbb{R}\,, \tag{6.4}$$

there exists an everywhere continuous and bounded second derivative $M''(x)$, and

$$c_t = ct^{-\gamma} \quad \text{for some } c > 0, \; 0 < \gamma < \tfrac{1}{2}. \tag{6.5}$$

Then we have $\lim_{t\to\infty} X_t = \theta$ and $\lim_{t\to\infty}[|W_t|] = m$ with probability 1.

If moreover $\frac{1}{4} < \gamma < \frac{1}{2}$, $\lim_{t\to\infty, x\to\theta} Ee^2(t, x) = \sigma^2$, $\lim_{t\to\infty, x\to\theta} E(e^2(t, x) I_{[e^2(t,x)\geqq rt]}) =$ 0,

$$E(e(t, x + c_t)e(t, x - c_t)) = 0 \; \forall t \geqq 2, \; x \in \mathbb{R}\,,$$

then the random variables $t^{1/2}(X_t - \theta)$ are asymptotically normally distributed with parameters $(0, \sigma^2/(2m^2))$.

Apparently, the latter asymptotic variance is half its minimal possible value. The explanation is that the adaptive procedure needs twice as many observations as the standard Robbins–Monro procedure. So we have to compare the precision of the t-th approximation of the adaptive procedure and the $2t$-th approximation of the standard procedure; both of them coincide then.

Recently, Lai and Robbins (1979) have shown that at least for identically distributed random errors, there is no need of two observations at each step for estimating $M'(\theta)$; one observation at each step suffices, as well as the standard least squares estimate based on all observations up to the current one.

THEOREM 6.2. *Let $M(x) \gtreqless 0$ for $x \gtreqless \theta$, $M^2(x) \leqq K(1 + x^2)$ $\forall x \in \mathbb{R}$, $M'(x)$ exists and is continuous $\forall x \in \mathbb{R}$. Define X_t, $t \geqq 1$, formally as in (6.3), where now $Y_t = M(X_t) + e(t+1)$; $X_1, e(2), e(3) \ldots$ are independent random variables, $e(t)$, $t \geqq 2$, indentically distributed, $Ee(t) = 0$, $Ee^2(t) = \sigma^2 < +\infty$,*

$$W_t = \frac{\sum_{i=1}^t (X_i - \bar{X})(Y_i - \bar{Y})}{\sum_{i=1}^t (X_i - \bar{X})^2}, \qquad \bar{X} = \frac{1}{t}\sum_{i=1}^t X_i, \quad \bar{Y} = \frac{1}{t}\sum_{i=1}^t Y_i,$$

$[|W_t|] = [|W_t|]_{r_1}^{r_2}$ for some $0 < r_1 \leqq m \leqq r_2 < +\infty$, $m = |M'(\theta)|$. Then we have $\lim_{t\to\infty} X_t = \theta$, $\lim_{t\to\infty}[|W_t|] = m$ with probability 1 and $t^{1/2}(X_t - \theta)$ is asymptotically normally distributed with parameters $(0, \sigma^2/m^2)$.

If in addition the continuous $M''(x)$ exists in a neighborhood of θ, then also $(\log t)^{1/2}([|W_t|] - m)$ is asymptotically normally distributed with parameters $(0, m^2)$.

The adaptive method – with an obvious modification – can be made use of, even if we don't know which of both cases occurs: whether $M(x) \gtreqless 0$ for $x \gtreqless \theta$ or $M(x) \lesseqgtr 0$ for $x \gtreqless \theta$. The method itself adapts to the true alternative. There are also multivariate analogs of the adaptive procedure. See Nevel'son and Has'minskiĭ (1973) and Pantel (1979) for both topics.

7. Asymptotic efficiency

Under assumptions listed in Section 5, the normed Robbins–Monro sequence $t^{1/2}(X_t - \theta)$ is asymptotically normally distributed with parameters $(0, a^2\sigma^2/(2ma - 1))$; the asymptotic variance can be minimized by choosing $a = 1/m$ or by making use of the adaptive method.

In some situations, the asymptotic variance can be further reduced by a proper transformation of observations, i.e. by replacing Y through $g(Y)$ in the approximation scheme. We confine ourselves to the case, when the errors $e(t, x) = e(t)$ do not depend on x and are all identically distributed, according to a probability density f possessing a positive and finite Fisher information

$$I(f) = \int_{-\infty}^{+\infty} \left(\frac{f'(y)}{f(y)}\right)^2 f(y)\,\mathrm{d}y .$$

At the same time, we consider only such transformations g, for which both the transformed regression function

$$M_g(x) = Eg(M(x) + e(t))$$

and the errors corresponding to it

$$e_g(t, x) = g(M(x) + e(t)) - M_g(x)$$

fulfil again the assumptions of the Asymptotic Normality Theorem, with possibly different constants K_g, $m_g = |M_g'(\theta)|$ and σ_g^2, but with the same $\theta = \theta_g$. Moreover, g should satisfy some regularity assumptions (not listed here) and the inequality $\int_{-\infty}^{+\infty} gf'\,\mathrm{d}y < +\infty$. The class of transformations g satisfying all the above requirements, will be denoted as $\mathcal{G}$.

Then the transformed procedure

$$X_{t+1} = X_t + \frac{a}{t}\, g(Y_t),$$

where $Y_t = M(X_t) + e(t+1)$, with $2K_g a > 1$, retains the convergence property $\lim_{t\to\infty} X_t = \theta$ with probability 1, and $t^{1/2}(X_t - \theta)$ is asymptotically normally distributed with parameters $(0, a^2\sigma_g^2/(2m_g a - 1))$. The asymptotic variance can again be minimized by choosing $a = 1/m_g$ or by making use of the adaptive procedure, the minimal value (for fixed g) of the asymptotic variance being σ_g^2/m_g^2.

If the function $g_0 = -f'/f$ belongs to the class $\mathcal{G}$, then g_0 realizes the $\min_{g\in\mathcal{G}} \sigma_g^2/m_g^2$, which equals $1/(m^2 I(f))$; Anbar (1973).

The above property of the g_0-transformed Robbins–Monro procedure can be called its asymptotic efficiency, in the following sense: If the regression function is linear, $M(x) = -m(x - \theta)$, then $1/(m^2 I(f))$ is exactly the Cramér–Rao lower bound for variances of regular unbiased estimates of θ.

If, for instance, the underlying density f is that of the normal $(0, \sigma^2)$ distribution, then the optimal transformation g_0 is the identity and the corresponding minimal asymptotic variance is σ^2/m^2. If f is the density of the double exponential distribution, again with 0 mean and variance σ^2, then the transformation $g_0(y) = \operatorname{sgn} y$ is optimal, the corresponding minimal asymptotic variance being $\sigma^2/(2m^2)$.

If the density f is specified only as a member of the class of ε-contaminated normal distributions, then the optimal transformation (this time in the minimax sense with respect to the mentioned class) is

$$g_0(y) = \begin{cases} y & \text{for } |y| \leqq K, \\ K \operatorname{sgn} y & \text{for } |y| > K, \end{cases}$$

K being uniquely determined by ε and σ^2.

8. Robbins–Monro procedure restricted to a bounded set

All the convergence and other asymptotic properties of the Robbins–Monro procedure listed up to now have been derived under conditions of the type $|M(x)|^2 + E|e(t, x)|^2 = K(1 + |x|^2)$, limiting the increase of both the regression function and the error variances for $|x| \to +\infty$. In almost all real situations, however, we can indicate in advance a bounded set containing the unknown θ. The above condition reduces then to the boundedness of M and $E|e(\cdot,\cdot)|^2$ on this set. It remains only to modify the approximation procedure so that it doesn't leave the bounded set, retaining at the same time all its convergence properties. What follows is a mathematical formulation of this idea.

Let $C \subset \mathbb{R}^p$ be a compact convex set with a nonempty interior C^0; let π_C denote the projection onto C. Let $M(x)$ and $e(t, x)$ fulfil the overall assymptions stated at the beginning of Section 3, this time for $x \in C$ only. Let there be a unique $\theta \in C$ such that $M(\theta) = 0$, assume that this θ is in C^0. Let a_t, $t \geqq 1$, be positive constants, $\Sigma_{t=1}^{\infty} a_t = +\infty$, $\Sigma_{t=1}^{\infty} a_t^2 < +\infty$. Further assume

$$\sup_{x \in C \cap U_\varepsilon^c(\theta)} \langle M(x), x - \theta \rangle < 0 \quad \forall \varepsilon > 0\,, \tag{8.1}$$

where $U_\varepsilon(\theta)$ denotes the ε-neighborhood of θ;

$$|M(x)|^2 + E|e(t, x)|^2 \leqq K \quad \forall t \geqq 2,\ x \in C. \tag{8.2}$$

Choose X_1 arbitrarily; define recursively

$$X_{t+1} = \pi_C(X_t + a_t Y_t), \quad t \geqq 1\,, \tag{8.3}$$

where $Y_t = M(X_t) + e(t+1, X_t)$. Then we have $\lim_{t\to\infty} X_t = \theta$ with probability 1.

Also the mean square convergence and the asymptotic normality hold true for the procedure (8.3) under the same additional assumptions as in Sections 3 and 5.

Calculating projections at each step of the procedure might be uncomfortable. For other possibilities see Nevel'son and Has'minskiĭ (1972/76, Chapter 7), Dupač and Fiala (1983).

9. The dynamic Robbins–Monro procedure

In some situations, it is not realistic to assume that the regression function M remains fixed in time, and we have to admit that M changes even during the approximation process. We will consider only a location trend in the one-dimensional case; Dupač (1965).

Let at time t, $1 \leqq t < +\infty$, the regression function $M(t, x) = M^0(x - \theta_t)$ be valid, observable with an experimental error only, $M(t, x) + e(t, x)$, where $M^0(x) \gtreqless 0$ for $x \gtreqless 0$,

$$K|x| \leqq |M^0(x)| \leqq K_1|x| \quad \forall x \in \mathbb{R} ;$$

$$\theta_t = b + ct, \quad \text{with unknown } b, c ;$$

$(e(t, x), x \in \mathbb{R})_{t=2}^{\infty}$ are independent measurable random functions with $Ee(t, x) = 0$, $Ee^2(t, x) \leqq K_2$.

Choose X_1 arbitrarily, $EX_1^2 < +\infty$; define recursively

$$X_{t+1} = X_t^* + a_t Y_t^* , \tag{9.1}$$

where

$$X_t^* = \left(1 + \frac{1}{t}\right)X_t, \qquad Y_t^* = M(t+1, X_t^*) + e(t+1, X_t^*) \tag{9.2}$$

and a_t, $t \geqq 1$, are positive constants such that

$$\lim_{t\to\infty} ta_t = +\infty, \qquad \sum_{t=1}^{\infty} a_t^2 < +\infty .$$

Then we have $\lim_{t\to\infty}(X_t - \theta_t) = 0$ with probability 1 as $\lim_{t\to\infty} E(X_t - \theta_t)^2 = 0$ as well.

The procedure (9.1), (9.2) is called the dynamic Robbins–Monro procedure. We can describe it in the following way: At time $t+1$ we seek the approximation X_{t+1} to θ_{t+1}; we start from the preceding approximation X_t to θ_t, make a correction for trend, $X_t^* = (1 + t^{-1})X_t$, then we estimate the regression function $M(t+1, \cdot)$ at point X_t^* by means of an observation Y_t^* and add finally the correction term $a_t Y_t^*$.

For an alternative approach to the problem see Ruppert (1979).

10. The Kiefer–Wolfowitz approximation procedure

Let M be a real function of p real variables, attaining its maximal value at the unique point θ. We assume again that at any time-instant t, the function value of M at an arbitrarily chosen point x can be unbiasedly estimated by an observation Y. If M is smooth and if θ is its only stationary point, then to find it is equivalent to solving the system of equations $\nabla M(x) = 0$; apparently, this system can be solved by means of the Robbins–Monro procedure. However, there is a hitch in this reasoning, as we are able to estimate unbiasedly the function M only, not its gradient ∇M. Hence, the gradient ∇M is to be replaced by an approximate gradient $\nabla_c M$, which is defined as a p-dimensional vector, whose i-th coordinate is given by the ratio

$$(M(x + c\varepsilon_i) - M(x - c\varepsilon_i))/(2c),$$

where ε_i is the i-th vector of the standard orthonormal basis in $\mathbb{R}^p$.

Here $c = c_t$ is chosen as a sequence of positive constants tending to 0 for $t \to \infty$. The p-vector of random errors, with which the (half-)differences $\frac{1}{2}[M(x + c_t\varepsilon_i) - M(x - c_t\varepsilon_i)]$, $i \leqq i \leqq p$, are observed, will be denoted by $e(t+1, x)$. As to the sequence $(e(t, x), x \in \mathbb{R}^p)_{t=2}^{\infty}$, the same assumptions are made as at the beginning of Section 3.

The Kiefer–Wolfowitz stochastic approximation procedure (introduced in Kiefer and Wolfowitz, 1952), is described as follows: Choose X_1 arbitrarily; define recursively

$$X_{t+1} = X_t + a_t Y_t,$$

where $Y_t = \nabla_{c_t} M(X_t) + c_t^{-1} e(t+1, X_t)$ and a_t, $t \geqq 1$, c_t, $t \geqq 1$ are bounded sequences of positive constants.

Analogs of convergence theorems for the Robbins–Monro procedure are valid for the Kiefer–Wolfowitz procedure; let us especially state analogs to Theorems 3.2 and 3.3.

THEOREM 10.1 *Let the second order derivatives of M exist and be (globally) Lipschitz continuous. Let there exist a positive definite matrix C such that*

$$\langle \nabla M(x), C(x - \theta) \rangle < 0 \quad \forall x \neq \theta\,; \tag{10.1}$$

further assume

$$|\nabla M(x)|^2 + E|e(t, x)|^2 \leqq K_1(1 + |x|^2), \quad x \in \mathbb{R}^p, \tag{10.2}$$

$$\sum_{t=1}^{\infty} a_t = +\infty, \qquad \sum_{t=1}^{\infty} a_t c_t^2 < +\infty, \qquad \sum_{t=1}^{\infty} a_t^2 c_t^{-2} < +\infty\,. \tag{10.3}$$

Then we have $\lim_{t\to\infty} X_t = \theta$ *with probability* 1.

Theorem 10.2. *Replace in the preceding theorem assumption* (10.1) *by the stronger*

$$\langle \nabla M(x), C(x-\theta)\rangle \leqq -K|x-\theta|^2 \quad \forall x \in \mathbb{R}^p, \tag{10.4}$$

and assumption (10.3) *by the weaker*

$$\sum_{t=1}^{\infty} a_t = +\infty, \quad a_t = o(c_t^2), \tag{10.5}$$

leaving the other assumptions unaltered. Then we have $\lim_{t\to\infty} E|X_t - \theta|^2 = 0$, *provided that* $E|X_1|^2 < +\infty$.

Practically all the asymptotic properties of the Robbins–Monro procedure (including the asymptotic normality) can be derived for the Kiefer–Wolfowitz procedure as well, with appropriate alterations. Let us mention only one problem here, which is specific for the Kiefer–Wolfowitz procedure, namely the optimal choice of constants a_t, c_t. We shall confine ourselves to constants of the form

$$\begin{aligned} a_t &= at^{-\alpha}, \quad a>0,\ 0<\alpha\leqq 1, \\ c_t &= ct^{-\gamma}, \quad c>0,\ 0<\gamma<\alpha/2, \end{aligned} \tag{10.6}$$

in the case $\alpha = 1$ we shall require additionally that $2Ka > 1 - 2\gamma$, where K is the constant from (10.4).

Theorem 10.3. *Under assumptions as in Theorem* 10.2 *and an additional assumption of existence of continuous 3rd order derivatives of* M *in a neighborhood of* θ, *with constants* a_t, c_t *satisfying* (10.6), *we have*

$$E|X_t-\theta|^2 = \begin{cases} O(t^{-4\gamma}) & \text{for } \gamma < \alpha/6, \\ O(t^{-\alpha+2\gamma}) & \text{for } \gamma \geqq \alpha/6. \end{cases}$$

In particular we have

$$E|X_t-\theta|^2 = O(t^{-2/3)} \quad \textit{for } a_t = at^{-1},\ c_t = ct^{-1/6},\ 3Ka>1,$$

and this choice of constants is optimal in the following sense: If in (10.6) *we have either* $\alpha \neq 1$ *or* $\gamma \neq \frac{1}{6}$, *then there exists a Kiefer–Wolfowitz sequence* X_t, $t \geqq 1$, *satisfying all the assumptions of Theorem* 10.3 *and such that* $t^{2/3}E|X_t-\theta|^2 \to +\infty$.

Theorem 10.3 is a slight modification of a result in Dupač (1957).

11. The Robbins–Monro procedure without the independence assumption

The assumption of independence of the random functions $e(t,x)$ in the Robbins–Monro approximation method is often unrealistic; hence, the behavior of the procedure has been studied also without this assumption. Let us consider a special case only, where the regression function M is (p-dimensional) linear; see Györfi (1980).

We have to solve the system of equations $Ax = b$, where the matrix $A \in \mathbb{R}^{p\times p}$ is negative definite, $b \in \mathbb{R}^p$; denote the solution as θ $(=A^{-1}b)$. Neither the matrix A now the vector b are known to us; at the t-th step of the approximation procedure, however, we have a matrix $A(t) = A + e_A(t)$ and a vector $b(t) = b + e_b(t)$ at our disposal, such that

$$\lim_{t\to\infty} \frac{1}{t}\sum_{k=1}^{t} e_A(k) = 0, \qquad \lim_{t\to\infty} \frac{1}{t}\sum_{k=1}^{t} e_b(k) = 0\,, \tag{11.1}$$

$$\text{and that a finite } \lim_{t\to\infty} (1/t)\, \textstyle\sum_{k=1}^{t} \|A(k)\|^2 \text{ exists.} \tag{11.2}$$

($\|\cdot\|$ means Euclidean norm of a matrix).

Choose X_1 arbitrarily and define recursively

$$X_{t+1} = X_t + \frac{1}{t}(A(t)X_t - b(t)), \quad t \geqq 1\,. \tag{11.3}$$

Then we have $\lim_{t\to\infty} X_t = \theta$.

A series of similar and partially more general results can be found in Ljung (1977).

Observe that the errors $e_A(t)$ and $e_b(t)$ in the just formulated theorem are not assumed to be random variables; hence, nor the X_t's are. Nevertheless, the theorem can be applied to stochastic problems. We only need to verify that the errors satisfy (11.1) and (11.2) with probability 1; then the assertion $X_t \to \theta$ will also hold with probability 1.

As an example, let us seek the best linear prediction $\eta^* = \langle \theta, \xi \rangle$ of a random variable η, based on a p-dimensional random vector ξ, i.e., let us seek the solution of the problem

$$E(\eta - \langle x, \xi \rangle)^2 = \min, \qquad x \in \mathbb{R}^p\,. \tag{11.4}$$

Assume that there is one realization of a stationary ergodic random sequence $((\xi_t, \eta_t), 1 \leqq t < +\infty)$ at our disposal, its stationary distribution being that of the pair (ξ, η). The solution θ of the problem (11.4) is equal to the solution of the system of normal equations

$$E(\xi\xi^{\mathrm{T}})x = E(\eta\xi)\,,$$

provided that the covariance matrix of the random vector ξ is positive definite and that $E\eta^2 < +\infty$. Denoting

$$-E(\xi\xi^{\mathrm{T}}) = A, \qquad -E(\eta\xi) = b\,,$$

$$-\xi\xi^{\mathrm{T}} = A(t), \qquad -\eta\xi = b(t)\,,$$

we are in a situation described by Györfi's theorem; the condition (11.2) reads now

$$\lim_{t\to\infty} \frac{1}{t} \sum_{k=1}^{t} |\xi_k|^4 < +\infty \tag{11.5}$$

with probability 1, as $\|\xi_t \xi_t^{\mathrm{T}}\| = |\xi_t|^2$. The condition (11.5) is, however, equivalent to $E|\xi|^4 < +\infty$, owing to the assumed ergodicity. The sequence X_t, $t \geqq 1$, defined by

$$X_{t+1} = X_t - \frac{1}{t}(\xi_t \xi_t^{\mathrm{T}} X_t - \eta_t \xi_t) \tag{11.6}$$

then converges to θ with probability 1. (11.6) can be rewritten as

$$X_{t+1} = X_t - \frac{1}{t}(\langle X_t, \xi_t\rangle - \eta_t)\xi_t\,,$$

so that we don't actually need matrix multiplication when using the iterative formula.

References

Anbar, D. (1973). On optimal estimation methods using stochastic approximation procedures. *Ann. Statist.* **1**, 1175–1184.

Dupač, V. (1957). On the Kiefer–Wolfowitz approximation method. (In Czech). *Časopis Pěst. matem.* **82**, 47–75.

Dupač, V. (1965). A dynamic stochastic approximation. *Ann. Math. Statist.* **36**, 1695–1702.

Dupač, V. and Fiala T. (1983). Stochastic approximation on a bounded convex set. In: *Proceedings of the Conference "Mathematical Learning Models-Theory and Algorithms"*, Bad Honnef, May 3–7, 1982. Lecture Notes in Statistics 20. Springer, Berlin, pp. 26–32.

Fabian, V. (1968). On asymptotic normality in stochastic approximation. *Ann. Math. Statist.* **39**, 1327–1332.

Györfi, L. (1980). Stochastic approximation from ergodic sample for linear regression. *Z. Wahrsch. Verw. Geb.* **54**, 47–55.

Kiefer, J. and Wolfowitz, J. (1952). Stochastic estimation of the maximum of a regression function. *Ann. Math. Statist.* **23**, 462–466.

Lai, T. L. and Robbins, H. (1979). Adaptive design and stochastic approximation. *Ann. Statist.* **7**, 1196–1221.

Ljung, L. (1977). Analysis of recursive stochastic algorithms. *IEEE Trans. Autom. Control* **22**, 551–575.

Nevel'son, M. B. and Has'minskii, R. Z. (1972/76). *Stochastičeskaja Approksimacija i Rekurrentnoe Ocenivanie.* Nauka, Moscow. English translation: *Stochastic Approximation and Recursive Estimation.* Translations of Math. Monographs, vol. 47, American Math. Society, Providence.

Nevel'son, M. B. and Has'minskii, R. Z. (1973). An adaptive Robbins–Monro procedure. *Autom. Rem. Contr.* **34**, 1594–1607.

Pantel, M. (1979). Adaptive Verfahren der stochastischen Approximation. Dept. of Mathem. Univ. of Essen. Thesis.

Révész, P. (1973). Robbins–Monro procedure in a Hilbert space and its application in the theory of learning processes I. *Stud. Sci. Math. Hungar.* **8**, 351–398.

Robbins, H. and Monro, S. (1951). A stochastic approximation method. *Ann. Math. Statist.* **22**, 400–407.

Ruppert, D. (1979). A new dynamic stochastic approximation procedure. *Ann. Statist.* **7**, 1179–1195.

Venter, J. H. (1967). An extension of the Robbins–Monro procedure. *Ann. Math. Statist.* **38**, 181–190.

Wasan, M. T. (1969). *Stochastic Approximation.* Cambridge Univ. Press.

P. R. Krishnaiah and P. K. Sen, eds., *Handbook of Statistics, Vol. 4*
© Elsevier Science Publishers (1984) 531–549

24

Density Estimation

P. Révész

1. Introduction

Let $X_1, X_2, \ldots$ be a sequence of i.i.d.r.v.'s with density function $f(x)$. Let $K_n(a, b)$ be the number of elements of the sample lying in the interval $[a, b)$, i.e.

$$K_n(a, b) = \sum_{i=1}^{n} I_{[a,b)}(X_i)$$

where

$$I_{[a,b)}(x) = \begin{cases} 1 & \text{if } a \leqslant x < b\,, \\ 0 & \text{otherwise}\,, \end{cases}$$

is the indicator function of the interval $[a, b)$. Then probability $P(a \leqslant X_1 < b) = \int_a^b f(t)\,\mathrm{d}t$ can be estimated by the relative frequency $n^{-1}K_n(a, b)$ (if n is big enough) and the value of a continuous $f(x)$ $(a \leqslant x < b)$ can be estimated by $(b-a)^{-1}\int_a^b f(t)\,\mathrm{d}t$ (if the interval is short enough), i.e.

$$f(x) \sim (b-a)^{-1}\int_a^b f(t)\,\mathrm{d}t \sim (n(b-a))^{-1}K_n(a, b) \quad (a \leqslant x < b)\,.$$

(Here the sign $\sim$ does not stand for any precise mathematical statement; it only indicates an intuitive near equality.)

Hence an empirical density function $f_n(x)$ in the interval $[a, b)$ can be defined as

$$f_n(x) = (n(b-a))^{-1}K_n(a, b) \quad (a \leqq x < b)\,.$$

Roughly speaking this is the idea behind most of the definitions of an empirical density function. When attempting a definition of this type one makes two errors. The first one occurs when the function $f(x)$ is estimated by its integral mean $(b-a)^{-1}\int_a^b f(t)\,\mathrm{d}t$ and the second error is made when the probability $\int_a^b f(t)\,\mathrm{d}t$ is estimated by the relative frequency $n^{-1}K_n(a, b)$. The first error is small if the interval is short while the second one is small if the interval $[a, b)$ contains a large enough number of elements of the sample, that is when $[a, b)$ is not too short. Hence one of the main problems is how to find a good compromise between these two opposing tendencies.

Now we give a very general definition of the empirical density function and later on we show how the most frequently used empirical densities can be obtained from this general concept.

General definition (Földes and Révész, 1974; Walter and Blum, 1979). Suppose that $f(x)$ is vanishing outside the interval $-\infty \leqslant C < D \leqslant +\infty$ and let $\Psi = \{\psi_k(x, y)\}$ be a sequence of Borel-measurable functions defined on the square $(A, B)^2$ where $(C, D) \subseteq (A, B)$. Then an empirical density function can be defined as follows:

$$f_n(x) = f_n^{(\phi)} = n^{-1} \sum_{k=1}^{n} \psi_n(x, X_k) = \int_C^D \psi_n(x, y)\, \mathrm{d}F_n(y) \tag{1}$$

where

$$F_n(y) = n^{-1} \sum_{i=1}^{\infty} I_{(-\infty, y)}(X_i)$$

is the empirical distribution function based on the sample $X_1, X_2, \ldots, X_n$.

Replacing ψ in this definition by some concrete sequence of functions one obtains different forms of the empirical density as special cases.

1. *Histogram*. Let

$$\cdots < x_{-1}(n) < x_0(n) < x_1(n) < \cdots$$

be a partition of the real line such that

$$x_{i+1}(n) - x_i(n) = h_n \quad (i = 0, \pm 1, \pm 2, \ldots)$$

where $\{h_n\}$ is a sequence of positive numbers tending to 0 as $n \to \infty$. Further let

$$\alpha_j^{(n)}(x) = \begin{cases} 1 & \text{if } x_{j-1}(n) \leqslant x < x_j(n)\,, \\ 0 & \text{otherwise,} \end{cases}$$

and

$$\psi_n(x, y) = h_n^{-1} \sum_{j=-\infty}^{+\infty} \alpha_j^{(n)}(x) \alpha_j^{(n)}(y)\,. \tag{2}$$

By this choice of Ψ the empirical density $f_n^{(\Psi)}$ becomes the well-known histogram. In fact

$$f_n^{(\Psi)}(x) = (nh_n)^{-1} K_n(x_i(n), x_{i+1}(n)) \quad \text{if } x_i(n) \leqslant x < x_{i+1}(n)\,. \tag{3}$$

2. *Smoothed histogram*. Using the notations introduced in the definition of the histogram we define a $\tilde{\psi}$ as follows. Let

$$\tilde{x}_i(n) = \frac{x_{i-1}(n) + x_i(n)}{2}$$

and

$$\tilde{\psi}_n(x, y) = \psi_n(\tilde{x}_i(n), y) + \frac{\psi_n(\tilde{x}_{i+1}(n), y) - \psi_n(\tilde{x}_i(n), y)}{h_n}(x - \tilde{x}_i^{(n)})$$

if $\tilde{x}_i(n) \leq x < \tilde{x}_{i+1}(n)$ where ψ_n is defined by (2). Then

$$f_n = f_n^{(\Psi)}(\tilde{x}_i^{(n)}) + \frac{f_n^{(\Psi)}(\tilde{x}_{i+1}^{(n)}) - f_n^{(\Psi)}(\tilde{x}_i^{(n)})}{h_n}(x - \tilde{x}_i^{(n)})$$

if $\tilde{x}_i^{(n)} \leq x < \tilde{x}_{i+1}^{(n)}$ where $f_n^{(\Psi)}$ is defined by (3).

3. *Kernel-type empirical density* (Rosenblatt 1956; Parzen 1962). Let $\lambda(x)$ be an arbitrary density function vanishing outside an interval $(A, B) \supset (C, D)$. Let $\{h_n\}$ be a sequence of positive numbers tending to 0 as $n \to \infty$ and define ψ by

$$\psi_n(x, y) = h_n^{-1}\lambda((x - y)h_n^{-1}).$$

Then

$$f_n^{(\Psi)} = f_n^{(\lambda)} = (nh_n^{-1})\sum_{i=1}^{n}\lambda((x - X_i)h_n^{-1}) = h_n^{-1}\int_{-\infty}^{+\infty}\lambda((x - y)h_n^{-1})\,\mathrm{d}F_n(y).$$

The density function λ of this definition is called kernel or window and h_n is called window-width.

4. *Orthogonal expansion* (Čencov 1962; van Ryzin 1966; Schwartz 1967). Let $\varphi = \{\varphi_k(x)\}_{k=1}^{\infty}$ be a complete orthonormal sequence defined on (A, B) and define ψ by

$$\psi_n(x, y) = \sum_{j=1}^{l_n}\varphi_j(x)\varphi_j(y)$$

where $\{l_n\}$ is a sequence of positive integers tending to $+\infty$. Then

$$f_n^{(\psi)} = f_n^{(\varphi)} = \sum_{k=1}^{l_n}\hat{c}_k\varphi_k(x) \quad (A < x < B)$$

where

$$\hat{c}_k = n^{-1}\sum_{j=1}^{n}\varphi_k(X_j).$$

Now we turn to some further widely used definitions of the empirical density function which are not special cases of the general definition (1).

5. *A sequential definition.* By a little modification of the kernel-type definition of the empirical density Wolverton and Wagner (1969) (see also Rejtö and Révész, 1973) introduced the following definition:

$$f_n(x) = n^{-1}\sum_{k=1}^{n}h_k^{-1}\lambda((x - X_k)h_k^{-1}) = \frac{n-1}{n}f_{n-1}(x) + (nh_n)^{-1}\lambda\left(\frac{x - X_n}{h_n}\right)$$

where λ is an arbitrary density function and h_n is a sequence of positive numbers tending to 0.

A great computational advantage of this definition is that f_n can be evaluated making use of f_{n-1} and X_n only (not using of sample elements $X_1, X_2, \ldots, X_{n-1}$ directly).

6. *Nearest neighbour type definitions.* Here we present two kinds of definitions.

6.a. *Nearest neighbour type histogram* (Révész, 1972). Define the random sequence $x_i(n)$ by

$$x_i(n) = X^*_{ib_n+1} \quad \left(i = 0, 1, 2, \ldots, \left[\frac{n-1}{b_n}\right]\right)$$

where $X_1^* < X_2^* < \cdots < X_n^*$ are the ordered statistics of the sample $X_1, X_2, \ldots, X_n$, and $\{b_n\}$ is an increasing sequence of positive integers tending to infinity. Then the empirical density function can be defined by

$$f_n(x) = \begin{cases} \dfrac{b_n}{n(X^*_{(i+1)b_n+1} - X^*_{ib_n+1})} & \text{if } X^*_{ib_n+1} \leq x < X^*_{(i+1)b_n+1}\,, \\ 0 & \text{if } x < X_1 \text{ or } x \geq X^*_{([(n-1)/b_n]+1)b_n}\,. \end{cases}$$

6.b. *Nearest neighbour idea in the kernel-type definition* (Loftsgarden and Quesenberry, 1965; Mack and Rosenblatt, 1979). Let λ be an arbitrary density function and $\{k_n\}$ be a sequence of positive integers tending to infinity. Then we give the following definition.

$$f_n(x) = (nR_n(x))^{-1} \sum_{k=1}^{n} \lambda\left(\frac{x - X_k}{R_n(x)}\right)$$

where $R_n(x)$ is the smallest positive number for which the interval $[x - R_n(x)/2, x + R_n(x)/2]$ contains k_n elements of the sample $X_1, X_2, \ldots, X_n$.

A lot of other definitions were introduced and investigated by different authors. Here we do not intend to give any further details. We refer only to the survey of Wertz (1978) and the bibliography of Wertz and Schneider (1979).

2. Formulation of the problems

In the Introduction we presented quite a few possible definitions of the empirical density function. In fact each definition gives a great freedom to the user in the choice of the parameters. In case of the histogram and smoothed histogram we have to specify the sequence $\{h_n\}$. Using the kernel-type empirical density or the sequential definition the density λ and the sequence $\{h_n\}$ should be specified. Having the orthogonal expansion the sequence $\{l_n\}$ and the orthonormal sequence $\{\varphi_n\}$ can be chosen relatively freely. In case of the nearest neighbour type histogram the sequence $\{b_n\}$ is to be chosen while in case of 6.b. λ and $\{k_n\}$ should be specified.

The practitioner certainly wants to know which of these definitions is the best for his or her purpose and how the parar· ,rs should be chosen. In other words we have to know in which case the empirical density is close to the real density functions. In order to answer this question we have to say precisely the meaning of 'close'.

The simplest way to characterize this closeness is the

1. *Pointwise distance*

$$\xi_n(x) = |f_n(x) - f(x)|$$

where x $(-\infty < x < +\infty)$ is any fixed real number and f_n is any empirical density.

A much more sophisticated and useful problem is the study of the global distances. Among the possible global distances we present here the three most popular ones.

2. *Uniform distance*

$$\sup_{-\infty<x<+\infty} \xi_n(x) = u_n .$$

3. *L^2-distance*

$$\int_{-\infty}^{+\infty} \xi_n^2(x)\,\mathrm{d}x = v_n .$$

4. *L^1-distance*

$$\int_{-\infty}^{+\infty} \xi_n(x)\,\mathrm{d}x = w_n .$$

Hence our problem is to minimize these distances. Since ξ_n, u_n, v_n, w_n are random variables we have several different possibilities to give a precise meaning of the expresion 'to minimize'. A possible way to minimize the

1. *Second moment.* Et_n^2 where t_n can be equal to any of ξ_n, u_n, v_n, w_n.

From a practical point of view one of the most important questions is to find

2. *Limit theorems.* Here the question is to find numerical sequences $\{a_n\}$ and $\{b_n\}$ $(b_n > 0)$ such that the sequence $\{b_n^{-1}(t_n - a_n)\}$ should have a nondegenerate limit distribution. (Here t_n is again any of ξ_n, u_n, v_n, w_n). In this case the goodness of the estimator f_n can be characterized by the largeness of b_n. If b_n is 'small' the estimator f_n is 'close' to f.

Largely from the theoretical point of view there is importance attached to

3. *Strong theorems.* Here the problem is to find numerical sequences $\{a_n\}$

and $\{b_n\}$ $(b_n > 0)$ such that one of the following relation should hold with probability one:

$$\lim_{n\to\infty} b_n^{-1}(t_n - a_n) = 0, \quad \limsup_{n\to\infty} b_n^{-1}(t_n - a_n) = 1, \quad \lim_{n\to\infty} b_n^{-1}(t_n - a_n) = 1.$$

Studying the above given characteristics of the different empirical density functions one realizes that in the limit (for large n) they (the densities obtained by the different definitions) behave very similarly. The parameters of these empirical densities have a great influence on the rate of convergence. However the different definitions produce nearly the same results if the parameters are chosen correctly. If we want to compare the different definitions for a fixed finite n (instead of studying the limit behaviour) then computer studies show some differences but these differences cannot be distinguished by theoretical methods.

3. The general method of studying the differences

In the Introduction we mentioned that in estimating the density function $f(x)$ $(a \leqq x < b)$ by $f_n(x) = (n(b-a))^{-1}K_n(a, b)$ one made two errors. The first error,

$$f(x) - (b-a)^{-1} \int_a^b f(t)\,\mathrm{d}t = f(x) - Ef_n(x),$$

is deterministic and depends on the continuity properties of f. This error is called bias. Its evaluation is based on approximation theory and it does not require any stochastic technique. The second error

$$f_n(x) - Ef_n(x)$$

is the random error. The study of its properties requires the rich apparatus of the limit theorems of probability theory. In general the 'distance' between f and f_n is estimated by the sum of the distances between f and Ef_n and between Ef_n and f_n.

As an example in case of the General Definition of the Introduction we present some details of the evaluation of the bias. Clearly in this case

$$Ef_n^{(\Psi)} = E\int_C^D \psi_n(x, y)\,\mathrm{d}F_n(y) = \int_C^D \psi_n(x, y)\,\mathrm{d}F(y).$$

Hence the bias is

$$f(x) - \int_C^D \psi_n(x, y)\,\mathrm{d}F(y) = f(x) - \int_C^D \psi_n(x, y) f(y)\,\mathrm{d}y.$$

This difference can be evaluated by the method of singular integrals (see e.g.,

Alexits, 1961). One can easily see that the concrete examples (given in the Introduction) of the General Definition produce asymptotically unbiased estimators under very mild conditions.

Applying this general method in some concrete cases the following results can be obtained:

1. *Histogram* (Révész, 1972). Assume that $f(x)$ is uniformly continuous on an interval $-\infty \leq A < B \leq +\infty$. Then for any $\varepsilon > 0$ we have

$$\sup_{A+\varepsilon \leq x \leq B-\varepsilon} |Ef_n(x) - f(x)| = O(h_n)$$

where $f_n = f_n^{(\psi)}$ is defined by (3). (Here we mean $-\infty + \varepsilon = -\infty$, $+\infty - \varepsilon = +\infty$.)

2. *Smoothed histogram*. Assume that $f(x)$ has a bounded second derivative on an interval $-\infty \leq A < B \leq +\infty$. Then for any $\varepsilon > 0$ we have

$$\sup_{A+\varepsilon \leq x \leq B-\varepsilon} |Ef_n(x) - f(x)| = O(h_n^2).$$

3. *Kernel-type empirical density*. Assume that λ is a bounded density function for which

(i) $$\lim_{|x|\to\infty} x^4 \lambda(x) = 0,$$

(ii) $$\int_{-\infty}^{+\infty} x\lambda(x)\,dx = 0.$$

Also assume that $f(x)$ has a bounded second derivative on an interval $-\infty \leq A < B \leq +\infty$. Then for any $\varepsilon > 0$ we have

$$\sup_{A+\varepsilon \leq x \leq B-\varepsilon} |Ef_n(x) - f(x)| = O(h_n^2).$$

4. *Orthogonal expansion*. Suppose that $f(x)$ is differentiable on $(0, \pi)$ and its derivative is of bounded variation. Define the orthonormal sequence φ on $(0, \pi)$ by $\varphi_k(x) = \sqrt{2/\pi}\cos kx$. Then for any $\varepsilon > 0$ we have

$$\sup_{\varepsilon \leq x \leq \pi-\varepsilon} |Ef_n - f| = O(l_n^{-3}).$$

Applying special but not very different methods one can see that the result formulated for the kernel-type empirical density remains true in the same form in the case of our sequential definition.

4. Negative results

As it is well-known the empirical distribution function is a uniformly consistent estimator of the distribution function with probability 1, i.e.

$$\lim_{n\to\infty} \sup_{-\infty<x<+\infty} |F_n(x) - F(x)| = 0 \quad \text{a.s.}$$

In the case of density estimation one cannot obtain such a complete result. This fact is shown by the following example; let the density function $f(x)$ be defined as follows:

$$f(x) = h_2(x) + h_3(x) + \cdots$$

where

$$h_i(x) = \begin{cases} 2^i(|x| - (2^i - 2^{-i})) & \text{if } 2^i - 2^{-i} \leq |x| \leq 2^i, \\ 2^i(2^i + 2^{-i} - |x|) & \text{if } 2^i \leq |x| \leq 2^i + 2^{-i}, \\ 0 & \text{otherwise}. \end{cases}$$

Note that for any n the sample $X_1, X_2, \ldots, X_n$ belongs to a finite interval and that for any adequate definition of the empirical density $f_n(x)$, $\lim_{|x|\to\infty} f_n(x) = 0$ a.s. (for any fixed n). Therefore we obtain

$$\lim_{n\to\infty} \sup_{-\infty<x<\infty} |f_n(x) - f(x)| = 1 \quad \text{a.s.}$$

The reader can see that in this example the reason for the nonconsistency is the bias, not the random error. Realizing that the reason for the bias is the irregular (nonsmooth) behaviour of the density function one cannot hope that without a condition about the smoothness of $f(x)$ one can find a uniformly consistent estimator.

A more surprising negative example was given by Devroye (1982). He has shown that for any sequence $\{f_n\}$ of empirical densities and for any sequence $\{a_n\}$ of positive numbers tending to 0, among the relatively smooth densities one can find an f such that

$$E\left(\int_{-\infty}^{+\infty} |f_n(x) - f(x)|\, \mathrm{d}x\right) > a_n \quad \text{infinitely often},$$

i.e. the rate of convergence can be arbitrarily slow.

Farrell (1967; for a more sophisticated version see Farrell, 1972) has shown that a sequence of estimators uniformly consistent at a fixed point cannot be obtained for a (relatively wide) class of densities. In fact let C_α be the set of densities $f(x)$ having a continuous derivative and satisfying the condition $\sup_{-\infty<x<\infty} f(x) \leq \alpha$ ($\alpha \geq 3$). Then for any sequence $\{f_n\}$ of empirical densities we have

$$\sup_{f\in C_\alpha} E(f_n(0) - f(0))^2 \geq \tfrac{1}{16}.$$

5. On the random error

After studying the bias $Ef_n - f$ we have to study the random error $f_n - Ef_n$. We list the known results in the most important cases.

1–2. *Histogram and smoothed histogram.* Let $f(x)$ be a bounded density function. Then

$$E(f_n - Ef_n)^2 = O\left(\frac{1}{nh_n}\right).$$

3. *Kernel-type empirical density.* Suppose that f and λ are bounded density functions. Then

$$E(f_n - Ef_n)^2 = O\left(\frac{1}{nh_n}\right).$$

4. *Orthogonal expansion.* Let $\varphi = \{\varphi_k\}$ be a uniformly bounded orthonormal system and assume that f is also bounded. Then

$$E(f_n - Ef_n)^2 = O(l_n n^{-1}).$$

6. The mean-square error

Combining the results of Sections 3 and 5 by the trivial formula

$$V_n = E(f_n - f)^2 = E(f_n - Ef_n)^2 + (Ef_n - f)^2$$

we obtain (under the conditions of Sections 3 and 5)

1. *Histogram*

$$V_n = E(f_n - f)^2 = O\left(h_n^2 + \frac{1}{nh_n}\right) \quad (A + \varepsilon \leqslant x \leqslant B - \varepsilon).$$

2–3. *Smoothed histogram and kernel-type empirical density*

$$V_n = E(f_n - f)^2 = O\left(h_n^4 + \frac{1}{nh_n}\right) \quad (A + \varepsilon \leqslant x \leqslant B - \varepsilon).$$

4. *Orthogonal expansion*

$$V_n = E(f_n - f)^2 = O\left(l_n^{-3} + \frac{l_n}{n}\right) \quad (A + \varepsilon \leqslant x \leqslant B - \varepsilon).$$

Having these formulas one can see that V_n takes its smallest possible value if

$h_n \sim n^{-1/3}$ in case 1,
$h_n \sim n^{-1/5}$ in cases 2 and 3,
$l_n \sim n^{1/4}$ in case 4

where the sign $\sim$ means that the corresponding terms are equal to each other asymptotically.

Making use of these choices we get

$$V_n = O(n^{-2/3}) \quad \text{in case } 1\,,$$
$$V_n = O(n^{-4/5}) \quad \text{in cases 2 and } 3\,,$$
$$V_n = O(n^{-3/4}) \quad \text{in case } 4\,.$$

REMARK. Assuming more regularity conditions of f one cannot get better rates in cases 1, 2 and 3. In case 4 assuming more differentiability conditions we get better rates, in fact the exponent of n can be made arbitrarily close to -1. In case 3 a better rate can be also achieved if we do not assume that λ is nonnegative and we do assume further regularity conditions on f. A slight disadvantage of 4 and 3 (when λ can be negative) is that the obtained empirical density f_n can be negative.

By a slight modification of the definition of the smoothed histogram one can also obtain a better rate. In fact using a polynomial approximation of higher order (instead of the linear approximation) and assuming more differentiability conditions of f the exponent of n in V_n can be arbitrary close to -1. This method is called spline approximation (cf. Boneva, Kendall and Stefanov, 1971).

7. The integral of the mean-square error

The results of Section 6 imply that the integral

$$I_n^2 = \int_C^D E(f_n - f)^2 \, dx$$

can be estimated by the estimators of V_n if the interval (C, D) is finite and by some further regularity conditions hold. The same results can be obtained for non-bounded support (C, D). However in some special cases more precise information can be obtained on I_n. Here we investigate the case of histogram and kernel-type density estimation.

1. *Histogram* (Freedman and Diaconis, 1981a). Assume that f and f' are absolutely continuous on (C, D) with $f \in L^2(C, D)$, $f' \in L^2(C, D)$, $f'' \in L^p(C, D)$ $(1 \leqq p \leqq 2)$. In the case when C and/or D are finite we also assume that for each n these endpoints belong to the partition $\{x_i(n)\}$ $(i = 0, \pm 1, \pm 2, \ldots)$. Introduce the notation:

$$\gamma = \int_A^B (f'(x))^2 \, dx > 0, \qquad \beta = \tfrac{1}{4} 6^{2/3} \gamma^{1/3}, \qquad \alpha = 6^{1/3} \gamma^{-1/3}\,.$$

Then, the cell width h_n which minimizes I_n is $\alpha n^{-1/3} + O(n^{-1/2})$, and at such h_n's, $I_n^2 = \beta n^{-2/3} + O(n^{-1})$.

The practical application of this result is hard because in most cases we do not have any information on the actual value of γ (and that of α). Hence we cannot find the best possible value of the cell width h_n. Studying this problem Freedman and Diaconis (1981a) say:

"In principle, γ can be estimated from the data, as in Woodroofe (1968). However, numerical computations suggest that the following simple, robust rule for choosing the cell width h_n often gives quite reasonable results.

Rule. Choose the cell width as twice the interquartile range of the data, divided by the cube root of the sample size."

2. *Smoothed histogram.* Freedman and Diaconis (1981b) also say that their method can be extended to this case. However they do not give any details.

3. *Kernel-type empirical density* (Rosenblatt, 1971). Assume that f is bounded, twice continuously differentiable on (C, D), $f \in L^2(C, D)$ and $f'' \in L^2(C, D)$. Also assume that λ is bounded and symmetric with $\int_A^B u^2\lambda(u)\,du < \infty$. Then

$$I_n^2 = \frac{1}{nh_n}\int_A^B \lambda^2(x)\,dx + \tfrac{1}{4}h_n^4 \int_C^D (f''(x))^2\,dx\left(\int_A^B u^2\lambda(u)\,du\right)^2 + o\left(\frac{1}{nh_n} + h_n^4\right).$$

This shows that the best possible choice in an asymptotic sense for the window width is $h_n = Kn^{-1/5}$ where

$$K = \frac{(\int_A^B \lambda^2(x)\,dx)^{1/5}}{(\int_C^D (f''(x))^2\,dx)^{1/2}(\int_A^B u^2\,\lambda(u)\,du)^{2/5}}.$$

With this choice of h_n we obtain

$$I_n^2 = \tfrac{5}{4}\left(\int_A^B \lambda^2(x)\,dx\right)^{4/5}\left(\int_A^B (f''(x))^2\,dx\left(\int_A^B u^2\lambda(u)\,du\right)^2\right)^{1/5} n^{-4/5} + o(n^{-4/5}).$$

In the practical application of this result we have again the problem that the value of K is unknown, unless we estimate the value of $\int_C^D (f''(x))^2\,dx$ from the sample.

Another problem which comes into the picture is the optimal choice of λ. Epanechnikov (1969) proved that we can minimize I_n^2 by the choice

$$\lambda(x) = \begin{cases} \tfrac{3}{4}5^{-1/2}\left(1 - \dfrac{x^2}{5}\right) & \text{if } |x| \leq 5^{1/2}, \\ 0 & \text{otherwise}. \end{cases}$$

8. Limit theorems

To prove a limit distribution theorem for $f_n(x) - f(x)$ (when x is fixed) is relatively easy. In fact we have the following result (Parzen, 1962). Let $f_n(x)$ be the kernel-type empirical density and assume that f is twice differentiable in a neighbourhood of a fixed x_0 (with bounded second derivative in the neighbourhood). About the kernel assume that it is bounded with $\lim_{|x|\to\infty} x^4\lambda(x)=0$. Finally, about the window-width we assume that $h_n = n^{-1/5-\varepsilon}$ $(0<\varepsilon<\frac{4}{5})$. Then if $f(x_0)>0$ we have

$$(nh_n)^{1/2}\Lambda^{-1}(f(x_0))^{-1/2}(f_n(x_0)-f(x_0))\xrightarrow{\mathscr{D}} N(0,1)$$

where $N(0, 1)$ is the standard normal law and $\Lambda^2=\int_A^B \lambda^2(x)\,\mathrm{d}x$. Similar results can be obtained using other definitions of the empirical density.

A much more complicated problem is to evaluate the limit distributions of the different global distances. In order to obtain such theorems in general we prove that the process $f_n(x)-f(x)$ (or its normalized version) can be approximated by a sequence of suitable Gaussian processes. To evaluate the limit distributions of the corresponding functionals of these Gaussian processes is somewhat easier.

At the first stage we give results for the L^2 distance $v_n = \int_C^D (f_n(x)-f(x))^2\,\mathrm{d}x$.

8.a. *Limit theorems of v_n* (Csörgő and Révész, 1981). Here we only mention some results regarding to a few concrete definitions of the Introduction.

1. *Histogram*. Suppose that f vanishes outside a finite interval $-\infty<C<D<+\infty$ and has a bounded derivative there. Then

$$\left(h_n\int_C^D f^2(x)\,\mathrm{d}x\right)^{-1/2}[nh_nv_n-1]\xrightarrow{\mathscr{D}} N(0,1)$$

provided that $n^{-1}h_n^{-3/2}\log^2 n\to 0$ and $h_n^{5/2}n\to 0$. For example if $h_n=n^{-\alpha}$ then it is assumed that $\frac{2}{5}<\alpha<\frac{2}{3}$.

3. *Kernel-type empirical density* (Bickel and Rosenblatt, 1973). Suppose that $f(x)$ vanishes outside a finite interval $-\infty<C<D<+\infty$ and has a bounded second derivative there. Also assume that λ vanishes outside a finite interval (A, B) with $\operatorname{var}_{x\in(A,B)}\lambda(x)\leqq \text{Const.}$ and $\int_A^B x\lambda(x)\,\mathrm{d}x=0$. Then

$$h_n^{-1/2}\sigma^{-1}[nh_nv_n-\Lambda^2]\xrightarrow{\mathscr{D}} N(0,1)$$

where

$$\sigma^2=\int_C^D f^2(x)\,\mathrm{d}x\int_A^B\left(\int_A^B\lambda(x+y)\lambda(x)\,\mathrm{d}x\right)^2\mathrm{d}y$$

and

$$\Lambda^2=\int_A^B\lambda^2(x)\,\mathrm{d}x$$

provided that $(\log^2 n)n^{-1}h_n^{-3/2}\to 0$ and $nh_n^{9/2}\to 0$. For example if $h_n=n^{-\alpha}$ then it is assumed $\frac{2}{9}<\alpha<\frac{2}{3}$.

4. *Orthogonal expansion* (Csörgő and Révész, 1981, p. 229). Suppose that f vanishes outside the interval $[0,\pi]$ and is absolutely continuous inside. Suppose

also that $f'(x)$ $(0<x<\pi)$ is of bounded variation. Then

$$l_n^{1/2}\left(\int_0^\pi f^2(x)\,\mathrm{d}x\right)^{-1}[nl_n^{-1}v_n-1]\xrightarrow{\mathscr{D}} N(0,1)$$

provided that $(\log^2 n)n^{-1}l_n^{3/2}\to 0$ and $nl_n^{-7/2}\to 0$. For example if $l_n = n^\beta$ then it is assumed that $\frac{2}{7}<\beta<\frac{2}{3}$.

8.b. *Limit theorems for the sup-distance.* Here we study only two concrete examples, the case of kernel-type empirical density and that of the nearest neighbour idea in the kernel-type definition. Both theorems contain a number of regularity conditions which are not too strong for practical applications but are too lengthy to be listed here. Hence we formulate the results without taking care of the formulation of these regularity conditions.

5. *Kernel-type empirical density* (Bickel and Rosenblatt, 1973). Let $h_n = n^{-\alpha}$ with $\frac{1}{5}<\alpha<\frac{1}{2}$. Then

$$\begin{aligned}\lim_{n\to\infty}\mathbb{P}\Big\{f_n(x)-\left(\frac{\Lambda f_n(x)}{nh_n}\right)^{1/2}&\left(\frac{z}{(2\alpha\log n)^{1/2}}+d_n\right)\leqslant f(x)\\ &\leqslant f_n(x)+\left(\frac{\Lambda f_n(x)}{nh_n}\right)^{1/2}\left(\frac{z}{(2\alpha\log n)^{1/2}}+d_n\right)\text{ for all }0\leqslant x\leqslant 1\Big\}\\ &=\exp(-2\mathrm{e}^{-z}),\end{aligned}$$

where $\Lambda = \int_A^B \lambda^2(x)\,\mathrm{d}x$,

$$d_n = (2\alpha\log n)^{1/2}+\frac{1}{(2\alpha\log n)^{1/2}}\left[\log\frac{1}{\pi}\left(\frac{K(\lambda)}{2}\right)^{1/2}\right]$$

and

$$K(\lambda)=\frac{1}{2\Lambda}\int_A^B(\lambda'(x))^2\,\mathrm{d}x\,.$$

6.a. *Nearest neighbour idea in the kernel-type definition* (Csörgő and Révész, 1982). Let $k_n=[n^\alpha]$ with $\frac{1}{2}<\alpha<\frac{4}{5}$ and define the r.v.'s

$$A_n = X^*_{[n^\beta]},\qquad B_n = X^*_{n-[n^\beta]}$$

where $X_1^*<X_2^*<\cdots<X_n^*$ are the order statistics of the sample $X_1, X_2, \ldots, X_n$ and

$$\max\left(\frac{2+\alpha}{3},\frac{8+5\alpha}{12}\right)<\beta<1\,.$$

Then

$$\lim_{n\to\infty}\mathbb{P}\left\{(2\log n/k_n)^{1/2}\left[\sup_{A_n\leqslant x\leqslant B_n}\Lambda^{-1}k_n^{1/2}\left|\frac{f_n(x)-f(x)}{f(x)}\right|-b\left(\frac{n}{k_n}\right)\right]<z\right\} =\exp(-2\mathrm{e}^{-z}),$$

where

$$b(u) = (2\log u)^{1/2} + (2\log u)^{-1/2}\log(2\pi)^{-1/2}\left[\frac{1}{2\pi\Lambda^2}\int_A^B \lambda(x)\lambda''(x)\,dx\right]^{1/2}.$$

REMARK. In the practical application of the results of this Section, the hardest problem is the choice of the parameters of h_n or l_n. In Section 5 we presented some ideas about the optimal choice of these parameters. Here we must realize that the theorems of the present section are not valid when the parameter of the estimator is chosen in the optimal way. Hence applying these results we can get a confidence band only whenever our estimator is certainly not the optimal one and we cannot evaluate the confidence band in the case when the estimator is as close to the unknown f as it is possible.

Beside the two special definitions studied in this Section, similar theorems can be obtained using other definitions. The most widely investigated case is that of the histogram (Smirnov, 1944; Tumanjan, 1955; Woodroofe, 1967; Révész, 1972). The best result in the histogram-case was obtained by Freedman and Diaconis (1981). They studied the limiting joint distribution of the location and size of the maximum deviation between the histogram and the underlying density.

9. Strong theorems

Under very mild regularity conditions one can prove that the estimators introduced in Section 1 are uniformly strongly consistent. That is, a number of theorems say that

$$\lim_{n\to\infty} u_n = \lim_{n\to\infty}\sup_x |f_n(x) - f(x)| = 0,$$

$$\lim_{n\to\infty} v_n = \lim_{n\to\infty}\int (f_n(x) - f(x))^2\,dx = 0,$$

$$\lim_{n\to\infty} w_n = \lim_{n\to\infty}\int |f_n(x) - f(x)|\,dx = 0$$

with probability 1. A much more interesting question is to characterize the rate of convergence in the above three statements. The results of Section 8.b suggest that order of magnitude of u_n is around $(nh_n)^{-1/2}$. Here we formulate only one result of this type what gives a very precise result under strong enough conditions (Révész, 1982; cf. also Silverman, 1976, 1978).

Let $f_n(x)$ be the kernel-type empirical density function. Assume that $f(x)$ is vanishing outside the interval [0, 1] and strictly positive and twice differentiable inside ($|f''| \leq C, f \geq \alpha > 0$ $(0 \leq x \leq 1)$). Also assume that λ is a twice differentiable, bounded, even function with $\lim_{x\to\infty} x^4\lambda(x) = 0$, $|\lambda''| \leq C$. Then for any $\varepsilon > 0$ we have

$$\lim_{n\to\infty}\left[(nh_n)^{1/2}\sup_{\varepsilon\leq x\leq 1-\varepsilon}\left|\frac{f_n(x)-f(x)}{f^{1/2}(x)}\right|-2\Lambda^2\log h_n^{-1}\right]=0\quad\text{a.s.}$$

(where $\Lambda^2=\int\lambda^2(x)\,\mathrm{d}x$) provided that

$$nh_n\nearrow\infty,\qquad h_n\searrow 0,\qquad \frac{\log^4 n}{nh_1\log h_n^{-1}}\to 0,\qquad \frac{nh_n^5}{\log h_n^{-1}}\to 0\,.$$

Studying the pointwise behaviour of f_n a law of iterated logarithm can be proved. Under some regularity conditions for several different definitions of the empirical density function P. Hall (1981; cf. also Wegman and Davies, 1979) proved that for any fixed x

$$\limsup_{n\to\infty}\left(\frac{nh}{2\log\log n}\right)^{1/2}|f_n(x)-f(x)|=C(f(x))^{1/2}$$

where the constant C depends on the underlying definition. For example using the kernel-type empirical density $C=(\int\lambda^2(x)\,\mathrm{d}x)^{1/2}$.

10. Multivariate density

Let $X_1, X_2, \ldots$ be a sequence of i.i.d.r.v.'s taking values from the d-dimensional Euclidean space $\mathbb{R}^d$ with density function $f(x)$ $(x\in\mathbb{R}^d)$. The different definitions of the empirical density functions f_n presented in the Introduction can be extended without any difficulty to the multivariate case. The results presented in the previous Sections can be mostly generalized without any essentially new idea. The hardest task in this generalization is the problem of limit theorems. This difficulty comes from two questions. The first one (and the harder one) is to approximate the process f_n-f by a suitable Gaussian process, the second one is to evaluate the distributions of the required functionals of the approximating Gaussian process. Especially the multivariate problem of sup-distance (except in the histogram case) looks hopeless.

Here we present only one theorem giving the limit distribution of the square integral $v_n=\int(f_n-f)^2$ in case of the kernel-type empirical density (Rosenblatt, 1975; Révész, 1976).

Let $\lambda(x)$ $(x\in\mathbb{R}^2)$ be a density function of bounded variation vanishing outside a bounded, convex, open set. Define the empirical density by

$$f_n(x)=(nh_n^2)^{-1}\sum_{i=1}^{n}\lambda\left(\frac{x-X_i}{h_n}\right).$$

Assume that $f(x)$ is vanishing outside a bounded, convex open set $\mathscr{L}$. Let

$$\mathscr{L}_\varepsilon = \{x: U_\varepsilon(x) \subset \mathscr{L}\}$$

where

$$U_\varepsilon(x) = \{u: |x-u| \leq \varepsilon\}$$

($|x-u|$ is the Euclidean distance). Assume further that there exist $\varepsilon > 0$ and $\delta > 0$ such that

$$f(x) \geq \delta \quad \text{if } x \in \mathscr{L}_\varepsilon$$

and the second partial derivatives of f are bounded on $\mathscr{L}$. Then

$$(h_n\sigma)^{-1}\left[nh_n^2 \int_{\mathscr{L}} (f_n(x) - f(x))^2 \, dx - \int \lambda^2(x) \, dx\right] \xrightarrow{\mathscr{D}} N(0, 1)$$

where

$$\sigma^2 = 2 \int \left(\int \lambda(x+y)\lambda(x) \, dx\right)^2 dy \int f^2(x) \, dx$$

provided that $n^{-1/4+\varepsilon} \leq h_n \leq n^{-1/5-\varepsilon}$ for some $\varepsilon > 0$.

Similar results can be obtained in the higher dimension case. Rosenblatt (1975) also proposed using this result to test the hypothesis of independence.

11. Goodness-of-fit (Statistical applications)

The results of Section 8 are useful in constructing confidence bands for an unknown density function and also in constructing goodness-of-fit tests for a completely specified f. To test a goodness-of-fit hypothesis most statisticians use the well-known Kolmogrov–Smirnov or the Cramér–von Mises statistics instead of statistics based on the limit distributions of the empirical density functions. The reason for this fact is the opinion that the tests based on the empirical distribution are stronger than the ones based on the empirical density. This opinion seems correct, although there is not any exact mathematical theory proving it. In spite of this fact there are several advantages of using empirical densities. Here we mention a few of them.

(1) The view of the graph of the density function says more to a practitioner than that of the distribution. Hence the graph of an empirical density with its confidence band aids statisticians in the study of the unknown distribution.

(2) There is no multivariate analogue of tests based on the empirical distribution. The most easily available (distribution-free) goodness-of-fit tests in the multivariate case are the ones based on the empirical densities.

(3) Most goodness-of-fit problems arising in practice do not usually specify F (or f) completely and, instead of one specific F, we are frequently given a whole parametric family of distribution functions $\{F(x, \vartheta); \vartheta \in \Theta \subseteq \mathbb{R}^p\}$. From a goodness-of-fit point of view the unknown parameter ϑ is a nuisance. Replacing ϑ in F by an estimation $\hat{\vartheta}_n$ of it the Kolmogorov–Smirnov and the Cramér–von

Mises limit theorems will not be valid anymore. However, most of the limit theorems of Section 8 remain valid after the mentioned replacing (Bickel and Rosenblatt, 1973). Hence the results of Section 8 can be used for composite goodness-of-fit problems.

(4) Some parameters, especially the mode of the distribution can be easily estimated via the empirical density. In fact the mode of the empirical density is a good estimator of the real mode (Parzen, 1962; Chernoff, 1964; Eddy, 1982).

12. Application in pattern recognition

Let $X_1^{(1)}, X_2^{(1)}, \ldots$ $(X_1^{(2)}, X_2^{(2)}, \ldots)$ be a sequence of i.i.d. random vectors in $\mathbb{R}^d$ having a common probability density $f_1(x)$ $(f_2(x))$. We then think of $\{X_i^{(1)}\}$ as a sequence of samples from Class 1 and $\{X_i^{(2)}\}$ as a sequence of samples from Class 2. For each i either we independently observe $X_i^{(1)}$ with probability q_1, or we observe $X_i^{(2)}$ with probability $q_2 = 1 - q_1$. If we define $\rho_i = 1$ when $X_i^{(1)}$ is observed and $\rho_i = 0$ when $X_i^{(2)}$ is observed, then $\rho_1, \rho_2, \ldots$ is a sequence of i.i.d.r.v.'s such that $\mathbb{P}(\rho_i = 1) = q_1$ and $\mathbb{P}(\rho_i = 0) = q_2$. We will assume that $X_i^{(1)}, X_j^{(2)}, \rho_k$ are independent for all i, j, k. If we define $X_i = \rho_i X_i^{(1)} + (1 - \rho_i) X_i^{(2)}$ then $\{X_i\}$ is the sequence of observed random vectors.

The problem is to find a decision procedure for classifying X_i, $i > n$ which is based on the sequence $(X_1, \rho_1), (X_2, \rho_2), \ldots, (X_n, \rho_n)$. When f_1, f_2, q_1, q_2 are known, the following procedure minimizes the probability of misclassification. Let

$$D_0(x) = q_1 f_1(x) - q_2 f_2(x) \quad \text{and} \quad \mathcal{D}_0 = \{x\colon D_0(x) \geqq 0\}\,.$$

Then decide: X_i is from Class 1 if $X_i \in \mathcal{D}_0$ and X_i is from Class 2 otherwise. In this case the probability of misclassification is

$$P_0 = q_1 \int_{\bar{\mathcal{D}}_0} f_1(x)\,\mathrm{d}x + q_2 \int_{\mathcal{D}_0} f_2(x)\,\mathrm{d}x\,.$$

In the case when q_1, f_1, f_2 are unknown, instead of $D_0(x)$ one can use the decision function

$$D^{(n)}(x) = \frac{\mu_n}{n} f_1^{(n)}(x) - \frac{n - \mu_n}{n} f_2^{(n)}(x)$$

where $\mu_n = \rho_1 + \rho_2 + \cdots + \rho_n$, $f_1^{(n)}$ (resp. $f_2^{(n)}$) is an empirical density function based on the nonzero elements of the sequence $\{\rho_i X(i)\}_{i=1}^n$ (resp. $\{(1-\rho_i)X_i^{(2)}\}_{i=1}^{(n)}$). Using the decision function $D^{(n)}$ instead of D_0 the probability of misclassification P_n will be larger than P_0. However, one can prove (Wolverton and Wagner, 1969; Rejtö and Révész, 1973) that the probability that P_n is much larger than P_0 is very small if n is big enough.

References

Alexits, G. (1961). *Convergence Problems of Orthogonal Series.* Akadémiai Kiadó, Budapest.

Bickel, P. J. and Rosenblatt, M. (1973). On some global measures of the deviations of density function estimates. *Ann. Statist.* **1**, 1071–1095.

Boneva, L., Kendall, D. and Stefanov, I. (1971). Spline transformations. Three new diagnostic aids for the statistical data-analyst. *J. Roy. Statist. Soc. Ser. B* **33**, 1–71.

Čencov, N. N. (1962). Evaluation of an unknown density from observations. *Soviet Math.* **3**, 1559–1569.

Chernoff, H. (1964). Estimation of the mode. *Ann. Inst. Statist. Math.* **16**, 31–41.

Csörgő, M. and Révész, P. (1981). *Strong Approximations in Probability and Statistics.* Akadémiai Kiadó, Budapest.

Csörgő, M. and Révész, P. (1982). An invariance principle for N. N. empirical density functions. (To appear.)

Devroye, L. P. and Wagner, T. J. (1977). The strong uniform consistency of nearest neighbor density estimates. *Ann. Statist.* **5**, 536–540.

Devroye, L. (1982). On arbitrary slow rates of global convergence in density estimation. (To appear.)

Eddy, W. F. (1982). The Asymptotic Distributions of Kernel Estimators of the Mode. *Z. Wahrsch. Verw. Geb.* **59**, 279–290.

Epanechnikov, V. A. (1969). Nonparametric estimates of a multivariate probability density. *Theor. Probability Appl.* **14**, 153–158.

Farrell, R. (1967). On the lack of a uniformly consistent sequence of estimators of a density function in certain cases. *Ann. Math. Statist.* **38**, 471–474.

Farrell, R. (1972). On the best obtainable asymptotic notes of convergence in estimation of a density function at a point. *Ann. Math. Statist.* **43**, 170–180.

Földes, A. and Révész, P. (1974). A general method for density estimation. *Studia Sci. Math. Hungar.* **9**, 81–92.

Freedman, D. and Diaconis, P. (1981a). On the Histogram as a Density Estimator: L_2 Theory. *Z. Wahrsch. Theorie* **57**, 453–476.

Freedman, D. and Diaconis, P. (1981b). On the Maximum Deviation between the Histogram and the Underlying Density. *Z.* Wahrsch. *Verw. Geb.* **58**, 139–167.

Hall, P. (1981). Laws of the Iterated Logarithm for Nonparametric Density Estimators. *Z. Wahrsch. Verw Geb.* **56**, 47–61.

Loftsgarden, D. O. and Quesenberry, C. P. (1965). A nonparametric estimate of a multivariate density function. *Ann. Math. Statist.* **36**, 1049–1051.

Mack, Y. P. and Rosenblatt, M. (1979). Multivariate k-nearest neighbor density estimates. *J. Multivariate Analysis* **9**, 1–15.

Parzen, E. (1962). On estimation of a probability density function and mode. *Ann. Math. Statist.* **33**, 1065–1076.

Rejtő, L. and Révész, P. (1973). Density estimation and pattern classification. *Problems of Control and Information Theory* **2**, 67–80.

Révész, P. (1972). On empirical density function. *Period. Math. Hungar.* **2**, 85–110.

Révész, P. (1976). On multivariate empirical density functions. *Sankhya Ser. A.* **38**, 212–220.

Révész, P. (1982). On the increments of Wiener and related processes. *Ann. Probability* **10**, 613–622.

Rosenblatt, M. (1956). Remarks on some nonparametric estimates of density function *Ann. Math. Statist.* **27**, 832–837.

Rosenblatt, M. (1971). Curve estimates. *Ann. Math. Statist.* **42**, 1815–1842.

Rosenblatt, M. (1975). A quadratic measure of deviation of two-dimensional density estimates and a test of independence. *Ann. Statist.* **3**, 1–14.

Schwartz, S. C. (1967). Estimation of probability density by an orthogonal series. *Ann. Math. Statist.* **38**, 1261–1265.

Silverman, B. W. (1976). On a Gaussian process related to multivariate probability density estimation. *Math. Proc. Cambridge Philos. Soc.* **80**, 185–199.

Silverman, B. W. (1978). Weak and strong uniform consistency of the kernel estimate and its derivatives. *Ann. Statist.* **6**, 177–184.

Smirnov, N. N. (1944). Approximate laws of distribution of random variables from empirical data (in Russian). *Uspehi Mat. Nauk* **10**, 179–206.

Tumansan, S. H. (1955). On the maximal deviation of the empirical density of a distribution (in Russian). *Nauču. Trudy Erevensk. Univ.* **48**, 3–48.

van Ryzin, I. (1966). Bayes risk consistency of classification procedures using density estimation. *Sankhya Ser. A.* **28**, 261–270.

Walter, G. and Blum, I. (1979). Probability density estimation using delta sequences. *Ann. Statist.* **7**, 328–340.

Wegman, E. J. and Davies, H. I. (1979). Remarks on some recursive estimators of a probability density. *Ann. Statist.* **7**, 316–317.

Wertz, W. (1978). Statistical density estimation – A survey.

Wertz, W. and Schneider, B. (1979). Statistical density estimation: a bibliography. *Int. Stat. Review* **47**, 155–175.

Wolverton, C. T. and Wagner, T. I. (1969). Asymptotically optimal discriminant functions for pattern classification. *IEEE Transactions on Information Theory* **15**, 258–266.

Woodroofe, M. (1966). On the maximum deviation of the sample density. *Ann. Math. Statist.* **38**, 475–481.

PART VI

LIFE TESTING AND RELIABILITY

P. R. Krishnaiah and P. K. Sen, eds., *Handbook of Statistics*, *Vol. 4*
© Elsevier Science Publishers (1984) 551–578 25

Censored Data

Asit P. Basu

1. Introduction and summary

In this chapter we present a survey of nonparametric methods for censored data. The literature in this field is quite extensive. In fact for almost every nonparametric method available for complete data there are some modifications available for censored data. Here we present a survey of some recent developments in the area.

In Section 2 we define the various types of censoring considered. We define Type I, Type II, arbitrarily censored and randomly censored data. The connection between random censoring and the theory of competing risks is pointed out. Section 3 considers the one sample problem of estimating the population distribution function. The Kaplan–Meier estimator for censored data and its properties are discussed in detail. Then some references for Bayesian and nonparametric Bayesian approaches for studying censored data are given. Sections 4 and 5 consider the two-sample and k-sample problems respectively. In Section 4 we primarily consider the modifications of the Wilcoxon–Mann–Whitney statistic and the Savage statistic. A unified approach to derive LMP rank tests is also given. Section 5 primarily considers modifications of the Kruskal–Wallis, Jonckheere and Friedman tests. Nonparametric regression is considered in Section 6 and problems of independence in Section 7. Finally, in Section 8, we consider a number of topics useful in problems of reliability and survival analysis. These include the classification problem, problem of accelerated life testing and isotonic regression. Some concepts useful in reliability are also given.

2. Types of censoring

Censored data arise naturally in a number of fields, particularly in problems of reliability and survival analysis. Let $X_1, X_2, \ldots, X_n$ be the life times of n

This research has been supported by the ONR Grant N00014-78-C-0655.

items put on test. Assume X_i's to be independent and identically distributed all having a common distribution function $F(x)$. F is usually assumed to be absolutely continuous. We may want to terminate the test before complete information on all n items is available for several reasons. The underlying test may be a destructive one so that items on test cannot be reused or, because of time and or cost constraint, we cannot afford to wait indefinitely for all items to fail.

In survival analysis, often we do not have complete control on the experiment. Patients may enter a hospital or clinic at arbitrary points of time for treatment and leave (before completion of treatment), or die from a cause different from the one under investigation. In many cases we may be forced to terminate the experiment at a given time (end of budget year, say) and try to develop appropriate inference procedure based on the available data. Depending on the nature of the test, we usually are led to the following types of censored data.

(a) *Type I censoring.* Here we assume n items are put on test and we terminate our test at a predetermined time T, so that complete information on the first k ordered observations

$$X_{(1)} < X_{(2)} < \cdots < X_{(k)}$$

is available. Here k is an integer valued random variable with

$$X_{(k)} < T < X_{(k+1)} .$$

Each of the remaining unobserved lifetimes is known to be greater than T.

(b) *Type II censoring.* As in Type I censoring, n items are simultaneously put on test and we terminate our test after a predetermined number (or fraction) of failures are obtained. In this case we have complete information on the first r ordered observations

$$X_{(1)} < X_{(2)} < \cdots < X_{(r)}$$

and the remaining observations are known to be greater than $X_{(r)}$. Here r (or r/n) is a fixed constant.

(c) *Arbitrary censoring and Random censoring.* In (a) and (b) it is assumed that all items are put on test simultaneously (or that the ordered observations are available). There may be situations, however, where all items cannot be tested simultaneously. For example each item may require certain time to set up the test and it may not be feasible to install all items simultaneously for test.

Similar situations may arise in clinical trials where patients enter a clinic for treatment at different points of time. A possible situation is illustrated in Figure 1.

Let X_i denote the survival time for patient i $(i = 1, 2, \ldots, n)$ where all the patients are being treated for the same disease, say cancer. The study begins at

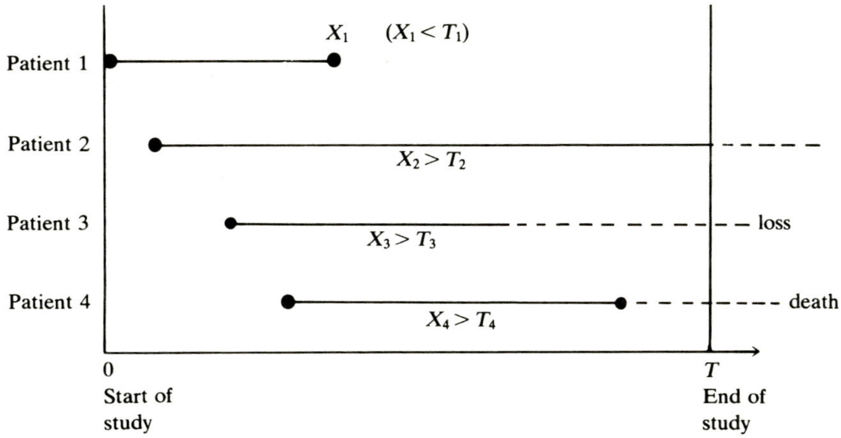

Fig. 1.

time 0 and ends at time T. Since the patients are entering the study at different points of time, the i-th patient can be observed for a given period, say T_i. Thus X_i is observed if $X_i \leq T_i$. Otherwise, it is censored. In the picture above $X_1 < T_1$. However $X_2 > T_2$ so that X_2 is not totally observed. For both patients 3 and 4, $X_i > T_i$ ($i = 3, 4$). X_3 is censored because the patient withdrew from the study (is lost so far as the current study is concerned), X_4 is censored because here the cause of death is different from cancer (say heart disease). However, no distinction is made between these two causes of censoring.

In general, the i-th patient is observed up to time T_i and we observe $\min(X_i, T_i)$. We also know whether $X_i \leq T_i$ (uncensored) or $X_i > T_i$ (censored).

In many studies T_i's are considered as given constants. In this case we say we have arbitrarily censored data. Note, Type I censoring is a special case of arbitrary censoring, where $T_i = T$ ($i = 1, 2, \ldots, n$).

Sometimes it is convenient to regard the T_i's to be random also, in that case we call the above censoring to be *random censoring*. In this case X and T are assumed independent. The assumption of randomness may be quite reasonable in many cases. Note, in this case we are essentially observing the minimum of two random variables X and T along with an indicator denoting which of the two is the minimum. In such a case we say we observe the *identified minimum* and use it to draw inference about the distribution function $F(x)$. Such a problem is called the problem of competing risks and many nonparametric problems can be handled using the techniques developed there. See Basu (1981), Basu and Ghosh (1980) and Basu and Klein (1982) for a bibliography of this related area.

(d) *Progressive censoring.* In the literature several versions of progressive censoring have been considered. Here the goal is to further reduce the testing time. In connection of parametric theory, the following version is usually used. See Klein and Basu (1981).

Let n items be put on test in a life testing experiment. At a predesignated censoring time T_i a fixed number $c_i \geq 0$ of items are removed from the test $(i = 1, 2, \ldots, m,\ T_1 < T_2 < \cdots < T_m)$. At time T_m either a fixed number c_m are removed from the test or testing is terminated with a random number c_m still functioning. Instead of fixed times $T_1, \ldots, T_m$ one could also carry out progressive censoring after specified number of failures. Thus after the first n_1 failures an additional number c_1 may be removed, and test be continued. After n_2 additional failures c_2 further items may be removed and so on such that at m steps we remove $c_1 + \cdots + c_m$ items from the test.

In connection of nonparametric methods, Chatterjee and Sen (1973) considered a special case where $m = r$, $c_i = 0$ and $n_i = 1$ $(i = 1, 2, \ldots, r)$.

In the next few sections we shall present a survey of some nonparametric methods available for the various type of censoring considered here.

3. One sample problem

Let $X_1, X_2, \ldots, X_n$ be a random sample of size n from a population with distribution function $F(x)$. $S(x) = 1 - F(x) = \bar{F}(x)$ is called the survival function or the reliability function. Let

$$F_n(x) = (\#\text{ of } X_i\text{'s} \leq x)/n \tag{3.1}$$

be the empirical distribution function. The problem of estimating the distribution based on the estimator $\hat{F}(x) = F_n(x)$ is well known. Properties of $F_n(x)$ have been studied extensively in standard test books. See for example, Wilks (1962) and Conover (1980). Durbin (1973) presents a survey of related distribution theory for tests based on the empirical distribution function. In particular, he considered the one-sample Kolmogorov–Smirnov and Cramér–von Mises tests.

3.1. *Kaplan–Meier (KM) estimator*

Considerable results for censored data have also been obtained based on the appropriate modification of $F_n(x)$. Substantial results have been obtained specially for random and arbitrarily censored data. Here the most important and widely studied result is the product limit (PL) estimator of the survival function $S(x)$ proposed by Kaplan and Meier (1958). Although this estimator can be defined for both arbitrarily censored and randomly censored data, for simplicity we shall consider the randomly censored model. In the case of randomly censored data we observe the following pair

$$(U_i, \delta_i), \quad i = 1, 2, \ldots, n,$$

where X_i's are observations on the variable under interest (lifetimes or survival times), T_i's are censoring variables, $U_i = \min(X_i, T_i)$,

and

$$\delta_i = \begin{cases} 1 & \text{if } U_i \text{ is uncensored } (X_i \leqslant T_i)\,, \\ 0 & \text{if } U_i \text{ is censored } (X_i > T_i)\,, \end{cases}$$

$(i = 1, 2, \ldots, n)$. Let $u_{(1)} \leqslant u_{(2)} \leqslant \cdots \leqslant u_{(r)}$ denote the ordered u's and $\delta_{[i]}$ be the value of δ_i associated with $u_{(i)}$. That is

$$\delta_{[i]} = \begin{cases} 1 & \text{if } u_{(i)} \text{ is uncensored}\,, \\ 0 & \text{otherwise}\,. \end{cases}$$

Let $u'_{(1)} < u'_{(2)} < \cdots < u'_{(r)}$ denote the distinct values of u_i, and let $\delta'_{[i]}$ be the value of δ_i associated with $U'_{(i)}$. The KM estimator is given by

$$\hat{S}(x) = \prod_{j : u'_{(j)} \leqslant x} \left(1 - \frac{d_j}{n_j}\right)^{\delta'_{[j]}} \tag{3.2}$$

where

n_j = number surviving (not failing) at time $u'_{(j)}$ (just before $u'_{(j)}$),
d_j = # died (failed at time $u'_{(j)}$).

Notice if there is no censoring, $\hat{S}(x)$ reduces to $S_n(x) = 1 - F_n(x)$, the empirical survival function.

The following example will illustrate the computation.

EXAMPLE. Ten items were put on life test. The data obtained is given below in a summary form.

Item no. (i)	1	2	3	4	5	6	7	8	9	10
u_i	4	90	55	15	20	35	9	9	45	100
δ_i	1	0	1	1	0	0	1	0	1	1

Here the ordered $u_{(i)}$'s with corresponding $\delta_{[i]}$'s are

$u_{(i)}$	4	9	9	15	20	35	45	55	90	100
$\delta_{[i]}$	1	1	0	1	0	0	1	1	0	1

The Kaplan–Meier PL estimator of the reliability function $S(x)$ is then given by

$$\hat{S}(0) = 1, \qquad \hat{S}(4) = \hat{S}(0)(1 - \tfrac{1}{10}) = 0.90\,,$$
$$\hat{S}(9) = \hat{S}(4)(1 - \tfrac{1}{9}) = 0.90 \times \tfrac{8}{9} = 0.80, \qquad \hat{S}(15) = \hat{S}(9)(1 - \tfrac{1}{7}) = 0.8 \times \tfrac{6}{7} = 0.69\,,$$
$$\hat{S}(45) = 0.69 \times \tfrac{3}{4} = 0.52, \qquad \hat{S}(55) = 0.52 \times \tfrac{2}{3} = 0.35, \qquad \hat{S}(100) = 0$$

Note, if the last item is censored the estimator would remain undefined for values >100. In that case, although it is bounded by $\hat{S}(100)$ and 0, one sometimes uses the convention that $\hat{S}(x) = 0$ for $x \geqslant u_{(n)}$.

Properties of Kaplan–Meier estimator (KME) have been studied extensively. KME has been shown by Kaplan and Meier (1958) and Johansen (1978) to be the generalized maximum likelihood estimator of F as defined by Kiefer and Wolfowitz (1956). Large sample properties of the KME have been obtained by many. Efron (1967), Breslow and Crowley (1974), Meier (1975) and Gill (1981) have established the weak convergence of the KME (regarded as a stochastic process). Breslow and Crowley give the following.

Let

$$H(t) = P(T \leqslant t) \quad \text{and} \quad T_F = \inf\{t \geqslant 0: F(t) = 1\},$$

$$\bar{G} = 1 - G = P(U > u) = (1 - H)(1 - F).$$

THEOREM 3.1. *If F and H are continuous and $T < T_G$ with $G(T) < 1$, then the process*

$$Z_N^*(t) = \sqrt{n}\{\hat{S}_n(t) - S(t)\}, \quad 0 \leqslant t \leqslant T,$$

converges weakly to a zero mean Gaussian process $Z^*(t)$ with covariance function

$$\operatorname{Cov}\{Z^*(s), Z^*(t)\} = C(s)S(s)S(t) \quad (s \leqslant t). \tag{3.3}$$

where

$$C(s) = \int_0^s \frac{\mathrm{d}F}{(1-F)^2(1-H)} \quad (s < T_F).$$

The theorem remains true also for the case of arbitrary censoring.

In particular, asymptotic variance of $\hat{S}(t)$, assuming no ties, can be estimated by

$$\widehat{\operatorname{Var}}(\hat{S}(t)) = \hat{S}^2(t) \sum_{i:u_{(i)} \leqslant t} \frac{\delta_{[i]}}{(n-i)(n-i+1)} \tag{3.4}$$

which is well known as the Greenwood's formula for the variance. Peterson (1977) has proven the strong consistency of KME whereas Földes and Rejtö (1981) have established the strong uniform consistency of the KME.

Asymptotic confidence bands for $S(t) = \bar{F}(t)$ have been obtained by Hall and Wellner (1980) for the randomly censored model based on KME.

Let

$$K(t) = C(t)\{1 + C(t)\}^{-1}, \qquad C_n(t) = n \sum_{(i:X_i < t)} (N-i)^{-1}(N-i+1)^{-1}\delta_i$$

and $\bar{K}_n(t) = \{1 + C_n(t)\}^{-1}$. Let B denote a Brownian bridge process on $[0, 1]$, and for $0 < a < 1$, $0 \leqslant \lambda < \infty$ set

$$G_a^+(\lambda) = \Pr\{\sup_{0\leq t\leq a} B(t) \leq \lambda\}$$
$$= 1 - \bar{\Phi}[\lambda\{a(1-a)\}^{-1/2}] - \bar{\Phi}[\lambda(1-2a)\{a(1-a)\}^{-1/2}],$$
$$G_a(\lambda) = \Pr\{\sup_{0\leq t\leq a} |B(t)| \leq \lambda\} \tag{3.5}$$
$$= 1 - 2\bar{\Phi}[\lambda\{a(1-2a)\}^{-1/2}] + 2\sum_{i=1}^{\infty}(-1)^k$$
$$\times e^{-2k^2\lambda^2}[\bar{\Phi}\{r(2k-d)\} - \bar{\Phi}\{r(2k+d)\}],$$

where Φ is the standard normal distribution, $\bar{\Phi} = 1 - \phi$, $r = \lambda\{(1-a)/a\}^{1/2}$ and $d = (1-a)^{-1}$.

THEOREM 3.2. *If $T < T_G$ so that $G(T) < 1$ and F and H are continuous, then, as $n \to \infty$,*

$$\Pr\{S(t) \leq \hat{S}_n(t) + \lambda D_n(t) \text{ for all } 0 \leq t \leq T\} \to G_a^+(\lambda) > G^+(\lambda),$$
$$\Pr\{\hat{S}_n(t) - \lambda D_n(t) \leq S(t) \leq \hat{S}_n(t) + \lambda D_n(t) \text{ for all } 0 \leq t \leq T\}$$
$$\to G_a(\lambda) > G(\lambda),$$

where

$$D_n(t) = n^{-1/2}\hat{S}_n(t)/\bar{K}_n(t), \quad a = K(T) = C(T)/(1 + C(T)) \quad \text{and}$$
$$G^+(\lambda) = G_1^+(\lambda).$$

Gillespie and Fisher (1979) and Nair (1981) also provide alternative large sample confidence bands based on the KME. However, unlike the Hall–Wellner bands their bands do not reduce to the standard bands in the uncensored case.

In case of Type I and Type II censored sample, confidence bands have also been obtained by Barr and Davidson (1973), Koziol and Byar (1975), Dufour and Maag (1978). One can, of course, carry out the corresponding goodness of fit test based on the modified Kolmogorov–Smirnov (KS) statistic. This has been discussed further in Koziol and Byar (1975).

Similar modifications of the modified Cramér–Von Mises statistics for Type II censored data have been considered by Petitt and Stephens (1978).

Dvoretzky, Kiefer and Wolfowitz (1956) have shown that the sample distribution function is asymptotically minimax over a wide class of loss functions. Wellner (1982) gives similar asymptotic minimax properties of the KME under a wide variety of loss functions.

There have been further extensions related to the Kaplan–Meier estimator. Turnbull (1974) has considered the problem when observations are doubly-censored. That is, he considers the case when some of the observations are censored on the right and some on the left. Meier (1975) has extended the theory to more general censoring mechanisms. Campbell (1981) and Langberg and Shaked (1982) have considered bivariate extensions of the survival function for censored data.

3.2. Bayesian nonparametric inference in reliability

Life test data often are grouped and presented in the form of a life table as used in actuarial methods. A Bayesian approach, which is weakly nonparametric, to analyze these data have been considered by Lochner and Basu (1969, 1972, 1976) using a Dirichlet prior. Ferguson (1973) has refined this process by proposing the Dirichlet process prior on the space of distribution functions. For randomly censored data Susarla and Van Ryzin (1976) obtain an estimate of the distribution function using the Dirichlet prior proposed by Ferguson. Ferguson and Phadia (1979) have further extended the results of Susarla and Van Ryzin to the class of prior distributions 'neutral to the right'. In reliability context Dykstra and Laud (1981) define a stochastic process whose sample paths are assumed to be increasing hazard rates or failure rates $r(x)$, where $r(x) = f(x)/\bar{F}(x)$. The posterior distribution of hazard rates is obtained and the Bayes estimates of $F(x)$ are obtained for both exact and censored data.

4. Two-sample problems

4.1. Linear rank statistic

Let $X_1, X_2, \ldots, X_m$ and $Y_1, Y_2, \ldots, Y_n$ be two independent random samples from two populations with distribution functions $F(x)$ and $G(y)$ respectively. Considerable work has been done to test the hypothesis

$$H_0:\quad F(x) = G(x) \text{ for all } x,$$

against the alternative that they are different. The two cases of primary interest are the location alternative

$$H_L:\quad G(x) = F(x - \theta),\ \theta \neq 0 \text{ (or } \theta > 0 \text{ or } \theta < 0)$$

and the scale alternative

$$H_S:\quad G(x) = F(x/\theta),\ \theta > 0 \text{ with } \theta \neq 1 \text{ (or } >1 \text{ or } <1).$$

Order all the $N + m + n$ observations and let

$$W_1 < W_2 < \cdots < W_N$$

denote the combined ordered sample.

Let

$$Z_i = \begin{cases} 1, & \text{if } W_i \text{ is from } F, \\ 0, & \text{otherwise}. \end{cases}$$

A general class of nonparametric tests, based on rank order statistics, to test H_0 agains H_L or H_S is of the type

$$T_N = \sum_{i=1}^{N} a_i Z_i , \tag{4.1}$$

where the a_i's are given numbers. Thus in the particular case of the Wilcoxon statistic $a_i = i$. See Gibbons (1971), Hájek and Sidák (1967) and Puri and Sen (1970) for detailed discussion of various properties of T_N.

Suppose, instead of a complete sample, we have a Type II (or Type I) censored sample. That is, we want to make a decision based on (at most) the first r of the combined set of N observations.

Initially, many of the tests proposed were of ad hoc nature. For example, in case of Type II censoring, Sobel (1957) proposed the statistic

$$V_r^{(N)} = \sum_{i=1}^{r} (nm_i - mn_i) = mn \sum_{i=1}^{r} \left(\frac{m_i}{m} - \frac{n_i}{n}\right), \tag{4.2}$$

where $m_i = \sum_{j=1}^{i} Z_j$ and $n_i = i - m_i$. The justification in this case is that under H_0, proportions of X and Y failures, among i failures ($i = 1, 2, \ldots, r$), should be same. Basu (1967a) showed $V_r^{(N)}$ is related to the Wilcoxon statistic and modified it to

$$V_r = V_r^{(N)} + \tfrac{1}{2}(N - r - 1)(nm_r - mn_r) . \tag{4.3}$$

It can be shown that

$$-\frac{V_r}{N^2} = \sum_{i=1}^{N} l_i Z_i ,$$

where

$$\begin{aligned} l_i &= (2i - N - 1)/2N, & 1 \leq i \leq r, \\ &= r/2N, & r + 1 \leq i \leq N. \end{aligned}$$

Thus V_r is a linear rank statistic and is related to the Wilcoxon rank test. Basu (1968) and Lochner (1968) considered a more general approach for modifying a statistic

$$T_N = \sum_{i=1}^{N} a_i Z_i$$

by

$$T_{Nr} = \sum_{1}^{r} a_i Z_i + \sum_{r+1}^{N} a_i E_0(Z_i \mid Z_1, \ldots, Z_r) \tag{4.4}$$

where $E_0(Z_i \mid Z_1, \ldots, Z_r)$ is the conditional expectation of Z_i $(i \geq r+1)$ given $(Z_1, Z_2, \ldots, Z_r)$. This is equivalent to the convention of replacing *unobserved* rank scores by average rank scores. Basu (1968) studied the generalized Savage statistic, based on exponential scores, and Lochner (1968) considered general linear rank statistics. From the results of Gastwirth (1965) it follows that, for the censored case, the test based on T_{Nr} (where $r/N \to p$ as $N \to \infty$, $0<p<1$) has maximum asymptotic efficiency relative to the T_N test. Asymptotic normality of the tests follows under general conditions.

For the Type I censoring, let the test be terminated at time T, and let $W_{r(T)} \leq T < W_{r(T)+1}$, so that $r(T)$ is a random variable. Under H_0

$$P(r(T)=r)=\binom{N}{r}[F(T)]^r[1-F(T)]^{N-r}, \quad r=0,1,\ldots,N. \tag{4.5}$$

In this case, appropriate modification of a rank test, as given for Type II censoring, is given by $T_{Nr(T)}$. Halperin (1960) has considered such a modification of the Wilcoxon–Mann–Whitney statistic. He has, however, expressed his statistic as a U-statistic using the Mann–Whitney form

$$\begin{aligned} U_{Nr(T)} = [\#\text{ of pairs } (X_i, Y_j) \text{ with } Y_j < X_i \text{ among} \\ (W_1, W_2, \ldots, W_{r(T)}] + \tfrac{1}{2}m - m_{r(T)})(n + n_{r(T)})\,. \end{aligned} \tag{4.6}$$

Rao, Savage and Sobel (1960) consider a number of other censoring schemes, each of which can be considered variations of Type I or Type II censoring. For each such variation one can consider suitable rank tests. Thus Young (1970) has considered the case when one continues observing until m_r observations from $F(x)$ are obtained. Halperin and Ware (1974) have considered a variation of the Wilcoxon statistic when one waits until m_rX's or n_rY's have been obtained. Spurrier (1981) has considered the case when m_rX's and n_rY's are obtained. Let the first r ordered observations $-\infty < W_1 < W_2 < \cdots < W_r < \infty$ be given and let the Z_i's be defined as before. A useful result to study rank statistics for censored data is,

$$\begin{aligned} &P(Z_1=z_1, Z_2=z_2, \ldots, Z_r=z_r) \\ &= \frac{m!n!}{(m-m_r)!(n-n_r)!} \int_{-\infty<w_1<w_2<\cdots<w_r<\infty} \cdots \int \prod_{i=1}^{r} [f^{Z_i}(w_i) g^{1-Z_i}(w_i)] \\ &\quad \times [1-F(w_r)]^{m-m_r}[1-G(w_r)]^{n-n_r}\, \mathrm{d}w_1 \ldots \mathrm{d}w_r\,. \end{aligned} \tag{4.7}$$

If the null hypothesis is true, above reduces to

$$P(Z_1=z_1, \ldots, Z_r=z_r) = \binom{N-r}{m-m_r} \Big/ \binom{N}{m}.$$

4.2. Locally most powerful rank tests

Basu, Ghosh and Sen (1983) have given a unified approach of obtaining locally most powerful (LMP) rank tests under censoring. Their method justifies why the heuristic methods used by Basu (1967a, 1968) and Lochner (1968) worked.

LEMMA 4.1 (Basu, Ghosh and Sen). *Let T_1, T_2 be two (possibly vector valued) statistics, and for each θ, let $L_\theta(T_1, T_2)$, $L_{1\theta}(T_1)$ and $L_{2\theta}(T_2)$ by respectively the joint and marginal densities of T_1 and T_2 (with respect to some σ-finite measure μ). Then*

$$\frac{L_{1\theta}(t_1)}{L_{1\theta_0}(t_1)} = E_{\theta_0}\left\{\frac{L_\theta(T_1, T_2)}{L_{\theta_0}(T_1, T_2)} \,\middle|\, T_1 = t_1\right\} \tag{4.8}$$

where the expectation at $(\theta = \theta_0)$ is with respect to T_2, given $T_1 = t_1$.

As a corollary we have

$$\psi_1(T_1) = E_{\theta_0}\{\psi(T_1, T_2) \mid T_1 = t_1\}\,, \tag{4.9}$$

where

$$\psi(T_1, T_2) = \frac{\partial}{\partial\theta} \log L_\theta((T_1, T_2) \mid \theta = \theta_0),$$

$$\psi_1(T_1) = \frac{\partial}{\partial\theta} \log L_{1\theta}(T_1 \mid \theta = \theta_0)$$

are the test statistics corresponding to the LMP tests based on (T_1, T_2) and T_1 alone. A variant of this is due to Sen (1981).

As an application, consider the two sample rank tests for H_0 against H_L based on a Type II censored sample, considered by Sobel (1957) and Basu (1967a, 1968). Here

$$T_1 = (Z_1, Z_2, \ldots, Z_r), \qquad T_2 = (Z_{r+1}, Z_{r+2}, \ldots, Z_N), \qquad Z = (T_1, T_2)\,.$$

It follows that the LMP rank test is given by

$$T_r^N = \sum_{i=1}^{r} a_i Z_i + \frac{n - n_r}{N - r}\left(\sum_{i=r+1}^{N} a_i\right) \tag{4.10}$$

where $T_N = \Sigma_{i=1}^N a_i Z_i$ is the LMP rank test when complete sample is available. Similarly one can obtain the LMP for scalar alternative.

For the Type I censored sample, considered by Halperin (1960), one similarly obtains the LMP test which is conditional given $r(T) = r$. Basu, Ghosh and Sen (1982) show that, under certain general conditions the two statistics corresponding to Type I and Type II censoring, are asymptotically equivalent in probability.

LMP rank tests for scalar alternative, rank tests for the cases considered by Young (1970), Halperin and Ware (1974) and Mehrotra, Johnson and Bhattacharyya (1977) can be similarly derived.

The Hodges–Lehmann estimator of the location and the ratio of scale parameters have been considered by Doksum (1967) and Spurrier (1981) respectively in some special cases for censored data.

4.3. *Arbitrary censored data and progressive censoring*

Considerable work has been done for the two-sample case when the data are arbitrarily (randomly) censored. See recent books by Johnson and Johnson (1980), Kalbfleisch and Prentice (1980), and Miller (1981) for some of these references. The two most studied tests are the modifications of the Mann–Whitney statistic and the Savage statistic (1956). We describe them below. Let

$$U_{ij} = U(X_i, Y_j) = \begin{cases} 1 & \text{if } Y_j < X_i \ (X_i \text{ censored or uncensored, } Y_i \text{ uncensored}), \\ -1 & \text{if } Y_j > X_i \ (Y_j \text{ censored or uncensored, } X_i \text{ uncensored}), \\ 0 & \text{otherwise}. \end{cases}$$

Then

$$U = \sum_{i=1}^{m} \sum_{j=1}^{n} U_{ij} \tag{4.11}$$

is a modification of the Mann–Whitney statistic for arbitrary censored data. U has been considered by Gehan (1965a), Gilbert (1962), Efron (1967) and Mantel (1967) among others. Halperin, Ware and Wu (1980) have considered the problem where censored data have been analyzed by distinguishing between administrative censoring, where participants in a clinical study participate at a different point of time, and censoring due to loss of follow-up. Wei (1980) considered a statistic similar to (4.11) for paired observations. Woolson and Lachenbruch (1980) have also considered rank tests for matched pairs.

The second test is the modification of the Savage statistic (1956) considered by Basu (1968) for Type II censored data. Crowley and Thomas (1975) show that log rank test can be defined by

$$T_{Nr} = \sum_{j=1}^{N} \left(\delta_j - \sum_{i=1}^{j} \frac{\delta_i}{N - i + 1} \right) (1 - Z_j). \tag{4.12}$$

T_{Nr} has been considered by Mantel (1966), Thomas (1975), Cox (1972) and Peto and Peto (1972). Note, if there is no censoring the above reduces to the Savage test based on exponential scores.

Chatterjee and Sen (1973) have considered the two-sample problem under progressive censoring, under the more general alternative hypothesis of regression. Here the alternative hypothesis is given by

$$H_R: \quad F_i(x) = F(x - \alpha - \beta d_i), \ i = 1, 2, \ldots, N, \ \beta \neq 0 \ \text{(or } \beta > 0 \text{ or } \beta < 0).$$

Here α is a nuisance parameter and $d_1, d_2, \ldots, d_N$ are known constants. Here test statistics are of the form

$$M^+_{N,r} = \max_{0 \leq k \leq r} T_{Nk} \quad \text{and} \quad M_{N,r} = \max_{0 \leq k \leq r} |T_{Nk}| \tag{4.13}$$

where T_{Nk} is of the form (4.4). See Sen (1981) for a review of two sample and other nonparametric tests under progressive censoring.

4.4. Tests based on empirical distribution function

Let $X_{(1)} < X_{(2)} < \cdots < X_{(m)}$ and $Y_{(1)} < Y_{(2)} < \cdots < Y_{(n)}$ be two independent ordered samples from two populations, $F(x)$ and $G(y)$ respectively. The two-sample Kolmogorov–Smirnov test for testing

$$H_0: \quad F(x) = G(x) \ \text{for all } x,$$

against the alternative that they are different, is given by

$$T_{m,n} = \sup_{-\infty < x < \infty} |F_m(x) - G_n(x)| \tag{4.14}$$

where $F_m(x)$ and $G_n(x)$ are the empirical distributions of the two independent samples. There have been several modifications of this test corresponding to the different types of censoring. Tsao (1954) and Conover (1967) studied the following censoring plans.

Plan 1. Continue testing until $X_{(r)}$ is observed. Here the statistic proposed is

$$T_r = \max_{x \leq X_{(r)}} |F_m(x) - G_n(x)|, \quad r \leq m. \tag{4.15}$$

Plan 2. Continue testing until $X_{(r)}$ and $Y_{(r)}$ are observed. Here the statistic proposed is

$$T'_r = \max_{x \leq \max(X_{(r)}, Y_{(r)})} |F_m(x) - G_n(x)|, \quad r \leq \min(m, n). \tag{4.16}$$

Plan 3. Continue testing until either $X_{(r)}$ or $Y_{(r)}$ is observed. Here the test proposed is

$$T''_r = \max_{x \leq \min(X_{(r)}, Y_{(r)})} |F_m(x) - G_n(x)|, \quad r \leq \min(m, n). \tag{4.17}$$

Tsao (1954) has provided tables of percentage points for T_r and T'_r. Conover (1967) has studied the properties of these.

Each of the above modifications were for Type II censoring. Similar modifications for Type I censoring has been considered by Koziol and Byar (1975). Suppose tests are continued until times T_1 (for X population) and T_2 (for Y population). Let $T = \min(T_1, T_2)$. Then the statistic considered is

$$D_T = \sup_{-\infty<x<T} |F_m(x) - G_n(x)| . \tag{4.18}$$

This statistic is related to the corresponding one sample problem considered by Barr and Davidson (1973), and Koziol and Byar (1975). Percentage points for the distribution of D_T is also given.

Pettitt and Stephens (1976) have considered similar modifications of the Cramér–Von Mises statistic for censored data.

Fleming, O'Fallon, O'Brien and Harrington (1980) have considered extension of the Kolmogorov–Smirnov statistic for arbitrarily censored data. We describe their statistic below.

The null hypothesis can be restated as

$$H_0: \quad S_1(t) = S_2(t) \text{ for all } t \tag{4.19}$$

where $S_1(t) = 1 - F(t)$ and $S_2(t) = 1 - G(t)$ are the respective survival functions. In context of survival studies the random variables of interest are the survival times (or times to death).

For convenience, let us charge the notation and let T_{ij} denote the j-th survival time from i-th population ($j = 1, 2, \ldots, N_i$, $i = 1, 2$). Let τ_{ij} denote the censoring time corresponding to the j-th observation in i-th sample, so that we observe

$$\min(T_{ij}, \tau_{ij}) .$$

Denote the survival distributions for the censoring variables by $C_1(\cdot)$ and $C_2(\cdot)$ respectively. The cumulative hazard function $\beta_i(t)$ corresponding to a survival function $S_i(t)$ is given by $\beta_i(t) = -\ln S_i(t)$. Similarly, define $\alpha_i(t) = -\ln H_i(t)$. Combine all the $N_1 + N_2$ observations. Let $\{t_j, j = 1, 2, \ldots, m\}$ be the distinct observations, $\{T_j, i = 1, 2, \ldots, d\}$ the distinct death (survival times) and $\{c_j, i = 1, 2, \ldots, c\}$ the distinct censoring times ($m \leqslant d + c$).

As in Section 3, let $N_i(t)$ be the number of individuals from sample i at risk at time t. Similarly, let $D_i(t)$ and $L_i(t)$ represent the number of deaths and censorships at time t from some i. Define $T = \max\{t_j\colon N_1(t_j)N_2(t_j) > 0\}$. The $\beta_i(t)$'s and $\alpha_i(t)$'s are estimated using Nelson's (1969) estimators and are given by

$$\hat{\beta}_i(t) = \sum_{T_j \leq t} \sum_{k=0}^{D_i(T_j)-1} \{N_i(T_j) - k\}^{-1},$$
$$\hat{\alpha}_i(t) = \sum_{\tau_j \leq t} \sum_{k=0}^{L_i(\tau_j)-1} \{N_i(\tau_j) - D_i(\tau_j) - k\}^{-1}. \tag{4.20}$$

Hence estimates of the survival functions are given by

$$\hat{S}_i(t) = \exp\{-\hat{\beta}_i(t)\} \quad \text{and} \quad \hat{H}_i(t) = \exp\{-\hat{a}_i(t)\}.$$

Then the statistics proposed to test above hypothesis are given by

$$D^{+}_{N_1,N_2}(T) = \sup_{0 \leq t \leq T} Y_{N_1,N_2}(t), \tag{4.21}$$

for the one-sided alternative and

$$D_{N_1,N_2}(T) = \sup_{0 \leq t \leq T} |Y_{N_1,N_2}(t)| \tag{4.22}$$

for the two-sided alternative. Here

$$Y_{N_1,N_2}(t) = \tfrac{1}{2}\{\hat{S}_1(t) + \hat{S}_2(t)\} \int_0^t \left\{\frac{N_1\hat{C}_1(S^-)N_2\hat{C}_2(S^-)}{N_1\hat{C}_1(S^-) + N_2\hat{C}_2(S^-)}\right\}^{1/2} \times I_{[N_1(s)N_2(s)>0]} d\{\hat{\beta}_1(s) - \hat{\beta}_2(s)\}. \tag{4.23}$$

Fleming et al. (1980) have given an algorithm for computing the above statistics and size and power of the test for small and moderate sample sizes have been computed using Monte–Carlo simulations.

The problem of estimating the ratio of scale parameters in the two-sample case with arbitrary right censorship has been considered by Padgett and Wei (1982). To this end, they considered a two-sample version of the Cramér–Von Mises statistic.

5. K-sample problem

K-sample extensions have been considered for a number of two-sample problems. Let X_{ij} ($j = 1, 2, \ldots, n_i$; $i = 1, 2, \ldots, k$) be k independent samples of sizes $n_1, n_2, \ldots, n_k$ respectively from k populations with continuous cumulative distribution functions $F_1, F_2, \ldots, F_k$ respectively. We assume that the F_i's belong to a family $\mathscr{F}$ of distribution functions indexed by a parameter θ.

We want to test the null hypothesis

$$H_0: \quad F_1(x) = F_2(x) = \cdots = F_k(x) \tag{5.1}$$

against the alternative hypothesis that they are different. Two important cases are

$$H_1:\quad F_i(x) = F(x, \theta_i),\ i = 1, 2, \ldots, k. \tag{5.2}$$

and

$$H_2:\quad F_1(x) < F_2(x) < \cdots < F_k(x) \quad \text{for all } x. \tag{5.3}$$

Basu (1967b) considers the following test statistics for the above hypothesis based on a Type II censored sample. That is, when the first r ordered observations among all the $N = \Sigma_1^k n_i$ observations are available. Let

$$\begin{aligned} Z_\alpha^{(i)} &= 1 \quad \text{if the } \alpha\text{-th ordered observation is from the } i\text{-th population},\\ &= 0 \quad \text{otherwise } (\alpha = 1, 2, \ldots, N), \end{aligned} \tag{5.4}$$

$$S_i = \sum_{\alpha=1}^{r} ((2\alpha - N - r - 1)/2N) Z_\alpha^{(i)} \quad (i = 1, 2, \ldots, k). \tag{5.5}$$

Then to test (5.1) against (5.2), the statistic $B_r^{(N)}$, which is a generalization of the Kruskal–Wallis statistic, is proposed. Here

$$B_r^{(N)} = \frac{12N^2(N-1)}{r[(r^2-1) + 3N(N-r)]} \sum_{i=1}^{k} \left(\frac{1}{n_i}\right)(S_i + rn_i/2N)^2. \tag{5.6}$$

Similarly, for testing (5.1) against the ordered alternative (5.3), the statistic proposed is

$$V(N, r) = \sum_{i<j} V_{ij}, \tag{5.7}$$

where

$$n_{ir} = \sum_{\alpha=1}^{r} Z_\alpha^{(i)},$$

and

$$V_{ij} = \sum_{\alpha=1}^{r} (n_j n_{i\alpha} - n_i n_{j\alpha}) + \tfrac{1}{2}(N - r - 1)(n_j n_{ir} - n_i n_{jr})$$
$$(i, j = 1, 2, \ldots, k;\ i < j).$$

Note $V(n, r)$ is a generalization of the well known Jonckheere statistic when complete samples are available. Properties of $B_r^{(N)}$ and $V(N, r)$ have been studied by Basu (1967b). Basu (1968) also considered a k-sample extension of the two sample Savage statistic defined in Section 4.

Similarly, for arbitrarily censored data, Breslow (1970) considers a generalization of the k-sample Kruskal–Wallis test, Patel and Hoel (1973) consider a generalization of the Jonckheere test, Patel (1975) considers an extension of the Friedman test, and Crowley and Thomas (1975) consider a generalization of the Savage test.

Let Z_{ij} be the censoring variable corresponding to X_{ij} so that we observe $Y_{ij} = \min(X_{ij}, Z_{ij})$ along with the indicator variable

$$\delta_{ij} = \begin{cases} 1 & \text{if } Y_{ij} = X_{ij}\,, \\ 0 & \text{if } Y_{ij} = Z_{ij}\,. \end{cases}$$

Let I_i be the common distribution of the censoring variables Z_{ij}, $j = 1, 2, \ldots, n_i$.

Let the score function ψ for comparing between two observations $Y_{i\alpha}$ and $Y_{j\beta}$ be

$$\psi(Y_{i\alpha}, \delta_{i\alpha}, Y_{j\beta}, \delta_{j\beta}) = \begin{cases} -1, & Y_{i\alpha} < Y_{j\beta},\ \delta_{i\alpha} = 1\,, \\ +1, & Y_{j\beta} < Y_{i\alpha},\ \delta_{j\beta} = 1\,, \\ 0, & \text{otherwise}\,. \end{cases} \tag{5.8}$$

Let

$$W_i = \sum_{\alpha=1}^{n_i} \sum_{j=1}^{k} \sum_{\beta=1}^{n_j} \psi(Y_{i\alpha}, \delta_{i\alpha}, Y_{j\beta}, \delta_{j\beta}), \qquad \boldsymbol{W} = (W_1, W_2, \ldots, W_k)', \tag{5.9}$$

($\boldsymbol{\Sigma}$ = Asymptotic covariance matrix of $N^{-3/2}\boldsymbol{W}$),

$$\boldsymbol{S}_i = N^{-3/2}\boldsymbol{\xi}_i'\boldsymbol{W} \text{ and,} \qquad \hat{\boldsymbol{\Sigma}} = \text{estimator of } \boldsymbol{\Sigma}', \qquad S = \sum_{i=1}^{k-1} S_i^2\,, \tag{5.10}$$

where $\boldsymbol{\xi}_i = (\xi_{i1}, \xi_{i2}, \ldots, \xi_{ik})$ $(i = 1, 2, \ldots, k-1)$ is such that

$$\boldsymbol{\xi}_i'\hat{\boldsymbol{\Sigma}}\boldsymbol{\xi}_j = \begin{cases} 1 & \text{if } i = j\,, \\ 0 & \text{if } i \neq j\,. \end{cases}$$

Breslow (1970) proposed the statistic S for testing (5.1) against (5.2) under the assumption that all censoring variables have the same distribution and studied its properties. Patel and Hoel (1973) and Patel (1975) proposed similar statistics for generalization of the Jonckheere test and the Friedman test respectively. Woolson and Lachenbruch (1981) have also considered extension of the Friedman test.

6. Regression

Several regression techniques are currently available for analyzing censored data. We describe them below.

6.1. Cox's model

Cox's (1972) approach is based on the proportional hazard model. Let $F(t; \boldsymbol{x})$ be the underlying distribution function and $f(t; \boldsymbol{x})$ the corresponding density function for the survival time T when the independent variables are $\boldsymbol{x}$. Then

$$\lambda(t; \boldsymbol{x}) = f(t; \boldsymbol{x})/(1 - F(t, \boldsymbol{x}))$$

denotes the hazard rate. Cox's model is then given by

$$\lambda(t; \boldsymbol{x}) = \lambda_0(t)\, \mathrm{e}^{\boldsymbol{x\beta}} \tag{6.1}$$

where $\boldsymbol{\beta}$ is the vector of regression coefficients and $\lambda_0(t)$ is the hazard rate when $\boldsymbol{x} = \boldsymbol{0}$. The above model is equivalent to the assumption of Lehmann alternatives

$$1 - F(t; \boldsymbol{x}) = [1 - F(t; \boldsymbol{0})]^{\exp(\boldsymbol{x\beta})}. \tag{6.2}$$

Since $\lambda_0(t)$ is unknown, Cox (1972, 1975) suggests a partial likelihood approach. Let $R(t)$ denote the risk set, the units that are still in test at time t- and let τ_i denote the censoring variable, so that we observe $Y_i = \min(T_i, \tau_i)$ and $\delta_i = I(T_i \leqslant \tau_i)$ and $\boldsymbol{x}'_i = (x_{i1}, \ldots, x_{ip})$, where I is the indicator function ($i = 1, 2, \ldots, n$). Let $y_{(1)} < y_{(2)} < \cdots < y_{(n)}$ be the ordered observations, assumed all distinct. Then the partial likelihood is

$$L(\boldsymbol{\beta}) = \prod_{i=1}^{n} \left[\frac{\mathrm{e}^{\boldsymbol{x}_i\boldsymbol{\beta}}}{\sum_{j\in R(t_{(i)})} \mathrm{e}^{\boldsymbol{x}_j\boldsymbol{\beta}}}\right]^{\delta_{[i]}}. \tag{6.3}$$

Here $\delta_{[i]} = 1$ if $y_{(i)}$ is censored, and 0, otherwise. The estimate $\hat{\boldsymbol{\beta}}$ of β is obtained by solving the set of equations

$$\frac{\partial L(\boldsymbol{\beta})}{\partial \boldsymbol{\beta}} = \boldsymbol{0}. \tag{6.4}$$

In case there are ties, (6.3) is appropriately modified.

Bailey (1979) and Tsiatis (1981) showed that $\hat{\boldsymbol{\beta}}$ is asymptotically normal. Efron (1977) and Oakes (1977) have shown that the Cox estimator is nearly fully efficient.

6.2. Standard linear model

Unlike Cox and Miller (1976), Buckley and James (1979) and Koul, Susarla and Van Ryzin (1981) have considered the usual linear model with

$$E(T \mid \boldsymbol{x}) = \alpha + \boldsymbol{x\beta}, \tag{6.5}$$

where α is the intercept and $\boldsymbol{\beta}$ is the vector of regression coefficients for the independent variables $\boldsymbol{x}$.

Miller (1976) suggests minimizing the sum of squares

$$n \int \mathrm{e}^2\, \mathrm{d}\hat{F}(e; a, \boldsymbol{b}), \tag{6.6}$$

with respect to a and $\boldsymbol{b}$ where $\hat{F}(e; a, b)$ is the Kaplan–Meier product limit estimator based on δ_i and $\hat{e}_i = \hat{e}_i(a, \boldsymbol{b}) = y_i - a - \boldsymbol{x}_i\boldsymbol{b}$ $(i = 1, 2, \ldots, n)$. Because of the difficulty in locating the infimum of (6.6), Miller proposes an iterative method to calculate the estimators. Let

$$\boldsymbol{y} = (y_1, \ldots, y_n)', \qquad \boldsymbol{x} = (x_{ij}), \quad i = 1, 2, \ldots, n, \; j = 1, 2, \ldots, p,$$
$$\bar{\boldsymbol{X}}^W = \left(\left(\sum_{i=1}^n w_i(\hat{\boldsymbol{\beta}}_k)x_{ij}\right)\right),$$

and

$$\boldsymbol{W}(\hat{\beta}_k) = \text{diagonal matrix } ((w_i(\hat{\boldsymbol{\beta}}_k))).$$

Then

$$\hat{\boldsymbol{\beta}}_{k+1} = [(\boldsymbol{X} - \bar{\boldsymbol{X}}^W)\boldsymbol{W}(\hat{\boldsymbol{\beta}}_k)(\boldsymbol{X} - \bar{\boldsymbol{X}}^W)]^{-1}(\boldsymbol{X} - \bar{\boldsymbol{X}}^W)\boldsymbol{W}(\hat{\boldsymbol{\beta}}_k)\boldsymbol{y}. \tag{6.7}$$

The limit of the sequence $\hat{\boldsymbol{\beta}}_k$, $k = 0, 1, \ldots,$ is the estimator $\hat{\boldsymbol{\beta}}$.

Buckley and James (1979), unlike Miller who minimizes (6.6), modifies the normal equations by replacing each censored observation by computing its conditional expectation computed from the Kaplan–Meier estimator for the residuals. Since

$$E(\delta_i Y_i + (1 - \delta_i)E(T_i \mid T_i > Y_i) \mid x_i) = \boldsymbol{\alpha} + \boldsymbol{x}_i\boldsymbol{\beta}, \tag{6.8}$$

y_i is replaced by

$$\hat{y}_i = \delta_i y_i + (1 - \delta_i)\hat{E}(T_i \mid T_i > y_i)$$

and then the usual least squares normal equations are solved. Buckley–James estimators are also computed iteratively.

Koul, Susarla and Van Ryzin (1981) propose an estimator based on the following relationship. Assume that the censoring distributions $G(t, \boldsymbol{x})$ are independent of $\boldsymbol{x}_i$. That is let $G(t, \boldsymbol{x}_i) = G(t)$ for all $\boldsymbol{x}_i$. Then

$$E(\delta_i Y_i(1 - G(Y_i))^{-1} \mid \boldsymbol{X}_i) = \boldsymbol{\alpha} + \boldsymbol{X}_i\boldsymbol{\beta}. \tag{6.9}$$

In their method $G(t)$ is estimated and the dependent variable is replaced by

$$\hat{y}_i = \delta_i y_i(1 - \hat{G}(y_i))^{-1}$$

and then the usual least square estimator normal equations are solved.

Unlike the Miller and Buckley–James estimators, Koul–Susarla–van Ryzin estimator is explicitly defined and does not require iterative methods – which may not always converge. Large sample methods are established more rigorously here. However, the assumption on the censoring distribution is not considered realistic in many cases.

Miller and Halpern (1981) have compared the above four regression methods.

7. Independence

Bivariate data in many situations may be censored, in one or both variables. The nonparametric statistics, usually considered for estimating and testing for independence, when complete samples are available, are the Spearman's rank correlation coefficient ρ and Kendall's rank correlation coefficient τ.

Let (X_i, Y_i), $i = 1, 2, \ldots, n$ be n independent and identically distributed random vectors with continuous bivariate distribution function $H(x, y; \theta)$ having marginal df's $F(x)$ and $G(y)$, respectively. Let θ be the measure of dependence so that under null hypothesis H_0: $\theta = 0$, we have $H(x, y; 0) = F(x)G(y)$. Let $h(x, y; \theta)$ be the density function corresponding to $H(x, y; \theta)$.

Consider the case for Type II censoring for each coordinate. Let R_i (Q_j) be the rank of X_i (Y_j) among $X_1, \ldots, X_n$ ($Y_1, \ldots, Y_n$), for $i = 1, 2, \ldots, r$ ($j = 1, 2, \ldots, s$). Then Shirahata (1975) and Basu, Ghosh and Sen (1982) show that under certain conditions, the asymptotically locally most powerful rank test for testing H_0: $\theta = 0$, against the alternative $\theta > 0$, using above censored data is based on the statistic

$$T = \sum_{i=1}^{n} a_{rs}(R_i, Q_i) \tag{7.1}$$

where

$$a_n(i, j) = n^2 \binom{n-1}{i-1}\binom{n-1}{j-1} \int_{-\infty}^{\infty}\int_{-\infty}^{\infty} \left\{ \frac{\partial_h(x, y; \theta)}{\partial\theta}\bigg|_{\theta=0} \right\}$$
$$[F(x)]^{i-1}[1 - F(x)]^{n-i}[G(y)]^{j-1}[1 - G(y)]^{n-j} f(x) g(y)\, dx\, dy\,, \tag{7.2}$$

and

$$a_{rs}(i, j) = \begin{cases} a_n(i, j) & \text{if } i \leq r,\ j \leq s\,, \\ \dfrac{1}{n-r} \displaystyle\sum_{l=r+1}^{n} a_n(l, j) & \text{if } i > r,\ j \leq s\,, \\ \dfrac{1}{n-s} \displaystyle\sum_{l=s+1}^{n} a_n(i, l) & \text{if } i \leq r,\ j > s\,, \\ \dfrac{1}{(n-r)(n-s)} \displaystyle\sum_{l=r+1}^{n} \sum_{l'=s+1}^{n} a_n(l, l') & \text{if } i > r,\ j > s\,. \end{cases} \tag{7.3}$$

The above test, based on scores, generalizes the Spearman rank correlation coefficient.

For arbitrary censored samples, generalization of Kendall's τ and Spearman's ρ have been considered by several. Brown, Hollander and Korwar (1974) discuss modifications of Kendall's τ and apply them to data consisting of survival time of Stanford heart transplant patients. Latta (1977b) modifies Spearman's ρ and applies it to the same data for a comparison to the statistics of Brown et al. (1974). Weier and Basu (1980) consider several other modifications where each censored observation is replaced by its conditional expectation. Thus if X_i is censored at time u_i, it is replaced by $E_{\hat{P}}(X_i \mid X_i > u_i)$ where $E_{\hat{P}}$ denotes the expected value under the Kaplan–Meier estimate.

Similarly expectation under the assumption of normal and exponential distributions are also considered. Computer simulation is then used to study power properties of various modifications. Under alternative hypothesis (X, Y) is assumed to have bivariate normal or the bivariate exponential proposed by Block and Basu (1974). Other works on independence have been considered by O'Brien (1978), Oakes (1982) and Cuzick (1982).

8. Other NP methods

In this section we consider a number of other ad hoc nonparametric methods for censored data. Most of these have been motivated by problems in reliability and survival analysis.

8.1. Classification problem

Basu (1977) has considered the following classification problem based on a Type II censored data. Let π_i, $i = 0, 1, 2$ denote three populations with distribution functions $F_i(x)$. It is known that $\pi_0 = \pi_i$ $(i = 1, 2)$ for exactly one i. Let X_{ij}, $j = 1, 2, \ldots, n_i$ denote a random sample of size n_i from $F_i(x)$ $(i = 0, 1, 2)$. Let the $N = n_0 + n_1 + n_2$ observations be combined and only the first r ordered observations among the combined sample of size N be available. To classify π_0 to one of the two populations π_1 and π_2 based on the first r ordered observations, let

$$Z_\alpha^{(i)} = \begin{cases} 1 & \text{if the } \alpha\text{-th ordered observation among the combined} \\ & N \text{ observations is from the } i\text{-th population,} \\ 0 & \text{otherwise}, \end{cases}$$

$$n_i T_i = \sum_{\alpha=1}^{r} \frac{2\alpha - N - 1}{2N} Z_\alpha^{(i)} + \frac{r(n_i - n_{ir})}{2N}, \qquad n_{ir} = \sum_{\alpha=1}^{r} Z_\alpha^{(i)}, \tag{8.1}$$

$$V = V(T) = 2T_0 - T_1 - T_2 . \tag{8.2}$$

Basu proposes the following rule.

$$R: \quad \text{classify } F_0 = F_1 \text{ if } V(T) > 0, \qquad F_0 = F_2 \text{ if } V(T) < 0 . \tag{8.3}$$

The above classification rule has been proven to be consistent and the asymptotic relative efficiency of this rule with respect to a classification rule for the normal distribution is computed.

8.2. Nonparametric concepts and methods in reliability

Nonparametric methods are used extensively in reliability theory. A comprehensive survey has been given by Hollander and Proschan (1982). Definitions of IFR, IFRA, NBU, NBUE classes of life distributions and their duals have been given in Hollander and Proschan. Rolski (1975) introduced a new class of life distributions called harmonic new better than used in expectation (HNBUE) with dual (HNWUE). We define them first. Let X be a nonnegative random variable with distribution function F and survival function $\bar{F} = 1 - F$ with finite mean $\mu = \int_0^\infty \bar{F}(x)\,dx$. F and $\bar{F}$ are said to be harmonic new better than used in expectation (HNBUE) if

$$\int_t^\infty \bar{F}(x)\,dx \leq \mu \exp(-t/\mu) \quad \text{for } t \geq 0 . \tag{8.4}$$

If the reversed inequality is true, F and $\bar{F}$ are said to have harmonic new worse than used in expectation (HNWUE). Klefsjö (1982) has studied different properties of the HNBUE and HNWUE classes. The following chain of implication holds between the six classes.

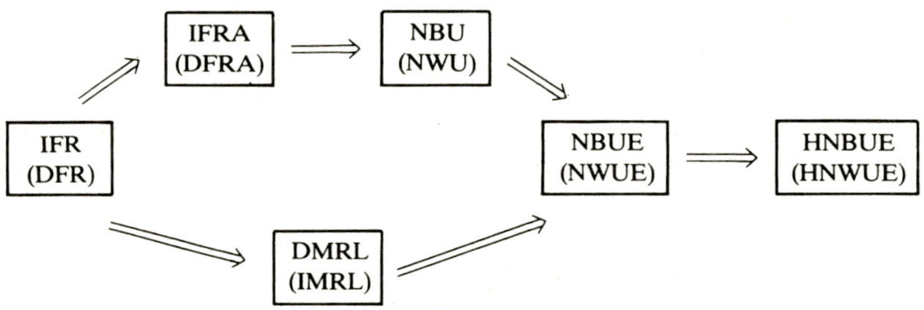

Multivariate extension of HNBUE class has been considered by Basu, Ebrahimi and Klefsjö (1982). Basu and Ebrahimi (1982b) have proposed a nonparametric test for testing whether F is HNBUE.

8.3. Accelerated life testing

Accelerated life testing of a product is commonly used to reduce test time and costs. Shaked (1979), Shaked and Singpurwalla (1982), and Shaked, Zimmer and Ball (1979) have considered a nonparametric version of accelerated testing when there is a single cause of failure and the samples are not censored. Basu and Ebrahimi (1982a) have extended the results to the case of arbitrarily (or randomly) censored data and to the case when there are competing causes of failures. We explain the procedures briefly. First, consider the case when there is only a single cause of failure.

Consider a system with survival function $\bar{F}_{10}$ under use condition, say V_0. Let $V_1, V_2, \ldots, V_J$ denote the accelerated stress levels and let $\bar{F}_{11}, \bar{F}_{12}, \ldots, \bar{F}_{1J}$ denote the corresponding survival functions. We assume all survival functions belong to a common but unknown family and that

$$\bar{F}_{1j}(t) = \bar{F}(AV_j^\alpha t), \quad j = 0, 1, \ldots, J, \tag{8.5}$$

where $A > 0$ and $\alpha > 0$ are unknown constants.

At stress level j, n_j items are put on test. Let X_{jl} and X_{jl}^c denote the lifetime and censoring time of l-th item at stress level V_j, $l = 1, 2, \ldots, n_j$, $j = 1, 2, \ldots, J$. Although X_{jl}^c's could be constants, here X_{jl} and X_{jl}^c are assumed independent variables. Thus the data available are

$$T_{jl} = \min(X_{jl}, X_{jl}^c) \quad \text{and} \quad \delta_{jl} = I(X_{jl} \leq X_{jl}^c).$$

For each pair (i, j), the scale factor between S_{1i} and S_{1j} is

$$\theta_{ij} = \frac{AV_i^\alpha}{AV_j^\alpha} = (V_i/V_j)^\alpha,$$

and

$$\alpha = \log \theta_{ij}/\log(V_i/V_j). \tag{8.6}$$

For each pair (i, j), we obtain estimator $\hat{\theta}_{ij}$ using the method of Padgett and Wei (1982), and all these are combined to obtain a 'pooled' estimate $\hat{\alpha}$ of α, where

$$\hat{\alpha} \frac{\sum_{j=1}^s \sum_{k=j+1}^s (\log \hat{\theta}_{jk}) \log(V_k/V_j)}{\sum_{j=1}^s \sum_{k=j+1}^s (\log(V_k/V_j))^2}. \tag{8.7}$$

We next rescale all the variables so that the rescaled value of all follow the same distribution F_0, by using the transformation

$$Z_{jl} = (V_j/V_0)\hat{\alpha}_{T_{jl}}. \tag{8.8}$$

We can then use (Z_{jl}, δ_{jl}) to obtain the Kaplan–Meier estimator of $\bar{F}_0$.

The above procedure is further extended by Basu and Ebrahimi (1982) to include the cases of competing risks.

8.4. *MLE of stochastically ordered survival functions*

Many times one wishes to estimate survival functions of populations from censored data when it is known that one population of survival times must be stochastically greater (or smaller) than those from a second population. For this problem Dykstra (1982) has considered the problem of finding MLE. He has considered two cases (a) one survival function being fixed in advance and (b) estimating two survival functions when the data include censored observations. Feltz (1982) has extended the result to the case of k stochastically ordered populations.

References

Anderson, P. K., Borgan, O., Gill, R. and Keiding, N. (1982). Linear nonparametric tests for comparison of counting processes, with applications to censored survival data. *Inter. Statistics Inst. Rev.* **50**, 219–258.

Bailey, K. R. (1979). The general maximum likelihood approach to the Cox regression model. Ph.D. Thesis, University of Chicago, Chicago, Illinois.

Barr, D. R. and Davison, T. (1973). A Kolmogorov–Smirnov test for censored samples. *Technometrics* **15**, 739–757.

Basu, A. P. (1967a). On the large sample properties of a generalized Wilcoxon–Mann–Whitney statistic. *Ann. Math. Statistics* **38**, 905–915.

Basu, A. P. (1967b). On two k-sample rank tests for censored data. *Ann. Math. Statistics* **38**, 1520–1535.

Basu, A. P. (1968). On a generalized Savage statistic with applications to life testing. *Ann. Math. Statistics* **39**, 1591–1604.

Basu, A. P. (1977). A generalized Wilcoxon–Mann–Whitney statistic with some applications in reliability. In: Tsokos and Shimi, eds., *The Theory and Applications of Reliability Vol. 1*. Academic Press, New York, pp. 131–149.

Basu, A. P. (1981). Identifiability problems in the theory of competing and complementary risks – A survey. In: Taillie et al., eds., *Statistical Distributions in Scientific Work 5*. Reidel, Dordrecht, Holland, pp. 335–348.

Basu, A. P. and Ebrahimi, N. (1982a). Nonparametric accelerated life testing. Technical Report No. 120, Department of Statistics, University of Missouri-Columbia. *IEEE Trans. Reliability* **31**, 432–435.

Basu, A. P. and Ebrahimi, N. (1982b). Testing whether survival function is harmonic new better than used in expectation. Technical Report No. 122, University of Missouri–Columbia.

Basu, A. P., Ebrahimi, N. and Klefsjö, B. (1983). Multivariate harmonic new better than used in expectation distributions. *Skand. J. Statistics* **10**, 19–25.

Basu, A. P. and Ghosh, J. K. (1980). Identifiability of distributions under competing risks and complementary risks models. *Communications in Statistics Theory Method A* **9**, 1515–1525.

Basu, A. P., Ghosh, J. K. and Sen, P. K. (1982). A unified way of deriving LMP rank tests from censored data. Technical Report No. 118, University of Missouri-Columbia.

Basu, A. P. and Klein, John (1982). Some recent results in competing risks theory. *Survival Analysis, IMS Monograph* **2**, 21–29.

Block, H. and Basu, A. P. (1974). A continuous bivariate exponential extension. *Jour. Amer. Stat. Assn.* **69**, 1031–1037.

Breslow, N. (1970). A generalized Kruskal–Wallis test for comparing k samples subject to unequal patterns of censorship. *Biometrika* **57**, 579–593.

Breslow, N. E. and Crowley, J. (1974). A large sample study of the life table and product limit estimates under random censorship. *Ann. Statistics* **2**, 437–453.

Brown, W. B., Jr., Hollander, M. and Korwar, R. M. (1974). Nonparametric tests of independence for censored data with applications to heart transplant studies. In: Proschan and Serfling, eds., *Reliability and Biometry: Statistical Analysis of Life Length*. SIAM, Philadelphia, pp. 327–354.

Buckley, J. and James, I. (1979). Linear regression with censored data. *Biometrika* **66**, 429–436.

Campbell, G. (1981). Nonparametric bivariate estimation with randomly censored data. *Biometrika* **68**, 417–422.

Chatterjee, S. K. and Sen, P. K. (1973). Nonparametric testing under progressive censoring. *Calcutta Statistical Association Bulletin* **22**, 13–50.

Conover, W. J. (1967). The distribution function of Tsao's truncated Smirnov tests. *Ann. Math. Statistics.* **38**, 1208–1215.

Conover, W. J. (1980). *Practical Nonparametric Statistics.* Wiley, New York.

Cox, D. R. (1972). Regression models and life tables (with discussion). *J. Roy. Stat. Soc. Ser. B* **34**, 187–220.

Cox, D. R. (1975). Partial likelihood. *Biometrika* **62**, 269–276.

Crowley, J. and Thomas, D. R. (1975). Large sample theory for the log rank test. Technical report No. 415, University of Wisconsin-Madison, Madison, Wisconsin.

Csörgő, S. and Horváth, L. (1981). On the Koziol–Green model for random censorship. *Biometrika* **68**, 391–401.

Cuzick, J. (1982). Rank tests for association with right censored data. *Biometrika* **69**, 351–364.

Doksum, K. (1972). Decision theory for some nonparametric models. *Proc. Sixth Berk. Symp. Math. Stat. Prob.* **1**, 331–341.

Dufour, R. and Maag, U. R. (1978). Distribution results for modified Kolmogorov–Smirnov statistics for truncated or censored samples. *Technometrics* **20**, 29–32.

Durbin, J. (1973). *Distribution Theory for Tests Based on the Sample Distribution Function.* SIAM, Philadelphia.

Dvoretzky, A., Kiefer, J. and Wolfowitz, J. (1956). Asymptotic minimax character of the sample distribution function and of the classical multinomial estimator. *Ann. Math. Statistics* **27**, 642–669.

Dykstra, R. L. (1982). Maximum likelihood estimation of the survival functions of stochastically ordered random variables. *Jour. Am. Stat. Assn.* **77**, 621–628.

Dykstra, R. and Laud, P. (1981). A Bayesian nonparametric approach to reliability. *Annals of Statistics* **9**, 356–367.

Efron, B. (1967). The two sample problem with censored data. In: *Proc. Fifth Berkeley Symposium in Mathematical Statistics, IV.* Prentice-Hall, New York, 831–853.

Efron, B. (1977). The efficiency of Cox's likelihood function for censored data. *J. Amer. Stat. Assoc.* **72**, 557–565.

Feltz, C. J. (1982). Nonparametric maximum likelihood estimation of stochastically ordered survival functions. Ph.D. Thesis, University of Missouri-Columbia.

Ferguson, T. S. (1973). A Bayesian analysis of some nonparametric problems. *Ann. Statistics* **1**, 209–230.

Ferguson, T. S. and Phadia, E. G. (1979). Bayesian nonparametric estimation based on censored data. *Ann. Statistics* **7**, 163–186.

Fleming, T. R., O'Fallon, J. R., O'Brien, P. C. and Harrington, D. P. (1980). Modified Kolmogorov–Smirnov test procedures with application to arbitrarily right censored data. *Biometrics* **36**, 607–625.

Földes, A. and Rejtö, L. (1981). Strong uniform consistency for nonparametric survival curve estimators from randomly censored data. *Ann. Statistics* **9**, 122–129.

Földes, A. and Rejtö, L. (1981). A LIL type result for the product limit estimator. *Z. Wahrsch. Verw. Geb.* **56**, 75–86.

Gail, M. H. and Ware, J. H. (1979). Comparing observed life table data with a known survival curve in the presence of random censorship. *Biometrics* **35**, 385–391.

Gastwirth, J. L. (1965). Asymptotically most powerful rank tests for the two-sample problem with censored data. *Ann. Math. Statistics* **36**, 1243–1247.

Gehan, Edmund A. (1965a). A generalized Wilcoxon test for comparing arbitrarily singly-censored samples. *Biometrika* **52**, 203–223.

Gehan, Edmund A. (1965b). A generalized two-sample Wilcoxon test for doubly censored data. *Biometrika* **52**, 650–652.

Gibbons, J. D. (1971). *Nonparametric Statistical Inference.* McGraw-Hill, New York.

Gilbert, J. P. (1962). Random censorship. Ph.D. Thesis, University of Chicago.

Gill, R. D. (1981). Large sample behavior of the product limit estimator on the whole line. *Mathematical Centre Report*, Amsterdam.

Gillespie, M. J. and Fisher, L. (1979). Confidence bands for the Kaplan–Meier survival curve estimate. *Ann. Statistics* **7**, 920–924.

Hájek, J. and Šidák, Z. (1967). *Theory of Rank Tests.* Academic Press, New York.

Hall, W. J. and Wellner, J. A. (1980). Confidence bands for a survival curve from censored data. *Biometrika* **67**, 133–143.

Halperin, M. (1960). Extension of the Wilcoxon–Mann–Whitney test to samples censored at the same fixed point. *J. Amer. Statist. Assoc.* **55**, 125–138.

Halperin, M. and Ware, J. (1974). Early decision in a censored Wilcoxon two-sample test for accumulating survival data. *J. Amer. Statist. Assoc.* **69**, 414–422.

Halperin, M., Ware, J. H. and Wu, M. (1980). Conditional distribution-free tests for the two-sample problem in the presence of right censoring. *J. Amer. Statist. Assoc.* **75**, 638–645.

Hollander, Myles and Proschan, F. (1972). Testing whether new is better than used. *Ann. Math. Statistics* **43**, 1136–1146.

Hollander, M. and Proschan, F. (1979). Testing to determine the underlying distribution using randomly censored data. *Biometrics* **35**, 393–401.

Hollander, M. and Proschan, F. (1982). Nonparametric concepts and methods in reliability. Technical Report No. 77, Stanford University.

Hyde, J. (1977). Testing survival under right censoring and left truncation. *Biometrika* **64**, 225–230.

Joe, H. and Proschan, F. (1981). Estimating a decreasing mean residual life distribution from complete or incomplete data. *Scand. J. Statist.* (to appear).

Johansen, S. (1978). The product limit estimator as maximum likelihood estimator. *Scand. J. Statist.* **5**, 195–199.

Johnson, R. C. E. and Johnson, N. L. (1980). *Survival Models and Data Analysis.* Wiley, New York.

Johnson, R. A. and Mehrotra (1972). Locally most powerful rank tests for the two-sample problem with censored data. *Ann. Math. Statistics* **43**, 823–831.

Kalbfleisch, J. D. and Prentice, R. L. (1980). *The Statistical anal Analysis of Failure Time Data.* Wiley, New York.

Kaplan, E. L. and Meier, P. (1958). Nonparametric estimation from incomplete observations. *J. Amer. Statist. Assoc.* **53**, 457–481.

Kiefer, J. and Wolfowitz, J. (1956). Consistency of the maximum likelihood estimator in the presence of infinitely many incidental parameters. *Ann. Math. Statistics* **27**, 887–906.

Kitchin, J. (1980). A new method for estimating life distributions from incomplete data. Ph.D. Dissertation, Florida State University.

Kitchin, J., Langberg, A. and Proschan, F. (1980). A new method for estimating life distributions from incomplete data. Florida State University, Department of Statistics Technical Report No. M-548.

Klein, J. P. and Basu, A. P. (1981). Weibull accelerated life tests when there are competing causes of failure. *Commun. Statist. Theor. Meth. A* **10**, 2073–2100.

Klefsjö, B. (1982). The HNBUE and HNWUE classes of life distributions. *Naval Research Logistics Quarterly* **29**, 331–344.

Koul, H. L. and Susarla, V. (1980). Testing for new better than used in expectation with incomplete data. *J. Amer. Statist. Assoc.* **75**, 952–956.

Koul, H. L., Susarla, V. and Ryzin, J. Van. (1981). Regression analysis with randomly right-censored data. *Ann. Statistics* **6**, 1276–1288.

Koziol, J. A. (1980). Goodness-of-fit tests for randomly censored data. *Biometrika* **67**, 693–696.

Koziol, J. A. and Byar, D. P. (1975). Percentage points of the asymptotic distributions of one and two sample *K–S* statistics for truncated or censored data. *Technometrics* **17**, 507–510.

Koziol, J. A. and Green, S. B. (1976). A Cramer–von Mises statistic for randomly censored data. *Biometrika* **63**, 465–474.

Langberg, N. A. and Shaked, M. (1982). On the identifiability of multivariate life distribution functions. *Ann. Prob.* **10**, 773–779.

Latta, R. B. (1977a). Generalized Wilcoxon statistics for the two sample problem with censored data. *Biometrika* **64**, 633–635.

Latta, R. B. (1977b). Rank tests for censored data. Technical Report No. 112, University of Kentucky.

Lochner, R. H. (1968). On some two-sample problems in life testing. Ph.D. Dissertation, University of Wisconsin, Madison.

Lochner, R. H. and Basu, A. P. (1969). Bayesian analysis of the time truncated samples. Technical Report No. 201, Department of Statistics, University of Wisconsin.

Lochner, R. H. and Basu, A. P. (1972). Bayesian analysis of the two-sample problem with incomplete data. *J. Amer. Stat. Assoc.* **67**, 432–438.

Lochner, R. H. and Basu, A. P. (1976). A Bayesian test for increasing failure rate. In: Tsokos and Shimi, eds., *The Theory and Applications of Reliability, Vol. 1.* Academic Press, New York, pp. 67–83.

Mantel, N. (1966). Evaluation of the survival data and two new rank order statistics arising in its consideration. *Cancer Chemotherapy Reports* **50**, 163–170.

Mantel, N. (1967). Ranking procedures for arbitrarily restricted observation. *Biometrics* **23**, 65–78.

Mehrotra, K. G., Johnson, R. A. and Bhattacharyya, G. (1977). Locally most powerful rank tests for multicensored data. *Communications in Statistics – Theory and Methods A* **6**, 459–469.

Meier, P. (1975). Estimation of a distribution function from incomplete observations. In: J. Gani, ed., *Perspectives in Probability and Statistics.* Academic Press, New York, 67–68.

Miller, R. G. (1976). Least squares regression with censored data. *Biometrika* **63**, 449–464.

Miller, R. G. Jr. (1981). *Survival Analysis.* Wiley, New York.

Miller, R. G. and Halpern, J. (1981). Regression with censored data. Stanford University Division of Biostatistics Technical Report No. 66. *Biometrika* **69**, 521–531.

Nair, V. N. (1981). Plots and tests for goodness of fit with randomly censored data. *Biometrika* **68**, 99–103.

Nelson, W. (1969). Hazard Plotting for incomplete failure data. *Journal of Quality Technology* **1**, 27–52.

Oakes, D. (1977). The asymptotic information in censored survival data. *Biometrika* **64**, 441–448.

Oakes, D. (1982). A concordance test for independence in the presence of censoring. *Biometrics* **38**, 451–455.

O'Brien, P. C. (1978). A nonparametric test for association for censored data. *Biometrics* **34**, 243–250.

Padgett, W. J. and Wei, L. J. (1980). Maximum likelihood estimation of a distribution function with increasing failure rate based on censored observations. *Biometrika* **67**, 470–474.

Padgett, W. J. and Wei, L. J. (1982). Estimation of the ratio of scale parameters in the two-sample problem with arbitrary right censorship. *Biometrika* **69**, 252–256.

Patel, K. M. (1975). A generalized Friedman test for randomized block designs when observations are subject to arbitrary right censorship. *Communications in Statistics* **4**, 389–395.

Patel, K. M. and Hoel, D. G. (1973). A generalized Jonckheere k-sample test against ordered alternatives when observations are subject to arbitrary right censorship. *Communications in Statistics* **2**, 373–380.

Peterson, A. V. (1977). Expressing the Kaplan–Meier estimator as a function of empirical subsurvival functions. *J. Amer. Statist. Assoc.* **72**, 854–858.

Peto, R. and Peto, J. (1972). Asymptotically efficient rank invariant test procedures (with discussion). *J. Roy. Stat. Soc. A* **135**, 185–206.

Pettitt, A. N. (1976). Cramér–von Mises statistics for testing normality with censored samples. *Biometrika* **63**, 475–481.

Pettitt, A. N. and Stephens, M. A. (1976). Modified Cramér–von Mises statistics for censored data. *Biometrika* **63**, 291–298.

Phadia, E. G. (1973). Minimax estimation of a cumulative distribution function. *Ann. Statistics* **1**, 1149–1157.

Prentice, R. L. (1978). Linear rank tests with right censored data. *Biometrika* **65**, 167–179.

Puri, M. L. and Sen, P. K. (1971). *Nonparametric Methods in Multivariate Analysis.* Wiley, New York.

Rao, U. V. R., Savage, I. R. and Sobel, M. (1960). Contributions to the theory of rank order statistics: Two sample censored case. *Ann. Math. Statistics* **31**, 415–426.

Rolski, T. (1975). Mean residual life. *Bulletin of the International Statistical Institute* **46**, 266–270.

Savage, I. R. (1956). Contributions to the theory of rank statistics: Two sample case. *Ann. Math. Statistic* **27**, 590–616.

Sen, P. K. (1981). *Sequential nonparametrics.* Wiley, New York.

Sen, P. K. (1981). On invariance principles for LMP conditional test statistics. *Calcutta Statist. Assoc. Bull.* **30**, 41–56.

Shaked, M. (1979). An estimator for generalized hazard rate function. *Communications in Statistics, Theory and Methods*, pp. 17–33.

Shaked, M. and Singpurwalla, N. D. (1982). Nonparametric estimation and goodness of fit testing of hypothesis for distribution in accelerated life testing. *IEEE Trans. Reliability* **31**, 69–74.

Shaked, M., Zimmer, W. J. and Ball, C. A. (1979). A nonparametric approach to accelerated life testing. *J. Amer. Statist. Assoc.* **74**, 696–699.

Shirahata, S. (1975). Locally most powerful rank tests for independence with censored data. *Annals of Statistics* **3**, 241–245.

Sobel, M. (1957). On a generalized Wilcoxon statistic for life testing. In: *Proc. Working Conference on the Theory of Reliability.* New York University and the RCA, pp. 8–13.

Spurrier, J. D. (1981). Comparison of two independent life tests subject to type II censoring. Technical report No. 78, Department of Mathematics and Statistics, The University of South Carolina, Columbia, South Carolina.

Susarla, V. and Van Ryzin, J. (1976). Nonparametric Bayesian estimation of survival curves from incomplete observations. *J. Amer. Stat. Assoc.* **71**, 897–902.

Tarone, R. E. and Ware, J. (1977). On distribution-free tests for equality of survival distributions. *Biometrika* **64**, 156–160.

Thomas, D. R. (1975). On a generalized Savage statistic for comparing two arbitrarily censored samples. Technical Report, Department of Statistics, Oregon State University.

Tsao, C. K. (1954). An extension of Massey's distribution of the maximum deviation between two-sample cumulative step functions. *Ann. Math. Statistics* **25**, 587–592.

Tsiatis, A. A. (1981). A large sample study of Cox's regression model. *Ann. Statistics* **9**, 93–108.

Turnbull, B. W. (1974). Nonparametric estimation of a survivorship function with doubly censored data. *J. Amer. Statist. Assoc.* **69**, 169–173.

Turnbull, B. W. and Weiss, L. (1978). A likelihood ratio statistic for testing goodness of fit with randomly censored data. *Biometrics* **34**, 367–375.

Wei, L. J. (1980). A generalized Gehan and Gilbert test for paired observations that are subject to arbitrary right censorship. *J. Amer. Statist. Assoc.* **75**, 364–367.

Weier, D. R. and Basu, A. P. (1980a). An investigation of Kendall's τ modified for censored data with applications. *Journal of Statistical Planning and Inference* **4**, 381–390.

Weier, D. R. and Basu, A. P. (1980b). Testing for independence in multivariate exponential distributions. *Australian Journal of Statistics* **22**, 276–288.

Weier, D. R. and Basu, A. P. (1981). On tests of independence under bivariate exponential models. In: Taillie et al., eds., *Statistical Distributions under Scientific Work 5.* Reidel, Dordrecht, Holland, pp. 169–180.

Wellner, J. A. (1982). Asymptotic optimality of the product limit estimator. *Ann. Statistics* **10**, 595–602.

Wilks, S. S. (1962). *Mathematical Statistics.* Wiley, New York.

Woolson, R. F. and Lachenbruch, P. A. (1980). Rank tests for censored matched pairs. *Biometrika* **67**, 597–606.

Woolson, R. F. and Lachenbruch, P. A. (1981). Rank tests for censored randomized block designs. *Biometrika* **68**, 427–435.

Young, D. H. (1970). Consideration of power for some two-sample tests with censoring based on a given order statistic. *Biometrika* **57**, 595–604.

P. R. Krishnaiah and P. K. Sen, eds., *Handbook of Statistics, Vol. 4*
© Elsevier Science Publishers (1984) 579–611

26

Tests for Exponentiality

Kjell A. Doksum and Brian S. Yandell

1. Introduction

The exponential hypothesis is important because of its implications concerning the random mechanism operating in the experiment being considered. In reliability, the exponential assumption may apply when one is dealing with failure times of items or equipment without any moving parts, such as for instance transistors, fuses, air monitors, car fenders, etc. In these examples, failure is not brought about by wear, but by a random shock, and the exponential assumption corresponds to assuming that this shock follows a Poisson process distribution. Thus testing the exponential assumption about the failure time distribution is equivalent to testing the Poisson assumption about the process producing the shock that causes failure.

Tests for exponentiality are subject to the usual dilemma concerning goodness of fit tests, namely, only when the hypothesis is rejected do we have a significant result. Thus, if the significance level of a test of exponentiality is $\alpha = 0.05$ and the true underlying model is Weibull and not exponential, the probability of falsely accepting H_0 can be nearly $1 - \alpha = 0.95$.

On the other hand, when a test rejects exponentiality it justifies the use of other more complicated models and the probabilistic and statistical methods that go along with these models. Such models and methods can be found in the books by Barlow and Proschan (1975) and Kalbfleisch and Prentice (1980).

In this paper we present some of the tests available for testing exponentiality. It is not a complete treatment of the topic and reflects the author's interests and biases.

In Section 2, we introduce some of the common parametric and nonparametric alternatives to exponentiality. The next section discusses tests designed for parametric models and it is shown that one of these tests is appropriate in a nonparametric setting. Spacings tests are discussed in Section 4, and their isotonic properties for increasing failure rate alternatives are

This research was supported in part by National Science Foundation Grant MCS-8102349. Brian Yandell is now at the departments of Statistics and Horticulture, University of Wisconsin, Madison.

developed. In Section 5, tests based on the total time on test transform are considered, while in Section 6 the nonparametric optimality of the total time on test statistic is developed.

Some of the common distance type statistics are discussed in Section 7 and graphical methods based on Q–Q plots and the total time on test transforms are given in Section 8. Section 9 gives some tests designed to detect 'New Better than Used' alternatives.

The rest of the paper concerns testing for exponentiality in the presence of right censoring. Censoring arises in many practical problems when individuals under study cannot be observed until failure. Section 10 presents several common types of censoring and details the notation used for this part. Estimates and their properties are reviewed in Section 11. Very few small-sample results exist for censored data problems (Chen, Hollander and Langberg, 1980). We therefore briefly review in Section 12 the weak convergence of the survival curve and of the cumulative failure rate, or hazard function, to a Gaussian process. This is used in later sections to examine asymptotic properties of several generalizations of tests without censoring. These include maximal deviation (Kolmogorov–Smirnov) tests which can be inverted to yield simultaneous confidence bands (Section 13); tests based on spacings and the total time on test (14); and others including average deviation, or Cramér–von Mises tests, and linear rank tests (15); Monte Carlo simulation results are summarized in Section 16. The question of exponentiality is explored using several tests on data from a prostate cancer study in Section 17.

2. Some alternatives to exponentiality

2a. Parametric models

It is often useful to study properties of nonparametric methods at certain important parametric models. Two such alternatives to the exponential model are the gamma and Weibull models whose probability densities are respectively

$$
\begin{aligned}
f_{\mathrm{G}}(t;\theta,\lambda) &= \lambda(\lambda t)^{\theta-1}\,\mathrm{e}^{-\lambda t}/\Gamma(\theta), \quad t>0\,,\\
f_{\mathrm{W}}(t;\theta,\lambda) &= \lambda\theta(\lambda t)^{\theta-1}\,\mathrm{e}^{-(\lambda t)^{\theta}}, \quad t>0\,,
\end{aligned}
\tag{2.1}
$$

where θ, $\lambda>0$. Their properties are discussed in Barlow and Proschan (1975) and Kalbfleisch and Prentice (1980). When $\theta=1$ both reduce to the exponential density so the exponential hypothesis can be written H_θ: $\theta=1$.

Properties of alternatives to exponentiality are most conveniently expressed through the *failure* (or *hazard*) *rate* function defined by

$$\lambda(t)=f(t)/[1-F(t)], \quad t>0\,,$$

where F is the failure distribution defined by $F(t)=P(T\leqslant t)$ and $f(t)$ is its density. The failure times of equipment or components with moving parts are modelled to have an increasing failure rate distribution since wear would increase the rate of failure.

In the case of gamma and Weibull distributions, we find that the failure rate is monotone increasing if $\theta > 1$ and monotone decreasing if $\theta < 1$. In fact, for the Weibull density, we have

$$\lambda_W(t) = \theta\lambda^\theta t^{\theta-1}, \quad t > 0\,.$$

Two other interesting, but less familiar parametric models are

$$\begin{aligned} f_L(t;\, \theta, \lambda) &= \lambda(1 + \theta\lambda t)\exp\{-(\lambda t + \tfrac{1}{2}\theta\lambda^2 t^2)\}\,, \\ f_M(t;\, \theta, \lambda) &= \lambda[1 + \theta K(\lambda t))]\exp\{-[\lambda t + \theta(\lambda t - K(\lambda t)]\} \end{aligned} \tag{2.2}$$

where $\theta \geqslant 0$, $t > 0$, and $K(x) = 1 - \exp(-x)$, $x > 0$. We refer to these as the linear failure rate density and Makeham (type) density, respectively. They were introduced by Bickel and Doksum (1969). Their failure rates are

$$\lambda_L(t) = \lambda(1 + \theta\lambda t), \qquad \lambda_M(t) = \lambda[1 + \theta(1 - e^{-\lambda t})]\,.$$

These densities reduce to the exponential density when $\theta = 0$, the failure rates are increasing when $\theta > 0$.

2b. Nonparametric models

It is usually hard to determine exactly which parametric family of densities is appropriate in a given experiment. Thus it is useful to turn to nonparametric classes of distributions that arise naturally from physical considerations of aging and wear. Three such natural classes of nonparametric models are listed below.

(1) The class of all IFR (Increasing Failure Rate) distributions. This is the class of distribution functions F that have failure rate $\lambda(t)$ nondecreasing for $t > 0$.

(2) The class of IFRA (IFR Average) distributions which is the class of F where the failure rate average

$$a(t) = t^{-1}\int_0^t \lambda(x)\,dx = -t^{-1}\log[1 - F(t)] \tag{2.3}$$

is nondecreasing. This class has nice closure properties: It is the smallest class of F's which includes the exponential distribution and is closed under the formation of coherent systems (Birnbaum, Esary and Marshall, 1966) and it is closed under convolution (Block and Savits, 1976).

(3) The class of NBU (New Better than Used) distributions F is the class with

$$S(s+t) \leq S(s)S(t), \quad s \geq 0,\ t \geq 0\,, \tag{2.4}$$

where $S(t) = 1 - F(t)$ is the survival function. Note that (2.4) is equivalent to stating that the conditional survival probability $S(s+t)/S(s)$ of a unit of age s is less than the corresponding survival probability of a new unit.

The three above classes satisfy

$$\text{IFR} \subset \text{IFRA} \subset \text{NBU}\,.$$

Thus the gamma and Weibull distributions with $\theta > 1$ are examples of F's for all three classes as are the F_L and F_M distributions when $\theta > 0$.

For further results on these nonparametric classes, see Barlow and Proschan (1975), and Hollander and Proschan (1984, this volume).

3. Parametric tests

In this section we consider tests that are asymptotically (approximately, for large sample size) optimal for parametric alternatives in the sense that in the class of all level α tests (assuming scale λ unknown) they maximize the asymptotic power. We will find that one of these tests is consistent for the nonparametric class of all IFRA alternatives.

Let $T_1, \ldots, T_n$ denote n survival or failure times assumed to be independent and to follow a continuous distribution F satisfying $F(0) = 0$. The exponential hypothesis H_0 is that $F(t) = K_\lambda(t)$, some λ, where

$$K_\lambda(t) = 1 - e^{-\lambda t}, \quad t > 0,\ \lambda > 0\,.$$

Suppose we have a parametric alternative with density $f(t; \theta, \lambda)$ in mind, where θ is a real shape parameter, λ is a real scale parameter, and $\theta = \theta_0$ corresponds to the exponential hypothesis.

With this setup, it is natural to apply the likelihood ratio test which is based on the likelihood ratio statistic

$$R(\boldsymbol{t}) = \frac{\sup_{\theta,\lambda} L(\boldsymbol{t}; \theta, \lambda)}{\sup_\lambda L(\boldsymbol{t}; \theta_0, \lambda)}$$

where $\boldsymbol{t} = (t_1, \ldots, t_n)$ is the observed sample vector, $L(\boldsymbol{t}; \theta, \lambda) = \Pi_{i=1}^n f(t_n; \theta, \lambda)$ is the likelihood function, and the sup is over $\lambda > 0$ and $\theta \in \Theta$, where Θ is the parameter set for θ. In the examples of Section 2, $\Theta = [0, \infty]$.

Note that since the maximum likelihood estimate of λ in the exponential model is $\hat{\lambda} = 1/\bar{t}$, then

$$\sup_\lambda L(\boldsymbol{t}; \theta_0, \lambda) = L\left(\boldsymbol{t}; \theta_0, \frac{1}{\bar{t}}\right) = \frac{1}{\bar{t}} e^{-n}\,.$$

For smooth models, as in Section 2(a), the value of $R(t)$ can be computed on a computer. The test rule based on $R(t)$ is to reject exponentiality when $R(t) \geq k_\alpha$, where k_α is the $(1-\alpha)$-th quartile of a χ^2 distribution with one degree of freedom (e.g., Bickel and Doksum, 1977, p. 229).

Another test suitable for a parametric alternative $f(t; \theta, \lambda)$ is Neyman's (1959) asymptotically most powerful $C(\alpha)$ test. This test is asymptotically most powerful in the class of all similar tests, that is, in the class of all tests that have level α no matter what the value of the unknown parameter λ is.

Let

$$h(t) = \frac{\partial}{\partial \theta} \log f(t; \theta, 1)\Big|_{\theta = \theta_0^+},$$

then it can be easily shown that in our setup the $C(\alpha)$ test reduces to a test which rejects exponentiality for large values of the test statistic

$$T(h) = (1/\sqrt{n}) \sum_{i=1}^{n} h(t_i/\bar{t})/\tau(h) \tag{3.1}$$

where

$$\bar{t} = \frac{1}{n} \sum_{i=1}^{n} t_i$$

and

$$\tau^2(h) = \int_0^\infty h^2(t)\, e^{-t}\, dt - \left[\int_0^\infty th(t)\, e^{-t}\, dt\right]^2. \tag{3.2}$$

The test rule is to reject H_0 when $T(h) \geq c_\alpha$, where c_α is the upper α critical value from a standard normal distribution, i.e. $c_{0.05} = 1.645$. For the four parametric models f_G, f_W, f_L and f_M of the previous section, we find, after some simplification,

$$T_G = \frac{1}{\sqrt{n}} \sum_{i=1}^{n} [\log(t_i/\bar{t}) + E]/\sqrt{\tfrac{1}{6}\pi^2 - 1},$$

$$T_W = \frac{1}{\sqrt{n}} \sum_{i=1}^{n} \{1 + [1 - (t_i/\bar{t})] \log(t_i/\bar{t})\}/\sqrt{\tfrac{1}{6}\pi^2},$$

$$T_L = \frac{1}{\sqrt{n}} \sum_{i=1}^{n} [1 - \tfrac{1}{2}(t_i/\bar{t})^2], \qquad T_M = \frac{1}{\sqrt{n}} \sum_{n=1}^{n} [2K(t_n/\bar{t}) - 1]/\sqrt{\tfrac{1}{12}}$$

respectively, where E = Euler's constant = 0.5772 and K is the standard exponential distribution function $1 - e^{-x}$.

Next, we consider the question of whether any of these four test statistics will have desirable properties not only for the parametric alternative they were derived for but also for nonparametric classes of failure distributions. We find that

THEOREM 3.1. *The test that rejects H_0 when $T_L \geq c_\alpha$ is consistent for any alternative F in the class of IFRA distributions.*

PROOF. Rewrite T_L as

$$T_L = \tfrac{1}{2}\sqrt{n}\left[1 - \frac{\hat{\sigma}^2}{\bar{t}^2}\right],$$

where $\hat{\sigma}^2 = n^{-1}\sum_{i=1}^n (t_i - \bar{t})^2$. Under H_0, T_L converges (in law) to a standard normal random variable. For IFRA alternatives, $\sigma^2 = \mathrm{Var}(T)$ exists, thus $(\hat{\sigma}^2/\bar{t}^2) \to \sigma^2/\mu^2$ a.s. as $n \to \infty$, where $\mu = E(T)$. If F is an IFRA distribution different from $K_\lambda(t)$, then from Barlow and Proschan (1975, p. 118), $(\sigma/\mu) < 1$; thus $T_L \to \infty$ (a.s.). Thus the power of the test converges to one as $n \to \infty$.

Note that this test is equivalent to rejecting H_0 for large values of the sample coefficient of variation $\bar{t}/\hat{\sigma}$ and that it can be carried out on any calculator that computes $\bar{t}$ and $\hat{\sigma}^2$.

EXAMPLE 3.1. In Table 3.1 we give 107 failure times for right rear breaks on D9G-66A Caterpillar tractors. These numbers are reproduced from Barlow and Campo (1975). We find $\bar{t} = 2024.26$ and $\hat{\sigma} = 1404.35$, thus $T_L = 2.68$ and the level $\alpha = 0.01$ test based on T_L rejects the hypothesis. The p-value is $p_L = 0.0037$. By comparison we find $T_M = 4.20$, so the test based on this statistic rejects H_0 with negligible p-value.

Shorack (1972) derived the uniformly most powerful invariant test for gamma alternatives. It is equivalent to the $C(\alpha)$ test based on T_G. Spiegelhalter (1983)

Table 3.1
Failure data for right rear brake on D9G-66A caterpillar tractor

56	806	1253	1927	2325	3185
83	834	1313	1957	2337	3191
104	838	1329	2005	2351	3439
116	862	1347	2010	2437	3617
244	897	1454	2016	2454	3685
305	904	1464	2022	2546	3756
429	981	1490	2037	2565	3826
452	1007	1491	2065	2584	3995
453	1008	1532	2096	2624	4007
503	1049	1549	2139	2675	4159
552	1069	1568	2150	2701	4300
614	1107	1574	2156	2755	4487
661	1125	1586	2160	2877	5074
673	1141	1599	2190	2879	5579
683	1153	1608	2210	2922	5623
685	1154	1723	2220	2986	6869
753	1193	1769	2248	3092	7739
763	1201	1795	2285	3160	

derived the locally most powerful test for Weibull alternatives and obtained the $C(\alpha)$ test based on T_W.

4. Tests based on spacings

Let $T_1, \ldots, T_n$ denote n survival or failure times assumed to be independent and to follow a continuous distribution F satisfying $F(0) = 0$. The exponential hypothesis is that

$$F(x) = 1 - e^{-\lambda x}, \quad x > 0, \ \lambda \geq 0 . \tag{4.1}$$

We look for a simple transformation of $T_1, \ldots, T_n$ that will yield new variables $D_1, \ldots, D_n$ with a distribution which is sensitive to IFR deviations from the exponential assumption. Such a transformation is defined by

$$D_i = (n + 1 - i)(T_{(i)} - (T_{(i-1)}), \quad i = 1, \ldots, n, \tag{4.2}$$

where $T_{(0)} = 0$ and $T_{(1)} < \cdots < T_{(n)}$ are the ordered T's. Using the Jacobian result on transformations of random variables, (e.g., Bickel and Doksum, 1977, p. 46), we find that under the exponential hypothesis, $D_1, \ldots, D_n$ are independent and each has the exponential distribution (4.1).

The D's are called the *normalized sample spacings*, or just *spacings* for short. They are useful since for the important class of IFR alternatives, there will be a stochastic downward trend in the spacings and tests that are good for trend will be good for IFR alternatives. To make this claim precise, we define a distribution F to be *more IFR* than G, written $F <_c G$, if $G^{-1}F$ is convex, where $G^{-1}F$ is defined by $P_F(T^{-1}F(T) \leq t) = G(t)$, $t \geq 0$ (Van Zwet, 1964; Bickel and Doksum, 1969). With this definition, 'F is IFR' is equivalent to '$F <_c K$', where $K(x)$ denotes the standard exponential distribution $1 - e^{-x}$. Moreover, for the gamma and Weibull families $F_{G,\theta}$ and $F_{W,\theta}$ of Section 2; $F_{G,\theta_2} <_c F_{G,\theta_1}$ and $F_{W,\theta_2} <_c F_{W,\theta_1}$ are both equivalent to $\theta_1 < \theta_2$.

We say that there is a *stronger downward trend* in $D_1, \ldots, D_n$ than $D'_1, \ldots, D'_n$ if D'_i/D_i is nondecreasing in i.

Now we can make precise the notion that the more increasing the failure rate, the stronger the stochastic downward trend in the spacings.

LEMMA 4.1. *Suppose F is more IFR than G. Let $T_1, \ldots, T_n$ be a sample from F with corresponding spacings $D_1, \ldots, D_n$. Then there is a sample $T'_1, \ldots, T'_n$ with distribution G and spacings $D'_1, \ldots, D'_n$ such that there is a stronger downward trend in $D_1, \ldots, D_n$ than in $D'_1, \ldots, D'_n$.*

PROOF. Let $T_{(1)} < \cdots < T_{(n)}$ be the ordered failure times and let $T'_{(i)} = G^{-1}F(T_{(i)})$, $i = 1, \ldots, n$. Then $T'_{(1)}, \ldots, T'_{(n)}$ are distributed as order statistics from G. Next let $D'_i = (n - i + 1)(T'_{(i)} - T'_{(i-1)})$. Since the function $G^{-1}F$ is convex, its slope is increasing and thus

$$(T'_{(i)} - T'_{(i-1)})/(T_{(i)} - T_{(i-1)}) \leq (T'_{(j)} - T'_{(j-1)})/(T_{(j)} - T_{(j-1)})$$

for $i < j$. It follows that $(D'_i/D_i) \leq (D'_j/D_j)$, $i < j$.

As an application of this Lemma, we note that there is a stronger downward trend in spacings from an IFR population than in spacings from an exponential population. Since spacings from an exponential population form a sample from an exponential distribution, there is no trend in these spacings.

Figure 4.1 shows a downward trend in the spacings for the tractor data of Example 3.1. The spacings D_i are plotted against $i/(n+1)$.

We consider two types of test statistics appropriate for testing no trend vs. downward trend. The first is the class of *linear rank statistics* of the form

$$\frac{1}{n}\sum_{i=1}^{n} c\left(\frac{i}{n+1}\right)J\left(\frac{R_i}{n+1}\right)$$

where R_i is the rank of D_i, and $c(1/(n+1)), \ldots, c(n/(n+1))$, $J(1/(n+1)), \ldots, J(n/(n+1))$ are constants to be chosen subject to the condition that $-c(i/(n+1))$ and $J(i/(n+1))$ are nondecreasing in i. Proschan and Pyke (1967) proposed $J(i/(n+1)) = i/(n+1)$, while Bickel and Doksum (1969) showed that it is both better and asymptotically optimal for all alternatives $f(t;\theta,\lambda)$ to choose $J(i/(n+1)) = -\log(1-i/(n+1))$. Thus we will from now on consider

$$W = \frac{1}{n}\sum_{i=1}^{n} c\left(\frac{i}{n+1}\right)\left[-\log\left(1-\frac{R_i}{n+1}\right)\right].$$

The choice of c depends on the alternative, and for the parametric alternatives f_G, f_W, f_L, f_M of Section 2 the respective asymptotically optimal choices of c are (Bickel and Doksum, 1969)

$$c_G(u) = (1-u)^{-1}\int_{-\log(1-u)}^{\infty} x^{-1}e^{-x}\,dx, \qquad c_W(u) = -\log[-\log(1-u)], \tag{4.3}$$
$$c_L(u) = \log(1-u), \qquad c_M(u) = -u.$$

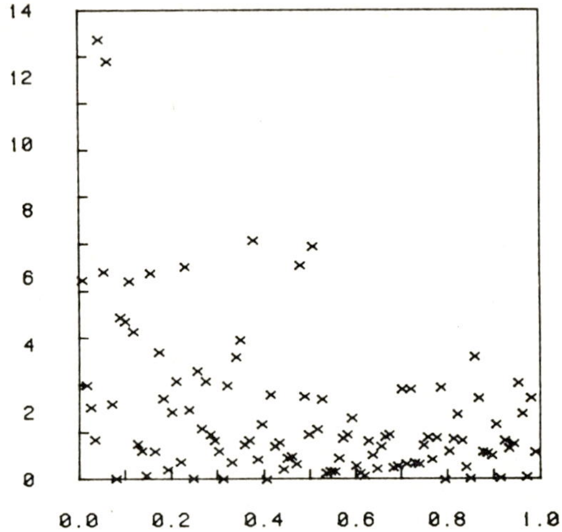

Fig. 4.1. Plot of spacings D_i vs. $i/(n+1)$ for the tractor data of Table 3.1.

The second class of statistics is the class of (*standardized*) *linear spacings statistics* which are of the form

$$S = \sum_{n=1}^{n} c\left(\frac{i}{n+1}\right) D_i \Big/ \sum_{i=1}^{n} D_i$$

where $c(i/(n+1))$ again is nonincreasing in i. This class was considered by Barlow and Prochan (1966). For the four parametric alternatives of Section 2, the optimal c to use in S is precisely as in (4.3) above (Bickel and Doksum, 1969; Bickel, 1969). We denote these asymptotically optimal spacings statistics by S_G, S_W, S_L and S_M respectively.

Next we turn to nonparametric properties of these two classes of statistics. We say that a statistic $T = T(D_1, \ldots, D_n)$ is *trend monotonic* if $T(D_1, \ldots, D_n) \geq T(D'_1, \ldots, D'_n)$ when there is a stronger downward trend in $D_1, \ldots, D_n$ than in $D'_1, \ldots, D'_n$.

From Lehmann (1966) and Bickel and Doksum (1969) we can conclude:

THEOREM 4.1. *If* $-c(i/(n+1))$ *and* $J(i/(n+1))$ *are nondecreasing in* i*, then the linear rank and spacings statistics* W *and* S *are trend monotonic.*

Recall that a similar test is one where the probability of rejecting H_0 when H_0 is true is the same for all values of the scale parameters λ. This probability is the significance level α. Tests that reject H_0 when $T \geq k$, where k is a critical constant and T is trend monotonic, are similar. This is because the downward trend in $\lambda D_1, \ldots, \lambda D_n$ is the same as that of $D_1, \ldots, D_n$, thus $T(D_1, \ldots, D_n) = T(\lambda D_1, \ldots, \lambda D_n)$.

From Lemma 4.1 and Theorem 4.1 we get the following important result.

COROLLARY 4.1. *Let* $\beta(T, F)$ *denote the power of the test that rejects* H_0 *when* $T \geq k$*, where* T *is trend monotonic. Then the test is unbiased and has isotonic power with respect to the IFR ordering, i.e. if* F *is in the IFR class, then the power* $\beta(T, F)$ *is greater than the significance level* $\alpha = \beta(T, K)$*, and if* F *is more IFR than* G*, then* $\beta(T, F) \geq \beta(T, G)$.

At this point, we have two classes of tests that are good for the nonparametric IFR class in the sense of being unbiased and having isotonic power. In each of the two classes of tests, we can obtain the asymptotically optimal test for a parametric alternative $f(t; \theta, \lambda)$, by choosing

$$c_h(u) = \frac{1}{1-u} \int_{-\log(1-u)}^{\infty} h'(t)\, e^{-t}\, dt \tag{4.4}$$

where

$$h(t) = \frac{\partial}{\partial \theta} \log f(t; \theta, 1) \Big|_{\theta = \theta_0}$$

as in Section 3. The resulting statistics W_h and S_h can be shown to be asymptotically equivalent (in the sense of having the same asymptotic power) to the $C(\alpha)$ test based on $T(h)$ given in Section 3. Thus the rank and spacings tests with c given by (4.4) are asymptotically most powerful for $f(t; \theta, \lambda)$ in the sense of maximizing the asymptotic power. See also Bickel (1969). Formula (4.4) was used to compute the examples given (in (4.3)).

Let

$$\bar{c} = n^{-1}\sum_{i=1}^{n} c(i/(n+1)) \quad \text{and} \quad s_c^2 = n^{-1}\sum_{i=1}^{n} (c_i - \bar{c})^2, \; c_i = c\left(\frac{i}{n+1}\right),$$

then the distribution of both $\sqrt{n}(W - \bar{c})/s_c$ and $\sqrt{n}(S - \bar{c})/s_c$ converge to a standard normal distribution under H_0. Thus approximate level α tests based on W and S reject H_0 when these quantities exceed the upper level α critical value c_α of a standard normal distribution.

Note that using integral approximations to sums, $\bar{c}$ and s_c^2 can be approximated by $\mu(c)$ and $\sigma^2(c)$ where

$$\mu(c) = \int_0^1 c(u)\,du \quad \text{and} \quad \sigma^2(c) = \int_0^1 c^2(u)\,du - \mu^2(c).$$

For the four examples c_G, c_W, c_L and c_E of (4.3), we find

$$\mu(c_G) = 1, \quad \sigma^2(c_G) = \tfrac{1}{6}\pi^2 - 1, \qquad \mu(c_W) = 1 - E, \quad \sigma^2(c_W) = \tfrac{1}{6}\pi^2,$$

$$\mu(c_L) = -1, \quad \sigma^2(c_L) = 1, \qquad \mu(c_E) = -\tfrac{1}{2}, \quad \sigma^2(c_E) = \tfrac{1}{12},$$

where $E = 0.5772$.

In the case of $c_M(u) = -u$, we have $\bar{c}_M = -\frac{1}{2}$ and $s_M^2 = \frac{1}{12}(n-1)/(n+1)$.

EXAMPLE 4.1. For the tractor data of Example 3.1, we find $S_L = -0.689$, $\bar{c}_L = -0.979$, $s_L = 0.931$ and $\sqrt{n}(S_L - \bar{c}_L)/s_L = 3.22$ which should be compared with the 'asymptotic equivalent' value $T_L = 2.68$ of Example 3.1. Similarly, $S_M = -0.370$ and $\sqrt{n}(S_M - \bar{c}_M)/s_M = 4.69$ as compared with $T_M = 4.20$ is Example 3.1. Clearly, S_L and S_M both reject exponentiality. Note that, in this example, the spacings tests appear to do better than the $C(\alpha)$ tests.

Finally, we remark that in terms of finite sample size Monte Carlo power, the spacing tests were shown in Bickel and Doksum (1969) to do better than the rank tests.

5. Test based on the total time on test transform

In this section, we introduce another transformation and other test statistics whose distributions are sensitive to IFR models. Suppose we put n independent items on test at the same time. Let $T_{(1)} < \cdots < T_{(n)}$ denote their ordered failure times. At time $T_{(i)}$, the total time the n items have spent on test is

$$TT_i = nT_{(1)} + (n-1)(T_{(2)} - T_{(1)}) + \cdots + (n+1-i)(T_{(i)} - T_{(i-1)})$$

$$= \sum_{i=1}^{i} (n+1-j)(T_{(j)} - T_{(j-1)}) = \sum_{j=1}^{i} D_j$$

where $T_{(0)} = 0$. Note that $TT_n = \Sigma_{j=1}^n D_j = \Sigma_{i=1}^n T_{(i)}$.

The transformation considered in this section is the one that transforms the survival or failure times $T_1, \ldots, T_n$ into $W_1, \ldots, W_{n-1}$, where

$$W_i = \frac{\Sigma_{j=1}^i D_j}{\Sigma_{j=1}^n D_j}.$$

We call $W_1, \ldots, W_{n-1}$ *total time on test transforms*, or *total time transforms* for short. Under H_0, $W_1, \ldots, W_{n-1}$ are distributed as the order statistics in a sample of size $n-1$ from a distribution uniform on (0, 1) (Epstein, 1960).

This transformation is useful since W_i tends to be larger for an IFR distribution than it is for an exponential distribution, more precisely:

THEOREM 5.1. *Suppose F is more IFR than G. Let $T_1, \ldots, T_n$ be a sample from F with corresponding total time transforms $W_1, \ldots, W_{n-1}$. Then there is a sample $T'_1, \ldots, T'_n$ with distribution G and total time transforms $W'_1, \ldots, W'_n$ with*

$$W_i \geq W'_i, \quad i = 1, \ldots, n-1 .$$

The proof can be found in Barlow and Proschan (1966), Barlow and Doksum (1972), and Barlow, Bartholomew, Bremner and Brunk (1972).

The result suggests using tests based on statistics that are *monótonic* in the W's in the sense that they are coordinate-wise increasing, i.e.

$$T(W_1, \ldots, W_{n-1}) \geq T(W'_1, \ldots, W'_{n-1})$$

whenever $W'_n \geq W'_i$, $i = 1, \ldots, n-1$.

For such tests we find

THEOREM 5.2. *Let $\beta(T, F)$ denote the power of the test that rejects H_0 when $T \geq k$, where T is monotonic. Then the test is monotonic and has isotonic power with respect to the IFR ordering, i.e. if F is in the IFR class, then the power $\beta(T, F)$ is greater than the significance level α, and if F is more IFR than G, then $\beta(T, F) \geq \beta(T, G)$.*

One important monotonic statistic is the total time on the test statistic which is defined by

$$V = \sum_{i=1}^{n-1} W_i \,.$$

Since V is distributed as the sum of uniform variables under H_0, its distribution is very close to normal. The exact distribution is tabled in Barlow et al. (1972) for $n \leq 12$. For $n > 9$,

$$\sqrt{12(n-1)}\left[\frac{V}{n-1} - \tfrac{1}{2}\right]$$

has practically a standard normal distribution.

A little algebra shows that

$$V = (n+1)\left(\frac{n}{n+1} + S_M\right)$$

where $S_{\mathrm{M}} = -\Sigma_{i=1}^{n} iD_i / \Sigma_{i=1}^{n} D_i$ as in Section 4. Thus V is equivalent to S_{M}, asymptotically equivalent to T_{M}, and asymptotically most powerful for the Makeham alternative $f_{\mathrm{M}}(t;\, \theta, \lambda)$.

Barlow and Doksum (1972) investigated a more general class of monotonic statistics, namely

$$V_J = \sum_{i=1}^{n-1} J(W_i)$$

where J is some nondecreasing function on $(0, 1)$. They found that for a given parametric alternative $f(t;\, \theta, \lambda)$, the test based on V_J will be asymptotically most powerful if $J(u)$ is chosen to equal $-c(u)$ where $c(u)$ is the function given in (4.1) and (4.2). Thus for the linear failure rate alternative $f_{\mathrm{L}}(t, \theta, \lambda)$, $-\Sigma_{i=1}^{n-1} \log(1 - W_i)$ is asymptotically optimal, while for the Weibull alternative $f_{\mathrm{W}}(t;\, \theta, \lambda)$, $\Sigma_{i=1}^{n-1} \log[-\log(1 - W_i)]$ is asymptotically optimal.

Other tests based on the spacings D_i or total time transforms W_i, have been considered by Störmer (1962), Seshadri, Csörgő and Stephens (1969), Csörgő, Seshadri and Yalovsky (1975), Koul (1978), Azzam (1978), Parzen (1979) and Csörgő and Révész (1981b), among others. An excellent source for results on spacings is the paper by Pyke (1965).

6. Nonparametric optimality

In Sections 3, 4 and 5, we have seen that different IFR parametric alternatives lead to different asymptotically optimal tests. Thus we have no basis on which to choose one test as being better than the others.

In this section, we outline the development of a theory that leads to one test, namely the one based on the total time on test statistc V, as being asymptotically optimal. These results are from Barlow and Doksum (1972).

We define the *total time on test transform* $\mathscr{H}_F^{-1}$ of the distribution function F as

$$\mathscr{H}_F^{-1}(u)=\int_0^{F^{-1}(u)}[1-F(v)]\,dv,\quad 0<u<1,$$

and the standardized total time on test transform as

$$H_F^{-1}(u)=\mathscr{H}_F^{-1}(u)/\mathscr{H}_F^{-1}(1),\quad 0<u<1.$$

Note that $H_F^{-1}(1)=E_F(T_i)=$ mean of T_i. The reason for the inverse notation is that H_F^{-1} can be regarded as the inverse of a distribution in (0, 1). We let H or H_F denote this distribution. Note that W_i of the previous section can be regarded as $H_{F_n}^{-1}(i/n)$ where F_n is the empirical distribution function of $T_1,\ldots,T_n$.

It is easy to check that when F is exponential, $H(u)=u$, $0\leqslant u\leqslant 1$; while F is IFR iff $H(t)$ is convex and $H(t)\leqslant t$ on [0, 1]. Thus the problem of testing for exponentiality can be formulated in terms of H as testing

H_0: H is uniform on [0, 1]

vs.

H_1: H is convex, $H(t)\leqslant t$ and H is not uniform on [0, 1].

The optimality criteria we are going to consider is the minimax criteria, i.e. we want to find the test that maximizes (asymptotically) the minimum power over a nonparametric class Ω. The term minimax is used since in decision theory terminology, risk = 1 − power.

We cannot take Ω to be the whole IFR class since then the minimum power would always be α. The total time on test transform H_F^{-1} gives us a convenient way of separating alternatives from H_0. We let $\Omega(\Delta)$, $0<\Delta<1$, be the class of all distributions F where H is convex, $H(t)\leqslant t$, and

$$\sup[t-H(t)]\geqslant\Delta.$$

If Δ is fixed, the minimum power over $\Omega(\Delta)$ will tend to one, thus we must allow $\Delta=\Delta_n$ to depend on n, in fact the interesting cases have

$$\Delta_n=O(n^{-1/2}).$$

Let $\beta(\varphi_T,F)$ denote the power of the level α Total Time on Test test which rejects H_0 when $V\geqslant k_\alpha$, then

LEMMA 6.1. *Assume that* $\lim_{n\to\infty}(\sqrt{n}\Delta_n)$ *exists and equals* c *where* c *is some number in* $[0,\infty]$, *then*

$$\lim_{n\to\infty}[\inf_{F\in\Omega(\Delta_n)}\beta(\varphi_T,F)]\leqslant\Phi(-k_\alpha+\sqrt{3}c).$$

Now suppose that $\beta(\varphi_J, F)$ denotes the power of the test which rejects H_0 when $V_J = \Sigma_{i=1}^{n-1} J(W_i)$ is greater than the appropriate critical constant. We want to choose J to maximize the limiting minimum power. This is achieved by choosing $J(w) = w$; thus the Total Time on Test test φ_{T} is optimal in the sense of being asymptotically minimax. The result follows from the fact that if $\lim_{n\to\infty}(\sqrt{n}\Delta_n) = c$, $c \in [0, \infty]$, then

$$\lim_{n\to\infty}[\inf_{F\in\Omega(\Delta_n)} \beta(\varphi_J, F)] \leq \Phi(-k_\alpha + \sqrt{3}c).$$

The proof can be obtained (under appropriate conditions) from Barlow and Doksum (1972) and Koul and Staudte (1976).

7. Distance statistics

If there is no natural alternative class of distributions (such as the IFR class), one can use statistics based on the distance between the exponential distribution $K_\lambda(t)$ and the empirical distribution $F_n(t)$ defined as $F_n(t) = n^{-1}[\#T_i \leq t]$. If $\lambda = \lambda_0$ is specified, the Kolmogorov statistic is given by

$$D_n(\lambda_0) = \max_t |F_n(t) - K_{\lambda_0}(t)|.$$

For tables, see Owen (1962).

In the more realistic case with λ unknown, we replace λ in K_κ by $\hat{\lambda} = 1/\bar{t}$ and use

$$D_n^* = \max_t |F_n(t) - K_{\hat{\lambda}}(t)|$$

where $K_{\hat{\lambda}}(t) = 1 - \exp(-\hat{\lambda}t)$.

The distribution of D_n^* has been studied by Lilliefors (1969), Stephens (1974) and Durbin (1975), among others. A very good approximation to the level α critical values k_α of D_n^* for $\alpha = 0.01$, 0.05 and 0.10 are given by

$$k_\alpha = \frac{0.2}{n} + \frac{d_\alpha}{\sqrt{n} + 0.26 + (0.5/\sqrt{n})}$$

where $d_\alpha = 1.308$, 1.094, 0.990 for $\alpha = 0.01$, 0.05, 0.10, respectively.

An alternative approach to estimating λ in $D_n(\lambda)$ is to first make a transformation of $T_1, \ldots, T_n$ to obtain new variables whose distribution does not depend on λ. Thus we could use the distance between the empirical distribution of $W_1, \ldots, W_{n-1}$ and the uniform distribution on (0, 1). The distribution of the resulting statistic is the same as that of the one-sample Kolmogorov statistic. Tables can be found in Owen (1962, p. 423).

For other distance statistics and their properties, see Seshadri, Csörgő and Stephens (1969), Durbin (1973, 1975), Csörgő, Seshadri and Yalovsky (1975), Sarkadi and Tusnady (1977), and Csörgő and Révész (1981(a)).

8. Graphical method in the uncensored case

8a. The Q–Q (Quantile–Quantile) Plot

The exponential quantile function evaluated at the population distribution function is

$$Q_F(t) = K^{-1}[F(t)]$$

where $K^{-1}(u) = -\log(1-u)$ is the inverse of the exponential distribution. If the exponential hypothesis is satisfied and in fact $F(t) = 1 - e^{-\lambda t} = K(\lambda t)$, then we find

$$Q_K(t) = \lambda t .$$

Thus a graphical method for checking exponentiality is to plot

$$Q_{F_n}(t) = K^{-1}[F_n(t)] = -\log[1 - F_n(t)]$$

and check if this plot falls close to a straight line through the origin. Since we cannot use the log of zero, we use the modification

$$\hat{Q}(t) = K^{-1}\left[F_n(t) - \frac{1}{2n}\right]$$

and plot $\hat{Q}(t)$ for $t = t_{(i)}$, $i = 1, \ldots, n$, where $\{t_{(i)}\}$ are the order statistic of the sample. Since the $t_{(i)}$ are sample quantiles, the resulting plot of $(t_{(i)}, K^{-1}[i - \frac{1}{2}/n])$ is called a *Q–Q* plot.

The reliability of $\hat{Q}(t)$ can be judged by giving the simultaneous level α confidence band

$$[K^{-1}(\hat{F}_n(t) - k_\alpha),\ K^{-1}(\hat{F}_n(t) + k_\alpha)] ,$$

where k_α is the level α critical value for the D_n^* test of Section 7. We reject exponentiality if the line $t/\bar{t}$ does not fall entirely within the band. This graphical test is equivalent to the D_n^* test of the previous section.

Note that, using Section 4, a convex shape for $\hat{Q}(t)$ indicates an IFR alternative.

8b. The total time on test plot

Barlow and Campo (1975) demonstrated that

$$H_n^{-1}(u) = H_{F_n}^{-1}(u), \quad 0 < u < 1,$$

where H_F^{-1} is the standardized total time on test transform of F defined in Section 6, gives a useful plot for checking exponentiality. Under exponentiality, H_n^{-1} should fall close to the identity function on (0, 1), while for IFR alternatives, we would expect $H_n^{-1}(t) \geq t$ and $H_n^{-1}(t)$ concave (see Section 6). Figure 8.1 shows this plot for the tractor data of Example 3.1. An IFR distribution is strongly indicated for this data.

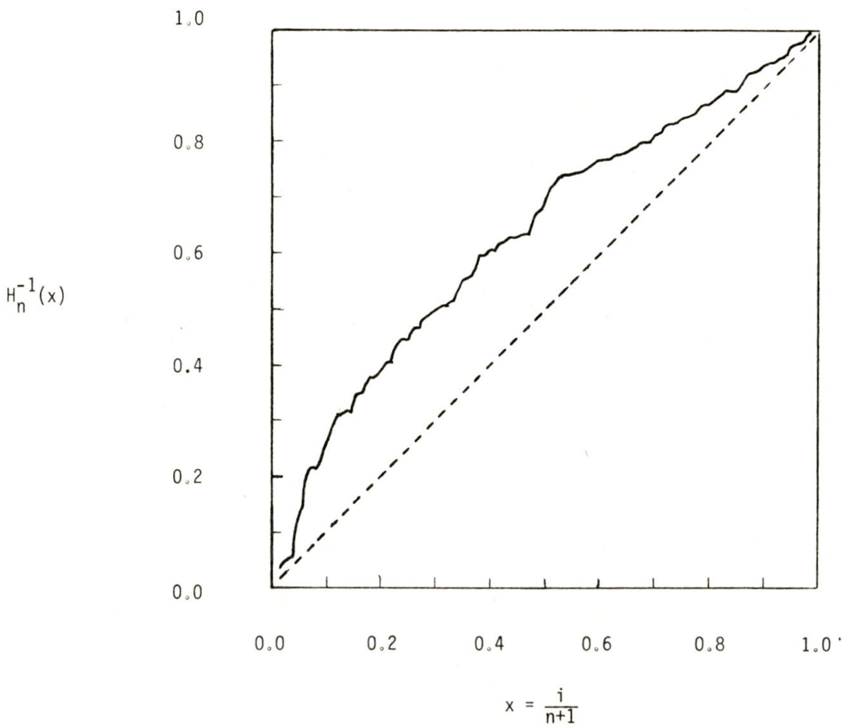

Fig. 8.1. Total time on test plot for the tractor data of Table 3.1.

The reliability of $H_n^{-1}(u)$ can be judged by using the asymptotic simultaneous level α confidence band

$$\left[H_n^{-1}(u) - \frac{b_\alpha}{\sqrt{n}},\ H_n^{-1}(u) + \frac{b_\alpha}{\sqrt{n}}\right], \quad 0 < u < 1,$$

where b_α is the critical value of the maximum of the Brownian Bridge on [0, 1]. Thus b_α is given in Owen (1962, p. 439).

9. NBU alternatives

Tests designed to detect NBU alternatives are motivated by measures of the deviation of F from exponentiality towards NBU alternatives. One such measure, considered by Hollander and Proschan (1972), is

$$\gamma(F) = \int_0^\infty \int_0^\infty [S(t)S(v) - S(t+v)]\, \mathrm{d}F(t)\, \mathrm{d}F(v)\,.$$

When F is exponential, $\gamma(F) = 0$, while it is positive when F is NBU. Thus an intuitive rule is to reject exponentiality for large values of $\gamma(F_n)$. Hollander and Proschan (1972) give the appropriate critical values and prove consistency of this test rule.

Koul (1977) considered

$$\alpha(F) = \inf_{t,v \geqslant 0} \{S(t+v) - S(t)S(v)\}$$

as a measure of NBUness. $\alpha(F)$ is 0 when F is exponential and negative when F is NBU. Koul (1977) gave critical values of the test based on $\alpha(F_n)$ for selected values of α and n.

Deshpande (1983) measures NBUness through

$$\xi(F) = \int_0^\infty [S^2(t) - S(2t)]\, \mathrm{d}F(t)$$

and considers the corresponding test statistic $\xi(F_n)$. He develops the asymptotic distribution and gives the Pitman asymptotic relative efficiencies 0.931, 1.006 and 0.946 of $\xi(F_n)$ to $\gamma(F_n)$ for linear failure rate, Weibull and Makeham alternatives, respectively.

For further results on measures of NBU alternatives, see Koul (1978) and Hollander and Proschan (this volume, Chapter 27).

10. Types of censoring

Censoring may arise in a variety of ways, leading to several possible assumptions about the form of censoring. Here we consider primarily right censoring. For individual i, $i = 1, \ldots, n$, the observed length of life, or time on test, is $Y_i = \min(T_i, C_i)$, in which T_i is the failure time with survival curve $S(t) = P(T_i \geqslant t)$, and C_i is the censoring time with censoring curve $G(t) = P(C_i \geqslant t)$. T_i and C_i are assumed to be independent.

'Type I' censoring concerns experiments in which observation is terminated at a predetermined time $C_i = C$, $i = 1, \ldots, n$. Thus a random number of failures are observed. For 'type II' censoring, observation continues until $r \leqslant n$ failures occur, with r fixed. Type II censoring may arise when one wants at least r failure times, for reasons of power, but cannot afford to wait until all individuals fail.

In many clinical trials, the beginning and end of the observation period is fixed, but individuals may enter the study at any time. This is an example of 'fixed' or 'progressive type I' censoring, in which the C_i, $i = 1, \ldots, n$, are fixed but not necessarily equal.

'Random censorship' refers to experiments in which the censoring times are randomly distributed. This may occur when censoring is due to competing risks, such as loss to follow-up or accidental death. However, T_i and C_i may be dependent, as is the case when individuals are removed from study based on mid-term diagnosis. The lack of independence brings problems of identifiability and interpretation (Horvath, 1980; see Prentice et al. (1978) for review).

Several other possible assumptions deserve mention. Hyde (1977) and Mihalko and Moore (1980) considered left truncation with right censoring. Left truncation may correspond to birth or to entering the risk stage of a disease (Chiang, 1979). Mantel (1967), Aalen (1978), Gill (1980) and others generalize this to arbitrary censoring.

Various authors (Koziol and Green, 1976; Hollander and Proschan, 1979; Koziol 1980; Chen, Hollander and Langberg, 1982) assumed a 'proportional hazards' model for censoring. That is, $G = S^{\beta}$ with β the 'censoring parameter'.

All these types of censoring are special cases of the multiplicative intensity model (Aalen, 1975, 1976, 1978; Gill, 1980). For our purposes, let $N(t)$, $t \geq 0$, be the number of failures in $[0, t]$ and $R(t)$ be the number at risk of failure at time $t \geq 0$. If we are only concerned with right censorship, then $R(t) = \#(Y_i \geq t)$. More generally $R(t)$ must be predictable, that is left-continuous with right-hand limits and depending only on the history of the process $\{N(u), R(u); 0 \leq u \leq t\}$. We assume that for each $t > 0$, the jump $\mathrm{d}N(t)$ is a zero–one random variable with expectation $R(t)\,\mathrm{d}H(t)$, in which $H(t)$ is the cumulative rate, or hazard function. Aalen (1975, 1978) and Gill (1980) and later authors use the fact that

$$N(t) - \int_0^t R(u)\,\mathrm{d}H(u), \quad t \geq 0,$$

is a square-integrable martingale to derive asymptotic properties of the estimators and tests presented below. Note that one does not need to assume continuity of the survival S or censoring G curve.

The remainder of this paper concerns right censorship unless otherwise noted.

11. Estimates in the censored case

The tests presented in later sections embody estimates of the survival curve, the censoring curve, and/or the hazard function. The survival curve is usually estimated by the Kaplan–Meier product limit estimator

$$S_n(t)\begin{cases} = \prod_{\{i|Y_i\leq t\}} \left(1-\frac{1}{R(Y_i)}\right) I(T_i \leq C_i) & \text{if } 0\leq t< Y_{(n)}, \\ = 0 & \text{if } t > Y_{(n)}, \end{cases}$$

with the Efron (1967) convention that the last event is considered a failure. The censoring curve may be estimated in a similar fashion, with the relation

$$G_n(t)S_n(t) = 1 - R(t^+)/n .$$

S_n and G_n are biased but consistent and self-consistent (Efron, 1967). If S is continuous and G is left-continuous, then S_n is asymptotically normal (Breslow and Crowley, 1974). If S and G are both continuous then S_n is strongly uniformly consistent on any finite interval in the support of both S and G (Földes and Rejtö, 1981).

The hazard function is estimated by the Nelson (1969) estimator

$$H_n(t) = \sum_{\{i|Y_i\geq t\}} \left(\frac{I(t_i \leq C_i)}{R(Y_i)}\right) = \int_0^t R^{-1}\, \mathrm{d}N .$$

H_n is biased, consistent and asymptotically normal (Aalen, 1978) under the same conditions as those for S_n. It is also strongly uniformly consistent (Yandell, 1983).

Some tests rely on a survival curve estimator based on H_n, namely

$$\hat{S}(t) = \exp(-H_n(t)) > t \geq 0 .$$

$\hat{G}(t)$ is defined in an analogous manner. The properties of these estimates are presented in Fleming and Harrington (1979).

The asymptotic variance V of $\sqrt{n}(H_n - H)$ and of $\sqrt{n}(S_n - S)/S$ has the form (Breslow and Crowley, 1974; Gill, 1980)

$$V(t) = \int_0^t S^{-1}G^{-1}\, \mathrm{d}H .$$

It can be estimated consistently by

$$V_n(t) = n \int_0^t R^{-1}(R-1)^{-1}\, \mathrm{d}N = n \sum_{\{i|Y_i\leq t\}} \left(\frac{I(T_i \leq C_i)}{R(Y_i)(R(Y_i)-1)}\right) .$$

12. Weak convergence

Several asymptotic tests for censored survival data are based on the weak convergence of the survival curve S_n or the hazard function to a Gaussian process. Throughout this section we assume that S is continuous, G is left-continuous, and censoring and survival (failure) act independently.

Breslow and Crowley (1974) first proved the weak convergence for S_n and H_n with G continuous. Meier (1975) handled the case of fixed censorship for S_n. Aalen (1978) and Gill (1980, 1983) considered the case of G left-continuous. The results can be stated in terms of Brownian motion B on $[0, \infty)$ or a Brownian bridge B^0 on $[0, 1]$ with a time change (Efron, 1967; Gillespie and Fisher, 1979; Hall and Wellner, 1980). Let $\circ$ denote composition.

THEOREM 12.1. *Let* $Z_n = \sqrt{n}(H_n - H)$ *or* $Z_n = \sqrt{n}(S_n - S)/S$. *Then*

$$Z_n \Rightarrow B \circ V, \qquad Z_n/(1+V) \Rightarrow B^0 \circ (V/(1+V))$$

in $D[0, T]$ *for* $T < T_{SG} = \inf\{t; S(t)G(t) > 0\}$.

Gill (1981) extended this result to the whole line:

THEOREM 12.2. *Let* $Z_n = \sqrt{n}(S_n - S)/S$. *Then*

$$Z_n/(1+V) \Rightarrow B^0 \circ (V/(1+V))$$

in $D[0, T_{SG}]$. *In addition,*

$$Z_n/(1+V_n) \Rightarrow B^0 \circ (V/(1+V))$$

in $D[0, T_{SG}]$ *provided that*

$$\int_0^{T_{SG}} S^2 \, \mathrm{d}V = \int_0^{T_{SG}} SG^{-1} \, \mathrm{d}H < \infty .$$

Nair (1980, 1981) and Gill (1983) introduced weight functions which allow weak convergence to weighted versions of B and B^0.

THEOREM 12.3 (Nair 1980). *Let* $Z_n = \sqrt{n}(H_n - H)$ *or* $Z_n = \sqrt{n}(S_n - S)/S$. *Let* q *be continuous and nonnegative on* $[0, 1]$, *and* $T_n \xrightarrow{\mathrm{p}} T < T_{GS}$. *Then*

$$Z_n V_n^{-1/2}(T_n) q \circ (V_n / V_n(T_n)) \Rightarrow (Bq) \circ (V/V(T)), \tag{12.1}$$

$$(Z_n/(1+V_n)) q \circ (V_n/(1+V_n)) \Rightarrow (B^0 q) \circ (V/(1+V)) \tag{12.2}$$

on $D[0, T]$.

Gill (1983) proved a similar result on the whole line for a restricted class of weight functions.

THEOREM 12.4 (Gill 1983). *Let* $Z_n = \sqrt{n}(S_n - S)/S$. *Let* q *be continuous on* $[0, 1]$, *symmetric at* $\frac{1}{2}$, *nondecreasing on* $(0, \frac{1}{2})$,

$$\int_0^1 q^{-2}(t)\, \mathrm{d}t < \infty$$

and $(1-t)q^{-1}(t)$ *nonincreasing near* $t = 1$. *Then*

$$(Z_n/(1+V))q \circ (V/(1+V)) \Rightarrow (B^0 q) \circ (V/(1+V)) .$$

These results will be used with various weight functions in later sections.

Csörgő and Horváth (1982a, 1982b) showed that Z_n (for the survival curve or hazard function) can be strongly approximated on $[0, T]$ by a Brownian bridge process. They required continuity of G, but mention in Remark 3.3 (Csörgő and Horvath, 1982) that continuity and independence of competing risks may not be needed (see Horvath, 1980). Their results yield the same test statistics as those available from the weak convergence results. In addition they provide the rate of convergence, and Chung and Strassen type laws of the iterated logarithm (Csörgő and Horvath, 1983).

13. Maximal deviation tests

One class of goodness-of-fit tests relies on the Kolmogorov–Smirnov metric of the maximal deviation of the empirical from the theoretical distribution. Here we exhibit results for a completely specified null distribution (S or H). For the exponential family, $S(x) = \mathrm{e}^{-\lambda x}$ or $H(x) = \lambda x$, one may view these as conservative in the sense that if no choice of λ yields a curve close enough to the empirical curve, then the hypothesis of exponentiality is rejected. In other words, if one cannot place a straight line completely within the $1-\alpha$ confidence bands for $H(t)$, $a_n \leq t \leq T_n$, then the exponential hypothesis is rejected at level α.

The basic result (Aalen, 1976; Gillespie and Fisher, 1979; Hall and Wellner, 1980; Nair, 1980, 1981; Gill, 1980) is for $Z_n = \sqrt{n}(H_n - H)$ or $Z_n = \sqrt{n}(S_n - S)/S$,

$$\sup_{a_n \leq t \leq T_n} \left| \frac{Z_n(t)}{\sqrt{V_n(T_n)}} q\left(\frac{V_n(t)}{V_n(T_n)}\right) \right| \Rightarrow \sup_{a \leq x \leq b} |q(x)B(x)| ,$$

$$\sup_{a_n \leq x \leq T_n} \left| \frac{Z_n(t)}{1 + V_n(t)} q\left(\frac{V_n(t)}{1 + V_n(t)}\right) \right| \Rightarrow \sup_{a \leq x \leq b} |q(x)B^0(x)| .$$

The cited authors restrict attention to the finite intervals $[a_n, T_n]$ with $T_n \xrightarrow{\mathrm{p}} T < T_{SG}$ and $a_n \geq 0$. The limiting distribution then depends upon S and G, with

$$a = V(a_n)/V(T_n) \quad \text{and} \quad b = 1$$

for the convergence to Brownian motion, and

$$a = \frac{V(a_n)}{1+V(a_n)} \quad \text{and} \quad b = \frac{V(T_n)}{1+V(T_n)}$$

for the convergence to B^0. In the latter case, if q satisfies the conditions of Theorem 12.4, then one can extend the sup to the whole line for $Z_n = \sqrt{n}(S_n - S)/S$.

The maximal deviation statistics can be inverted to yield simultaneous confidence bands. More precisely, the $1-\alpha$ confidence band for $H(t)$ is

$$H_n(t) \pm K_{q,\alpha} n^{-1/2}(1+V_n(t))q^{-1}\left(\frac{V_n(t)}{1+V_n(t)}\right)$$

with $K_{q,\alpha}$ the $1-\alpha$ point of $\sup|qB^0|$.

Consider several choices of q for this band. If $q(u) = 1/(1-u)$, one gets bands proportional to 1, with asymptotic distribution that of $\sup|B|$. This has distribution (Feller, 1971)

$$\Pr\{\sup_{0 \leq u \leq 1} |B(u)| < x\} = \frac{4}{\pi}\sum_{n=0}^{\infty}\frac{(-1)^n}{2n+1}\exp\left(-\frac{(2n+1)^2\pi^2}{8x^2}\right) \approx 4\phi(x) - 3$$

in which $\phi(x)$ is the standard normal distribution. With the choice $q(u) = u^{-1/2}(1-u)^{-1/2}$, the bands are proportional to $[V_n(t)]^{1/2}$, with asymptotic distribution

$$2\Pr\left\{\sup_{a<u<1}\left|\frac{B(u)}{\sqrt{u}}\right| \leq x\right\} = 2\Pr\left\{\sup_{a/(1+a)<u<1/2}\frac{B^0(u)}{\sqrt{u(1-u)}} \leq x\right\}$$

tabled by Borovkov and Sycheva (1968). The choice $q(u) = 1$ yields bands proportional to $(1+V_n(t))$ with asymptotic distribution equivalent to the Kolmogorov–Smirnov distribution tabled by Pearson and Hartley (1976, Table 54) (see Hall and Wellner (1980) for the case $T < T_{SG}$).

Useful approximations to the distributions of $\sup_{a \leq x \leq b}|q(x)B(x)|$ and $\sup_{a \leq x \leq b}|q(x)B^0(x)|$ can be found in the papers by Jennen and Lerche (1981) and Jennen (1981).

The above tests are consistent but biased against continuous alternatives. They are distribution-free asymptotically, up to the choice of interval end points. The choice of $q(\cdot)$ is open, with the obvious remark that different choices emphasize different intervals of the survival or hazard function. The Borovkov–Sycheva (1968) type choice is appealing as the bands are then proportionally wider than pointwise confidence intervals. These bands also have equal variance at every point.

Fleming and Harrington (Fleming, O'Fallon, O'Brien and Harrington, 1980; Fleming and Harrington, 1981) introduced a class of Kolmogorov–Smirnov type tests which differ in an important manner from those considered above. They point out (Fleming and Harrington, 1981) that the asymptotic distribution of tests based on (12.1) with $q = 1$ depend on the maximum of a Gaussian process with variance function which depends on the censoring curve,

$$V(t)/V(T) = \int_0^t G^{-1}S^{-1}\,\mathrm{d}H \Big/ \int_0^T G^{-1}S^{-1}\,\mathrm{d}H\,.$$

They claim that such a test 'has the undesirable property that its probability of rejection of [the null hypothesis] based upon information up to time t systematically tends to zero when censorship of data after time t is increased'. The tests based on (12.1) are asymptotically distribution-free (Nair, 1980), but the power against alternatives will certainly depend on the choice of T and the degree of censoring.

Fleming and Harrington (1981) propose instead to examine

$$Z_{n,a}(t) = \int_0^t \tfrac{1}{2}[S^a(u) + \hat{S}^a(u)]\hat{G}^{-1/2}(u)\,\mathrm{d}(H^*(u) - H_u(u))$$

with $a \geq 0$. This converges weakly to a zero mean independent increment Gaussian process with variance

$$V_a(t) = \int_0^t S^{(2a-1)}\,\mathrm{d}H$$

which does not depend on the censoring G. Their statistic is

$$K_{n,a} = \sup_{0\leq t\leq T} (V_{n,a}(T))^{-1/2}|Z_{n,a}(t)|$$

with $V_{n,a}(t) = \int_0^t \hat{S}^{(2a-1)}\,\mathrm{d}H_n$ and $T < T_{SG}$. K_+^a and K^a converge in distribution to $\sup B$ and $\sup|B|$, respectively. The parameter $a > 0$ acts as the weighting factor; the early part of the distribution S is more emphasized if $0 < a < 1$ while the tail is more heavily weighted if $a > 1$. This can be seen by noting that

$$-\mathrm{d}(S^a) = -S^{a-1}\,\mathrm{d}S = S^a\,\mathrm{d}H\,.$$

See Fleming and Harrington (1981). Note that T may be replaced by $T_n \xrightarrow{\mathrm{P}} T$ and the weights and transformations discussed in Section 12 may be used here, with the obvious modifications.

One-sided maximal deviation tests and simultaneous confidence bands arise in an analogous manner. See the above references for details.

14. Spacings and total time on test

Barlow and Proschan (1969) first derived the distribution of the total time on test plots under the exponential hypothesis for censored data. Barlow and Campo (1975) considered several types of censoring, showing the form of the total time on test and indicating how censoring may affect the stochastic ordering of scaled total time on test plots. Others (Lurie, Hartley and Stroud, 1974; Mehrotra, 1982) considered weighted spacings tests under type II censoring. Aalen and Hoem (1978) considered the multiplicative intensity model of Aalen (1978), generalizing earlier results to arbitrary censorship. The Aalen–Hoem approach will be considered here.

We construct a random time change on the counting process of failures to derive a stationary Poisson process under the null hypothesis of exponentiality. The total time on test transform, based on this random time change, has the same distribution as that in the noncensored case. Define

$$\psi(t)=\int_0^t R(u)\,\mathrm{d}u$$

in which $R(u)$ is defined as in Section 10. If $t_0=0$ and $t_1<t_2<\cdots<t_k$ are the k distinct failure times, assuming no tied failures, then

$$D_i=\int_{t_{i-1}}^{t_i} R(u)\,\mathrm{d}u=\psi(t_i)-\psi(t_{i-1})$$

is the i-th spacing. Aalen and Hoem (1978) show that if the survival curve is $S(t)=\exp(-H(t))$ then

$$N^*(t)=N(\psi^{-1}(t))$$

is a Poisson process with parameter $h(\cdot)=H'(\cdot)$ (their results are more general). If $S(\cdot)$ is exponential then N^* is a stationary process. Hence

$$N^*(\psi(t_i))=i,\quad i=1,\ldots,k,$$

and $(D_1,\ldots,D_k)$ has the same distribution as a random sample from $S(\cdot)$. For exponential $S(x)=\mathrm{e}^{-\lambda x}$,

$$\begin{aligned}\Pr\{D_1>x\}&=\Pr\{t_1>\psi^{-1}(x)\}=\Pr\{N(\psi^{-1}(x))=0\}\\&=\Pr\{N^*(x)=0\}=\mathrm{e}^{-\lambda x}.\end{aligned}$$

Thus many results for noncensored data apply to $(D_1,\ldots,D_k)$. The scaled total time on test transform is

$$\psi(t_i)/\psi(t_k), \quad i = 0, 1, \ldots, k .$$

This is plotted for some censored data on prostate cancer in Section 17. The tests based on spacings presented in the first part of this paper generalize in a natural way. In particular the cumulative total time on test statistic of Barlow et al. (1972) becomes, for fixed k,

$$V_k = \sum_{i=1}^{k-1} \psi(t_i)/\psi(t_k) .$$

15. Other tests

Several other goodness of fit tests have been proposed in the literature for censored data. These include tests based on contingency tables (Mihalko and Moore, 1980), average deviations (Koziol and Green, 1976; Csörgő and Horvath, 1981; Nair, 1980, 1981), generalized ranks (Breslow, 1975; Hyde, 1977; Hollander and Proschan, 1979; Gill, 1980; Anderson et al., 1981; Harrington and Fleming, 1981), and kernel density or failure rate estimators (Yandell, 1983; see Bickel and Rosenblatt, 1973). We briefly present general forms of the average deviation and generalized linear rank tests.

The average deviation, or Cramér–von Mises, tests are based on weighted average deviations, from the null distribution. Let $K_n(x) = V_n(x)/(1 + V_n(x))$ and

$$Z_n(x) = \sqrt{n}q(K_n(x))(1 - K_n(x))(S_n(x) - S(x))/S(x) .$$

Then the statistics are of the form

$$W_n^i = \int_0^{T_n} |Z_n|^i \, \mathrm{d}K_n, \quad i = 1, 2 ,$$

for specified weight function q. Similar statistics obtain for the hazard function and for the transform based on equation (12.1). Asymptotic distribution of W_n^2 for $q = 1$ corresponds to that of the classical Cramér–von Mises test

$$\psi_n^2 = -n \int (S_n - S)^2 \, \mathrm{d}S_n$$

in the case of no censoring, and is tabled in Pearson and Hartley (1976, Table 54). Koziol and Green (1976) show that ψ_n^2 converges to a distribution which depends on the censoring parameter β of the proportional hazards model (Koziol and Green, 1976). Clearly, the choice of weights $q(\cdot)$ will force emphasis on different aspects of the distribution S.

Generalized linear rank tests take the form

$$\int_0^{T_n} K(s)(\mathrm{d}H_n(s) - \mathrm{d}H^*(s))$$

in which $H^*(t) = \int_0^t I[R(s) > 0]\, \mathrm{d}H(s)$ is the estimable portion of $H(\cdot)$. $K(t)$ is some function of the history of the survival process, $\{(N(u), R(u)),\ u \in [0, t]\}$. If $K(t) = R(t)$, this becomes, with $T_n = Y_{(n)}$,

$$N(T_n) - \int_0^{T_n} T\, \mathrm{d}H = N(T_n) + \sum_{i=1}^{n} \log(S(Y_i))$$

which is equivalent to Breslow's (1975)

$$\left(N(T_n) - \int_0^{T_n} R\, \mathrm{d}H\right)^2 \Big/ \int_0^{T_n} R\, \mathrm{d}H$$

which converges to chi square with one degree of freedom. Hyde's (1977) statistic is a modification of this to allow left truncation. The asymptotic theory for general $K(\cdot)$ is presented in Anderson et al. (1981) and Gill (1980). Finally we mention that Burke (1982) has constructed a test for the hypothesis that both T and C have exponential distributions.

16. Simulation results

A few Monte Carlo results concerning goodness of fit tests with censored data are available. Koziol (1980) compared the censored Kolmogorov–Smirnov test $\hat{D}_n$ of Hall and Wellner (1980), the Cramér–von Mises statistic

$$\hat{W}_n^2 = n \int_0^T (S_n - S)^2/(S^2(1 + V_n)^2)\, \mathrm{d}(V_n/(1 + V_n))$$

and a 'traditional analogue' of the Cramér–von Mises statistic

$$\psi_n^2 = -n \int_0^T (S_n - S)^2\, \mathrm{d}S_n\,.$$

He considered scale $(S(t) = \mathrm{e}^{-\lambda t})$ and Weibull $(S(t) = \exp(-t^\gamma))$ alternatives to the unit exponential in the Koziol–Green (1976) proportional hazards model, with 1000 trials and sample sizes 20 and 50. At level 0.05, ψ_n^2 had the right size, with $\hat{W}_n^2$ a close second. The size of $\hat{D}_n$ was between 0.066 and 0.107, depending on the degree of censoring ($\beta = 0.5, 1$). ψ_n^2 and W_n^2 had better power against Weibull alternatives, but D_n and W_n^2 had more power than ψ_n^2 against scale alternatives. One may be surprised that D_n performed as well as it did, since the alternatives represent small changes along the whole distribution rather than marked change at any one point. The power of D_n against Weibull alternatives dropped from 0.904 to 0.576 as the censoring parameter increased from 0.5 to 1. This suggests looking at the statistics of Fleming and Harrington (1981).

Unfortunately (for our situation), the simulations of Fleming et al. (1980) (Fleming and Harrington, 1981; Harrington and Fleming, 1981) were only done for 2-sample situations. Further, these simulations concern statistics which differ from those considered here. They examine variations on Kolmogorov–Smirnov tests and several linear rank tests.

Hollander and Proschan (1979) compare the Cramér–von Mises statistic ψ_n^2 with two linear rank statistics.

17. Data analysis

Data were obtained from Hollander and Proschan (1979) on 211 patients with stage IV prostate cancer who were treated with estrogen in a Veterans Administration Cooperative Urological Research Group (1967) study. The observations span the years 1967 through March, 1977. Ninety patients died of prostate cancer, 105 died of other diseases and 16 were alive in March, 1977. The live patients and deaths from other causes were counted as censored.

Koziol and Green (1976) failed to reject the hypothesis of exponentiality with parameter $\lambda = 1/100$. Using the ψ_n^2 Cramér–von Mises statistic with the data truncated at an earlier date, Hollander and Proschan (1979) could not reproduce the earlier value of ψ_n^2, but their value and those of the Hyde (1977) and their own test were not significant at $\alpha = 0.10$. The significance probabilities of the tests varied considerably (0.86, 0.49, 0.14, respectively). Csörgő and Horvath (1981) state that Koziol has computed the Cramér–von Mises W_n^2, the Kolmogorov–Smirnov D_n, and the Kuiper statistic with p-values of 0.15, 0.1, and 0.04, respectively. The ordering of p-values reflects the deviation of S_n from S in Figure 1 of Hollander and Proschan (1979). Csörgő and Horvath's (1981) version of the Cramér–von Mises test is somewhat more significant ($p = 0.0405$).

Our graphical tests indicate that the data may not be exponential, or may at least be a borderline situation. Figure 17.1 is the total time on test plot, showing the same criss-cross of the exponential case curve as seen in Figure 1 of Hollander and Proschan (1979). The hazard function plot of Figure 17.2 suggests that the data may be exponential over most of its range, but the rate appears to taper off. Figures 17.3 and 17.4 are both transformations of the survival curve (see Nair, 1981). Confidence bands are 80% based on the Borokov–Sycheva (1968) weights.

The P–P plot in Figure 17.3 shows some discrepancy with the exponential. This is a plot of

$$(u, S_n(S^{-1}(u)), \quad 0 \leq u \leq 1\,,$$

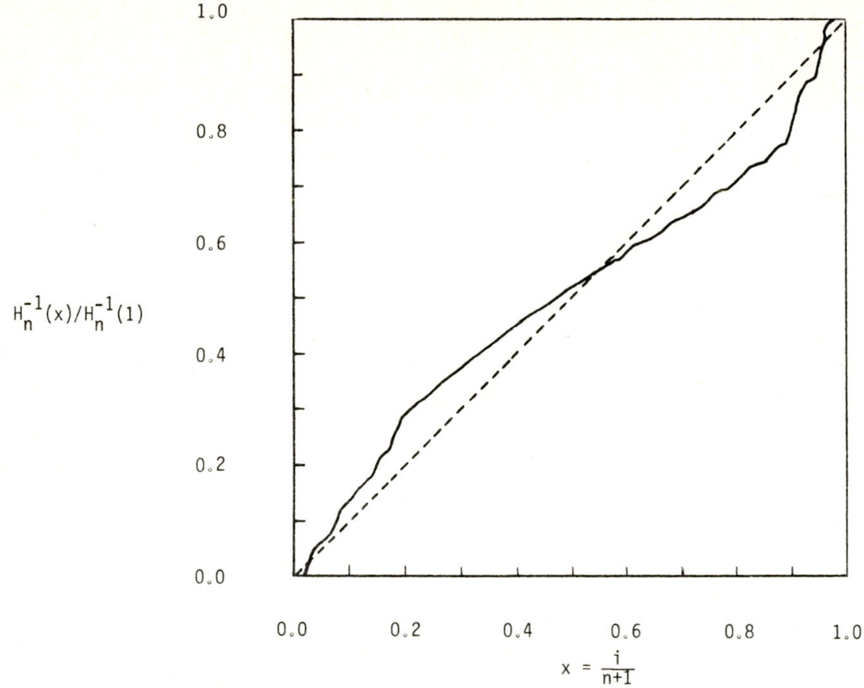

Fig. 17.1. Total time on test plot for the prostate data.

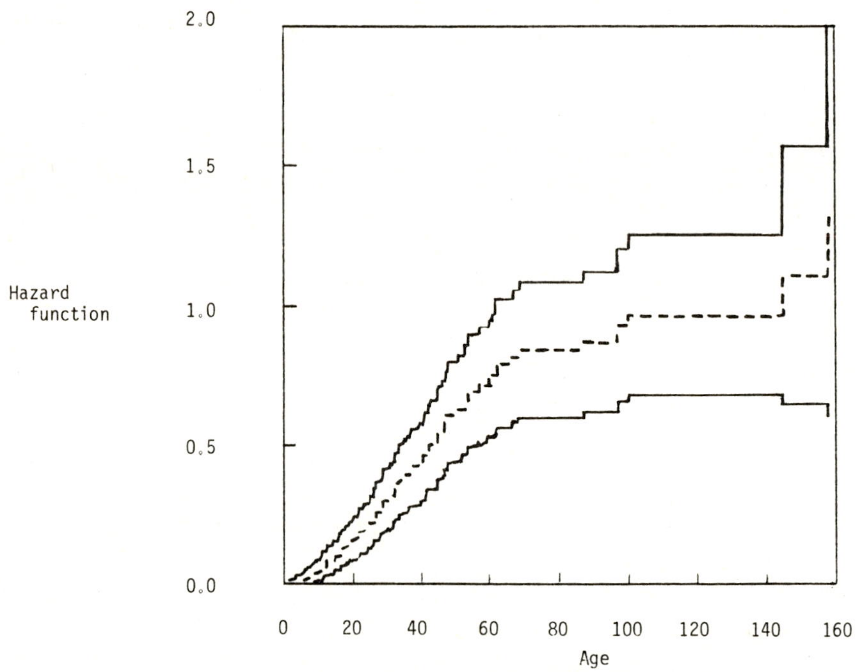

Fig. 17.2. Hazard function plot for the prostate data.

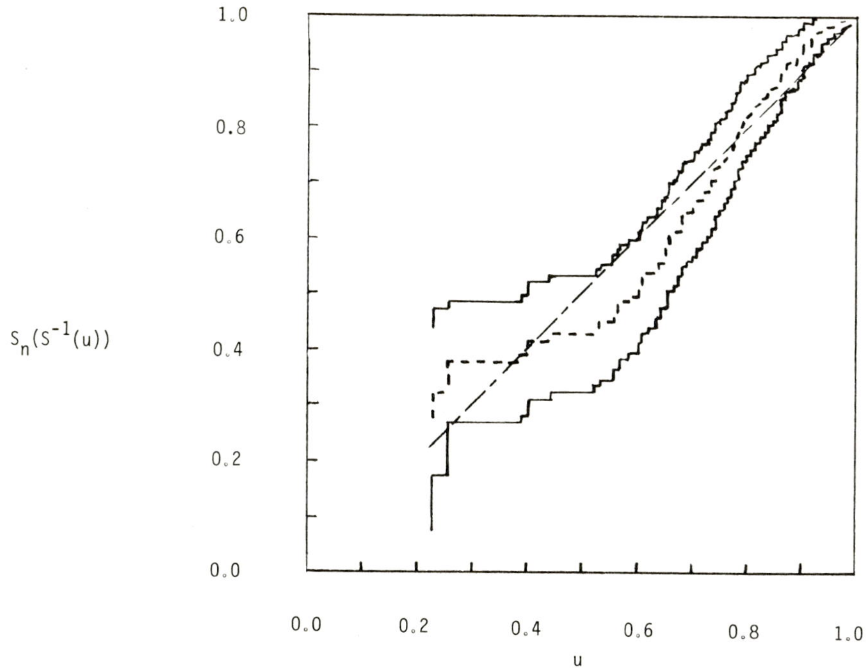

Fig. 17.3. *P–P* plot with 80% simultaneous confidence band for the prostate data. The straight line (diagonal) represents the exponential hypothesis.

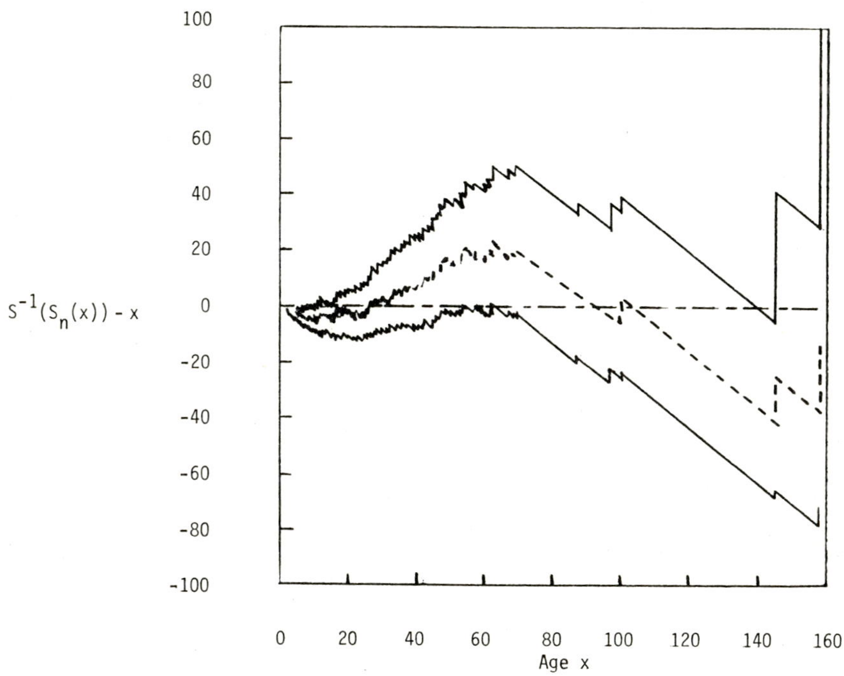

Fig. 17.4. The shift function with 80% simultaneous confidence band for the prostate data. The horizontal line (axis) represents the exponential hypothesis.

along with appropriately transformed 80% simultaneous confidence bands. Note that the confidence bands cross over the diagonal again near the tail ($u \to 0$). Figure 17.4 is a plot of the shift function, which is a version of the *Q–Q* plot (see Doksum and Sievers, 1976; Nair, 1981). The curve is

$$(x, S^{-1}(S_n(x)) - x), \quad x \geqslant 0 .$$

Since $S(u) = e^{-u\lambda}$ is continuous, the shift function, or *Q–Q*, plot and the *P–P* plot contain the same information. The rate parameter for these two plots was estimated from the data as $\hat{\lambda} = 0.00939$.

References

Aalen, O. (1975). Statistical inference for a family of counting processes. Ph.D. dissertation, Department of Statistics, Univ. California, Berkeley.

Aalen, O. (1976). Nonparametric inferences in connection with multiple decrement models. *Scand. J. Statistic.* **3**, 15–27.

Aalen, O. (1978). Nonparametric inference for a family of counting processes. *Ann. Statist.* **6**, 701–726.

Aalen, O. and Hoem, J. (1978). Random time changes for multivariate counting processes. *Scand. Actuarial J. 1978*, 81–101.

Andersen, P. K., Borgan, O., Gill, R. and Kieding, N. (1981). Linear nonparametric tests for comparison of counting processes, with applications to censored survival data. *Inter. Statist. Rev.*, to appear.

Azzam, M. M. (1978). Tests for increasing failure rate and convex ordering. Ph.D. dissertation, Department of Statistics, Univ. California, Berkeley.

Barlow, R. E., Bartholomew, D. J., Bremner, J. M. and Brunk, H. D. (1972). *Statistical Inference under Order Restrictions.* Wiley, New York.

Barlow, R. E. and Campo, R. (1975). Total time on test processes and applications to failure data analysis. In: *Reliability and Fault Tree Analysis*, SIAM, pp. 451–481.

Barlow, R. E. and Doksum, K. A. (1972). Isotomic tests for convex orderings. *Proc. 6th Berkeley Symp. Math. Statist. Probab.* **I**, 293–323.

Barlow, R. E. and Proschan, F. (1966). Inequalities for linear combinations of order statistics from restricted families. *Ann. Math. Statist.* **37**, 1574–1592.

Barlow, R. E. and Proschan, F. (1969). A note on tests for monotone failure rate based on incomplete data. *Ann. Math. Statist.* **40**, 595–600.

Barlow, R. E. and Proschan, F. (1975). *Statistical Theory of Reliability and Life Testing.* Holt, Rinehart and Winston, New York.

Bickel, P. J. (1969). Tests for monotone failure rate II. *Ann. Math. Statist.* **40**, 1250–1260.

Bickel, P. J. and Doksum, K. A. (1969). Tests for monotone failure rate based on normalized spacings. *Ann. Math. Statist.* **40**, 1216–1235.

Bickel, P. J. and Doksum, K. A. (1977). *Mathematical Statistics: Basic Ideas and Selected Topics.* Holden-Day, San Francisco.

Bickel, P. J. and Rosenblatt, M. (1973). On some global measures of the deviations of density function estimates. *Ann. Statist.* **1**, 1071–1095. Correction note (1975), *Ann. Statist.* **3**, 1370.

Birnbaum, Z. W., Esary, J. D. and Marshall, A. W. (1966). Stochastic characterization of wearout for components and systems. *Ann. Math. Statist.* **37**, 816–825.

Block, H. W. (1975) and Savits, T. H. (1976). The IFRA closure problem. *Ann. Probab.* **4**, 1030–1032.

Borovkov, A. A. and Sycheva, N. M. (1968). On asymptotically optimal non-parametric criteria. *Theory Probab. Appl.* **13**, 359–393.

Breslow, N. and Crowley, J. (1974). A large sample study of the life table and product limit estimates under random censorship. *Ann. Statist.* **2**, 437–453.

Burke, M. D. (1982). Tests for exponentiality based on randomly censored data. *Colloquia Math. Soc. J. Bolyai* **32**, 89–101.

Chen, Y. Y., Hollander, M. and Langberg, N. A. (1982). Small-sample results for the Kaplan–Meier estimator. *J. Amer. Statist. Assoc.* **77**, 141–144.

Csörgő, M. and Révész, P. (1981a). *Strong Approximations in Probability and Statistics.* Academic Press, New York.

Csörgő, M. and Révész, P. (1981b). Quantile processes and sums of weighted spacings for composite goodness-of-fit. In: M. Csörgő, D. A. Dawson, J. N. K. Rao and A. K. Md. E. Saleh, eds., *Statistics and Related Topics.* North-Holland, Amsterdam, pp. 69–87.

Csörgő, M., Seshadri, V. and Yalovsky, M. (1975). Applications of Characterizations in the Area of Goodness-of-Fit. In: C. P. Patil, K. Kotz and J. K. Ord, eds., *Statistical Distributions in Scientific Work, Vol. 2.* Reidel, Boston, pp. 79–90.

Csörgő, S. and Horváth, L. (1981). On the Koziol–Green model for random censorship. *Biometrika* **68**, 391–401.

Csörgő, S. and Horváth, L. (1982a). On cumulative hazard processes under random censorship. *Scand. J. Statist.* **9**, 13–21.

Csörgő, S. and Horváth, L. (1982b). On random censorship from the right. *Acta Sci. Math.* (Szeged), **44**, 23–34.

Csörgő, S. and Horváth, L. (1983). The rate of strong uniform consistency for the product-limit estimator. *Z. Warsch. Verw. Geb.* **62**, 411–462.

Doksum, K. A. and Sievers, G. L. (1976). Plotting with confidence: Graphical comparisons of two populations. *Biometrika* **63**, 421–434.

Durbin, J. (1973). *Distribution Theory for Tests Based on the Sample Distribution Function.* Society for Industrial and Applied Mathematics.

Durbin, J. (1975). Kolmogrov–Smirnov tests when parameters are estimated with applications to tests of exponentiality and tests of spacings. *Biometrika* **62**, 5–22.

Efron, B. (1967). The two-sample problem with censored data. *Proc. Fifth Berkeley Symp. Math. Statist. Probabl.* **4**, 831–853.

Epstein, B. (1960). Tests for the validity of the assumption that the underlying distribution of life is exponential. *Technometrics* **2**, 83–101.

Feller, W. (1971). *An Introduction to Probability Theory and its Applications.* Wiley, New York.

Fleming, T. R., O'Fallon, J. R., O'Brien, P. C. and Harrington, D. P. (1980). Modified Kolmogorov–Smirnov test procedures with application to arbitrarily right censored data. *Biometrics* **36**, 607–625.

Fleming, T. R. and Harrington, D. P. (1979). Nonparametric estimation of the survival distribution in censored data. Technical Report Series No. 8, Mayo Clinic.

Fleming, T. R. and Harrington, D. P. (1981). A class of hypothesis tests for one and two sample censored survival data. *Commun. Statist. A* **10**, 763–794.

Földes, A. and Rejtö, L. (1981). Strong uniform consistency for nonparametric survival curve estimators from randomly censored data. *Ann. Statist.* **9**, 122–129.

Gill, R. D. (1980). Censoring and stochastic integrals. Mathematical Centre Tracts 124, Mathematisch Centrum, Amsterdam.

Gill, R. D. (1983). Large sample behaviour of the product-limit on the whole line. *Ann. Statist.* **11**, 59–67.

Gillespie, M. J. and Fisher, L. (1979). Confidence bands for Kaplan–Meier survival curve estimate. *Ann. Statist.* **7**, 920–924.

Hall, W. J. and Wellner, J. A. (1980). Confidence bands for a survival curve from censored data. *Biometrika* **67**, 133–143.

Harrington, D. P. and Fleming, T. R. (1982). A class of rank test procedures for censored survival data. *Biometrika* **69**, 553–566.

Hollander, M. and Proschan, F. (1972). Testing whether new is better than used. *Ann. Math. Stat.* **43**, 1136–1146.

Hollander, M. and Proschan, F. (1979). Testing to determine the underlying distribution using randomly censored data. *Biometrics* **35**, 393–401.

Hollander, M. and Proschan, F. (1984). Nonparametric concepts and methods in reliability. In: P. R. Krishnaiah and P. K. Sen, eds., *Handbook of Statistics, Vol. 4, Nonparametric Methods,* this volume.

Horváth, L. (1980). Dropping continuity and independence assumptions in random censorship models. *Studia Sci. Math. Hung.* **15**, 381–389.

Hyde, J. (1977). Testing survival under right censoring and left truncation. *Biometrika* **64**, 225–230.

Jennen, C. (1981). Asymptotishe Bestimmung von Kenngrössen sequentieller verfahren. Doctorial Dissertation. University of Heidelberg.

Jennen, C. and Lerche, H. R. (1981). First exit densities of Brownian motion through one-sided moving boundaries. *Z. Wahrsch. Verw. Gebiete* **55**, 133–148.

Kalbfleisch, J. D. and Prentice, R. L. (1980). *The Statistical Analysis of Failure Time Data.* Wiley, New York.

Koul, H. L. (1977). A test for new is better than used. *Comm. Statist. A* **6**, 563–573.

Koul, H. L. (1978a). A class of tests for testing "new is better than used". *Canad. J. Statist.* **6**, 249–271.

Koul, H. L. (1978b). Testing for new is better than used in expectation. *Comm. Statist. A*, **7**, 685–701.

Koul, H. L. and Staudte, Jr., R. G. (1976). Power bounds for a Smirnov statistic in testing the hypothesis of symmetry. *Ann. Statist.* **4**, 924–935.

Koziol, J. A. (1980). Goodness-of-fit tests for randomly censored data. *Biometrika* **67**, 693–696.

Koziol, J. A. and Green, S. B. (1976). A Cramer–von Mises statistic for randomly censored data. *Biometrika* 63, 465–474.

Lilliefors, H. W. (1969). On the Kolmogorov–Smirnov test for the exponential distribution with mean unknown. *J. Amer. Statist. Assoc.* **64**, 387–389.

Lurie, D., Hartley, H. O. and Stroud, M. R. (1974). A goodness of fit test for censored data. *Commun. Statist.* **3**, 745–753.

Mantel, N. (1967). Ranking procedures for arbitrarily restricted observations. *Biometrics* **65**, 311–317.

Mehrotra, K. G. (1982). On goodness of fit tests based on spacings for type II censored samples. *Commun. Statist.* **11**, 869–878.

Meier, P. (1975). Estimation of a distribution from incomplete observations. In: J. Gani, ed., *Perspectives in Probab. and Statistic.: Papers in Honour of M. S. Bartlett.* Academic Press, New York, pp. 67–82.

Mihalko, D. P. and Moore, D. S. (1980). Chi square tests of fit for type II censored data. *Ann. Statist.* **8**, 625–644.

Nair, V. N. (1980). Goodness of fit test for multiply right censored data. Tech. Report, 1 Sept. 1980, Bell Tele. Lab., Holmdel, NJ, 22 pp.

Nair, V. N. (1981). Plots and tests for goodness of fit with randomly censored data. *Biometrika* **68**, 99–103.

Nelson, W. (1969). Hazard plotting for incomplete failure data. *J. Qual. Tech.* **1**, 27–52.

Neyman, J. (1959). Optimal asymptotic tests of composite statistical hypothesis. *Probab. and Statist.* The Harold Cramer Volume, Almquist and Wiksells, Uppsala, Sweden, pp. 213–234.

Owen, D. B. (1962). *Handbook of Statistical Tables.* Addison-Wesley, Reading, MA.

Parzen, E. (1979). Nonparametric statistical data modeling. *J. Amer. Statist. Assoc.* **74**, 105–131.

Pearson, E. S. and Hartley, H. O. (1975). *Biometrika Tables for Statisticians, Vol. 2.* Griffin, London.

Prentice, R. L., Kalbfleisch, J. D., Peterson, A. V., Jr., Fournoy, N., Farewell, V. T. and Breslow, N. E. (1978). The analysis of failure times in the presence of competing risks. *Biometrics* **34**, 541–554.

Proschan, F. and Pyke, R. (1967). Tests for monotone failure rate. *Proc. Fifth Berkeley Symp. Math. Statist. Probab.* **3**, 293–312.

Pyke, R. (1965). Spacings. *J. Roy. Statist. Soc. Ser. B* **27**, 395–436.

Sarkadi, K. and Tusnady, G. (1977). Testing for normality and the exponential distribution. *Proc. Fifth Conf. Probab. Theory*. Brasov, Romania, 99–118.

Seshadri, V., Csörgő, M. and Stephens, M. A. (1969). Tests for the exponential distribution using Kolmogorov-type statistics. *J. R. Statist. Soc. B* **31**, 499–509.

Shorack, G. R. (1972). The best test of exponentiality against gamma alternatives. *J. Amer. Statist. Assoc.* **67**, 213–214.

Spiegelhalter, D. J. (1983). Diagnostic tests of distributional shape. *Biometrika* **70**, 401–410.

Stephens, M. A. (1974). EDF statistics for goodness of fit and some comparisons. *J. Amer. Statist. Assoc.* **69**, 730–737.

Störmer, H. (1962). On a test of the exponential distribution. *Metrika* **5**, 128–137.

Veterans Administration Cooperative Urological Research Group. (1967). Treatment and survival of patients with cancer of the prostate. *Surgery, Gynecology, Obstetrics* **124**, 1011–1017.

Yandell, B. S. (1983). Nonparametric inference for rates with censored survival data. *Ann. Statist.* **11**, 1119–1135.

van Zwet, W. (1964). *Convex Transformations of Random Variables*. Math. Centrum, Amsterdam.

P. R. Krishnaiah and P. K. Sen, eds., *Handbook of Statistics, Vol. 4*
© Elsevier Science Publishers (1984) 613–655

27

Nonparametric Concepts and Methods in Reliability

Myles Hollander and Frank Proschan

1. Introduction and summary

In this chapter we survey the use of nonparametric methods in reliability theory. Our survey is not complete, but focuses on classes of life distributions corresponding to various notions of aging. We have chosen this topic because a great deal of the nonparametric analysis in reliability is devoted to these classes of life distributions.

Classes of life distributions based on notions of aging afford nonparametric statisticians an opportunity to consider problems of a character somewhat different from the usual. Instead of assuming that he knows nothing about the underlying life distribution, the statistician assumes that he does not know the parametric form of the life distribution, but that he does know, for example, that the failure rate is increasing. (See Section 2 for definitions.) More generally, he knows that some type of aging property holds for the life distribution; this aging property gives rise to corresponding geometric property for the life distribution.

The chapter is divided into two basic parts. Part A summarizes the probabilistic properties of and logical relationships among these classes of life distributions. Part B surveys nonparametric inference for these classes. We do not give proofs but rather present results and some discussion of them.

Part A. In Section 2, for the classes of life distributions, we present definitions, physical interpretations, useful geometric characterizations, and implications among the classes. In Section 3 we then list for each life distribution class, those reliability operations which lead to closure of the class and those which do not. The reliability operations considered are:

(a) formation of coherent systems,

(b) addition of independent lifelengths (convolution of life distributions),

Research supported by the Air Force Office of Scientific Research, under Grant AFOSR 82-K-0007 to Florida State University and by the National Institute of General Medical Sciences under Grant R01 GM21215 to Stanford University. Part of this research was done while M. Hollander was on sabbatical leave visiting Stanford University.

(c) selection of a lifelength observation from one of a set of distributions (mixture of distributions), and

(d) subjecting a device to shocks.

In Section 4 we describe sample applications of results concerning the life distribution classes. These applications are in maintenance, spares provisioning, checking procedures, bounds for survival function and moment inequalities.

Section 5 considers briefly generalizations to multivariate models and to multistate systems.

Section 6 gives references and comments on the preceding material.

Part B. In Section 7 we consider nonparametric estimators of the distribution function F (or equivalently, the survival function $\bar{F} = 1 - F$). Estimators discussed include the classical empirical distribution function, the maximum likelihood estimator of an F (and its associated failure rate) that is known to be in the 'increasing failure rate' class, the maximum likelihood estimator of an F known to be in the 'increasing failure rate average' class, an estimate of an F that is known to be in the 'new better than used' class, a Bayesian nonparametric estimator, and an empirical Bayes nonparametric estimator.

In Section 8 we consider tests of exponentiality versus the various life distribution classes introduced in Section 2. Also briefly discussed are total-time-on-test plots and empirical mean residual life functions.

Finally, Section 9 contains references to some censored-data generalizations of the inferential procedures considered in Sections 7 and 8, and also references to some other important techniques for censored data.

Part A. Probabilistic Aspects

2. Life distribution classes

We formulate a variety of classes of life distributions based on notions of aging. We define some basic concepts first.

2.1. DEFINITIONS. (a) A life distribution F is a distribution satisfying $F(x) = 0$ for $x < 0$. (Note that $F(0)$ need not be 0. Throughout Part A, distributions will be life distributions unless otherwise stated or clear from the context.)

(b) The corresponding *survival function* (*survival probability, survival distribution*) is $\bar{F} \stackrel{\text{def}}{=} 1 - F$.

(c) The corresponding *lifelength* X satisfies $P[X \leqslant x] = F(x)$ and $P[X > x] = \bar{F}(x)$ for $0 \geqslant x < \infty$.

2.2. DEFINITION . Let F have density f. Then the *failure rate function* $r(x)$ is defined as

$$r(x) = f(x)/\bar{F}(x) \tag{2.1}$$

for x such that $\bar{F}(x) > 0$.

Physically, we may interpret $r(x)\,dx$ as the probability that a unit alive at age x will fail in $(x, x+dx)$, where dx is small. $r(x)$ is also called the *conditional failure rate function*, the *hazard rate function*, and the *intensity function.* WARNING: In some of the literature, $r(x)$ is called the hazard function; this conflicts with our usage, as defined in Definition 2.3.

From (2.1), we obtain by integrating and exponentiating, the well-known identity:

$$\bar{F}(x) = e^{-\int_0^x r(u)\,du} \tag{2.2}$$

for $0 \leq x < \infty$.

2.3. Definition. Assume the failure rate function exists. Then $\int_0^x r(u)\,du$ is called the *hazard function.*

More generally, from (2.2), the hazard function $R(x)$ is given by

$$R(x) = -\ln \bar{F}(x) \tag{2.3}$$

for $x \geq 0$ satisfying $\bar{F}(x) > 0$. Throughout, $R(x)$ will denote the hazard function.

Classes of life distributions based on aging properties

2.4. Definition. Suppose $r(x)$ exists. Then the distribution F has *increasing failure rate* (IFR) if $r(x)\uparrow$ for $0 \leq x < \infty$. More generally, F is IFR if $F(0) = 0$ and

$$\bar{F}(x+y)/\bar{F}(x) \quad \downarrow \text{ in } x \tag{2.4}$$

for $0 \leq x < \infty$ and fixed $y > 0$.

2.5. Notation. $\uparrow$ means nondecreasing; $\downarrow$ means nonincreasing.

2.6. Definition. Let F have failure rate r. Then F has *increasing failure rate average* (IFRA) if $F(0) = 0$ and $(1/x)\int_0^x r(u)\,du \uparrow$ for all $0 < x < \infty$. More generally, F is IFRA if $F(0) = 0$ and $(1/x)R(x) \uparrow$ for $0 < x < \infty$.

2.7. Definition. F is new better than used (NBU) if for all $0 \leq x < \infty$, $0 \leq y < \infty$:

$$\bar{F}(x+y) \leq \bar{F}(x)\bar{F}(y)\,. \tag{2.5}$$

2.8. Definition. F is *new better than used in expectation* (NBUE) if the mean μ of F is finite and

$$\int_0^\infty \bar{F}(x+t)\,\mathrm{d}t \leqslant \mu\bar{F}(x) \tag{2.6}$$

for all $0 \leqslant x < \infty$.

2.9. Definition. F has *decreasing mean residual life* (DMRL) if $F(0) = 0$ and

$$\int_0^\infty \frac{\bar{F}(x+t)}{\bar{F}(x)}\,\mathrm{d}t \quad \downarrow \text{ in } x \tag{2.7}$$

for all $x \geqslant 0$ for which $\bar{F}(x) > 0$.

For the classes of life distributions defined above (and for their duals) we will apply the acronym (say IFR) to the distribution function, survival function, hazard function, random variable, etc., as needed; e.g., we will say 'an IFR random variable'.

A dual class may be defined for each of the life distribution classes defined above by reversing the inequality or direction of monotonicity, and by making appropriate adjustments at end points of support. E.g., we define the DFR distribution by:

2.10. Definition. Suppose $r(x)$ exists. Then the distribution F has *decreasing failure rate* (DFR) if $r(x)$ $\downarrow$ for $0 \leqslant x < \infty$. More generally, F is DFR if

$$\bar{F}(x+y)/\bar{F}(x) \quad \downarrow \text{ in } x\,, \tag{2.8}$$

$0 \leqslant x < \infty$ for fixed $y > 0$.

Note that F may have mass at the origin.

The remaining dual classes are DFRA, NWU (new worse than used), NWUE, IMRL. In the case of NWUE, we permit μ to be infinite.

For the results given below for the IFR, IFRA, NBU, NBUE and DMRL classes, there exist corresponding results for the dual classes; we do not present these explicitly.

Statistical interpretation of life distribution classes

Each of the distribution classes has a simple physical interpretation; this feature is crucial in correctly modelling real-life problems. Thus

2.11. IFR. If the failure rate exists, it is increasing with age. More generally, the conditional probability of completing a mission of fixed duration successfully decreases with initial age of the device.

2.12. IFRA. It has been shown (Birnbaum, Esary and Marshall, 1966) that a coherent system of independent IFR components has an IFRA system life. See

Barlow and Proschan (BP) (1981) for the definition and properties of a coherent system.

It has also been shown that a device subject to shocks governed by a Poisson process, which fails when accumulated damage exceeds a fixed threshold has an IFRA distribution (Esary, Marshall and Proschan, 1973).

2.13. NBU. A used NBU device of any fixed age has stochastically smaller residual lifelength than does a new device. This interpretation of (2.5) follows after dividing both sides by $\bar{F}(x)$.

2.14. NBUE. A weaker comparison of this type is obtained by dividing both sides of (2.6) by $\bar{F}(x)$: A used NBUE device of any fixed age has a smaller mean residual lifelength than does a new device.

2.15. DMRL. The older the device is, the smaller is the mean residual life. This represents a weaker version of the following IFR property: The older the device, the smaller *stochastically* is the residual life.

The remarks above give probabilistic and statistical interpretations of the definitions of the various classes of life distributions. To obtain further mathematical and statistical results, it is helpful to understand basic geometric properties of the life distribution classes.

Geometric properties of the life distribution classes

2.16. IFR. Since $r(x) \uparrow$, it follows that the hazard function $R(x) = \int_0^x r(u)\,du$ is convex. Equivalently, the survival function $\bar{F}(x) = e^{-R(x)}$ is log concave. This, in turn, is equivalent to $\bar{F}(x)$ is a Pólya frequency function of order 2.

2.17. Definition. A function $f(x) \geq 0$ for all real x is a *Pólya frequency function of order 2* (PF_2) if $x_1 < x_2$, $y_1 < y_2$ implies

$$\begin{vmatrix} f(x_1 - y_1) & f(x_1 - y_2) \\ f(x_2 - y_1) & f(x_2 - y_2) \end{vmatrix} \geq 0 .$$

Functions of the PF_2 class enjoy many desirable geometric properties: unimodality, absolute continuity (except possibly at the endpoints of the nonzero domain), monotone likelihood ratio property, etc. Also, the PF_2 class is in the broader class of totally positive functions, for which an extensive theory has been developed. (See Karlin, 1968.)

2.18. IFRA. From Definition 2.6 we see that the hazard function $R(x)$ is starshaped.

2.19. Definition. A function $f(x) \geq 0$ defined on $[0, \infty)$ is *starshaped* if $f(0) = 0$ and $(1/x)f(x) \uparrow$ for $x > 0$.

A useful characterization of starshaped functions is given in

2.20. Characterizations of starshaped functions. $f(x)$ is starshaped iff

$$f(\alpha x) \leq \alpha f(x) \tag{2.9}$$

for $0 \leq x < \infty$, $0 \leq \alpha \leq 1$.

From the definition of IFRA, we obtain a useful geometric characterization:

2.21. IFRA survival function. Let $F(0) = 0$. Then F is IFRA iff $\bar{F}^{1/t}(t) \downarrow$ for $t > 0$.

Another useful geometric characterization is given in

2.22. Single crossing property of IFRA. F is IFRA iff $\bar{F}$ crosses any exponential survival function at most once and if a crossing does occur, $\bar{F}$ crosses the exponential from above. Moreover, if F and the exponential have the same mean, then a single crossing *must* occur.

See BP (1981, Chapter 4), for the proof and applications of this single crossing property in shock models and in the derivation of bounds and inequalities.

NBU Geometric Characterization. From Definition 2.5 we may obtain a geometric characterization for NBU distributions.

2.23. Let life distribution F have hazard function R. Then F is NBU iff R is superadditive.

2.24. Definition. A function $f(x) \geq 0$ defined on $[0, \infty)$ is *superadditive* iff

$$f(x+y) \geq f(x) + f(y) \tag{2.10}$$

for all $x, y \geq 0$.

An interesting characterization of the DMRL distributions may be obtained by reasoning as follows. From Definition 2.9 we have

$$\int_0^\infty \frac{\bar{F}(x+t)}{\bar{F}(x)} \mathrm{d}t \quad \downarrow \text{ in } x,$$

which implies

$$\int_0^\infty \left[\frac{1}{\mu}\bar{F}(x+t) \Big/ \frac{1}{\mu}\bar{F}(x)\right] \mathrm{d}t \quad \downarrow \text{ in } x.$$

This monotonicity, in turn, implies

$$\frac{1}{\mu}\bar{F}(x)\Big/\int_0^\infty \frac{1}{\mu}\bar{F}(x+t)\,\mathrm{d}t \quad \uparrow \text{ in } x\,. \tag{2.11}$$

We identify the numerator $(1/\mu)\bar{F}(x)$ as the density of the forward recurrence time random variable (say Z) in a renewal process in the stationary state, where the underlying distribution is F. (See Cox, 1962.) The ratio in (2.11) then represents the failure rate function of Z. We thus have

2.25. DMRL CHARACTERIZATION. Let F have mean μ. Then F is DMRL iff the survival function $(1/\mu)\int_x^\infty \bar{F}(y)\,\mathrm{d}y$ is IFR.

Some useful applications of this DMRL characterization are:
(1) To test whether F is DMRL, it suffices to test whether $(1/\mu)\int_x^\infty \bar{F}$ is IFR.
(2) To estimate a DMRL F, it suffices to estimate an IFR $(1/\mu)\int_x^\infty \bar{F}$.
In a similar fashion we may characterize the NBUE distribution:

2.26. NBUE CHARACTERIZATION. Let F have mean μ. Then F is NBUE iff the survival function $\bar{F}_1(x) =^{\text{def}} (1/\mu)\int_x^\infty \bar{F}(y)\,\mathrm{d}y$ has a failure rate function r_1 that satisfies

$$r_1(0) \le r_1(x) \tag{2.12}$$

for $0 \le x < \infty$.

Implications among life distribution classes

From the geometric characterizations of the life distribution classes, we readily obtain the implications shown in Figure 2.1. Where no implication is shown, no implication exists, as may be demonstrated by counterexample.

For example, F IFR $\Rightarrow R(0) = 0$ and R convex $\Rightarrow R$ starshaped $\Rightarrow R$ superadditive. This proves the first line of implications in Figure 2.1. The other implications are proved just as easily.

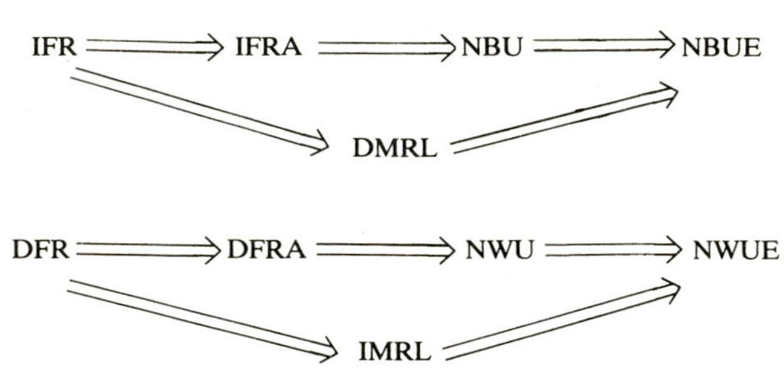

Fig. 2.1. Implications among life distribution classes.

3. Closure and inheritance properties of life distribution classes under reliability operations

The reliability operations that we consider are:

(a) Formation of coherent structures using independent components.

(b) Addition of independent lifelengths; this corresponds to convolution of the corresponding life distributions.

(c) Selection of lifelengths from various distributions in random fashion; this corresponds to mixture of distributions.

(d) Subjecting a device to random shocks; the time of occurrence of the shock and the effect of the shock are random.

In this section, we shall list for each class of life distributions and each of reliability operations (a), (b) and (c), whether closure of the class occurs under the reliability operation. For example, the convolution of IFR distributions is an IFR distribution: we say that the IFR class *is closed under convolution.* Another example: A coherent structure of independent IFR components does not necessarily have an IFR life distribution: we say that the IFR class *is not closed under the formation of coherent structures.*

To understand reliability operation (c), we present:

3.1. Definition. Let $\{F_\alpha\}$ be a set of life distributions indexed by α. Let μ be a probability measure on α. Then

$$F = \int F_\alpha \, \mathrm{d}\mu(\alpha) \tag{3.1}$$

is the *mixture* of life distributions $\{F_\alpha\}$; μ is the *mixing distribution.*

An important special case occurs when α takes on only a finite number of values, say n. Then

$$F = \sum_{i=1}^{n} p_i F_i\,, \tag{3.2}$$

where the mixing distribution places probability p_i on the index i, $i = 1, \ldots, n$. The physical interpretation of (3.2) is that with probability p_i, the distribution F_i is selected and then a lifelength governed by F_i is observed.

As for reliability operation (d), we use the following shock model.

3.2. Shock model. A device is subject to shocks occurring randomly over time according to a Poisson process with event rate λ. The probability that the device survives k shocks is $\bar{P}_k$, where $\bar{P}_0 \geq \bar{P}_1 \geq \bar{P}_2 \geq \cdots$. Then it follows immediately that the probability $\bar{H}(t)$ of survival of the device until time t is given by:

$$\bar{H}(t) = \sum_{k=0}^{\infty} e^{-\lambda t} \frac{(\lambda t)^k}{k!} \bar{P}_k \tag{3.3}$$

for $0 \leq t < \infty$.

Before we state the basic problem, we need to define discrete versions of the classes of continuous life distributions. In Definition 3.3, we define classes corresponding to adverse aging.

3.3. DEFINITION. Let X be a (discrete) lifelength taking on integer values $0, 1, 2, \ldots$. Let $\bar{P}_k = P[X > k]$, $k = 0, 1, 2, \ldots$. Assume $\bar{P}_0 = 1$. (We relax this assumption for the dual classes of discrete life distributions.) We say that:

(a) $\bar{P}_k$ is IFR if $\bar{P}_{k+1}/\bar{P}_k \downarrow$ in $k = 0, 1, 2, \ldots$.

(b) $\bar{P}_k$ is IFRA if $\bar{P}_k^{1/k} \downarrow$ in $k = 1, 2, \ldots$.

(c) $\bar{P}_k$ is NBU if $\bar{P}_{k+l} \leq \bar{P}_k \bar{P}_l$ for $k = 0, 1, 2, \ldots$; $l = 0, 1, 2, \ldots$.

(d) $\bar{P}_k$ is NBUE if $\bar{P}_k$ has finite mean and $\bar{P}_k \sum_{j=0}^{\infty} \bar{P}_j \geq \sum_{j=k}^{\infty} \bar{P}_j$ for $k = 0, 1, 2, \ldots$.

(e) $\bar{P}_k$ is DMRL if $\bar{P}_k$ has finite mean and $\sum_{j=k}^{\infty} \bar{P}_j/\bar{P}_k \downarrow$ in $k = 0, 1, 2, \ldots$.

As in the continuous case, dual classes may be defined corresponding to beneficial aging. For these dual classes, $\bar{P}_0 < 1$ is permitted.

We now present the main result, which essentially states that if the discrete survival function $\bar{P}_k$ is in a given class of discrete life distributions, then the continuous survival function $\bar{H}(t)$ given by (3.3) is in the corresponding class of continuous life distributions.

3.4. THEOREM (Esary, Marshall and Proschan, 1973). *Suppose that* (3.3) *holds. Then*

(a) $\bar{P}_k$ *is discrete IFR* $\Rightarrow$ $\bar{H}(t)$ *is IFR.*

(b) $\bar{P}_k$ *is discrete IFRA* $\Rightarrow$ $\bar{H}(t)$ *is IFRA.*

(c) $\bar{P}_k$ *is discrete NBU* $\Rightarrow$ $\bar{H}(t)$ *is NBU.*

(d) $\bar{P}_k$ *is discrete NBUE* $\Rightarrow$ $\bar{H}(t)$ *is NBUE.*

(e) $\bar{P}_k$ *is discrete DMRL* $\Rightarrow$ $\bar{H}(t)$ *is DMRL.*

A similar theorem holds for the dual classes representing beneficial aging.

Note that Theorem 3.4 does not represent a closure theorem but rather a preservation or inheritance theorem. That is, given a class of discrete life distributions of a certain type, under Shock model 3.2, the corresponding class of continuous life distributions is engendered. Thus the type of life distribution describing discrete $\bar{P}_k$ is inherited by continuous $\bar{H}(t)$.

We may now summarize the closure and inheritance properties of the life distribution classes corresponding to adverse and beneficial aging.

'Mixture of noncrossing distributions' in Table 3.1 refers to the subclass of mixtures (3.1) in which for every $\alpha_1 \neq \alpha_2$, either F_{α_1} is stochastically smaller than F_{α_2} or F_{α_1} is stochastically larger than F_{α_2}.

An interesting feature is shown in Table 3.1. With the exception of the

Table 3.1
Closure and inheritance of classes of life distributions under reliability operations

Class of Life Distribution	Reliability Operation	Formation of Coherent Structure	Convolution of Life Distributions	Mixture of Life Distributions	Mixture of Noncrossing Life Distributions	Survival of Shocks as Described In Model 3.2
Adverse aging	IFR	Not Closed	Closed	Not Closed	Not Closed	Discrete → Continuous
	IFRA	Closed	Closed	Not Closed	Not Closed	Discrete → Continuous
	NBU	Closed	Closed	Not Closed	Not Closed	Discrete → Continuous
	NBUE	Not Closed	Closed	Not Closed	Not Closed	Discrete → Continuous
	DMRL	Not Closed	Not Closed	Not Closed	Not Closed	Discrete → Continuous
Beneficial aging	DFR	Not Closed	Not Closed	Closed	Closed	Discrete → Continuous
	DFRA	Not Closed	Not Closed	Closed	Closed	Discrete → Continuous
	NWU	Not Closed	Not Closed	Not Closed	Closed	Discrete → Continuous
	NWUE	Not Closed	Not Closed	Not Closed	Closed	Discrete → Continuous
	IMRL	Not Closed	Not Closed	Closed	Closed	Discrete → Continuous

DMRL class, each of the classes of life distributions representing adverse aging is closed under convolution of distributions. On the other hand, none of these classes is closed under mixture of distributions. For the classes representing beneficial aging, the reverse is true: Each is closed under mixture of distribution (in the NWU and NWUE cases, the distributions being mixed are noncrossing). On the other hand, none of these beneficial aging life distribution classes is closed under convolution of distributions.

4. Applications

How are these physically motivated classes of life distributions used in reliability applications? Space limitation does not permit detailed descriptions of applications. Instead we content ourselves with sample applications for each of which we sketch the key idea.

4.1. BOUNDS. Knowing that the life distribution is in a certain class permits us to develop a bound for survival probability (or other parameters of interest) given the mean, a specified percentile, or other limited information.

Example:
IFR Survival Function Bound. Let F be IFR with mean μ. Then

$$\bar{F}(x) \geq \begin{cases} e^{-x/\mu} & \text{for } 0 \leq x \leq \mu\,, \\ 0 & \text{for } x \geq \mu\,. \end{cases} \tag{4.1}$$

The bound is sharp.
Similar bounds exist for each class.

4.2. INEQUALITIES. A beautiful example of the inheritance of geometric properties occurs in comparing certain normalized moments.

Moment Comparisons. Let $\lambda_s = \int (x^s/\Gamma(s+1))\,\mathrm{d}F(x)$. Then

$$\bar{F} \text{ IFR } \left(\text{i.e., } \frac{\bar{F}(x+y)}{\bar{F}(x)} \downarrow \text{ in } x \geq 0 \text{ for } y > 0\right)$$

$$\Rightarrow \frac{\lambda_{s+t}}{\lambda_s} \downarrow \text{ in } s \geq 0 \text{ for } t > 0;$$

$\bar{F}$ IFRA (i.e., $\bar{F}^{1/x} \downarrow$ in $x > 0$) $\Rightarrow \lambda_s^{1/s} \downarrow$ in $s > 0$;

$\bar{F}$ NBU (i.e., $\bar{F}(x+y) \leq \bar{F}(x)\bar{F}(y)$ for $x \geq 0, y \geq 0$)
$\Rightarrow \lambda_{s+t} \leq \lambda_s \lambda_t$ for $s \geq 0, t \geq 0$;

$\bar{F}$ NBUE (i.e., $\int \bar{F}(x+y)\,\mathrm{d}y \leq \bar{F}(x) \int \bar{F}(y)\,\mathrm{d}y$ for $x \geq 0$)
$\Rightarrow \lambda_{s+1} \leq \lambda_s \lambda_1$ for $s \geq 0$.

4.3. Optimum spare parts provisioning. We wish to determine the composition of a minimum cost spare parts kit of essential components so as to provide specified assurance of no system shutdown because of spares shortage during a specified period. We know the life distribution and cost of a single unit of each component type.

If each component is IFR, the total spares cost function being minimized is concave in n_i, the number of spares of component type i for all i. A simple algorithm yields the optimum spares kit.

4.4. Optimum checking procedure. A device must be inspected at intervals of time to see if it is still functioning, since otherwise failure may not be known until too late. (Examples of devices of this type are stored batteries, safety systems such as fire detection systems, nuclear radiation monitors, and anti-freeze protection.) Undetected failure costs c_1 per unit of time undetected and each check costs c_2. The failure distribution F is known. We can show that when the density f is PF_2, a simple algorithm yields the optimum sequence of check intervals (not necessarily of equal length) as a function of c_1, c_2 and F.

4.5. Comparison of maintenance policies. Well-known maintenance policies are:

(a) *Age replacement* – Replace at failure *or* at *age* T (a constant), whichever comes first.

(b) *Block replacement* – Replace at failure *and* at chronological times T, $2T$, $3T, \ldots$.

(c) *Replace at failure only.*

Note that Policy (c) generates a renewal process with underlying distribution F, where F is the life distribution of the device. Policy (a) also generates a renewal process, but with underlying distribution F_T, where $F_T(x) = F(x)$ for $x < T$ and $F_T(T) = 1$.

To make comparisons, we need notation:
$N(t)$ = number of renewals in $[0, T]$ under policy (c),
$N_A(t, T)$ = number of failures in $[0, t]$ under policy (a),
$N_B(t, T)$ = number of failures in $[0, t]$ under policy (b).
Comparisons and related results are of the following type:
(1) $N(t) \geq_{st} N_B(t, T)$ for all $t \geq 0$, $T > 0$, $T > 0 \Leftrightarrow F$ is NBU.
(2) $N_A(t, T)$ is stochastically $\uparrow$ in $T > 0$ for each fixed $t \Leftrightarrow F$ is IFR.
(3) $N_A(t, kT) \geq_{st} N_A(t, T)$ for all $t \geq 0$; $T > 0$; $k = 1, 2, \ldots \Leftrightarrow F$ is NBU.
(4) Let $M(t) = EN(t)$, the renewal function for a renewal process with underlying NBUE distribution F having mean μ. Then

$$\frac{t}{\mu} - 1 \leq M(t) \leq \frac{t}{\mu} \quad \text{for } 0 \leq t < \infty . \tag{4.2}$$

This bound on the renewal function holds uniformly on the whole positive line. Note that the error in approximating $M(t)$ by $(t/\mu)-\frac{1}{2}$ is at most $\frac{1}{2}$ no matter how large t is. For many practical reliability problems, such a small error for the expected number of failures under policy (c) is acceptable, making the tedious exact calculation of $M(t)$ unnecessary.

Many additional applications of the classes of life distributions based on notions of aging may be found in the literature. One particular application is referenced at this point because it combines several of the ideas reviewed in this chapter and applies them to a real-life problem: Proschan (1963).

5. Extensions to multivariate life distributions and to multivariate performance

The development of classes of life distributions based on aging has been extended to the multivariate case. Multivariate versions of IFR, IFRA, NBU, NBUE, and of their duals have been defined and their properties have been developed. Actually several multivariate versions of each univariate class of life distributions have been defined. In this brief survey, we dare not go into detail; references are given in Section 6. The chief problem at this stage with these multivariate extensions is that although the multivariate versions are mathematically attractive, there seems to have been no applications of these multivariate life classes, to our knowledge.

Another direction of generalization is based on the fact that for many components and systems, we are interested in levels of performance rather than only on functioning and failure. This has lead naturally to the development of models in which both component performance and system performance are described by *levels of performance*, the levels being indexed by elements of $\{0, 1, \ldots, M\}$ or of $\{[0, \infty)\}$.

The theory is still in the early stages of development. Here too, several competing models have been formulated. Thus far, no applications of the multistate reliability theory have been made to actual systems, as far as we know.

6. Notes and references on probabilistic part

The first systematic treatment of the IFR class appears in Barlow, Marshall and Proschan (1963). The paper proves the closure of the IFR class under convolution, closure of the DFR class under mixture (see Table 3.1), the bound on IFR survival (see 4.1), and a number of other fundamental properties of the IFR class and of the failure rate function. Actually, DFR mixture closure is a consequence of an earlier result of Artin (1964), in which he shows that the sum of log convex functions is log convex.

The IFRA class was introduced in Birnbaum, Esary and Marshall (1966) as the class of life distributions of coherent structures of IFR components. IFRA life distributions have been shown to arise much more generally as first passage time distributions (Brown and Chaganty, 1983). IFRA distributions also describe lifelengths experiencing damage from random shocks under fairly general assumptions (Esary, Marshall and Proschan, 1973). This paper also shows that the renewal random variable $N(t)$ (see 4.5) has a discrete IFRA distribution. The IFRA convolution closure result listed in Table 3.1 is due to Block and Savits (1976).

The NBU and NBUE classes are treated in a systematic way and their fundamental roles in maintenance analysis are shown in Marshall and Proschan (1972).

A convenient single reference for IFR, IFRA, NBU, NBUE is BP (1981). It also contains references which supplement the results presented in the book. See also Mehrotra (1981) for a discussion of mixtures of NWUE.

The DMRL class is receiving a good deal of current attention, especially in biostatistics. (See Chen, Hollander and Langberg, 1983b; and Hollander and Proschan, 1975.) In medical research, one measure of effectiveness of treatment is the mean residual life (MRL) of the patient; MRL, in general, is also a measure of great importance in demography, life insurance, communication between doctor and patient, and comparison of diseases. IMRL mixture closure is discussed in Haines and Singpurwalla (1974) and Brown (1981). Bondesson (1983) showed that the DMRL class is not closed under convolution.

The optimum spares provisioning application of 4.3 is discussed in BP (1965, Chapter 6). A more general treatment appears in BP (1981, Chapter 7). The optimum checking application of 4.4 is described in BP (1965, Chapter 4). Application 4.5 is developed in detail originally in BP (1964) and then under weaker hypotheses in Marshall and Proschan (1972).

Research on multivariate versions of life distribution classes based on notions of aging is currently being vigorously pursued. A survey is presented in Block and Savits (1981).

Systematic treatment of multistate performance of components and systems is in its infancy, but the infant is growing rapidly. A survey is presented in El-Neweihi and Proschan (1980). Additional research is discussed and references are given in Ross (1979), Butler (1979a,b), Griffith (1980), Natvig (1980), and Fardis and Cornell (1981).

Part B. Statistical Aspects

7. Some nonparametric estimators of the distribution function

7.1. THE EMPIRICAL DISTRIBUTION FUNCTION. In the fully nonparametric case when nothing is known about the underlying c.d.f. F, the empirical distribution function (e.d.f.) $\hat{F}_n(x)$ (alternatively called the sample distribution function) is known to possess many desirable properties as an estimator of $F(x)$. For a random sample $X_1, \ldots, X_n$ from F, the e.d.f. is defined as

$$\hat{F}_n(x) = n^{-1} \sum_{i=1}^{n} I\{X_i \leq x\}. \tag{7.1}$$

Since the r.v. $n\hat{F}_n(x)$ has a binomial distribution with parameters n and $p = F(x)$, it is easily seen that

$$E(\hat{F}_n(x)) = F(x), \quad \mathrm{Var}(\hat{F}_n(x)) = n^{-1}F(x)\bar{F}(x)$$

and

$$n^{1/2}(\hat{F}_n(x) - F(x)) \to N(0, \sigma_x^2)$$

where $\sigma_x^2 = F(x)\bar{F}(x)$. Furthermore, the Glivenko–Cantelli theorem states that, as $n \to \infty$, $\sup_{-\infty<x<\infty} |\hat{F}_n(x) - F(x)| \to 0$ almost surely. Regarded as a stochastic process in x, $n^{1/2}(\hat{F}_n(x) - F(x))$ converges to a Gaussian process. Consider the group $\mathcal{G}$ of transformations ϕ, where ϕ is a continuous strictly increasing function from the real line into the real line. For F continuous, Aggarwal (1955) showed that for the group $\mathcal{G}$, $\hat{F}_n$ is a minimax invariant estimator of F under the loss function

$$L(F, \hat{F}) = \int \{[F(x) - \hat{F}(x)]^2/[F(x)\bar{F}(x)]\} \, \mathrm{d}F(x).$$

Dvoretzky, Kiefer and Wolfowitz (1956) have shown that the sample distribution function is asymptotically minimax over a wide class of loss functions. Phadia (1973) proves minimaxity of $\hat{F}_n$ under the loss function

$$L(F, \hat{F}) = \int \{[F(x) - \hat{F}(x)]^2/[F(x)\bar{F}(x)]\} \, \mathrm{d}W(x),$$

where W is a given finite measure (a weight function) on $(\mathbb{R}, \mathcal{B})$, the real line with the σ-field $\mathcal{B}$ of Borel subsets of $\mathbb{R}$. See Read (1971) for a setting in which $\hat{F}_n$ is asymptotically inadmissible.

Confidence bands for $F(x)$ can be obtained by inverting the Kolmogorov goodness-of-fit statistic $D = \sup_{-\infty<x<\infty} |\hat{F}_n(x) - F_0(x)|$; D is used to test the hypothesis H_0: $F(x) = F_0(x)$, where $F_0(x)$ is completely specified. From tables of the distribution of D under H_0 (cf. Birnbaum, 1952) we can find a critical value c_α such that $P\{D \leq c_\alpha \mid H_0\} = 1 - \alpha$. Then the 'curves' $\{\hat{F}_n(x) - c_\alpha, \hat{F}_n(x) + c_\alpha\}$ form a simultaneous confidence band for $F(x)$ with confidence coefficient $1 - \alpha$.

The e.d.f. plays a central role in estimating parameters (other than $F(\cdot)$ itself). For example if $\theta(F)$ is a parameter of interest such as the mean, median, or standard deviation of F, $\theta(\hat{F}_n)$ is a natural estimator of $\theta(F)$.

The e.d.f. also plays a useful role in Efron's (1979) method of *bootstrapping*. Bootstrapping uses the e.d.f. to estimate the sampling distribution of some prespecified r.v. $R(\boldsymbol{X}, F)$ (for example, one may be interested in $R(\boldsymbol{X}, F) = t(\boldsymbol{X}) - \theta(F)$, where $t(\boldsymbol{X})$ is an estimator of $\theta(F)$), where $\boldsymbol{X} = (X_1, \ldots, X_n)$.

Consider the observed realization $\boldsymbol{x} = (x_1, \ldots, x_n)$ of $\boldsymbol{X}$. $\hat{F}_n$ is a d.f. which puts mass n^{-1} at each observed value $x_1, \ldots, x_n$. So Efron suggests drawing, with $\hat{F}_n$ fixed, a random sample of size n from $\hat{F}_n$. This sample $\boldsymbol{X}^* = (X_1^*, \ldots, X_n^*)$ with realization $\boldsymbol{x}^* = (x_1^*, \ldots, x_n^*)$ is called the 'bootstrap sample'. Efron then suggests approximating the sampling distribution of $R(\boldsymbol{X}, F)$ by the 'bootstrap distribution' of $R^* = R(\boldsymbol{X}^*, \hat{F}_n)$. The practical difficulty comes in the actual calculation of the bootstrap distribution. When direct calculation is intractable, Efron suggests generating repeated realizations of $\boldsymbol{X}^*$ by taking random samples of size n from $\hat{F}_n$. Denoting the samples by $\boldsymbol{x}^{*1}, \boldsymbol{x}^{*2}, \ldots, \boldsymbol{x}^{*N}$, the histogram of the corresponding values $R(\boldsymbol{x}^{*1}, \hat{F}_n), R(\boldsymbol{x}^{*2}, \hat{F}_n), \ldots, R(\boldsymbol{x}^{*N}, \hat{F}_n)$ is taken as an approximation to the actual bootstrap distribution. For more on bootstrapping, see Efron (1979), Singh (1981), Bickel and Freedman (1981), Freedman (1981), and Efron and Gong (1983).

7.2. The maximum likelihood estimator when F is known to be IFR (Grenander, Marshall and Proschan). Suppose $X_1, \ldots, X_n$ is a random sample from the life distribution F with density f having increasing failure rate $r(t)$. Using the relationship $\bar{F}(x) = \exp\{-\int_0^x r(t)\,dt\}$, $0 \leq x \leq \infty$, the log of the likelihood function can be written as

$$\log L = \sum_{i=1}^{n} \log r(X_{(i)}) - \sum_{i=1}^{n} \int_0^{X_{(i)}} r(t)\,dt\,, \tag{7.2}$$

where $X_{(1)} \leq \cdots \leq X_{(n)}$ are the ordered X's. $\log L$ can be made arbitrarily large by taking $f(X_{(n)})$ arbitrarily large, so first consider the (constrained) class $\mathcal{F}^M$ of IFR distributions with failure rate bounded by M. Grenander (1956) and Marshall and Proschan (1965) (also see Barlow et al., 1972) first find the unique $\hat{F}_n^M$ (say) in $\mathcal{F}^M$ which maximizes (7.2). Letting $M \to \infty$, it is then shown that $\hat{F}_n^M$ converges in distribution to an estimator $\hat{G}_n$. The latter is called the MLE for F in the class of IFR distributions.

Grenander shows that $\log L$ is maximized over $\mathcal{F}^M$ by a d.f. with failure rate constant between observations, and the estimator for r (corresponding to F in $\mathcal{F}^M$) is

$$r_n^M(X_{(i)}) = \min\left\{\min_{v \geq i+1} \max_{u \leq i} \left[\frac{1}{v-u}(r_u^{-1} + \cdots + r_{v-1}^{-1})\right]^{-1}, M\right\} \tag{7.3}$$

where $r_n = M$ and

$$r_j = \{(n-j)(X_{(j+1)} - X_{(j)})\}^{-1} \quad \text{for } j = 1, 2, \ldots, n-1\,. \tag{7.4}$$

Letting $M \to \infty$, the estimator for r, corresponding to F in the IFR class, is

$$\begin{aligned} \hat{r}_n(X_{(i)}) = \min_{v \geq i+1} \max_{u \leq i} [v-u]^{-1}[(n-u)(X_{(u+1)} - X_{(u)}) + \cdots \\ + (n-v+1)(X_{(v)} - X_{(v-1)})]^{-1}, \quad i = 1, \ldots, n-1\,, \end{aligned} \tag{7.5}$$

$$\hat{r}_n(X_{(n)}) = \infty\,.$$

For the remaining values of x, $\hat{r}_n(x)$ is 0 for $0 \leq x < X_{(1)}$, ∞ for $x \geq X_{(n)}$, constant between observations, and right continuous.

The MLE $\hat{G}_n$ for the d.f. is obtained via

$$\hat{G}_n(x) = 1 - \exp\left[-\int_{-\infty}^{x} \hat{r}_n(u)\,du\right], \tag{7.6}$$

where $\hat{r}_n$ is given by (7.5). Marshall and Proschan (1965) show that if F is IFR with continuous failure rate r, then for all x, $\hat{G}_n(x) \to F(x)$ almost surely. Barlow et al. (1972) point out that $\hat{r}_n$ and $\hat{G}_n$ perform badly in the tails of the distribution.

Rather than compute $\hat{r}_n$ directly via (7.5), a more intuitive (and easier computational) method proceeds as follows. Note that the values $r_j = 1/\{(n-j)(X_{(j+1)} - X_{(j)})\}$ may be considered naive estimates for the constant failure rate on the interval $[X_{(j)}, X_{(j+1)})$, for on that interval the observed 'time on test' is $(n-j)(X_{(j+1)} - X_{(j)})$ and there has been 1 failure. To obtain $\hat{r}_n$, first calculate the naive estimators given by (7.4). If these estimators are in the 'right' order, that is if $r_1 \leq r_2 \leq \cdots \leq r_{n-1}$, then take $r_n(X_{(i)}) = r_i$, $i = 1, \ldots, n-1$. If there is a 'reversal' in that $r_i > r_{i+1}$, then replace both r_i and r_{i+1} by their harmonic mean

$$\{2^{-1}(r_i^{-1} + r_{i+1}^{-1})\}^{-1} = 2^{-1}\{(n-i)(X_{(i+1)} - X_{(i)}) + (n-i+1)(X_{(i+2)} - X_{(i+1)})\}^{-1}.$$

Keep averaging this manner just to the point necessary to eliminate all reversals.

Similar calculations will yield the MLE of a failure rate known to be decreasing over time. See Marshall and Proschan (1965) for details.

Dykstra and Laud (1981) obtain Bayesian estimates of the failure rate and the distribution function for both complete and censored data models.

EXAMPLE 7.1. In an experiment at Florida State University to study the effect of methylmercury poisoning on the lifelengths of fish, goldfish were subjected to various dosages of methylmercury. At one dosage level, the ordered times to death in days were 42, 43, 51, 61, 66, 69, 71, 81, 82, 82 (van Belle, 1972). For these data, we show how to obtain, via pooling, the estimator $\hat{r}_{10}$ given by (7.5). See Table 7.1.

Since the fourth-pool r's are nondecreasing, they yield the MLE. That is,

$$\begin{aligned}
\hat{r}_{10}(x) &= 0 && \text{for } 0 \leq x < 42; \\
&= 0.0210 && \text{for } 42 \leq x < 61; \\
&= 0.0333 && \text{for } 61 \leq x < 66; \\
&= 0.0566 && \text{for } 66 \leq x < 81; \\
&= 0.5000 && \text{for } 81 \leq x < 82; \\
&= \infty && \text{for } x \geq 82.
\end{aligned}$$

Table 7.1
Pooling to obtain $\hat{r}$

j	$X_{(j)}$	$X_{(j+1)} - X_{(j)}$	$n-j$	r_j (naive)	r_j (first pool)	r_j (second pool)	r_j (third pool)	r_j (fourth pool)
1	42	1	9	$1/\{1(9)\} = 0.1111$	$2/(9+64) = 0.0274$	$3/143 = 0.0210$	$3/143 = 0.0210$	$3/143 = 0.0210$
2	43	8	8	$1/\{8(8)\} = 0.0156$	$2/(9+64) = 0.0274$	$3/143 = 0.0210$	$3/143 = 0.0210$	$3/143 = 0.0210$
3	51	10	7	$1/\{10(7)\} = 0.0143$	$1/70 = 0.0143$	$3/143 = 0.0210$	$3/143 = 0.0210$	$3/143 = 0.0210$
4	61	5	6	$1/\{5(6)\} = 0.0333$	$1/30 = 0.0333$	$1/30 = 0.0333$	$1/30 = 0.0333$	$1/30 = 0.0333$
5	66	3	5	$1/\{3(5)\} = 0.0667$	$1/15 = 0.0667$	$1/15 = 0.0667$	$1/15 = 0.0667$	$3/53 = 0.0566$
6	69	2	4	$1/\{2(4)\} = 0.1250$	$1/8 = 0.1250$	$1/8 = 0.1250$	$2/38 = 0.0588$	$3/53 = 0.0566$
7	71	10	3	$1/\{10(3)\} = 0.0333$	$1/30 = 0.0333$	$1/30 = 0.0333$	$2/38 = 0.0588$	$3/53 = 0.0566$
8	81	1	2	$1/\{1(2)\} = 0.5000$	$1/2 = 0.5000$	$1/2 = 0.5000$	$1/2 = 0.5000$	$1/2 = 0.5000$
9	82	0	1	$1/\{0(1)\} = \infty$	∞	∞	∞	∞
10	82							

7.3. THE MAXIMUM LIKELIHOOD ESTIMATOR WHEN F IS KNOWN TO BE IFRA (Barlow, Marshall and Proschan). The happy situation for the IFR class, where the MLE estimators of the d.f. and the failure rate function are strongly consistent (when the underlying d.f. is known to be IFR), is not the case for the IFRA class. Marshall and Proschan (1975) show that the MLE of the d.f. and of the average failure rate function of an IFRA life distribution converge almost surely to functions other than the true ones. (Marshall and Proschan (1975) also show that in the DFRA case, the MLE principle fails completely to produce an estimator.) Marshall and Proschan (1975) (see also Barlow, 1968) first show that the MLE of $\bar{F}$ assumes the form

$$-\ln \hat{\bar{F}}(x) = \begin{cases} 0, & 0 \leq x < X_{(1)}, \\ \lambda_i x, & X_{(i)} \leq x < X_{(i+1)},\ i = 1, \ldots, n-1, \\ \infty, & X_{(n)} \leq x, \end{cases} \tag{7.7}$$

with $\lambda_1 \leq \lambda_2 \leq \cdots \leq \lambda_{n-1}$. Note that $-x^{-1} \ln \hat{\bar{F}}(x)$ is increasing on $[0, \infty)$, so that $\hat{F}$ is IFRA. Furthermore, by choosing $-\ln \hat{\bar{F}}$ linear between ordered observations, no probability is wasted. Rather, probability is assigned to the fullest extent at the order statistics, where it makes the maximum contribution to the likelihood function. Thus, the likelihood function can be expressed as

$$L = n!(1 - e^{-\lambda_1 X_{(1)}})(e^{-\lambda_1 X_{(2)}} - e^{-\lambda_2 X_{(2)}}) \cdots (e^{-\lambda_{n-2} X_{(n-1)}} - e^{-\lambda_{n-1} X_{(n-1)}})\, e^{-\lambda_{n-1} X_{(n)}}.$$

Marshall and Proschan show that L is maximized by the choice

$$\hat{\lambda}_j = X_{(1)}^{-1}\Big\{\log \sum_{i=1}^{n} X_{(i)} - \log \sum_{i=2}^{n} X_{(i)}\Big\} + \cdots + X_{(j)}^{-1}\Big\{\log \sum_{i=j}^{n} X_{(i)} - \log \sum_{i=j+1}^{n} X_{(i)}\Big\}. \tag{7.8}$$

Thus the MLE of $\bar{F}(X_{(j)})$, $j = 1, \ldots, n-1$, is

$$\hat{\bar{F}}(X_{(j)}) = e^{-\lambda_j X_{(j)}} = \left\{\frac{\sum_{i=2}^{n} X_{(i)}}{\sum_{i=1}^{n} X_{(i)}}\right\}^{X_{(j)}/X_{(1)}} \left\{\frac{\sum_{i=3}^{n} X_{(i)}}{\sum_{i=2}^{n} X_{(i)}}\right\}^{X_{(j)}/X_{(2)}} \cdots \left\{\frac{\sum_{i=j+1}^{n} X_{(i)}}{\sum_{i=j}^{n} X_{(i)}}\right\}. \tag{7.9}$$

For values of the MLE of $\bar{F}(x)$ for x between the ordered observations, replace λ_i in (7.7) by $\hat{\lambda}_i$ given in (7.8).

Marshall and Proschan prove that, as $j/n \to u$, $0 < u < 1$,

$$\hat{\lambda}_j \xrightarrow{\text{a.s.}} \lambda_G(u) = \int_0^{F^{-1}(u)} \Big[\int_y^{\infty} x \,\mathrm{d}F(x)\Big]^{-1} \mathrm{d}F(y) \tag{7.10}$$

whereas the true average failure rate $-t^{-1} \log \bar{F}(t)$ can be written as

$$\lambda(u) = \frac{1}{F^{-1}(u)} \int_0^{F^{-1}(u)} \Big[\int_y^{\infty} \mathrm{d}F(x)\Big]^{-1} \mathrm{d}F(y). \tag{7.11}$$

From (7.10) it follows that the MLE of an IFRA F is a strongly consistent estimator of

$$G_F(x) = 1 - \exp\left\{-x \int_0^x \left[\int_y^\infty z \, \mathrm{d}F(z)\right]^{-1} \mathrm{d}F(y)\right\}, \tag{7.12}$$

rather than a consistent estimator of

$$F(x) = 1 - \exp\left\{-x \frac{1}{x} \int_0^x \left[\int_y^\infty \mathrm{d}F(z)\right]^{-1} \mathrm{d}F(y)\right\}.$$

Marshall and Proschan give various examples where G_F differs from F. In the exponential case, where $\bar{F}(x) = \exp(-\lambda x)$, $\hat{\bar{F}}(x) \xrightarrow{\text{a.s.}} (1+\lambda x)\exp(-\lambda x)$ rather than to the (correct) survival function $\exp(-\lambda x)$. In the uniform case, where $F(x) = x$, $0 \leq x \leq 1$, $G_F(x) = 1 - \{(1-x)/(1+x)\}^x$, which differs from the correct d.f. $F(x) = x$.

7.4. An estimator when F is known to be NBU (Samaniego and Boyles). Samaniego and Boyles (1980) addressed the problem of finding, when it is known that the underlying F is NBU, a sequence $\{\tilde{\bar{F}}_n\}$ of NBU distributions such that $\tilde{\bar{F}}_n \xrightarrow{\text{a.s.}} \bar{F}$. As they point out, the estimator $\tilde{\bar{F}}_n$ defined by (7.13) below, only partially succeeds. Samaniego and Boyles study the function

$$\tilde{S}_n(x) = \sup_y \frac{\hat{\bar{F}}_n(x+y)}{\hat{\bar{F}}_n(y)}, \tag{7.13}$$

as an estimator of the survival function $\bar{F}(x)$, where $\hat{\bar{F}}_n$ is the empirical survival function (that is, $\hat{\bar{F}}_n(x) = 1 - \hat{F}_n(x)$ where $\hat{F}_n(x)$ is given by (7.1)) and where the supremum is over nonnegative y for which $\hat{\bar{F}}_n(y) > 0$. Their motivation is as follows. If $\bar{F}_n$ were an NBU survival function (of course in general it is not), then it would satisfy for all appropriate x and y the inequality

$$\hat{\bar{F}}_n(x) \geq \hat{\bar{F}}_n(x+y)/\hat{\bar{F}}_n(y). \tag{7.14}$$

The survival function estimator $\tilde{S}_n$ is constructed to overcome violations of (7.14) by taking for each fixed x, a supremum over all violations. Samaniego and Boyles show that $\tilde{S}_n$ is an NBU survival curve and that it is relatively easy to compute. Computation is simplified since $\tilde{S}_n$ can be rewritten as $\tilde{S}_n(x) = \sup_i \{\hat{\bar{F}}_n(x + X_{(i)})/\hat{\bar{F}}_n(X_{(i)})\}$. Furthermore, $\tilde{S}_n(x)$ is a step function and if it has a jump at x, then $x = X_{(r)} - X_{(s)}$ for some r and s, where $0 \leq s < r \leq n$, with $X_{(0)} \stackrel{\text{def}}{=} 0$. This means that to compute $\tilde{S}_n$, at most $\binom{n+1}{2}$ points need to be checked for possible jumps and at each potential jump point x, comparison of the values $\hat{\bar{F}}_n(x + X_{(i)})/\hat{\bar{F}}_n(X_{(i)})$ yields the value of $\tilde{S}_n$.

Samaniego and Boyles show that, in general, $\tilde{S}_n$ is not a consistent estimator of $\bar{F}$ (in fact, it is not even consistent when F is exponential). The lack of consistency is due to the fact that the tail behavior of $\hat{\bar{F}}_n$ plays a critical role in the asymptotic behavior of $\tilde{S}_n$. Samaniego and Boyles do show that $\tilde{S}_n$ is a strongly uniformly consistent estimator of $\bar{F}$ when F is NBU and $T = \inf\{M: F(M) = 1\}$ is well-defined and finite.

7.5. A Bayes nonparametric estimator of the distribution function (Ferguson). Ferguson (1973) has introduced the Dirichlet process prior on space of distribution functions. This provides a method for treating F as random, and obtaining Bayesian nonparametric estimates of a d.f., of the mean of a d.f., of percentiles of a d.f., and of other parameters. Here we will informally describe the Dirichlet process (the reader should refer to Ferguson, 1973, for further details) and Ferguson's Bayesian nonparametric estimator of a d.f.

Let $\alpha(\cdot)$ be a nonnegative finite measure on the real line $\mathbb{R}$ with the Borel σ-field $\mathcal{B}$. Then P is a Dirichlet process on $(\mathbb{R}, \mathcal{B})$ with parameter α if, for every $m = 1, 2, \ldots,$ and every measurable partition $B_1, \ldots, B_m$ of $\mathbb{R}$, $(P(B_1), \ldots, P(B_m))$ has a Dirichlet distribution with parameter $(\alpha(B_1), \ldots, \alpha(B_m))$. See Wilks (1962) and Ferguson (1973) for the definition and basic properties of the Dirichlet distribution.

A sample from the randomly chosen F selected by the Dirichlet process can be viewed as follows. First, a d.f. F is chosen according to the Dirichlet process prior, and then, given F, a random sample is obtained from F. Ferguson (1973, Theorem 1) shows that the posterior distribution of the Dirichlet process P with parameter α, given a sample $X_1, \ldots, X_n$ from F, is again a Dirichlet process with updated parameter $\alpha + \Sigma_{i=1}^n \delta_{X_i}$, where δ_z is the measure which concentrates a unit mass at the point z.

Now consider the problem of estimating $F(x) = P((-\infty, x])$ with loss function $L(P, \hat{F}) = \int (F(x) - \hat{F}(x))^2 \,\mathrm{d}W(x)$, where the action space is the space of all d.f.'s on $\mathbb{R}$, and where W is a given finite measure on $(\mathbb{R}, \mathcal{B})$. If P is chosen according to the Dirichlet process, then, for each x, the r.v. $F(x)$ has a Beta distribution with parameters $\alpha((-\infty, x])$, $\alpha((x, \infty))$. The Bayes risk for the no-sample problem ($n = 0$) is

$$EL(P, \hat{F}) = \int E(F(x) - \hat{F}(x))^2 \,\mathrm{d}W(x), \tag{7.15}$$

and the right-hand-side of (7.15) is minimized by choosing $\hat{F}(x)$ for each x to minimize $E(F(x) - \hat{F}(x))^2$. That is, by taking

$$\hat{F}(x) = EF(x) = \frac{\alpha((-\infty, x])}{\alpha((-\infty, x]) + \alpha((x, \infty))} = \frac{\alpha((-\infty, x])}{\alpha(\mathbb{R})} \overset{\text{def}}{=} F_0(x). \tag{7.16}$$

The d.f. $F_0(x)$ can be viewed as the prior guess at the unknown $F(x)$. Using the result that the posterior distribution of P, given a sample $X_1, \ldots, X_n$, is

Dirichlet with parameter $\alpha + \Sigma_{i=1}^{n} \delta_{X_i}$, in conjunction with (7.16), gives the Bayes estimator for a sample of size n as

$$\hat{F}_B(x) = p_n F_0(x) + (1 - p_n)\hat{F}_n(x) \tag{7.17}$$

where

$$p_n = \alpha(\mathbb{R})/(\alpha(\mathbb{R}) + n), \tag{7.18}$$

$\hat{F}_n$ is the e.d.f., and F_0 is given by (7.16).

Ferguson's Bayes estimate is a mixture of the prior choice F_0 and the e.d.f. Note that if $\alpha(\mathbb{R})$ is small relative to n, then little weight is given to the prior choice, but if $\alpha(\mathbb{R})$ is large relative to n, little weight is given to the observations. Thus Ferguson views $\alpha(\mathbb{R})$ as a measure of belief in the prior choice F_0, measured in units of numbers of observations. As $\alpha(\mathbb{R}) \to 0$, the case Ferguson calls the noninformative prior, $\hat{F}_B$ converges to the e.d.f. (Korwar and Hollander (1976) present a different context and motivation for viewing $\alpha(\mathbb{R})$ as a measure of belief in the prior choice, but see Sethuraman and Tiwari (1983) for a probabilistic setting in which very small values of $\alpha(\mathbb{R})$ correspond to very definite information concerning the unknown true distribution.)

The estimator $\hat{F}_B$ converges to the true F uniformly almost surely. This is a consequence of the Glivenko–Cantelli theorem and the fact that $p_n \to 0$ as $n \to \infty$.

The Bayes risk $R(\alpha)$ of $\hat{F}_B$ with respect to the Dirichlet prior using weighted squared-error loss is

$$R(\alpha) = E_{\boldsymbol{X}}\left[\int \{E_{F(x)|\boldsymbol{X}}(F(x) - \hat{F}_B(x))^2\}\, \mathrm{d}W(x)\right],$$

where $\boldsymbol{X} = (X_1, \ldots, X_n)$. Korwar and Hollander (1976) and Goldstein (1975b) show that

$$R(\alpha) = [\alpha(\mathbb{R})/\{(\alpha(\mathbb{R}) + 1)(\alpha(\mathbb{R}) + n)\}] \int F_0(x)(1 - F_0(x))\, \mathrm{d}W(x). \tag{7.19}$$

More general classes of nonparametric Bayes estimators of F, which include $\hat{F}_B$ as a special case, are considered by Doksum (1972), Ferguson (1974), Doksum (1974) and Goldstein (1975a, 1975b).

7.6. An empirical Bayes nonparametric estimator of the distribution function (Hollander and Korwar). Motivated by Ferguson's Bayes estimator (7.17), Korwar and Hollander (1976) and Hollander and Korwar (1977) proposed an empirical Bayes estimator of F which requires less information about the prior choice $\alpha(\cdot)/\alpha(\mathbb{R})$ than does Ferguson's estimator. The empirical Bayes model is as follows. Let $(F_i, \boldsymbol{X}_i)$, $i = 1, 2, \ldots,$ be a sequence of pairs of independent elements. The F's are random probability measures which have common prior distribution given by a Dirichlet process on $(\mathbb{R}, \mathcal{B})$. Given $F_i = F'$ (say), $\boldsymbol{X}_i = (X_{i1}, \ldots, X_{im_i})$ is a random sample of size m_i from F'. For the problem of estimating F_{n+1} on the basis of $\boldsymbol{X}_1, \ldots, \boldsymbol{X}_{n+1}$, Hollander and Korwar (1977) propose a sequence $H = \{H_{n+1}\}$ of estimators which for weighted squared-error loss is asymptotically optimal in the sense of Robbins (1964). The proposed sequence is for $n = 1, 2, \ldots,$

$$H_{n+1}(x) = p_{n+1} \sum_{i=1}^{n} \hat{F}_i(x)/n + (1 - p_{n+1})\hat{F}_{n+1}(x), \tag{7.20}$$

where $p_{n+1} = \alpha(\mathbb{R})/\{\alpha(\mathbb{R}) + m_{n+1}\}$ and $\hat{F}_i$ is the e.d.f. of $\boldsymbol{X}_i$. Hollander and Korwar (1977) compare the performance of H_{n+1} with that of the e.d.f. $\hat{F}_{n+1}$ and show that the inequality

$$m_{n+1}^{-1} > \frac{\{\alpha(\mathbb{R}) \sum_{i=1}^{n} m_i^{-1}\} + n}{n^2\{\alpha(\mathbb{R}) + m_{n+1}\}} \tag{7.21}$$

is a necessary and sufficient condition for the Bayes risk of $\hat{F}_{n+1}$, with respect to the Dirichlet process prior, to be larger than the overall expected loss using H_{n+1}. A sufficient condition for H_{n+1} to be better than $\hat{F}_{n+1}$ is

$$n \min(m_1, m_2, \ldots, m_{n+1}) > \max(m_1, m_2, \ldots, m_{n+1}). \tag{7.22}$$

Another sufficient condition for H_{n+1} to be better than $\hat{F}_{n+1}$ is

$$(2n-1) \min(\alpha(\mathbb{R}), m_1, m_2, \ldots, m_{n+1}) > \max(\alpha(\mathbb{R}), m_1, m_2, \ldots, m_{n+1}). \tag{7.23}$$

The sequence of estimators given by (7.20) can also be used for the problem of simultaneously estimating $n + 1$ distribution functions. For note that by interchanging the roles of samples 1 and $n + 1$, the H estimator defined by (7.20) becomes an estimator of F_1 based on $\boldsymbol{X}_1$ and the 'past' samples $\boldsymbol{X}_2, \ldots, \boldsymbol{X}_{n+1}$. More generally, an estimator of F_j based on all the samples is

$$H_j(x) = p_j \sum_{\substack{i=1 \\ i \neq j}}^{n} \hat{F}_i(x)/n + (1 - p_j)\hat{F}_j(x), \quad j = 1, 2, \ldots, n+1, \tag{7.24}$$

where

$$p_j = \alpha(\mathbb{R})/\{\alpha(\mathbb{R}) + m_j\}. \tag{7.25}$$

Note that if all the sample sizes are equal condition (7.22) reduces to $n > 1$ (the latter condition was given in Theorem 3.1 of Korwar and Hollander (1976)). Their result is reminiscent of, though much weaker than, the famous James and Stein (1961) result (see also Stein, 1955, and Efron and Morris, 1975) for simultaneous estimation of k normal means. The James–Stein estimator does better, for each point in the parameter space, when $k \geqslant 3$, in terms of mean squared error, than the classical rule which estimates each population mean by its sample mean. The Korwar–Hollander result says that in the equal sample size case, if there are at least three distribution functions to be estimated, one can do better (not pointwise for each point in the parameter space but on the average where the average is with respect to the Dirichlet prior) than using, for each distribution, the corresponding sample distribution function.

EXAMPLE 7.2. The data of Table 7.2, adapted from Proschan (1963), give the intervals between successive failures of the air conditioning systems of three '720' jet airplanes. We use these data to illustrate the estimators defined by (7.24).

Table 7.2
Intervals between failures of air conditioning systems

Plane		
7912	7913	7914
23	97	50
261	51	44
87	11	102
7	4	72
120	141	22
14	18	39
62	142	3
47	68	15
225	77	197
71	80	188
246	1	79
21	16	88
42	106	46
20	206	5
5	82	5
12	54	36
120	31	22
11	216	139
3	46	210
14	111	97
71	39	30
11	63	23
14	18	13
11	191	14
16	18	
90	163	
1	24	
16		
52		
95		

For the data in Table 7.2, $n+1=3$, $m_1=30$, $m_2=27$ and $m_3=24$, and note that in this case, inequality (7.22) is satisfied. To compute the H estimators given by (7.24) we need only specify $\alpha(\mathbb{R})$, whereas to utilize Ferguson's Bayes estimator we must fully specify the prior measure $\alpha(\cdot)/\alpha(\mathbb{R})$. Ferguson (1973) gives a justification for interpreting $\alpha(\mathbb{R})$ as the 'prior sample size' of the process. Note that as $\alpha(\mathbb{R})$ decreases, the estimator H_j of F_j puts more weight on the observations from the j-th sample, and less weight on the observations from the other samples. For purposes of illustration, we take $\alpha(\mathbb{R})=7$ and from (7.25) we find $p_1=7/(7+30)=0.19$, $p_2=0.21$, $p_3=0.23$, so that from (7.24) we obtain

$$H_1(x)=0.19\{\hat{F}_2(x)+\hat{F}_3(x)\}/2+0.81\hat{F}_1(x)\,,$$
$$H_2(x)=0.21\{\hat{F}_1(x)+\hat{F}_3(x)\}/2+0.79\hat{F}_2(x)\,,$$
$$H_3(x)=0.23\{\hat{F}_1(x)+\hat{F}_2(x)\}/2+0.77\hat{F}_3(x)\,.$$

8. Tests for nonparametric classes of life distributions

8.1. An IFR test motivated by the total-time-on-test-transform (Klefsjö). The total-time-on-test (TTT)-transform has been advocated by Barlow and Campo (1975) and others as a useful method for graphical analysis of life data, and it has also been used by Klefsjö (1983) and others to derive tests of exponentiality versus various nonparametric alternatives.

For a life distribution F with finite mean μ, the *TTT-transform* H_F^{-1} of F is defined as $H_F^{-1}(x)=\int_0^{F^{-1}(x)}\bar{F}(u)\,\mathrm{d}u$ for $0\leq x\leq 1$, where $F^{-1}(x)=\inf\{u\colon F(u)\geq x\}$. Since there is a 1–1 correspondence between life distributions and their TTT-transforms, the TTT-transform can be viewed as a central device for studying properties of F (in the same way as are the characteristic function, the failure rate function, the mean residual life function, etc.). The transform $\phi_F(x)=\mu^{-1}\int_0^{F^{-1}(x)}\bar{F}(t)\,\mathrm{d}t$, where $\mu=H_F^{-1}(1)=\int_0^\infty\bar{F}(t)\,\mathrm{d}t$, is scale invariant and is called the *scaled TTT-transform*. The *empirical scaled TTT-transform* is defined by

$$\phi_n(x)=H_n^{-1}(x)/H_n^{-1}(1),\quad 0\leq x\leq 1\,, \tag{8.1}$$

where

$$H_n^{-1}(x)=\int_0^{\hat{F}_n^{-1}(x)}\bar{\hat{F}}_n(t)\,\mathrm{d}t,\quad 0\leq x\leq 1\,,$$

and $\hat{F}_n$ is the e.d.f. It can be shown that $H_n^{-1}(j/n)=S_j/n$, $j=0,\ldots,n$, where

$$S_j=\sum_{i=1}^{j}(n-i+1)(X_{(i)}-X_{(i-1)})\,,$$

and $X_{(1)}<\cdots<X_{(n)}$ are the order statistics of a random sample $X_1,\ldots,X_n$ from F, $X_{(0)}\overset{\text{def}}{=}0$ and $S_0=0$. S_j is called the total time on test at $X_{(j)}$. If we set

$$U_j = S_j/S_n, \quad j = 0, 1, \ldots, n, \tag{8.2}$$

then $\phi_n(j/n) = U_j$. The *TTT-plot* is a plot of U_j versus j/n for $j = 0, 1, \ldots, n$, with plotted adjacent points connected by straight lines. Since $\phi_n(j/n)$ converges to $\phi_F(x)$ a.s. uniformly in $[0, 1]$ as $n \to \infty$ and $j/n \to x$, Barlow and Campo (1975) suggest comparing TTT-plots with graphs of scaled TTT-transforms for making subjective inferences about models governing the underlying F.

Motivated by the result that F is IFR (DFR) if and only if the scaled TTT-transform ϕ_F is concave (convex) (results due to Barlow and Campo, 1975), Klefsjö suggests a statistic A that should reflect evidence of concavity (convexity) in the TTT-plot. His statistic is

$$A = \sum_{j=0}^{n-2} \sum_{k=2}^{n-j} \sum_{i=1}^{k-1} \{k(U_{j+i} - U_j) - i(U_{j+k} - U_j)\}. \tag{8.3}$$

Significantly large values of A lead to rejection of

$$H_0: \quad F(x) = 1 - \exp(-\lambda x), \quad x \geq 0, \ \lambda > 0 \ (\lambda \text{ unspecified}),$$

in favor of

$$H_1: \quad F \text{ is IFR (and not exponential)}.$$

H_0 is rejected in favor of

$$H_1': \quad F \text{ is DFR (and not exponential)}$$

if A is significantly small. Klefsjö shows that A can be written in terms of the normalized spacings $D_i = (n - i + 1)(X_{(i)} - X_{(i-1)})$, $i = 1, \ldots, n$, as

$$A = \sum_{j=1}^{n} \alpha_j D_j / S_n \tag{8.3'}$$

where

$$\alpha_j = 6^{-1}\{(n+1)^3 j - 3(n+1)^2 j^2 + 2(n+1) j^3\}. \tag{8.4}$$

The null distribution of A can be determined by using the result that under H_0, $D_1, D_2, \ldots, D_n$ are iid according to $F(x) = 1 - \exp(-\lambda x)$. Klefsjö provides null distribution tables of A^* (given by (8.5)) for the sample sizes $n = 5(5)75$, giving the upper and lower 0.01, 0.05, 0.10 percentiles. Klefsjö also shows that under H_0 the statistic

$$A^* = A(7560/n^7)^{1/2} \tag{8.5}$$

can be treated asymptotically as an $N(0, 1)$ random variable. Klefsjö also shows that the test which rejects for large values of A is consistent against the class of continuous IFR distributions.

Other IFR tests. Many tests of H_0 versus H_1 have been based on the normalized spacings, while other tests utilize only the ranks of the normalized spacings. Bickel and Doksum (1969) study both types of tests based on spacings. Their tests are partially motivated by a result of Proschan and Pyke (1967) that shows that when F is IFR, the D's exhibit a decreasing trend in that $P(D_i \geq D_j) < \frac{1}{2}$ whenever $i > j$. This led Bickel and Doksum to define a test function $\phi = \phi(D_1, \ldots, D_n)$ (the probability of rejecting H_0 in favor of H_1, given the D's) to be monotone in the D's if

$$\phi(D'_1, \ldots, D'_n) \leq \phi(D_1, \ldots, D_n) \quad \text{for all } (D_1, \ldots, D_n)$$

and

$$(D'_1, \ldots, D'_n) \text{ such that } i < j \text{ and } D'_i \geq D'_j \text{ implies } D_i \geq D_j .$$

Bickel and Doksum show that all monotone tests are rank tests. Furthermore, letting R_i denote the rank of D_i in the joint ranking from least to greatest of $D_1, \ldots, D_n$, they showed that the rank test which rejects H_0 in favor of H_1 for large values of $W_1 = \sum_{i=1}^{n} i \log(1 - R_i/(n+1))$ is asymptotically most powerful for IFR Makeham alternatives in the class of linear rank statistics. Bickel and Doksum also found that the Pitman asymptotic relative efficiency $e(W_1, M)$, of W_1 with respect to the Proschan–Pyke (1967) rank statistic M, is equal to $\frac{4}{3}$ for all sequences of alternatives $\{F_{\theta_n}\}$ tending to H_0. The Proschan–Pyke statistic is $M = \sum_{i<j} \psi(R_i, R_j)$ where $\psi(a, b) = 1$ if $a > b, 0$ otherwise, and H_0 is to be rejected in favor of H_1 for large values of M.

Although W_1 dominates M with respect to Pitman asymptotic relative efficiency, that result does not hold for finite n and fixed IFR alternatives. Furthermore, although null distribution tables for W_1 are easily generated (using the fact that under H_0 all $n!$ possible values of $(R_1, \ldots, R_n)$ are equally likely) we are unaware of such tables, whereas the M-statistic can easily be referred to published tables of Kendall's rank correlation coefficient. (Specifically, refer $2M - n(n-1)/2$ to the null distribution of the statistic K as given in Table A.21 of Hollander and Wolfe (1973).) A normal approximation treats

$$M^* = \{M - [n(n-1)/4]\}\{n(n-1)(2n+5)/72\}^{-1/2}$$

as a standard normal r.v. under H_0.

Other rank statistics considered by Bickel and Doksum are

$$W_0 = -\sum_{i=1}^{n} iR_i, \qquad W_2 = \sum_{i=1}^{n} \log(1 - i/(n+1)) \log(1 - R_i/(n+1)),$$

$$W_3 = \sum_{i=1}^{n} \log\{-\log(1 - i/(n+1))\} \log(1 - R_i/(n+1)),$$

$$W_4 = -\sum_{i=1}^{n} g(i/(n+1)) \log(1 - R_i/(n+1)),$$

where $g(t) = (1-t)^{-1} \int_{-\log(1-t)}^{\infty} x^{-1} e^{-x} dx$. Large values lead to rejection of H_0 in favor of H_1. Bickel and Doksum show that W_0 is asymptotically equivalent to Proschan and Pyke's M, and that, in the class of linear rank statistics, W_2 is asymptotically most powerful for linear failure rate alternatives, W_3 is asymptotically most powerful for Weibull alternatives, and W_4 is asymptotically most powerful for gamma alternatives.

Bickel and Doksum also consider test statistics, such as the total-time-on-test statistic, based on studentized spacings statistics. (We discuss the total-time-on-test statistic in the context of a test for NBUE alternatives in Section 8.5.) In particular, Bickel and Doksum show that the rank tests, despite being as good in terms of Pitman asymptotic relative efficiency as their counterparts based on studentized linear spacings, are less powerful than their counterparts based on the studentized linear spacings.

8.2. An IFRA test motivated by the total-time-on-test-transform (Klefsjö). Barlow and Campo (1975) proved that if F is a life distribution which is IFRA (DFRA), then $\phi_F(t)/t$ is decreasing (increasing) for $0<t<1$. Thus, since $\phi_F(t)/t$ being decreasing is a necessary (it is not sufficient) condition for F to be IFRA, Klefsjö (1983) proposes a statistic which investigates whether the analogous property tends to hold for the TTT-plot. If F is IFRA, we would expect $U_i/(i/n) > U_j/(j/n)$ for all $j>i$, and $i = 1, 2, \dots, n-1$. This suggests the statistic

$$B = \sum_{i=1}^{n-1} \sum_{j=i+1}^{n} (jU_i - iU_j). \tag{8.6}$$

Significantly large values of B lead to rejection of

H_0: $F(x)$ is exponential

in favor of

H_2: F is IFRA (and not exponential)

H_0 is rejected in favor of

H_2': F is DFRA (and not exponential)

if B is significantly small. Klefsjö shows that B can be written in terms of the normalized spacings as

$$B = \sum_{j=1}^{n} \beta_j D_j / S_n, \tag{8.6$'$}$$

where

$$\beta_j = 6^{-1}\{2j^3 - 3j^2 + j(1-3n-3n^2) + 2n + 3n^2 + n^3\}. \tag{8.7}$$

Klefsjö provides null distribution tables of B^* (given by (8.8)) for the sample sizes $n = 5(5)75$, giving the upper and lower 0.01, 0.05, 0.10 percentiles. He also shows that, under H_0, the statistic

$$B^* = B(210/n^5)^{1/2} \tag{8.8}$$

can be treated (asymptotically) as an $N(0, 1)$ variable.

Klefsjö also shows that the test that rejects for large values of B is consistent against the class of continuous IFRA distributions.

Other IFRA tests. Barlow (1968) derives a likelihood ratio statistic, lower percentiles of which are for testing exponentially versus IFRA, upper percentiles of which are intended for testing IFRA versus DFRA. He gives tables for $n = 2(1)10$, and percentiles 0.01, 0.05, 0.10, 0.90, 0.95, 0.99. Other tests of exponentiality versus IFRA alternatives, motivated by total-time-on-test processes, are given by Barlow and Campo (1975) and Bergman (1977). Specifically, Barlow and Campo suggest the statistic L = 'number of crossings between the TTT-plot and the 45° line'. Bergman compares L with the TTT-statistic (discussed here in Section 8.5 as a test for NBUE alternatives) and finds the latter statistic superior to L.

Motivated by the fact that F is IFRA if and only if, for $x > 0$, $0 < b < 1$, $\bar{F}(bx) \geq \{\bar{F}(x)\}^b$, Deshpande (1983) defines a class of tests based on the statistics $J_b = n(n-1)^{-1} \Sigma^* h_b(X_i, X_j)$ where Σ^* denotes summation over $1 \leq i \leq n$, $1 \leq j \leq n$ such that $i \neq j$, and

$$h_b(X_1, X_2) = 1 \quad \text{if } X_1 > bX_2\,,$$
$$\qquad\quad = 0 \quad \text{otherwise}\,.$$

J_b is the U-statistic based on the kernel h_b, and $E(J_b) = \int_0^\infty \bar{F}(bx)\,dF(x)$. Significantly large values of J_b lead to rejection of H_0 in favor of H_2. The value b, $b \in (0, 1)$, must be specified by the user. For the choice $b = \frac{1}{2}$, Deshpande shows that the Pitman asymptotic relative efficiencies of $J_{1/2}$, with respect to the Hollander–Proschan J^* statistic presented in the next section, are 0.931 for linear failure rate alternatives, 0.946 for Makeham alternatives, and 1.006 for Weibull alternatives. (The statistic $J_{1/2}$ has also been proposed as an NBU test by Kumazawa (1981) but in a private communication Dr. S. C. Kochar has pointed out to us that Kumazawa's reported asymptotic efficiency values of $J_{1/2}$ versus J^* for Weibull and linear failure rate alternatives are incorrect.)

8.3. An NBU test (Hollander and Proschan). Hollander and Proschan (1972) developed a test of

$$H_0: \quad F(x) = 1 - \exp(-\lambda x), \quad x \geq 0,\ \lambda > 0\ (\lambda \text{ unspecified})$$

versus

$$H_3: \quad F \text{ is NBU (and not exponential)}\,,$$

based on a random sample $X_1, \ldots, X_n$ from a continuous life distribution F.

Their test is motivated by considering the parameter

$$\begin{aligned}\gamma(F) &= \int_0^\infty \int_0^\infty \{\bar{F}(x)\bar{F}(y) - \bar{F}(x+y)\}\, \mathrm{d}F(x)\, \mathrm{d}F(y) \\ &= \tfrac{1}{4} - \int_0^\infty \int_0^\infty \bar{F}(x+y)\, \mathrm{d}F(x)\, \mathrm{d}F(y) \\ &\overset{\text{def}}{=} \tfrac{1}{4} - \Delta(F)\end{aligned} \tag{8.9}$$

as a measure of the deviation of F from exponentiality towards NBU (or NWU) alternatives. Note that $\Delta(F) = P(X_1 > X_2 + X_3)$ where X_1, X_2, X_3 are iid according to F, that $\Delta(F) = \frac{1}{4}$ when F is exponential, and that

$$P(X_1 > X_2 + X_3 \mid X_1 > X_2) = 2\Delta(F)\,. \tag{8.10}$$

Looking at the left-hand-side of (8.10), we see that $\Delta(F)$ is less than $\frac{1}{4}$ when the conditional chance "that a used item (which has already survived past the random time X_2) will survive an additional random time X_3" is less than the chance "that a new item will survive a random time X_3" (the latter chance of course being $\frac{1}{2}$).

Replacing F by the e.d.f. $\hat{F}_n$, Hollander and Proschan suggest rejecting H_0 in favor of H_3 if J, defined by (8.11), is too small. (J is asymptotically equivalent to $\Delta(\hat{F}_n)$ and is more convenient to work with.) H_0 is rejected in favor of

$$H_3'\colon\quad F \text{ is NWU (and not exponential)}$$

if J is significantly large. The statistic J can be written as

$$J = 2[n(n-1)(n-2)]^{-1} \textstyle\sum' \psi(X_{\alpha_1}, X_{\alpha_2} + X_{\alpha_3})$$

where $\psi(a, b) = 1$ if $a > b$, 0 otherwise and the Σ' is over all $n(n-1)(n-2)/2$ triples $(\alpha_1, \alpha_2, \alpha_3)$ of three integers such that $1 \leq \alpha_i \leq n$, $\alpha_1 \neq \alpha_2$, $\alpha_1 \neq \alpha_3$ and $\alpha_2 < \alpha_3$.

Hollander and Proschan give upper and lower critical values of

$$T = n(n-1)(n-2)J/2 = \sum_{i>j>k} \psi(X_{(i)}, X_{(j)} + X_{(k)}) \tag{8.11}$$

in the $\alpha = 0.01$, 0.025, 0.05, 0.075 and 0.10 regions for $n = 4(1)20(5)50$. The normal approximation treats

$$J^* = \{n(432/5)\}^{1/2}(J - \tfrac{1}{4}) \tag{8.12}$$

as an $N(0, 1)$ random variable under H_0.

Hollander and Proschan also show that the test which rejects for large values of J is consistent against the class of continuous NBU distributions. (For more on this NBU test, see Hollander and Wolfe (1973) and Cox and Hinkley (1974).)

Other NBU tests. Koul (1977) suggests rejecting H_0 in favor of H_3 for significantly small values of $S = \min_{1 \le k \le j \le n} T_{kj}$, where for $1 \le k \le j \le n$, $T_{kj} = nS_{kj} - (n-k)(n-j)$, and $S_{kj} = \sum_{i=1}^{n} \psi(X_{(i)}, X_{(k)} + X_{(j)})$. The motivation for the S statistic is that $n^{-2}S$ estimates the parameter $\alpha(F) = \inf_{x,y \ge 0}\{\bar{F}(x+y) - \bar{F}(x)\bar{F}(y)\}$, and $\alpha(F)$ is a measure of the deviation of F from H_0 towards H_3, being 0 when F is exponential and negative when F is NBU. Koul gives critical values of S for $\alpha = 0.005, 0.01, 0.025, 0.05, 0.10, 0.20$, $n = 3(1)30(5)50$. Koul's test is not as readily implementable as the Hollander–Proschan test based on J as Koul does not provide a convenient large sample approximation for critical values of S and he also does not provide a dual test of H_0 versus NWU alternatives. Koul (1978a) suggests a class of tests of H_0 versus H_3 based on $\int\int \psi(\hat{\bar{F}}_n(x+y))\,d\hat{F}_n(x)\,d\hat{F}_n(y)$, where $\hat{F}_n$ is the e.d.f. and ψ is a nondecreasing right continuous function from $[0, 1)$ to $[0, \infty)$ with $\psi(0) = 0$. The Hollander–Proschan J statistic corresponds to the choice $\psi(u) = u$. Koul studies in detail and advocates the choice $\psi(u) = u^{1/2}$.

8.4. A DMRL TEST (Hollander and Proschan). For the problem of testing

$$H_0: \quad F(x) = 1 - \exp(-\lambda x), \quad x \ge 0,\ \lambda > 0\ (\lambda \text{ unspecified})$$

versus

$$H_4: \quad F \text{ is DMRL (and not exponential)},$$

Hollander and Proschan considered the following integral as a measure of deviation, for a given F, from H_0 to H_4. Let

$$\nu(F) = \iint_{x<y} \bar{F}(x)\bar{F}(y)\{\varepsilon_F(x) - \varepsilon_F(y)\}\,dF(x)\,dF(y)$$

where $\varepsilon_F(x) = \{\int_x^\infty \bar{F}(u)\,du\}/\bar{F}(x)$ is the mean residual life at time x for $x \ge 0$ (and $\varepsilon_F(x) \equiv 0$ whenever $\bar{F}(x) = 0$). The parameter $\nu(F)$ is an average value of the deviation $\bar{F}(x)\bar{F}(y)\{\varepsilon_F(x) - \varepsilon_F(y)\}$, with the weights $\bar{F}(x)$ and $\bar{F}(y)$ representing the proportions of the population still alive at times x and y respectively, and thus furnishing comparisons concerning the mean residual lifelengths from x and y respectively. Hollander and Proschan replace F by $\hat{F}_n$, the e.d.f., and a statistic asymptotically equivalent to $\nu(\hat{F}_n)$ is

$$V = n^{-4} \sum_{i=1}^{n} c_i X_{(i)} \tag{8.13}$$

where

$$c_i = \tfrac{4}{3}i^3 - 4ni^2 + 3n^2 i - \tfrac{1}{2}n^3 + \tfrac{1}{2}n^2 - \tfrac{1}{2}i^2 + \tfrac{1}{6}i\,. \tag{8.14}$$

In order to make the test scale invariant, Hollander and Proschan utilize $V^* = V/\bar{X}$, significantly large values suggesting DMRL alternatives and signficantly small values suggesting IMRL alternatives. They give critical values (obtained by Monte Carlo sampling) of $V' = \{(210)n\}^{1/2} V^*$ for $n = 2(1)20(5)50$ and α in the upper and lower 0.01, 0.05, 0.10 regions. Exact tables of V' are given by Langenberg and Srinivasan (1979) for α in the upper and lower 0.01, 0.05, 0.10 regions and $n = 2(1)20(5)60$. Hollander and Proschan show that, under H_0,

$$V' = \{(210)n\}^{1/2} V^* \tag{8.15}$$

can be treated (asymptotically) as an $N(0, 1)$ random variable. They also show that the test that rejects for large values of V' is consistent against the class of continuous DMRL distributions.

Klefsjö (1983) derives the V^* statistic using the total-time-on-test plot. Bryson (1974) has suggested a test for IMRL alternatives and gives critical values for $n = 10$, 15, 20, 25 and 30 but does not derive the asymptotic distribution of his test statistic.

The mean residual life function. There is an inversion formula giving the survival function $\bar{F}(x)$ in terms of the mean residual life function $\varepsilon_F(x)$. The formula is often attributed to Cox (1962, Exercise 1, p. 128), but see Kotz and Shanbhag (1980) for a general result and related references. Hall and Wellner (1981) use the inversion formula and show how knowledge of mean residual life functions can be used in modelling and model identification. Bryson and Siddiqui (1969) point out that the mean residual life function lends itself readily to graphical analysis; Figure 2 of Bryson and Siddiqui plots the empirical mean residual life function $\hat{\varepsilon}_F(x) = \{\int_x^\infty \hat{\bar{F}}_n(u)\,du/\hat{\bar{F}}_n(x)\} I\{x < X_{(n)}\}$ for survival times of patients suffering from chronic granulocytic leukemia, with $x = 0$ taken as the date of diagnosis. Yang (1978) established strong consistency of $\hat{\varepsilon}_F$ on a finite interval $[0, T]$ and also showed that the associated process $n^{1/2}\{\hat{\varepsilon}_F(x) - \varepsilon_F(x)\}$ converges weakly to a Gaussian process. Hall and Wellner (1979) strengthened Yang's results; in particular they extended her weak convergence result to the positive real line, and they also derived nonparametric simultaneous confidence bands for $\varepsilon_F(x)$. Hall and Wellner (1981) show that the empirical mean residual life function is a useful addition to the arsenal of techniques (histograms, empirical survival functions, failure rate estimators, total-time-on-test plots, etc.) for analysis of survival data.

8.5. An NBUE test using the total-time-on-test statistic (Hollander and Proschan). The total-time-on-test statistic $\Sigma_{i=1}^{n-1} U_i$, where the U's are defined by equation (8.2) in Section 8.1, has been considered by many authors including Barlow (1968), Bickel and Doksum (1969), Barlow and Proschan (1969), Barlow et al. (1972) and Barlow and Doksum (1972) as a test statistic for testing exponentiality against IFR (or DFR) alternatives. However, Hollander and

Proschan (1975), and later Klefsjö (1983), show that K arises in a natural way as a test against NBUE (or NWUE) alternatives. Hollander and Proschan consider the parameter $\eta(F) = \int_0^\infty \bar{F}(x)\{\varepsilon_F(0) - \varepsilon_F(x)\}\, \mathrm{d}F(x)$ as a measure of deviation for a given F from

$$H_0: \quad F(x) = 1 - \exp(-\lambda x), \quad x \geq 0, \ \lambda > 0 \ (\lambda \text{ unspecified})$$

to

$$H_5: \ F \text{ is NBUE (and not exponential)}.$$

The sample counterpart to $\eta(F)$, obtained by replacing F by the e.d.f. $\hat{F}_n$, is

$$K = n^{-2} \sum_{i=1}^{n} d_i X_{(i)},$$

where

$$d_i = \frac{3n}{2} - 2i + \tfrac{1}{2}. \tag{8.16}$$

Dividing K by $\bar{X}$ to make it scale invariant, Hollander and Proschan proposed $K^* = K/\bar{X}$ as a statistic for testing exponentiality against NBUE alternatives and pointed out that

$$\sum_{i=1}^{n-1} U_i = nK^* + \frac{(n-1)}{2}. \tag{8.17}$$

Significantly large values of K^* suggest NBUE alternatives; significantly small values suggest NWUE alternatives. Thus the total-time-on-test statistic, originally proposed to detect IFR (DFR) alternatives, can be more suitably viewed as a test statistic designed to detect the larger NBUE (NWUE) class. Barlow (1968) tables percentile points of $\Sigma_{i=1}^{n-1} U_i$ for $n = 2(1)10$ and α in the lower and upper 0.01, 0.05 and 0.10 regions. The large sample approximatión under H_0 treats (asymptotically)

$$K' = \{(12)n\}^{1/2} K^* \tag{8.18}$$

as an $N(0, 1)$ random variable.

Klefsjö (1983), by considering the TTT-plot, is also led to the derivation of the K^* statistic as a test statistic for exponentiality versus NBUE alternatives.

Barlow and Doksum (1972) advocated the statistic $D^+ = \max_{1 \leq i \leq n} [U_i - i/n]$ for testing H_0 versus H_1, large values being significant. Koul (1978b) shows that rejecting H_0 for large values of D^+ can be more appropriately viewed as a test against the (larger) H_5 class of alternatives. The null distribution of D^+ as tabled by Birnbaum and Tingey (1951) is also appropriate in this testing context. Asymptotically, under H_0, $P\{n^{1/2} D^+ \leq x\} = 1 - \exp(-2x^2)$. Borges, Proschan and Rodrigues (1982) develop a test against NBUE alternatives using the parameter $\eta^*(F) = \int_0^\infty \bar{F}(x)\{\varepsilon_F(0) - \varepsilon_F(x)\}\, \mathrm{d}x$. Note that $\eta^*(F)$ has the integrand in common with the parameter $\eta(F)$ suggested by Hollander and Proschan with the essential difference being that $\eta^*(F)$ integrates with respect to '$\mathrm{d}x$' rather than '$\mathrm{d}F(x)$'. Borges, Proschan and Rodrigues are led to consider the sample coefficient of variation $s/\bar{X}$, where $s^2 = n^{-1}\Sigma_{i=1}^{n}(X_i - \bar{X})^2$. They compute Pitman asymptotic relative efficiencies of $s/\bar{X}$ with respect to K^* for Weibull alternatives and for the 'Barlow–Doksum' (1972) distribution

$$F_{\mu,\theta}(x) = 1 - \exp(-a_1 x), \quad 0 \leq x \leq x_0,$$
$$= 1 - \exp\{-a_1 x_0 - a_2(x - x_0)\}, \quad x_0 \leq x < \infty$$

where $x_0 = -u \ln(1 - u + \theta)/\{u - \theta\}$, $a_1 = 1 - (\theta/u)$, $a_2 = 1 + \{\theta/(1-u)\}$, and $0 \leq \theta < u$, for a fixed number u such that $0 < u < 1$. Barlow and Doksum (1972) showed that $\{F_{u,\theta}\}$ is IFR (and hence NBUE). Borges, Proschan and Rodrigues show that (in terms of Pitman asymptotic relative efficiency) $s/\bar{X}$ is better than K^* for the Bickel–Doksum distribution with large u. For the Bickel–Doksum distribution with small u, and for the Weibull distribution, K^* outperforms $s/\bar{X}$. The test proposed by Borges, Proschan and Rodrigues is equivalent to a test based on $(\Sigma_{i=1}^{n} X_i^2/n)/\bar{X}^2$ studied by Lee, Locke and Spurrier (1980).

8.6. Some Pitman efficiency values. One commonly used measure for comparing two competing test sequences $\{T_{1,n}\}$, $\{T_{2,n}\}$ satisfying certain regularity conditions is the Pitman asymptotic relative efficiency (A.R.E.) (cf. Lehmann, 1975). Let $\{F_{\theta_n}\}$ be a sequence of alternatives with $\theta_n = \theta_0 + bn^{-1/2}$, where b is an arbitrary positive constant and F_{θ_0} satisfies the null hypothesis. The Pitman efficacy for a test statistic sequence $\{T_n\}$ (say) that is asymptotically normal with mean $E_\theta(T_n)$ and standard deviation $\sigma_\theta(T_n)/n$ against the sequence of alternatives $\{F_{\theta_n}\}$ is, subject to suitable regularity (cf. Lehmann, 1975),

$$c_T(F) = \lim_{n\to\infty} n^{-1/2}\{(\mathrm{d}/\mathrm{d}\theta)E_\theta(T_n) \mid \theta = \theta_0\}/\sigma_{\theta_0}(T_n).$$

For two such competing test sequences $\{T_{1,n}\}$, $\{T_{2,n}\}$, the Pitman A.R.E. $e_F(T_1, T_2)$ of T_1 with respect to T_2 is $e_F(T_1, T_2) = \{c_{T_1}(F)/c_{T_2}(F)\}^2$. We may roughly say that the Pitman A.R.E. of T_1 with respect to T_2 is the limiting ratio of samples sizes n_2/n_1 such that both tests achieve equal power against equal alternatives that are 'close to' the null hypothesis.

Here F_{θ_0} is exponential and we consider the linear failure rate, Makeham, Pareto, Weibull, and gamma distributions (F_1, F_2, F_3, F_4, F_5 say) with corresponding densities

$$f_\theta^{(1)}(x) = (1 + \theta x)\exp\left\{-\left(x + \frac{\theta x^2}{2}\right)\right\},$$

$$f_\theta^{(2)}(x) = [1 + \theta(1 - \mathrm{e}^{-x})]\exp\{-[x + \theta(x + \mathrm{e}^{-x} - 1)]\},$$

$$f_\theta^{(3)}(x) = (1 + \theta x)^{-\{(1+\theta)/\theta\}}, \qquad f_\theta^{(4)}(x) = (1 + \theta)x^\theta \exp\{-x^{(1+\theta)}\},$$

$$f_\theta^{(5)}(x) = [x^\theta \mathrm{e}^{-x}]/\Gamma(1 + \theta).$$

For each density, $x \geq 0$, $\theta \geq 0$ and $\theta = \theta_0 = 0$ corresponds to the exponential. Each of f_1, f_2, f_4, f_5 is IFR for $\theta > 0$, whereas f_3 is DFR for $\theta > 0$. The entries in Table 8.1 (extracted from Klefsjö, 1983, and based on efficiency calculations reported in Bickel and Doksum, 1969, Hollander and Proschan, 1975, and Klefsjö, 1983) give for each F_1, F_2, F_3, F_4, F_5 the Pitman A.R.E.'s of A^*, B^*, V^*, K^* relative to the statistic (among A^*, B^*, V^*, K^*) having the largest efficacy for that particular F. The 'c^2_{MAX}' column of Table 8.1 gives, for a given F, the largest squared efficacy for the four included statistics.

Table 8.1
Pitman A.R.E.

	A^*	B^*	V^*	K^*	c^2_{MAX}
F_1 (linear failure rate)	0.44	0.31	1.00	0.91	0.820
F_2 (Makeham)	0.70	0.70	0.70	1.00	0.083
F_3 (Pareto)	0.44	0.31	1.00	0.91	0.820
F_4 (Weibull)	0.51	0.87	0.49	1.00	1.441
F_5 (gamma)	0.39	1.00	0.28	0.90	0.498

For the J^* statistic given by (8.12), Hollander and Proschan (1972) show that $e_{F_4}(J^*, K^*) = 0.937$ and $e_{F_1}(J^*, K^*) = 0.45$. Other efficiency values for J^* are given by Koul (1978b) and Deshpande (1983). Other efficiency values for K^* are given by Bickel and Doksum (1969), and Borges, Proschan and Rodrigues (1982).

EXAMPLE 8.1. We use the methylmercury poisoning data of Table 7.1 to illustrate the IFR test based on A^*, the IFRA test based on B^*, the NBU test based on J^*, the DMRL test based on V^*, and the NBUE test based on K^*. Table 8.2 expedites the calculation of these statistics by giving the ordered sample, the normalized spacings, and the α's, β's, c's and d's defined by (8.4), (8.7), (8.14) and (8.16), respectively.

Table 8.2
Calculation of A^*, B^*, V^*, K^*

i	$X_{(i)}$	D_i	α_i	β_i	c_i	d_i
1	42	420	165	165	−189	13.5
2	43	9	231	111	−1	11.5
3	51	64	220	60	122	9.5
4	61	70	154	14	188	7.5
5	66	30	55	−25	205	5.5
6	69	15	−55	−55	181	3.5
7	71	8	−154	−74	124	1.5
8	81	30	−220	−80	42	−0.5
9	82	2	−231	−71	−57	−2.5
10	82	0	−165	−45	−165	−4.5

Although the tied values at 82 are not consistent with the assumption that F is continuous, we use the null distribution tables based on that assumption.

For A we find using (8.3)′, (8.4), (8.5) and Table 8.2, $A^* = 3.77$, with $P < 0.01$ from Klefsjö's (1983) table for $n = 10$. For B we find using (8.6)′, (8.7), (8.8) and Table 8.2, $B^* = 4.98$ with $P < 0.01$ from Klefsjö's (1983) table for $n = 10$. For J we find, using (8.11), $J = 0$ (since $X_{(10)} < X_{(1)} + X_{(2)}$) and from Hollander and Proschan (1972), $P = 1/\binom{18}{10} = 1/43758 = 0.00002$. For V we find using (8.13), (8.14), (8.15) and Table 8.2, $V' = 2.10$ and $0.01 < P < 0.05$ from Langenberg and Srinivasan's (1979) table (the 0.01 critical value is $V' = 2.14$). For K, we find using (8.16), (8.17), and Table 8.2, $(10)K^* + 4.5 = 7.74$ with a $P < 0.01$ from Barlow's (1968) table.

9. Generalizations to censored data

There has been vigorous research in the area of survival analysis for censored data. Recent books covering portions of the research are Lee (1980), Elandt–Johnson and Johnson (1980), Kalbfleisch and Prentice (1980), Miller (1981), and Lawless (1982). These are many types of censoring including Type I censoring, Type II censoring, and random censoring (cf. Miller, 1981). Many of the inferential procedures discussed in Sections 7 and 8 have been generalized to accommodate the various types of censoring. Space limitations prohibit a comprehensive account here, and instead we will simply reference some of the generalizations in the randomly censored model.

In the randomly censored model, instead of observing a complete sample $X_1, \ldots, X_n$, one is able to observe only the pairs $Z_i = \min(X_i, T_i)$, $\delta_i = 1$ if $Z_i = X_i$ (i-th observation is uncensored) and $\delta_i = 0$ if $Z_i = T_i$ (i-th observation is censored). We assume that $X_1, \ldots, X_n$ are iid according to the continuous life distribution F, $T_1, \ldots, T_n$ are iid according to the continuous censoring distribution H, and the T's and X's are mutually independent. The censoring distribution H is typically, though not necessarily, unknown and is treated as a nuisance parameter.

The Kaplan–Meier (1958) estimator (KME) can be viewed as a nonparametric MLE (see Kaplan and Meier, 1958) and when there is no censoring it reduces to the e.d.f. of Section 7.1. Under our continuity assumptions, the KME $\bar{F}_{kn}(t)$ can be written as

$$\bar{F}_{kn}(t) = \prod_{i=1}^{nK_n(t)} c_{in}^{\delta_{(i)}} I\{Z_{(n)} \geq t\}, \quad t \in [0, \infty)\,, \tag{9.1}$$

where $c_{in} = (n-i)(n-i+1)^{-1}$, $Z_{(1)} < \cdots < Z_{(n)}$ are the ordered Z's, $\delta_{(i)}$ is the δ corresponding to $Z_{(i)}$, $K_n(t) = n^{-1} \Sigma_{i=1}^n I\{Z_i \leq t\}$ is the empirical distribution function of the Z's, and where a product over an empty set of indices is defined to be 1.

Large sample properties of the KME have been studied by many authors. In particular, weak convergence of the KME (regarded as a stochastic process) has been established by Efron (1967), Breslow and Crowley (1974), Meier (1975), and Gill (1983). Strong consistency of the KME is proved by Peterson (1977) and Langberg, Proschan and Quinzi (1981), and strong uniform consistency by Földes and Rejtö (1981). Hall and Wellner (1980) provide asymptotic confidence bands for $\bar{F}$ for the randomly censored model based on the KME; competing asymptotic bands also based on the KME are given by Gillespie and Fisher (1979) and Nair (1981).

Chen, Hollander and Langberg (1982) give exact small-sample results, including the exact mean and variance of the KME, under a model of proportional hazards where $\bar{H} = \bar{F}^{\beta}$. Additional discussion of the KME can be found in the following books: Brown and Hollander (1977), Lee (1980), Elandt–Johnson and Johnson (1980), Kalbfleisch and Prentice (1980), Miller (1981), and Lawless (1982).

Kitchin, Langberg and Proschan (1980) introduce the Piecewise Exponential Estimator (PEXE) as a competitor of the KME. The two estimators have the same asymptotic properties but Kitchin (1980) shows some advantages of the PEXE in small samples.

Susarla and Van Ryzin (1976) derive a Bayes estimator of F in the randomly censored model using Ferguson's Dirichlet process. Their result is further generalized, when the prior distribution of F is a process 'neutral to the right', by Ferguson and Phadia (1979). Dykstra and Laud (1981) define a stochastic process whose sample paths are failure rates and use the process to derive Bayes estimators of the failure rate and distribution for both complete and censored data models.

Tests of exponentiality versus the classes IFR, IFRA, NBU, DMRL and NBUE are also available for the incomplete data case of random censoring. IFR and IFRA tests for incomplete data are proposed by Barlow and Proschan (1969). Chen, Hollander and Langberg (1983a) generalize the NBU test of Section 7.3 to accommodate randomly censored data and Chen, Hollander and Langberg (1983b) generalize the DMRL test of Section 7.4 to accommodate randomly censored data. Koul and Susarla (1980) modify the total-time-on-test statistic to provide an NBUE test for incomplete data.

Thus far in this section we have referenced censored-data analogues of the methods for complete data discussed in Sections 7 and 8. However, there is a large body of censored-data techniques for other important nonparametric problems. For example, a two-sample censored data test of the hypothesis that two population distributions F, G are equal, based on a generalization of the Wilcoxon–Mann–Whitney two-sample rank sum test, was proposed by Gehan (1965) and Gilbert (1962). Efron (1967), in a fundamental paper that was a catalyst for future work in censoring, showed that the Gehan–Gilbert test could be adversely affected by unequal censoring distributions, proposed a test based on $\int F_{km}\, dG_{kn}$ where F_{km} and G_{kn} are the KME's for samples 1 and 2, respectively (this test was a forerunner of many proposals where test statistics and estimators of parameters of the form $\Delta(F, G, \ldots)$ were obtained via substituting the respective KME's for $F, G, \ldots$), and showed how an appropriate transformation of $n^{1/2}\{F_{kn}(t) - F(t)\}$ yielded the standard Weiner process. There are many two-sample competitors; see also Mantel (1966), Peto and Peto (1972), Latta (1977a), Aalen (1978), Fleming et al. (1980) and the survey papers of Oakes (1981) and Andersen et al. (1982).

One of the earliest K-sample censored data tests of the hypothesis that K population distributions $F_1, F_2, \ldots, F_K$ are equal is Breslow's (1970) generalization of the Kruskal–Wallis test. See also Tarone and Ware (1977), Prentice (1978), Brookmeyer and Crowley (1980), and Andersen et al. (1982) for competitors.

A paired-sample test for censored data is proposed by Wei (1980).

Sign tests and confidence intervals for the median survival time when the data are right censored are given by Brookmeyer and Crowley (1982) and Emerson (1982).

Tests, using randomly censored data, that the underlying distribution is a specified distribution F_0 (say), include those proposed by Breslow (1975), Koziol and Green (1976), Hyde (1977), Turnbull and Weiss (1978), Hollander and Proschan (1979), Gail and Ware (1979), Koziol (1980), Fleming et al. (1980), Csörgő and Horváth (1981), Woolson (1981), and Andersen et al. (1982). Turnbull and Weiss (1978) and Chen (1981) have devised procedures for the goodness-of-fit problem where the null hypothesis is composite.

Tests of the independence of X, Y, where one (or both) variables are subject to censoring include a generalization of Kendall's τ due to Brown, Hollander and Korwar (1973) and a generalization of Spearman's ρ due to Latta (1977b).

Two-sample tests, K-sample tests, goodness-of-fit tests and tests of independence can also be developed from the general regression methods for censored data proposed by Cox (1972) in a landmark paper. Competing regression methods are due to Miller (1976), Buckley and James (1979), and Koul, Susarla and Van Ryzin (1981). Miller (1981) and Miller and Halpern (1981) contrast the advantages and disadvantages of the different regression approaches; approaches that can be used to determine the effects of covariates on survival. See Miller (1981), Birnbaum (1979), Basu (1984), and Doksum and Yandell (1984) for concise descriptions of additional techniques for censored data.

References

Aalen, O. O. (1978). Nonparametric inference for a family of counting processes. *Ann. Statist.* **6**, 701–726.

Aggarwal, O. P. (1955). Some minimax invariant procedures for estimating a cumulative distribution function. *Ann. Math. Statist.* **26**, 450–463.

Andersen, P. K., Borgan, O., Gill, R. and Keiding, N. (1982). Linear nonparametric tests for comparison of counting processes, with applications to censored survival data. *Inter. Statist. Inst. Rev.* **50**, 219–258.

Artin, E. (1964). *The Gamma Function.* Holt, New York.

Barlow, R. E. (1968). Likelihood ratio tests for restricted families. *Ann. Math. Statist.* **39**, 547–560.

Barlow, R. E., Bartholomew, D. J., Bremner, J. M. and Brunk, H. D. (1972). *Statistical Inference Under Order Restrictions.* Wiley, New York.

Barlow, R. E. and Campo, R. (1975). Total time on test processes and applications to failure data analysis. In: R. E. Barlow, J. Fursell and N. Singpurwalla, eds., *Reliability and Fault Tree Analysis*. SIAM, Philadelphia, pp. 451–481.

Barlow, R. E. and Doksum, K. (1972). Isotonic tests for convex orderings. *Proc. Sixth Berk. Symp. Math. Statist. Prob.* **1**, 293–323.

Barlow, R. E., Marshall, A. W. and Proschan, F. (1963). Properties of probability distributions with monotone hazard rate. *Ann. Math. Statist.* **34**, 375–389.

Barlow, R. E. and Proschan, F. (1964). Comparison of replacement policies, and renewal theory implications. *Ann. Math. Statist.* **35**, 577–589.

Barlow, R. E. and Proschan, F. (1965). *Mathematical Theory of Reliability*. Wiley, New York.

Barlow, R. E. and Proschan, F. (1966). Inequalities for linear combinations of order statistics from restricted families. *Ann. Math. Statist.* **37**, 1574–1592.

Barlow, R. E. and Proschan, F. (1969). A note on tests for monotone failure rate based on incomplete data. *Ann. Math. Statist.* **40**, 595–600.

Barlow, R. E. and Proschan, F. (1981). *Statistical Theory of Reliability and Life Testing*. Second Printing Publisher: To Begin With, 1137 Hornell Drive, Silver Spring, Maryland 20904.

Basu, A. P. (1984). Censored data. Chapter 25 in this volume.

Bergman, B. (1977). Crossings in the total time on test plot. *Scand. J. Statist.* **4**, 171–177.

Bickel, P. J. and Doksum, K. (1969). Tests for monotone failure rate based on normalized spacings. *Ann. Math. Statist.* **40**, 1216–1235.

Bickel, P. J. and Freedman, D. A. (1981). Some asymptotic theory for the bootstrap. *Ann. Statist.* **9**, 1196–1217.

Birnbaum, Z. W. (1952). Numerical tabulation of the distribution of Kolmogorov's statistic for finite sample size. *J. Amer. Statist. Assoc.* **47**, 425–441.

Birnbaum, Z. W. (1979). *On the Mathematics of Competing Risks*. DHEW Publication No. PHS 79–1351.

Birnbaum, Z. W., Esary, J. D. and Marshall, A. W. (1966). Stochastic characterization of wearout for components and systems. *Ann. Math. Statist.* **37**, 816–825.

Birnbaum, Z. W. and Tingey, F. H. (1951). One-sided confidence contours for probability distribution functions. *Ann. Math. Statist.* **22**, 592–596.

Block, H. W. and Savits, T. H. (1976). The IFRA closure problem. *Ann. Prob.* **4**, 1030–1032.

Block, H. W. and Savits, T. H. (1981). Multivariate classes in reliability theory. *Mathematics of Operations Research* **6**, 453–461.

Bondesson, L. (1983). On preservation of classes of life distributions under reliability operations: Some complementary results. *Naval Res. Logist. Quarterly* **30**, 443–447.

Borges, W., Proschan, F. and Rodrigues, J. (1982). A simple test for new better than used in expectation. Florida State University Department of Statistics Report No. M-604.

Breslow, N. E. (1970). A generalized Kruskal–Wallis test for comparing K samples subject to unequal patterns of censorship. *Biometrika* **57**, 579–594.

Breslow, N. E. and Crowley, J. (1974). A large sample study of the life table and product limit estimates under random censorship. *Ann. Statist.* **2**, 437–453.

Brookmeyer, R. and Crowley, J. (1982a). A k-sample median test for censored data. *J. Amer. Statist. Assoc.* **77**, 433–440.

Brookmeyer, R. and Crowley, J. (1982b). A confidence interval for the median survival time. *Biometrics* **38**, 29–41.

Brown, B. Wm. Jr. and Hollander, M. (1977). *Statistics: A Biomedical Introduction*. Wiley, New York.

Brown, B. W., Hollander, M. and Korwar, R. M. (1974). Nonparametric tests of independence for censored data with applications to heart transplant studies. In: F. Proschan and R. J. Serfling, eds., *Reliability and Biometry*. SIAM, Philadelphia, pp. 327–354.

Brown, M. (1981). Further monotonicity properties for specialized renewal processes. *Ann. Prob.* **9**, 891–895.

Brown, M. and Chaganty, N. R. (1983). On the first passage time distribution for a class of Markov chains. *Ann. Probab.* **11**, 1000–1008.

Bryson, M. C. (1974). Heavy-tailed distributions: Properties and tests. *Technometrics* **16**, 61–68.

Bryson, M. C. and Siddiqui, M. M. (1969). Some criteria for aging. *J. Amer. Statist. Assoc.* **64**, 1472–1483.

Buckley, J. and James, I. (1979). Linear regression with censored data. *Biometrika* **66**, 429–436.

Butler, D. A. (1979a). A complete importance ranking for components of binary coherent systems, with extensions to multistate systems. *Naval Res. Log. Quart.* **26**, 565–578.

Butler, D. A. (1979b). Bounding the reliability of multistate systems. Stanford University Technical Report No. 193.

Chen, C. (1981). Correlation-type goodness-of-fit tests for randomly censored data. Stanford University Division of Biostatistics Technical Report No. 73.

Chen, Y. Y., Hollander, M. and Langberg, N. A. (1982). Small-sample results for the Kaplan–Meier estimator. *J. Amer. Statist. Assoc.* **77**, 141–144.

Chen, Y. Y., Hollander, M. and Langberg, N. A. (1983a). Testing whether new is better than used with randomly censored data. *Ann. Statist.* **11**, 267–274.

Chen, Y. Y., Hollander, M. and Langberg, N. A. (1983b). Tests for monotone mean residual life using randomly censored data. *Biometrics* **39**, 119–127.

Cox, D. R. (1962). *Renewal Theory.* Methuen, London.

Cox, D. R. (1972). Regression models and life tables (with discussion). *J. R. Statist. Soc. B* **34**, 187–220.

Cox, D. R. and Hinkley, D. V. (1974). *Theoretical Statistics.* Chapman and Hall, London.

Csörgő, S. and Horváth, L. (1981). On the Koziol–Green model for random censorship. *Biometrika* **68**, 391–401.

Deshpande, J. V. (1983). A class of tests for exponentiality against increasing failure rate average alternatives. *Biometrika* **70**, 514–518.

Doksum, K. (1972). Decision theory for some nonparametric models. *Proc. Sixth. Berk. Symp. Math. Statist. Prob.* **1**, 331–341.

Doksum, K. (1974). Tailfree and neutral random probabilities and their posterior distributions. *Ann. Prob.* **2**, 183–201.

Doksum, K. and Yandell, B. S. (1984). Tests for exponentiality. Chapter 26 in this volume.

Dvoretzky, A., Kiefer, J. and Wolfowitz, J. (1956). Asymptotic minimax character of the sample distribution function and of the classical multinomial estimator. *Ann. Math. Statist.* **27**, 642–669.

Dykstra, R. L. and Laud, P. (1981). A Bayesian nonparametric approach to reliability. *Ann. Statist.* **9**, 356–367.

Efron, B. (1967). The two sample problem with censored data. *Proc. Fifth Berk. Symp. Math. Statist. Prob.* **4**, 831–854.

Efron, B. (1979). Bootstrap methods: Another look at the jackknife. *Ann. Statist.* **7**, 1–26.

Efron, B. and Gong, G. (1983). A leisurely look at the bootstrap, the jackknife and cross-validation. *The American Statistician* **37**, 36–48.

Efron, B. and Morris, C. (1975). Data analysis using Stein's estimator and its generalizations. *J. Amer. Statist. Assoc.* **70**, 311–319.

Elandt-Johnson, R. C. and Johnson, N. L. (1980). *Survival Models and Data Analysis.* Wiley, New York.

El-Neweihi, E. and Proschan, F. (1980). Multistate reliability models: A survey. *Multivariate Analysis V*, 523–541.

Emerson, J. D. (1982). Nonparametric confidence intervals for the median in the presence of right censoring. *Biometrics* **38**, 17–27.

Esary, J. D., Marshall, A. W. and Proschan, F. (1973). Shock models and wear processes. *Ann. Prob.* **1**, 627–649.

Fardis, M. N. and Cornell, C. A. (1981). Analysis of coherent multistate systems. *IEEE Trans. Reliability* **30**, 117–122.

Ferguson, T. S. (1973). A Bayesian analysis of some nonparametric problems. *Ann. Statist.* **1**, 209–230.

Ferguson, T. S. (1974). Prior distributions on spaces of probability measures. *Ann. Statist.* **2**, 615–629.

Ferguson, T. S. and Phadia, E. G. (1979). Bayesian nonparametric estimation based on censored data. *Ann. Statist.* **7**, 163–186.

Fleming, T. R., O'Fallon, J. R., O'Brien, P. C. and Harrington, D. P. (1980). Modified Kolmogorov–Smirnov test procedures with application to arbitrarily right censored data. *Biometrics* **36**, 607–625.

Földes, A. and Rejtö, L. (1981). Strong uniform consistency for nonparametric survival curve estimators from randomly censored data. *Ann. Statist.* **9**, 122–129.

Freedman, D. A. (1981). Bootstrapping regression models. *Ann. Statist.* **9**, 1218–1228.

Gail, M. H. and Ware, J. H. (1979). Comparing observed life table data with a known survival curve in the presence of random censorship. *Biometrics* **35**, 385–391.

Gehan, E. A. (1965). A generalized Wilcoxon test for comparing arbitrarily singly-censored samples. *Biometrika* **52**, 203–223.

Gilbert, J. P. (1962). Ph.D. dissertation. University of Chicago.

Gill, R. D. (1983). Large sample behaviour of the product limit estimator on the whole line. *Ann. Statist.* **11**, 49–58.

Gillespie, M. J. and Fisher, L. (1979). Confidence bands for the Kaplan–Meier survival curve estimate. *Ann. Statist.* **7**, 920–924.

Goldstein, M. (1975a). Approximate Bayes solutions to some nonparametric problems. *Ann. Statist.* **3**, 512–517.

Goldstein, M. (1975b). A note on some Bayesian nonparametric estimates. *Ann. Statist.* **3**, 736–740.

Grenander, U. (1956). On the theory of mortality measure, Part II. *Skan. Aktuarietidskr* **39**, 125–153.

Griffith, W. S. (1980). Multistate reliability methods. *J. Appl. Prob.* **17**, 735–744.

Haines, A. L. and Singpurwalla, N. D. (1974). Some contributions to the stochastic characterization of wear. In: F. Proschan and R. J. Serfling, eds., *Reliability and Biometry.* SIAM, Philadelphia, pp. 47–80.

Hall, W. J. and Wellner, J. A. (1979). Estimation of mean residual life. University of Rochester Department of Statistics Technical Report.

Hall, W. J. and Wellner, J. A. (1981). Mean residual life. In: M. Csörgő, D. A. Dawson, J. N. K. Rao and A. K. Md. E. Saleh, eds. *Statistics and Related Topics.* North-Holland, Amsterdam, pp. 169–184.

Hall, W. J. and Wellner, J. A. (1980). Confidence bands for a survival curve from censored data. *Biometrika* **67**, 133–143.

Halperin, M., Ware, J. H. and Wu, M. (1980). Conditional distribution-free tests for the two-sample problem in the presence of right censoring. *J. Amer. Statist. Assoc.* **75**, 638–695.

Hollander, M. and Korwar, R. M. (1977). Nonparametric estimation of distribution functions. In: C. P. Tsokos and I. N. Shimi, eds., *The Theory and Applications of Reliability I.* Academic Press, New York, pp. 85–107.

Hollander, M. and Proschan, F. (1979). Testing the determine the underlying distribution using randomly censored data. *Biometrics* **35**, 393–401.

Hollander, M. and Wolfe, D. A. (1973). *Nonparametric Statistical Methods.* Wiley, New York.

Hyde, J. (1977). Testing survival under right censoring and left truncation. *Biometrika* **64**, 225–230.

James, W. and Stein, C. (1961). Estimation with quadratic loss. *Proc. Fourth Berk. Symp. Math. Statist. Prob.* **1**, 361–379.

Kalbfleisch, J. D. and Prentice, R. L. (1980). *The Statistical Analysis of Failure Time Data.* Wiley, New York.

Kaplan, E. L. and Meier, P. (1958). Nonparametric estimation from incomplete observations. *J. Amer. Statist. Assoc.* **53**, 457–481.

Karlin, S. (1968). *Total Positivity. No. I.* Stanford University Press, Stanford, California.

Kitchin, J. (1980). A new method for estimating life distributions from incomplete data. Ph.D. dissertation. Florida State University.

Kitchin, J., Langberg, N. A. and Proschan, F. (1980). A new method for estimating life distributions from incomplete data. Florida State University Department of Statistics Technical Report No. M-548.

Klefsjö, B. (1983). Some tests against aging based on the total time on test transform. *Comm. Statist. A-Theory Methods* **12**, 907–927.

Korwar, R. M. and Hollander, M. (1976). Empirical Bayes estimation of a distribution function. *Ann. Statist.* **4**, 580–587.

Kotz, S. and Shanbhag, D. N. (1980). Some new approaches to probability distributions. *Advances in Applied Probability* **12**, 903–921.

Koziol, J. A. (1980). Goodness-of-fit tests for randomly censored data. *Biometrika* **67**, 693–696.

Koziol, J. A. and Green, S. B. (1976). A Cramér–von Mises statistic for randomly censored data. *Biometrika* **63**, 465–474.

Koul, H. L. (1977). A test for new is better than used. *Comm. Statist. A-Theory Methods* **6**, 563–573.

Koul, H. L. (1978a). A class of tests for testing 'new is better than used'. *Canad. J. Statist.* **6**, 249–271.

Koul, H. L. (1978b). Testing for new is better than used in expectation. *Comm. Statist. A-Theory Methods* **7**, 685–701.

Koul, H. L. and Susarla, V. (1980). Testing for new better than used in expectation with incomplete data. *J. Amer. Statist. Assoc.* **75**, 952–956.

Koul, H., Susarla, V. and Van Ryzin, J. (1981). Regression analysis with randomly right-censored data. *Ann. Statist.* **6**, 1276–1288.

Kumazawa, Y. (1981). A test of whether new is better than used. Osaka University Report.

Langberg, N. A., Proschan, F. and Quinzi, A. J. (1981). Estimating dependent lifelengths with applications to the theory of competing risks. *Ann. Statist.* **9**, 157–167.

Langenberg, P. and Srinivasan, R. (1979). Null distribution of the Hollander–Proschan statistic for decreasing mean residual life. *Biometrika* **66**, 679–680.

Latta, R. B. (1977a). Generalized Wilcoxon statistics for the two-sample problem with censored data. *Biometrika* **64**, 633–635.

Latta, R. B. (1977b). Rank tests for censored data. University of Kentucky Department of Statistics Technical Report No. 112.

Lawless, J. F. (1981). *Statistical Models and Methods for Lifetime Data.* Wiley, New York.

Lee, E. T. (1980). *Statistical Methods for Survival Data Analysis.* Lifetime Learning Publications, Belmont, California.

Lee, S. C. S., Locke, C. and Spurrier, J. D. (1980). On a class of tests of exponentiality. *Technometrics* **22**, 547–554.

Lehmann, E. L. (1975). *Nonparametrics: Statistical Methods Based on Ranks.* Holden-Day, San Francisco.

Mantel, N. (1966). Evaluation of survival data and two new rank order statistics arising in its consideration. *Cancer Chemotherapy Reports* **50**, 163–170.

Marshall, A. W. and Proschan, F. (1965). Maximum likelihood estimation for distributions with monotone failure rate. *Ann. Math. Statist.* **36**, 69–77.

Marshall, A. W. and Proschan, F. (1972). Classes of distributions applicable in replacement, with renewal theory implications. *Proc. Sixth Berk. Symp. Math. Statist. Prob.* **1**, 395–415.

Marshall, A. W. and Proschan, F. (1975). Inconsistency of maximum likelihood estimator of distributions having increasing failure rate average. Florida State University Department of Statistics Technical Report No. M-350.

Mehrotra, K. G. (1981). A note on the mixture of new worse than used in expectation. *Naval Res. Log. Quart.* **28**, 181–184.

Meier, P. (1975). Estimation of a distribution function from incomplete observations. *Perspectives in Probability and Statistics* (J. Gani, ed.) 67–87. Academic Press, New York.

Miller, R. G. (1976). Least squares regression with censored data. *Biometrika* **63**, 449–464.

Miller, R. G. (1981). *Survival Analysis.* Wiley, New York.

Miller, R. G. and Halpern, J. (1982). Regression with censored data. *Biometrika* **69**, 521–531.

Nair, V. N. (1981). Plots and tests for goodness of fit with randomly censored data. *Biometrika* **68**, 99–103.

Natvig, B. (1980). Two suggestions of how to define a multistate coherent system. University of Oslo Statistical Research Report No. 4.

Oakes, D. (1981). Survival times: Aspects of partial likelihood (with discussion). *Inter. Statist. Inst. Rev.* **49**, 235–264.

Peterson, A. V. (1977). Expressing the Kaplan–Meier estimator as a function of empirical subsurvival functions. *J. Amer. Statist. Assoc.* **72**, 854–858.

Peto, R. and Peto, J. (1972). Asymptotically efficient rank invariant test procedures (with discussion). *J. R. Statist. Soc. A* **135**, 185–206.

Phadia, E. G. (1973). Minimax estimation of a cumulative distribution function. *Ann. Statist.* **1**, 1149–1157.

Prentice, R. L. (1978). Linear rank tests with right censored data. *Biometrika* **65**, 167–179.

Proschan, F. (1963). Theoretical explanation of observed decreasing failure rate. *Technometrics* **5**, 375–383.

Proschan, F. and Hollander, M. (1972). Testing whether new is better than used. *Ann. Math. Statist.* **43**, 1136–1146.

Proschan, F. and Hollander, M. (1975). Tests for the mean residual life. *Biometrika* **62**, 585–593.

Proschan, F. and Pyke, R. (1967). Tests for nomotone failure rate. *Proc. Fifth Berk. Symp. Math. Statist. Prob.* **3**, 293–312.

Read, R. R. (1971). The asymptotic inadmissibility of the sample distribution function. *Ann. Math. Statist.* **42**, 89–95.

Robbins, H. E. (1964). The empirical Bayes approach to statistical decision problems. *Ann. Math. Statist.* **35**, 1–20.

Ross, S. M. (1979). Multivalued state component systems. *Ann. Prob.* **7**, 379–383.

Samaniego, F. J. and Boyles, R. A. (1980). Estimating a distribution function when new is better than used. University of California at Davis Division of Statistics Technical Report No. 22.

Sethuraman, J. and Tiwari, R. C. (1982). Convergence of Dirichlet measures and the interpretation of their parameters. In: S. S. Gupta and J. O. Berger, eds., *Statistical Decision Theory and Related Topics III*, Vol. 2. Academic Press, New York, pp. 305–315.

Singh, K. (1981). On the asymptotic accuracy of Efron's bootstrap. *Ann. Statist.* **9**, 1187–1195.

Stein, C. (1955). Inadmissibility of the usual estimator for the mean of a multivariate normal distribution. *Proc. Third Berk. Symp. Math. Statist. Prob.* **1**, 197–206.

Susarla, V. and Van Ryzin, J. (1976). Nonparametric Bayesian estimation of survival curves from incomplete observations. *J. Amer. Statist. Assoc.* **71**, 897–902.

Tarone, R. E. and Ware, J. (1977). On distribution-free tests for equality of survival distributions. *Biometrika* **64**, 156–160.

Turnbull, B. W. and Weiss, L. (1978). A likelihood ratio statistic for testing goodness of fit with randomly censored data. *Biometrics* **34**, 367–375.

van Belle, G. (1972). Personal communication.

Wei, L. J. (1980). A generalized Gehan and Gilbert test for paired observations that are subject to arbitrary right censorship. *J. Amer. Statist. Assoc.* **75**, 634–637.

Wilks, S. S. (1962). *Mathematical Statistics.* Wiley, New York.

Woolson, R. F. (1981). Rank tests and a one-sample logrank test for comparing observed survival data to a standard population. *Biometrics* **37**, 687–696.

Yang, G. L. (1978). Estimation of a biometric function. *Ann. Statist.* **6**, 112–116.

PART VII

MISCELLANEOUS TOPICS

P. R. Krishnaiah and P. K. Sen, eds., *Handbook of Statistics, Vol. 4*
© Elsevier Science Publishers (1984) 657–698

28

Sequential Nonparametric Tests

Ulrich Müller-Funk

1. General background and basic types of statistical hypotheses

1.1. Generalities

In this article we are going to survey sequential probability ratio tests (SPRT) and related procedures, tests with power one as well as truncated sequential tests designed for situations that are somewhat vaguely classified as nonparametric. There will be no mentioning of life testing problems, which are dealt with in a separate chapter. As to the basic facts concerning sequential tests, the reader is referred to B. K. Ghosh 'Sequential Tests of Statistical Hypotheses' (1970). A complete and detailed treatment of the sequential nonparametric procedures proposed up to now can be found in a recent monograph by P. K. Sen 'Sequential Nonparametrics' (1981). The field partly consists of ad hoc tests, a more or less coherent statistical justification only existing for sequential rank tests. The latter, therefore, will somewhat predominate.

Methodically, nonparametric theory largely constitutes a chapter of asymptotic statistics and all the more this is true in the sequential case. Accordingly, the tools employed include functional limit theorems and 'nonlinear' renewal arguments but are rather different from the ones that shape the exact (finite) sequential theory. In most cases the validity of these probabilistic limit theorems rests on a decomposition of the statistics under consideration into an average (or, more generally, a U-statistic) of independent variates and into a remainder term approaching zero sufficiently fast. Once a representation of this kind has been established, the distribution theory required has no longer anything specifically nonparametric about it but solely hinges on analytic and probabilistic standard techniques. A closer look at these things would lead us too far afield. Instead, we shall mostly call for a limit theorem without specifying exactly the conditions on which it is valid. Incidentally, most of these tools are comparatively recent and that is why this part of statistics is still in a state of flux despite its early outset, cf. Noether (1954), Tsao (1954) and Romani (1956).

Throughout the paper we shall adhere to the following assumptions and notations. Let $X, X_1, X_2, \ldots$ be a sequence of $\mathbb{R}^d$-valued random variables (r.v.), defined on a common basic space, which are independent and identically distributed (i.i.d.) under each probability measure (p.m.) belonging to either of two classes $\mathfrak{P}_0$ and $\mathfrak{P}_1$. The sets of p.m. F induced by X under $\mathfrak{P}_0$ and $\mathfrak{P}_1$ are denoted by $\mathfrak{H}_0$ and $\mathfrak{H}_1$, respectively. We shall specify the (disjoint) hypotheses $\mathfrak{H}_i$ more precisely in the testing problems to be discussed below. For the sake of convenience it is supposed w.l.o.g. that $\mathfrak{P}_i$ can be indexed by $\mathfrak{H}_i$, i.e. $\mathfrak{P}_i = \{P_F : F \in \mathfrak{H}_i\}$, $i = 0, 1$. Borel measures on $\mathbb{R}^d$ and their distribution functions (d.f.) will be identified. In order to exclude ties let us assume that $\mathfrak{H}_0, \mathfrak{H}_1$ are contained in the class of continuous d-dimensional distributions, $\mathfrak{F}_d$ say. We shall mostly come across the cases $d = 1$ or $d = 2$. In the latter we shall denote the components of X by Y and Z and corresponding marginals by G and H, respectively. The corresponding empirical distributions pertaining to the first n observations are labelled $\hat{F}_n, \hat{G}_n, \hat{H}_n$. $\mathfrak{A}_n$ is to denote the information contained in $X_1, \ldots, X_n$, i.e. the σ-field $\sigma(X_1, \ldots, X_n)$ generated by these r.v. on the basic space. Moreover, we shall meet with filtrations $\mathfrak{C} = (\mathfrak{C}_n)_{n\geq 1}$ etc. contained in $\mathfrak{A} = (\mathfrak{A}_n)_{n\geq 1}$, i.e. nondecreasing sequences of sub-σ-fields $\mathfrak{C}_n \subseteq \mathfrak{A}_n$. $\mathfrak{C}$ is to be thought of as the information on which the nonparametric procedure is based. It will be specified in each problem separately. Sometimes, however, we shall deal with the nonincreasing σ-fields $\mathfrak{D} = (\mathfrak{D}_n)_{n\geq 1}$ instead, $\mathfrak{D}_n = \sigma(\hat{F}_n, X_{n+1}, X_{n+2}, \ldots)$. Our 'generic' sequence of test statistics $Q = (Q_n)_{n\geq 1}$ will always be assumed to be $\mathfrak{C}(\mathfrak{D})$-measurable, i.e. adapted to these σ-algebras. A sequential test is specified by a possibly extended stopping time N w.r.t $\mathfrak{A}$ and a decision rule $\delta = (A_n, B_n)_{n\geq 1}$, where $\{A_n, B_n\}$ is an $\mathfrak{A}_n$-measurable partition of $\{N = n\}$. The test is said to be $\mathfrak{C}(\mathfrak{D})$-measurable, of course, if all quantities involved are.

In nonparametric theory, more often than not, curves $\{F_\vartheta : \vartheta \in \Theta \subset \mathbb{R}^1\}$ are singled out, where Θ is an interval containing zero, $F_0 \in \mathfrak{H}_0$ and $F_\vartheta \in \mathfrak{H}_1$ if $\vartheta \neq 0$. The probability ratio (p.r.) $L_n(F_\vartheta : F_0)$ of $\mathfrak{L}_\vartheta(X_1, \ldots, X_n)$ relative to $\mathfrak{L}_0(X_1, \ldots, X_n)$ will be written as

$$L_n(F_\vartheta : F_0) = L_n(\vartheta) = \prod_{j=1}^{n} l_\vartheta(X_i), \qquad l_\vartheta(x) = \frac{\mathrm{d}F_\vartheta}{\mathrm{d}F_0}(x)\,.$$

If that curve is smoothly parametrized (e.g. L_p-differentiable, $p \geqq 1$) we are interested in the slopes as well

$$\dot{L}_n = \frac{\partial}{\partial\vartheta} L_n(\vartheta)\bigg|_{\vartheta=0} = \sum_{j=1}^{n} \dot{l}(X_j)\,,$$

$$\ddot{L}_n = \frac{\partial^2}{\partial\vartheta^2} L_n(\vartheta)\bigg|_{\vartheta=0} = \dot{L}_n^2 + \sum_{j=1}^{n} \ddot{l}(X_j)\,,$$

where $\dot{l}$ and $\ddot{l}$, of course, stand for the first and the second log-derivative of l_ϑ.

From the local point of view, $\dot{L}_n$ resp. $\ddot{L}_n$ are the test statistics to be used in the one-sided testing problem $\vartheta = 0$ vs. $\vartheta > 0$ (i.e. $\Theta =]\underline{\vartheta}, \bar{\vartheta}[, \bar{\vartheta} > 0$) resp. the two-sided testing problem $\vartheta = 0$ vs. $\vartheta \neq 0$ (i.e. $\Theta =]\underline{\vartheta}, \bar{\vartheta}[, \underline{\vartheta} < 0 < \bar{\vartheta}$) within this parametric submodel. Note that $n^{-1}\ddot{L}_n$ asymptotically behaves like $n^{-1}\dot{L}_n^2$ (up to a constant) which is why two-sided tests are usually simply based on $\dot{L}_n$. In the remaining parts of this section we are going to list some statistical hypotheses that have undergone a more intensive study within a sequential context and to introduce the statistics related to them.

1.2. Hypotheses leading to a set of ranks

The structure of such problems may be summarized as follows. The pair of hypotheses is invariant with respect to a group of transformations Γ on $\mathbb{R}^d$. For every $n \geq 1$ a reduction by invariance relative to

$$(x_1, \ldots, x_n) \mapsto (\gamma(x_1), \ldots, \gamma(x_n)) \quad (\gamma \in \Gamma, x_j \in \mathbb{R}^d)$$

and a subsequent reduction by sufficiency yield an invariantly sufficient statistic generating a σ-field $\mathfrak{C}_n$. This statistic takes values in a finite set over which it induces the (discrete) uniform distribution under every $F \in \mathfrak{H}_0$. In particular, $\mathfrak{H}_0$ is reduced to a simple hypothesis. The sequence of these invariantly sufficient statistics is transitive and, accordingly, carries all the essential information. If $\{F_\vartheta: \vartheta \in \Theta\}$ is a parametric submodel of the kind mentioned above, then the invariant p.r. $L_{\mathfrak{C}n}(F_\vartheta: F_0)$ (i.e. the p.r. of $\mathfrak{L}_\vartheta(X_1, \ldots, X_n) \mid \mathfrak{C}_n$ relative to $\mathfrak{L}_0(X_1, \ldots, X_n) \mid \mathfrak{C}_n$) takes on the form

$$L_{\mathfrak{C}n}(F_\vartheta: F_0) = L_{\mathfrak{C}n}(\vartheta) = E_0(L_n \mid \mathfrak{C}_n). \tag{1.1}$$

Taking for granted a smooth parametrization, we also get

$$\dot{L}_{\mathfrak{C}n} = E_0(\dot{L}_n \mid \mathfrak{C}_n) = \sum_{j=1}^{n} E_0(\dot{l}(X_j) \mid \mathfrak{C}_n), \tag{1.2}$$

$$\ddot{L}_{\mathfrak{C}n} = E_0(\ddot{L}_n \mid \mathfrak{C}_n) = E_0(\dot{L}_n^2 \mid \mathfrak{C}_n) + \sum_{j=1}^{n} E_0(\ddot{l}(X_j) \mid \mathfrak{C}_n).$$

In what follows, we shall be concerned with three standard problems that fit into the foregoing scheme. As it turns out that $\mathfrak{C}_n$ is generated by a vector of ranks in these cases, we shall prefer the more suggestive notations L_{Rn}, $\dot{L}_{Rn}$ to $L_{\mathfrak{C}n}$, $\dot{L}_{\mathfrak{C}n}$ etc. Here, the two-sided testing problems require a few extra considerations as it has to be shown that $n^{-1}\ddot{L}_{Rn}$ and $n^{-1}\dot{L}_{Rn}^2$ again behave in essentially the same way and hence that the commonly proposed tests employing $\dot{L}_{Rn}$ are asymptotically justifiable. In order not to overburden the exposition we shall mainly deal with one-sided alternatives only and refer the reader to Müller-Funk et al. (1983) for some comments concerning the two-sided case.

I. The two-sample problem $(d = 2)$

Here we are testing for the equality of the d.f. G and H of two independent populations $Y_1, Y_2, \ldots$ and $Z_1, Z_2, \ldots$. Formally, we have $X_j = (Y_j, Z_j)$, which means that we draw an equal number of Y- and Z-observations. This sampling scheme is by no means indispensable. As long as no allocation rule is available, however, which is more convincing and truly sequential, we regard this to be a natural device. In the present case, the hypotheses to be tested are

$$\begin{aligned}\mathfrak{H}_0 &= \{F \in \mathfrak{F}_2 : F = G \otimes G\ \exists G \in \mathfrak{F}_1\},\\ \mathfrak{H}_1 &= \{F \in \mathfrak{F}_2 : F = G \otimes H,\ G(x) \geqq H(x)\ \forall x \in \mathbb{R}^1,\ G \neq H\},\end{aligned} \tag{1.3}$$

where $G \otimes H$ denotes the product measure. The problem is invariant with respect to all transformations $\gamma : (y, z) \mapsto (\gamma_0(y), \gamma_0(z))$, γ_0 continuous and strictly isotonic. Accordingly,

$$\mathfrak{C}_n = \sigma(R_{n1}, \ldots, R_{nn}), \qquad R_{nk} = \sum_{j=1}^{n} \{1(Y_j \leq Z_k) + 1(Z_j \leq Z_k)\},$$

$1 \leq k \leq n$, i.e. R_{nk} is the rank of Z_k in the pooled sample. There are two standard ways by which one-parameter families $\{F_\vartheta : 0 \leq \vartheta < \bar{\vartheta}\}$ as above can be generated.

Translation alternatives: $F_\vartheta(y, z) = G(y)G(z - \vartheta)\ \forall (y, z)$, where $G \in \mathfrak{F}_1$ is assumed to have an absolutely continuous Lebesgue density g the Fisher information of which is finite. If $(R_{n1}, \ldots, R_{nn}) = (r_1, \ldots, r_n)$, we get

$$L_{Rn}(\vartheta) = E_0\Big(\prod_{j=1}^{n} (1 + \vartheta\psi_\vartheta(U_{2n(r_j)}))\Big), \qquad \psi_\vartheta(\cdot) = \frac{g(G^{-1}(\cdot) - \vartheta) - g(G^{-1}(\cdot))}{\vartheta g(G^{-1}(\cdot))}, \tag{1.4}$$

$$\dot{L}_{Rn} = \sum_{j=1}^{n} E_0(\psi_0(U_{2n(r_j)})), \qquad \psi_0(\cdot) = -\frac{g'(G^{-1}(\cdot))}{g(G^{-1}(\cdot))}, \tag{1.5}$$

where $G^{-1}(\cdot)$ stands for the usual pseudo-inverse and where $U_{2n(1)} < \cdots < U_{2n(2n)}$ denotes an ordered sample drawn from the rectangular distribution.

Nonparametric alternatives: $F_\vartheta(y, z) = G(y)\Psi_\vartheta(G(z))\ \ \forall (y, z)$, where $\{\Psi_\vartheta : 0 \leq \vartheta < \bar{\vartheta}\}$ is a family of d.f. on the unit interval such that $\Psi_\vartheta(t) \leq t = \Psi_0(t)$ for all $0 < t < 1$, $0 < \vartheta < \bar{\vartheta}$ but $\Psi_\vartheta(t) \not\equiv t$. Lehmann's alternatives, which will be considered in Section 2.1, form a prominent example of this kind. Another choice of $\Psi_\vartheta(\cdot)$ is due to Behnen (1972), who simply defined $\Psi_\vartheta(\mathrm{d}t) = (1 + \vartheta\psi(t))\,\mathrm{d}t$, $\psi \in L_1(]0, 1[)$ being bounded from below and satisfying

$$0 \not\equiv \int_0^s \psi(t)\,\mathrm{d}t \leq 0 \quad \forall 0 < s < 1, \qquad \int_0^1 \psi(t)\,\mathrm{d}t = 0\,. \tag{1.6}$$

With this choice, of course, the identities (1.4), (1.5) remain valid if we replace

ψ_ϑ, ψ_0 by ψ throughout. At first sight, this approach may appear rather restricted. It can be shown, however, that every $F \in \mathfrak{H}_1$ can be represented in this way. As for the connexion with translation alternatives, the reader is referred to Hajek and Sidak (1967, p. 21).

II. The one-sample problem for testing symmetry ($d = 1$)

The discussion runs fairly analogous to the previous one and for that reason we shall merely list the terminology needed. The statistical testing problem and the quantities related to the reduction by invariance and sufficiency are,

$$\mathfrak{H}_0 = \{F \in \mathfrak{F}_1 \colon F = F_-\}, \qquad \mathfrak{H}_1 = \{F \in \mathfrak{F}_1 \colon F(x) \leq F_-, F \neq F_-\},$$

where $F_-(x) = 1 - F(-x)$, $x \in \mathbb{R}^1$. The problem is invariant under all continuous, strictly isotonic and uneven transformations. Here,

$$\mathfrak{C}_n = \sigma(\operatorname{sgn}(X_j)R^+_{nj}, 1 \leq j \leq n), \qquad R^+_{nk} = \sum_{j=1}^{n} 1(|X_j| \leq |X_k|).$$

For *translation alternatives* $F_\vartheta(x) = F(x - \vartheta)$, $F \in \mathfrak{H}_0$, where F possesses an absolutely continuous Lebesgue density f. Then, (1.4), (1.5) become

$$L_{Rn}(\vartheta) = E_0\Big(\prod_{j=1}^{n} (1 + \vartheta\psi_\vartheta(\varepsilon_j U_{n(r_j)}))\Big), \tag{1.7}$$

$$\dot{L}_{Rn} = \sum_{j=1}^{n} \varepsilon_j E_0(\psi_0(U_{n(r_j)} + 1/2)), \quad \psi_0(t) = -f'(F^{-1}(t))/f(F^{-1}(t)), \tag{1.8}$$

if $(\operatorname{sgn}(X_1)R^+_{n1}, \ldots, \operatorname{sgn}(X_n)R^+_{nn}) = (\varepsilon_1 r_1, \ldots, \varepsilon_n r_n)$. ($\psi_\vartheta(\cdot)$ is defined in the same manner as before but the explicit formula is tedious and hence omitted.) In order to define Behnen-style *nonparametric alternatives* $\Psi_\vartheta(F)$, $F \in \mathfrak{H}_0$, we choose ψ and Ψ_ϑ as before but require in addition to the previous assumptions that $\psi(t) + \psi(1 - t) = 1$. With ψ and ψ_0 replaced by ψ, (1.7) and (1.8) also provide us with the expressions for the p.r. and its slope in this case as well.

III. The bivariate independence problem ($d = 2$)

We frame the hypotheses

$$\mathfrak{H}_0 = \{F \in \mathfrak{F}_2 \colon F(y, z) = G(y)H(z)\ \forall (y, z)\},$$
$$\mathfrak{H}_1 = \{F \in \mathfrak{F}_2 \colon F(y, z) \geq G(y)H(z)\ \forall (y, z),\ F \neq G \otimes H\}.$$

This pair of hypotheses is invariant with respect to all transformations $(y, z) \mapsto (\gamma_1(y), \gamma_2(z))$, where again γ_1, γ_2 are continuous and strictly isotonic. If R^Y_{nk} and R^Z_{nk} denote the ranks of the Y- and Z-observations, respectively, and $(\kappa_1, \ldots, \kappa_n)$ the random permutation that orders the first vector of ranks, then

$$\mathfrak{C}_n = (\{R^Y_{n1}, \ldots, R^Y_{nn}\}, (R^Z_{n\kappa_1}, \ldots, R^Z_{n\kappa_n})).$$

No equivalent exists for translation families and therefore we are entirely dependent on Behnen's *nonparametric alternatives.* We shall be contented with the following form. Let $\psi_1, \psi_2 \in L_1(]0, 1[)$ be bounded from below and suppose that each of them fulfills (1.6). Put

$$\Psi_\vartheta(\mathrm{d}s, \mathrm{d}t) = (1 + \vartheta\psi_1(s)\psi_2(t))\, \mathrm{d}s\, \mathrm{d}t\,. \tag{1.9}$$

The classes $F_\vartheta(y, z) = \{\Psi_\vartheta(G(y), H(z)), 0 \leq \vartheta < \vartheta\}$, $G, H \in \mathfrak{F}_1$, will serve our purposes. On the atoms $(R^Y_{n1}, \ldots, R^Y_{nn}, R^Z_{n1}, \ldots, R^Z_{nn}) = (r_1, \ldots, r_n, S_1, \ldots, s_n)$ the p.r. and its slope take on the form

$$L_{Rn}(\vartheta) = E_0\Big(\prod_{j=1}^{n} (1 + \vartheta\psi_1(U_{n(r_j)})\psi_2(U_{n(s_j)}))\Big)\,, \tag{1.10}$$

$$\dot{L}_{Rn} = \sum_{j=1}^{n} E_0(\psi_1(U_{n(r_j)}))E(\psi_2(U_{n(s_j)})) \tag{1.11}$$

where the variates $U^{(1)}_{nk}$ and $U^{(2)}_{nk}$ belong to independent, ordered random samples with a rectangular parent distribution.

In defining Behnen's alternatives we had to require in all cases that the function ψ involved is bounded from below in order to ensure that $1 + \vartheta\psi(\cdot)$ is nonnegative for small values of ϑ and, hence is a density. Within the asymptotic approach of Section 2.2 to come, we shall dispense with this assumption by means of a simple truncation technique.

The slopes $\dot{L}_{Rn}$ are but special cases of linear rank statistics. As for the general theory of such statistics we refer the reader to Hajek and Sidak (1967), Puri and Sen (1971) or Witting and Nölle (1970). Here, we shall only recall their definitions together with some basic facts. Let

$$\varphi_n(t) = \sum_{j=1}^{n} \varphi_{nj} 1_{]j-1,j]}(nt), \qquad 0 < t < 1, \quad (\phi_{nj})_{1 \leq j \leq n} \subset \mathbb{R}^1\,,$$

be a simple score function tending towards some limiting function φ in an appropriate sense (e.g. in mean square). The general linear rank statistic and its asymptotic functional take on the following form:

(i) In the two-sample problem:

$$T_n = \sum_{j=1}^{n} \varphi_{nR_{nj}} = n \int_{\mathbb{R}} \varphi_n(n(n+1)^{-1}\hat{F}_n(x))\hat{H}_n(\mathrm{d}x)\,,$$
$$\theta_\varphi(F) = \int_{\mathbb{R}} \varphi(2^{-1}(G+H)(x))H(\mathrm{d}x)\,. \tag{1.12}$$

(ii) In the univariate one-sample-case $(H(x) = H_F(x) = (F(x) - F(-x)) \times 1_{\mathbb{R}_+}(x))$:

$$T_n = \sum_{j=1}^{n} \operatorname{sgn}(X_j)\varphi_{nR^+_{nj}} = n \int_{\mathbf{R}} \operatorname{sgn}(x)\varphi_n(n(n+1)^{-1}\hat{H}_n(|x|))\hat{F}_n(\mathrm{d}x), \tag{1.13}$$

$$\theta_\varphi(F) = \int_{\mathbf{R}} \operatorname{sgn}(x)\varphi(H(|x|))F(\mathrm{d}x)\,.$$

(iii) In the bivariate independence problem

$$T_n = \sum_{j=1}^{n} \varphi_{1nR^Y_{nj}}\varphi_{2nR^Z_{nj}}$$

$$= n \iint_{\mathbf{R}^2} \varphi_{1n}(n(n+1)^{-1}\hat{G}_n(y))\varphi_{2n}(n(n+1)^{-1}\hat{H}_n(z))\hat{F}_n(\mathrm{d}y, \mathrm{d}z)\,, \tag{1.14}$$

$$\theta_\varphi(F) = \iint_{\mathbf{R}^2} \varphi_1(G(y))\varphi_2(H(z))F(\mathrm{d}y, \mathrm{d}z)\,.$$

In what follows, we shall restrict our attention to either of the following types of scores:

$$\varphi^*_{nj} = E(\varphi(U_{n(j)})), \qquad \text{`exact scores'}\,,$$

$$\varphi^\circ_{nj} = \varphi(E(U_{n(j)})) = \varphi(j(n+1)^{-1}), \qquad \text{`approximative scores'}\,.$$

For a smooth φ, it does not make any difference asymptotically, which of the two is employed. If φ is nondecreasing and if its Lebesgue integral over the unit interval is zero, then $\theta_\varphi(\cdot)$ reflects the stochastic ordering by means of which we introduced the hypotheses, i.e.

$$F_1 \leq F_2 \Rightarrow \theta_\varphi(F_1) \geq \theta_\varphi(F_2), \qquad \theta_\varphi(F) = 0,\ F \in \mathfrak{H}_0\,,$$

$$\theta_\varphi(F) \geq 0,\ F \in \mathfrak{H}_1\,.$$

Next, let us come to some fundamental results concerning the above quantities. Although we shall state some of them a bit informally, we like to put them into the shape of theorems.

THEOREM A ('SLLN', Müller-Funk, 1983b). *Suppose that $\varphi_n \to \varphi$ in the sense L_1 resp. that $\varphi_{in} \to \varphi_i$ in the sense L_2 $(i = 1, 2)$. For all $\varepsilon > 0$, there are $c(\varepsilon) > 0$, $n(\varepsilon) \in \mathbb{N}$ so that for every $F \in \mathfrak{F}_1$ resp. $F \in \mathfrak{F}_2$*

$$|n^{-1}T_n - \theta_\varphi(F)| \leq c(\varepsilon)d_K(\hat{F}_n, F) + \varepsilon \quad \forall n \geq n(\varepsilon)\,,$$

where $d_K(F_1, F_2) = \sup_x|F_1(x) - F_2(x)|$ denotes the Kolmogorov distance between d.f.

The above theorem generalizes SLLN due to Sen (1970), Hajek (1974), Sen and Ghosh (cf. Sen, 1981, p. 120), Rieder (1981, 1982), etc. It will be useful in connexion with a well-known inequality obtained by Dvoretzky et al. (1956): For some $d \geq 0$,

$$P_F(d_K(\hat{F}_n, F) > t) \leq d \exp(-2nt^2) \quad \forall n \in \mathbb{N}, \quad F \in \mathfrak{F}_1 . \tag{1.15}$$

The distribution theory under the null-hypothesis is largely facilitated by the following observation due to Sen (1981, p. 90, 127, 152).

THEOREM B. *If T_n^* denotes a linear rank statistic corresponding to exact scores, then $(T_n^*, \mathfrak{C}_n)$ is a martingale sequence under $\mathfrak{H}_0$.*

This result makes it easy to establish the invariance principle for linear rank statistics under the null-hypothesis and alternatives close to it. Moreover, Sen and Ghosh (1972), (1973b), (1974b) used it to prove strong embedding theorems which (slightly modified) read as follows: If φ is normalized in the sense L_2, then (switching to a new basic probability space if necessary) there are i.i.d. standard normal variables $W_1, W_2, \ldots$ so that under $\mathfrak{H}_0$ a.s.

$$T_n - \sum_{j=1}^{n} W_j = o(c_n) \tag{1.16}$$

where the growth of the positive constants c_n depends on the score function. For a broad class of φ (including the inverse Φ^{-1} of the standard normal d.f.) we may choose $c_n = (n \log\log n)^{1/2}$. In that case, (1.16) implies the LIL under the null-hypothesis.

Invariance principles under arbitrary, fixed d.f. $F \in \mathfrak{F}_1$ (resp. $\in \mathfrak{F}_2$) were obtained by various authors using different methods. We refer to Sen (1981), Chapters 4, 5, 6 for references. Here, we are going to reproduce a somewhat stronger assertion (requiring more restrictive assumptions on the score functions φ), cf. Sen and Ghosh (1971), (1973a), Lai (1975a), Müller-Funk (1979), Bönner et al. (1980), Sen (1981, p. 135, 164) for proofs in each of the three cases.

THEOREM C. ('Chernoff–Savage-representation'). *Suppose that the score function φ (resp. the pair (φ_1, φ_2)) satisfies a suitable Chernoff–Savage-type growth condition. Then for all $F \in \mathfrak{F}_1$ (resp. $F \in \mathfrak{F}_2$) there is a Borel function h_F,*

$$E_F|h_F(X_1)|^{2+\iota} < \infty \quad \exists \iota > 0, \qquad E_F(h_F(X_1)) = \theta_\varphi(F),$$

so that for all $\kappa > 1$ there are $0 < \gamma = \gamma(\kappa) < 2^{-1}$, $n(\kappa) \in \mathbb{N}$ for which

$$\sup_F P_F(|T_n - S_{nF}| > n^\gamma) \leq n^{-\kappa} \quad \forall n \geq n(\kappa),$$

where $S_{nF} = \sum_{j=1}^{n} h_F(X_j)$, $n \geq 1$.

The formulation of this result requires further comments. Firstly, the functions h_F in question are not just any but quite specific ones and we dropped their definition for the mere sake of space. Secondly, the regularity conditions referred to above are fulfilled for any φ with a continuous second derivative so that φ resp. φ', φ'' do not increase faster near zero and one than Φ^{-1} and its first two derivatives. Especially, the normal scores statistics are included. The foregoing Theorem yields the invariance principle as well as the LIL. Over and above that it will be our main tool in obtaining approximations to average sample numbers (ASN). For that purpose, the uniformity statement in the above result turns out to be crucial.

In the one- and the two-sample location model it is more convenient to vary the statistics by a translation parameter than the distributions, that is to make use of the identity $\mathfrak{L}_\vartheta(T) = \mathfrak{L}_0(T(\vartheta))$, where $T_n(\vartheta)$ are the linear rank statistics based on the shifted observations $X_j + \vartheta$ resp. $(Y_j, Z_j + \vartheta)$. When dealing with local alternatives, $T_n(\vartheta)$ can be expanded around $\vartheta = 0$. To make this more precise, we associate the following quantities with every d.f. F which possesses an absolutely continuous Lebesgue density f and a finite Fisher information $I(F)$:

$$\begin{aligned} I(F) &= \int_{\mathbb{R}} f'^2(x)/f(x)\,\mathrm{d}x\,, \\ \psi_F(u) &= -I^{-1/2}(F)f'(F^{-1}(u))/f(F^{-1}(u)), \quad 0<u<1\,, \\ \rho_\varphi(F) &= \int_0^1 \varphi(u)\psi_F(u)\,\mathrm{d}u = I^{-1/2}(F)\mu_\varphi(F)\,, \end{aligned} \tag{1.17}$$

as well as the remainder terms $D_n = D_n(F, \varphi, r)$, $r > 0$, where

$$D_n = n^{-1/2} \sup_{|\eta|\leq r} |T_{\varphi n}(\eta n^{-1/2}) - T_{\varphi n}(0) - \eta(nI(F))^{1/2}\rho_\varphi(F)|\,.$$

Asymptotic linearity results were first proved by Jurečková (1969) and van Eeden (1972), who showed that $(D_n)_{n\geq 1}$ tends to zero in probability. Almost sure versions are due to Sen and Ghosh (1971), Jurečková (1973) and Sen (1980). At the same time, van Eeden's paper supplements earlier work by Lehmann (1966) concerning orderings of vectors of ranks. In case of a non-decreasing score function, it can be drawn from these sources that

$$F_0 \in \mathfrak{H}_0,\, F_1 \in \mathfrak{H}_1 \Rightarrow P_{F_0}(T_n > t) \leq P_{F_1}(T_n > t) \quad \forall n \geq 1, \quad t \in \mathbb{R}^1\,, \tag{1.18}$$

i.e. T_n is stochastically larger under the alternative than under the null-situation.

1.3. Testing for a functional of a distribution

Suppose that all of the unknown d.f. we are really interested in can be expressed by means of a real-valued functional $\nu(\cdot)$. Let us assume that every uniform distribution over a finite set of points is in the domain of $\nu(\cdot)$, i.e. that $V_n = \nu(\hat{F}_n)$ is always defined. Finally, we require that this functional is continuous in the sense that $V = (V_n)_{n\geq 1}$ is a strongly consistent estimator for $\nu(F)$ in case F is the true d.f. It has been common usage for a long time to turn estimators into test statistics for the parameter involved. In order to frame a pair of hypotheses we fix some ν_0 in the range of $\nu(\cdot)$ and put, for instance,

$$\begin{aligned}\mathfrak{H}_0 &= \{F \in \mathfrak{H}_1 \cap \text{domain}(\nu)\colon \nu(F) = \nu_0\}\,,\\ \mathfrak{H}_1 &= \{F \in \mathfrak{H}_1 \cap \text{domain}(\nu)\colon \nu(F) > \nu_0\}\,.\end{aligned} \tag{1.19}$$

Generally speaking, V is not distribution free over either of these classes. We shall mostly encounter functionals having some kind of integral representation. For instance, $\nu(\cdot)$ may be a 'regular functional' i.e.

$$\nu(F) = \nu_g(F) = \int \cdots \int g(x_1, \ldots, x_p) \prod_{j=1}^{p} F(\mathrm{d}x_j), \quad p \geq 1\,.$$

where the kernel g is at least square integrable and symmetric in its arguments. Besides, we are then also interested in the related U-statistics $U = (U_n)_{n\geq p}$,

$$U_n = E(g(X_1, \ldots, X_p) \mid \mathfrak{D}_n) = \binom{n}{p}^{-1} \sum_{1\leq j_1 < \cdots < j_p \leq n} g(X_{j_1}, \ldots, X_{j_p})\,.$$

Among the standard examples of regular functionals are the moments of a distribution, e.g. the variance which corresponds to the kernel $g(Y, Z) = (Y - Z)^2/2$. The quantile $\nu(F) = F^{-1}(n)$, $0 < n < 1$, on the other hand, is a familiar functional which fails to be regular in the above sense.

Other functionals of interest can be written in the form

$$\nu(F) = \int h(x, F(x))\,\mathrm{d}x \quad \text{or} \quad \nu(F) = \int h(x, F^{-1}(x))\,\mathrm{d}x\,.$$

(In the latter case, of course, we have to assume that $d = 1$.) The pertaining V-statistics comprise statistics of the Cramér–von Mises-type in the first case and certain linear combinations of order statistics (in short: L-statistics) in the second one. As for the general theory of all these sorts of statistics we refer the reader to Puri and Sen (1971), Serfling (1980) and Sen (1981). As almost all statistics of interest can be made arbitrarily close to a U-statistic we shall somewhat concentrate on that type of statistics. Let us recall those structural results concerning U-statistics and their V-statistics counterparts that will turn out to be crucial in the course of the following discussion of sequential tests. To begin with, $(U_n, \mathfrak{D}_n)_{n\geq p}$ is a reversed martingale and a.s. converges towards $\nu(F)$ under F. Next, we have

THEOREM D (Hoeffding). *Suppose that $\nu(\cdot) = \nu_g(\cdot)$ is regular and that F is in the domain of $\nu(\cdot)$. Then*

(a) *There are U-statistics $U_n^{(j)}$ corresponding to kernels g_j (depending on F) of length j, $1 \leq j \leq p$, so that*

$$U_n = \nu(F) + \sum_{j=1}^{p} \binom{p}{j} U_n^{(j)}, \qquad E_F(U_n^{(j)} U_n^{(k)}) = 0 \quad \text{if } j \neq k,$$
$$E_F(U_n^{(j)}) = 0, \qquad \mathrm{Var}_F(U_n^{(j)}) = \binom{n}{j}^{-1} \mathrm{Var}_F(g_j(X_1, \ldots, X_j)). \tag{1.20}$$

Particularly, $\mathrm{Var}_F(U_n) = p^2\sigma^2(F)n^{-1} + \mathrm{O}(n^{-2})$, $\sigma^2(F) = \mathrm{Var}_F(g_1(X_i))$.
(b) *If g possesses moments up to order k, then*

$$E_F|U_n - V_n|^k = \mathrm{O}(n^{-k}).$$

The martingale property and the Hoeffding decomposition (1.20) are the main tools for proving limit theorems (CLT, invariance principle, LIL) and inequalities for U-statistics. For instance,

THEOREM E. *Let $\nu(\cdot)$ and F be as in the preceding theorem.*
(a) (Grams and Serfling) *If g possesses moments up to order 2k, then*

$$E_F|U_n - pU_n^{(1)} - \nu(F)|^{2k} = \mathrm{O}(n^{-2k}). \tag{1.21}$$

(b) (Grams and Serfling) *If g possesses moments up to order 2k, then*

$$P_F(\sup_{k \geq n} |U_n - \nu(F)| > \varepsilon) = \mathrm{O}(n^{-2k+1}) \quad \forall \varepsilon > 0. \tag{1.22}$$

(c) (Berk) *If g possesses a moment generating function which is finite in a neighbourhood of the origin under F, then, for all $\varepsilon > 0$ there are constants $c(\varepsilon) > 0$, $1 > u(\varepsilon) > 0$, so that*

$$P_F(|U_n - \nu(F)| > \varepsilon) \leq c(\varepsilon) u(\varepsilon)^n, \quad n \geq p. \tag{1.23}$$

The exact references as well as extensions and refinements may be found in Serfling's monograph (1980). The gist of the above theorems is, of course, that U/V-statistics not only generalize but in fact behave like an average of i.i.d. variates. This, however, remains true for V-statistics corresponding to a functional that is not regular but allows for a one-step von Mises expansion. For the quantile e.g., such an expansion was established by Bahadur (cf. Serfling, 1980, Section 2.5): If F has a strictly positive second derivative at $\nu(F) = F^{-1}(u)$, $0 < u < 1$, then P_F-a.s.

$$V_n = \nu(\hat{F}_n) = \nu(F) - (\hat{F}_n(F^{-1}(u)) - u)/F'(F^{-1}(u)) + \mathrm{O}((n^{-1} \log\log n)^{3/4}).$$

A similar behaviour is to be expected for functionals that are expressible as weighted averages of quantiles, i.e. functionals of the form

$$\nu(F) = \int_0^1 F^{-1}(u)\varphi(u)\,\mathrm{d}u + \sum_{j=1}^{m} c_j F^{-1}(u_j) = \int x\varphi(F(x))F(\mathrm{d}x) + \sum_{j=1}^{m} c_j F^{-1}(u_j).$$

As for an analogue to Theorem C for that type of statistics confer Sen (1977a) or Govindarajulu and Mason (1980).

Apart from a trivial modification, linear rank statistics, too, can be regarded as V-statistics. In fact

$$T_n = (1 + n^{-1})\theta_\varphi(n\hat{F}_n/(n+1)).$$

As is widely known, special rank statistics even happen to be (generalized) U-statistics, e.g. the Wilcoxon statistic and certain rank correlations. Statistically, however, this point of view is hardly rewarding.

Besides integral-type functionals we shall also come across 'distance-type' functionals, e.g.

$$\nu(F) = \nu_{F_0}(F) = \sup_x (F_0(x) - F(x)),$$

where F_0 is a fixed d.f. In this particular case and if $d = 1$ the testing problem (1.19) simply boils down to

$$\begin{aligned} \mathfrak{H}_0 &= \mathfrak{H}_0(F_0) = \{F \in \mathfrak{F}_1 : F \geq F_0\}, \\ \mathfrak{H}_1 &= \mathfrak{H}_1(F_0) = \{F \in \mathfrak{F}_1 : F \leq F_0, F \not\equiv F_0\}. \end{aligned} \tag{1.24}$$

This problem as well as the two-sample problem (1.3) will be treated, accordingly, by means of the Kolmogorov–Smirnov type statistics $K^{(1)}$ and $K^{(2)}$, respectively, where

$$K_n^{(1)} = \sup_x (F_0(x) - \hat{F}_n(x)), \qquad K_n^{(2)} = \sup_x (\hat{G}_n(x) - \hat{H}_n(x)).$$

2. Nonparametric Wald tests

As before, let $Q = (Q_n)_{n \geq 1}$ denote our generic sequence of test statistics adapted to $\mathfrak{A} = (\mathfrak{A}_n)_{n \geq 1}$. By a Wald test based on Q and on constants $b < 0 < a$ we mean a test that stops sampling as soon as Q_n crosses one of the horizontal barriers b or a and decides in favour of $\mathfrak{H}_1$ if Q_n is no less than a and accepts $\mathfrak{H}_0$ otherwise. To be in line with our formal definition of a sequential test, we symbolize the procedure by (N_Q^*, δ_Q^*), where

$$\begin{aligned} N_Q^* &= N_Q^*(b, a) = \inf\{n \geq 1 : Q_n \notin\,]b, a[\}, \\ \delta_Q^* &= \delta_Q^*(b, a) = (\{Q_n \leq b\}, \{Q_n \geq a\})_{n \geq 1} \end{aligned} \tag{2.1}$$

($\inf \emptyset = 0$). Such a test, of course, is nothing else but a familiar SPRT, if $Q_n = \log L_n$ is chosen to be the log-p.r. corresponding to a fixed pair of alternatives. Another well-known type of parametric Wald tests is built upon slopes $Q_n = \dot{L}_n$ of p.r. pertaining to suitably parametrized families of distributions. Confer Berk (1975a) for details, who established their local optimum character. Nonparametric analogues to both types of Wald tests result, e.g. If we replace L_n resp. $\dot{L}_n$ by $L_{Rn} = E_0(L_n \mid \mathfrak{C}_n)$ resp. $\dot{L}_{Rn} = E_0(\dot{L}_n \mid \mathfrak{C}_n)$, where $\mathfrak{C}_n$ is again induced by a vector of ranks. Moreover, it is near at hand to construct Wald tests from U/V-statistics. Unlike $\log L_n$, the latter statistics fail to be sums of i.i.d. variables, generally speaking, and only $E_0(L_n \mid \mathfrak{C}_n)$ is again a p.r. Accordingly, we have to put anew all questions concerning the basic properties of the ensuing procedures.

(1) Is N_Q^* a.s. finite (integrable, exponentially bounded)?

(2) What approximations to operating characteristic functions (OC) and ASN can be given?

(3) How to justify these tests? Which optimum properties can be guaranteed?

2.1. Rank SPRT

Let us look at any of the testing problems described in Section 1.2. We fix alternatives $F_i \in \mathfrak{H}_1$, $i = 0, 1$, and denote the corresponding p.r. again by $L_n(F_1 : F_0)$, $L_{Rn}(F_1 : F_0) = E_0(L_n(F_1 : F_0) \mid \mathfrak{C}_n)$. A rank SPRT (N_R^*, δ_R^*), i.e. a Wald test built upon $Q_n = \log L_{Rn}(F_1 : F_0)$, and is but a special case of an invariant SPRT, for which various results are available in the literature. Savage and Savage (1965), for instance, stated sufficient conditions that ensure the finiteness of the stopping times. Wijsman (1977a, 1977b) and Lai (1975b, c) examined the properties of these random times more closely. We shall not reproduce their findings but refer the reader to Wijsman's (1979) excellent survey paper on the subject. Little can be said on behalf of the other questions raised above without appealing to some kind of asymptotics. At least Wald's approximations to the stopping bounds remain valid, i.e.

$$a \simeq \log \frac{1-\beta}{\alpha} \geq a, \qquad b \simeq \log \frac{\beta}{1-\alpha} \leq b, \tag{2.2}$$

where α and β are the error probabilities under F_0 resp. F_1. The Wald–Wolfowitz theorem is no longer in force but Eisenberg et al. (1976) established some sort of weak admissibility. Perhaps the most convincing argument supporting the use of a rank SPRT, however, is asymptotic in nature. First, we are going to quote a somewhat stripped version of a general result due to Lai (1981). Let $\mathfrak{E} = (\mathfrak{E}_n)_{n \geq 1}$ be any filtration contained in $\mathfrak{A}$ and denote by $\mathfrak{T}(\alpha, \beta)$ the class of $\mathfrak{E}$-measurable sequential tests (N, δ) such that

$$P_{F_0}((N, \delta) \text{ rejects } \mathfrak{H}_0) \leq \alpha, \qquad P_{F_1}((N, \delta) \text{ rejects } \mathfrak{H}_1) \leq \beta. \tag{2.3}$$

THEOREM F. *For* $0 < \alpha, \beta, \alpha + \beta < 1$, *let* $b = b(\alpha, \beta) < a = a(\alpha, \beta)$ *be chosen so that* (2.3) *holds for* $(N_Q^*(b, a), \delta_Q^*(b, a))$, $Q_n = \log L_{\mathfrak{E}n}$, *and that, moreover,*

$$a \sim \log \alpha^{-1}, \quad b \sim \log \beta \qquad as\ \alpha + \beta \to 0\,. \tag{2.4}$$

Suppose there are finite constants $I_0 < 0 < I_1$ such that for all $\zeta > 0$

$$M_i(\zeta) = \sup\{m \geqslant 1 : |m^{-1} \log L_{\mathfrak{E}m} - I_i| \geqslant \zeta\}$$

has a finite expectation under P_{F_i} $(i = 0, 1)$. Then, as $\alpha + \beta \to 0$.

$$\inf\{E_{F_0}(N) : (N, \delta) \in \mathfrak{T}(\alpha, \beta)\} \sim E_{F_0}(N^*_{\mathfrak{Q}}) \sim I_0^{-1} \log \beta\,,$$

$$\inf\{E_{F_1}(N) : (N, \delta) \in \mathfrak{T}(\alpha, \beta)\} \sim E_{F_1}(N^*_{\mathfrak{Q}}) \sim I_1^{-1} \log \alpha^{-1}\,.$$

The assertion above figures as an asymptotic substitute for the Wald–Wolfowitz theorem in a context where the latter is no longer applicable. As for its assumptions, note that (2.4) merely requires the excess over the boundaries being negligible in the limit; compare with (2.2). Integrability of $M_i(\zeta)$ defines a mode of convergence of $n^{-1} \log L_{\mathfrak{E}n}$ towards I_i which is stronger than a.s. convergence ('1-quick convergence', cf. Lai, 1976a, and Chow and Teicher, 1978, p. 368). In case $E_{F_i}(\log L_1^2)$ is finite, the foregoing result can be applied to $\mathfrak{A}$ as well. Obviously, the factor at which the costs increase cannot be smaller if only the partial information $\mathfrak{E}$ is available. In the special situation in which $\mathfrak{E} = \mathfrak{C}$ is generated by a maximal invariant vector of ranks, however, it is possible that $\mathfrak{C}$ is asymptotically fully informative. Thus, it can be shown in the two-sample case that for a broad class of $F_1 = G \otimes H \in \mathfrak{F}_1$ and a suitably selected $F_0 = J \otimes J \in \mathfrak{H}_0$

$$\lim_n n^{-1} \log L_n(F_1 : F_0) = \lim_n n^{-1} \log L_{Rn}(F_1 : F_0) = I_1 \tag{2.5}$$

under F_1 (in the sense of 1-quick convergence). Here, suitably selected means that F_0 minimizes the Kullback–Leibler numbers $K - L(F_1 : \cdot)$ over $\mathfrak{H}_0$. Besides, I_1 coincides with $K - L(F_1 : F_0)$, i.e.

$$I_1 = \int \log\left(\frac{\mathrm{d}F_1}{\mathrm{d}F_0}\right) \mathrm{d}F_1 = \min_{F \in \mathfrak{H}_0} \int \log\left(\frac{\mathrm{d}F_1}{\mathrm{d}F}\right) \mathrm{d}F_1\,. \tag{2.6}$$

A simple variational argument shows that $J = \frac{1}{2}(G + H)$ provides the solution to this minimization problem. As for a discussion in detail, the reader is referred to Berk and Savage (1968), Bahadur and Raghavachari (1972), and Hajek (1974). So far, however, a more complete treatment of this sort of 'asymptotic sufficiency' seems to be lacking.

Irrespective of the foregoing remarks, there is always one advantage which is gained if a rank SPRT is employed instead of an ordinary SPRT. The reduction by invariance shrinks $\mathfrak{H}_0$ and, more generally, subclasses of the form $\{\Psi(F) : F \in \mathfrak{H}_0\}$ to simple hypotheses. (Again, $\Psi(\cdot)$ denotes an appropriate d.f.

on $]0,1[^{(d)}$.) Hence the optimality of the classical SPRT can be extended to composite hypotheses, at least asymptotically, if L_n is replaced by L_{Rn} throughout. The entire approach, however, suffers from a serious drawback. There are hardly any interesting classes of alternatives that lead to simple expressions for $\log L_{Rn}$. The examples to come and two related problems treated by Govindarajulu (1975, pp. 281, 283) seem to comprise all rank-SPRT which have been investigated in detail.

We are going to take up the two-sample testing problem specified in Section 1.2. For this problem, Wilcoxon et al. (1963) first proposed a SPRT based on ranks. These authors treated grouped data and ranking within groups, a topic we shall turn to only later on. Savage and Sethuraman (1966) suggested a rank-SPRT within the meaning of the present paper, i.e. a SPRT which makes more effective use of the data by a complete reranking of the observations at each stage. Their procedure was further investigated by Sethuraman (1970), Savage and Sethuraman (1972), Govindarajulu (1975) among others. Research concerning generalizations of Stein's lemma partly originated from the Savage–Sethuraman paper. In all of the afore-mentioned articles the p.r. are built upon $\mathfrak{H}_0$ and Lehmann alternatives $\mathfrak{R}(\eta)$,

$$\mathfrak{R}(\eta)=\{F(y,z)=G(y)H(z)=J(y)J^{1+\eta}(z)\colon J\in\mathfrak{F}_1\}\subseteq\mathfrak{H}_1,\quad 0<\eta\,.$$

These alternatives lack intuitive meaning but are popular for their analytical tractability. It has been pointed out that both $\mathfrak{H}_0$ and $\mathfrak{R}(\eta)$ turn into simple hypotheses after the reduction by invariance. With those alternatives, (1.7) becomes

$$L_{Rn}(\eta)=\frac{(1+\eta)^n(2n)!}{n^{2n}}\prod_{j=1}^{n}\frac{1}{W_n(Y_j)W_n(Z_j)}\Rightarrow n^{-1}\log L_{Rn}(\eta)$$
$$=\log(4(1+\eta))-2-n^{-1}\sum_{j=1}^{n}\{\log W_n(Y_j)+\log W_n(Z_j)\}+O(n^{-1}\log n)\,, \tag{2.7}$$

where $W_n(\cdot)=\hat{G}_n(\cdot)+(1+\eta)\hat{H}_n(\cdot)$ and $O(n^{-1}\log n)$ is deterministic (cf. Govindarajulu, 1975, p. 261, for a simple direct proof of (2.7)). In view of the well-known fact that $\hat{G}_n$, $\hat{H}_n$ approach their theoretical counterparts G, H exponentially fast, it is tempting to substitute $W(\cdot)=G(\cdot)+(1+\eta)H(\cdot)$ for $W_n(\cdot)$ in (2.7). Having in mind Stein's lemma and the fact that an average of i.i.d. r.v. approaches the mean 1-quickly if second moments exist (cf. once more Chow and Teicher, 1978, p. 368) we 'conjecture' the following

Theorem G. *For all $F=G\otimes H\in\mathfrak{H}_1\cup\mathfrak{H}_0$ and $0<\eta$ put*

$$I(\eta\mid F)=\log(4(1+\eta))-2$$
$$-\int\log(G(u)+(1+\eta)H(u))\{G(\mathrm{d}u)+H(\mathrm{d}u)\}\,.$$

To be in agreement with our previous notation we write $I_0(\eta)$ resp. $I_1(\eta)$ instead of $I(\eta \mid F)$ if $F \in \mathfrak{H}_0$ resp. $F \in \mathfrak{R}(\eta)$.

(i) (*Savage and Sethuraman*, 1966). *For all $F \in \mathfrak{H}_1 \cup \mathfrak{H}_0$ for which $I(\eta \mid F) = 0$, there exists some $0 < \zeta < 1$ so that for $Q_n = \log L_{Rn}(\eta)$ and all n sufficiently large*:

$$P_F(N_Q^*(b, a) > n) < \zeta^n \quad (\text{'}N_Q^*(b, a) \text{ exponentially bounded'}) .$$

(ii) (*confer Berk and Savage*, 1968). *For all $F_0 \in \mathfrak{H}_0$, $F_1 \in \mathfrak{R}(\eta)$*:

$$n^{-1} \log L_{Rn}(\eta) \to I_1(\eta) \text{ exponentially fast under } F_1.$$

A refined version of part (i) appears in Sethuraman (1970) and Wijsman (1979). Part (ii) is but a special case of the main result in the Berk–Savage (1968) paper which deals with a broad class of nonparametric alternatives.

The validity of the functional limit theorem and other probabilistic statements concerning $\log L_{Rn}(\eta)$ can be drawn from Lai (1975a). In a recent paper Woodroofe (1982a) obtained a Chernoff–Savage theorem that allows for an application of the nonlinear renewal theory developed by Lai and Siegmund (1977, 1979). This approach yields refined approximations to both error probabilities.

In many practical situations, observations only become available in groups or the evaluation of an item requires such an effort of time that grouping seems advisable. Two SPRT based on ranks were proposed for experiments wherein groups of m observations are taken sequentially. The test presented by Wilcoxon et al. (1983) is based on $\Sigma_1^n \log L_{Rn}^{(k)}(\eta)$, where $L_{Rm}^{(k)}(\eta)$ is the p.r. computed from the k-th group. The statistics $(L_{Rm}^{(k)}(\eta))_{k\geq 1}$ form an i.i.d. sequence whence all interesting features of this test can be drawn from the standard theory. As this device merely joins together independent experiments but neglects the information that can be gained by comparing observations from different groups, it is suspected to be somewhat inefficient. In order to meet this objection, Bradley et al. (1966) suggested to maintain the above sampling scheme but to rerank the whole data collected once a new group of observations is obtained, i.e. suggested the use of Wald tests based on $(\log L_{Rn\cdot m})_{n\geq 1}$. These authors, too, assume Lehmann alternatives and mention some elementary properties of their procedure (a.s. termination, adequacy of the Wald approximations (2.2)). It should be added that these papers also treat samples of Y- and Z-observations which are not necessarily of the same size.

The discussion of the one-sample problem for testing symmetry is somewhat less complete but largely parallels the one of the two-sample case. For that reason we shall not enter into details but refer the reader to Weed et al. (1974) and Weed and Bradley (1971). Choi (1973) considered the corresponding sequential tests for independence.

2.2. *Wald tests based on linear rank statistics and U/V-statistics*

A rather natural modification of the rank-SPRT discussed so far leads to a class of Wald tests that are much simpler to carry out: Instead of using rank-p.r., which are awkward to handle for almost all classes of alternatives, we can employ their slopes, i.e. linear rank statistics. It is easy to conjecture that the resulting tests will asymptotically enjoy some kind of optimum property within a local approach. To corroborate this and, at the same time, to obtain approximations to OC and ASN, we shall rely on an invariance principle. This tool as well as other results needed to answer the questions posed at the beginning are valid, however, for many other statistics, too. Accordingly, it seems economical first to discuss properties of Wald tests based on more general statistics $Q=(Q_n)_{n\geq 1}$ and subsequently to turn to those aspects that are characteristic of linear rank statistics resp. U/V-statistics.

Termination properties of Wald tests. We shall mainly come across statistics Q that not only obey the SLLN, i.e. for all $F \in \mathfrak{H}_0 \cup \mathfrak{H}_1$ there is some $\mu(F) \in \mathbb{R}$ so that for all $\varepsilon > 0$

$$P_F(\sup_{k\geq n}|k^{-1}Q_k - \mu(F)| > \varepsilon) = p_{Fn}(\varepsilon) = o(1)\,,$$

but for which, in addition, some information concerning the rate of this convergence is available. In case $\mu(F) \neq 0$, the a.s. finiteness (integrability, exponentially boundedness) of $N_Q^*(b, a)$ can be concluded from this by means of some crude bounds. If, for instance, $p_{Fn}(\varepsilon) = O(n^{-r})$, then

$$\begin{aligned} P_F(N_Q^*(b,a) > n) &\leq P_F(-\mu(F) + bn^{-1} < n^{-1}Q_n - \mu(F) < -\mu(F) + an^{-1}) \\ &\leq p_{Fn}(\varepsilon) \quad (|\mu(F)| > \varepsilon > 0,\ n \text{ large}) \end{aligned} \tag{2.8}$$

implies the existence of all moments $E_F(N_Q^*(b,a))^s$, $s<r$. In case $\mu(F)=0$ the behaviour of Q is reminiscent of that of a recurrent random walk. In fact, we shall reduce this case to the discussion of the corresponding situation known from the classical SPRT when the log-p.r. becomes driftless under the 'exceptional' parameter point. To this end, we assume that for these $F \in \mathfrak{H}_0 \cup \mathfrak{H}_1$ we can find statistics $D_F = (DFn)_{n\geq 1}$ and a random walk $S_F = (S_{Fn})_{n\geq 1}$, $E_F(S_{F1}) = 0$, so that, for some $0 < \gamma \leq \frac{1}{2}$,

$$0 < D_{Fn} \to 1, \qquad D_{Fn}Q_n - S_{Fn} = o(n^\gamma) \qquad [P_F]\,. \tag{2.9}$$

(Both D_F and S_F are supposed to be $\mathfrak{A}$-measurable.) Define

$$M_1(\varepsilon) = \sup\{m \geq 1 : |D_{Fn} - 1| > \varepsilon\}\,,$$

$$M_2(\varepsilon, \gamma) = \sup\{m \geq 1 : |D_{Fn}Q_n - S_{Fn}| > \varepsilon n^\gamma\}\,.$$

Fix $k \geq 1$ (to be specified later on). With every $n \geq 1$ we associate the numbers $n_j = [jk^{-1}n]$, $1 \leq j \leq k$ ([·] denoting the interger part). Then,

$$
\begin{aligned}
P_F(N_Q^* > n) &\leq P_F(Q_{n_j} \in]b, a[\forall 1 \leq j \leq k) \\
&\leq P_F((1+\varepsilon)b - \varepsilon n_j^\gamma \leq S_{Fn_j} \leq (1+\varepsilon)a + \varepsilon n_j^\gamma \cdot \forall 1 \leq j \leq k) \\
&\quad + P_F(M_1(\varepsilon) \geq n_1) + P_F(M_2(\varepsilon, \gamma) \geq n_1) \\
&\leq \prod_{j=1}^{k} P_F(|S_{Fn_j} - S_{Fn_{j-1}}| \leq (1+\varepsilon)(a-b) + 4\varepsilon n^\gamma) \\
&\quad + P_F(2kM_1(\varepsilon) > n) + P_F(2kM_2(\varepsilon, \gamma) > n)
\end{aligned}
\tag{2.10}
$$

for every $n > 2k$. These estimates make it easy to derive sufficient conditions for the finiteness of N_Q^* and its moments. Essentially following Lai (1975b) we arrive at

THEOREM H. *Stick to the above notations and the assumption* $E_F(S_{F1}) = 0$.

(i) *Suppose that* (2.9) *is fulfilled with* $\gamma = \frac{1}{2}$ *and that* $E_F(S_{F1}^2)$ *is positive and finite. Then,* $N_Q^*(b, a)$ *is* P_F*-a.s. finite.*

(ii) *In addition to the assumptions made in* (i), *suppose that* $\gamma < \frac{1}{2}$ *and that, for some* $r, \varepsilon, \iota > 0$,

$$E_F(|S_{F1}|^{2+\iota}) < \infty, \qquad E_F(M_1^r(\varepsilon)) < \infty, \qquad E_F(M_2^r(\varepsilon, \gamma)) < \infty,$$

then

$$E_F((N_Q^*(b, a))^r) < \infty .$$

SKETCH OF PROOF. To prove the first part, it suffices to choose $k = 1$ and to apply the central limit theorem to the first summand in the last line of (2.10). The other two terms are negligible because of (2.9). In order to show the second assertion, we apply the Berry–Esseen theorem to each of the k factors appearing in (2.10). If k is suitably selected (depending on r, γ, ι), then the whole product can be shown to be of the order n^{-s}, $s > r$. Multiplying (2.10) by n^{r-1} and summing up we arrive at the result.

We dwelled a little bit on details because the argument used above is quite typical of the way in which classical results are carried over to a nonparametric set up. Mutatis mutandis, this remark applies as well to the asymptotic theory to come.

Local approximations to OC and ASN. Let us briefly recapitulate the necessary distribution theory. For every $\Delta_0 > \Delta > 0$, let F_Δ, $\tilde{F}_\Delta$ be p.m. and let $S_\Delta = (S_{\Delta n})_{n \geq 1}$ be the partial sums of variables that are independent under both F_Δ and $\tilde{F}_\Delta$. With S_Δ we associate the Donsker processes $\mathcal{S}_\Delta(\cdot)$ which result from linearly interpolating the values $\Delta S_{\Delta n}$ at the epochs $\Delta^2 n$, $n \geq 0$ ($S_{\Delta 0} \equiv 0$). (The definition of these processes is, in essence, a matter of scaling and depends on the nature in which F_Δ, $\tilde{F}_\Delta$ are parametrized. We are going to deal exclusively with that case that has been labelled the 'regular' one.) Processes $\mathcal{Q}_\Delta(\cdot)$ are defined in a like manner.

Assume that the random functions $\mathcal{S}_\Delta(\cdot)$ weakly approach $\mathcal{B}(\cdot\,|\,0,1)$ under $\tilde{F}_\Delta$, $\Delta\downarrow 0$, where $\mathcal{B}(\cdot\,|\,\zeta,\sigma^2)$ denotes a Brownian motion with drift ζt and variance $\sigma^2 t$. If $\mathfrak{L}_\Delta(\mathcal{S}_\Delta(t)_{0\le t\le r})$ is contiguous[1] to $\tilde{\mathfrak{L}}_\Delta(\mathcal{S}_\Delta(t)_{0\le t\le r})$, then there exists an equicontinuous family of centering functions $\zeta_\Delta(\cdot)$, so that the processes $\mathcal{S}_\Delta(\cdot)-\zeta_\Delta(\cdot)$ tend towards $\mathcal{B}(\cdot\,|\,0,1)$ under F_Δ as well; cf. Müller-Funk (1980). This includes that every weak limit of $\mathcal{S}_\Delta(\cdot)$ under F_Δ is necessarily Gaussian and that the covariance structure remains unchanged. In most applications, moreover, $\mathcal{S}_\Delta(\cdot)$ behaves like a process with independent and stationary increments and, correspondingly, $\zeta_\Delta(t)=\zeta_\Delta t$ is linear, ζ_Δ bounded. Typically, F_Δ is defined in such a way that these drift coefficients actually converge.

Now, suppose that $\mathcal{Q}_\Delta(\cdot)$ can be approximated under $\tilde{F}_\Delta$ by the above kind of partial sums processes $\mathcal{S}_\Delta(\cdot)$, i.e. that, for all $r,\ \varepsilon>0$,

$$\tilde{P}_\Delta(\sup_{0\le t\le r}|\mathcal{Q}_\Delta(t)-\mathcal{S}_\Delta(t)|>\varepsilon)=o(1)\quad\text{as }\Delta\to 0\,. \tag{2.11}$$

Under contiguity, (2.11) is valid with P_Δ instead of $\tilde{P}_\Delta$. The foregoing discussion, therefore, indicates that some $\mathcal{B}(\cdot\,|\,\zeta,1)$ will emerge as a limit process of $\mathcal{Q}_\Delta(\cdot)$. If $\tilde{F}_\Delta\equiv F_0$ and if Q is a (reversed) martingale sequence under F_0, then an even simpler derivation of that limiting behaviour is possible; confer Sen (1975). In this case, the tightness of $\mathfrak{L}_0((\mathcal{Q}_\Delta(t))_{0\le t\le r})$ is implied by the convergence of the finite dimensional marginals (Brown, 1971; Loynes, 1970) and, accordingly, it suffices to verify that, for all $0<t_1<\cdots<t_k$ and $\varepsilon>0$,

$$P_0(\max_{1\le j\le k}|\mathcal{Q}_\Delta(t_j)-\mathcal{S}_\Delta(t_j)|>\varepsilon)=o(1)\,. \tag{2.11$'$}$$

Such weak convergence results under local alternatives form the basis of approximations to OC and of Sen and Ghosh's (1980) approach to the sequential asymptotic relative Pitman efficiency (ARPE). In order to obtain the approximate ASN under F_Δ as well, we need a stronger result (requiring more restrictive assumptions). In varying forms, it has been known for a long time.

To be more precise, we select some $F_0\in\mathfrak{H}_0$ together with alternatives $\{F_\Delta:\Delta_0>\Delta>0\}$. The formulation of the invariance principle below calls for Borel functions $h_\Delta(\cdot)$ so that for some constant $c>0$:

$$\begin{aligned}&\exists\iota>0:\ E_\Delta|h_\Delta(X_1)|^{2+\iota}\le c\,,\\&\exists\zeta>0:\ E_\Delta(h_\Delta(X_1))=\Delta\zeta(1+o(1))\,,\\&\exists\sigma>0:\ \mathrm{Var}_\Delta(h_\Delta(X_1))=\sigma^2+o(1)\qquad(\text{as }\Delta\to 0)\end{aligned} \tag{2.12}$$

Put $S_{\Delta n}=\Sigma_1^n h_\Delta(X_j)$. Next, the Wald test $(N_Q^*(b,a),\ \delta_Q^*(b,a))$ is turned into an asymptotic sequential test by setting $N_N^*(\Delta)=N_Q^*(b\Delta^{-1},a\Delta^{-1})$, and $\delta_Q^*(\Delta)=$

[1]As for the definition and basic facts, confer Hájek and Šidák (1967, p. 202).

$\delta_Q^*(b\Delta^{-1}, a\Delta^{-1})$. Furthermore, we define

$$\tau^*(x) = \tau^*_{b,a}(x) = \inf\{t \geq 0: x(t) \notin]b, a[\},$$

where $x(\cdot)$ belongs to the space $C([0,\infty[)$ of all real valued continuous functions. Note that

$$0 \leq \Delta^2 N_Q^*(\Delta) - \tau^*(\mathcal{Q}_\Delta) \leqq \Delta^2 . \tag{2.13}$$

THEOREM I. *Suppose that* $\mathfrak{L}_\Delta(X_1) \to \mathfrak{L}_0(X_1)$ *in total variation, that* (2.12) *holds true, and that there are constants* $d > 0$, $\frac{1}{2} > \gamma > 0$, $\kappa > 2$ *so that*

$$P_\Delta(|Q_n - S_{\Delta n}| > n^\gamma) \leq dn^{-\kappa} \quad \text{for all } \Delta . \tag{2.14}$$

Then, as $\Delta \to 0$,

$$\mathfrak{L}_\Delta(\mathcal{Q}_\Delta(\cdot)) \xrightarrow{w} \mathfrak{L}(\mathcal{B}(\cdot \mid \zeta, \sigma^2)), \qquad \mathfrak{L}_\Delta(\Delta^2 N_Q^*(\Delta)) \xrightarrow{w} \mathfrak{L}(\tau^*(\mathcal{B}(\cdot \mid \zeta, \sigma^2))),$$

$$\Delta^2 E_\Delta(N_Q^*(\Delta)) \to E(\tau^*(\mathcal{B}(\cdot|\zeta, \sigma^2))),$$

where $\xrightarrow{w}$ *denotes the usual weak convergence on* $C([0,\infty[)$.

SKETCH OF PROOF. The first assertion is but a reformulation of Donsker's theorem for triangular arrays, the second one follows from the first one, (2.13), and the a.s. continuity of τ^* with respect to Wiener measure. In order to show the uniform integrability of the stopping variables involved, one can use a variant of the argument that is at the core of (2.10).

We shall also need a version of Theorem I concerning the modified stopping times

$$\bar{N}_Q(\Delta) = \inf\{n \geq n_0(\Delta): \Delta Q_n \notin]b, a[\},$$

where the initial sample size $n_0(\Delta)$ tends to infinity, but where $\Delta^2 n_0(\Delta)$ converges to zero. Sen (1981, p. 258, 267) essentially proved

THEOREM I′. *Suppose that all assumptions of Theorem* I *are satisfied except that in* (2.14) *we replace* Q_n *by* D_nQ_n, *where* $D = (D_n)_{n\geq 1}$ *are* ($\mathfrak{A}$-*measurable*) *statistics for which some* $\kappa' > 1$ *exists so that for all* $\varepsilon > 0$ *there are constants* $d' > 0$, $n_1 \geq 1$ *for which*

$$P_\Delta(|D_n - 1| > \varepsilon) \leq d'n^{-\kappa'} \quad \forall n \geq n_1 \ \forall \Delta > 0 . \tag{2.15}$$

Then the assertion of Theorem I *holds true for* $\bar{N}_Q(\Delta)$ *instead of* $N_Q^*(\Delta)$.

We shall not reproduce the familiar formulas for the statistically relevant limiting expressions

$$\chi_{b,a}(\zeta,\sigma^2)=P(\mathscr{B}(\tau^*\mid\zeta,\sigma^2)=b),\qquad \Lambda_{b,a}(\zeta,\sigma^2)=E(\tau^*(\mathscr{B}(\cdot\mid\zeta,\sigma^2)))\,,$$

but refer to Dvoretzky et al. (1953).

Locally optimal Wald tests, ARPE. In order to justify Wald tests based upon slopes of p.r. we want to model such tests on the optimal procedure of the matching limit problem. Consequently, let us first look at some continuous time problems. Suppose that we are observing $\mathscr{S}(\cdot)=\mathscr{B}(\cdot\mid\xi,\sigma^2)$, where σ^2 is, of course, known. We are interested in testing

$$H_0:\ \xi=0\qquad\text{versus}\qquad H_1:\ \xi>0\quad\text{resp. } H_1':\ \xi=1\,. \tag{2.16}$$

As is generally known, the p.r. of $\mathfrak{L}(\mathscr{B}(t\mid\xi,\sigma^2)_{0\leq t\leq u})$ relative to $\mathfrak{L}(\mathscr{B}(t\mid 0,\sigma^2)_{0\leq t\leq u})$ is just the exponential martingale

$$\mathscr{L}(u\mid\xi,\sigma^2)=\exp(\sigma^{-2}(\xi\mathscr{S}(u)-\xi^2u/2)),\quad u\geq 0\,. \tag{2.17}$$

By definition, a LMP sequential test for H_0 versus H_1 maximizes the slope of the power function at $\xi=0$ among all tests the error-I-type of which is exactly α and the ASN at zero of which does not exceed a prescribed upper bound. Irle (1980), paralleling Berk's (1975a) work in discrete time, showed that the solution is but a Wald test based on $\mathscr{S}(\cdot)$. Accordingly, this LMP test coincides with the SPRT for the conjugate pair of parameters $\xi_0=-\frac{1}{2}$ versus $\xi_1=\frac{1}{2}$. If, moreover, the boundaries happen to be symmetrical, i.e. $a=-b$, then this test may be thought of as a minimax sequential test, too; confer DeGroot (1960), Schmitz (1982). Alternatively, one can orientate oneself at the SPRT for H_0 versus H_1' which is based on $\mathscr{S}(t)-t/2$. As for SPRT in continuous time the reader is referred to Dvoretzky et al. (1953), Ghosh (1970), Irle and Schmitz (1981).

Now, suppose that Theorem I is applicable to $F_0\in\mathfrak{H}_0$, $Q=(Q_n)_{n\geq1}$, and a suitable family $\{F_\Delta:\Delta_0>\Delta>0\}\subset\mathfrak{H}_1$. Introducing a local parameter $0\leq\eta\leq 1$ by putting $F_{\Delta\eta}=F_{\Delta\cdot\eta}$ we can express the asymptotic problem on hand in the form $\eta=0$ against $\eta>0$. If $\zeta=1$ in (2.12), then the Wald test based on ΔQ, i.e. $(N_Q^*(\Delta),\delta_Q^*(\Delta))$, formally acts within the local model like the sequential LMP for H_0 versus H_1. This, of course, does not yet imply any kind of asymptotic optimality unless we can relate Q to the p.r. To illustrate what is meant by that, we let $\mathfrak{C}=(\mathfrak{C}_n)_{n\geq1}$ denote the information which is available in the nonparametric formulation of the problem and assume that $Q_n=E_0(\dot{L}_n\mid\mathfrak{C}_n)=\Sigma_1^n E_0(\dot{l}(X_j)\mid\mathfrak{C}_n)$, where $\dot{L}_n$ is again the slope of $L_n(F_{\Delta\eta}:F_0)$. Heuristic considerations suggest to look for the following kind of expansion under F_0:

$$\begin{aligned}
\mathscr{L}_{\Delta\eta}(t) &= E_0\Big(\prod_{j=1}^{n} l_{\Delta\eta}(X_j) \mid \mathfrak{C}_n\Big), \quad t = \Delta^2 n, \\
&= E_0\Big(\prod_{j=1}^{n} (1 + \eta\Delta \dot{l}(X_j) + \cdots) \mid \mathfrak{C}_n\Big) \\
&= \prod_{j=1}^{n} (1 + \eta\Delta E_0(\dot{l}(X_j) \mid \mathfrak{C}_n) + \cdots) \\
&\simeq \exp\Big(\eta\Delta \sum_{j=1}^{n} E_0(\dot{l}(X_j) \mid \mathfrak{C}_n) - \eta^2\Delta^2 n E_0\Big(n^{-1}\sum_{j=1}^{n} \dot{l}^2(X_j) \mid \mathfrak{C}_n\Big)\Big/2\Big) \\
&\simeq \exp(\eta \mathscr{Q}_\Delta(t) - \eta^2 \upsilon^2 t/2), \quad \upsilon^2 = E_0(\dot{l}^2(X_1)), \\
&\to \exp(\eta\upsilon \mathscr{B}(t \mid 0, 1) - \eta^2\upsilon^2 t/2) \quad \text{as } \Delta \downarrow 0 .
\end{aligned}$$

The validity of such an expansion essentially amounts to the following: $\{F_\Delta : \Delta_0 > \Delta \geq 0\}$ is locally asymptotically normal and $\mathfrak{C}$ is asymptotically fully informative. LeCam (1979) showed how this approximation device can then be used to obtain asymptotically optimal tests. Unfortunately, the implementation of the whole programme turns out to be rather complicated in the present situation. For that reason, we shall be contented with a weaker form of asymptotic optimality in the special cases to come. There, we shall merely verify that the optimal parametric procedure and the nonparametric Wald test based on the conditioned statistics have the same asymptotic characteristics which, moreover, coincide with those of the optimal test for the limit problem. The same remarks apply to Wald tests patterned after the SPRT for H_0 versus H_1', i.e. to Wald tests based on $\Delta(Q_n - n\Delta/2)$, where Q is asymptotically equivalent to the slopes of the p.r. In many cases $\upsilon^2 = \upsilon^2(F_0)$ depends on the choice of $F_0 \in \mathfrak{H}_0$ and has to be estimated. The guiding rule for constructing sequential tests involving unknown parameters dates back to Bartlett (1946) and Cox (1963), who considered parametric testing problems in the presence of nuisance parameters which were taken care of with the help of ML-estimates. In essence, the reasoning is but a variant of the preceding discussion. Incorporating estimators into the test statistics requires a modification of the stopping rule. For technical reasons we have to allow for an initial sample size tending to infinity at a moderately large rate. Theorem I′ now provides the basis for handling that case.

Theorems I and I′ can also be used to judge the relative performances of Wald tests based on the same stopping boundaries but on different statistics. The ARPE was investigated by various authors at varying levels of generality; confer Hall (1974), Sen (1973a), Sen and Ghosh (1974a, 1980), Ghosh and Sen (1976, 1977), Braun (1976), Lai (1975a, 1978), Müller-Funk (1979), among others. Simplistically, we may summarize their results in a blanket assertion: The nonsequential and the sequential ARPE numerically coincide. This claim can be verified, for instance, by a modification of the argument which is used in the nonsequential theory in order to relate the limiting shift of the test statistics to the ratios of sample sizes. We shall not touch on the more delicate question of comparing sequential tests that have different types of stopping boundaries. For the difficulties connected with that confer Berk (1975b, 1976) and the references given there. Another topic that is left out is the problem how to obtain correction terms for the foregoing approximations or how to ensure second (higher) order properties of nonparametric Wald tests. So far, nobody seems to have made efforts in that direction.

Asymptotics under a fixed distribution. In case Q behaves like a random walk with mean zero and variance $\sigma^2(F)$ per observation, Lai (1975b) determined the limit of $N_Q^*(b, a)$ and its moments as $\min\{a, |b|\}$ tends to infinity and $|b|(a+|b|)^{-1}$ approaches $0<\omega<1$. Omitting regularity conditions, his result reads as follows:

$$\sigma^2(F)(a+|b|)^{-2}N_Q^*(b,a) \xrightarrow{w} \tau^*_{\omega,1-\omega}(\mathcal{B}(\cdot\mid 0,1)),$$

$$E_F(N_Q^*(b,a))^\kappa \sim E(\tau^*_{\omega,1-\omega}(\mathcal{B}(\cdot\mid 0,1)))^\kappa \sigma^{-2\kappa}(F)(a+|b|)^{2\kappa}.$$

Typically, this case corresponds to the null-situation. If, however, $n^{-1}Q_n$ converges towards a positive constant $\mu(F)$, then a standard argument from classical renewal theory shows that

$$N_Q^*(b,a) \sim a/\mu(F) \qquad [P_F] \quad \text{as } \min\{a,|b|\}\to\infty.$$

Of course, we would like to replace $N_Q^*(b, a)$ by its expectation. The uniform integrability needed for that has to be established separately. For linear rank statistics e.g., this can be achieved by means of Theorem A. Better approximations can be obtained along the lines suggested by Lai and Siegmund (1977, 1979).

I. The use of linear rank statistics

Let T, or T_φ if the dependence on the score function φ respectively the pair $\varphi = (\varphi_1, \varphi_2)$ is to be stressed, be one of the three types of linear rank statistics already introduced. Throughout, we shall assume that $\varphi, \varphi_1, \varphi_2$ are at least square integrable with L_2-norm equal to one and that all of them are non-decreasing.

Wald tests based on ΔT, or simply T (as the factor Δ can be absorbed into the boundaries), were investigated by several authors. Let us only mention the papers by Wilcoxon et al. (1963), Bradley et al. (1965, 1966), Holm (1973a, b, 1975), Sen and Ghosh (1974a, b), Lai (1975a), Braun (1976), Ghosh and Sen (1977), Hall and Loynes (1977), Müller-Funk (1979), Bönner et al. (1980).

First, let us somewhat telegraphically deal with those points that are immediate consequences of the previous discussion.

Termination properties. Theorem A and (1.15) entail that N_T^* is exponentially bounded for every $F \in \mathfrak{H}_1$ for which $\theta_\varphi(F)$ is strictly positive. In fact, for all $\varepsilon > 0$ and n large enough,

$$
\begin{aligned}
P_F(N_T^* > n) &\leq P_F(|n^{-1}T_{\varphi n} - \theta_\varphi(F)| > \varepsilon) \\
&\leq P_F(d_K(\hat{F}_n, F) > \varepsilon/2cd) \leq \exp(-n\varepsilon^2/2c^2d^2)\,.
\end{aligned}
$$

The case $\theta_\varphi(F) = 0$ can be treated by means of Theorem H. If $F \in \mathfrak{H}_0$, for instance, then the first part of this result and the CLT for T under $\mathfrak{H}_0$ (cf. Hajek and Sidak, 1967, pp. 160, 163, 166, 168) ensure that N_T^* is a.s. finite in the null-situation.

Unbiasedness. Put

$$
\bar{N}_T^* = \inf\{n \geq 1\colon T_n \geq a\Delta^{-1}\}, \qquad \underline{N}_T^* = \inf\{n \geq 1\colon T_n \leq b\Delta^{-1}\}\,.
$$

Obviously, $N_T^* = \min\{\bar{N}_T^*, \underline{N}_T^*\}$ and

$$
\mathrm{OC}(F) = P_F((N_T^*, \delta_T^*) \text{ accepts } \mathfrak{H}_0) = P_F(\underline{N}_T^* \leq \bar{N}_T^*)\,.
$$

We know from (1.18), that $\mathfrak{L}_{F_1}(T_n)$ is stochastically larger than $\mathfrak{L}_{F_0}(T_n) = \mathfrak{L}_{\mathfrak{H}_0}(T_n)$ for every choice of $F_0 \in \mathfrak{H}_0$ and $F_1 \in \mathfrak{H}_1$ and all n. With the help of an i.i.d. sample drawn from the rectangular d.f. we can construct random sequences $T^{(i)} = (T_n^{(i)})_{n \geq 1}$, $i = 0, 1$, so that (i) $\mathfrak{L}(T^{(i)})$ and $\mathfrak{L}_{F_i}(T)$ are equal and (ii) $T^{(0)} \leq T^{(1)}$ (in each component). Because of

$$
T^{(0)} \leq T^{(1)} \Rightarrow \begin{Bmatrix} \underline{N}_{(0)}^* \leq \underline{N}_{(1)}^* \\ \bar{N}_{(0)}^* \geq \bar{N}_{(1)}^* \end{Bmatrix} \Rightarrow \{\underline{N}_{(1)}^* \leq \bar{N}_{(1)}^*\} \subset \{\underline{N}_{(0)}^* \leq \bar{N}_{(0)}^*\}\,,
$$

where $\underline{N}_{(0)}^* = \underline{N}_T^*(0)$, etc., we realize that (N_T^*, δ_T^*) is in fact unbiased.

Approximations to OC and ASN under local alternatives. It has been mentioned that the invariance principle under $\mathfrak{H}_0$ and alternatives close to it is implied by the martingale property (cf. Theorem B) and the CLT. The latter, however, holds true without further conditions on the score function(s). Hence, every class of contiguous alternatives under which $\mathcal{T}_\Delta(\cdot)$ is weakly convergent (and not just stochastically bounded) leads to a limit process $\mathcal{B}(\cdot \mid \zeta, 1)$. Accordingly, the limiting OC is given by $\chi_{b,a}(\zeta, 1)$ and our first task is accomplished once we can compute ζ.

To arrive at approximate formulas for the ASN we have to rely on Theorem I. This result is particularly easy to apply if we assume nonparametric alternatives. The reasoning is exemplified by the univariate one-sample problem the obvious changes in the other cases being left to the reader. Select some square integrable ψ_0 defined on the unit interval which satisfies (1.6) and which is skew, i.e. $\psi_0(t) + \psi_0(1-t) \equiv 0$. If, in addition, ψ fulfills Chernoff–Savage type smoothness and growth conditions, then ψ suitably truncated leads to a family $\{\psi_\Delta\colon \Delta_0 > \Delta > 0\}$ each member of which enjoys the same properties as ψ_0 and, moreover, so that (i) $\psi_\Delta \to \psi_0$ in mean square, (ii) $\Delta \sup\{|\psi_\Delta(t)|: 1 > t > 0\} \leq 1$, (iii) $\Delta^2 \sup\{|\psi_\Delta(t)|: 1 > t > 0\} = o(1)$ as $\Delta \downarrow 0$.

Fix $F_0 \in \mathfrak{H}_0$, $\eta > 0$ and put $\Psi_{\Delta\eta}(\mathrm{d}t) = (1 + \eta\Delta\psi_\Delta(t))\,\mathrm{d}t$, $F_{\Delta\eta} = \Psi_{\Delta\eta}(F_0)$. Let $h_{\Delta\eta}$ be the corresponding Borel functions in the Chernoff–Savage representation. We shall find it convenient to express the score function in the form $\varphi(t) = \varphi_0((t+1)/2)$, where φ_0 is again skew. Somewhat tedious calculations show that (2.12) is fulfilled with $\zeta = \eta\rho$ and $\sigma^2 = 1$, where

$$\rho = \rho(\varphi_0, \psi_0) = \int_0^1 \varphi_0(t)\psi_0(t)\,\mathrm{d}t \in [-1, 1]\,. \tag{2.18}$$

According to Theorem I, the asymptotic Wald test $(N_T^*(\Delta), \delta_T^*(\Delta))$ has the asymptotic OC $\chi_{b,a}(\eta\rho, 1)$ and the asymptotic ASN $\Lambda(\eta\rho, 1)$ under these local alternatives.

Asymptotic optimality within a local model. Again, we shall only discuss the univariate one-sample case by way of example. As before, we consider nonparametric alternatives $F_{\Delta\eta} = \Psi_{\Delta\eta}(F_0)$. The pertaining p.r. relative to F_0 are termed $L_{\Delta n}(\eta)$. For every fixed Δ, the LMP sequential test for $\eta = 0$ versus $\eta > 0$ is based on $\partial/\partial\eta\, L_{\Delta n}(\eta)|_{\eta=0} = \dot{L}_{\Delta n} = \Sigma_1^n \psi_\Delta(X_j)$. Putting $h_\Delta = \psi_\Delta$, $\zeta = \eta$, $\sigma^2 = 1$, we can apply Theorem I (or a direct argument) in order to realize that the limiting characteristics of these optimal tests are $\chi_{b,a}(\eta, 1)$ and $\Lambda_{b,a}(\eta, 1)$. If we choose $\varphi(t) = \psi(t) = \psi_0((t+1)/2)$ (implying $\rho(\varphi_0, \psi_0) = 1$), then the asymptotic Wald test based on T_ψ yields the same limiting OC and ASN within this model. (Note, that these limits do not depend on the starting point $F_0 \in \mathfrak{H}_0$.) In that weak sense, T_ψ induces an asymptotic LMP sequential test for $\mathfrak{H}_0$ against local ψ_0-alternatives. This statement, of course, corresponds to the familiar fact that linear rank statistics with score function ψ asymptotically yield an optimal nonsequential test if ψ_0 is the direction in which we leave the null-hypothesis (irrespective of the starting point). A more precise meaning can be attached to this kind of optimality if we only compare Wald tests based on linear rank statistics. As already pointed out, the ARPE of any Wald test built upon T_φ relative to the T_ψ-Wald test is given by the quantity

$$\mathrm{ARPE}(T_\varphi : T_\psi \mid \Psi_{\Delta\eta}(F_0)) = \rho^2(\varphi_0, \psi_0) \quad \forall 0 < \eta < 1, \qquad F_0 \in \mathfrak{H}_0\,.$$

Obviously, the RHS takes on the maximal value 1 iff $\varphi = \psi$.

Variations on the theme. (1) The situation becomes slightly different if we find it natural to assume a location model where the precise form of the underlying d.f. is not known. For the sake of simplicity let us again consider the one-sample problem. When we now select local alternatives $F_0(\cdot - \eta\Delta)$, $F_0 \in \mathfrak{H}_0$, then the tangent ψ_0 changes with every F_0 and the limiting shift, accordingly, depends on this starting point. Clearly, $n^{-1/2}T_{\varphi n}$ has the asymptotic mean $\eta\mu(F_0)$ under $F_0(\cdot - \eta n^{-1/2})$ (cf. (1.17) for the definition of the quantities involved). Correspondingly, $\mathcal{T}_{\varphi\Delta}(\cdot)$ converges towards $\mathcal{B}(\cdot \mid \eta\mu(F_0), 1)$. If the drift coefficient were known, the optimal procedure for H_0 versus H_1 would suggest the use of the statistics $\mu(F_0)T_{\varphi n}$. The natural thing to do, of course, is to replace $\mu(F_0)$ by an estimate $(\hat{\mu}_n)_{n\geq 1}$, which meets the requirements of Theorem I′. Sen and Ghosh (1974a) proposed the Hodges–Lehmann type estimator

$$\hat{\mu}_n = \frac{2u_{n,\alpha/2}}{n(\hat{\vartheta}_{u,n} - \hat{\vartheta}_{l,n})}, \quad n \geq 1,$$

$$\hat{\vartheta}_{u,n} = \inf\{\vartheta: T_n(\vartheta) > -u_{n,\alpha/2}\}, \qquad \hat{\vartheta}_{l,n} = \sup\{\vartheta: T_n(\vartheta) > u_{n,\alpha/2}\}$$

where $n^{-1/2}u_{n,\alpha/2}$ tends to the upper $\alpha/2$-point of the standard normal distribution. The demonstration that $(\hat{\mu}_n)_{n\geq 1}$ satisfies all conditions required by Theorem I′ seems to be rather laborious, however; confer Sen (1981; p. 265) for a sketch of proof. Under alternatives $F_0(\cdot - \eta\Delta)$, the limiting formulas for OC and ASN again become $\chi_{b,a}(\eta\mu(F_0), 1)$ resp. $\Lambda_{b,a}(\eta\mu(F_0), 1)$ if $(\hat{\mu}_n T_{\varphi n})_{n\geqq 1}$ is employed. Let ψ_0 be defined as in (1.5) and, as before, denote its right-hand tail by ψ. The above rank procedure is (weakly) asymptotically LMP with respect to this model in case $\varphi = \psi$. Here, of course, the starting point F_0 matters.

(2) Wald tests modelled on the SPRT for H_0 against H_1' (cf. (2.16)) can be treated in a similar fashion. If such SPRT-type tests are constructed with the help of $\Delta(T_n - n\Delta/2)$, the changes that have to be made go without saying. In the one-sample location model, (2.17) and the asymptotic linearity result would suggest the use of the quantiy

$$\mu(F_0)T_n(\Delta/2) \simeq \mu(F_0)(T_n - n\Delta\mu(F_0)/2).$$

Inserting the estimator $(\hat{\mu}_n)_{n\geq 1}$, we arrive at the test statistics $(\hat{\mu}_n T_n(\Delta/2))_{n\geq 1}$ the use of which was proposed by Sen and Ghosh (1974a).

(3) A different kind of justification was provided by Holm (1973a, b). Assuming a suitable loss structure, this author established the asymptotic minimax character of the sign test.

The theory presented so far can be modified and generalized in various directions. For instance, we can employ linear rank statistics into which only the sequential ranks R_{jj}, R_{jj}^+ etc. enter. This type of statistics is particularly well suited for the sequential approach as T_{n+1} is easily computed from T_n and the additional observation. Confer Lombard (1981) and Section 3.2.

II. The use of U/V-statistics

As already mentioned, U- and V-statistics are in general not distribution-free under $\mathfrak{H}_0$ or similar classes of d.f. Accordingly, no matter how we set up sequential tests with the help of U/V-statistics, we shall have to rely on some invariance principle in order to deduce anything at all about their error probabilities etc. To render this remark technically workable, we make the following assumptions on the (not necessarily regular) functional $\nu(\cdot)$:

(A1) $\nu(\cdot)$ is 'differentiable at the true d.f.', i.e. for all $F \in \mathfrak{H}_0$ there is a function h_F which is at least square integrable and is centered at its expectation (under F) so that for some $\gamma \geqslant \frac{1}{2}$

$$V_n - \nu(F) - \int_{\mathbb{R}} h_F \, \mathrm{d}(\hat{F}_n - F) = \mathrm{o}(n^{-\gamma})[P_F] .$$

(A2) There is an ($\mathfrak{A}$-measurable) estimator $\hat{\sigma}^2 = (\hat{\sigma}_n^2)_{n\geqslant 1}$ for $\sigma^2(F) = \mathrm{Var}_F(h_F(X_1))$ that is strongly consistent under every $F \in \mathfrak{H}_0$.

Formula (2.17) then suggests to base Wald tests on $\tilde{V} = (\tilde{V}_n)_{n\geqslant 1}$,

$$\tilde{V}_n = \hat{\sigma}_n^{-2} n(V_n - \nu_0), \quad n \geqslant 1 ,$$

or, more precisely, on $\Delta \tilde{V}$. (We have $\hat{\sigma}_n^2$ instead of $\hat{\sigma}_n$ in the denominator because $\mathfrak{H}_0$ specifies $\nu(F)$ and not $\nu(F)/\sigma(F)$.)

Examples of such functionals are provided by smoothly weighted averages of quantiles, i.e.

$$\nu(F) = \int_{\mathbb{R}} x\varphi(F(x))F(\mathrm{d}x) \tag{2.19}$$

where φ is assumed to satisfy Chernoff–Savage-type conditions. Formally expanding the L-statistic $\nu(\hat{F}_n)$ around F and recalling the well-known formula for the asymptotic variance of $\nu(\hat{F}_n)$ we have

$$h_F(x) = x\varphi(F(x)) + \int_{\mathbb{R}} u\varphi'(F(u))(1_{[0,\infty[}(u - x) - F(u))F(\mathrm{d}u) ,$$

$$\sigma^2(F) = \iint_{\mathbb{R}^2} (F(x \wedge y) - F(x)F(y))\varphi(F(x))\varphi(F(y)) \, \mathrm{d}x \, \mathrm{d}y .$$

It has been pointed out earlier that (A1), i.e. the Chernoff–Savage representation, is valid on regularity conditions (cf. the references in 1.2). Under suitable assumptions, moreover, the natural candidate for $\hat{\sigma}^2$, and that is $\sigma^2(\hat{F}_n)$, is strongly consistent as required by (A2). In case of a regular functional, both (A1) and (A2) can easily be verified by means of Theorems D and E. That goes for (A2) as well because the limiting variance of U/V-statistics is itself a regular functional as observed by Sen (1960) and Sproule (1969). Sen proposed the following estimator: Let U_{nj} be the 'U-statistic corresponding to the kernel $g_j(y_1, \ldots, y_{j-1}) = g(X_j, y_1, \ldots, y_{j-1})$ and the random sample $X_1, \ldots, X_{j-1}, X_{j+1}, \ldots, X_n$' and put

$$\hat{\sigma}_n^2 = (n-1)^{-1} \sum_{j=1}^{n} (U_{nj} - U_n)^2, \quad n > p . \tag{2.20}$$

In this case we can define $\tilde{U}_n$ in analogy to $\tilde{V}_n$. With regard to Theorem D(b) both, of course, are asymptotically equivalent.

Wald tests based on $\tilde{V}$, $\tilde{U}$ or related statistics were investigated by Sen (1973a, b), Ghosh and Sen (1976). Lin (1981) studied the moments of the stopping times of Wald tests based on U and V.

Finiteness of the stopping variables and their moments, etc. The stopping times $N^*_{\tilde{V}}(v, a)$, $\bar{N}_{\tilde{V}}(b, a)$ as well as $N^*_{\tilde{U}}(b, a)$, $\bar{N}_{\tilde{U}}(b, a)$ are amenable to our previous methods. In fact, Theorem H – in connexion with Theorem E(a, b) – entails that all these stopping rules together with their expectations are finite if the kernel g possesses $(2+\iota)$ moments for some $\iota > 0$. Theorem E(c) can be used to discuss the exponential boundedness of these quantities. Confer also Lin (1981), who, in addition, provided approximations under a fixed distribution.

Asymptotics under $\mathfrak{H}_0$ and alternatives close to it. The random sequence $(\hat{\sigma}_n^{-1} n(V_n - \nu_0))_{n\geq 1}$, properly scaled, behaves like a standard Brownian motion under $\mathfrak{H}_0$ and, in this sense, is asymptotically distribution-free. Note, however, that the invariance principle on which we depend will in general not be uniformly valid. Accordingly, we cannot really circumvent the annoying lack of distribution-freeness by simply passing on to asymptotic methods. $\tilde{V}$ is not asymptotically distribution-free anyhow, and thus every assertion concerning Wald tests based on $\tilde{V}$ depends on the choice of a class $\{F_\Delta : \Delta_0 > \Delta \geq 0\}$ allowing for an expansion of the form $\nu(F_{\Delta\eta}) = \nu_0 + \eta\Delta\zeta(F_0) + \cdots$. Theorem I′ then provides us with the limiting OC and ASN. In the case of a regular $\nu(\cdot)$, simple sufficient conditions guaranteeing its applicability can be deduced with the help of Theorems D, E.

Even if we arrange these Wald tests after the example of a limiting optimal procedure, we are not trying to adapt our former reasoning in the case of rank statistics to the present situation. Otherwise, we would have to fix $F_0 \in \mathfrak{H}_0$ and to specify a class of alternatives $\{F_\Delta : \Delta_0 > \Delta > 0\}$ as above, so that the slopes of the p.r. and the U/V-statistics agree asymptotically. To put families of the form $F_\Delta(\mathrm{d}x) = \exp(\Delta h_{F_0}(x) - w_{F_0}(\Delta))F_0(\mathrm{d}x)$ ($\in \mathfrak{H}_1$?) or similar classes to trial may be reasonable in special cases but seems to be unrewarding more generally.

The foregoing remarks make it clear that the present derivation of Wald tests built upon U/V-statistics can only be regarded as a guiding rule that has to be adjusted to the actual problem.

A supplement. Mukhopadhyay (1981) proposed sequential tests based on U that are in the spirit of the Chow–Robbins type sequential bounded length confidence intervals. It seems to be difficult to compare these tests with Wald tests.

3. Nonparametric sequential significance tests

In this section we encounter sequential procedures for testing problems into which the hypotheses enter into a nonsymmetrical way. In their outward appearance they differ from the tests of the foregoing section as they only have one-sided stopping regions. Such sequential tests may be thought of as a sequence of one-sided (nonsequential) tests $1(Q_n \geq a(n))$ based on an increasing number of observations. We stop sampling once the first of these tests becomes significant in which case we reject the null-hypothesis. If this does not occur, we continue sampling either indefinitely or until some target sample size is reached. In the subsections to come we are going to deal with the open-ended as well as the truncated versions.

Open-ended versions: *Nonparametric tests with power one.* Tests for one-sided hypotheses that have a type-I-error not exceeding a prescribed level $0 < \alpha < 1$ and a type-II-error zero were first investigated by Fabian (1956), Farrell (1964) as well as Darling and Robbins (1967a). Such tests were proposed for some sort of quality control problems. A more detailed discussion of the logic underlying these procedures appears in Darling and Robbins (1968a). Apart from rather artificial examples, of course, only sequential tests will meet both requirements on the error probabilities. In mathematical terms, a test with power one (TWP 1) based on statistics $Q = (Q_n)_{n\geq 1}$, an initial sample size n_0, and a stopping boundary $a(\cdot): [n_0, \infty[\to \mathbb{R}_+$ is defined by the random time

$$\tilde{N}_Q = \tilde{N}_Q(a) = \inf\{n \geq n_0 \colon Q_n \geq a(n)\}$$

and the decision rule $\tilde{\delta}_Q(a) = (\emptyset, \{\tilde{N}_Q(a) = n\})_{n\geq 1}$. In other words, the test decides in favor of $\mathfrak{H}_1$ whenever sampling stops and is subject to the conditions

$$P_F(\tilde{N}_Q(a) < \infty) \leq \alpha \quad \forall F \in \mathfrak{H}_0, \qquad P_F(\tilde{N}_Q(a) < \infty) = 1 \quad \forall F \in \mathfrak{H}_1 . \tag{3.1}$$

Note, that the test is completely determined by $\tilde{N}_Q$. Typically, the existence of a TWP 1 based on Q is due to the mere fact that the LIL is valid under $\mathfrak{H}_0$ and that the SLLN holds true under $\mathfrak{H}_1$, i.e.

$$\exists c > 0 \ \forall F \in \mathfrak{H}_0 \quad \overline{\lim_n}\, (n \log_2 n)^{-1/2} Q_n \leq c \quad [P_F] , \tag{3.2}$$

$$\forall F \in \mathfrak{H}_1 \ \exists \zeta > 0 \quad n^{-1} Q_n \to \zeta \quad [P_F] , \tag{3.3}$$

where $\log_2 t = \log((\log t)^+)$. In this case we can choose the stopping curve in such a way that $a(t) \sim r(t \log_2 t)^{1/2}$, $r > c$, and such that

$$P_F(Q_n \geq a(n) \text{ for some } n \geq m) = q(m) = o(1) \quad \text{as } m \to \infty . \tag{3.4}$$

If the convergence in (3.4) is uniform with respect to $F \in \mathfrak{H}_0$, then $a(\cdot)$ together with a suitably selected n_0 induces a TWP 1. In all cases of interest, moreover, the $\overline{\lim}$ in (3.2) is actually equal to c $[P_F]$ for some particular $F \in \mathfrak{H}_0$. If this happens, a boundary leading to a TWP 1 cannot tend to infinity at a rate less than $(n \log_2 n)^{1/2}$, vice versa. Accordingly, this rate is the best one can strive for in order to render the ASN under $\mathfrak{H}_1$ small. The foregoing remarks, however, only enable us to carry out the test in practice if we can derive explicit bounds for $q(m)$. Such 'iterated logarithm inequalities' for the sample mean were derived by Darling and Robbins (1967a, 1967b). If Q is a (reversed) martingale, then one can try to mimic the technique in the last mentioned paper, i.e. to apply the Chow–Hajek–Renyi inequality to appropriate blocks of statistics Q_n. The best we can hope for with that device, however, is a boundary which increases to infinity at some rate $t^{1/2}(\log t)^\kappa$ and which, accordingly, leads to a test with a comparatively large stopping variable. Corresponding inequalities for Brownian motion were obtained by Robbins (1970), Robbins and Siegmund (1970, 1973). Hence it is near at hand to shift the problem to the limit by means of a suitable invariance principle. If the Q_n happen to be sums of i.i.d. variates, the necessary limit theorems can be found in Robbins and Siegmund (1970) as well as in Lai (1976b). Extensions to these results to 'disturbed' random walks yield TWP 1 that approximately fulfill the first relation in (3.1). Technically, we can handle the oncoming remainder terms and random coefficients by the same method that has been employed in the discussion of the termination properties of Wald tests in Subsection 2.2. Proceeding that way, we arrive at the following result, which is but a corollary to Lai (1976b, Theorem 5).

THEOREM J. *Fix $F \in \mathfrak{H}_0$ and suppose that there are i.i.d. variables $W_1, W_2, \ldots, E_F(W_1^2) = 1$ and $E_F(W_1) = 0$, and statistics $D = (D_n)_{n\geq 1}$ (both depending on F) so that for some $0 < \gamma < \frac{1}{2}$*

$$D_n Q_n - \sum_{j=1}^{n} W_j = o(n^\gamma) \qquad [P_F] .$$

Assume that $a(\cdot)$ is continuous and that $t^{-1/2}(a(t) - t^\gamma)$ is ultimately nondecreasing.

(i) *If $D_n \equiv 1$ and $a(\cdot)$ is an upper class function for $\mathcal{B}(\cdot \mid 0, 1)$*, i.e.

$$\int_{s_0}^{\infty} a(t) t^{-3/2} \exp(-a^2(t)/(2t))\, \mathrm{d}t < \infty \quad \exists s_0 \geq 0 , \tag{3.5}$$

then for all $s \geq s_0$

$$P_F(Q_n \geq m^{1/2} a(n/m)\, \exists n \geq sm) \rightarrow P(\mathcal{B}(t \mid 0, 1) \geq a(t)\, \exists t \geq s) . \tag{3.6}$$

(ii) *If $0 < D_n \rightarrow 1$ $[P_F]$ and if there is some $0 < d < 1$ so that* (3.5) *is fulfilled with* $\mathrm{d}a(\cdot)$ *instead of* $a(\cdot)$, *then* (3.6) *remains true.*

Suitable rate $(t \log_2 t)^{1/2}$ boundaries were made fairly explicit by Robbins and Siegmund (1970, 1973).

Truncated versions: *Nonparametric repeated significance tests.* Armitage being concerned with medical trials introduced certain sequential three-decision procedures into the statistical literature; confer his monograph (1975) for a more recent account. The related two-decision procedures have been termed repeated significance tests (RST). In short, their procession can be described as follows. A target sample size m is specified and the incoming observations are constantly scrutinized. Sampling stops and the null-hypothesis is rejected as soon as the accumulated data shows enough evidence to do so. If this does not happen to be the case up to (and including) time m, then we stick to $\mathfrak{H}_0$. Tests of this kind were proposed in an ad hoc fashion and motivated on ethical grounds as well as on practical considerations.

Armitage only allowed for models involving the binomial or the normal distribution. The first model, of course, corresponds to a crude 0–1 classification of the data by the experimenter while the second one actually requires that the observations can be measured on a physical scale. In order to permit an assessment of the data that is somewhere between these extremes, it is near at hand to use ranks. Miller (1970) started research on such RST by proposing a test based on the Wilcoxon statistics. A full account of further developments in this area is contained in a survey paper by Sen (1978).

Formally, a RST is but a TWP 1 truncated at the time m. More precisely, it is determined by a stopping time min $\tilde{N}_Q(a)$, m and a terminal decision rule $(A_n, B_n) = (\emptyset, \{\tilde{N}_Q(a) = n\})$ if $n < m$ and $(A_m, B_m) = (\{\tilde{N}_Q(a) > m\}, \{\tilde{N}_Q(a) = m\})$, where $N_Q(a)$ is defined as before but where $a(\cdot)$ is now subject to the requirement that

$$P_F(\tilde{N}_Q(a) \leq m) \leq \alpha \quad \forall F \in \mathfrak{H}_0 \qquad (0 < \alpha < 1)\,. \tag{3.7}$$

With no optimal procedure lurking in the background, neither for a suitable parametric model nor for a limiting continuous time problem, after which we could pattern the shape of the stopping region, we may simply put $a(t) \equiv a > 0$. This choice is common and has the advantage that the law of $\tilde{N}_Q(a)$ can be expressed in a simple manner,

$$P_F(\tilde{N}_Q(a) \leq n) = P_F(\max_{1 \leq j \leq n} Q_j \geq a) \quad \forall n \geq 1\,. \tag{3.8}$$

Finiteness of the ASN, integrability of the stopping rules $\tilde{N}_Q(a)$. Obviously, the answer to the problem becomes trivial in the case of a RST and turns out to be negative under the null-hypothesis for a TWP 1. It remains to derive conditions which ensure the finiteness of the moments of $\tilde{N}_Q(a)$ under an alternative F satisfying (3.3). No general result seems to exist that embraces all cases of interest. Many examples, however, are covered by the following device. Suppose that $a(t) \leq rt^{1-q}$ for all t sufficiently large, $r > 0$, $1 > q \geq 0$. For the time being, moreover, we assume that $Q_n = nU_n$, $n \geq p$, where $U = (U_n)_{n \geq p}$ is again a sequence of U-statistics corresponding to some kernel g. Fix $1 > \omega > 0$, $\varepsilon > 0$ and argue as follows:

$$P_F(\tilde{N}_Q(a) > n) \leq P_F(U_j < rj^{-q} \ \ \forall \omega n \leqq j \leqq n)$$
$$\leq P_F(\max_{\omega n \leq j \leq n} j^q(pU_j^{(1)} + \nu(F)) < r + \varepsilon)$$
$$+ P_F(\max_{\omega n \leq j \leq n} j^q |U_j - \nu(F) - pU_j^{(1)}| > \varepsilon) ; \tag{3.9}$$

confer Theorem D for the quantities $\nu(F)$ $(=\zeta)$ and $U_k^{(1)}$. The second summand above can further be treated by means of Theorem E(b):

$$P_F(\max_{\omega n < j \leq n} j^q |U_j - \nu(F) - pU_j^{(1)}| > \varepsilon)$$
$$\leq P_F(\sup_{j \geq \omega n} |U_j - \nu(F) - pU_j^{(1)}| > n^{-q}\varepsilon)$$
$$\leq \varepsilon^{-2k} n^{2qk} E_F(\sup_{j \geq \omega n} |U_j - \nu(F) - pU_j^{(1)}|^{2k})$$
$$\leq \text{const. } n^{-2k(1-q)} ,$$

provided g possesses moments of order $2k$. Putting $S_j = j(pU_j^{(1)} + \nu(F))$, $c(t) = (r + \varepsilon)t^{1-q}$, the first summand in (3.9) becomes

$$P_F(\max_{\omega n \leq j \leq n} j^q(U_j^{(1)} + \nu(F)) < r + \varepsilon) = P_F(\omega^{-1}\tilde{N}_S(c) > n) + O(n^{-2k+1}) .$$

According to Gut (1974), $\tilde{N}_S(c)$ has moments up to order $2k$. Multiplying across (3.9) with n^{s-1}, $0 < s < 2k(1-q) - 1$, and adding up, we realize that $E_F(\tilde{N}_Q(a))^s$ is finite. Because of Theorem D(b), the above reasoning extends to V-statistics. After that we can move on to statistics that can be bounded or approximated from below by U/V-statistics. This is true, for instance, for L-statistics with a smooth weight function, confer Helmers (1981), or for signed linear rank statistics with an increasing, convex score function. As for the latter case, simply note that

$$T_{\varphi n} \geq \int_0^\infty 2(\varphi(H) + (n\hat{H}_n/(n+1) - H)\varphi'(H))\, \mathrm{d}\hat{F}_n - \int_0^1 \varphi_n(u)\, \mathrm{d}u ,$$

where the first summand is essentially a V-statistic and where the second term is bounded.

Asymptotic OC and ASN, the limiting distribution of $\tilde{N}_Q(a)$. First let us consider horizontal barriers $\Delta^{-1}a$ the convenience of which has been mentioned in connexion with RST; confer (3.8). In this case, approximate expressions for OC with respect to local alternatives $\{F_\Delta : \Delta_0 > \Delta > 0\}$ starting from some $F_0 \in \mathfrak{H}_0$ are provided by the well-known Bartlett formula if Q again behaves like a linearly shifted Brownian motion, $\mathcal{B}(\cdot \mid \zeta, 1)$ say. Expressed in formulas,

$$P_\Delta(\tilde{N}_Q(a\Delta^{-1}) \leqslant r\Delta^{-2}) \simeq P_\Delta(\Delta \max_{1\leqslant j\leqslant [r\Delta^{-2}]} Q_j \geqslant a)$$
$$\simeq P(\mathscr{B}(t \mid \zeta, 1) \geqslant a \quad \text{for some } 0 \leqslant t \leqslant r)$$
$$= \Phi(\zeta r^{1/2} - ar^{-1/2}) + e^{2a\zeta}(1 - \Phi(\zeta r^{1/2} + ar^{-1/2})) . \tag{3.10}$$

Clearly, (3.10) also yields an approximation to the ASN because of

$$\Delta^2 E_\Delta(\tilde{N}_Q(a\Delta^{-1}) \wedge r\Delta^{-2}) = \int_0^r P_\Delta(\Delta^2\tilde{N}_Q(a\Delta^{-1}) > u)\, du$$
$$\simeq \int_0^r P(\mathscr{B}(s \mid \zeta, 1) < a,\ \forall 0 \leqslant s \leqslant u)\, du .$$

It is tempting to take (3.10) as a basis for an asymptotic comparison of RST based on different kinds of statistics as well. In a formal manner, it is in fact possible to do so. As pointed out by Sen (1981, p. 251), however, the way in which the OC of any two RST are forced to agree in the limit makes it difficult to attach any meaning to such numbers. Sen, for that reason took recourse to some form of the Bahadur concept. Here, we shall be contented with a rough comparison under a fixed alternative by means of the theorem to come.

In order to get the limiting distribution of $\tilde{N}_Q(ra(\cdot))$, $r\to\infty$, for a class of regularly varying boundaries under a fixed alternative we state a result that partly generalizes works by Siegmund (1968), Bhattacharya and Mallik (1973), Gut (1974), Chow et al. (1981).

THEOREM K. *Fix a d.f.* $F \in \mathfrak{H}_1$. *Assume that there are constants* $c > 0$, $0 \leqslant p < 1$, $\mu = \mu(F) > 0$ *as well as a real-valued continuous function* $q(\cdot)$ *vanishing at infinity so that, as* $n\to\infty$, $\Delta\to 0$,

$$n^{-1}Q_n \to \mu \quad [P_F], \qquad \mathscr{Q}_\Delta(t) - \mu\Delta^{-1}t \xrightarrow{w} \mathscr{A}(t) \quad \text{under } F,$$

$$a(t) = ct^p \exp\Bigl(\int_{t_0}^t q(s)\big/ s\, ds\Bigr), \quad t \geqslant t_0 ,$$

where $\mathscr{A}(\cdot) = \mathscr{A}_F(\cdot)$ *is a centered Gaussian process almost all sample paths of which are supposed to be continuous. Let* $b(\cdot)$ *be an asymptotic inverse of* $t/a(t)$ *and set* $n_r = b(r/\mu)$, $r > 0$. *Then, as* $r\to\infty$,

$$P_F(\mu(1-p)(\tilde{N}_Q(ra) - n_r) \leqq xn_r^{1/2}) \xrightarrow{w} P(\mathscr{A}_F(1) \leqq x), \quad x \in \mathbb{R} .$$

Asymptotic expressions for the ASN are easy to conjecture. To this end, suppose again that $a(t) = t^p l(t)$, $0 \leqslant p < 1$, where $l(\cdot)$ is slowly varying, and that $b(t)$ is an asymptotic inverse of $t/a(t)$. When applied to the positively drifting sequence Q, the SLLN and the renewal argument at which we already alluded in Section 2.2 imply that $\tilde{N}_Q(ra) \sim b(r/\mu(F))$ P_F-a.s. This fact suggests the heuristic formula

$$E_F(\tilde{N}_Q^\kappa(ra)) \sim b^\kappa(r/\mu(F)) \quad \text{as } r \to \infty .$$

The uniform integrability of the stopping times that is necessary for its validity remains to be checked in every special case. No such condition is required, of course, if we deal with the truncated random times $\min\{\tilde{N}_Q(ra), m_r\}$, $m_r = m_r(\Delta) = b(r/\Delta)$, $\Delta > 0$. In this case, as $r \uparrow \infty$ (and Δ remains fixed),

$$m_r^{-1} E_F(\tilde{N}_Q(ra) \wedge m_r) = \begin{cases} 1 + o(1), & \text{if } \mu(F) < \Delta , \\ (\Delta/\mu(F))^{1-p} + o(1), & \text{if } \mu(F) > \Delta . \end{cases} \tag{3.11}$$

If we consider two different sequences of such test statistics, Q and R say, then

$$E_F(\tilde{N}_Q(ra) \wedge m_r)/E_F(\tilde{N}_R(ra) \wedge m_r) = (\mu_R(F)/\mu_Q(F))^{1-p} + o(1) . \tag{3.12}$$

If, moreover, both Q and R satisfy the assumptions of Theorem K, then

$$P_F(\tilde{N}_Q(ra) \leq m_r) = \begin{cases} o(1), & \text{if } \mu_Q(F) < \Delta , \\ 1 + o(1), & \text{if } \mu_Q(F) > \Delta , \end{cases} \tag{3.13}$$

and the same limiting relation holds true if Q is replaced by R. If Δ ('small') is thought of to determine an indifference zone, i.e. only alternatives F are considered to be important for which $\mu_Q(F) > \Delta$ resp. $\mu_R(F) > \Delta$, then (3.13) expresses that the limiting OC of the Q- and the R-based RST agree 'by consistency'. Accordingly, the RHS of (3.12) can be interpreted as a measure of asymptotic efficiency. With that interpretation, (3.11) is just the efficiency of the RST as compared with the corresponding fixed-sample size. It is noteworthy, that these numbers still reflect the asymptotic behaviour of the stopping boundaries.

I. The use of linear rank statistics

Throughout, we suppose that the (pair of) score function(s) φ is nondecreasing and normalized in the sense L_2. To be able to refer to those alternatives that can be distinguished by the limiting functional from a d.f. in $\mathfrak{H}_0$ we introduce the classes

$$\mathfrak{H}_1(\varphi) = \{F \in \mathfrak{H}_1 \colon \theta_\varphi(F) > 0\} .$$

Tests with power one. TWP 1 for $\mathfrak{H}_0$ versus $\mathfrak{H}_1(\varphi)$ based on T_φ were first considered by Sen and Ghosh (1973b) in the univariate one-sample problem. These authors used a strong embedding theorem instead of Theorem J in order to specify suitable boundaries; confer (1.16). (In the present case, this device is more convenient as only milder conditions have to be imposed.) The two-sample as well as the independence case can be dealt with along the same lines, cf. Sen (1981, p. 241), or with the help of Theorem J. So far, most efforts have been concentrated on the determination of the quantities $a(\cdot)$, n_0, but little has been done otherwise.

Repeated significance tests. Research on rank RST was initiated by Miller (1970) and pursued by Lombard (1976, 1977), Sen (1977b) and others. The unbiasedness of these procedures can be shown in the same manner as with Wald tests in Section 2.2. As the invariance principle holds true under $\mathfrak{H}_0$ and alternatives close to it, it is clear that (3.10) is valid with the familiar shift parameters ζ. If, for example, we consider nonparametric alternatives $\mathrm{d}F_{\Delta\eta} = (1 + \eta\Delta\psi_\Delta(F_0))$, $F_0 \in \mathfrak{H}_0$, as in the preceding sections, then ζ takes on the form $\zeta = \eta\rho(\varphi_0, \psi_0)$, etc. By means of Theorems C and K and the previous discussion, we obtain the efficiency numbers $(\theta_{\varphi_2}(F)/\theta_{\varphi_1}(F))^{1-p}$ for any two RST based on linear rank statistics using score functions φ_1 and φ_2, respectively. Alternatively, we can compare these tests with RST based on sequential rank statistics. The latter were first proposed by Reynolds (1975), who considered the Wilcoxon type. Later on, statistics of this form were employed by Müller-Funk (1980) and Lombard (1981); confer also Mason (1981). In the univariate one-sample case, which we are going to treat again by way of example, the general form of these statistics is

$$T_n^0 = \sum_{j=1}^n c_j \operatorname{sgn}(X_j)\varphi(R_{jj}^+/(j+1)), \quad 0 < c_j \to c_\infty .$$

Under the null-situation, these rank statistics are sums of independent variables for which the invariance principle yields a standard Brownian motion in the limit. Switching to local alternatives, we obtain the same limiting shift as with the corresponding ordinary signed rank statistics. Under a fixed alternative F, however, $\mathcal{T}_\Delta^0(\cdot)$ centered at $\Delta^{-1}\theta_\varphi(F)$ tends towards a mean zero Gaussian process $\mathcal{S}(\cdot)$, which is representable in the form

$$\mathcal{S}(t) = a\mathcal{B}(t \mid 0, 1) + b \int_0^t u^{-1}\mathcal{B}(u \mid 0, 1)\,\mathrm{d}u, \quad t \geq 0$$

(cf. Müller-Funk, 1983b, and also Lombard and Mason, 1983).

Both a, b depend on the underlying d.f. as well as on the score function. If F belongs to $\mathfrak{H}_0$ or if φ is the identity (Wilcoxon case), then b vanishes. T and T^0 both have the same asymptotic mean $\theta_\varphi(F)$ and, accordingly, the two RST based on them are asymptotically equivalent (provided φ is smooth enough to allow for an application of Theorem K). It has already been mentioned that the statistics T^0 offer some computational advantage over their counterparts T where all ranks and scores have to be calculated anew as long as sampling goes on.

II. The use of U/V-statistics

As in the case of Wald tests we have to require that all functionals $\nu(\cdot)$ considered fulfill the assumptions (A1), (A2) of Section 2.2. Then, sequential significance tests based on

$$\check{V}_n = \hat{\sigma}_n^{-1} n(V_n - \nu_0), \quad n \geq 1,$$

or, in the case of a regular functional, on $\check{U}$ derived analogously from the related U-statistics, perfectly fit into our general discussion. As all essential tools and aspects have already been talked about, it only remains to bring forward the relevant papers in the area.

Tests with power one. Altogether, there are only few examples of non-parametric TWP 1 in the literature. The sample mean being the prime example of a U/V-statistic has been investigated within a location model by Lai (1977; F_0 known) and Sen (1981; F_0 known and unknown). The Wilcoxon statistic (looked upon as a special U-statistic) was treated by Strauch (1982). TWP 1 for the median and, more generally, for quantiles appear in Sen (1981; p. 238, 239). Further functionals, e.g. the variance, can be handled without difficulties by means of Theorem J, however.

Repeated significance tests. Sen's (1978, 1981) works, which deal with U- as well as with L-statistics, seem to be the only sources.

III. The use of Kolmogorov–Smirnov-type statistics

Tests with power one. The earliest nonparametric TWP 1 considered at all were proposed by Darling and Robbins (1968b) for the problems (1.3), (1.24) and based on the statistics $K^{(i)}$, $i = 1, 2$. These authors did not rely on invariance principles etc. but made use of the familiar fixed sample size distributions (the derivation of which is combinatorial in its nature). The complete specification of these tests requires a boundary $a(\cdot)$ and an initial sample size n_0 of the following kind:

(i) $a(t)$ is concave, increasing and strictly positive,

(ii) $a(t)/t$ strictly decreases to zero (as $t \to \infty$) and is bounded by 1,

(iii) $$\sum_{n \geqq n_0} \exp(-a^2(n)/(n+1)) \leq \alpha .$$

No boundary increasing at a rate essentially slower than $(t \log t)^{1/2}$ comes within the range of such curves. No attempts seem to have been made so far to arrive at better boundaries by other methods. On the other hand, Darling and Robbins were able to derive explicit upper bounds for the ASN.

Burdick (1973) tackled the one-sample problem for testing symmetry by a method that is similar to the one used by Darling and Robbins. His test procedure is based on the quantity $\sup\{|M_n^+(x) - M_n^-(x)| : x \in \mathbb{R}\}$, where $M_n^+(x)$ resp. $M_n^-(x)$ is the number of positive resp. negative observations among $X_1, \ldots, X_n$ the absolute value of which does not exceed x.

Repeated significance tests of Kolmogorov–Smirnov type do not seem to have been investigated in detail up to now.

IV. The use of rank likelihood ratio statistics

A *repeated significance test* recently proposed by Woodroofe (1982b) seems to be the sole specimen of this kind in the literature. He deals with the two-sample case and assumes Lehmann alternatives, i.e. $F(x, y) = J(x)J^{\eta}(y)$, $0 < \eta < \infty$. In contrast to Section 2.1, however, the parameter η is no longer considered to be known but taken into account by means of the rank ML-estimator. Asymptotic properties of this test procedure are determined (under a fixed alternative).

Some related procedures. We mentioned earlier that the model underlying a TWP 1 is meant to describe a type of quality control problem. Sometimes, however, the sequential detection (disruption) problem more realistically reflects the situation. To formulate it in mathematical terms, let $X_1, X_2, \ldots$ be a sequence of independent r.v. with d.f. $F_1, F_2, \ldots, F_i \in \mathscr{F}_1$. Consider the pair of statistical hypotheses

$$\begin{aligned} \mathscr{H}_0&: \ \exists \text{ d.f. } J \text{ so that } F_i = J\ \forall 1 \leq i \leq m\,, \\ \mathscr{H}_1&: \ \exists \text{ d.f. } J_1, J_2 (J_1 \neq J_2)\ \exists 1 \leq k < m,\ k \text{ finite, so that} \\ &\quad\ F_i = J_1\ \forall 1 \leq i < k,\ F_i = J_2\ \forall k \leq i\,, \end{aligned}$$

where $1 < m \leq \infty$. The detection problem corresponds to the case $m = \infty$. Of course, one would like to choose a sequential test with a stopping rule that stops as soon as possible if a change in distribution has occurred but keeps the probability of a false alarm under control. Up to now there does not seem to be a paper that explores this problem within a nonparametric set-up. Recently, however, the case of a finite m, commonly referred to as the change-point problem, has attracted the attention of people working in sequential nonparametrics. The ultimate goal remains the same as in the unrestricted case. As for details we refer to Bhattacharya and Frierson (1981), Sen (1980b, 1983a, 1983b) and Lombard (1983).

4. Other types of nonparametric sequential tests

Our coverage of nonparametric sequential tests is still rather incomplete and certainly biased as we stressed some topics at the cost of others which may be regarded as equally or even more important. Let us briefly discuss some types of models and procedures that have not been mentioned so far.

(1) For the sake of simplicity we only considered one- and two-sample problems but did not deal with the k-sample case or, more generally, with the regression case. Let us only mention the papers by Govindarajulu (1977; rank SPRT for the k-sample problem), Ghosh and Sen (1977; Wald test for regression based on linear rank statistics), and Jurečková and Sen (1981a, b; Wald tests for regression based on M-statistics). Sen's monograph (1981, Chapter 9) provides further examples. Ahmad and Al-Mutair (1979) considered the multivariate k-sample problem.

(2) The previous discussion has entirely been restricted to genuinely sequential tests whereas two- and multi-stage tests may be preferrable from the

practical point of view. Nonparametric procedures of this type were investigated by Hewett and Spurrier (1979), Spurrier (1978) as well as Spurrier and Hewett (1976) among others. We also skipped nonparametric tests for the two-sample case where only one sample size is allowed to be random while the other one is kept fixed. For details confer Randles and Wolfe (1979), Orban and Wolfe (1980), or Bandyopadhyay (1980).

(3) Many of the probabilistic results on which we have relied have versions that allow for observations that are dependent on each other in a specific way and/or are nonstationary. Theorem A, for instance, does not make any use of the underlying stochastic structure and hence continues to hold. Theorem B is no longer true but may be replaced by a suitable version of Theorem C. Chernoff–Savage representations under nonstandard assumptions were obtained by Sen and Ghosh (1973a) for linear rank statistics and by Ruymgaart and van Zuijlen (1978) for L-statistics, for example. Accordingly, our previous reasoning remains valid in these cases.

References

Ahmad, R. and Al-Mutair, M. A. (1979). The univariate and multivariate k-sample problem: A nonparametric sequential contrast method approach. *Trans. 8-th Prague Conference on Information Theory, vol. C*, Reidel, Dordrecht, pp. 17–29.

Armitage, P. (1975). *Sequential Medical Trials.* Blackwell, Oxford.

Bartlett, M. S. (1946). The large sample theory of sequential tests. *Proc. Cambr. Phil. Soc.* **42**, 239–244.

Bahadur, R. R. and Raghavachari, M. (1972). Some asymptotic properties of likelihood ratios on general sample spaces. *Proc. VI-th Berkeley Symp. Math. Stat. Prob.*

Bandyopadhyay, U. (1980). Semi-sequential nonparametric tests for two populations with one sample fixed. *Calcutta Stat. Ass. Bull.* **29**, 45–64. Correction: *Calcutta Stat. Ass. Bull* **30**, 95.

Behnen, K. (1972). A characterization of certain rank-order tests with bounds for the asymptotic relative efficiency. *Ann. Math. Stat.* **43**, 1122–1135.

Berk, R. H. (1973). Some asymptotic aspects of sequential analysis. *Ann. Stat.* **1**, 1126–1138.

Berk, R. H. (1975a). Locally most powerful sequential tests. *Ann. Stat.* **3**, 373–381.

Berk, R. H. (1975b). Comparing sequential and non-sequential tests. *Ann. Stat.* **3**, 991–996.

Berk, R. H. (1976). Asymptotic efficiencies of sequential tests. *Ann. Stat.* **4**, 891–911.

Berk, R. H. and Savage, I. R. (1968). The information in a rank order and stopping the stopping time of some associated SPRT's. *Ann. Math. Stat.* **39**, 1661–1674.

Bhattacharya, P. K. and Mallik, A. (1973). Asymptotic normality of the stopping times of some sequential procedures. *Ann. Stat.* **1**, 1203–1211.

Bhattacharya, P. K., Frierson, D. (1981). A nonparametric control chart for detecting small disorders. *Ann. Stat.* **9**, 544–554.

Bönner, N., Müller-Funk, U. and Witting, H. (1980). A Chernoff–Savage representation for correlation rank statistics with applications to sequential tests. In: I. M. Chakarvarti ed., *Asymptotic Theory of Statistical Tests and Estimation.* Academic Press, New York.

Bradley, R. A. (1967). Topics in rank-order statistics. *Proc. V-th Berkeley Symp. Math. Stat. Prob.*

Bradley, R. A., Martin, D. C. and Wilcoxon, F. (1965). Sequential rank tests. I. Monte Carlo studies in the two-sample case. *Technometrics* **7**, 463–484.

Bradley, R. A., Merchant, S. D. and Wilcoxon, F. (1966). Sequential rank tests. II. Modified two sample procedures. *Technometrics* **8**, 615–623.

Braun, H. (1976). Weak convergence of sequential rank statistics. *Ann. Stat.* **4**, 554–575.

Brown, B. M. (1971). Martingale central limit theorems. *Ann. Math. Statist.* **42**, 59–66.

Burdick, D. L. (1973). A best sequential test for symmetry when the probability of termination is not one. *Ann. Stat.* **1**, 1195–1199.

Choi, S. C. (1973). On nonparametric sequential tests for independence. *Technometrics* **15**, 625–629.

Chow, Y. S., Hsiung, C. A. and Yu, K. F. (1980). Limit theorems for a positively drifting process and its related first passage times. *Bull. Inst. Math. Acad. Sin.* **8**, 141–172.

Chow, Y. S. and Teicher, H. (1978). *Probability Theory*. Springer, Heidelberg.

Cox, D. R. (1963). Large sample sequential tests of composite hypotheses. *Sankhya Ser. A* **25**, 5–12.

Darling, D. A. and Robbins, H. (1967a). Iterated logarithm inequalities. *Proc. Nat. Acad. Sci.* **57**, 1188–1192.

Darling, D. A. and Robbins, H. (1967b). Inequalities for the sequence of sample means. *Proc. Nat. Acad. Sci.* 57, 1577–1580.

Darling, D. A. and Robbins, H. (1968a). Some further remarks on inequalities for sample sums. *Proc. Nat. Acad. Sci.* **60**, 1175–1182.

Darling, D. A. and Robbins, H. (1968b). Some nonparametric sequential tests with power 1. *Proc. Nat. Acad. Sci.* **61**, 804–809.

DeGroot, M. H. (1960). Minimax sequential tests of some composite hypotheses. *Ann. Math. Statist.* **31**, 1193–1200.

Dvoretzky, A., Kiefer, J. and Wolfowitz, J. (1953). Sequential decision problems for processes with continuous time parameter. Testing hypotheses. *Ann. Math. Statist.* **24**, 254–264.

Dvoretzky, A., Kiefer, J. and Wolowitz, J. (1956). Asymptotic minimax character of the sample distribution function and the classical multinomial estimator. *Ann. Math. Stat.* **27**, 642–669.

Eisenberg, B., Ghosh, B. K. and Simons, G. (1976). Properties of generalized sequential probability ratio tests. *Ann. Stat.* **4**, 237–251.

Fabian, V. (1956). A decision function. *Czechoslovak Math. J.* **6**, 31–41.

Farrell, R. H. (1964). Asymptotic behavior of expected sample size in certain one sided tests. *Ann. Math. Statist.* **35**, 36–72.

Ghosh, B. K. (1970). *Sequential Tests of Statistical Hypotheses*. Addison-Wesley, Reading.

Ghosh, M. and Sen, P. K. (1976). Asymptotic theory of sequential tests based on linear functions of order statistics. In: Ikeda et al., eds., *Essays in Probability and Statistics*. Tokyo.

Ghosh, M. and Sen, P. K. (1977). Sequential rank tests for regression. *Sankhya Ser. A* **39**, 45–62.

Govindarajulu, Z. (1975). *Sequential Statistical Procedures*. Academic Press, New York.

Govindarajulu, Z. (1977). Stopping time of a c-sample rank-order sequential probability ratio test. *Trans. 7th Prague conference* 1974, *Vol. B*. Reidel, Dordrecht, pp. 163–174.

Govindarajulu, Z. and Mason, D. M. (1980). A strong representation for linear combinations of order statistics with applications to fixed-width confidence intervals for location and scale parameters. Tech. Rep.

Gut, A. (1974). On the moments and limit distributions of some first passage times. *Ann. Prob.* **2**, 277–308.

Hajek, J. (1974). Asymptotic sufficiency of the vector of ranks in the Bahadur sense. *Ann. Stat.* **2**, 75–83.

Hajek, J. and Sidak, Z. (1967). *Theory of Rank Tests*. Academic Press, New York.

Hall, P. (1974). Two asymptotically efficient sequential t-tests. In: J. Hajek, ed., *Proc. Prague Conf. Asympt. Methods*. Academia, Prague.

Hall, W. J. and Loynes, R. M. (1977). Weak convergence of processes related to likelihood ratios. *Ann. Stat.* **5**, 330–341.

Helmers, R. (1981). A Berry–Essen theorem for linear combinations of order statistics. *Ann. Stat.* **9**, 342–347.

Hewett, J. E. and Spurrier, J. D. (1979). Some two-stage k-sample tests. *J. Amer. Stat. Ass.* **74**, 398–404.

Holm, S. (1973a). A sequential rank test. In: *Proc. Prague Symp. Asymp. Stat.*, pp. 157–172.

Holm, S. (1973b). The asymptotic minimax character of sequential binomial and sign tests. *Ann. Stat.* **1**, 1139–1148.

Holm, S. (1975). Sequential inversion sum test. *Scand. J. Stat.* **2**, 1–10.

Irle, A. (1980). Locally best tests for Gaussian processes. *Metrika* **27**, 15–28.

Irle, A. and Schmitz, N. (1981). On the optimality of the SPRT for processes with continuous time parameter. Unpublished manuscript.

Jurečková, J. (1969). Asymptotic linearity of a rank statistic in regression parameter. *Ann. Math. Stat.* **40**, 1889–1900.

Jurečková, J. (1973). Almost sure uniform asymptotic linearity of rank statistics in regression parameter. Trans. VI Prague Conf.

Jurečková, J. and Sen, P. K. (1981a). Invariance principles for some stochastic processes relating to M-estimators and their role in sequential statistical inference. To appear in *Sankhya Ser. A.*

Jurečková, J. and Sen, P. K. (1981b). Sequential procedures based on M-estimators with discontinuous score functions. *J. Stat. Planning and Inference* **5**, 253–266.

Lai, T. L. (1975a). On Chernoff–Savage statistics and sequential rank tests. *Ann. Stat.* **3**, 825–845.

Lai, T. L. (1975b). A note on first exit time with applications to sequential analysis. *Ann. Stat.* **3**, 999–1005.

Lai, T. L. (1975c). Termination, moments and exponential boundedness of the stopping rule of certain invariant sequential probability ratio tests. *Ann. Stat.* **3**, 581–598.

Lai, T. L. (1976a). On *r*-quick convergence and a conjecture of Strassen. *Ann. Prob.* **4**, 612–627.

Lai, T. L. (1976b). Boundary crossing probabilities for sample sums and confidence sequences. *Ann. Prob.* **4**, 299–312.

Lai, T. L. (1977). Power-one tests based on sample sums. *Ann. Stat.* **5**, 866–880.

Lai, T. L. (1981). Asymptotic optimality of invariant sequential probability ratio tests. *Ann. Stat.* **9**, 318–333.

Lai, T. L. and Siegmund, D. (1977). A nonlinear renewal theory with applications to sequential analysis I. *Ann. Stat.* **5**, 946–954.

Lai, T. L. and Siegmund, D. (1979). A nonlinear renewal theory with applications to sequential analysis II. *Ann. Stat.* **7**, 60–76.

LeCam, L. (1976). A reduction theorem for certain sequential experiments II. *Ann. Stat.* **7**, 847–859.

Lehmann, E. L. (1966). Some concepts of dependence. *Ann. Math. Statist.* **37**, 1137–1153.

Lin, K.-H. (1981). Convergence rate and the first exit time for U-statistics. *Bull. Inst. Math. Acad. Sin.* **9**, 129–143.

Lombard, F. (1976). Truncated sequential tests based on one sample rank order statistics. *South Afr. J.* **10**, 177–185.

Lombard, F. (1977). Sequential procedures based on Kendall's tau statistics. *South Afr. J.* **11**, 79–87.

Lombard, F. (1981). An invariance principle for sequential nonparametric test statistics under contiguous alternatives. *South Afr. J.* **15**, 129–152.

Lombard, F. (1983). Asymptotic distributions of rank statistics in the change-point problem. Unpublished manuscript.

Lombard, F. and Swanepoel, J. W. H. (1979). An asymptotic sequential test based on confidence sequences. *Comm. Stat. A* **8**, 107–116.

Lombard, F. and Mason, D. M. (1983). Limit theorems for generalized sequential rank statistics. Unpublished manuscript.

Loynes, R. M. (1970). An invariance principle for reverse martingales. *Proc. Amer. Math. Soc.* **25**, 56–64.

Mason, D. M. (1981). On the use of a statistic based on sequential ranks to prove limit theorems for simple linear rank statistics. *Ann. Stat.* **9**, 424–436.

Miller, R. G. (1970). Sequential signed-rank tests. *J. American Stat. Ass.* **65**, 1554–1561.

Mukhopadhyay, N. (1981). Convergence rates of sequential confidence intervals and tests for the mean of a U-statistic. *Comm. Stat.-Theor. Meth. A* **10**, 2231–2244.

Müller-Funk, U. (1979). Non-parametric sequential tests for symmetry. *Z. Wahrsch. Verw. Geb.* **46**, 325–342.

Müller-Funk, U. (1980). On contiguity and weak convergence with an application to sequential analysis. *Proc. Coll. Math. Soc. Janos Bolyai*, 619–636.

Müller-Funk, U. (1983a). Sequential signed rank statistics. *Comm. Stat. C-Sequential Analysis* **2**, 123–148.

Müller-Funk, U. (1983b). A quantitative SLLN for linear rank statistics. *Stat. and Dec.* **1**, 371–378. *Dec.* **1**.

Müller-Funk, U., Pukelsheim, F. and Witting, H. (1983). Locally most powerful tests for two-sided hypotheses. To appear in *Proc. IV Pann. Symp. Math. Stat.*

Noether, G. E. (1954). A sequential test of randomness against linear trend. *Ann. Math. Stat.* **25**, 176 (abstract).

Orban, J. and Wolfe, D. A. (1980). Distribution-free partially sequential placement procedures. *Comm. Stat. A* **9**, 883–904.

Puri, M. L. and Sen, P. K. (1971). *Nonparametric Methods in Multivariate Analysis.* Wiley, New York.

Randles, R. H. and Wolfe, D. A. (1979). *Introduction to the Theory of Nonparametric Statistics.* Wiley, New York.

Reynolds, M. (1975). A sequential signed-rank test for symmetry. *Ann. Stat.* **3**, 382–400.

Rieder, H. (1981). Robustness of one- and two-sample rank tests against gross errors. To appear in *Ann. Stat.* **9**.

Rieder, H. (1982). A general robustness property of rank correlations. Unpublished manuscript.

Robbins, H. (1970). Statistical methods related to the law of the iterated logarithm. *Ann. Math. Statist.* **41**, 1397–1409.

Robbins, H. and Siegmund, D. (1970). Boundary crossing probabilities for the Wiener process and sample sums. *Ann. Math. Statist.* **41**, 1410–1429.

Robbins, H. and Siegmund, D. (1973). Statistical tests of power one and the integral representation of solutions of certain partial differential equations. *Bull. Inst. Math. Acad. Sin.* **1**, 93–120.

Romani, J. (1956). Tests no parametricos en forma secuencial. *Trabajos de Estadistica* **7**, 43–96.

Ruymgaart, F. and van Zuijlen, M. C. A. (1978). On convergence of the remainder term in linear combinations of functions of order statistics in the non-i.i.d. case. *Sankhya Ser. A* **40**, 369–387.

Savage, I. R. and Savage, L. J. (1965). Finite stopping time and finite expected stopping time. *J. Roy. Stat. Soc.* **27**, 284–289.

Savage, I. R. and Sethuraman, J. (1966). Stopping time of a rank-order sequential probability ratio test based on Lehmann alternatives. *Ann. Math. Stat.* **37**, 1154–1160; Correction *Ann. Math. Stat.* **38**, 1309 (1967).

Savage, I. R. and Sethuraman, J. (1972). Asymptotic distribution of the log likelihood ratio based on ranks in the two sample problem. In: *Proc. VI Berkeley Symp. Math. Stat. Prob.*

Schmitz, N. (1982). Minimax sequential tests for the drift of a Wiener process. To appear in: *Mathematische Systeme in der Ökonometrie.*

Sen, P. K. (1960). On some convergence properties of U-statistics. *Calcutta Stat. Ass. Bull.* **10**, 1–18.

Sen, P. K. (1970). On some convergence properties of one-sample rank-order statistics. *Ann. Math. Stat.* **41**, 2140–2143.

Sen, P. K. (1973a). Asymptotic sequential tests for regular functionals of distribution functions. *Teoria Veroyatnostey i ee Primenyia* **18**, 235–249.

Sen, P. K. (1973b). An asymptotically optimal test for the bundle strength of filaments. *Ann. Stat.* **1**, 526–537.

Sen, P. K. (1975). Rank statistics, martingales, and limit theorems. In: M. L. Puri, ed., *Stat. Inf. and Rel. Topics.* Academic Press, New York.

Sen, P. K. (1977a). On Wiener process embedding for linear combinations of order statistics. *Sankhya Ser. A* **39**, 138–143.

Sen, P. K. (1977b). Some invariance principles relating to jack-knifing and their role in sequential analysis. *Ann. Stat.* **5**, 316–329.

Sen, P. K. (1977c). Tied-down Wiener process approximations for aligned rank order statistics and some applications. *Ann. Stat.* **5**, 1107–1123.

Sen, P. K. (1978). Nonparametric repeated significance tests. In: P. P. Krishnaiah, ed., *Developments in Statistics, vol. I.* Academic Press, New York.

Sen, P. K. (1980a). On almost sure linearity theorems for signed rank order statistics. *Ann. Stat.* **8**, 313–321.

Sen, P. K. (1980b). Asymptotic theory of some tests for a possible change in the regression slope occurring at an unknown time point. *Z. Wahrsch. Verw. Geb.* **52**, 203–218.

Sen, P. K. (1981). *Sequential Nonparametrics.* Wiley, New York.

Sen, P. K. (1983a). Some recursive residual rank tests for change-points. In: M. H. Rizvi et al., eds., *Recent Advances in Statistics*: Papers in Honor of Herman Chermoff's Sixtieth Birthday. Academic Press, New York.

Sen, P. K. (1983b). Tests for change-points based on recursive U-statistics. *Comm. Stat.-Sequ. Analysis* **1**, 263–284.

Sen, P. K. and Ghosh, M. (1971). On bounded length sequential confidence intervals based on one-sample rank-order statistics. *Ann. Math. Stat.* **42**, 189–203.

Sen, P. K. and Ghosh, M. (1972). On strong convergence of regression rank statistics. *Sankhya Ser. A* **34**, 335–348.

Sen, P. K. and Ghosh, M. (1973a). A Chernoff–Savage representation of rank order statistics for ϕ-mixing processes. *Sankhya Ser. A* **35**, 153–172.

Sen, P. K. and Ghosh, M. (1973b). A law of the iterated logarithm for one-sample rank order statistics and an application. *Ann. Stat.* **1**, 568–576.

Sen, P. K. and Ghosh, M. (1974a). On sequential rank tests for location. *Ann. Stat.* **2**, 540–552.

Sen, P. K. and Ghosh, M. (1974b). Some invariance principles for rank statistics for testing independence. *Z. Wahrsch. Verw. Geb.* **29**, 93–108.

Sen, P. K. and Ghosh, M. (1980). On the Pitman efficiency of sequential tests. *Calcutta Stat. Ass. Bull.* **29**, 65–72.

Serfling, R. (1980). *Approximation Theorems of Mathematical Statistics.* Wiley, New York.

Sethuraman, J. (1970). Stopping time of a rank-order sequential probability ratio test based on Lehmann alternatives II. *Ann. Math. Stat.* **41**, 1322–1333.

Siegmund, D. (1968). On the asymptotic normality of one sided stopping rules. *Ann. Math. Stat.* **39**, 1493–1497.

Sproule, R. N. (1969). A sequential fixed-width confidence interval for the mean of a U-statistic. Ph.D. dissertation, Chapel Hill, North Carolina.

Spurrier, J. D. (1978). Two stage sign tests that allow first stage acceptance and rejection. *Comm. Stat. Ser. A* **7**, 399–408.

Spurrier, J. D. and Hewett, J. E. (1979). Two-stage Wilcoxon tests of hypotheses. *J. Amer. Stat. Ass.* **71**, 982–987.

Strauch, J. (1982). A nonparametric interminable test for symmetry of power one. *Comm. Stat. C-Sequential Analysis* **2**, 87–97.

Tsao, C. K. (1954). Sequential rank sums tests. *Ann. Math. Stat.* **25**, 177 (abstract).

van Eeden, C. (1972). An analogue of signed-rank statistics of Jureckova's asymptotic linearity theorem for rank statistics. *Ann. Math. Stat.* **43**, 791–802.

Weed, Jr., H. D. and Bradley, R. A. (1971). Sequential one-sample grouped signed rank tests for symmetry. *J. Amer. Stat. Assoc.* **66**, 321–326.

Weed, Jr., H. D., Bradley, R. A. and Govindarajulu, Z. (1974). Stopping times of two rank order sequential probability ratio tests for symmetry based on Lehmann alternatives. *Ann. Stat.* **2**, 1314–1322.

Wijsman, R. A. (1977a). A general theorem with applications on exponentially bounded stopping time without moment conditions. *Ann. Stat.* **5**, 292–315.

Wijsman, R. A. (1977b). Obstructive distributions in a sequential rank-order test based on Lehmann alternatives. In: S. S. Gupta and D. S. Moore, eds., *Statistical Decision Theory and Related Topics II.* Academic Press, New York.

Wijsman, R. A. (1979). Stopping time of invariant sequential probability ratio tests. In: P. Krishnaiah, ed., *Developments in Statistics II.* Academic Press, New York.

Wilcoxon, F., Rhodes, L. J. and Bradley, R. A. (1963). Two sequential two-sample grouped rank tests with applications to screening experiments. *Biometrics* **19**, 58–84.

Witting, H. and Nölle, G. (1970). *Angewandte Mathematische Statisik.* Teubner, Stuttgart.

Woodroofe, M. (1982a). On sequential rank tests. Univ. Michigan preprint.

Woodroofe, M. (1982b). Likelihood ratio tests with ranks. Univ. Michigan preprint.

P. R. Krishnaiah and P. K. Sen, eds., *Handbook of Statistics, Vol. 4*
© Elsevier Science Publishers (1984) 699–739

29

Nonparametric Procedures for some Miscellaneous Problems

Pranab Kumar Sen

1. Introduction

A variety of nonparametric procedures pertaining to univariate as well as multivariate and fixed-sample sizes as well as sequential schemes has been discussed in this volume. In this context, a few of the basic problems has either been barely mentioned or treated very briefly. A more complete treatment of these will be considered in this chapter.

Nonparametric methods for the *change-point problem* have been discussed in Chapter 5 (by G. K. Bhattacharyya). Some recent developments in this area (including recursive residuals, recursive U-statistics and recursively aligned rank statistics) are systematically reviewed in Section 2 of this chapter. Various types of *censoring* and appropriate nonparametric methods for such censored data have been considered in Chapter 25 (by A. P. Basu). In the context of life testing, survival analysis and reliability theory, some related results have also been presented in Chapters 27 (by Hollander and Prochan), 26 (by Doksum and Yandell) and 32 (by Wieand). In follow-up studies, *progressively censoring schemes* are quite useful, and an up-to-date and informative discussion of some recent developments in this area is presented in Section 3. *Rank analysis of covariance* constitutes another area of real interest. Since this topic has not been covered in detail in earlier chapters, an adequate description of some nonparametric procedures is given in Section 4. *Growth curve models* (both in the parametric and nonparametric setups) are commonly encountered in many practical problems. Section 5 is devoted to a systematic account of some nonparametric procedures for such models (and is based on the S. N. Roy memorial lectures given by the author in 1979 at the Calcutta University). The final section deals with some nonparametric procedures in *biological assays*; nonparametric estimates of *relative potency* and tests for the basic assumption are considered and their relative merits and demerits studied.

2. Nonparametric tests for change-points

Let $X_1, \ldots, X_n$ be n independent random vectors (r.v.) taken at ordered time points $t_1, \ldots, t_n$, respectively. Let F_i be the distribution function (d.f.) of

X_i, defined on the p ($\geqslant$1-)dimensional Euclidean space E^p, for $i = 1, \ldots, n$. Though, conventionally, one assumes that the F_i are all the same (and then proceeds to draw inference on some characteristics (or parameters) of this common d.f.), in the context of *continuous inspection schemes* (viz., Page (1957)) and in some other problems of practical importance, a change of the d.f. may occur at an unknown time point (τ), where $\tau \in (t_1, t_n)$. As such, it may be of interest to test for such a possible change, i.e., to test for the null hypothesis

$$H_0: \quad F_1 = \cdots = F_n = F \text{ (unknown)}, \tag{2.1}$$

against the composite alternative

$$H: \quad F_1 = \cdots = F_q \neq F_{q+1} = \cdots = F_n \text{ for some } q: 1 \leqslant q \leqslant n-1, \tag{2.2}$$

that is, $\tau \in (t_q, t_{q+1}]$, for some unknown $q: 1 \leqslant q \leqslant n-1$. A similar situation arises in the *sequential detection problem* (viz., Shirayayev (1963, 1978)), where one may conceive of an infinite sequence $\{X_i; i \geqslant 1\}$ of independent r.v.'s, gathered over a sequence $\{t_i; i \geqslant 1\}$ of ordered time points, such that for some integer q ($\geqslant$1), $\{X_i; i \leqslant q\}$ are i.i.d.r.v. with a d.f. F_1 and $\{X_i; i \geqslant q+1\}$ are i.i.d.r.v. with a d.f. F_{q+1}; here q may even be equal to ∞. The problem is to raise an alarm if $q < \infty$, while, for $q = \infty$, the process should not be unnecessarily stopped. Thus, one would like to choose a *stopping number* N, such that if $q < \infty$, then, $E\{\max(N-q, 0)\} = E(N-q)^+$ (or some other measure of the excess of N over q) should be as small as possible, while, for $q = \infty$, the probability of a *false alarm* (i.e., $N < \infty$) should be small, i.e., EN should be large. Thus, in a sequential detection problem, we have a genuine sequential scheme, while, in a change-point problem, n is specified in advance, though the test for (2.1) against (2.2) may be done sequentially or not. Though, in the context of statistical quality control, the sequential detection problem seems to be more appropriate, it may be remarked that in view of the customary adjustments and inspections at regular intervals, the change-point problem remains equally appropriate. However, within each inspection period (i.e., for a maximum n specified in advance), the test for (2.1) against (2.2) may be made recursively, so that an early termination may be recommended whenever desired. This may call for some quasi-sequential schemes in a change-point problem. Some of these procedures are discussed here.

In the parametric case, the simplest change-point problem relates to the *shift alternative* where one takes $F_i(x) = F(x - \theta_i)$, $i \geqslant 1$, F defined on E and the θ_i real, and then one wants to test for H_0: $\theta_1 = \cdots = \theta_n = \theta$ (unknown) against H: $\theta_1 = \cdots = \theta_q \neq \theta_{q+1} = \cdots = \theta_n$, for some unknown $q \in [1, n-1]$. In a somewhat more general setup, one may consider a linear model:

$$F_i(x) = F(x - \boldsymbol{\beta}_i' \boldsymbol{c}_i), \quad i = 1, \ldots, n, \quad x \in E, \tag{2.3}$$

where the $\boldsymbol{c}_i$ are specified vectors of real elements, the $\boldsymbol{\beta}_i$ are unknown regression coefficients (vectors), and one wants to test for the *constancy of regression relationships over time* i.e., for H_0: $\boldsymbol{\beta}_1 = \cdots = \boldsymbol{\beta}_n = \boldsymbol{\beta}$ (unknown) against H: $\boldsymbol{\beta}_1 = \cdots = \boldsymbol{\beta}_q \neq \boldsymbol{\beta}_{q+1} = \cdots = \boldsymbol{\beta}_n$ for some unknown $q \in [1, n-1]$. The shift alternative is a particular case of (2.3) when the c_i are all real elements, equal to 1 and $\beta_i = \theta_i$, $i \geq 1$. Tests for the change-point problem relating to the model (2.3) have been studied by a host of workers. Most of these parametric tests are non-sequential in nature, and they are based on the residuals based on the terminal estimate of the hypothesized common $\boldsymbol{\beta}$. Some quasi-sequential tests based on recursive estimates of $\boldsymbol{\beta}$ are also available; the monograph due to Hackl (1980) contains a good account of these parametric procedures. *Recursive residuals* can be used with advantage in this problem. For normal F, a detailed account of some cumulative sum (CUSUM) procedures based on recursive residuals is due to Brown, Durbin and Evans (1975). The theory has been extended to the nonparametric case (of unknown F belonging to some suitable family) by Sen (1982a). Chernoff and Zacks (1964) have considered some Bayes procedures (for normal F) having a good impact on the nonparametric case too. In this article, we confine ourselves to the developments in the nonparametric case only. Also, in view of the fact that the case of the shift alternative has been treated in the article by G. K. Bhattacharyya (viz., Chapter 5), we shall deal with this model only very briefly, and then proceed on to the case of the general model in (2.3).

Assuming that the observations are initially from a symmetric (unknown) distribution with a specified median θ_0, Page (1955) proposed a test for a change-point based on the cumulative sums of the $\operatorname{sgn}(X_i - \theta_0)$, $i \geq 1$. Bhattacharyya and Johnson (1968) considered a general class of locally optimal rank tests for the same problem, where a Bayesian setup of Chernoff and Zacks (1964) has been incorporated. If we assume that for the change-point τ, $P\{\tau = t_i\} = d_i$, $i = 1, \ldots, n$, and set $D_i = \sum_{j \geq i} d_j$, $i = 1, \ldots, n$, and if R^+_{ni} stands for the rank of $|X_i - \theta_0|$ among $|X_1 - \theta_0|, \ldots, |X_n - \theta_0|$, for $i = 1, \ldots, n$, then, for an assumed form f of the density function corresponding to the d.f. F (having a finite Fisher information), the statistic

$$S_n = \sum_{i=1}^{n} D_i \operatorname{sgn}(X_i - \theta_0) a^+_n(R^+_{ni}) \tag{2.4}$$

leads to a locally most powerful rank (LMPR) test when

$$a^+_n(k) = E\phi^+(U_{nk}), \quad k = 1, \ldots, n, \qquad \phi^+(u) = \phi((1+u)/2), \quad 0 < u < 1, \tag{2.5}$$

where $U_{n1} < \cdots < U_{nn}$ are the ordered r.v. of a sample of size n from the uniform $(0, 1)$ d.f. and $\phi(u) = -f'(F^{-1}(u))/f(F^{-1}(u))$, $0 < u < 1$ is the Fisher score function. In particular, if we take for F a double exponential, logistic or normal d.f., then $a^+_n(k) = 1$, $k/(n+1)$ or the expected value of the k-th order statistic of a sample of size n from the chi distribution with 1 degree of freedom $(1 \leq k \leq n)$, and these correspond to the weighted sign, signed rank and normal scores statistics. Note that, often, in the absence of any knowledge on the d_i, they are all taken to be equal (i.e., $D_i = i/n$, $i = 1, \ldots, n$), and this leads to a simplified version of (2.4). When the initial level θ_0 is not specified, as is mostly the case, the corresponding locally optimal invariant test statistic is

$$L_n = \sum_{i=1}^{n} D_i a_n(R_{ni}), \qquad a_n(k) = E\phi(U_{nk}), \quad k = 1, \ldots, n, \tag{2.6}$$

where R_{ni} = rank of X_i among $X_1, \ldots, X_n$, for $i = 1, \ldots, n$. For small values of n, the exact null distribution of S_n (or L_n) can be obtained by direct enumeration of the $2^n(n!)$ (or $n!$) equally likely realizations of the vectors of ranks and signs (or ranks), while the asymptotic normality results hold under very general conditions on the D_i and the score function ϕ. In the context of equal d_i and for the specific scores $\phi(u) = \operatorname{sgn}(u - \frac{1}{2})$ and $\phi(u) = u - \frac{1}{2}$, $0 < u < 1$, more simplified expressions for the S_n and L_n are available. If we define a two-sample median (or Wilcoxon) statistic by $M_{k,n-k}$ (or $W_{k,n-k}$) when $X_1, \ldots, X_k$ constitute the first sample and $X_{k+1}, \ldots, X_n$ the second sample, then L_n may be expressed as the sum (over $k: 1 \leq k \leq n-1$) of these pseudo two-sample statistics. The pseudo two sample approach has further been considered by A. Sen and Srivastava (1975). They considered the statistics

$$\max_{1 \leq k \leq n-1} \{[M_{k,n-k} - E_0 M_{k,n-k}]/(\operatorname{var}_0(M_{k,n-k}))^{1/2}\}, \tag{2.7}$$

and

$$\max_{1 \leq k \leq n-1} \{[U_{k,n-k} - k(n-k)/2]/[k(n-k)(n+1)/12]^{1/2}\}, \tag{2.8}$$

where E_0 and var_0 stand for the expectation and variance under H_0 and $U_{k,n-k}$ is the Mann–Whitney form of the Wilcoxon statistic. Two-sided versions of (2.7) and (2.8) were also considered by them. Pettitt (1979) considered a variant form of (2.8), viz.,

$$\max_{1 \leq k \leq n-1} \{U_{k,n-k} - k(n-k)/2\} \tag{2.9}$$

and its two-sided version, and (2.8) and (2.9) differ only with respect to the weights $\sqrt{k(n-k)/12}$. Schechtman and Wolfe (1981) have also considered these statistics and provided some simulation studies of the allied distribution theory for specific values of n. For some asymptotic theory, we may refer to Sen (1978, Section 6). P. K. Bhattacharya and Frierson (1981) have considered some sequential procedures (for the control chart setup) for detecting a possible shift in location; some allied work is also due to Lombard (1981, 1983). These procedures are genuinely distribution-free. Some alternative asymptotically distribution-free procedures for more general models have been considered by Sen (1977, 1980, 1982b).

For parametric linear models, tests for change-points based on recursive residuals have been considered by a host of workers; a detailed account is given in Brown, Durbin and Evans (1975). Only very recently, such recursive residual tests have been developed for the nonparametric case. Use of recursive residuals in a nonparametric setup has been considered in Sen (1982a), wherein some invariance principles have been developed, and these provide the desired asymptotic distribution theory of CUSUM test statistics under very general conditions. Recursive ranking scheme has been employed by Bhattacharya and Frierson (1981) for the simple shift alternative change-point problem – some description of their procedure is given in the article by G. K. Bhattacharyya (see Chapter 5). For general linear models, the recursive ranking scheme of Bhattacharya and Frierson (1981) may not workout validly, and some aligned recursive ranking schemes have been considered by Sen (1983a) which provide valid and efficient solutions. Sen (1982c) has also considered the recursive procedure for general estimable parameters.

With the same setup as in (2.1)–(2.2), consider an estimable parameter $\theta = \theta(F) = \int \cdots \int \phi(x_1, \ldots, x_m)\,\mathrm{d}F(x_1), \ldots, \mathrm{d}F(x_m)$ of *degree* m (≥ 1). Under (2.1), whenever $k \geq m$, a symmetric, unbiased and optimal estimator of θ is given by

$$U_k = \binom{k}{m}^{-1} \sum_{1 \leq i_1 < \cdots < i_m \leq k} \phi(X_{i_1}, \ldots, X_{i_m}) . \tag{2.10}$$

U_k is termed a U-statistic. Side by side, we may introduce the *recursive* U-statistics $\{U_k^*;\, k \geq m\}$ by letting

$$U_k^* = \binom{k-1}{m-1}^{-1} \sum_{1 \leq i_1 < \cdots < i_{m-1} \leq k-1} \phi(X_{i_1}, \ldots, X_{i_{m-1}}, X_k), \quad k \geq m . \tag{2.11}$$

Thus, at the k-th stage ($k \geq m+1$), the *recursive residual* based on U_k^* is

$$w_k = U_k^* - U_{k-1}, \text{ for } k \geq m+1 \text{ and } w_k = 0 \text{ for } k \leq m . \tag{2.12}$$

We also define the CUSUMS for these residuals by

$$W_r = \sum_{k \leq r} w_k, \text{ for } r \geq 0 . \tag{2.13}$$

(Note that $W_r = 0$, for $r \leq m$).

To estimate the variance function of these CUSUMS, we define, for every $k \geq m$,

$$U_{ki} = \binom{k-1}{m-1}^{-1} \sum_{k,i} \phi(X_i, X_{i_2}, \ldots, X_{i_m}), \quad i = 1, \ldots, k , \tag{2.14}$$

where the summation $\Sigma_{k,i}$ extends over all possible $1 \leq i_2 < \cdots < i_m \leq k$ with

$i_j \neq i$, for $j = 2, \ldots, m$; $i = 1, \ldots, k$. Further, let

$$s_k^2 = (k-1)^{-1} \sum_{i=1}^{k} \{U_{ki} - U_k\}^2, \qquad \tilde{s}_k = \max\{s_k, k^{-1/2}\}, \quad k \geq m+1\,. \tag{2.15}$$

The recursive tests for change-points (based on the CUSUMS in (2.13)) rest on the statistic $\tilde{D}_n^+$ (for one-sided alternatives) and $\tilde{D}_n$ (for two-sided alternatives) where

$$\tilde{D}_n^+ = (n-m)^{-1/2} \max\{W_k/\tilde{s}_k : m < k \leq n\}, \tag{2.16}$$

$$\tilde{D}_n = (n-m)^{-1/2} \max\{|W_k|/\tilde{s}_k : m < k \leq n\}. \tag{2.17}$$

Nonrecursive versions of (2.16)–(2.17) may be obtained by replacing the $\tilde{s}_k$ by the terminal s_n. It follows from Sen (1982c) that under quite general regularity conditions, as $n \to \infty$, for every $x \geq 0$,

$$P\{\tilde{D}_n^+ \leq x \mid H_0\} \to 2\Phi(x) - 1\,, \tag{2.18}$$

$$P\{\tilde{D}_n \leq x \mid H_0\} \to \sum_{k=-\infty}^{\infty} (-1)^k [\Phi((2k+1)x) - \Phi((2k-1)x)]\,, \tag{2.19}$$

where $\Phi(x)$ is the standard normal d.f. Thus, with the (asymptotic) critical values determined from (2.18)–(2.19), a control chart type procedure may be based on the normalized CUSUMS $W_k/\tilde{s}_k$, rejecting the null hypothesis whenever the graph crosses the critical level(s). The tests have been shown to be consistent under very general regularity conditions. Also, for local alternatives, the limiting nonnull distribution theory of $\tilde{D}_n^+$ and $\tilde{D}_n$ has been obtained in terms of the boundary crossing probabilities of some drifted Brownian motions, with nonlinear drift functions). It may be remarked that for $\phi(x) = x$ ($m = 1$), we have

$$U_k = \bar{X}_k = k^{-1} \sum_{i=1}^{k} X_i \ \forall k \geq 1, \qquad U_k^* = X_k \ \forall k \geq 1\,,$$

$$w_k = X_k - \bar{X}_{k-1} = k(\bar{X}_k - \bar{X}_{k-1})/(k-1), \quad k \geq 2$$

and s_k^2 is the sample variance $((k-1)^{-1} \Sigma_{i=1}^k (X_i - \bar{X}_k)^2)$ for sample size $k \geq 2$. The corresponding $\tilde{D}_n^+$ (or $\tilde{D}_n$) is then suitable for the shift alternative. Similarly, for $m = 2$, $\phi(x_1, x_2) = \frac{1}{2}(x_1 - x_2)^2$, $\theta(F)$ reduces to the variance of the distribution F and the corresponding $\tilde{D}_n^+$ (or $\tilde{D}_n$) is suitable for any change in dispersion.

Let us now consider some recursive residual rank tests for change-points relating to the general linear model in (2.3), where the $\boldsymbol{\beta}_i$ and $\boldsymbol{c}_i$ are t-vectors for some $t \geq 1$. Assuming that $\boldsymbol{\beta}_1 = \cdots = \boldsymbol{\beta}_k = \boldsymbol{\beta}$, let $\hat{\boldsymbol{\beta}}_k$ be some suitable estimator or $\boldsymbol{\beta}$ and consider the (recurvise) residuals as

$$\hat{X}_{k+1,i} = X_i - \hat{\boldsymbol{\beta}}_k' \boldsymbol{c}_i \quad \text{for } i = 1, \ldots, k+1, \qquad 1 \leq k \leq n-1. \tag{2.20}$$

(Generally, for $k < t$, the residuals $\hat{X}_{k+1,i}$ may be taken to be equal to 0.)

Note that the $\hat{\boldsymbol{\beta}}_k$ may be quite arbitrary (i.e., least squares estimators, M-estimators, R-estimators, etc.), and the only regularity condition assumed for such a sequence of estimators is that under H_0: $\boldsymbol{\beta}_1 = \cdots = \boldsymbol{\beta}_n = \boldsymbol{\beta}$, for every $\varepsilon > 0$, there exists an interger k_0 (≥ 1) such that

$$P\{\max_{k_0 \leq k \leq n} (\log k)^{-1} k^{1/2} \|\hat{\boldsymbol{\beta}}_k - \boldsymbol{\beta}\| \geq 1\} < \varepsilon \quad \forall n \geq k_0. \tag{2.21}$$

(It has been shown by Sen (1983a) that (2.21) holds (under quite general regularity conditions) for the least squares, M- and R-estimators.) Now, for every k ($\geq t+1$), let $\hat{R}_{ki}^+$ be the rank of $|\hat{X}_{k,i}|$ among $|\hat{X}_{k,1}|, \ldots, |\hat{X}_{k,k}|$, for $i = 1, \ldots, k$. Also, define the scores $\{a_k^+(i), i = 1, \ldots, k; k \geq 1\}$ as in (2.5). Then, the *residual signed-rank scores are defined by*

$$\hat{u}_k = \operatorname{sgn}(\hat{X}_{k,k}) a_k^+(\hat{R}_{kk}^+) \quad \text{for } k = t+1, \ldots, n; \tag{2.22}$$

conventionally, we let $\hat{u}_k = 0$ for $k \leq t$. Then, the CUSUM's for the residual rank scores in (2.22) are defined by

$$\hat{U}_r = \sum_{k \leq r} \hat{u}_k \quad \text{for } r = 0, 1, \ldots, n. \tag{2.23}$$

We also define

$$A^2 = A_\phi^2 = \int_0^1 \phi^2(u)\, du = \int_0^1 \{\phi^+(u)\}^2\, du, \tag{2.24}$$

and assume that $0 < A < \infty$. Let then

$$D_n^+ = (n-t)^{-1/2} A^{-1} \max\{U_r \colon r \leq n\}, \tag{2.25}$$

$$D_n = (n-t)^{-1/2} A^{-1} \max\{|U_r| \colon r \leq n\}. \tag{2.26}$$

The recursive tests for the change-point problem are based on D_n^+ and D_n. Unlike the location (shift) model, here, the use of D_n^+ may only be advocated when under the alternative hypothesis, the $\boldsymbol{\beta}_k' \boldsymbol{c}_k$ are monotone. Since this may not generally be the case, the two-sided test statistic in (2.26) is generally adopted. It may be noted that the recursive residual scores $\hat{u}_k$, $k \geq t$ are, generally, neither strictly independent nor marginally identically distributed. Hence, the tests based on D_n^+ or D_n may not be genuinely distribution-free.

However, they are asymptotically distribution-free under quite general conditions. For density functions with a finite Fisher information and for $\{\boldsymbol{c}_k\}$, for which

$$n^{-1}\boldsymbol{C}_n = n^{-1}\sum_{k=1}^{n} \boldsymbol{c}_k\boldsymbol{c}_k' \to \boldsymbol{C}_0 \quad \text{(p.d.)} \qquad \text{as } n \to \infty, \tag{2.27}$$

and

$$\max\{\boldsymbol{c}_k'\boldsymbol{C}_0^{-1}\boldsymbol{c}_k : 1 \leq k \leq n\} = \mathrm{O}((\log n)^2), \tag{2.28}$$

it follows from Sen (1983a) that, under H_0, D_n^+ and D_n have respectively the limiting distributions given by the right hand sides of (2.18) and (2.19). In this context, we may, of course, allow the score function $\phi = \{\phi(u), 0 < u < 1\}$ to be quite arbitrary, such that on letting $\phi^{(r)}(u) = (\mathrm{d}^r/\mathrm{d}u^r)\phi(u)$, $r = 0, 1, 2$, there exist a generic positive constant K $(<\infty)$ and a δ $(<\frac{1}{6})$ for which

$$|\phi^{(r)}(u)| \leq K[u(1-u)]^{-r-\delta}, \quad 0 < u < 1, \qquad r = 0, 1, 2. \tag{2.29}$$

All the commonly adapted score functions satisfy (2.29). Asymptotic nonnull distribution theory of D_n^+ and D_n, for local alternatives, has also been studied in Sen (1983a). For the simple shift alternative scheme, for which the Bhattacharya–Frierson (1981) sequential ranking scheme works out well, Lombard (1981, 1983) has considered some weighted quadratic sums of rank statistics for the change point problem. His procedure, however, may not apply to the linear models treated above.

3. Nonparametric testing under progressive censoring

Sequential nonparametric procedures have been discussed in Chapters 28 (by Müller-Funk) and 22 (by Sen). In the context of clinical trials, life testing problems and in survival analysis, a somewhat different type of sequential procedure is generally adapted. Typically, in a follow-up study, where the responses (failures) occur sequentially (over time), because of the limitations of cost and time of experimentations, the study may have to be curtailed at an intermediate stage, and based on the partial (censored) experimental outcome, statistical conclusions have to be made. A curtailment may occur after a prefixed time period (*Truncation* or *Type I censoring*) or after a given number of failures (responses) have been recorded (*Type II censoring*). In the first case, the number of failures occurring during the given time period is random, while, in the second case, the study period needed to record a predetermined number of failures is random. In either case, statistical conclusions are drawn only at the end of the study period. In many clinical trials, interim (or repeated) analysis of experimental data as they accumulate (over time) is common. From the point of view of medical ethics, this is often desirable. For example, in a (randomized) clinical trial, if the patients are allocated (at random) to a control group and a treatment group, and the response relates to the time of occurrence of an event (e.g., heart attack/death), then, instead of waiting till the study period is over, one may like to monitor the study from the very beginning so that if one of the two groups performs significantly better than the other one, then the surviving patients may all be switched to that group. This may also call for a curtailment of the study at an early stage. On the other hand, if there is no real difference between the two groups then the continuation of the follow-up study does not pose any extra risk to any particular group, and an early termination is really not that needed. This basic motivation underlies the formulation of *progressive censoring schemes* (PCS). In passing, we may remark that a repeated analysis scheme on the accumulating data may lead to a higher risk of making incorrect decisions, unless proper statistical analysis schemes are validly formulated.

We may refer to Chapter 25 (by Basu) where the concepts of various types of censoring have been introduced and the related nonparametric methods have also been discussed. With these in mind, we conceive of a set $\{X_1, \ldots, X_n\}$ of independent random variables with continuous distribution functions (d.f.) $\{F_1, \ldots, F_n\}$. Suppose that we want to test for the null hypothesis H_0: $F_1 = \cdots = F_n = F$ (unknown), against some alternative (where the F_i are not all the same). Let $Z_{n1} \leqslant \cdots \leqslant Z_{nn}$ be the order statistics associated with $X_1, \ldots, X_n$, and let us define

$$R_i: X_i = Z_{nR_i} \quad \text{and} \quad S_i: X_{S_i} = Z_{ni} \qquad \text{for } i = 1, \ldots, n\,. \tag{3.1}$$

Thus, $\boldsymbol{R} = (R_1, \ldots, R_n)$ and $\boldsymbol{S} = (S_1, \ldots, S_n)$ stand for the vectors of the *ranks* and *anti-ranks* of the observations. In a life-testing problem, the failures occur spatially, so that, one observes the following sequence in order

$$(Z_{n1}, S_1), (Z_{n2}, S_2), \ldots, (Z_{nk}, S_k), \ldots, (Z_{nn}, S_n)\,. \tag{3.2}$$

If the experiment is conducted for a fixed period of time $[0, T]$ and if $r = r(T)$ is defined by $r(T) = \max\{k: Z_{nk} \leqslant T \text{ and } k \leqslant n\}$, then, for the Type I censoring scheme, one observes the data set $\{(Z_{n1}, S_1), \ldots, (Z_{nr(T)}, S_{r(T)})\}$ along with the complementary set $(S_{r(T)+1}, \ldots, S_n)$, though the exact permutation of this later vector is not known at time-point T. In a Type II censoring scheme, for some prefixed r $(1 \leqslant r \leqslant n)$, one waits upto a (random) time-point Z_{nr} and obtains the data set $\{(Z_{n1}, S_1), \ldots, (Z_{nr}, S_r)\}$ along with the complementary set $(S_{r+1}, \ldots, S_n)$, again without the precise knowledge on the exact permutation of this vector. Nonparametric tests based on such censored data have been discussed in Chapters 25 (by Basu), 26 (by Doksum and Yandell) and 32 (by Wieand). Generally, in a life testing problem, continuous monitoring is needed to record the set in (3.2), and, as such, it seems unreasonable to wait until the time-point T or Z_{nr} has been reached and then to test for H_0. It is more natural to test for the null hypothesis as more information becomes available at the successive failures (Z_{ni}), so that if at any early stage the null hypothesis becomes untenable, the study may be curtailed along with the rejection of H_0. Basically, this calls for a repeated significance testing procedure (at each failure point), and, as the accumulating data may not have independent or stationary increments, special care needs to be taken so that the Type I or II error rates for the procedure are under control. This is done through the formulation of some suitable *time-sequential procedures*, which are described below.

Consider a typical linear rank statistic

$$T_n = \sum_{i=1}^{n} (c_i - \bar{c}_n) a_n(R_i), \tag{3.3}$$

where $c_1, \ldots, c_n$ are given constants, $\bar{c}_n = n^{-1} \sum_{i=1}^{n} c_i$ and $a_n(1), \ldots, a_n(n)$ are the *scores*. Using (3.1), we may rewrite (3.3) as $T_n = \sum_{i=1}^{n} (c_{S_i} - \bar{c}_n) a_n(i)$. With this form, we may consider the *censored rank statistics* $\{T_{nr}\}$ by letting, for every $r \leq n$,

$$T_{nr} = E\{T_n \mid S_1, \ldots, S_r; H_0\} = \sum_{i=1}^{r} (c_{S_i} - \bar{c}_n)[a_n(i) - a_n^*(r)], \tag{3.4}$$

where

$$a_n^*(r) = \begin{cases} (n-r)^{-1} \sum_{j=r+1}^{n} a_n(j), & 0 \leq r \leq n-1, \\ 0, & r = n. \end{cases} \tag{3.5}$$

Note that at the k-th failure Z_{nk}, one can compute the censored rank statistic T_{nk}, for $k \geq 1$; conventionally, we let $T_{n0} = 0$. Operationally, in a progressive censoring scheme, at the k-th failure Z_{nk}, one computes T_{nk} (or $|T_{nk}|$): If, for the first time, for some k (≥ 1), T_{nk} (or $|T_{nk}|$) exceeds a critical value τ_n^+ (or τ_n), experimentation is stopped at that point of time (along with the rejection of H_0), while, otherwise, one proceeds on to the next failure point. In this setup, one may either prefix a maximum number r ($\leq n$), such that at Z_{nr}, the study is curtailed (if it has not been so earlier), or even, one may take $r = n$. Thus, the probability of the Type I error for this procedure is given by

$$\alpha_{nr}^+ = P\{T_{nk} > \tau_n^+, \text{ for some } k: 1 \leq k \leq r \mid H_0\} \tag{3.6}$$

and a similar expression holds for the two-sided case. If we define, for every r, $n: 1 \leq r \leq n$, $n \geq 2$,

$$A_{nr}^2 = (n-1)^{-1} \left\{ \sum_{i=1}^{r} a_n^2(i) + (n-r)(a_n^*(r))^2 - n^{-1} \left(\sum_{i=1}^{n} a_n(i) \right)^2 \right\}, \tag{3.7}$$

and let

$$C_n^2 = \sum_{i=1}^{n} (c_i - \bar{c}_n)^2, \tag{3.8}$$

then, if follows from the results of Chatterjee and Sen (1973) that under H_0 and some very mild regularity conditions on the scores, for r: $r/n \to p$: $0 < p \leqslant 1$,

$$\max_{0 \leqslant k \leqslant r} T_{nk}/(A_{nr}C_n) \xrightarrow{\mathscr{D}} \sup\{W(t): 0 \leqslant t \leqslant 1\}, \tag{3.9}$$

$$\max_{0 \leqslant k \leqslant r} |T_{nk}|/(A_{nr}C_n) \xrightarrow{\mathscr{D}} \sup\{|W(t)|: 0 \leqslant t \leqslant 1\}, \tag{3.10}$$

where $W = \{W(t); 0 \leqslant t \leqslant 1\}$ is a standard Brownian motion on $[0, 1]$. Note that, for every $\lambda \geqslant 0$,

$$P\{\sup_{0 \leqslant t \leqslant 1} W(t) > \lambda\} = (2/\pi)^{1/2} \int_\lambda^\infty \exp(-\tfrac{1}{2}t^2)\, dt = 2[1 - \Phi(\lambda)], \tag{3.11}$$

$$P\{\sup_{0 \leqslant t \leqslant 1} |W(t)| > \lambda\} = 1 - \sum_{k=-\infty}^{\infty} (-1)^k [\Phi((2k+1)\lambda) - \Phi((2k-1)\lambda)], \tag{3.12}$$

where $\Phi(\cdot)$ is the standard normal d.f. If λ_α^+ and λ_α stand for the solutions for the right hand sides of (3.11) and (3.12) being equated to a given α $(0 < \alpha < 1)$ then, for large n, $\tau_n^* \simeq A_{nr}C_n\lambda_\alpha^+$ and a similar approximation holds for the two-sided case. For small n, the next value of τ_n^* may be obtained by direct enumeration of the permutation distributions of $(S_1, \ldots, S_n)$ (over the set of permutations of $(1, \ldots, n)$). Asymptotic power properties of this PCS testing procedure have been studied by Chatterjee and Sen (1973), Sen (1976) and others, and a detailed account of these is given in Chapter 11 of Sen (1981a). In the particular case of two-sample statistics, we have, for some n_1: $1 \leqslant n_1 < n$, $n_2 = n - n_1$,

$$c_1 = \cdots = c_{n_1} = 0, \quad c_{n_1+1} = \cdots = c_n = 1 \quad \text{and} \quad C_n^2 = n_1 n_2/n .$$

Also, for Wilcoxon scores, i.e., $a_n(k) = k/(n+1)$, $k = 1, \ldots, n$, we have

$$A_{nr}^2 = [n/12(n+1)][1 - \{(n-r)^3 - (n-r)\}/(n^3 - n)] .$$

For some numerical studies in this case, we may refer to Davis (1978). For the case of log-rank scores, i.e., $a_n(k) = -1 + \sum_{i=1}^{k} (n - i + 1)^{-1}$, $k = 1, \ldots, n$, we may refer to Majumdar and Sen (1978b) and Koziol and Petkau (1978).

Majumdar and Sen (1978a) considered a vector generalization of the above PCS procedure. In (3.3), replace the c_i (and $\bar{c}_n$) by q-vectors $\boldsymbol{c}_i$ (and $\bar{\boldsymbol{c}}_n$), for some $q \geqslant 1$. This typically arises in a multi-sample model. A similar change is needed in (3.4), while, in (3.8), replace C_n^2 by a $q \times q$ matrix $\boldsymbol{C}_n = \sum_{i=1}^n (\boldsymbol{c}_i - \bar{\boldsymbol{c}}_n)(\boldsymbol{c}_i - \bar{\boldsymbol{c}}_n)'$. Then, at the k-th failure, we consider the test statistic

$$\mathscr{L}_{nk} = A_{nr}^{-2}(\boldsymbol{T}_{nk}'\boldsymbol{C}_n^- \boldsymbol{T}_{nk}) \quad \text{for } k \leqslant r, \tag{3.13}$$

where C_n^- is a (reflexible) generalized inverse of C_n. With this extension, in (3.9)–(3.10), we need to replace the Brownian motion W by a q-dimensional *Bessel process* $B_q = \{B_q(t) = \{\Sigma_{j=1}^q W_j^2(t)\}^{1/2}; 0 \le t \le 1\}$, where the $W_j = \{W_j(t); 0 \le t \le 1\}$, $j \ge 1$, are independent copies of a standard Brownian motion. Then, under H_0, as n increases,

$$\max\{\mathscr{L}_{nk}^{1/2} : 0 \le k \le r\} \xrightarrow{\mathscr{D}} \sup\{B_q(t) : 0 \le t \le 1\} = B_q^*, \tag{3.14}$$

say, and the percentile points of the distribution of B_q^* have been tabulated by De Long (1981). Asymptotic power properties of this test, for local alternatives, have been studied by Majumdar and Sen (1978a) and De Long (1980).

For the specific two-sample problem, Koziol and Byar (1975) proposed the use of truncated versions of the classical Kolmogorov–Smirnov statistic and tabulated their percentage points too; for some one-sided tests, we may refer to Schey (1977). Sinha and Sen (1979a) considered a weighted empirical process which includes the two-sample problem as a particular case. Let

$$S_n(x) = n^{-1} \sum_{i=1}^n I(X_i \le x)$$

and

$$H_n(x) = C_n^{-1} \sum_{i=1}^n (c_i - \bar{c}_n) I(X_i \le x), \qquad x \ge 0, \tag{3.15}$$

where C_n^2 is defined by (3.8). Also, let $\omega = \{\omega(t), 0 < t < 1\}$ be a nonnegative, smooth weight function. Consider then the statistics

$$K_{nr}^{+(\omega)} = \sup_{x \le Z_{nr}} H_n(x)/\omega(S_n(x)) \quad \text{and} \quad K_{nr}^{(\omega)} = \sup_{x \le Z_{nr}} |H_n(x)|/\omega(S_n(x)), \tag{3.16}$$

where Z_{nr} is defined as in (3.1)–(3.2). The choice of the uniform weight function $[\omega(t) = 1]$ leads to the unweighted statistics. In this special case, under H_0, $K_{nr}^{+(1)}$ and $K_{nr}^{(1)}$ have asymptotically the same distribution as of $\sup\{W^0(t): 0 \le t \le p\}$ and $\sup\{|W^0(t)|: 0 \le t \le p\}$, where $W^0 = \{W^0(t) = W(t) - tW(1), 0 \le t \le 1\}$ is a standard *Brownian bridge* and $p = \lim_{n\to\infty} r/n: 0 < p \le 1$. For the weighted case where $\omega(t) \to 0$ with $t \downarrow 0$, allowing a small truncation at 0, the distribution theory for the boundary crossing of a Brownian bridge over a curvilinear boundary has been incorporated in the study of the distribution theory of $K_{nr}^{+(\omega)}$ and $K_{nr}^{(\omega)}$. Sinha and Sen (1979b) have also extended their procedure to the q-vector case by letting

$$\boldsymbol{H}_n(x) = \boldsymbol{C}_n^{-1/2} \sum_{i=1}^n (\boldsymbol{c}_i - \bar{\boldsymbol{c}}_n) I(X_i \le x), \quad x \ge 0, \tag{3.17}$$

$$K_{nr}^* = \sup\{[\boldsymbol{H}_n(x)]'[\boldsymbol{H}_n(x)]/\omega^2(S_n(x)): 0 < x \le Z_{nr}\}. \tag{3.18}$$

Note that like the $\mathscr{L}_{nk}$, K^*_{nr} remains invariant under a reparameterization: $X_i - \boldsymbol{\beta}'\boldsymbol{c}_i \to X_i - \boldsymbol{\gamma}'\boldsymbol{d}_i$ where $\boldsymbol{d}_i = \boldsymbol{D}\boldsymbol{c}_i$ and $\boldsymbol{D}$ is nonsingular. For the asymptotic distribution theory of K^*_{nr}, the standardized Bessel process may be called for, and, percentile points for the relevant distributions can be obtained from De Long (1981).

PCS tests for the analysis of covariance (ANOCOVA) models based on appropriate rank statistics have also been considered by Sen (1979, 1981b) and others. These will be discussed briefly in the next section. In the rest of this section, we consider some PCS procedures relating to the Cox (1972) *proportional hazard model.* Since the proportional hazard models have already been discussed in Chapters 32 (by Wieand) and 26 (Doksum and Yandell), we shall treat the same only briefly and stress mainly on the relevant PCS procedures. Note that these models are quasi-nonparametric in character: the hazard function is of nonparametric nature, but, the dependence on the covariates is of a specified structure.

For the i-th subject having survival time Y_i and a set of concomitant variates $\boldsymbol{d}_i = (d_{i1}, \ldots, d_{iq})'$, for some $q \geq 1$, consider the model that the conditional hazard rate, given $\boldsymbol{d}_i$, is of the form:

$$h_i(t) = (\mathrm{d}/\mathrm{d}t) \log P\{Y_i \geq t \mid \boldsymbol{d}_i\} = h_0(t) \cdot \exp\{\boldsymbol{\beta}'\boldsymbol{d}_i\}, \quad i = 1, \ldots, n, \tag{3.19}$$

where $h_0(t)$ is the hazard function for $\boldsymbol{d}_i = \boldsymbol{0}$ and is quite arbitrary in nature, while $\boldsymbol{\beta} = (\beta_1, \ldots, \beta_q)'$ parameterizes the regression of the survival time on the covariates. Thus, the model is nonparametric with respect to $h_0(t)$ but parametric with respect to the covariates through the proportionality assumption in (3.19). The null hypothesis of interest is H_0: $\boldsymbol{\beta} = \boldsymbol{0}$ against $\boldsymbol{\beta} \neq \boldsymbol{0}$. To incorporate possible withdrawals of subjects from the scheme, we take $Y_i^0 = \min\{Y_i, W_i\}$ and $\delta_i = I(Y_i = Y_i^0)$, $i = 1, \ldots, n$, where $W_1, \ldots, W_n$ stand for the withdrawal times, assumed to be independent of the Y_i. Let then $T = \{t_1 < \cdots < t_m\}$ be the set of ordered failure times among the Y_i^0 (i.e., $m = \sum_{i=1}^n \delta_i$ and the t_j correspond to the points for which $\delta_i = 1$). Then, at time $t_j - 0$, there is a *risk set* $\mathscr{R}_j$ of r_j subjects who are surviving upto that point and have not dropped out yet, for $j = 1, \ldots, m$. Note that $\mathscr{R}_m \subseteq \mathscr{R}_{m-1} \subseteq \cdots \subseteq \mathscr{R}_1$. Now, at the k-th point t_k, one has the picture for the partial set $\{t_j: j \leq k\}$, and the corresponding *partial* (*log*)-*likelihood function* may be defined by

$$\log L^*_{nk} = \sum_{j=1}^{k} \Big\{\boldsymbol{\beta}'\boldsymbol{d}_{Q_j} - \log\Big(\sum_{i \in \mathscr{R}_j} \exp(\boldsymbol{\beta}'\boldsymbol{d}_i)\Big)\Big\}, \quad 1 \leq k \leq m, \tag{3.20}$$

where $Q_1, \ldots, Q_m$ stand for the anti-ranks corresponding to the failure points $t_1, \ldots, t_m$, respectively. The partial likelihood scores are then defined by

$$\boldsymbol{U}^*_{nk} = (\partial/\partial\boldsymbol{\beta}) \log L^*_{nk}\,|_{\boldsymbol{\beta}=\boldsymbol{0}} = \sum_{j=1}^{k} \Big\{\boldsymbol{d}_{Q_j} - r_j^{-1} \sum_{i \in \mathbf{R}} \boldsymbol{d}_i\Big\}, \quad k = 1, \ldots, m. \tag{3.21}$$

Also, let

$$J^*_{nk} = \sum_{j=1}^{k} \left\{ \sum_{i \in \mathcal{R}_j} (d_i - \bar{d}^*_j)(d_i - \bar{d}^*_j)' \right\}, \quad k = 1, \ldots, m, \tag{3.22}$$

where $\bar{d}^*_j = r_j^{-1} \sum_{i \in \mathcal{R}_j} d_i$, $j = 1, \ldots, m$. At the k-th failure point, one may then consider the Cox (1972) form of the partial likelihood ratio test statistic

$$\mathcal{L}^{*2}_{nk} = (U^*_{nk})'(J^*_{nk})^-(U^*_{nk}). \tag{3.23}$$

In a PCS setup, we look at the data at each failure point, i.e., at each k $(=1, \ldots, m)$, we compute $\mathcal{L}^*_{nk}$. If, for the first time, for some k, $\mathcal{L}^*_{nk}$ exceeds a critical value $\tau^*_{n\alpha}$, we stop at that point of time (t_k) along with the rejection of the null hypothesis H_0: $\boldsymbol{\beta} = \mathbf{0}$. If no such k occurs during the tenure of the study, the null hypothesis is accepted. It follows from the results of Sen (1981) that for the sequence $\{\mathcal{L}^*_{nk}: k = 1, \ldots, m\}$, the standardized q-dimensional Bessel process may be used to study the asymptotic null distribution of $\max\{\mathcal{L}^*_{nk}: k \leq m\}$, and hence, the procedure is similar in nature to the one based on the rank statistics. For null hypotheses other than $\boldsymbol{\beta} = \mathbf{0}$, in the scores in (3.21) and the information matrix in (3.22), unknown parameters (i.e., the unspecified part of $\boldsymbol{\beta}$) may enter and substitution of their estimates may be necessary. This will result in a nondistribution-free nature of the resulting test statistic. However, the asymptotic distribution-free character still retains, and parallel PCS procedures work out well. We may refer to Tsiatis (1981a,b) and Anderson and Gill (1982), among others, where these procedures have been worked out.

4. Rank analysis of covariance

Rank procedures for the analysis of variance (ANOVA) and multivariate ANOVA (MANOVA) problems have been discussed in Chapters 2 (by Bhapkar), 11 (Adichie) and 12 (Aubuchon and Hettmansperger). For some simple (viz., one-way ANOVA) models, rank analysis of convariance (RANOCOVA) procedures have also been developed by Quade (1967), Puri and Sen (1969) and Sen and Puri (1970), among others. Essentially, the multivariate approach prescribed by these authors goes through neatly in the general cases too. We discuss some of these procedures in this section.

Let $\boldsymbol{X}^*_i = (\boldsymbol{X}'_{i0}, \boldsymbol{X}'_i)'$, $i = 1, \ldots, n$, be n independent r.v.'s with $(p+q)$-variate continuous distribution functions F^*_i, $i = 1, \ldots, n$, where $p \geq 1$ and $q \geq 1$. The $\boldsymbol{X}_{i0}$ are the primary variates (p-vectors) and the $\boldsymbol{X}_i$ are the covariates (q-vectors). We assume that the covariates are not affected by the treatments, so that the $\boldsymbol{X}_i$ are i.i.d.r.v.'s with a common q-variate d.f. F. Let then $F^0_i(\boldsymbol{y} \mid \boldsymbol{x})$ be the conditional d.f. of $\boldsymbol{X}_{i0}$, given $\boldsymbol{X}_i = \boldsymbol{x}$, for $i = 1, \ldots, n$. Basically, we want to test for the null hypothesis

$$H_0:\quad F_1^0 = \cdots = F_n^0 = F^0 \ \text{(unknown)}\,, \tag{4.1}$$

against the set of alternatives that they are not all equal. It may be convenient to conceive of the model

$$F_i^0(y \mid \boldsymbol{x}) = F^0(y - \boldsymbol{\beta}(\boldsymbol{c}_i - \bar{\boldsymbol{c}}_n) \mid \boldsymbol{x}), \quad i \geq 1\,, \tag{4.2}$$

where the $\boldsymbol{c}_i$ are specified r (≥ 1)-vectors, not all equal, $\bar{\boldsymbol{c}}_n = n^{-1}\sum_{i=1}^n \boldsymbol{c}_i$ and $\boldsymbol{\beta}$ parameterizes the regression of the primary variates on the $\boldsymbol{c}_i$. For the particular case of one-way RANOCOVA model, the $\boldsymbol{c}_i$ can only assume the realizations $(1, 0, \ldots, 0)$, $(0, 1, \ldots, 0), \ldots, (0, \ldots, 0, 1)$. In this more general setup in (4.2), we like to test for H_0: $\boldsymbol{\beta} = \boldsymbol{0}$ against $\boldsymbol{\beta} \neq \boldsymbol{0}$.

Let $\boldsymbol{E}_n^* = (\boldsymbol{X}_1^*, \ldots, \boldsymbol{X}_n^*)$ be the $(p+q) \times n$ matrix of the sample observations. For each row, we adapt a separate ranking scheme. This leads us to the following *rank collection matrix*

$$\boldsymbol{R}_n^* = \begin{pmatrix} \boldsymbol{R}_n^0 \\ \boldsymbol{R}_n \end{pmatrix} \begin{matrix} {\scriptstyle p\times n} \\ {\scriptstyle q\times n} \end{matrix} \tag{4.3}$$

where each row of $\boldsymbol{R}_n^*$ consists of the numbers $1, \ldots, n$, permuted in some order; ties are neglected, with probability one, by virtue of the assumed continuity of the F_i^*. Also, we consider a $(p+q) \times n$ matrix of *scores*

$$\boldsymbol{A}_n = \begin{pmatrix} a_{n1}^0(1) \cdots a_{n1}^0(n) \\ \cdots\cdots\cdots\cdots\cdots \\ a_{np}^0(1) \cdots a_{np}^0(n) \\ a_{n1}(1) \cdots a_{n1}(n) \\ \cdots\cdots\cdots\cdots\cdots \\ a_{nq}(1) \cdots a_{nq}(n) \end{pmatrix} \tag{4.4}$$

where the scores are defined as in (2.5) and (2.6) with possibly different score generating functions for the different variates. As in (3.3), we define the rank statistics

$$\begin{aligned} \underset{r\times(p+q)}{\boldsymbol{T}_n^*} &= (\underset{r\times p}{\boldsymbol{T}_n^0}, \underset{r\times q}{\boldsymbol{T}_n}) \\ &= \sum_{i=1}^n (\boldsymbol{c}_i - \bar{\boldsymbol{c}}_n)[a_{n1}^0(R_{1i}^0), \ldots, a_{np}^0(R_{ip}^0), a_{n1}(R_{1i}), \ldots, a_{nq}(R_{qi})]\,. \end{aligned} \tag{4.5}$$

The within row averages of the scores in (4.4) are denoted by $\bar{a}_{n1}^0, \ldots, \bar{a}_{np}^0$ and $\bar{a}_{n1}, \ldots, \bar{a}_{nq}$, respectively. Let then

$$v_{njj'}^{00} = (n-1)^{-1}\Big\{\sum_{i=1}^n (a_{nj}^0(R_{ni}^0) - \bar{a}_{nj}^0)(a_{nj'}^0(R_{j'i}^0) - \bar{a}_{nj'}^0)\Big\}, \quad j, j' = 1, \ldots, p\,, \tag{4.6}$$

$$v^0_{njj'} = (n-1)^{-1}\left\{\sum_{i=1}^{n} (a^0_{nj}(R^0_{ji}) - \bar{a}^0_{nj})(a_{nj'}(R_{j'i}) - \bar{a}_{nj'})\right\},$$
$$j = 1, \ldots, p,\ j' = 1, \ldots, q, \tag{4.7}$$

$$v_{njj'} = (n-1)^{-1}\left\{\sum_{i=1}^{n} (a_{nj}(R_{ji}) - \bar{a}_{nj})(a_{nj'}(R_{j'i}) - \bar{a}_{nj'})\right\},$$
$$j, j' = 1, \ldots, q. \tag{4.8}$$

We denote

$$\boldsymbol{V}^*_n = \begin{pmatrix} \boldsymbol{V}^{00}_n & \boldsymbol{V}^0_n \\ \boldsymbol{V}^{0\prime}_n & \boldsymbol{V}_n \end{pmatrix} \quad \text{where } \boldsymbol{V}^{00}_n = ((v^{00}_{njj'})),\ \boldsymbol{V}^0_n = ((v^0_{njj'})) \text{ and } \boldsymbol{V}_n = ((v_{njj'}))\,. \tag{4.9}$$

Further, we define

$$\boldsymbol{C}_n = \sum_{i=1}^{n} (\boldsymbol{c}_i - \bar{\boldsymbol{c}}_n)(\boldsymbol{c}_i - \bar{\boldsymbol{c}}_n)'\,. \tag{4.10}$$

Then, the first step is to use the fact that the permutational dispersion matrix of $\boldsymbol{T}^*_n$ is equal to $\boldsymbol{C}_n \otimes \boldsymbol{V}^*_n$, and hence, the fitted value of the primary variate rank statistics on the covariate rank statistics yields the following residuals:

$$\boldsymbol{T}^{0*}_n = \boldsymbol{T}^0_n - \boldsymbol{T}_n(\boldsymbol{V}_n)^- \boldsymbol{V}^{0\prime}_n\,. \tag{4.11}$$

Also, let

$$\boldsymbol{V}^{0*}_n = \boldsymbol{V}^{00}_n - \boldsymbol{V}^0_n(\boldsymbol{V}_n)^- \boldsymbol{V}^{0\prime}_n\,. \tag{4.12}$$

Finally, let $\tilde{\boldsymbol{T}}^{0*}_n$ be the rolled out rp-vector from $\boldsymbol{T}^{0*}_n$, and let

$$\mathscr{L}^*_n = (\tilde{\boldsymbol{T}}^{0*}_n)'(\boldsymbol{C}_n \otimes \boldsymbol{V}^{0*}_n)^-(\tilde{\boldsymbol{T}}^{0*}_n)\,. \tag{4.13}$$

Then, $\tilde{\boldsymbol{T}}^{0*}_n$ is the $r \times p$ matrix of covariate-adjusted rank order statistics and $\mathscr{L}^*_n$ is the test statistic based on these adjusted statistics. For small values of n, the exact permutational (conditional) distribution of $\mathscr{L}^*_n$ can be obtained by direct enumeration of the $n!$ equally likely column permutation of the matrix $\boldsymbol{R}^*_n$ in (4.3), and, for large n, $\mathscr{L}^*_n$ has closely the chi square distribution with pr degrees of freedom, when the rank of $\boldsymbol{C}_n = r$ and the null hypothesis holds.

The multivariate approach developed in Sen and Puri (1970) also yields the asymptotic distribution theory under alternative hypotheses. For suitable sequences of local alternatives, the asymptotic distribution of $\mathscr{L}^*_n$ turns out to be a noncentral chi square with rp degrees of freedom and an appropriate noncentrality parameter $\Delta^*_{\mathscr{L}}$. In passing, we remark that for the MANOVA problem of testing the equality of the p-variate (marginal) d.f.'s F^{00}_i (of the $\boldsymbol{X}_{i0}$), one actually ignores the concomitant variates ($\boldsymbol{X}_i$) and, based on the $\boldsymbol{T}^0_n$ and

V_n^{00} in (4.5) and (4.6), one considers the test statistic

$$\mathscr{L}_n^0 = (\tilde{T}_n^0)'(C_n \otimes V_n^{00})^-(\tilde{T}_n^0), \tag{4.14}$$

where $\tilde{T}_n^0$ is the rolledout rp-vector from T_n^0. For small values of n, the exact permutation distribution of $\mathscr{L}_n^0$ can be obtained by direct enumeration, while, for large n, under H_0, $\mathscr{L}_n^0$ has closely the central chi square distribution with pr degrees of freedom. For local alternatives, similar to the MANOCOVA model, the asymptotic distribution of $\mathscr{L}_n^0$ is noncentral chi square with pr degrees of freedom and noncentrality parameter $\Delta_{\mathscr{L}}$, where $\Delta_{\mathscr{L}}$ depends on the marginal df's F_i^{00}, $i = 1, \ldots, n$. It can be shown that for a sequence of common alternatives,

$$\Delta_{\mathscr{L}}^* \geq \Delta_{\mathscr{L}} \quad \text{for all admissible alternatives}, \tag{4.15}$$

where the equality sign holds only when $V_n^{00} - V_n^{0*}$ $(= V_n^0 V_n^- V_n^{0\prime})$ converges to a null matrix, in probability, when $n \to \infty$, i.e., the concomitant variates ranks are not associated with those of the primary variates. This explains the asymptotic power superiority of the rank ANOCOVA procedure over the corresponding ANOVA procedure. From the computational point of view, we may have an alternative look at (4.13). Let $\tilde{T}_n^*$ be the rolledout $r(p+q)$-vector from T_n^* in (4.5), and let $\tilde{T}_n$ be the rolledout form of T_n in (4.5). Then, it follows that

$$\mathscr{L}_n^* = \tilde{T}_n^{*\prime}(C_n \otimes V_n^*)^- \tilde{T}_n^* - \tilde{T}_n'(C_n \otimes V_n)^- \tilde{T}_n. \tag{4.16}$$

Thus, $\mathscr{L}_n^*$ is the difference between the classical rank MANOVA statistic for the entire set of $p+q$ variates and the one for the set of q covariates alone. This formula avoids the computation of the residuals in (4.11) and the adjusted matrix in (4.12). Gerig (1975) has used this characterization of the RANOCOVA statistics for the two-way layout problem. He used the multivariate generalization of the Friedman (intra-block) rank statistics (viz., Gerig, 1969) and computed the two test statistics for the entire set of $p+q$ characters and the subset of q covariates only, and their difference provides the desired test statistic for the MONOCOVA problem. MANOCOVA models incorporating aligned ranking are also considered in Puri and Sen (1971, Ch. 7). These are conditionally distribution-free and the results run parallel to the ones in the one-way layout case.

Let us now consider the case of censored data relating to ANOCOVA. For simplicity, we consider the case of $p = 1$ and $q \geq 1$. Corresponding to the rank vector R_n^0 in (4.3), we define the anti-rank vector S_n^0 for the primary variate by letting

$$R_{S_i^0}^0 = S_{R_i^0}^0 = i \quad \text{for } i = 1, \ldots, n. \tag{4.17}$$

We also assume that the covariates (and hence the rank matrix R_n) are observable at the beginning. Thus, in a Type II censoring scheme, for some k ($\leq n$), one has the knowledge about the rank collection matrix

$$\begin{pmatrix} 1 & 2 & \cdots & k \\ R_{1S_1^0} & R_{1S_2^0} & \cdots & R_{1S_k^0} \\ \cdots & \cdots & \cdots & \cdots \\ R_{qS_1^0} & R_{qS_2^0} & \cdots & R_{qS_k^0} \end{pmatrix}; \tag{4.18}$$

we also know about the remaining $(n-k)$ columns of $\boldsymbol{R}_n$, but, without the specific order in which they would have appeared in (4.18) if the censoring were not made. Let $R_i^0 = R_{0i}$, $i = 1, \ldots, n$, and, for each j $(=0, 1, \ldots, q)$, define

$$\begin{aligned} a_{nj}^*(k) &= (n-k)^{-1}\Big\{\sum_{r=1}^{n} a_{nj}(r) - \sum_{i=1}^{k} a_{nj}(R_{js_i^0})\Big\}, \quad 1 \leq k \leq n-1, \\ &= 0, \qquad\qquad k = n, \end{aligned} \tag{4.19}$$

where $a_{n0}(\cdot) = a_{n1}^0(\cdot)$. Also, for each k $(\leq n)$, define

$$v_{njj'}^{(k)} = (n-1)^{-1}\Big\{\sum_{i=1}^{k} a_{nj}(R_{js_i^0})a_{nj'}(R_{j's_i^0}) + (n-k)a_{nj}^*(k)a_{nj'}^*(k) - n\bar{a}_{nj}\bar{a}_{nj'}\Big\}, \tag{4.20}$$

for $j, j' = 0, 1, \ldots, q$. Then, as in (4.9), we let

$$\boldsymbol{V}_n^{*(k)} = \begin{pmatrix} v_{n00}^{(k)} & \boldsymbol{v}_{n0}^{(k)} \\ \boldsymbol{v}_{n0}^{(k)\prime} & \boldsymbol{V}_n^{(k)} \end{pmatrix} \quad \text{for } k = 1, \ldots, n. \tag{4.21}$$

Finally, as in (3.4), we write, for each k $(=1, \ldots, n)$,

$$\begin{aligned} \boldsymbol{T}_{n^*}^{(k)} &= \sum_{i=1}^{k} (\boldsymbol{c}_{s_i^0} - \bar{\boldsymbol{c}}_n)[a_{n0}(R_{0s_i^0}) - a_{n0}^*(k), \ldots, a_{nq}(R_{qs_i^0}) - a_{nq}^*(k)] \\ &= (\boldsymbol{T}_{n0}^{(k)}, \boldsymbol{T}_n^{(k)}). \end{aligned} \tag{4.22}$$

Then, proceeding as in (3.13) and using the adjustment for the covariates as in (4.16), we may define a censored test statistic (at the k-th failure) by

$$\mathscr{L}_{nk}^* = (\tilde{\boldsymbol{T}}_{n^*}^{(k)})'(\boldsymbol{C}_n \otimes \boldsymbol{V}_n^{*(k)})^-(\tilde{\boldsymbol{T}}_{n^*}^{(k)}) - (\tilde{\boldsymbol{T}}_n^{(k)})'(\boldsymbol{C}_n \otimes \boldsymbol{V}_n^{(k)})^-(\tilde{\boldsymbol{T}}_n^{(k)}), \tag{4.23}$$

for $k = 1, \ldots, n$, where $\tilde{A}$ stands for the rolledout vector form of a matrix A. Note that $\mathscr{L}_{nk}^*$ is the rank analysis of covariance test statistic when the covariates are observable at the beginning while the first k failures among the primary variables are observed and the rest censored (in a Type II censoring scheme). Here also, under the null hypothesis in (4.1), for small values of n, the exact permutation distribution of $\mathscr{L}_{nk}^*$ can be obtained by direct enumeration, while, for large n whenever k/n is away from 0, $\mathscr{L}_{nk}^*$ has closely the central chi square distribution with r degrees of freedom (under H_0). The asymptotic distribution theory for local alternatives also follows on parallel lines. In the case of Type I censoring, as in Section 3, $r(T)$, the number of failures (with respect to the primary variate) occurring in the prescribed duration of the study, will be a nonnegative integer valued random variable. Here also, under H_0, the permutational distribution of $\mathscr{L}^*_{nr(T)}$, conditioned on $r(T) = r$, can be used to construct a conditionally distribution-free test. For large n, again this conditional (permutational) null distribution as well as the corresponding unconditional one may be closely approximated by the chi square distribution with r degrees of freedom. Sen (1979, 1981b) has employed the triangular array of censored ANOCOVA statistics $\{\mathscr{L}^*_{nk}: 0 \leq k \leq n; n \geq 1\}$ for testing the null hypothesis H_0 in (4.1) under a PCS. Operationally, the procedure is the same as in the case of rank ANOVA under PCS, treated in Section 3 (see (3.5) through (3.13)), and the basic task is to find out a critical value l^*_n which also provides a stopping rule for the PCS. It follows that if we define the Bessel process as in before (3.14), and, if for every ε': $0 < \varepsilon \leq 1$, we let

$$B_{r,\varepsilon} = \{t^{-1/2}B_r(t);\ \varepsilon \leq t \leq 1\},$$

then, under mild regularity conditions, when H_0 in (4.1) holds,

$$\max_{k:\varepsilon n \leq k \leq n} \mathscr{L}^{*1/2}_{nk} \xrightarrow{\mathscr{D}} \sup\{t^{-1/2}B_r(t):\ \varepsilon \leq t \leq 1\} = B^{**}_{r\varepsilon}\,, \tag{4.24}$$

say. For various ε (>0), the percentile points of the distribution of $B^{**}_{r\varepsilon}$ $(r \geq 1)$ have been studied by De Long (1981). Note that as $t \downarrow 0$, $t^{-1/2}B_r(t)$ may not behave smoothly (and, in fact, it may almost surely bounces to $\pm\infty$), and hence, in (4.24), the weak convergence result may not be appropriate if we let $\varepsilon = 0$. Also, from the practical point of view, statistical monitoring may not be very useful unless we have gathered at least some data to enable us to use the estimators $\boldsymbol{V}^*_{nk}$ etc. in a consistent manner. Thus, it seems quite appropriate to start the repeated significance testing not from the very first failure, but, after a given number of failures have occurred. Technically, we may also replace the upper limit (n) for k in (4.24) by any number r: $r/n \to p$: $0 < p \leq 1$, and in that case, in the right hand side, the upper limit (1) for t has also to be replaced by p. Then, given a choice of p $(0 < p \leq 1)$, one may choose a small ε $(0 < \varepsilon < p)$ and start the repeated significance testing scheme at the n_0-th failure point, where $n_0 = [n\varepsilon] + 1$. With this choice, one may use the PCS scheme mentioned before and use the partial sequence $\{\mathscr{L}^*_{nk};\ k \geq n_0\}$ for this purpose. Note that for any $p \in (0, 1]$,

$$\sup\{t^{-1/2}B_r(t):\ \varepsilon \leq t \leq p\} \overset{\mathscr{D}}{=} \sup\{t^{-1/2}B_r(t):\ \varepsilon/p \leq t \leq 1\} \tag{4.25}$$

so that the DeLong (1981) tables remain applicable with ε replaced by ε/p.

5. Nonparametrics in growth curve models

In a typical MANOVA model, one assumes that

$$Y = (Y_1, \ldots, Y_n) = \beta X + e, \quad e = (e_1, \ldots, e_n), \tag{5.1}$$

where the e_i are i.i.d.r.v. with a distribution function F defined on the $p(\geq 1)$-dimensional Euclidean space E^p, β is a $p \times q$ matrix of unknown parameters (for some $q \geq 1$) and X is a $q \times n$ matrix of known constants. In the parametric case, one assumes that F is a multivariate normal distribution, while in the nonparametric case, the d.f. F may be of quite arbitrary (but continuous) form. Rank based tests for linear hypotheses involving the β has already been considered in Chapters 11 (by Adichie) and 12 (Aubuchon and Hettmansperger). We write $Y_i = (Y_{i1}, \ldots, Y_{ip})'$, for $i = 1, \ldots, n$. The p characteristics on the same individual, in general, need not relate to a common character and may not all be measured on a common metric scale. In a longitudinal study, however, one has typically a set of repeated measurements (on a common metric scale) on the same unit or individual over differing conditions or periods of time. In such a case, often, the matrix β can be expressed in terms of another matrix involving lesser number of unknown parameters, viz.,

$$\beta = G\theta, \quad G(p \times r) \text{ known and } \theta(r \times q) \text{ unknown, with } r \leq p. \tag{5.2}$$

For example, if Y_{ij} refers to the measurements on the i-th individual at time point t_j $(1 \leq j \leq p)$, then, conceivably, the growth (change) of the systematic component may well be described in terms of a polynomial or some other smooth function of the time points (which are given) so that the system may be described in terms of a lesser number of unknown parameters. This may be particularly advantageous if p is not very small while the degree of the polynomial function is small compared to p. In such a case, the avoidance of the redundant parameters through reparameterization may lead to increased efficiency of the statistical inference procedures, and, this is the main idea in a growth curve analysis. In the parametric case, this reparameterization has been incorporated in a reduced dimensional MANOVA model, and, more generally, in a MANOCOVA model. We shall consider here nonparametric analogues of both of these models. The MANOVA procedures are comparatively simpler, but, may entail some loss of efficiency due to dimension-reduction, and the MANOCOVA procedures may fare well in this respect.

5.1. *Nonparametric* MANOVA *for growth curve models*

With respect to (5.2), we assume that the known coefficient matrix G is a rank r $(\leq p)$, so that $G'G$ is an $r \times r$ matrix of full rank. This is no loss of generality, as otherwise, a further reparameterization may be induced to achieve the full rank model. In the parametric case, Potthoff and Roy (1964) advocated the use of the following transformation:

$$Z = (Z_1, \ldots, Z_n) = (G'G)^{-1}G'Y. \tag{5.3}$$

Note that by (5.1), (5.2) and (5.3),

$$Z = \theta X + e^*, \quad e^* = (e_1^*, \ldots, e_n^*) = (G'G)^{-1}G'e. \tag{5.4}$$

Note that the columns of Z are still independent (vectors) and the e_i^* are i.i.d.r.v. with a d.f. F^*, defined on E^r, where F^* depends on the underlying F and the matrix G. In any case, whenever F is continuous, so will be F^*. Now, (5.4) represents a reduced dimensional MANOVA model, and, for this model, rank tests may be applied in the same fashion as in the classical case.

Let us first consider the case where the design matrix X can be partitioned as $X' = (1_n, X^{*\prime})$ where $1_n = (1, \ldots, 1)'$ and X^* is a $(q-1)\times n$ matrix. This is typically the case in the one-way layout model. We also write $\theta = (\theta_0, \theta^*)$, where θ_0 is of order $r\times 1$ and θ^* of order $r\times(q-1)$. Then, we may rewrite (5.4) as

$$Z = \theta_0 1_n' + \theta^* X^* + e^*, \tag{5.5}$$

where we are interested in testing suitable linear hypotheses on θ^*, treating θ_0 as a nuisance parameter. Suppose that we want to test for

$$H_0\colon\ \theta^* = 0 \quad \text{against} \quad H_1\colon\ \theta^* \neq 0. \tag{5.6}$$

For each row of the $r\times n$ matrix Z, we adopt a coordinate-wise ranking and as in (4.3), we denote this $r\times n$ rank matrix by R_n. Also, as in (4.4)–(4.5), we define a matrix T_n, of order $r\times(q-1)$, of linear rank statistics based on R_n and X^*. Note that the c_i in (4.5) are to be replaced by the columns of X^*. With a similar change in (4.10), but, still denoting the resulting matrix by C_n, we conclude that $\mathcal{L}_n^0$, defined by (4.14) with the modifications of C_n, V_n^{00}, etc. mentioned earlier, is the desired rank MANOVA test statistics for testing H_0 against H_1 in (5.6). For small values of n, here also, the permutational (conditional) null distribution of $\mathcal{L}_n^0$ may be obtained by direct enumeration, while, for large n, this can safely be approximated by the central chi square distribution with $r(q-1)$ degrees of freedom. For suitable local alternatives, the asymptotic nonnull distribution of $\mathcal{L}_n^0$ is noncentral chi square with $r(q-1)$ degrees of freedom and some appropriate noncentrality parameter. In passing, we may remark that for the particular testing problem in (5.6), we might as well have considered the original matrix Y (of order $p\times n$) and constructed (as in (4.14)) a rank MANOVA test statistic based on the Y-rankings. This rank statistic would have asymptotically, under H_0 in (5.6), chi square distribution with $p(q-1)$ degrees of freedom, where $p \geqslant r$. Also, under the common sequence of local alternatives, this statistic will have a noncentral chi square distribution with $p(q-1)$ degrees of freedom and an appropriate noncentrality parameter. It is well known that for tests based on chi square statistics, an increase in the degrees of freedom without an increase in the noncentrality parameter reduces the power of the test, so that unless an increased degrees of freedom is accompanied by an adequately increased noncentrality parameter, the resulting test may be less powerful. On this criterion, if we compare the two asymptotic noncentral chi square distributions for the two rank MANOVA tests based on the Z-rankings and Y-rankings, we may note that for $r < p$, under (5.4), there may not be any significant difference between the two noncentrality parameters, so that the test based on the Z-rankings would generally perform better than the other one. This really provides the justification for the data-reduction in (5.4). The larger is the difference between p and r, the better would be the relative performance of the rank MANOVA test based on the Z-rankings. On the other hand, r should be so chosen that the model in (5.2) is adequately testable. A value of r less than the minimum adequate value may lead to loss of efficiency through smaller noncentrality parameters. It may also lead to considerable bias through the inadequate fit and the resulting test may not perform that well. On the other hand, a value of r chosen larger than the minimum adequate value may unnecessarily increase the degrees of freedom without, possibly, any further increment in the noncentrality parameter, so that the resulting test may be less efficient than desired. Thus, with respect to (5.2), choice of a proper value of r and $\boldsymbol{G}$ is of crucial importance for the growth curve analysis to be valid and efficient. One of the technical problems in this context is to compare the relative performance of two competing test statistics with different values of r, by a single numerical measure (such as the Pitman efficiency). While comparing two noncentral chi square distributions with different degrees of freedoms and possibly nonproportional noncentrality parameters, the classical concept of the Pitman efficiency is not generally adaptable; the two asymptotic power functions may not be made equal by simply adjusting the sample sizes (as is done in the Pitman efficiency case). One may, of course, use some other measures, such as the approximate Bahadur efficiency, but, they may not also perform that well in the current context. Another measure of local efficiency (comparing the curvature of the asymptotic power functions at the null value) has been considered for this growth curve model by Woolson and Sen (1974), and, in the light of this, some studies have also been made on the impact of an erroneous choice of r in (5.2). In passing, we may also remark that in a majority of practical problems, polynomial equations for (5.2) work out well, and, in this setup, a quadratic or cubic function generally performs well. Thus, we may be able to use, in many cases, a value of r as small as 2 or 3, and this may lead to substantial gain in efficiency when p is not very small. For some specific problems of this type, we may refer to Ghosh, Grizzle and Sen (1973), Sen (1973) and Koziol et al. (1981), among others.

Let us now proceed on to a sub-hypothesis testing problem in a growth curve model. Here, the rank based tests may not be distribution-free, even permutationally (conditionally), so that we may have to be satisfied with asymptotically distribution-free tests. In the setup of (5.5), consider now the partition $\boldsymbol{\theta}^* = (\boldsymbol{\theta}_1^*, \boldsymbol{\theta}_2^*)$ where $\boldsymbol{\theta}_j^*$ is of the order $r \times s_j$, $j = 1, 2$, $s_1 \geq 0$, $s_2 \geq 0$ and $s_1 + s_2 =$

$q-1$. Suppose that we want to test for

$$H_0\colon\ \boldsymbol{\theta}_1^* = \mathbf{0} \quad \text{vs.} \quad H_1\colon\ \boldsymbol{\theta}_1^* \neq \mathbf{0}\,, \tag{5.7}$$

where $\boldsymbol{\theta}_0$ and $\boldsymbol{\theta}_2^*$ are treated as nuisance parameters. In this testing problem, we may use the theory of aligned rank tests, developed in Sen and Puri (1977), along with the data reduction (from $\boldsymbol{Y}$ to $\boldsymbol{Z}$) made earlier. Note that under H_0 in (5.5), we have from (5.5),

$$\boldsymbol{Z} = \boldsymbol{\theta}_0\mathbf{1}_n' + \boldsymbol{\theta}_2^*\boldsymbol{X}_2^* + \boldsymbol{e}^* \quad \text{where } \boldsymbol{X}^{*\prime} = (\boldsymbol{X}_1^{*\prime}, \boldsymbol{X}_2^{*\prime})\,. \tag{5.8}$$

Since the ranks are shift invariant, we may not have to be worried with the nuisance parameter $\boldsymbol{\theta}_0$, but, we need to estimate $\boldsymbol{\theta}_2^*$ and align the observations through the estimate before using ranks. This is what is called an *aligned ranking procedure*. To estimate $\boldsymbol{\theta}_2^*$, we use, as in Jurečková (1971), suitable rank statistics, and obtain the estimators by similar alignment schemes. Let $\boldsymbol{B}$ be an $r \times s_2$ matrix of real elements (which we may allow to vary over the space E^{rs_2}), and for every $\boldsymbol{B}$, define $\boldsymbol{Z}(\boldsymbol{B}) = \boldsymbol{Z} - \boldsymbol{B}\boldsymbol{X}_2^*$. On the $r \times n$ residual matrix $\boldsymbol{Z}(\boldsymbol{B})$, we adopt the coordinatewise ranking scheme (as in after (5.6)), denote this rank collection matrix by $\boldsymbol{R}_n(\boldsymbol{B})$ and the corresponding matrix of linear rank statistics (based on the regressor $\boldsymbol{X}_2^*$ and $\boldsymbol{Z}(\boldsymbol{B})$), of order $r \times s_2$, by $\boldsymbol{T}_{n2}(\boldsymbol{B})$. This is defined for all $\boldsymbol{B} \in E^{rs_2}$. Note that $\boldsymbol{B} = \boldsymbol{\theta}_2^*$, the coordinates of $\boldsymbol{T}_{n2}(\boldsymbol{\theta}_2^*)$ all have the mean 0, and hence, we may estimate $\boldsymbol{\theta}_2^*$ by equating $\boldsymbol{T}_{n2}(\boldsymbol{B})$ to $\mathbf{0}$. In view of the fact that $\boldsymbol{T}_{n2}(\boldsymbol{B})$ may not have the exact value $\mathbf{0}$ admissible (note that it has elements having discrete distributions), we define

$$B_n = \{\boldsymbol{B}\colon \|\boldsymbol{T}_{n2}(\boldsymbol{B})\| = \text{minimum}\}\,, \tag{5.9}$$

where $\|\cdot\|$ stands for the maximum norm. It follows from Jurečková (1971) that under suitable regularity conditions, the set B_n is a closed convex set with a maximum diameter converging to 0, in probability, as $n \to \infty$. We take the center of gravity of set B_n as our estimate of $\boldsymbol{\theta}_2^*$, and denote this estimator by $\hat{\boldsymbol{\theta}}_2^*$. Note that $\hat{\boldsymbol{\theta}}_2^*$ is the so-called R-estimator of $\boldsymbol{\theta}_2^*$, (see Chapter 21 (by Jurečková)), and as in Chapter 12 (Aubuchon and Hettmansperger) and 11 (Adichie), we employ this R-estimator for testing H_0 in (5.7). For this purpose, we define

$$\hat{\boldsymbol{Z}} = \boldsymbol{Z} - \hat{\boldsymbol{\theta}}_2^*\boldsymbol{X}_2^* = (\hat{\boldsymbol{Z}}_1, \ldots, \hat{\boldsymbol{Z}}_n)\,. \tag{5.10}$$

Let $\hat{\boldsymbol{R}}_n$ be the $r \times n$ matrix of ranks of $\hat{\boldsymbol{Z}}$, where for each row, a separate ranking of the elements is made. Now, in (4.5), we replace the $\boldsymbol{c}_i$ by the column vectors of $\boldsymbol{X}_1^*$ and the matrix $\boldsymbol{R}_n$ by $\hat{\boldsymbol{R}}_n$; we denote the resulting matrix (of order $r \times s_1$) of aligned linear rank statistics by $\hat{\boldsymbol{T}}_{n1}$. Further, in (4.10), we replace the $\boldsymbol{c}_i$ by the columns of $\boldsymbol{X}^*$ (and similarly for $\bar{\boldsymbol{c}}_n$), and denote the resulting $(q-1) \times (q-1)$ matrix by $\boldsymbol{C}_n$. We partition $\boldsymbol{C}_n$ as

$$C_n = \begin{pmatrix} C_{n11} & C_{n12} \\ C_{n21} & C_{n22} \end{pmatrix}, \quad C_{nij} \text{ of order } s_i \times s_j, \text{ for } i, j = 1, 2 . \tag{5.11}$$

Finally, let

$$C_{n11.2} = C_{n11} - C_{n12} C_{n22}^{-} C_{n21} . \tag{5.12}$$

The aligned rank test statistic for testing H_0 in (5.7) is ten given by

$$\hat{\mathscr{L}}_n = (\bar{\bar{T}}_{n1})'(C_{n11.2} \otimes V_n)^{-}(\bar{\bar{T}}_{n1}) , \tag{5.13}$$

where $\bar{\bar{T}}_{n1}$ is the rolledout form of $\hat{T}_{n1}$ and V_n is the rank covariance matrix in (4.9) based on the $r \times n$ matrix Z. Under the null hypothesis H_0 in (5.7), $\hat{\mathscr{L}}_n$ has asymptotically central chi square distribution with rs_1 degrees of freedom. Hence, an asymptotic test may be made by using the appropriate percentile point of this latter distribution as the critical value of $\hat{\mathscr{L}}_n$. Under suitable sequence of local alternatives, $\hat{\mathscr{L}}_n$ has asymptotically noncentral chi square distribution with rs_1 degrees of freedom and an appropriate noncentrality parameter. Hence, the discussions following (5.6) all pertain to the case of this aligned rank test too.

Finally, we consider the case where for (5.4), X' may not be expressible as $(\mathbf{1}_n, X^{*\prime})$ or in (5.6) or (5.7), we want to test for $\boldsymbol{\theta}$ instead of $\boldsymbol{\theta}^*$. For this model, we may note that the ordinary ranks are invariant under shift and hence will not be useful in testing for $\boldsymbol{\theta}_0$. Thus, to overcome this problem, we may have to use signed rank statistics instead of linear rank statistic, and, this, in turn, demands the extra assumption that the d.f. F^* of e^* in (5.4) is *diagonally symmetric about* $\mathbf{0}$ (i.e., both e^* and $(-1)e^*$ have the same d.f. F^*). This assumption, though less restrictive than the multinormality of F^*, was not needed for the rank tests for (5.6) and (5.7). We may define (a vector of) signed rank statistics as in (2.4) wherein we replace the D_i by the columns of X and for the scores $a_n^+(k)$ in (2.5) we may use the score generating function ϕ_j^+ for the j-th coordinate of Z $(1 \leq j \leq r)$. The definition of the R_{ij}^+ $(1 \leq i \leq n)$ is the same as in (2.4), but restricted to the j-th row of Z, for $j = 1, \ldots, r$. With these adjustments, we are now in a position to use the general theory developed by Adichie (1978) for rank tests for linear hypotheses (based on aligned signed rank statistics), where at the beginning we replace the data set Y by Z. Since this theory has been discussed in Chapter 11 (by Adichie), we refer to the details there. One technical advantage of using the aligned signed rank statistics is that instead of using specifically the R-estimators (viz., (5.10)) for the alignment process, one may also use any other estimator having the 'root n' consistency property. This may make the computations relatively simpler, though at the cost of the more stringent assumption of diagonal symmetry of F^*.

5.2. *Nonparametric* MANOCOVA *for growth curve models*

Whenever G in (5.2) is chosen properly, the theory developed in Section 5.1

works out well. However, there are certain drawbacks of the MANOVA approach in Section 5.1 and some of these can be avoided in the alternative MANOCOVA approach of this section. It has been pointed out by Rao (1965) (see also Khatri (1966)) that the dimension reducing transformation in (5.3) may throw away some information contained in $\boldsymbol{Y}$ through the complementary part of $\boldsymbol{Z}$, which is dropped out in the subsequent analysis. Though these criticisms were mainly aimed at the normal theory analysis, nevertheless, they remain pertinent in the nonparametric case too. By considering this complementary part of $\boldsymbol{Z}$ as a matrix of (stochastic) covariates, information contained in this matrix can be at least partially recovered by using the multivariate ANOCOVA procedures. Since this MANOCOVA procedure based on rank statistics has already been discussed in Section 4, we may take advantage of that and consider parallel procedures for the growth curve models.

In (5.3), side by side, consider another matrix $\boldsymbol{W}$ of order $(p-r)\times n$

$$\boldsymbol{W} = \boldsymbol{H}'\boldsymbol{Y} \quad \text{where } \boldsymbol{H} \text{ is a } p\times(p-r) \text{ matrix of known constants}, \tag{5.14}$$

and, corresponding to a given $\boldsymbol{G}$ in (5.2), we choose $\boldsymbol{H}$ such that

$$\boldsymbol{G}'\boldsymbol{H} = \boldsymbol{0}(r\times(p-r)) \tag{5.15}$$

Note that by (5.1), (5.2), (5.14) and (5.15), we have

$$\begin{pmatrix}\boldsymbol{Z}\\ \boldsymbol{W}\end{pmatrix} = \begin{pmatrix}(\boldsymbol{G}'\boldsymbol{G})^{-1}\boldsymbol{G}'\boldsymbol{Y}\\ \boldsymbol{H}'\boldsymbol{Y}\end{pmatrix} = \begin{pmatrix}\boldsymbol{\theta}\boldsymbol{X}+\boldsymbol{e}^{*}\\ \boldsymbol{e}^{**}\end{pmatrix}, \quad \boldsymbol{e}^{**} = \boldsymbol{H}'\boldsymbol{e}. \tag{5.16}$$

Thus, the marginal ($(p-r)$-variate) distributions of the columns of $\boldsymbol{W}$ are not dependent on $\boldsymbol{\theta}$ and they are i.i.d.r.v.'s too. Hence, $\boldsymbol{W}$ satisfies the basic requirements for being qualified as a matrix of concomitant vectors. The left hand side of (5.16) is $p\times n$ matrix where the first r rows only contain information on $\boldsymbol{\theta}$, while the last $p-r$ rows do not. This makes it possible to use the model in (4.1)–(4.2) where the $\boldsymbol{c}_i$ need to be replaced by the column vectors of $\boldsymbol{X}$, $p+q$ by p, p by r and q by $p-r$. These changes are to be made in (4.3) through (4.12). With these modifications, we replace the test statistics ($\mathscr{L}_n^0$ and $\hat{\mathscr{L}}_n$) in Section 5.1 (analogous to the one in (4.14)) by the corresponding ones obtained from (4.16), i.e., we compute the same statistics for the entire set and then for the set of covariates only, and finally, take their differences. By virtue of the discussions made after (4.14) (including the inequality in (4.15)), it becomes clear that the incorporation of the ranks of $\boldsymbol{W}$ leads to a generally larger noncentrality parameter (without changing the degrees of freedom) for the test statistic (over the MANOVA form in Section 5.1), and this accounts for the increased efficiency of the MANOCOVA approach over the corresponding MANOVA approach. We conclude this section with the following remarks. For the results in both Sections 5.1 and 5.2, the matrix $\boldsymbol{G}$ plays a vital role. As in

Potthoff and Roy (1964), one may also choose a symmetric and positive definite matrix $\boldsymbol{Q}$ (of order $p \times p$) and define $\boldsymbol{Z} = (\boldsymbol{G}'\boldsymbol{Q}^{-1}\boldsymbol{G})^{-1}\boldsymbol{G}'\boldsymbol{Q}^{-1}\boldsymbol{Y}$. Then, by (5.1) and (5.2), we would have $\boldsymbol{Z} = \boldsymbol{\theta X} + (\boldsymbol{G}'\boldsymbol{Q}^{-1}\boldsymbol{G})^{-1}\boldsymbol{G}'\boldsymbol{Q}^{-1}\boldsymbol{e}$. The solutions in the normal theory case as well as in the nonparametric case all work out for any positive definite $\boldsymbol{Q}$, and hence, one may like to know whether there is any optimal choice of $\boldsymbol{Q}$ in this context. In the normal theory case, from the point of view of best linear unbiased estimation of a linear (estimable) function of $\boldsymbol{\theta}$, Potthoff and Roy (1964) showed that for $\boldsymbol{Q}$ the optimal choice is the true covariance matrix of the $\boldsymbol{e}_i$. Since this covariance matrix is unknown, they suggested the use of the sample mean (residual) product matrix, and as it turns out that this choice is isomorphic to the alternatives suggested by Rao (1965) and Khatri (1966) from the ANOCOVA point of view. The situation is different in the nonparametric case. First, in the normal theory case, the residual sum of product matrix is stochastically independent of the estimates of linear combinations of $\boldsymbol{\theta}$. This result is not generally true for the nonnormal case. Secondly, the normal theory estimates are linear ones, while the rank based estimates are nonlinear functions of $\boldsymbol{Y}$. Further, the invariance of the normal theory estimates, under nonsingular transformations on the observation vectors, may not apply to the rank statistics and the derived estimates, where a coordinatewise ranking is made. For different choices of $\boldsymbol{Q}$, we may not have therefore the desired linear relations, and hence, locating an optimal $\boldsymbol{Q}$ may be a harder problem in the theory of nonlinear programming. However, these are, to some extent, only pathological points: In most of the practical applications, the $\boldsymbol{G}$ matrix in (5.2) may be selected in a very natural way and the specific choice of $\boldsymbol{Q} = \boldsymbol{I}$, made in this section, would work out well. We may also note that the form of $\boldsymbol{G}$ becomes quite simple in many longitudinal studies where the time points at which the repeated measurements are made are the same for each unit or individual. However, in practice, missing observations are not uncommon. These missing values may occur systematically (which is easier to handle) or may occur haphazardly. The patients may drop off at some point of time, so that the observations at time points beyond that point would be missing. Alternatively, the patients may not show up on some specific dates, thus skipping some appointments, which results in missing observations at random points of times. In the parametric case, to accommodate such missing patterns, more general linear models (MGLM) have been introduced. The analysis schemes for such MGLM's are generally complicated: Kleinbaum (1973) suggested the use of BAN (best asymptotically normal) estimates for such an analysis. However, the exactness of the solutions are affected and only asymptotic solutions are available. Nonparametric procedures for such MGLM's have not been developed to the full generality, and, more research on this line is needed to bring it down to the practical users' level. Finally, the solutions to the nonparametric growth curve analysis considered in this section relate specifically to the multivariate one-way layout designs. Sen (1973) has shown that similar solutions can be worked out in some higher order factorial designs. For complete block designs, ANOCOVA rank procedures are available in the literature and these would provide the desired solutions for the corresponding growth curve analysis based on rank statistics (after a suitable dimensional reduction transformation is used). In principle, the methodological approach is the same, and hence, these details will not be considered here. Growth curve models may also be quite appropriate for many clinical trials: Here, one may have some additional complications due to censoring of various types. Since rank analysis of covariance procedures have also been developed for such censored data (see Section 4), modifications in the scheme for growth curve analysis due to censoring may be formulated in the same manner.

6. Nonparametric methods in bio-assays

In a biological assay, typically, a new (test) preparation and an old (standard) one are compared by means of the reactions that follow their applications to living matter, and, the *relative potency* of the test preparation with respect to the standard one is the main item under investigation. Thus, in a typical case, a *stimulus* is applied to a *subject* and this induces some change in some measurable characteristic of the subject (termed, *response*). The magnitude of this response may depend on the intensity of the stimulus (termed *dose*). The relationship between the dose and the response may not be exact and is subject to random variation either in the application of the doses or in the manifestation of the responses or both. We shall confine ourselves mainly to *quantitative assays* which may be classified into three broad types: (i) Direct assays, (ii) indirect assays with quantitative responses, and (iii) indirect assays with quantal (i.e., all or nothing) responses. In a direct assay, the dose of a given preparation needed to produce a specified response is recorded, so that the response is certain while the dose is a random variable. In an indirect assay, for given levels of doses, the responses are measured. Thus, the response is a random variable whose distribution may well depend on the dose. In an indirect assay with quantitative responses, the response variables are quantitative, while in a quantal one, the response is of the type all or nothing. For example, if several doses of a blood pressure reducing drug are used in an assay where for each dose, on several subjects, the changes in the blood pressure are recorded after a week of application of the drug, then it will be an indirect assay of the quantitative type. On the other hand, if some toxic drug is used in several doses, and for each dose, a set of animals are given the preparation following which the number of animals killed are recorded, we have an indirect assay of the quantal (death or no death) type. Alternatively, if a toxic preparation is injected into the blood stream of a cat until the heart stops beating, then the dose is random in nature while the response (death) is not, so that we have a direct assay. In this section, we shall mainly review the progress of nonparametric methods in direct assays and indirect assays with quantitative responses.

First, consider a typical direct assay involving a standard preparation (S) and a test preparation (T). Let there be m subjects on which the standard preparation has been administered, and let $X_1, \ldots, X_m$ be the respective doses to yield a common response. These doses are then nonnegative random variables, and we assume that $X_1, \ldots, X_m$ are independent and identically distributed random variables (i.i.d.r.v.) with a continuous distribution function (d.f.) $F_S(x)$, $x \geq 0$ (where $F_S(0) = 0$). Similarly, let $Y_1, \ldots, Y_n$ be the doses for the n subjects on which the test preparation has been administered; these are assumed to be i.i.d.r.v. with a continuous d.f. $F_T(x)$, $x \geq 0$ (where $F_T(0) = 0$). In a typical direct assay model, one assumes that the test preparation behaves as if it is a *dilution* (or concentration) of the standard one. In statistical terms, this may be represented by

$$F_T(x) = F_S(\rho x), \quad 0 \leq x < \infty, \quad \text{where } \rho > 0 . \tag{6.1}$$

The positive constant ρ is termed the relative potency of the test preparation with respect to the standard one. Also, the model in (6.1) is termed the *fundamental assumption* of a direct (-dilution) assay. Our main interest lies in estimating the relative potency ρ and verifying this fundamental assumption. Parametric procedures for these inference problems are usually based on the assumption that F_T is normal or lognormal (or sometimes, logistic or log-logistic). These procedures are discussed in detail in Finney (1964). The form of the estimator of ρ depends explicitly on the assumed form of the d.f. F_T, and these parametric estimates are generally not very robust against departures from the assumed form of the tolerance d.f. For example, if we assume that F_T is normal (only justified if the standardized mean is sufficiently large so that $F_T(0)$ is very small), then the estimate of ρ comes out as the ratio of the sample means for the two preparations, while, if F_T is taken as the log-normal d.f., then the estimator is the ratio of the two geometric means, and these are generally not the same. Rank based estimates of relative potency have been considered by Sen (1963), Shorack (1966) and Rao and Littell (1976), among others. These estimates are invariant under the choice of any monotone transformation on the dose (called *dosage*), and, besides being robust, are generally quite efficient for normal, log-nominal, logistic or other common forms of the tolerance distributions.

For convenience of description, we choose the dosage as equal to log-dose. Let $X_i^* = \log X_i$, $i = 1, \ldots, m$, be the dosages for the standard preparation and let $F_S^*(x)$ be the d.f. of the X_i^*. Then,

$$F_S^*(x) = P\{X_i^* \leq x\} = P\{X_i \leq e^x\} = F_S(e^x) \quad \text{for } x \in (-\infty, \infty) .$$

Similarly, let $Y_i^* = \log Y_i$, $i = 1, \ldots, n$, be the dosages for the test preparation and $F_T^*(x) = P\{Y_i^* \leq x\} = F_T(e^x)$ be the corresponding d.f. Then, by (6.1), we conclude that for the dilution-model,

$$F_T^*(x) = F_T(e^x) = F_S(\rho\, e^x) = F_S(e^{x + \log \rho}) = F_S^*(x + \log \rho) , \tag{6.2}$$

for every $x \in (-\infty, \infty)$. Thus, the two d.f.'s F_S^* and F_T^* differ only by a shift $\Delta = \log \rho$, and for this shift model, efficient rank based estimators have already been considered in Chapter 21 (by Jurečková). As in Section 5, we consider the two-sample linear rank statistic $T_N(b) = \Sigma_{i=1}^m a_N(R_{Ni}(b))$, $b \in (-\infty, \infty)$, where $N = m + n$, $a_N(1), \ldots, a_N(N)$ are monotone ($\nearrow$) scores and $R_{Ni}(b)$ is the rank of $X_i^* - b$ among $X_1^* - b, \ldots, X_m^* - b$, $Y_1^*, \ldots, Y_n^*$, for $i = 1, \ldots, m$. Then, $T_N(b)$ is $\searrow$ in b, and we define an R-estimator $\hat{\Delta}_{N(R)}$ of Δ by

$$\hat{\Delta}_{N(R)} = (\sup\{b\colon T_N(b) > \bar{a}_N\} + \inf\{b\colon T_N(b) < \bar{a}_N\})/2\,, \tag{6.3}$$

where $\bar{a}_N = N^{-1}\Sigma_{i=1}^N a_N(i)$, and, we may set, without any loss of generality, $\bar{a}_N = 0$. In particular, if we choose the two-sample Wilcoxon–Mann–Whitney statistic, i.e., $a_N(k) = (k - (N+1)/2)/(N+1)$, $k = 1, \ldots, N$, then, (6.3) simplifies to the median of the mn differences $X_i^* - Y_j^*$, $1 \le i \le m$, $1 \le j \le n$. This simple estimator has some optimal properties when F_S^* is a logistic d.f. Even otherwise, it is a very simple, robust and efficient estimator of Δ; the estimator of ρ can simply be obtained by taking the anti-log of this estimator. For scores other than the ones relating to the median and Wilcoxon statistics (worked out by Sen, 1963), an explicit solution for (6.3) may not be available. But, starting with the Wilcoxon scores estimator, an iteration procedure may be employed and a few iteration steps should yield the desired estimator upto a given level of accuracy. Let us next consider the confidence intervals for the relative potency. In this respect, the lack of robustness with the parametric procedure is more noticeable. Attainment of the specified coverage probability by the prescribed confidence interval, in the parametric case, may be seriously affected by an incorrect assumption on the form of F_S (or F_S^*). The picture is quite different for the nonparametric case. Distribution-free confidence interval for Δ can again be obtained by using linear rank statistics, and this, in turn, provides the confidence interval for the relative potency ρ. Note that when ρ obtains, $T_N(\Delta)$ has the same distribution as of $T_N(0)$ under $H_0\colon \Delta = 0$, and the latter is independent of the underlying F_S^*. Hence, we can always find two constants $t_N^{(1)}$ and $t_N^{(2)}$, depending on m, n, α and the scores $a_N(1), \ldots, a_N(N)$, such that

$$P\{T_N(0) \in [t_N^{(1)}, t_N^{(2)}] \mid H_0\} = 1 - \alpha_N \ge 1 - \alpha\,,$$

where α_N is a known function of N and it converges to α, as $N \to \infty$. Let then

$$\hat{\Delta}_{N,L} = \sup\{b;\ T_N(b) > t_N^{(1)}\}\,, \tag{6.4}$$

$$\hat{\Delta}_{N,U} = \inf\{b\colon T_N(b) < t_N^{(2)}\}\,, \tag{6.5}$$

and

$$I_N = [\hat{\Delta}_{N,L}, \hat{\Delta}_{N,U}]\,. \tag{6.6}$$

The desired confidence interval for Δ is given by I_N in (6.6), and the confidence interval for ρ is obtained from (6.6) by replacing $\hat{\Delta}_{N,L}$ and $\hat{\Delta}_{N,U}$ by their anti-logs. If we use the Wilcoxon scores, these confidence limits can be expressed in terms of appropriate sample quantiles of the differences $X_i^* - Y_j^*$, $1 \leq i \leq m$, $1 \leq j \leq n$, and for the median scores, in terms of similar quantities. We may refer to Shorack (1966) for some graphical procedures for obtaining these estimates (both the point and interval ones) for such simple scores. For scores other than these two, again, an iterative solution to (6.4)–(6.5) is generally needed. These confidence intervals are robust, distribution-free and generally quite efficient for the common forms of tolerance d.f.'s encountered in practice.

Rao and Littell (1976) have advocated the use of the two-sample Kolmogorov–Smirnov statistic (instead of the two-sample linear rank statistic $T_N(b)$), and obtained (point as well as interval) estimators by similar alignment procedures. These estimates are somewhat more complicated and may not be fully efficient for the common types of tolerance d.f.'s.

With respect to the basic model in (6.1), a nonparametric test for a null hypothesis H_0: $\rho = \rho_0$, ρ_0 (>0) specified, can easily be constructed by considering the two-sample rank statistic $T_N(\Delta_0)$, where $\Delta_0 = \log \rho_0$. Note that under H_0, $T_N(\Delta_0)$ has the same distribution as of $T_N(0)$ under the hypothesis that $\Delta = 0$, where the latter does not depend on the underlying d.f. F_S. Hence, the two-sample rank tests, discussed in detail in Chapter 2 (by Bhapkar), as adapted to $T_N(\Delta_0)$, remain applicable in this context. A more pertinent question in this context is the validity of the fundamental assumption in (6.1). With respect to the log-dose transformation, the problem reduces to that of testing for the null hypothesis that the two d.f.'s F_S^* and F_T^* differ only by some unknown shift. This has been treated in Sen (1964), and we present an updated account of the same.

These tests are not, in general, distribution-free, but are asymptotically distribution-free under quite general regularity conditions. First, if we assume that the two d.f.'s F_S^* and F_T^* are both symmetric (around their respective medians), then, we may use some standard aligned rank test for scale when locations are not necessarily the same. Let $\tilde{X}_m^*$ and $\tilde{Y}_n^*$ be respectively the sample medians of the X_i^* ($1 \leq i \leq m$) and Y_j^* ($1 \leq j \leq n$). Let then

$$\hat{X}_i^* = X_i^* - \tilde{X}_m^*, \quad i = 1, \ldots, m, \qquad \hat{Y}_j^* = Y_j^* - \tilde{Y}_n^*, \quad i = 1, \ldots, n. \tag{6.7}$$

Also, let $\hat{R}_{Ni}$ be the rank of $\hat{X}_i^*$ among the N aligned observations in (6.7), for $i = 1, \ldots, m$. Consider then a two-sample rank statistic $\hat{T}_N^* = m^{-1} \sum_{i=1}^m a_N^*(\hat{R}_{Ni})$, where the scores are symmetric in the sense that $a_N^*(k) = a_N^*(N - k + 1)$, for every k ($\leq N$), as is typically the case with rank tests for scale. Then, using the linearity results of Jurečková (1969), it follows that whatever be the unknown ρ, under (6.1), $(Nm/n)^{1/2}(\hat{T}_N^* - \bar{a}_N^*)/A_N^*$ has asymptotically normal distribution with 0 mean and unit variance, where $\bar{a}_N^* = N^{-1} \sum_{i=1}^N a_N^*(i)$ and $A_N^{*2} = (N-1)^{-1} \sum_{i=1}^N (a_N^*(i) - \bar{a}_N^*)^2$. Hence, $\hat{T}_N^*$ is asymptotically distribution-free, under (6.1), and an asymptotic test for the fundamental assumption in (6.1) can be based on $\hat{T}_N^*$, using its asymptotic normality. The assumption that F_S^* in (6.2) is a symmetric d.f. may not always be very realistic in bio-assays, where the tolerance d.f.'s may sometimes be so skewed that even the log-dose transformation may not render symmetry to the transformed d.f. In such a case, use of the aligned rank test (as in above) may not be very suitable on the ground of robustness (against asymmetry of the d.f.'s). We may, however, use some alternative nonparametric tests which do not require the symmetry of the d.f.'s; these tests may not be otherwise full efficient for some specific models. Thus, we may have to make a compromise between high efficiency for specific alternatives and validity for a broad class of tolerance d.f.'s.

Let $\phi(a, b; c, d)$ be equal to $\text{sign}(|a-b|-|c-d|)$, and consider the two-sample (generalized) U-statistic

$$U_N = \binom{m}{2}^{-1}\binom{n}{2}^{-1} \sum_{1\leq i<j\leq m} \sum_{1\leq r<s\leq n} \phi(X_i^*, X_j^*; Y_r^*, Y_s^*), \tag{6.8}$$

which has been considered by Lehmann (1951) as a suitable nonparametric test statistic (for scale alternatives). Though U_N unbiasedly estimates 0 when (6.2) holds, it is not a genuine rank statistic, nor it has a distribution independent of the underlying d.f.'s when (6.2) holds. Hence, an exact nonparametric test based on U_N may not be worked out. However, jackknife estimator of the variance of U_N has been worked out by Sen (1960), and using this estimator, one can construct a studentized statistic which has asymptotically (under (6.2)) normal distribution with 0 mean and unit variance, even when F_S^* is not symmetric. This provides an asymptotic test for the validity of the fundamental assumption in (6.1) without necessarily assuming that F_S^* is symmetric. Use of the Mood's two sample scale test based on the residuals in (6.7) may also be recommended on the same ground. However, the Lehmann test may perform better in the majority of cases.

Note that the model (6.2) relates to the hypothesis that the two d.f.'s differ only in locations. If we are in a position to assume that F_S^* is a symmetric d.f., then an aligned Kolmogorov–Smirnov type test (viz., Sen, 1984c) may be used. Let $\tilde{X}_m^*$ and $\tilde{Y}_n^*$ be respectively the sample medians of the X_i^* and the Y_j^*. Then define $\hat{X}_{mi}^{*+} = |X_i - \tilde{X}_m^*|$, $i = 1, \ldots, m$, and let $F_{Sm}^*(x) = m^{-1}\sum_{i=1}^m I(\hat{X}_{mi}^{*+} \leq x)$, $x \geq 0$, be the corresponding empirical d.f. Similarly, let $\hat{Y}_{ni}^{*+} = |Y_i - \tilde{Y}_n^*|$, $i = 1, \ldots, n$, and $F_{Tn}^*(x) = n^{-1}\sum_{i=1}^n I(\hat{Y}_{ni}^{*+} \leq x)$, $x \geq 0$, be the empirical d.f. for the test preparation. Consider the usual Kolmogorov–Smirnov statistic for these aligned observations:

$$K_{mn}^* = (mn/N)^{1/2} \sup\{|F_{Sm}^*(x) - F_{Tn}^*(x)|: x \geq 0\}. \tag{6.9}$$

Then, under (6.2), whatever be the value of ρ, when F_S^* is symmetric and has a continuous density function almost everywhere, we have, for every $d \geq 0$,

$$\lim_{N\to\infty} P\{K_{nm}^* \geq d \mid H_0^*\} = 2\sum_{r=1}^{\infty} (-1)^{r-1} \exp(-2r^2d^2), \tag{6.10}$$

where H_0^* relates to the model in (6.2). Hence, an asymptotically distribution-free test, with critical value obtained from (6.10), can be based on K_{mn}^* in (6.9). This test, like the Kolmogorov–Smirnov test, is consistent against a broad class of alternatives where (6.2) does not hold.

For more than one dilution direct assays (relating to the same standard and test preparations) conducted under varying conditions, there also remains the question of a desirable way of combining the estimates of the relative potency from the different assays. This problem, in a nonparametric setup, has been studied by Sen (1965) where a linear combination of the individual assay linear rank statistics has been used (in place of (6.3)) to derive a robust and efficient rank based estimator of the common value of the relative potency. Related efficiency results are also studied there.

Let us next consider the case of indirect quantitative assays. In an indirect quantitative assay, specified *doses* are given, each to several subjects, and their *responses* are recorded. The response is quantitative in nature. For a dose z, the d.f. of the response $U(z)$ is denoted by $F_z(u)$. An average response $\mu(z)$ (such as the median, mean or some other measure of the central tendency of F_z) expressed as a function of z is known as the *dose-response regression.* In practice, often, *log-dose transformation* along with a suitable *response-metameter* yield a linearized dosage-response regression $Y = \alpha + \beta x + e$ where e represents the chance variation component (with a d.f. $G(e)$), x the dosage and (α, β) represents the vector of unknown parameters. Parametric procedures for the estimation of these parameters and for testing suitable hypotheses are discussed in detail in Finney (1964). Here also, these parametric procedures may be attacked on the ground of lack of robustness, and, we discuss here some alternative nonparametric procedures which are robust and efficient for broad class of d.f.

Suppose that we have a standard and a test preparation with respective (linearized) dosage-response regressions

$$Y_S = \alpha_S + \beta_S x + e_S \quad \text{and} \quad Y_T = \alpha_T + \beta_T x + e_T, \tag{6.11}$$

where the errors e_S and e_T both have the common (unknown) d.f. G. If the test preparation behaves as a dilution (or concentration) of the standard one, we have then

$$\beta_S = \beta_T = \beta \text{ (unknown)} \quad \text{and} \quad \alpha_T - \alpha_S = \beta \log \rho, \tag{6.12}$$

where ρ (>0) is the *relative potency* of the test preparation with respect to the standard one, and, the equality of the regression coefficients constitutes the *fundamental assumption* of this *parallel line assay*. We consider some non-parametric tests for the validity of the fundamental assumption and some nonparametric estimates of the relative potency. These were studied earlier by Sen (1971). Consider a symmetrical $2k$-point design (for some $k \geqslant 2$) with k doses of each preparation such that the successive doses bear a constant ratio

D (>0) to one-another, and n_0 (≥ 1) subjects are used for each dose. For the standard preparation, the k doses are denoted by $Z_{1j} = aD^{j-1}$, $a > 0$, for $j = 1, \ldots, k$, while, for the test preparation, these are abD^{j-1}, $b > 0$, for $j = 1, \ldots, k$. Note that each preparation is administered to $kn_0 = n$ subjects. The dosages for the standard and test preparations are

$$x_{1j} = \log_D Z_{1j} = (j-1) + \log_D a, \qquad x_{2j} = \log_D Z_{2j} = (j-1) + \log_D(ab), \tag{6.13}$$

for $j = 1, \ldots, k$. If we write $x_j = (j - (k+1)/2)$, $j = 1, \ldots, k$, then, we may rewrite (6.11) as

$$Y_S = \alpha_S^* + \beta_S x_j + e_S \quad \text{and} \quad Y_T = \alpha_T^* + \beta_T x_j + e_T, \tag{6.14}$$

where

$$\alpha_S^* = \alpha_S + \beta_S[\log_D a + (k-1)/2],$$

$$\alpha_T^* = \alpha_T + \beta_T[\log_D ab + (k-1)/2]. \tag{6.15}$$

With this change in the scale and origin of the dosage, we thus obtain the following two sets of responses:

	Standard Preparation				Test Preparation			
Dosage	x_1	x_2	$\cdots$	x_k	x_1	x_2	$\cdots$	x_k
	$Y_{11}^{(1)}$	$Y_{21}^{(1)}$	$\cdots$	$Y_{k1}^{(1)}$	$Y_{11}^{(2)}$	$Y_{21}^{(2)}$	$\cdots$	$Y_{k1}^{(2)}$
	$\cdots$	$\cdots$	$\cdots$	$\cdots$	$\cdots$	$\cdots$	$\cdots$	$\cdots$
	$Y_{1n_0}^{(1)}$	$Y_{2n_0}^{(1)}$	$\cdots$	$Y_{kn_0}^{(1)}$	$Y_{1n_0}^{(2)}$	$Y_{2n_0}^{(2)}$	$\cdots$	$Y_{kn_0}^{(2)}$

First, to test the validity of the fundamental assumption i.e., the parallelism of the two regression lines in (6.14), we proceed as in Sen (1971) and define the set of divided differences as follows: Let

$$W_{lj,rs}^{(i)} = (Y_{ls}^{(i)} - Y_{jr}^{(i)})/(l-j), \quad r, s = 1, \ldots, n_0, \ 1 \leq j < l \leq k, \quad i = 1, 2. \tag{6.16}$$

Then, as in Sen (1968), the median of these $k(k-1)n_0^2$ entries is the pooled sample estimator of the hypothesized common value β, and is denoted by β_n^*. Let then

$$U_n^{(i)}(\beta_n^*) = \sum_{1 \leq j < l \leq k} \sum_{r=1}^{n_0} \sum_{s=1}^{n_0} \operatorname{sign}(Y_{ls}^{(i)} - Y_{jr}^{(i)} - \beta_n^*(x_l - x_j)), \quad i = 1, 2. \tag{6.17}$$

Also, let

$$V_n = kn_0[(kn_0 - 1)(2kn_0 + 5) - (n_0 - 1)(2n_0 + 5)]/18. \tag{6.18}$$

Then, the test statistic is

$$S_n = \{(U_n^{(1)}(\beta_n^*))^2 + (U_n^{(2)}(\beta_n^*))^2\}/V_n . \tag{6.19}$$

Under the null hypothesis of the equality of β_S and β_T, S_n has closely the chi square distribution with 1 degree of freedom (viz., Sen, 1971) and the test is quite robust in character. Instead of using this simple test statistic based on the aligned Kendall tau statistics, one may consider alternative tests based on general linear rank statistics aligned in a similar manner; these tests are worked out in detail in Sen (1969) and are easily adaptable here.

To estimate the relative potency ρ, we note that when $\beta_S = \beta_T = \beta$,

$$\log_D(b\rho) = (\alpha_T^* - \alpha_S^*)/\beta = \delta/\beta, \quad \text{say}, \tag{6.20}$$

say, where D and b are specified positive constants. As an estimator of β, we use β_n^*, defined after (6.16). Let then

$$\hat{Y}_{jr}^{(i)} = Y_{jr}^{(i)} - \beta_n^* x_j \quad \text{for } r = 1, \ldots, n_0,\ j = 1, \ldots, k, \qquad i = 1, 2. \tag{6.21}$$

Then, we use the following estimator of δ:

$$\delta_n = \text{median of the } n^2 \text{ differences } \hat{Y}_{ls}^{(2)} - \hat{Y}_{jr}^{(1)}:$$

$$1 \leq r \leq n_0, \quad 1 \leq s \leq n_0,\ 1 \leq j \leq k,\ 1 \leq l \leq k. \tag{6.22}$$

Finally, the estimator of ρ is defined by the solution:

$$\log_D(b\rho_n^*) = \delta_n^*/\beta_n^* . \tag{6.23}$$

Here also, instead of using the simple Wilcoxon scores estimator in (6.22) one may use a general R-estimator based on the aligned observations in (6.21).

Asymptotic properties of the estimator ρ_n^* have been studied by Sen (1971). In the above development, for simplicity, we have considered a symmetric $2k$-point design. The picture becomes little more computationally involved when the number of subjects for each dosage is not the same or the design is not a symmetric one. However, looking at (6.11), we may note that essentially the problem is to test for parallelism of two regression lones and to estimate the relative potency by a ratio formula as in (6.12). Hence, the R-estimators for linear models, see for example Chapter 11 (by Adichie), are generally adaptable for this specific problem. In planned bio-assays, of course, one can adopt a symmetric design, and, the solutions considered here remain useful.

So far, we have considered the case where the dosage is given by log-dose. Often, the dosage is taken as $(\text{dose})^\lambda$, for some $\lambda > 0$. In such a case, referred to the dosage-response regressions in (6.11), we have

$$\alpha_S = \alpha_T = \alpha \text{ (unknown)} \quad \text{and} \quad \beta_T = \rho^\lambda \beta_S . \tag{6.24}$$

Since the relative potency ρ is expressible in terms of the ratio of the two slopes β_T and β_S, the assay is termed a *slope-ratio assay*, and the equality of the two intercepts α_S and α_T constitutes the *fundamental assumption* of the slope-ratio assay.

First, we consider a nonparametric test for the validity of the fundamental assumption. For each i (=1, 2), consider the $\binom{k}{2}n_0^2$ divided differences in (6.16), and the median of these is taken as the point estimator of the respective slope. We denote these estimates by β^*_{Sn} and β^*_{Tn}, respectively. Consider then the residuals, defined as in (6.21), where β^*_n is replaced by β^*_{Sn} (for $i = 1$) and β^*_{Tn} (for $i = 2$). Then, for each i (=1, 2), consider the $\binom{kn_0+1}{2}$ midranges

$$(\hat{Y}^{(i)}_{jr} + \hat{Y}^{(i)}_{ls})/2, \quad 1 \leq j < l \leq k, \; 1 \leq r, s \leq n_0 \quad \text{and}$$

$$1 \leq j = l \leq k, \; 1 \leq r < s \leq n_0 .$$

The median of these $2\binom{kn_0+1}{2}$ midranges (for the combined sample) is denoted by α^*_n and is taken as the pooled sample estimator of the hypothesized common value α. Subtracting α^*_n further from the $\hat{Y}^{(i)}_{jr}$, we denote the ultimate residuals as $\bar{Y}^{(i)}_{jr}$, for $r = 1, \ldots, n_0$, $j = 1, \ldots, k$ and $i = 1, 2$. Let $\bar{W}_{nS}$ and $\bar{W}_{nT}$ be respectively the Wilcoxon signed rank statistics based on the ultimate residuals for the standard and the test preparation. Then, as in Sen (1972a), we consider the test statistic

$$Q_N = [3(k-1)/(2k+1)N(N+1)(2N+1)](\bar{\bar{W}}^2_{NS} + \bar{\bar{W}}^2_{nT}) \tag{6.25}$$

where $N = kn_0$. Under the null hypothesis is equal intercepts, Q_N has closely chi square distribution with 1 degree of freedom. It is a robust statistic and the test based on Q_N is asymptotically distribution-free. In the above discussion, we have confined ourselves to a $2k$-point design. An analogous statistic can be worked out for the $(2k+1)$-point design. Also, instead of the Wilcoxon scores procedures, one may use a general rank procedure based on R-estimates of the two regression slopes and a general R-estimator of the intercept. These are worked out in Sen (1972b) and are comparatively more involved. However, in principle, these run on parallel lines, and hence, we omit the details. We proceed on to the estimation of the relative potency. Since in a slope-ratio assay, the constant λ (>0) is specified, we may define $\tilde{\rho} = \rho^\lambda$, and, we may proceed to estimate $\tilde{\rho}$ as well. We define

$$\tilde{Y}^{(i)}_j = \text{median}\{(Y^{(i)}_{jr} + Y^{(i)}_{js})/2 \colon 1 \leq r \leq s \leq n_0\},$$
$$\text{for } j = 1, \ldots, k, \; i = 1, 2 . \tag{6.26}$$

Minimizing the usual sum of squares

$$\sum_{j=1}^{k}\{(\tilde{Y}_j^{(1)}-\alpha-\beta_S x_j)^2+(\tilde{Y}_j^{(2)}-\alpha-\beta_T x_j)^2\} \tag{6.27}$$

with respect to the unknown α, β_S and β_T, we obtain the following estimator

$$\tilde{\rho}_n^* = (a/b)[f_1Q_2^* - f_2Q_1^*]/[f_1Q_1^* - f_2Q_2^*], \tag{6.28}$$

where

$$Q_i^* = k^{-1}\sum_{j=1}^{k} j\tilde{Y}_j^{(i)} - \frac{k+1}{2}\tilde{Y}, \qquad \tilde{Y} = (2k)^{-1}\sum_{i=1}^{2}\sum_{j=1}^{k}\tilde{Y}_j^{(i)}, \tag{6.29}$$

$$f_1 = (k+1)(5k+1)/24k \quad \text{and} \quad f_2 = -(k+1)^2/8k\,; \tag{6.30}$$

a and b are the scale factors appearing in the doses for the standard and test preparation. A similar formula works out well for the $(2k+1)$-point design (viz., Sen, 1972a). Asymptotic properties of the estimator $\tilde{\rho}_n^*$ (of $\tilde{\rho}$) have been studied by Sen (1972a). In this context also, one may use some general R-estimators of within cell locations and use them in (6.27) for deriving the corresponding estimator of $\tilde{\rho}$. The procedure remains the same.

We conclude this section with some remarks on the nonparametric procedures for the indirect quantal assays. In this indirect assay, the response is quantal (i.e., all or nothing) in nature, so that for each preparation and each dose, among the subjects administered, some manifest a certain reaction and the others not. For a given dose z, if $P(z)$ stands for the probability of a response, then, one is interested in knowing the dependence of $P(z)$ on the dose z. In particular, if we assume that $P(z)$ is a monotone function of z, and there exists a unique ξ_α such that

$$P(\xi_\alpha) = \alpha \quad (\text{where } 0<\alpha<1), \tag{6.31}$$

then, ξ_α is called the 100α % effective dose; for $\alpha = \frac{1}{2}$, it is called the *median effective dose.* If the quantal response relates to the death, $\xi_{1/2}$ is termed the *median lethal dose.* Estimation of the median effective dose is the main task in a quantal assay. A detailed account of the parametric theory (based on specific forms of $P(\cdot)$, such as the (log)-normal, logistic etc.) is available with Finney (1964, Chapter 17). There are a few nonparametric estimates not fully explored, and, we shall comment briefly on them.

Let $x_1, \ldots, x_k$ be the k ($\geq$2) doses which we assume to be equally spaced (either on the linear or logarithmic scale). Suppose that each dose is administered to n ($\geq$1) subjects. Let U_{ij} be the response of the j-th subject at the dose level x_i, for $j = 1, \ldots, n$ and $i = 1, \ldots, k$. The U_{ij}, $1 \leq j \leq n$ are i.i.d.r.v. having a binomial distribution $P(U_{ij}=0) = 1 - P(U_{ij}=1) = 1-\pi_i$, for $i = 1, \ldots, k$. Note that $\bar{U}_i = n^{-1}\sum_{j=1}^{n} U_{ij}$ is an unbiased estimator of π_i, for $i = 1, \ldots, k$. We write

$$\pi_i = \Pi(x_i), \quad i = 1, \ldots, k, \tag{6.32}$$

where Π is the unknown tolerance d.f., and, in a nonparametric setup, excepting possibly the symmetry of Π, nothing is assumed about this d.f. Among the ad hoc estimators, we may mention the following:

(i) The *Spearman–Kärber estimator.* If d (>0) be the common inter-dose spacing, then, the estimator is

$$\xi_{1/2}^{(1)} = x_k + d/2 - d(\bar{U}_1 + \cdots + \bar{U}_k), \tag{6.33}$$

which is a linear function of the individual averages $\bar{U}_1, \ldots, \bar{U}_k$, and hence,

$$E\hat{\xi}_{1/2}^{(1)} = x_k + d/2 - d\sum_{j=1}^{k} \Pi(x_j), \quad \mathrm{Var}(\hat{\xi}_{1/2}^{(1)}) = n^{-1}d^2 \sum_{j=1}^{k} \pi_j(1-\pi_j). \tag{6.34}$$

(ii) The *Reed–Muench estimator.* Note that $x_1 < \cdots < x_k$ and by assumption the π_i are therefore ordered too. In such a case, the sample counterparts $\bar{U}_i$ are stochastically ordered. If there exists a positive interger m $(<k)$, such that $(\bar{U}_1 + \cdots + \bar{U}_m) = (k-m+1) - (\bar{U}_k + \cdots + \bar{U}_k)$, then, the Reed–Muench estimator is

$$\hat{\xi}_{1/2}^{(2)} = x_k - d(k-m). \tag{6.35}$$

If for no m, the strict equality sign holds (for the $\bar{U}_i$), then, one may get two consecutive values of m for which opposite inequalities hold, and, one may obtain the estimator (parallel to (6.35)) by linear interpolation. This situation is more likely to occur as the $\bar{U}_i$ are r.v.'s.

(iii) The *Dragstedt–Behrens estimator.* Consider the partial sequence

$$\tilde{U}_j = (\bar{U}_1 + \cdots + \bar{U}_j)/\{(\bar{U}_1 + \cdots + \bar{U}_j) + (k-j+1) - (\bar{U}_j + \cdots + \bar{U}_k)\},$$
$$j = 1, \ldots, k, \tag{6.36}$$

and define

$$m = \max\{r: \tilde{U}_r \leq \tfrac{1}{2}, r \leq k-1\}. \tag{6.37}$$

Then, the estimator is defined by

$$\hat{\xi}_{1/2}^{(3)} = x_k - d(k-m) + d(\tfrac{1}{2} - \tilde{U}_m)/(\tilde{U}_{m+1} - \tilde{U}_m). \tag{6.38}$$

Note that if d is small and k, the number of dose is large, then

$$x_k + d/2 - d\sum_{j=1}^{k} \Pi(x_j) \to \int_0^\infty x \, \mathrm{d}\Pi(x),$$

the mean of the tolerance distribution. Thus, if the tolerance distribution is symmetric, the Spearman–Kärber estimator estimates the median effective dose closely. Otherwise, it may estimate some other characteristic of the d.f. Π. Miller (1973) has studied the relative performance of the three nonparametric estimators when k is large and d small. It casts light on the bias of these estimators when Π is not symmetric. These three estimators behave very similarly in this asymptotic case, though, for small values of k, the Spearman–Kärber estimator may behave a bit better than the other ones.

From the practical point of view, these estimators are of not much importance. Because of limitations of cost and other factors, usually, in a quantal assay, only a small number of doses are prescribed, and, for each dose, a number of subjects are administered. In this setup, neither n nor k is usually large. Hence, the small sample behaviour of the estimates is of prime importance. In the parametric case, the assumed functional form of Π provides more information for a comparatively more efficient analysis. On the other hand, the parametric procedures may not be very robust against any departure from the assumed form of Π. In this respect, suitable M-procedures having good robustness properties (for local departures from the assumed model) may work out as a good compromise when k or n is not large.

References

[1] Adichie, J. N. (1978). Rank tests of subhypotheses in the general linear regression. *Ann. Statistist.* **6**, 1012–1026.

[2] Anderson, P. K. and Gill, R. D. (1982). Cox's regression model for counting processes: a large sample study. *Ann. Statist.* **10**, 1100–1120.

[3] Bhattacharya, P. K. and Frierson, D. (1981). A nonparametric control chart for detecting small disorders. *Ann. Statist.* **9**, 544–554.

[4] Bhattacharyya, G. K. and Johnson, R. A. (1968). Nonparametric tests for shift at unknown time point. *Ann. Math. Statist.* **39**, 1731–1743.

[5] Brown, R. L., Durbin, J. and Evans, J. M. (1975). Techniques for testing constancy of regression relationship over time. (With discussion). *J. Roy. Statist. Soc. Ser. B* **37**, 149–192.

[6] Chatterjee, S. K. and Sen, P. K. (1973). Nonparametric testing under progressive censoring. *Calcutta Statistical Assoc. Bull.* **22**, 13–50.

[7] Chernoff, H. and Zacks, S. (1964). Estimating the current mean of a normal distribution which is subjected to change in time. *Ann. Math. Statist.* **35**, 999–1018.

[8] Cox, D. R. (1972). Regression models and life tables. (With discussion). *J. Roy. Statist. Soc. Ser. B.* **34**, 187–220.

[9] Davis, C. E. (1978). A two-sample Wilcoxon test for progressively censored data. *Comm. Statist. Ser. A* **7**, 389–398.

[10] DeLong, D. M. (1980). Some asymptotic properties of a progressively censored nonparametric test for multiple regression. *J. Multivar. Anal.* **10**, 363–370.

[11] DeLong, D. M. (1981). Crossing probabilities for a square root boundary by a Bessel process. *Comm. Statist. Ser. A* **10**, 2197–2213.

[12] Finney, D. J. (1964). *Statistical Method in Biological Assay.* Charles Griffin, London, 2nd ed.

[13] Gerig, T. M. (1969). A multivariate extension of Friedman's χ^2-test. *J. Amer. Statist. Assoc.* **64**, 1595–1608.

[14] Gerig, T. M. (1975). A multivariate extension of Friedman's χ^2-test with random covariates. *J. Amer. Statist. Assoc.* **70**, 443–447.

[15] Ghosh, M., Grizzle, J. E. and Sen, P. K. (1973). Nonparametric methods in longitudinal studies. *J. Amer. Statist. Assoc.* **68**, 29–36.

[16] Hackl, P. (1980). *Testing the Constancy of Regression Relationships Over Time.* Vandenhoeck and Ruprecht, Gottingen.

[17] Jurečková, J. (1969). Asymptotic linearity of a rank statistic in regression parameter. *Ann. Math. Statist.* **40**, 1889–1900.

[18] Jurečková, J. (1971). Nonparametric estimates of regression coefficients, *Ann. Math. Statist.* **42**, 1328–1338.

[19] Khatri, C. G. (1966). A note on a MANOVA model applied to problems in growth curves. *Ann. Inst. Statist. Math.* **18**, 75–86.

[20] Kleinbaum, D. G. (1973). A generalization of the growth curve model which allows missing data. *J. Multivar. Anal.* **3**, 117–124.

[21] Koziol, J. A. and Byar, D. P. (1975). Percentage points of the asymptotic distributions of one and two-sample K–S statistics for truncated or censored data. *Technometrics* **17**, 507–510.

[22] Koziol, J. A., Maxwell, D. A., Fukushima, M., Colmerauer, M. E. M. and Pilch, Y. H. (1981). A distribution-free test for tumor growth curve analysis with applications to an animal tumor immunotherapy experiment. *Biometrics* **37**, 383–390.

[23] Koziol, J. A. and Petkau, A. J. (1978). Sequential testing of equality of two survival distributions using modified Savage statistics. *Biometrika* **65**, 615–623.

[24] Lehmann, E. L. (1951). Consistency and unbiasedness of certain nonparametric tests. *Ann. Math. Statist.* **22**, 165–179.

[25] Lombard, F. (1981). An invariance principle for sequential nonparametric test statistics under contiguous alternatives. *South African Statist. J.* **15**, 129–152.

[26] Lombard, F. (1983). Asymptotic distributions of rank statistics in the change-point problem. *South African Statist. Jour.* **17**.

[27] Majumdar, H. and Sen, P. K. (1977). Rank order tests for grouped data under progressive censoring. *Comm. Statist. Ser. A.* **6**, 507–524.

[28] Majumdar, H. and Sen, P. K. (1978a). Nonparametric tests for multiple regression under progressive censoring. *J. Multivar. Anal.* **8**, 73–95.

[29] Majumdar, H. and Sen, P. K. (1978b). Nonparametric testing for simple regression under progressive censoring with staggering entry and random withdrawal. *Comm. Statist. Ser. A.* **7**, 349–371.

[30] Miller, R. G., Jr. (1973). Nonparametric estimators of the mean tolerance in bioassay. *Biometrika* **60**, 535–542.

[31] Page, E. S. (1955). A test for a change in a parameter occurring at an unknown point. *Biometrika* **42**, 523–526.

[32] Page, E. S. (1957). On problems in which a change of parameters occurs at an unknown time point. *Biometrika* **44**, 248–252.

[33] Pettitt, A. N. (1979). A nonparametric approach to the change point problem. *Appl. Statist.* **28**, 126–135.

[34] Potthoff, R. F. and Roy, S. N. (1964). A generalized multivariate analysis of variance model especially useful for growth curve problems. *Biometrika* **51**, 313–326.

[35] Puri, M. L. and Sen, P. K. (1969). Analysis of covariance based on general rank scores. *Ann. Math. Statist.* **40**, 610–618.

[36] Puri, M. L. and Sen, P. K. (1971). *Nonparametric Methods in Multivariate Analysis.* Wiley, New York.

[37] Quade, D. (1967). Rank analysis of covariance. *J. Amer. Statist. Assoc.* **62**, 1187 ff.

[38] Rao, C. R. (1965). The theory of least squares when the parameters are stochastic and its application to the analysis of growth curves. *Biometrika* **52**, 447–458.

[39] Rao, P. V. and Littell, R. C. (1976). An estimator of relative potency. *Comm. Statist. Ser. A* **5**, 183–189.

[40] Schechtman, E. and Wolfe, D. A. (1981). Distribution-free tests for the change-point problem. Tech. Report, Ohio State Univ.

[41] Schey, H. M. (1977). The asymptotic distribution of the one-sided Kolmogorov–Smirnov statistic for truncated data. *Comm. Statist. Ser. A* **6**, 1361–1366.

[42] Sen, A. and Srivastava, M. S. (1975). On tests for detecting changes in mean. *Ann. Statist.* **3**, 98–108.

[43] Sen, P. K. (1960). On some convergence properties of U-statistics. *Calcutta Statist. Assoc. Bull.* **10**, 1–18.

[44] Sen, P. K. (1963). On the estimation of relative potency in dilution (-direct) assays by distribution-free methods. *Biometrics* **19**, 532–552.

[45] Sen, P. K. (1964). Tests for the validity of fundamental assumption in dilution (-direct) assays. *Biometrics* **20**, 770–784.

[46] Sen, P. K. (1965). Some further applications of nonparametric methods in dilution (-direct) assays. *Biometrics* **21**, 799–810.

[47] Sen, P. K. (1968). Estimates of the regression coefficient based on Kendall's tau. *J. Amer. Statist. Assoc.* **63**, 1379–1389.

[48] Sen, P. K. (1969). On a class of rank order tests for the parallelism of several regression lines. *Ann. Math. Statist.* **40**, 1668–1683.

[49] Sen, P. K. (1971). Robust statistical procedures in problems of linear regression with special reference to quantitative bio-assays. I. *Internat. Statist. Rev.* **39**, 21–38.

[50] Sen, P. K. (1972a). Robust statistical procedures in problems of linear regression with special reference to quantitative bio-assays, II. *Internat. Statist. Rev.* **40**, 161–172.

[51] Sen, P. K. (1972b). On a class of aligned rank order tests for the identity of the intercepts of several regression lines. *Ann. Math. Statist.* **43**, 2004–2012.

[52] Sen, P. K. (1973). Some aspects of nonparametric procedures in multivariate statistical analysis. In *Multivariate Statistical Inference* (ed: D. G. Kabe and R. P. Gupta), North-Holland, Amsterdam, pp. 230–240.

[53] Sen, P. K. (1976). Asymptotically optimal rank order tests for progressive censoring. *Calcutta Statist. Assoc. Bull.* **25**, 65–78.

[54] Sen, P. K. (1977). Tied-down Wiener process approximations for aligned rank order statistics and some applications. *Ann. Statist.* **5**, 1107–1123.

[55] Sen, P. K. (1978). Invariance principles for rank statistics revisited. *Sankhya Ser. A* **40**, 215–236.

[56] Sen, P. K. (1979). Rank analysis of covariance under progressive censoring. *Sankhya Ser. A* **41**, 147–169.

[57] Sen, P. K. (1980). Asymptotic theory of some tests for a possible change in the regression slope occurring at an unknown time point. *Z. Wahrsch. Verw. Geb.* **52**, 203–218.

[58] Sen, P. K. (1981a). *Sequential Nonparametrics: Invariance Principles and Statistical Inference.* Wiley, New York.

[59] Sen, P. K. (1981b). Rank analysis of covariance under progressive censoring, II. In: M. Csörgő et al., eds., *Statistics and Related Topics.* North-Holland, Amsterdam, pp. 285–295.

[60] Sen, P. K. (1982a). Invariance principles for recursive residuals. *Ann. Statistist.* **10**, 307–312.

[61] Sen, P. K. (1982b). Asymptotic theory of some tests for constancy of regression relationship over time. *Math. Operat. Statist., Ser. Statist.* **13**, 21–32.

[62] Sen, P. K. (1982c). Tests for changepoints based on recursive U-statistics. *Sequential Anal.* **1**, 263–284.

[63] Sen, P. K. (1983). Some recursive residual rank tests for change points. In: M. H. Rizvi et al., eds., *Recent Advances in Statistics: Papers in Honor of Herman Chernoff's Sixtieth Birthday.* Academic Press, New York, pp. 371–391.

[64] Sen, P. K. (1984a). Subhypotheses testing against restricted alternatives for the Cox regression model. *J. Statist. Plann. Infer.* **10**, 31–42.

[65] Sen, P. K. (1984b). The Cox regression model, random censoring and locally optimal rank tests. *J. Statist. Plann. Infer.* **9**, 355–366.

[66] Sen, P. K. (1984c). On a Kolmogorov–Smirnov type aligned test. To appear in *Statist. Probability Letters* **2**.

[67] Sen, P. K. and Puri, M. L. (1970). Asymptotic theory of likelihood ratio and rank order tests in some multivariate linear models. *Ann. Math. Statist.* **41**, 87–100.

[68] Sen, P. K. and Puri, M. L. (1977). Asymptotically distribution-free aligned rank order tests for composite hypotheses for general multivariate linear models. *Z. Wahrsch. Verw. Geb.* **39**, 175–186.

[69] Sinha, A. N. and Sen, P. K. (1979a). Progressively censored tests for clinical experiments and life testing problems based on weighted empirical distributions. *Comm. Statist. Ser. A* **8**, 871–898.

[70] Sinha, A. N. and Sen, P. K. (1979b). Progressively censored tests for multiple regression based on weighted empirical distributions. *Calcutta Statist.* Assoc. Bull. **28**, 57–82.

[71] Sinha, A. N. and Sen, P. K. (1982). Tests based on empirical processes for progressive censoring schemes with staggering entry and random withdrawal. *Sankhya Ser. B* **44**, 1–18.

[72] Shiryayev, A. N. (1963). On optimum methods in quickest detection problems. *Theor. Probability Appl.* **8**, 22–46.

[73] Shiryayev, A. N. (1978). *Optimal Stopping Rules.* Springer-Verlag, New York.

[74] Shorack, G. R. (1966). Graphical procedures for using distribution-free methods in the estimation of relative potency in dilution (-direct) assays. *Biometrics* **22**, 610–619.

[75] Tsiatis, A. (1981a). A large sample study of Cox's regression model. *Ann. Statist.* **9**, 93–108.

[76] Tsiatis, A. (1981b). The asymptotic distribution of the efficient score test for the proportional hazard model calculated over time. *Biometrika* **68**, 311–315.

[77] Woolson, R. F. and Sen, P. K. (1974). Asymptotic comparison of a class of multivariate multi-parameter tests. *Comm. Statist.* **3**, 813–828.

P. R. Krishnaiah and P. K. Sen, eds., *Handbook of Statistics, Vol. 4*
© Elsevier Science Publishers (1984) 741–754 30

Minimum Distance Procedures

Rudolf Beran

1. Introduction

Of lasting influence in statistics is the idea that a fitted statistical model should summarize plausibly the data under consideration and that adequacy of the fitted model can be assessed, in part, by calculating the distance between the data and the fitted model. Minimum distance procedures for parametric models are the most direct expression of this notion. By minimum distance procedures, we mean especially minimum distance estimates and minimum distance tests. Minimum distance estimates are estimates of the parameters chosen to minimize the distance between the data and the fitted model. Minimum distance tests are based upon the shortest distance between the data and members of the null hypothesis parametric model.

Suppose that, under the parametric model, the observations $X_1, X_2, \ldots, X_n$ are i.i.d. with distribution function F_θ, where $\theta \in \Theta \subset \mathbb{R}^k$ is unknown. Let $\hat{F}_n$ denote the empirical distribution function. Examples of minimum distance estimates include:

(a) The least squares estimate, which minimizes $\int [y - \int x\, dF_\theta(x)]^2\, d\hat{F}_n(y)$ by choice of θ;

(b) The minimum Kolmogorov–Smirnov distance estimate, which minimizes $\sup_x |\hat{F}_n(x) - F_\theta(x)|$;

(c) The minimum Cramér–von Mises distance estimate, which minimizes $\int [\hat{F}_n(x) - F_\theta(x)]^2\, d\mu$, where μ is a probability measure;

(d) The minimum Hellinger distance estimate, which minimizes $\int [\hat{f}_n^{1/2}(x) - f_\theta^{1/2}(x)]^2\, dx$, where f_θ is the Lebesgue density of F_θ and $\hat{f}_n$ is a nonparametric density estimate (the discrete distribution version is obvious);

(e) The minimum chi-square estimate.

Associated with each of these estimates is a minimum distance test of the hypothesis that the parametric model fits the data.

Maximum likelihood estimates are related to minimum distance procedures because the maximum likelihood estimates of θ minimizes $-\int \log[f_\theta(x)]\, d\hat{F}_n(x)$,

This research was supported in part by National Science Foundation Grant MCS 80-02648.

a quantity possessing some of the properties of a distance. Most notably, $-\int \log[f(x)]g(x)\,dx$ is minimized over all densities f if and only if $f = g$.

Parametric models are a simple approximation to the real world. In the past three decades, statisticians have increasingly realized that procedures optimal for a specific parametric model may perform poorly on actual data. Least squares estimates for normal models are overly sensitive to outliers. Maximum likelihood estimates in the gamma model are strongly affected by observations near zero as well as by large positive outliers. F-tests are robust in neither level nor power. It has become apparent that statistical theory should consider performance of procedures not only at the parametric model being fitted but also over small, pertinent neighborhoods of this parametric model.

What are realistic, mathematically tractable, neighborhoods? What is being estimated or tested when the parametric model does not hold strictly? Distance ideas are helpful in resolving these questions. For instance, in the i.i.d. case, to say that the actual distribution function G is close to the parametric model distribution function F_{θ_0} in Kolmogorov–Smirnov distance permits a small percentage of outliers, or other troublesome observations, in the sample. When G differs from F_{θ_0}, the least squares estimate of θ estimates not θ_0 but $T(G)$, the value of θ which minimizes $\int [y - \int x\,dF_\theta(x)]^2\,dG(y)$. Closeness of G to F_{θ_0} in Kolmogorov–Smirnov distance need not entail closeness of $T(G)$ to θ_0. A similar phenomenon underlies the nonrobustness of maximum likelihood estimates.

A plausible remedy for such nonrobustness is to use minimum distance estimates based on the Kolmogorov–Smirnov metric or on a distance weaker than the Kolmogorov–Smirnov metric, such as the Cramér–von Mises distance. The general principle here is that the estimation distance be no stronger than the distance describing the contamination neighborhood about the parametric model. A small departure of the actual distribution from distributions in the parametric model, as measured in the contamination distance, should then have a relatively small effect on the minimum distance estimate of θ, provided the estimation distance between F_t and G is a continuous function of t.

How should we assess the performance of estimates, given that deviations from the parametric model occur? The traditional comparisons of efficiency at the parametric model are insufficient, indeed irrelevant, since it is most unlikely that the parametric model contains the actual distribution of the data. A better way of assessing performance is to calculate, at least asymptotically, the maximum risk incurred over the entire contamination neighborhood. This approach is not excessively cautious when applied to sufficiently local contamination neighborhoods; that is, to departures from the parametric model which cannot be detected reliably by goodness-of-fit tests. In performance comparisons over such neighborhoods, minimum Cramér–von Mises or Hellinger distance estimates can be much superior to maximum likelihood or other classical estimates.

Global performance of estimates, under large, detectable departures from the parametric model, can be assessed less formally. The main requirement is that the functional being estimated remain sensible and that the estimate be precise. Minimum Cramér–von Mises, or Kolmogorov–Smirnov, or Hellinger distance estimates share this property.

The robustness of certain minimum distance estimates is reflected by the sensitivity of the corresponding minimum distance tests to most departures from the parametric model. For tests based on the Kolmogorov–Smirnov or Cramér–von Mises distances, this broad sensitivity was recognized relatively early. Unfortunately, finding critical values for these tests is not easy; even their asymptotic theory, under the null hypothesis that the parametric model holds, is complex. Despite some analytical progress on the problem, this circumstance has greatly hampered the use of these otherwise appealing tests.

Recent research suggests a new approach to finding critical values: estimate the null distribution of the test statistic by parametric bootstrapping. One version of the procedure runs as follows. Let T_n denote the minimum distance estimate of θ, calculated from the observed sample of size n. Draw m pseudo-random samples of size n from the distribution F_{T_n}. For each sample, evaluate the minimum distance between sample and parametric model. The empirical distribution of the m values of the minimum distance so realized is an estimate of the null distribution of the test statistic. It yields estimated critical values for the minimum distance goodness-of-fit test. Current experience with bootstrapping suggests taking m between 100 and 1000.

In what follows, we will examine more closely the minimum distance estimates and tests based on the Cramér–von Mises metric

$$d(G, F) = \left\{\int [G(x) - F(x)]^2\right\} \mathrm{d}\mu\,, \tag{1.1}$$

where, F, G are distribution functions and μ is a probability measure. The theory for these procedures is relatively simple, yields interesting results, and most importantly, illustrates general features of minimum distance procedures. Arguments that work for the Cramér–von Mises metric often have counterparts for other metrics, even when substantial technical differences occur (as is the case for the Hellinger and Kolmogorov–Smirnov metrics).

Section 2 develops asymptotic distribution theory for minimum Cramér–von Mises distance estimates and establishes their robustness under local perturbations from the parametric model. Asymptotic performance of the minimized Cramér–von Mises distance is studied in Section 3. The discussion includes theoretical justification of the bootstrap technique for selecting critical values and asymptotic power calculations for the corresponding goodness-of-fit tests.

2. Minimum Cramér–von Mises distance estimates

Suppose the observations $X_1, X_2, \ldots, X_n$ are i.i.d. random variables whose distribution function is G. Suppose we fit to this data the parametric model with distribution function F_θ, where $\theta \in \Theta \subset \mathbb{R}^k$. The minimum Cramér–von Mises distance estimate T_n is defined as the value of θ which minimizes $d(\hat{F}_n, F_\theta)$ for d defined by (1.1). If T_n is not unique, we pick one of its values arbitrarily. Existence of T_n will be discussed below. Intuitively, T_n estimates $T(G)$, the value(s) of θ which minimize $d(G, F_\theta)$. Since $T_n = T(\hat{F}_n)$, we begin our study of the estimate T_n by studying the functional T.

2.1. Existence and continuity of T

Suppose that the parameter space Θ is a compact subset of $\mathbb{R}^k$ and that the parametric model has the following two properties:

(A) For every distribution function G, $d(G, F_t)$ is a continuous function of t.

(B) $d(F_t, F_\theta) = 0$ if and only if $t = \theta$.

Then $T(G)$ exists for every distribution function G and $T(F_\theta) = \theta$ uniquely. The functional T is continuous at F_θ in the sense that $\lim_{n\to\infty} d(G_n, F_\theta) = 0$ implies $\lim_{n\to\infty} T(G_n) = \theta$.

We will check these assertions in order. Since $d(G, F_t)$ is continuous in t over a compact space Θ, there exists a minimizing value $T(G)$ of t. The uniqueness of $T(F_\theta)$ is ensured by assumption (B).

Write $h(t, G) \equiv d(G, F_t)$. Suppose $\lim_{n\to\infty} d(G_n, F_\theta) = 0$. Then

$$\sup |h(t, G_n) - h(t, F_\theta)| \le d(G_n, F_\theta) \to 0\,, \tag{2.1}$$

which implies that $\min_t h(t, G_n) \to \min_t h(t, F_\theta) = 0$ or, equivalently, $h(T(G_n), G_n) \to 0$. Since (2.1) also implies $|h(T(G_n), G_n) - h(T(G_n), F_\theta)| \to 0$, we conclude that

$$\lim_{n\to\infty} h(T(G_n), F_\theta) = 0\,. \tag{2.2}$$

Hence $\lim_{n\to\infty} T(G_n) = \theta$. Otherwise, since Θ is compact, there would exist a subsequence $\{m\} \subset \{n\}$ such that $T(G_m) \to \theta_1 \ne \theta$. But then $h(T(G_m), F_\theta) \to h(\theta_1, F_\theta)$ by assumption (A) and the limit is positive by assumption (B). This contradicts (2.2).

2.2. Differentiability of T

Let us regard the elements of Θ as $k \times 1$ vectors. Suppose the parametric model has three additional properties for every θ in the interior of Θ:

(C) There exists a $k \times 1$ vector function δ_θ in $L_2^k(\mu)$ such that

$$\int [F_t - F_\theta - (t-\theta)'\delta_\theta]^2 \, \mathrm{d}\mu = \mathrm{o}(|t-\theta|^2) \quad \text{whenever } t \to \theta\,. \tag{2.3}$$

(D) The function δ_θ is $L_2^k(\mu)$ continuous in that

$$\lim_{t\to\theta} \int |\delta_t - \delta_\theta|^2 \, d\mu = 0 . \tag{2.4}$$

(E) The matrix $\int \delta_\theta \delta'_\theta \, d\mu$ is nonsingular.
Then the minimum distance functional $T(G)$ is differentiable at F_θ in the sense that, for every θ in the interior of Θ and for every G in a d-neighborhood of F_θ,

$$T(G) = \theta + \int \gamma_\theta [G - F_\theta] \, d\mu + o[d(G, F_\theta)] , \tag{2.5}$$

where

$$\gamma_\theta(x) = \left[\int \delta_\theta \delta'_\theta \, d\mu \right]^{-1} \delta_\theta(x) . \tag{2.6}$$

Indeed, assumption (C) implies

$$\begin{aligned} d^2(G, F_t) - d^2(G, F_\theta) &= \int (F_t - F_\theta)(F_t + F_\theta - 2G) \, d\mu \\ &= 2(t - \theta)' \int \delta_\theta (F_\theta - G) \, d\mu + o(|t - \theta|) . \end{aligned} \tag{2.7}$$

Since $\theta = T(G)$ minimizes $d(G, F_\theta)$, it follows from (2.7) that

$$\int \delta_{T(G)}(F_{T(G)} - G) \, d\mu = 0 . \tag{2.8}$$

On the other hand, from (2.3),

$$\begin{aligned} \int \delta_{T(G)}(F_{T(G)} - G) \, d\mu = \int \delta_{T(G)}(F_\theta - G) \, d\mu + \left[\int \delta_{T(G)} \delta'_\theta \, d\mu \right] \\ \times (T(G) - \theta) + o(|T(G) - \theta|) . \end{aligned} \tag{2.9}$$

Continuity of the functional T at F_θ and (2.3) imply that, for every G in a d-neighborhood of F_θ,

$$|T(G) - \theta| \leq 2\lambda_\theta^{-1/2} d(F_{T(G)}, F_\theta) \leq 4\lambda_\theta^{-1/2} d(G, F_\theta) , \tag{2.10}$$

where λ_θ is the smallest eigenvalue of $\int \delta_\theta \delta'_\theta \, d\mu$. Assumption (E) ensures the strict positivity of λ_θ. The second inequality in (2.10) rests on the triangle inequality and the fact $d(G, F_{T(G)}) \leq d(G, F_\theta)$. Combining (2.8), (2.9), (2.10) yields (2.5), in view of assumption (D).

Equation (2.5) can be rewritten as

$$T(G) = \theta + \int \rho_\theta \, dG + o[d(G, F_\theta)] , \tag{2.11}$$

where

$$\rho_\theta(x) = \int I(t \geq x)\gamma_\theta(t)\,\mathrm{d}\mu(t) - \int\int I(t \geq x)\gamma_\theta(t)\,\mathrm{d}\mu(t)\,\mathrm{d}F_\theta(x). \quad (2.12)$$

This form is convenient for discussing asymptotic behavior of the minimum distance estimate $T_n = T(\hat{F}_n)$.

2.3. Asymptotics and qualitative local robustness

Asymptotic performance of the estimate $T_n = T(\hat{F}_n)$ over neighborhoods of the parametric model F_θ can be deduced from the behavior of the empirical distribution function $\hat{F}_n$ over the same neighborhoods. Let $B_n(\theta, c)$ be the set of all distribution functions G such that $d(G, F_\theta) \leq n^{-1/2}c$, where c is positive. The Dvoretzky, Kiefer, Wolfowitz (1956) inequality on the Kolmogorov–Smirnov norm of the empirical distribution function process implies that

$$\limsup_{t\to\infty} \sup_n \sup_{G\in B_n(\theta,c)} P_G[n^{1/2}d(\hat{F}_n, G) > t] = 0\,, \quad (2.13)$$

for every positive c. Hence, using the definition of $B_n(\theta, c)$ and the triangle inequality,

$$\limsup_{t\to\infty} \sup_n \sup_{G\in B_n(\theta,c)} P_G[n^{1/2}d(\hat{F}_n, F_\theta) > t] = 0\,. \quad (2.14)$$

Equation (2.14) implies in particular that for every positive ε,

$$\lim_{n\to\infty} \sup_{G\in B_n(\theta,c)} P_G[d(\hat{F}_n, F_\theta) > \varepsilon] = 0\,. \quad (2.15)$$

Since the functional T is continuous at F_θ and $T(F_\theta) = \theta$, it follows from (2.15) that

$$\lim_{n\to\infty} \sup_{G\in B_n(\theta,\,c)} P_G[|T_n - \theta| > \varepsilon] = 0 \quad (2.16)$$

for every positive ε and c. On the other hand,

$$\lim_{n\to\infty} \sup_{G\in B_n(\theta,\,c)} |T(G) - \theta| = 0\,. \quad (2.17)$$

Combining (2.16) and (2.17) establishes

$$\lim_{n\to\infty} \sup_{G\in B_n(\theta,\,c)} P_G[|T_n - T(G)| > \varepsilon] = 0 \quad (2.18)$$

for every positive ε and c.

Equation (2.18) is a simple qualitative robustness property, which asserts that the minimum distance estimate T_n converges locally uniformly to the functional $T(G)$ being estimated, if the actual distribution function G is near F_θ; all values of $T(G)$ are close to θ in this case. A stronger qualitative robustness property holds under assumptions (C), (D) and (E) when $\theta \in \text{int}(\Theta)$: the limiting distribution of $n^{1/2}[T_n - T(G_n)]$ under any sequence of distributions $G_n \in B_n(\theta, c)$ is $N(0, \Sigma_\theta)$ where

$$\Sigma_\theta = \int \rho_\theta \rho'_\theta \, \mathrm{d}F_\theta \,. \tag{2.19}$$

Indeed, equations (2.11) and (2.14) imply

$$n^{1/2}(T_n - \theta) = n^{1/2} \int \rho_\theta \, \mathrm{d}\hat{F}_n + \mathrm{o}_p(1) \tag{2.20}$$

under $G_n \in B_n(\theta, c)$; moreover

$$n^{1/2}[T(G_n) - \theta] = n^{1/2} \int \rho_\theta \, \mathrm{d}G_n + \mathrm{o}(1) \,. \tag{2.21}$$

Combining (2.20) and (2.21) yields

$$n^{1/2}[T_n - T(G_n)] = n^{1/2} \int \rho_\theta \, \mathrm{d}(\hat{F}_n - G_n) + \mathrm{o}_p(1) \tag{2.22}$$

under $G_n \in B_n(\theta, c)$. This implies the asserted locally uniform asymptotic normality.

2.4. *Quantitative local robustness*

Performance of the minimum distance estimate T_n may be assessed quantitatively by calculating its maximum risk over all distribution functions G in a small, realistic neighborhood of F_θ. An asymptotic version of such a calculation proves tractable and provides an approximation to the finite sample size situation.

Let $u : \mathbb{R}^+ \to \mathbb{R}^+$ be a bounded, monotone increasing function with $u(0) = 0$. Let T_n^* be any estimate of $T(G)$ and let

$$R_n(T_n^*, G) = E_G u[n^{1/2}|T_n^* - T(G)|] \tag{2.23}$$

be the risk associated with T_n^*. Since $T(G)$ is differentiable at F_θ, in the sense of equation (2.11), an argument based on Hájek's asymptotic minimax theorem yields the following lower bound on maximum risk over $B_n(\theta, c)$ when $\theta \in \text{int}(\Theta)$:

$$\lim_{c\to\infty} \liminf_{n\to\infty} \inf_{T_n^*} \sup_{G\in B_n(\theta,c)} R_n(T_n^*, G) \geq R_0(\theta) \tag{2.24}$$

where

$$R_0(\theta) = Eu(|\Sigma_\theta^{1/2} Z|) \tag{2.25}$$

and Z is a standard k-dimensional normal random vector. (For a similar argument, see Koshevnik and Levit (1976).) Note that the infimum in (2.24) is taken over all possible estimates T_n^* of $T(G)$.

Since the loss function u is monotone increasing and bounded, it is continuous almost everywhere. The locally uniform asymptotic normality of T_n, established in Section 2.3, implies that

$$\lim_{n\to\infty} E_{G_n} u[n^{1/2}|T_n - T(G_n)|] = R_0(\theta) \tag{2.26}$$

for every sequence $G_n \in B_n(\theta, c)$ and every positive c. Hence,

$$\lim_{n\to\infty} \sup_{G\in B_n(\theta,c)} R_n(T_n, G) = R_0(\theta) \tag{2.27}$$

for every positive c. Thus, the minimum distance estimate T_n attains the lower bound (2.24) on maximum risk over the contamination neighborhood $B_n(\theta, c)$.

We have replaced the classical problem of estimating θ in the parametric model F_θ by the more realistic problem of estimating the minimum Cramér–von Mises distance functional $T(G)$ for underlying distributions G near F_θ. If c and n are large, the minimum distance estimate $T_n = T(\hat{F}_n)$ is approximately minimax for $T(G)$ over all distribution functions G in the contamination neighborhood $B_n(\theta, c)$ about F_θ. This property may be interpreted as quantitative robustness of the estimate T_n. Similar results are available for minimum Hellinger distance estimates. Whether minimum Kolmogorov–Smirnov distance estimates are asymptotically minimax is not known, primarily because the asymptotic distributions of these estimates are not normal.

3. Minimum Cramér–von Mises distance goodness-of-fit tests

Suppose the observations $X_1, X_2, \ldots, X_n$ are i.i.d. random variables. To test the null hypotheses that the common distribution function of the observations belongs to the parametric model $\{F_\theta: \theta \in \Theta\}$, it is natural to consider the statistic

$$S_n = nd^2(\hat{F}_n, F_{T_n}), \tag{3.1}$$

where T_n is the minimum distance estimate discussed in Section 2. The statistic S_n is the shortest distance between the empirical distribution function $\hat{F}_n$ and members of the parametric model. We will pursue this idea by finding the asymptotic distribution of S_n under null hypotheses and local alternatives. The asymptotics suggests two ways to estimate critical values of S_n. One way is analytic and complex; the other is the parametric bootstrap.

3.1. *Asymptotic distribution of* S_n

Consider a hypothetical sequence of experiments. In the n-th experiment, the observations $\{X_i: 1 \leq i \leq n\}$ are i.i.d. with common distribution function G_n; the problem is to test

$$H_n: \quad G_n = F_{\theta_0} \text{ for some } \theta_0 \in \operatorname{int}(\Theta)$$

versus the local alternative

$$K_n: \quad G_n \text{ is such that } \lim_{n\to\infty} \int [n^{1/2}(G_n - F_{\theta_0}) - \xi_{\theta_0}]^2 \, d\mu = 0 \,,$$

where ξ_{θ_0} is a function in $L_2(\mu)$.

The parameter value θ_0 is unknown. The assumptions made in Section 2 on the parametric model are retained.

By (2.20) and the definition (2.12) of ρ_θ,

$$\begin{aligned} n^{1/2}(T_n - \theta_n) &= n^{1/2} \int \rho_{\theta_0} \, d(\hat{F}_n - G_n) + n^{1/2} \int \rho_{\theta_0} \, d(G_n - F_{\theta_0}) + o_p(1) \\ &= n^{1/2} \int \rho_{\theta_0} \, d(\hat{F}_n - G_n) + \int \gamma_{\theta_0} \xi_{\theta_0} \, d\mu + o_p(1) \end{aligned} \tag{3.2}$$

under K_n. Since the random variables $\{n^{1/2}(T_n - \theta_0)\}$ are tight, assumption (C) yields

$$\int [n^{1/2}(F_{T_n} - F_{\theta_0}) - n^{1/2}(T_n - \theta_0)' \delta_{\theta_0}]^2 \, d\mu = o_p(1) \,. \tag{3.3}$$

Thus, under K_n,

$$\begin{aligned} S_n &= \int [n^{1/2}(\hat{F}_n - G_n) + n^{1/2}(G_n - F_{\theta_0}) + n^{1/2}(F_{\theta_0} - F_{T_n})]^2 \, d\mu \\ &= \int \Big[n^{1/2} \Big\{ \hat{F}_n - G_n - \delta'_{\theta_0} \int \rho_{\theta_0} \, d(\hat{F}_n - G_n) \Big\} + b_{\theta_0} \Big]^2 \, d\mu + o_p(1) \end{aligned} \tag{3.4}$$

where

$$b_{\theta_0} = \xi_{\theta_0} - \delta'_{\theta_0} \int \gamma_{\theta_0} \xi_{\theta_0} \, d\mu \,. \tag{3.5}$$

A weak convergence argument in $L_2(\mu)$ now shows that S_n converges weakly, under K_n, to the random variable

$$S(b)=\int [Y(x,\theta_0)+b_{\theta_0}(x)]^2\,\mathrm{d}\mu\,, \tag{3.6}$$

where $Y(x, \theta_0)$ is a gaussian process with mean zero and covariance function

$$C(x,y,\theta_0)=\int d_{\theta_0}(x,z)d_{\theta_0}(y,z)\,\mathrm{d}F_{\theta_0}(z)\,; \tag{3.7}$$

here

$$d_{\theta_0}(x,z)=I(z\leqslant x)-F_{\theta_0}(x)-\delta'_{\theta_0}(x)\rho_{\theta_0}(z)\,. \tag{3.8}$$

In particular, under the null hypotheses H_n, the S_n converge weakly to the random variable

$$S=\int [Y(x,\theta_0)]^2\,\mathrm{d}\mu\,. \tag{3.9}$$

More explicit representations are available for the random variables S and $S(b)$. Let the $\{\lambda_k(\theta_0); k\geqslant 1\}$ denote the distinct, nonzero eigenvalues of $C(x, y, \theta_0)$, ordered so that $\lambda_1(\theta_0)>\lambda_2(\theta_0)>\cdots>0$. Let $r_k(\theta_0)$ be the multiplicity of $\lambda_k(\theta_0)$ and let $\lambda_k(\theta_0)b_k^2(\theta_0)$ be the squared length in $L_2(\mu)$ of the projection of b_{θ_0} onto the eigenspace of $\lambda_k(\theta_0)$. Then

$$S(b)=\sum_{k=1}^{\infty}\lambda_k\chi^2(r_k,b_k^2)\,, \tag{3.10}$$

where the $\{\chi^2(r_k, b_k^2)\}$ are independent random variables with noncentral chi-square distributions, degrees-of-freedom $\{r_k\}$, and noncentrality parameters $\{b_k\}$. Similarly,

$$S=\sum_{k=1}^{\infty}\lambda_k\chi^2(r_k) \tag{3.11}$$

where $\chi^2(r_k)=\chi^2(r_k, 0)$ has chi-square distribution with r_k degrees-of-freedom.

While the characteristic functions of S and $S(b)$ are readily found from (3.10) and (3.11), computable expressions for the distribution functions of S or $S(b)$ exist only in special cases. Hoeffding (1964) gives an asymptotic expansion for $P[S>x]$ which is valid for large x. In particular,

$$\lim_{x\to\infty}\{P[S>x]/P[\lambda_1\chi^2(r_1)>x]\}=A \tag{3.12}$$

where

$$A=\prod_{k\geqslant 2}[1-\lambda_k/\lambda_1]^{r_k/2}<\infty\,. \tag{3.13}$$

A similar, though more complicated, argument (Beran, 1975) establishes

$$\lim_{x\to\infty} \{P[S(b) > x]/P[\lambda_1\chi^2(r_1, b_1^2) > x]\}$$
$$= A \exp\left[2^{-1} \sum_{k\geq 2} b_k^2\lambda_k \Big/ (\lambda_1 - \lambda_k)^2\right]. \tag{3.14}$$

Durbin and Knott (1972) discuss numerical inversion of the characteristic functions for S and $S(b)$.

3.2. The goodness-of-fit test

The null hypothesis H_n becomes implausible when S_n is relatively large. How large is 'relatively large'? A traditional answer is to consult the asymptotic distribution of S_n under H_n. Suppose α is the desired test level. Let $c(\alpha, \theta_0)$ be such that $P[S > c(\alpha, \theta_0)] = \alpha$. The strict monotonicity and continuity of the distribution function of S on R^+ ensures existence and uniqueness of $c(\alpha, \theta_0)$. Since θ_0 is unknown, $c(\alpha, \theta_0)$ cannot be used as a critical value for the test. It is plausible, however, that the test which rejects H_n if $S_n > c(\alpha, T_n)$, where T_n is the minimum distance estimate of θ_0, will have approximate level α when n is large. Justifying this claim rests on showing that $c(\alpha, T_n) \xrightarrow{P} c(\alpha, \theta_0)$ under H_n. For then

$$\lim_{n\to\infty} P_{H_n}[S_n > c_n(\alpha, T_n)] = P[S > c(\alpha, \theta_0)] = \alpha. \tag{3.15}$$

From a practical viewpoint, this approach is not very appealing. To perform the test requires evaluation of $c(\alpha, T_n)$, which involves finding the eigenvalues $\{\lambda_k(T_n); k \geq 1\}$ and then approximately inverting the estimated characteristic function of S, either numerically or by use of Hoeffding's expansion. The calculations have to be redone for every sample and every parametric model.

More intuitive is the parametric bootstrap approximation to $c(\alpha, \theta_0)$, which is obtained as follows. Let $J_n(x, \theta_0)$ denote the exact distribution function of S_n under the assumption that the $\{X_i; 1 \leq i \leq n\}$ are i.i.d. with distribution function F_{θ_0}. The bootstrap estimate of $J_n(x, \theta_0)$ is $J_n(x, T_n)$ and the bootstrap critical value estimate is

$$c_n(\alpha) = \inf\{x: J_n(x, T_n) \geq 1 - \alpha\}. \tag{3.16}$$

While exact calculation of $J_n(x, T_n)$ is usually impractical, Monte Carlo approximations are fairly straightforward. For instance: Draw m pseudorandom samples of size n from the distribution F_{T_n}. For each sample, evaluate S_n. The empirical distribution of the m values of S_n so realized is an approximation to $J_n(x, T_n)$ which readily yields an approximation to $c_n(\alpha)$. Current experience with bootstrapping suggests taking m between 100 and 1000.

The corresponding goodness-of-fit test ϕ_n is to reject H_n if $S_n > c_n(\alpha)$ and to accept H_n otherwise. We will show in the next two paragraphs that $c_n(\alpha) \xrightarrow{\text{P}} c(\alpha, \theta_0)$ under both H_n and K_n. Consequently, the asymptotic level of ϕ_n is α. The asymptotic power of ϕ_n will be analyzed in Section 3.3.

Let $J_n(x, \theta_0)$ be the distribution function of S. Let $\{h_n \in \mathbb{R}^k; n \geq 1\}$ be any sequence of $k \times 1$ vectors converging to some $h \in \mathbb{R}^k$, $|h| \leq c$. Let $\theta_n = \theta_0 + n^{-1/2} h_n$. For $G_n = F_{\theta_n}$ in the asymptotics of Section 3.1, the function $\xi_{\theta_0} = h' \delta_{\theta_0}$ by assumption (C) and therefore b_{θ_0} vanishes identically. Hence, $J_n(x, \theta_n)$ converges to $J(x, \theta_0)$, uniformly in x since the limit distribution function is continuous. This implies

$$\lim_{n\to\infty} \sup_{|h| \leq c} \sup_x |J_n(x, \theta_0 + n^{-1/2} h) - J(x, \theta_0)| = 0 \tag{3.17}$$

for every positive c.

Since the random variables $\{n^{1/2}(T_n - \theta_0)\}$ are tight under both H_n and K_n (see Section 2.3), it follows from (3.17) that

$$\sup_x |J_n(x, T_n) - J(x, \theta_0)| \xrightarrow{\text{P}} 0 \tag{3.18}$$

under both H_n and K_n. In particular, (3.18) entails $c_n(\alpha) \xrightarrow{\text{P}} c(\alpha, \theta_0)$ under both H_n and K_n, because $J(x, \theta_0)$ is continuous and strictly monotone for all positive x.

3.3. *Asymptotic power of the test*

The asymptotic power of the goodness-of-fit test ϕ_n under the local alternatives K_n is

$$\lim_{n\to\infty} P_{K_n}[S_n > c_n(\alpha)] = P[S(b) > c(\alpha, \theta_0)]\,, \tag{3.19}$$

because $c_n(\alpha) \xrightarrow{\text{P}} c(\alpha, \theta_0)$ under K_n as well as under H_n. The random variable $S(b)$ was defined in (3.6) and (3.10). Since a noncentral chi-square distribution increases stochastically with its noncentrality parameter, it follows that ϕ_n is asymptotically unbiased against K_n whenever $\int b_{\theta_0}^2 \, d\mu = \lim_{n\to\infty} nd(G_n, F_{T(G_n)}) > 0$. Thus, the test has some sensitivity to every alternative whose minimum distance from the parametric model is positive.

Further analysis, based on (3.12) and (3.14), reveals that for small levels α, the asymptotic power of ϕ_n is largely determined, to a surprising extent, by the first term $\lambda_1 \chi^2(r_1, b_1^2)$ in (3.14). If K_n is such that $b_1^2 = 0$ but $b_k^2 > 0$ for some $k \geq 2$, the test ϕ_n is not very efficient. Details appear in Beran (1975). Numerical studies by Durbin and Knott (1972) for special cases support these conclusions when $\alpha = 0.05$.

It is the qualitative robustness of the estimates T_n, expressed as tightness of the $\{n^{1/2}(T_n - \theta_0)\}$ under every sequence of alternatives K_n, which ensures that $c_n(\alpha) \xrightarrow{P} c(\alpha, \theta_0)$ under K_n and thereby justifies (3.19). Bootstrap critical values based on a nonrobust estimate of θ_0 are not recommended.

How well does the goodness-of-fit test ϕ_n perform under alternatives in which the $\{X_i\}$ are i.i.d. with fixed distribution function G? In this case, the minimum distance estimate converges in probability to $T(G)$, if $T(G)$ is unique. By arguments similar to those in Section 3.2, the bootstrap critical values $c_n(\alpha)$ converge in probability to $c(\alpha, T(G))$, which is finite. The test ϕ_n is consistent, provided $d(G, F_{T(G)}) > 0$.

Strictly speaking, the null hypothesis H_n is too narrow, virtually certain to be false. We can weaken the force of this objection by enlarging the null hypothesis to contain all contributions in the ball $B_n(\theta, c)$, for some specified positive c. Requiring ϕ_n to have level α over this augmented null hypothesis amounts to requiring a smaller nominal level over H_n.

4. Sources

Neyman (1949) studied minimum chi-squared estimates and tests. Wolfowitz (1957) considered minimum distance procedures more abstractly, establishing consistency of minimum distance estimates under general conditions. Asymptotic distributions of particular minimum distance estimates were derived by Blackman (1955), Parr and Schucany (1980), Millar (1981) for the Cramér–von Mises distance; by Rao, Schuster and Littel (1975) for the Kolmogorov–Smirnov distance; by Beran (1977) for the Hellinger distance. Kac, Kiefer and Wolfowitz (1955) developed asymptotics for some modified minimum distance tests.

Robustness of minimum distance estimates was discussed, in various ways, by Holm (1976), Beran (1977), Parr and Schucany (1980), and Millar (1981). The analysis of robustly modified maximum likelihood estimates in Beran (1981) extends to minimum Hellinger distance estimates. Parr and Schucany's (1980) paper contains Monte Carlo results for some minimum distance location estimates.

The consistency of bootstrap estimates was studied by Efron (1979), Bickel and Freedman (1981).

Section 2 rederives some of the results in Millar (1981). Also pertinent is Bolthausen (1977). The asymptotics for the statistics S_n in Section 3.1 are related to Kac, Kiefer and Wolfowitz (1955) and to Pollard (1980); the analysis of the bootstrap critical values $c_n(\alpha)$ is new and solves an old problem.

Parr (1980) has compiled a bibliography of the extensive literature on minimum distance estimates. Our exposition here covers only parts of this growing subject.

References

Beran, R. (1975). Tail probabilities of noncentral quadratic forms. *Ann. Statist.* **3**, 969–974.

Beran, R. (1977). Minimum Hellinger distance estimates for parametric models. *Ann. Statist.* **5**, 445–463.

Beran, R. (1981). Efficient robust estimates in parametric models. *Z. Wahrsch. Verw. Gebiete* **55**, 91–108.

Bickel, P. J. and Freedman, D. A. (1981). Some asymptotic theory for the bootstrap. *Ann. Statist.* **9**, 1196–1217.

Blackman, J. (1955). On the approximation of a distribution function by an empirical distribution. *Ann. Math. Statist.* **26**, 256–267.

Bolthausen, E. (1977). Convergence in distribution of minimum distance estimators. *Metrika* **24**.

Durbin, J. and Knott, M. (1972). Components of Cramér–von Mises statistics I. *J. Roy. Statist. Soc. Ser. B* **34**, 290–307.

Dvoretzky, A., Kiefer, J. and Wolfowitz, J. (1956). Asymptotic minimax character of the sample distribution function and of the classical multinomial estimator. *Ann. Math. Statist.* **27**, 642–669.

Efron, B. (1979). Bootstrap methods: another look at the jackknife. *Ann. Statist.* **7**, 1–26.

Hoeffding, W. (1964). On a theorem of V. M. Zolotarev. *Th. Probab. Appl.* **9**, 89–92.

Holm, S. (1976). Discussion to a paper by P. J. Bickel. *Scand. J. Statist.* **3**, 158–161.

Kac, M., Kiefer, J. and Wolfowitz, J. (1955). On tests of normality and other tests of goodness-of-fit based on distance methods. *Ann. Math. Statist.* **26**, 189–211.

Koshevnik, Yu. A. and Levit, B. Ya. (1976). On a nonparametric analog of the information matrix. *Th. Probab. Appl.* **21**, 738–753.

Millar, P. W. (1981). Robust estimation via minimum distance methods. *Z. Wahrsch. Verw. Gebiete* **55**, 73–84.

Neyman, J. (1949). Contributions to the theory of the χ^2 test. *Proc. First Berkeley Symp. Math. Statist. Probab.* 239–273, University of California Press.

Parr, W. C. and Schucany, W. R. (1980). Minimum distance and robust estimation. *J. Amer. Statist. Assoc.* **75**, 616–637.

Parr, W. C. (1980). Minimum distance estimation: a bibliography. Unpublished preprint.

Pollard, D. (1980). The minimum distance method of testing. *Metrika* **27**, 43–70.

Rao, P. V., Schuster, E. F. and Littel, R. C. (1975). Estimation of shift and center of symmetry based on Kolmogorov–Smirnov statistics. *Ann. Statist.* **3**, 862–873.

Wolfowitz, J. (1957). The minimum distance method. *Ann. Math. Statist.* **28**, 75–88.

P. R. Krishnaiah and P. K. Sen, eds., *Handbook of Statistics, Vol. 4*

31

Nonparametric Methods in Directional Data Analysis

S. Rao Jammalamadaka

1. Introduction

In many natural and physical sciences the observations are in the form of directions – directions either in plane or in three-dimensional space. Such is the case when a biologist investigates the flight directions of birds or a geologist measures the paleomagnetic directions or an ecologist records the directions of wind or water. A convenient sample frame for two-dimensional directions is the circumference of a unit circle centered at the origin with each point on the circumference representing a direction; or, equivalently, since magnitude has no relevance, each direction may be represented by a unit vector. Such data on two-dimensional directions will be called 'circular data'. Similarly the surface of a unit sphere in three-dimensions may be used as the sample space for directions in space, with each point on the surface representing a three-dimensional direction; or alternatively, such a direction may be represented by a unit vector in three-dimensions. Such data is referred to as the 'spherical data'. Also, studies on any periodic phenomena with a known period (such as circadian rhythms in animals) can be represented as circular data, for instance by identifying each cycle or period with points on the circumference, pooling observations over several such periods, if necessary.

The analysis of directional data gives rise to a host of novel statistical problems and does not fit into the usual methods of statistical analysis which one employs for observations on the real line or Euclidean space. Since there is no natural zero-direction, any method of numerically representing a direction depends on the arbitrary choice of this zero direction. It is important that the statistical analyses and conclusions remain independent of this arbitrary zero direction. Unfortunately, however, usual statistics like the arithmetic mean and standard deviation (and all the higher moments) which one employs in linear statistical analyses fail to have this required rotational invariance so that one is forced to seek alternate statistics for describing directional data. To do this, we treat each direction as a unit vector in plane or space. One computes the resultant vector, whose direction provides a meaningful measure of the average direction in unimodal populations. The length of this vector resultant measures the concentration of the data since observations closer together lead to a longer resultant.

One of the basic parametric models for unimodal directional data is called the von Mises–Fisher distribution and is discussed briefly in Section 2. This plays as prominent a role in a directional data analysis as does the normal distribution in the linear case. Sections 3 and 4 review nonparametric methods for circular (two-dimensional) and spherical (three-dimensional) data, respectively. Section 3 is considerably larger since more distribution-free methods have been developed for circular data. The reader may consult the books by Mardia (1972), Batschelet (1981) and Watson (1983) for a more complete introduction to this novel area of statistics.

2. The von Mises–Fisher model for directional data

A parametric model which plays a central role in directional data analysis is called the von Mises–Fisher distribution. In general, if $\boldsymbol{x}$ is a unit vector in p-dimensions ($p \geq 2$) or equivalently, represents a point on S_p, the surface of a unit ball in p dimensions, then the probability density of the von Mises–Fisher distribution is of the form

$$C_p(\kappa)\exp(\kappa \cdot \boldsymbol{x}'\boldsymbol{\mu}) \tag{2.1}$$

where $\kappa > 0$ is a concentration parameter and the unit vector $\boldsymbol{\mu}$ denotes the mean direction. Here the normalizing constant

$$C_p(\kappa) = \kappa^{(1/2)p-1}/[(2\pi)^{p/2} I_{p/2-1}(\kappa)] \tag{2.2}$$

where $I_r(\kappa)$ is the modified Bessel function of the first kind and order r. When $p = 2$, this density reduces to

$$f(\alpha \mid \kappa, \mu) = [2\pi I_0(\kappa)]^{-1}\exp[\kappa \cdot \cos(\alpha - \mu)] \tag{2.3}$$

where $0 \leq \alpha < 2\pi$ and $0 \leq \mu < 2\pi$ are the angles (in polar coordinates) corresponding to $\boldsymbol{x}$ and $\boldsymbol{\mu}$ in (2.1). This density was introduced by von Mises (1918) to test the hypothesis that the atomic weights are integers. When $p = 3$, Fisher (1953) studied the pdf with zero mean direction,

$$f(\alpha, \beta \mid \kappa) = \frac{\kappa}{4\pi \sinh \kappa} e^{\kappa \cos \alpha} \cdot \sin \alpha \tag{2.4}$$

where $0 \leq \alpha < \pi$ and $0 \leq \beta < 2\pi$ are the polar coordinates of $\boldsymbol{x}$. Fisher's 1953 paper is the first comprehensive treatment of the sampling distributions and statistical inference for the spherical model (2.4).

If the concentration parameter $\kappa = 0$ in (2.1), this reduces to the uniform (isotropic) distribution on S_p. For $\kappa > 0$, this is a unimodal distribution with mode at $\boldsymbol{x} = \boldsymbol{\mu}$. Since the likelihood for a sample of n observations is given by

$$[C_p(\kappa)]^n \cdot \exp(\kappa \cdot \boldsymbol{R}'\boldsymbol{\mu})$$

where $\boldsymbol{R}$ is the vector resultant of the sample, $\boldsymbol{R}$ is a sufficient statistic for this family of distributions. The Maximum Likelihood Estimator of μ is given by $(1/|R|)\cdot \boldsymbol{R}$, where $|R|$ is the length of the resultant. The concentration parameter κ is estimated by solving the equation

$$\frac{I_1(\kappa)}{I_0(\kappa)} = \frac{|R|}{n}$$

for $p = 2$, and the equation

$$\coth \kappa - \frac{1}{\kappa} = \frac{|R|}{n}$$

for $p = 3$. Various one- and two-sample tests on the parameters can be performed using $\boldsymbol{R}$. See, for instance, Chapters 6 and 8 of Mardia (1972). One important preliminary test is to verify if indeed there is a preferred direction, i.e., H_0: $\kappa = 0$. The UMP invariant test for this is based on the length of the resultant, $|R|$, whose density under the null hypothesis of uniform distribution (or random walk model) is given (for $r > 0$) by

$$r\int_0^\infty J_0(rt)J_0^n(t)t\,\mathrm{d}t$$

when $p = 2$, and

$$\frac{r}{2^{n-1}(n-2)!}\sum_{j=0}^{n}(-1)^j\binom{n}{j}\langle n-r-2j\rangle^{n-2}$$

when $p = 3$. Here $\langle x\rangle = x$ if $x > 0$ and 0 otherwise. This test of 'no preferred direction', i.e., of H_0: $\kappa = 0$ which is based on $|R|$, is known as Rayleigh's test.

3. Nonparametric methods for circular data

Though considerable statistical theory has been developed for the von Mises–Fisher distribution and to a much lesser extent for some of the other parametric models for directions, these models may not provide an adequate description of the data or the distributional information may be imprecise. For instance, information about the unimodality or axial symmetry that a particular parametric model assumes may be lacking or might be inappropriate for a given data set. The search for methods which are robust leads naturally, as in linear statistical inference, to techniques which are nonparametric or model-free. In linear inference, there are a number of considerations on which one can justify for instance an assumption of normality as for example when one deals with averages, or when the samples are large enough. Unfortunately, there is no corresponding rationale for invoking the von Mises–Fisher distribution and thus the need for model-free methods might indeed be stronger in directional data analysis.

This section will be subdivided into three subsections dealing with one-, two- and multi-sample nonparametric techniques.

3.1. One-sample tests and the goodness-of-fit problem

For simplicity, let us assume that the circle has unit circumference and that the circular data is presented in terms of angles $(\alpha_1, \ldots, \alpha_n)$ with $0 \leqslant \alpha_i < 1$ with respect to some arbitrary zero direction. Given such a random sample, one of the fundamental problems in circular data is to test if there is no preferred direction against the alternative of one (or more) preferred direction(s). Since having no preferred direction corresponds to a uniform (or isotropic) distribution, the null hypothesis to test is

$$H_0: \quad \alpha \sim \text{uniform distribution on } [0, 1)\,. \tag{3.1}$$

As in the linear case, the goodness-of-fit problem of testing whether the sample came from a specified circular distribution can also be reduced to testing uniformity on the circle.

We seek rotationally invariant tests, i.e., tests invariant under changes in zero direction as well as the sense of rotation (clockwise or anticlockwise). There are three broad groups of tests for this problem, which are described below.

(i) Tests based on sample arc lengths or spacings. If $\alpha_{(1)} \leqslant \cdots \leqslant \alpha_{(n)}$ denote the order statistics in the linear sense, the differences

$$\alpha_i^* = (\alpha_{(i)} - \alpha_{(1)}), \quad i = 2, \ldots, n\,, \tag{3.2}$$

form a maximal invariant. But if one defines

$$D_i = (\alpha_{(i)} - \alpha_{(i-1)}), \quad i = 1, \ldots, n\,, \tag{3.3}$$

with $\alpha_{(0)} = (\alpha_{(n)} - 1)$, these are the lengths of the arcs into which the sample partitions the unit circumference and are called the sample spacings. Clearly $\alpha_i^* = \Sigma_{j=2}^{i} D_j$. Any symmetric function of the sample spacings will have the rotational invariance property, and Rao (1969) suggested the use of such a class of spacings tests for testing H_0 in (3.1). See Rao (1976) and the references contained there. In particular the statistic

$$\frac{1}{2}\sum_{i=1}^{n}\left|D_i - \frac{1}{n}\right| = \sum_{i=1}^{n}\max\left(D_i - \frac{1}{n}, 0\right) \tag{3.4}$$

corresponds to the uncovered portion of the circumference when n arcs of length $(1/n)$ are placed to cover the circumference starting at each of the observations. Its exact and asymptotic distributions and a table of percentage points are given in Rao (1976) and reproduced in Batschelet (1981). Among all such symmetric test statistics, the one based on $\Sigma_{i=1}^{n}(D_i - 1/n)^2$, which is referred to as the Greenwood statistic, has asymptotically maximum local power. Burrows (1979), Currie (1981) and Stephens (1981) discuss computational methods for obtaining the percentage points of Greenwood's statistic. See also Rao and Kuo (1984) for a discussion of some variants of this statistic which are asymptotically better. Another group of spacings statistics are based on ordered spacings. In particular, if $D_{(n)} = \max_{1\leq i\leq n} D_i$, then $R_n = (1 - D_{(n)})$ is referred to as the 'circular range', the shortest arc on the circumference containing all the observations. This is discussed in Rao (1969) and Laubscher and Rudolph (1968).

(ii) Tests based on empirical distribution functions. Given the random sample $\alpha_1, \ldots, \alpha_n$ on the circumference $[0, 1)$, one can define the empirical distribution function (in the usual linear sense) as

$$F_n(x) = \frac{\text{number of } \alpha_i \leq x}{n} \tag{3.5}$$

for $0 \leq x < 1$. The usual test statistics like the Kolmogorov–Smirnov statistic

$$K_n = \sqrt{n} \sup_{0\leq x<1} |F_n(x) - x| \tag{3.6}$$

or the Cramer–von Mises statistic

$$W_n^2 = n\int_0^1 (F_n(x) - x)^2\,\mathrm{d}x \tag{3.7}$$

do not have the required invariance property. Kuiper (1960) suggested the following variation of (3.6) which is rotationally invariant and hence usable with circular data. Let

$$D_n^+ = \sup_{0\leq x<1} (F_n(n) - x) \quad \text{and} \quad D_n^- = \sup_{0\leq x<1} (x - F_n(x))\,. \tag{3.8}$$

While the Kolmogorov–Smirnov statistic $K_n = \max(D_n^+, D_n^-)$, Kuiper's statistic

$$V_n = \sqrt{n}(D_n^+ + D_n^-)\,. \tag{3.9}$$

Its asymptotic null distribution (under the hypothesis (3.1) of uniformity) is given by (cf. Kuiper, 1960)

$$P(V_n \geq x) = \sum_{m=1}^{\infty} 2(4m^2x^2 - 1)\,\mathrm{e}^{-2m^2x^2} - \frac{8x}{3\sqrt{n}} \sum_{m=1}^{\infty} m^2(4m^2x^2 - 3)\,\mathrm{e}^{-2m^2x^2} + \mathrm{O}\Big(\frac{1}{n}\Big) \tag{3.10}$$

for $x \geq 0$. Stephens (1965) provides upper percentage points for small n. Watson (1961) defined an invariant version of (3.7), namely

$$U_n^2 = n \int_0^1 \Big[F_n(x) - x - \int_0^1 (F_n(y) - y)\,\mathrm{d}y\Big]^2 \mathrm{d}x \tag{3.11}$$

for use with circular data. Observe that U_n^2 is of the form of a variance while W_n^2 is like the second moment. The asymptotic null distribution is given by (refer to Watson, 1961)

$$\lim_{n\to\infty} P(U_n^2 > x) = 2 \sum_{m=1}^{\infty} (-1)^{m-1}\,\mathrm{e}^{-2m^2\pi^2 x} \tag{3.12}$$

for $x \geq 0$.

(iii) Scan statistics and chi-square-type tests. Ajne (1968) suggested two test statistics based on the number of observations in a half-circle

$$N(\alpha) = \text{number of observations in } [\alpha, \alpha + \tfrac{1}{2})$$

for $0 \leq \alpha < 1$. One of them is to take

$$N = \sup_{0 \leq \alpha < 1} N(\alpha)\,, \tag{3.13}$$

the maximum number in any half-circle. As Rao (1969) and Bhattacharya and Johnson (1969) pointed out, this is related to a bivariate sign test suggested earlier by Hodges (1955). The exact null distribution of N is given by (cf. Ajne, 1968)

$$P(N \geq k) = \frac{(2k - n)}{2^{n+1}} \sum_{j=0}^{\infty} \binom{n}{k + j(2k - n)} \tag{3.14}$$

for $k \geq [n/2] + 1$ which reduces for $k > \frac{2}{3}n$ to the simpler expression

$$P(N \geq k) = \frac{2k - n}{2^{n+1}} \binom{n}{k}\,.$$

The asymptotic null distribution of

$$N^* = \frac{2}{\sqrt{n}}\left(N - \frac{n}{2}\right)$$

is given by

$$\lim_{n\to\infty} P(N^* > c) = 2c\sqrt{\frac{2}{\pi}} \sum_{j=0}^{\infty} e^{-(2j+1)^2c^2/2} \tag{3.15}$$

for $c \geq 0$. Rothman (1972) considered the maximum number of observations in any arc of length p, $0 < p < 1$. These are related to the scan statistics. See, for instance, Naus (1982).

Another statistic for testing uniformity is obtained by averaging, i.e., by considering

$$A_n = \int_0^1 \left(N(\alpha) - \frac{n}{2}\right)^2 d\alpha . \tag{3.16}$$

The asymptotic null distribution of A_n is given by

$$\lim_{n\to\infty} P(A_n > x) = \sum_{m=1}^{\infty} \frac{4(-1)^{m-1}}{\pi(2m-1)} e^{-\pi^2(2m-1)^2x/2}$$

for $x \geq 0$. The statistic in (3.16) has been generalized in two different directions by Beran (1969a) and Rao (1972b). Rao (1972b) considered dividing the unit circumference into m (≥ 2) equal class intervals with the i-th interval being

$$I_i(\alpha) = \left[\alpha + \left(\frac{i-1}{m}\right), \alpha + \frac{i}{m}\right), \quad i = 1, \ldots, m .$$

Using the observed class frequencies $N_i(\alpha)$, $i = 1, \ldots, m$ in these intervals, one can construct a χ^2 statistic $\chi_n^2(\alpha) = \Sigma_{i=1}^m (N_i(\alpha) - n/m)^2/(n/m)$. This can be made invariant with respect to the choice of origin α by taking either the supremum over α or alternately by averaging as in (3.16), namely,

$$\chi_n^2 = \int_0^1 \chi_n^2(\alpha)\, d\alpha . \tag{3.17}$$

The statistic A_n corresponds to χ_n^2 with $m = 2$. The asymptotic null distribution of χ_n^2 in (3.17) and a computational form for it are provided in Rao (1972b).

Beran (1969a) proposed the class of test statistics of the form

$$T_n = \int_0^1 \left[\sum_{i=1}^n (f(\alpha + \alpha_i) - 1)\right]^2 d\alpha \tag{3.18}$$

where f is any probability density function on the circle. It can be verified that

A_n and U_n^2 defined in (3.16) and (3.11) are of this form. Beran (1969a) obtains the asymptotic distribution of this statistic under the null hypothesis (3.1) as well as under fixed alternatives and derives the approximate Bahadur slope. Tests based on T_n are best invariant against local alternatives (i.e., for small k) of the form

$$g(\alpha; k) = 1 + k(f(\alpha + \mu_0) - 1), \quad 0 \leq \alpha < 1, \; 0 \leq k \leq 1 .$$

If we define

$$h(\theta) = 2 \sum_{p=1}^{\infty} \rho_p^2 \cos p\theta$$

where

$$\rho_p^2 = \left| \int_0^1 \mathrm{e}^{\mathrm{i}p\alpha} f(\alpha)\, \mathrm{d}\alpha \right|^2 ,$$

then T_n can be rewritten as

$$T_n = \sum_{i=1}^{n} \sum_{j=1}^{n} h(\theta_i - \theta_j) . \tag{3.19}$$

See for example Mardia (1975, pp. 190–191).

Some comparisons. Rao (1972a) compared the various tests of uniformity through the Bahadur efficiencies using von Mises–Fisher alternatives (cf. Equation (2.3)) with small concentration. For this situation Rayleigh's test based on the length $|R|$ of the resultant is uniformly most powerful invariant. These comparisons based on Bahadur efficiencies show that Watson's U_n^2 (cf. (3.11)) and Ajne's A_n (cf. (3.16)) tests have the same asymptotic efficiencies as the Rayleigh test, while Kuiper's test (cf. (3.9)) and the Hodges–Ajne N test (cf. (3.13)) have a lower asymptotic efficiency of $8/\pi$ or about 81% compared to the former group of test statistics. Symmetric spacings tests have poor asymptotic efficiencies but Monte Carlo comparisons show that for small samples, they have reasonable power compared to Rayleigh's test. Stephens (1969) compares Kuiper's V_n, Watson's U_n^2 and Ajne's A_n using Monte Carlo powers. His conclusions indicate that while these three tests are about equal in performance against unimodal alternatives, differences show up in testing uniformity against multimodal alternatives with Kuiper's V_n faring the best and U_n^2 and A_n following in that order.

Other one-sample tests. Schach (1969b) applies the Wilcoxon signed rank test for testing the hypothesis of symmetry. Rothman (1971) and Puri and Rao (1977) consider the problem of testing coordinate independence given bivariate data on a torus. There are a number of papers on the topic of developing measures of association for angular–angular or angular–linear data, some of them nonparametric. See Jupp and Mardia (1980) and other references contained there for measures of correlation for the angular–angular case, i.e., for observations on a torus. Fisher and Lee (1981) develop a U-statistic which is analogous to Kendall's τ for measuring angular linear association and derive its distribution.

Lenth (1981) utilizes a periodic version of the commonly used ψ functions to adapt robust M-estimation for use with directional data.

3.2. Two-sample tests

The usual nonparametric theory for the two-sample observations in the combined sample. See, e.g., Hajek and Sidak (1967). Since rank values on a circle depend on the starting point as well as the sense of rotation, such rank tests cannot be used with circular data. Schach (1969a) defines what may be called the 'circular ranks', which remain invariant under the following group of transformations. Let $(\alpha_1, \ldots, \alpha_n)$ and $(\beta_1, \ldots, \beta_n)$ denote the two independent random samples from F and G respectively, the hypothesis of interest being

$$H_0: \quad F(\alpha) = G(\alpha), \; 0 \leq \alpha < 1 . \tag{3.20}$$

Let $(r_1, \ldots, r_m)$ denote the (linear) ranks of the first sample in the combined sample of $N = (m + n)$ observations in the usual fashion, and let

$$R = \{(r_1, \ldots, r_N) : (r_1, \ldots, r_N) \text{ is a permutation of integers} (1, \ldots, N)\}$$

be the space of rank vectors for the combined sample. Define groups of transformations $\{g\}$ (corresponding to changes in zero direction) and $\{h\}$ (corresponding to changes in sense of rotation), of R onto itself by

$$g: (r_1, \ldots, r_N) \to (r_1 + 1, \ldots, r_N + 1)$$

and

$$h: (r_1, \ldots, r_N) \to (N + 1 - r_N, \ldots, N + 1 - r_1)$$

where the components of the transformed vector are defined modulo N. Let $\mathcal{G}$ be the group of transformations $R \to R$ generated by $\{g\}$ and $\{h\}$. We may define circular ranks $(C_1, \ldots, C_m)$ of $(\alpha_1, \ldots, \alpha_m)$ as an equivalence class of $(r_1, \ldots, r_m)$ under the group $\mathcal{G}$. One can then define, corresponding to any linear rank test $T(\boldsymbol{r})$ based on linear ranks $\boldsymbol{r}$, a circular rank test

$$T(\boldsymbol{c}) = \sup_{g^* \in \mathcal{G}} T(g^*(\boldsymbol{r}))$$

which will then possess the required invariance. Batschelet (1965) suggested such an invariant version of the Wilcoxon–Mann–Whitney statistic and provided a short table of critical values. Epplett (1982) pursues this further and obtains its asymptotic null distribution.

Let $F_m(x)$ and $G_n(x)$ denote the empirical distribution functions of the α's and β's, respectively. Define

$$D^+_{m,n} = \sup_{0\le x<1} [F_m(x) - G_n(x)] \tag{3.21}$$

and

$$D^-_{m,n} = \sup_{0\le x<1} [G_n(x) - F_m(x)] . \tag{3.22}$$

Let $\boldsymbol{r} = (r_1, \ldots, r_m)$ denote the linear ranks and

$$W_{m,n}(\boldsymbol{r}) = \sum_{i=1}^{m} r_i \tag{3.23}$$

the Wilcoxon test statistic. Then Epplett (1982) shows that the circular version

$$\begin{aligned} W_{m,n}(\boldsymbol{c}) &= \sup_{g^*\in\mathcal{G}} W_{m,n}(g^*(\boldsymbol{r})) \\ &= \max\{W_{m,n}(\boldsymbol{r}) + mnD^+_{m,n},\, m(N+1) - W_{m,n}(\boldsymbol{r}) + mnD^-_{m,n}\} \end{aligned} \tag{3.24}$$

and establishes that $\sqrt{mnN}\{W_{m,n}(\boldsymbol{c}) - \frac{1}{2}m(N+1)\}$ converges in distribution to that of $\sup_t|S(t)|$ where $S(t)$ is a Gaussian process with zero mean and covariance kernel

$$K(s, t) = \tfrac{1}{12} - \tfrac{1}{2}(t-s)(1-(t-s))$$

for $0 \le s \le t \le 1$. This test is shown to compare favorably with the two-sample Kuiper test (see Equation (3.25)) in terms of Bahadur efficiency. Through inclusion–exclusion, Epplett (1979) relates the exact probabilities for the circular statistic to those of the linear Wilcoxon statistic and provides a recurrence relation.

(i) Tests based on empirical distribution functions. Since the two-sample versions of the Kolmogorov–Smirnov and Cramer–von Mises statistics are not rotationally invariant, they are inappropriate for testing the hypothesis (3.20). Kuiper (1960) suggested the following two-sample variation of the Kolmogorov–Smirnov statistic:

$$V_{m,n} = (D^+_{m,n} + D^-_{m,n}) \tag{3.25}$$

where $D^+_{m,n}$ and $D^-_{m,n}$ are as defined in (3.21) and (3.22). Its asymptotic null distribution, properly normalized, is the same as that given in (3.10). Barr and Shudde (1973) show that

$$V_{m,n} = \sup_{g^*\in\mathcal{G}} D_{m,n}(g^*(\boldsymbol{r})) \tag{3.26}$$

where $D_{m,n} = \max(D^+_{m,n}, D^-_{m,n})$ is the usual two-sample Kolmogorov–Smirnov statistic. Similarly, a two-sample version of (3.11) was proposed by Watson (1962), namely

$$U^2_{m,n} = \frac{mn}{N}\int_0^1 \left[F_m(x) - G_n(x) - \int_0^1 (F_m(y) - G_n(y))\,\mathrm{d}H_N(y)\right]^2 \mathrm{d}H_N(x) \tag{3.27}$$

where $F_m(x)$ and $G_n(x)$ are the empirical distribution functions of α's and β's respectively and $H_N(x) = [mF_m(x) + nG_n(x)]/N$. The asymptotic null distribution of $U^2_{m,n}$ is again the same as that given in Equation (3.12).

(ii) Tests based on uniform scores. Beran (1969b) pointed out that two-sample tests for the hypothesis (3.20) can be obtained from tests of uniformity as follows: If $(r_1, \ldots, r_m)$ denote the (linear) ranks of the first sample in the combined sample, then define

$$u_i = r_i/N, \quad i = 1, \ldots, m, \tag{3.28}$$

called the 'uniform scores'. Under the null hypothesis $F = G$, these scores must be uniformly distributed on the circle of unit circumference. Thus any test of uniformity discussed in Section 3.1 can then be used on $\{u_i\}$ to test the hypothesis (3.20). A test which rejects this hypothesis for large values of $|\boldsymbol{R}_1|$, the length of the resultant of $\{u_i, i = 1, \ldots, m\}$ was proposed by Wheeler and Watson (1964). Mardia (1967) considered the statistic based on $|\boldsymbol{R}_1|$ in connection with a bivariate location problem. Mardia (1969) and Schach (1969a) discuss the asymptotic power and consistency properties of the Wheeler and Watson statistic. Schach (1969a) considers a general class of statistics of the form

$$T_{m,n} = \sum_{i=1}^{m}\sum_{j=1}^{m} h_N(u_i - u_j) \tag{3.29}$$

which corresponds to the two-sample adaptation of Beran's statistic T_n (cf. (3.19)) and shows that the asymptotic null distribution of $((N-1)T_{m,n}/mn)$ is the same as that of the one-sample statistic T_n if $n/N \to \lambda$, $0 < \lambda < 1$.

(iii) Tests based on spacing-frequencies. For the general two-sample problem, Holst and Rao (1980) investigate families of statistics based on the 'spacing-frequencies'. These are the frequencies of one sample, say β_j's, that fall in between the spacings made by the other sample. Thus the spacing-frequencies are defined by

$$S_i = \text{number of } \beta_j\text{'s in } [\alpha_{(i-1)}, \alpha_{(i)}), \quad i = 1, \ldots, m, \tag{3.30}$$

where $\{\alpha_{(i)}\}$ are ordered. Statistics based symmetrically on $\{S_i, i = 1, \ldots, m\}$ are clearly rotation invariant. Thus one may use test statistics of the form

$$T_{m,n} = \sum_{i=1}^{m} h_N(S_i) \tag{3.31}$$

for 'reasonable' functions $h_N(\cdot)$. The circular run test (cf. David and Barton, 1962) and a test suggested by Dixon (1940) based on $\Sigma_1^m S_i^2$, are special cases of this form. Holst and Rao (1980) show that under mild conditions on $h_N(\cdot)$, the statistics $T_{m,n}$ are asymptotically normal and that the Dixon test based on $\Sigma_1^m S_i^2$ is asymptotically locally most powerful among this class. Some further power comparisons and the special relevance of this class (3.31) to circular data problems are discussed in Rao and Mardia (1980). More recently, tests based on k-th order spacing-frequencies (for fixed finite k), i.e., on $S_i^{(k)}$ = the number of observations in $[\alpha_{(i-1)}, \alpha_{(i+k-1)})$, are considered in Rao and Schweitzer (1982) where it is shown that among tests symmetric in $\{S_i^{(k)}\}$, which can be used for circular distributions, $\Sigma_{i=1}^m S_i^{(k)2}$ is asymptotically locally most powerful.

3.3. *Multi-sample tests for circular data*

Given a random sample of size n_i, say $(\alpha_{i1}, \ldots, \alpha_{in_i})$ from the i-th population, $i = 1, \ldots, k$, one is interested in tests of homogeneity of these k populations. If these populations are unimodal, such tests of homogeneity (a) with respect to mean directions and (b) with respect to concentrations are proposed for large samples in Rao (1966) and are further discussed in Yoshimura (1978).

A k-sample analogue of Watson's $U_{m,n}^2$ (see Equation (3.27)) and its asymptotic null distribution are discussed in Maag (1966). Multisample analogues of other statistics like $V_{m,n}$ (Equation (3.25)) do not appear to have been considered in the literature. A test based on multiple runs on the circle is discussed in David and Barton (1962, pp. 119–136) and this has a long history. One can also construct tests based on 'uniform scores'

$$u_{ij} = r_{ij}/N \tag{3.32}$$

where $\{r_{ij}, u = 1, \ldots, n_i\}$ are the ranks of the i-th sample observations among the combined sample of $N = (n_1 + \cdots + n_k)$ observations. Mardia (1972b) considers a test based on the statistic

$$2\sum_{i=1}^{k} (R_i^2/n_i) \tag{3.33}$$

where

$$R_i^2 = \Big(\sum_{j=1}^{n_i} \cos 2\pi u_{ij}\Big)^2 + \Big(\sum_{j=1}^{n_i} \sin 2\pi u_{ij}\Big)^2$$

is the squared length of the resultant for the uniform scores of the i-th sample. The statistic in (3.33) corresponds to the log likelihood ratio for testing homogeneity of mean directions of k von Mises–Fisher distributions and may also be thought of as an extension of the Wheeler and Watson test mentioned earlier. Mardia (1972b) also shows that the test in (3.33) when compared with the parametric competitor for the von Mises–Fisher distributions, has a Bahadur efficiency approaching one as the concentration parameter of the von Mises–Fisher distributions approaches zero.

4. Nonparametric methods for spherical data

Most of statistical methods developed for one-, two- or multi-sample spherical problems assume parametric models like the von Mises–Fisher distribution in (2.4) or other distributions with different properties. There has not been, in general, as much progress in developing useful nonparametric tests for spherical data. Though this is somewhat analogous to the situation in regard to nonparametric procedures for multivariate statistical analyses, it seems that there is scope for progress here along the lines of Puri and Sen (1971).

4.1. One-sample problem-testing uniformity

Beran (1968) considered a class of statistics similar to (3.18) for the more general problem of testing uniformity on a compact homogeneous space. He shows that this test is locally most powerful invariant and derives the asymptotic null distribution of the statistic. Specifically for the sphere, an analogue of (3.19) is based on

$$T_n = \frac{n}{4} - \frac{1}{\pi n} \sum_{i<j} \psi_{ij} \tag{4.1}$$

where ψ_{ij} is the smaller angle between the i-th and j-th observations (in polar coordinates) (α_i, β_i) and (α_j, β_j). Gates and Westcott (1980) discuss bounds on the distribution of the minimum interpoint angular distance $\psi = \min_{i,j} \psi_{ij}$ under the hypothesis of uniformity. Giné (1975) considers a class of invariant tests for uniformity based on Sobolev norms, which contains as special cases tests for uniformity on the circle, the sphere and the hemisphere (where the antipodes are identified) introduced earlier by Rayleigh, Watson (1961), Ajne (1968), Rao (1972b), Beran (1968) and Bingham (1964). Prentice (1978) follows along the lines of Giné (1975) and Beran (1968) to obtain a class of invariant tests for spheres and hemispheres in any dimension $p \geqslant 1$. Stephens (1966) tabulates the percentage points of three statistics which are useful in testing uniformity on the sphere against specified alternatives listed.

4.2. Two-sample tests for spherical data

Wellner (1979) considers a class of permutation tests for the two-sample problem when the data come from any arbitrary compact Riemannian manifold. Special cases of interest include tests for comparing two samples from the unit sphere in three dimensions or the hemisphere or the torus. The test statistics are the two-sample analogues of Gine's (1975) tests of uniformity. As with any permutation test, the idea is to calculate an invariant statistic $T_{m,n}$ for all $\binom{m+n}{n}$ choices of the first sample relabellings and reject the null hypothesis of identical distributions if the observed $T_{m,n}$ is 'too big' relative to the resulting conditional distribution (conditional on the pooled sample). Specified consistency properties and asymptotic distributions under the null and fixed alternatives are derived. Two-sample versions of the various test statistics of uniformity are discussed as examples.

References

Ajne, B. (1968). A simple test for uniformity of a circular distribution. *Biometrika* **55**, 343–354.

Barr, D. and Shudde, R. H. (1973). A note on Kuiper's V_n statistics. *Biometrika* **60**, 663–664.

Batschelet, E. (1965). *Statistical Methods for the Analysis of Problems in Animal Orientation and Certain Biological Rhythms*, Am. Inst. of Bio. Sciences, Washington.

Batschelet, E. (1981). *Circular Statistics in Biology*, Academic Press, London.

Beran, R. J. (1968). Testing for uniformity on a compact homogeneous space. *J. Appl. Prob.* **5**, 177–195.

Beran, R. J. (1969a). Asymptotic theory of a class of tests for uniformity of a circular distribution. *Ann. Math. Statist.* **40**, 1196–1206.

Beran, R. J. (1969b). The derivation of nonparametric two sample tests from tests for uniformity of a circular distribution. *Biometrika* **56**, 561–570.

Bhattacharya, G. K. and Johnson, R. A. (1969). On Hodges' bivariate sign test and a test for uniformity of a circular distribution. *Biometrika* **56**, 446–449.

Bingham, C. (1964). Distributions on the sphere and on the projective plane, Ph.D. thesis, Yale University.

Burrows, P. M. (1979). Selected percentage points of Greenwood's statistic. *J. R. Statist. Soc. Ser. A* **142**, 256–258.

Currie, F. D. (1981). Further percentage points of Greenwood's statistic. *J. R. Statist. Soc. Ser. A* **144**, 360–363.

David, F. N. and Barton, D. E. (1962). *Combinatorial Chance*, Griffin and Co., London.

Dixon, W. J. (1940). A criterion for testing the hypothesis that two samples are from the same population. *Ann. Math. Statist.* **11**, 199–204.

Eplett, W. J. R. (1979). The small sample distribution of a Mann–Whitney type statistic for circular data. *Ann. Statist.* **7**, 446–453.

Eplett, W. J. R. (1982). Two Mann–Whitney type rank tests. *J. R. Statist. Soc. Ser. B* **44**, 270–286.

Fisher, N. I. and Lee, A. J. (1981). Nonparametric measures of angular–linear association. *Biometrika* **68**, 629–636.

Fisher, R. A. (1953). Dispersion on a sphere. *Proc. Roy. Soc. Lond. A* **217**, 295–305.

Gates, D. J. and Westcott, M. (1980). Further bounds for the distribution of minimum interpoint distance on a sphere. *Biometrika* **67**, 466–469.

Giné, E. (1975). Invariant tests for uniformity on compact Riemannian manifolds based on Sobolev norms. *Ann. Statist.* **3**, 1243–1266.

Hajek, J. and Sidak, Z. (1967). *Theory of Rank Tests*, Academic Press, New York.

Hodges, J. L., Jr. (1955). A bivariate sign test. *Ann. Math. Statist.* **26**, 523–527.

Holst, L. and Rao, J. S. (1980). Asymptotic theory for families of two-sample nonparametric statistics. *Sankhya Ser. A* **42**, 19–52.

Jupp, P. E. and Mardia, K. V. (1980). A general correlation coefficient for directional data and related regression problems. *Biometrika* **67**, 163–173.

Kuiper, N. H. (1960). Tests concerning random points on a circle. *Proc. Koninkl. Nederl. Akad. Van Wetenschappen Ser. A*, **63**, 38–47.

Laubscher, N. F. and Rudolph, G. J. (1968). A distribution arising from random points on the circumference of a circle. *Nat. Res. Inst. Math. Sci. Rep.* **268**, Pretoria, South Africa, pp. 1–15.

Lenth, R. V. (1981). Robust measures of location for directional data. *Technometrics* **23**, 77–81.

Maag, U. R. (1966). A k-sample analogue of Watson's U^2 statistic. *Biometrika* **53**, 579–583.

Mardia, K. V. (1967). A nonparametric test for the bivariate two-sample location problem. *J. R. Statist. Soc. Ser. B* **29**, 320–342.

Mardia, K. V. (1969). On Wheeler and Watson's two-sample test on a circle. *Sankhya Ser. A.* **31**, 177–190.

Mardia, K. V. (1972a). *Statistics of Directional Data*, Academic Press, London.

Mardia, K. V. (1972b). A multisample uniform scores test on a circle and its parametric competitor. *J. R. Statist. Soc. Ser. B* **34**, 102–113.

Mardia, K. V. (1975). Statistics of directional data. *J. R. Statist. Soc. Ser. B* **37**, 349–393.

Naus, J. I. (1982). Approximations for distributions of scan statistics. *J. Amer. Statist. Assoc.* **77**, 177–183.

Prentice, M. J. (1978). On invariant tests of uniformity for directions and orientations. *Ann. Statist.* **6**, 169–176.

Puri, M. L. and Rao, J. S. (1977). Problems of association for bivariate circular data and a new test of independence. In: P. R. Krishnaiah, ed., *Multivariate Analysis, Vol. 4.* North-Holland Publishing Co., Amsterdam.

Puri, M. L. and Sen, P. K. (1971). *Nonparametric Methods in Multivariate Analysis*, Wiley, New York.

Rao, J. S. (1966). Large sample tests for homogeneity of angular data (Appendix to a paper by S. Sengupta and J. S. Rao). *Sankhya Ser. B* **28**, 172–174.

Rao, J. S. (1969). Some contributions to the analysis of circular data, Ph.D. thesis, Indian Statistical Institute, Calcutta.

Rao, J. S. (1972a). Bahadur efficiencies of some tests for uniformity on the circle. *Ann. Math. Statist.* **43**, 468–479.

Rao, J. S. (1972b). Some variants of chi-square for testing uniformity on the circle. *Z. Wahrsch. Verw. Geb.* **22**, 33–44.

Rao, J. S. (1976). Some tests based on arc lengths for the circle. *Sankhya Ser. B* **38**, 329–338.

Rao, J. S. and Kuo, M. (1984). Asymptotic results on the Greenwood statistic and some of its generalizations. *J. Roy. Statist. Soc. Ser. B* **46**.

Rao, J. S. and Mardia, K. V. (1980). Pitman efficiencies of some two-sample nonparametric tests. In: K. Matusita, ed., *Recent Developments in Statistical Inference and Data Analysis.* North-Holland, Amsterdam.

Rao, J. S. and Schweitzer, R. L. (1982). On tests for the two sample problem based on higher order spacing-frequencies. To appear.

Rothman, E. D. (1971). Tests of coordinate independence for a bivariate sample on a torus. *Ann. Math. Statist.* **42**, 1962–1969.

Rothman, E. D. (1972). Tests of uniformity for circular distributions. *Sankhya Ser. A* **34**, 23–32.

Schach, S. (1969a). On a class of nonparametric two-sample tests for circular distributions. *Ann. Math. Statist.* **40**, 1791–1800.

Schach (1969b). Nonparametric symmetry tests for circular distributions. *Biometrika* **56**, 571–577.

Stephens, M. A. (1965). The goodness of fit statistic V_n: distribution and significance points. *Biometrika* **52**, 309–321.

Stephens, M. A. (1966). Statistics connected with the uniform distribution: percentage points and application to testing for randomness of directions. *Biometrika* **53**, 235–240.

Stephens, M. A. (1969). A goodness of fit statistic for the circle with some comparisons. *Biometrika* **56**, 161–168.

Stephens, M. A. (1981). Further percentage points of Greenwood's statistic. *J. R. Statist. Soc. Ser. A* **144**, 364–366.

von Mises, R. (1981). Über die 'Ganzzahligkeit' der Atomgewicht und verwandte fragen. *Physikal Z.* **19**, 490–500.

Watson, G. S. (1961). Goodness of fit tests on a circle. *Biometrika* **48**, 109–114.

Watson, G. S. (1962). Goodness of fit tests on a circle II. *Biometrika* **49**, 57–63.

Watson, G. S. (1983). *Statistics on the sphere.* Wiley, New York.

Wellner, J. A. (1979). Permutational tests for directional data. *Ann. Statist.* **7**, 929–943.

Wheeler, S. and Watson, G. S. (1964). A distribution free two sample test on a circle. *Biometrika* **51**, 256–257.

Yoshimura, I. (1978). On a test of homogeneity hypothesis for directional data. *Sankhya Ser. A* **40**, 310–312.

PART VIII

APPLICATIONS

P. R. Krishnaiah and P. K. Sen, eds., *Handbook of Statistics, Vol. 4*

32

Application of Nonparametric Statistics to Cancer Data

H. S. Wieand

1. Introduction

This chapter deals with applications of nonparametric statistics to cancer data. The approach is to present data sets and to use nonparametric statistics in the analysis of the data, giving considerable detail when computations are involved. Appropriate references are given for the reader who wants to learn more about the theory and derivations of the statistics.

In Section 2, a single data set consisting only of times to 'failure' or 'censor' for patients with high-grade ovarian carcinoma is analyzed. The Kaplan–Meier estimated survival curve is used to obtain estimates of the mean, median and cumulative hazard function. The actuarial approach is also considered. In Section 3, data are presented for a second set of patients with low-grade carcinoma and tests are performed to see if difference in grade is associated with time to progression of disease. As before, the only information available is time to 'failure' or 'censor'. The statistics used in the analysis include the Mantel–Haenszel, Gehan–Wilcoxon, Prentice–Wilcoxon and Kolmogorov–Smirnov statistics. The section ends with a discussion of the extension of Mantel–Haenszel and Gehan–Wilcoxon to more than two samples. In Section 4, a data set for patients with some known covariates, such as age at entry, is presented. The importance of identifying important covariates before comparing treatments is demonstrated. Special emphasis is given to the Cox proportional hazards model and stratified Mantel–Haenszel statistics.

Throughout the chapter, T will denote a positive random variable with continuous distribution function

$$F(t) = P(T \leq t), \tag{1.1}$$

survival function

$$S(t) = P(T \geq t), \tag{1.2}$$

and density function

$$f(t)=\frac{\mathrm{d}}{\mathrm{d}t}F(t)\,. \tag{1.3}$$

We will define the hazard function by

$$\lambda(t)=-\frac{\mathrm{d}}{\mathrm{d}t}\log S(t)=\frac{f(t)}{S(t)} \tag{1.4}$$

and the cumulative hazard function by

$$\Lambda(t)=\int_0^t \lambda(u)\,\mathrm{d}u\,. \tag{1.5}$$

When observation times are presented, censored values will be followed by a + sign. For example, $t=100$ will mean that a failure occurred at time 100, while $t=100+$ will mean that a censoring occurred at time 100. In any analysis which follows, 100+ will be considered to be greater than 100.

In general, sample values will be distinguished from population values by a ˆ. For example, $\hat{S}(t)$ is the sample value of the survival function at time t and it is an estimate of the population value $S(t)$.

2. The one-sample problem

The data set considered in this section is taken from an example appearing in a paper by Fleming et al. (1980). It consists of observations of time to progression of disease for patients with high-grade or undifferentiated Stage II or Stage III ovarian carcinoma who were followed at Mayo Clinic. The observation times were 34, 88, 137, 199, 280, 291, 299+, 300+, 309, 351, 358, 369, 369, 370, 375, 382, 392, 429+, 451 and 1119+ days.

We will begin our analysis by estimating the survival function, using the Kaplan–Meier (1958) product limit estimator. Let k be the number of distinct failure times. We will let t_i represent the i-th ordered failure time, n_i the number of patients at risk at time t_i, and d_i the number of deaths at time t_i. The Kaplan–Meier estimate is then defined by

$$\hat{S}(t)=\prod_{i:t_i\leq t}\left(1-\frac{d_i}{n_i}\right). \tag{2.1}$$

Table 2.1 gives the values of i, t_i, n_i, d_i and $\hat{S}(t)$ in the interval $[t_i, t_{i+1})$ for the ovarian carcinoma data. Notice that the statistic does not change values at 299+ since there is no failure. Furthermore, if 299+ had been a 291+ or 308+, none of the entries in the table would have changed. However, if the 299+ had been a 309+, n_7 would have been 13 and other changes would have resulted. Although it is not mentioned in the table, we define $\hat{S}(t)=1$ in the interval

Table 2.1
Mayo Clinic data. Values of the Kaplan–Meier estimator for the high-grade ovarian carcinoma patients

i	t_i	n_i	d_i	$\hat{S}(t)$
1	34.0	20	1	0.9500
2	88.0	19	1	0.9000
3	137.0	18	1	0.8500
4	199.0	17	1	0.8000
5	280.0	16	1	0.7500
6	291.0	15	1	0.7000
7	309.0	12	1	0.6417
8	351.0	11	1	0.5833
9	358.0	10	1	0.5250
10	369.0	9	2	0.4083
11	370.0	7	1	0.3500
12	375.0	6	1	0.2917
13	382.0	5	1	0.2333
14	392.0	4	1	0.1750
15	451.0	2	1	0.0875

$[0, t_1)$ and $\hat{S}(t) = \hat{S}(t_k) = 0.0875$ in the interval $[t_k, \infty)$. A plot of $\hat{S}(t)$ versus t (Figure 2.1) gives an overall survival picture. Note that from the plot one can see that a patient has roughly an 85% chance of surviving (without progression of the disease) for six months and a 50% chance of surviving without progression for one year.

It has been shown under fairly general conditions (Breslow and Crowley, 1974), that $\hat{S}(t)$ is asymptotically normal with mean $S(t)$ and a variance which

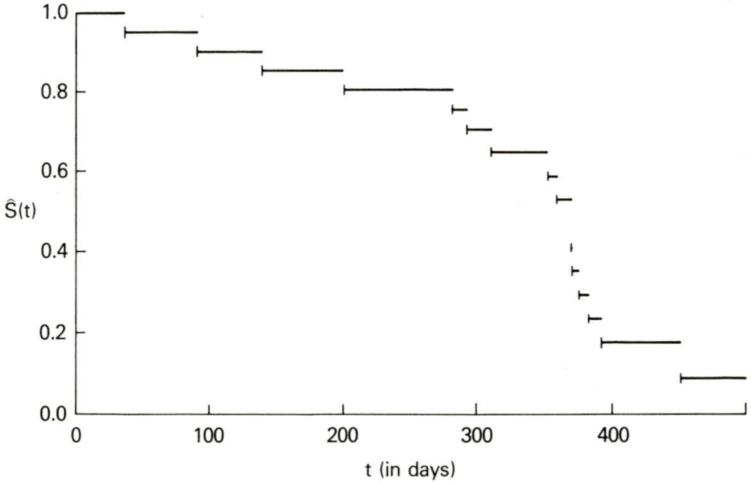

Fig. 2.1. Estimated survival function for the 20 high grade ovarian carcinoma patients.

can be approximated by

$$\hat{\sigma}_S^2(t) = \hat{S}^2(t) \sum_{i:t_i \le t} \frac{d_i}{n_i(n_i - d_i)} . \tag{2.2}$$

(2.2) is sometimes referred to as Greenwood's formula (Greenwood (1926)). Hence, an asymptotic $1-\alpha$ level confidence interval for $S(t)$ is

$$(\hat{S}(t) - z_{\alpha/2}\hat{\sigma}_S(t), \hat{S}(t) + z_{\alpha/2}\hat{\sigma}_S(t)) \tag{2.3}$$

where z_α satisfies $\alpha = \int_{z_\alpha}^{\infty} (2\pi)^{-0.5} \exp(-x^2/2)\,dx$. To obtain a 90% confidence interval for the probability that a patient with high-grade cancer will survive at least one year, we note that $z_{0.05} = 1.645$, $\hat{S}(365) = \hat{S}(358) = 0.5250$ and $(\hat{\sigma}_S(358)) = 0.11646$, which when used in (2.3) yield the interval (0.33, 0.72), i.e., we are 90% sure that the probability a patient will survive at least a year is greater than 0.33 but less than 0.72.

We can also use $\hat{S}(t)$ to help us estimate the mean (μ) and median (θ) of the survival time. The natural estimator of the median is:

$$\hat{\theta} = \hat{S}^{-1}(0.5) . \tag{2.4}$$

In our example, we cannot evaluate $\hat{S}^{-1}(0.5)$, but we know that $\hat{S}(358) = 0.525$ and $\hat{S}(369) = 0.4083$, so our estimate should be between 358 and 369. A linear smoothing would yield $\hat{\theta} = 358 + (369 - 358)(0.5250 - 0.5000)/(0.5250 - 0.4083) = 360.4$. An interpretation is that half the patients with high-grade cancer will have progression when 360 days have elapsed and half will not. This method of smoothing is often employed (see Miller, 1981) and is convenient, although others have been considered. Rai, Susarla and Van Ryzin (1980) discuss Bayes estimators which result in a different smoothing mechanism and which have smaller mean square errors in many cases.

Estimation of the mean is not as intuitive as that of the median, particularly if there are censoring times after the last failure. One approach, discussed in Miller (1981) is to let:

$$\hat{\mu} = \sum_{i=0}^{k} \hat{S}(t_i)(t_{i+1} - t_i) \tag{2.5}$$

where t_0 is assumed to be zero and t_{k+1} is the last observation time, regardless of whether it is a failure time or censored time. If the last time is a failure, $t_{k+1} = t_k$ and the last term on the RHS of (2.5) is zero. For the Mayo Clinic data, the value of the mean is 380 days (see Table 2.2). The variance of $\hat{\mu}$ can be estimated by

$$\hat{\sigma}_{\hat{\mu}}^2 = \sum_{i=1}^{k} \left(\sum_{j=i}^{k} \hat{S}(t_i)(t_{i+1} - t_i) \right)^2 \frac{d_i}{n_i(n_i - d_i)} . \tag{2.6}$$

Table 2.2
Mayo Clinic data. Values needed for computation of the mean survival time for high grade carcinoma patients

i	$\hat{S}(t_i)$	$t_{i+1}-t_i$	$\hat{S}(t_i)(t_{i+1}-t_i)$	$\sum_{j=i}^{K} \hat{S}(t_i)(t_{i+1}-t_i)$	$\left(\sum_{j=i}^{K} \hat{S}(t_i)(t_{i+1}-t_i)\right)^2 \frac{d_i}{n_i(n_i-d_i)}$
0	1.0000	34	34.00		
1	0.9500	54	51.30	345.86	314.78
2	0.9000	49	44.10	294.56	253.70
3	0.8500	62	52.70	250.46	205.00
4	0.8000	81	64.80	197.76	143.78
5	0.7500	11	8.25	132.96	73.66
6	0.7000	18	12.60	124.71	74.06
7	0.6417	42	26.95	112.11	95.22
8	0.5833	7	4.08	85.16	65.93
9	0.5250	11	5.78	81.08	73.04
10	0.4083	1	0.41	75.31	180.05
11	0.3500	5	1.75	74.90	133.57
12	0.2917	7	2.04	73.15	178.36
13	0.2333	10	2.33	71.11	252.83
14	0.1750	59	10.33	68.78	394.22
15	0.0875	668	58.45	58.45	1708.20
			379.87		4146.40

In this case, if the last time is a failure the term corresponding to $i = k$ is 0. For the Mayo Clinic data, $\hat{\sigma}^2_{\hat{\mu}}$ is 4146 and the standard error is 64.

Estimation of the density function or hazard function is quite involved and will not be addressed here, however, estimation of the cumulative hazard function is straightforward. If one has already found the Kaplan–Meier estimator, an easily obtained estimator (see Peterson, 1977) is:

$$\hat{\Lambda}_p(t) = -\log \hat{S}(t)\,. \tag{2.7}$$

A second estimator (Nelson, 1969), which has proved to be useful in hypothesis testing and will be discussed further in Section 3, is:

$$\hat{\Lambda}_N(t) = \sum_{i:t_i \leq t} \sum_{l=0}^{d_i-1} \frac{1}{n_i - l}\,. \tag{2.8}$$

The two estimators are generally very close. For example, for the Mayo Clinic data, $\hat{\Lambda}_p(365) = -\log(0.525) = 0.644$ while $\hat{\Lambda}_N(365) = (1/20) + (1/19) + (1/18) + (1/17) + (1/16) + (1/15) + (1/12) + (1/11) + (1/10) = 0.620$. Note than to get $\hat{\Lambda}_N(369)$, we would have to add $(1/9) + (1/8)$ to $\hat{\Lambda}_N(368)$ since there were two failures at $t = 369$.

As shown above, once the Kaplan–Meier estimate of the survival function is obtained, one can very easily obtain estimates of quantiles (including the median), the mean, variance of the mean, and hazard function. For large samples, or samples where specific failure times are hard to obtain, it is often convenient to replace the Kaplan–Meier estimator by the actuarial life table estimator, which is the classical method. Details can be found in virtually any survival analysis textbook, however a short discussion may be worthwhile.

We divide the time axis into intervals $I_1, I_2, \ldots, I_M$ and let w_j be the number of censored values in the interval I_j and d_j the number of failures. We define $n'_j = n_j - 0.5w_j$ (essentially the effective risk set during the interval). Then, for t in I_k,

$$\hat{S}_A(t) = \prod_{j<k} \left(1 - \frac{d_j}{n'_j}\right) \tag{2.9}$$

is the actuarial estimate with variance estimate

$$\hat{\sigma}^2_{\hat{S}_A}(t) = \hat{S}^2_A(t) \sum_{j<k} \frac{d_j}{n'_j(n'_j - d_j)} \,. \tag{2.10}$$

As the lengths of the intervals approach 0, $\hat{S}_A(t) \to \hat{S}(t)$.

The Kaplan–Meier estimator is more precise than the actuarial estimator. However, for large data sets, the actuarial estimator is easier to compute and differs little from the Kaplan–Meier estimate. For example, if we divide the time into 6-month intervals (Table 2.3) we obtain an estimate of $\hat{S}_A(365) = 0.5313$ for survival at 1 year which is quite close to the Kaplan–Meier value of $\hat{S} = 0.5250$.

Table 2.3
Mayo Clinic data. Values of the actuarial estimator for the high-grade carcinoma patients

i	I_i	n_i	d_i	i	n'_i	$\hat{S}_k(t)$, $t \in I$
1	[0–6)	20	3	0	20	1.00
2	[6–12)	17	6	2	16	0.8500
3	[12–18)	9	7	1	8.5	0.5313
4	[18–24)	1	0	0	1	0.0937
5	[24–30)	1	0	0	1	0.0937
6	[30–36)	1	0	0	1	0.0937
7	[36–42)	1	0	1	0.5	0.0937
	[42–∞)					0.0937

3. The two-sample problem

In this section we will compare the group of patients considered in Section 2 with another group of Stage II or III ovarian carcinoma patients followed at

Mayo Clinic, who differ from the first group in that they had low-grade or well differentiated cancer. We want to test to see if progression of disease is different for patients with high-grade cancer than it is for patients with low-grade cancer.

The observation times for the low-grade cancer patients were 28, 89, 175, 195, 309, 377+, 393+, 421+, 447+, 462, 709+, 744+, 770+, 1106+ and 1206+ days. Kaplan–Meier plots (Figure 3.1) indicate that if there is a difference in time to progression of disease in the two groups, it is in favor of the low-grade patients. However, more analysis is required to determine whether such a difference is likely to occur by random chance or if it truly represents an improved prognosis of low-grade patients.

The notation given below will be used in the analyses which follow. Let k be the number of distinct failure times in the combined sample, and $t_1 < t_2 < \cdots < t_k$ be the ordered failure times. n_{ji} will be the number of patients from group j at risk at time t_i. A patient is considered to be at risk at time t_i if his time to failure or censor is at least t_i. d_{ji} will be the number of patients from group j who fail at time t_i. n_i and d_i will be the corresponding values in the entire sample. Define

$$E_{ji} = \frac{n_{ji} d_i}{n_i} \tag{3.1}$$

Table 3.1
Mayo Clinic data. Values needed for computation of the Tarone–Ware statistics

n_i	d_i	n_{1i}	d_{1i}	n_{2i}	d_{2i}	E_{1i}	$d_{1i}-E_{1i}$	V_{1i}	$n_i(d_{1i}-E_{1i})$	$n_i^2 V_{1i}$	$\hat{F}(t_i)(d_{1i}-E_{1i})$	$\hat{F}^2(t_i)V_{1i}$
35	1	20	0	15	1	0.57	−0.57	0.24	−20	300	−0.56	0.23
34	1	20	1	14	0	0.59	0.41	0.24	14	280	0.39	0.22
33	1	19	1	14	0	0.58	0.42	0.24	14	266	0.39	0.20
32	1	18	0	14	1	0.56	−0.56	0.25	−18	252	−0.50	0.19
31	1	18	1	13	0	0.58	0.42	0.24	13	234	0.36	0.18
30	1	17	0	13	1	0.57	−0.57	0.25	−17	221	−0.47	0.17
29	1	17	0	12	1	0.59	−0.59	0.24	−17	204	−0.47	0.16
28	1	17	1	11	0	0.61	0.39	0.24	11	187	0.30	0.14
27	1	16	1	11	0	0.59	0.41	0.24	11	176	0.30	0.13
26	1	15	1	11	0	0.58	0.42	0.24	11	165	0.30	0.12
23	2	12	1	11	1	1.04	−0.04	0.48	−1	252	−0.03	0.20
21	1	11	1	10	0	0.52	0.48	0.25	10	110	0.30	0.10
20	1	10	1	10	0	0.50	0.50	0.25	10	100	0.30	0.09
19	2	9	2	10	0	0.95	1.05	0.47	20	170	0.56	0.13
17	1	7	1	10	0	0.41	0.59	0.24	10	70	0.29	0.06
16	1	6	1	10	0	0.39	0.63	0.23	10	60	0.29	0.05
14	1	5	1	9	0	0.36	0.64	0.23	9	45	0.28	0.04
13	1	4	1	9	0	0.31	0.69	0.21	9	36	0.28	0.03
8	1	2	1	6	0	0.25	0.75	0.19	6	12	0.26	0.02
7	1	1	0	6	1	0.14	−0.14	0.12	−1	6	−0.04	0.01
							5.33	5.11	84	3146	2.53	2.48

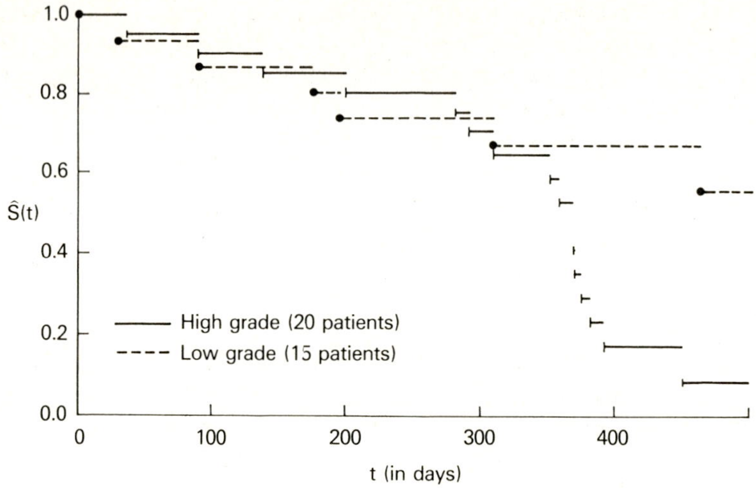

Fig. 3.1. Estimated survival functions for the high and low grade ovarian carcinoma patients.

and

$$V_{ji} = \frac{n_{1i}n_{2i}d_i(n_i - d_i)}{n_i^2(n_i - 1)}. \tag{3.2}$$

E_{ji} and V_{ji} can be thought of as the expected value and variance, respectively, of d_{ji}, $j = 1, 2$ if there is no difference in prognosis for the two groups. Table 3.1 gives these values for the Mayo Clinic data. For example, at time $t = 34$, $n_{1i} = 20$, $d_{1i} = 1$, $n_{1i} = 14$ and $d_{2i} = 0$. Hence $n_i = 34$, $d_i = 1$, $E_{1i} = (20 \times 1)/34 = 0.59$ and $V_{1i} = (20 \times 14 \times 1 \times 33)/(34^2 \times 33) = 0.24$.

The statistics used most frequently to form two-sample tests are special cases of a class of statistics discussed by Tarone and Ware (1977). These statistics are of the form

$$T = \frac{\Sigma_{i=1}^k w_i(d_{1i} - E_{1i})}{(\Sigma_{i=1}^k w_i^2 V_{1i})^{1/2}}. \tag{3.3}$$

If $w_i = 1$ for all i, T is the well-known Mantel–Haenszel (1959) statistic. This statistic was also considered by Cox (1972) and Peto and Peto (1972) and is sometimes referred to as the Mantel–Cox statistic or the logrank statistic. If $w_i = n_i$, T is the asymptotic version of the Gehan generalized Wilcoxon (Gehan, 1965) statistic. Another generalization of the Wilcoxon statistic, considered by Prentice (1978) is formed by letting $w_i = \hat{F}(t_i)$ where $\hat{F}$ is the Kaplan–Meier estimator over the combined sample. Under fairly general conditions, it can be shown that if there is no difference in prognosis between groups 1 and 2, T is asymptotically normal with mean 0 and variance 1 (see Tarone and Ware, 1977). Hence one may test the hypothesis of no difference by computing T (using any of the above weight functions) and comparing its value

rather complex, but if we let

$$R = 1 - 0.5 \times (\exp\{-\Lambda_{1k}\} + \exp\{-\Lambda_{2K}\})$$

and

$$p = 2 \times \left(1 - \Phi\left(\frac{A}{(R - R^2)^{0.5}}\right) + \Phi\left(\frac{A(2R-1)}{(R - R^2)^{0.5}}\right) \exp\{-2A^2\}\right),$$

p is a close approximation to the p-value for the two-sided test.

To illustrate, we again consider the ovarian carcinoma data. The entries given in Table 3.2 (which were given in the Fleming et al. (1980) paper) are useful when computing A.

For example, in the line corresponding to

$$t_i = 309, \qquad \Lambda_{1i} = \Lambda_{1(i-1)} + \Delta\Lambda_{1i} = 0.346 + \tfrac{1}{12} = 0.430\,,$$

$$\Lambda_{2i} = 0.298 + 1/11 = 0.389\,, \quad \alpha_{1i} = 0 + \tfrac{1}{14} + \tfrac{1}{13} = 0.148, \text{ and } \alpha_{2i} = 0\,.$$

Furthermore,

$$w_i = (20 \exp\{-0.148\})^{-1} + (15 \exp\{0\})^{-1} = 0.058 + 0.067 = 0.125$$

so $\eta_i = (1/0.125)^{0.5} = 2.832$. Now,

Table 3.2
Mayo Clinic data. Calculation of the statistics needed to evaluate the two-sided generalized Smirnov statistic

t_i	n_{1i}	d_{1i}	$\hat{\Lambda}_{1i}$	$\hat{\alpha}_{1i}$	n_{2i}	d_{2i}	$\hat{\Lambda}_{2i}$	$\hat{\alpha}_{2i}$	$\hat{\eta}_i$	U_i	Y_i
28	20	0	0.000	0.000	15	1	0.067	0.000	2.928	−0.195	−0.189
34	20	1	0.050	0.000	14	0	0.067	0.000	2.928	−0.049	−0.046
88	19	1	0.103	0.000	14	0	0.067	0.000	2.928	0.105	0.097
89	18	0	0.103	0.000	14	1	0.138	0.000	2.928	−0.104	−0.092
137	18	1	0.158	0.000	13	0	0.138	0.000	2.928	0.059	0.051
175	17	0	0.158	0.000	13	1	0.215	0.000	2.928	−0.166	−0.138
195	17	0	0.158	0.000	12	1	0.298	0.000	2.928	−0.410	−0.327
199	17	1	0.217	0.000	11	0	0.298	0.000	2.928	−0.238	−0.184
280	16	1	0.280	0.000	11	0	0.298	0.000	2.928	−0.055	−0.041
291	15	1	0.346	0.000	11	0	0.298	0.000	2.928	0.140	0.101
309	12	1	0.430	0.148	11	1	0.389	0.000	2.832	0.119	0.079
351	11	1	0.520	0.148	10	0	0.389	0.000	2.832	0.376	0.239
358	10	1	0.620	0.148	10	0	0.389	0.000	2.832	0.659	0.401
369	9	2	0.857	0.148	10	0	0.389	0.000	2.832	1.328	0.732
370	7	1	0.999	0.148	10	0	0.389	0.000	2.832	1.733	0.906
375	6	1	1.166	0.148	10	0	0.389	0.000	2.832	2.205	1.090
382	5	1	1.366	0.148	9	0	0.389	0.100	2.756	2.756	1.285
392	4	1	1.616	0.148	9	0	0.389	0.100	2.756	3.445	1.509
451	2	1	2.116	0.482	6	0	0.389	0.479	2.303	4.596	1.834
462	1	0	2.116	0.482	6	1	0.556	0.479	2.303	4.212	1.462

$$U_i = U_{i-1} + \eta_i \times (\Delta\Lambda_{1i} - \Delta\Lambda_{2i}) = 0.140 + 2.832(\tfrac{1}{12} - \tfrac{1}{11}) = 0.119$$

and

$$Y_i = (0.5)(\exp\{-\Lambda_{1i}\} + \exp\{-\Lambda_{2i}\}) \times U_i = (0.5)(0.651 + 0.678)(0.119) = 0.079\,.$$

Using similar computations for each i, we obtain $A = \max|Y_i| = 1.834$.

Finally $R = 1 - 0.5 \times (\exp\{-\Lambda_{1k}\} + \exp\{-\Lambda_{2k}\}) = 1 - 0.5 \times (0.121 + 0.573) = 0.653$ and. from (3.16) $p = 2 \times (1 - \phi(3.85) + \Phi(1.179) \times 0.00120) = 2 \times (1 - 0.99994 + (0.881)(0.00120)) = 0.00223$.

We note that this p-value is considerably smaller than those obtained for the Mantel–Haenszel ($p = 0.018$), Gehan ($p = 0.13$), and Prentice ($p = 0.11$) statistics. Fleming et al. (1980) do several Monte Carlo studies which indicate that this generalized Kolmogorov–Smirnov statistic is often able to detect difference which the Tarone–Ware class of statistics would fail to detect.

Each of the statistics discussed in this section can be superior to the others in certain situations and there is no set rule for deciding which statistic would be best to use. However, the following guidelines may be helpful. If one is comparing two treatments and one suspects that if there is a difference between the two, it will remain roughly the same throughout (in the sense that the ratio of the hazards will not cross), then a Tarone–Ware type of statistic should be

Table 3.3
Data for breast cancer patients from the NSABP study

Placebo				Treatment A				Treatment B			
Time	Status	Age	Nodes	Time	Status	Age	Nodes	Time	Status	Age	Nodes
846	0	58	1	737	0	44	1	576	0	64	4
642	0	68	1	162	0	40	1	688	0	46	1
687	0	50	1	671	0	57	1	717	0	55	1
814	0	69	1	759	0	38	1	681	0	47	1
574	0	48	1	697	0	71	1	710	0	29	1
799	0	46	1	596	0	60	1	606	0	42	1
840	0	70	1	621	0	43	1	729	0	62	1
1001	0	42	1	632	0	64	1	648	0	50	1
987	0	52	1	707	0	39	1	632	0	64	1
662	0	50	1	692	0	33	1	709	0	58	1
888	0	38	1	667	0	41	1	694	0	48	1
835	1	42	1	692	0	71	1	641	0	27	1
716	1	57	1	668	0	57	1	51	1	54	1
171	0	68	1	647	0	57	1	376	0	41	1
143	1	39	1	621	0	58	1	813	0	33	2
774	0	42	2	692	0	68	1	758	0	45	2
988	0	46	2	633	0	67	1	646	0	28	2
897	0	68	2	769	0	48	1	654	0	54	2
845	0	56	2	481	1	49	1	654	0	49	2
410	1	68	2	465	1	41	1	680	1	51	2
483	0	51	2	272	1	40	1	447	0	59	2

Table 3.3 (Continued)

Placebo				Treatment *A*				Treatment *B*			
Time	Status	Age	Nodes	Time	Status	Age	Nodes	Time	Status	Age	Nodes
175	1	43	2	459	1	58	1	205	1	55	2
98	1	54	2	159	1	32	1	586	0	38	3
119	1	44	2	218	1	43	1	651	0	70	3
28	0	39	2	234	1	57	1	683	1	54	3
845	0	50	3	1	0	49	1	353	1	61	3
955	0	57	3	760	0	37	2	181	1	48	3
1006	0	56	3	704	0	55	2	712	0	55	4
851	0	65	3	654	0	58	2	722	0	50	4
624	1	51	3	781	0	40	2	697	0	61	4
62	0	66	3	738	0	46	2	656	0	59	4
24	1	28	3	669	0	48	2	728	0	40	4
96	1	58	3	782	0	59	2	682	0	49	4
96	1	62	3	719	0	50	2	609	1	50	4
942	0	63	4	631	0	56	2	386	1	53	4
785	0	41	4	748	0	55	2	416	1	34	5
364	1	39	4	704	0	53	2	498	1	61	5
291	1	63	4	646	0	50	2	585	1	50	5
52	1	37	4	31	0	39	2	345	1	62	5
72	1	68	4	752	0	53	2	149	1	46	5
183	1	49	4	290	0	64	2	405	0	39	5
136	1	43	4	478	1	52	2	728	0	48	6
889	0	60	5	765	0	59	3	642	1	58	6
812	0	63	5	626	1	45	3	67	1	32	6
810	1	45	5	91	1	61	3	385	0	38	6
113	1	56	5	759	0	54	4	743	0	51	7
33	1	55	5	283	1	53	4	146	1	64	7
903	0	47	6	493	0	44	5	522	1	63	7
638	1	60	6	730	0	54	5	574	1	73	7
137	1	44	6	326	1	58	5	767	0	51	8
62	1	47	7	48	1	53	5	56	1	53	8
110	1	56	7	354	1	50	5	674	1	62	9
136	1	70	7	803	0	51	6	638	0	51	10
76	1	60	7	687	0	45	6	338	1	74	10
960	0	50	8	627	1	61	6	382	1	60	12
397	1	54	9	71	1	36	7	258	1	47	12
830	0	59	10	92	1	62	7	763	0	53	13
422	1	47	11	719	0	37	8	262	1	39	13
405	1	34	11	283	1	52	8	667	0	51	14
678	1	48	13	299	1	35	9	177	1	58	15
35	1	59	13	766	0	52	10	324	1	54	18
117	1	35	13	326	1	65	13	433	1	63	18
841	1	39	15	69	1	51	14	381	1	46	21
36	1	36	15	88	1	31	17	187	1	48	22
258	1	71	16	31	1	39	19	89	1	58	22
508	1	49	18	205	1	45	19	315	0	46	24
172	1	49	20	61	1	32	24	274	1	53	24
32	1	57	32					370	1	54	25

adequate. (This is often the case in clinical trials.) The Mantel–Haenszel form is easy to compute and can be used in most cases. However if one expects the group to differ the most at an early stage, the Gehan–Wilcoxon or Prentice–Wilcoxon would be more appropriate. Finally, if one is expecting a difference that might include 'crossing hazards' or has no idea what type of differences might occur, the generalized Kolmogorov–Smirnov statistic would be preferable.

A set of data which will be analyzed in the next section is given in Table 3.3. The three groups of patients considered were participants in a study conducted by the National Surgical Adjuvant Project for Breast and Bowel Cancers (NSABP). The patients all had a radical mastectomy for mammary carcinoma and histologically positive axillary nodes. In the actual studies, patients were randomized to various treatment groups, however for this analysis subsets have been selected in a nonrandom fashion. Sixty-eight patients were selected from a placebo group, 67 from a group which received a chemotherapy combination (referred to as treatment A), and 68 from another group receiving a different chemotherapy combination (referred to as treatment B).

We will begin analyzing this data as if the patients had been randomized into the three groups by using a 3-sample version of a Tarone–Ware statistic. The method of computation is similar to that described earlier in the section for the two-sample problem, however some additional notation is required. For simplicity, let $j = 1$ refer to the placebo group and $j = 2$ and 3 correspond to treatment groups A and B respectively. d_{ji}, n_{ji}, etc. will have the same meanings as before. The definitions which parallel those given by (3.1) and (3.2) are

$$E_{ji} = \frac{n_{ji}d_i}{n_i}, \quad j = 1, 2, 3, \tag{3.15}$$

$$V_{jji} = \frac{n_{ji}(n_i - n_{ji})d_i(n_i - d_i)}{n_i^2(n_i - 1)}, \tag{3.16}$$

$$V_{jj'i} = \frac{-n_{ji}n_{j'i}d_i(n_i - d_i)}{n_i^2(n_i - 1)}, \quad j \neq j'. \tag{3.17}$$

Let

$$U = v_w V^{-1} v_w' \tag{3.18}$$

where

$$v_w = \left(\sum_{i=1}^{k} w_i(d_{1i} - E_{1i}), \sum_{i=1}^{k} w_i(d_{2i} - E_{2i})\right) \tag{3.19}$$

and V is a 2×2 matrix with

$$V_{jj'} = \sum_{i=1}^{k} w_i^2 V_{jj'i}, \quad j = 1, 2 \text{ and } j' = 1, 2. \tag{3.20}$$

Under the hypothesis that all three groups have the same survival prognosis, U

will have a chi-square distribution with 2 degrees of freedom. It should be noted that the above notation is only applicable for a three group comparison, but if there were N_G groups, the above equations still are valid, but $j = 1, \ldots, N_G$ and v_w and V are extended to have dimensions $N_G - 1$ and $N_G - 1 \times N_G - 1$, respectively, and the final statistic has $N_G - 1$ degrees of freedom.

Applying the definitions to the NSABP data, it can be shown that at $t = 20.5$ months $n_{1i} = 41$, $d_{1i} = 0$, $n_{2i} = 55$, $d_{2i} = 1$, $n_{3i} = 59$ and $d_{3i} = 1$. Then

$$E_{1i} = (41 \times 2)/155 = 0.529, \qquad E_{2i} = 0.710\,,$$

$$V_{11i} = (41 \times (155 - 41) \times 2 \times (155 - 2))/(155^2 \times 154) = 0.387\,,$$

$$V_{12i} = V_{21i} = -(41 \times 55 \times 2 \times 153)/(155^2 \times 154) = 0.187\,,$$

and $V_{22i} = 0.455$. Similar computations are made at every failure time.

To compute the 3-sample Mantel–Haenszel statistic, we recall that $w_i = 1$ so

$$v_w = \left(\sum_{i=1}^{k} (d_{1i} - E_{1i}), \sum_{i=1}^{k} (d_{2i} - E_{2i})\right) = (38 - 30.11, 25 - 31.70)$$

and

$$V = \begin{pmatrix} 19.07 & -9.33 \\ -9.33 & 20.74 \end{pmatrix} \quad \text{where each } V_{jj'} = \sum_{i=1}^{k} V_{jj'i}\,.$$

This leads to

$$V^{-1} = \begin{pmatrix} 0.067 & 0.030 \\ 0.030 & 0.062 \end{pmatrix}$$

and $U = 3.76$ ($p = 0.15$). Hence, one would not reject the hypothesis that the groups differ in terms of survival at $\alpha = 0.05$ or even $\alpha = 0.10$.

4. Analysis with covariates included

The methods described in Section 3 are the standard procedures for use in randomized trials or any situation where patients in different groups are considered to have 'similar' characteristics. However, in practice, one occasionally must compare groups where this is not the case. For example, the data in Table 3.3 were obtained by choosing the first patients to appear in a file which had not been created randomly. Table 4.1 indicates that the patients are fairly well balanced according to age, but not according to the number of positive nodes found at surgery. Hence a comparison of treatments should take this imbalance into account. If we don't adjust for this imbalance and use the Mantel–Haenszel statistic to compare placebo to treatment A, we obtain $T = 1.86$ (one-sided $p = 0.03$) and if we compare placebo to treatment B, we obtain $T = 1.15$ ($p = 0.12$), indicating that treatment A is significantly better than placebo while treatment B is not.

To adjust for nodes when computing the Mantel–Haenszel statistic, the patients may be divided into subsets (by number of positive nodes). Then

Table 4.1
Summary tables of characteristics for 203 NSABP patients[a]

Characteristic		Treatment arm		
Age	Placebo	Treatment *A*	Treatment *B*	Total
<50	29 (42.6)	29 (43.3)	26 (38.2)	84 (41.4)
50–59	21 (30.9)	27 (40.3)	27 (39.7)	75 (36.9)
⩾60	18 (26.5)	11 (16.4)	15 (22.1)	44 (21.7)
Total	68	67	68	203
Number of Positive Nodes	Placebo	Treatment *A*	Treatment *B*	Total
1–3	34 (50.0)	45 (67.2)	26 (38.2)	105 (51.7)
4–9	22 (32.4)	15 (22.4)	26 (38.2)	63 (31.0)
⩾10	12 (17.6)	7 (10.4)	16 (23.5)	35 (17.2)
Total	68	67	68	203

[a]The values in parentheses are %'s over columns.

within each subset the $\Sigma_i d_{1i} - E_{1i}$ and $\Sigma_i V_{1i}$ are found, and the statistic $T = (\Sigma_S \Sigma_i (d_{1i} - E_{1i}))/\Sigma_S \Sigma_i (V_{1i})^{1/2})$ computed where Σ_S refers to the sum over all the different subsets. Again T is asymptotically normal with mean 0 and variance 1 under the hypothesis of no difference.

To illustrate using the NSABP data, we divide the patients into subsets with 1–3, 4–9 and 10 or more positive nodes. Comparing placebo to treatment *A*, we first rank the failure times of patients with 1–3 nodes from the combined (placebo and treatment *A*) groups and use the method of computation discussed in Section 3 to obtain $\Sigma_i (O_{1i} - E_{1i}) = 2.15$ and $\Sigma_i V_{1i} = 4.72$.

We repeat this process for the patients with 4–9 positive nodes to obtain $\Sigma_i (O_{1i} - E_{1i}) = 1.63$ and $\Sigma_i V_{1i} = 5.84$ and for the patients with 10 or more nodes to obtain $\Sigma_i (O_{1i} - E_{1i}) = -1.45$ and $\Sigma_i V_{1i} = 3.17$. We then sum these results and get $T = (2.15 + 1.63 - 1.45)/(4.72 + 5.84 + 3.17) = 0.63$ $(p = 0.26)$. The same technique applied to placebo versus treatment *B* yields $T = 1.44$ $(p = 0.08)$. Using this adjusted method, the results indicate no significant improvement using treatment *A*, but borderline significance with treatment *B*, quite contrary to the results obtained in the unadjusted case.

There is no set rule for when adjustments are appropriate, but if the patients are randomized to treatment groups, unadjusted statistics are usually preferable. In this case adjusting may introduce as many problems as it solves. If the patients are not randomized, but a characteristic is balanced across groups, it still need not be adjusted for but, if there is an imbalance in some characteristic which might influence survival, use an adjusted statistic. In the above case, adjustment is clearly required.

Although the Mantel–Haenszel statistic was used in the above discussion, the same technique can be used with any Tarone–Ware type of statistic.

Another topic of interest when covariates (patient characteristics) are given is trying to determine which covariates are important in predicting treatment response. The standard nonparametric method for addressing this problem uses the Cox (1972) proportional hazard model. This model assumes that the hazard function for every patient is of the form

$$\lambda_z(t) = \lambda_0(t)\, e^{z\beta} \tag{4.1}$$

where $\lambda_0(t)$ is called the base-line hazard, z is a vector of covariates for an individual patient, and β is a vector of regression parameters. A particularly thorough discussion of the model is given in Kalbfleisch–Prentice (1980). The discussion involves some reasonably complex mathematics which is beyond the scope of this chapter, however an outline of how the model can be used (to include preparation for and interpretation of computer packages) follows.

The starting point is usually to obtain estimators for the values of β which will indicate whether individual covariates influence prognosis, whether the effect is favorable or unfavorable, and the magnitude of the effect. It is possible to standardize each component of β and determine whether the corresponding covariate has a statistically significant effect. The first step in running a program using the proportional hazards model is to arrange the data so that one column has the time to event (failure or censor), another has status (fail or censor), and each succeeding column has a value for a covariate. For the NSABP data, we let z_1 represent age, z_2 represent number of positive nodes, and z_3 represent treatment ($0 =$ placebo and $1 =$ treated). Hence the file has 5 columns and the entry for the first patient is '846 0 58 1 0', where the first 0 indicates the patient is censored and the second 0 indicates the patient is a placebo patient. This file was used as input for the BMDP Cox regression program (using the placebo and treatment *A* patients) and some of the results obtained are given in Table 4.2.

The values of the standardized coefficients are helpful for a preliminary analysis since under the hypothesis that $\beta_i = 0$, i.e., covariate z_i has no effect on the hazard, these represent observations of a standardized (mean 0 and variance 1) normal random variable. In this example the standardized coefficient for β_1 is -1.328 (two-sided p-value $= 0.19$), that of nodes is 5.75 ($p \ll 0.001$) and treatment is 0.81 ($p = 0.42$). This would indicate that a highly significant prognostic factor is the number of positive nodes, while age and treatment may not be significant.

The values of β_i are used to estimate the effect of a particular covariate. For example, $\beta_2 = 0.122$ implies that if two patients are the same age and are receiving the same treatment, but the second patient has one more positive node than the first, the second patient's estimated hazard function will always be $\exp\{0.122\} = 1.13$ times that of the first patient. If the second patient had 5

Table 4.2
Summary of output from a BMDP run using the Cox regression model applied to the NSABP data

Variable	Coefficient Z	Standardized Coefficient	EXP(Coeff)
	Placebo and Treatment A		
1. Age	−0.0180	−1.32	0.98
2. Nodes	0.1220	5.75	1.13
3. Treatment[a]	−0.2208	−0.81	0.80
	Placebo and Treatment B		
1. Age	0.0029	0.23	1.00
2. Nodes	0.0982	5.58	1.10
3. Treatment	−0.4612	−1.83	0.63

[a]0 = placebo, 1 = Treatment A (resp. B).

more positive nodes than the first patient, his estimated hazard would be $\exp\{5\times 0.122\} = 1.84$ times that of the first patient. $\beta_3 = -0.221$ implies that if two patients have the same number of positive nodes and are the same age, a treated patient's estimated hazard will be $\exp\{-0.221\} = 0.80$ times that of a placebo patient.

A similar run was made on the placebo and treatment B patients and these results are also shown in Table 4.2. Again the number of positive nodes appears to be highly significant while age is not at all significant and treatment is of borderline significance (which is consistent with the findings using the adjusted Mantel–Haenszel statistic). In this case the estimated hazard of a greated patient is $\exp\{-0.461\} = 0.63$ times that of a placebo patient.

The above example illustrates some uses of the proportional hazards model, but ignores topics such as covariate interactions and time-dependent covariates. These topics are discussed in some detail in Kalbfleisch–Prentice (1981), Breslow (1975), and Byar and Green (1980).

5. Additional references

As stated previously, my intent was to give considerable detail in examples. This necessitated the omission of some topics. Furthermore, several articles and books have appeared since this chapter was initiated. Hence, a short summary of other references seems appropriate.

Many nonparametric techniques in survival analysis are discussed in Miller (1981) and an extensive bibliography is provided. A review of partial likelihood theory and the Cox model is given in Oakes (1981). Sequential forms of the Tarone–Ware statistics are discussed by Koziol and Petkau (1978), Jones and Whitehead (1979), Rubinstein and Gail (1982), Slud and Wei (1982), and Tsiatis (1981a, 1982) among others. Chatterjee and Sen (1973) introduced a slightly different modification of linear rank statistics for use in sequential analysis which is summarized in Sen (1981).

Considerable progress has been made in the asymptotic theory using two different approaches. A weak convergence (Billingsley, 1968) approach is used by Breslow and Crowley (1974), Crowley (1974), and Tsiatis (1981b) and is discussed in Miller (1981). A martingale theory approach, which has received considerably more attention, was introduced by Aalen (1976, 1978) with refinements in Gill (1981). The present state of the theory is summarized in Helland (1982). Applications to two and k-sample statistics are discussed in Fleming and Harrington (1981, 1982) and Anderson et al. (1982), to the Cox proportional hazards model by Anderson and Gill and (1982), and to sequential analysis by Sen (1979, 1981).

Acknowledgement

I would like to thank David Henry, Richard Sass, and P. K. Sen for suggestions leading to improvements in this chapter. I would also like to thank Drs. Bernard Fisher and Carol Redmond for permission to use the data files of the National Surgical Adjuvant Breast and Bowel Projects for several examples. Finally, I want to thank The Biometric Society for permission to use the Mayo Clinic data as it was presented in Fleming et al. (1980).

References

Aalen, O. O. (1976). Nonparametric inference in connection with multiple decrement models. *Scand. J. Statist.* **3**, 15–27.

Aalen, O. O. (1978). Nonparametric inference for a family of counting processes. *Ann. Statist.* **6**, 701–726.

Anderson, P. K. and Gill, R. D. (1982). Cox's regression model for counting processes: A large sample study. *Ann. Statist.* **10**, 1100–1120.

Anderson, P. K., Borgan, O., Gill, R. and Keiding, N. (1982). Linear nonparametric tests for comparison of counting processes, with applications to censored survival data. *Internat. Statist. Rev.* **50**, 219–259.

Billingsley, P. (1968). *Convergence of Probability Measures.* Wiley, New York.

Breslow, N. (1975). Analysis of survival data under the proportional hazards model. *Internat. Statist. Rev.* **43**, 45–57.

Breslow, N. and Crowley, J. (1974). A large sample study of the life table and product limit estimates under random censorship. *Ann. Statist.* **2**, 437–453.

Byar, D. P. and Green, S. B. (1980). The choice of treatment for cancer patients based on covariate information: Application to prostate cancer. *Bulletin du Cancer* (Paris) **67**, 477–490.

Chatterjee, S. K. and Sen, P. K. (1973). Nonparametric testing under progressive censoring. *Calcutta Statist. Assoc. Bull.* **22**, 13–50.

Cox, D. R. (1972). Regression models and life tables. *J. Roy. Statist. Soc. Ser. B* **34**, 187–202.

Cramer, H. (1928). On the composition of elementary errors. *Skandinavisk Aktuarietidskrift* **11**, 13–74.

Crowley, J. (1974). Asymptotic normality of a new nonparametric statistic for use in organ transplant studies. *J. Amer. Statist. Assoc.* **69**, 1006–1011.

Fleming, T. R. and Harrington, D. P. (1981). A class of hypothesis tests for one and two sample censored survival data. *Comm. Statist. A* **10**, 763–794.

Fleming, T. R. and Harrington, D. P. (1982). A class of rank test proceduces for censored survival data. *Biometrika* **69**, 553–566.

Fleming, T., O'Fallon, J., O'Brien, P. and Harrington, D. (1980). Modified Kolmogorov–Smirnov test procedures with application to arbitrarily right-censored data. *Biometrics* **36**, 607–625.

Gehan, E. A. (1965). A generalized Wilcoxon test for comparing arbitrarily singly-censored samples. *Biometrika* **52**, 203–223.

Gill, R. D. (1980). *Censoring and Stochastic Integrals*. Mathematical Centre Tracts 124, Mathematisch Centrum, Amsterdam.

Greenwood, M. (1926). The natural duration of cancer. In: *Reports on Public Health and Medical Subjects*, No. 33.

Helland, I. S. (1982). Central limit theorems for martingales with discrete or continuous time. *Scand. J. Statist.* **9**, 79–94.

Jones, D. and Whitehead, J. (1979). Sequential forms of the log rank and modified Wilcoxon tests for censored data. *Biometrika* **66**, 105–113.

Kalbfleisch, J. and Prentice, R. L. (1980). *The Standard Analysis of Failure Time Data*. Wiley, New York.

Kaplan, E. L. and Meier, P. (1958). Nonparametric estimation from incomplete observations. *J. Amer. Statist. Assoc.* **53**, 457–481.

Koziol, J. A. (1978). A two sample Cramér–von Mises test for randomly censored data. *Biom. J.* **20**, 603–608.

Koziol, J. A. and Petkau, A. J. (1978). Sequential testing of the equality of two survival distributions using the modified Savage statistic. *Biometrika* **65**, 615–623.

Mantel, N. and Haenszel, W. (1959). Statistical aspects of the analysis of data from retrospective studies of disease. *J. Nat. Cancer Inst.* **22**, 719–748.

Miller, R. (1981). *Survival Analysis*. Wiley, New York.

Nelson, W. (1969). Hazard plotting for incomplete failure data. *J. Qual. Tech.* **1**, 27–52.

Oakes, D. (1981). Survival times: Aspects of partial likelihood. *Int. Statist. Rev.* **49**, 235–264.

Peterson, A. V. (1977). Expressing the Kaplan–Meier estimator as a function of empirical subsurvival functions. *J. Amer. Statist. Assoc.* **72**, 854–858.

Peto, P. and Peto, J. (1972). Asymptotically efficient rank invariant test procedures. *J. Roy. Statist. Soc. Ser. A* **135**, 185–198.

Prentice, R. L. (1978). Linear rank tests with right censored data. *Biometrika* **65**, 167–179.

Prentice, R. L. and Marek, P. (1979). A qualitative discrepancy between censored data rank tests. *Biometrics* **35**, 861–869.

Rai, K., Susarla, V. and Van Ryzin, J. (1979). Shrinkage estimation in nonparametric Bayesian survival analysis. *Comm. Statist. B* **9**, 271–298.

Rubinstein, L. V. and Gail, M. H. (1982). Monitoring rules for stopping accrual in comparative survival studies. *Controlled Clinical Trials* **3**, 325–343.

Schey, H. M. (1977). The asymptotic distribution of the one-sided Kolmogorov–Smirnov statistic for truncated data. *Comm. Statist. A* **6**, 1361–1365.

Sen, P. K. (1979). Weak convergence of some quantile processes arising in progressively censored tests. *Ann. Statist.* **7**, 414–431.

Sen, P. K. (1981). The Cox regression model, invariance principles for some induced quantile processes and some repeated significance tests. *Ann. Statist.* **9**, 109–121.

Slud, E. and Wei, L. J. (1982). Two-sample repeated significance tests based on the modified Wilcoxon statistic. *J. Amer. Statist. Assoc.* **77**, 862–868.

Smirnov, N. V. (1939). Estimate of deviation between empirical distribution functions in two independent samples (Russian). *Bulletin Moscow Univ.* **2**(2), 3–16.

Tarone, R. E. and Ware, J. (1977). On distribution free tests for equality of survival distributions. *Biometrika* **64**, 156–160.

Tsiatis, A. A. (1981a). The asymptotic joint distribution of the efficient scores test for the proportional hazards model calculated over time. *Biometrika* **68**, 311–315.

Tsiatis, A. A. (1981b). A large sample study of Cox's regression model. *Ann. Statist.* **9**, 93–108.

Von Mises, R. (1931). *Wahrscheinlichkeitsrechnung*. Leipzig-Wein.

P. R. Krishnaiah and P. K. Sen, eds., *Handbook of Statistics, Vol. 4*

33

Nonparametric Frequentist Proposals for Monitoring Comparative Survival Studies

Mitchell Gail

1. Introduction

Clinical trials to compare survival on two treatment groups are often costly and time-consuming, and they may impose psychological and physical burdens on participating patients. Yet these trials offer a uniquely sound methodology for acquiring new medical knowledge. To obtain convincing scientific evidence from such a trial requires that one follow a substantial number of patients for a sufficient time interval. Over the past fifteen years, a number of methods for monitoring the results of clinical trials have been developed in an effort to shorten the trials and reduce the numbers of patients required without greatly diminishing the quality of the scientific information gained.

The question of how to monitor a clinical trial is inevitably determined by one's view of the central purpose and meaning of such an experiment. Armitage (1960, 1975) presents classical frequentist methods for a variety of endpoints. These methods emphasize hypothesis tests and proper control of the significance level despite repeated looks at the data, in hypothetical repetitions of the entire clinical trial. The *frequentist viewpoint* has been vigorously challenged by Cornfield (1966), who presents a *Bayesian alternative.* Anscombe (1963), too, criticized the frequentist approach, and presented a *decision theoretic* approach to monitoring. This formulation, which was developed simultaneously by Colton (1963), is to regard the monitoring of a clinical trial as an effort to minimize the total number of patients who receive the inferior treatment; patients other than those participating in the trial must be included in such calculations. Yet another viewpoint is that of the *selection* theorist (e.g. Bechhoffer, 1954), who aims to select the superior treatment with prespecified high probability in hypothetical repetitions of the trial. Gail (1982) has discussed these several approaches and gives further references, as do DeMets and Lan (1984). In this paper we concentrate on recent developments and adaptations of the frequentist paradigm, and we confine attention to survival tests based on ranks.

It is assumed that the main response of interest is survival time and that decisions to stop the trial or to declare a treatment preference, if any, depend primarily on the survival comparisons. This is certainly an oversimplification of the activities of real monitoring committees, who must take into account other factors, such as administrative and financial problems, developments in the medical literature, credibility of the present findings, and treatment toxicities.

Appropriate statistical techniques for the sequential monitoring of survival data are relatively new. Survival experiments typically enter patients over a protracted accrual period, and follow-up usually continues for a period of continued observation after accrual. A patient's survival time is the interval from his date of entry to his date of death, and, if observation ceases before death occurs, his experimental observation time is the interval from entry to termination of observation. In most experiments it is necessary to distinguish between the experimental time and the real calendar time at which an event occurs. In the case of *simultaneous entry*, however, the two time scales are superimposable, and the experiment unfolds in an orderly manner. In particular, the ranks of earlier death times are not changed by the accumulating information on subsequent death times. This simplicity is exhibited and exploited by Chatterjee and Sen (1973), who present a general theory for monitoring linear rank statistics with simultaneous entry. For clinical trials, however, the usual pattern of entry over a period of accrual is called *staggered entry*. With staggered entry, at any real time there may be a complicated pattern of observed and unobserved experimental survival times. The survival times may be unobserved either because the patient has not yet entered the study, or, because, having entered, his survival time exceeds his experimental follow-up. This latter case is termed right censorship. The theory of two-sample rank tests for survival data with staggered entry is well understood if the analysis is to be performed only once, as discussed by Kalbfleisch and Prentice (1980) and Gill (1980). However, the distribution theory for monitoring a trial with staggered entry at several points in real time remains a topic of active research, as outlined in Section 2.

We review several frequentist innovations for monitoring clinical trials in this paper. We use the term *continuous monitoring* to describe plans that calculate rank statistics at each death time, or, as proposed by Majumdar and Sen (1978b), at death times and entry times. It is often impractical to monitor trials continuously, and recent attempts have been made to adapt the notion of '*group sequential methods*' (Pocock, 1977) to survival data. One can either examine the data at fixed real time intervals, or one can perform the analysis at the random real times when fixed increments of statistical information have been obtained. In the former case, the boundary defining the critical region will be data dependent, because required covariances will depend on random events such as the number of patients accrued. Prespecified boundaries can be used if the analyses are performed at fixed increments of statistical information and if increments are uncorrelated. A special case is the analysis of survival data using the logrank statistic at intervals defined by fixed numbers of deaths. Yet another approach to monitoring is termed *stochastic curtailment*. This procedure stops the experiment when it is clear that, with high probability, further observation will not change the current opinion as to whether to accept or reject the null hypothesis. Finally, we shall discuss a strategy that is applicable to many cancer chemotherapy trials, namely *accrual monitoring*. The idea is to monitor the survival data only to determine when accrual should be stopped, with the understanding that the final data analysis and judgment as to treatment preference will be performed only once, at the time when a prespecified amount of statistical information has been obtained.

As we review this potpourri of frequentist innovations for monitoring survival data, it is well to bear in mind that any procedure that reaches a treatment decision on the basis of early experimental follow-up data leaves open the possibility that long term follow-up may reverse an earlier opinion, as can happen if the two hazard functions cross. Another difficulty in using sequential methods is that data used for interim analyses are more likely to contain errors than is a carefully edited final dataset.

2. Null distribution theory for two-sample rank tests used for monitoring

As in Tsiatis (1982), we consider a class of two-sample rank tests computed at a real time t. Define the following variables for each patient: E is the real time of entry into the study, Y is the experimental survival time, measured from E, W is the experimental time to censorship, measured from E, and $Z=1$ or 0 according as the patient is in treatment group 1 or 2. We wish to test the null hypothesis that the distribution of Y, which is assumed to be independent of entry time and experimental censoring time, is the same in both treatment groups. At time t, we consider only those $N(t)$ patients with entry times $E<t$, and we order their $d(t)$ distinct experimental death times $Y(1)<Y(2)<\cdots<Y(d(t))$. At experimental time $Y(i)$, there are $n_1(t, Y(i))$ patients with $E<t$, $Y \geqslant Y(i)$, $t-E \geqslant Y(i)$, $W \geqslant Y(i)$ and $Z=1$. These patients are said to be 'at risk in group 1'. Likewise $n_2(t, Y(i))$ are at risk in group 2 and $n(t, Y(i))=n_1(t, Y(i))+n_2(t, Y(i))$ are the total at risk at experimental time $Y(i)$ and real time t. The proportion at risk in group 1 is

$$p(t, Y(i))=n_1(t, Y(i))/n(t, Y(i)),$$

and we use rank statistics of the form

$$S(t)=\sum_{1}^{d(t)} Q(t, Y(i))\{Z(t, Y(i))-p(t, Y(i))\}. \tag{2.1}$$

The summation in (2.1) extends only over the $d(t)$ distinct death times, $Y(i)$, and $Z(t, Y(i))$ indicates the group membership of the patient who died at $Y(i)$. The positive function $Q(t, Y(i))$ defines the particular rank test of interest. For example $Q(t, Y(i))=1$ defines the logrank statistic, which was first proposed by Mantel (1966), and $Q(t, Y(i))=n(t, Y(i))$ defines the modified Wilcoxon test given by Gehan (1965). The class of tests represented by (2.1) has been discussed by Tarone and Ware (1977), Prentice and Marek (1979), Gill (1980), and Harrington and Fleming (1982).

As mentioned by Lagakos (1982), it is revealing to think of (2.1) as the score statistic associated with the partial likelihood of Cox (1972). Suppose the hazard ratio $h_1(t, Y(i))/h_2(t, Y(i)) = \exp\{\alpha Q(t, Y(i))\}$. The partial likelihood represents the survival experiment as a product of independent Bernoulli trials with log likelihood

$$L \equiv \sum \alpha Z(t, Y(i))Q(t, Y(i)) - \log[n_1(t, Y(i)) \exp\{\alpha Q(t, Y(i))\} + n_2(t, Y(i))] . \tag{2.2}$$

The quantity $S(t)$ in (2.1) is the score statistic $\dot{L}(\alpha = 0)$. The observed Fisher information is

$$-\ddot{L}(\alpha = 0) \equiv \hat{V}(d) = \sum_1^d Q^2(t, Y(i))p_i(1 - p_i) , \tag{2.3}$$

where the shortened notation $Z_i \equiv Z(t, Y(i))$ and $p_i \equiv p(t, Y(i))$ will be used.

If (2.2) had indeed arisen from independent Bernoulli trials with variates Z_i having mass functions $p_i^{Z_i}(1 - p_i)^{1-Z_i}$ under H_0: $\alpha = 0$, then $S(t)$ would be the sum of independent centered, weighted variates with expectations 0 and variances $Q^2(t, Y(i))p_i(1 - p_i)$. Hence

$$S(t)\hat{V}(d)^{-1/2} \tag{2.4}$$

would converge to a standard normal deviate, provided only that Liapounov conditions held for $Q(t, Y(i))$. For fixed t, this weak convergence has been demonstrated rigorously by Gill (1980), who identified $S(t)$ as a martingale. It should be emphasized that this result pertains to a single analysis at the end of the trial, t, and not to multiple analyses. For a single t, one can unambiguously rank all patients' experimental observation times.

The heuristic of regarding $S(t)$ as arising from independent Bernoulli trials is also a guide to available results for sequential monitoring of S at several real times $t_1 < t_2 < \cdots < t_K$. Suppose, first, that all patients enter simultaneously. Then, for $t_1 < t_2$, the increment $S(t_2) - S(t_1)$ corresponds to new Bernoulli variates in the partial likelihood, and terms in the partial likelihood associated with $S(t_1)$ are unaltered by the new information obtained between t_1 and t_2. This suggests that with *simultaneous entry*, statistics of the form (2.1) will have uncorrelated increments. Chatterjee and Sen (1973) prove that for $S(t)$ computed at the random times at which deaths occur, $S(t)$ is a martingale, and a normalized version of $S(t)$ converges to a Wiener process. Note that relationships given by Mehrotra, Michalek and Mihalko (1982) may be used to translate Chatterjee and Sen's results for linear rank statistics into the form

(2.1). These martingale and convergence results hold whether one calculates S at fixed real times or at the random times of observed deaths. Majumdar and Sen (1978a) and Sinha and Sen (1979) extend these results to several samples.

The distribution theory is also simple if entry is *purely sequential*; that is, if the entry of the next patient occurs only after the death of the previous patient. Sen and Ghosh (1972) demonstrate that a normed version of $S(t)$, computed after each observed death, again converges to a Wiener process. However, example 3.1 of Slud (1983) shows that the martingale property may fail in the presence of loss to follow-up represented by $W < \infty$.

Results are more complicated for *staggered entry*. Suppose first that $S(t_1)$ and $S(t_2)$ are computed at two predetermined real times, $t_1 < t_2$. The previous heuristic device can be extended in the following way. Let $Y(i, t_1)$, $i = 1, 2, \ldots, d(t_1)$, denote the ordered experimental death times as of real time t_1, and let $Y(i, t_2)$, $i = 1, 2, \ldots, d(t_2)$, denote such death times as of real time t_2. Note $Y(i, t_1)$ need not equal $Y(i, t_2)$. Express $S(t_2)$ as

$$\begin{aligned} S(t_2) &= \sum_1^{d(t_2)} Q(t_2, Y(i, t_2))\{Z(t_2, Y(i, t_2)) - p(t_2, Y(i, t_2))\} \\ &\equiv \sum\nolimits^{*} Q(t_2, Y^*)\{Z^* - p(t_2, Y^*)\} + \sum\nolimits^{**} Q(t_2, Y^{**})\{Z^{**} - p(t_2, Y^{**})\} \\ &= \sum\nolimits^{*} Q(t_2, Y^*)\{Z - p(t_1, Y^*)\} + \sum\nolimits^{**} Q(t_2, Y^{**})\{Z^{**} - p(t_2, Y^{**})\} \\ &\quad + \sum\nolimits^{*} Q(t_2, Y^*)\{p(t_1, Y^*) - p(t_2, Y^*)\} \end{aligned} \tag{2.5}$$

where Σ^* represents a sum over those patients who had already died in $[0, t_1]$, $\{Y^*\}$ are their experimental death times, Σ^{**} represents a sum over those patients who die in $(t_1, t_2]$, and $\{Y^{**}\}$ are their experimental death times. At real time t_2, the first of the three sums in (2.5) might be regarded as a sum of independent weighted, centered Bernoulli variates with centering $p(t_1, Y^*)$ determined by the data at t_1 and with weights $Q(t_2, Y^*)$ depending on t_2. The second sum might be regarded as another independent set of independent weighted, centered Bernoulli variates with centerings $p(t_2, Y^{**})$ dependent on all the information available at t_2. For randomized trials, the third sum can be shown to be asymptotically negligible compared to the first two, because the entry time distributions are the same on the two treatment arms. Indeed, from Tsiatis (1982, equation 3.3), it follows that $p(t_1, Y^*) - p(t_2, Y^*)$ converges to zero and that the third sum is asymptotically negligible. Because $S(t_1)$ only depends on the set of Bernoulli variates in Σ^*, we have

$$\widehat{\mathrm{Cov}}(S(t_1), S(t_2)) = \sum_1^{d(t_1)} Q(t_1, Y(i, t_1))Q(t_2, Y(i, t_1))p_i(1 - p_i) \tag{2.6}$$

where $p_i = p(t_1, Y(i, t_1))$ and where the summation is over the times $Y(i, t_1)$.

This relationship was proved by Tsiatis (1982). In particular, Tsiatis showed that, at fixed real times $t_1, t_2, \ldots, t_K$, a normed version of $S(t_1), \ldots, S(t_k)$ converges to a multivariate normal distribution with covariance estimated from (2.6). Importantly, the covariance (2.6) implies that increments of $S(t)$ are correlated unless $Q(t_2, Y(i, t_1))$ is asymptotically independent of real time t_2. In particular, increments of Gehan's modified Wilcoxon test are asymptotically correlated, whereas, the increments of the logrank test, with $Q(t, Y(i)) = 1$, are asymptotically uncorrelated. The G^ρ statistics of Harrington and Fleming (1982) and the generalization of the Wilcoxon statistic discussed by Prentice and Marek (1979) also have asymptotically uncorrelated increments. Slud and Wei (1982) were the first to show that Gehan's modified Wilcoxon statistic had asymptotically correlated increments. Harrington, Fleming and Green (1982) obtained results similar to those of Tsiatis (1982) for repeatedly testing the null hypothesis that the hazard ratio was an arbitrary constant, not necessarily one. An unpublished manuscript by Slud (1983) extends Tsiatis' results for the logrank test, computed at fixed times. He shows with staggered entry that increments of the logrank test are exactly uncorrelated, and that the logrank statistic converges to a Wiener process. Indeed, he proves these results whenever $Q(t, Y(i))$ is independent of real time t, provided certain technical restrictions on patient entry are satisfied.

If one computes the statistic S at the random times of observed deaths, $T_1, T_2, \ldots, T_k$ rather than at fixed real times, then with staggered entry $S(T_j)$ is not in general a martingale, as shown by example 3.3 of Slud (1983). However, Sellke and Siegmund (1983) prove that the logrank test, while not a martingale, converges to a Wiener process, nonetheless. It seems that their argument can also be extended to other statistics for which $Q(t, Y(i))$ is independent of t.

To summarize, under the null hypothesis that the lifetimes X have the same distribution in the two treatment groups, the previous heuristic based on imagined sets of independent centered Bernoulli variates

$$\{Z^* - p(t_1, Y^*)\}, \{Z^{**} - p(t_2, Y)^{**})\}$$

and so forth, may be used to reconstruct the available asymptotic theory of $S(t_1), S(t_2), \ldots, S(t_k)$. If $t_1, t_2, \ldots, t_k$ are pre-specified real times, the statistics tend to joint normality with covariance estimated from (2.6). When $Q(t, Y(i))$ is asymptotically independent of t, convergence to a Wiener process is obtained. Similar results appear to hold for analyses performed at the random times at which the Fisher information (2.3) increases. No similar results are available for fixed alternative distributions of Y in the two treatment groups, and, indeed, simulations by Gail, DeMets and Slud (1982) prove that the logrank increments are correlated under a fixed alternative. However, for local alternatives with fixed t, the Bernoulli model leads to correct calculations of noncentrality parameter for proportional hazards, in agreement with Gill (1980, Chapter 5) and Schoenfeld (1981).

3. Continuous monitoring

Armitage (1975, p. 142) suggested that one can monitor the logrank test by plotting $S(T_d)$, the value of S at the random time of death d, against $\hat{V}(d) = \Sigma_1^d p_i(1-p_i)$. He asserted under the null hypothesis that, $S(T_d)$ would behave approximately like the cumulative sum of $\hat{V}(d)$ independent normal deviates, each with mean 0 and variance $\frac{1}{4}$ if patients were assigned with equal probability to each treatment. He therefore suggested using classical boundaries for normal variates, in conjunction with the logrank statistic, to monitor survival studies. For proportional hazards alternatives $h_1(t, Y(i)) = h_2(t, Y(i)) \exp(\theta)$, with small θ, a Taylor series expansion argument shows that the mean of one of these normal deviates is $\mu \doteq p_i(1-p_i)\theta \doteq \theta/4$ and the variance is $\sigma^2 \doteq p_i(1-p_i) \doteq \frac{1}{4}$. Thus boundaries appropriate for monitoring the cumulative sum of d independent $N(\mu, \sigma)$ variates may be used to monitor a logrank plot of $S(T_d)$ against $\hat{V}(d)$, where the noncentrality parameter $\delta = \mu/\sigma = \theta/2$ is used to help select a boundary. Since $\hat{V}(d)$ is only slightly less than $d/4$, one might choose to plot $S(T_d)$ against $d/4$ or $2S(T_d)$ against d, instead.

Chapter 5 of Armitage (1975) contains two-sided boundaries for testing $\theta = 0$. A 'restricted boundary' for the logrank test is expressible in terms of the design constants a, b and V_f. One continues observation so long as

$$-a\sigma - b\sigma\hat{V}(d) < S(T_d) < a\sigma + b\sigma\hat{V}(d) \tag{3.1}$$

and $\hat{V}(d) < V_f$. If the boundary (3.1) is infringed with $\hat{V}(d) < V_f$, one rejects the null hypothesis. Otherwise, one stops monitoring when $\hat{V}(d) \geq V_f$ and accepts the hypothesis. To use this plan one must specify a, b and V_f in advance; these constants are determined by the alternative $\delta = \theta/2$ one wishes to detect and by the desired size and power of the test. Armitage (Chapter 5, 1975) also presents a 'repeated significance test' boundary of the type: continue observation as long as

$$|S(T_d)| < k(\alpha, V_f)\hat{V}(d)^{1/2} \tag{3.2}$$

and $\hat{V}(d) < V_f$. Reject the null hypothesis as soon as the boundary (3.2) is infringed. Otherwise, accept the hypothesis as soon as $\hat{V}(d) \geq V_f$. The constants V_f and $k(\alpha, V_f)$ are again determined by $\delta = \theta/2$ and the desired size and power. The boundary (3.2) is a parabola, and it results from repeatedly comparing the standardized deviate (2.4) with the constant $k(\alpha, V_f)$. The calculations needed to construct repeated significance test boundaries for independent normal variates are given by Armitage, McPherson and Rowe (1969) and McPherson and Armitage (1971). The operating characteristics of restricted boundaries (3.1) and repeated significance test boundaries (3.2) are similar. Under the alternative, both result in substantial average reductions in the numbers of deaths which must be observed to terminate a trial, compared with a fixed sample design that is analyzed only when a prespecified number of deaths have been observed. To illustrate, suppose one wished to detect a relative hazard of $e^{\theta} = 2$ with power 0.95 using a fixed sample two-sided $\alpha = 0.05$ level logrank test. Lininger, Gail, Green and Byar (1978) showed by simulation that the required total number of deaths for a fixed sample design is approximately

$$(Z_\alpha + Z_\beta)^2(\sigma/\mu)^2 = 4(Z_\alpha + Z_\beta)^2/\theta^2\,, \tag{3.3}$$

where Z_α and Z_β are two-sided standard normal deviates corresponding to size α and power $1-\beta$. In this example, $Z_\alpha = Z_\beta = 1.96$, $\theta = \log 2$, and a total of 128 deaths are required for the fixed sample design. By interpolation in Table 5.5 of Armitage (1975), one should choose $V_f = 158/4$, so that a maximum of 158 deaths are required for the repeated significance test boundary. On average, the repeated significance test boundary would stop after 153 deaths if $\theta = 0$ and after only 68 deaths if $\theta = \log 2$. This example shows that the savings in numbers of deaths required can be substantial under the alternative, but that sequential trials can run on for longer than a fixed sample trial of equivalent power if the null hypothesis is true.

Jones and Whitehead (1979) and Whitehead and Jones (1979) also proposed plotting $S(T_d)$ against $\hat{V}(d)$, and they developed approximations for the average number of deaths required for straight line boundaries with continuation regions

$$S(T_d) \in (-a + b\hat{V}(d),\ a + c\hat{V}(d))\,. \tag{3.4}$$

Equation (3.4) encompasses two-sided horizontal boundaries with $b = c = 0$, classical Wald (1947, Chapter 3) boundaries with $b = c$, and triangular boundaries as proposed by Anderson (1960). The work of Whitehead and Jones (1979) permits one to state a significance level, based on the number of deaths observed when the boundary was crossed, and to form a valid interval estimate of the log relative hazard, θ, after stopping. Although they present results for the modified Wilcoxon test as well as for the logrank test, the distribution theory in Section 2 shows that these techniques can only be used for the Wilcoxon test with simultaneous entry. For staggered entry, these methods should only be used with the logrank test, or, possibly, with other tests for which the weight function $Q(t, Y(i))$ in (2.1) is asymptotically independent of t. Whitehead (1983) has summarized work in this area.

Horizontal boundaries with rejection regions of the type

$$\max |S(T_d)| > c \tag{3.5}$$

were also proposed by Chatterjee and Sen (1973) for simultaneous entry, and this procedure was specialized for the logrank test by Koziol and Petkau (1978) and for the modified Wilcoxon test by Davis (1978). Since the total sample size is known with simultaneous entry, the uncensored permutational variance of

$S(\infty)$, namely V_f, is known, and c can be determined by noting that $S(T_d)/V_f^{1/2}$ converges to a Wiener process on $[0, 1]$ with transformed 'time' scale $\omega = \hat{V}(d)/V_f$.

More recently, Majumdar and Sen (1978b) adapted boundaries like (3.5) to staggered entry. However, instead of restricting analyses to the times at which deaths occur, they proposed to monitor the data at the random real times either of a death or of a new entry into the study. At each real time of analysis, they consider a sequence of type I censored survival problems, one corresponding to each member of a finite set of potential follow-up times. For each member of the sequence, they find the maximum of $S(T_d)$, and then they maximize over these maxima. The resulting distribution theory is complex, and rejection occurs when one of the computed statistics exceeds a critical value appropriate for a two-dimensional 'Brownian sheet'. Sinha and Sen (1982) proposed closely related procedures. A disadvantage of these techniques is the need to specify a 'target sample size', which is the putative number of patients to be entered in the event monitoring doesn't lead to rejection. This quantity, which is hypothetical, is needed to calculate the monitoring statistics. No corresponding parameter is needed to calculate $S(T_d)$ and $\hat{V}(d)$ for the methods discussed previously.

Sen (1979) developed an adjustment procedure for covariates that permits continuous monitoring for simultaneous entry, and Whitehead (1983) presents covariate adjustment procedures for the case of staggered entry.

It is often impractical to attempt continuous monitoring because of the difficulty of maintaining accurate up-to-date data. This is a particular problem for cooperative clinical trials involving several institutions. To alleviate this difficulty, group sequential plans have been proposed for examining the data only a few times (Pocock, 1977). Group sequential plans capture most of the efficiencies of continuous monitoring, and they greatly reduce, but do not eliminate, practical problems (see DeMets, Williams and Brown, 1982). In the next two sections, we therefore describe group sequential plans for survival data.

4. Group sequential plans with pre-specified boundaries: Analyses at fixed increments of information

Rather than attempt to monitor the data continuously, one might choose to look only a few times and to stop the trial if early evidence against the null hypothesis is strong. Pocock (1977) proposed this 'group sequential' approach for a variety of clinical trial response variables, and he concluded that "In general, a group sequential design with even a quite small number of groups provides a substantial reduction in average sample size when treatment differences exist. In fact, such a reduction may often be close to or even better than that achieved by standard sequential designs". Further work by Pocock (1981) suggested that it was rarely worthwhile to examine accumulating information more than five times if a repeated significance test boundary is used.

The following simple normal model has been used to construct group sequential boundaries. After k groups of g observations each, the standardized statistic

$$Z(k) = \left(\sum_{i=1}^{k} \sum_{j=1}^{q} C_{ij}\right)(\sigma^2 kg)^{-1/2} \tag{4.1}$$

is computed and compared with symmetric, two-sided group sequential boundaries b_k for $n = 1, 2, \ldots, K$, where K is a prespecified maximum number of looks at the data. The null hypothesis rejected at the smallest $k \leq K$ for which $|Z(k)| > b_k$. One-sided rejection regions, with rejection at the smallest $k \leq K$ for which $Z(k) > b_k$ have also been proposed. To compute boundaries of appropriate size it is assumed that:

(1) the group increments, $\Sigma_{j=1}^{q} C_{ij}$, are normally distributed,
(2) uncorrelated, and
(3) homoscedastic with known variance $g\sigma^2$.

Power is computed under the alternative that the group increments have mean δg.

Pocock (1981) suggested that boundaries computed under this simple normal model would be applicable to the logrank test provided the analyses were performed at equally spaced numbers of deaths. The results in Section 2 indeed suggest that increments of the logrank numerator that are computed after each g deaths will be approximately normally distributed, uncorrelated, and homoscedastic with variance $g\sigma^2 \doteq g/4$. More generally, the boundaries we discuss next might be used for any rank statistic (2.1) with asymptotically uncorrelated increments provided analyses are performed at nearly equally spaced values of Fisher information, $\hat{V}(d)$. The proposal is to use group sequential boundaries with the standardized deviate

$$Z(k) = S(T_{kg})\{\hat{V}(kg)\}^{-1/2} . \tag{4.2}$$

Gail, DeMets and Slud (1982) compared the simulated performance of the logrank test, computed after each eighteen deaths, with the theoretical predictions of the simple normal model. They chose $K = 5$ maximum looks, and considered four two-sided boundaries. The Haybittle (1971) boundary (H) has $b_k = 3.0$ for $k = 1, 2, 3, 4$ and $b_5 = 1.96$. This boundary is conservative and only detects extreme early differences. Its size is 0.053, slightly in excess of the nominal 0.05 level. The Pocock boundary (P) is a repeated significance test boundary with $b_k = 2.413$ for $k = 1, 2, 3, 4, 5$. Note that b_k exceeds 1.96 in order to assure proper size; had 1.96 been used instead, the size of this test would be about 0.142. The O'Brien–Fleming boundary (O) is $b_k = (4.149 \times 5/k)^{1/2}$ for $k = 1, 2, 3, 4, 5$. A fixed sample size boundary (F) was also defined as $b_k = 100$ for $k = 1, 2, 3, 4$, and $b_5 = 1.96$.

If the increments of the logrank score statistic S, computed after each g

deaths, satisfied the assumptions (1)–(3) above, then the properties of these boundaries could be calculated theoretically. For proportional hazards alternatives with hazard ratio $\exp(\theta)$, the expectation of an increment based on g deaths is approximately $g\delta = g\theta/4$ for small θ, and the variance is $g\sigma^2 \doteq g/4$. Hence the required noncentrality parameter is

$$\Delta = (g\theta/4)(g/4)^{-1/2} = \theta(g/4)^{1/2}\,.$$

The quantity Δ is used for tabulations in Pocock (1977), and theoretical quantities such as the size, power and average sample number of groups, $\bar{k}$, may be computed from Δ as in Armitage, McPherson and Rowe (1969), McPherson and Armitage (1971), and DeMets and Ware (1980).

The theoretical properties of these boundaries show that the Pocock boundary, P, has the greatest potential for early stopping, especially under the alternative $\exp(\theta) = 2$ (Table 1). On the other hand, the fixed boundary F has the greatest power among tests with size 0.050. An attractive feature of O and H is that they seldom stop when the trial is just beginning, and the final boundary point b_k is close to 1.96. This conservative behavior is often acceptable to the monitoring committees that decide when to stop a trial (DeMets and Ware, 1982).

The simulations by Gail, DeMets and Slud (1982) show that the simple normal model indeed describes the null case behavior of the logrank statistic quite accurately, both for simultaneous entry and for staggered entry, as is expected from the theoretical results in 2. Under the alternative $\exp(\theta) = 2$, increments of the logrank score are correlated, and the simple normal model is not strictly valid. Nonetheless, the observed operating characteristics of these boundaries are close enough to the values in Table 1 for most applications. For example, the power of P is about 3% less than the predicted 0.845, and the average number of groups required for P exceeds the theoretical value 3.083 by about 4%. Discrepancies for the other boundaries in Table 1 are even less. Robustness studies demonstrate that the size of P may increase to 0.07 if

Table 1
Theoretical properties of four group sequential boundaries with $K = 5$ and $g = 18$ deaths for each increment of the logrank score[a]

	Pocock (P)	O'Brien–Fleming (O)	Haybittle (H)	Fixed (F)
null case				
size	0.050	0.050	0.053	0.050
$\bar{k}$	4.876	4.964	4.977	5.000
relative hazard $\exp(\theta) = 2$				
power	0.845	0.901	0.909	0.907
$\bar{k}$	3.083	3.648	3.864	5.000

[a] $\bar{k}$ is the average number of groups required.

healthier patients enter later in the trial, and that the other boundaries are less sensitive to trends in the life distribution of entering patients. Altogether, these results suggest that the simple normal model may be used to design and study group sequential boundaries for the logrank test with fixed numbers of deaths in each group.

It is a strength of methods based on analyses at fixed increments of $\hat{V}(d)$ that the properties of proposed boundaries can be evaluated prior to the experiment, either by use of the simple normal model or through simulations. Moreover, standardized designs are available for the practitioner, who needs no special facilities to use them. Recent work on the selection of group sequential boundaries includes that of DeMets and Ware (1980, 1982), who proposed one-sided rejection regions, Whitehead and Stratton (1983), who propose an asymmetric triangular continuation region, Gould and Pecore (1982), who insert an inner wedge to permit earlier stopping under the null hypothesis, and Pocock (1981) and McPherson (1982), who consider how many repeated looks at the data are useful.

Standard confidence intervals and point estimates of the log relative risk, θ, based either on the partial likelihood of Cox (1972) or on likelihoods from parametric models, do not have their intended frequentist properties in hypothetical repetitions of the group sequential trial. However, valid frequentist confidence intervals have been constructed for group sequential plans with predetermined boundaries. Jennison and Turnbull (1983a) obtained valid confidence intervals for the binomial parameter following group sequential tests by defining an ordering on the outcomes, and Tsiatis, Rosner and Mehta (1983) have extended these ideas to the standard normal model. They obtained confidence intervals, which can be applied to the log relative hazard, θ. For our problem, the outcomes are ordered according to how much the data favor survival curve G_2 over survival curve $G_1 = G_2^{\exp(\theta)}$. Thus G_2 is most favored if $Z(1)$ rejects at a large positive value, somewhat less favored if $Z(2)$ rejects at a large positive value and so forth, with G_1 most favored if $Z(1)$ has a large negative value. With this ordering of the group sequential outcome space, one can compute the probability that the results would favor G_2 as much or more than did the observed outcome. These probabilities are inverted to produce confidence intervals on θ. Similar ideas are found in the work of Fairbanks and Madsen (1982) and Madsen and Fairbanks (1983), who define P values according to an implicit ordering of the possible outcomes.

The confidence intervals of Tsiatis, Rosner and Mehta (1983) are designed for the analysis *after* a group sequential boundary has been used to test the null hypothesis. Jennison (1982) and Jennison and Turnbull (1983b) construct repeated confidence intervals based on the logrank test that can be calculated as the trial proceeds. These confidence intervals can be used for estimation of the log relative hazard θ and for hypothesis tests which reject whenever a sequentially computed confidence interval excludes the null value $\theta = 0$. To use the method of Jennison and Turnbull (1983b), one must pre-specify a group sequential boundary, b_k. Based on the fact that the logrank statistic, $S(T_{kg})$, is approximately normally distributed with mean $kg\theta/4$ and variance $kg/4$, compute the confidence interval after k groups of g deaths from

$$[4S(T_{kg})/kg - 2b_k(kg)^{-1/2}, 4S(T_{kg})/kg + 2b_k(kg)^{-1/2}]\,. \tag{4.3}$$

By construction, confidence intervals $k = 1, 2, \ldots, K$ are simultaneously valid; that is, each covers θ with probability $\geq 1 - \alpha$. Hence, one can reject H_0 as soon as any such interval excludes the null value of θ. As might be expected, confidence intervals based on the boundary O are quite wide for the first few looks, compared to a fixed sample boundary. These simultaneous confidence intervals remain valid regardless of when or how the trial is stopped. This property may be useful to a monitoring committee that decides to stop the trial because of undue toxicity or other factors and that is mainly concerned with the magnitude of the treatment effect, as discussed by Meier (1979).

There are practical difficulties in using pre-specified boundaries designed for tests after fixed increments of $\hat{V}(d)$. The monitoring committee must be prepared to establish the maximum number of looks, K, in advance and to be satisfied with analyses performed when the prespecified increments in $\hat{V}(d)$ have occurred, rather than when the committee plans to meet.

5. Group sequential plans with data dependent boundaries: Analyses at designated calendar times

Two methods have been developed to allow the monitoring committee to control the times of analysis. The method of Slud and Wei (1982) permits analyses at pre-specified calendar times $t_1, t_2, \ldots, t_k$, where K, the total number of planned looks, is set in advance. A second proposal by Lan and DeMets (1983), permits one to perform any number of analyses at any desired calendar time, but interim analyses require specification of a hypothetical final total information $\hat{V}(D)$ and of a function which determines how fast the size of the design is 'used up'. For both methods, if the times of analysis are set by the meeting schedule of the monitoring committee, the incremental information between analyses will vary, depending on such factors as patient accrual and chance variations in the numbers of observed deaths. Thus the estimated covariance (2.6) will not exhibit the regularities that are required for the construction of prespecified boundaries defined in Section 4. Instead, the boundaries are determined adaptively, based on the data in hand, to preserve overall size α.

Slud and Wei (1982) proposed the following procedure for monitoring survival data at pre-specified calendar times $t_1 < t_2 < \cdots < t_K$, where K, the maximum number of looks, is fixed in advance. At each t_k, use the variance estimate $\hat{V}(d(t_k)) = \widehat{\text{Cov}}(S(t_k), S(t_k))$ to compute a standardized deviate $Z(t_k) = S(t_k)\hat{V}(d(t_k))^{-1/2}$. Under the null hypothesis, $Z(t_1), Z(t_2), \ldots, Z(t_K)$ converges to a multivariate normal distribution with means zero, unit variances, and correlations given by the limit of $\widehat{\mathrm{Cov}}(S(t_i), S(t_j))\{\hat{V}(d(t_i))\hat{V}(d(t_j))\}^{-1/2}$. For a fixed size $\alpha > 0$, define K values $\alpha_k \geq 0$, $\Sigma_1^K \alpha_k = \alpha$. At time t_1, compute a two-sided boundary point $b(t_1)$ from $P[|Z(t_1)| > b(t_1)] = \alpha_1$. Define subsequent boundary points recursively from

$$P[|Z(t_1)| < b(t_1), |Z(t_2)| \geq b(t_2)] = \alpha_2\,,$$

$$P[|Z(t_1)| < b(t_1), |Z(t_2)| < b(t_2), |Z(t_3)| \geq b(t_3)] = \alpha_3$$

and so forth for $b(t_4), \ldots, b(t_K)$. One-sided boundaries can be calculated similarly.

From this construction, the overall size of the procedure will be α. It is clear that the choice of the pre-specified sequence $\alpha_1, \alpha_2, \ldots, \alpha_K$ determines the boundary points $b(t_k)$. A practical disadvantage of this procedure is the need to fix K, and α_k in advance. If the times of analysis, t_k, are also prespecified, a valuable look at the data may be 'used up' even for intervals which provide little new information. A further disadvantage is that the boundary points $b(t_k)$ are random variables. Thus the operating characteristics of the procedure cannot be studied in advance unless one makes strong and questionable assumptions about the accrual rates and common unknown survival distribution G.

Lan and DeMets (1983) suggest a procedure that allows one to perform analyses whenever one wants and still preserve overall significance level α. Their ideas can be applied easily to any survival process with independent increments, but we use the logrank statistic for illustration. First one must specify a final Fisher information $V(D)$, which, for the logrank test is $D/4$; here the trial is to be carried out until exactly D deaths have been observed, unless stopped earlier by the procedure outlined below. Next one defines a Brownian motion time scale $\omega = \hat{V}(d(t))/V(D) = d(t)/D$ which satisfies $0 \leq \omega \leq 1$. Lan and DeMets develop one-sided boundaries and show how to convert them to symmetric two-sided tests. Consider a standard Brownian motion $B(\omega)$ on $[0, 1]$ and a boundary $b(\omega)$ which satisfies $P[B(\omega) \geq b(\omega), 0 \leq \omega \leq 1] = \alpha$. The time to first exit, τ, is a random variable with a sub-distribution function $\alpha^*(\omega) = P[\tau \leq \omega]$ satisfying $\alpha^*(0) = 0$, $\alpha^*(1) = \alpha$. This function is determined implicitly by the boundary $b(\omega)$. Conversely, an increasing positive function $\alpha^*(\omega)$ that satisfies $\alpha^*(0) = 0$ and $\alpha^*(1) = \alpha$ implicitly defines a boundary $b(\omega)$. Suppose such a function, $\alpha^*(\omega)$ has been predetermined. At a calendar time t_1, we have observed $d(t_1)$ deaths, which we convert to process time $\omega_1 = d(t_1)/D$. The appropriate Brownian motion boundary value is determined from

$$P[B(\omega_1) \geq b(\omega_1)] = \alpha^*(\omega_1)\,.$$

One rejects the null hypothesis if $Z(t_1) \geq b(\omega_1)\omega_1^{-1/2}$. If no rejection occurs at t_1, one tests again at t_2 and determines $b(\omega_2)$ from $P[B(\omega_1) < b(\omega_1), B(\omega_2) \geq b(\omega_2)] = \alpha^*(\omega_2) - \alpha^*(\omega_1)$. Again, one rejects if $Z(t_2) \geq b(\omega_2)\omega_2^{-1/2}$. The procedure continues, based on such recursive calculations, either until the null hypothesis has been rejected or until D deaths have been observed ($\omega = 1$). The function $\alpha^*(\omega)$ determines how rapidly the size of the test is to be 'used up'. Lan and DeMets define functions $\alpha^*(\omega)$ that correspond to the O'Brien–Fleming (O) and Pocock (P) boundaries described in Section 4 if $t_1, t_2, \ldots, t_k$ are chosen to correspond to equal increments $\omega_k - \omega_{k-1} = 1/K$ for $k = 1, 2, \ldots, K$ and $\omega_0 = 0$. Indeed, the procedures in Section 4 can be regarded as special cases of the Lan–DeMets procedure. However, when analyses are performed at the times of meetings of the monitoring committee, the corresponding boundaries will be unpredictable. Therefore, as for the procedure of Slud and Wei, it is not possible to study the properties of these boundaries in advance without making strong assumptions about accrual rates and survival.

Canner (1976, 1977) discusses group sequential monitoring of survival data at designated calendar times, but, for staggered entry, he studies a parametric test of the equality of two exponential survival distributions (Canner, 1977).

6. Curtailed experiments

Suppose a fixed sample clinical trial is planned, and that the trial has size $\alpha = P[R \mid H_0]$ and type 2 error $\beta = P[\bar{R} \mid H_a]$ where R is the event that the test statistic falls in the rejection region at completion of the planned experiment and $\bar{R}$ is the complementary event. If one reaches a point in the trial from which it is known with certainty whether the final result will be in R or in $\bar{R}$, irregardless of any future observations, one could stop the trial and either accept or reject the null hypothesis. Such a procedure is called *curtailment.*

If all patients enter simultaneously, curtailment is feasible for survival statistics like (2.1), because a sequential path may be so extreme that all possible arrangements of later ranks lead to R, for example. Halperin and Ware (1974) studied such curtailment for Gehan's modification of the Wilcoxon test, and Verter (1983) extended their results to other statistics like (2.1), including the logrank test.

The procedures above are very conservative because they require that a foregone conclusion be reached before curtailment. To permit somewhat earlier stopping one might require only that the *probability* that the present decision will be reversed is small. Such a procedure is called *stochastic curtailment.* To illustrate this idea, suppose one plans a 'fixed sample' logrank analysis after D deaths, with rejection if $S(T_D) \in R$. The quantity D and the rejection region R are chosen to give size α and power $1-\beta$ for a hazard ratio e^{θ}. Suppose, however, one examines the logrank score after each death and plans to reject H_0 at the smallest $d \leqslant D$ such that

$$P[S(T_D) \in R \mid S(T_d); H_0] \geqslant \gamma \tag{6.1}$$

where γ is near 1. Also, if the new treatment is not doing well, one might wish to stop at the smallest $d \leq D$ such that

$$P[S(T_D) \in \bar{R} \mid S(T_d); H_a] \geq \gamma' . \tag{6.2}$$

Lan, Simon and Halperin (1982) proved the very general result that the overall size of this procedure is α/γ and the power is $1 - \beta/\gamma'$. For example if $\gamma = 0.95$, stochastic curtailment only increases the size from $\alpha = 0.05$ to $\alpha/\gamma = 0.0526$. To attain adequate power, a value of γ' as small as 0.5 may be used. For example, if the original fixed sample power is $1 - \beta = 0.9$, the stochastically curtailed power is $1 - 0.10/0.5 = 0.8$, which is often acceptable.

There are practical problems with this procedure. Early stopping to reject H_0 according to (6.1) is feasible, because the probability in (6.1) can be calculated under H_0. For the logrank statistic, the independent increments property makes the calculation tractable. Early stopping to accept the null hypothesis according to (6.2) is more problematic. The probability in (6.2) depends on the unknown alternative. For proportional hazards, one must specify θ. If the true θ is larger than anticipated, one will overestimate the left hand side of (6.2), leading to inappropriate early stopping and loss of power. There is the further technical problem that, under H_a, the logrank increments are correlated, and no theory is yet available for the computation of (6.2). The calculation of (6.2) is also problematic in situations which require that one guess at the future form of the two survival distributions. Halperin, Lan, Ware, Johnson and DeMets (1982) discuss this difficulty in connection with stochastic curtailment of a comparison of the proportions dead on the two treatments.

7. Monitoring to stop accrual

All the methods for monitoring that we have discussed so far are based on the premise that a final treatment comparison will be made at the moment the trial is stopped. In some circumstances, it may be desirable to separate the decision to stop the accrual of new patients from the final analysis for the comparison of treatments. Rubinstein and Gail (1982) have studied the following procedure and have used it in the design and monitoring of lung cancer trials.

It is agreed that the final two-sample treatment comparison, based on the proportional hazards model, will be made after $D = 90$ deaths have occurred. This assures a power 0.9 against relative hazard $e^{\theta} = 2$ for the two-sided $\alpha = 0.05$ level logrank test. The standard deviation of the estimate of θ will be approximately $V(D)^{-1/2} \doteq (4/90)^{1/2} = 0.21$. In essence, this is a fixed sample experiment, based on D deaths, as regards the final treatment comparison. However, Rubinstein and Gail (1982) show that one can monitor the accumulating survival data with the logrank statistic and stop accrual early if an important treatment difference begins to emerge, without perturbing the final statistical comparison of treatments, so long as one waits until D deaths have been observed before performing the final analysis of treatment effect. This strategy takes advantage of the fact that, at the moment accrual is stopped, there are a number of patients still on study who will, in time, provide the required additional survival information. The rules for stopping accrual are quite flexible and usually contain the proviso that the time to death D will not be unduly prolonged by early termination of accrual. The technique should not be used with statistics that have correlated increments.

A major advantage of this procedure is that one can prespecify the precision of the final estimate of treatment effect and that the final fixed sample analysis is simple and noncontroversial. Also, the frequentist analysis and the relative likelihood analysis will agree, because for such fixed sample experiments, the likelihood ratio is a monotone function of the observed p value. Furthermore, the data may be monitored whenever it is convenient, and rigid rules for stopping accrual are not necessary.

This technique is not universally applicable, because it is assumed that all accrued patients will remain on their originally assigned treatment until $D = 90$ deaths are observed. This is often appropriate in a cancer trial in which the treatment is given early in the course of follow-up, because there is usually no compelling evidence that delayed treatment with the apparently better regime could benefit those initially given the other treatment. However, in a trial of a chronically administered treatment, such as a beta-blocker to prevent sudden death from cardiac arrhythmia, there may be an urgent need to switch all patients to the apparently preferable treatment as soon as substantial evidence of a treatment difference arises. In such cases, one cannot usefully separate the decision to stop accrual from the final analysis of treatment effect.

8. Discussion

Several themes emerge from this survey. First, good progress has been made in developing a null distribution theory for sequential application of survival statistics with staggered entry. Convergence to a Gaussian process has been proved for a wide class of statistics for analysis in real time (see Slud, 1983, Tsiatis, 1982 and Harrington, Fleming and Green, 1982), and Sellke and Siegmund (1983) have proved convergence to a Brownian motion for the logrank test in ‘process time’, $V(d)$. It would be useful to have similar results for other statistics in process time. No rigorous theory has been published for fixed alternatives, even for the logrank test with proportional hazards. This deficit impedes the evaluation of group sequential designs, and it prevents the precise evaluation of probabilities like (6.2) needed for stochastic curtailment. Nonetheless, at least for the logrank statistic with relative hazards in the range 0.5 to 2.0, a Brownian motion model with drift leads to results that are sufficiently accurate for planning group sequential trials (Gail, DeMets and Slud, 1982).

A second theme is flexibility. The procedure of Slud and Wei (1982) permits analyses at arbitrary real times, though the number of looks and allocation of α level must be prespecified. If one is willing to specify a final intended amount of process information $V(D)$, and a function for allocating α level, then the methods of Lan and DeMets (1983) may be used to look at the data whenever desired. Stochastic curtailment (Lan, Halperin and Simon, 1982) may also be employed at any time in the trial, provided the required probabilities can be calculated. The method of accrual monitoring (Rubinstein and Gail, 1982) offers investigators great freedom in the monitoring process, provided they agree to defer a treatment decision until the prespecified final information, $V(D)$, has been obtained.

Implicit in the variety of proposed boundaries and techniques are underlying judgments as to the relative importance of early stopping and the need for compelling medical evidence. Accrual monitoring, stochastic curtailment, and conservative group sequential boundaries such as those proposed by Haybittle (1971) and O'Brien and Fleming (1979) generally result in more clinical data and less potential for early stopping. Additional clinical data and longer follow-up times can be especially helpful in survival studies, because later survival experience on the two treatments may not resemble the early relative performance.

Those procedures which offer the greatest potential for early stopping pose special problems at the time of analysis, because a frequentist analysis that is based on hypothetical repetitions of the experiment with its stopping procedure can differ from a likelihood based analysis that ignores the stopping rule (Cornfield, 1966). A medical investigator, who was asked to write up the results of a clinical trial that stopped early, was frustrated by the advice of the following three statistical consultants. The first statistician, who had planned the original group sequential study with the Pocock boundary, presented a confidence interval for the hazard ratio. The second statistician, who had replaced the first during the course of the study, presented a different confidence interval, because he had decided to switch to an O'Brien–Fleming boundary. The third statistician, who had not had access to the information on experimental design and who said he didn't want it anyhow, based his analysis on the observed likelihood and produced yet a third confidence interval. Relieved, the medical investigator chose the last confidence interval, since it agreed most closely with his own fixed sample analysis. This story, which I hope never happened, is a reminder that there is disagreement among statisticians as to whether the strict frequentist analysis is meaningful or appropriate for a trial which may never be repeated and for which it is unlikely that any pre-specified stopping procedure will be strictly observed.

Acknowledgements

I would like to thank K. K. G. Lan, D. P. Harrington, B. Turnbull, E. V. Slud and A. A. Tsiatis for their generous provision of preprints and helpful comments, the reviewer for pointing out important references, and Julie Paolella for typing the manuscript. This paper was used as the basis of a presentation to the *1983 Conference on Biostatistics* in Philadelphia, sponsored by Temple University and the Merck Company in Philadelphia.

References

Anderson, T. W. (1960). A modification of the sequential probability ratio test to reduce sample size. *Anals of Mathematical Statistics* **31**, 165–197.

Anscombe, F. J. (1963). Sequential medical trials. *Journal of the American Statistical Association* **58**, 365–383.

Armitage, P. (1960). *Sequential Medical Trials*, Blackwell, Oxford.

Armitage, P. (1975). *Sequential Medical Trials*, Wiley, New York.

Armitage, P., McPherson, C. K. and Rowe, B. C. (1969). Repeated significance tests on accumulating data. *Journal of the Royal Statistical Society A* **132**, 235–244.

Bechhofer, R. E. (1954). A single-sample multiple decision procedure for ranking means of normal populations with known variances. *Annals of Mathematical Statistics* **25**, 16–39.

Canner, P. L. (1976). Repeated analysis of clinical trial data. In: *Proceedings of the Ninth International Biometric Conference*, Volume 1, Biometric Society, Boston.

Canner, P. L. (1977). Monitoring treatment differences in long-term clinical trials. *Biometrics* **33**, 603–615.

Chatterjee, S. K. and Sen, P. K. (1973). Nonparametric testing under progressive censorship. *Calcutta Statistical Association Bulletin* **22**, 13–50.

Colton, T. (1963). A model for selecting one of two medical treatments. *Journal of the American Statistical Association* **58**, 388–400.

Cornfield, J. (1966). Sequential trials, sequential analysis, and the likelihood principle. *American Statistician* **20**, 18–23.

Cox, D. R. (1972). Regression models and life tables (with discussion). *Journal of the Royal Statistical Society B* **34**, 187–202.

Davis, C. E. (1978). A two-sample Wilcoxon test for progressively censored survival data. *Communication Statistical Theory and Methods A* **7**, 389–398.

DeMets, D. L. and Lan, K. K. G. (1984). An overview of sequential methods and their applications in clinical trials. *Communications in Statistics Series A* (to appear).

DeMets, D. L. and Ware, J. H. (1980). Group sequential methods in clinical trials with a one-sided hypothesis. *Biometrika* **67**, 651–660.

DeMets, D. L. and Ware, J. H. (1982). Asymmetric group sequential boundaries for monitoring clinical trials. *Biometrika* **69**, 661–663.

DeMets, D. L., Williams, G. W. and Brown, B. W. Jr. and the NOTT Research Group (1982). A case report of data monitoring experience: the nocturnal oxygen therapy trial. *Controlled Clinical Trials* **3**, 113–124.

Fairbanks, K. and Madsen, R. (1982). *P* values for tests using a repeated significance test design. *Biometrika* **69**, 69–74.

Gail, M. H. (1982). Monitoring and stopping clinical trials. In: V. Miké and K. E. Stanley, eds., *Statistics in Medical Research: Methods and Issues with Applications in Cancer Research*. Wiley, New York, pp. 455–484.

Gail, M. H., DeMets, D. L. and Slud, E. V. (1982). Simulation studies on increments of the two-sample logrank score test for survival time data, with application to group sequential boundaries. In: J. Crowley and R. A. Johnson, eds., *Survival Analysis*, Institute of Mathematical Statistics, Hayward, California, pp. 287–301.

Gehan, E. A. (1965). A generalized Wilcoxon test for comparing arbitrarily singly-censored samples. *Biometrika* **52**, 203–223.

Gill, R. D. (1980). *Censoring and Stochastic Integrals.* Mathematical Centre Tracts 124, Mathematische Centre, Amsterdam.

Gould, A. L. and Pecore, V. J. (1982). Group sequential methods for clinical trials allowing early acceptance of H_0 and incorporating costs. *Biometrika* **69**, 75–80.

Halperin, M. and Ware, J. (1974). Early decision in a censored Wilcoxon two-sample test for accumulating survival data. *Journal of the American Statistical Association* **69**, 414–422.

Halperin, M., Lan, K. K. G., Ware, J. H., Johnson, N. J. and DeMets, D. L. (1982). An aid to data monitoring in long-term clinical trials. *Controlled Clinical Trials* **3**, 311–323.

Harringron, D. P. and Fleming, T. R. (1982). A class of rank test procedures for censored survival data. *Biometrika* **69**, 553–566.

Harrington, D. P., Fleming, T. R. and Green, S. J. (1982). Procedures for serial testing in censored survival data. In: J. Crowley and R. A. Johnson, eds., *Survival Analysis,* Institute of Mathematical Statistics Monograph Series, Volume 2, Hayward, California.

Haybittle, J. L. (1971). Repeated assessment of results in clinical trials of cancer treatment. *British Journal of Radiology* **44**, 793–797.

Jennison, C. (1982). Sequential methods for medical experiments. Ph.D. Dissertation, Cornell University.

Jennison, C. and Turnbull, B. W. (1983a). Confidence intervals for a binomial parameter following a multistage test with application to MIL-STD 105D and medical trials. *Technometrics* **25**, 49–58.

Jennison, C. and Turnbull, B. W. (1983b). Repeated confidence intervals for sequential clinical trials. *Controlled Clinical Trials* **5**, 33–45.

Jones, D. and Whitehead, J. (1979). Sequential forms of the logrank and modified Wilcoxon tests for censored data. *Biometrika* **66**, 105–113.

Jones, D. R., Newman, C. E. and Whitehead, J. (1982). The design of a sequential clinical trial for the comparison of two lung cancer treatments. *Statistics in Medicine* **1**, 73–82.

Kalbfleisch, J. D. and Prentice, R. L. (1980). *The Statistical Analysis of Failure Time Data,* Wiley, New York.

Koziol, J. and Petkau, A. J. (1978). Sequential testing of the equality of two survival distributions using the modified Savage statistic. *Biometrika* **65**, 615–623.

Lagakos, S. W. (1982). Inference in Survival Analysis: nonparametric tests to compare survival distributions. In: V. Miké and K. E. Stanley, eds., *Statistics in Medical Research: Methods and Issues with Applications in Cancer Research.* Wiley, New York, pp. 340–364.

Lan, K. K. and DeMets, D. L. (1983). Discrete sequential boundaries for clinical trials. *Biometrika* **70**, 659–663.

Lan, K. K. G., Simon, R. and Halperin, M. (1982). Stochastically curtailed tests in long-term clinical trials. *Communications in Statistics: Sequential Analysis* **1**, 207–219.

Lininger, L., Gail, M. H., Green, S. B. and Byar, D. P. (1979). Comparison of four tests for equality of survival curves in the presence of stratification and censoring. *Biometrika* **66**, 419–428.

Madsen, R. W. and Fairbanks, K. B. (1983). *P* values for multistage and sequential tests. *Technometrics* **25**, 285–293.

Majumdar, H. and Sen, P. K. (1978a). Nonparametric tests for multiple regression under progressive censoring. *J. Multivariate Analysis* **8**, 73–95.

Majumdar, H. and Sen, P. K. (1978b). Nonparametric testing for simple regression under progressive censoring with staggering entry and random withdrawal. *Communications in Statistics Series A* **7**, 349–371.

Mantel, N. (1966). Evaluation of survival data and two new rank tests arising in its consideration. *Cancer Chemotherapy Reports* **50**, 163–170.

McPherson, K. (1982). On choosing the number of interim analyses in clinical trials. *Statistics in Medicine* **1**, 25–36.

McPherson, C. K. and Armitage, P. (1971). Repeated significance tests on accumulating data when the null hypothesis is not true. *Journal of the Royal Statistical Association A* **134**, 15–65.

Mehrotra, K. G., Michalek, J. E. and Mihalko, D. (1982). A relationship between two forms of linear rank procedures for censored data. *Biometrika* **69**, 674–676.

Meier, P. (1979). Terminating a trial – the ethical problem. *Clinical Pharmacology and Therapeutics* **25**, 633–640.

Morton, R. (1978). Regression analysis of life tables and related nonparametric tests. *Biometrika* **65**, 329–333.

O'Brien, P. C. and Fleming, T. R. (1979). A multiple testing procedure for clinical trials. *Biometrics* **35**, 549–556.

Pocock, S. J. (1977). Group sequential methods in the design and analysis of clinical trials. *Biometrika* **64**, 191–199.

Pocock, S. J. (1981). Interim analyses and stopping rules for clinical trials. In: J. F. Bithell and R. Coppi, eds., *Perspectives in Medical Statistics*. Grune and Stratton, New York, pp. 191–223.

Prentice, R. (1978). Linear rank tests and right censored data. *Biometrika* **65**, 167–179.

Prentice, R. and Marek, P. (1979). A qualitative discrepancy between censored data rank tests. *Biometrics* **35**, 861–867.

Rubinstein, L. V. and Gail, M. H. (1982). Monitoring rules for stopping accrual in comparative survival studies. *Controlled Clinical Trials* **3**, 325–343.

Schoenfeld, D. (1981). The asymptotic properties of nonparametric tests for comparing survival distributions. *Biometrika* **68**, 316–319.

Sellke, T. and Siegmund, D. (1983). Sequential analysis of the proportional hazards model. *Biometrika* **70**, 315–326.

Sen, P. K. (1979). Rank analysis of covariance under progressive censoring. *Sankhya Series A* **41**, 147–169.

Sen, P. K. (1981). The Cox regression model, invariance principles for some induced quantile processes and some repeated significance tests. *The Annals of Statistics* **9**, 109–121.

Sen, P. K. and Ghosh, M. (1972). On strong convergence of regression rank statistics. *Sankhya A* **34**, 335–348.

Sinha, A. N. and Sen, P. K. (1979). Progressively censored tests for clinical experiments and life testing based on weighted empirical distributions. *Communications in Statistics Series A* **8**, 871–897.

Sinha, A. N. and Sen, P. K. (1982). Tests based on empirical processes for progressive censoring schemes with staggering entry and random withdrawal. *Sankhya Series B* **44**, 1–18.

Slud, E. V. (1983). Sequential linear rank tests for two-sample censored survival data. Submitted for publication.

Slud, E. V. and Wei, L. J. (1982). Two-sample repeated significance tests based on the modified Wilcoxon statistic. *Journal of the American Statistical Association* **77**, 862–868.

Tarone, R. and Ware, J. (1977). On distribution-free tests for equality of survival distributions. *Biometrika* **64**, 156–160.

Tsiatis, A. A. (1982). Repeated significance testing for a general class of statistics used in censored survival analysis. *Journal of the American Statistical Association* **77**, 855–861.

Tsiatis, A. A., Rosner, G. L. and Mehta, C. R. (1983). Exact confidence intervals following a group sequential test. Dana–Farber Cancer Institute Technical Report 310Z.

Tsiatis, A. A. and Tritchler, D. L. (1982). Application of grouped sequential tests to survival analysis in clinical trials. Technical Report 275 ERCZ, Sidney Farber Cancer Institute Department of Biostatistics.

Verter, J. (1983). Early decision using simple rank statistics for accumulating survival data. Ph.D. Dissertation, University of North Carolina at Chapel Hill.

Wald, A. (1947). *Sequential Analysis*. Wiley, New York.

Whitehead, J. (1983). *Design and Analysis of Sequential Clinical Trials*. Wiley, New York.

Whitehead, J. and Jones, D. (1979). The analysis of sequential clinical trials. *Biometrika* **66**, 443–452.

Whitehead, J. and Stratton, I. (1983). Group sequential clinical trials with triangular continuation regions. *Biometrics* **39**, 227–236.

P. R. Krishnaiah and P. K. Sen, eds., *Handbook of Statistics, Vol. 4*
© Elsevier Science Publishers (1984) 813–830

34

Meteorological Applications of Permutation Techniques based on Distance Functions

Paul W. Mielke, Jr.

1. Introduction

Recent applications of classical linear rank tests in meteorological investigations (Mielke et al., 1981c, 1982) have raised concerns regarding the interpretation of P-values arising from such tests. The basis for these concerns is that the analysis space associated with classical linear rank test statistics is almost always nonmetric and consequently incomprehensible to any meteorologist (perhaps investigators in any field once these concerns are recognized). These concerns arise from the fact that classical linear rank test statistics mimic statistics associated with parametric tests based on normality (e.g., one-way analysis of variance, one-sample t test, and analysis of variance involving randomized blocks). This paper describes recently developed permutation techniques which include both some suggested alternative nonparametric rank tests which are free of these concerns and the classical linear rank tests for comparison.

Section 2 describes multi-response permutation procedures (MRPP). Included in this description is the relation with well-known permutation techniques and some presently known asymptotic results. Recently developed univariate rank tests based on MRPP (Mielke et al., 1981b) coupled with median regression are used in an example which involves recent reanalyses of the Climax wintertime orographic cloud seeding experiments (Mielke et al., 1981c, 1982). Other examples involving multivariate applications of MRPP are also described (Mielke et al., 1981a).

Section 3 describes one-sample permutation techniques (Mielke and Berry, 1982) and multivariate permutation techniques for randomized blocks (Mielke and Iyer, 1982) which are closely related to MRPP. The relation of these recent techniques with well-known permutation tests is described. Also a multivariate

This work has been supported by the Division of Atmospheric Resources Research, Bureau of Reclamation, U.S. Department of the Interior, under Contract 8-07-83-V0009 and Cooperative Agreement 2-07-83-V0273, by the National Science Foundation, under Grant ATM-81-07056, and by National Environmental Satellite Service, National Oceanic and Atmospheric Administration, U.S. Department of Commerce, under Contract NA81RA-H-00001.

measure of agreement is discussed in an example to compare different precipitation estimates for large areas which are based on satellite data with corresponding precipitation estimates which are based either on conventional surface gauge network data or radar data.

2. Comparisons without blocking

The techniques and examples presented in this section are based on multi-response permutation procedures (MRPP). Since these techniques are relatively new, a description of MRPP along with their relation to well known techniques will precede some recent meteorological applications involving weather modification and tropical storms.

2.1. *Methodological description* (*MRPP*)

Let $\Omega = \{\omega_1, \ldots, \omega_N\}$ be a finite population of N objects, let $x_I' = (x_{1I}, \ldots, x_{rI})$ denote r commensurate response measurements (these might be functions of response measurements or residuals adjusted by predictors) for object ω_I $(I = 1, \ldots, N)$, and let $S_1, \ldots, S_{g+1}$ designate an exhaustive partitioning of the N objects comprising Ω into $g+1$ disjoint subgroups. Also let $\Delta_{I,J}$ be a symmetric distance function value of the response measurements associated with the objects ω_I and ω_J. The statistic underlying MRPP is given by

$$\delta = \sum_{i=1}^{g} C_i \xi_i$$

where

$$\xi_i = \binom{n_i}{2}^{-1} \sum_{I<J} \Delta_{I,J} \Psi_i(\omega_I) \Psi_i(\omega_J)$$

is the average distance function value for all distinct pairs of objects in subgroup S_i $(i = 1, \ldots, g)$, $n_i \geq 2$ is the number of *a priori* classified objects in subgroup S_i $(i = 1, \ldots, g)$, $K = \Sigma_{i=1}^{g} n_i$, $n_{g+1} = N - K$ is the number of remaining (unclassified) objects in the excess subgroup S_{g+1} (this is an empty subgroup in many applications), $\Sigma_{I<J}$ is the sum over all I and J such that $1 \leq I < J \leq N$, $\Psi_i(\omega_I)$ is 1 if $\omega_I \in S_i$ and 0 otherwise, $C_i > 0$ $(i = 1, \ldots, g)$, and $\Sigma_{i=1}^{g} C_i = 1$. The underlying permutation distribution of δ (the null hypothesis) assigns equal probabilities to the

$$M = N! \Big/ \left(\prod_{i=1}^{g+1} n_i! \right)$$

possible allocations of the N objects to the $g+1$ subgroups. The mean, variance and skewness of δ under the null hypothesis are denoted respectively by μ_δ, σ_δ^2 and γ_δ. Under the null hypothesis, preliminary findings (Mielke et al., 1976; Mielke, 1978, 1979b) indicated some situations when the asymptotic distribution of $N(\delta - \mu_\delta)$ is nondegenerate with γ_δ being substantially negative. Based on results due to Sen (1970, 1972), O'Reilly and Mielke (1980) presented general theorems for the multivariate case of MRPP which characterize situations in which the distribution of $N^{1/2}(\delta - \mu_\delta)$ is asymptotically normal under the null hypothesis. More recently Brockwell et al. (1982) presented theorems for the univariate case of MRPP which delineate distributions for situations (probably the most important situations) in which the nondegenerate distribution of $N(\delta - \mu_\delta)$ is *not* asymptotically normal under the null hypothesis (with rare exceptions, invariance principles fail for these situations). The multivariate generalizations of the results by Brockwell et al. (1982) and special situations analogous to those considered by Mielke and Sen (1981) for linear rank statistics are open questions which require further attention.

The symmetric distance function ($\Delta_{I,J}$) is extremely important since it defines the structure of the underlying analysis space of MRPP. The form of the symmetric distance functions considered in this paper will be confined to

$$\Delta_{I,J} = \left(\sum_{k=1}^{r} |x_{kI} - x_{kJ}|^p \right)^{v/p}$$

where $p \geq 1$ and $v > 0$ (p is not relevant when $r = 1$). In particular, the underlying analysis space of MRPP is nonmetric when $v > 1$ (i.e., the triangle inequality property of a metric space fails) and is metric when $v \leq 1$ (a distorted metric space when $v < 1$). The analysis space of MRPP is a Euclidean space when $p = 2$ and $v = 1$. While the validity of a permutation test is not affected by these geometric considerations, the rejection region of any test is highly dependent on the underlying geometry. Thus a geometry problem (i.e., either a nonmetric or a distorted metric space) will affect the power of a permutation test. The results of a permutation test will surely be misleading if the rejection region is incomprehensible. As a consequence, the choice of $p = 2$ and $v = 1$ is recommended for routine applications. This may be a controversial conclusion since, as subsequently demonstrated, the majority of permutation tests presently used in routine applications are based on $v = 2$.

Since the symmetric distance functions considered in this presentation are of the type described in the previous paragraph, small values of δ imply a concentration of response measurements within the g subgroups. Thus the P-value for a realized value of δ (say δ_0) is the probability statement given by $P(\delta \leq \delta_0)$. Although an efficient algorithm for calculating the exact P-value for an observed value of δ has been developed (Berry, 1982), this procedure is extremely expensive when M is large (e.g., $M > 10^6$). Since $n_1 = n_2 = 12$ and $N = 24$ yields $M \doteq 2.7 \times 10^6$ or $n_1 = n_2 = n_3 = 6$ and $N = 18$ yields $M \doteq 1.7 \times 10^7$, the need for an approximation procedure becomes essential for cases involving even relatively small sample sizes. Because the distribution of δ under the null hypothesis may differ substantially from a normal distribution for either small, moderate or extremely large sample sizes (Mielke, 1978, 1979b; O'Reilly and Mielke, 1980; Brockwell et al., 1982), approximate P-values are based on the Pearson type III distribution which compensates for the fact that the underlying permutation distribution is often substantially skewed (Harter, 1969; Mielke et al., 1981a). In particular, the standardized test statistic given by

$$T = (\delta - \mu_\delta)/\sigma_\delta$$

is presumed to follow the Pearson type III distribution with the density function given by

$$f(y) = \frac{(-2\gamma)^{4/\gamma^2}}{\Gamma(4/\gamma^2)} [-(2+y\gamma)/\gamma]^{(4-\gamma^2)/\gamma^2} \, \mathrm{e}^{-2(2+y\gamma)/\gamma^2}$$

where $-\infty < y < -2/\gamma$. If $T_0 = (\delta_0 - \mu_\delta)/\sigma_\delta$ and $\gamma = \gamma_\delta \leq -0.001$, then

$$P(\delta \leq \delta_0) \doteq \int_{-\infty}^{T_0} f(y)\,\mathrm{d}y$$

is the approximate P-value (an approximate P-value based on the standard normal distribution is reported if $\gamma_\delta > -0.001$). The approximate P-value is evaluated with Simpson's rule over the interval $(T_0 - 9, T_0)$. Efficient computational expressions for μ_δ, σ_δ^2 and γ_δ are given by

$$\mu_\delta = D(1)\,,$$

$$\sigma_\delta^2 = 2\left\{\sum_{i=1}^{g} C_i^2 [n_i^{(2)}]^{-1} - [N^{(2)}]^{-1}\right\}[D(2) - 2D(2') + D(2'')] + 4\left[\sum_{i=1}^{g} C_i^2 n_i^{-1} - N^{-1}\right][D(2') - D(2'')]\,,$$

$$\gamma_\delta = \{E[\delta^3] - 3\mu_\delta\sigma_\delta^2 - \mu_\delta^3\}/\sigma_\delta^3\,,$$

and

$$\begin{aligned} E[\delta^3] = {} & 4\sum_{i=1}^{g} C_i^3 [n_i^{(2)}]^{-2} D(3) + 8\sum_{i=1}^{g} C_i^3 n_i^{(3)} [n_i^{(2)}]^{-3} [3D(3') + D(3^*)] \\ & + 8\sum_{i=1}^{g} C_i^3 n_i^{(4)} [n_i^{(2)}]^{-3} [3D(3^{**}) + D(3^{***})] \\ & + 6\sum_{i=1}^{g} C_i^2 \{1 - C_i + C_i n_i^{(4)} [n_i^{(2)}]^{-2}\} [n_i^{(2)}]^{-1} D(3'') \\ & + 12\sum_{i=1}^{g} C_i^2 \{(1 - C_i) n_i^{(3)} + C_i n_i^{(5)} [n_i^{(2)}]^{-1}\} [n_i^{(2)}]^{-2} D(3''') \\ & + \sum_{i=1}^{g} C_i \{(1 - C_i)(1 - 2C_i) + 3C_i(1 - C_i) n_i^{(4)} [n_i^{(2)}]^{-2} \\ & + C_i^2 n_i^{(6)} [n_i^{(2)}]^{-3}\} D(3'''') \end{aligned}$$

where $N^{(c)} = N!/(N-c)!$ and the symmetric function model parameters are given by

$$D(1) = \frac{1}{N^{(2)}} \sum \Delta_{J_1,J_2},$$

$$D(2) = \frac{1}{N^{(2)}} \sum \Delta^2_{J_1,J_2}, \qquad D(2') = \frac{1}{N^{(3)}} \sum \Delta_{J_1,J_2}\Delta_{J_1,J_3},$$

$$D(2'') = \frac{1}{N^{(4)}} \sum \Delta_{J_1,J_2}\Delta_{J_3,J_4},$$

$$D(3) = \frac{1}{N^{(2)}} \sum \Delta^3_{J_1,J_2}, \qquad D(3') = \frac{1}{N^{(3)}} \sum \Delta^2_{J_1,J_2}\Delta_{J_1,J_3},$$

$$D(3'') = \frac{1}{N^{(4)}} \sum \Delta^2_{J_1,J_2}\Delta_{J_3,J_4}, \qquad D(3''') = \frac{1}{N^{(5)}} \sum \Delta_{j_1,J_2}\Delta_{j_1,J_3}\Delta_{J_4,J_5},$$

$$D(3'''') = \frac{1}{N^{(6)}} \sum \Delta_{J_1,J_2}\Delta_{J_3,J_4}\Delta_{J_5,J_6}, \qquad D(3^*) = \frac{1}{N^{(3)}} \sum \Delta_{J_1,J_2}\Delta_{J_1,J_3}\Delta_{J_2,J_3},$$

$$D(3^{**}) = \frac{1}{N^{(4)}} \sum \Delta_{J_1,J_2}\Delta_{J_1,J_3}\Delta_{J_2,J_4}$$

and

$$D(3^{***}) = \frac{1}{N^{(4)}} \sum \Delta_{J_1,J_2}\Delta_{J_1,J_3}\Delta_{J_1,J_4}$$

where J_1, J_2, J_3, J_4, J_5 and J_6 denote distinct integers from 1 through N and the sums are over all permutations of distinct indices. Efficient computations to obtain these parameters are given in the appendix of Mielke et al. (1976). The Pearson type III distribution has been investigated and appears to be an excellent approximation (Mielke and Berry, 1982; Mielke et al., 1982).

A number of different choices for C_i have been mentioned and/or used. An inefficient choice of C_i when the n_i's are not equal is $C_i = n_i^{(2)}/[\sum_{j=1}^{g} n_j^{(2)}]$. This choice was used by Mantel and Valand (1970) and Mielke et al. (1976) and discussed by Mielke (1978) and O'Reilly and Mielke (1980). Another inefficient choice of C_i when the n_i's are not equal is $C_i = 1/g$ (O'Reilly and Mielke, 1980). An efficient choice of C_i when the n_i's are not equal is $C_i = n_i/K$. This choice was suggested in Mielke (1979b) and used in Mielke et al. (1981a, 1981b, 1982). Because the choice of $C_i = n_i/K$ causes the second of the two expressions comprising σ^2_δ (the potentially dominant expression when large subgroup sizes are involved) to vanish when $N = K$, this is the recommended choice. An asymptotically equivalent choice of C_i which is also efficient when the n_i's are not equal is $C_i = (n_i - 1)/(K - g)$. As will be shown, this last choice of C_i underlies the one-way analysis of variance and many classical linear rank statistics since the permutation tests associated with these statistics are also special cases of MRPP. Note that all of the choices for C_i mentioned above are equivalent if the n_i's are equal $(i = 1, \ldots, g)$.

2.2. Relation to well-known methods

The relation between the permutation version of one-way analysis of variance (two-sample t test when $g = 2$) and MRPP is first described. Let $F = MS_A/MS_W$ be the ordinary one-way analysis of variance statistic. If $g \geq 2$, $v = 2$, $r = 1$, $A = \Sigma_{I=1}^{N} x_{1I}$, $B = \Sigma_{I=1}^{N} x_{1I}^2$, $N = K$ and $C_i = (n_i - 1)/(N - g)$ for $i = 1, \ldots, g$, then the identity relating F and δ is given by

$$N\delta = 2(NB - A^2)/[N - g + (g - 1)F] .$$

Because F is based on $v = 2$, the previously mentioned geometry problem of the underlying analysis space is a relevant concern for the permutation version of one-way analysis of variance.

The relation between simple two-sample linear rank tests and MRPP is now described. Let $H = \Sigma_{I=1}^{N} f_I U_I$ be a two-sample linear rank test statistic where $U_I = 1$ if $\omega_I \in S_1$ and 0 otherwise, $n_1 = n$, $n_2 = N - n$, $g = 2$, $v = 2$, $r = 1$, f_I is a score function of the rank order value of x_{1I} from below relative to the finite population of N response measurements, $A = \Sigma_{I=1}^{N} f_I$, $B = \Sigma_{I=1}^{N} f_I^2$ and $C_i = (n_i - 1)/(N - 2)$ for $i = 1$ and 2. Here the identity relating H and δ is given by

$$N(N - 2)\delta = 2\{NB - A^2 - (NH - nA)^2/[n(N - n)]\} .$$

Since H is also based on $v = 2$, the geometry problem of the underlying analysis space is again a relevant concern.

Whaley (1983) demonstrates the equivalence of (i) the multidimensional runs statistic developed by Friedman and Rafsky (1979), (ii) the spatial autocorrelation statistic introduced by Cliff and Ord (1973), and (iii) the special case of the MRPP statistic when $n_1 = n$, $n_2 = N - n$, $g = 2$, $C_i = n_i^{(2)}/[n_1^{(2)} + n_2^{(2)}]$ for $i = 1$ and 2, and $\Delta_{I,J}$ is 1 if ω_I and ω_J are linked and 0 otherwise. When $n_1 \neq n_2$, this interesting observation by Whaley (1983) suggests that the performance of this statistic might be improved if the present choice of C_i is replaced with $C_i = n_i/N$ for $i = 1$ and 2 (the simple structure of $\Delta_{I,J}$ eliminates the geometry problem).

2.3. Weather modification evaluations

This example is based on two recent papers (Mielke et al., 1981c, 1982) which were prompted by concerns (Mielke, 1979a) involving earlier evaluations (Mielke et al., 1971) of the Climax I and II wintertime orographic cloud seeding experiments. The concerns resulted from suggestions that the positive results of Mielke et al. (1971) might have been the consequences of natural regional increases during treated experimental units (24 hour periods) rather than being primarily attributed to a cloud seeding treatment (i.e., a type I statistical error). Since data from specified National Weather Service control stations were not used by Mielke et al. (1971), Mielke et al. (1981c) obtained even stronger statistical results when the control data were used to adjust for the concern regarding natural regional increases during the treated experimental units.

Although the results of Mielke et al. (1982) are in general agreement with those of Mielke et al. (1981c), the methodology of Mielke et al. (1981c) is attacked in Mielke et al. (1982). The following discussion describes the reason for this attack and also compares selected results of Mielke et al. (1981c) with results of the alternative methodology suggested in Mielke et al. (1982).

Before describing the attack on the methodology of Mielke et al. (1981c), the methodology of that paper is presented. Let $(x_1, y_1), \ldots, (x_N, y_N)$ denote N pairs of control and target precipitation amount observations, respectively, associated with N experimental units. A treatment (cloud seeding with silver iodide) which should only affect target observations is applied to a randomly obtained subset of n experimental units and no treatment is applied to the remaining subset of $N-n$ experimental units. The statistical analyses are based on residual data ($e_I = y_I - bx_I$ for $I = 1, \ldots, N$) based on a least squares regression line, i.e., the estimate of b minimizes

$$\sum_{I=1}^{N} (y_I - bx_I)^2 .$$

The regression line is forced through the origin since $x_I = y_I = 0$ is a common situation when the forecasts specifying the experimental units are faulty (no precipitation occurs in either the control or target vicinity). The analyses are based on five specific two-sample linear rank tests of the type described in the second paragraph of Subsection 2.2. The score functions of these five tests are given by

$$f_I = \begin{cases} -|I-(N+1)/2|^w & \text{if } I < (N+1)/2 , \\ 0 & \text{if } I = (N+1)/2 , \\ |I-(N+1)/2|^w & \text{if } I > (N+1)/2 , \end{cases}$$

for $w = -\frac{1}{2}, 0, 1, 2$ and 4. Asymptotic properties of these tests including the power and the null distribution for any $w > -1$ are given elsewhere (Mielke, 1972, 1974; Mielke and Sen, 1981). Since the larger residuals roughly correspond to the larger precipitation amounts, the tests associated with large and small values of w respectively place emphasis on changes in the large and small precipitation amounts. The reason for the five choices of w is that any effect due to a cloud seeding treatment is complex (certainly not a simple location or scale shift) and the five tests emphasize different types of changes. Even crude statements regarding the analytic form of either the underlying distribution or the change in question are not feasible (the science of weather modification is very infantile).

The rationale prompting Mielke et al. (1982) to attack the methodology of Mielke et al. (1981c) included (1) the disproportionately high influence on least squares regression residual data by a very few large values and (2) the incomprehensible underlying analysis space associated with two-sample linear rank tests. The alternative methodology used by Mielke et al. (1982) involved (1) utilization of median regression residual data, i.e., the estimate of b minimizes

$$\sum_{I=1}^{N} |y_I - bx_I|,$$

and (2) the two-sample rank tests introduced by Mielke et al. (1981b) where $n_1 = n$, $n_2 = N - n$, $g = 2$, $v = 1$, $C_i = n_i/N$ for $i = 1$ and 2, and f_I is the previously defined score function with $w = 0$, 1 and 2. As pointed out by Huber (1974), a median regression line is resistant against arbitrary residuals with large magnitudes. In addition, an extremely efficient algorithm for obtaining median regression residual data has been recently developed by Bloomfield and Steiger (1980). As previously emphasized, the underlying analysis space associated with $v = 1$ is a simple Euclidean space (i.e., does not involve either the non-metric or distorted metric spaces associated with other choices of v).

The primary motivation for preferring the methodology of Mielke et al. (1982) over that of Mielke et al. (1981c) is the physical interpretation. Meteorologists generally believe that cloud seeding will be most effective only when relatively small precipitation amounts are involved (i.e., cases associated with small residuals). The basis for this belief is that the large precipitation amounts are associated with synoptic storms which are endowed with an abundance of natural ice crystals (i.e., there is no need for additional ice crystals induced by cloud seeding). Since a few large residuals may severely influence the position of a least squares regression line, the immediate consequence is that the residual data considered most important by meteorologists (small values) may be distorted in a totally unreasonable manner by a very small subset of the residual data that is considered least important. With respect to analysis techniques, it is intrinsically assumed by investigators in most any discipline that the underlying analysis space of a chosen statistical technique is congruent with the perceived Euclidean space of the data being analyzed. Since very few investigators recognize the complexity of the underlying analysis space associated with either linear rank tests or closely related techniques based on $v = 2$, a reasonable demand is that a statistical technique must possess an underlying analysis space (i.e., $v = 1$) which is compatible with the perceived data space in question.

Table 1 presents selected two-sided P-value comparisons, involving experimental units with warm 500 mb temperatures, between the *old* methodology of Mielke et al. (1981c) and the new methodology of Mielke et al. (1982). The comparisons of Table 1 are presented separately for the independent Climax I and II experiments. The major differences in the results of the Climax II experiment are caused by a few very large residuals. Further discussion and details which include physical interpretations and the availability of data associated with these and other comparisons are given in Mielke et al. (1981c, 1982).

Table 1
Two-sided P-value comparisons between the old and new methodologies involving warm 500 mb temperatures (−20 to −11°C) for the independent Climax I and II experiments

	Climax I		Climax II	
w	old	new	old	new
0	0.150	0.113	0.006	0.002
1	0.049	0.068	0.022	0.010
2	0.032	0.048	0.066	0.031

The discussion up to now has been confined to wintertime orographic cloud seeding experiments. Weather modification experiments associated with cumulous clouds present many interesting problems which were not encountered with the Climax I and II experiments. The experimental units associated with many cumulous cloud experiments are the individual clouds. Also the response variables of recent cumulous cloud seeding experiments usually involve specific collections of in situ cloud measurements relative to the time of the seeding treatment (double blind experiments are used to insure that a seeding aircraft is not aware of the treatment: seeding material or a placebo). The responses of such experiments involve all levels of measurements (viz. nominal, ordinal, interval and ratio). Any combination of the responses is easily included with MRPP since a nominal response (e.g., presence or absence of an attribute) enters as a simple dichotomous response (0, 1). Though the number of dimensions used with MRPP is of essentially no consequence computationally, the physical interpretation of results associated with many responses may be far from trivial.

2.4. *Tropical storm forecasts*

The following application of MRPP is concerned with forecasting phenomena such as direction and intensity changes of tropical storms (e.g. hurricanes and typhoons). This application is based on a composited rawinsonde data set obtained from the west Atlantic during a 14 year period (Gray, 1979). The composited rawinsonde data consists of individual rawinsonde reports at specific locations relative to a tropical storm's center and direction of movement. The data associated with each rawinsonde involves many responses such as temperature, humidity, tangential and radial wind velocities, and heights of pressure levels (e.g. 900 mb, 500 mb or 200 mb levels) as the rawinsonde balloon carries its instrumented package up through the atmosphere. The compositing is essential since the island and ship sites used to loft each rawinsonde balloon are essentially random locations relative to the storm's center and direction. As a consequence, many observations over long periods of time are needed to fill in various positions (radial belts and octants) relative to a 'conceptual composited' storm's center and direction. The radial belts are annuli of two degrees thickness (1 to 3°, 3 to 5°, 5 to 7°, etc.) about the storm's center and the octants (north, northwest, west, etc.) correspond to the storm's forward direction passing through the center of the north octant.

Comparisons of one or more responses (made commensurate by appropriate scaling) between developing and nondeveloping storms are considered. The classification of a developing and nondeveloping storm at a given time is determined from the known history of each storm. Substantial differences between one or more responses for given radial belt, octant and pressure level combinations would yield a basis for future forecasts of development by obtaining essential information with appropriate aircraft penetrations. Some preliminary results based on the west Atlantic data set are given in Table 2. The P-values of Table 2 are based on MRPP with $v = 1$. The results suggest that tangential winds and height responses may be very important for subsequent forecast criteria. This same procedure will also be used to develop forecast criteria for storm direction changes and other phenomena.

The present example is one of an apparent endless number of multi-response situations which are routinely encountered in meteorology. Another typical example involving seasonal mean sea-level pressure pattern changes over broad

Table 2
P-values for testing response differences between developing and nondeveloping storms with responses restricted to the 5 to 7° radial belt

Response	Octant	Pressure level 900 mb	500 mb	200 mb
Height	North	0.002	0.002	0.06
	West	0.05	0.01	0.03
	South	0.77	0.65	0.41
	East	1.00	0.58	0.57
Temperature	North	0.32	0.06	0.02
	West	0.39	0.09	0.81
	South	0.09	0.65	0.66
	East	0.17	0.02	0.32
Tangential wind velocity	North	0.0002	0.02	0.46
	West	0.00004	0.003	0.18
	South	0.11	0.03	0.36
	East	0.19	0.06	0.03
Radial wind velocity	North	0.16	0.01	0.17
	West	0.81	0.91	0.76
	South	0.72	0.77	0.31
	East	0.07	0.64	1.00

geographical areas is given by Mielke et al. (1981a). Meteorologists generally scoff at the suggestion that their data can be adequately described by a simply defined multivariate distribution (certainly not a multivariate normal distribution, with or without transformations, whose only virtue is that many existing inference techniques depend on this unrealistic assumption). Because of the seemingly impossible task required in providing realistic and tractable multivariate distributions for the numerous applications, permutation techniques based on empirical data appear to be the only reasonable solution. Since the results of any statistical analysis must meaningfully represent the physical phenomena being investigated (even when based on a permutation approach), the use of a statistical inference technique whose underlying analysis space is not compatible with the data space (a Euclidean space) is unacceptable. The fact that so many commonly used statistical techniques are based on an incomprehensible nonmetric analysis space (the symmetric distance function being *squared* Euclidean distance) is both disturbing and unnecessary. While these concerns are stated for meteorologists, they apply equally well to investigators in any other discipline.

3. Comparisons with blocking

The techniques and examples of this section involve recently developed analogs of MRPP which utilize blocking (Mielke and Berry, 1982; Mielke and Iyer, 1982). Following a description of these techniques along with their relation to well known techniques, a summary is given for recent location shift power comparisons between nonparametric tests based on a Euclidean analysis space and classical nonparametric tests for matched pairs. This section is then concluded with a meteorological application involving comparisons of satellite derived precipitation estimates with ordinary precipitation estimates based on conventional surface gauge network data and radar data.

3.1. Methodological description

Let b blocks and g treatments be associated with a randomized block design. Let $x'_{ij} = (x_{1ij}, \ldots, x_{rij})$ denote r commensurate response measurements corresponding to treatment i and block j (the response measurements might again be functions of response measurements or residuals adjusted by predictors). The modified MRPP statistic for this situation is given by

$$\delta = \left[g\binom{b}{2}\right]^{-1} \sum_{i=1}^{g} \sum_{j<k} \Delta(x_{ij}, x_{ik})$$

where $\Delta(x, y)$ is a symmetric distance function value of the points $x' = (x_1, \ldots, x_r)$ and $y' = (y_1, \ldots, y_r)$ in the r-dimensional Euclidean space. The underlying permutation distribution of δ (the null hypothesis) assigns equal probabilities to the

$$M = (g!)^b$$

possible allocations of the g r-dimensional response measurements to the g treatment positions within each of the b blocks. The mean, variance and skewness of δ under the null hypothesis are again denoted by μ_δ, σ_δ^2 and γ_δ, respectively. Except for subsequently described special cases when δ is equivalent to well known statistics, little is presently known about the asymptotic distribution of δ. However, under the null hypothesis and fairly reasonable conditions, it is conjectured that (1) $g^{1/2}(\delta - \mu_\delta)$ is asymptotically a normal random variable when $b \geqslant 2$ is fixed and $g \to \infty$, and (2) $b(\delta - \mu_\delta)$ is asymptotically a nondegenerate and nonnormal random variable with $\gamma_\delta < 0$ when $g \geqslant 2$ is fixed and $b \to \infty$.

The symmetric distance function is again confined to

$$\Delta(x, y) = \left(\sum_{h=1}^{r} |x_h - y_h|^p \right)^{v/p}$$

where $p \geqslant 1$ and $v > 0$. Since the choice of the symmetric distance function defines the structure of the underlying analysis space of these procedures, the discussion in Section 2 concerning this choice is equally pertinent here.

In a manner analogous to MRPP, small values of δ imply a concentration of the response measurements associated with each of the g treatments (i.e., over blocks). Therefore $P(\delta \leqslant \delta_0)$ is again the P-value associated with δ_0 (the realized value of δ). Though an efficient algorithm for calculating the exact P-value for an observed value of δ exists, this approach becomes prohibitively expensive when M is large (e.g., greater than 10^6). Noting that $b = 6$ and $g = 4$ yields $M \doteqdot 1.9 \times 10^8$ or that $b = 4$ and $g = 6$ yields $M \doteqdot 2.7 \times 10^{11}$, the necessity for an approximation technique is obvious for even relatively small randomized block configurations.

As in Section 2, approximate P-values are again based on the Pearson type III distribution to compensate for the commonly encountered substantial skewness of the underlying permutation distribution of δ. Thus the standardized test statistic given by

$$T = (\delta - \mu_\delta)/\sigma_\delta$$

is again presumed to follow the Pearson type III distribution and the approximate P-value is obtained by the previously described approach in Subsection 2.1. To obtain T and the P-value for a realized value of δ, the determination of μ_δ, σ_δ^2 and γ_δ is again essential. If

$$\Delta(i, r; j, s) = \Delta(x_{ir}, x_{js})$$

and

$$D(i, r; j, s) = \Delta(i, r; j, s) - g^{-1}\sum_{i=1}^{g}\Delta(i, r; j, s) - g^{-1}\sum_{j=1}^{g}\Delta(i, r; j, s)$$
$$+ g^{-2}\sum_{i=1}^{g}\sum_{j=1}^{g}\Delta(i, r; j, s),$$

then μ_δ, σ_δ^2 and γ_δ are conveniently expressed as

$$\mu_\delta = \left[g^2\binom{b}{2}\right]^{-1}\sum_{r<s}\sum_{i=1}^{g}\sum_{j=1}^{g}\Delta(i, r; j, s),$$

$$\sigma_\delta^2 = \left[g\binom{b}{2}\right]^{-2}\frac{1}{g-1}\sum_{r<s}\sum_{i=1}^{g}\sum_{j=1}^{g}[D(i, r; j, s)]^2,$$

$$\gamma_\delta = \kappa_3(\delta)/\sigma_\delta^3$$

and

$$\kappa_3(\delta) = \left[g\binom{b}{2}\right]^{-3}\frac{1}{g-1}[H(g) + L(b)]$$

where $g \geq 2$, $b \geq 2$,

$$H(g) = \begin{cases} 0 & \text{if } g = 2, \\ \dfrac{g}{g-2}\displaystyle\sum_{r<s}\sum_{i=1}^{g}\sum_{j=1}^{g}[D(i, r; j, s)]^3 & \text{if } g \geq 3, \end{cases}$$

$$L(b) = \begin{cases} 0 & \text{if } b = 2, \\ \dfrac{6}{g-1}\displaystyle\sum_{r<s<t}\sum_{i=1}^{g}\sum_{j=1}^{g}\sum_{k=1}^{g} D(i, r; j, s)D(i, r; k, t)D(j, s; k, t) & \text{if } b \geq 3, \end{cases}$$

and $\Sigma_{r<s<t}$ denotes the sum over all r, s and t such that $1 \leq r < s < t \leq b$.

Efficient computational expressions to obtain μ_δ, σ_δ^2 and γ_δ are described in detail by Mielke and Iyer (1982). For the special case of these techniques involving matched pairs (Mielke and Berry, 1982), empirical results indicate that P-values based on the Pearson type III distribution are excellent approximations.

3.2. *Relation to well-known methods*

The relation between δ and the classical F statistic for testing the null hypothesis of a randomized block design is initially described. If $v = 2$ and $r = 1$, then the functional relation between F and δ is given by

$$F = (b-1)[2SS_T - g(b-1)\delta]/[g(b-1)\delta - 2SS_B]$$

where the corrected total sum of squares is given by $SS_T = (\Sigma_{i=1}^{g}\Sigma_{j=1}^{b} x_{ij}^2) - SS_M$, the block sum of squares is given by $SS_B = \{\Sigma_{j=1}^{b}[(\Sigma_{i=1}^{g} x_{ij})^2/g]\} - SS_M$, and $SS_M = (\Sigma_{i=1}^{g}\Sigma_{j=1}^{b} x_{ij})^2/bg$. Thus F and δ are equivalent under the null hypothesis since SS_T and SS_B are invariant relative to the $(g!)^b$ permutations of the response measurements. (For this and other cases involving univariate responses, the response measurement subscript is omitted, i.e., $x_{1ij} = x_{ij}.$) Incidentally, δ is equivalent to Cochran's Q test statistic if $r = 1$ and each x_{ij} is either 0 or 1.

Let R denote the ordinary Pearson correlation coefficient. If $v = 2$, $b = 2$ and $r = 1$, then the functional relation between R and δ is given by

$$R = (\mu_\delta - \delta)/(2S_1S_2)$$

where $R = [\Sigma_{i=1}^g (x_{i1} - \bar{x}_1)(x_{i2} - \bar{x}_2)]/(gS_1S_2)$, $\mu_\delta = S_1^2 + S_2^2 + (\bar{x}_1 - \bar{x}_2)^2$, $\bar{x}_j = (\Sigma_{i=1}^g x_{ij})/g$ and $S_j^2 = [\Sigma_{i=1}^g (x_{ij} - \bar{x}_j)^2]/g$ for $j = 1$ and 2. Then R and δ are equivalent under the null hypothesis because $\bar{x}_1$, $\bar{x}_2$, S_1 and S_2 are invariant relative to the $(g!)^2$ response measurement permutations.

If $g = 2$, $r = 1$, $x_{1j} = -x_{2j} = x_j$ and $|x_j| > 0$ for $j = 1, \ldots, b$, then the test based on δ is equivalent to an extended class of permutation techniques for matched pairs (Mielke and Berry, 1982).

If $v = 2$, $r = 1$ and the response measurements for each block are replaced by their corresponding ranks, then the test based on δ is equivalent to the Friedman two-way analysis of variance (Kendall coefficient of concordance). The values of μ_δ, σ_δ^2 and γ_δ for this case are

$$\mu_\delta = (g^2 - 1)/6, \qquad \sigma_\delta^2 = [(g+1)(g^2-1)]/[18b(b-1)],$$

and

$$\gamma_\delta = -\{8(b-2)^2/[(g-1)b(b-1)]\}^{1/2}.$$

Furthermore $\rho = 1 - \delta/\mu_\delta$ is Spearman's rho (a measure of correlation) when $b = 2$. In the present context, the 'correlation' measure $\rho = 1 - \delta/\mu_\delta$ can be interpreted in a much broader setting.

If $v = r = 1$ and the response measurements for each block are again replaced by their corresponding ranks, then the test based on δ is the Euclidean space analog of the Friedman two-way analysis of variance (Kendall coefficient of concordance). The values of μ_δ, σ_δ^2 and γ_δ for this case are

$$\mu_\delta = (g^2 - 1)/(3g), \qquad \sigma_\delta^2 = \frac{2(g+1)(2g^2+7)}{45g^2b(b-1)},$$

and

$$\gamma_\delta = -\frac{(g+2)(2g^2+31)\theta(g) + (8g^4+29g^2+71)(b-2)/(g-1)}{[49(g+1)(2g^2+7)^3b(b-1)/40]^{1/2}}$$

where $\theta(g) = 0$ or 1 if $g = 2$ or $g \geq 3$, respectively. In this case δ is the Spearman footrule statistic when $b = 2$ (cf. Diaconis and Graham, 1977).

3.3. Power simulations for location alternatives

The simulated power comparisons presented here involve four specific permutation tests for matched pairs. If $g = 2$, $r = 1$, $x_{1j} = -x_{2j} = x_j$ and $|x_j| > 0$

for $j = 1, \ldots, b$, then δ may be expressed as

$$\delta = \binom{b}{2}^{-1} \sum_{i<j} |x_i - x_j|^v$$

where $v > 0$. An equivalent representation (the usual matched-pairs model) is to let $x_i = |x_i| Z_i$ for $i = 1, \ldots, b$ where $|x_i|$ is a fixed positive score and Z_i is a random variable specified by $P(Z_i = 1) = P(Z_i = -1) = \frac{1}{2}$ under the null hypothesis. This class of permutation techniques for matched pairs has been considered by Mielke and Berry (1982). The four tests which will be compared include the sign test, the Wilcoxon signed-ranks test, and two rank tests for matched pairs which depend on a Euclidean space. Let δ^* denote the specific case with $|x_i| = \frac{1}{2}$ for $i = 1, \ldots, b$ (note that δ^* does not depend on v). The test associated with δ^* is equivalent to the two-sided version of the sign test. Also let $\delta_{s,v}$ denote the specific case with $|x_i| = r_i^s$ for $i = 1, \ldots, b$ and $r_1, \ldots, r_b$ are the rank order statistics from below. The test associated with $\delta_{1,2}$ is equivalent to the two-sided version of the Wilcoxon signed-ranks test. Also the tests associated with $\delta_{1,1}$ and $\delta_{2,1}$ are rank tests for matched pairs which depend on a Euclidean space. The simulated power comparisons presented in Tables 3 and 4 both involve the pooled results of three independent comparisons given by

Table 3
Estimated power against a location shift of size 0.3σ where σ is the standard deviation of the distribution specified and $b = 80$

	α	δ^*	$\delta_{1,1}$	$\delta_{2,1}$	$\delta_{1,2}$
Laplace	0.10	0.957	0.960	0.937	0.950
	0.02	0.807	0.857	0.783	0.830
	0.002	0.553	0.593	0.467	0.530
	0.0002	0.280	0.283	0.187	0.247
Logistic	0.10	0.803	0.907	0.893	0.920
	0.02	0.523	0.720	0.713	0.727
	0.002	0.240	0.363	0.360	0.373
	0.0002	0.070	0.143	0.127	0.140
Normal	0.10	0.680	0.850	0.860	0.873
	0.02	0.400	0.613	0.663	0.663
	0.002	0.167	0.253	0.287	0.283
	0.0002	0.047	0.107	0.100	0.103
Uniform	0.10	0.460	0.800	0.917	0.843
	0.02	0.217	0.513	0.723	0.600
	0.002	0.077	0.177	0.370	0.233
	0.0002	0.007	0.057	0.127	0.100
U-shaped	0.10	0.093	1.000	1.000	0.977
	0.02	0.007	0.987	1.000	0.927
	0.002	0.000	0.807	0.990	0.673
	0.0002	0.000	0.450	0.953	0.427

Table 4
Estimated power against a location shift of size 0.6σ where σ is the standard deviation of the distribution specified and $b = 20$

	α	δ^*	$\delta_{1,1}$	$\delta_{2,1}$	$\delta_{1,2}$
Laplace	0.10	0.900	0.913	0.870	0.910
	0.02	0.567	0.723	0.660	0.700
	0.002	0.350	0.377	0.333	0.370
	0.0002	0.053	0.107	0.077	0.100
Logistic	0.10	0.800	0.857	0.857	0.863
	0.02	0.387	0.610	0.610	0.617
	0.002	0.207	0.277	0.273	0.277
	0.0002	0.027	0.063	0.043	0.057
Normal	0.10	0.720	0.797	0.827	0.833
	0.02	0.293	0.543	0.577	0.570
	0.002	0.143	0.203	0.230	0.227
	0.0002	0.017	0.043	0.030	0.043
Uniform	0.10	0.490	0.727	0.830	0.780
	0.02	0.163	0.420	0.593	0.467
	0.002	0.070	0.120	0.203	0.170
	0.0002	0.007	0.017	0.013	0.017
U-shaped	0.10	0.137	0.640	0.877	0.630
	0.02	0.027	0.317	0.603	0.360
	0.002	0.013	0.053	0.220	0.070
	0.0002	0.000	0.020	0.017	0.017

Mielke and Berry (1982). The power comparisons of δ^*, $\delta_{1,1}$, $\delta_{2,1}$ and $\delta_{1,2}$ in Table 3 involve (1) a fixed size of $b = 80$, (2) five origin symmetric distributions including the Laplace (double exponential), logistic, normal, uniform, and a U-shaped distribution with density $(3y^2)/2$, $-1 < y < 1$, and (3) a location shift of 0.3σ for the distribution specified. Table 4 differs from Table 3 in that the fixed size is $b = 20$ and the location shift is 0.6σ for the distribution specified. Each power estimate in Table 3 or Table 4 for δ^*, $\delta_{1,1}$, $\delta_{2,1}$ and $\delta_{1,2}$ (corresponding to each significance level, α, and each distribution) depends on 300 P-values associated with the same collection of 300 independent random samples of 80 or 20 values, respectively, from the uniform (0, 1) distribution. Complete details concerning these comparisons are given by Mielke and Berry (1982). The purpose of these comparisons is to demonstrate that specific advantages can be gained when classical tests are replaced with tests based on a Euclidean space (in addition to the geometric appeal stressed in Section 2). The results of Tables 3 and 4 indicate that $\delta_{1,1}$ is a good choice when heavy-tailed distributions are encountered and that $\delta_{2,1}$ is a seemingly outstanding choice when light-tailed (including uniform and U-shaped) distributions are encountered. Similar comparisons involving two-sample analogs of $\delta_{1,1}$ and $\delta_{1,2}$ (two-sample analogs of δ^* and $\delta_{2,1}$ not included) are given by Mielke et al. (1981b).

3.4. Verifying satellite precipitation estimates

Recent meteorological investigations are concerned with the development of useful precipitation estimates derived from satellite data. If satellite precipitation estimates are suitable, then precipitation estimates will be available from various parts of the world where no reliable precipitation estimates are currently available.

In order to verify the suitability and also compare distinct types of satellite precipitation estimates, agreement must be established between satellite precipitation estimates and conventional precipitation estimates based on surface gauge network data and/or radar data. The correspondence between this application and the procedures of this section involves (1) the distinct types of precipitation estimates (i.e., surface gauge network, radar, and satellite) are associated with blocks, (2) the time periods (i.e., hours, days, weeks or months) are associated with treatments and (3) the specific regions which yield the precipitation estimates (i.e., a number of well defined geographical areas) are the associated multi-responses. If a specific type of satellite precipitation estimate is compared to surface gauge network and radar precipitation estimates, seven specific weeks are considered, and five geographical areas are included, then $b = 3$, $g = 7$ and $r = 5$. Also the symmetric distance function is Euclidean distance ($p = 2$ and $v = 1$) since any other choice would not have a realistic physical interpretation. A descriptive measure of agreement for this situation is $\rho = 1 - \delta/\mu_\delta$ (i.e., the previously suggested broader interpretation of Spearman's rho). The inferential measure of agreement is the P-value based on the permutation procedure for randomized blocks. If two or more satellite precipitation estimation techniques are compared, then the one yielding the smallest P-value would be preferred. Since many meteorologists routinely question radar precipitation estimates (questions regarding surface gauge network precipitation estimates should have the same relevance), comparisons should also be made between (1) the agreement between specific satellite and surface gauge network precipitation estimates and (2) the agreement between radar and surface gauge network precipitation estimates.

References

Berry, K. J. (1982). Algorithm AS 179: Enumeration of all permutations of multi-sets with fixed repetition numbers. *Appl. Statist.* **31**, 169–173.

Bloomfield, P. and Steiger, W. L. (1980). Least absolute deviations curve-fitting. *SIAM J. Sci. Statist. Comput.* **1**, 290–301.

Brockwell, P. J., Mielke, P. W. and Robinson, J. (1982). On non-normal invariance principles for multi-response permutation procedures. *Austral. J. Statist.* **24**, 33–41.

Cliff, A. D. and Ord, J. K. (1973). *Spatial Autocorrelation.* Pion Limited, London, England.

Diaconis, P. and Graham, R. L. (1977). Spearman's footrule as a measure of disarray. *J. Roy. Statist. Soc. Ser. B* **39**, 262–268.

Friedman, J. H. and Rafsky, L. C. (1979). Multivariate generalizations of the Wald–Wolfowitz and Smirnov two-sample tests. *Ann. Statist.* **7**, 697–717.

Gray, W. M. (1979). Hurricanes: Their formation, structure and likely role in the tropical circulation. In: D. B. Shaw, ed., *Meterology Over the Tropical Oceans*. Royal Meteorological Society, James Glaisher House, Bracknell, England, pp. 155–218.

Harter, H. L. (1969). A new table of percentage points of the Pearson type III distribution. *Technometrics* **11**, 177–187.

Huber, P. J. (1974). Comment on adaptive robust procedures. *J. Amer. Statist. Assoc.* **69**, 926–927.

Mantel, N. and Valand, R. S. (1970). A technique of nonparametric multivariate analysis. *Biometrics* **26**, 547–558.

Mielke, P. W. (1972). Asymptotic behavior of two-sample tests based on powers of ranks for detecting scale and location alternatives. *J. Amer. Statist. Assoc.* **67**, 850–854.

Mielke, P. W. (1974). Squared rank test appropriate to weather modification cross-over design. *Technometrics* **16**, 13–16.

Mielke, P. W. (1978). Clarification and appropriate inferences for Mantel and Valand's nonparametric multivariate analysis technique. *Biometrics* **34**, 277–282.

Mielke P. W. (1979a). Comment on field experimentation in weather modification. *J. Amer. Statist. Assoc.* **74**, 87–88.

Mielke P. W. (1979b). On asymptotic non-normality of null distributions of MRPP statistics. *Commun. Statist. A* **8**, 1541–1550. Errata: *A* **10**, 1795; *A* **11**, 847.

Mielke P. W. and Berry, K. J. (1982). An extended class of permutation techniques for matched pairs. *Commun. Statistic. A* **11**, 1197–1207.

Mielke, P. W., Berry, K. J. and Brier, G. W. (1981a). Application of multiresponse permutation procedures for examining seasonal changes in monthly mean sea-level pressure patterns. *Mon. Wea. Rev.* **109**, 120–126.

Mielke, P. W., Berry, K. J., Brockwell, P. J. and Williams, J. S. (1981b). A class of nonparametric tests based on multiresponse permutation procedures. *Biometrika* **68**, 720–724.

Mielke, P. W., Berry, K. J. and Johnson, E. S. (1976). Multi-response permutation procedures for *a priori* classifications. *Commun. Statist. A* **5**, 1409–1424.

Mielke, P. W., Berry, K. J. and Medina, J. G. (1982). Climax I and II: Distortion resistant residual analyses. *J. Appl. Meteor.* **21**, 788–792.

Mielke, P. W., Brier, G. W., Grant, L. O., Mulvey, G. J. and Rosenzweig, P. N. (1981c). A statistical reanalysis of the replicated Climax I and II wintertime orographic cloud seeding experiments. *J. Appl. Meteor.* **20**, 643–659.

Mielke, P. W., Grant, L. O. and Chappell, C. F. (1971). An independent replication of the Climax wintertime orographic cloud seeding experiment. *J. Appl. Meteor.* **10**, 1198–1212. Corrigendum: **15**, 801.

Mielke, P. W. and Iyer, H. K. (1982). Permutation techniques for analyzing multiresponse data from randomized block experiments. *Commun. Statist. A* **11**, 1427–1437.

Mielke, P. W. and Sen, P. K. (1981). On asymptotic non-normal null distributions for locally most powerful rank test statistics. *Commun. Statist. A* **10**, 1079–1094.

O'Reilly, F. J. and Mielke, P. W. (1980). Asymptotic normality of MRPP statistics from invariance principles of *U*-statistics. *Commun. Statist. A* **9**, 629–637.

Sen, P. K. (1970). The Hájek–Rényi inequality for sampling from a finite population. *Sankhyā Ser. A* **32**, 181–188.

Sen, P. K. (1972). Finite population sampling and weak convergence to a Brownian bridge. *Sankhyā Ser. A* **34**, 85–90.

Whaley, F. S. (1983). The equivalence of three independently derived permutation procedures for testing the homogeneity of multidimensional samples. *Biometrics* **39**, 741–745.

P. R. Krishnaiah and P. K. Sen, eds., *Handbook of Statistics, Vol. 4*

35

Categorical Data Problems Using Information Theoretic Approach

S. Kullback and J. C. Keegel

1. Introduction

Concepts of statistical information theory have been applied in a very general mathematical formulation to problems of statistics involving continuous and discrete variables (Kullback, 1959, 1983). The discussions and applications in this chapter will however be limited to considerations of frequency or count data dealing with categorical variables. Such data is usually cross-classified into multi-way contigency tables, see for example, Bishop et al. (1975), Fienberg (1977), Gokhale and Kullback (1978a, 1978b), Haberman (1974), Kullback (1959, Chapter 8). The reader will observe later that our formulation and approach does not necessarily require that the cross-classified count data be arrayed in a contingency table. The impracticability of studying the simultaneous effects of large numbers of variables by contemplation of a multiple cross-classification is now generally recognized. We shall assume that the reader has some familiarity with elementary contingency tables and the usual notation, including the dot notation to represent summation over an index. We shall first present the underlying theory and then illustrate and amplify the analytic procedures with examples. The data in most of the examples have not been previously published.

Suppose there is a multinomial experiment resulting in count data distributed into Ω cells. Let $x(\omega)$ denote the observed frequency of occurrence associated with a typical cell $\omega \in \Omega$. The symbol ω will be used to denote cells like (ij) in a two-way table, $(hijk)$ in a four-way table, and so on. For example, in a $5\times3\times2$ table, the symbol ω will replace (ijk) one of the $5\times3\times2=30$ cells. The symbol ω here corresponds to the triplet (ijk) and takes on values in lexicographic order $(1,1,1)$, $(1,1,2)$, $(1,2,1)$, $(1,2,2)$, $(1,3,1)$, $(1,3,2)$, $(2,1,1), \ldots,$ $(2,3,2), \ldots, (5,3,2)$. The lexicographic order, a form of numerical alphabetizing, is especially useful in organizing categorical data for computer programs.

Depending on the design of the data collection procedure we shall look upon the cross-classification of the count data as a single sample from a multinomial distribution or as a collection of samples from many multinomial distributions. In particular, in observations with a dichotomous dependent variable and several explanatory variables, we shall treat the data as many binomials. The examples at the end of this chapter illustrate the preceding discussion.

2. Discrimination information

Within the statistical literature there are a number of possible measures for pseudo-distances between probability distributions (Rao, 1965, pp. 288–289). One such measure is the discrimination information function which we shall now define. Suppose there are two probability distributions $p(\omega)$, $\pi(\omega)$, defined over the cells ω of the space Ω where $\Sigma_\Omega\, p(\omega) = 1$, $\Sigma_\Omega\, \pi(\omega) = 1$. The discrimination information function is

$$I(p : \pi) = \sum_{\Omega} p(\omega) \ln(p(\omega)/\pi(\omega)) . \tag{1}$$

Supporting arguments for, properties of, and applications of $I(p : \pi)$ in statistical inference may be found, among others, in Johnson (1979), Kullback (1959, 1983). At this point we merely cite that $I(p : \pi) \geq 0$, with equality if and only if $p = \pi$, that for fixed π, $I(p : \pi)$ is a convex function of p, and that it can be taken as a measure of closeness between p and π. See Kotz and Johnson (1983).

We use the discrimination information function $I(p : \pi)$ as the basic criterion for the minimum discrimination information (MDI) approach using the principle of MDI estimation. As we shall see, this approach provides a *unified* treatment for categorical or count data and can treat univariate and multivariate logit analysis and quantal response analysis as particular cases. The use of the principle of MDI estimation leads naturally to exponential families, which result in *multiplicative* or *loglinear* models.

3. MDI estimation

Suppose the statistician initially holds the viewpoint that the appropriate distribution is $\pi(\omega)$. However, given a judgement that the distribution is a member of a family $\mathbb{P}$ of distributions satisfying the linearly independent moment constraints

$$\boldsymbol{Cp} = \boldsymbol{\theta}, \qquad \boldsymbol{C} \text{ is } (r+1)\times \Omega, \quad \boldsymbol{p} \text{ is } \Omega \times 1, \quad \boldsymbol{\theta} \text{ is } (r+1)\times 1 , \tag{2}$$

where we use the symbol Ω to represent the space as well as the number of cells and the rank of the model matrix $\boldsymbol{C}$ is $r + 1 \leq \Omega$. The matrix relation (2) may be written as

$$\sum_{\Omega} c_i(\omega)p(\omega) = \theta_i, \quad i = 0, 1, \ldots, r, \tag{3}$$

where the elements of $\boldsymbol{C}$ are $c_i(\omega)$, $i = 0, 1, \ldots, r$, $\omega = 1, \ldots, \Omega$. We take $c_0(\omega) = 1$ for all ω, and $\theta_0 = 1$ to satisfy the natural constraint $\Sigma_\Omega p(\omega) = 1$. The adjusted viewpoint of the statistician is then $p^*(\omega)$ where $p^*(\omega)$ is the unique distribution minimizing $I(p:\pi)$ for $p \in \mathbb{P}$, that is, $p^*(\omega)$ also satisfies the constraints (2) or (3). It has been shown by Shore and Johnson (1980) that the use of any separator other than (1) above for inductive inference when new information is in the form of expected values leads to a violation of one or more reasonable consistency axioms.

4. Loglinear representation

Straightforward application of the calculus yields the fact that the MDI estimate $p^*(\omega)$ is

$$\begin{aligned} p^*(\omega) &= \exp(\tau_0 + \tau_1 c_1(\omega) + \tau_2 c_2(\omega) + \cdots + \tau_r c_r(\omega))\pi(\omega), \quad \omega \in \Omega, \\ &= \exp(\tau_0)\exp(\tau_1 c_1(\omega))\exp(\tau_2 c_2(\omega)) \cdots \exp(\tau_r c_r(\omega))\pi(\omega), \\ &\qquad \omega \in \Omega, \quad (4) \end{aligned}$$

or

$$\ln(p^*(\omega)/\pi(\omega)) = \tau_0 + \tau_1 c_1(\omega) + \tau_2 c_2(\omega) + \cdots + \tau_r c_r(\omega), \quad \omega \in \Omega, \quad (5)$$

where the exponential or natural parameters τ_i are related to the moment parameters θ_i by the relations

$$\sum_\Omega c_i(\omega) p^*(\omega) = \theta_i, \quad i = 0, 1, \ldots, r. \quad (6)$$

It may be readily determined from (4) that $\tau_0 = -\ln M(\tau_1, \tau_2, \ldots, \tau_r)$ where

$$M(\tau_1, \tau_2, \ldots, \tau_r) = \sum_\Omega \exp(\tau_1 c_1(\omega) + \tau_2 c_2(\omega) + \cdots + \tau_r c_r(\omega))\pi(\omega).$$

The fact that $I(p:\pi)$ in (1) is a convex function of $p(\omega)$ insures a unique MDI estimate $p^*(\omega)$. The representation (4) is an exponential family which has also been represented as a multiplicative model. The representation (5) is known as a loglinear model. Loglinear models are particularly appropriate for the analysis of contingency tables or more generally, categorical count data. Extensive bibliographies and applications may be found among others in Bishop et al. (1975), Fienberg (1977), Gokhale and Kullback (1978a, 1978b), Haberman (1974), Ku and Kullback (1974), Plackett (1974).

In loglinear models the logarithm of cell estimates is expressed as a linear combination of main effect and interaction parameters associated with various characteristics (variables) and their levels. The partial association between a pair of characteristics for all values of associated variables (covariates) is a sum and difference of the logarithms of appropriate cell estimates and thus a linear combination of the parameters. The average value of these partial associations is also a linear combination of the parameters. Since, as we shall see, it is possible to determine the covariance matrix of the parameters, the variances of the partial associations and of the average partial association may be calculated and confidence intervals determined. We illustrate these ideas in some of the examples at the end of this chapter.

It may be shown that for any $p \in \mathbb{P}$, that is, satisfying (3) the Pythagorean type property

$$I(p:\pi) = I(p^*:\pi) + I(p:p^*) \tag{7}$$

holds. The property (7) is the basis for the Analysis of Information tables we shall use.

5. Internal constraint and external constraint problems

We establish two broad classes of problems according to the genesis of the values of the moment constraints θ_i in (3). The first class of problems has a data fitting or smoothing or model building objective. In this class of problems which we call internal constraints problems, ICP, the values of the moment constraints are derived from the observed data. The second class of problems is one in which the investigator is concerned with testing various hypotheses about the probabilities $p(\omega)$. In this class of problems which we call external constraint problems, ECP, the values of the moment constraints are determined as a result of the hypotheses in question and are not derived from the observed data (Gokhale and Kullback, 1978a, 1978b).

We have already introduced the designation $x(\omega)$ for the observed cell counts and now define $x^*(\omega) = Np^*(\omega)$ where $N = \Sigma_\Omega x(\omega)$. For the ICP the constraints (3) are

$$\sum_\Omega c_i(\omega)x^*(\omega) = \sum_\Omega c_i(\omega)x(\omega), \quad i = 0, 1, \ldots, r, \tag{8}$$

and in particular $\Sigma_\Omega x^*(\omega) = \Sigma_\Omega x(\omega)$. The goodness-of-fit or MDI statistic for ICP is

$$2I(x:x^*) = 2\sum_\Omega x(\omega)\ln(x(\omega)/x^*(\omega)) \tag{9}$$

which is asymptotically distributed as chi-square with degrees of freedom equal to the difference of the dimensions of the model matrix $\boldsymbol{C}$. For the ICP case the results of the MDI estimation procedure are the same as the maximum likelihood estimates and the MDI statistic in (9) is the log-likelihood ratio statistic. This is however not true for the ECP case, although the MDI estimates are BAN.

6. Analysis of information

The analysis of information is based on the Pythagorean type property (7) applied to nested models. Specifically if $x_a^*(\omega)$ is the MDI estimate corresponding to a set of moment constraints H_a, and $x_b^*(\omega)$ is the MDI estimate corresponding to a set of moment constraints H_b where the set H_a is nested in H_b, that is, every constraint in H_a is explicitly or implicitly contained in the set H_b, then

$$2I(x:x_a^*) = 2I(x_b^*:x_a^*) + 2I(x:x_b^*)\,, \tag{10}$$
$$\Omega - r_a - 1 = r_b - r_a + \Omega - r_b - 1, \quad \text{degrees of freedom}\,.$$

The analysis in (10) is an additive analysis into components which are MDI statistics with additivity relations for the associated degrees of freedom. The component

$$2I(x_b^*:x_a^*) = 2\sum_{\Omega} x_b^*(\omega)\ln(x_b^*(\omega)/x_a^*(\omega)), \quad r_b - r_a \text{ DF}\,, \tag{11}$$

measures the *effect* of the constraints in $x_b^*(\omega)$ which are not included in $x_a^*(\omega)$. In the algorithms used in the computer programs to determine the MDI estimate and other associated values the arbitrary distribution $\pi(\omega)$ in ICP is usually taken as the uniform distribution. For the ICP case, since measures of the form $2I(x:x_a^*)$ may also be interpreted as measures of the variation unexplained by the MDI estimate x_a^* the additive relationship (10) leads to the interpretation of the ratio

$$(2I(x:x_a^*) - 2I(x:x_b^*))/2I(x:x_a^*) = 2I(x_b^*:x_a^*)/2I(x:x_a^*) \tag{12}$$

as the fraction of the variation unexplained by x_a^* accounted for by the additional moment constraints defining x_b^*. The ratio (12) is thus similar to the squared correlation coefficient associated with normal distributions (Goodman, 1970).

For the ECP case the constraints are considered in the form $\boldsymbol{C}\boldsymbol{x}^* = \boldsymbol{N}\boldsymbol{\theta}$, and the MDI statistic to test the hypothesis is

$$2I(x^*:x) = 2\sum_{\Omega} x^*(\omega)\ln(x^*(\omega)/x(\omega)) \tag{13}$$

which is asymptotically distributed as chi-square with r degrees of freedom. In the ECP cases the distribution $\pi(\omega)$ is usually taken so that $x(\omega) = N\pi(\omega)$. In

ECP if $C_2 p = \theta_2$ implies $C_1 p = \theta_1$, where C_2 is $(r_2+1)\times \Omega$ and C_1 is $(r_1+1)\times \Omega$, $r_2 > r_1$, then the analysis of information is

$$2I(x_2^*:x) = 2I(x_2^*:x_1^*) + 2I(x_1^*:x) \tag{14}$$

with the associated degrees of freedom

$$r_2 = (r_2 - r_1) + r_1 . \tag{15}$$

7. Covariance matrices

Since the MDI estimate $x^*(\omega)$ is a member of an exponential family the estimated asymptotic covariances of the variables $c_i(\omega)$ and the associated natural parameters τ_i, $i = 1, 2, \ldots r$ are related. Compute $\boldsymbol{S} = \boldsymbol{CDC'}$ where $\boldsymbol{C}$ is the model matrix and $\boldsymbol{D}$ is a diagonal matrix with entries the estimates $x^*(\omega)$ in lexicographic order over the cells and $\boldsymbol{C'}$ is the transpose of $\boldsymbol{C}$. Partition the matrix $\boldsymbol{S}$ as

$$\boldsymbol{S} = \begin{Vmatrix} \boldsymbol{S}_{11} & \boldsymbol{S}_{12} \\ \boldsymbol{S}_{21} & \boldsymbol{S}_{22} \end{Vmatrix}$$

where $\boldsymbol{S}_{11}$ is 1×1, $\boldsymbol{S}_{12} = \boldsymbol{S}'_{21}$ is $1\times r$, $\boldsymbol{S}_{22}$ is $r\times r$. The estimated asymptotic covariance matrix of the $c_i(\omega)$ is $\boldsymbol{S}_{22.1}$ and the estimated asymptotic covariance matrix of the τ_i is $\boldsymbol{S}_{22.1}^{-1}$ where $\boldsymbol{S}_{22.1} = \boldsymbol{S}_{22} - \boldsymbol{S}_{21}\boldsymbol{S}_{11}^{-1}\boldsymbol{S}_{12}$. These matrices along with the values of the τ_i are available as part of the computer output for some of the programs in use.

8. Confidence intervals

Since the asymptotic covariance matrix of the natural parameters, the τ_i, is available one can determine asymptotic simultaneous sets of confidence intervals for a set of estimates of the natural parameters, or for their linear combinations, in the model selected for detailed analysis. We describe joint confidence intervals based on the Bonferroni inequality (Miller, 1966). With probability $\geqslant 1-\alpha$, joint confidence intervals for k variables are given by

$$x_i - z(1-\alpha/2k)\sigma_i \leqslant \mu_i \leqslant x_i + z(1-\alpha/2k)\sigma_i, \quad i = 1, 2, \ldots, k,$$

where σ_i is the standard deviation of the i-th variable and the function $z(1-\alpha/2k)$ is the $100(\alpha/2k)$ percentile of the standard normal distribution, that is,

$$1-\alpha/2k = \int_{-\infty}^{z} (1/2\pi)^{1/2} \exp(-u^2/2)\,\mathrm{d}u$$

Table 1

k	z	k	z	k	z
1	1.96	8	2.74	15	2.93
2	2.2425	9	2.77	16	2.94
3	2.395	10	2.81	17	2.96
4	2.495	11	2.83	18	2.98
5	2.575	12	2.86	19	3.0
6	2.635	13	2.90	20	3.0
7	2.69	14	2.91		

e.g., $\alpha = 0.05$, $k = 1$, $1-0.025 = 0.975$, $z = 1.96$. Note that $(z(1-\alpha/2k))^2 = \chi^2_{1,1-\alpha/k}$ where $1-\alpha/k = \text{Prob}(\chi^2 \leq \chi^2_{1,1-\alpha/k})$. For $\alpha = 0.05$ we list for values of k the values of the function $z(1-\alpha/2k)$; see Table 1.

9. Outliers

In the ICP case various procedures have been suggested for measuring the adequacy of a loglinear model with respect to specific cells in a contingency table. A cell which does not fit the model is called an OUTLIER. These OUTLIERS may lead one to reject a model which fits the other cells. In other cases even though a model seems to fit, the OUTLIERS contribute much more than reasonable to the measure of deviation between the observed data and the fitted value of the model.

Many writers have used the difference between the test statistic computed for some fitted loglinear model and the test statistic using the same model but ignoring several preselected cells. A separate OUTLIER computation for each cell is time consuming. We now indicate a quick and easy approximation using MDI techniques (Ireland, 1972). The output of various computer programs we use for estimating models includes a listing for each cell called OUTLIER. The value of OUTLIER given for each cell is a lower bound for the decrease in the corresponding $2I(x:x^*)$ if that cell were not included in the estimation procedure. Large values of OUTLIER are those which are at least as large as very significant chi-square values for one degree of freedom. The basis for the OUTLIER interpretation follows. Let x^*_a denote the MDI estimate subject to certain internal constraints. Let x^*_b denote the MDI estimate subject to the same internal constraints as x^*_a except that the value $x(\omega_1)$, say, is not included, so that $x^*_b(\omega_1) = x(\omega_1)$. The basic additive property of the MDI statistics is (10) or

$$2I(x:x^*_a) - 2I(x:x^*_b) = 2I(x^*_b:x^*_a)\,. \tag{16}$$

Using the convexity property of the information function it may be shown that

$$2I(x_b^*:x_a^*) \geq 2(x(\omega_1)\ln(x(\omega_1)/x_a^*(\omega_1)) + (N - x(\omega_1))\ln((N - x(\omega_1))/(N - x_a^*(\omega_1)))) . \quad (17)$$

The last value can be computed and is listed as the OUTLIER entry for each cell of the complete computer output for the MDI estimate x_a^*. The ratio as in (12) may also be used to indicate the percentage of the unexplained variation due to the OUTLIER cell (Gokhale and Kullback, 1978a, 1978b).

Note that if $x(\omega_1) = 0$ the OUTLIER value for that cell as computed from the right side of (17) is $2N\ln(1/(1 - x_a^*(\omega_1)/N))$ and for large N is approximately $2x_a^*(\omega_1)$. We have Table 2 to help interpretation.

Table 2

$x_a^*(\omega_1)$	OUTLIER	Probab. zero count under Poisson: $\exp(-x_a^*(\omega_1))$
1.0	2.00	0.36788
1.5	3.00	0.22313
2.0	4.00	0.13534
2.5	5.00	0.08208
3.0	6.00	0.04979
3.5	7.00	0.03020
4.0	8.00	0.01832
4.5	9.00	0.01111
5.0	10.00	0.00674

Similarly if $x(\omega_1) = 1$, the OUTLIER value for that cell is

$$-2\ln x_a^*(\omega_1) + 2(N - 1)\ln(1/(1 - (x_a^*(\omega_1) - 1)/(N - 1))) ,$$

and for large N is approximately

$$-2\ln x_a^*(\omega_1) + 2x_a^*(\omega_1) .$$

Thus we obtain Table 3. See Larntz (1978). Applications of OUTLIERS in model building and selection will be found in some of the examples at the end of this chapter. At this point it may be useful to illustrate with actual values obtained from a real set of data. The observed values are in a $3 \times 3 \times 3$ contingency table found in Agresti (1977, p. 42), and taken from an article by Smith (1976). The data presents attitudes toward abortion by years of schooling completed with ideal number of children controlled for a sample of size 1425. We shall denote the observed data by $x(ijk)$ where the index i represents number of children, the index j represents years of schooling, and the index k represents attitude, $i = 1, 2, 3$, $j = 1, 2, 3$, $k = 1, 2, 3$. It should be noted that the indices simply represent the levels of the three variables and are not dummy variables. For the MDI estimate $x_a^*(ijk)$ fitting the two-way marginals, that is,

Table 3

		Poisson probability	
$x_a^*(\omega_1)$	OUTLIER	zero or one count	one count
0.5	0.39	0.90980	0.30327
2.0	0.61	0.40601	0.27067
2.5	1.17	0.28730	0.20522
3.0	1.80	0.19915	0.14936
3.5	2.49	0.13589	0.10569
4.0	3.23	0.09158	0.07326
4.5	3.99	0.06110	0.04999
5.0	4.78	0.04043	0.03369
5.5	5.59	0.02657	0.02248
6.0	6.42	0.01735	0.01487
6.5	7.26	0.01127	0.00977
7.0	8.11	0.00730	0.00638

subject to the constraints $x_a^*(ij\cdot) = x(ij\cdot)$, $x_a^*(\cdot jk) = x(\cdot jk)$, $x_a^*(i\cdot k) = x(i\cdot k)$, it is found that $2I(x:x_a^*) = 20.953$ with 8 degrees of freedom. The tabulated value of chi-square for 8 degrees of freedom, at the 0.01 level is 20.1. An examination of the computer output for the MDI estimate x_a^* shows that the cell (131) has an OUTLIER value of 6.280. For the MDI estimate x_b^* using the same constraints as x_a^* but omitting the cell (131) from the estimation, that is, $x_b^*(ij\cdot) = x(ij\cdot)$, $x_b^*(\cdot jk) = x(\cdot jk)$, $x_b^*(i\cdot k) = x(i\cdot k)$, $x_b^*(131) = x(131)$, it is found that $2I(x:x_b^*) = 9.499$ with 7 degrees of freedom. The tabulated value of chi-square for 7 degrees of freedom at the 0.1 level is 12.0. We can summarize these values in Table 4. Note that the actual reduction due to omitting cell (131) from the estimation procedure is 11.454 and the OUTLIER value indicated a lower bound of 6.280. The percentage reduction omitting the cell (131) is $11.454/20.953 = 0.55$ or 55 percent.

Table 4
Analysis of information

Component due to	Information	D.F.
$x(ij\cdot), x(\cdot jk), x(i\cdot k)$	$2I(x:x_a^*) = 20.953$	8
	$2I(x_b^*:x_a^*) = 11.454$	1
$x(ij\cdot), x(\cdot jk), x(i\cdot k), x(131)$	$2I(x:x_b^*) = 9.499$	7

10. k-Sample problem

Earlier we had stated that depending on the design of the data collection procecure we shall look upon the cross-classification of the count data as a single sample from a multinomial distribution or as a collection of samples from many multinomial distributions. Our discussion thus far has been in terms of a single multinomial distribution. We shall now consider a canonical reduction of the k-sample problem to the single sample problem already discussed (Gokhale, 1973; Gokhale and Kullback, 1978a, 1978b). The basic idea is to convert the k samples into one pseudo sample.

Consider the k discrete spaces Ω_i, $i = 1, 2, \ldots, k$, where we denote the points or cells of Ω_i by $\omega(ij)$, $j = 1, 2, \ldots, \Omega_i$, $\boldsymbol{\omega}'(i) = (\omega(i1), \ldots, \omega(i\Omega_i))$. We use Ω_i both for the i-th space and the number of cells in it. Let $\boldsymbol{p}'_i = (p_i(\omega(i1)), \ldots, p_i(\omega(i\Omega_i)))$, $i = 1, 2, \ldots, k$, be k sets of probability distributions defined respectively over Ω_i, $i = 1, 2, \ldots, k$. Let $\boldsymbol{p}' = (\boldsymbol{p}'_1, \ldots, \boldsymbol{p}'_k)$ be a $1 \times \Omega$ matrix where $\Omega = \Omega_1 + \Omega_2 + \cdots + \Omega_k$. Let $\mathscr{P}$ be the family of all such matrices $\boldsymbol{p}'$. For a given $\pi' = (\pi'_1, \ldots, \pi'_k) \in \mathscr{P}$ and $\boldsymbol{p}' \in \mathscr{P}$ the generalized discrimination information is given by

$$I(p:\boldsymbol{\pi}) = \sum w_i \sum_{j=1}^{\Omega_i} p_i(\omega(ij)) \ln(p_i(\omega(ij))/\pi_i(\omega(ij))), \tag{18}$$

where w_i are known weights such that $\Sigma_i w_i = 1$, $0 < w_i < 1$. Denote the elements (points or cells) of Ω by $\omega(ij)$, $i = 1, \ldots, k$, $j = 1, \ldots, \Omega_i$, so that $\omega(i1), \ldots, \omega(i\Omega_i)$ are the cells of Ω belonging to Ω_i. The number of cells in the Ω_i need not be the same.

The MDI estimate is the value of $\boldsymbol{p}$ which minimizes $I(\boldsymbol{p}:\boldsymbol{\pi})$ in (18) over the family of $\boldsymbol{p}$'s which satisfy the linearly independent moment constraints

$$\boldsymbol{Bp} = \boldsymbol{\theta}, \qquad \boldsymbol{B} \text{ is } (k+r)\times\Omega, \quad \boldsymbol{p} \text{ is } \Omega\times 1, \quad \boldsymbol{\theta} \text{ is } (k+r)\times 1,$$
$$\text{rank of } \boldsymbol{B} \text{ is } (k+r) \leqslant \Omega. \tag{19}$$

11. Canonical transformation

To transform to a canonical form so that one may utilize essentially the procedures and algorithms of the single sample case we proceed as follows. Let W_i be an $\Omega_i \times \Omega_i$ diagonal matrix with diagonal elements w_i. Define

$$\boldsymbol{W} = \begin{Vmatrix} \boldsymbol{W}_1 & \boldsymbol{0} & \cdots & \boldsymbol{0} \\ \boldsymbol{0} & \boldsymbol{W}_2 & \cdots & \boldsymbol{0} \\ & & & \\ \boldsymbol{0} & \boldsymbol{0} & \cdots & \boldsymbol{W}_k \end{Vmatrix}, \tag{20}$$

$$\boldsymbol{P} = \boldsymbol{Wp}, \quad \boldsymbol{\Pi} = \boldsymbol{W\pi}, \quad \boldsymbol{C} = \boldsymbol{BW}^{-1}, \quad \boldsymbol{C} \text{ is } (k+r)\times\Omega, \quad \boldsymbol{W}^{-1} = \boldsymbol{V}.$$

Note that $\Sigma_\Omega P(\omega) = 1$, $\Sigma_\Omega \boldsymbol{\Pi}(\omega) = 1$, $I(\boldsymbol{p}:\boldsymbol{\pi}) = \Sigma_\Omega P(\omega) \ln(P(\omega)/\Pi(\omega)) = I(P:\Pi)$, $\boldsymbol{Bp} = \boldsymbol{BW}^{-1}\boldsymbol{Wp} = \boldsymbol{CP} = \boldsymbol{\theta}$. In terms of the canonical transformation the k-sample problem may now be formulated as finding the MDI estimate $P^*(\omega)$ minimizing

$$I(P:\Pi)=\sum_{\Omega} P(\omega)\ln(P(\omega)/\boldsymbol{\Pi}(\omega)), \tag{21}$$

subject to the linearly independent constraints

$$\begin{aligned} \boldsymbol{CP}=\boldsymbol{\theta}, \quad & \boldsymbol{C} \text{ is } (k+r)\times\Omega, \quad \boldsymbol{P} \text{ is } \Omega\times 1, \quad \boldsymbol{\theta} \text{ is } (k+r)\times 1, \\ & \text{rank of } \boldsymbol{C} \text{ is } (k+r)\leqq\Omega . \end{aligned} \tag{22}$$

Denote the elements of $\boldsymbol{C}$ by $c_i(\omega)$, $i=1,\dots,k,k+1,\dots,k+r$, $\omega=11,\dots,1\Omega_1,\dots,k1,\dots,k\Omega_k$. We may write (22) as

$$\sum_{\Omega} c_i(\omega)P(\omega)=\theta_i, \quad i=1,\dots,k,k+1,\dots,k+r. \tag{23}$$

We usually take the elements of the B-matrix $b_i(\omega(ij))=1$, $j=1,\dots,\Omega_i$, $i=1,\dots,k$, and zero otherwise, that is, the natural constraint for each sample. Thus

$$\begin{aligned} c_i(\omega)&=v_i \quad \text{for } \omega=i1,\dots,i\Omega_i,\ v_i=1/w_i, \\ &=0 \quad \text{otherwise, } i=1,2,\dots,k, \\ \theta_i&=1, \quad i=1,2,\dots,k. \end{aligned} \tag{24}$$

12. MDI estimate

The exponential family or multiplicative representation for the MDI estimate is

$$\begin{aligned} P^*(\omega)&=\exp(\lambda_1 c_1(\omega)+\cdots+\lambda_k c_k(\omega)+\tau_1 c_{k+1}(\omega)+\cdots+\tau_r c_{k+r}(\omega))\Pi(\omega) \\ &=\exp(\lambda_1 c_1(\omega))\cdots\exp(\lambda_k c_k(\omega))\exp(\tau_1 c_{k+1}(\omega))\cdots\exp(\tau_r c_{k+r}(\omega))\Pi(\omega), \end{aligned} \tag{25}$$

and the loglinear representation is

$$\ln(P^*(\omega)/\Pi(\omega))=\lambda_1 c_1(\omega)+\cdots+\lambda_k c_k(\omega)+\tau_1 c_{k+1}(\omega)+\cdots+\tau_r c_{k+r}(\omega), \quad \omega=11,\dots,k\Omega_k. \tag{26}$$

The λ's are scaling constants and the τ's are the exponential or natural parameters of interest. Applications of the k-sample procedure will be found in the examples at the end of this chapter. We remark that one usually takes the weights $w_i=N_i/N$, where the sum of the observations in the i-th sample is N_i

and $N = N_1 + N_2 + \cdots + N_k$, although the analyst may select any values consistent with his view of the problem.

13. Marginals

The marginals of contingency tables have long played a critical role in their analysis. In the case of smoothing or fitting the aim of the analysis usually is to get a good fit to the observed table using a minimal or parsimonious number of natural parameters in the loglinear model depending only on some of the observed marginals. This shows how much of the total information is contained in a summary consisting only of sets of marginals. The observed distribution can then be said to be explainable in terms of a smaller number of linear functions of cell frequencies. The observed distribution, the full or complete model requires a maximum number of parameters for its description, whereas the uniform distribution requires the least. When sets of marginals are fitted the values of the $c_i(\omega)$ in (8) are either 0's or 1's since the sum $\Sigma_\Omega c_i(\omega)x(\omega)$ is a marginal value if and only if $c_i(\omega)$ is a 1 for every cell that enters into the marginal and is a 0 otherwise.

14. Notation

To relate the tau or natural parameters with the associated marginal value notationally and avoid a possible need for record keeping, we shall use superscripts and subscripts on the taus. The superscripts indicate the variables or factors involved and the subscripts refer to the levels or categories of the variables. Thus τ_{11}^{AB} will correspond to the two-way marginal for variables A and B each at level 1, τ_{213}^{BCD} will correspond to the three-way marginal for variables B, C, D, respectively at level 2, level 1, and level 3. To insure linear independence in the C-matrix we follow the convention of setting every tau with any subscript equal to the last level or category value as zero. Any level could have been selected for the reference value. Other parameterizations use one similar to the analysis of variance in which sums of the parameters over all levels are set equal to zero.

15. Computer algorithms

When the moment constraints in (8) involve only sets of marginals the Deming–Stephan iterative proportional fitting algorithm may be used to determine the MDI estimate $x^*(\omega)$ satisfying (8) and then the natural parameters are determined from the loglinear representation. The proportional fitting algorithm may be described as successively cycling through adjustments of the marginals of interest starting with the marginals of the $\pi(\omega)$ distribution until a desired accuracy of agreement between the set of observed marginals of interest and the computed marginals is attained (Gokhale and Kullback, 1978a, pp. 214–216; Ireland and Kullback, 1968). Fitting sets of marginals leads to so called hierarchical models since any higher order marginal such as $x(ijk\cdot\cdot)$ for example implies the lower order marginals $x(ij\cdot\cdot\cdot)$, $x(i\cdot k\cdot\cdot)$, $x(\cdot jk\cdot\cdot)$, $x(i\cdot\cdot\cdot\cdot)$, $x(\cdot j\cdot\cdot\cdot)$, $x(\cdot\cdot k\cdot\cdot)$, $x(\cdot\cdot\cdot\cdot\cdot)$. Thus along with the interaction parameters involving superscripts of the three variables there must also appear the parameters involving all combinations of the three variables two at a time and singly.

When the model matrix C contains entries other than 0 or 1 arising from sets of marginals, the Deming–Stephan iteration is not applicable. For any model matrix C there is available a Newton–Raphson type iterative algorithm that in effect solves (6) for the values of the taus and then determines the values of $x^*(\omega)$ therefrom. The algorithm in fact iteratively solves linearized versions of the equations $\theta_i = (\partial/\partial\tau_i) \ln M(\tau_1, \tau_2, \ldots, \tau_r)$, where $M(\tau_1, \tau_2, \ldots, \tau_r)$ is defined following (6). The relation is a consequence of the exponential family property of the MDI estimate (Gokhale and Kullback, 1978a, pp. 193–195; Keegel, 1981).

16. Fitting strategies

In seeking a good fit to data by fitting a nested sequence of marginals the tests are carried out sequentially. Hence they become conditional on the outcome of the previous ones. A balance should be achieved between the amount of variation explained and the number of parameters associated with the marginals fitted. The information number of the first estimate in the fitting sequence will usually be the basis to determine the variation explained by subsequent estimates. Accordingly, the first model is determined by the nature of the problem under consideration. If the contingency table involves classifications all of which are random variables and the problem is essentially one of determining mutual associations and interactions similar to a correlation analysis, the first model will be that for mutual independence and the estimate is obtained by fitting all the one-way marginals. If the contingency table involves a dependent classification and a number of explanatory classifications, and the problem is that of determining the relationship of the dependent classification on the explanatory classifications, similar to a regression analysis, the first model will be that for homogeneity of the dependent classification over the combinations of the explanatory classifications. The estimate fits the one-way marginal of the dependent classification and the joint multiway marginal of the explanatory classifications. In this last case it is convenient for examining the output of the computer programs to use the last index for the dependent classification in the lexicographic ordering.

The order in which the other marginals are chosen for fitting may sometimes be indicated by the objective of the analysis, fixed marginal totals, the nature of the classifications, such as time of occurrence, cause-effect phenomena, etc. Otherwise, the choice of order is arbitrary and may be governed by considerations such as previous experience, comparison with an earlier analysis by different methods and so on. The value of the information numbers will, of course, depend on the order of fitting. The fit of the model selected for analysis will depend on the set of marginals finally selected for fitting and not on their order. It is our experience that one usually arrives at the same final model, based on a set of marginals having large effects, even following different orders of fitting, barring extreme unusual cases of strong correlation. When in doubt several different orderings may be tried. Large effects for marginals tend to remain large, and small effects for marginals tend to remain small, independent of the sequence of fitting.

17. Structural simplicity

It should be noted that in the case of a single sample the complete contingency table can be specified in terms of $\Omega - 1$ probabilities. In the k-sample case the complete table can be specified in terms of $\Sigma_{i=1}^{k} (\Omega_i - 1)$ probabilities. The numbers $\Omega - 1$ or $\Sigma (\Omega_i - 1)$ in practical situations are large. An objective of the fitting procedure and the analysis is to determine whether the observed data are consistent with a simpler meaningful structure satisfied by the large number of individual probabilities.

Such a structure is imposed by a loglinear model. The smaller the number of parameters in the model, the simpler is the structure. In the one-sample case, since the contingency table is completely determined by $\Omega - 1$ probabilities, a model with $\Omega - 1$ linearly independent parameters should yield any given set of probabilities of a multinomial distribution. Simplicity of structure is indicated when the observed distribution can be satisfactorily estimated by a model containing less than $\Omega - 1$ parameters.

As a simple example consider the following loglinear model for a 3×2 contingency table with classifications A and B:

$$\begin{aligned}
\ln(p(11)/\pi(11)) &= \tau_0 + \tau_1^A + \tau_1^B + \tau_{11}^{AB}, \\
\ln(p(12)/\pi(12)) &= \tau_0 + \tau_1^A, \\
\ln(p(21)/\pi(21)) &= \tau_0 + \tau_2^A + \tau_1^B + \tau_{21}^{AB}, \\
\ln(p(22)/\pi(22)) &= \tau_0 + \tau_2^A, \\
\ln(p(31)/\pi(31)) &= \tau_0 \qquad + \tau_1^B, \\
\ln(p(32)/\pi(32)) &= \tau_0
\end{aligned} \tag{27}$$

where $\pi(ij)$, $i = 1, 2, 3$, $j = 1, 2$, is an arbitrary but known probability distribution over the six cells. Note that in accordance with earlier comments we have set every tau parameter with subscript $i = 3$ and/or $j = 2$ equal to zero. Observe that there are five linearly independent parameters τ_1^A, τ_2^A, τ_1^B, τ_{11}^{AB}, τ_{21}^{AB}, and that τ_0 is a scaling factor so that the probabilities $p(ij)$ sum to one.

The associated C-matrix is

$$
\begin{array}{c|cccccc|}
\omega & 1 & 2 & 3 & 4 & 5 & 6 \\
A & 1 & 1 & 2 & 2 & 3 & 3 \\
B & 1 & 2 & 1 & 2 & 1 & 2
\end{array}
$$

$$
\boldsymbol{C} = \left\| \begin{array}{cccccc}
1 & 1 & 1 & 1 & 1 & 1 \\
1 & 1 & 0 & 0 & 0 & 0 \\
0 & 0 & 1 & 1 & 0 & 0 \\
1 & 0 & 1 & 0 & 1 & 0 \\
1 & 0 & 0 & 0 & 0 & 0 \\
0 & 0 & 1 & 0 & 0 & 0
\end{array} \right\| . \qquad (28)
$$

Given any particular values to the five parameters a probability distribution $p(ij)$, $i = 1, 2, 3$, $j = 1, 2$, is uniquely determined. Conversely, for any distribution $p(ij)$, say an observed distribution with positive entries in each cell one can solve the equations (27) to find the values of $\tau_1^A, \tau_2^A, \tau_1^B, \tau_{11}^{AB}, \tau_{21}^{AB}$.

In the present example, simplicity of structure will be introduced if we assume say that $\tau_{11}^{AB} = \tau_{21}^{AB} = 0$. The model then has only three linearly independent parameters $\tau_1^A, \tau_2^A, \tau_1^B$. In fact under this simpler structure if we assume that $\pi(ij) = \frac{1}{6}$, $i = 1, 2, 3$, $j = 1, 2$, then it can be shown that $p(ij) = p(i\,\cdot)p(\cdot\,j)$, $i = 1, 2, 3$, $j = 1, 2$, corresponding to mutual independence of the classifications A and B. For this simpler structure the appropriate C-matrix will consist only of the first four rows of the matrix in (28).

18. Nonhierarchical models

In current practice it is conventional to use hierarchical loglinear models. These models are those in which whenever we include an interaction term, we must also include all lower-order interactions involving variables in the higher-order term (Bishop et al., 1975, pp. 34, 68; Fienberg, 1977, p. 60). This is primarily a consequence of the fact that estimates are obtained by the Deming–Stephan iterative proportional fitting procedure using sets of marginal values. Thus, since a marginal, say $x(hij\,\cdot)$ implies the lower order marginals $x(hi\,\cdot\,\cdot)$, $x(h\,\cdot\,j\,\cdot)$, $x(\cdot\,ij\,\cdot)$, $x(h\,\cdot\,\cdot\,\cdot)$, $x(\cdot\,i\,\cdot\,\cdot)$, $x(\cdot\,\cdot\,j\,\cdot)$, the above restriction applies.

However there do arise data for which nonhierarchical loglinear models are appropriate. These models "can be related to the concept of synergism where a response occurs when two factors are present together but not when either occurs alone" (Bishop et al., 1975, p. 38; Worcester, 1971). In the fitting procedure using the Deming–Stephan proportional fitting iteration one may infer that a certain set of marginals say $x(hij\,\cdot)$ introduces significant effects. Essentially this means that we reject a null hypothesis that *all* interaction parameters of a set are zero. However, this does not rule out the possibility that *some* of them may be zero. Thus to complete the examination of the model we should obtain the output using the Newton–Raphson type iteration, because this output includes the values of the taus for the model and their covariance matrix. One can then infer whether certain individual taus are not significantly different from zero and get a new estimate using this fact. We may therefore determine a final model that provides an acceptable estimate with a simpler structure and fewer parameters. Such models may be nonhierarchical in structure. Walls and Weeks (1969) have given proofs that the variance of least square estimates of regression coefficients increases when variables are added to a regression equation. Although a similar result has not been proven for loglinear models, this property has been observed in enough cases to lead one to believe it to be so. Also Bishop et al. (1975, p. 313) state: "A model with fewer parameters may improve the precision of the estimates. Suppose we have two models for predicting the cell frequencies in a table of counts, both of which are compatible with the observed data, one model being a special case of and having fewer parameters than the other. Then the overall variability of the estimates from the simpler model about the 'true' values for the cells is smaller than the 'overall variability' for the model with more parameters requiring estimation. We have no general proof of this theorem, although it is specifically true for many regression problems and we believe it is true in a rather general way". The model building philosophy in the examples at the end of this chapter is in accordance with the preceding empirical theorems.

We quote some of the negative comments about nonhierarchical loglinear models because we believe that our proposed approach and methodology and the examples tend to refute these comments. "In larger tables with non-hierarchical structure a possible strategy is to partition and look at smaller sections of data .... We can either consider the structure of each set of tables separately or rearrange the cells to form new compound variables" (Bishop et al., 1975, p. 38). "As in conventional analysis of variance, interpretation of parameters in non-hierarchical models appears difficult; consequently, the usefulness of non-hierarchical models is not clear" (Haberman, 1974, p. 200). "It is possible to consider fitting nonhierarchical models to data, but we cannot then compute the estimated expected values directly via our iterative proportional fitting procedure. Rather, we need to transform the table, interchanging cells, so that the non-hierarchical model for the original table becomes a hierarchical model for the transformed table" (Fienberg, 1977, p. 39).

The examples introduce some additional concepts related to the previous discussion but in our opinion such that a meaningful presentation is more easily accomplished in terms of specific data rather than theory.

19. Example 1: Case-control study

This example considers the MDI analysis of a case-control study of two types of exposure. The count data were cross-classified into a $3\times2\times2\times2$ contingency table. The log-odds or logit representation for the observed data is indicated. A nonhierarchical model is derived and its interpretation indicated.

The data are from a study made by Professor Julius Schachter and used as an illustration by Heilbron (1981) for an analysis of ratios of odds ratios. The study relates to the possible role of the herpes simplex virus Type 2 (HSV-2), and of another common sexually-transmitted agent, *Chlamydia trachomatis*, in the etiology of cervical dysplasia in women. Exposure to HSV-2 is here taken to be indicated by

$$\frac{\log(\text{neutralizing antibody titer for HSV-2})}{\log(\text{neutralizing antibody titer for HSV-1})} \geqslant 0.85\,,$$

while exposure to *Chlamydia* is taken as indicated by a micro-immuno-fluorescent antibody titer $\geqq 1{:}8$. The present results are stratified by three levels, the total number of sexual partners.

The observed occurrences of the entries in the cells of the $3\times2\times2\times2$ cross-classification $x(hijk)$ are listed in lexicographic order in Table 5. Also listed are the values of the cross-product ratios or odds ratios

$$x(hi11)x(hi22)/x(hi12)x(hi21) \quad \text{for } h = 1, 2, 3,\ i = 1, 2\,.$$

We shall treat the data as 12 samples of the binomial variable Group over the twelve combinations of the explanatory variables Partners $\times$ Titer $\times$ H.Ratio. The log-odds or logit representation for the observed data (complete

Table 5
Observed and estimated data

hijk	*x*(*hijk*)	*x**(*hijk*)	*hijk*	*x*(*hijk*)	*x**(*hijk*)	*hijk*	*x*(*hijk*)	*x**(*hijk*)
1111	56	53.612	2111	47	42.460	3111	43	50.217
1112	42	44.388	2112	57	61.540	3112	80	72.783
1121	54	53.892	2121	46	48.992	3121	44	40.827
1122	78	78.108	2122	74	71.008	3122	56	59.173
1211	14	16.388	2211	8	9.904	3211	4	5.354
1212	28	25.612	2212	29	27.096	3212	16	14.646
1221	34	31.319	2221	18	17.667	3221	12	9.369
1222	83	85.681	2222	48	48.333	3222	23	25.631

$x(1111)x(1122)/x(1112)x(1121) = 1.926$	$x^*(1111)x^*(1122)/x^*(1112)x^*(1121) = 1.751$
$x(1211)x(1222)/x(1212)x(1221) = 1.221$	$x^*(1211)x^*(1222)/x^*(1212)x^*(1221) = 1.750$
$x(2111)x(2122)/x(2112)x(2121) = 1.326$	$x^*(2111)x^*(2122)/x^*(2112)x^*(2121) = 1.000$
$x(2211)x(2222)/x(2212)x(2221) = 0.736$	$x^*(2211)x^*(2222)/x^*(2212)x^*(2221) = 1.000$
$x(3111)x(3122)/x(3112)x(3121) = 0.684$	$x^*(3111)x^*(3122)/x^*(3112)x^*(3121) = 1.000$
$x(3211)x(3222)/x(3212)x(3221) = 0.479$	$x^*(3211)x^*(3222)/x^*(3212)x^*(3221) = 1.000$

Characteristic		Index	1	2	3
Partners	*P*	*h*	1–3	4–10	>10
Titer	*T*	*i*	⩾1:8	<1:8	
H.Ratio	*R*	*j*	⩾0.85	<0.85	
Group	*G*	*k*	Dysplasia	Control	

model) is

$$\ln(x(hij1)/x(hij2)) = \tau_1^G + \tau_{h1}^{PG} + \tau_{i1}^{TG} + \tau_{j1}^{RG} + \tau_{hi1}^{PTG} + \tau_{hj1}^{PRG} + \tau_{ij1}^{TRG} + \tau_{hij1}^{PTRG},$$
$$h = 1, 2, \quad i = 1, \quad j = 1, \quad k = 1, \tag{1}$$

where the tau parameters represent the interactions of the explanatory variables P, T, and R with the dependent variable G. All parameters with only superscripts P, T, R and combinations thereof in the loglinear representation of $x(hijk)$ drop out of the log-odds representation. Note that all parameters with subscript $h = 3$ and/or $i = 2$ and/or $j = 2$ and/or $k = 2$ are zero by convention to insure linear independence in the C-matrix.

To determine a parsimonious loglinear (or multiplicative) model which is a good fit to the data, a sequence of nested MDI models using various marginals as constraints was fitted to the data. The marginal constraints require the MDI estimate to have the same values for the fitted marginals as the original data. The results of this first overview of the data are summarized in the Analysis of Information Table 6a. The values were obtained by using the Deming–Stephan iterative proportional fitting algorithm. The use of $x(hij\cdot)$ as one of the constraints in all nine models reflects our treating the data as 12 samples of the binomial variable Group over the twelve combinations of the explanatory variables Partners × Titer × H.Ratio.

The first model in Table 6a, x_a^*, which has the explicit representation $x_a^*(hijk) = x(hij\cdot)x(\cdot\cdot\cdot k)/N$ is used to test the null hypothesis that the dependent binomial variable is homogeneous over the twelve experimental situa-

Table 6a
Analysis of information

Component due to		Information	D.F.
(a) $x(hij\cdot)$, $x(\cdot\cdot\cdot k)$		$2I(x:x_a^*) = 35.053$	11
	$x(h\cdot\cdot k)$	$2I(x_b^*:x_a^*) = 1.567$	2
(b) $x(hij\cdot)$, $x(h\cdot\cdot k)$		$2I(x:x_b^*) = 33.486$	9
	$x(\cdot i\cdot k)$	$2I(x_c^*:x_b^*) = 21.906$	1
(c) $x(hij\cdot)$, $x(h\cdot\cdot k)$, $x(\cdot i\cdot k)$		$2I(x:x_c^*) = 11.580$	8
	$x(\cdot\cdot jk)$	$2I(x_d^*:x_c^*) = 0.667$	1
(d) $x(hij\cdot)$, $x(h\cdot\cdot k)$, $x(\cdot i\cdot k)$, $x(\cdot\cdot jk)$		$2I(x:x_d^*) = 10.913$	7
	$x(hi\cdot k)$	$2I(x_e^*:x_d^*) = 0.694$	2
(e) $x(hij\cdot)$, $x(hi\cdot k)$, $x(\cdot\cdot jk)$		$2I(x:x_e^*) = 10.219$	5
	$x(h\cdot jk)$	$2I(x_f^*:x_d^*) = 8.345$	2
(f) $x(hij\cdot)$, $x(h\cdot jk)$, $x(\cdot i\cdot k)$		$2I(x:x_f^*) = 2.568$	5
	$x(\cdot ijk)$	$2I(x_g^*:x_d^*) = 1.000$	1
(g) $x(hij\cdot)$, $x(h\cdot\cdot k)$, $x(\cdot ijk)$		$2I(x:x_g^*) = 9.913$	6
	$x(h\cdot jk)$	$2I(x_h^*:x_e^*) = 7.874$	2
(h) $x(hij\cdot)$, $x(hi\cdot k)$, $x(h\cdot jk)$		$2I(x:x_h^*) = 2.345$	3
	$x(\cdot ijk)$	$2I(x_i^*:x_h^*) = 2.274$	1
(i) $x(hij\cdot)$, $x(hi\cdot k)$, $x(h\cdot jk)$, $x(\cdot ijk)$		$2I(x:x_i^*) = 0.071$	2

Table 6b
Analysis of information

Component due to		Information	D.F.
(a) $x(hij\cdot), x(\cdots k)$		$2I(x:x_a^*)=35.053$	11
	$x(\cdot 1\cdot 1), x(1\cdot 11)$	$2I(x^*:x_a^*)=28.619$	2
(*) $x(hij\cdot), x(\cdots 1), x(\cdot 1\cdot 1), x(1\cdot 11)$		$2I(x:x^*)=6.434$	9
	$x(2\cdot 11)$	$2I(x_j^*:x^*)=0.257$	1
(j) $x(hij\cdot), x(\cdots 1), x(\cdot 1\cdot 1), x(1\cdot 11), x(2\cdot 11)$		$2I(x:x_j^*)=6.177$	8

tions. This null hypothesis is rejected, and we seek to account for the variation. An examination of the various effect and goodness-of-fit values in Table 6a, in particular

$$\begin{array}{lll} x(\cdot\, i\cdot k) & 2I(x_c^*:x_b^*)=21.906 & 1\text{ D.F.} \\ x(h\cdot jk) & 2I(x_f^*:x_d^*)=8.345 & 2\text{ D.F.} \end{array}$$

suggests that as an initial model we consider the nonhierarchical model x_j^* with fitting constraints $x(hij\cdot)$, $x(\cdots 1)$, $x(\cdot 1\cdot 1)$, $x(h\cdot 11)$. The log-odds or logit representation for this nonhierarchical model is

$$\ln(x_j^*(hij1)/x_j^*(hij2))=\tau_1^G+\tau_{11}^{TG}+\tau_{h11}^{PRG},\quad h=1,2\,. \tag{2}$$

To obtain the MDI estimate $x_j^*(hijk)$, the tau parameters and their covariance matrix, the Newton–Raphson type algorithm was employed. We do not list the estimated values $x_j^*(hijk)$, but list in Table 7 the values of the tau parameters in the log-odds representation (2), as well as their standardized values. Since the standardized value of τ_{211}^{PRG} is 0.5081 we obtained a final

Table 7

Parameters of x_j^*			
	TAU	EXPTAU	TAUSTANDARD
τ_1^G	−1.020230	0.360512	−7.7702
τ_{11}^{TG}	0.631706	1.880817	4.2805
τ_{111}^{PRG}	0.576277	1.779401	3.0636
τ_{211}^{PRG}	0.097223	1.102106	0.5081
Parameters of x^*			
τ_1^G	−1.006426	0.365523	−7.8420
τ_{11}^{TG}	0.635305	1.887597	4.3101
τ_{111}^{PRG}	0.559924	1.750539	3.0219

nonhierarchical model x^* fitting the constraints $x(hij\cdot)$, $x(\cdot\cdot\cdot 1)$, $x(\cdot 1 \cdot 1)$, $x(1\cdot 11)$. The log-odds or logit representation for the model x^* is

$$\ln(x^*(hij1)/x^*(hij2)) = \tau_1^G + \tau_{11}^{TG} + \tau_{111}^{PRG} . \tag{3}$$

Using the Newton–Raphson type algorithm there were obtained the values of $x^*(hijk)$ listed in Table 5, along with the cross-product ratios or odds ratios

$$x^*(hi11)x^*(hi22)/x^*(hi12)x^*(hi21) \quad \text{for } h = 1, 2, 3,\ i = 1, 2 .$$

The tau parameters in the log-odds or logit representation (3) are listed in Table 7, and their covariance matrix is listed in Table 8.

Table 8
Covariance matrix, parameters of x^*

τ_1^G	τ_{11}^{TG}	τ_{111}^{PRG}
0.016489	−0.015695	−0.005370
−0.015695	0.021743	0.000291
−0.005370	0.000291	0.034339

We note from Analysis of Information Table 6b that $2I(x:x^*) = 6.434$, 9 D.F., implies that x^* is a very good fit to the original data. We also note that the square of the standardized value of τ_{211}^{PRG} as given in Table 7 among the parameters of x_j^*, that is, $(0.5081)^2 = 0.2582$, is the tau approximation to $2I(x_j^*:x^*) = 0.257$, 1 D.F. The results in Table 6b confirm our use of x^* as the final model.

We convert the log-odds representation in (3) to the odds representation

$$x^*(hij1)/x^*(hij2) = (\exp \tau_1^G)(\exp \tau_{11}^{TG})(\exp \tau_{111}^{PRG}) . \tag{4}$$

The values of the odds factors for x^* are given in Table 9. From Table 9 we see that other things being equal the odds of Dysplasia to Control are 1.89 to 1 for Titer ≥ 1:8. There is a synergistic effect between Partners and H.Ratio and

Table 9
Odds factors, x^*

BASE	TITER		PARTNERS × H.RATIO		
				$j = 1$	$j = 2$
0.365523	$i = 1$	1.887597	$h = 1$	1.750539	1
	$i = 2$	1	$h = 2$	1	1
			$h = 3$	1	1

there is no difference among Partners for values four or more. Other things being equal the odds of Dysplasia to Control are 1.75 to 1 for Partners × H.Ratio (1 − 3) × (≥0.85) as compared to any other combination.

We may use (3) to get

$$\ln(x^*(hi11)/x^*(hi12)) - \ln(x^*(hi21)/x^*(hi22)) = \tau_{111}^{PRG}, \quad h = 1, \qquad (5)$$
$$= 0, \qquad h = 2, 3$$

or

$$x^*(hi11)x^*(hi22)/x^*(hi12)x^*(hi21) = \exp(\tau_{111}^{PRG}), \quad h = 1, \qquad (6)$$
$$= 1, \qquad h = 2, 3.$$

From the values for $x^*(hijk)$, which is an excellent parsimonious fit to the original data, we may infer that the ratios of the odds ratios in each stratum of the original data may be taken as one.

20. Example 2: School support from various community sources

This example considers in parallel information analyses of five similar three-way contingency tables. Parsimonious loglinear models are derived. The use of odds factors is illustrated. The square of the standarized value of a tau parameter which is inferred to be zero is shown to be a very good approximation to the MDI statistic comparing estimates with and without the parameter in question. Treatment of an OUTLIER is also illustrated.

The particular data we analyze in this study are derived from a mail survey conducted by the National Center for Education Statistics, presented in Table B-6.6, page B-43, Volume I of Violent Schools – Safe Schools, The Safe School Study Report to the Congress, U.S. Department of HEW, National Institute of Education. Table B-6.6 lists the percentage of schools reporting 'very much' support from five community sources:

PARENTS, LOCAL POLICE, LOCAL COURTS,
SCHOOL BOARD, and SCHOOL SYSTEM CENTRAL OFFICE,

by level and location in the handling of discipline problems. The sample size upon which the percentages were based was also given.

We shall present our statistical analysis as an application of the principle of *minimum discrimination information estimation* (MDIE).

Accordingly the data in Table B-6.6 were converted to the form given as our Table 1. We indicate the number of YES responses to 'very much' support and the number of NO responses.

Since the joint responses to the question of support from the community sources are not available, we must examine the data in Table 10 as five contingency tables, one for each of the community sources.

Table 10
Numbers of schools receiving 'very much' support from various community sources in the handling of discipline problems, by level and location

		ELEMENTARY				JUNIOR HIGH				SENIOR HIGH			
COMMUNITY SOURCE		LARGE CITIES	SMALL CITIES	SUB. AREAS	RURAL AREAS	LARGE CITIES	SMALL CITIES	SUB. AREAS	RURAL AREAS	LARGE CITIES	SMALL CITIES	SUB. AREAS	RURAL AREAS
1. PARENTS	YES	135	148	278	135	124	147	312	155	119	113	292	114
	NO	140	99	186	152	165	131	265	205	178	138	291	222
		275	247	464	287	289	278	577	360	297	251	583	336
2. LOCAL POLICE	YES	62	80	174	91	110	133	261	155	129	131	326	142
	NO	195	135	221	168	180	138	283	205	170	116	257	188
		257	215	395	259	290	271	544	360	299	247	583	330
3. LOCAL COURTS	YES	16	31	67	42	19	27	63	69	29	42	68	51
	NO	214	163	285	192	259	239	465	276	260	205	502	267
		230	194	352	234	278	266	528	345	289	247	570	318
4. SCHOOL BOARD	YES	54	94	242	123	76	146	343	242	86	130	381	230
	NO	191	119	162	79	207	120	210	109	199	115	196	99
		245	213	404	202	283	266	553	351	285	245	577	329
5. SCHOOL SYSTEM CENTRAL OFFICE	YES	74	121	272	186	96	172	366	268	110	146	400	243
	NO	182	108	146	84	187	101	197	89	188	101	179	85
		256	229	418	270	283	273	563	357	298	247	579	328

Characteristic	Index	1	2	3	4
Level	i	Elementary	Junior High	Senior High	
Location	j	Large Cities	Small Cities	Suburbs	Rural
Support	k	Yes	No		

The different values for the Level × Location total for the different community sources are a consequence of missing reports. In most cases the differences are small and did not affect the analysis. However the total 352 for Elementary × Suburban area for LOCAL COURTS implies about 25% ((464 − 352)/464) not reporting. This seems to have affected the analysis for LOCAL COURTS as compared with other community sources. We shall comment on this later.

For each of the community sources we denote the observed occurrences in Table 10 by $x(ijk)$.

To eliminate some of the 'random noise' from the original observations, and obtain a simple structural model relating the response, Support, with the explanatory variables, Level, Location, we fitted a loglinear model to the data for each of the community sources by MDIE. The models are derived by fitting certain marginals or combinations of marginals, that is, the estimates are constrained to have some set of marginals or combination of marginals equal to those of the original observed values for each community source. The model is not necessarily the same in detail for each community source. We present later in Section 21 details about the fitting procedure but at this point we shall consider some of the implications of the models.

We denote the estimated occurrences by $x^*(ijk)$ and list the estimated values for each community source in Table 11.

The loglinear model can be reformulated as a multiplicative model and the odds (YES/NO) expressed as a product of three factors: a base value factor relative to each community source, a factor depending on Level, and a factor depending on Location. For LOCAL COURTS there is also a factor for ELEM × SUBURBS interaction. The magnitudes of the factors are an indication of the relative importance of the various effects and interactions. The odds can also be obtained as the ratio of YES/NO in Table 11 but the gross odds give no indication of the relative importance of the component effects. The odds factors for each community source are given in Table 12.

We calculate from Table 12 that the odds (YES/NO) for community source, PARENTS, for ELEMENTARY, SMALL CITIES is the product $0.5886 \times 1.5878 \times 1.4993 = 1.4012$. From Table 11 we see that the corresponding ratio is $144.131/102.869 = 1.4011$. The odds (YES/NO) for community source, SCHOOL SYSTEM CENTRAL OFFICE, for JUNIOR HIGH, RURAL is the product $2.8867 \times 1.0000 \times 1.0000 = 2.8867$. From Table 11 we see that the corresponding ratio is $265.149/91.851 = 2.8867$. The odds (YES/NO) for community source, LOCAL COURTS, for ELEMENTARY, SUBURBS is the product $0.2204 \times 1.0000 \times 0.6659 \times 1.6019 = 0.2351$. From Table 11 we see that the corresponding ratio is $67.000/285.000 = 0.2351$.

For community source PARENTS we note by examining the odds factors that the best odds for support are for ELEMENTARY, SUBURB. The poorest odds are the same for SENIOR HIGH, LARGE CITIES and SENIOR HIGH, RURAL.

For community source LOCAL POLICE we note by examining the odds

Table 11
Estimated values based on appropriate loglinear models

COMMUNITY SOURCE		ELEMENTARY LARGE CITIES	SMALL CITIES	SUBURBS	RURAL	JUNIOR HIGH LARGE CITIES
1. PARENTS	YES	132.844	144.131	280.384	138.641	122.899
	NO	142.156	102.869	183.616	148.359	166.101
		275.000	247.000	464.000	287.000	289.000
2. LOCAL POLICE	YES	73.188	83.971	164.075	85.767	105.364
	NO	183.812	131.029	230.925	173.233	184.636
		257.000	215.000	395.000	259.000	290.000
3. LOCAL COURTS	YES	18.469	24.828	67.000	42.261	22.324
	NO	211.531	169.172	285.000	191.739	255.676
		230.000	194.000	352.000	234.000	278.000
4. SCHOOL BOARD	YES	55.559	97.810	233.674	125.957	79.938
	NO	189.441	115.190	170.326	76.043	203.062
		245.000	213.000	404.000	202.000	283.000
5. SCHOOL SYSTEM CENTRAL OFFICE	YES	76.836	125.444	262.480	188.240	98.960
	NO	179.164	103.556	155.520	81.760	184.040
		256.000	229.000	418.000	270.000	283.000

factors that the best odds for support are for SENIOR HIGH, SUBURBS. The poorest odds are for ELEMENTARY, LARGE CITIES.

For community source LOCAL COURTS we see from the odds factor table that the factors are the same for all levels. Small cities and Surburbs also have the same factor. Since $0.6659 \times 1.6019 = 1.0667$ we also have the alternative version of the odds factors in the two-way table. The best odds for support are for ELEM, SUBURBS and the poorest odds are for LARGE CITIES.

For community source SCHOOL BOARD we note by examining the odds factors that the best odds are the same for JUNIOR HIGH, RURAL and SENIOR HIGH, RURAL. The poorest odds are for ELEMENTARY, LARGE CITIES.

For community source SCHOOL SYSTEM CENTRAL OFFICE we note by examining the odds factors that the best odds are the same for JUNIOR HIGH, RURAL and SENIOR HIGH, RURAL. The poorest odds are for ELEMENTARY, LARGE CITIES.

We see from Table 12 that the support of Parents is apparently similarly stimulated for large cities and rural areas but less than for suburbs and small cities. This seems to be a surprising pattern for parents although the decreasing support from elementary to junior high to senior high seems understandable as the children grow older. The official and administrative support areas all show an increase from large cities to small cities to surburbs to rural areas except for

Table 11 (continued)

				SENIOR HIGH			
	SMALL CITIES	SUBURBS	RURAL	LARGE CITIES	SMALL CITIES	SUBURBS	RURAL
1.	146.207	315.802	153.092	110.037	117.662	285.814	124.487
	131.793	261.198	206.908	186.963	133.338	297.186	211.513
	278.000	577.000	360.000	297.000	251.000	583.000	336.000
2.	129.743	274.469	149.423	122.448	130.286	322.456	152.809
	141.257	269.531	210.577	176.552	116.714	260.544	177.191
	271.000	544.000	360.000	299.000	247.000	583.000	330.000
3.	34.042	67.572	62.308	23.207	31.611	72.947	57.432
	231.958	460.428	282.692	265.793	215.389	497.053	260.568
	266.000	528.000	345.000	289.000	247.000	570.000	318.000
4.	141.688	358.386	242.109	80.503	130.502	373.940	226.934
	124.312	194.614	108.891	204.497	114.498	203.060	102.066
	266.000	553.000	351.000	285.000	245.000	577.000	329.000
5.	164.617	382.327	265.149	104.205	148.939	393.193	243.610
	108.383	180.673	91.851	193.795	98.061	185.807	84.390
	273.000	563.000	357.000	298.000	247.000	579.000	328.000

local police, and local courts for elementary schools. Schools in large cities seem to lack the support of community sources.

We make no attempt to delve into the results in Table 12 from a Social or Sociological point of view, though there seem to be interesting implications.

21. Technical appendix

We summarize in this technical appendix the bases for our preceding analysis.

As an initial overview of the data we obtained for each community source an Analysis of Information based on fitting the marginals:

(a) $x(ij\cdot)$, $x(\cdot\cdot k)$

(b) $x(ij\cdot)$, $x(i\cdot k)$

(c) $x(ij\cdot)$, $x(i\cdot k)$, $x(\cdot jk)$

(d) $x(ij\cdot)$, $x(\cdot jk)$

As suggested by the Analysis of Information for each community source, we then obtained for each of the models x_c^*, that is, the model fitting all the two-way marginals, implying no second-order interaction, complete details including estimated values, the tau parameters and their covariance matrix.

An examination of this detail showed that some of the tau parameters were

Table 12
Odds Factors

BASE	LEVEL i		LOCATION j	
	PARENTS			
0.5886	1. ELEM	1.5878	1. LG. CITIES	1.0000
	2. JR. HI	1.2572	2. SM.CITIES	1.4993
	3. SR. HI	1.0000	3. SUBURBS	1.6341
			4. RURAL	1.0000
	LOCAL POLICE			
0.8624	1. ELEM.	0.5741	1. LG. CITIES	0.8042
	2. JR. HI	0.8228	2. SM. CITIES	1.2944
	3. SR. HI	1.0000	3. SUBURBS	1.4351
			4. RURAL	1.0000
	LOCAL COURTS			
0.2204	1. ELEM.	1.0000	1. LG. CITIES	0.3961
	2. JR. HI	1.0000	2. SM.CITIES	0.6659
	3. SR. HI	1.0000	3. SUBURBS	0.6659
			4. RURAL	1.000
	ELEM × SUBURBS	1.6019		

		j			
		1. LG. CITIES	2. SM. CITIES	3. SUBURBS	4. RURAL
	1. ELEM	0.3961	0.6659	1.0667	1.0000
0.2204	$i =$ 2. JR. HI	0.3961	0.6659	0.6659	1.0000
	3. SR. HI	0.3961	0.6659	0.6659	1.0000

	SCHOOL BOARD			
2.2234	1. ELEM	0.7450	1. LG. CITIES	0.1771
	2. JR. HI	1.0000	2. SM. CITIES	0.5126
	3. SR. HI	1.0000	3. SUBURBS	0.8282
			4. RURAL	1.0000
	SCHOOL SYSTEM CENTRAL OFFICE			
2.8867	1. ELEM	0.7976	1. LG. CITIES	0.1863
	2. JR. HI	1.0000	2. SM. CITIES	0.5261
	3. SR. HI	1.0000	3. SUBURBS	0.7331
			4. RURAL	1.0000

not significantly different from zero or from each other. For the community source LOCAL COURTS second-order interaction also seems to be present. We summarize these results in Table 13. (Note that we have used the indices as superscripts to represent variables.)

The estimates listed in Table 11 and the values in Table 12 were obtained by

Table 13

Community Source	Values of tau	Standardized Value
PARENTS	$\tau^{jk}_{11} = 0.109957$	1.1592
LOCAL COURTS	$\tau^{ik}_{11} = 0.181121$	1.5404
	$\tau^{ik}_{21} \doteq -0.081328$	−0.7238
	$\tau^{jk}_{21} - \tau^{jk}_{31} = 0.040872$	0.3090
	OUTLIER Cell $(ij) = (13)$	OUTLIER 131 = 6.289
SCHOOL BOARD	$\tau^{ik}_{21} = -0.93768$	−1.1905
SCHOOL SYSTEM CENTRAL OFFICE	$\tau^{jk}_{21} = -0.056957$	−0.7180

Table 14

Community Source	Information	D.F.
PARENTS	$2I(x:x^*) = 3.901$	7
	$2I(x^*_c:x^*) = 1.345$	1
	$2I(x:x^*_c) = 2.556$	6
Note that $(1.1592)^2 = 1.344$		
LOCAL COURTS Component due to	**Information**	**D.F.**
$x(ij\cdot)$, $(x\cdot 11)$, $x(\cdot 21)+x(\cdot 31)$	$2I(x:x^*_f) = 20.892$	9
	$2I(x^*_e:x^*_f) = 0.094$	1
$x(ij\cdot)$, $x(\cdot jk)$	$2I(x:x^*_e) = 20.798$	8
	$2I(x^*_c:x^*_e) = 4.949$	2
$x(ij\cdot)$, $x(\cdot jk)$, $x(i\cdot k)$	$2I(x:x^*_c) = 15.849$	6
$x(ij\cdot)$, $x(\cdot 11)$, $x(\cdot 21)+x(\cdot 31)$	$2I(x:x^*_f) = 20.892$	9
	$2I(x^*:x^*_f) = 8.939$	1
$x(ij\cdot)$, $x(\cdot 11)$, $x(\cdot 21)+x(\cdot 31)$, $x(131)$	$2I(x:x^*) = 11.953$	8
SCHOOL BOARD	$2I(x:x^*) = 4.665$	7
	$2I(x^*_c:x^*) = 1.417$	1
Note that $(-1.1905)^2 = 1.417$	$2I(x:x^*_c) = 3.248$	6
SCHOOL SYSTEM CENTRAL OFFICE	$2I(x:x^*) = 5.769$	7
	$2I(x^*_c:x^*) = 0.515$	1
	$2I(x:x^*_c) = 5.254$	6
Note that $(-0.7180)^2 = 0.516$		

rerunning the data with the tau parameters as above set equal to zero, or to each other. For LOCAL COURTS the constraint $x^*(131) = x(131)$ was also used.

We give in Table 14 the Analysis of Information values comparing the x_c^* models with the final x^* models all of which fit their respective data sets well.

22. Example 3: Coronary heart disease

The data set for this example originates from a study on the incidence of coronary heart disease (CHD) done by the Medical Bureau for Occupational Diseases in Johannesburg. We are grateful to the director of the bureau, Dr. F. J. Wiles for permission to use the data, and to Dr. T. J. Hastie and Dr. June Juritz for making the data available to us.

1. The data

In a sample of 2012 miners studied by the Medical Bureau for Occupational Diseases of the Chamber of Mines, there were 108 who suffered from coronary heart disease (CHD). The problem was to relate the occurrence of coronary heart disease on the factors serum cholesterol, systolic blood pressure, and smoking habits. The cross-classification of the observed data is represented by a $3 \times 3 \times 3 \times 2$ contingency table $x(jklm)$ where the values of the cell occurrences $x(jklm)$ are given in lexicographic order in Table 15. Note that the ages of the miners was not furnished with the data.

2. The analysis

In order to obtain a first overview of the possible relationship of the dependent variable Coronary Heart Disease (D) on the explanatory variables Serum Cholesterol (C), Systolic Blood Pressure (P), Smoking Habits (H), a sequence of nested marginals was fitted using the Deming–Stephan algorithm or iterative proportional fitting procedure (Gokhale and Kullback, 1978a, pp. 214–216; Ku and Kullback, 1974, p. 116). Summary results for the initial model x_a^* and the set of marginals selected as a potential final model x_b^* are shown in the Analysis of Information Table 16.

The log-odds (logit) representation for the estimate x_b^* is

$$\ln(x_b^*(jkl1)/x_b^*(jkl2)) = \tau_1^D + \tau_{j1}^{CD} + \tau_{k1}^{PD} + \tau_{jk1}^{CPD} + \tau_{l1}^{HD} . \qquad (1)$$

We recall that any parameter with subscript $m = 2$, and/or $j = 3$, and/or $k = 3$, and/or $l = 3$, is zero by convention. The values of the eleven parameters in (1) are listed in Table 17. Also given in Table 17 are the standardized values of the parameters, that is, the ratio of each tau to its standard deviation.

Although the statistics indicated that x_b^* was a good fit to the observed data there was an anomalous behavior of the odds of No CHD to CHD across levels

Table 15
Observed and estimated occurrences

$jklm$	x	x_c^*	$jklm$	x	x_c^*	$jklm$	x	x_c^*
1111	39	38.782	2111	31	31.288	3111	33	33.243
1112	0	0.218	2112	1	0.712	3112	1	0.757
1121	75	74.569	2121	78	73.310	3121	66	67.742
1122	1	1.431	2122	1	5.690	3122	7	5.258
1131	224	224.688	2131	178	172.604	3131	133	137.341
1132	5	4.312	2132	8	13.396	3132	15	10.659
1211	40	39.176	2211	35	35.199	3211	36	36.176
1212	0	0.824	2212	1	0.801	3212	1	0.824
1221	47	46.656	2221	74	69.598	3221	78	75.166
1222	3	3.344	2222	1	5.402	3222	3	5.834
1231	136	137.168	2231	127	126.205	3231	125	128.061
1232	11	9.832	2232	9	9.795	3232	13	9.939
1311	11	10.938	2311	15	14.666	3311	20	20.532
1312	0	0.062	2312	0	0.334	3312	1	0.468
1321	28	28.454	2321	32	32.479	3321	29	31.551
1322	1	0.546	2322	3	2.521	3322	5	2.449
1331	75	74.569	2331	66	66.814	3331	73	77.022
1332	1	1.431	2332	6	5.186	3332	10	5.978

Characteristic		Index	1	2	3
Serum Cholesterol	C	j	<220	220–260	>260
Blood Pressure	P	k	<130	130–155	>155
Smoking Habits	H	l	Nonsmoker	Ex-smoker	Current smoker
CHD	D	m	No	Yes	

Table 16
Analysis of information

Component due to	Information	D.F.
(a) $x(jkl\cdot), x(\cdots m)$	$2I(x:x_a^*) = 58.627$	26
	$2I(x_b^*:x_a^*) = 47.953$	10
(b) $x(jkl\cdot), x(jk\cdot m), x(\cdot\cdot lm)$	$2I(x:x_b^*) = 10.674$	16
(a) $x(jkl\cdot), x(\cdots m)$	$2I(x:x_a^*) = 58.627$	26
	$2I(x_c^*:x_a^*) = 29.501$	3
(c) $x(jkl\cdot), x(\cdots 1)$, $x(1\cdot\cdot 1), x(12\cdot 1), x(\cdot\cdot 11)$	$2I(x:x_c^*) = 29.126$	23
	$2I(x_b^*:x_c^*) = 18.452$	7
(b) $x(jkl\cdot), x(jk\cdot m), x(\cdot\cdot lm)$	$2I(x:x_b^*) = 10.674$	16

of the characteristics. These odds did not show the monotonic behavior one would expect. An examination of the standardized values in Table 17 seemed to imply that only four of the parameters were significantly different from zero, that is, τ_1^D, τ_{11}^{CD}, τ_{121}^{CPD}, τ_{11}^{HD}. The nonhierarchical parsimonious model x_c^* was obtained by fitting the marginals x_c^*: $x(jkl\cdot)$, $x(\cdots 1)$, $x(1\cdot\cdot 1)$, $x(12\cdot 1)$,

Table 17
Parameters of x_b^*

Tau	Standardized	Tau	Standardized
(1) $\tau_1^D = 1.803170$	6.5634	(6) $\tau_{111}^{CPD} = -0.302631$	−0.3385
(2) $\tau_{11}^{CD} = 2.074424$	2.7205	(7) $\tau_{121}^{CPD} = -1.929345$	−2.2711
(3) $\tau_{21}^{CD} = 0.515967$	1.1759	(8) $\tau_{211}^{CPD} = 0.567675$	0.9670
(4) $\tau_{11}^{PD} = 0.285422$	0.8242	(9) $\tau_{221}^{CPD} = -0.086671$	−0.1462
(5) $\tau_{21}^{PD} = 0.596931$	1.6228	(10) $\tau_{11}^{HD} = 1.369462$	2.9240
		(11) $\tau_{21}^{HD} = 0.389440$	1.6321

$x(\cdot\cdot 11)$. The estimate x_c^* has the log-odds (logit) representation

$$\ln(x_c^*(jkl1)/x_c^*(jkl2)) = \tau_1^D + \tau_{11}^{CD} + \tau_{121}^{CPD} + \tau_{11}^{HD}. \tag{2}$$

We consider the data as 27 binomials of the binary variable CHD (D). To use the Newton–Raphson type k-samples iterative algorithm (Gokhale and Kullback, 1978a, pp. 199–205, 211–212, 245) the appropriate 31×54 B-design-matrix is set up as follows (cf. Gokhale and Kullback, 1978b, 1002):

(a) The 54 columns correspond respectively to the 54 cells in lexicographic order as in Table 15;

(b) Rows 1 to 27 each contain two ones, one each respectively in the columns corresponding to the cells $jkl1$ and $jkl2$ and zeros elsewhere for $j = 1, 2, 3$, $k = 1, 2, 3$, $l = 1, 2, 3$, in lexicographic order;

(c) Row 28 has a one in every column in which $m = 1$ and zeros elsewhere;

(d) Row 29 has a one in every column in which $j = 1$ and $m = 1$, and zeros elsewhere;

(e) Row 30 has a one in every column in which $j = 1$ and $k = 2$ and $m = 1$, and zeros elsewhere;

(f) Row 31 has a one in every column in which $l = 1$ and $m = 1$, and zeros elsewhere. The first 27 rows of the matrix correspond to the constraints $x_c^*(jkl\cdot) = x(jkl\cdot)$, and the next four rows of the matrix correspond to the respecitve constraints

$$x_c^*(\cdots 1) = x(\cdots 1), \quad \tau_1^D; \qquad x_c^*(1\cdot\cdot 1) = x(1\cdot\cdot 1), \quad \tau_{11}^{CD};$$

$$x_c^*(12\cdot 1) = x(12\cdot 1), \quad \tau_{121}^{CPD}; \qquad x_c^*(\cdot\cdot 11) = x(\cdot\cdot 11), \quad \tau_{11}^{HD}.$$

The values of the MDI estimate x_c^* are also listed in Table 15. In accordance with the entries in Analysis of Information Table 16 we note that $2I(x:x_c^*) = 29.126$ with 23 D.F. implies that x_c^* is a good fit to the observed data (the 0.1 significance level of the tabulated chi-square for 23 D.F. is 32.0). The values of the parameters for x_c^* are listed in Table 18 and the covariance matrix of these parameters is given in Table 19.

Table 18
Parameters of x_c^*

Tau	Standardized	Exp(tau)
(1) $\tau_1^D = 2.556049$	22.4290	12.8848
(2) $\tau_{11}^{CD} = 1.397269$	3.7359	4.0441
(3) $\tau_{121}^{CPD} = -1.317753$	−2.9188	0.2677
(4) $\tau_{11}^{HD} = 1.226255$	2.6453	3.4084

Table 19
Covariance matrix x_c^*

	τ_1^D (1)	τ_{11}^{CD} (2)	τ_{121}^{CPD} (3)	τ_{11}^{HD} (4)
(1)	0.012987	−0.012625	0.000267	−0.010232
(2)		0.139883	−0.127419	0.002613
(3)			0.203824	−0.005616
(4)				0.214888

The log-odds representation in (2) can be expressed as the multiplicative representation of the odds

$$x_c^*(jkl1)/x_c^*(jkl2) = \exp(\tau_1^D)\exp(\tau_{11}^{CD})\exp(\tau_{121}^{CPD})\exp(\tau_{11}^{HD}) . \tag{3}$$

The odds factors are listed in Table 20 in which for convenience, because of the interaction we have combined the factors $\exp(\tau_{j1}^{CD})\exp(\tau_{jk1}^{CPD})$. Thus the best

Table 20
Odds factors $x_c^*(jkl1)/x_c^*(jkl2)$

BASE			CHOL × SBP			SHAB	
			k = 1: <130	k = 2: 130–155	k = 3: >155		
	j = 1:	<220	4.0441	1.0828	1.0000	l = 1	3.4084
12.8848	j = 2:	220–260	1.0000	1.0000	1.0000	l = 2	1.0000
	j = 3:	>260	1.0000	1.0000	1.0000	l = 3	1.0000

Table 21
ODDS $x_c^*(jkl1)/x_c^*(jkl2)$ No CHD/CHD

	SYSTOLIC BLOOD PRESSURE								
	k = 1			k = 2			k = 3		
SMOKING	CHOLESTEROL			CHOLESTEROL			CHOLESTEROL		
HABIT	j = 1	j = 2	j = 3	j = 1	j = 2	j = 3	j = 1	j = 2	j = 3
l = 1	177.60	43.92	43.92	47.55	43.92	43.92	43.92	43.92	43.92
l = 2	52.11	12.88	12.88	13.95	12.88	12.88	12.88	12.88	12.88
l = 3	52.11	12.88	12.88	13.95	12.88	12.88	12.88	12.88	12.88

odds for No CHD to CHD correspond to $x_c^*(1111)/x_c^*(1112) = 177.60$ to 1; the next best to $x_c^*(11l1)/x_c^*(11l2) = 52.11$ to 1, $l = 2, 3$. The smallest odds correspond to $x_c^*(jkl1)/x_c^*(jkl2) = 12.88$ to 1, $j = 2, 3$, $k = 2, 3$, $l = 2, 3$. The odds in the original data $x(\cdots 1)/x(\cdots 2) = 1904/108 = 17.63$ to 1. Other things being equal the odds of No CHD to CHD for a nonsmoker are 3.4 times those for an ex-smoker or current smoker and the odds for the latter two are equal.

The odds of No CHD to CHD, $x_c^*(jkl1)/x_c^*(jkl2)$, are given in Table 21. We note that the odds now show a monotonic progression across the levels of the explanatory variables.

23. Example 4: Driver records

1. Introduction

In an illuminating paper Fuchs (1979) presented an example of real data in which biased inferences about the relationship between two variables, while controlling for the effect of one or several covariables, may result from the use of insufficient covariables in the analysis. Fuchs (1979) notes: "One of the basic assumptions in testing the average partial association is that the investigator is aware of the relevant covariables that may influence the response profiles of the dependent variable within the subpopulations". Fuchs (1979) cites tests proposed by Cochran (1954), Mantel and Haenszel (1959), Hopkins and Gross (1971), Sugiura and Otake (1974), and Landis, Heyman and Koch (1977) for testing the average partial association between the dependent variable and the subpopulations.

Fuchs (1979) computed average partial associations by collapsing the original table over various combinations of the covariables. He noted discrepancies in the assessments of the average partial association when different covariables are used. We shall use the same data as Fuchs (1979) and find a suitable loglinear model fitting the data. We shall compute the average partial association using the fitted model.

2. The data

As indicated by Fuchs (1979) two groups of drivers (D) are compared in the analysis. One group is a simple random sample of the entire population of drivers in Wisconsin (Control). The other group includes drivers with known cardiovascular deficiencies (Cardiovascular) and was obtained by subdividing, according to the type of existing condition, a simple random sample of drivers with several medical conditions. The dependent variable is the number of traffic violations (V) within a one-year period (1974). For each driver, the available data also include information on the age interval (A), sex (S), and place of residence (R). The cross-classification of the observed data is represented by a $2 \times 5 \times 3 \times 2 \times 2$ contingency table $x(ghijk)$. The values of the cell occurrences $x(ghijk)$ are given in Table 22. *In this analysis it is important to assess which group of drivers has a better record.*

Table 22
Observed and estimated occurrences

ghijk	*x*	*x**	*ghijk*	*x*	*x**	*ghijk*	*x*	*x**	*ghijk*	*x*	*x**
11111	3	2.976	14111	8	7.092	22111	64	63.601	25111	53	56.029
11112	2	2.024	14112	1	1.908	22112	20	20.399	25112	21	17.971
11121	6	4.987	14121	5	4.628	22121	58	57.497	25121	87	83.963
11122	0	1.013	14122	0	0.372	22122	5	5.503	25122	5	8.037
11211	42	40.162	14211	66	65.640	22211	42	39.444	25211	40	39.444
11212	8	9.838	14212	6	6.360	22212	2	4.556	25212	4	4.556
11221	10	11.183	14221	12	12.635	22221	36	38.667	25221	47	46.401
11222	2	0.817	14222	1	0.365	22222	4	1.333	25222	1	1.599
11311	113	112.725	14311	193	193.926	22311	23	22.470	25311	54	51.495
11312	16	16.275	14312	12	11.074	22312	1	1.530	25312	1	3.505
11321	22	23.967	14321	54	54.078	22321	20	19.602	25321	24	23.522
11322	3	1.033	14322	1	0.922	22322	0	0.398	25322	0	0.478
12111	9	8.870	15111	5	4.812	23111	31	31.800			
12112	2	2.130	15112	2	2.188	23112	11	10.200			
12121	3	3.733	15121	3	2.642	23121	62	62.060			
12122	1	0.267	15122	0	0.358	23122	6	5.940			
12211	37	37.737	15211	58	59.292	23211	36	34.066			
12212	4	3.263	15212	11	9.708	23212	2	3.934			
12221	17	16.572	15221	13	12.394	23221	37	37.701			
12222	0	0.428	15222	0	0.606	23222	2	1.299			
12311	127	126.552	15311	142	140.447	23311	18	18.725			
12312	6	6.448	15312	12	13.553	23312	2	1.275			
12321	31	30.536	15321	19	20.412	23321	15	14.701			
12322	0	0.464	15322	2	0.588	23322	0	0.299			
13111	2	2.011	21111	58	57.543	24111	69	70.415			
13112	1	0.989	21112	18	18.457	24112	24	22.585			
13121	2	1.744	21121	60	59.322	24121	68	70.274			
13122	0	0.256	21122	5	5.678	24122	9	6.726			
13211	42	45.031	21211	39	35.858	24211	60	61.856			
13212	11	7.969	21212	1	4.142	24212	9	7.144			
13221	13	12.348	21221	41	40.601	24221	58	58.969			
13222	0	0.652	21222	1	1.399	24222	3	2.032			
13311	91	89.650	21311	26	28.088	24311	52	55.240			
13312	8	9.350	21312	4	1.912	24312	7	3.760			
13321	26	25.215	21321	23	22.542	24321	45	44.104			
13322	0	0.785	21322	0	0.458	24322	0	0.896			

Characteristic		Index	Level				
			1	2	3	4	5
Driver Group	*D*	*g*	Cardiovascular	Control			
Residence[a]	*R*	*h*	Urban 1	Urban 2	Urban 3	Urban 4	Rural
Age[b]	*A*	*i*	16–35	36–55	≥56		
Sex	*S*	*j*	Male	Female			
Violations	*V*	*k*	0	≥1			

[a]Urban 1: ≥150 000 Habitants, Urban 2: 39 000–149 999, Urban 3: 10 000–38 999 Urban 4: <10 000.

[b]16–35: Birth years 1938–57, 36–55: Birth years 1918–37, ≥56: Birth years ≤1917.

3. The analysis

In order to obtain a first overview of the possible relationship of the dependent variable Violations (V) on the explanatory variables Driver Group (D), Residence (R), Age (A), Sex (S), a sequence of nested marginals was fitted using the Deming–Stephan algorithm or iterative proportional fitting procedure (Gokhale and Kullback, 1978a, pp. 214–216; Ku and Kullback, 1974, p. 116). The fitting constrains the estimated tables denoted by x_a^*, x_b^*, etc., to have the same marginal values, for the fitted set, as does the observed table. Summary results for the first six sets of fitted marginals are shown in the Analysis of Information Table 23. The information numbers for the subsequent sets of fitted marginals are not included since they implied no further significant interactions. We remark that information numbers of the form

$$2I(x:x_a^*) = 2\sum x(ghijk)\ln(x(ghijk)/x_a^*(ghijk))$$

and

$$2I(x_b^*:x_a^*) = 2\sum x_b^*(ghijk)\ln(x_b^*(ghijk)/x_a^*(ghijk))$$

Table 23
Analysis of information

Component due to	Information	D.F.
(a) $x(ghij\cdot), x(\cdot\cdot\cdot\cdot k)$	$2I(x:x_a^*) = 190.211$	59
	$2I(x_b^*:x_a^*) = 4.439$	1
(b) $x(ghij\cdot), x(g\cdots k)$	$2I(x:x_b^*) = 185.772$	58
	$2I(x_c^*:x_b^*) = 3.031$	4
(c) $x(ghij\cdot), x(g\cdots k), x(\cdot h\cdot\cdot k)$	$2I(x:x_c^*) = 182.741$	54
	$2I(x_d^*:x_c^*) = 60.165$	2
(d) $x(ghij\cdot), x(g\cdots k), x(\cdot h\cdot\cdot k), x(\cdot\cdot i\cdot k)$	$2I(x:x_d^*) = 122.576$	52
	$2I(x_e^*:x_d^*) = 61.894$	1
(e) $x(ghij\cdot), x(g\cdots k), x(\cdot h\cdot\cdot k), x(\cdot\cdot i\cdot k)$, $x(\cdots jk)$	$2I(x:x_e^*) = 60.682$	51
	$2I(x_f^*:x_e^*) = 13.811$	4
(f) $x(ghij\cdot), x(gh\cdot\cdot k), x(\cdot\cdot i\cdot k), x(\cdots jk)$	$2I(x:x_f^*) = 46.871$	47

Component due to	Information	D.F.
(a) $x(ghij\cdot), x(\cdot\cdot\cdot\cdot k)$	$2I(x:x_a^*) = 190.211$	59
	$2I(x^*:x_a^*) = 140.198$	8
(*) $x(ghij\cdot), x(\cdot\cdot\cdot\cdot 1), x(1\cdots 1), x(\cdot\cdot 1\cdot 1)$, $x(\cdot\cdot 2\cdot 1), x(\cdots 11), x(11\cdot\cdot 1), x(12\cdot\cdot 1)$, $x(13\cdot\cdot), x(14\cdot\cdot 1)$	$2I(x:x^*) = 50.013$	51

Information	D.F.
$x(g\cdots k)$: $2I(x_b^*:x_a^*) = 4.439$	1
$x(\cdot\cdot i\cdot k)$: $2I(x_d^*:x_c^*) = 60.165$	2
$x(\cdots jk)$: $2I(x_e^*:x_d^*) = 61.894$	1
$x(gh\cdot\cdot k)$: $2I(x_f^*:x_e^*) = 13.811$	4

for this case are the same as log-likelihood-ratio statistics. Also the MDI estimate $x_a^*(ghijk)$ has the explicit representation in terms of the marginals as $x_a^*(ghijk) = x(ghij\,\cdot)x(\cdot\cdot\cdot\cdot\,k)/N$, $N = x(\cdot\cdot\cdot\cdot\cdot)$ (Bishop et al., 1975; Gokhale and Kullback, 1978a, 1978b). In Table 23 the value $2I(x:x_a^*) = 190.211$, 59 D.F., implies that the binary variable Violations (V) is not homogeneous over the 60 binomials of the combinations of the explanatory variables (Kullback, 1959, Chapter 8) and we seek a model to account for the behavior. The conditional effects in Table 23, last part, suggest that main effects (two-factor interactions) of Driver Group (D), Age (A), Sex (S), on Violations (V) are significant and that the interaction Driver Group×Residence on Violations (three-factor interactions) is significant. Note that the results suggest no Residence (R) main effect but a Driver Group×Residence interaction.

To obtain the MDI estimates of the nonhierarchical parsimonious loglinear model implied by the preceding analysis, its parameters, and their covariance matrix, we consider the data as 60 binomials of the binary variable Violations (V). If we denote the MDI estimate satisfying the suggested constraints by $x^*(ghijk)$, then we have the log-odds (logit) representation for $x^*: x(ghij\,\cdot)$, $x(g\cdot\cdot\cdot\,k)$, $x(\cdot\cdot\,i\,\cdot\,k)$, $x(\cdot\cdot\cdot\,jk)$, $x(gh\cdot\cdot\,k)$,

$$\ln(x^*(ghij1)/x^*(ghij2)) = \tau_1^V + \tau_{g1}^{DV} - \tau_{i1}^{AV} + \tau_{j1}^{SV} + \tau_{gh1}^{DRV}\,. \tag{1}$$

Since by convention, to insure linear independence, any parameter with a subscript corresponding to a last level or category index value is zero, there are nine parameters to be determined:

$$\tau_1^V;\qquad \tau_{g1}^{DV} \text{ for } g=1;\qquad \tau_{i1}^{AV} \text{ for } i=1,2;$$
$$\tau_{j1}^{SV} \text{ for } j=1;\qquad \tau_{gh1}^{DRV} \text{ for } g=1 \text{ and } h=1,2,3,4\,.$$

To use the Newton–Raphson type k-Samples iterative algorithm (Gokhale and Kullback, 1978a, pp. 199–205, 211–212, 245) the appropriate 69×120 B-design-matrix is set up as follows (cf. Gokhale and Kullback, 1978b, p. 1002):

(a) The 120 columns correspond respectively to the 120 cells in lexicographic order as in Table 22;

(b) Rows 1 to 60 each contain two ones, one each respectively in the columns corresponding to the cells $ghij1$ and $ghij2$ and zeros elsewhere for $g = 1, 2$, $h = 1, 2, 3, 4, 5$, $i = 1, 2, 3$, $j = 1, 2$, in lexicographic order;

(c) Row 61 has a one in every column in which $k = 1$, and zeros elsewhere;

(d) Row 62 has a one in every column in which $g = 1$ and $k = 1$, and zeros elsewhere;

(e) Row 63 has a one in every column in which $i = 1$ and $k = 1$, and zeros elsewhere;

(f) Row 64 has a one in every column in which $i = 2$ and $k = 1$, and zeros elsewhere;

(g) Row 65 has a one in every column in which $j = 1$ and $k = 1$, and zeros elsewhere;

(h) Row 66 has a one in every column in which $g = 1$ and $h = 1$ and $k = 1$, and zeros elsewhere;

(i) Row 67 has a one in every column in which $g = 1$ and $h = 2$ and $k = 1$, and zeros elsewhere;

(j) Row 68 has a one in every column in which $g = 1$ and $h = 3$ and $k = 1$, and zeros elsewhere;

(k) Row 69 has a one in every column in which $g = 1$ and $h = 4$ and $k = 1$, and zeros elsewhere.

The first 60 rows of the matrix correspond to the constraints $x^*(ghij\cdot) = x(ghij\cdot)$ and the next nine rows of the matrix correspond to the respective constraints

$$\begin{array}{llll} x^*(\cdot\cdot\cdot\cdot 1) = x(\cdot\cdot\cdot\cdot 1), & \tau_1^V; & x^*(1\cdot\cdot\cdot 1) = x(1\cdot\cdot\cdot 1), & \tau_{11}^{DV}, \\ x^*(\cdot\cdot 1\cdot 1) = x(\cdot\cdot 1\cdot 1), & \tau_{11}^{AV}; & x^*(\cdot\cdot 2\cdot 1) = x(\cdot\cdot 2\cdot 1), & \tau_{21}^{AV}, \\ x^*(\cdot\cdot\cdot 11) = x(\cdot\cdot\cdot 11), & \tau_{11}^{SV}; & x^*(11\cdot\cdot 1) = x(11\cdot\cdot 1), & \tau_{111}^{DRV}, \\ x^*(12\cdot\cdot 1) = x(12\cdot\cdot 1), & \tau_{121}^{DRV}; & x^*(13\cdot\cdot 1) = x(13\cdot\cdot 1), & \tau_{131}^{DRV}, \\ x^*(14\cdot\cdot 1) = x(14\cdot\cdot 1), & \tau_{141}^{DRV}. & & \end{array}$$

The MDI estimate $x^*(ghijk)$ is a good fit to the original table since $2I(x:x^*) = 50.013$, 51 D.F. There is a summary analysis of information in Table 23. The values of the estimated cell entries $x^*(ghijk)$ are listed in Table 22. The values of the parameters in the log-odds representation (1) are given in Table 24, and their covariance matrix in Table 25. It is worth remarking that the use of the Newton–Raphson type algorithm with appropriate design matrix enables us to estimate a nonhierarchical loglinear model with no Residence (R) main effects but a Driver Group $\times$ Residence interaction.

The object of this analysis is to assess which group of drivers has a better record. We see from (1) that the difference of the log-odds for the two driver groups, or the partial association, is

$$\begin{aligned} \beta h &= \ln(x^*(1hij1)/x^*(1hij2)) - \ln(x^*(2hij1)/x^*(2hij2)) \\ &= \ln \frac{x^*(1hij1)x^*(2hij2)}{x^*(1hij2)x^*(2hij1)} = \tau_{11}^{DV} + \tau_{1h1}^{DRV}, \quad h = 1, 2, 3, 4, 5\,. \end{aligned} \tag{2}$$

We recall that parameters with subscript $g = 2$ are zero. Note that for the model x^* the partial association between Driver Group (D) and Violations (V) is related to the Residence (R) categories but not with Age (A) or Sex (S).

Using the values of the parameters in Table 24 we compute the values of the partial associations. Their variances may be obtained from Table 25, using the fact that

$$\mathrm{Var}(\beta_h) = \mathrm{Var}(\tau_{11}^{DV}) + \mathrm{Var}(\tau_{1h1}^{DRV}) + 2\,\mathrm{Cov}(\tau_{11}^{DV}, \tau_{1h1}^{DRV})\,. \tag{3}$$

It is found that

Table 24
Parameters of x^*

(1) $\tau_1^V = 3.896484$	(6) $\tau_{111}^{DRV} = -0.402919$
(2) $\tau_{11}^{DV} = -0.348965$	(7) $\tau_{121}^{DRV} = 0.638561$
(3) $\tau_{11}^{AV} = -1.550100$	(8) $\tau_{131}^{DRV} = -0.077727$
(4) $\tau_{21}^{AV} = -0.528708$	(9) $\tau_{141}^{DRV} = 0.524661$
(5) $\tau_{11}^{SV} = -1.209265$	

Table 25
Covariance matrix

	(1)	(2)	(3)	(4)	(5)	(6)	(7)	(8)	(9)
(1)	0.051353	−0.019399	−0.033689	−0.021332	−0.022729	−0.001708	0.000265	−0.000595	−0.001123
(2)		0.061672	0.020198	0.006695	−0.004367	−0.042175	−0.043543	−0.042155	−0.042318
(3)			0.039323	0.020004	0.003122	0.000890	−0.001505	0.000219	0.000528
(4)				0.032236	0.001014	0.001371	0.001232	−0.000906	0.001567
(5)					0.028218	0.001116	0.000723	0.000877	0.000595
(6)						0.080928	0.042335	0.042246	0.042340
(7)							0.125993	0.042178	0.042368
(8)								0.099454	0.042212
(9)									0.093941

$$\begin{aligned}
\beta_1 &= -0.751884, & \mathrm{Var}(\beta_1) &= 0.058250\,, \\
\beta_2 &= 0.289596, & \mathrm{Var}(\beta_2) &= 0.100579\,, \\
\beta_3 &= -0.426692, & \mathrm{Var}(\beta_3) &= 0.076816\,, \\
\beta_4 &= 0.175696, & \mathrm{Var}(\beta_4) &= 0.070977\,, \\
\beta_5 &= -0.348965, & \mathrm{Var}(\beta_5) &= 0.061672\,.
\end{aligned} \tag{4}$$

Using the Bonferroni inequality (Miller, 1966, p. 8) we can obtain joint 95% confidence intervals for the β's in (4) as

$$\beta_h \pm 2.575(\mathrm{Var}(\beta_h))^{1/2}\,. \tag{5}$$

We compute the confidence intervals for the β's as

$$\begin{aligned}
&\beta_1:\ -1.373361,\ -0.130407, \qquad && \beta_4:\ -0.510323,\ 0.861715\,, \\
&\beta_2:\ -0.527044,\ 1.105944, && \beta_5:\ -0.988437,\ 0.290507\,. \\
&\beta_3:\ -1.140371,\ 0.286987,
\end{aligned} \tag{6}$$

From the results in (6) we may infer that the Control Group has a better driving record in Residence category Urban 1 but that there is no significant difference in driving records for the other Residence categories. The average partial association between Driver Group and Violations is

$$\beta = (\beta_1 + \beta_2 + \beta_3 + \beta_4 + \beta_5)/5 = \tau_{11}^{DV} + (\tau_{111}^{DRV} + \tau_{121}^{DRV} + \tau_{131}^{DRV} + \tau_{141}^{DRV})/5\,. \tag{7}$$

Using the values of the β's in (4) we compute $\beta = -0.212450$. The variance of β is given by

$$\begin{aligned}
\mathrm{Var}(\beta) = {} & \mathrm{Var}(\tau_{11}^{DV}) + \tfrac{1}{25}\sum_{h=1}^{4} \mathrm{Var}(\tau_{1h1}^{DRV}) + \tfrac{2}{5}\sum_{h=1}^{4} \mathrm{Cov}(\tau_{11}^{DV}, \tau_{1h1}^{DRV}) \\
& + \tfrac{2}{25}\sum_{m=1}^{4}\sum_{n=1}^{4} \mathrm{Cov}(\tau_{1m1}^{DRV}, \tau_{1n1}^{DRV}), \quad m \neq n\,.
\end{aligned} \tag{8}$$

Using the values of the variances and covariances in Table 25 we compute $\mathrm{Var}(\beta) = 0.029903$. The 95% confidence interval for β is

$$\beta:\ -0.551383,\ 0.126483\,.$$

We may conclude that there is no significant average difference in the driving records of the two Driver Groups.

We note that Fuchs (1979, p. 123) computed an average partial association and a 95% confidence interval as −0.21, and −0.55, 0.12, controlled on Sex (S) and Age (A), and −0.21, and −0.54, 0.11, controlled on Age (A), Sex (S), and Residence (R). We have changed the sign as given by Fuchs for comparison since he considered the odds of Violations $(V) \geqq 1$ to 0.

We remark that the log-odds representation in (1) can be expressed as the multiplicative representation of the odds

$$x^*(ghij1)/x^*(ghij2) = \exp(\tau_1^V)\exp(\tau_{g1}^{DV})\exp(\tau_{i1}^{AV})\exp(\tau_{j1}^{SV})\exp(\tau_{gh1}^{DRV}). \tag{9}$$

We summarize (9) in the odds factors Table 26 in which we have combined $\exp(\tau_{g1}^{DV})\exp(\tau_{gh1}^{DRV})$ because of the interaction.

From Table 26 we see that the largest odds of Violations (0 to ≥1) occur for the cell $ghij = 1232$, that is, Cardiovascular Group in Urban 2 for Females of Age ≥56. The odds are $49.2291 \times 1.3359 = 65.7652$. The smallest odds of Violations (0 to ≥1) occur for the cell $ghij = 1111$, that is, Cardiovascular Group in Urban 1 for Males of Age 16–35. The odds are $49.2291 \times 0.4715 \times 0.2122 \times 0.2984 = 1.4698$.

Table 26
Odds factors

Base		Driver Group × Residence					Age	Sex
		$h=1$	$h=2$	$h=3$	$h=4$	$h=5$		
							$i=1$ 0.2122	$j=1$ 0.2984
49.2291	$g=1$	0.4715	1.3359	0.6526	1.1921	0.7054	$i=2$ 0.5894	$j=2$ 1.0000
	$g=2$	1.0000	1.0000	1.0000	1.0000	1.0000	$i=3$ 1.0000	

Acknowledgements

We are grateful to Dr. Carlyle E. Maw for his interest, support, and making available data tapes provided in conjunction with Violent Schools – Safe Schools, The Safe School Study Report to the Congress, U.S. Department of HEW, National Institute of Education. Some of the examples are based on these data tapes. The work for Section 19 was supported under grant NIE-9-76-0091 which we gratefully acknowledge.

Computations for the examples were performed using programs on file at the Computer Center of The George Washington University.

Information regarding the procurement of these computer programs and appropriate instructions may be obtained by writing to Dr. J. C. Keegel, Department of Statistics, The George Washington University, Washington D.C. 20052, U.S.A.

References

Agresti, A. (1977). Considerations in measuring partial association for ordinal categorical data. *J. Amer. Statist. Assoc.* **72** (357), 37–45.

Bishop, Y. M. M., Fienberg, S. E. and Holland, P. W. (1975). *Discrete Multivariate Analysis.* MIT Press, Cambridge, MA.

Cochran, W. G. (1954). Some methods for strenghtening the common χ^2 test. *Biometrics* **10**, 417–451.

Fienberg, S. E. (1977). *The Analysis of Cross-Classified Categorical Data.* MIT Press, Cambridge, MA.

Fuchs, C. (1979). Possible biased inferences in tests for average partial association. *The American Statistician* **33**, 120–126.

Gokhale, D. V. (1973). Approximating discrete distributions with applications. *J. Amer. Statist. Assoc.* **68**, (344), 1009–1012.

Gokhale, D. V. and Kullback, S. (1978a). *The Information in Contingency Tables.* Marcel Dekker Inc., New York.

Gokhale, D. V. and Kullback, S. (1978b). The minimum discrimination information approach in analyzing categorical data. *Communications in Statistics – Theory and Methods* **A7**, 987–1005.

Goodman, L. A. (1970). The multivariate analysis of qualitative data: Interaction among multiple classifications, *J. Amer. Statist. Assoc.*, **65**, 226–256.

Haberman, S. J. (1974). *The Analysis of Frequency Data.* University of Chicago Press, Chicago, IL.

Heilbron, D. C. (1981). The analysis of ratios of odds ratios in stratified contingency tables. *Biometrics* **37**, 55–66.

Hopkins, C. E. and Gross, A. J. (1971). A generalization of Cochran's procedure for the combining of $r \times c$ contingency tables. *Statistica Neerlandica* **25**, 57–62.

Ireland, C. T. (1972). Sequential cell deletion in contingency tables. Statistics Department, The George Washington University.

Ireland, C. T. and Kullback, S. (1968). Contingency tables with given marginals. *Biometrika* **55**, 179–188.

Johnson, R. W. (1979). Axiomatic characterization of the directed divergences and their linear combinations. *IEEE Trans. Inf. Theory* **25** (6), 709–716.

Keegel, J. C. (1981). Iterative fitting procedures for minimum discrimination information estimation in contingency table analysis. In: G. A. Fleischer, ed., *Contingency Table Analysis for Road Safety Studies.* Sijthoff & Noordhoff, NATO Advanced Study Institute Series, pp. 107–117.

Kotz, S. and Johnson, N. L. (1983), editors. *Encyclopedia of Statistical Sciences, Vol. IV.* Article "Kullback Information". Wiley, New York, pp. 421–425.

Ku, H. H. and Kullback, S. (1974). Loglinear models in contingency table analysis. *The American Statistician* **28**, 115–122.

Kullback, S. (1959). *Information Theory and Statistics.* Wiley, New York, 1968; Dover, Inc., New York, 1978; Peter Smith Publisher, Inc., Magnolia, MA.

Kullback, S. (1983). Kullback Information. In: S. Kotz and N. L. Johnson, eds., *Encyclopedia of Statistical Sciences, Vol. IV.* Wiley, New York, pp. 421–425.

Landis, J. R., Heyman, E. R. and Koch, G. G. (1978). Average partial association in three-way contingency tables: A review and discussion of alternative tests. *International Statistical Review* **46**, 237–254.

Larntz, K. (1978). Small-sample comparisons of exact levels for chi-squared goodness-of-fit statistics. *J. Amer. Statist. Assn.*, **73**, 253–263.

Mantel, N. and Haenszel, W. M. (1959). Statistical aspects of the analyses of data from retrospective studies of disease. *Journal of the National Cancer Institute* **22**, 719–748.

Miller, R. G., Jr. (1966). *Simultaneous Statistical Inference.* McGraw-Hill, New York, p. 8.

Plackett, R. L. (1974). *The Analysis of Categorical Data*, Griffins Statist. Monographs and Courses, No. 35.

Rao, C. R. (1965). *Linear Statistical Inference and Its Applications.* Wiley, New York, 2nd Edition, 1973.

Shore, J. E. and Johnson, R. W. (1980). Axiomatic derivation of the principle of maximum entropy and the principle of minimum cross-entropy. *IEEE Trans. Inf. Theory*, **26** (1), 26–37.

Smith, K. W. (1976). Marginal standardization and table shrinking: Aids inthe traditional analysis of contingency tables. *Social Forces* **54**, 669–693.

Sugiura, N. and Otake, M. (1974). An extension of the Mantel–Haenszel procedure to k $2 \times c$ contingency tables and the relation to the logit model. *Communications in Statistics* **3**, 829–842.

Walls and Weeks (1969). *The American Statistician* **23**, 24–26.

Worcester, J. (1971). The relative odds in the 2^3-contingency table. *American Journal of Epidemiology* **93** (3), 145–149.

PART IX

TABLES

P. R. Krishnaiah and P. K. Sen, eds., *Handbook of Statistics, Vol. 4*
© Elsevier Science Publishers (1984) 873–935

36

Tables for Order Statistics

P. R. Krishnaiah and P. K. Sen

1. Introduction

For various nonparametric procedures, moments (usually the first and second order) of order statistics play a vital role. For example, in the nonparametric tests for symmetry, for the so called normal scores statistics (cf. Chapter 3 by M. Huskova), one needs to compute the expected values of the ordered random variables of a sample of size n from the chi distribution with 1 degree of freedom (i.e., a gamma population with $\theta = \frac{1}{2}$, where θ is the shape parameter). Similarly, for the log-rank test in the two-sample problem (or in the simple regression problem), one needs to compute the expected order statistics from the exponential distribution (i.e., a gamma population with $\theta = 1$). Also, expected values of the order statistics from a normal distribution are needed for the two-sample normal scores statistics (as well as the linear rank statistics based on normal scores.) In Section 2 of this chapter, the moments of order statistics for gamma distributions are presented. In Section 3, parallel results for the normal distribution are considered. In the context of multiple comparisons (in the parametric as well as nonparametric cases), one usually needs to study the distribution of the maximum of a set of correlated normal variables. This also arises in the context of tests against restricted alternatives. Some of the relevant tables on these topics are presented in Section 4. In the practical use of finite intersection tests for simultaneous inference, one similarly confronts the problem of studying the distribution of the maximum of several correlated chi-square variables with one degree of freedom. These tables are considered in Section 5.

The tables in Section 2 are adapted from Breiter and Krishnaiah (1968) and Gupta (1960). The ones in Section 3 are taken from Harter (1961) and Sarhan and Greenberg (1956). Tables in Section 4 are from Gupta (1963). Finally, the tables in Section 5 are adapted from Krishnaiah and Armitage (1965).

2. Order statistics from gamma population

In this section, we discuss the computation of the moments of gamma order statistics. Let $x_1, x_2, \ldots, x_n$ be n independent observations from a population

The work of Krishnaiah was sponsored by the Air Force Office of Scientific Research under Contract F49620-82-K-0001. Reproduction in whole or in part is permitted for any purpose of the United States Government.

with density

$$g(x;\theta)=\frac{\exp(-x)x^{\theta-1}}{\Gamma(\theta)},\quad x\geqslant 0\,. \tag{2.1}$$

Also, let $x_{k,n}$ denote k-th order statistic when the x_i's are arranged in ascending order, that is, $x_{1,n}\leqslant\cdots\leqslant x_{k,n}$. The r-th moment of $x_{k,n}$ is given by

$$\mu_r'(k,n)=\frac{n!}{(k-1)!(n-k)!}\int_{-\infty}^{\infty}x^r\{G(x;\theta)\}^{k-1}\{1-G(x;\theta)\}^{n-k}g(x;\theta)\,\mathrm{d}x \tag{2.2}$$

where

$$G(x;\theta)=\int_0^x g(z;\theta)\,\mathrm{d}z=\exp(-x)x^\theta\sum_{i=0}^{\infty}\{x^i/(\theta+i)!\}\,. \tag{2.3}$$

Gupta (1960) obtained explicit expressions for $\mu_r'(k,n)$ when θ is an integer whereas Krishnaiah and Rizvi (1967) obtained corresponding expressions when θ is a half-integer. But, these explicit expressions are not convenient for practical purposes. Breiter and Krishnaiah (1968) computed percentage points by using Gauss–Legendre quadrature formula. A brief description of the method used by them is given below.

The r-th moment of $x_{k,n}$ is approximated as

$$\mu_r(k,n)\cong(c/2)\int_{-1}^{1}\psi(k,n,r,\theta,c(w+1)^2/4)(w+1)\,\mathrm{d}w \tag{2.4}$$

where c is a properly chosen constant,

$$\psi(k,n,r,\theta,v)=\frac{n!}{\Gamma(\theta)(k-1)!(n-k)!}\{G(v;\theta)\}^{k-1}\times\{1-G(v;\theta)\}^{n-k}g(v;\theta)v^r$$

and

$$G(v,\theta)=\exp(-v)v^\theta\sum_{j=0}^{\infty}\frac{v^j}{(\theta+j)!}\,. \tag{2.5}$$

The right-hand side of (2.4) can be computed by using Gauss Legendre quadrature formula. In computing the tables, Breiter and Krishnaiah (1968) chose c to be equal to 100 and used 48 point Gauss–Legendre quadrature formula. They checked the accuracy of these tables by using 96 point Gauss–Legendre quadrature formula also. Using the above method, they computed

Table 1
Moments of gamma order statistics; $\theta = 0.5$

n	k	$\mu'_1(k,n)$	$\mu'_2(k,n)$	$\mu'_3(k,n)$	$\mu'_4(k,n)$
1	1	0.50000E–00	0.75000E 00	0.18750E 01	0.65625E 01
2	1	0.18169E–00	0.11338E–00	0.12430E–00	0.19630E–00
	2	0.81831E 00	0.13866E 01	0.36257E 01	0.12929E 02
3	1	0.96399E–01	0.34687E–01	0.22748E–01	0.21954E–01
	2	0.35227E–00	0.27077E–00	0.32739E–00	0.54500E 00
	3	0.10513E 01	0.19445E 01	0.52749E 01	0.19121E 02
4	1	0.60361E–01	0.14416E–01	0.64561E–02	0.43276E–02
	2	0.20454E–00	0.95498E–01	0.71625E–01	0.74838E–01
	3	0.50000E–00	0.44604E–00	0.68815E 00	0.10152E 01
	4	0.12351E 01	0.24444E 01	0.67338E 01	0.25156E 02
5	1	0.41539E–01	0.71387E–02	0.23531E–02	0.11774E–02
	2	0.13560E–00	0.43525E–01	0.22868E–01	0.16928E–01
	3	0.30798E–00	0.17846E–00	0.14476E–00	0.16169E–00
	4	0.62803E 00	0.62776E 00	0.87542E 00	0.15841E 01
	5	0.13869E 01	0.28931E 01	0.83296E 01	0.31049E 02
6	1	0.30417E–01	0.39642E–02	0.10099E–02	0.39523E–03
	2	0.97145E–01	0.23011E–01	0.90698E–02	0.50882E–02
	3	0.21251E–00	0.84552E–01	0.50467E–01	0.40808E–01
	4	0.40340E–00	0.26236E–00	0.23906E–00	0.26377E–00
	5	0.74034E 00	0.81045E 00	0.11986E 01	0.22348E 01
	6	0.15162E 01	0.33157E 01	0.97518E 01	0.36811E 02
7	1	0.23273E–01	0.23877E–02	0.48658E–03	0.15395E–03
	2	0.73287E–01	0.13428E–01	0.41494E–02	0.18429E–02
	3	0.15679E–00	0.46981E–01	0.21369E–01	0.13202E–01
	4	0.28681E–00	0.13465E–00	0.89263E–01	0.77149E–01
	5	0.49085E–00	0.35815E–00	0.35140E–00	0.43699E–00
	6	0.84014E 00	0.99137E 00	0.15306E 01	0.29540E 01
	7	0.16289E 01	0.37030E 01	0.11128E 02	0.42454E 02
8	1	0.18400E–01	0.15283E–02	0.25567E–03	0.67026E–04
	2	0.57384E–01	0.84031E–02	0.21028E–02	0.76242E–03
	3	0.12100E–00	0.28484E–01	0.10289E–01	0.50845E–02
	4	0.21645E–00	0.77810E–01	0.39836E–01	0.26730E–01
	5	0.35717E–00	0.19149E–00	0.13869E–00	0.12757E–00
	6	0.57106E 00	0.45814E–00	0.47902E–00	0.62264E 00
	7	0.92983E 00	0.11691E 01	0.18810E 01	0.37311E 01
	8	0.17287E 01	0.40650E 01	0.12449E 02	0.47896E 02
9	1	0.14922E–01	0.10257E–02	0.14873E–03	0.31824E–04
	2	0.46217E–01	0.55491E–02	0.11512E–02	0.34864E–03
	3	0.90469E–01	0.18392E–01	0.54334E–02	0.22106E–02
	4	0.17005E–00	0.48668E–01	0.20000E–01	0.10832E–01
	5	0.27445E–00	0.11424E–00	0.64639E–01	0.46603E–01
	6	0.42335E–00	0.25829E–00	0.19794E–00	0.19234E–00
	7	0.64491E 00	0.56057E 00	0.61066E 00	0.83779E 00
	8	0.10112E 01	0.13480E 01	0.22414E 01	0.45577E 01
	9	0.18184E 01	0.44063E 01	0.13725E 02	0.53415E 02

Table 1 (continued)
$\theta = 1.5$

n	k	$\mu_1'(k,n)$	$\mu_2'(k,n)$	$\mu_3'(k,n)$	$\mu_4'(k,n)$
1	1	0.15000E 01	0.37500E 01	0.13125E 02	0.59062E 02
2	1	0.86338E 00	0.12035 E01	0.23025E 01	0.55864E 01
	2	0.21366E 01	0.62965E 01	0.23948E 02	0.11254E 03
3	1	0.63154E 00	0.63350E 00	0.86257E 00	0.14776 E01
	2	0.13271E 01	0.23436E 01	0.51822E 01	0.13804E 02
	3	0.25414E 01	0.82729E 01	0.33330E 02	0.16191E 03
4	1	0.50801E 00	0.40564E 00	0.43647E–00	0.58758E 00
	2	0.10021E 01	0.13171E 01	0.21409E 01	0.41476E 01
	3	0.16520E 01	0.33700E 01	0.82236E 01	0.23461E 02
	4	0.28379E 01	0.99072E 01	0.41699E 02	0.20805E 03
5	1	0.43003E–00	0.28852E–00	0.25946E–00	0.29074E–00
	2	0.81991E 00	0.87414E 00	0.11445E 01	0.17749E 01
	3	0.12755E 01	0.19815E 01	0.36355E 01	0.77065E 01
	4	0.19030E 01	0.42958E 01	0.11282E 02	0.33964E 02
	5	0.30716E 01	0.11310E 02	0.49303E 02	0.25158E 03
6	1	0.37579E–00	0.21909E–00	0.17050E–00	0.16482E–00
	2	0.70125E 00	0.63569E 00	0.70427E 00	0.92039E 00
	3	0.10573E 01	0.13510E 01	0.20249E 01	0.34840E 01
	4	0.14037E 01	0.26119E 01	0.52462E 01	0.11029E 02
	5	0.21076E 01	0.51377E 01	0.14300E 02	0.44981E 02
	6	0.32644E 01	0.12545E 02	0.56304E 02	0.29290E 03
7	1	0.33559E–00	0.17395E–00	0.11996E–00	0.10249E–00
	2	0.61697E 00	0.48993E–00	0.47377E–00	0.53877E 00
	3	0.91195E 00	0.10001E 01	0.12805E 01	0.18744E 01
	4	0.12510E 01	0.18190E 01	0.30174E 01	0.56300E 01
	5	0.16757E 01	0.32066E 01	0.69177E 01	0.16653E 02
	6	0.22804E 01	0.59101E 01	0.17253E 02	0.56312E 02
	7	0.34284E 01	0.13650E 02	0.62812E 02	0.33233E 03
8	1	0.30445E–00	0.14264E–00	0.88676E–01	0.68149E–01
	2	0.55359E 00	0.39309E–00	0.33895E–00	0.34288E–00
	3	0.80710E 00	0.78044E 00	0.87821E 00	0.11264E 01
	4	0.10867E 01	0.13662E 01	0.19510E 01	0.31211E 01
	5	0.14153E 01	0.22717E 01	0.40838E 01	0.81390E 01
	6	0.18320E 01	0.37676E 01	0.86181E 01	0.21762E 02
	7	0.24298E 01	0.66243E 01	0.20132E 02	0.67829E 02
	8	0.35710E 01	0.14654E 02	0.68909E 02	0.37011E 03
9	1	0.27951E–00	0.11986E–00	0.68048E–01	0.47670E–01
	2	0.50395E 00	0.32485E–00	0.25370E–00	0.23198E–00
	3	0.72732E 00	0.63195E 00	0.63735E 00	0.73102E 00
	4	0.96667E 00	0.10774E 01	0.13599E 01	0.19173E 01
	5	0.12367E 01	0.17271E 01	0.26899E 01	0.46259E 01
	6	0.15581E 01	0.27074E 01	0.51989E 01	0.10949E 02
	7	0.19690E 01	0.42976E 01	0.10328E 02	0.27168E 02
	8	0.25615E 01	0.72890E 01	0.22933E 02	0.79446E 02
	9	0.36972E 01	0.15575E 02	0.74656E 02	0.40645E 03

Table 1 (continued)
$\theta = 2.5$

n	k	$\mu'_1(k,n)$	$\mu'_2(k,n)$	$\mu'_3(k,n)$	$\mu'_4(k,n)$
1	1	0.25000E 01	0.87500E 01	0.39375E 02	0.21656E 03
2	1	0.16512E 01	0.36570E 01	0.10090E 02	0.33216E 02
	2	0.33488E 01	0.13843E 02	0.68659E 02	0.39991E 03
3	1	0.13156E 01	0.22744E 01	0.48213E 01	0.12039E 02
	2	0.23224E 01	0.64224E 01	0.20629E 02	0.75570E 02
	3	0.38621E 01	0.17553E 02	0.92675E 02	0.56208E 03
4	1	0.11265E 01	0.16464E 01	0.29215E 01	0.60567E 01
	2	0.18830E 01	0.41581E 01	0.10521E 02	0.29985E 02
	3	0.27617E 01	0.86867E 01	0.30737E 02	0.12116E 03
	4	0.42288E 01	0.20509E 02	0.11332E 03	0.70905E 03
5	1	0.10018E 01	0.12907E 01	0.20048E 01	0.36170E 01
	2	0.16251E 01	0.30693E 01	0.65883E 01	0.15815E 02
	3	0.22697E 01	0.57913E 01	0.16419E 02	0.51240E 02
	4	0.30898E 01	0.10617E 02	0.40283E 02	0.16777E 03
	5	0.45136E 01	0.22982E 02	0.13158E 03	0.84437E 03
6	1	0.91192E 00	0.10625E 01	0.14845E 01	0.23986E 01
	2	0.14511E 01	0.24319E 01	0.46063E 01	0.97090E 01
	3	0.19732E 01	0.43442E 01	0.10552E 02	0.28028E 02
	4	0.25662E 01	0.72385E 01	0.22286E 02	0.74452E 02
	5	0.33516E 01	0.12306E 02	0.49282E 02	0.21442E 03
	6	0.47460E 01	0.25117E 02	0.14804E 03	0.97036E 03
7	1	0.84327E 00	0.90386E 00	0.11569E 01	0.17066E 01
	2	0.13238E 01	0.20143E 01	0.34500E 01	0.65510E 01
	3	0.17695E 01	0.34758E 01	0.74971E 01	0.17604E 02
	4	0.22448E 01	0.55021E 01	0.14626E 02	0.41926E 02
	5	0.28073E 01	0.85408E 01	0.28030E 02	0.98846E 02
	6	0.35693E 01	0.13812E 02	0.57782E 02	0.26065E 03
	7	0.49421E 01	0.27001E 02	0.16308E 03	0.10886E 04
8	1	0.78866E 00	0.78726E 00	0.93531E 00	0.12769E 01
	2	0.12255E 01	0.17200E 01	0.27084E 01	0.47144E 01
	3	0.16186E 01	0.28972E 01	0.56747E 01	0.12061E 02
	4	0.20210E 01	0.44400E 01	0.10534E 02	0.26843E 02
	5	0.24685E 01	0.65641E 01	0.18717E 02	0.57009E 02
	6	0.30106E 01	0.97269E 01	0.33618E 02	0.12395E 03
	7	0.37555E 02	0.15174E 02	0.65837E 02	0.30622E 03
	8	0.51117E 01	0.28691E 02	0.17697E 03	0.12004E 04
9	1	0.74389E 00	0.69797E 00	0.77721E 00	0.99210E 00
	2	0.11468E 01	0.15016E 01	0.22001E 01	0.35550E 01
	3	0.15011E 01	0.24845E 01	0.44874E 01	0.87725E 01
	4	0.18535E 01	0.37227E 01	0.80493E 01	0.18637E 02
	5	0.22302E 01	0.53367E 01	0.13640E 02	0.37099E 02
	6	0.26592E 01	0.75461E 01	0.22779E 02	0.72938E 02
	7	0.31863E 01	0.10817E 02	0.39038E 02	0.14945E 03
	8	0.39181E 01	0.16419E 02	0.73494E 02	0.35101E 03
	9	0.25609E 01	0.30225E 02	0.18991E 03	0.13066E 04

Table 1 (continued)
$\theta = 3.5$

n	k	$\mu_1'(k,n)$	$\mu_2'(k,n)$	$\mu_3'(k,n)$	$\mu_4'(k,n)$
1	1	0.35000E 01	0.15750E 02	0.86625E 02	0.56306E 03
2	1	0.24814E 01	0.76013E 01	0.27547E 02	0.11488E 03
	2	0.45186E 01	0.23899E 02	0.14570E 03	0.10112E 04
3	1	0.20610E 01	0.51507E 01	0.14987E 02	0.49531E 02
	2	0.33205E 01	0.12502E 02	0.52666E 02	0.24558E 03
	3	0.51177E 01	0.29597E 02	0.19222E 03	0.13941E 04
4	1	0.18190E 01	0.39636E 01	0.99647E 01	0.28223E 02
	2	0.27904E 01	0.87119E 01	0.30055E 02	0.11345E 03
	3	0.38506E 01	0.16293E 02	0.75277E 02	0.37771E 03
	4	0.55400E 01	0.34031E 02	0.23120E 03	0.17329E 04
5	1	0.16557E 01	0.32582E 01	0.73492E 01	0.18570E 02
	2	0.24724E 01	0.67852E 01	0.20427E 02	0.66835E 02
	3	0.32674E 01	0.11602E 02	0.44497E 02	0.18338E 03
	4	0.42393E 01	0.19420E 02	0.95797E 02	0.50727E 03
	5	0.58652E 01	0.37684E 02	0.26506E 03	0.20393E 04
6	1	0.15361E 01	0.27880E 01	0.57721E 01	0.13330E 02
	2	0.22540E 01	0.56090E 01	0.15235E 02	0.44769E 02
	3	0.29090E 01	0.91375E 01	0.30810E 02	0.11097E 03
	4	0.36258E 01	0.14066E 02	0.58184E 02	0.25580E 03
	5	0.45461E 01	0.22097E 02	0.11460E 03	0.63300E 03
	6	0.61290E 01	0.40802E 02	0.29515E 03	0.23205E 04
7	1	0.14434E 01	0.24507E 01	0.47280E 01	0.10141E 02
	2	0.20920E 01	0.48121E 01	0.12037E 02	0.32465E 02
	3	0.26591E 01	0.76014E 01	0.23229E 02	0.75527E 02
	4	0.32423E 01	0.11186E 02	0.40917E 02	0.15823E 03
	5	0.39134E 01	0.16227E 02	0.71134E 02	0.32898E 03
	6	0.47902E 01	0.24446E 02	0.13199E 03	0.75461E 03
	7	0.63506E 01	0.43528E 02	0.32234E 03	0.25815E 04
8	1	0.13688E 01	0.21959E 01	0.39904E 01	0.80404E 01
	2	0.19655E 01	0.42342E 01	0.98907E 01	0.24845E 02
	3	0.24716E 01	0.65459E 01	0.18476E 02	0.55325E 02
	4	0.29716E 01	0.93606E 01	0.31151E 02	0.10920E 03
	5	0.35130E 01	0.13011E 02	0.50684E 02	0.20725E 03
	6	0.41536E 01	0.18157E 02	0.83404E 02	0,40201E 03
	7	0.50144E 01	0.26542E 02	0.14819E 03	0.87214E 03
	8	0.65415E 01	0.45954E 02	0.34722E 03	0.28257E 04
9	1	0.13070E 01	0.19960E 01	0.34442E 01	0.65746E 01
	2	0.18680E 01	0.37945E 01	0.83606E 01	0.19767E 02
	3	0.23240E 01	0.57728E 01	0.15246E 02	0.42620E 02
	4	0.27667E 01	0.80920E 01	0.24936E 02	0.80736E 02
	5	0.32277E 01	0.10946E 02	0.38919E 02	0.14477E 03
	6	0.37412E 01	0.14662E 02	0.60095E 02	0.25724E 03
	7	0.43599E 01	0.19905E 02	0.95058E 02	0.47440E 03
	8	0.52014E 01	0.28438E 02	0.16337E 03	0.98579E 03
	9	0.67090E 01	0.48144E 02	0.37020E 03	0.30557E 04

Table 1 (continued)
$\theta = 4.5$

n	k	$\mu'_1(k,n)$	$\mu'_2(k,n)$	$\mu'_3(k,n)$	$\mu'_4(k,n)$
1	1	0.45000E 01	0.24750E 02	0.16087E 03	0.12066E 04
2	1	0.33359E 01	0.13109E 02	0.59016E 02	0.29856E 03
	2	0.56641E 01	0.36391E 02	0.26273E 03	0.21146E 04
3	1	0.28441E 01	0.93719E 01	0.34878E 02	0.14405E 03
	2	0.43195E 01	0.20583E 02	0.10729E 03	0.60759E 03
	3	0.63364E 01	0.44295E 02	0.34046E 03	0.28681E 04
4	1	0.25548E 01	0.74877E 01	0.24572E 02	0.88811E 02
	2	0.37119E 01	0.15025E 02	0.65796E 02	0.30976E 03
	3	0.49270E 01	0.26141E 02	0.14879E 03	0.90542E 03
	4	0.68062E 01	0.50346E 02	0.40434E 03	0.35223E 04
5	1	0.23579E 01	0.63348E 01	0.18945E 02	0.62074E 02
	2	0.33425E 01	0.12099E 02	0.47080E 02	0.19576E 03
	3	0.42662E 01	0.19413E 02	0.93869E 02	0.48076E 03
	4	0.53675E 01	0.30627E 02	0.18540E 03	0.11885E 04
	5	0.71659E 01	0.55276E 02	0.45908E 03	0.41057E 04
6	1	0.22123E 01	0.55485E 01	0.15423E 02	0.46791E 02
	2	0.30861E 01	0.10266E 02	0.36553E 02	0.13849E 03
	3	0.38551E 01	0.15765E 02	0.68135E 02	0.31028E 03
	4	0.46773E 01	0.23061E 02	0.11960E 03	0.65124E 03
	5	0.57126E 01	0.34410E 02	0.21830E 03	0.14572E 04
	6	0.74566E 01	0.59449E 02	0.50724E 03	0.46354E 04
7	1	0.20986E 01	0.49735E 01	0.13019E 02	0.37084E 02
	2	0.28943E 01	0.89986E 01	0.29847E 02	0.10503E 03
	3	0.35657E 01	0.13436E 02	0.53317E 02	0.22215E 03
	4	0.42410E 01	0.18869E 02	0.87892E 02	0.42779E 03
	5	0.50045E 01	0.26205E 02	0.14339E 03	0.81882E 03
	6	0.59959E 01	0.37692E 02	0.24826E 03	0.17125E 04
	7	0.77000E 01	0.63076E 02	0.55040E 03	0.51226E 04
8	1	0.20065E 01	0.45321E 01	0.11276E 02	0.30456E 02
	2	0.27435E 01	0.80631E 01	0.25218E 02	0.83481E 02
	3	0.33469E 01	0.11805E 02	0.43733E 02	0.16968E 03
	4	0.39303E 01	0.16154E 02	0.69292E 02	0.30961E 03
	5	0.45516E 01	0.21584E 02	0.10649E 03	0.54597E 03
	6	0.52762E 01	0.28978E 02	0.16552E 03	0.98254E 03
	7	0.62358E 01	0.40596E 02	0.27585E 03	0.19558E 04
	8	0.79092E 01	0.66287E 02	0.58962E 03	0.55749E 04
9	1	0.19297E 01	0.41810E 01	0.99561E 01	0.25684E 02
	2	0.26206E 01	0.73409E 01	0.21838E 02	0.68633E 02
	3	0.31735E 01	0.10591E 02	0.37047E 02	0.13545E 03
	4	0.36935E 01	0.14233E 02	0.57105E 02	0.23813E 03
	5	0.42263E 01	0.18556E 02	0.84526E 02	0.39896E 03
	6	0.48119E 01	0.24007E 02	0.12407E 03	0.66357E 03
	7	0.55084E 01	0.31464E 02	0.18625E 03	0.11420E 04
	8	0.64430E 01	0.43206E 02	0.30144E 03	0.21883E 04
	9	0.80924E 01	0.69172E 02	0.62564E 03	0.59983E 04

Table 1 (continued)
$\theta = 5.5$

n	k	$\mu_1'(k,n)$	$\mu_2'(k,n)$	$\mu_3'(k,n)$	$\mu_4'(k,n)$
1	1	0.55000E 01	0.35750E 02	0.26812E 03	0.22791E 04
2	1	0.42065E 01	0.20229E 02	0.10903E 03	0.64932E 03
	2	0.67934E 01	0.51271E 02	0.42722E 03	0.39088E 04
3	1	0.36504E 01	0.15011E 02	0.68292E 02	0.33931E 03
	2	0.53188E 01	0.30664E 02	0.19051E 03	0.12693E 04
	3	0.75308E 01	0.61575E 02	0.54558E 03	0.52286E 04
4	1	0.33196E 01	0.12306E 02	0.50082E 02	0.22105E 03
	2	0.46427E 01	0.23125E 02	0.12292E 03	0.69410E 03
	3	0.59950E 01	0.38202E 02	0.25809E 03	0.18445E 04
	4	0.80427E 01	0.69366E 02	0.64140E 03	0.63566E 04
5	1	0.30927E 01	0.10618E 02	0.39809E 02	0.16110E 03
	2	0.42275E 01	0.19060E 02	0.91177E 02	0.46084E 03
	3	0.52654E 01	0.29224E 02	0.17054E 03	0.10440E 04
	4	0.64814E 01	0.44187E 02	0.31647E 03	0.23782E 04
	5	0.84330E 01	0.75661E 02	0.72264E 03	0.73511E 04
6	1	0.29237E 01	0.94480E 01	0.33213E 02	0.12559E 03
	2	0.39374E 01	0.16467E 02	0.72787E 02	0.33869E 03
	3	0.48076E 01	0.24246E 02	0.12796E 03	0.70513E 03
	4	0.57232E 01	0.34202E 02	0.21312E 03	0.13828E 04
	5	0.68605E 01	0.49180E 02	0.36814E 03	0.27259E 04
	6	0.87475E 01	0.80957E 02	0.79353E 03	0.82462E 04
7	1	0.27912E 01	0.85815E 01	0.28616E 02	0.10236E 03
	2	0.37191E 01	0.14647E 02	0.60797E 02	0.26494E 03
	3	0.44832E 01	0.21016E 02	0.10276E 03	0.52307E 03
	4	0.52402E 01	0.28553E 02	0.16155E 03	0.94787E 03
	5	0.60855E 01	0.38439E 02	0.25179E 03	0.17091E 03
	6	0.71705E 01	0.53477E 02	0.41468E 03	0.33427E 04
	7	0.90103E 01	0.85537E 02	0.85668E 03	0.90634E 04
8	1	0.26832E 01	0.79091E 01	0.25223E 02	0.86105E 02
	2	0.35466E 01	0.13288E 02	0.52361E 02	0.21615E 03
	3	0.42367E 01	0.18723E 02	0.86107E 02	0.41131E 03
	4	0.48941E 01	0.24837E 02	0.13052E 03	0.70934E 03
	5	0.55862E 01	0.32269E 02	0.19258E 03	0.11864E 04
	6	0.63851E 01	0.42140E 02	0.28731E 03	0.20227E 04
	7	0.74323E 01	0.57255E 02	0.45714E 03	0.37827E 04
	8	0.92357E 01	0.89578E 02	0.91375E 03	0.98178E 04
9	1	0.25930E 01	0.73691E 01	0.22614E 02	0.74149E 02
	2	0.34055E 01	0.12229E 02	0.46099E 02	0.18175E 03
	3	0.40405E 01	0.16997E 02	0.74277E 02	0.33654E 03
	4	0.46291E 01	0.22174E 02	0.10977E 03	0.56085E 03
	5	0.52255E 01	0.28166E 02	0.15646E 03	0.89497E 03
	6	0.58748E 01	0.35551E 02	0.22148E 03	0.14196E 04
	7	0.66402E 01	0.45435E 02	0.32023E 03	0.23242E 04
	8	0.76585E 01	0.60633E 02	0.49626E 03	0.41994E 04
	9	0.94329E 01	0.93196E 02	0.96594E 03	0.10520E 05

Table 1 (continued)
$\theta = 6.5$

n	k	$\mu_1'(k,n)$	$\mu_2'(k,n)$	$\mu_3'(k,n)$	$\mu_4'(k,n)$
1	1	0.65000E 01	0.48750E 02	0.41437E 03	0.39366E 04
2	1	0.50890E 01	0.28995E 02	0.18226E 03	0.12499E 04
	2	0.79110E 01	0.68505E 02	0.64649E 03	0.66232E 04
3	1	0.44742E 01	0.22121E 02	0.11923E 03	0.69366E 03
	2	0.63184E 01	0.42745E 02	0.30831E 03	0.23625E 04
	3	0.87073E 01	0.81384E 02	0.81558E 03	0.87535E 04
4	1	0.41057E 01	0.18483E 02	0.90079E 02	0.47084E 03
	2	0.55799E 01	0.33035E 02	0.20669E 03	0.13621E 04
	3	0.70569E 01	0.52454E 02	0.40993E 03	0.33629E 04
	4	0.92575E 01	0.91028E 02	0.95079E 03	0.10550E 05
5	1	0.38513E 01	0.16178E 02	0.73227E 02	0.35394E 03
	2	0.51233E 01	0.27702E 02	0.15749E 03	0.93843E 03
	3	0.62649E 01	0.41035E 02	0.28050E 03	0.19977E 04
	4	0.75849E 01	0.60067E 02	0.49622E 03	0.42731E 04
	5	0.96756E 01	0.98768E 02	0.10644E 04	0.12120E 05
6	1	0.36610E 01	0.14563E 02	0.62203E 02	0.28282E 03
	2	0.48025E 01	0.24254E 02	0.12835E 03	0.70954E 03
	3	0.57647E 01	0.34596E 02	0.21577E 03	0.13962E 04
	4	0.67651E 01	0.47474E 02	0.34523E 03	0.25992E 04
	5	0.79948E 01	0.66363E 02	0.57172E 03	0.51100E 04
	6	0.10012E 02	0.10525E 03	0.11630E 04	0.13522E 05
7	1	0.35112E 01	0.13355E 02	0.54400E 02	0.23530E 03
	2	0.45601E 01	0.21809E 02	0.10902E 03	0.56793E 03
	3	0.54086E 01	0.30366E 02	0.17667E 03	0.10636E 04
	4	0.62396E 01	0.40237E 02	0.26789E 03	0.18397E 04
	5	0.71592E 01	0.52902E 02	0.40323E 03	0.31687E 04
	6	0.83290E 01	0.71748E 02	0.63912E 03	0.58865E 04
	7	0.10292E 02	0.11083E 03	0.12503E 04	0.14794E 05
8	1	0.33888E 01	0.12411E 02	0.48568E 02	0.20143E 03
	2	0.43679E 01	0.19968E 02	0.95225E 02	0.47236E 03
	3	0.51368E 01	0.27333E 02	0.15041E 03	0.85464E 03
	4	0.58615E 01	0.35421E 02	0.22045E 03	0.14118E 04
	5	0.66178E 01	0.45054E 02	0.31534E 03	0.22677E 04
	6	0.74840E 01	0.57610E 02	0.45597E 03	0.37093E 04
	7	0.86107E 01	0.76460E 02	0.70016E 03	0.66122E 04
	8	0.10532E 02	0.11574E 03	0.13289E 04	0.15963E 05
9	1	0.32861E 01	0.11647E 02	0.44031E 02	0.17613E 03
	2	0.42102E 01	0.18521E 02	0.84861E 02	0.40383E 03
	3	0.49199E 01	0.25032E 02	0.13150E 03	0.71222E 03
	4	0.55707E 01	0.31936E 02	0.18822E 03	0.11395E 04
	5	0.62250E 01	0.39776E 02	0.26073E 03	0.17521E 04
	6	0.69320E 01	0.49276E 02	0.35903E 03	0.26802E 04
	7	0.77600E 01	0.61778E 02	0.50444E 03	0.42239E 04
	8	0.88537E 01	0.80655E 02	0.75608E 03	0.72946E 04
	9	0.10742E 02	0.12013E 03	0.14005E 04	0.17047E 05

Table 1 (continued)
$\theta = 7.5$

n	k	$\mu_1'(k,n)$	$\mu_2'(k,n)$	$\mu_3'(k,n)$	$\mu_4'(k,n)$
1	1	0.75000E 01	0.63750E 02	0.60562E 03	0.63591E 04
2	1	0.59804E 01	0.39437E 02	0.28347E 03	0.22015E 04
	2	0.90196E 01	0.88063E 02	0.92778E 03	0.10517E 05
3	1	0.53116E 01	0.30742E 02	0.19186E 03	0.12808E 04
	2	0.73181E 01	0.59826E 02	0.46671E 03	0.40429E 04
	3	0.98703E 01	0.10368E 03	0.11583E 04	0.13753E 05
4	1	0.49080E 01	0.26067E 02	0.14833E 03	0.89760E 03
	2	0.65222E 01	0.44769E 02	0.32244E 03	0.24303E 04
	3	0.81141E 01	0.68882E 02	0.61098E 03	0.56556E 04
	4	0.10456E 02	0.11528E 03	0.13408E 04	0.16453E 05
5	1	0.46282E 01	0.23071E 02	0.12269E 03	0.69113E 03
	2	0.60272E 01	0.38051E 02	0.25089E 03	0.17235E 04
	3	0.72646E 01	0.54846E 02	0.42975E 03	0.34905E 04
	4	0.86084E 01	0.78239E 02	0.73180E 03	0.70990E 04
	5	0.10990E 02	0.12454E 03	0.14930E 04	0.18791E 05
6	1	0.44183E 01	0.20953E 02	0.10567E 03	0.56290E 03
	2	0.56782E 01	0.33662E 02	0.20778E 03	0.13323E 04
	3	0.67253E 01	0.46827E 02	0.33710E 03	0.25060E 04
	4	0.78038E 01	0.62866E 02	0.52241E 03	0.44750E 04
	5	0.91187E 01	0.85926E 02	0.83649E 03	0.84110E 04
	6	0.11256E 02	0.13227E 03	0.16243E 04	0.20867E 05
7	1	0.42524E 01	0.19357E 02	0.93486E 02	0.47579E 03
	2	0.54135E 01	0.30525E 02	0.17880E 03	0.10856E 04
	3	0.63399E 01	0.41506E 02	0.28023E 03	0.19490E 04
	4	0.72392E 01	0.53921E 02	0.41292E 03	0.32487E 04
	5	0.82272E 01	0.69575E 02	0.60453E 03	0.53947E 04
	6	0.94753E 01	0.92466E 02	0.92927E 03	0.96175E 04
	7	0.11552E 02	0.13890E 03	0.17401E 04	0.22742E 05
8	1	0.41166E 01	0.18102E 02	0.84285E 02	0.41283E 03
	2	0.52031E 01	0.28146E 02	0.15789E 03	0.91648E 03
	3	0.60449E 01	0.37662E 02	0.24154E 03	0.15928E 04
	4	0.68315E 01	0.47912E 02	0.34472E 03	0.25425E 04
	5	0.76470E 01	0.59930E 02	0.48112E 03	0.39548E 04
	6	0.85753E 01	0.75362E 02	0.67858E 03	0.62587E 04
	7	0.97753E 01	0.98167E 02	0.10128E 04	0.10737E 05
	8	0.11806E 02	0.14472E 03	0.18440E 04	0.24457E 05
9	1	0.40024E 01	0.17081E 02	0.77067E 02	0.36523E 03
	2	0.50300E 01	0.26266E 02	0.14203E 03	0.79363E 03
	3	0.58088E 01	0.34726E 02	0.21340E 03	0.13465E 04
	4	0.65171E 01	0.43533E 02	0.29784E 03	0.20856E 04
	5	0.72245E 01	0.53386E 02	0.40332E 03	0.31137E 04
	6	0.79849E 01	0.65164E 02	0,54335E 03	0,46277E 04
	7	0.88706E 01	0.80462E 02	0.74619E 03	0.70742E 04
	8	0.10034E 02	0.10323E 03	0.10890E 04	0.11784E 05
	9	0.12028E 02	0.14990E 03	0.19384E 04	0.26041E 05

Table 1 (continued)
$\theta = 8.5$

n	k	$\mu_1'(k,n)$	$\mu_2'(k,n)$	$\mu_3'(k,n)$	$\mu_4'(k,n)$
1	1	0.85000E 01	0.80750E 02	0.84787E 03	0.97506E 04
2	1	0.68791E 01	0.51574E 02	0.41753E 03	0.36236E 04
	2	0.10121E 02	0.10993E 03	0.12782E 04	0.15878E 05
3	1	0.61597E 01	0.40908E 02	0.29045E 03	0.21903E 04
	2	0.83179E 01	0.72906E 02	0.67169E 03	0.64903E 04
	3	0.11022E 02	0.12844E 03	0.15815E 04	0.20571E 05
4	1	0.57235E 01	0.35097E 02	0.22876E 03	0.15750E 04
	2	0.74683E 01	0.58339E 02	0.47552E 03	0.40360E 04
	3	0.91674E 01	0.87474E 02	0.86786E 03	0.89446E 04
	4	0.11641E 02	0.14209E 03	0.18194E 04	0.24447E 05
5	1	0.54200E 01	0.31340E 02	0.19186E 03	0.12363E 04
	2	0.69377E 01	0.50127E 02	0.37634E 03	0.29300E 04
	3	0.82643E 01	0.70658E 02	0.62431E 03	0.56950E 04
	4	0.97695E 01	0.98684E 02	0.10302E 04	0.11111E 05
	5	0.12109E 02	0.15294E 03	0.20166E 04	0.27781E 05
6	1	0.51915E 01	0.28664E 02	0.16709E 03	0.10224E 04
	2	0.65623E 01	0.44718E 02	0.31574E 03	0.23056E 04
	3	0.76886E 01	0.60945E 02	0.49752E 03	0.41788E 04
	4	0.88390E 01	0.80370E 02	0.75109E 03	0.72111E 04
	5	0.10234E 02	0.10784E 03	0.11698E 04	0.13016E 05
	6	0.12483E 02	0.16196E 03	0.21860E 04	0.30724E 05
7	1	0.50106E 01	0.26638E 02	0.14918E 03	0.87516E 03
	2	0.62768E 01	0.40825E 02	0.27456E 03	0.19059E 04
	3	0.72759E 01	0.54450E 02	0.41870E 03	0.33051E 04
	4	0.82389E 01	0.69605E 02	0.60263E 03	0.53439E 04
	5	0.92907E 01	0.88444E 02	0.86243E 03	0.86116E 04
	6	0.10612E 02	0.11560E 03	0.12927E 04	0.14841E 05
	7	0.12795E 02	0.16969E 03	0.23349E 04	0.33372E 05
8	1	0.48622E 01	0.25035E 02	0.13555E 03	0.76756E 03
	2	0.60494E 01	0.37857E 02	0.24458E 03	0.16284E 04
	3	0.69592E 01	0.49729E 02	0.36451E 03	0.27383E 04
	4	0.78036E 01	0.62320E 02	0.50901E 03	0.42497E 04
	5	0.86743E 01	0.76891E 02	0.69625E 03	0.64380E 04
	6	0.96605E 01	0.95376E 02	0.96214E 03	0.99158E 04
	7	0.10929E 02	0.12234E 03	0.14029E 04	0.16482E 05
	8	0.13062E 02	0.17645E 03	0.24680E 04	0.35785E 05
9	1	0.47373E 01	0.23727E 02	0.12478E 03	0.68541E 03
	2	0.58620E 01	0.35501E 02	0.22167E 03	0.14247E 04
	3	0.67052E 01	0.46104E 02	0.32476E 03	0.23414E 04
	4	0.74672E 01	0.56978E 02	0.44402E 03	0.35319E 04
	5	0.82242E 01	0.68997E 02	0.59024E 03	0.51470E 04
	6	0.90343E 01	0.83206E 02	0.78105E 03	0.74709E 04
	7	0.99737E 01	0.10146E 03	0.10527E 04	0.11138E 05
	8	0.11202E 02	0.12831E 03	0.15030E 04	0.18009E 05
	9	0.13294E 02	0.18247E 03	0.25887E 04	0.38006E 05

Table 1 (continued)
$\theta = 9.5$

n	k	$\mu_1'(k,n)$	$\mu_2'(k,n)$	$\mu_3'(k,n)$	$\mu_4'(k,n)$
1	1	0.95000E 01	0.99750E 02	0.11471E 04	0.14339E 05
2	1	0.78838E 01	0.65425E 02	0.58935E 03	0.56549E 04
	2	0.11216E 02	0.13407E 03	0.17049E 04	0.23023E 05
3	1	0.70168E 01	0.52644E 02	0.41939E 03	0.35282E 04
	2	0.93177E 01	0.90987E 02	0.92926E 03	0.99083E 04
	3	0.12165E 02	0.15562E 03	0.20927E 04	0.29581E 05
4	1	0.65498E 01	0.45607E 02	0.33540E 03	0.25916E 04
	2	0.84178E 01	0.73756E 02	0.67137E 03	0.63380E 04
	3	0.10218E 02	0.10822E 03	0.11872E 04	0.13479E 05
	4	0.12815E 02	0.17142E 03	0.23946E 04	0.34948E 05
5	1	0.62239E 01	0.41022E 02	0.28454E 03	0.20667E 04
	2	0.78535E 01	0.63948E 02	0.53886E 03	0.46914E 04
	3	0.92641E 01	0.88469E 02	0.87015E 03	0.88078E 04
	4	0.10853E 01	0.12138E 03	0.13985E 04	0.16593E 05
	5	0.13305E 02	0.18393E 03	0.26436E 04	0.39537E 05
6	1	0.59780E 01	0.37738E 02	0.25005E 03	0.17307E 04
	2	0.74533E 01	0.57442E 02	0.45696E 03	0.37465E 04
	3	0.86541E 01	0.76959E 02	0.70264E 03	0.65812E 04
	4	0.98740E 01	0.99979E 02	0.10377E 04	0.11034E 05
	5	0.11343E 02	0.13209E 03	0.15789E 04	0.19372E 05
	6	0.13698E 02	0.19429E 03	0.28565E 04	0.43570E 05
7	1	0.57830E 01	0.35238E 02	0.22492E 03	0.14969E 04
	2	0.71483E 01	0.52734E 02	0.40081E 03	0.31338E 04
	3	0.82157E 01	0.69211E 02	0.59736E 03	0.52782E 04
	4	0.92387E 01	0.87289E 02	0.84302E 03	0.83186E 04
	5	0.10350E 02	0.10950E 03	0.11836E 04	0.13071E 05
	6	0.11740E 02	0.14112E 03	0.17370E 04	0.21892E 05
	7	0.14024E 02	0.20316E 03	0.30431E 04	0.47183E 05
8	1	0.56227E 01	0.33254E 02	0.20568E 03	0.13244E 04
	2	0.69048E 01	0.49129E 02	0.35962E 03	0.27042E 04
	3	0.78787E 01	0.63550E 02	0.52436E 03	0.44227E 04
	4	0.87775E 01	0.78647E 02	0.71901E 03	0.67039E 04
	5	0.97000E 01	0.95931E 02	0.96703E 03	0.99334E 04
	6	0.10741E 02	0.11764E 03	0.13136E 04	0.14954E 05
	7	0.12073E 02	0.14895E 03	0.18781E 04	0.24205E 05
	8	0.14302E 02	0.21090E 03	0.32096E 04	0.50465E 05
9	1	0.54875E 01	0.31629E 02	0.19040E 03	0.11917E 04
	2	0.67039E 01	0.46256E 02	0.32796E 03	0.23859E 04
	3	0.76079E 01	0.59185E 02	0.47044E 03	0.38179E 04
	4	0.84202E 01	0.72280E 02	0.63222E 03	0.56324E 04
	5	0.92240E 01	0.86607E 02	0.82749E 03	0.80433E 04
	6	0.10081E 02	0.10339E 03	0.10787E 04	0.11445E 05
	7	0.11071E 02	0.12476E 03	0.14311E 04	0.16708E 05
	8	0.12360E 02	0.15587E 03	0.20059E 04	0.26346E 05
	9	0.14545E 02	0.21778E 03	0.33600E 04	0.53480E 05

Table 1 (continued)
$\theta = 10.5$

n	k	$\mu_1'(k,n)$	$\mu_2'(k,n)$	$\mu_3'(k,n)$	$\mu_4'(k,n)$
1	1	0.10500E 02	0.12075E 03	0.15094E 04	0.20377E 05
2	1	0.86934E 01	0.81006E 02	0.80391E 03	0.84533E 04
	2	0.12307E 02	0.16049E 03	0.22148E 04	0.32300E 05
3	1	0.78814E 01	0.65974E 02	0.58316E 03	0.54175E 04
	2	0.10318E 02	0.11107E 03	0.12454E 04	0.14525E 05
	3	0.13301E 02	0.18521E 03	0.26995E 04	0.41187E 05
4	1	0.73852E 01	0.57623E 02	0.47240E 03	0.40512E 04
	2	0.93699E 01	0.91028E 02	0.91543E 03	0.95162E 04
	3	0.11265E 02	0.13111E 03	0.15754E 04	0.19533E 05
	4	0.13980E 02	0.20324E 03	0.30743E 04	0.48405E 05
5	1	0.70381E 01	0.52147E 02	0.40462E 03	0.32739E 04
	2	0.87739E 01	0.79527E 02	0.74352E 03	0.71606E 04
	3	0.10264E 02	0.10828E 03	0.11733E 03	0.13050E 05
	4	0.11933E 02	0.14633E 03	0.18435E 04	0.23856E 05
	5	0.14491E 02	0.21747E 03	0.33819E 04	0.54543E 05
6	1	0.67757E 01	0.48206E 02	0.35829E 03	0.27707E 04
	2	0.83500E 01	0.71853E 02	0.63625E 03	0.57899E 04
	3	0.96215E 01	0.94874E 02	0.95807E 03	0.99019E 04
	4	0.10906E 02	0.12169E 03	0.13885E 04	0.16197E 05
	5	0.12446E 02	0.15865E 03	0.20710E 04	0.27685E 05
	6	0.14900E 02	0.22923E 03	0.36441E 04	0.59914E 05
7	1	0.65672E 01	0.45195E 02	0.32432E 03	0.24172E 04
	2	0.80265E 01	0.66275E 02	0.56213E 03	0.48910E 04
	3	0.91588E 01	0.85799E 02	0.82155E 03	0.80356E 04
	4	0.10239E 02	0.10697E 03	0.11401E 04	0.12390E 05
	5	0.11407E 02	0.13272E 03	0.15748E 04	0.19053E 05
	6	0.12862E 02	0.16902E 03	0.22605E 04	0.31138E 05
	7	0.15240E 02	0.23927E 03	0.38732E 04	0.64710E 05
8	1	0.63957E 01	0.42796E 02	0.29816E 03	0.21545E 04
	2	0.77679E 01	0.61987E 02	0.50744E 03	0.42562E 04
	3	0.88024E 01	0.79139E 02	0.72622E 03	0.67979E 04
	4	0.97527E 01	0.96900E 02	0.98043E 03	0.10098E 05
	5	0.10724E 02	0.11705E 03	0.12998E 04	0.14682E 05
	6	0.11817E 02	0.14213E 03	0.17390E 04	0.21675E 05
	7	0.13210E 02	0.17798E 03	0.24460E 04	0.34293E 05
	8	0.15530E 02	0.24802E 03	0.40771E 04	0.69055E 05
9	1	0.62508E 01	0.40825E 02	0.27728E 03	0.19511E 04
	2	0.75542E 01	0.58558E 02	0.46517E 03	0.37822E 04
	3	0.85157E 01	0.73985E 02	0.65537E 03	0.59155E 04
	4	0.93759E 01	0.89446E 02	0.86792E 03	0.85627E 04
	5	0.10224E 02	0.10622E 03	0.11211E 04	0.12018E 05
	6	0.11125E 02	0.12571E 03	0.14427E 04	0.16813E 05
	7	0.12163E 02	0.15034E 04	0.18884E 04	0.24106E 05
	8	0.13509E 02	0.18588E 03	0.26053E 04	0.37203E 05
	9	0.15783E 02	0.25579E 03	0.42611E 04	0.73037E 05

Table 1 is reproduced from *Sankhyā* with the kind permission of the Sankhyā's editors.

Table 1a
Moments of order statistics from exponential population

N	k	$\mu'_1(k,N)$	$\mu'_2(k,N)$	$\mu'_3(k,N)$	$\mu'_4(k,N)$	$\mu_2(k,N)$	$\mu_3(k,N)$	$\mu_4(k,N)$
1	1	1.00000	2.00000	6.00000	24.0000	1.00000	2.00000	9.00000
2	1	0.500000	0.500000	0.750000	1.50000	0.250000	0.250000	0.562500
	2	1.50000	3.50000	11.2500	46.5000	1.25000	2.25000	11.0625
3	1	0.333333	0.222222	0.222222	0.296296	0.111111	0.0740741	0.111111
	2	0.833333	1.05556	1.80556	3.90741	0.361111	0.324074	0.840278
	3	1.83333	4.72222	15.9722	67.7963	1.36111	2.32407	1.20069
4	1	0.250000	0.125000	0.0937500	0.0937500	0.0625000	0.0312500	0.0351563
	2	0.583333	0.513889	0.607639	0.903935	0.173611	0.105325	0.187934
	3	1.08333	1.59722	3.00347	6.91088	0.423611	0.355324	1.01085
	4	2.08333	5.76389	20.2951	88.0914	1.42361	2.35532	12.5525
5	1	0.200000	0.0800000	0.0478000	0.0384000	0.0400000	0.0160000	0.0144000
	2	0.450000	0.305000	0.276750	0.315150	0.102500	0.0472500	0.0645563
	3	0.783333	0.827222	1.10397	1.78711	0.213611	0.121324	0.244001
	4	1.28333	2.11056	4.26981	10.3267	0.463611	0.371324	1.12692
	5	2.28333	6.67722	24.3015	107.533	1.46361	2.37132	12.9086
6	1	0.166667	0.0555556	0.0277778	0.0185185	0.0277778	0.00925926	0.00694444
	2	0.366667	0.202222	0.149111	0.137807	0.0677778	0.0252593	0.0280111
	3	0.616667	0.510556	0.532028	0.669835	0.130278	0.0565093	0.0885840
	4	0.950000	1.14369	1.67592	2.90439	0.130583	0.286547	
	5	1.45000	2.59389	5.56675	14.0379	0.491389	0.380583	1.21113
	6	2.45000	7.49389	28.0484	126.232	1.49139	2.38058	13.1595
7	1	0.142857	0.0408163	0.0174927	0.00999583	0.0204082	0.00583090	0.00374844
	2	0.309524	0.143991	0.0894882	0.0696546	0.0481859	0.0150902	0.0140992
	3	0.509524	0.347800	0.298168	0.308189	0.0881859	0.0310902	0.0400589
	4	0.759524	0.727563	0.843840	1.15203	0.150686	0.0623401	0.108285
	5	1.09286	1.45613	2.29997	4.21866	0.261798	0.136414	0.319854
	6	1.59286	3.04899	6.87346	17.9656	0.511796	0.386416	1.27504
	7	2.59286	8.23471	31.5776	144.276	1.51180	2.38641	13.3458
8	1	0.125000	0.0312500	0.117188	0.00585938	0.0156250	0.00390625	0.00219727
	2	0.267857	0.107781	0.0579104	0.0389511	0.0360332	0.00973716	0.00785897
	3	0.434524	0.252622	0.184221	0.161765	0.0638109	0.0189964	0.0208089
	4	0.634524	0.506431	0.488080	0.552229	0.103811	0.0349965	0.0505235
	5	0.884524	0.948694	1.19960	1.75183	0.166311	0.0662462	0.124609
	6	1.21786	1.76060	2.96020	5.69876	0.277422	0.140321	0.346592
	7	1.71786	3.47846	8.17788	22.0545	0.527422	0.390323	1.32522
	8	2.71786	8.91417	34.9204	161.736	1.52742	2.39031	13.4898
9	1	0.111111	0.0246914	0.00823045	0.00365793	0.0123457	0.00274348	0.00137174
	2	0.236111	0.0837191	0.0396251	0.0234705	0.0279707	0.00664973	0.00472642
	3	0.378988	0.191996	0.121909	0.0931328	0.0483789	0.0124806	0.0118998
	4	0.545635	0.373874	0.038846	0.299030	0.0761564	0.0217399	0.0269073
	5	0.745634	0.672127	0.712122	0.868728	0.116158	0.0377396	0.0595857
	6	0.995636	1.16995	1.58958	2.45831	0.178657	0.0669887	0.138302
	7	1.32897	2.05592	3.64550	7.31698	0.289774	0.143057	0.368531
	8	1.82897	3.88469	9.47285	26.2647	0.539762	0.393073	1.36564
	9	2.82897	9.54283	38.1013	178.670	1.53977	2.39306	13.6043
10	1	0.100000	0.0200000	0.00600000	0.00240000	0.0100000	0.00200000	0.000900000
	2	0.211111	0.0669136	0.0283045	0.0149798	0.0223457	0.00474348	0.00301248
	3	0.336111	0.150941	0.0849075	0.0574336	0.0379707	0.00864973	0.00730466
	4	0.478968	0.287789	0.208246	0.176431	0.0583790	0.0144806	0.0157026
	5	0.645637	0.503002	0.459747	0.482929	0.0861552	0.0237405	0.0323758
	6	0.845632	0.841253	0.964498	1.25453	0.126159	0.0397384	0.0674573
	7	1.09564	1.38908	2.00631	3.26083	0.188650	0.0709957	0.149908
	8	1.42896	2.34171	4.34802	9.05819	0.299783	0.145045	0.386844
	9	1.92897	4.27069	10.7541	30.5663	0.549752	0.395096	1.39885
	10	2.92897	10.1286	41.1399	195.126	1.54977	2.39504	13.6977
11	1	0.0909091	0.0165289	0.00450789	0.00163923	0.00826446	0.00150263	0.000614712
12	1	0.0833333	0.0138889	0.00347222	0.00115741	0.00694444	0.00115741	0.000434028
13	1	0.0769231	0.0118343	0.00273100	0.000840307	0.00591716	0.000910332	0.000315115
14	1	0.0714286	0.0102041	0.00218659	0.000624740	0.00510204	0.000728863	0.000234277
15	1	0.0666667	0.00888889	0.00177778	0.000474074	0.00444444	0.000592593	0.000177778

the values of $\mu_1'(k, n)$, $\mu_2'(k, n)$, $\mu_3'(k, n)$ and $\mu_4'(k, n)$ when $k = 1(1)n$, $n = 1(1)9$ and $\theta = 0.5(1.0)10.5$. These values are given in Table 1. Here, we note that Gupta (1960) constructed tables for the moments of gamma order statistics when θ is a positive integer whereas Harter (1964) constructed tables for $\mu_1'(k, n)$ when $\theta = 0.5(0.5)4.0$. In Table 1a, we give the values of $\mu_1'(k, N)$, $\mu_2'(k, N)$, $\mu_3'(k, N)$, $\mu_4'(k, N)$, $\mu_2(k, N)$, $\mu_3(k, N)$ and $\mu_4(k, N)$ when $\theta = 1$ and

$$\mu_i(k, N) = E[\{x_{k,N} - E(x_{k,N})\}^i] .$$

The entries in Table 1 are reproduced from Breiter and Krishnaiah (1968) whereas the entries in Table 1a are reproduced from Gupta (1960).

3. Moments of order statistics from normal population

Let $x_1, x_2, \ldots, x_n$ be a sample from a normal population with mean zero and variance one. Also, let $x_{(1)} \geq \cdots \geq x_{(n)}$ be order statistics from the above population when the observations are arranged in descending order of magnitude. The r-th moment of $x_{(k)}$ is given by

$$\mu_r'(k, n) = \frac{n!}{(k-1)!(n-k)!} \int_{-\infty}^{\infty} x[1 - F(x)]^{k-1}[F(x)]^{n-k} f(x)\, dx . \tag{3.1}$$

The expected values of the order statistics satisfy the relation

$$\mu_1'(k, n) = -\mu_1'(n - k + 1, n) . \tag{3.2}$$

Also,

$$E(x_{(i)}x_{(j)}) = \frac{n!}{(i-1)!(j-i-1)!(n-j)!} \int_{-\infty}^{\infty} \int_{-\infty}^{y} xyf(x)f(y) \times [F(x)]^{i-1}[1 - F(y)]^{n-j}[F(y) - F(x)]^{j-i-1}\, dx\, dy \tag{3.3}$$

$$E(x_{(i)}x_{(j)}) = E(x_{(j)}x_{(i)}) = E(x_{(n-i+1)}x_{(n-j+1)}) . \tag{3.4}$$

Harter (1961) computed the expected values of order statistics for $n = 2(1)100(25)250(50)400$. In Table 2, we reproduce the values of order statistics for $n = 2(1)79$ and $k = 1(1)[n/2]$ where $[n/2]$ is the integral value of $n/2$. The moments of order statistics for the remaining values of k can be obtained using the relation (3.2).

Tables of $E(x_{(i)})$ and $E(x_{(i)}x_{(j)})$ were given in Teichroew (1956) for $n = 2(1)20$. Using the above tables, Sarhan and Greenberg (1956) computed the values of σ_{ii} and σ_{ij} where

$$\sigma_{ii} = E[(x_{(i)} - E(x_{(i)})]^2 \tag{3.5}$$

$$\sigma_{ij} = E[\{(x_{(i)} - E(x_{(i)})\}\{x_{(j)} - E(x_{(j)})\}] . \tag{3.6}$$

The values given by Sarhan and Greenberg are reproduced in Table 3.

Table 2
Expected values of order statistics from normal populations

k \ n			2	3	4	5	6	7	8	9
1			0·56419	0·84628	1·02938	1·16296	1·26721	1·35218	1·42360	1·48501
2			—	·00000	0·29701	0·49502	0·64176	0·75737	0·85222	0·93230
3			—	—	—	·00000	·20155	·35271	·47282	·57197
4			—	—	—	—	—	·00000	·15251	·27453
5			—	—	—	—	—	—	—	·00000

k \ n	10	11	12	13	14	15	16	17	18	19
1	1·53875	1·58644	1·62923	1·66799	1·70338	1·73591	1·76599	1·79394	1·82003	1·84448
2	1·00136	1·06192	1·11573	1·16408	1·20790	1·24794	1·28474	1·31878	1·35041	1·37994
3	0·65606	0·72884	0·79284	0·84983	0·90113	0·94769	0·99027	1·02946	1·06573	1·09945
4	·37576	·46198	·53684	·60285	·66176	·71488	·76317	0·80738	0·84812	0·88586
5	·12267	·22489	·31225	·38833	·45557	·51570	·57001	·61946	·66479	·70661
6	—	0·00000	0·10259	0·19052	0·26730	0·33530	0·39622	0·45133	0·50158	0·54771
7	—	—	—	·00000	0·08816	·16530	·23375	·29519	·35084	·40164
8	—	—	—	—	—	·00000	·07729	·14599	·20774	·26374
9	—	—	—	—	—	—	—	·00000	·06880	·13072
10	—	—	—	—	—	—	—	—	—	·00000

k \ n	20	21	22	23	24	25	26	27	28	29
1	1·86748	1·88917	1·90969	1·92916	1·94767	1·96531	1·98216	1·99827	2·01371	2·02852
2	1·40760	1·43362	1·45816	1·48137	1·50338	1·52430	1·54423	1·56326	1·58145	1·59888
3	1·13095	1·16047	1·18824	1·21445	1·23924	1·26275	1·28511	1·30641	1·32674	1·34619
4	0·92098	0·95380	0·98459	1·01356	1·04091	1·06679	1·09135	1·11471	1·13697	1·15822
5	·74538	·78150	·81527	0·84697	0·87682	0·90501	0·93171	0·95705	0·98115	1·00414
6	0·59030	0·62982	0·66667	0·70115	0·73354	0·76405	0·79289	0·82021	0·84615	0·87084
7	·44833	·49148	·53157	·56896	·60399	·63690	·66794	·69727	·72508	·75150
8	·31493	·36203	·40559	·44609	·48391	·51935	·55267	·58411	·61385	·64205
9	·18696	·23841	·28579	·32965	·37047	·40860	·44436	·47801	·50977	·53982
10	·06200	·11836	·16997	·21755	·26163	·30268	·34105	·37706	·41096	·44298
11	—	0·00000	0·05642	0·10813	0·15583	0·20006	0·24128	0·27983	0·31603	0·35013
12	—	—	—	·00000	·05176	·09953	·14387	·18520	·22389	·26023
13	—	—	—	—	—	·00000	·04781	·09220	·13361	·17240
14	—	—	—	—	—	—	—	·00000	·04442	·08588
15	—	—	—	—	—	—	—	—	—	·00000

k \ n	30	31	32	33	34	35	36	37	38	39
1	2·04276	2·05646	2·06967	2·08241	2·09471	2·10661	2·11812	2·12928	2·14009	2·15059
2	1·61560	1·63166	1·64712	1·66200	1·67636	1·69023	1·70362	1·71659	1·72914	1·74131
3	1·36481	1·38268	1·39985	1·41637	1·43228	1·44762	1·46244	1·47676	1·49061	1·50402
4	1·17855	1·19803	1·21672	1·23468	1·25196	1·26860	1·28466	1·30016	1·31514	1·32964
5	1·02609	1·04709	1·06721	1·08652	1·10509	1·12295	1·14016	1·15677	1·17280	1·18830
6	0·89439	0·91688	0·93841	0·95905	0·97886	0·99790	1·01624	1·03390	1·05095	1·06741
7	·77666	·80066	·82359	·84555	·86660	·88681	0·90625	0·92496	0·94300	0·96041
8	·66885	·69438	·71875	·74204	·76435	·78574	·80629	·82605	·84508	·86343
9	·56834	·59545	·62129	·64596	·66954	·69214	·71382	·73465	·75468	·77398
10	·47329	·50206	·52943	·55552	·58043	·60427	·62710	·64902	·67009	·69035
11	0·38235	0·41287	0·44185	0·46942	0·49572	0·52084	0·54488	0·56793	0·59005	0·61131
12	·29449	·32686	·35755	·38669	·41444	·44091	·46620	·49042	·51363	·53592
13	·20885	·24322	·27573	·30654	·33582	·36371	·39032	·41576	·44012	·46348
14	·12473	·16126	·19572	·22832	·25924	·28863	·31663	·34336	·36892	·39340
15	·04148	·08037	·11695	·15147	·18415	·21515	·24463	·27272	·29954	·32520
16	—	0·00000	0·03890	0·07552	0·11009	0·14282	0·17388	0·20342	0·23159	0·25849
17	—	—	—	·00000	·03663	·07123	·10399	·13509	·16469	·19292
18	—	—	—	—	—	·00000	·03461	·06739	·09853	·12817
19	—	—	—	—	—	—	—	·00000	·03280	·06395
20	—	—	—	—	—	—	—	—	—	·00000

Table 2 (continued)

k \ n	40	41	42	43	44	45	46	47	48	49
1	2·16078	2·17068	2·18032	2·18969	2·19882	2·20772	2·21639	2·22486	2·23312	2·24119
2	1·75312	1·76456	1·77571	1·78654	1·79707	1·80733	1·81732	1·82706	1·83655	1·84582
3	1·51702	1·52964	1·54188	1·55377	1·56533	1·57658	1·58754	1·59820	1·60860	1·61874
4	1·34368	1·35728	1·37048	1·38329	1·39574	1·40784	1·41962	1·43108	1·44224	1·45312
5	1·20330	1·21782	1·23190	1·24556	1·25881	1·27170	1·28422	1·29641	1·30827	1·31983
6	1·08332	1·09872	1·11364	1·12810	1·14213	1·15576	1·16899	1·18186	1·19439	1·20658
7	0·97722	0·99348	1·00922	1·02446	1·03924	1·05358	1·06761	1·08104	1·09420	1·10701
8	·88114	·89825	0·91480	0·93082	0·94634	0·96139	0·97599	0·99018	1·00396	1·01737
9	·79259	·81056	·82792	·84472	·86097	·87673	·89201	·90684	0·92125	0·93525
10	·70988	·72871	·74690	·76448	·78148	·79705	·81391	·82939	·84442	·85902
11	0·63177	0·65149	0·67052	0·68880	0·70666	0·72385	0·74049	0·75663	0·77228	0·78748
12	·55736	·57709	·59788	·61707	·63661	·65353	·67088	·68768	·70397	·71978
13	·48591	·50749	·52827	·54830	·56763	·58631	·60438	·62186	·63881	·65523
14	·41688	·43944	·46114	·48204	·50220	·52166	·54046	·55865	·57625	·59331
15	·34978	·37337	·39604	·41784	·43885	·45912	·47868	·49759	·51588	·53360
16	0·28423	0·30890	0·33257	0·35533	0·37723	0·39833	0·41868	0·43834	0·45734	0·47573
17	·21988	·24569	·27043	·29418	·31701	·33898	·36016	·38060	·40034	·41942
18	·15644	·18345	·20931	·23411	·25792	·28081	·30285	·32410	·34460	·36441
19	·09362	·12192	·14897	·17488	·19972	·22358	·24652	·26862	·28992	·31049
20	·03117	·06085	·08917	·11625	·14219	·16707	·19097	·21396	·23610	·25746
21	—	0·00000	0·02969	0·05803	0·08513	0·11109	0·13600	0·15993	0·18296	0·20514
22	—	—	—	·00000	·02835	·05546	·08144	·10637	·13033	·15338
23	—	—	—	—	—	·00000	·02712	·05311	·07805	·10203
24	—	—	—	—	—	—	—	·00000	·02599	·05095
25	—	—	—	—	—	—	—	—	—	·00000

k \ n	50	51	52	53	54	55	56	57	58	59
1	2·24907	2·25678	2·26432	2·27169	2·27891	2·28598	2·29291	2·29970	2·30635	2·31288
2	1·85487	1·86371	1·87235	1·88080	1·88906	1·89715	1·90506	1·91282	1·92041	1·92786
3	1·62863	1·63829	1·64773	1·65695	1·66596	1·67478	1·68340	1·69185	1·70012	1·70822
4	1·46374	1·47409	1·48420	1·49407	1·50372	1·51315	1·52237	1·53140	1·54024	1·54889
5	1·33109	1·34207	1·35279	1·36326	1·37348	1·38346	1·39323	1·40278	1·41212	1·42127
6	1·21846	1·23003	1·24132	1·25234	1·26310	1·27361	1·28387	1·29391	1·30373	1·31334
7	1·11948	1·13162	1·14347	1·15502	1·16629	1·17729	1·18804	1·19855	1·20882	1·21886
8	1·03042	1·04312	1·05550	1·06757	1·07934	1·09083	1·10205	1·11300	1·12371	1·13419
9	0·94887	0·96213	0·97504	0·98762	0·99988	1·01185	1·02352	1·03493	1·04607	1·05695
10	·87321	·88701	·90045	·91354	·92629	0·93873	0·95086	0·96271	0·97427	0·98557
11	0·80225	0·81661	0·83058	0·84417	0·85742	0·87033	0·88292	0·89520	0·90719	0·91890
12	·73513	·75004	·76455	·77866	·79240	·80578	·81883	·83155	·84397	·85609
13	·67117	·68666	·70170	·71633	·73057	·74444	·75794	·77111	·78396	·79649
14	·60986	·62592	·64152	·65668	·67143	·68578	·69976	·71337	·72665	·73960
15	·55077	·56742	·58358	·59928	·61455	·62940	·64385	·65793	·67164	·68502
16	0·49354	0·51080	0·52755	0·54380	0·55960	0·57495	0·58989	0·60444	0·61860	0·63241
17	·43789	·45578	·47312	·48995	·50629	·52217	·53761	·55263	·56725	·58150
18	·38357	·40211	·42007	·43749	·45439	·47080	·48675	·50226	·51736	·53205
19	·33036	·34957	·36818	·38621	·40369	·42065	·43713	·45314	·46872	·48388
20	·27807	·29799	·31726	·33592	·35400	·37154	·38856	·40510	·42117	·43681
21	0·22653	0·24719	0·26716	0·28648	0·30518	0·32331	0·34090	0·35797	0·37456	0·39068
22	·17559	·19702	·21772	·23772	·25708	·27583	·29400	·31163	·32875	·34538
23	·12511	·14735	·16880	·18953	·20957	·22896	·24774	·26595	·28362	·30078
24	·07494	·09803	·12029	·14177	·16252	·18259	·20201	·22082	·23906	·25677
25	·02496	·04896	·07206	·09434	·11584	·13661	·15669	·17614	·19498	·21325
26	—	0·00000	0·02400	0·04712	0·06940	0·09091	0·11170	0·13180	0·15127	0·17013
27	—	—	—	·00000	·02312	·04541	·06693	·08773	·10785	·12733
28	—	—	—	—	—	·00000	·02229	·04382	·06463	·08476
29	—	—	—	—	—	—	—	·00000	·02153	·04234
30	—	—	—	—	—	—	—	—	—	·00000

Table 2 (continued)

k \ n	60	61	62	63	64	65	66	67	68	69
1	2·31928	2·32556	2·33173	2·33778	2·34373	2·34958	2·35532	2·36097	2·36652	2·37199
2	1·93516	1·94232	1·94934	1·95624	1·96301	1·96965	1·97618	1·98260	1·98891	1·99510
3	1·71616	1·72394	1·73158	1·73906	1·74641	1·75363	1·76071	1·76767	1·77451	1·78122
4	1·55736	1·56567	1·57381	1·58180	1·58963	1·59732	1·60487	1·61228	1·61955	1·62670
5	1·43023	1·43900	1·44760	1·45603	1·46430	1·47241	1·48036	1·48817	1·49584	1·50338
6	1·32274	1·33195	1·34097	1·34982	1·35848	1·36698	1·37532	1·38351	1·39154	1·39942
7	1·22869	1·23832	1·24774	1·25698	1·26603	1·27490	1·28360	1·29213	1·30051	1·30873
8	1·14443	1·15445	1·16427	1·17388	1·18329	1·19252	1·20157	1·21044	1·21915	1·22769
9	1·06760	1·07802	1·08821	1·09819	1·10797	1·11754	1·12693	1·13613	1·14516	1·15401
10	0·99662	1·00742	1·01799	1·02833	1·03846	1·04838	1·05810	1·06762	1·07696	1·08612
11	0·93034	0·94153	0·95247	0·96317	0·97365	0·98391	0·99395	1·00380	1·01345	1·02291
12	·86793	·87950	·89081	·90187	·91270	·92329	·93367	0·94383	0·95379	0·96355
13	·80873	·82068	·83237	·84379	·85496	·86590	·87660	·88708	·89735	·90741
14	·75224	·76459	·77665	·78843	·79996	·81123	·82226	·83306	·84364	·85400
15	·69807	·71081	·72324	·73540	·74727	·75889	·77025	·78138	·79226	·80293
16	0·64587	0·65901	0·67183	0·68436	0·69659	0·70856	0·72025	0·73170	0·74290	0·75387
17	·59538	·60893	·62214	·63504	·64764	·65996	·67200	·68377	·69529	·70657
18	·54637	·56033	·57395	·58723	·60020	·61288	·62526	·63737	·64921	·66080
19	·49864	·51303	·52705	·54073	·55408	·56712	·57985	·59230	·60447	·61638
20	·45202	·46685	·48129	·49537	·50911	·52252	·53561	·54841	·56091	·57314
21	0·40637	0·42164	0·43652	0·45101	0·46515	0·47894	0·49240	0·50555	0·51839	0·53095
22	·36155	·37729	·39260	·40752	·42207	·43625	·45009	·46360	·47680	·48969
23	·31745	·33366	·34944	·36480	·37976	·39435	·40857	·42245	·43601	·44925
24	·27396	·29066	·30691	·32272	·33812	·35312	·36775	·38201	·39594	·40953
25	·23098	·24820	·26494	·28122	·29706	·31249	·32753	·34219	·35649	·37045
26	0·18842	0·20618	0·22343	0·24019	0·25650	0·27237	0·28784	0·30290	0·31759	0·33192
27	·14621	·16452	·18230	·19957	·21636	·23269	·24859	·26408	·27917	·29389
28	·10425	·12315	·14148	·15927	·17656	·19337	·20973	·22565	·24116	·25627
29	·06248	·08198	·10089	·11923	·13704	·15435	·17118	·18755	·20349	·21902
30	·02081	·04096	·06047	·07938	·09774	·11556	·13288	·14972	·16611	·18207
31	—	0·00000	0·02014	0·03966	0·05858	0·07694	0·09478	0·11211	0·12896	0·14536
32	—	—	—	·00000	·01952	·03844	·05681	·07465	·09199	·10885
33	—	—	—	—	—	·00000	·01893	·03730	·05514	·07249
34	—	—	—	—	—	—	—	·00000	·01837	·03622
35	—	—	—	—	—	—	—	—	—	·00000

k \ n	70	71	72	73	74	75	76	77	78	79
1	2·37736	2·38265	2·38785	2·39298	2·39802	2·40299	2·40789	2·41271	2·41747	2·42215
2	2·00120	2·00720	2·01310	2·01890	2·02462	2·03024	2·03578	2·04124	2·04662	2·05191
3	1·78783	1·79432	1·80071	1·80699	1·81317	1·81926	1·82525	1·83115	1·83696	1·84268
4	1·63373	1·64063	1·64742	1·65410	1·66067	1·66714	1·67350	1·67976	1·68592	1·69200
5	1·51078	1·51805	1·52520	1·53223	1·53914	1·54594	1·55263	1·55921	1·56569	1·57207
6	1·40717	1·41478	1·42226	1·42961	1·43684	1·44395	1·45094	1·45782	1·46459	1·47125
7	1·31680	1·32473	1·33252	1·34017	1·34770	1·35510	1·36237	1·36953	1·37657	1·38350
8	1·23608	1·24431	1·25240	1·26034	1·26815	1·27583	1·28338	1·29080	1·29810	1·30529
9	1·16270	1·17123	1·17961	1·18784	1·19592	1·20387	1·21168	1·21936	1·22691	1·23434
10	1·09511	1·10393	1·11259	1·12110	1·12945	1·13766	1·14572	1·15365	1·16145	1·16912
11	1·03220	1·04130	1·05024	1·05902	1·06764	1·07610	1·08442	1·09260	1·10063	1·10854
12	0·97313	0·98252	0·99173	1·00078	1·00966	1·01838	1·02695	1·03537	1·04364	1·05178
13	·91728	·92695	·93644	0·94576	0·95490	0·96387	0·97269	0·98135	0·98986	0·99822
14	·86416	·87412	·88388	·89346	·90286	·91209	·92115	·93005	·93880	·94739
15	·81338	·82362	·83366	·84351	·85317	·86265	·87196	·88110	·89008	·89890
16	0·76462	0·77514	0·78546	0·79558	0·80550	0·81524	0·82480	0·83418	0·84339	0·85244
17	·71761	·72843	·73903	·74942	·75960	·76960	·77940	·78903	·79848	·80776
18	·67214	·68325	·69413	·70480	·71526	·72551	·73557	·74544	·75512	·76463
19	·62803	·63943	·65060	·66155	·67227	·68279	·69310	·70322	·71314	·72289
20	·58510	·59681	·60827	·61950	·63050	·64128	·65185	·66222	·67239	·68237
21	0·54323	0·55525	0·56701	0·57852	0·58980	0·60085	0·61168	0·62230	0·63272	0·64294
22	·50230	·51463	·52669	·53850	·55006	·56138	·57248	·58336	·59403	·60449
23	·46219	·47484	·48721	·49932	·51117	·52277	·53414	·54528	·55621	·56692
24	·42281	·43579	·44848	·46089	·47304	·48493	·49657	·50798	·51917	·53013
25	·38404	·39739	·41041	·42313	·43558	·44777	·45970	·47138	·48283	·49404

Table 2 (continued)

k \ n	70	71	72	73	74	75	76	77	78	79
26	0·34591	0·35958	0·37292	0·38597	0·39873	0·41122	0·42343	0·43540	0·44711	0·45859
27	·30825	·32227	·33596	·34934	·36242	·37521	·38772	·39997	·41196	·42371
28	·27102	·28540	·29945	·31317	·32657	·33968	·35250	·36504	·37731	·38934
29	·23416	·24893	·26333	·27740	·29114	·30457	·31770	·33055	·34311	·35542
30	·19762	·21277	·22756	·24199	·25608	·26984	·28329	·29645	·30931	·32190
31	0·16134	0·17690	0·19208	0·20688	0·22133	0·23543	0·24922	0·26269	0·27586	0·28875
32	·12527	·14125	·15683	·17202	·18684	·20130	·21543	·22923	·24272	·25591
33	·08936	·10579	·12178	·13737	·15257	·16740	·18188	·19602	·20983	·22334
34	·05357	·07045	·08688	·10289	·11848	·13370	·14854	·16303	·17718	·19101
35	·01785	·03520	·05209	·06852	·08453	·10014	·11536	·13021	·14471	·15888
36	—	0·00000	0·01736	0·03424	0·05068	0·06670	0·08231	0·09754	0·11240	0·12691
37	—	—	—	·00000	·01689	·03333	·04935	·06497	·08020	·09507
38	—	—	—	—	—	·00000	·01644	·03247	·04809	·06333
39	—	—	—	—	—	—	—	·00000	·01602	·03165
40	—	—	—	—	—	—	—	—	—	·00000

Table 2 is reproduced from *Biometrika* with the kind permission of the Biometrika Trust.

4. Probability integral of the maximum of correlated normal variables

Let $x_1, x_2, \ldots, x_N$ be distributed as multivariate normal with mean vector $\mathbf{0}$ and covariance matrix $\Omega = (\omega_{ij})$ where $\omega_{ii} = 1$ and $\omega_{ij} = c_i c_j$, $i \neq j$. In this case, Dunnett and Sobel (1954) showed that the above distribution can be generated by using the relations

$$x_i = \sqrt{1 - c_i^2}\, y_i - c_i y_0 \tag{4.1}$$

for $i = 1, 2, \ldots, N$ where $y_0, y_1, \ldots, y_N$ are distributed independent of each other as normal with mean zero and variance one. From (4.1), it follows easily that

$$P[\max(x_1, \ldots, x_N) \leq H] = \int_{-\infty}^{\infty} \left\{ F\left(\frac{z\rho^{1/2} + H}{(1-\rho)^{1/2}} \right) \right\}^N f(x)\, \mathrm{d}z \tag{4.2}$$

where $c_{ij} = \rho$,

$$f(x) = \frac{1}{\sqrt{2\pi}} \exp(-x^2/2), \qquad F(a) = \int_{-\infty}^{a} f(x)\, \mathrm{d}x .$$

Using (4.2), Gupta (1963) computed the values of the right side of (4.2) for $\rho = 0.100, 0.125, 0.200, 0.250, 0.300, \frac{1}{3}, 0.375, 0.400, 0.500, 0.600, \frac{2}{3}, 0.700, 0.750, 0.800, 0.875, 0.900$, $N = 1(1)12$, and $H = -3.50(0.10)3.50$. In Table 4, we reproduce the values for $\rho = 0.125, \frac{1}{3}, 0.500, \frac{2}{3}, 0.800, 0.900$. Gupta, Nagel and Panchapakesan (1973) constructed the percentage points of the maximum of equi-correlated normal variables.

Table 3
Variances and covariances of order statistics in samples of sizes up to 20 from a normal population

n	i	j	Value	n	i	j	Value	n	i	j	Value
2	1	1	.6816901139	7				9			
		2	.3183098861		3	3	.2197215626		4	4	.1706588464
	2	2	.6816901139			4	.1655598429			5	.1369913669
3	1	1	.5594672038			5	.1290048425			6	.1126671842
		2	.2756644477		4	4	.2104468615		5	5	.1661012814
		3	.1648683485	8	1	1	.3728071434	10	1	1	.3443438233
	2	2	.4486711046			2	.1863073997			2	.1712629030
4	1	1	.4917152369			3	.1259660300			3	.1162600989
		2	.2455926930			4	.0947230277			4	.0882494247
		3	.1580080701			5	.0747650242			5	.0707413677
		4	.1046840000			6	.0602075169			6	.0583987134
	2	2	.3604553434			7	.0482085508			7	.0489206279
		3	.2359438035			8	.0368353073			8	.0410844589
5	1	1	.4475340691		2	2	.2394010458			9	.0340406470
		2	.2243309506			3	.1631958727			10	.0266989351
		3	.1481477252			4	.1232633317		2	2	.2145241430
		4	.1057710776			5	.0975647193			3	.1466226180
		5	.0742152685			6	.0787224662			4	.1117015961
	2	2	.3115189521			7	.0632468118			5	.0897428245
		3	.2084354440		3	3	.2007687900			6	.0741996414
		4	.1499420668			4	.1623584312			7	.0622278486
	3	3	.2868336616			5	.1209637555			8	.0523067222
6	1	1	.4159271090			6	.0978171355			9	.0433711561
		2	.2085030023		4	4	.1871862195		3	3	.1760032834
		3	.1394352565			5	.1491754908			4	.1338022448
		4	.1024293940	9	1	1	.3573533264			5	.1077445338
		5	.0773637839			2	.1781434240			6	.0892254012
		6	.0563414544			3	.1207454442			7	.0749183943
	2	2	.2795777392			4	.0913071400			8	.0630332449
		3	.1889859560			5	.0727422354		4	4	.1570389144
		4	.1396640604			6	.0594831125			5	.1275089296
		5	.1059054582			7	.0490764081			6	.1057868169
	3	3	.2462125354			8	.0400936927			7	.0889462026
		4	.1832727978			9	.0310552188		5	5	.1510639039
7	1	1	.3919177761		2	2	.2256968778			6	.1255969678
		2	.1961990246			3	.1541163526	11	1	1	.3332474428
		3	.1321155811			4	.1170056918			2	.1653647712
		4	.0984868607			5	.0934477394			3	.1123684351
		5	.0765598346			6	.0765461431			4	.0855170698
		6	.0599187124			7	.0632354695			5	.0688483064
		7	.0448022105			8	.0517146001			6	.0572007588
	2	2	.2567328862		3	3	.1863826133			7	.0483754063
		3	.1744833274			4	.1420779776			8	.0412423472
		4	.1307298658			5	.1137080176			9	.0351103357
		5	.1019550089			6	.0933625386			10	.0294198503
		6	.0799811748			7	.0772351806			11	.0233152868

Table 3 (continued)

n	i	j	Value	n	i	j	Value	n	i	j	Value
11				12	3			13	3		
	2	2	.2051975798			4	.1212063211			8	.0589221432
		3	.1403096511			5	.0982806602			9	.0514460445
		4	.1071492595			6	.0822228461			10	.0449637542
		5	.0864430257			7	.0701213964			11	.0390643799
		6	.0719305024			8	.0604384621		4	4	.1330111820
		7	.0608869662			9	.0522825611			5	.1082512667
		8	.0519504506			10	.0450357615			6	.0909855605
		9	.0442549455		4	4	.1398109405			7	.0780173339
		10	.0371029077			5	.1135687821			8	.0677217143
	3	3	.1657242880			6	.0951645279			9	.0591628729
		4	.1269672025			7	.0812419810			10	.0517328050
		5	.1026407291			8	.0700795832		5	5	.1232503256
		6	.0855178832			9	.0606620874			6	.1037367701
		7	.0724741050		5	5	.1306137359			7	.0890434754
		8	.0618873278			6	.1096212247			8	.0773552864
		9	.0527550069			7	.0936951520			9	.0676230904
	4	4	.1479546565			8	.0808972960		6	6	.1183175325
		5	.1198752861		6	6	.1266377911			7	.1016824204
		6	.1000346585			7	.1083945831			8	.0884194610
		7	.0848765182	13	1	1	.3152053842		7	7	.1167989950
		8	.0725451434			2	.1557272904	14	1	1	.3077301026
	5	5	.1396410804			3	.1058908842			2	.1517203662
		6	.1167449805			4	.0808649736			3	.1031719531
		7	.0991935960			5	.0654634490			4	.0788715916
	6	6	.1371624335			6	.0548221797			5	.0639657428
12	1	1	.3236363870			7	.0468833088			6	.0537064714
		2	.1602373762			8	.0406132548			7	.0460899180
		3	.1089309641			9	.0354226462			8	.0401141688
		4	.0830686767			10	.0309322744			9	.0352141760
		5	.0670884464			11	.0268537250			10	.0310371163
		6	.0559933694			12	.0228858068			11	.0273362865
		7	.0476620974			13	.0184348220			12	.0239061001
		8	.0410208554		2	2	.1904130721			13	.0205080257
		9	.0354439060			3	.1302055829			14	.0166279801
		10	.0305012591			4	.0997262696		2	2	.1844200252
		11	.0257945392			5	.0808785938			3	.1260791989
		12	.0208221233			6	.0678145832			4	.0966524633
	2	2	.1972846039			7	.0580457285			5	.0785202981
		3	.1349020328			8	.0503167946			6	.0660028340
		4	.1031959206			9	.0439095087			7	.0566896715
		5	.0835045822			10	.0383601798			8	.0493708148
		6	.0697859658			11	.0333147765			9	.0433617156
		7	.0594590652			12	.0284018130			10	.0382337404
		8	.0512113198		3	3	.1513917013			11	.0336863221
		9	.0442747124			4	.1162698131			12	.0294681314
		10	.0381191478			5	.0944566603			13	.0252863928
		11	.0322507340			6	.0792922993		3	3	.1457045665
	3	3	.1579786877			7	.0679282354			4	.1119816877

Table 3 (continued)

n	i	j	Value	n	i	j	Value	n	i	j	Value
14	3			15	2			16	1		
		5	.0911181271			6	.0643390895			3	.0985009764
		6	.0768754957			7	.0554074400			4	.0754040023
		7	.0659084825			8	.0484238833			5	.0613086724
		8	.0574341188			9	.0427294113			6	.0516624963
		9	.0504677802			10	.0379177516			7	.0445503705
		10	.0445169192			11	.0337151721			8	.0390194716
		11	.0392352316			12	.0299152347			9	.0345378158
		12	.0343322071			13	.0263303885			10	.0307810003
	4	4	.1272273070			14	.0227213594			11	.0275353612
		5	.1036931108		3	3	.1407322502			12	.0246479007
		6	.0873562483			4	.1082138452			13	.0219956755
		7	.0751519909			5	.0881605755			14	.0194585037
		8	.0655310936			6	.0743268436			15	.0168710289
		9	.0576120957			7	.0640558183			16	.0138287378
		10	.0508402240			8	.0560136122		2	2	.1743940788
		11	.0448243469			9	.0494485109			3	.1191409287
	5	5	.1171012461			10	.0438960670			4	.0914350918
		6	.0987747550			11	.0390426915			5	.0744591145
		7	.0850536546			12	.0346513382			6	.0628003900
		8	.0742181416			13	.0305060359			7	.0542033941
		9	.0652867776		4	4	.1222328270			8	.0475009760
		10	.0576401464			5	.0997323941			9	.0420638230
	6	6	.1115324579			6	.0841705696			10	.0375018250
		7	.0961405595			7	.0725946869			11	.0335574012
		8	.0839617110			8	.0635175007			12	.0300461208
		9	.0739069221			9	.0560990511			13	.0268180579
	7	7	.1090269480			10	.0498187836			14	.0237301562
		8	.0953087256			11	.0443247452			15	.0205785433
15	1	1	.3010415703			12	.0393501820		3	3	.1363385612
		2	.1481297708		5	5	.1118698986			4	.1048706756
		3	.1007223449			6	.0945206004			5	.0855189036
		4	.0770594060			7	.0815891122			6	.0722075087
		5	.0625845851			8	.0714331681			7	.0623568515
		6	.0526530129			9	.0631224388			8	.0546749107
		7	.0453078886			10	.0560795005			9	.0484366006
		8	.0395736873			11	.0499127743			10	.0431079377
		9	.0349035905		6	6	.1058606366			11	.0386652905
		10	.0309614122			7	.0914683204			12	.0346277256
		11	.0275211039			8	.0801407559			13	.0309149135
		12	.0244126313			9	.0708582090			14	.0273595378
		13	.0214819828			10	.0629824402		4	4	.1178657554
		14	.0185333263		7	7	.1026916923			5	.0962513413
		15	.0151137071			8	.0900499964			6	.0813480448
	2	2	.1791215291			9	.0796738323			7	.0703000911
		3	.1224176953		8	8	.1016946521			8	.0616728900
		4	.0939067144	16	1	1	.2950098090			9	.0546595026
		5	.0763912337			2	.1448881089			10	.0487647746

Table 3 (continued)

n	i	j	Value	n	i	j	Value	n	i	j	Value
16	4			17	2			17	6		
		11	.0436607328			9	.0413928192			12	.0478122500
		12	.0391112669			10	.0370349110		7	7	.0920031780
		13	.0349253749			11	.0332940802			8	.0818194607
	5	5	.1073517089			12	.0299082825			9	.0728154074
		6	.0908232622			13	.0270170379			10	.0652667274
		7	.0785480532			14	.0242386812			11	.0587626219
		8	.0689488802			15	.0215459396		8	8	.0907361650
		9	.0611364182			16	.0187658306			9	.0808000287
		10	.0545638941		3	3	.1324207975			10	.0724509963
		11	.0488684327			4	.1018792434		9	9	.0900465814
		12	.0437882959			5	.0831421716	18	1	1	.2845301297
	6	6	.1010461906			6	.0702850403			2	.1392501620
		7	.0874627156			7	.0607964413			3	.0946172637
		8	.0768239668			8	.0534208202			4	.0724851730
		9	.0681545540			9	.0474555487			5	.0590304274
		10	.0608534805			10	.0424726884			6	.0498600635
		11	.0545210724			11	.0381025587			7	.0431302310
	7	7	.0974026613			12	.0344194567			8	.0370260195
		8	.0856181910			13	.0310047771			9	.0337388141
		9	.0760015577			14	.0278210708			10	.0302610667
		10	.0678931922			15	.0247342095			11	.0272938041
	8	8	.0957213007		4	4	.1140068197			12	.0247002471
		9	.0850291218			5	.0931620330			13	.0223801573
17	1	1	.2895330037			6	.0788266621			14	.0202537421
		2	.1419424629			7	.0682298909			15	.0182488619
		3	.0964748737			8	.0599826092			16	.0162850441
		4	.0738849615			9	.0533057575			17	.0142368875
		5	.0601272302			10	.0477239973			18	.0117719054
		6	.0507326948			11	.0429261816		2	2	.1662929294
		7	.0438236491			12	.0386042630			3	.1135058132
		8	.0384672834			13	.0348624030			4	.0871597604
		9	.0341441055			14	.0312881041			5	.0710825990
		10	.0305380548		5	5	.1034004377			6	.0600975754
		11	.0274465527			6	.0875720930			7	.0520217423
		12	.0247237144			7	.0758534534			8	.0457683625
		13	.0222620771			8	.0667204245			9	.0407317967
		14	.0199600651			9	.0593187706			10	.0365451034
		15	.0177476891			10	.0531257771			11	.0329704894
		16	.0154552071			11	.0477987292			12	.0298442464
		17	.0127264751			12	.0430070793			13	.0270462281
	2	2	.1701426762			13	.0388375657			14	.0244806350
		3	.1161866734		6	6	.0968824669			15	.0220607111
		4	.0891982557			7	.0839811738			16	.0196804667
		5	.0726970385			8	.0739130200			17	.0172154925
		6	.0613908459			9	.0657442736		3	3	.1288098043
		7	.0530761573			10	.0589030403			4	.0991828539
		8	.0466140918			11	.0530137275			5	.0809890792

Table 3 (continued)

n	i	j	Value	n	i	j	Value	n	i	j	Value
18	3			18				19	3		
		6	.0685324700		8	8	.0864960639			8	.0511541418
		7	.0593598602			9	.0771762286			9	.0456228816
		8	.0522488413			10	.0693891332			10	.0410365629
		9	.0465162123			11	.0627116906			11	.0371346427
		10	.0417473296		9	9	.0853127880			12	.0337391171
		11	.0376730987			10	.0767442321			13	.0307215918
		12	.0341080171	19	1	1	.2799358050			14	.0279835020
		13	.0309157650			2	.1367768168			15	.0254424108
		14	.0279875014			3	.0929061763			16	.0230195063
		15	.0252244786			4	.0711902425			17	.0206214645
		16	.0225161109			5	.0580094835		4	4	.1074740839
	4	4	.1105660331			6	.0490405678			5	.0879061905
		5	.0903973787			7	.0424705246			6	.0745033878
		6	.0765579277			8	.0374006329			7	.0646406188
		7	.0663522086			9	.0333319395			8	.0570032284
		8	.0584310521			10	.0299634144			9	.0508572608
		9	.0520394281			11	.0271011338			10	.0457576598
		10	.0467183404			12	.0246129452			11	.0414165001
		11	.0421694861			13	.0224037540			12	.0376368753
		12	.0381869632			14	.0204007370			13	.0342765540
		13	.0346192645			15	.0185431530			14	.0312262549
		14	.0313452497			16	.0167731147			15	.0283944527
		15	.0282548286			17	.0150223067			16	.0256935148
	5	5	.0999084321			18	.0131789994		5	5	.0967944745
		6	.0846879168			19	.0109382527			6	.0821055695
		7	.0734460811		2	2	.1627856651			7	.0712796742
		8	.0647101858			3	.1110590145			8	.0628870005
		9	.0576543520			4	.0852931063			9	.0561272025
		10	.0517756675			5	.0695970759			10	.0505141639
		11	.0467468133			6	.0588910196			11	.0457330144
		12	.0423415563			7	.0510351093			12	.0415681234
		13	.0383932046			8	.0449652247			13	.0378636088
		14	.0347682770			9	.0400891754			14	.0344995261
	6	6	.0932407331			10	.0360490040			15	.0313752928
		7	.0800202644			11	.0326137544		6	6	.0900218693
		8	.0713338046			12	.0296258236			7	.0782029063
		9	.0635829688			13	.0269716592			8	.0690294360
		10	.0571197288			14	.0245641909			9	.0616336896
		11	.0515868552			15	.0223306885			10	.0554877905
		12	.0467370896			16	.0202017247			11	.0502493169
		13	.0423879846			17	.0180952193			12	.0456834841
	7	7	.0890167025			18	.0158767294			13	.0416203596
		8	.0785179677		3	3	.1257138904			14	.0379290224
		9	.0700199026			4	.0967367097		7	7	.0856172981
		10	.0629269074			5	.0790298792			8	.0756153413
		11	.0568501034			6	.0669273696			9	.0675433161
		12	.0515199092			7	.0580336124			10	.0608297030

Table 3 (continued)

n	i	j	Value	n	i	j	Value	n	i	j	Value
19	7			20	2			20	5		
		11	.0551032224			11	.0322405467			7	.0693175756
		12	.0501089625			12	.0293684960			8	.0612251429
		13	.0456621835			13	.0268315105			9	.0547222526
	8	8	.0828339961			14	.0245479493			10	.0493374275
		9	.0740273546			15	.0224526609			11	.0447662310
		10	.0666958229			16	.0204888032			12	.0408014074
		11	.0604372723			17	.0185994024			13	.0372948400
		12	.0549752083			18	.0167136502			14	.0341351571
	9	9	.0812876330			19	.0147107671			15	.0312332040
		10	.0732703911		3	3	.1228134687			16	.0285109200
		11	.0664202898			4	.0945049010		6	6	.0871511254
	10	10	.0807900751			5	.0772355098			7	.0757703360
20	1	1	.2756066156			6	.0654510179			8	.0669555780
		2	.1344941714			7	.0568056677			9	.0598659769
		3	.0913234064			8	.0501310269			10	.0539910639
		4	.0699870991			9	.0447763202			11	.0490008080
		5	.0570566384			10	.0403482354			12	.0446702771
		6	.0482701093			11	.0365934287			13	.0408385549
		7	.0418437826			12	.0333397949			14	.0373845194
		8	.0368937058			13	.0304645792			15	.0342111024
		9	.0329296302			14	.0278756579		7	7	.0826123055
		10	.0296562523			15	.0254904381			8	.0730383676
		11	.0268838808			16	.0232716371			9	.0653307665
		12	.0244839567			17	.0211277373			10	.0589387428
		13	.0223640803			18	.0189874448			11	.0535056766
		14	.0204584277		4	4	.1046766243			12	.0487882257
		15	.0187096782			5	.0856442356			13	.0446121000
		16	.0170711408			6	,0726321560			14	.0408450989
		17	.0154951854			7	.0630731775		8	8	.0796309757
		18	.0139227072			8	.0556855081			9	.0712591607
		19	.0122530117			9	.0497539273			10	.0643103375
		20	.0102047204			10	.0448455403			11	.0583997310
	2	2	.1505731636			11	.0406811669			12	.0532644495
		3	.1088143707			12	.0370700493			13	.0487159834
		4	.0835758044			13	.0338793392		9	9	.0778118317
		5	.0682247554			14	.0310045146			10	.0702526464
		6	.0577699656			15	.0283650517			11	.0638176734
		7	.0501109523			16	.0258897454			12	.0582229133
		8	.0442041191			17	.0235070343		10	10	.0769474356
		9	.0394693443		5	5	.0939960007			11	.0699266108
		10	.0355565554			6	.0797773755				

Table 3 is reproduced from the *Annals of Mathematical Statistics* with the kind permission of the Institute of Mathematical Statistics.

Table 4
Probability that N standard normal random variables with common correlation ρ are simultaneously less than or equal to H; $\rho = 0.125$

$-H$ \ N	1	2	3	4	5	6	7	8	9	10	11	12
3.50	.00023	.00000	.00000	.00000	.00000	.00000	.00000	.00000	.00000	.00000	.00000	.00000
3.40	.00034	.00000	.00000	.00000	.00000	.00000	.00000	.00000	.00000	.00000	.00000	.00000
3.30	.00048	.00000	.00000	.00000	.00000	.00000	.00000	.00000	.00000	.00000	.00000	.00000
3.20	.00069	.00000	.00000	.00000	.00000	.00000	.00000	.00000	.00000	.00000	.00000	.00000
3.10	.00097	.00000	.00000	.00000	.00000	.00000	.00000	.00000	.00000	.00000	.00000	.00000
3.00	.00135	.00001	.00000	.00000	.00000	.00000	.00000	.00000	.00000	.00000	.00000	.00000
2.90	.00187	.00001	.00000	.00000	.00000	.00000	.00000	.00000	.00000	.00000	.00000	.00000
2.80	.00256	.00002	.00000	.00000	.00000	.00000	.00000	.00000	.00000	.00000	.00000	.00000
2.70	.00347	.00003	.00000	.00000	.00000	.00000	.00000	.00000	.00000	.00000	.00000	.00000
2.60	.00466	.00006	.00000	.00000	.00000	.00000	.00000	.00000	.00000	.00000	.00000	.00000
2.50	.00621	.00009	.00000	.00000	.00000	.00000	.00000	.00000	.00000	.00000	.00000	.00000
2.40	.00820	.00016	.00001	.00000	.00000	.00000	.00000	.00000	.00000	.00000	.00000	.00000
2.30	.01072	.00025	.00001	.00000	.00000	.00000	.00000	.00000	.00000	.00000	.00000	.00000
2.20	.01390	.00040	.00002	.00000	.00000	.00000	.00000	.00000	.00000	.00000	.00000	.00000
2.10	.01786	.00064	.00004	.00000	.00000	.00000	.00000	.00000	.00000	.00000	.00000	.00000
2.00	.02275	.00098	.00007	.00001	.00000	.00000	.00000	.00000	.00000	.00000	.00000	.00000
1.90	.02872	.00149	.00012	.00001	.00000	.00000	.00000	.00000	.00000	.00000	.00000	.00000
1.80	.03593	.00224	.00021	.00003	.00000	.00000	.00000	.00000	.00000	.00000	.00000	.00000
1.70	.04457	.00330	.00036	.00005	.00001	.00000	.00000	.00000	.00000	.00000	.00000	.00000
1.60	.05480	.00480	.00059	.00010	.00002	.00000	.00000	.00000	.00000	.00000	.00000	.00000
1.50	.06681	.00686	.00097	.00018	.00004	.00001	.00000	.00000	.00000	.00000	.00000	.00000
1.40	.08076	.00967	.00155	.00031	.00008	.00002	.00001	.00000	.00000	.00000	.00000	.00000
1.30	.09680	.01343	.00244	.00054	.00014	.00004	.00001	.00001	.00000	.00000	.00000	.00000
1.20	.11507	.01838	.00375	.00093	.00027	.00009	.00003	.00001	.00001	.00000	.00000	.00000
1.10	.13567	.02479	.00566	.00154	.00048	.00017	.00006	.00003	.00001	.00001	.00000	.00000
1.00	.15866	.03295	.00838	.00250	.00085	.00032	.00013	.00006	.00003	.00001	.00001	.00000
0.90	.18406	.04318	.01217	.00396	.00145	.00058	.00025	.00012	.00006	.00003	.00002	.00001
0.80	.21186	.05580	.01734	.00614	.00241	.00103	.00047	.00023	.00012	.00006	.00004	.00008
0.70	.24196	.07112	.02425	.00930	.00392	.00178	.00087	.00045	.00024	.00013	.00008	.00005
0.60	.27425	.08942	.03330	.01377	.00620	.00300	.00154	.00083	.00047	.00027	.00016	.00010
0.50	.30854	.11096	.04491	.01997	.00958	.00490	.00264	.00149	.00088	.00053	.00033	.00021
0.40	.34458	.13589	.05951	.02833	.01444	.00779	.00441	.00260	.00159	.00100	.00065	.00043
0.30	.38209	.16432	.07749	.03936	.02124	.01205	.00714	.00439	.00279	.00182	.00122	.00083
0.20	.42074	.19623	.09923	.05356	.03050	.01817	.01125	.00720	.00474	.00320	.00221	.00155
0.10	.46017	.23152	.12499	.07144	.04281	.02670	.01722	.01144	.00780	.00544	.00387	.00280
0.00	.50000	.26995	.15492	.09344	.05875	.03825	.02566	.01766	.01244	.00894	.00654	.00486

H \ N	1	2	3	4	5	6	7	8	9	10	11	12
0.00	.50000	.26995	.15492	.09344	.05875	.03825	.02566	.01766	.01244	.00894	.00654	.00486
0.10	.53983	.31117	.18905	.11992	.07885	.05346	.03721	.02650	.01925	.01424	.01070	.00815
0.20	.57926	.35475	.22724	.15106	.10360	.07296	.05258	.03865	.02892	.02198	.01694	.01322
0.30	.61791	.40014	.26919	.18692	.13332	.09729	.07242	.05486	.04220	.03291	.02598	.02074
0.40	.65542	.44673	.31443	.22733	.16815	.12685	.09734	.07583	.05987	.04783	.03863	.03150
0.50	.69146	.49388	.36235	.27192	.20804	.16185	.12777	.10217	.08265	.06755	.05573	.04636
0.60	.72575	.54092	.41221	.32011	.25264	.20223	.16390	.13430	.11114	.09279	.07809	.06619
0.70	.75804	.58719	.46321	.37115	.30142	.24769	.20568	.17239	.14570	.12407	.10637	.09177
0.80	.78814	.63209	.51449	.42414	.35357	.29764	.25274	.21629	.18641	.16167	.14103	.12368
0.90	.81594	.67506	.56519	.47811	.40814	.35121	.30441	.26555	.23300	.20554	.18220	.16223
1.00	.84134	.71564	.61450	.53204	.46404	.40737	.35973	.31935	.28487	.25524	.22961	.20733
1.10	.86433	.75346	.66170	.58495	.52013	.46493	.41756	.37663	.34105	.30995	.28263	.25850
1.20	.88493	.78824	.70618	.63593	.57531	.52264	.47659	.43610	.40032	.36856	.34023	.31487
1.30	.90320	.81983	.74746	.68419	.62853	.57929	.53549	.49637	.46127	.42966	.40108	.37517
1.40	.91924	.84816	.78520	.72912	.67891	.63376	.59299	.55603	.52241	.49174	.46366	.43790
1.50	.93319	.87325	.81920	.77025	.72574	.68512	.64792	.61377	.58230	.55325	.52636	.50141
1.60	.94520	.89520	.84940	.80731	.76850	.73261	.69935	.66843	.63963	.61276	.58762	.56407
1.70	.95543	.91417	.87585	.84018	.80688	.77573	.74653	.71911	.69331	.66900	.64605	.62435
1.80	.96407	.93038	.89871	.86890	.84077	.81418	.78902	.76516	.74252	.72099	.70050	.68097
1.90	.97128	.94406	.91821	.89363	.87022	.84790	.82658	.80621	.78671	.76804	.75013	.73295
2.00	.97725	.95548	.93463	.91463	.89542	.87697	.85921	.84212	.82565	.80976	.79443	.77962
2.10	.98214	.96491	.94827	.93221	.91667	.90164	.88708	.87298	.85931	.84605	.83317	.82067
2.20	.98610	.97260	.95948	.94673	.93433	.92226	.91051	.89907	.88791	.87703	.86642	.85606
2.30	.98928	.97880	.96858	.95858	.94881	.93925	.92990	.92075	.91179	.90302	.89442	.88600
2.40	.99180	.98376	.97587	.96813	.96052	.95305	.94572	.93851	.93142	.92446	.91761	.91087
2.50	.99379	.98768	.98165	.97572	.96987	.96411	.95843	.95283	.94731	.94187	.93649	.93119
2.60	.99534	.99073	.98618	.98169	.97725	.97286	.96852	.96422	.95998	.95578	.95163	.94753
2.70	.99653	.99310	.98970	.98633	.98299	.97968	.97641	.97316	.96994	.96675	.96359	.96046
2.80	.99744	.99491	.99239	.98989	.98741	.98495	.98251	.98008	.97767	.97528	.97290	.97054
2.90	.99813	.99628	.99443	.99260	.99078	.98897	.98716	.98537	.98359	.98182	.98005	.97830
3.00	.99865	.99731	.99597	.99464	.99331	.99199	.99068	.98937	.98807	.98677	.98548	.98419
3.10	.99903	.99807	.99711	.99615	.99519	.99424	.99330	.99235	.99141	.99047	.98954	.98860
3.20	.99931	.99863	.99794	.99726	.99658	.99590	.99523	.99455	.99388	.99321	.99254	.99187
3.30	.99952	.99903	.99855	.99807	.99759	.99711	.99664	.99616	.99568	.99521	.99473	.99426
3.40	.99966	.99933	.99899	.99865	.99832	.99799	.99765	.99732	.99699	.99665	.99632	.99599
3.50	.99977	.99953	.99930	.99907	.99884	.99861	.99838	.99815	.99792	.99768	.99745	.99722

Table 4 (continued)

$\rho = \frac{1}{3}$

$-H$ \ N	1	2	3	4	5	6	7	8	9	10	11	12
3.50	.00023	.00000	.00000	.00000	.00000	.00000	.00000	.00000	.00000	.00000	.00000	.00000
3.40	.00034	.00000	.00000	.00000	.00000	.00000	.00000	.00000	.00000	.00000	.00000	.00000
3.30	.00048	.00001	.00000	.00000	.00000	.00000	.00000	.00000	.00000	.00000	.00000	.00000
3.20	.00069	.00001	.00000	.00000	.00000	.00000	.00000	.00000	.00000	.00000	.00000	.00000
3.10	.00097	.00002	.00000	.00000	.00000	.00000	.00000	.00000	.00000	.00000	.00000	.00000
3.00	.00135	.00003	.00000	.00000	.00000	.00000	.00000	.00000	.00000	.00000	.00000	.00000
2.90	.00187	.00005	.00000	.00000	.00000	.00000	.00000	.00000	.00000	.00000	.00000	.00000
2.80	.00256	.00008	.00001	.00000	.00000	.00000	.00000	.00000	.00000	.00000	.00000	.00000
2.70	.00347	.00013	.00001	.00000	.00000	.00000	.00000	.00000	.00000	.00000	.00000	.00000
2.60	.00466	.00020	.00002	.00000	.00000	.00000	.00000	.00000	.00000	.00000	.00000	.00000
2.50	.00621	.00031	.00004	.00001	.00000	.00000	.00000	.00000	.00000	.00000	.00000	.00000
2.40	.00820	.00048	.00007	.00002	.00000	.00000	.00000	.00000	.00000	.00000	.00000	.00000
2.30	.01072	.00072	.00011	.00003	.00001	.00000	.00000	.00000	.00000	.00000	.00000	.00000
2.20	.01390	.00108	.00018	.00005	.00002	.00001	.00000	.00000	.00000	.00000	.00000	.00000
2.10	.01786	.00159	.00029	.00008	.00003	.00001	.00001	.00000	.00000	.00000	.00000	.00000
2.00	.02275	.00231	.00046	.00014	.00005	.00002	.00001	.00001	.00000	.00000	.00000	.00000
1.90	.02872	.00331	.00072	.00023	.00009	.00004	.00002	.00001	.00001	.00000	.00000	.00000
1.80	.03593	.00469	.00111	.00037	.00015	.00007	.00004	.00002	.00001	.00001	.00001	.00000
1.70	.04457	.00656	.00168	.00059	.00026	.00013	.00007	.00004	.00003	.00002	.00001	.00001
1.60	.05480	.00906	.00251	.00093	.00042	.00022	.00012	.00007	.00005	.00003	.00002	.00002
1.50	.06681	.01234	.00368	.00145	.00068	.00036	.00021	.00013	.00008	.00006	.00004	.00003
1.40	.08076	.01661	.00532	.00220	.00108	.00059	.00035	.00022	.00015	.00010	.00007	.00006
1.30	.09680	.02207	.00758	.00330	.00168	.00095	.00058	.00038	.00026	.00018	.00013	.00010
1.20	.11507	.02896	.01063	.00486	.00257	.00150	.00094	.00063	.00043	.00031	.00023	.00018
1.10	.13567	.03754	.01469	.00705	.00386	.00232	.00150	.00102	.00072	.00053	.00040	.00031
1.00	.15866	.04807	.02002	.01005	.00571	.00353	.00233	.00162	.00117	.00087	.00066	.00052
0.90	.18406	.06083	.02688	.01411	.00829	.00528	.00357	.00253	.00186	.00141	.00109	.00086
0.80	.21186	.07607	.03558	.01948	.01185	.00776	.00537	.00388	.00290	.00223	.00175	.00140
0.70	.24196	.09402	.04644	.02649	.01664	.01119	.00792	.00583	.00443	.00346	.00275	.00223
0.60	.27425	.11489	.05978	.03547	.02298	.01586	.01147	.00860	.00665	.00526	.00425	.00349
0.50	.30854	.13881	.07591	.04578	.03123	.02209	.01631	.01246	.00978	.00785	.00642	.00533
0.40	.34458	.16586	.09512	.06076	.04176	.03024	.02280	.01772	.01413	.01150	.00952	.00799
0.30	.38209	.19604	.11763	.07778	.05495	.04072	.03130	.02475	.02003	.01651	.01383	.01174
0.20	.42074	.22928	.14360	.09813	.07119	.05393	.04224	.03395	.02787	.02320	.01972	.01692
0.10	.46017	.26538	.17311	.12206	.09082	.07028	.05604	.04576	.03808	.03221	.02760	.02392
0.00	.50000	.30409	.20613	.14974	.11413	.09012	.07311	.06061	.05113	.04375	.03790	.03318

H \ N	1	2	3	4	5	6	7	8	9	10	11	12
0.00	.50000	.30409	.20613	.14974	.11413	.09012	.07311	.06061	.05113	.04375	.03790	.03318
0.10	.53983	.34504	.24251	.18121	.14133	.11377	.09384	.07893	.06744	.05838	.05111	.04516
0.20	.57926	.38780	.28201	.21642	.17250	.14144	.11853	.10109	.08744	.07654	.06768	.06036
0.30	.61791	.43187	.32423	.25516	.20763	.17324	.14739	.12737	.11149	.09863	.08805	.07922
0.40	.65542	.47670	.36871	.29711	.24653	.20913	.18049	.15796	.13983	.12497	.11260	.10217
0.50	.69146	.52173	.41489	.34180	.28887	.24891	.21777	.19288	.17258	.15574	.14157	.12950
0.60	.72575	.56638	.46212	.38866	.33418	.29223	.25897	.23200	.20971	.19100	.17508	.16140
0.70	.75804	.61010	.50974	.43703	.38187	.33858	.30371	.27501	.25100	.23061	.21310	.19789
0.80	.78814	.65236	.55707	.48618	.43124	.38733	.35140	.32143	.29605	.27426	.25536	.23880
0.90	.81594	.69271	.60344	.53536	.48150	.43771	.40135	.37062	.34429	.32145	.30144	.28377
1.00	.84134	.73076	.64824	.58381	.53186	.48892	.45275	.42180	.39499	.37150	.35073	.33223
1.10	.86433	.76621	.69093	.63084	.58150	.54009	.50473	.47412	.44731	.42360	.40245	.38346
1.20	.88493	.79882	.73104	.67581	.62968	.59039	.55641	.52665	.50033	.47683	.45570	.43657
1.30	.90320	.82847	.76822	.71818	.67570	.63902	.60691	.57849	.55310	.53024	.50951	.49062
1.40	.91924	.85509	.80223	.75753	.71900	.68529	.65545	.62876	.60471	.58286	.56291	.54458
1.50	.93319	.87873	.83292	.79354	.75912	.72863	.70134	.67670	.65430	.63380	.61493	.59748
1.60	.94520	.89946	.86026	.82604	.79573	.76858	.74403	.72166	.70115	.68224	.66471	.64841
1.70	.95543	.91743	.88431	.85497	.82866	.80484	.78310	.76312	.74466	.72752	.71153	.69656
1.80	.96407	.93283	.90518	.88036	.85785	.83727	.81831	.80076	.78441	.76913	.75479	.74129
1.90	.97128	.94588	.92307	.90235	.88336	.86583	.84956	.83437	.82013	.80674	.79409	.78213
2.00	.97725	.95681	.93821	.92114	.90533	.89062	.87686	.86392	.85172	.84017	.82921	.81878
2.10	.98214	.96586	.95088	.93698	.92401	.91184	.90037	.88952	.87922	.86942	.86007	.85113
2.20	.98610	.97327	.96134	.95017	.93966	.92973	.92031	.91134	.90279	.89460	.88676	.87922
2.30	.98928	.97927	.96988	.96102	.95261	.94462	.93699	.92969	.92269	.91595	.90947	.90322
2.40	.99180	.98408	.97677	.96982	.96319	.95684	.95074	.94488	.93923	.93378	.92851	.92341
2.50	.99379	.98789	.98226	.97688	.97171	.96673	.96193	.95729	.95280	.94845	.94423	.94012
2.60	.99534	.99088	.98659	.98247	.97849	.97463	.97090	.96728	.96376	.96034	.95701	.95376
2.70	.99653	.99319	.98997	.98684	.98381	.98087	.97801	.97522	.97250	.96984	.96725	.96471
2.80	.99744	.99497	.99256	.99023	.98795	.98573	.98356	.98144	.97937	.97734	.97535	.97339
2.90	.99813	.99632	.99454	.99281	.99112	.98947	.98784	.98625	.98469	.98316	.98165	.98017
3.00	.99865	.99733	.99604	.99477	.99353	.99231	.99111	.98993	.98877	.98763	.98650	.98540
3.10	.99903	.99808	.99715	.99623	.99533	.99444	.99357	.99270	.99185	.99101	.99018	.98937
3.20	.99931	.99863	.99797	.99731	.99666	.99602	.99539	.99477	.99415	.99354	.99294	.99234
3.30	.99951	.99904	.99857	.99810	.99764	.99718	.99673	.99629	.99585	.99541	.99498	.99455
3.40	.99966	.99933	.99900	.99867	.99835	.99803	.99771	.99739	.99708	.99677	.99647	.99616
3.50	.99977	.99953	.99931	.99908	.99885	.99863	.99841	.99819	.99797	.99775	.99754	.99732

Table 4 (continued)
$\rho = 0.500$

-H \ N	1	2	3	4	5	6	7	8	9	10	11	12
3.50	.00023	.00001	.00000	.00000	.00000	.00000	.00000	.00000	.00000	.00000	.00000	.00000
3.40	.00034	.00001	.00000	.00000	.00000	.00000	.00000	.00000	.00000	.00000	.00000	.00000
3.30	.00048	.00002	.00000	.00000	.00000	.00000	.00000	.00000	.00000	.00000	.00000	.00000
3.20	.00069	.00003	.00001	.00000	.00000	.00000	.00000	.00000	.00000	.00000	.00000	.00000
3.10	.00097	.00005	.00001	.00000	.00000	.00000	.00000	.00000	.00000	.00000	.00000	.00000
3.00	.00135	.00008	.00002	.00000	.00000	.00000	.00000	.00000	.00000	.00000	.00000	.00000
2.90	.00187	.00013	.00003	.00001	.00000	.00000	.00000	.00000	.00000	.00000	.00000	.00000
2.80	.00256	.00020	.00004	.00001	.00001	.00000	.00000	.00000	.00000	.00000	.00000	.00000
2.70	.00347	.00030	.00007	.00002	.00001	.00001	.00000	.00000	.00000	.00000	.00000	.00000
2.60	.00466	.00045	.00011	.00004	.00002	.00001	.00001	.00000	.00000	.00000	.00000	.00000
2.50	.00621	.00067	.00017	.00006	.00003	.00002	.00001	.00001	.00000	.00000	.00000	.00000
2.40	.00820	.00098	.00027	.00011	.00005	.00003	.00002	.00001	.00001	.00001	.00000	.00000
2.30	.01072	.00143	.00041	.00017	.00009	.00005	.00003	.00002	.00001	.00001	.00001	.00001
2.20	.01390	.00204	.00062	.00027	.00014	.00008	.00005	.00004	.00003	.00002	.00001	.00001
2.10	.01786	.00289	.00093	.00041	.00022	.00013	.00009	.00006	.00004	.00003	.00003	.00002
2.00	.02275	.00405	.00137	.00063	.00035	.00021	.00014	.00010	.00007	.00006	.00004	.00004
1.90	.02872	.00561	.00201	.00096	.00054	.00034	.00023	.00016	.00012	.00009	.00007	.00006
1.80	.03593	.00767	.00289	.00143	.00083	.00053	.00037	.00026	.00020	.00016	.00012	.00010
1.70	.04457	.01037	.00411	.00210	.00125	.00082	.00057	.00042	.00032	.00025	.00020	.00017
1.60	.05480	.01386	.00576	.00305	.00185	.00124	.00088	.00066	.00051	.00040	.00033	.00027
1.50	.06681	.01832	.00799	.00436	.00272	.00185	.00133	.00101	.00079	.00063	.00052	.00043
1.40	.08076	.02394	.01093	.00616	.00393	.00272	.00200	.00153	.00121	.00098	.00081	.00068
1.30	.09680	.03094	.01476	.00858	.00560	.00395	.00294	.00228	.00182	.00149	.00124	.00105
1.20	.11507	.03955	.01970	.01179	.00788	.00565	.00427	.00335	.00270	.00223	.00188	.00160
1.10	.13567	.04999	.02596	.01600	.01092	.00797	.00611	.00485	.00395	.00329	.00279	.00241
1.00	.15866	.06251	.03380	.02142	.01494	.01109	.00861	.00692	.00570	.00479	.00410	.00355
0.90	.18406	.07734	.04347	.02832	.02015	.01521	.01197	.00973	.00809	.00686	.00591	.00516
0.80	.21186	.09469	.05526	.03695	.02683	.02058	.01641	.01348	.01133	.00969	.00841	.00739
0.70	.24196	.11472	.06941	.04762	.03525	.02746	.02218	.01842	.01562	.01348	.01179	.01043
0.60	.27425	.13757	.08619	.06060	.04570	.03614	.02957	.02481	.02124	.01847	.01627	.01449
0.50	.30854	.16332	.10580	.07617	.05850	.04694	.03887	.03296	.02847	.02495	.02214	.01984
0.40	.34458	.19198	.12841	.09458	.07393	.06017	.05041	.04318	.03762	.03323	.02969	.02677
0.30	.38209	.22349	.15414	.11606	.09227	.07612	.06451	.05579	.04902	.04363	.03924	.03561
0.20	.42074	.25771	.18303	.14075	.11374	.09508	.08147	.07113	.06302	.05650	.05115	.04668
0.10	.46017	.29442	.21503	.16873	.13850	.11727	.10156	.08948	.07991	.07215	.06574	.06034
0.00	.50000	.33333	.25000	.20000	.16667	.14286	.12500	.11111	.10000	.09091	.08333	.07692

H \ N	1	2	3	4	5	6	7	8	9	10	11	12
0.00	.50000	.33333	.25000	.20000	.16667	.14286	.12500	.11111	.10000	.09091	.08333	.07692
0.10	.53983	.37408	.28772	.23446	.19823	.17192	.15193	.13620	.12350	.11302	.10422	.09673
0.20	.57926	.41623	.32788	.27192	.23308	.20445	.18240	.16487	.15058	.13869	.12863	.12000
0.30	.61791	.45931	.37006	.31206	.27104	.24032	.21637	.19713	.18129	.16801	.15669	.14692
0.40	.65542	.50282	.41379	.35450	.31177	.27930	.25368	.23287	.21559	.20099	.18845	.17757
0.50	.69146	.54624	.45855	.39874	.35486	.32104	.29403	.27187	.25332	.23750	.22384	.21190
0.60	.72575	.58906	.50376	.44425	.39981	.36509	.33704	.31380	.29417	.27732	.26266	.24977
0.70	.75804	.63079	.54885	.49041	.44605	.41091	.38220	.35820	.33775	.32006	.30458	.29089
0.80	.78814	.67098	.59323	.53661	.49293	.45788	.42894	.40451	.38353	.36525	.34915	.33483
0.90	.81594	.70922	.63638	.58224	.53983	.50536	.47660	.45211	.43090	.41231	.39582	.38107
1.00	.84134	.74520	.67778	.62670	.58608	.55257	.52450	.50030	.47920	.46056	.44394	.42898
1.10	.86433	.77865	.71701	.66944	.63107	.59913	.57195	.54839	.52770	.50930	.49279	.47786
1.20	.88493	.80941	.75373	.71000	.67424	.64414	.61828	.59568	.57569	.55780	.54166	.52698
1.30	.90320	.83734	.78766	.74799	.71510	.68713	.66287	.64151	.62248	.60535	.58980	.57559
1.40	.91924	.86243	.81864	.78309	.75326	.72753	.70519	.68529	.66744	.65127	.63652	.62297
1.50	.93319	.88471	.84656	.81513	.78843	.76525	.74480	.72652	.71001	.69498	.68119	.66847
1.60	.94520	.90427	.87143	.84398	.82040	.79973	.78134	.76479	.74975	.73598	.72328	.71151
1.70	.95543	.92124	.89331	.86964	.84908	.83090	.81459	.79982	.78631	.77388	.76236	.75163
1.80	.96407	.93581	.91233	.89218	.87448	.85870	.84444	.83144	.81948	.80841	.79811	.78848
1.90	.97128	.94817	.92867	.91172	.89669	.88317	.87087	.85958	.84914	.83943	.83036	.82183
2.00	.97725	.95855	.94253	.92845	.91585	.90442	.89395	.88429	.87531	.86691	.85902	.85159
2.10	.98214	.96717	.95416	.94260	.93217	.92264	.91385	.90569	.89806	.89090	.88415	.87776
2.20	.98610	.97424	.96380	.95443	.94590	.93805	.93077	.92397	.91758	.91156	.90586	.90045
2.30	.98928	.97996	.97169	.96419	.95730	.95092	.94497	.93938	.93411	.92911	.92437	.91984
2.40	.99180	.98459	.97809	.97215	.96665	.96153	.95673	.95220	.94790	.94382	.93992	.93619
2.50	.99379	.98825	.98321	.97856	.97423	.97017	.96635	.96272	.95927	.95597	.95282	.94979
2.60	.99534	.99113	.98726	.98366	.98030	.97712	.97411	.97125	.96851	.96588	.96337	.96094
2.70	.99653	.99336	.99043	.98768	.98509	.98264	.98030	.97807	.97592	.97386	.97188	.96996
2.80	.99744	.99509	.99288	.99081	.98884	.98697	.98517	.98345	.98180	.98020	.97865	.97716
2.90	.99813	.99640	.99476	.99321	.99173	.99032	.98876	.98765	.98639	.98517	.98398	.98283
3.00	.99865	.99738	.99618	.99504	.99394	.99288	.99187	.99089	.98994	.98901	.98812	.98724
3.10	.99903	.99812	.99724	.99641	.99560	.99483	.99408	.99335	.99254	.99195	.99128	.99063
3.20	.99931	.99866	.99803	.99743	.99684	.99628	.99573	.99520	.99468	.99417	.99367	.99319
3.30	.99952	.99905	.99861	.99817	.99776	.99735	.99695	.99657	.99619	.99582	.99546	.99511
3.40	.99966	.99934	.99902	.99872	.99842	.99813	.99785	.99757	.99730	.99704	.99678	.99652
3.50	.99977	.99954	.99932	.99911	.99890	.99870	.99850	.99830	.99811	.99792	.99774	.99756

Table 4 (continued)

$\rho = \frac{2}{3}$

H \ N	1	2	3	4	5	6	7	8	9	10	11	12
3.50	.00023	.00002	.00001	.00000	.00000	.00000	.00000	.00000	.00000	.00000	.00000	.00000
3.40	.00034	.00003	.00001	.00000	.00000	.00000	.00000	.00000	.00000	.00000	.00000	.00000
3.30	.00048	.00005	.00002	.00001	.00000	.00000	.00000	.00000	.00000	.00000	.00000	.00000
3.20	.00069	.00008	.00003	.00001	.00001	.00000	.00000	.00000	.00000	.00000	.00000	.00000
3.10	.00097	.00013	.00004	.00002	.00001	.00001	.00001	.00000	.00000	.00000	.00000	.00000
3.00	.00135	.00020	.00007	.00003	.00002	.00001	.00001	.00001	.00001	.00000	.00000	.00000
2.90	.00187	.00029	.00010	.00005	.00003	.00002	.00002	.00001	.00001	.00001	.00001	.00000
2.80	.00256	.00043	.00016	.00008	.00005	.00003	.00002	.00002	.00001	.00001	.00001	.00001
2.70	.00347	.00063	.00024	.00013	.00008	.00005	.00004	.00003	.00002	.00002	.00002	.00001
2.60	.00466	.00091	.00037	.00020	.00013	.00009	.00006	.00005	.00004	.00003	.00003	.00002
2.50	.00621	.00131	.00055	.00030	.00019	.00014	.00010	.00008	.00006	.00005	.00004	.00004
2.40	.00820	.00185	.00080	.00045	.00030	.00021	.00016	.00012	.00010	.00008	.00007	.00006
2.30	.01072	.00259	.00116	.00067	.00044	.00032	.00024	.00019	.00016	.00013	.00011	.00010
2.20	.01390	.00360	.00167	.00099	.00066	.00048	.00037	.00030	.00025	.00021	.00018	.00016
2.10	.01786	.00493	.00236	.00143	.00098	.00072	.00056	.00045	.00038	.00032	.00028	.00024
2.00	.02275	.00669	.00331	.00204	.00142	.00106	.00083	.00068	.00057	.00048	.00042	.00037
1.90	.02872	.00897	.00459	.00289	.00204	.00154	.00122	.00100	.00085	.00073	.00063	.00056
1.80	.03593	.01192	.00630	.00405	.00290	.00222	.00177	.00147	.00124	.00107	.00094	.00084
1.70	.04457	.01567	.00854	.00560	.00407	.00315	.00254	.00212	.00181	.00157	.00139	.00124
1.60	.05480	.02039	.01144	.00765	.00564	.00441	.00360	.00302	.00259	.00227	.00201	.00180
1.50	.06681	.02626	.01517	.01035	.00773	.00611	.00503	.00425	.00368	.00323	.00288	.00259
1.40	.08076	.03347	.01990	.01383	.01047	.00837	.00694	.00592	.00515	.00455	.00407	.00368
1.30	.09680	.04224	.02581	.01827	.01402	.01133	.00948	.00813	.00712	.00632	.00569	.00516
1.20	.11507	.05277	.03313	.02387	.01856	.01515	.01278	.01105	.00973	.00869	.00785	.00716
1.10	.13567	.06528	.04208	.03085	.02430	.02003	.01704	.01483	.01313	.01179	.01070	.00980
1.00	.15866	.07497	.05287	.03943	.03145	.02618	.02245	.01967	.01752	.01581	.01441	.01325
0.90	.18406	.09702	.06575	.04985	.04025	.03383	.02924	.02579	.02310	.02095	.01919	.01771
0.80	.21186	.11659	.08092	.06234	.05094	.04323	.03765	.03342	.03011	.02744	.02523	.02338
0.70	.24196	.13880	.09859	.07715	.06378	.05462	.04793	.04282	.03878	.03551	.03280	.03051
0.60	.27425	.16371	.11892	.09448	.07899	.06825	.06033	.05424	.04939	.04543	.04214	.03935
0.50	.30854	.19136	.14203	.11451	.09679	.08436	.07510	.06793	.06218	.05746	.05351	.05015
0.40	.34458	.22161	.16797	.13737	.11736	.10315	.09248	.08413	.07741	.07185	.06718	.06319
0.30	.38209	.25448	.19677	.16314	.14082	.12479	.11264	.10307	.09536	.08885	.08340	.07871
0.20	.42074	.28970	.22835	.19185	.16727	.14941	.13575	.12491	.11605	.10866	.10238	.09696
0.10	.46017	.32702	.26257	.22343	.19669	.17705	.16189	.14977	.13981	.13145	.12431	.11812
0.00	.50000	.36614	.29921	.25775	.22902	.20769	.19109	.17771	.16665	.15732	.14931	.14234

H \ N	1	2	3	4	5	6	7	8	9	10	11	12
0.00	.50000	.36614	.24921	.25775	.22902	.20769	.19109	.17771	.16665	.15732	.14931	.14234
0.10	.53983	.43068	.33798	.29459	.26411	.24124	.22328	.20871	.19654	.18630	.17749	.16969
0.20	.57926	.44822	.37852	.33166	.30173	.27751	.25833	.24266	.22955	.21837	.20868	.20018
0.30	.61791	.49031	.42041	.37459	.34154	.31622	.29601	.27937	.26537	.25336	.24292	.23372
0.40	.65542	.51248	.46319	.41696	.38317	.35703	.33599	.31856	.30380	.29108	.27996	.27014
0.50	.69146	.57426	.50837	.46027	.42616	.39950	.37788	.35985	.34449	.33119	.31952	.30916
0.60	.72575	.61520	.54944	.50403	.47001	.44317	.42123	.40281	.38704	.37331	.36121	.35043
0.70	.75804	.65487	.59190	.54770	.51418	.48750	.46551	.44694	.43095	.41697	.40459	.39351
0.80	.78814	.69288	.63320	.59076	.55815	.53194	.51020	.49171	.47570	.46164	.44913	.43790
0.90	.81594	.72890	.67313	.63272	.60137	.57596	.55471	.51656	.52072	.50676	.49424	.48305
1.00	.84134	.76266	.71106	.67311	.64335	.61900	.59851	.58087	.56544	.55175	.53948	.52839
1.10	.86433	.79394	.74676	.71154	.68362	.66059	.64106	.62416	.60929	.59605	.58414	.57332
1.20	.88493	.82263	.77946	.74766	.72179	.70026	.68189	.66590	.65176	.63911	.62767	.61728
1.30	.90320	.85063	.81049	.78122	.75753	.73766	.72059	.70565	.69237	.68044	.66963	.65976
1.40	.91924	.87196	.83824	.81202	.79059	.77248	.75681	.74302	.73072	.71961	.70950	.70023
1.50	.93319	.89264	.86317	.83997	.82081	.80450	.79030	.77774	.76647	.75626	.74691	.73835
1.60	.94520	.91079	.88533	.86502	.84810	.83359	.82088	.80957	.79939	.79012	.78163	.77378
1.70	.95543	.92654	.90478	.88722	.87246	.85971	.84847	.83842	.82933	.82102	.81338	.80630
1.80	.96407	.94006	.92168	.90667	.89394	.88287	.87305	.86423	.85621	.84886	.84207	.83577
1.90	.97128	.95154	.93618	.92350	.91266	.90316	.89469	.88704	.88006	.87363	.86768	.86213
2.00	.97725	.96119	.94849	.93790	.92877	.92072	.91350	.90695	.90095	.89541	.89025	.88543
2.10	.98213	.96920	.95883	.95009	.94249	.93574	.92967	.92413	.91903	.91430	.90990	.90576
2.20	.98610	.97579	.96741	.96027	.95402	.94844	.94339	.93876	.93448	.93050	.92678	.92328
2.30	.98928	.98115	.97445	.96869	.96361	.95905	.95489	.95107	.94753	.94422	.94112	.93819
2.40	.99180	.98546	.98016	.97557	.97149	.96780	.96443	.96131	.95841	.95570	.95314	.95072
2.50	.99379	.98889	.98475	.98113	.97789	.97494	.97224	.96973	.96738	.96518	.96310	.96113
2.60	.99534	.99159	.98839	.98557	.98302	.98070	.97855	.97656	.97468	.97292	.97125	.96966
2.70	.99653	.99370	.99125	.98907	.98710	.98529	.98361	.90203	.98056	.97916	.97783	.97657
2.80	.99744	.99532	.99347	.99181	.99029	.98890	.9876	.98638	.98522	.98411	.98309	.98210
2.90	.99813	.99656	.99517	.99392	.99277	.99171	.99071	.98978	.98889	.98805	.98724	.98647
3.00	.99865	.99750	.99647	.99554	.99467	.99387	.99312	.99241	.99173	.99109	.99047	.98988
3.10	.99903	.99819	.99744	.99675	.99612	.99552	.99496	.99442	.99391	.99343	.99296	.99252
3.20	.99931	.99871	.99816	.99766	.99720	.99676	.99634	.99595	.99557	.99521	.99486	.99452
3.30	.99952	.99909	.99870	.99833	.99800	.99768	.99737	.99708	.99681	.99654	.99628	.99603
3.40	.99966	.99936	.99908	.99883	.99858	.99835	.99813	.99792	.99772	.99753	.99734	.99716
3.50	.99977	.99956	.99936	.99918	.99901	.99884	.99869	.99854	.99839	.99825	.99812	.99799

Table 4 (continued)
$\rho = 0.800$

-M \ N	1	2	3	4	5	6	7	8	9	10	11	12
3.50	.00023	.00005	.00002	.00001	.00001	.00001	.00001	.00000	.00000	.00000	.00000	.00000
3.40	.00034	.00008	.00004	.00002	.00001	.00001	.00001	.00001	.00001	.00001	.00000	.00000
3.30	.00048	.00011	.00005	.00003	.00002	.00002	.00001	.00001	.00001	.00001	.00001	.00001
3.20	.00069	.00017	.00008	.00005	.00004	.00003	.00002	.00002	.00002	.00001	.00001	.00001
3.10	.00097	.00025	.00013	.00008	.00006	.00004	.00004	.00003	.00003	.00002	.00002	.00002
3.00	.00135	.00037	.00019	.00012	.00009	.00007	.00006	.00005	.00004	.00004	.00003	.00003
2.90	.00187	.00054	.00028	.00018	.00013	.00011	.00009	.00007	.00006	.00006	.00005	.00004
2.80	.00256	.00078	.00042	.00028	.00020	.00016	.00013	.00011	.00010	.00009	.00008	.00007
2.70	.00347	.00110	.00060	.00041	.00030	.00024	.00020	.00017	.00015	.00013	.00012	.00011
2.60	.00466	.00156	.00087	.00059	.00045	.00036	.00030	.00025	.00022	.00020	.00018	.00016
2.50	.00621	.00217	.00124	.00086	.00065	.00052	.00044	.00038	.00033	.00030	.00027	.00024
2.40	.00820	.00299	.00175	.00122	.00094	.00076	.00064	.00056	.00049	.00044	.00040	.00037
2.30	.01072	.00409	.00244	.00173	.00134	.00110	.00093	.00081	.00072	.00064	.00059	.00054
2.20	.01390	.00553	.00337	.00242	.00190	.00156	.00133	.00116	.00104	.00094	.00085	.00079
2.10	.01786	.00741	.00461	.00336	.00265	.00220	.00189	.00166	.00148	.00134	.00123	.00113
2.00	.02275	.00983	.00624	.00460	.00367	.00307	.00264	.00233	.00209	.00190	.00175	.00162
1.90	.02872	.01290	.00836	.00624	.00502	.00423	.00367	.00325	.00293	.00267	.00246	.00228
1.80	.03593	.01678	.01109	.00838	.00680	.00576	.00503	.00448	.00405	.00370	.00342	.00319
1.70	.04457	.02162	.01455	.01114	.00912	.00778	.00682	.00610	.00554	.00509	.00471	.00440
1.60	.05480	.02758	.01891	.01465	.01209	.01039	.00916	.00823	.00750	.00691	.00642	.00601
1.50	.06681	.03486	.02434	.01907	.01580	.01373	.01217	.01098	.01005	.00929	.00866	.00813
1.40	.08076	.04363	.03101	.02458	.02064	.01796	.01600	.01450	.01332	.01235	.01155	.01087
1.30	.09680	.05411	.03913	.03136	.02655	.02325	.02082	.01896	.01747	.01626	.01524	.01438
1.20	.11507	.06648	.04890	.03962	.03381	.02979	.02682	.02452	.02268	.02117	.01990	.01882
1.10	.13567	.08093	.06052	.04957	.04263	.03779	.03419	.03139	.02914	.02728	.02572	.02439
1.00	.15866	.09764	.07419	.06141	.05322	.04746	.04315	.03978	.03705	.03480	.03290	.03127
0.90	.18406	.11673	.09010	.07535	.06580	.05902	.05391	.04990	.04664	.04394	.04165	.03968
0.80	.21186	.13833	.10841	.09158	.08056	.07267	.06669	.06197	.05813	.05492	.05219	.04984
0.70	.24196	.16250	.12925	.11025	.09768	.08862	.08170	.07621	.07172	.06796	.06475	.06198
0.60	.27425	.18926	.15271	.13151	.11733	.10703	.09912	.09281	.08762	.08326	.07954	.07630
0.50	.30854	.21856	.17883	.15543	.13962	.12804	.11910	.11193	.10601	.10102	.09674	.09301
0.40	.34458	.25031	.20758	.18205	.16462	.15176	.14177	.13371	.12734	.12139	.11652	.11227
0.30	.38209	.28434	.23890	.21134	.19234	.17822	.16718	.15824	.15079	.14447	.13901	.13423
0.20	.42074	.32042	.27262	.24321	.22274	.20740	.19535	.18553	.17733	.17034	.16428	.15895
0.10	.46017	.35828	.30854	.27750	.25569	.23923	.22621	.21556	.20663	.19898	.19234	.18649
0.00	.50000	.39758	.34638	.31194	.29100	.27354	.25965	.24823	.23861	.23034	.22314	.21678

M \ N	1	2	3	4	5	6	7	8	9	10	11	12
0.00	.50000	.39758	.34638	.31399	.29100	.27354	.25965	.24823	.23861	.23034	.22314	.21678
0.10	.53983	.43794	.38580	.35238	.32844	.31011	.29545	.28335	.27311	.26428	.25657	.24973
0.20	.57926	.47894	.42643	.39232	.36766	.34866	.33337	.32068	.30990	.30059	.29242	.28516
0.30	.61791	.52016	.46765	.43342	.40831	.38882	.37306	.35990	.34870	.33898	.33042	.32281
0.40	.65542	.56115	.50960	.47524	.44995	.43019	.41411	.40065	.38913	.37910	.37028	.36235
0.50	.69146	.60149	.55124	.51732	.49214	.47233	.45613	.44250	.43079	.42057	.41151	.40342
0.60	.72575	.64075	.59230	.55920	.53440	.51477	.49862	.48498	.47328	.46292	.45377	.44586
0.70	.75804	.67858	.63236	.60040	.57626	.55702	.54112	.52763	.51594	.50568	.49653	.48831
0.80	.78814	.71462	.67102	.64051	.61727	.59863	.58314	.56994	.55847	.54836	.53933	.53119
0.90	.81594	.74861	.70792	.67911	.65698	.63912	.62421	.61145	.60032	.59047	.58166	.57369
1.00	.84134	.78033	.74275	.71585	.69502	.67811	.66391	.65171	.64103	.63156	.62305	.61534
1.10	.86433	.80960	.77528	.75044	.73105	.71521	.70188	.69032	.68019	.67117	.66306	.65568
1.20	.88493	.83634	.80534	.78264	.76480	.75012	.73769	.72692	.71742	.70894	.70129	.69432
1.30	.90320	.86051	.83279	.81229	.79605	.78261	.77117	.76122	.75242	.74454	.73740	.73089
1.40	.91924	.88212	.85761	.83930	.82467	.81251	.80210	.79302	.78498	.77770	.77113	.76511
1.50	.93319	.90124	.87901	.86362	.85060	.83971	.83035	.82215	.81484	.80826	.80227	.79677
1.60	.94520	.91708	.89943	.88529	.87383	.86419	.85587	.84854	.84199	.83608	.83069	.82573
1.70	.95543	.93249	.91861	.90438	.89440	.88597	.87865	.87218	.86638	.86113	.85633	.85191
1.80	.96407	.94492	.93147	.92102	.91243	.90512	.89876	.89312	.88804	.88343	.87921	.87530
1.90	.97128	.95547	.94420	.93536	.92805	.92179	.91632	.91145	.90706	.90306	.89938	.89598
2.00	.97725	.96412	.95498	.94759	.94143	.93614	.93149	.92733	.92357	.92014	.91698	.91404
2.10	.98214	.97168	.96402	.95790	.95277	.94834	.94443	.94093	.93775	.93483	.93216	.92964
2.20	.98610	.97772	.97151	.96650	.96227	.95861	.95536	.95244	.94978	.94733	.94507	.94296
2.30	.98928	.98264	.97765	.97359	.97015	.96715	.96448	.96207	.95987	.95785	.95597	.95421
2.40	.99180	.98660	.98263	.97939	.97661	.97418	.97201	.97005	.96825	.96659	.96504	.96360
2.50	.99379	.98975	.98663	.98406	.98185	.97991	.97816	.97658	.97512	.97378	.97252	.97135
2.60	.99534	.99223	.98981	.98780	.98608	.98451	.98313	.98186	.98070	.97962	.97861	.97767
2.70	.99653	.99417	.99231	.99074	.98939	.98818	.98709	.98610	.98518	.98432	.98352	.98277
2.80	.99744	.99567	.99425	.99305	.99200	.99107	.99028	.98948	.98873	.98806	.98743	.98684
2.90	.99813	.99681	.99574	.99483	.99403	.99332	.99267	.99207	.99151	.99099	.99051	.99004
3.00	.99865	.99767	.99688	.99619	.99559	.99508	.99456	.99410	.99367	.99328	.99290	.99255
3.10	.99903	.99832	.99773	.99723	.99678	.99637	.99600	.99565	.99533	.99503	.99475	.99448
3.20	.99931	.99880	.99837	.99800	.99767	.99736	.99709	.99683	.99659	.99637	.99615	.99596
3.30	.99952	.99918	.99884	.99867	.99833	.99811	.99790	.99771	.99754	.99737	.99721	.99706
3.40	.99966	.99940	.99918	.99899	.99881	.99865	.99850	.99837	.99824	.99811	.99800	.99789
3.50	.99977	.99958	.99943	.99929	.99916	.99905	.99894	.99884	.99875	.99866	.99858	.99850

Table 4 (continued)
$\rho = 0.900$

−H \ N	1	2	3	4	5	6	7	8	9	10	11	12
3.50	.00823	.00009	.00006	.00004	.00003	.00003	.00002	.00002	.00002	.00002	.00002	.00002
3.40	.00034	.00014	.00009	.00006	.00005	.00004	.00004	.00003	.00003	.00003	.00003	.00002
3.30	.00048	.00020	.00013	.00010	.00008	.00007	.00006	.00005	.00005	.00004	.00004	.00004
3.20	.00069	.00029	.00019	.00014	.00012	.00010	.00009	.00008	.00007	.00007	.00006	.00006
3.10	.00097	.00043	.00028	.00021	.00018	.00015	.00013	.00012	.00011	.00010	.00009	.00009
3.00	.00135	.00061	.00041	.00031	.00026	.00022	.00020	.00018	.00016	.00015	.00014	.00013
2.90	.00187	.00087	.00059	.00046	.00038	.00033	.00029	.00026	.00024	.00023	.00021	.00020
2.80	.00256	.00122	.00084	.00066	.00055	.00048	.00043	.00039	.00036	.00033	.00031	.00029
2.70	.00347	.00170	.00119	.00094	.00079	.00069	.00062	.00056	.00052	.00048	.00045	.00043
2.60	.00466	.00235	.00166	.00132	.00112	.00098	.00088	.00080	.00074	.00069	.00065	.00062
2.50	.00621	.00322	.00230	.00184	.00157	.00138	.00124	.00114	.00106	.00099	.00093	.00088
2.40	.00820	.00436	.00315	.00255	.00218	.00193	.00174	.00160	.00149	.00140	.00132	.00125
2.30	.01072	.00585	.00428	.00349	.00300	.00266	.00242	.00223	.00208	.00195	.00185	.00176
2.20	.01390	.00778	.00576	.00473	.00409	.00365	.00332	.00307	.00287	.00270	.00256	.00244
2.10	.01786	.01024	.00768	.00635	.00552	.00494	.00452	.00418	.00392	.00370	.00351	.00335
2.00	.02275	.01336	.01013	.00844	.00738	.00663	.00608	.00565	.00530	.00501	.00477	.00456
1.90	.02872	.01727	.01325	.01111	.00976	.00882	.00811	.00755	.00710	.00673	.00642	.00615
1.80	.03593	.02211	.01715	.01450	.01280	.01160	.01071	.01000	.00943	.00895	.00855	.00820
1.70	.04457	.02806	.02201	.01873	.01662	.01513	.01400	.01311	.01239	.01179	.01128	.01084
1.60	.05480	.03527	.02797	.02397	.02138	.01953	.01814	.01703	.01613	.01538	.01473	.01418
1.50	.06681	.04395	.03522	.03039	.02724	.02499	.02327	.02191	.02079	.01986	.01906	.01837
1.40	.08076	.05427	.04395	.03818	.03439	.03166	.02957	.02791	.02655	.02541	.02443	.02358
1.30	.09680	.06641	.05434	.04752	.04300	.03974	.03723	.03523	.03359	.03220	.03101	.02998
1.20	.11507	.08056	.06659	.05860	.05329	.04942	.04644	.04406	.04209	.04043	.03900	.03776
1.10	.13567	.09687	.08086	.07162	.06543	.06090	.05739	.05458	.05225	.05028	.04858	.04710
1.00	.15866	.11549	.09734	.08675	.07961	.07436	.07028	.06700	.06427	.06196	.05996	.05822
0.90	.18406	.13652	.11615	.10415	.09600	.08998	.08529	.08149	.07833	.07565	.07333	.07130
0.80	.21186	.16002	.13740	.12395	.11474	.10791	.10257	.09823	.09462	.09153	.08886	.08652
0.70	.24196	.18602	.16116	.14624	.13595	.12829	.12226	.11736	.11326	.10976	.10672	.10405
0.60	.27425	.21448	.18746	.17107	.15970	.15118	.14447	.13899	.13439	.13046	.12703	.12402
0.50	.30854	.24533	.21624	.19844	.18601	.17665	.16925	.16318	.15808	.15370	.14989	.14652
0.40	.34458	.27841	.24743	.22829	.21484	.20467	.19659	.18995	.18435	.17954	.17533	.17161
0.30	.38209	.31352	.28088	.26051	.24611	.23517	.22645	.21926	.21318	.20794	.20335	.19929
0.20	.42074	.35040	.31636	.29492	.27967	.26803	.25871	.25100	.24447	.23883	.23388	.22949
0.10	.46017	.38875	.35361	.33129	.31530	.30304	.29319	.28502	.27808	.27207	.26680	.26210
0.00	.50000	.42822	.39233	.36931	.35274	.33996	.32967	.32110	.31380	.30747	.30189	.29693

H \ N	1	2	3	4	5	6	7	8	9	10	11	12
0.00	.50000	.42822	.39233	.36931	.35274	.33996	.32967	.32110	.31380	.30747	.30189	.29693
0.10	.53983	.46841	.43213	.40866	.39165	.37848	.36783	.35894	.35135	.34475	.33892	.33373
0.20	.57926	.50892	.47263	.44894	.43168	.41825	.40734	.39822	.39040	.38359	.37758	.37220
0.30	.61791	.54934	.51341	.48976	.47241	.45886	.44782	.43855	.43059	.42364	.41749	.41198
0.40	.65542	.58925	.55405	.53068	.51344	.49991	.48884	.47953	.47151	.46450	.45827	.45269
0.50	.69146	.62825	.59412	.57127	.55431	.54095	.52998	.52072	.51273	.50572	.49950	.49390
0.60	.72575	.66597	.63323	.61112	.59462	.58155	.57079	.56160	.55301	.54609	.54072	.53517
0.70	.75804	.70209	.67099	.64982	.63393	.62130	.61086	.60200	.59431	.58754	.58151	.57607
0.80	.78814	.73631	.70708	.68703	.67189	.65980	.64978	.64124	.63383	.62728	.62143	.61615
0.90	.81594	.76840	.74122	.72241	.70814	.69670	.68718	.67905	.67197	.66570	.66009	.65502
1.00	.84134	.79818	.77317	.75572	.74241	.73169	.72274	.71508	.70839	.70246	.69714	.69232
1.10	.86433	.82554	.80276	.78674	.77446	.76452	.75620	.74905	.74280	.73725	.73226	.72773
1.20	.88493	.85042	.82988	.81533	.80411	.79500	.78734	.78075	.77497	.76982	.76519	.76098
1.30	.90320	.87281	.85449	.84141	.83127	.82300	.81603	.81001	.80472	.80000	.79575	.79189
1.40	.91924	.89275	.87658	.86494	.85587	.84845	.84217	.83674	.83195	.82767	.82380	.82028
1.50	.93319	.91033	.89620	.88596	.87793	.87134	.86574	.86089	.85660	.85276	.84929	.84611
1.60	.94520	.92568	.91345	.90453	.89751	.89171	.88678	.88249	.87868	.87527	.87218	.86936
1.70	.95543	.93893	.92847	.92078	.91469	.90965	.90534	.90159	.89826	.89526	.89254	.89005
1.80	.96407	.95025	.94139	.93483	.92962	.92528	.92156	.91831	.91542	.91282	.91045	.90828
1.90	.97128	.95984	.95241	.94687	.94245	.93875	.93558	.93280	.93032	.92808	.92604	.92417
2.00	.97725	.96786	.96170	.95708	.95336	.95025	.94757	.94521	.94310	.94120	.93946	.93786
2.10	.98214	.97451	.96946	.96563	.96255	.95995	.95771	.95573	.95396	.95236	.95090	.94955
2.20	.98610	.97997	.97586	.97273	.97020	.96806	.96620	.96456	.96309	.96176	.96054	.95941
2.30	.98928	.98440	.98110	.97857	.97650	.97476	.97324	.97189	.97068	.96958	.96858	.96764
2.40	.99180	.98796	.98533	.98331	.98165	.98024	.97901	.97791	.97693	.97603	.97521	.97445
2.50	.99379	.99080	.98873	.98712	.98580	.98467	.98368	.98281	.98201	.98129	.98063	.98001
2.60	.99534	.99303	.99142	.99015	.98911	.98822	.98744	.98674	.98611	.98553	.98500	.98451
2.70	.99653	.99477	.99352	.99254	.99173	.99103	.99042	.98987	.98937	.98892	.98850	.98811
2.80	.99744	.99611	.99516	.99441	.99378	.99324	.99276	.99234	.99195	.99159	.99127	.99096
2.90	.99813	.99714	.99642	.99585	.99537	.99495	.99459	.99426	.99396	.99368	.99343	.99319
3.00	.99865	.99791	.99737	.99694	.99658	.99627	.99599	.99574	.99551	.99530	.99510	.99492
3.10	.99903	.99849	.99809	.99777	.99750	.99727	.99706	.99687	.99669	.99654	.99639	.99625
3.20	.99931	.99892	.99863	.99839	.99819	.99802	.99786	.99772	.99759	.99747	.99736	.99726
3.30	.99952	.99923	.99902	.99885	.99870	.99857	.99846	.99836	.99826	.99817	.99809	.99801
3.40	.99966	.99946	.99931	.99919	.99908	.99899	.99890	.99883	.99876	.99869	.99863	.99858
3.50	.99977	.99963	.99952	.99943	.99935	.99929	.99923	.99917	.99912	.99907	.99903	.99899

Table 4 is reproduced from the *Annals of Mathematical Statistics* with the kind permission of the Institute of Mathematical Statistics.

5. Probability integral of the maximum of correlated chi-square variables with one degree of freedom

Let $x_1, x_2, \ldots, x_N$ be distributed as a multivariate normal with mean vector $\mathbf{0}$ and covariance matrix $\Omega = (\rho_{ij})$ where $\rho_{ii} = 1$ $(i = 1, 2, \ldots, N)$ and $\rho_{ij} = \rho$ $(\neq j)$. Also, let $y_i = x_i^2$ for $i = 1, 2, \ldots, N$. Then, the joint distribution of $y_1, y_2, \ldots, y_N$ is known to be a multivariate chi-square distribution with one degree of freedom and with Ω as the covariance matrix of the "accompanying" multivariate normal. But

$$P[y_i \leq a;\ i = 1, 2, \ldots, N] = P[-\sqrt{a} \leq x_i \leq \sqrt{a};\ i = 1, 2, \ldots, N]$$
$$= \int_{-\infty}^{\infty} \psi(\rho, a, z) f(z)\, \mathrm{d}z \tag{5.1}$$

where

$$\psi(\rho, a, z) = F\left(\frac{\sqrt{a} + z\rho^{1/2}}{\sqrt{1-\rho}}\right) - F\left(\frac{-\sqrt{a} + z\rho^{1/2}}{\sqrt{1-\rho}}\right)$$

and $f(z)$ and $F(a)$ are defined in Section 4. Using (5.1), Krishnaiah and Armitage (1965) computzd the values of α for $N = 1(1)10$, $a = 0.1(0.1)11.5$, $\rho = 0.0(0.0125)0.85$. They also give values of a for $\alpha = 0.10, 0.05, 0.025, 0.01$ and the above values of N and ρ. Table 5 gives the values of α for $a = 0.1(0.1)11.5$, $N = 1(1)10$ and $\rho = 0.000(0.100)0.800, 0.850$, and these values are reproduced from Krishnaiah and Armitage (1965).

Table 5
Probability integral of the maximum of correlated chi-square variables with one degree of freedom;
$\rho = 0.0$

a \ N	1	2	3	4	5	6	7	8	9	10
0.1	0.248170	0.061589	0.015284	0.003793	0.000941	0.000234	0.000058	0.000014	0.000004	0.000001
0.2	0.345279	0.119218	0.041163	0.014213	0.004907	0.001694	0.000585	0.000202	0.000070	0.000024
0.3	0.416117	0.171154	0.072052	0.029982	0.012476	0.005192	0.002160	0.000899	0.000374	0.000156
0.4	0.472911	0.223645	0.105764	0.050017	0.023654	0.011186	0.005290	0.002502	0.001183	0.000559
0.5	0.520500	0.270920	0.141014	0.073398	0.038203	0.019885	0.010350	0.005387	0.002804	0.001460
0.6	0.561422	0.315195	0.176957	0.099348	0.055776	0.031314	0.017580	0.009870	0.005541	0.003111
0.7	0.597216	0.356667	0.213007	0.127212	0.075973	0.045372	0.027097	0.016183	0.009665	0.005772
0.8	0.628906	0.395523	0.248747	0.156439	0.098385	0.061875	0.038914	0.024473	0.015391	0.009680
0.9	0.657218	0.431936	0.283876	0.186568	0.122616	0.080586	0.052962	0.034808	0.022876	0.015035
1.0	0.682689	0.466065	0.318178	0.217216	0.148291	0.101237	0.069113	0.047183	0.032211	0.021990
1.1	0.705734	0.498060	0.351498	0.248064	0.175067	0.123551	0.087194	0.061536	0.043428	0.030648
1.2	0.726678	0.528061	0.383731	0.278849	0.202633	0.147249	0.107003	0.077757	0.056504	0.041060
1.3	0.745787	0.556198	0.414805	0.309356	0.230714	0.172063	0.128322	0.095701	0.071373	0.053229
1.4	0.763276	0.582591	0.444678	0.339412	0.259065	0.197738	0.150929	0.115201	0.087930	0.067115
1.5	0.779328	0.607353	0.473328	0.368878	0.287477	0.224039	0.174600	0.136071	0.106044	0.082643
1.6	0.794097	0.630590	0.500749	0.397643	0.315767	0.250750	0.199119	0.158120	0.125563	0.099709
1.7	0.807712	0.652398	0.526950	0.425624	0.343781	0.277676	0.224283	0.181156	0.146322	0.118186
1.8	0.820287	0.672871	0.551948	0.452756	0.371390	0.304646	0.249898	0.204988	0.168149	0.137931
1.9	0.831921	0.692093	0.575767	0.478993	0.398485	0.331508	0.275789	0.229435	0.190872	0.158790
2.0	0.842701	0.710144	0.598439	0.504305	0.424978	0.358129	0.301796	0.254324	0.214319	0.180607
2.1	0.852701	0.727099	0.619997	0.528672	0.450799	0.384397	0.327776	0.279494	0.238325	0.203220
2.2	0.861989	0.743025	0.640480	0.552087	0.475893	0.410214	0.353600	0.304800	0.262734	0.226474
2.3	0.870626	0.757989	0.659925	0.574548	0.500216	0.435501	0.379159	0.330105	0.287398	0.250216
2.4	0.878665	0.772052	0.678374	0.596064	0.523740	0.460192	0.404354	0.355292	0.312182	0.274304
2.5	0.886154	0.785268	0.695868	0.616646	0.546443	0.484233	0.429104	0.380252	0.336962	0.298600
2.6	0.893136	0.797692	0.712448	0.636313	0.568314	0.507582	0.453340	0.404894	0.361626	0.322981
2.7	0.899652	0.809373	0.728154	0.655085	0.589348	0.530208	0.477002	0.429136	0.386073	0.347331
2.8	0.905735	0.820357	0.743026	0.672985	0.609547	0.552088	0.500046	0.452909	0.410216	0.371547
2.9	0.911420	0.830687	0.757105	0.690041	0.628917	0.573208	0.522433	0.476156	0.433979	0.395537
3.0	0.916735	0.840404	0.770428	0.706278	0.647470	0.593559	0.544137	0.498829	0.457294	0.419218
3.1	0.921706	0.849545	0.783032	0.721726	0.665221	0.613139	0.565135	0.520889	0.480108	0.442519
3.2	0.926362	0.858146	0.794953	0.736414	0.682186	0.631951	0.585415	0.542306	0.502371	0.465378
3.3	0.930720	0.866240	0.806227	0.750371	0.698386	0.650001	0.604969	0.563057	0.524048	0.487742
3.4	0.934803	0.873857	0.816885	0.763627	0.713841	0.667301	0.623796	0.583126	0.545108	0.509569
3.5	0.938631	0.881028	0.826960	0.776211	0.728575	0.683863	0.641895	0.602503	0.565528	0.530822
3.6	0.942220	0.887779	0.836483	0.788152	0.742613	0.699705	0.659276	0.621183	0.585291	0.551473
3.7	0.945587	0.894135	0.845483	0.799478	0.755977	0.714842	0.675945	0.639165	0.604387	0.571501
3.8	0.948747	0.900121	0.853988	0.810218	0.768693	0.729295	0.691917	0.656454	0.622809	0.590888

Table 5 (continued)
$\rho = 0.0$

3.9	0.951714	0.905759	0.862023	0.820400	0.780786	0.743084	0.707204	0.673055	0.640556	0.609626
4.0	0.954500	0.911069	0.869615	0.830048	0.792280	0.756231	0.721822	0.688979	0.657630	0.627708
4.1	0.957117	0.916072	0.876788	0.839189	0.803201	0.768757	0.735791	0.704237	0.674037	0.645133
4.2	0.959576	0.920786	0.883564	0.847847	0.813573	0.780685	0.749127	0.718844	0.689785	0.661901
4.3	0.961887	0.925228	0.889965	0.856046	0.823420	0.792038	0.761851	0.732815	0.704886	0.678021
4.4	0.964061	0.929413	0.896011	0.863809	0.832765	0.802836	0.773983	0.746167	0.719350	0.693498
4.5	0.966105	0.933359	0.901723	0.871159	0.841631	0.813104	0.785544	0.758918	0.733194	0.708343
4.6	0.968028	0.937078	0.907118	0.878115	0.850340	0.822862	0.796554	0.771086	0.746433	0.722568
4.7	0.969837	0.940584	0.912214	0.884699	0.858014	0.832134	0.807034	0.782692	0.759084	0.736188
4.8	0.971540	0.943890	0.917027	0.890929	0.865573	0.840939	0.817006	0.793754	0.771164	0.749217
4.9	0.973143	0.947008	0.921574	0.896823	0.872738	0.849299	0.826489	0.804292	0.782692	0.761671
5.0	0.974652	0.949948	0.925869	0.902400	0.879527	0.857233	0.835504	0.814327	0.793686	0.773568
5.1	0.976074	0.952721	0.929926	0.907676	0.885959	0.864762	0.844072	0.823877	0.804165	0.784924
5.2	0.977413	0.955336	0.933758	0.912667	0.892053	0.871904	0.852210	0.832961	0.814147	0.795758
5.3	0.978674	0.957804	0.937378	0.917388	0.897824	0.878677	0.859939	0.841601	0.823653	0.806088
5.4	0.979863	0.960132	0.940798	0.921853	0.903290	0.885100	0.867277	0.849813	0.832700	0.815932
5.5	0.980983	0.962328	0.944028	0.926076	0.908465	0.891189	0.874242	0.857617	0.841308	0.825309
5.6	0.982039	0.964401	0.947080	0.930070	0.913365	0.896961	0.880851	0.865030	0.849494	0.834236
5.7	0.983035	0.966358	0.949963	0.933847	0.918004	0.902430	0.887121	0.872071	0.857276	0.842732
5.8	0.983974	0.968204	0.952687	0.937419	0.922396	0.907613	0.893068	0.878755	0.864672	0.850814
5.9	0.984859	0.969947	0.955261	0.940798	0.926553	0.912524	0.898707	0.885100	0.871699	0.858500
6.0	0.985694	0.971593	0.957693	0.943992	0.930488	0.917176	0.904055	0.891121	0.878373	0.865807
6.1	0.986482	0.973146	0.959991	0.947013	0.934211	0.921582	0.909124	0.896834	0.884711	0.872751
6.2	0.987225	0.974613	0.962162	0.949870	0.937735	0.925756	0.913929	0.902253	0.890727	0.879348
6.3	0.987926	0.975998	0.964214	0.952572	0.941071	0.929708	0.918483	0.907393	0.896438	0.885614
6.4	0.988588	0.977306	0.966153	0.955127	0.944227	0.933451	0.922798	0.912267	0.901856	0.891564
6.5	0.989212	0.978541	0.967985	0.957543	0.947213	0.936995	0.926887	0.916889	0.906998	0.897213
6.6	0.989802	0.979708	0.969717	0.959828	0.950039	0.940351	0.930761	0.921269	0.911874	0.902575
6.7	0.990359	0.980810	0.971354	0.961988	0.952714	0.943528	0.934431	0.925422	0.916500	0.907663
6.8	0.990884	0.981851	0.972901	0.964032	0.955244	0.946536	0.937907	0.929357	0.920885	0.912491
6.9	0.991380	0.982835	0.974363	0.965964	0.957638	0.949383	0.941200	0.933087	0.925044	0.917070
7.0	0.991849	0.983764	0.975745	0.967792	0.959903	0.952079	0.944318	0.936621	0.928986	0.921414
7.1	0.992291	0.984642	0.977052	0.969520	0.962047	0.954631	0.947272	0.939970	0.932724	0.925534
7.2	0.992709	0.985472	0.978287	0.971155	0.964075	0.957046	0.950069	0.943143	0.936267	0.929441
7.3	0.993104	0.986256	0.979455	0.972701	0.965994	0.959333	0.952718	0.946148	0.939624	0.933145
7.4	0.993477	0.986997	0.980560	0.974164	0.967810	0.961497	0.955226	0.948995	0.942805	0.936656
7.5	0.993830	0.987698	0.981604	0.975547	0.969528	0.963546	0.957601	0.951692	0.945820	0.939985
7.6	0.994163	0.988360	0.982591	0.976856	0.971154	0.965485	0.959850	0.954247	0.948677	0.943140
7.7	0.994478	0.988986	0.983525	0.978093	0.972692	0.967321	0.961979	0.956667	0.951384	0.946130

7.8	0.994775	0.989578	0.984407	0.979264	0.974148	0.969058	0.963995	0.958958	0.953948	0.948964
7.9	0.995056	0.990137	0.985242	0.980372	0.975525	0.970702	0.965904	0.961129	0.956377	0.951649
8.0	0.995322	0.990666	0.986032	0.981419	0.976828	0.972259	0.967711	0.963184	0.958678	0.954194
8.1	0.995573	0.991166	0.986779	0.982410	0.978062	0.973732	0.969422	0.965130	0.960858	0.956605
8.2	0.995811	0.991639	0.987485	0.983348	0.979229	0.975127	0.971042	0.966974	0.962923	0.958889
8.3	0.996035	0.992086	0.988153	0.984235	0.980333	0.976447	0.972575	0.968720	0.964879	0.961054
8.4	0.996248	0.992509	0.988785	0.985075	0.981378	0.977696	0.974027	0.970372	0.966731	0.963103
8.5	0.996448	0.992909	0.989383	0.985869	0.982358	0.978879	0.975402	0.971938	0.968486	0.965046
8.6	0.996638	0.993288	0.989948	0.986620	0.983303	0.979998	0.976703	0.973420	0.970147	0.966886
8.7	0.996818	0.993646	0.990484	0.987332	0.984190	0.981058	0.977936	0.974824	0.971722	0.968629
8.8	0.996987	0.993984	0.990993	0.988004	0.985028	0.982061	0.979102	0.976153	0.973212	0.970280
8.9	0.997148	0.994304	0.991469	0.988641	0.985822	0.983010	0.980207	0.977411	0.974624	0.971845
9.0	0.997300	0.994607	0.991922	0.989244	0.986573	0.983909	0.981253	0.978603	0.975961	0.973326
9.1	0.997444	0.994894	0.992351	0.989814	0.987284	0.984760	0.982243	0.979732	0.977227	0.974729
9.2	0.997580	0.995165	0.992757	0.990354	0.987957	0.985566	0.983180	0.980801	0.978427	0.976059
9.3	0.997708	0.995422	0.993141	0.990865	0.988594	0.986328	0.984068	0.981813	0.979563	0.977318
9.4	0.997830	0.995665	0.993504	0.991348	0.989197	0.987050	0.984909	0.982771	0.980639	0.978511
9.5	0.997945	0.995894	0.993848	0.991806	0.989768	0.987734	0.985704	0.983679	0.981658	0.979640
9.6	0.998054	0.996112	0.994174	0.992239	0.990308	0.988381	0.986458	0.984538	0.982622	0.980710
9.7	0.998157	0.996318	0.994482	0.992649	0.990820	0.988994	0.987171	0.985352	0.983536	0.981724
9.8	0.998255	0.996512	0.994773	0.993037	0.991304	0.989574	0.987847	0.986123	0.984402	0.982684
9.9	0.998347	0.996697	0.995049	0.993405	0.991763	0.990123	0.988487	0.986853	0.985221	0.983593
10.0	0.998434	0.996871	0.995310	0.993752	0.992196	0.990643	0.989092	0.987544	0.985997	0.984454
10.1	0.998517	0.997036	0.995558	0.994082	0.992608	0.991136	0.989666	0.988198	0.986733	0.985270
10.2	0.998595	0.997193	0.995792	0.994393	0.992997	0.991602	0.990209	0.988818	0.987429	0.986042
10.3	0.998669	0.997341	0.996014	0.994689	0.993365	0.992043	0.990724	0.989406	0.988089	0.986774
10.4	0.998740	0.997481	0.996224	0.994968	0.993714	0.992462	0.991211	0.989962	0.988714	0.987468
10.5	0.998806	0.997613	0.996422	0.995233	0.994044	0.992858	0.991672	0.990488	0.989306	0.988124
10.6	0.998869	0.997739	0.996611	0.995483	0.994357	0.993233	0.992109	0.990987	0.989866	0.988747
10.7	0.998928	0.997858	0.996789	0.995721	0.994654	0.993588	0.992524	0.991460	0.990398	0.989336
10.8	0.998985	0.997971	0.996958	0.995945	0.994934	0.993924	0.992915	0.991908	0.990901	0.989895
10.9	0.999038	0.998077	0.997117	0.996158	0.995200	0.994243	0.993287	0.992332	0.991377	0.990424
11.0	0.999089	0.998178	0.997269	0.996360	0.995452	0.994545	0.993638	0.992733	0.991828	0.990924
11.1	0.999136	0.998274	0.997412	0.996551	0.995690	0.994830	0.993971	0.993113	0.992256	0.991399
11.2	0.999182	0.998364	0.997543	0.996731	0.995916	0.995101	0.994287	0.993474	0.992661	0.991849
11.3	0.999225	0.998450	0.997676	0.996903	0.996130	0.995358	0.994586	0.993815	0.993044	0.992275
11.4	0.999265	0.998531	0.997798	0.997065	0.996332	0.995600	0.994869	0.994138	0.993408	0.992678
11.5	0.999304	0.998608	0.997913	0.997218	0.996524	0.995830	0.995137	0.994444	0.993752	0.993060

Table 5 (continued)
$\rho = 0.1000$

a \ N	1	2	3	4	5	6	7	8	9	10
0.1	0.248170	0.061879	0.015489	0.003890	0.000983	0.000247	0.000063	0.000016	0.000004	0.000001
0.2	0.345279	0.119742	0.041679	0.014553	0.005095	0.001788	0.000629	0.000221	0.000078	0.000028
0.3	0.416118	0.173866	0.072897	0.030653	0.012923	0.005460	0.002311	0.000980	0.000416	0.000177
0.4	0.472911	0.224503	0.106923	0.051065	0.024447	0.011728	0.005637	0.002714	0.001308	0.000632
0.5	0.520500	0.271890	0.142458	0.074837	0.039403	0.020788	0.010986	0.005815	0.003083	0.001636
0.6	0.561422	0.316248	0.178650	0.101169	0.057415	0.032646	0.018593	0.010605	0.006057	0.003464
0.7	0.597216	0.357779	0.214910	0.129392	0.078061	0.047179	0.028559	0.017313	0.010509	0.006386
0.8	0.628906	0.396672	0.250821	0.158944	0.100915	0.064181	0.040880	0.026074	0.016651	0.010645
0.9	0.657218	0.433105	0.286384	0.189358	0.125563	0.083394	0.055467	0.036940	0.024630	0.016440
1.0	0.682689	0.467240	0.320485	0.220249	0.151621	0.104536	0.072171	0.049888	0.034524	0.023916
1.1	0.705734	0.499229	0.353875	0.251295	0.178738	0.127314	0.090802	0.064837	0.046346	0.033161
1.2	0.726678	0.529215	0.386148	0.282237	0.206600	0.151438	0.111141	0.081658	0.060056	0.044210
1.3	0.745787	0.557328	0.417239	0.312869	0.234928	0.176634	0.132959	0.100189	0.075569	0.057049
1.4	0.763276	0.583692	0.447108	0.342995	0.263479	0.202641	0.156022	0.120249	0.092762	0.071619
1.5	0.779328	0.608420	0.475735	0.372506	0.292043	0.229222	0.180101	0.141640	0.111489	0.087827
1.6	0.794097	0.631619	0.503118	0.401285	0.320441	0.256159	0.204973	0.164162	0.131585	0.105553
1.7	0.807712	0.653388	0.529269	0.429252	0.348522	0.283261	0.230433	0.187615	0.152874	0.124655
1.8	0.820287	0.673820	0.554206	0.456347	0.376159	0.310356	0.256287	0.211807	0.175176	0.144979
1.9	0.831921	0.692999	0.577956	0.482527	0.403248	0.337295	0.282360	0.236551	0.198313	0.166363
2.0	0.842701	0.711007	0.600553	0.507765	0.429705	0.363950	0.308495	0.261676	0.222110	0.188642
2.1	0.852701	0.727918	0.622031	0.532043	0.455462	0.390211	0.334551	0.287023	0.246401	0.211651
2.2	0.861989	0.743802	0.642430	0.555357	0.480470	0.415987	0.360404	0.312446	0.271029	0.235232
2.3	0.870626	0.758724	0.661790	0.577709	0.504687	0.441201	0.385948	0.337815	0.295850	0.259232
2.4	0.878665	0.772745	0.680152	0.599108	0.528089	0.465791	0.411088	0.363014	0.320729	0.283509
2.5	0.886154	0.785922	0.697560	0.619569	0.550657	0.489707	0.435749	0.387941	0.345548	0.307930
2.6	0.893136	0.798307	0.714053	0.639111	0.572383	0.512913	0.459864	0.412507	0.370198	0.332373
2.7	0.899652	0.809951	0.729674	0.657756	0.593264	0.535379	0.483381	0.436636	0.394584	0.356728
2.8	0.905735	0.820899	0.744464	0.675530	0.613304	0.557086	0.506256	0.460264	0.418623	0.380895
2.9	0.911420	0.831195	0.758462	0.692459	0.632513	0.578023	0.528457	0.483339	0.442243	0.404788
3.0	0.916735	0.840879	0.771706	0.708572	0.650907	0.598184	0.549959	0.505816	0.465384	0.428330
3.1	0.921708	0.849989	0.784234	0.723897	0.668488	0.617569	0.570745	0.527661	0.487994	0.451453
3.2	0.926361	0.858561	0.796083	0.738465	0.685290	0.636183	0.590805	0.548847	0.510031	0.474103
3.3	0.930720	0.866627	0.807286	0.752305	0.701329	0.654036	0.610133	0.569357	0.531463	0.496231
3.4	0.934803	0.874219	0.817878	0.765448	0.716626	0.671138	0.628730	0.589175	0.552262	0.517799
3.5	0.938631	0.881365	0.827890	0.777923	0.731207	0.687505	0.646600	0.608296	0.572410	0.538775
3.6	0.942220	0.888092	0.837352	0.789760	0.745094	0.703153	0.663751	0.626718	0.591894	0.559135
3.7	0.945587	0.894427	0.846295	0.800987	0.758314	0.718103	0.680194	0.644441	0.610705	0.578861
3.8	0.948747	0.900392	0.854745	0.811631	0.770890	0.732373	0.695942	0.661471	0.628840	0.597941

3.9	0.951714	0.906011	0.862729	0.821721	0.782849	0.745985	0.711011	0.677817	0.646300	0.616366
4.0	0.954500	0.911303	0.870273	0.831283	0.794215	0.758960	0.725417	0.693489	0.663090	0.634135
4.1	0.957117	0.916289	0.877400	0.840342	0.805014	0.771322	0.739179	0.708502	0.679216	0.651248
4.2	0.959576	0.920987	0.884133	0.848922	0.815269	0.783092	0.752316	0.722870	0.694688	0.667708
4.3	0.961887	0.925413	0.890494	0.857049	0.825005	0.794293	0.764848	0.736610	0.709519	0.683524
4.4	0.964061	0.929586	0.896502	0.864743	0.834245	0.804948	0.776797	0.749738	0.723723	0.698704
4.5	0.966105	0.933518	0.902178	0.872027	0.843011	0.815079	0.788181	0.762274	0.737314	0.713260
4.6	0.968028	0.937225	0.907540	0.878923	0.851326	0.824707	0.799024	0.774237	0.750309	0.727204
4.7	0.969837	0.940721	0.912605	0.885449	0.859212	0.833855	0.809345	0.785645	0.762725	0.740553
4.8	0.971540	0.944016	0.917390	0.891625	0.866687	0.842544	0.819164	0.796519	0.774580	0.753321
4.9	0.973143	0.947124	0.921910	0.897470	0.873774	0.850794	0.828504	0.806878	0.785893	0.765524
5.0	0.974652	0.950055	0.926180	0.903000	0.880489	0.858625	0.837383	0.816742	0.796682	0.777181
5.1	0.976074	0.952820	0.930213	0.908232	0.886853	0.866056	0.845822	0.826131	0.806966	0.788308
5.2	0.977413	0.955428	0.934024	0.913181	0.892882	0.873107	0.853840	0.835064	0.816764	0.798924
5.3	0.978674	0.957888	0.937624	0.917864	0.898593	0.879794	0.861455	0.843559	0.826094	0.809047
5.4	0.979863	0.960210	0.941025	0.922293	0.904002	0.886137	0.868686	0.851636	0.834976	0.818695
5.5	0.980983	0.962400	0.944238	0.926483	0.909125	0.892151	0.875550	0.859312	0.843428	0.827886
5.6	0.982039	0.964468	0.947273	0.930446	0.913976	0.897852	0.882065	0.866606	0.851466	0.836637
5.7	0.983035	0.966419	0.950142	0.934195	0.918569	0.903256	0.888247	0.873534	0.859110	0.844968
5.8	0.983974	0.968260	0.952852	0.937740	0.922918	0.908377	0.894111	0.880113	0.866376	0.852894
5.9	0.984859	0.969999	0.955413	0.941094	0.927035	0.913231	0.899674	0.886360	0.873281	0.860433
6.0	0.985694	0.971640	0.957833	0.944266	0.930933	0.917830	0.904950	0.892289	0.879841	0.867602
6.1	0.986482	0.973190	0.960120	0.947265	0.934623	0.922186	0.909952	0.897915	0.886072	0.874417
6.2	0.987225	0.974653	0.962281	0.950103	0.938115	0.926314	0.914695	0.903254	0.891989	0.880894
6.3	0.987926	0.976035	0.964323	0.952786	0.941421	0.930224	0.919191	0.908319	0.897606	0.887047
6.4	0.988588	0.977340	0.966253	0.955324	0.944550	0.933927	0.923452	0.913124	0.902937	0.892891
6.5	0.989212	0.978573	0.968078	0.957725	0.947511	0.937434	0.927491	0.917680	0.907998	0.898442
6.6	0.989802	0.979737	0.969802	0.959995	0.950314	0.940756	0.931319	0.922000	0.912799	0.903712
6.7	0.990358	0.980837	0.971432	0.962143	0.952966	0.943901	0.934945	0.926097	0.917354	0.908715
6.8	0.990884	0.981876	0.972973	0.964174	0.955477	0.946880	0.938382	0.929981	0.921675	0.913463
6.9	0.991380	0.982857	0.974429	0.966095	0.957852	0.949700	0.941637	0.933662	0.925773	0.917969
7.0	0.991849	0.983785	0.975806	0.967912	0.960101	0.952371	0.944722	0.937152	0.929660	0.922244
7.1	0.992291	0.984661	0.977108	0.969631	0.962229	0.954900	0.947644	0.940459	0.933345	0.926300
7.2	0.992709	0.985489	0.978339	0.971257	0.964242	0.957294	0.950411	0.943593	0.936839	0.930147
7.3	0.993104	0.986272	0.979503	0.972795	0.966148	0.959561	0.953033	0.946563	0.940151	0.933796
7.4	0.993477	0.987012	0.980603	0.974250	0.967951	0.961707	0.955516	0.949378	0.943292	0.937257
7.5	0.993830	0.987711	0.981644	0.975626	0.969658	0.963739	0.957868	0.952044	0.946268	0.940538
7.6	0.994163	0.988372	0.982628	0.976928	0.971273	0.965662	0.960095	0.954571	0.949089	0.943650
7.7	0.994478	0.988997	0.983558	0.978160	0.972802	0.967483	0.962205	0.956965	0.951763	0.946600
7.8	0.994775	0.989588	0.984438	0.979325	0.974248	0.969208	0.964202	0.959232	0.954297	0.949396
7.9	0.995056	0.990147	0.985271	0.980428	0.975617	0.970840	0.966094	0.961381	0.956698	0.952047
8.0	0.995322	0.990675	0.986058	0.981471	0.976913	0.972385	0.967886	0.963415	0.958973	0.954560
8.1	0.995573	0.991174	0.986802	0.982458	0.978140	0.973848	0.969583	0.965343	0.961129	0.956941

Table 5 (continued)
$\rho = 0.1000$

8.2	0.995811	0.991646	0.987507	0.983391	0.979300	0.975233	0.971189	0.967169	0.963172	0.959198
8.3	0.996035	0.992093	0.988173	0.984275	0.980399	0.976544	0.972711	0.968899	0.965108	0.961338
8.4	0.996248	0.992515	0.988803	0.985111	0.981439	0.977786	0.974152	0.970537	0.966942	0.963365
8.5	0.996448	0.992915	0.989400	0.985902	0.982423	0.978961	0.975516	0.972089	0.968679	0.965286
8.6	0.996638	0.993293	0.989964	0.986651	0.983354	0.980073	0.976808	0.973559	0.970325	0.967107
8.7	0.996818	0.993650	0.990498	0.987360	0.984236	0.981127	0.978032	0.974951	0.971885	0.968832
8.8	0.996987	0.993988	0.991003	0.988030	0.985071	0.982124	0.979191	0.976270	0.973362	0.970466
8.9	0.997148	0.994308	0.991481	0.988665	0.985861	0.983068	0.980288	0.977519	0.974761	0.972015
9.0	0.997300	0.994611	0.991933	0.989265	0.986609	0.983962	0.981327	0.978702	0.976087	0.973483
9.1	0.997444	0.994897	0.992361	0.989834	0.987317	0.984809	0.982311	0.979822	0.977343	0.974873
9.2	0.997580	0.995168	0.992766	0.990372	0.987987	0.985610	0.983243	0.980884	0.978533	0.976191
9.3	0.997708	0.995425	0.993149	0.990881	0.988621	0.986369	0.984125	0.981889	0.979660	0.977439
9.4	0.997830	0.995667	0.993512	0.991363	0.989222	0.987[illegible]	0.984961	0.982841	0.980728	0.978621
9.5	0.997945	0.995897	0.993855	0.991820	0.989791	0.987768	0.985752	0.983742	0.981739	0.979742
9.6	0.998054	0.996114	0.994180	0.992252	0.990329	0.988413	0.986502	0.984596	0.982697	0.980803
9.7	0.998157	0.996320	0.994487	0.992661	0.990839	0.989023	0.987211	0.985406	0.983605	0.981809
9.8	0.998255	0.996514	0.994779	0.993048	0.991322	0.989600	0.987884	0.986172	0.984465	0.982762
9.9	0.998347	0.996698	0.995054	0.993414	0.991779	0.990147	0.988520	0.986898	0.985279	0.983665
10.0	0.998434	0.996873	0.995315	0.993761	0.992211	0.990665	0.989123	0.987585	0.986050	0.984520
10.1	0.998517	0.997038	0.995562	0.994090	0.992621	0.991156	0.989694	0.988236	0.986781	0.985330
10.2	0.998595	0.997194	0.995796	0.994401	0.993009	0.991621	0.990235	0.988853	0.987474	0.986098
10.3	0.998669	0.997342	0.996017	0.994695	0.993377	0.992061	0.990748	0.989437	0.988130	0.986825
10.4	0.998740	0.997482	0.996227	0.994974	0.993724	0.992477	0.991233	0.989990	0.988751	0.987514
10.5	0.998806	0.997614	0.996425	0.995239	0.994054	0.992872	0.991692	0.990515	0.989340	0.988167
10.6	0.998869	0.997740	0.996613	0.995489	0.994366	0.993246	0.992128	0.991011	0.989897	0.988785
10.7	0.998928	0.997859	0.996791	0.995725	0.994662	0.993600	0.992540	0.991482	0.990426	0.989372
10.8	0.998985	0.997971	0.996960	0.995950	0.994942	0.993935	0.992931	0.991928	0.990926	0.989927
10.9	0.999038	0.998078	0.997119	0.996162	0.995207	0.994253	0.993300	0.992350	0.991401	0.990453
11.0	0.999089	0.998179	0.997270	0.996363	0.995458	0.994554	0.993651	0.992750	0.991850	0.990951
11.1	0.999136	0.998274	0.997413	0.996554	0.995696	0.994839	0.993983	0.993129	0.992275	0.991424
11.2	0.999182	0.998365	0.997549	0.996734	0.995921	0.995109	0.994298	0.993488	0.992679	0.991871
11.3	0.999225	0.998450	0.997677	0.996905	0.996134	0.995364	0.994595	0.993828	0.993061	0.992295
11.4	0.999265	0.998532	0.997799	0.997067	0.996336	0.995607	0.994878	0.994150	0.993423	0.992697
11.5	0.999304	0.998608	0.997914	0.997221	0.996528	0.995836	0.995145	0.994455	0.993766	0.993078

Table 5 (continued)
$\rho = 0.2000$

a \ N	1	2	3	4	5	6	7	8	9	10
0.1	0.248170	0.062774	0.016091	0.004165	0.001386	0.000285	0.000075	0.000020	0.000005	0.000001
0.2	0.345279	0.121359	0.043194	0.015518	0.005616	0.002044	0.000747	0.000274	0.000101	0.000037
0.3	0.416118	0.176054	0.075374	0.032558	0.014161	0.006193	0.002721	0.001200	0.000531	0.000236
0.4	0.472911	0.227137	0.110319	0.054035	0.026542	0.013206	0.006574	0.003285	0.001646	0.000827
0.5	0.520500	0.274864	0.146684	0.078909	0.042717	0.023243	0.012700	0.006964	0.003830	0.002112
0.6	0.561422	0.319469	0.183599	0.106318	0.061937	0.036259	0.021312	0.012569	0.007435	0.004408
0.7	0.597216	0.361173	0.220467	0.135548	0.083816	0.052070	0.032473	0.020318	0.012749	0.008018
0.8	0.628906	0.400175	0.256872	0.166011	0.107875	0.070411	0.046129	0.030316	0.019978	0.013196
0.9	0.657218	0.436664	0.292520	0.197224	0.133662	0.090972	0.062137	0.042571	0.029241	0.020131
1.0	0.682689	0.470812	0.327208	0.228791	0.160764	0.113424	0.080297	0.057010	0.040579	0.028946
1.1	0.705734	0.502779	0.360792	0.260393	0.188811	0.137437	0.100369	0.073503	0.053958	0.039694
1.2	0.726678	0.532713	0.393181	0.291771	0.217476	0.162698	0.122097	0.091873	0.069291	0.052366
1.3	0.745787	0.560752	0.424316	0.322718	0.246477	0.188908	0.145217	0.111917	0.086445	0.066901
1.4	0.763276	0.587024	0.454167	0.353071	0.275570	0.215797	0.169470	0.133415	0.105254	0.083194
1.5	0.779328	0.611646	0.482725	0.382705	0.304548	0.243121	0.194608	0.156141	0.125533	0.101108
1.6	0.794097	0.634729	0.509996	0.411522	0.333241	0.270661	0.220399	0.179873	0.147085	0.120483
1.7	0.807712	0.656375	0.535998	0.439451	0.361503	0.298227	0.246630	0.204394	0.169709	0.141143
1.8	0.820287	0.676678	0.560757	0.466443	0.389218	0.325654	0.273106	0.229502	0.193205	0.162906
1.9	0.831922	0.695726	0.584306	0.492463	0.416293	0.352802	0.299655	0.255006	0.217381	0.185589
2.0	0.842701	0.713602	0.606683	0.517492	0.442653	0.379550	0.326125	0.280736	0.242057	0.209010
2.1	0.852701	0.730382	0.627929	0.541524	0.468242	0.405800	0.352384	0.306533	0.267063	0.232997
2.2	0.861989	0.746136	0.648086	0.564559	0.493019	0.431471	0.378316	0.332261	0.292243	0.257385
2.3	0.870626	0.760930	0.667199	0.586606	0.516954	0.456497	0.403827	0.357797	0.317458	0.282023
2.4	0.878665	0.774827	0.685311	0.607682	0.540030	0.480826	0.428834	0.383034	0.342583	0.306771
2.5	0.886153	0.787883	0.702468	0.627807	0.562236	0.504421	0.453271	0.407883	0.367505	0.331502
2.6	0.893136	0.800152	0.718714	0.647002	0.583572	0.527253	0.477084	0.432266	0.392129	0.356105
2.7	0.899652	0.811684	0.734091	0.665297	0.604043	0.549303	0.500231	0.456119	0.416370	0.380478
2.8	0.905736	0.822525	0.748641	0.682718	0.623657	0.570560	0.522679	0.479389	0.440159	0.404535
2.9	0.911420	0.832719	0.762406	0.699296	0.642430	0.591020	0.544407	0.502036	0.463436	0.428200
3.0	0.916735	0.842306	0.775424	0.715061	0.660379	0.610684	0.565397	0.524026	0.486151	0.451410
3.1	0.921707	0.851323	0.787734	0.730045	0.677523	0.629560	0.585642	0.545336	0.508266	0.474111
3.2	0.926362	0.859807	0.799373	0.744280	0.693886	0.647656	0.605139	0.565948	0.529750	0.496258
3.3	0.930720	0.867790	0.810375	0.757796	0.709491	0.664988	0.623889	0.585852	0.550580	0.517817
3.4	0.934803	0.875303	0.820775	0.770625	0.724362	0.681571	0.641899	0.605043	0.570740	0.538760
3.5	0.938631	0.882375	0.830603	0.782798	0.738526	0.697424	0.659179	0.623522	0.590220	0.559066
3.6	0.942220	0.889033	0.839892	0.794343	0.752008	0.712566	0.675740	0.641292	0.609013	0.578721
3.7	0.945587	0.895302	0.848669	0.805291	0.764835	0.727019	0.691598	0.658360	0.627121	0.597715
3.8	0.948747	0.901206	0.856962	0.815669	0.777033	0.740806	0.706770	0.674738	0.644545	0.616045

Table 5 (continued)
$\rho = 0.2000$

3.9	0.951714	0.906767	0.864799	0.825505	0.788628	0.753948	0.721274	0.690437	0.661293	0.633710
4.0	0.954500	0.912006	0.872203	0.834825	0.799645	0.766470	0.735128	0.705474	0.677374	0.650713
4.1	0.957117	0.916941	0.879198	0.843654	0.810110	0.778393	0.748355	0.719862	0.692799	0.667062
4.2	0.959576	0.921592	0.885807	0.852017	0.820047	0.789743	0.760974	0.733622	0.707583	0.682765
4.3	0.961887	0.925975	0.892051	0.859938	0.829479	0.800541	0.773006	0.746770	0.721740	0.697833
4.4	0.964061	0.930106	0.897951	0.867438	0.838431	0.810811	0.784474	0.759326	0.735286	0.712280
4.5	0.966105	0.934000	0.903525	0.874540	0.846925	0.820575	0.795398	0.771310	0.748239	0.726119
4.6	0.968028	0.937672	0.908791	0.881263	0.854982	0.829854	0.805799	0.782742	0.760618	0.739367
4.7	0.969837	0.941134	0.913766	0.887627	0.862623	0.838671	0.815699	0.793641	0.772439	0.752040
4.8	0.971540	0.944399	0.918467	0.893652	0.869869	0.847046	0.825118	0.804028	0.783723	0.764156
4.9	0.973143	0.947478	0.922909	0.899353	0.876738	0.854999	0.834077	0.813922	0.794488	0.775731
5.0	0.974652	0.950382	0.927106	0.904750	0.883250	0.862549	0.842596	0.823344	0.804753	0.786785
5.1	0.976074	0.953122	0.931071	0.909857	0.889422	0.869716	0.850693	0.832312	0.814537	0.797334
5.2	0.977413	0.955707	0.934819	0.914690	0.895271	0.876517	0.858387	0.840846	0.823859	0.807398
5.3	0.978674	0.958146	0.938363	0.919263	0.900814	0.882970	0.865698	0.848963	0.832738	0.816995
5.4	0.979863	0.960448	0.941706	0.923591	0.906065	0.889093	0.872642	0.856683	0.841191	0.826142
5.5	0.980983	0.962621	0.944868	0.927686	0.911040	0.894900	0.879236	0.864022	0.849237	0.834857
5.6	0.982039	0.964671	0.947856	0.931561	0.915754	0.900407	0.885496	0.870998	0.856892	0.843158
5.7	0.983035	0.966606	0.950681	0.935227	0.920219	0.905630	0.891440	0.877627	0.864174	0.851062
5.8	0.983974	0.968434	0.953350	0.938696	0.924447	0.910582	0.897081	0.883925	0.871099	0.858586
5.9	0.984859	0.970159	0.955873	0.941978	0.928453	0.915277	0.902434	0.889908	0.877682	0.865744
6.0	0.985694	0.971788	0.958258	0.945084	0.932246	0.919728	0.907514	0.895589	0.883940	0.872555
6.1	0.986482	0.973326	0.960512	0.948022	0.935839	0.923947	0.912333	0.900983	0.889887	0.879033
6.2	0.987225	0.974778	0.962643	0.950802	0.939241	0.927946	0.916904	0.906105	0.895537	0.885192
6.3	0.987926	0.976150	0.964657	0.953433	0.942463	0.931735	0.921240	0.910967	0.900905	0.891047
6.4	0.988588	0.977446	0.966562	0.955921	0.945513	0.935327	0.925352	0.915581	0.906004	0.896613
6.5	0.989212	0.978671	0.968362	0.958276	0.948402	0.938730	0.929252	0.919960	0.910845	0.901902
6.6	0.989802	0.979827	0.970065	0.960505	0.951138	0.941955	0.932950	0.924115	0.915442	0.906927
6.7	0.990358	0.980920	0.971674	0.962613	0.953728	0.945011	0.936456	0.928057	0.919807	0.911700
6.8	0.990884	0.981952	0.973196	0.964608	0.956180	0.947906	0.939781	0.931797	0.923950	0.916234
6.9	0.991380	0.982928	0.974635	0.966495	0.958502	0.950650	0.942932	0.935344	0.927882	0.920540
7.0	0.991849	0.983850	0.975996	0.968282	0.960701	0.953248	0.945919	0.938709	0.931614	0.924629
7.1	0.992291	0.984721	0.977283	0.969972	0.962783	0.955710	0.948751	0.941901	0.935155	0.928511
7.2	0.992709	0.985544	0.978500	0.971571	0.964753	0.958043	0.951435	0.944927	0.938515	0.932196
7.3	0.993104	0.986323	0.979651	0.973085	0.966620	0.960252	0.953979	0.947797	0.941703	0.935694
7.4	0.993477	0.987059	0.980740	0.974517	0.968386	0.962345	0.956390	0.950518	0.944727	0.939014
7.5	0.993830	0.987754	0.981769	0.975872	0.970059	0.964328	0.958675	0.953099	0.947596	0.942165
7.6	0.994163	0.988412	0.982744	0.977155	0.971643	0.966206	0.960841	0.955546	0.950318	0.945155
7.7	0.994478	0.989034	0.983665	0.978369	0.973143	0.967985	0.962893	0.957865	0.952899	0.947992

7.8	0.994775	0.989621	0.984536	0.979518	0.974563	0.969670	0.964838	0.960064	0.955347	0.950684
7.9	0.995056	0.990177	0.985361	0.980605	0.975907	0.971267	0.966681	0.962148	0.957668	0.953238
8.0	0.995322	0.990703	0.986141	0.981634	0.977181	0.972779	0.968427	0.964125	0.959870	0.955661
8.1	0.995573	0.991200	0.986879	0.982608	0.978386	0.974211	0.970082	0.965998	0.961957	0.957958
8.2	0.995811	0.991670	0.987577	0.983530	0.979527	0.975568	0.971650	0.967773	0.963936	0.960138
8.3	0.996035	0.992115	0.988238	0.984403	0.980608	0.976853	0.973136	0.969456	0.965813	0.962206
8.4	0.996248	0.992536	0.988863	0.985229	0.981631	0.978070	0.974543	0.971051	0.967593	0.964166
8.5	0.996448	0.992933	0.989454	0.986010	0.982600	0.979222	0.975877	0.972563	0.969280	0.966026
8.6	0.996638	0.993310	0.990014	0.986750	0.983517	0.980314	0.977141	0.973996	0.970879	0.967789
8.7	0.996818	0.993666	0.990544	0.987451	0.984386	0.981349	0.978338	0.975354	0.972395	0.969462
8.8	0.996987	0.994003	0.991045	0.988114	0.985209	0.982329	0.979473	0.976641	0.973833	0.971047
8.9	0.997148	0.994322	0.991520	0.988742	0.985988	0.983257	0.980548	0.977861	0.975196	0.972551
9.0	0.997300	0.994623	0.991969	0.989337	0.986726	0.984136	0.981566	0.979017	0.976487	0.973977
9.1	0.997444	0.994908	0.992394	0.989899	0.987424	0.984969	0.982532	0.980113	0.977712	0.975329
9.2	0.997580	0.995178	0.992796	0.990432	0.988086	0.985757	0.983446	0.981151	0.978873	0.976611
9.3	0.997708	0.995434	0.993177	0.990937	0.988713	0.986505	0.984312	0.982135	0.979973	0.977826
9.4	0.997830	0.995676	0.993537	0.991414	0.989306	0.987212	0.985133	0.983068	0.981016	0.978978
9.5	0.997945	0.995905	0.993879	0.991866	0.989868	0.987883	0.985911	0.983951	0.982005	0.980070
9.6	0.998054	0.996121	0.994202	0.992295	0.990400	0.988518	0.986648	0.984789	0.982942	0.981106
9.7	0.998157	0.996326	0.994507	0.992700	0.990904	0.989119	0.987346	0.985583	0.983830	0.982088
9.8	0.998255	0.996520	0.994797	0.993084	0.991382	0.989689	0.988007	0.986335	0.984672	0.983019
9.9	0.998347	0.996704	0.995071	0.993448	0.991834	0.990229	0.988634	0.987047	0.985470	0.983901
10.0	0.998434	0.996878	0.995331	0.993792	0.992262	0.990741	0.989228	0.987723	0.986226	0.984737
10.1	0.998517	0.997042	0.995576	0.994118	0.992668	0.991225	0.989790	0.988363	0.986943	0.985530
10.2	0.998595	0.997198	0.995809	0.994427	0.993052	0.991684	0.990324	0.988970	0.987623	0.986282
10.3	0.998669	0.997346	0.996029	0.994719	0.993416	0.992119	0.990829	0.989545	0.988267	0.986995
10.4	0.998740	0.997486	0.996238	0.994996	0.993761	0.992531	0.991307	0.990089	0.988877	0.987670
10.5	0.998806	0.997618	0.996435	0.995259	0.994087	0.992922	0.991761	0.990606	0.989456	0.988311
10.6	0.998869	0.997743	0.996623	0.995507	0.994397	0.993291	0.992191	0.991095	0.990004	0.988918
10.7	0.998928	0.997862	0.996800	0.995742	0.994690	0.993642	0.992598	0.991559	0.990524	0.989493
10.8	0.998985	0.997974	0.996968	0.995965	0.994968	0.993974	0.992984	0.991998	0.991017	0.990039
10.9	0.999038	0.998080	0.997127	0.996177	0.995231	0.994288	0.993350	0.992415	0.991484	0.990556
11.0	0.999089	0.998181	0.997277	0.996377	0.995480	0.994586	0.993696	0.992810	0.991926	0.991046
11.1	0.999136	0.998276	0.997420	0.996566	0.995716	0.994869	0.994025	0.993184	0.992346	0.991511
11.2	0.999182	0.998367	0.997555	0.996745	0.995939	0.995136	0.994336	0.993538	0.992743	0.991951
11.3	0.999225	0.998452	0.997683	0.996916	0.996151	0.995390	0.994631	0.993874	0.993120	0.992369
11.4	0.999265	0.998533	0.997804	0.997077	0.996352	0.995630	0.994910	0.994193	0.993477	0.992765
11.5	0.999304	0.998610	0.997918	0.997229	0.996542	0.995857	0.995175	0.994494	0.993816	0.993140

Table 5 (continued)
$\rho = 0.3000$

a \ N	1	2	3	4	5	6	7	8	9	10
0.1	0.248170	0.064357	0.017128	0.004637	0.001271	0.000351	0.000098	0.000027	0.000008	0.000002
0.2	0.345279	0.124204	0.045790	0.017160	0.006507	0.002489	0.000959	0.000371	0.000144	0.000056
0.3	0.416118	0.179891	0.079597	0.035776	0.016262	0.007455	0.003441	0.001596	0.000744	0.000348
0.4	0.472911	0.231738	0.116080	0.059021	0.030337	0.015722	0.008200	0.004299	0.002264	0.001196
0.5	0.520500	0.280037	0.153822	0.085706	0.048252	0.027382	0.015636	0.008974	0.005172	0.002991
0.6	0.561422	0.325055	0.191922	0.114864	0.069433	0.042293	0.025917	0.015960	0.009868	0.006123
0.7	0.597216	0.367037	0.229774	0.145711	0.093289	0.060166	0.039031	0.025440	0.016648	0.010930
0.8	0.628906	0.406209	0.266968	0.177619	0.119256	0.080637	0.054832	0.037456	0.025685	0.017670
0.9	0.657218	0.442776	0.303222	0.210081	0.146823	0.103310	0.073088	0.051937	0.037043	0.026504
1.0	0.682689	0.476929	0.338346	0.242693	0.175534	0.127786	0.093512	0.068725	0.050690	0.037501
1.1	0.705734	0.508840	0.372217	0.275136	0.204991	0.153682	0.115794	0.087608	0.066513	0.050646
1.2	0.726678	0.538672	0.404761	0.307159	0.234856	0.180643	0.139614	0.108335	0.084344	0.065854
1.3	0.745787	0.566571	0.435936	0.338570	0.264842	0.208348	0.164664	0.130638	0.103978	0.082987
1.4	0.763276	0.592673	0.465729	0.369220	0.294708	0.236512	0.190649	0.154246	0.125182	0.101866
1.5	0.779328	0.617104	0.494147	0.398999	0.324260	0.264886	0.217301	0.178894	0.147718	0.122288
1.6	0.794097	0.639980	0.521209	0.427828	0.353336	0.293253	0.244374	0.204332	0.171344	0.144037
1.7	0.807712	0.661409	0.546946	0.455653	0.381809	0.321433	0.271652	0.230326	0.195828	0.166894
1.8	0.820287	0.681487	0.571395	0.482441	0.409579	0.349273	0.298945	0.256665	0.220949	0.190640
1.9	0.831921	0.700309	0.594603	0.508172	0.436570	0.376645	0.326089	0.283158	0.246504	0.215069
2.0	0.842701	0.717957	0.616607	0.532843	0.462725	0.403449	0.352945	0.309639	0.272308	0.239985
2.1	0.852701	0.734511	0.637464	0.556458	0.488004	0.429602	0.379394	0.335961	0.298194	0.265209
2.2	0.861989	0.750043	0.657221	0.579031	0.512382	0.455041	0.405341	0.362000	0.324014	0.290577
2.3	0.870626	0.764621	0.675926	0.600582	0.535844	0.479718	0.430705	0.387650	0.349641	0.315945
2.4	0.878665	0.778307	0.693629	0.621135	0.558386	0.503598	0.455425	0.412823	0.374964	0.341183
2.5	0.886153	0.791160	0.710378	0.640719	0.580011	0.526659	0.479452	0.437445	0.399890	0.366180
2.6	0.893136	0.803234	0.726220	0.659364	0.600729	0.548887	0.502748	0.461459	0.424341	0.390839
2.7	0.899652	0.814578	0.741201	0.677102	0.620554	0.570277	0.525289	0.484819	0.448250	0.415080
2.8	0.905736	0.825239	0.755365	0.693967	0.639506	0.590831	0.547058	0.507490	0.471568	0.438832
2.9	0.911420	0.835262	0.768755	0.709994	0.657604	0.610554	0.568045	0.529447	0.494252	0.462042
3.0	0.916735	0.844686	0.781411	0.725217	0.674875	0.629459	0.588249	0.550674	0.516272	0.484663
3.1	0.921707	0.853550	0.793372	0.739669	0.691343	0.647560	0.607671	0.571159	0.537605	0.506661
3.2	0.926362	0.861888	0.804675	0.753386	0.707036	0.664876	0.626321	0.590901	0.558236	0.528009
3.3	0.930720	0.869733	0.815357	0.766400	0.721981	0.681427	0.644208	0.609900	0.578156	0.548688
3.4	0.934803	0.877116	0.825450	0.778743	0.736207	0.697235	0.661348	0.628163	0.597363	0.568687
3.5	0.938631	0.884066	0.834987	0.790448	0.749742	0.712322	0.677757	0.645698	0.615858	0.587997
3.6	0.942220	0.890609	0.843998	0.801544	0.762613	0.726713	0.693454	0.662519	0.633646	0.606619
3.7	0.945587	0.896770	0.852512	0.812062	0.774850	0.740433	0.708459	0.678639	0.650737	0.624554
3.8	0.948747	0.902573	0.860557	0.822030	0.786479	0.753506	0.722792	0.694075	0.667141	0.641808

3.9	0.951714	0.908038	0.868159	0.831474	0.797527	0.765957	0.736476	0.708847	0.682873	0.658390
4.0	0.954500	0.913188	0.875341	0.840422	0.808020	0.777811	0.749533	0.722971	0.697948	0.674311
4.1	0.957117	0.918040	0.882127	0.848898	0.817984	0.789092	0.761985	0.736470	0.712382	0.689586
4.2	0.959576	0.922613	0.888540	0.856927	0.827443	0.799824	0.773856	0.749362	0.726195	0.704228
4.3	0.961887	0.926924	0.894599	0.864531	0.836421	0.810031	0.785168	0.761670	0.739404	0.718254
4.4	0.964061	0.930987	0.900325	0.871733	0.844941	0.819736	0.795942	0.773414	0.752028	0.731681
4.5	0.966105	0.934818	0.905736	0.878552	0.853025	0.828961	0.806202	0.784615	0.764088	0.744528
4.6	0.968028	0.938431	0.910850	0.885010	0.860694	0.837728	0.815968	0.795294	0.775604	0.756812
4.7	0.969837	0.941838	0.915683	0.891124	0.867969	0.846058	0.825262	0.805471	0.786594	0.768551
4.8	0.971540	0.945052	0.920250	0.896914	0.874868	0.853970	0.834103	0.815168	0.797079	0.779766
4.9	0.973143	0.948084	0.924567	0.902395	0.881411	0.861486	0.842513	0.824403	0.807079	0.790475
5.0	0.974652	0.950944	0.928647	0.907584	0.887615	0.868622	0.850509	0.833196	0.816611	0.800696
5.1	0.976074	0.953643	0.932503	0.912497	0.893497	0.875398	0.858112	0.841566	0.825696	0.810448
5.2	0.977413	0.956190	0.936149	0.917149	0.899074	0.881831	0.865339	0.849532	0.834352	0.819749
5.3	0.978674	0.958593	0.939595	0.921552	0.904361	0.887936	0.872206	0.857111	0.842596	0.828618
5.4	0.979863	0.960862	0.942853	0.925720	0.909372	0.893731	0.878732	0.864320	0.850447	0.837071
5.5	0.980983	0.963004	0.945932	0.929667	0.914123	0.899231	0.884932	0.871176	0.857921	0.845127
5.6	0.982039	0.965026	0.948844	0.933403	0.918625	0.904449	0.890821	0.877696	0.865035	0.852802
5.7	0.983035	0.966935	0.951597	0.936939	0.922892	0.909400	0.896415	0.883895	0.871804	0.860111
5.8	0.983974	0.968738	0.954200	0.940287	0.926936	0.914098	0.901727	0.889787	0.878244	0.867071
5.9	0.984859	0.970440	0.956662	0.943456	0.930769	0.918554	0.906771	0.895387	0.884371	0.873698
6.0	0.985694	0.972048	0.958989	0.946456	0.934401	0.922781	0.911560	0.900708	0.890198	0.880005
6.1	0.986482	0.973567	0.961190	0.949297	0.937843	0.926790	0.916107	0.905765	0.895739	0.886008
6.2	0.987225	0.975002	0.963271	0.951986	0.941104	0.930593	0.920423	0.910568	0.901007	0.891719
6.3	0.987926	0.976357	0.965240	0.954531	0.944195	0.934200	0.924520	0.915131	0.906015	0.897152
6.4	0.988588	0.977637	0.967101	0.956941	0.947123	0.937620	0.928408	0.919466	0.910775	0.902320
6.5	0.989212	0.978847	0.968862	0.959222	0.949898	0.940863	0.932098	0.923582	0.915299	0.907235
6.6	0.989802	0.979990	0.970528	0.961382	0.952526	0.943939	0.935600	0.927491	0.919599	0.911908
6.7	0.990358	0.981071	0.972103	0.963426	0.955017	0.946855	0.938923	0.931203	0.923684	0.916351
6.8	0.990884	0.982092	0.973593	0.965362	0.957378	0.949621	0.942075	0.934728	0.927565	0.920575
6.9	0.991380	0.983057	0.975003	0.967195	0.959614	0.952243	0.945067	0.938074	0.931252	0.924591
7.0	0.991849	0.983969	0.976336	0.968930	0.961732	0.954728	0.947905	0.941250	0.934754	0.928407
7.1	0.992291	0.984831	0.977598	0.970573	0.963740	0.957085	0.950597	0.944265	0.938080	0.932033
7.2	0.992709	0.985646	0.978792	0.972128	0.965641	0.959320	0.953151	0.947128	0.941239	0.935479
7.3	0.993104	0.986417	0.979921	0.973600	0.967443	0.961438	0.955574	0.949844	0.944240	0.938754
7.4	0.993477	0.987146	0.980989	0.974995	0.969150	0.963446	0.957872	0.952422	0.947088	0.941864
7.5	0.993830	0.987835	0.982001	0.976315	0.970768	0.965349	0.960052	0.954869	0.949793	0.944819
7.6	0.994163	0.988486	0.982957	0.977565	0.972300	0.967154	0.962120	0.957191	0.952362	0.947626
7.7	0.994478	0.989102	0.983863	0.978749	0.973752	0.968865	0.964081	0.959394	0.954800	0.950292
7.8	0.994775	0.989685	0.984719	0.979869	0.975127	0.970486	0.965941	0.961485	0.957114	0.952824
7.9	0.995056	0.990236	0.985530	0.980931	0.976430	0.972024	0.967705	0.963469	0.959311	0.955229
8.0	0.995322	0.990757	0.986298	0.981936	0.977665	0.973481	0.969377	0.965351	0.961397	0.957513
8.1	0.995573	0.991250	0.987024	0.982887	0.978835	0.974862	0.970964	0.967137	0.963377	0.959681

Table 5 (continued)
$\rho = 0.3000$

8.2	0.995811	0.991717	0.987711	0.983788	0.979943	0.976171	0.972468	0.968831	0.965255	0.961740
8.3	0.996035	0.992158	0.988362	0.984642	0.980993	0.977412	0.973895	0.970438	0.967039	0.963694
8.4	0.996248	0.992575	0.988977	0.985450	0.981988	0.978588	0.975247	0.971962	0.968731	0.965550
8.5	0.996448	0.992970	0.989560	0.986215	0.982931	0.979703	0.976530	0.973409	0.970337	0.967312
8.6	0.996638	0.993344	0.990112	0.986940	0.983824	0.980760	0.977747	0.974781	0.971861	0.968984
8.7	0.996818	0.993697	0.990635	0.987626	0.984570	0.981762	0.978900	0.976082	0.973307	0.970571
8.8	0.996987	0.994032	0.991129	0.988277	0.985471	0.982711	0.979994	0.977317	0.974679	0.972078
8.9	0.997148	0.994348	0.991597	0.988892	0.986231	0.983611	0.981031	0.978488	0.975981	0.973508
9.0	0.997300	0.994648	0.992040	0.989475	0.986951	0.984464	0.982014	0.979598	0.977216	0.974865
9.1	0.997444	0.994931	0.992460	0.990028	0.987633	0.985273	0.982946	0.980652	0.978388	0.976154
9.2	0.997580	0.995199	0.992857	0.990551	0.988279	0.986039	0.983830	0.981651	0.979500	0.977376
9.3	0.997708	0.995453	0.993233	0.991046	0.988891	0.986765	0.984668	0.982598	0.980555	0.978536
9.4	0.997830	0.995694	0.993590	0.991516	0.989471	0.987454	0.985463	0.983497	0.981556	0.979638
9.5	0.997945	0.995921	0.993927	0.991961	0.990021	0.988106	0.986216	0.984350	0.982505	0.980682
9.6	0.998054	0.996137	0.994246	0.992382	0.990542	0.988725	0.986931	0.985158	0.983406	0.981674
9.7	0.998157	0.996340	0.994549	0.992781	0.991035	0.989311	0.987608	0.985925	0.984260	0.982615
9.8	0.998255	0.996533	0.994835	0.993159	0.991503	0.989867	0.988250	0.986652	0.985071	0.983507
9.9	0.998347	0.996716	0.995106	0.993517	0.991946	0.990394	0.988859	0.987341	0.985840	0.984354
10.0	0.998434	0.996889	0.995363	0.993856	0.992366	0.990893	0.989436	0.987995	0.986569	0.985158
10.1	0.998517	0.997053	0.995606	0.994177	0.992764	0.991366	0.989984	0.988615	0.987261	0.985920
10.2	0.998595	0.997208	0.995837	0.994481	0.993141	0.991815	0.990502	0.989203	0.987917	0.986643
10.3	0.998669	0.997355	0.996055	0.994770	0.993498	0.992240	0.990994	0.989761	0.988540	0.987330
10.4	0.998740	0.997494	0.996262	0.995043	0.993837	0.992643	0.991461	0.990290	0.989130	0.987981
10.5	0.998806	0.997625	0.996457	0.995302	0.994158	0.993025	0.991903	0.990791	0.989690	0.988598
10.6	0.998869	0.997750	0.996643	0.995547	0.994462	0.993387	0.992322	0.991267	0.990221	0.989184
10.7	0.998928	0.997868	0.996818	0.995779	0.994750	0.993730	0.992720	0.991718	0.990725	0.989740
10.8	0.998985	0.997980	0.996985	0.995999	0.995023	0.994056	0.993096	0.992146	0.991203	0.990268
10.9	0.999038	0.998086	0.997143	0.996208	0.995282	0.994364	0.993454	0.992551	0.991656	0.990768
11.0	0.999089	0.998186	0.997292	0.996406	0.995527	0.994656	0.993793	0.992936	0.992086	0.991243
11.1	0.999136	0.998281	0.997433	0.996593	0.995760	0.994933	0.994114	0.993301	0.992494	0.991693
11.2	0.999182	0.998371	0.997567	0.996770	0.995980	0.995196	0.994419	0.993647	0.992881	0.992120
11.3	0.999225	0.998456	0.997694	0.996939	0.996189	0.995445	0.994707	0.993975	0.993248	0.992525
11.4	0.999265	0.998537	0.997815	0.997098	0.996387	0.995681	0.994981	0.994286	0.993595	0.992910
11.5	0.999304	0.998613	0.997929	0.997249	0.996575	0.995905	0.995241	0.994581	0.993925	0.993274

Table 5 (continued)

$\rho = 0.4000$

a N	1	2	3	4	5	6	7	8	9	10
0.1	0.248170	0.066788	0.018716	0.005374	0.001569	0.000463	0.000138	0.000041	0.000012	0.000004
0.2	0.345279	0.128544	0.049733	0.019695	0.007924	0.003224	0.001323	0.000547	0.000227	0.000095
0.3	0.416117	0.185705	0.085952	0.040684	0.019552	0.009501	0.004655	0.002296	0.001138	0.000567
0.4	0.472911	0.238665	0.124675	0.066541	0.036039	0.019728	0.010887	0.006047	0.003375	0.001892
0.5	0.520500	0.287779	0.164385	0.095846	0.056678	0.033864	0.020393	0.012357	0.007526	0.004602
0.6	0.561422	0.333366	0.204143	0.127482	0.080696	0.051593	0.033238	0.021543	0.014032	0.009176
0.7	0.597216	0.375717	0.243343	0.160572	0.107345	0.072455	0.049268	0.033699	0.023160	0.015980
0.8	0.628906	0.415093	0.281581	0.194435	0.135946	0.095936	0.068187	0.048741	0.035003	0.025234
0.9	0.657218	0.451732	0.318608	0.228543	0.165937	0.121516	0.089619	0.066461	0.049509	0.037020
1.0	0.682689	0.485850	0.354262	0.262489	0.196724	0.148703	0.113156	0.086567	0.066516	0.051296
1.1	0.705734	0.517643	0.388448	0.295965	0.227973	0.177348	0.138385	0.108724	0.085783	0.067923
1.2	0.726678	0.547290	0.421121	0.328742	0.259309	0.206156	0.164914	0.132580	0.107024	0.086692
1.3	0.745787	0.574955	0.452268	0.360648	0.290450	0.235683	0.192381	0.157789	0.129932	0.107349
1.4	0.763276	0.600785	0.481903	0.391565	0.321173	0.265338	0.220461	0.184022	0.154193	0.129617
1.5	0.779328	0.624917	0.510053	0.421409	0.351302	0.294877	0.248869	0.210973	0.179508	0.153212
1.6	0.794097	0.647475	0.536761	0.450130	0.380703	0.324101	0.277359	0.238372	0.205593	0.177855
1.7	0.807712	0.668574	0.562074	0.477698	0.409275	0.352849	0.305724	0.265978	0.232190	0.203282
1.8	0.820287	0.688318	0.586045	0.504103	0.436945	0.380991	0.333790	0.293584	0.259069	0.229249
1.9	0.831921	0.706803	0.608733	0.529351	0.463664	0.408430	0.361416	0.321014	0.286026	0.255535
2.0	0.842701	0.724117	0.630194	0.553455	0.489400	0.435389	0.388486	0.348119	0.312887	0.281946
2.1	0.852701	0.743343	0.650486	0.576439	0.514138	0.460914	0.414910	0.374777	0.339503	0.308310
2.2	0.861989	0.755554	0.669668	0.598332	0.537874	0.485868	0.440618	0.400889	0.365750	0.334480
2.3	0.870626	0.769820	0.687795	0.619166	0.560613	0.509928	0.465560	0.426378	0.391524	0.360333
2.4	0.878665	0.783206	0.704923	0.638979	0.582368	0.533081	0.489699	0.451182	0.416742	0.385766
2.5	0.886153	0.795769	0.721104	0.657808	0.603159	0.555325	0.513010	0.475257	0.441340	0.410692
2.6	0.893136	0.807565	0.736389	0.675693	0.623009	0.576667	0.535480	0.498569	0.465265	0.435045
2.7	0.899652	0.818644	0.750827	0.692673	0.641944	0.597117	0.557104	0.521098	0.488482	0.458771
2.8	0.905735	0.829054	0.764464	0.708788	0.659993	0.616691	0.577885	0.542832	0.510962	0.481829
2.9	0.911420	0.838837	0.777344	0.724076	0.677187	0.635410	0.597830	0.563766	0.532690	0.504189
3.0	0.916735	0.848034	0.789510	0.738576	0.693557	0.653296	0.616953	0.583901	0.553655	0.525832
3.1	0.921708	0.856682	0.801000	0.752325	0.709136	0.670373	0.635270	0.603247	0.573857	0.546745
3.2	0.926361	0.864817	0.811853	0.765369	0.723954	0.686669	0.652799	0.621813	0.593296	0.566922
3.3	0.930720	0.872471	0.822105	0.777716	0.738045	0.702210	0.669562	0.639614	0.611982	0.586364
3.4	0.934803	0.879674	0.831789	0.789426	0.751439	0.717023	0.685582	0.656667	0.629925	0.605074
3.5	0.938631	0.886455	0.840938	0.800523	0.764168	0.731136	0.700883	0.672992	0.647139	0.623063
3.6	0.942220	0.892838	0.849580	0.811037	0.776261	0.744578	0.715489	0.688609	0.663641	0.640342
3.7	0.945587	0.898850	0.857745	0.820999	0.787748	0.757376	0.729424	0.703541	0.679448	0.656923

Table 5 (continued)
$\rho = 0.4000$

3.8	0.948747	0.904512	0.865460	0.830437	0.798657	0.769557	0.742716	0.717809	0.694581	0.672825
3.9	0.951714	0.909847	0.872753	0.839378	0.809015	0.781147	0.755387	0.731436	0.709059	0.688064
4.0	0.954500	0.914873	0.879638	0.847847	0.818849	0.792173	0.767463	0.744446	0.722904	0.702659
4.1	0.957117	0.919611	0.886147	0.855869	0.828183	0.802659	0.778969	0.756862	0.736136	0.716629
4.2	0.959576	0.924076	0.892299	0.863468	0.837043	0.812629	0.789928	0.768706	0.748779	0.729994
4.3	0.961887	0.928286	0.898114	0.870666	0.845451	0.822108	0.800363	0.780002	0.760852	0.742775
4.4	0.964061	0.932255	0.903609	0.877484	0.853430	0.831119	0.810298	0.790771	0.772378	0.754992
4.5	0.966105	0.935999	0.908804	0.883941	0.861001	0.839682	0.819754	0.801035	0.783378	0.766666
4.6	0.968028	0.939530	0.913715	0.890057	0.868184	0.847820	0.828752	0.810815	0.793873	0.777816
4.7	0.969837	0.942861	0.918357	0.895849	0.874998	0.855551	0.837314	0.820133	0.803883	0.788463
4.8	0.971540	0.946003	0.922746	0.901336	0.881463	0.862897	0.845459	0.829007	0.813428	0.798626
4.9	0.973143	0.948969	0.926896	0.906532	0.887595	0.869875	0.853206	0.837458	0.822527	0.808325
5.0	0.974652	0.951767	0.930820	0.911454	0.893412	0.876502	0.860573	0.845504	0.831200	0.817578
5.1	0.976074	0.954408	0.934530	0.916115	0.898930	0.882797	0.867578	0.853164	0.839464	0.826405
5.2	0.977413	0.956902	0.938039	0.920530	0.904163	0.888775	0.874239	0.860454	0.847338	0.834821
5.3	0.978674	0.959255	0.941357	0.924712	0.909126	0.894451	0.880571	0.867392	0.854838	0.842846
5.4	0.979863	0.961478	0.944495	0.928673	0.913834	0.899841	0.886590	0.873993	0.861982	0.850496
5.5	0.980983	0.963576	0.947463	0.932425	0.918298	0.904959	0.892311	0.880274	0.868784	0.857786
5.6	0.982039	0.965558	0.950271	0.935978	0.922531	0.909818	0.897748	0.886248	0.875260	0.864733
5.7	0.983035	0.967430	0.952926	0.939344	0.926546	0.914431	0.902915	0.891931	0.881426	0.871352
5.8	0.983974	0.969198	0.955439	0.942532	0.930354	0.918810	0.907824	0.897336	0.887294	0.877656
5.9	0.984859	0.970868	0.957815	0.945551	0.933964	0.922966	0.912489	0.902475	0.892879	0.883661
6.0	0.985694	0.972445	0.960063	0.948411	0.937387	0.926912	0.916921	0.907363	0.898194	0.889380
6.1	0.986482	0.973936	0.962190	0.951121	0.940634	0.930657	0.921131	0.912009	0.903251	0.894825
6.2	0.987225	0.975345	0.964203	0.953687	0.943712	0.934211	0.925130	0.916426	0.908063	0.900009
6.3	0.987926	0.976676	0.9661[illegible]7	0.956118	0.946631	0.937584	0.928929	0.920625	0.912640	0.904943
6.4	0.988588	0.977934	0.967909	0.958421	0.949398	0.940785	0.932537	0.924617	0.916994	0.909641
6.5	0.989212	0.979122	0.969614	0.960602	0.952022	0.943824	0.935964	0.928411	0.921134	0.914111
6.6	0.989802	0.980246	0.971227	0.962669	0.954511	0.946707	0.939219	0.932016	0.925073	0.918365
6.7	0.990358	0.981308	0.972754	0.964626	0.956870	0.949442	0.942310	0.935443	0.928817	0.922413
6.8	0.990884	0.982313	0.974199	0.966480	0.959107	0.952039	0.945245	0.938699	0.932378	0.926264
6.9	0.991380	0.983262	0.975567	0.968237	0.961227	0.954502	0.948032	0.941793	0.935764	0.929928
7.0	0.991849	0.984160	0.976861	0.969901	0.963238	0.956839	0.950678	0.944732	0.938983	0.933414
7.1	0.992291	0.985008	0.978087	0.971478	0.965145	0.959057	0.953191	0.947525	0.942043	0.936729
7.2	0.992709	0.985811	0.979246	0.972972	0.966952	0.961162	0.955577	0.950179	0.944952	0.939882
7.3	0.993104	0.986570	0.980344	0.974367	0.968066	0.963158	0.957842	0.952699	0.947716	0.942881
7.4	0.993477	0.987288	0.981383	0.975727	0.970291	0.965052	0.959992	0.955094	0.950344	0.945732
7.5	0.993839	0.987967	0.982367	0.976997	0.971832	0.966849	0.962033	0.957368	0.952841	0.948443
7.6	0.994163	0.988609	0.983298	0.978200	0.973292	0.968554	0.963971	0.959528	0.955215	0.951021

7.7	0.994478	0.989216	0.984180	0.979340	0.974677	0.970172	0.965810	0.961580	0.957470	0.953472
7.8	0.994775	0.989791	0.985014	0.980421	0.975990	0.971706	0.967556	0.963528	0.959613	0.955802
7.9	0.995056	0.990334	0.985804	0.981444	0.977235	0.973162	0.969213	0.965379	0.961649	0.958017
8.0	0.995322	0.990849	0.986553	0.982414	0.978415	0.974543	0.970787	0.967136	0.963584	0.960122
8.1	0.995573	0.991335	0.987261	0.983332	0.979534	0.975853	0.972280	0.968805	0.965422	0.962124
8.2	0.995811	0.991795	0.987932	0.984203	0.980595	0.977096	0.973697	0.970390	0.967168	0.964026
8.3	0.996035	0.992231	0.988567	0.985028	0.981600	0.978275	0.975042	0.971895	0.968827	0.965833
8.4	0.996248	0.992643	0.989168	0.985809	0.982554	0.979393	0.976318	0.973324	0.970403	0.967551
8.5	0.996448	0.993033	0.989738	0.986550	0.983458	0.980454	0.977530	0.974680	0.971900	0.969183
8.6	0.996638	0.993402	0.990277	0.987251	0.984315	0.981460	0.978680	0.975969	0.973322	0.970735
8.7	0.996818	0.993752	0.990788	0.987916	0.985127	0.982414	0.979771	0.977191	0.974672	0.972208
8.8	0.996987	0.994082	0.991272	0.988546	0.985898	0.983320	0.980806	0.978352	0.975954	0.973609
8.9	0.997148	0.994395	0.991730	0.989143	0.986623	0.984178	0.981789	0.979455	0.977173	0.974939
9.0	0.997300	0.994691	0.992164	0.989709	0.987321	0.984993	0.982721	0.980501	0.978330	0.976203
9.1	0.997444	0.994972	0.992575	0.990245	0.987977	0.985766	0.983606	0.981495	0.979429	0.977404
9.2	0.997580	0.995237	0.992964	0.990754	0.988600	0.986499	0.984446	0.982438	0.980472	0.978545
9.3	0.997708	0.995488	0.993333	0.991235	0.989190	0.987194	0.985243	0.983333	0.981463	0.979629
9.4	0.997830	0.995726	0.993682	0.991691	0.989750	0.987853	0.985999	0.984183	0.982403	0.980658
9.5	0.997945	0.995951	0.994013	0.992124	0.990280	0.988479	0.986716	0.984989	0.983297	0.981636
9.6	0.998054	0.996165	0.994326	0.992534	0.990783	0.989072	0.987396	0.985755	0.984145	0.982564
9.7	0.998157	0.996366	0.994623	0.992922	0.991260	0.989634	0.988042	0.986481	0.984950	0.983446
9.8	0.998255	0.996558	0.994904	0.993290	0.991712	0.990168	0.988655	0.987171	0.985715	0.984284
9.9	0.998347	0.996739	0.995170	0.993639	0.992141	0.990674	0.989236	0.987826	0.986441	0.985079
10.0	0.998434	0.996910	0.995423	0.993970	0.992548	0.991154	0.989788	0.988447	0.987130	0.985835
10.1	0.998517	0.997072	0.995662	0.994283	0.992933	0.991610	0.990312	0.989037	0.987784	0.986553
10.2	0.998595	0.997226	0.995888	0.994580	0.993298	0.992042	0.990808	0.989597	0.988406	0.987234
10.3	0.998669	0.997371	0.996103	0.994861	0.993645	0.992451	0.991279	0.990128	0.988995	0.987881
10.4	0.998740	0.997509	0.996306	0.995128	0.993973	0.992840	0.991726	0.990632	0.989555	0.988495
10.5	0.998806	0.997640	0.996499	0.995381	0.994285	0.993208	0.992150	0.991110	0.990087	0.989079
10.6	0.998869	0.997763	0.996681	0.995621	0.994580	0.993558	0.992553	0.991564	0.990591	0.989633
10.7	0.998929	0.997881	0.996854	0.995848	0.994860	0.993889	0.992935	0.991995	0.991070	0.990159
10.8	0.998985	0.997991	0.997018	0.996063	0.995125	0.994204	0.993297	0.992404	0.991525	0.990658
10.9	0.999038	0.998097	0.997173	0.996267	0.995377	0.994502	0.993641	0.992792	0.991956	0.991132
11.0	0.999089	0.998196	0.997321	0.996461	0.995616	0.994785	0.993967	0.993161	0.992366	0.991583
11.1	0.999137	0.998290	0.997460	0.996644	0.995843	0.995053	0.994276	0.993510	0.992755	0.992010
11.2	0.999182	0.998380	0.997592	0.996818	0.996057	0.995308	0.994570	0.993842	0.993124	0.992416
11.3	0.999225	0.998464	0.997717	0.996983	0.996261	0.995549	0.994848	0.994157	0.993475	0.992802
11.4	0.999265	0.998544	0.997836	0.997139	0.996454	0.995778	0.995112	0.994455	0.993807	0.993167
11.5	0.999304	0.998620	0.997949	0.997288	0.996637	0.995995	0.995363	0.994739	0.994123	0.993514

Table 5 (continued)
$\rho = 0.5000$

a \ N	1	2	3	4	5	6	7	8	9	10
0.1	0.248170	0.070359	0.021099	0.006527	0.002060	0.000659	0.000213	0.000069	0.000023	0.000007
0.2	0.345279	0.134851	0.055553	0.023582	0.010205	0.004476	0.001982	0.000885	0.000397	0.000179
0.3	0.416118	0.194069	0.095209	0.048074	0.024730	0.012889	0.006783	0.003596	0.001917	0.001027
0.4	0.472911	0.248534	0.137030	0.077667	0.044819	0.026194	0.015453	0.009183	0.005489	0.003296
0.5	0.520500	0.298708	0.179376	0.110599	0.069386	0.044068	0.028245	0.018233	0.011836	0.007719
0.6	0.561422	0.344997	0.221277	0.145551	0.097350	0.065890	0.044994	0.030938	0.021391	0.014856
0.7	0.597216	0.387764	0.262145	0.181535	0.127744	0.090927	0.065281	0.047185	0.034288	0.025027
0.8	0.628906	0.427330	0.301621	0.217823	0.159741	0.118447	0.088563	0.066652	0.050425	0.038312
0.9	0.657218	0.463981	0.339498	0.253879	0.192664	0.147771	0.114254	0.088899	0.069523	0.054599
1.0	0.682689	0.497972	0.375667	0.289317	0.225968	0.178299	0.141783	0.113436	0.091205	0.073630
1.1	0.705734	0.529532	0.410085	0.323864	0.259223	0.209524	0.170623	0.139767	0.115039	0.095061
1.2	0.726678	0.558865	0.442753	0.357332	0.292099	0.241022	0.200310	0.167424	0.140585	0.118501
1.3	0.745787	0.586158	0.473702	0.389597	0.324346	0.272453	0.230445	0.195984	0.167421	0.143552
1.4	0.763276	0.611574	0.502981	0.420587	0.355779	0.303543	0.260693	0.225073	0.195157	0.169824
1.5	0.779328	0.635264	0.530651	0.450261	0.386266	0.334082	0.290781	0.254374	0.223448	0.196960
1.6	0.794097	0.657365	0.556782	0.478609	0.415719	0.363910	0.320490	0.283621	0.251993	0.224638
1.7	0.807712	0.677997	0.581446	0.505640	0.444078	0.392908	0.349647	0.312596	0.280539	0.252577
1.8	0.820287	0.697274	0.604718	0.531376	0.471314	0.420993	0.378120	0.341124	0.308876	0.280536
1.9	0.831922	0.715296	0.626669	0.555850	0.497415	0.448107	0.405811	0.369068	0.336831	0.308316
2.0	0.842701	0.732157	0.647373	0.579103	0.522384	0.474218	0.432648	0.396324	0.364268	0.335752
2.1	0.852701	0.747940	0.666898	0.601178	0.546235	0.499307	0.458585	0.422812	0.391080	0.362710
2.2	0.861989	0.762723	0.685312	0.622123	0.568992	0.523373	0.483590	0.448477	0.417187	0.389086
2.3	0.870626	0.776576	0.702678	0.641987	0.590684	0.546422	0.507649	0.473281	0.442529	0.414799
2.4	0.878665	0.789566	0.719057	0.660818	0.611344	0.568471	0.530760	0.497203	0.467065	0.439791
2.5	0.886154	0.801751	0.734506	0.678664	0.631009	0.589541	0.552929	0.520233	0.490769	0.464017
2.6	0.893136	0.813186	0.749080	0.695575	0.649715	0.609658	0.574167	0.542369	0.513625	0.487451
2.7	0.899652	0.823923	0.762831	0.711595	0.667503	0.628851	0.594494	0.563620	0.535631	0.510074
2.8	0.905735	0.834007	0.775805	0.726770	0.684410	0.647152	0.613932	0.583997	0.556787	0.531881
2.9	0.911420	0.843482	0.788049	0.741144	0.700476	0.664592	0.632507	0.603518	0.577105	0.552871
3.0	0.916735	0.852388	0.799606	0.754757	0.715738	0.681206	0.650246	0.622205	0.596597	0.573051
3.1	0.921707	0.860762	0.810515	0.767651	0.730234	0.697026	0.667177	0.640079	0.615281	0.592433
3.2	0.926362	0.868638	0.820814	0.779861	0.744000	0.712085	0.683330	0.657168	0.633176	0.611032
3.3	0.930720	0.876048	0.830539	0.791425	0.757070	0.726417	0.698734	0.673495	0.650307	0.628865
3.4	0.934803	0.883022	0.839723	0.802377	0.769479	0.740053	0.713419	0.689089	0.666695	0.645953
3.5	0.938631	0.889586	0.848397	0.812749	0.781259	0.753024	0.727415	0.703977	0.682366	0.662318
3.6	0.942220	0.895768	0.856590	0.822572	0.792441	0.765361	0.740750	0.718185	0.697344	0.677981
3.7	0.945587	0.901589	0.864331	0.831876	0.803054	0.777093	0.753453	0.731740	0.711655	0.692967

3.8	0.948747	0.907074	0.871644	0.840688	0.813127	0.788248	0.765552	0.744669	0.725324	0.707298
3.9	0.951714	0.912241	0.878555	0.849034	0.822686	0.798854	0.777072	0.756998	0.738375	0.720999
4.0	0.954500	0.917112	0.885086	0.856939	0.831759	0.808936	0.788040	0.768753	0.750833	0.734092
4.1	0.957117	0.921702	0.891258	0.864427	0.840368	0.818519	0.798480	0.779956	0.762722	0.746602
4.2	0.959576	0.926031	0.897093	0.871520	0.848539	0.827628	0.808418	0.790634	0.774066	0.758551
4.3	0.961887	0.930112	0.902609	0.878240	0.856292	0.836285	0.817875	0.800808	0.784888	0.769961
4.4	0.964061	0.933962	0.907824	0.884605	0.863649	0.844511	0.826874	0.810501	0.795209	0.780854
4.5	0.966105	0.937593	0.912755	0.890635	0.870630	0.852329	0.835438	0.819735	0.805051	0.791252
4.6	0.968028	0.941019	0.917417	0.896348	0.877254	0.859758	0.843585	0.828529	0.814435	0.801175
4.7	0.969837	0.944251	0.921826	0.901760	0.883540	0.866816	0.851335	0.836905	0.823380	0.810643
4.8	0.971540	0.947302	0.925997	0.906888	0.889505	0.873523	0.858707	0.844881	0.831907	0.819676
4.9	0.973143	0.950182	0.929941	0.911746	0.895164	0.879895	0.865720	0.852475	0.840033	0.828293
5.0	0.974652	0.952900	0.933672	0.916350	0.900534	0.885948	0.872389	0.859705	0.847777	0.836510
5.1	0.976074	0.955466	0.937201	0.920711	0.905629	0.891699	0.878732	0.866588	0.855156	0.844347
5.2	0.977413	0.957889	0.940540	0.924845	0.910464	0.897162	0.884764	0.873140	0.862185	0.851819
5.3	0.978674	0.960178	0.943699	0.928761	0.915051	0.902352	0.890500	0.879376	0.868882	0.858942
5.4	0.979863	0.962339	0.946688	0.932472	0.919404	0.907281	0.895955	0.885311	0.875261	0.865732
5.5	0.980983	0.964380	0.949516	0.935989	0.923534	0.911964	0.901141	0.890959	0.881336	0.872204
5.6	0.982039	0.966309	0.952192	0.939322	0.927452	0.916411	0.906071	0.896334	0.887121	0.878372
5.7	0.983035	0.968130	0.954725	0.942480	0.931170	0.920636	0.910759	0.901447	0.892631	0.884250
5.8	0.983974	0.969852	0.957122	0.945473	0.934697	0.924648	0.915215	0.906313	0.897876	0.889851
5.9	0.984859	0.971478	0.959393	0.948309	0.938044	0.928458	0.919451	0.910942	0.902871	0.895187
6.0	0.985694	0.973016	0.961537	0.950997	0.941220	0.932078	0.923478	0.915346	0.907626	0.900270
6.1	0.986482	0.974468	0.963570	0.953545	0.944232	0.935515	0.927305	0.919536	0.912152	0.905112
6.2	0.987225	0.975842	0.965494	0.955960	0.947091	0.938779	0.930943	0.923520	0.916461	0.909724
6.3	0.987926	0.977140	0.967315	0.958248	0.949803	0.941879	0.934401	0.927311	0.920562	0.914117
6.4	0.988588	0.978367	0.969039	0.960417	0.952376	0.944822	0.937687	0.930916	0.924465	0.918300
6.5	0.989212	0.979527	0.970671	0.962473	0.954817	0.947618	0.940810	0.934344	0.928179	0.922283
6.6	0.989802	0.980624	0.972216	0.964422	0.957134	0.950272	0.943778	0.937604	0.931714	0.926076
6.7	0.990358	0.981661	0.973679	0.966269	0.959331	0.952793	0.946598	0.940705	0.935077	0.929688
6.8	0.990884	0.982642	0.975064	0.968020	0.961416	0.955186	0.949278	0.943653	0.938278	0.933126
6.9	0.991380	0.983569	0.976376	0.969679	0.963394	0.957459	0.951825	0.946456	0.941322	0.936399
7.0	0.991849	0.984446	0.977618	0.971253	0.965271	0.959617	0.954245	0.949122	0.944220	0.939515
7.1	0.992291	0.985276	0.978795	0.972744	0.967052	0.961665	0.956544	0.951656	0.946975	0.942480
7.2	0.992709	0.986061	0.979908	0.974157	0.968741	0.963611	0.958729	0.954065	0.949597	0.945302
7.3	0.993104	0.986803	0.980963	0.975497	0.970344	0.965458	0.960804	0.956356	0.952090	0.947988
7.4	0.993477	0.987506	0.981962	0.976767	0.971864	0.967211	0.962776	0.958533	0.954462	0.950545
7.5	0.993830	0.988170	0.982908	0.977971	0.973307	0.968876	0.964649	0.960603	0.956718	0.952977
7.6	0.994163	0.988799	0.983804	0.979113	0.974675	0.970457	0.966429	0.962571	0.958863	0.955292
7.7	0.994478	0.989393	0.984653	0.980195	0.975974	0.971957	0.968120	0.964441	0.960903	0.957494
7.8	0.994775	0.989956	0.985457	0.981220	0.977205	0.973382	0.969726	0.966218	0.962844	0.959589
7.9	0.995056	0.990489	0.986218	0.982193	0.978374	0.974734	0.971252	0.967908	0.964689	0.961583
8.0	0.995322	0.990993	0.986940	0.983115	0.979483	0.976018	0.972701	0.969513	0.966443	0.963479

Table 5 (continued)
$\rho = 0.5000$

8.1	0.995573	0.991470	0.987623	0.983989	0.980535	0.977237	0.974077	0.971039	0.968111	0.965283
8.2	0.995811	0.991921	0.988270	0.984817	0.981533	0.978394	0.975385	0.972490	0.969698	0.966999
8.3	0.996035	0.992348	0.988883	0.985603	0.982480	0.979493	0.976627	0.973868	0.971206	0.968631
8.4	0.996248	0.992753	0.989464	0.986347	0.983378	0.980536	0.977806	0.975177	0.972639	0.970183
8.5	0.996448	0.993135	0.990014	0.987054	0.984230	0.981525	0.978927	0.976422	0.974002	0.971660
8.6	0.996638	0.993498	0.990536	0.987723	0.985038	0.982465	0.979991	0.977604	0.975298	0.973064
8.7	0.996818	0.993841	0.991030	0.988358	0.985805	0.983357	0.981001	0.978728	0.976530	0.974399
8.8	0.996987	0.994165	0.991498	0.988960	0.986533	0.984204	0.981961	0.979796	0.977701	0.975669
8.9	0.997148	0.994473	0.991941	0.989530	0.987223	0.985007	0.982872	0.980810	0.978814	0.976877
9.0	0.997300	0.994764	0.992361	0.990071	0.987878	0.985770	0.983738	0.981774	0.979872	0.978026
9.1	0.997444	0.995040	0.992759	0.990584	0.988500	0.986495	0.984561	0.982690	0.980877	0.979117
9.2	0.997580	0.995300	0.993137	0.991071	0.989089	0.987182	0.985341	0.983560	0.981833	0.980155
9.3	0.997708	0.995547	0.993494	0.991532	0.989648	0.987834	0.986082	0.984386	0.982741	0.981142
9.4	0.997830	0.995781	0.993833	0.991969	0.990179	0.988454	0.986786	0.985172	0.983604	0.982081
9.5	0.997945	0.996003	0.994154	0.992383	0.990682	0.989041	0.987455	0.985917	0.984425	0.982973
9.6	0.998054	0.996213	0.994458	0.992777	0.991160	0.989599	0.988090	0.986626	0.985204	0.983821
9.7	0.998157	0.996411	0.994746	0.993149	0.991612	0.990129	0.988692	0.987299	0.985945	0.984627
9.8	0.998255	0.996599	0.995019	0.993503	0.992042	0.990631	0.989265	0.987938	0.986649	0.985393
9.9	0.998347	0.996778	0.995278	0.993838	0.992450	0.991108	0.989808	0.988546	0.987318	0.986121
10.0	0.998434	0.996946	0.995523	0.994156	0.992837	0.991561	0.990324	0.989123	0.987953	0.986814
10.1	0.998517	0.997106	0.995756	0.994457	0.993204	0.991991	0.990814	0.989671	0.988557	0.987472
10.2	0.998595	0.997258	0.995976	0.994743	0.993552	0.992399	0.991280	0.990191	0.989131	0.988097
10.3	0.998669	0.997401	0.996185	0.995014	0.993882	0.992786	0.991721	0.990686	0.989676	0.988692
10.4	0.998740	0.997537	0.996383	0.995271	0.994195	0.993153	0.992141	0.991155	0.990195	0.989257
10.5	0.998806	0.997666	0.996571	0.995514	0.994493	0.993502	0.992539	0.991601	0.990687	0.989794
10.6	0.998869	0.997788	0.996748	0.995745	0.994775	0.993833	0.992917	0.992025	0.991154	0.990304
10.7	0.998928	0.997903	0.996917	0.995965	0.995042	0.994147	0.993276	0.992427	0.991599	0.990789
10.8	0.998985	0.998013	0.997077	0.996172	0.995296	0.994445	0.993617	0.992809	0.992020	0.991250
10.9	0.999038	0.998116	0.997228	0.996369	0.995537	0.994728	0.993940	0.993172	0.992421	0.991688
11.0	0.999089	0.998214	0.997372	0.996556	0.995765	0.994997	0.994247	0.993517	0.992802	0.992104
11.1	0.999137	0.998308	0.997508	0.996734	0.995982	0.995251	0.994539	0.993844	0.993164	0.992499
11.2	0.999182	0.998396	0.997637	0.996902	0.996188	0.995494	0.994816	0.994155	0.993508	0.992875
11.3	0.999225	0.998479	0.997759	0.997061	0.996383	0.995723	0.995079	0.994450	0.993835	0.993232
11.4	0.999265	0.998558	0.997875	0.997213	0.996568	0.995941	0.995329	0.994730	0.994145	0.993572
11.5	0.999304	0.998633	0.997985	0.997356	0.996744	0.996148	0.995566	0.994997	0.994440	0.993894

Table 5 (continued)
$\rho = 0.6000$

a \ N	1	2	3	4	5	6	7	8	9	10
0.1	0.248170	0.075615	0.024767	0.008421	0.002929	0.001035	0.000369	0.000133	0.000048	0.000017
0.2	0.345279	0.143977	0.064310	0.029779	0.014099	0.006777	0.003292	0.001612	0.000794	0.000393
0.3	0.416118	0.205980	0.108830	0.059526	0.033268	0.018867	0.010814	0.006247	0.003630	0.002119
0.4	0.472911	0.262382	0.154836	0.094453	0.058829	0.037168	0.023727	0.015265	0.009880	0.006425
0.5	0.520500	0.313834	0.200566	0.132307	0.089045	0.060763	0.041885	0.029095	0.020331	0.014274
0.6	0.561422	0.360892	0.245065	0.171519	0.122377	0.088489	0.064616	0.047540	0.035183	0.026161
0.7	0.597216	0.404036	0.287816	0.211006	0.157574	0.119196	0.091027	0.070024	0.054180	0.042116
0.8	0.628906	0.443682	0.328555	0.250034	0.193664	0.151869	0.120192	0.095798	0.076787	0.061830
0.9	0.657218	0.480188	0.367172	0.288114	0.229915	0.185662	0.151257	0.124074	0.102333	0.084779
1.0	0.682689	0.513868	0.403651	0.324934	0.265794	0.219901	0.183483	0.154110	0.130124	0.110348
1.1	0.705734	0.544997	0.438031	0.360303	0.300921	0.254068	0.216262	0.185254	0.159502	0.137906
1.2	0.726678	0.573814	0.470385	0.394119	0.335035	0.287771	0.249109	0.216959	0.189886	0.166863
1.3	0.745787	0.600533	0.500806	0.426340	0.367965	0.320730	0.281649	0.248778	0.220780	0.196695
1.4	0.763276	0.625340	0.529395	0.456964	0.399609	0.352746	0.313601	0.280361	0.251779	0.226956
1.5	0.779328	0.648401	0.556255	0.486018	0.429911	0.383686	0.344760	0.311442	0.282563	0.257280
1.6	0.794097	0.669866	0.581492	0.513546	0.458853	0.413468	0.374984	0.341820	0.312880	0.287375
1.7	0.807712	0.689866	0.605204	0.539606	0.486442	0.442048	0.404178	0.371353	0.342545	0.317011
1.8	0.820287	0.708521	0.627490	0.564259	0.512701	0.469408	0.432284	0.399944	0.371421	0.346017
1.9	0.831922	0.725936	0.648440	0.587572	0.537666	0.495553	0.459274	0.427529	0.399413	0.374266
2.0	0.842701	0.742208	0.668140	0.609612	0.561382	0.520502	0.485139	0.454075	0.426459	0.401670
2.1	0.852701	0.757425	0.686670	0.630446	0.583898	0.544282	0.509886	0.479566	0.452523	0.428170
2.2	0.861989	0.771665	0.704107	0.650138	0.605265	0.566931	0.533534	0.504004	0.477587	0.453731
2.3	0.870626	0.784999	0.720520	0.668752	0.625534	0.588487	0.556111	0.527401	0.501650	0.478336
2.4	0.878665	0.797495	0.735974	0.686347	0.644759	0.608993	0.577647	0.549779	0.524720	0.501982
2.5	0.886154	0.809210	0.750530	0.702981	0.662991	0.628493	0.598179	0.571162	0.546815	0.524676
2.6	0.893136	0.820200	0.764245	0.718708	0.680279	0.647032	0.617743	0.591582	0.567958	0.546434
2.7	0.899652	0.830514	0.777171	0.733580	0.696671	0.664653	0.636380	0.611072	0.588174	0.567276
2.8	0.905735	0.840200	0.789357	0.747644	0.712215	0.681399	0.654127	0.629666	0.607495	0.587226
2.9	0.911420	0.849299	0.800849	0.760946	0.726953	0.697312	0.671023	0.647399	0.625950	0.606311
3.0	0.916735	0.857850	0.811689	0.773530	0.740928	0.712432	0.687106	0.664307	0.643573	0.624560
3.1	0.921708	0.865890	0.821916	0.785436	0.754180	0.726798	0.702413	0.680424	0.660395	0.642003
3.2	0.926362	0.873452	0.831568	0.796702	0.766748	0.740447	0.716980	0.695783	0.676448	0.658669
3.3	0.930720	0.880566	0.840679	0.807364	0.778666	0.753414	0.730842	0.710420	0.691764	0.674589
3.4	0.934803	0.887261	0.849281	0.817455	0.789969	0.765733	0.744031	0.724365	0.706376	0.689793
3.5	0.938631	0.893564	0.857404	0.827007	0.800691	0.777438	0.756580	0.737650	0.720312	0.704309
3.6	0.942220	0.899499	0.865076	0.836050	0.810860	0.788558	0.768519	0.750306	0.733602	0.718167
3.7	0.945587	0.905090	0.872324	0.844612	0.820506	0.799122	0.779877	0.762361	0.746275	0.731395
3.8	0.948747	0.910357	0.879172	0.852719	0.829656	0.809159	0.790682	0.773842	0.758358	0.744019

Table 5 (continued)
$\rho = 0.6000$

3.9	0.951714	0.915321	0.885643	0.860397	0.838337	0.818695	0.800961	0.784777	0.769878	0.756065
4.0	0.954499	0.920000	0.891758	0.867668	0.846572	0.827755	0.810740	0.795190	0.780859	0.767560
4.1	0.957117	0.924411	0.897539	0.874556	0.854385	0.836363	0.820042	0.805107	0.791327	0.778525
4.2	0.959576	0.928571	0.903005	0.881080	0.861799	0.844541	0.828890	0.814550	0.801304	0.788987
4.3	0.961887	0.932494	0.908172	0.887260	0.868832	0.852311	0.837306	0.823542	0.810813	0.798965
4.4	0.964061	0.936195	0.913058	0.893115	0.875506	0.859693	0.845311	0.832103	0.819875	0.808482
4.5	0.966105	0.939688	0.917679	0.898663	0.881839	0.866707	0.852926	0.840254	0.828511	0.817559
4.6	0.968028	0.942983	0.922050	0.903919	0.887849	0.873371	0.860169	0.848015	0.836740	0.826215
4.7	0.969837	0.946093	0.926185	0.908900	0.893551	0.879703	0.867057	0.855403	0.844581	0.834469
4.8	0.971540	0.949029	0.930396	0.913620	0.898963	0.885718	0.873609	0.862436	0.852051	0.842339
4.9	0.973143	0.951801	0.933796	0.918093	0.904099	0.891434	0.879840	0.869131	0.859168	0.849842
5.0	0.974652	0.954418	0.937298	0.922333	0.908973	0.896865	0.885767	0.875505	0.865948	0.856996
5.1	0.976074	0.956890	0.940611	0.926351	0.913598	0.902024	0.891403	0.881571	0.872407	0.863815
5.2	0.977413	0.959224	0.943746	0.930159	0.917988	0.906926	0.896763	0.887346	0.878559	0.870315
5.3	0.978674	0.961429	0.946714	0.933769	0.922154	0.911584	0.901861	0.892842	0.884420	0.876511
5.4	0.979863	0.963513	0.949522	0.937191	0.926109	0.916009	0.906709	0.898073	0.890002	0.882416
5.5	0.980983	0.965481	0.952181	0.940435	0.929862	0.920213	0.911319	0.903051	0.895318	0.888044
5.6	0.982039	0.967341	0.954697	0.943510	0.933424	0.924208	0.915702	0.907789	0.900381	0.893407
5.7	0.983035	0.969098	0.957383	0.946425	0.936804	0.928003	0.919871	0.912298	0.905202	0.898518
5.8	0.983974	0.970759	0.959335	0.949189	0.940013	0.931609	0.923835	0.916589	0.909793	0.903388
5.9	0.984859	0.972329	0.961471	0.951809	0.943059	0.935034	0.927604	0.920672	0.914165	0.908027
6.0	0.985694	0.973813	0.963493	0.954293	0.945950	0.938289	0.931188	0.924557	0.918329	0.912448
6.1	0.986482	0.975216	0.965408	0.956649	0.948694	0.941381	0.934595	0.928254	0.922292	0.916660
6.2	0.987225	0.976543	0.967221	0.958882	0.951298	0.944318	0.937836	0.931772	0.926066	0.920672
6.3	0.987926	0.977797	0.968938	0.961000	0.953771	0.947109	0.940916	0.935118	0.929659	0.924495
6.4	0.988588	0.978983	0.970565	0.963008	0.956117	0.949761	0.943845	0.938303	0.933080	0.928136
6.5	0.989212	0.980105	0.972105	0.964912	0.958344	0.952279	0.946630	0.941332	0.936336	0.931603
6.6	0.989802	0.981166	0.973563	0.966718	0.960459	0.954672	0.949277	0.944214	0.939436	0.934906
6.7	0.990359	0.982169	0.974945	0.968430	0.962466	0.956945	0.951794	0.946956	0.942386	0.938052
6.8	0.990884	0.983118	0.976254	0.970054	0.964370	0.959105	0.954187	0.949564	0.945195	0.941048
6.9	0.991380	0.984016	0.977494	0.971594	0.966179	0.961157	0.956462	0.952045	0.947868	0.943901
7.0	0.991849	0.984865	0.978669	0.973054	0.967895	0.963106	0.958625	0.954405	0.950413	0.946618
7.1	0.992291	0.985669	0.979781	0.974439	0.969524	0.964957	0.960680	0.956650	0.952834	0.949205
7.2	0.992709	0.986429	0.980835	0.975752	0.971071	0.966716	0.962634	0.958786	0.955139	0.951668
7.3	0.993104	0.987149	0.981834	0.976998	0.972539	0.968387	0.964492	0.960817	0.957332	0.954014
7.4	0.993477	0.987829	0.982780	0.978180	0.973932	0.969974	0.966258	0.962748	0.959419	0.956247
7.5	0.993830	0.988474	0.983677	0.979300	0.975255	0.971482	0.967936	0.964586	0.961404	0.958372
7.6	0.994163	0.989083	0.984526	0.980363	0.976511	0.972914	0.969531	0.966333	0.963294	0.960396
7.7	0.994478	0.989660	0.985331	0.981371	0.977702	0.974274	0.971048	0.967995	0.965092	0.962323

7.8	0.994775	0.990206	0.986094	0.982327	0.978834	0.975567	0.972489	0.969575	0.966803	0.964157
7.9	0.995056	0.990723	0.986817	0.983234	0.979908	0.976794	0.973859	0.971078	0.968431	0.965903
8.0	0.995322	0.991213	0.987502	0.984094	0.980927	0.977960	0.975161	0.972507	0.969980	0.967564
8.1	0.995573	0.991676	0.988151	0.984910	0.981895	0.979068	0.976399	0.973866	0.971453	0.969145
8.2	0.995811	0.992114	0.988766	0.985683	0.982813	0.980120	0.977575	0.975159	0.972853	0.970651
8.3	0.996035	0.992530	0.989349	0.986417	0.983685	0.981119	0.978693	0.976388	0.974189	0.972083
8.4	0.996248	0.992923	0.989902	0.987114	0.984513	0.982068	0.979755	0.977556	0.975457	0.973447
8.5	0.996448	0.993295	0.990425	0.987774	0.985299	0.982970	0.980765	0.978667	0.976664	0.974744
8.6	0.996638	0.993647	0.990922	0.988401	0.986045	0.983826	0.981724	0.979724	0.977812	0.975979
8.7	0.996818	0.993981	0.991392	0.988995	0.986753	0.984639	0.982636	0.980728	0.978904	0.977154
8.8	0.996987	0.994297	0.991838	0.989559	0.987425	0.985412	0.983503	0.981683	0.979942	0.978272
8.9	0.997148	0.994596	0.992261	0.990093	0.988063	0.986146	0.984326	0.982591	0.980930	0.979336
9.0	0.997300	0.994879	0.992661	0.990601	0.988668	0.986843	0.985109	0.983454	0.981870	0.980348
9.1	0.997444	0.995148	0.993041	0.991082	0.989243	0.987505	0.985852	0.984275	0.982764	0.981311
9.2	0.997580	0.995402	0.993401	0.991539	0.989789	0.988134	0.986559	0.985055	0.983614	0.982228
9.3	0.997708	0.995643	0.993743	0.991972	0.990307	0.988731	0.987231	0.985797	0.984422	0.983100
9.4	0.997830	0.995871	0.994066	0.992383	0.990799	0.989298	0.987869	0.986502	0.985191	0.983929
9.5	0.997945	0.996087	0.994373	0.992773	0.991265	0.989837	0.988475	0.987173	0.985922	0.984718
9.6	0.998054	0.996291	0.994664	0.993142	0.991708	0.990348	0.989051	0.987810	0.986617	0.985469
9.7	0.998157	0.996485	0.994939	0.993493	0.992129	0.990834	0.989599	0.988415	0.987278	0.986183
9.8	0.998255	0.996669	0.995201	0.993826	0.992528	0.991295	0.990119	0.988991	0.987907	0.986862
9.9	0.998347	0.996842	0.995448	0.994142	0.992907	0.991734	0.990613	0.989538	0.988505	0.987508
10.0	0.998434	0.997007	0.995683	0.994441	0.993267	0.992150	0.991083	0.990059	0.989073	0.988123
10.1	0.998517	0.997163	0.995906	0.994725	0.993608	0.992545	0.991529	0.990553	0.989614	0.988707
10.2	0.998595	0.997311	0.996117	0.994995	0.993933	0.992921	0.991953	0.991023	0.990128	0.989263
10.3	0.998670	0.997451	0.996317	0.995251	0.994240	0.993277	0.992355	0.991470	0.990616	0.989792
10.4	0.998740	0.997584	0.996507	0.995494	0.994532	0.993616	0.992738	0.991894	0.991081	0.990295
10.5	0.998806	0.997710	0.996687	0.995724	0.994810	0.993937	0.993102	0.992298	0.991523	0.990773
10.6	0.998869	0.997829	0.996858	0.995942	0.995073	0.994243	0.993447	0.992681	0.991943	0.991228
10.7	0.998928	0.997942	0.997020	0.996150	0.995323	0.994533	0.993775	0.993046	0.992342	0.991661
10.8	0.998985	0.998049	0.997173	0.996346	0.995560	0.994808	0.994087	0.993392	0.992722	0.992072
10.9	0.999038	0.998150	0.997319	0.996533	0.995785	0.995070	0.994383	0.993722	0.993082	0.992464
11.0	0.999089	0.998246	0.997457	0.996710	0.995999	0.995318	0.994665	0.994034	0.993426	0.992836
11.1	0.999137	0.998337	0.997588	0.996878	0.996202	0.995554	0.994932	0.994332	0.993752	0.993189
11.2	0.999182	0.998424	0.997712	0.997037	0.996394	0.995778	0.995186	0.994614	0.994062	0.993526
11.3	0.999225	0.998506	0.997829	0.997189	0.996577	0.995991	0.995427	0.994883	0.994356	0.993846
11.4	0.999265	0.998583	0.997941	0.997332	0.996751	0.996193	0.995656	0.995138	0.994637	0.994150
11.5	0.999304	0.998657	0.998047	0.997468	0.996915	0.996385	0.995874	0.995381	0.994903	0.994439

Table 5 (continued)
$\rho = 0.7000$

a \ N	1	2	3	4	5	6	7	8	9	10
0.1	0.248170	0.083658	0.030815	0.011847	0.004668	0.001867	0.000753	0.000306	0.000125	0.000051
0.2	0.345279	0.157564	0.078214	0.040465	0.021450	0.011546	0.006277	0.003435	0.001887	0.001040
0.3	0.416117	0.223270	0.129707	0.078405	0.048533	0.030511	0.019381	0.012399	0.007970	0.005140
0.4	0.472911	0.282027	0.181256	0.120992	0.082643	0.057318	0.040178	0.028374	0.020145	0.014355
0.5	0.520500	0.334850	0.231087	0.165329	0.120923	0.089772	0.067354	0.050925	0.038721	0.029563
0.6	0.561422	0.382566	0.278411	0.209647	0.161232	0.125792	0.099166	0.078781	0.062953	0.050528
0.7	0.597216	0.425857	0.322927	0.252902	0.202076	0.163702	0.133953	0.110448	0.091603	0.076323
0.8	0.628906	0.465286	0.364585	0.294501	0.242461	0.202253	0.170346	0.144536	0.123348	0.105754
0.9	0.657218	0.501323	0.403465	0.334140	0.281759	0.240570	0.207299	0.179900	0.157003	0.137645
1.0	0.682689	0.534362	0.439706	0.371689	0.319592	0.278066	0.244058	0.215662	0.191601	0.170978
1.1	0.705734	0.564739	0.473478	0.407131	0.355756	0.314365	0.280102	0.251181	0.226407	0.204941
1.2	0.726678	0.592736	0.504957	0.440508	0.390158	0.349245	0.315088	0.286007	0.260880	0.238919
1.3	0.745787	0.618598	0.534314	0.471904	0.422782	0.382587	0.348801	0.319838	0.294639	0.272461
1.4	0.763276	0.642536	0.561714	0.501417	0.453656	0.414348	0.381123	0.352484	0.327428	0.305252
1.5	0.779328	0.664732	0.587308	0.529157	0.482838	0.444528	0.411998	0.383833	0.359081	0.337074
1.6	0.794097	0.685346	0.611236	0.555232	0.510401	0.473162	0.441418	0.413832	0.389502	0.367790
1.7	0.807712	0.704519	0.633626	0.579750	0.536427	0.500303	0.469404	0.442469	0.418642	0.397317
1.8	0.820287	0.722375	0.654594	0.602811	0.560999	0.526013	0.495996	0.469759	0.446492	0.425618
1.9	0.831921	0.739023	0.674246	0.624512	0.584200	0.550360	0.521247	0.495737	0.473066	0.452686
2.0	0.842701	0.754562	0.692678	0.644943	0.606111	0.573416	0.545214	0.520449	0.498395	0.478536
2.1	0.852701	0.769079	0.709979	0.664187	0.626809	0.595248	0.567960	0.543946	0.522522	0.503199
2.2	0.861989	0.782654	0.726227	0.682320	0.646364	0.615925	0.589545	0.566283	0.545495	0.526716
2.3	0.870626	0.795359	0.741498	0.699415	0.664847	0.635509	0.610028	0.587516	0.567364	0.549133
2.4	0.878665	0.807257	0.755857	0.715538	0.682321	0.654061	0.629467	0.607699	0.588181	0.570497
2.5	0.886154	0.818407	0.769365	0.730750	0.698845	0.671640	0.647917	0.626884	0.607995	0.590857
2.6	0.893136	0.828863	0.782080	0.745108	0.714475	0.688298	0.665430	0.645121	0.626854	0.610259
2.7	0.899652	0.838674	0.794053	0.758663	0.729264	0.704087	0.682054	0.662456	0.644805	0.628746
2.8	0.905735	0.847884	0.805332	0.771466	0.743260	0.719055	0.697837	0.678936	0.661889	0.646362
2.9	0.911420	0.856534	0.815962	0.783562	0.756509	0.733247	0.712822	0.694603	0.678150	0.663146
3.0	0.916735	0.864663	0.825983	0.794993	0.769053	0.746705	0.727051	0.709497	0.693626	0.679137
3.1	0.921708	0.872304	0.835433	0.805798	0.780932	0.759469	0.740565	0.723658	0.708355	0.694370
3.2	0.926362	0.879490	0.844348	0.816014	0.792185	0.771578	0.753399	0.737121	0.722372	0.708881
3.3	0.930720	0.886250	0.852759	0.825675	0.802845	0.783065	0.765591	0.749924	0.735713	0.722703
3.4	0.934803	0.892612	0.860699	0.834814	0.812946	0.793966	0.777173	0.762098	0.748411	0.735870
3.5	0.938631	0.898601	0.868194	0.843460	0.822518	0.804310	0.788176	0.773676	0.760497	0.748411
3.6	0.942220	0.904241	0.875272	0.851642	0.831591	0.814128	0.798632	0.784688	0.772002	0.760357
3.7	0.945587	0.909553	0.881957	0.859385	0.840191	0.823447	0.808567	0.795162	0.782953	0.771738
3.8	0.948747	0.914558	0.888273	0.866714	0.848345	0.832293	0.818010	0.805125	0.793379	0.782579

3.9	0.951714	0.919276	0.894240	0.873652	0.856076	0.840692	0.826984	0.814604	0.803306	0.792909
4.0	0.954500	0.923722	0.899879	0.880221	0.863407	0.848666	0.835513	0.823622	0.812758	0.802751
4.1	0.957117	0.927915	0.905210	0.886442	0.870359	0.856237	0.843621	0.832201	0.821757	0.812129
4.2	0.959576	0.931869	0.910249	0.892334	0.876952	0.863427	0.851328	0.840364	0.830327	0.821065
4.3	0.961887	0.935599	0.915013	0.897914	0.883206	0.870254	0.858653	0.848130	0.838487	0.829580
4.4	0.964061	0.939118	0.919518	0.903200	0.889138	0.876737	0.865617	0.855519	0.846257	0.837695
4.5	0.966165	0.942438	0.923779	0.908207	0.894765	0.882895	0.872237	0.862550	0.853656	0.845427
4.6	0.968028	0.945572	0.927809	0.912952	0.900104	0.888742	0.878531	0.869239	0.860701	0.852795
4.7	0.969837	0.948530	0.931622	0.917447	0.905169	0.894297	0.884513	0.875603	0.867409	0.859814
4.8	0.971540	0.951323	0.935229	0.921707	0.909975	0.899572	0.890201	0.881658	0.873795	0.866503
4.9	0.973143	0.953960	0.938642	0.925744	0.914535	0.904583	0.895608	0.887418	0.879875	0.872874
5.0	0.974652	0.956451	0.941872	0.929571	0.918863	0.909342	0.900748	0.892899	0.885664	0.878945
5.1	0.976074	0.958803	0.944928	0.933197	0.922969	0.913864	0.905636	0.898114	0.891176	0.884728
5.2	0.977413	0.961025	0.947821	0.936635	0.926866	0.918159	0.910282	0.903075	0.896423	0.890236
5.3	0.978674	0.963124	0.950560	0.939894	0.930565	0.922239	0.914700	0.907796	0.901417	0.895482
5.4	0.979863	0.965108	0.953152	0.942983	0.934075	0.926115	0.918900	0.912287	0.906173	0.900480
5.5	0.980983	0.966982	0.955607	0.945912	0.937407	0.929798	0.922893	0.916560	0.910700	0.905240
5.6	0.982039	0.968754	0.957930	0.948689	0.940569	0.933297	0.926690	0.920625	0.915009	0.909774
5.7	0.983035	0.970429	0.960131	0.951321	0.943571	0.936620	0.930301	0.924493	0.919112	0.914093
5.8	0.983974	0.972011	0.962214	0.953818	0.946420	0.939778	0.933733	0.928174	0.923018	0.918206
5.9	0.984859	0.973508	0.964188	0.956185	0.949125	0.942779	0.936997	0.931676	0.926737	0.922124
6.0	0.985694	0.974923	0.966056	0.958430	0.951693	0.945630	0.940101	0.935008	0.930278	0.925857
6.1	0.986482	0.976261	0.967826	0.960559	0.954130	0.948338	0.943052	0.938178	0.933648	0.929411
6.2	0.987225	0.977526	0.969503	0.962578	0.956444	0.950912	0.945858	0.941194	0.936857	0.932798
6.3	0.987926	0.978722	0.971091	0.964493	0.958641	0.953357	0.948526	0.944064	0.939912	0.936023
6.4	0.988588	0.979854	0.972595	0.966309	0.960726	0.955681	0.951063	0.946795	0.942820	0.939095
6.5	0.989212	0.980925	0.974021	0.968032	0.962706	0.957888	0.953475	0.949393	0.945589	0.942021
6.6	0.989802	0.981937	0.975371	0.969666	0.964586	0.959986	0.955769	0.951865	0.948224	0.944808
6.7	0.990359	0.982895	0.976650	0.971216	0.966371	0.961979	0.957949	0.954217	0.950733	0.947462
6.8	0.990884	0.983802	0.977862	0.972686	0.968065	0.963873	0.960023	0.956454	0.953121	0.949990
6.9	0.991380	0.984660	0.979011	0.974080	0.969674	0.965672	0.961994	0.958582	0.955394	0.952398
7.0	0.991849	0.985471	0.980099	0.975403	0.971202	0.967382	0.963868	0.960607	0.957558	0.954690
7.1	0.992291	0.986239	0.981131	0.976659	0.972652	0.969007	0.965651	0.962534	0.959618	0.956874
7.2	0.992709	0.986966	0.982109	0.977849	0.974029	0.970550	0.967345	0.964366	0.961578	0.958952
7.3	0.993104	0.987654	0.983035	0.978979	0.975337	0.972017	0.968956	0.966109	0.963443	0.960932
7.4	0.993477	0.988305	0.983913	0.980050	0.976579	0.973411	0.970488	0.967768	0.965218	0.962816
7.5	0.993830	0.988921	0.984745	0.981067	0.977758	0.974735	0.971944	0.969345	0.966908	0.964610
7.6	0.994163	0.989505	0.985534	0.982032	0.978877	0.975994	0.973329	0.970846	0.968516	0.966318
7.7	0.994478	0.990057	0.986281	0.982947	0.979940	0.977189	0.974645	0.972273	0.970046	0.967945
7.8	0.994775	0.990580	0.986990	0.983816	0.980950	0.978326	0.975897	0.973631	0.971502	0.969493
7.9	0.995056	0.991075	0.987662	0.984640	0.981908	0.979405	0.977087	0.974922	0.972888	0.970967
8.0	0.995322	0.991544	0.988299	0.985421	0.982819	0.980431	0.978218	0.976151	0.974207	0.972370
8.1	0.995573	0.991987	0.988902	0.986163	0.983683	0.981406	0.979294	0.977319	0.975462	0.973706

Table 5 (continued)
$\rho = 0.7000$

8.2	0.995811	0.992408	0.989475	0.986867	0.984504	0.982332	0.980316	0.978431	0.976656	0.974977
8.3	0.996035	0.992806	0.990017	0.987535	0.985283	0.983212	0.981289	0.979488	0.977793	0.976188
8.4	0.996248	0.993182	0.990532	0.988169	0.986024	0.984049	0.982213	0.980494	0.978874	0.977340
8.5	0.996448	0.993539	0.991019	0.988771	0.986726	0.984843	0.983092	0.981451	0.979903	0.978437
8.6	0.996638	0.993877	0.991482	0.989342	0.987394	0.985599	0.983928	0.982361	0.980883	0.979482
8.7	0.996818	0.994198	0.991920	0.989883	0.988028	0.986316	0.984722	0.983226	0.981815	0.980476
8.8	0.996988	0.994501	0.992336	0.990397	0.988630	0.986998	0.985477	0.984049	0.982701	0.981422
8.9	0.997148	0.994788	0.992730	0.990885	0.989202	0.987646	0.986195	0.984833	0.983545	0.982323
9.0	0.997300	0.995060	0.993104	0.991348	0.989745	0.988262	0.986878	0.985577	0.984348	0.983180
9.1	0.997444	0.995318	0.993459	0.991788	0.990260	0.988847	0.987527	0.986286	0.985112	0.983997
9.2	0.997580	0.995562	0.993795	0.992205	0.990750	0.989403	0.988144	0.986959	0.985839	0.984773
9.3	0.997708	0.995793	0.994113	0.992600	0.991215	0.989931	0.988730	0.987600	0.986530	0.985513
9.4	0.997830	0.996013	0.994416	0.992976	0.991656	0.990433	0.989288	0.988209	0.987188	0.986217
9.5	0.997945	0.996220	0.994703	0.993333	0.992076	0.990910	0.989818	0.988789	0.987815	0.986887
9.6	0.998054	0.996417	0.994974	0.993671	0.992474	0.991363	0.990322	0.989341	0.988410	0.987524
9.7	0.998157	0.996603	0.995232	0.993992	0.992853	0.991793	0.990801	0.989865	0.988977	0.988131
9.8	0.998255	0.996780	0.995477	0.994297	0.993212	0.992203	0.991256	0.990363	0.989516	0.988708
9.9	0.998347	0.996947	0.995709	0.994586	0.993553	0.992591	0.991689	0.990837	0.990029	0.989258
10.0	0.998434	0.997106	0.995929	0.994861	0.993877	0.992961	0.992101	0.991288	0.990517	0.989781
10.1	0.998517	0.997256	0.996137	0.995121	0.994185	0.993312	0.992492	0.991717	0.990981	0.990278
10.2	0.998595	0.997399	0.996335	0.995369	0.994477	0.993645	0.992864	0.992124	0.991422	0.990752
10.3	0.998670	0.997534	0.996523	0.995603	0.994754	0.993962	0.993217	0.992512	0.991842	0.991202
10.4	0.998740	0.997662	0.996701	0.995826	0.995018	0.994263	0.993553	0.992881	0.992241	0.991631
10.5	0.998806	0.997783	0.996870	0.996038	0.995268	0.994549	0.993872	0.993231	0.992622	0.992039
10.6	0.998869	0.997898	0.997030	0.996239	0.995506	0.994821	0.994176	0.993565	0.992983	0.992427
10.7	0.998929	0.998007	0.997182	0.996429	0.995732	0.995079	0.994464	0.993882	0.993327	0.992796
10.8	0.998985	0.998110	0.997326	0.996610	0.995946	0.995325	0.994739	0.994183	0.993654	0.993148
10.9	0.999038	0.998208	0.997463	0.996782	0.996150	0.995558	0.995000	0.994470	0.993965	0.993482
11.0	0.999089	0.998300	0.997593	0.996945	0.996343	0.995779	0.995247	0.994742	0.994261	0.993800
11.1	0.999137	0.998388	0.997716	0.997099	0.996527	0.995990	0.995483	0.995001	0.994542	0.994102
11.2	0.999182	0.998472	0.997833	0.997246	0.996701	0.996190	0.995707	0.995248	0.994810	0.994390
11.3	0.999225	0.998551	0.997943	0.997386	0.996867	0.996380	0.995920	0.995482	0.995065	0.994664
11.4	0.999265	0.998625	0.998048	0.997518	0.997024	0.996561	0.996122	0.995705	0.995307	0.994925
11.5	0.999304	0.998696	0.998148	0.997644	0.997174	0.996732	0.996314	0.995917	0.995537	0.995173

Table 5 (continued)
$\rho = 0.8000$

a \ N	1	2	3	4	5	6	7	8	9	10
0.1	0.248170	0.097127	0.042161	0.019111	0.008834	0.004120	0.001929	0.000904	0.000424	0.000199
0.2	0.345279	0.179213	0.102629	0.061421	0.037588	0.023266	0.014484	0.009043	0.005653	0.003537
0.3	0.416118	0.249640	0.164206	0.112777	0.079313	0.056530	0.040595	0.029273	0.021158	0.015311
0.4	0.472911	0.310871	0.222595	0.166087	0.126921	0.098418	0.077003	0.060577	0.047812	0.037812
0.5	0.520500	0.364718	0.276597	0.218032	0.175897	0.144052	0.119148	0.099194	0.082934	0.069532
0.6	0.561422	0.412532	0.326104	0.267219	0.223817	0.190250	0.163389	0.141338	0.122882	0.107211
0.7	0.597216	0.455336	0.371394	0.313193	0.269556	0.235272	0.207445	0.184286	0.164623	0.147659
0.8	0.628906	0.493920	0.412854	0.355924	0.312691	0.278313	0.250119	0.226448	0.206192	0.188578
0.9	0.657218	0.528899	0.450881	0.395560	0.353138	0.319075	0.290892	0.267061	0.246555	0.228648
1.0	0.682689	0.560766	0.485842	0.432311	0.390962	0.357498	0.329594	0.305837	0.285286	0.267271
1.1	0.705734	0.589918	0.518063	0.466406	0.426289	0.393627	0.366208	0.342710	0.322265	0.304264
1.2	0.726678	0.616681	0.547832	0.498069	0.459267	0.427540	0.400768	0.377690	0.357493	0.339615
1.3	0.745787	0.641324	0.575398	0.527511	0.490053	0.459337	0.433327	0.410802	0.390986	0.373352
1.4	0.763276	0.664075	0.600980	0.554930	0.518807	0.489132	0.463949	0.442076	0.422756	0.405483
1.5	0.779328	0.685126	0.624766	0.580505	0.545689	0.517046	0.492715	0.471549	0.452809	0.435998
1.6	0.794097	0.704644	0.646921	0.604399	0.570852	0.543212	0.519723	0.499282	0.481167	0.464886
1.7	0.807712	0.722770	0.667589	0.626757	0.594438	0.567766	0.545088	0.525357	0.507874	0.492155
1.8	0.820287	0.739629	0.686895	0.647705	0.616581	0.590839	0.568933	0.549879	0.533007	0.517848
1.9	0.831922	0.755331	0.704950	0.667355	0.637396	0.612553	0.591382	0.572962	0.556665	0.542038
2.0	0.842701	0.769973	0.721853	0.685806	0.656987	0.633018	0.612551	0.594729	0.578963	0.564827
2.1	0.852701	0.783641	0.737690	0.703145	0.675441	0.652331	0.632548	0.615293	0.600020	0.586332
2.2	0.861989	0.796412	0.752540	0.719450	0.692837	0.670574	0.651463	0.634757	0.619949	0.606673
2.3	0.870626	0.808356	0.766476	0.734789	0.709242	0.687816	0.669372	0.653207	0.638848	0.625959
2.4	0.878665	0.819534	0.779562	0.749227	0.724718	0.704117	0.686338	0.670714	0.656800	0.644285
2.5	0.886154	0.830004	0.791857	0.762823	0.739320	0.719530	0.702414	0.687333	0.673869	0.661726
2.6	0.893136	0.839817	0.803416	0.775631	0.753100	0.734102	0.717643	0.703111	0.690102	0.678340
2.7	0.899652	0.849020	0.81[illegible]288	0.787702	0.766107	0.747878	0.732067	0.718083	0.705538	0.694167
2.8	0.905736	0.857656	0.824521	0.799083	0.778387	0.760900	0.745722	0.732282	0.720207	0.709237
2.9	0.911420	0.865765	0.834156	0.809819	0.789984	0.773211	0.758645	0.745740	0.734133	0.723573
3.0	0.916735	0.873381	0.843231	0.819951	0.800942	0.784851	0.770874	0.758489	0.747344	0.737194
3.1	0.921708	0.880538	0.851784	0.829517	0.811298	0.795861	0.782447	0.770561	0.759865	0.750122
3.2	0.926362	0.887267	0.859848	0.838552	0.821092	0.806279	0.793402	0.781992	0.771729	0.762381
3.3	0.930720	0.893596	0.867452	0.847090	0.830358	0.816142	0.803775	0.792818	0.782966	0.773998
3.4	0.934803	0.899551	0.874625	0.855159	0.839128	0.825484	0.813603	0.803076	0.793612	0.785004
3.5	0.938631	0.905155	0.881394	0.862788	0.847431	0.834337	0.822921	0.812800	0.803704	0.795434
3.6	0.942220	0.910432	0.887782	0.870003	0.855297	0.842732	0.831762	0.822027	0.813276	0.805323
3.7	0.945587	0.915401	0.893814	0.876828	0.862748	0.850695	0.840153	0.830788	0.822364	0.814709
3.8	0.948747	0.920082	0.899509	0.883284	0.869808	0.858251	0.848124	0.839113	0.831001	0.823625

Table 5 (continued)
$\rho = 0.8000$

3.9	0.951714	0.924493	0.904888	0.889392	0.876499	0.865421	0.855696	0.847028	0.839215	0.832105
4.0	0.954500	0.928650	0.909963	0.895171	0.882840	0.872226	0.862892	0.854558	0.847033	0.840178
4.1	0.957117	0.932569	0.914769	0.900641	0.888849	0.878685	0.869731	0.861722	0.854478	0.847870
4.2	0.959576	0.936265	0.919305	0.905817	0.894544	0.884815	0.876231	0.868540	0.861571	0.855203
4.3	0.961887	0.939751	0.923592	0.910716	0.899940	0.890631	0.882407	0.875027	0.868328	0.862196
4.4	0.964061	0.943039	0.927645	0.915353	0.905055	0.896149	0.888274	0.881197	0.874764	0.868865
4.5	0.966105	0.946142	0.931476	0.919743	0.909901	0.901383	0.893845	0.887065	0.880893	0.875223
4.6	0.968028	0.949070	0.935099	0.923900	0.914495	0.906349	0.899136	0.892643	0.886726	0.881283
4.7	0.969837	0.951834	0.938526	0.927836	0.918848	0.911058	0.904157	0.897943	0.892275	0.887056
4.8	0.971540	0.954444	0.941767	0.931564	0.922974	0.915525	0.908924	0.902977	0.897550	0.892550
4.9	0.973143	0.956908	0.944833	0.935095	0.926886	0.919762	0.913447	0.907757	0.902564	0.897777
5.0	0.974653	0.959235	0.947734	0.938441	0.930596	0.923781	0.917739	0.912295	0.907327	0.902746
5.1	0.976074	0.961433	0.950479	0.941611	0.934113	0.927595	0.921813	0.916604	0.911850	0.907467
5.2	0.977413	0.963509	0.953077	0.944615	0.937449	0.931214	0.925680	0.920694	0.916145	0.911951
5.3	0.978674	0.965470	0.955537	0.947462	0.940614	0.934648	0.929351	0.924577	0.920223	0.916210
5.4	0.979863	0.967324	0.957865	0.950161	0.943617	0.937910	0.932838	0.928265	0.924095	0.920255
5.5	0.980983	0.969076	0.960068	0.952719	0.946467	0.941006	0.936150	0.931769	0.927774	0.924096
5.6	0.982039	0.970731	0.962155	0.955145	0.949172	0.943948	0.939297	0.935099	0.931270	0.927745
5.7	0.983035	0.972296	0.964131	0.957445	0.951739	0.946742	0.942288	0.938265	0.934594	0.931214
5.8	0.983974	0.973775	0.966002	0.959625	0.954176	0.949397	0.945132	0.941276	0.937755	0.934512
5.9	0.984859	0.975174	0.967773	0.961693	0.956490	0.951920	0.947836	0.944140	0.940762	0.937651
6.0	0.985694	0.976497	0.969451	0.963654	0.958686	0.954317	0.950408	0.946866	0.943626	0.940639
6.1	0.986482	0.977747	0.971040	0.965513	0.960771	0.956596	0.952855	0.949460	0.946352	0.943484
6.2	0.987225	0.978930	0.972546	0.967276	0.962750	0.958761	0.955182	0.951930	0.948949	0.946196
6.3	0.987926	0.980049	0.973971	0.968948	0.964629	0.960818	0.957395	0.954282	0.951423	0.948781
6.4	0.988588	0.981107	0.975322	0.970534	0.966412	0.962772	0.959500	0.956520	0.953781	0.951245
6.5	0.989212	0.982108	0.976602	0.972037	0.968105	0.964629	0.961502	0.958650	0.956026	0.953594
6.6	0.989802	0.983055	0.977814	0.973463	0.969712	0.966393	0.963405	0.960678	0.958166	0.955834
6.7	0.990359	0.983951	0.978963	0.974816	0.971237	0.968069	0.965214	0.962607	0.960203	0.957970
6.8	0.990884	0.984799	0.980051	0.976098	0.972684	0.969660	0.966934	0.964443	0.962143	0.960005
6.9	0.991380	0.985601	0.981082	0.977315	0.974058	0.971172	0.968568	0.966188	0.963990	0.961943
7.0	0.991849	0.986361	0.982059	0.978469	0.975362	0.972607	0.970121	0.967848	0.965747	0.963790
7.1	0.992291	0.987079	0.982986	0.979564	0.976600	0.973970	0.971597	0.969426	0.967419	0.965548
7.2	0.992710	0.987760	0.983864	0.980602	0.977774	0.975265	0.972999	0.970926	0.969009	0.967221
7.3	0.993104	0.988403	0.984696	0.981587	0.978890	0.976494	0.974330	0.972351	0.970520	0.968813
7.4	0.993477	0.989013	0.985484	0.982522	0.979948	0.977661	0.975595	0.973705	0.971957	0.970326
7.5	0.993830	0.989590	0.986232	0.983409	0.980953	0.978770	0.976797	0.974992	0.973322	0.971765
7.6	0.994163	0.990136	0.986941	0.984250	0.981908	0.979823	0.977939	0.976214	0.974620	0.973133
7.7	0.994478	0.990654	0.987613	0.985049	0.982814	0.980824	0.979024	0.977376	0.975853	0.974433

7.8	0.994775	0.991143	0.988250	0.985807	0.983675	0.981774	0.980055	0.978481	0.977025	0.975668
7.9	0.995056	0.991607	0.988854	0.986526	0.984492	0.982678	0.981035	0.979530	0.978139	0.976843
8.0	0.995322	0.992046	0.989427	0.987208	0.985268	0.983536	0.981966	0.980528	0.979198	0.977959
8.1	0.995573	0.992462	0.989970	0.987856	0.986006	0.984352	0.982852	0.981477	0.980205	0.979020
8.2	0.995811	0.992856	0.990485	0.988471	0.986706	0.985127	0.983694	0.982379	0.981163	0.980030
8.3	0.996035	0.993229	0.990973	0.989055	0.987372	0.985864	0.984495	0.983238	0.982074	0.980990
8.4	0.996248	0.993583	0.991436	0.989609	0.988004	0.986565	0.985256	0.984054	0.982941	0.981904
8.5	0.996448	0.993917	0.991875	0.990135	0.988604	0.987231	0.985981	0.984832	0.983767	0.982773
8.6	0.996638	0.994235	0.992291	0.990634	0.989175	0.987864	0.986670	0.985571	0.984552	0.983601
8.7	0.996818	0.994535	0.992686	0.991108	0.989717	0.988466	0.987326	0.986275	0.985300	0.984390
8.8	0.996988	0.994819	0.993061	0.991557	0.990231	0.989038	0.987949	0.986945	0.986013	0.985141
8.9	0.997148	0.995089	0.993416	0.991984	0.990720	0.989582	0.988543	0.987583	0.986691	0.985857
9.0	0.997300	0.995344	0.993753	0.992389	0.991185	0.990099	0.989107	0.988190	0.987337	0.986539
9.1	0.997444	0.995586	0.994072	0.992774	0.991626	0.990591	0.989644	0.988768	0.987953	0.987189
9.2	0.997580	0.995816	0.994375	0.993139	0.992045	0.991058	0.990154	0.989318	0.988539	0.987808
9.3	0.997708	0.996033	0.994663	0.993485	0.992443	0.991502	0.990640	0.989842	0.989097	0.988398
9.4	0.997830	0.996239	0.994936	0.993814	0.992821	0.991924	0.991101	0.990340	0.989628	0.988960
9.5	0.997945	0.996434	0.995194	0.994126	0.993180	0.992325	0.991540	0.990813	0.990134	0.989496
9.6	0.998054	0.996619	0.995439	0.994423	0.993521	0.992706	0.991958	0.991264	0.990616	0.990006
9.7	0.998157	0.996794	0.995672	0.994704	0.993845	0.993068	0.992354	0.991693	0.991074	0.990492
9.8	0.998255	0.996960	0.995893	0.994971	0.994152	0.993411	0.992731	0.992100	0.991510	0.990954
9.9	0.998347	0.997118	0.996102	0.995225	0.994445	0.993738	0.993090	0.992488	0.991925	0.991395
10.0	0.998434	0.997267	0.996301	0.995465	0.994722	0.994049	0.993430	0.992856	0.992319	0.991813
10.1	0.998517	0.997408	0.996490	0.995694	0.994985	0.994343	0.993754	0.993207	0.992694	0.992212
10.2	0.998595	0.997542	0.996668	0.995911	0.995236	0.994624	0.994062	0.993540	0.993051	0.992591
10.3	0.998670	0.997669	0.996838	0.996117	0.995473	0.994890	0.994354	0.993856	0.993390	0.992951
10.4	0.998740	0.997790	0.996999	0.996312	0.995699	0.995143	0.994632	0.994157	0.993713	0.993294
10.5	0.998806	0.997904	0.997152	0.996498	0.995914	0.995383	0.994896	0.994443	0.994019	0.993620
10.6	0.998869	0.998012	0.997297	0.996674	0.996117	0.995612	0.995147	0.994715	0.994310	0.993929
10.7	0.998929	0.998114	0.997434	0.996841	0.996311	0.995829	0.995385	0.994973	0.994587	0.994224
10.8	0.998985	0.998212	0.997565	0.997000	0.996495	0.996035	0.995612	0.995219	0.994851	0.994504
10.9	0.999038	0.998304	0.997689	0.997151	0.996670	0.996231	0.995827	0.995452	0.995101	0.994770
11.0	0.999089	0.998391	0.997806	0.997294	0.996836	0.996418	0.996032	0.995674	0.995339	0.995023
11.1	0.999137	0.998474	0.997918	0.997430	0.996993	0.996595	0.996227	0.995885	0.995565	0.995264
11.2	0.999182	0.998553	0.998023	0.997560	0.997143	0.996763	0.996412	0.996086	0.995781	0.995493
11.3	0.999225	0.998627	0.998124	0.997682	0.997286	0.996923	0.996589	0.996277	0.995985	0.995710
11.4	0.999265	0.998698	0.998219	0.997799	0.997421	0.997075	0.996756	0.996459	0.996180	0.995918
11.5	0.999304	0.998765	0.998310	0.997910	0.997549	0.997220	0.996916	0.996632	0.996366	0.996115

Table 5 (continued)
$\rho = 0.8500$

a	N 1	2	3	4	5	6	7	8	9	10
0.1	0.248171	0.107937	0.052197	0.026162	0.013273	0.006761	0.003448	0.001760	0.000898	0.000458
0.2	0.345280	0.195597	0.122854	0.080418	0.053558	0.035923	0.024164	0.016273	0.010964	0.007388
0.3	0.416117	0.268571	0.190928	0.141708	0.107427	0.082282	0.063335	0.048864	0.037741	0.029165
0.4	0.472910	0.330684	0.252653	0.201306	0.164151	0.135594	0.112805	0.094208	0.078842	0.066055
0.5	0.520459	0.384534	0.307924	0.256398	0.218514	0.188935	0.164820	0.144566	0.127216	0.112169
0.6	0.561422	0.431901	0.357521	0.306644	0.268841	0.239185	0.214925	0.194419	0.176650	0.160973
0.7	0.597216	0.474036	0.402285	0.352508	0.315101	0.285614	0.261513	0.241206	0.223652	0.208158
0.8	0.628907	0.511840	0.442939	0.394553	0.357774	0.328535	0.304571	0.284431	0.267124	0.251954
0.9	0.657219	0.545993	0.479934	0.433236	0.397336	0.368481	0.344648	0.324559	0.307329	0.292309
1.0	0.682690	0.577019	0.513792	0.468891	0.434110	0.405859	0.382277	0.362238	0.344973	0.329917
1.1	0.705734	0.605337	0.544844	0.501779	0.468292	0.440883	0.417769	0.397920	0.380665	0.365523
1.2	0.726678	0.631288	0.573406	0.532132	0.500016	0.473632	0.451218	0.431785	0.414716	0.399593
1.3	0.745786	0.655152	0.599755	0.560177	0.529417	0.504148	0.482610	0.463813	0.447158	0.432256
1.4	0.763276	0.677162	0.624136	0.586143	0.556655	0.532493	0.511913	0.493907	0.477867	0.463407
1.5	0.779328	0.697513	0.646762	0.610249	0.581920	0.558785	0.539149	0.521994	0.506693	0.492845
1.6	0.794096	0.716371	0.667815	0.632700	0.605417	0.583193	0.564416	0.548082	0.533550	0.520400
1.7	0.807712	0.733877	0.687445	0.653674	0.627349	0.605918	0.587880	0.572273	0.558457	0.546001
1.8	0.820287	0.750151	0.705779	0.673323	0.647899	0.627166	0.609747	0.594743	0.581539	0.569702
1.9	0.831922	0.765301	0.722923	0.691765	0.667221	0.647126	0.630231	0.615712	0.602996	0.591664
2.0	0.842701	0.779419	0.738968	0.709097	0.685434	0.665957	0.649526	0.635402	0.623061	0.612114
2.1	0.852701	0.792589	0.753992	0.725395	0.702627	0.683776	0.667792	0.654011	0.641963	0.631298
2.2	0.861989	0.804886	0.768067	0.740720	0.718864	0.700665	0.685143	0.671693	0.659898	0.649446
2.3	0.870626	0.816378	0.781256	0.755127	0.734191	0.716678	0.701651	0.688553	0.677008	0.666742
2.4	0.878665	0.827127	0.793622	0.768666	0.748645	0.731845	0.717356	0.704650	0.693384	0.683313
2.5	0.886154	0.837189	0.805221	0.781386	0.762262	0.746188	0.732275	0.720009	0.709067	0.699228
2.6	0.893136	0.846615	0.816108	0.793337	0.775076	0.759727	0.746414	0.734631	0.724065	0.714507
2.7	0.899652	0.855451	0.826335	0.804568	0.787127	0.772483	0.759778	0.748509	0.738363	0.729140
2.8	0.905735	0.863741	0.835948	0.815128	0.798458	0.784486	0.772378	0.761635	0.751942	0.743097
2.9	0.911420	0.871522	0.844992	0.825068	0.809116	0.795772	0.784236	0.774011	0.764784	0.756348
3.0	0.916735	0.878829	0.853508	0.834431	0.819149	0.806387	0.795383	0.785655	0.776887	0.768871
3.1	0.921707	0.885696	0.861531	0.843262	0.828605	0.816376	0.805861	0.796593	0.788264	0.780660
3.2	0.926361	0.892151	0.869095	0.851600	0.837530	0.825791	0.815717	0.806869	0.798943	0.791729
3.3	0.930720	0.898221	0.876229	0.859479	0.845967	0.834680	0.825005	0.816530	0.808966	0.802107
3.4	0.934803	0.903931	0.882961	0.866931	0.853953	0.843090	0.833776	0.825632	0.818387	0.811843
3.5	0.938631	0.909304	0.889314	0.873982	0.861522	0.851061	0.842079	0.834230	0.827263	0.820991
3.6	0.942220	0.914362	0.895310	0.880655	0.868701	0.858628	0.849958	0.842376	0.835652	0.829613
3.7	0.945587	0.919123	0.900971	0.886972	0.875514	0.865821	0.857451	0.850116	0.843608	0.837770
3.8	0.948747	0.923608	0.906313	0.892950	0.881980	0.872664	0.864588	0.857489	0.851180	0.845518

3.9	0.951714	0.927832	0.911357	0.898607	0.888116	0.879174	0.871392	0.864526	0.858406	0.852905
4.0	0.954500	0.931812	0.916118	0.903958	0.893935	0.885367	0.877881	0.871249	0.865316	0.859969
4.1	0.957117	0.935563	0.920612	0.909017	0.899450	0.891253	0.884067	0.877674	0.871933	0.866739
4.2	0.959576	0.939099	0.924855	0.913800	0.904674	0.896844	0.889959	0.883812	0.878269	0.873234
4.3	0.961888	0.942433	0.928862	0.918320	0.909619	0.902147	0.895565	0.889670	0.884333	0.879466
4.4	0.964061	0.945577	0.932645	0.922592	0.914296	0.907173	0.900891	0.895252	0.890129	0.885440
4.5	0.966105	0.948543	0.936219	0.926629	0.918719	0.911931	0.905944	0.900561	0.895659	0.891157
4.6	0.968028	0.951342	0.939597	0.930445	0.922900	0.916432	0.910730	0.905601	0.900922	0.896614
4.7	0.969837	0.953984	0.942789	0.934053	0.926852	0.920687	0.915258	0.910377	0.905922	0.901812
4.8	0.971540	0.956478	0.945807	0.937465	0.930589	0.924709	0.919539	0.914896	0.910660	0.906748
4.9	0.973143	0.958833	0.948662	0.940694	0.934123	0.928510	0.923584	0.919168	0.915142	0.911426
5.0	0.974652	0.961056	0.951363	0.943751	0.937468	0.932104	0.927406	0.923202	0.919377	0.915850
5.1	0.976074	0.963157	0.953918	0.946646	0.940635	0.935505	0.931017	0.927011	0.923373	0.920026
5.2	0.977413	0.965141	0.956338	0.949390	0.943636	0.938724	0.934432	0.930608	0.927145	0.923965
5.3	0.978674	0.967016	0.958629	0.951991	0.946483	0.941776	0.937665	0.934008	0.930705	0.927679
5.4	0.979863	0.968788	0.960798	0.954458	0.949184	0.944671	0.940728	0.937226	0.934067	0.931183
5.5	0.980983	0.970462	0.962852	0.956798	0.951750	0.947420	0.943635	0.940274	0.937248	0.934491
5.6	0.982039	0.972044	0.964797	0.959019	0.954187	0.950034	0.946397	0.943168	0.940262	0.937620
5.7	0.983035	0.973540	0.966639	0.961126	0.956503	0.952520	0.949025	0.945918	0.943123	0.940584
5.8	0.983974	0.974953	0.968383	0.963125	0.958705	0.954886	0.951527	0.948536	0.945844	0.943400
5.9	0.984859	0.976290	0.970036	0.965022	0.960798	0.957139	0.953912	0.951032	0.948437	0.946079
6.0	0.985694	0.977553	0.971600	0.966821	0.962788	0.959284	0.956185	0.953413	0.950910	0.948634
6.1	0.986482	0.978748	0.973081	0.968527	0.964678	0.961326	0.958354	0.955687	0.953273	0.951074
6.2	0.987225	0.979878	0.974484	0.970145	0.966474	0.963270	0.960422	0.957859	0.955533	0.953409
6.3	0.987926	0.980946	0.975812	0.971679	0.968179	0.965120	0.962394	0.959933	0.957694	0.955644
6.4	0.988588	0.981957	0.977069	0.973133	0.969798	0.966879	0.964273	0.961914	0.959762	0.957785
6.5	0.989212	0.982912	0.978260	0.974511	0.971333	0.968551	0.966062	0.963805	0.961739	0.959837
6.6	0.989802	0.983816	0.979387	0.975816	0.972789	0.970138	0.967764	0.965607	0.963628	0.961800
6.7	0.990359	0.984671	0.980455	0.977052	0.974170	0.971645	0.969382	0.967324	0.965431	0.963679
6.8	0.990884	0.985481	0.981466	0.978224	0.975478	0.973074	0.970919	0.968957	0.967150	0.965474
6.9	0.991380	0.986246	0.982423	0.979334	0.976718	0.974428	0.972378	0.970510	0.968787	0.967186
7.0	0.991849	0.986971	0.983330	0.980386	0.977892	0.975712	0.973761	0.971984	0.970344	0.968817
7.1	0.992292	0.987657	0.984190	0.981382	0.979005	0.976929	0.975072	0.973382	0.971822	0.970369
7.2	0.992710	0.988306	0.985004	0.982327	0.980060	0.978081	0.976314	0.974706	0.973224	0.971843
7.3	0.993104	0.988921	0.985776	0.983223	0.981060	0.979173	0.977490	0.975961	0.974553	0.973241
7.4	0.993477	0.989502	0.986508	0.984072	0.982008	0.980208	0.978604	0.977149	0.975811	0.974565
7.5	0.993830	0.990053	0.987202	0.984878	0.982907	0.981189	0.979659	0.978274	0.977001	0.975819
7.6	0.994163	0.990575	0.987860	0.985643	0.983760	0.982119	0.980659	0.979339	0.978128	0.977004
7.7	0.994478	0.991069	0.988483	0.986369	0.984570	0.983001	0.981607	0.980347	0.979194	0.978126
7.8	0.994775	0.991536	0.989075	0.987058	0.985339	0.983839	0.982506	0.981303	0.980203	0.979186
7.9	0.995056	0.991979	0.989636	0.987712	0.986070	0.984635	0.983359	0.982209	0.981159	0.980190
8.0	0.995322	0.992399	0.990168	0.988333	0.986764	0.985391	0.984170	0.983069	0.982065	0.981140
8.1	0.995573	0.992796	0.990673	0.988923	0.987424	0.986110	0.984940	0.983886	0.982924	0.982040

Table 5 (continued)
$\rho = 0.8500$

8.2	0.995811	0.993172	0.991152	0.989483	0.988051	0.986794	0.985673	0.984662	0.983741	0.982894
8.3	0.996035	0.993529	0.991606	0.990016	0.988548	0.987445	0.986371	0.985401	0.984518	0.983705
8.4	0.996248	0.993866	0.992036	0.990521	0.989216	0.988065	0.987036	0.986105	0.985257	0.984477
8.5	0.996448	0.994186	0.992445	0.991001	0.989756	0.988656	0.987669	0.986776	0.985961	0.985212
8.6	0.996638	0.994489	0.992832	0.991457	0.990269	0.989218	0.988273	0.987416	0.986633	0.985912
8.7	0.996818	0.994776	0.993200	0.991890	0.990757	0.989753	0.988849	0.988027	0.987274	0.986581
8.8	0.996988	0.995048	0.993548	0.992302	0.991222	0.990262	0.989397	0.988610	0.987887	0.987220
8.9	0.997148	0.995305	0.993879	0.992692	0.991663	0.990747	0.989921	0.989166	0.988472	0.987830
9.0	0.997300	0.995549	0.994192	0.993062	0.992082	0.991209	0.990419	0.989696	0.989031	0.988414
9.1	0.997444	0.995780	0.994489	0.993414	0.992480	0.991648	0.990894	0.990203	0.989565	0.988972
9.2	0.997580	0.995999	0.994771	0.993747	0.992858	0.992066	0.991346	0.990686	0.990075	0.989507
9.3	0.997708	0.996207	0.995038	0.994064	0.993217	0.992463	0.991777	0.991146	0.990562	0.990017
9.4	0.997830	0.996403	0.995291	0.994364	0.993558	0.992840	0.992186	0.991585	0.991027	0.990505
9.5	0.997945	0.996590	0.995531	0.994648	0.993882	0.993198	0.992576	0.992003	0.991470	0.990971
9.6	0.998054	0.996766	0.995759	0.994919	0.994189	0.993538	0.992946	0.992400	0.991892	0.991416
9.7	0.998157	0.996933	0.995975	0.995175	0.994480	0.993860	0.993297	0.992778	0.992294	0.991840
9.8	0.998255	0.997092	0.996180	0.995418	0.994756	0.994167	0.993631	0.993137	0.992676	0.992244
9.9	0.998347	0.997242	0.996374	0.995648	0.995018	0.994457	0.993948	0.993478	0.993040	0.992628
10.0	0.998435	0.997385	0.996559	0.995867	0.995267	0.994733	0.994248	0.993801	0.993385	0.992994
10.1	0.998517	0.997520	0.996734	0.996075	0.995503	0.994994	0.994533	0.994108	0.993712	0.993341
10.2	0.998595	0.997648	0.996900	0.996272	0.995727	0.995242	0.994803	0.994399	0.994023	0.993670
10.3	0.998670	0.997769	0.997057	0.996459	0.995939	0.995477	0.995059	0.994675	0.994318	0.993982
10.4	0.998740	0.997884	0.997206	0.996636	0.996141	0.995700	0.995302	0.994936	0.994597	0.994278
10.5	0.998806	0.997993	0.997348	0.996805	0.996332	0.995912	0.995532	0.995184	0.994861	0.994559
10.6	0.998869	0.998097	0.997483	0.996965	0.996514	0.996113	0.995751	0.995419	0.995112	0.994824
10.7	0.998929	0.998195	0.997610	0.997117	0.996687	0.996304	0.995959	0.995642	0.995350	0.995076
10.8	0.998985	0.998288	0.997732	0.997261	0.996851	0.996486	0.996156	0.995854	0.995575	0.995315
10.9	0.999038	0.998376	0.997847	0.997398	0.997007	0.996658	0.996343	0.996055	0.995789	0.995541
11.0	0.999089	0.998459	0.997956	0.997529	0.997155	0.996822	0.996521	0.996245	0.995991	0.995755
11.1	0.999137	0.998539	0.998060	0.997653	0.997296	0.996978	0.996690	0.996427	0.996184	0.995958
11.2	0.999182	0.998614	0.998158	0.997770	0.997430	0.997126	0.996851	0.996599	0.996367	0.996151
11.3	0.999225	0.998685	0.998251	0.997882	0.997558	0.997267	0.997004	0.996763	0.996541	0.996335
11.4	0.999265	0.998753	0.998340	0.997988	0.997679	0.997401	0.997150	0.996919	0.996706	0.996509
11.5	0.999304	0.998817	0.998424	0.998089	0.997794	0.997529	0.997288	0.997068	0.996864	0.996675

References

[1] Breiter, M. C. and Krishnaiah, P. R. (1968). Tables for the moments of gamma order statistics. *Sankhyā Ser. B* **30**, 59–72.

[2] Dunnett, C. W. and Sobel, M. (1954). A bivariate generalization of student's t-distribution, with tables for certain cases. *Biometrika* **41**, 153–169.

[3] Gupta, S. S. (1960). Order statistics from the gamma distribution. *Technometrics* **2**, 243–262.

[4] Gupta, S. S. (1963). Probability integrals of multivariate normal and multivariate t. *Ann. Math. Statist.* **34**, 792–828.

[5] Gupta, S. S., Nagel, K. and Panchapakesan, S. (1973). On the order statistics from equally correlated normal random variables. *Biometrika* **60**, 403–413.

[6] Harter, H. L. (1961). Expected values of normal order statistics. *Biometrika* **48**, 151–165.

[7] Harter, H. L. (1964). Expected values of exponential, Weibull, and gamma order statistics. ARL 64-31. Wright-Patterson Air Force Base, Ohio.

[8] Krishnaiah, P. R. (1980). Computations of some multivariate distributions. In: P. R. Krishnaiah, ed., *Handbook of Statistics, Vol. 1.* North-Holland, Amsterdam.

[9] Krishnaiah, P. R. and Armitage, J. V. (1965). Tables for the distributions of the maximum of correlated chi-square variates with one degree of freedom. ARL 65-136. Wright-Patterson Air Force Base, Ohio.

[10] Krishnaiah, P. R. and Rizvi, M. H. (1967). A note on moments of gamma order statistics. *Technometrics* **9**, 315–318.

[11] Sarhan, A. E. and Greenberg, B. G. (1956). Estimation of location and scale parameters by order statistics from singly and doubly censored samples. *Ann. Math. Statist.* **27**, 427–451.

[12] Teichroew, D. (1956). Tables of expected values of order statistics and products of order statistics for samples of size twenty and less from the normal distribution. *Ann. Math. Statist.* **27**, 410–426.

P. R. Krishnaiah and P. K. Sen, eds., *Handbook of Statistics, Vol. 4*
© Elsevier Science Publishers (1984) 937–958

37

Selected Tables for Nonparametric Statistics

P. K. Sen and P. R. Krishnaiah

1. Introduction

For the classical nonparametric statistical inference problems, as has been discussed in the earlier chapters, under appropriate null hypotheses, there are genuinely distribution-free tests. In many simple cases, the exact (small sample) null distributions of such statistics have been tabulated by various workers, and these tables facilitate the use of these tests in practice. We shall present some of these tables for their potential use in the various contexts discussed in the preceding chapters. We may also note that in most of the cases, the asymptotic theory (discussed in the earlier chapters) provides simple distributional results for large sample sizes; some of these are also considered in this chapter and appropriate critical values are tabulated.

In Section 2, we consider the exact critical values relating to the Wilcoxon rank-sum and signed rank statistics. These are mainly adapted from the detailed tables provided by Wilcoxon, Katti and Wilcox (1968). Section 3 is devoted to the tables for the Kruskal–Wallis several sample rank-sum test statistics; these tables are adapted from Iman, Quade and Alexander (1975). Section 4 deals with the Friedman rank test related to the complete randomized block design. A greater part of these tabulated entries is adapted from Quade (1972) with some additional adaptations from Odeh (1977), Kendall and Smith (1939), Michaelis (1971) and Hollander and Wolfe (1973). Section 5 is devoted to the Kolmogorov-Smirnov type statistics, and the tabular entries are from Millar (1956), Birnbaum and Hall (1960) and others. Section 6 deals with the rank correlation statistics, and the critical values have kindly been provided by Professor Dana Quade and are mainly adapted from Owen (1962) and Otten (1973). Some additional tables, mainly adapted from Kiefer (1959), Schey (1977) and DeLong (1980, 1981), relate to the asymptotic cases, and are reported in Section 7.

We acknowledge with thanks the kind permissions granted by these authors

The work of Krishnaiah was sponsored by the Air Force Office of Scientific Research under Contract F4920-82-K-0001. Reproduction in whole or in part is permitted for any purpose of the United States Government.

to adopt some of their entries in this set of tables presented in this chapter. These adaptations from the original sources are made with the kind permissions granted by the following Societies and Publishers: Academic Press, American Mathematical Society, American Statistical Association, Biometrika Trustees, Institute of Mathematical Statistics, Marcel Dekker Inc. and Wiley.

2. Tables for the Wilcoxon statistics

For n observations $X_1, \dots, X_n$, the (one-sample) Wilcoxon signed rank statistic is defined as

$$W_n^* = (\text{sign } X_1)R_1^+ + \cdots + (\text{sign } X_n)R_n^+ \tag{2.1}$$

where R_i^+ is the rank of $|X_i|$ among $|X_1|, \dots, |X_n|$, for $i = 1, \dots, n$; by virtue of the assumed continuity of the underlying distribution, ties are neglected with probability 1. Note that under the null hypothesis that the distribution of X_i is symmetric about 0, W_n^* is also symmetrically distributed around 0. The mean of W_n^* is therefore 0 and its variance is equal to $n(n+1)(2n+1)/6$. For large values of n, one may use the normalized statistic

$$Z_n = W_n^*/\sqrt{n(n+1)(2n+1)/6} \tag{2.2}$$

and use the standard normal distribution tables for computing the critical values. In fact, this approximation works out quite well for n greater than 35. For this reason, we shall consider only the range: $5 \leq n \leq 35$. Note that if c_i is equal to 1 or 0 according as X_i is nonnegative or not, and if

$$W_n = c_1R_1^+ + \cdots + c_nR_n^+ , \tag{2.3}$$

then,

$$W_n^* = 2W_n - n(n+1)/2 . \tag{2.4}$$

Thus, for every k (≥ 0), under the null hypothesis,

$$P\{W_n^* \leq 2k - n(n+1)/2\} = P\{W_n^* \geq n(n+1)/2 - 2k\} = P\{W_n \leq k\} . \tag{2.5}$$

Consequently, specifying the value of k along with the probability level in (2.5) will suffice, for both the one and two-sided critical values. This is done for the probability levels (less than or equal to) 0.05, 0.025, 0.01 and 0.005, so that in the two-sided case, we have the corresponding levels 0.10, 0.05, 0.02 and 0.01. These entries for k and the corresponding probability levels in (2.5) are adapted from Wilcoxon, Katti and Wilcox (1968), and given in Table 2.1. Note that W_n has a discrete distribution with the mass-points $0, 1, \dots, n(n+1)/2$ and

Table 2.1
Selected critical values for the Wilcoxon signed rank statistics

n	k, $P\{W_n \leqslant k\}$			
5				0, 0.0313
6			0, 0.0156	1, 0.0313
7		0, 0.0078	2, 0.0234	3, 0.0391
8	0, 0.0039	1, 0.0078	3, 0.0195	5, 0.0391
9	1, 0.0039	3, 0.0098	5, 0.0195	8, 0.0488
10	3, 0.0049	5, 0.0098	8, 0.0244	10, 0.0420
11	5, 0.0049	7, 0.0093	10, 0.0210	13, 0.0415
12	7, 0.0046	9, 0.0081	13, 0.0212	17, 0.0461
13	9, 0.0040	12, 0.0085	17, 0.0239	21, 0.0471
14	12, 0.0043	15, 0.0083	21, 0.0247	25, 0.0453
15	15, 0.0042	19, 0.0090	25, 0.0240	30, 0.0473
16	19, 0.0046	23, 0.0091	29, 0.0222	35, 0.0467
17	23, 0.0047	27, 0.0087	34, 0.0224	41, 0.0492
18	27, 0.0045	32, 0.0091	40, 0.0241	47, 0.0494
19	32, 0.0047	37, 0.0090	46, 0.0247	53, 0.0478
20	37, 0.0047	43, 0.0096	52, 0.0242	60, 0.0487
21	42, 0.0045	49, 0.0097	58, 0.0230	67, 0.0479
22	48, 0.0046	55, 0.0095	66, 0.0250	75, 0.0492
23	54, 0.0046	62, 0.0098	73, 0.0242	83, 0.0490
24	61, 0.0048	69, 0.0097	81, 0.0245	91, 0.0475
25	68, 0.0048	76, 0.0094	89, 0.0241	100, 0.0479
26	75, 0.0047	84, 0.0095	98, 0.0247	110, 0.0497
27	83, 0.0048	93, 0.0100	107, 0.0246	119, 0.0477
28	91, 0.0048	101, 0.0096	116, 0.0239	130, 0.0496
29	100, 0.0049	110, 0.0095	126, 0.0240	140, 0.0482
30	109, 0.0050	120, 0.0098	137, 0.0249	151, 0.0481
31	118, 0.0049	130, 0.0099	147, 0.0239	163, 0.0491
32	127, 0.0047	140, 0.0097	159, 0.0249	175, 0.0492
33	138, 0.0049	151, 0.0099	170, 0.0242	187, 0.0485
34	148, 0.0048	162, 0.0098	182, 0.0242	200, 0.0488
35	159, 0.0048	174, 0.0100	195, 0.0247	213, 0.0484

the probability masses are multiples of 2^{-n}. As such, for small values of n, we may not have entries for all values of α (viz., 0.005, 0.01, 0.025 and 0.05). For example, for $n = 5$, the smallest nonzero significance level is 0.0313 corresponding to $k = 0$, while for $k = 1$, the significance value 0.0625 is greater than 0.05. Thus, some caution must be used in choosing a level of significance when n is small. It is recommended that instead of the conventional level, one should use the exact level provided by the tables. For $n = 35$, using continuity-correction and normal approximation, we have for $k = 213$, $P\{W_n \leqslant k\} = 0.0480$ compared to the exact value 0.0484. Thus, the error is less than 0.0005. This picture holds for all $n \geqslant 36$.

For two samples of sizes m and n, we denote the sample observations by $X_1, \ldots, X_m$ and $Y_1, \ldots, Y_n$, respectively. Let $R_1, \ldots, R_m$ be the ranks of $X_1, \ldots, X_m$ in the combined sample of size $N = m + n$; without any loss of generality, we take $m \leqslant n$. Here also, ties among the observations are neglected. Then, the Wilcoxon (Mann-Whitney-) statistic is based on the rank-sum

$$W_{mn} = R_1 + \cdots + R_m . \tag{2.6}$$

Under the null hypothesis (H_0) that the two samples are drawn independently from the same population, we have $EW_{mn} = m(N+1)/2$ and $\mathrm{Var}(W_{mn}) = mn(N+1)/12$. Also, for large values of (m, n), under H_0,

$$Z_n = (W_{mn} - m(N+1)/2)/\sqrt{mn(N+1)/12} \tag{2.7}$$

has closely the standard normal distribution, and this provides good approximation to the critical values of W_{mn}. Since the distribution of W_{mn} is discrete, for small values of (m, n), the following table adapted from Wilcoxon, Katti and Wilcox (1968) is recommended for choosing appropriate significance levels and the corresponding critical values. Here, for unequal sample sizes, the distribution of $W_{mn} - m(N+1)/2$ may not be symmetric about 0. As such, corresponding to each critical level, there is a need to provide both the lower and upper critical values, so that a two-sided test can be made using both of these values, while, for the one-sided test, the appropriate one should be used. As in the case of Table 2.1, here also, we confine ourselves to probability levels 0.005, 0.01, 0.025 and 0.05 (for the one-sided case). For each combination of (m, n), the entries in Table 2.2 relate to w_1 and w_2, such that

$$P\{W_{mn} \leq w_1\} = P\{W_{mn} \geq w_2\} = \alpha_{mn}(\leq \alpha) \tag{2.8}$$

Table 2.2
Critical values and probability levels for the rank sum statistics

n	m	w_1, w_2 and α_{mn}			
3	3				6, 15 (0.0500)
4	3				6, 18 (0.0286)
	4			10, 26 (0.0143)	11, 25 (0.0286)
5	3			6, 21 (0.0179)	7, 20 (0.0357)
	4		10, 30 (0.0079)	11, 29 (0.0159)	12, 28 (0.0317)
	5	15, 40 (0.0040)	16, 39 (0.0079)	17, 38 (0.0159)	19, 36 (0.0476)
6	3			7, 23 (0.0238)	8, 22 (0.0476)
	4	10, 34 (0.0048)	11, 33 (0.0095)	12, 32 (0.0190)	13, 31 (0.0333)
	5	16, 44 (0.0043)	17, 43 (0.0087)	18, 42 (0.0152)	20, 40 (0.0411)
	6	23, 55 (0.0043)	24, 54 (0.0076)	26, 52 (0.0206)	28, 50 (0.0465)
7	3		6, 27 (0.0083)	7, 26 (0.0167)	8, 25 (0.0333)
	4	10, 38 (0.0030)	11, 37 (0.0061)	13, 35 (0.0212)	14, 34 (0.0364)
	5	16, 49 (0.0025)	18, 47 (0.0088)	20, 45 (0.0240)	21, 44, (0.0366)
	6	24, 60 (0.0040)	25, 59 (0.0070)	27, 57 (0.0175)	29, 55 (0.0367)
	7	32, 73 (0.0035)	34, 71 (0.0087)	36, 69 (0.0189)	39, 66 (0.0487)
8	3		6, 30 (0.0061)	8, 28 (0.0242)	9, 27 (0.0424)
	4	11, 41 (0.0040)	12, 40 (0.0081)	14, 38 (0.0242)	15, 37 (0.0364)
	5	17, 53 (0.0031)	19, 51 (0.0093)	21, 49 (0.0225)	23, 47 (0.0466)
	6	25, 65 (0.0040)	27, 63 (0.0100)	29, 61 (0.0213)	31, 59 (0.0406)
	7	34, 78 (0.0047)	35, 77 (0.0070)	38, 74 (0.0200)	41, 71 (0.0469)

Table 2.2 (continued)

n	m	w_1, w_2 and α_{mn}			
8	8	43, 93 (0.0035)	45, 91 (0.0074)	49, 87 (0.0249)	51, 85 (0.0415)
9	3	6, 33 (0.0045)	7, 32 (0.0091)	8, 31 (0.0182)	10, 29 (0.0500)
	4	11, 45 (0.0028)	13, 43 (0.0098)	14, 42 (0.0168)	16, 40 (0.0378)
	5	18, 57 (0.0035)	20, 55 (0.0095)	22, 53 (0.0210)	24, 51 (0.0415)
	6	26, 70 (0.0038)	28, 68 (0.0088)	31, 65 (0.0248)	33, 63 (0.0440)
	7	35, 84 (0.0039)	37, 82 (0.0082)	40, 79 (0.0209)	43, 76 (0.0454)
	8	45, 99 (0.0039)	47, 97 (0.0076)	51, 93 (0.0232)	54, 90 (0.0464)
	9	56, 115 (0.0039)	59, 112 (0.0094)	62, 109 (0.0200)	66, 105 (0.0470)
10	3	6, 36 (0.0035)	7, 35 (0.0070)	9, 33 (0.0245)	10, 32 (0.0385)
	4	12, 48 (0.0040)	13, 47 (0.0070)	15, 45 (0.0180)	17, 43 (0.0380)
	5	19, 61 (0.0040)	21, 59 (0.0097)	23, 57 (0.0200)	26, 54 (0.0496)
	6	27, 75 (0.0037)	29, 73 (0.0080)	32, 70 (0.0210)	35, 67 (0.0467)
	7	37, 89 (0.0048)	39, 87 (0.0093)	42, 84 (0.0215)	45, 81 (0.0439)
	8	47, 105 (0.0043)	49, 103 (0.0078)	53, 99 (0.0217)	56, 96 (0.0416)
	9	58, 122 (0.0038)	61, 119 (0.0086)	65, 115 (0.0217)	69, 111 (0.0474)
	10	71, 139 (0.0045)	74, 136 (0.0093)	78, 132 (0.0216)	82, 128 (0.0446)
11	3	6, 39 (0.0027)	7, 38 (0.0055)	9, 36 (0.0192)	11, 34 (0.0440)
	4	12, 52 (0.0029)	14, 50 (0.0088)	16, 48 (0.0198)	18, 46 (0.0388)
	5	20, 65 (0.0043)	22, 63 (0.0096)	24, 61 (0.0190)	27, 58 (0.0449)
	6	28, 80 (0.0036)	30, 78 (0.0077)	34, 74 (0.0238)	37, 71 (0.0491)
	7	38, 95 (0.0041)	40, 93 (0.0077)	44, 89 (0.0221)	47, 86 (0.0427)
	8	49, 111 (0.0046)	51, 109 (0.0079)	55, 105 (0.0204)	59, 101 (0.0454)
	9	61, 128 (0.0048)	63, 126 (0.0079)	68, 121 (0.0232)	72, 117 (0.0476)
	10	73, 147 (0.0040)	77, 143 (0.0098)	81, 139 (0.0215)	86, 134 (0.0493)
	11	87, 166 (0.0042)	91, 162 (0.0098)	96, 157 (0.0237)	100, 153 (0.0440)
12	3	7, 41 (0.0044)	8, 40 (0.0088)	10, 38 (0.0242)	11, 37 (0.0352)
	4	13, 55 (0.0038)	15, 53 (0.0099)	17, 51 (0.0209)	19, 49 (0.0390)
	5	21, 69 (0.0047)	23, 67 (0.0097)	26, 64 (0.0242)	28, 62 (0.0409)
	6	30, 84 (0.0048)	32, 82 (0.0091)	35, 79 (0.0207)	38, 76 (0.0415)
	7	40, 100 (0.0049)	42, 98 (0.0085)	46, 94 (0.0225)	49, 91 (0.0416)
	8	51, 117 (0.0048)	53, 115 (0.0079)	58, 110 (0.0237)	62, 106 (0.0489)
	9	63, 135 (0.0046)	66, 132 (0.0092)	71, 127 (0.0245)	75, 123 (0.0477)
	10	76, 154 (0.0045)	79, 151 (0.0084)	84, 146 (0.0213)	89, 141 (0.0465)
	11	90, 174 (0.0043)	94, 170 (0.0094)	99, 165 (0.0219)	104, 160 (0.0454)
	12	105, 195 (0.0041)	109, 191 (0.0086)	115, 185 (0.0225)	120, 180 (0.0444)
13	3	7, 44 (0.0036)	8, 43 (0.0071)	10, 41 (0.0196)	12, 39 (0.0411)
	4	13, 59 (0.0029)	15, 57 (0.0076)	18, 54 (0.0223)	20, 52 (0.0395)
	5	22, 73 (0.0049)	24, 71 (0.0097)	27, 68 (0.0230)	30, 65 (0.0473)
	6	31, 89 (0.0046)	33, 87 (0.0084)	37, 83 (0.0231)	40, 80 (0.0437)
	7	41, 106 (0.0042)	44, 103 (0.0093)	48, 99 (0.0228)	52, 95 (0.0484)
	8	53, 123 (0.0050)	56, 120 (0.0099)	60, 116 (0.0223)	64, 112 (0.0445)
	9	65, 142 (0.0045)	68, 139 (0.0085)	73, 134 (0.0217)	78, 129 (0.0478)
	10	79, 161 (0.0049)	82, 158 (0.0089)	88, 152 (0.0247)	92, 148 (0.0441)
	11	93, 182 (0.0044)	97, 178 (0.0092)	103, 172 (0.0237)	108, 167 (0.0467)
	12	109, 203 (0.0048)	113, 199 (0.0094)	119, 193 (0.0229)	125, 187 (0.0488)
	13	125, 226 (0.0043)	130, 221 (0.0095)	136, 215 (0.0221)	142, 209 (0.0454)
14	3	7, 47 (0.0029)	8, 46 (0.0059)	11, 43 (0.0235)	13, 41 (0.0456)
	4	14, 62 (0.0039)	16, 60 (0.0088)	19, 57 (0.0232)	21, 55 (0.0395)
	5	22, 78 (0.0036)	25, 75 (0.0097)	28, 72 (0.0218)	31, 69 (0.0435)
	6	32, 94 (0.0044)	35, 91 (0.0100)	38, 88 (0.0204)	42, 84 (0.0457)
	7	43, 111 (0.0048)	46, 108 (0.0100)	50, 104 (0.0230)	54, 100 (0.0469)

Table 2.2 (continued)

n	m	w_1, w_2 and α_{mn}			
14	8	54, 130 (0.0041)	58, 126 (0.0098)	62, 122 (0.0211)	67, 117 (0.0475)
	9	67, 149 (0.0043)	71, 145 (0.0096)	76, 140 (0.0228)	81, 135 (0.0478)
	10	81, 169 (0.0044)	85, 165 (0.0093)	91, 159 (0.0242)	96, 154 (0.0478)
	11	96, 190 (0.0045)	100, 186 (0.0090)	106, 180 (0.0221)	112, 174 (0.0477)
	12	112, 212 (0.0046)	116, 208 (0.0087)	123, 201 (0.0232)	129, 195 (0.0475)
	13	129, 235 (0.0046)	134, 230 (0.0097)	141, 223 (0.0241)	147, 217 (0.0472)
	14	147, 259 (0.0046)	152, 254 (0.0093)	160, 246 (0.0249)	166, 240 (0.0469)
15	3	8, 49 (0.0049)	9, 48 (0.0086)	11, 46 (0.0196)	13, 44 (0.0380)
	4	15, 65 (0.0046)	17, 63 (0.0098)	20, 60 (0.0243)	22, 58 (0.0400)
	5	23, 82 (0.0039)	26, 79 (0.0097)	29, 76 (0.0209)	33, 72 (0.0491)
	6	33, 99 (0.0042)	36, 96 (0.0092)	40, 92 (0.0224)	44, 88 (0.0474)
	7	44, 117 (0.0043)	47, 114 (0.0086)	52, 109 (0.0233)	56, 105 (0.0455)
	8	56, 136 (0.0042)	60, 132 (0.0097)	65, 127 (0.0237)	69, 123 (0.0437)
	9	70, 155 (0.0050)	73, 152 (0.0089)	79, 146 (0.0238)	84, 141 (0.0478)
	10	84, 176 (0.0048)	88, 172 (0.0096)	94, 166 (0.0238)	99, 161 (0.0455)
	11	99, 198 (0.0046)	103, 194 (0.0088)	110, 187 (0.0236)	116, 181 (0.0486)
	12	115, 221 (0.0044)	120, 216 (0.0093)	127, 209 (0.0234)	133, 203 (0.0463)
	13	133, 244 (0.0048)	138, 239 (0.0097)	145, 232 (0.0232)	152, 225 (0.0489)
	14	151, 269 (0.0046)	156, 264 (0.0089)	164, 256 (0.0229)	171, 249 (0.0466)
	15	171, 294 (0.0049)	176, 289 (0.0093)	184, 281 (0.0227)	192, 273 (0.0488)
16	4	15, 69 (0.0037)	17, 67 (0.0078)	21, 63 (0.0250)	24, 60 (0.0497)
	5	24, 86 (0.0041)	27, 83 (0.0097)	31, 79 (0.0250)	34, 76 (0.0455)
	6	34, 104 (0.0040)	37, 101 (0.0085)	42, 96 (0.0244)	46, 92 (0.0490)
	7	46, 122 (0.0048)	49, 119 (0.0092)	54, 114 (0.0234)	58, 110 (0.0443)
	8	58, 142 (0.0044)	62, 138 (0.0096)	67, 133 (0.0224)	72, 128 (0.0463)
	9	72, 162 (0.0048)	76, 158 (0.0098)	82, 152 (0.0247)	87, 147 (0.0477)
	10	86, 184 (0.0043)	91, 179 (0.0099)	97, 173 (0.0234)	103, 167 (0.0487)
	11	102, 206 (0.0046)	107, 201 (0.0099)	114, 194 (0.0250)	120, 188 (0.0494)
	12	119, 229 (0.0049)	124, 224 (0.0099)	131, 217 (0.0236)	138, 210 (0.0500)
	13	137, 253 (0.0050)	142, 248 (0.0098)	150, 250 (0.0250)	156, 234 (0.0458)
	14	155, 279 (0.0045)	161, 273 (0.0097)	169, 265 (0.0236)	176, 258 (0.0463)
	15	175, 305 (0.0047)	181, 299 (0.0096)	190, 290 (0.0247)	197, 283 (0.0466)
	16	196, 332 (0.0048)	202, 326 (0.0095)	211, 317 (0.0234)	219, 309 (0.0469)
17	5	25, 90 (0.0043)	28, 87 (0.0096)	32, 83 (0.0238)	35, 80 (0.0425)
	6	36, 180 (0.0049)	39, 105 (0.0099)	43, 101 (0.0219)	47, 97 (0.0433)
	7	47, 128 (0.0043)	51, 124 (0.0097)	56, 119 (0.0236)	61, 114 (0.0497)
	8	60, 148 (0.0045)	64, 144 (0.0095)	70, 138 (0.0247)	75, 133 (0.0487)
	9	74, 169 (0.0046)	78, 165 (0.0091)	84, 159 (0.0223)	90, 153 (0.0476)
	10	89, 191 (0.0047)	93, 187 (0.0088)	100, 180 (0.0230)	106, 174 (0.0465)
	11	105, 214 (0.0047)	110, 209 (0.0096)	117, 202 (0.0235)	123, 196 (0.0453)
	12	122, 238 (0.0046)	127, 233 (0.0092)	135, 225 (0.0238)	142, 218 (0.0486)
	13	140, 263 (0.0046)	146, 257 (0.0098)	154, 249 (0.0240)	161, 242 (0.0472)
	14	159, 289 (0.0045)	165, 283 (0.0093)	174, 274 (0.0242)	182, 266 (0.0500)
	15	180, 315 (0.0050)	186, 309 (0.0099)	195, 300 (0.0243)	203, 292 (0.0485)
	16	201, 343 (0.0049)	207, 337 (0.0093)	217, 327 (0.0243)	225, 319 (0.0471)
	17	223, 372 (0.0047)	230, 365 (0.0098)	240, 355 (0.0243)	249, 346 (0.0493)
18	6	37, 113 (0.0047)	40, 110 (0.0091)	45, 105 (0.0236)	49, 101 (0.0448)
	7	49, 133 (0.0047)	52, 130 (0.0085)	58, 124 (0.0237)	63, 119 (0.0484)
	8	62, 154 (0.0046)	66, 150 (0.0094)	72, 144 (0.0235)	77, 139 (0.0452)
	9	76, 176 (0.0044)	81, 171 (0.0100)	87, 165 (0.0231)	93, 159 (0.0475)
	10	92, 198 (0.0050)	96, 194 (0.0090)	103, 187 (0.0226)	110, 180 (0.0493)

Table 2.2 (continued)

n	m	w_1, w_2 and α_{mn}			
18	11	108, 222 (0.0047)	113, 217 (0.0094)	121, 209 (0.0247)	127, 203 (0.0461)
	12	125, 247 (0.0044)	131, 241 (0.0097)	139, 233 (0.0239)	146, 226 (0.0474)
	13	144, 272 (0.0048)	150, 266 (0.0099)	158, 258 (0.0232)	166, 250 (0.0485)
	14	164, 298 (0.0050)	170, 292 (0.0100)	179, 283 (0.0247)	187, 275 (0.0495)
	15	184, 326 (0.0047)	190, 320 (0.0091)	200, 310 (0.0239)	208, 302 (0.0465)
	16	206, 354 (0.0049)	212, 348 (0.0092)	222, 338 (0.0231)	231, 329 (0.0473)
	17	228, 384 (0.0046)	235, 377 (0.0093)	246, 366 (0.0243)	255, 357 (0.0479)
	18	252, 414 (0.0048)	259, 407 (0.0094)	270, 396 (0.0235)	280, 386 (0.0485)
19	7	50, 139 (0.0042)	54, 135 (0.0090)	60, 129 (0.0238)	65, 124 (0.0471)
	8	64, 160 (0.0047)	68, 156 (0.0093)	74, 150 (0.0224)	80, 144 (0.0475)
	9	79, 182 (0.0050)	83, 178 (0.0093)	90, 171 (0.0239)	96, 165 (0.0474)
	10	94, 206 (0.0045)	99, 201 (0.0093)	107, 193 (0.0250)	113, 187 (0.0472)
	11	111, 230 (0.0047)	116, 225 (0.0092)	124, 217 (0.0233)	131, 210 (0.0468)
	12	129, 255 (0.0048)	134, 250 (0.0090)	143, 241 (0.0240)	150, 224 (0.0463)
	13	148, 281 (0.0049)	154, 275 (0.0099)	163, 266 (0.0247)	171, 258 (0.0497)
	14	168, 308 (0.0050)	174, 302 (0.0096)	183, 293 (0.0230)	192, 284 (0.0489)
	15	189, 336 (0.0050)	195, 330 (0.0093)	205, 320 (0.0235)	214, 311 (0.0482)
	16	211, 365 (0.0050)	218, 358 (0.0100)	228, 348 (0.0239)	237, 339 (0.0474)
	17	234, 395 (0.0050)	241, 388 (0.0097)	252, 377 (0.0243)	262, 367 (0.0499)
	18	258, 426 (0.0050)	265, 419 (0.0094)	277, 407 (0.0246)	287, 397 (0.0490)
	19	283, 458 (0.0050)	291, 450 (0.0099)	303, 438 (0.0248)	313, 428 (0.0482)
20	8	66, 166 (0.0048)	70, 162 (0.0092)	77, 155 (0.0244)	83, 149 (0.0495)
	9	81, 189 (0.0048)	85, 185 (0.0088)	93, 177 (0.0245)	99, 171 (0.0473)
	10	97, 213 (0.0048)	102, 208 (0.0095)	110, 200 (0.0245)	117, 193 (0.0498)
	11	114, 238 (0.0047)	119, 233 (0.0089)	128, 224 (0.0244)	135, 217 (0.0474)
	12	132, 264 (0.0046)	138, 258 (0.0094)	147, 249 (0.0241)	155, 241 (0.0493)
	13	151, 291 (0.0045)	158, 284 (0.0099)	167, 275 (0.0238)	175, 267 (0.0470)
	14	172, 318 (0.0049)	178, 312 (0.0092)	188, 302 (0.0235)	197, 293 (0.0484)
	15	193, 347 (0.0047)	200, 300 (0.0095)	210, 330 (0.0232)	220, 320 (0.0497)
	16	215, 377 (0.0046)	223, 369 (0.0098)	234, 358 (0.0247)	243, 349 (0.0475)
	17	239, 407 (0.0049)	247, 399 (0.0100)	258, 388 (0.0242)	268, 378 (0.0485)
	18	263, 439 (0.0047)	271, 431 (0.0094)	283, 419 (0.0238)	294, 408 (0.0495)
	19	289, 471 (0.0049)	297, 463 (0.0096)	310, 450 (0.0250)	320, 440 (0.0474)
	20	315, 505 (0.0047)	324, 496 (0.0098)	337, 483 (0.0245)	348, 472 (0.0482)
21	9	83, 196 (0.0047)	88, 191 (0.0095)	95, 184 (0.0225)	102, 177 (0.0472)
	10	99, 221 (0.0044)	105, 215 (0.0097)	113, 207 (0.0241)	120, 200 (0.0478)
	11	117, 246 (0.0047)	123, 240 (0.0098)	131, 232 (0.0230)	139, 224 (0.0480)
	12	136, 272 (0.0050)	142, 266 (0.0099)	151, 257 (0.0242)	159, 249 (0.0481)
	13	155, 300 (0.0047)	162, 293 (0.0099)	171, 284 (0.0231)	180, 275 (0.0481)
	14	176, 328 (0.0048)	183, 321 (0.0098)	193, 311 (0.0239)	202, 302 (0.0480)
	15	198, 357 (0.0050)	205, 350 (0.0097)	216, 339 (0.0247)	225, 330 (0.0478)
	16	220, 388 (0.0046)	228, 380 (0.0096)	239, 369 (0.0235)	249, 359 (0.0475)
	17	244, 419 (0.0047)	252, 411 (0.0095)	264, 399 (0.0242)	274, 389 (0.0473)
	18	269, 451 (0.0048)	277, 443 (0.0094)	290, 430 (0.0247)	301, 419 (0.0499)
22	9	85, 203 (0.0045)	90, 198 (0.0089)	98, 190 (0.0231)	105, 183 (0.0471)
	10	102, 228 (0.0047)	108, 222 (0.0099)	116, 214 (0.0237)	123, 207 (0.0459)
	11	120, 254 (0.0047)	126, 248 (0.0096)	135, 239 (0.0240)	143, 231 (0.0486)
	12	139, 281 (0.0048)	145, 275 (0.0092)	155, 265 (0.0242)	163, 257 (0.0471)
	13	159, 309 (0.0048)	166, 302 (0.0098)	176, 292 (0.0243)	185, 283 (0.0491)
	14	180, 338 (0.0048)	187, 331 (0.0094)	198, 320 (0.0243)	207, 311 (0.0475)
	15	202, 368 (0.0047)	210, 360 (0.0099)	221, 349 (0.0243)	231, 339 (0.0492)

Table 2.2 (continued)

n	m	w_1, w_2 and α_{mn}			
22	16	225, 399 (0.0047)	233, 391 (0.0095)	245, 379 (0.0242)	255, 369 (0.0476)
	17	250, 430 (0.0050)	258, 422 (0.0099)	270, 410 (0.0241)	281, 399 (0.0490)
23	10	105, 235 (0.0049)	111, 229 (0.0100)	119, 221 (0.0233)	127, 213 (0.0482)
	11	123, 262 (0.0047)	129, 256 (0.0093)	139, 246 (0.0250)	147, 238 (0.0490)
	12	142, 290 (0.0046)	149, 283 (0.0096)	159, 273 (0.0243)	168, 264 (0.0496)
	13	163, 318 (0.0049)	170, 311 (0.0098)	180, 301 (0.0236)	190, 291 (0.0500)
	14	184, 348 (0.0047)	192, 340 (0.0100)	203, 329 (0.0247)	212, 320 (0.0471)
	15	207, 378 (0.0049)	214, 371 (0.0092)	226, 359 (0.0239)	236, 349 (0.0474)
	16	230, 410 (0.0047)	238, 402 (0.0093)	251, 389 (0.0248)	261, 379 (0.0476)
24	10	107, 243 (0.0045)	113, 237 (0.0091)	122, 228 (0.0230)	130, 220 (0.0465)
	11	126, 270 (0.0047)	132, 264 (0.0091)	142, 254 (0.0237)	151, 245 (0.0495)
	12	146, 298 (0.0049)	153, 291 (0.0100)	163, 281 (0.0243)	172, 272 (0.0486)
	13	167, 327 (0.0050)	174, 320 (0.0098)	185, 309 (0.0247)	194, 300 (0.0476)
	14	188, 358 (0.0046)	196, 350 (0.0096)	208, 338 (0.0250)	218, 328 (0.0498)
	15	211, 389 (0.0047)	219, 381 (0.0094)	231, 369 (0.0235)	242, 358 (0.0486)
	16	235, 421 (0.0047)	244, 412 (0.0099)	256, 400 (0.0238)	267, 389 (0.0476)
25	10	110, 250 (0.0047)	116, 244 (0.0099)	126, 243 (0.0248)	134, 226 (0.0486)
	11	129, 278 (0.0047)	136, 271 (0.0099)	146, 261 (0.0246)	155, 252 (0.0499)
	12	149, 307 (0.0047)	156, 300 (0.0094)	167, 289 (0.0243)	176, 280 (0.0475)
	13	170, 337 (0.0047)	178, 329 (0.0098)	189, 318 (0.0240)	199, 308 (0.0485)
	14	193, 367 (0.0050)	200, 360 (0.0093)	212, 348 (0.0236)	223, 337 (0.0492)
	15	216, 399 (0.0049)	224, 391 (0.0095)	237, 378 (0.0248)	248, 367 (0.0499)
	16	240, 432 (0.0047)	249, 423 (0.0098)	262, 410 (0.0243)	273, 399 (0.0476)
26	11	132, 286 (0.0047)	139, 279 (0.0096)	149, 269 (0.0235)	158, 260 (0.0468)
	12	153, 315 (0.0050)	160, 308 (0.0097)	171, 297 (0.0243)	181, 287 (0.0498)
	13	174, 346 (0.0048)	182, 338 (0.0098)	193, 327 (0.0233)	204, 316 (0.0493)
	14	197, 377 (0.0049)	205, 369 (0.0097)	217, 357 (0.0239)	228, 346 (0.0488)
	15	220, 410 (0.0047)	229, 401 (0.0097)	242, 388 (0.0244)	253, 377 (0.0482)
	16	245, 443 (0.0048)	254, 434 (0.0096)	268, 420 (0.0249)	279, 409 (0.0475)
27	11	135, 294 (0.0047)	142, 287 (0.0094)	153, 276 (0.0243)	162, 267 (0.0473)
	12	156, 324 (0.0048)	163, 317 (0.0092)	175, 305 (0.0243)	185, 295 (0.0488)
	13	178, 355 (0.0049)	186, 347 (0.0097)	198, 335 (0.0243)	208, 325 (0.0471)
	14	201, 387 (0.0049)	209, 379 (0.0094)	222, 336 (0.0242)	233, 355 (0.0483)
	15	225, 420 (0.0049)	234, 411 (0.0098)	247, 398 (0.0241)	259, 386 (0.0493)
	16	250, 454 (0.0048)	259, 445 (0.0094)	273, 431 (0.0239)	285, 419 (0.0475)
28	12	159, 333 (0.0046)	167, 325 (0.0095)	179, 313 (0.0243)	189, 303 (0.0479)
	13	182, 364 (0.0050)	190, 356 (0.0097)	202, 344 (0.0236)	213, 333 (0.0479)
	14	205, 397 (0.0048)	214, 388 (0.0098)	227, 375 (0.0245)	238, 364 (0.0479)
	15	230, 430 (0.0050)	239, 421 (0.0099)	252, 408 (0.0237)	264, 396 (0.0477)
29	12	163, 341 (0.0049)	171, 333 (0.0098)	183, 321 (0.0243)	194, 310 (0.0500)
	13	186, 373 (0.0050)	194, 365 (0.0097)	207, 352 (0.0246)	218, 341 (0.0487)
	14	209, 407 (0.0047)	218, 398 (0.0095)	232, 384 (0.0248)	243, 373 (0.0474)
	15	234, 441 (0.0048)	243, 432 (0.0093)	258, 417 (0.0248)	270, 405 (0.0487)
30	13	189, 383 (0.0047)	198, 374 (0.0096)	211, 361 (0.0240)	223, 349 (0.0494)
	14	213, 417 (0.0047)	223, 407 (0.0099)	236, 394 (0.0235)	249, 381 (0.0496)
	15	239, 451 (0.0050)	248, 442 (0.0095)	263, 427 (0.0245)	276, 414 (0.0497)

For m and n larger than the tabulated entries, normal approximation holds quite well. Some of the extreme unequal sample size combinations have been left out here; these may be obtained from Wilcoxon, Katti and Wilcox (1968).

where α corresponds to the desired probability level and α_{mn} the actual one, being the largest possible value among all such solutions. The entries in Tables 2.1 and 2.2 are reproduced from Wilcoxson, Katti and Wilcox (1968) with the kind permission of the American Cyanamid Company and the Department of Statistics, Florida State University.

3. Tables for the Kruskal–Wallis statistics

For k (≥ 2) samples of sizes $n_1, \ldots, n_k$, respectively, let $R_{i1}, \ldots, R_{in_i}$ be the ranks of the ith sample observations in the combined sample of size $n =$

Table 3.1[a]
Selected critical values for the three sample Kruskal–Wallis statistics

n_1	n_2	n_3	h	$P(H \geq h)$		
2	2	2	4.571	0.0667		
3	2	2	4.714	0.0476		
3	3	2	5.139	0.0607		
3	3	3	5.600	0.0500		
4	2	1	4.821	0.0571		
4	2	2	5.125	0.0524		
4	3	1	5.208	0.0500		
4	3	2	5.400	0.0508		
4	3	3	5.727	0.0505	6.745	0.0100
4	4	1	4.867	0.0540	6.667	0.0095
4	4	2	5.236	0.0521	6.873	0.0108
4	4	3	5.576	0.0507	7.136	0.0107
4	4	4	5.692	0.0487	7.538	0.0107
5	2	1	5.000	0.0476		
5	2	2	5.040	0.0556	6.533	0.0079
5	3	1	4.871	0.0516	6.400	0.0119
5	3	2	5.251	0.0492	6.822	0.0103
5	3	3	5.515	0.0507	7.079	0.0087
5	4	1	4.860	0.0556	6.840	0.0111
5	4	2	5.268	0.0505	7.118	0.0101
5	4	3	5.631	0.0503	7.445	0.0097
5	4	4	5.618	0.0503	7.760	0.0095
5	5	1	4.909	0.0534	6.836	0.0108
5	5	2	5.246	0.0511	7.269	0.0103
5	5	3	5.626	0.0508	7.543	0.0102
5	5	4	5.643	0.0502	7.823	0.0098
5	5	5	5.660	0.0509	7.980	0.0105
6	2	1	4.822	0.0478		
6	3	1	4.855	0.0500	6.582	0.0119
6	3	2	5.227	0.0520	6.970	0.0091
6	3	3	5.615	0.0497	7.192	0.0102
6	4	1	4.947	0.0468	7.083	0.0104
6	4	2	5.263	0.0502	7.212	0.0108

Table 3.1. (continued)

n_1	n_2	n_3	h	$P(H \geqslant h)$		
6	4	3	5.604	0.0504	7.467	0.0101
6	4	4	5.667	0.0505	7.724	0.0101
6	5	1	4.836	0.0509	6.997	0.0101
6	5	2	5.319	0.0506	7.299	0.0102
6	5	3	5.600	0.0500	7.560	0.0102
6	5	4	5.661	0.0499	7.936	0.0100
6	5	5	5.729	0.0497	8.012	0.0100
6	6	1	4.857	0.0511	7.066	0.0103
6	6	2	5.410	0.0499	7.410	0.0102
6	6	3	5.625	0.0500	7.725	0.0099
6	6	4	5.721	0.0501	8.000	0.0100
6	6	5	5.765	0.0499	8.119	0.0100
6	6	6	5.719	0.0502	8.187	0.0102
7	7	7	5.766	0.0506	8.334	0.0101
8	8	8	5.805	0.0497	8.435	0.0101
Asymptotic value			5.991	0.0500	9.210	0.0100

[a]The entries in Tables 3.1–3.3 are reproduced from Iman, Quade and Alexander (1975) with the kind permission of the American Mathematical Society.

$n_1 + \cdots + n_k$, for $i = 1, \ldots, k$. Then, the Kruskal–Wallis statistic may be defined as

$$H = \frac{12}{n(n+1)} \left[\sum_{i=1}^{k} n_i^{-1} \left\{ \sum_{j=1}^{n_i} R_{ij} - n_i(n+1)/2 \right\}^2 \right]. \tag{3.1}$$

For $k = 2$, H in (3.1) reduces to Z_n^2, where Z_n is defined by (2.7).

Under the null hypothesis that all the k samples have been drawn independently from a common population, the (exact) distribution of H is generated by the $n!$ equally likely permutations of the ranks among themselves (for tied observations, the modifications are apparent). For large values of $n_1, \ldots, n_k$, this distribution can safely be approximated by a chi square distribution with $k-1$ degrees of freedom. The enumeration of the exact (permutation) distribution of H is by no means simple. For the special cases of $k = 3$, 4 and 5, and for some specific (small) values of the n_i, this distribution has been extensively tabulated by Iman, Quade and Alexander (1975). With the kind permission of the authors and the publishers, we have chosen some selected critical values, and these are presented in Tables 3.1 (three sample case), 3.2 (four sample case) and 3.3 (five sample case). As in Section 2, we have chosen the entries for which the exact probability levels are close to 0.05 and 0.01. These related to

$$P\{H \geqslant h\} \quad \text{for given } n_1, \ldots, n_k. \tag{3.2}$$

Table 3.2
Selected critical values for the four sample Kruskal–Wallis statistics

n_1	n_2	n_3	n_4	h	$P\{H \geqslant h\}$		
3	2	2	2	6.333	0.0476	7.133	0.0079
3	3	2	1	6.156	0.0560	7.044	0.0107
3	3	2	2	6.527	0.0492	7.636	0.0100
3	3	3	1	6.600	0.0493	7.400	0.0086
3	3	3	2	6.727	0.0495	8.015	0.0096
3	3	3	3	6.879	0.0502	8.436	0.0108
4	2	2	1	6.000	0.0566	7.000	0.0095
4	2	2	2	6.545	0.0492	7.391	0.0089
4	3	1	1	6.178	0.0492	7.067	0.0095
4	3	2	1	6.309	0.0494	7.455	0.0098
4	3	2	2	6.621	0.0495	7.871	0.0100
4	3	3	1	6.545	0.0495	7.758	0.0097
4	3	3	2	6.782	0.0501	8.333	0.0099
4	3	3	3	6.967	0.0503	8.659	0.0099
4	4	1	1	5.945	0.0495	7.500	0.0114
4	4	2	1	6.364	0.0500	7.886	0.0102
4	4	2	2	6.731	0.0487	8.308	0.0102
4	4	3	1	6.635	0.0498	8.218	0.0103
4	4	3	2	6.874	0.0498	8.621	0.0100
4	4	3	3	7.038	0.0499	8.867	0.0100
4	4	4	1	6.725	0.0498	8.571	0.0101
4	4	4	2	6.957	0.0496	8.857	0.0101
4	4	4	3	7.129	0.0502	9.075	0.0100
4	4	4	4	7.213	0.0507	9.287	0.0100
Asymptotic value				7.815	0.0500	11.345	0.0100

Table 3.3
Selected critical values for the five sample Kruskal–Wallis statistics

n_1	n_2	n_3	n_4	n_5	h	$P\{H \geqslant h\}$		
2	2	2	2	2	7.418	0.0487	8.291	0.0095
3	2	2	1	1	7.200	0.0500	7.600	0.0079
3	2	2	2	1	7.309	0.0489	8.127	0.0094
3	2	2	2	2	7.667	0.0508	8.682	0.0096
3	3	2	1	1	7.200	0.0500	8.055	0.0102
3	3	2	2	1	7.591	0.0492	8.576	0.0098
3	3	2	2	2	7.897	0.0505	9.103	0.0101
3	3	3	1	1	7.515	0.0538	8.424	0.0091
3	3	3	2	1	7.769	0.0489	9.051	0.0098
3	3	3	2	2	8.044	0.0492	9.505	0.0100
3	3	3	3	1	7.956	0.0505	9.451	0.0100
3	3	3	3	2	8.171	0.0504	9.848	0.0101
3	3	3	3	3	8.333	0.0496	10.200	0.0099
Asymptotic value					9.488	0.0500	13.277	0.0100

Without any loss of generality, one can take $n_1 \geq n_2 \geq n_3 \geq n_4$, and in the tables, this has been adapted. In general, the exact values are somewhat smaller than the asymptotic values (provided by the chi square distribution), so that the use of the asymptotic critical levels may lead to somewhat conservative tests.

4. Tables for the Friedman rank statistic

For $n(\geq 2)$ blocks of $p(\geq 2)$ plots each, let r_{ij}, $j = 1, \ldots, p$, be the within-block ranks of the p observations in the ith block, for $i = 1, \ldots, n$. Let $R_j = \sum_{i=1}^{n} r_{ij}$ be the rank-sum for the jth treatment, for $j = 1, \ldots, p$. Then, the Friedman rank statistic is defined by

$$\chi_F^2 = 12\left(\sum_{j=1}^{p} (R_j - n(p+1)/2)^2/\{np(p+1)\}\right). \tag{4.1}$$

Table 4.1
Selected critical values for the Friedman rank statistics

p	n	$c_{n,p}^{(\alpha)}$ and $\alpha_{n,p}$			
3	3	6.000	(0.0278)		
	4	6.500	(0.0417)	8.000	(0.0046)
	5	6.400	(0.0394)	8.400	(0.0085)
	6	7.000	(0.0289)	9.000	(0.0081)
	7	7.143	(0.0272)	8.857	(0.0084)
	8	6.250	(0.0469)	9.000	(0.0099)
	9	6.222	(0.0476)	8.667	(0.0103)
	10	6.200	(0.0456)	9.600	(0.0075)
	11	6.546	(0.0435)	9.456	(0.0065)
3	12	6.167	(0.0510)	8.667	(0.0107)
	13	6.000	(0.0501)	9.385	(0.0087)
	14	6.143	(0.0480)	9.000	(0.0101)
	15	6.400	(0.0468)	8.933	(0.0097)
	asymptotic	5.991	(0.0500)	9.210	(0.0100)
4	3	7.400	(0.0330)	9.000	(0.0017)
	4	7.800	(0.0364)	9.600	(0.0067)
	5	7.800	(0.0443)	9.960	(0.0087)
	6	7.600	(0.0433)	10.200	(0.0096)
	7	7.800	(0.0413)	10.543	(0.0090)
	8	7.650	(0.0488)	10.500	(0.0094)
	asymptotic	7.815	(0.0500)	11.345	(0.0100)
5	3	8.53	(0.0455)	10.13	(0.0078)
	4	8.8	(0.0489)	11.2	(0.0079)
	5	8.96	(0.049)	11.52	(0.010)
	6	9.067	(0.049)	11.867	(0.0099)
	7	9.143	(0.049)	12.114	(0.0100)
	8	9.200	(0.050)	12.300	(0.0099)
	asymptotic	9.488	(0.050)	13.277	(0.0100)

Table 4.1 (continued)

p	n	$c_{n,p}^{(\alpha)}$ and $\alpha_{n,p}$			
6	3	9.857	(0.046)	11.762	(0.0095)
	4	10.286	(0.047)	12.571	(0.0109)
	5	10.486	(0.048)	13.229	(0.0099)
	6	10.571	(0.049)	13.619	(0.0097)
	asymptotic	11.071	(0.050)	15.086	(0.0100)

Note that for various combinations of (p, n), the entries were computed by different workers and they have different degrees of accuracy; viz, the entries (5, 4) and (5, 5). We may note that across the table, the actual right hand tails of the exact distribution of χ_F^2 is dominated by that of the chi-square distribution (with the appropriate degrees of freedom), so that the use of the asymptotic critical values usually results in a more conservative test.

By virtue of the assumed continuity of the distributions of the responses, ties among the observations are neglected, with probability 1. Under the null hypothesis of interchangeability (within each block), the distribution of χ_F^2 is generated by the $(p!)^n$ equally likely within-block permutations of the ranks (over $1, \ldots, p$). For specific smaller values of n (and p) the exact critical values of χ_F^2 (though different from a preassigned level) have been tabulated by various workers, and some of these are reproduced here. These entries are mostly taken from Quade (1972), Odeh (1977), Kendall and Smith (1939), Michaelis (1971) and Hollander and Wolfe (1973). Specifically, for a given level of significance α $(0<\alpha<1)$ and (n, p), a critical value $c_{n,p}^{(\alpha)}$ is given for which, for every admissible $d > c_{n,p}^{(\alpha)}$,

$$P\{\chi_F^2 \geqslant c_{n,p}^{(\alpha)}\} = \alpha_{n,p} \leqslant \alpha < P\{\chi_F^2 \geqslant d\}. \tag{4.2}$$

The tabulated entries relate to $c_{n,p}^{(\alpha)}$ and $\alpha_{n,p}$ for $\alpha = 0.05$ and 0.01. For large values of n, under the null hypothesis, χ_F^2 has closely the central chi-square distribution with $p-1$ degrees of freedom, so that the approximate critical values can be obtained from the chi-square distributional tables.

5. Tables for the Kolmogorov–Smirnov type statistics

For n observations drawn from a continuous distribution F, defined on the real line $\mathbb{R}$, let $F_n(x) = n^{-1}$ (number of observations with values $\leqslant x$), $x \in \mathbb{R}$, be the sample (empirical) distribution function. Then, the one-sample Kolmogorov–Smirnov statistics are defined as

$$D_n^+ = \sup\{F_n(x) - F(x) : x \in R\} \tag{5.1}$$

and

$$D_n = \sup\{|F_n(x) - F(x)| : x \in R\} . \tag{5.2}$$

The probability distribution of D_n, for small values of n, has been tabulated by Birnbaum (1952). For D_n^+, some tabulations of the critical values are due to Millar (1956). We adopt these tables to provide the critical values for D_n^+ and D_n, for some specific values of n and for significance levels close to 0.05 and 0.01. Note that for significance levels not so large (i.e., less than 0.1), we have

$$\alpha_n = P\{D_n \geq \lambda\} \doteqdot 2P\{D_n^+ \geq \lambda\} . \tag{5.3}$$

Hence, we only provide the critical values for D_n for α_n close to 0.1, 0.05, 0.02 and 0.01. These are given in Table 5.1. In passing, we may note that for large n,

Table 5.1
Table for the critical values of the one-sample Kolmogorov–Smirnov statistic D_n

n	λ and α_n			
5			4/5 (0.031)	5/5 (0.0006)
6		4/6 (0.066)		5/6 (0.004)
7			5/7 (0.011)	6/7 (0.0004)
8			5/8 (0.023)	6/8 (0.0015)
9			5/9 (0.039)	6/9 (0.0039)
10		5/10(0.059)		6/10(0.0078)
11		5/11(0.083)	6/11(0.014)	7/11(0.0013)
12	5/12(0.109)		6/12(0.021)	7/12(0.0027)
13			6/13(0.031)	7/13(0.0047)
14			6/14(0.042)	7/14(0.0075)
15		6/15(0.055)	7/15(0.011)	8/15(0.0016)
16		6/16(0.069)	7/16(0.016)	8/16(0.0026)
17		6/17(0.085)	7/17(0.021)	8/17(0.0040)
18	6/18(0.102)		7/18(0.028)	8/18(0.0058)
19	6/19(0.120)		7/19(0.035)	8/19(0.0080)
20			7/20(0.043)	8/20(0.0107)
21		7/21(0.052)	8/21(0.014)	9/21(0.0030)
22		7/22(0.062)	8/22(0.018)	9/22(0.0041)
23		7/23(0.072)	8/23(0.022)	9/23(0.0054)
24		7/24(0.083)	8/24(0.027)	9/24(0.0070)
25		7/25(0.094)	8/25(0.032)	9/25(0.0089)
26	7/26(0.106)	8/26(0.037)	9/26(0.011)	10/26(0.0027)
27	7/27(0.119)	8/27(0.043)	9/27(0.013)	10/27(0.0035)
28		8/28(0.050)	9/28(0.016)	10/28(0.0045)
29		8/29(0.057)	9/29(0.020)	10/29(0.0056)
30		8/30(0.064)	9/30(0.023)	10/30(0.0068)

for every $t \geq 0$,

$$P\{n^{1/2}D_n^+ \geq t\} \to e^{-2t^2}, \quad (5.4)$$

$$P\{n^{1/2}D_n \geq t\} \to 2(e^{-2t^2} - e^{-8t2} + e^{-18t2} - \cdots) \quad (5.5)$$

and, actually, the right hand sides of (5.4) and (5.5) provide upper bounds for any finite sample size. These approximations are quite good for $n \geq 31$. Hence, we provide the entries only for $n \leq 30$.

For two samples of equal sizes n, if F_n and G_n stand for the empirical distributions, then one may define the one and two-sided Kolmogorov–Smirnov statistics as in (5.1) and (5.2) with F being replaced by G_n. In this case, (5.4) and (5.5) hold when we replace $n^{1/2}D_n^+$ and $n^{1/2}D_n$ by $(n/2)^{1/2}D_n^+$ and $(n/2)^{1/2}D_n$, respectively. For this two-sample case, Birnbaum and Hall (1960) have tabulated the distributions for specific values of n, and we adopt their tables to provide the critical values for specific level of significances. For the case of more than two samples, we refer to Section 7 for some tabulation of the asymptotic critical values, mostly due to Kiefer (1959). In Tables 5.1 and 5.2, the entries refer to λ and α_n, for (5.3) in the one and two-sample cases.

Table 5.2
Table for the critical values of two-sample D_n^+ and D_n for some specific n

n	D_n^+		D_n	
5	4/5 (0.040)	5/5 (0.004)	4/5 (0.079)	5/5 (0.008)
6	4/6 (0.061)	5/6 (0.013)	5/6 (0.026)	6/6 (0.002)
7	5/7 (0.027)	6/7 (0.004)	5/7 (0.053)	6/7 (0.008)
8	5/8 (0.044)	6/8 (0.009)	5/8 (0.087)	6/8 (0.019)
9	5/9 (0.063)	6/9 (0.017)	6/9 (0.034)	7/9 (0.006)
10	6/10(0.026)	7/10(0.006)	6/10(0.053)	7/10(0.012)
11	6/11(0.038)	7/11(0.010)	6/11(0.075)	7/11(0.020)
12	6/12(0.050)	7/12(0.016)	7/12(0.031)	8/12(0.008)
13	7/13(0.022)	8/13(0.006)	7/13(0.045)	8/13(0.013)
14	7/14(0.030)	8/14(0.009)	7/14(0.059)	8/14(0.019)
15	7/15(0.038)	8/15(0.013)	8/15(0.026)	9/15(0.008)
16	7/16(0.047)	8/16(0.017)	8/16(0.035)	9/16(0.011)
17	8/17(0.022)	9/17(0.008)	8/17(0.045)	9/17(0.016)
18	8/18(0.028)	9/18(0.010)	8/18(0.056)	9/18(0.021)
19	8/19(0.034)	9/19(0.013)	9/19(0.027)	10/19(0.009)
20	8/20(0.041)	10/20(0.006)	9/20(0.034)	10/20(0.012)
21	8/21(0.047)	10/21(0.008)	9/21(0.041)	11/21(0.006)
22	8/22(0.055)	10/22(0.010)	9/22(0.049)	11/22(0.007)
23	9/23(0.029)	11/23(0.005)	9/23(0.058)	11/23(0.009)
24	9/24(0.034)	11/24(0.006)	10/24(0.030)	11/24(0.012)
25	9/25(0.039)	11/25(0.007)	10/25(0.036)	12/25(0.006)
26	9/26(0.045)	11/26(0.009)	10/26(0.042)	12/26(0.007)
27	9/27(0.050)	11/27(0.011)	10/27(0.049)	12/27(0.009)
28	9/28(0.055)	12/28(0.005)	10/28(0.056)	12/28(0.011)
29	10/29(0.032)	12/29(0.007)	11/29(0.030)	13/29(0.005)
30	10/30(0.035)	12/30(0.008)	11/30(0.035)	13/30(0.007)

6. Tables for the Spearman rho and Kendall tau statistics

For a bivariate sample (X_1, Y_i), $i = 1, \ldots, n$, of size n, the Kendall rank correlation coefficient (tau) is defined by

$$t = T/\binom{n}{2} \quad \text{where } T = \sum_{1 \leqslant i \leqslant j \leqslant n} \text{sign}(X_i - X_j)\,\text{sign}(Y_i - Y_j)\,. \tag{6.1}$$

Let K be the number of concordant pairs, (i, j), such that $i < j$ and $(X_i - X_j)(Y_i - Y_j)$ is positive. Then, we have

$$K = n(n-1)(1+t)/4\,. \tag{6.2}$$

Thus, for every t^* $(-1 \leqslant t^* \leqslant 1)$, there exists a K^* $(0 \leqslant K^* \leqslant \binom{n}{2})$, such that

$$P\{t \leqslant t^*\} = P\{K \leqslant K^*\}; \qquad K^* = n(n-1)(1+t^*)/4\,. \tag{6.3}$$

In Table 6.1, the values of K^* for which the probabilities in (6.3) are close to typical significance levels, along with their exact levels, are presented. This table has kindly been provided by Professor Dana Quade who has used a program of his own to record these values up to four decimal places of accuracy.

The Spearman rank correlation coefficient (rho) is defined by

$$r^* = \left[\sum_{i=1}^{n} (R_i - (n+1)/2)(S_i - (n+1)/2)\right][12/n(n^2-1)] \tag{6.4}$$

where R_i (and S_i) are the ranks of X_i (and Y_i) among $X_1, \ldots, X_n$ (and $Y_1, \ldots, Y_n$) respectively, for $i = 1, \ldots., n$ (ties neglected). If, we define

$$S^* = (R_1 - S_1)^2 + \cdots + (R_n - S_n)^2\,, \tag{6.5}$$

then, we have

$$r^* = 1 - 6S^*/(n^3 - n)\,. \tag{6.6}$$

Thus, for every r $(-1 \leqslant r \leqslant 1)$, there exists an s, such that

$$P\{r^* \geqslant r\} = P\{S^* \leqslant s\} \quad \text{where } s = (n^3 - n)(1 - r)/6\,. \tag{6.7}$$

Table 6.1
Critical values and probability levels for the Kendall tau statistics

n	K^* ($P\{K \leq K^*\}$)							
4							0	0.0417
5			0	0.0083	0	0.0083	1	0.0417
6	0	0.0014	1	0.0083	1	0.0083	2	0.0278
7	1	0.0014	2	0.0054	3	0.0151	4	0.0345
8	3	0.0028	4	0.0071	5	0.0156	6	0.0305
9	5	0.0029	6	0.0063	8	0.0223	9	0.0376
10	8	0.0046	9	0.0083	11	0.0233	12	0.0363
11	11	0.0050	12	0.0083	14	0.0203	16	0.0433
12	14	0.0044	15	0.0069	18	0.0224	20	0.0432
13	17	0.0033	19	0.0075	22	0.0211	25	0.0500
14	22	0.0049	24	0.0096	27	0.0236	29	0.0397
15	26	0.0041	28	0.0078	32	0.0231	35	0.0463
16	31	0.0043	34	0.0099	37	0.0206	41	0.0480
17	36	0.0040	39	0.0086	43	0.0211	47	0.0457
18	42	0.0043	45	0.0086	50	0.0239	54	0.0479
19	48	0.0041	52	0.0097	57	0.0245	61	0.0466
20	55	0.0045	59	0.0099	64	0.0234	69	0.0492
21	62	0.0045	66	0.0093	72	0.0244	77	0.0485
22	70	0.0050	74	0.0097	80	0.0237	85	0.0454
23	77	0.0043	82	0.0094	89	0.0249	94	0.0456
24	86	0.0048	91	0.0099	98	0.0246	104	0.0484
25	95	0.0049	100	0.0098	107	0.0232	114	0.0488
26	104	0.0048	109	0.0092	117	0.0233	124	0.0470
27	113	0.0044	119	0.0092	128	0.0247	135	0.0478
28	124	0.0049	130	0.0099	139	0.0249	146	0.0466
29	134	0.0046	140	0.0090	150	0.0241	158	0.0476
30	145	0.0047	152	0.0097	162	0.0245	170	0.0468
31	157	0.0050	164	0.0099	174	0.0240	183	0.0480
32	168	0.0046	176	0.0098	187	0.0246	196	0.0475
33	181	0.0049	188	0.0092	200	0.0243	210	0.0488
34	193	0.0046	202	0.0100	214	0.0249	224	0.0485
35	207	0.0049	215	0.0095	228	0.0247	239	0.0499

In Table 6.2, corresponding to typical significance levels, the values of s (along with the exact significance levels) are presented for $n \leq 16$. This table has also been prepared by Professor Dana Quade. Values for $n = 4$ through 11 are adapted from Table 13.2 (pp. 400–406) of Owen (1962), though the three decimal places listings in Owen (1962) have been improved here to four decimal places. For values of n in between 12 and 16, the entries are adapted from Otten (1973), converted to four decimal places.

For large sample sizes, both t and r^*, when standardized, have closely the normal distribution, and hence, their critical levels can be computed by reference to the standard normal tables.

Table 6.2
Critical values and probability levels for the Spearman rho statistics

n	s $(P\{S^* \leq s\})$			
4				0, 0.0417
5		0, 0.0083	0, 0.0083	2, 0.0417
6	0, 0.0014	2, 0.0083	4, 0.0167	6, 0.0292
7	4, 0.0034	6, 0.0062	12, 0.0240	16, 0.0440
8	10, 0.0036	14, 0.0077	22, 0.0229	30, 0.0481
9	20, 0.0041	26, 0.0086	36, 0.0216	48, 0.0484
10	34, 0.0044	42, 0.0087	58, 0.0245	72, 0.0481
11	54, 0.0049	64, 0.0091	84, 0.0239	102, 0.0470
12	78, 0.0048	92, 0.0093	118, 0.0244	142, 0.0495
13	108, 0.0047	128, 0.0097	160, 0.0249	188, 0.0485
14	146, 0.0047	170, 0.0095	210, 0.0250	244, 0.0486
15	194, 0.0050	222, 0.0097	268, 0.0244	310, 0.0486
16	248, 0.0049	284, 0.0100	338, 0.0247	388, 0.0493

7. Tables for some statistics based on Bessel process approximations

As natural generalizations of the two-sample Kolmogorov–Smirnov tests, Kiefer (1959) considered some multi-sample tests. If we have k (≥ 2) independent samples of sizes $n_1, \ldots, n_k$, respectively, and if $S_{n_1}^{(1)}, \ldots, S_{n_k}^{(k)}$ stand for the k sample distributions, then, Kiefer (1959), among others, considered the following test statistics (for the equality of all the true distributions $F_1, \ldots, F_k$):

$$K_1 = \left\{ \sup_x \sum_{i=1}^{k} n_i [S_{n_i}^{(i)}(x) - \bar{S}_N(x)]^2 \right\}^{1/2}, \tag{7.1}$$

and

$$K_2 = \int_{-\infty}^{\infty} \sum_{i=1}^{k} n_i [S_{n_i}^{(i)} - \bar{S}_N(x)]^2 \, d\bar{S}_N(x), \tag{7.2}$$

where $N = n_1 + \cdots + n_k$ and $\bar{S}_N$ is the combined sample distribution function. If $W_i^0 = \{W_i^0(t),\ t \in (0, 1)\}$, $i = 1, \ldots, k$, are independent copies of a standard Brownian Bridge (so that $EW_i^0 = 0$ and $EW_i^0(s)W_i^0(t) = s \wedge t - st$, $s, t \in (0, 1)$), then, a k-parameter tied-down Bessel process $B_k^0 = \{B_k^0(t), t \in (0, 1)\}$ is defined by $(B_k^0(t))^2 = \Sigma_{i=1}^k (W_i^0(t))^2$, for $t \in (0, 1)$. Then let

$$\kappa_1^{(k)} = \sup_{0 \leq t \leq 1} B_k^0(t) \quad \text{and} \quad \kappa_2^{(k)} = \int_0^1 (B_k^0(t))^2 \, dt. \tag{7.3}$$

Under the null hypothesis that the F_i are all equal, for large sample sizes, K_1 has the same distribution as $\kappa_1^{(k-1)}$ and K_2 has the same one as $\kappa_2^{(k-1)}$. Thus, if we denote by

$$A_k(a) = P\{\kappa_1^{(k)} \leq a\} \quad \text{and} \quad B_k(a) = P\{\kappa_2^{(k)} \leq a\},\ a \geq 0\,, \tag{7.4}$$

then the percentile points of $A_{k-1}(\cdot)$ and $B_{k-1}(\cdot)$ provides the asymptotic critical values for K_1 and K_2, respectively. With this motivation, we provide in Table 7.1, the upper $100\alpha\%$ points of $A_k(\cdot)$ and $B_k(\cdot)$, for some typical values of α. These are taken from Kiefer (1959).

Note that for $k = 1$, the entries in Table 7.1 relate to the two-sample Kolmogorov–Smirnov and Cramér–von Mises' test statistics.

In the context of progressively censored tests, one typically encounters a test statistic which (under the null hypothesis) has asymptotically the same distribution as of

$$K_k^* = \sup\{B_k(x) \colon x \in (0, 1)\}\,, \tag{7.5}$$

where $B_k^2(t) = \Sigma_{i=1}^k W_i^2(t)$, $0 \leq t < \infty$, and the W_i are independent copies of a standard Brownian motion. The process $B_k = \{B_k(t), t \in (0, \infty)\}$ is termed a k-parameter Bessel process. The critical values of K_k^* in (7.5) are therefore of interest for suitable adaptation. DeLong (1980) has tabulated the percentile points of the distribution of K_k^*, for various values of k. In Table 7.2, we provide some of these typical entries.

Note that in the context of rank analysis of covariance under progressive censoring, and elsewhere, one has a test statistic whose asymptotic null distribution is given by that of the following

Table 7.1
Table for the percentile points of $A_k(\cdot)$ and $B_k(\cdot)$, $k \leq 5$

k	$A_k^{-1}(\alpha)$			$B_k^{-1}(\alpha)$		
	$\alpha = 0.01$	0.05	0.10	$\alpha = 0.01$	0.05	0.10
1	1.6276	1.3581	1.2239	0.7435	0.4614	0.3473
2	1.8427	1.5838	1.4540	1.0737	0.7475	0.6070
3	2.0009	1.7473	1.6196	1.3586	1.0002	0.8412
4	2.1326	1.8823	1.7559	1.6226	1.2373	1.0631
5	2.2480	2.0001	1.8746	1.8722	1.4647	1.2775

Table 7.2
Critical values for K_k^* for $k \leq 7$ and some typical levels of significance

α	k = 1	2	3	4	5	6	7
0.01	2.807	3.242	3.562	3.827	4.059	4.269	4.461
0.05	2.241	2.695	3.023	3.294	3.530	3.743	3.938
0.10	1.960	2.419	2.750	3.023	3.260	3.474	3.669

Table 7.3
Table for the critical values of $B_k^S(T)$ for specific values of k and T

	$k = 1$ (one-sided)			$k = 2$			$k = 3$			$k = 4$					
T	$P = 0.90$	$P = 0.95$	$P = 0.99$	$P = 0.90$	$P = 0.95$	$P = 0.99$	$P = 0.90$	$P = 0.95$	$P = 0.99$	$P = 0.90$	$P = 0.95$	$P = 0.99$	$P = 0.90$	$P = 0.95$	$P = 0.99$
1.10	1.54	1.89	2.57	1.88	2.19	2.81	2.38	2.68	3.26	2.73	3.03	3.60	3.02	3.31	3.87
1.20	1.61	1.97	2.64	1.96	2.27	2.88	2.46	2.76	3.34	2.82	3.11	3.67	3.11	3.39	3.95
1.25	1.64	2.00	2.67	1.99	2.31	2.91	2.49	2.79	3.37	2.85	3.14	3.70	3.14	3.42	3.98
1.50	1.75	2.10	2.77	2.10	2.41	3.01	2.60	2.90	3.47	2.95	3.24	3.80	3.24	3.53	4.07
1.75	1.82	2.17	2.83	2.17	2.48	3.07	2.67	2.96	3.53	3.02	3.31	3.86	3.31	3.59	4.13
2.00	1.87	2.22	2.88	2.22	2.52	3.12	2.72	3.01	3.57	3.07	3.35	3.90	3.36	3.64	4.18
3.00	2.00	2.34	2.99	2.34	2.64	3.22	2.84	3.12	3.67	3.19	3.46	4.00	3.47	3.75	4.27
4.00	2.07	2.41	3.05	2.41	2.70	3.28	2.90	3.18	3.73	3.25	3.52	4.06	3.54	3.81	4.33
5.00	2.12	2.45	3.08	2.45	2.74	3.32	2.94	3.22	3.76	3.29	3.56	4.09	3.58	3.84	4.36
6.00	2.15	2.49	3.11	2.49	2.78	3.34	2.98	3.25	3.79	3.33	3.59	4.12	3.61	3.87	4.39
7.00	2.18	2.51	3.14	2.51	2.80	3.37	3.00	3.27	3.81	3.35	3.62	4.14	3.64	3.90	4.41
8.00	2.20	2.53	3.15	2.53	2.82	3.38	3.02	3.29	3.83	3.37	3.63	4.15	3.66	3.92	4.43
9.00	2.22	2.55	3.17	2.55	2.84	3.40	3.04	3.31	3.84	3.39	3.65	4.17	3.67	3.93	4.44
10.00	2.24	2.57	3.18	2.57	2.85	3.41	3.05	3.32	3.85	3.40	3.66	4.18	3.69	3.94	4.45
12.50	2.27	2.60	3.21	2.60	2.88	3.44	3.08	3.35	3.88	3.43	3.69	4.20	3.71	3.97	4.47
15.00	2.30	2.62	3.23	2.62	2.90	3.45	3.10	3.37	3.90	3.45	3.71	4.22	3.74	3.99	4.49
17.50	2.32	2.64	3.24	2.64	2.92	3.47	3.12	3.39	3.91	3.47	3.73	4.24	3.75	4.01	4.50
20.00	2.33	2.65	3.26	2.65	2.93	3.48	3.14	3.40	3.92	3.48	3.74	4.25	3.77	4.02	4.52
30.00	2.38	2.69	3.29	2.69	2.97	3.52	3.18	3.44	3.96	3.52	3.78	4.28	3.81	4.06	4.55
40.00	2.41	2.72	3.32	2.72	3.00	3.54	3.20	3.46	3.98	3.55	3.80	4.30	3.83	4.08	4.57
50.00	2.43	2.74	3.33	2.74	3.01	3.55	3.22	3.48	3.99	3.57	3.82	4.32	3.85	4.10	4.58
60.00	2.45	2.76	3.35	2.76	3.03	3.57	3.24	3.50	4.01	3.58	3.83	4.33	3.86	4.11	4.59
70.00	2.46	2.77	3.36	2.77	3.04	3.58	3.25	3.51	4.01	3.59	3.84	4.34	3.88	4.12	4.60
80.00	2.47	2.78	3.37	2.78	3.05	3.59	3.26	3.52	4.02	3.60	3.85	4.34	3.89	4.13	4.61
100.00	2.49	2.80	3.38	2.80	3.07	3.60	3.28	3.53	4.04	3.62	3.87	4.36	3.90	4.14	4.62

$$K_{k,\varepsilon}^{0} = \sup\{t^{-1/2}B_k(t): \varepsilon \leq t \leq 1\}, \tag{7.6}$$

where, as before, B_k is a Bessel process, so that $\{t^{-1/2}B_k(t), t > 0\}$ is a standardized Bessel process, and $\varepsilon > 0$ is prefixed. Note that $K_{k,\varepsilon}^{0}$ has the same distribution as of

$$B_k^S(T) = \sup\{t^{-1/2}B_k(t): 1 \leq t \leq T\}, \quad T = \varepsilon^{-1}(>1). \tag{7.7}$$

Also, for the truncated version of the Kolmogorov–Smirnov test, one encounters a test statistic whose asymptotic null-distribution agrees with that of $\sup\{(t(1-t))^{-1/2}B_k^0(t): \varepsilon_1 \leq t \leq 1-\varepsilon_2\}$, where ε_1 and ε_2 are positive numbers, and B_k^0 is the tied-down Bessel process, defined as in before. For $k = 1$, we may also have a one-sided version, wherein $B_1^0(t)$ is replaced by $W_1^0(t)$. Note that if we let $T = (1-\varepsilon_2)/\varepsilon_1$ (>1), then, for this statistic too, the distribution is the same as in (7.7). For this reason, we use the tables in DeLong (1981) and provide some critical values of $B_k^S(T)$ for various T and k. For $k = 1$, the one-sided entries are also presented.

We conclude this section with the remark that for the unweighted one and two-sample Kolmogorov–Smirnov statistics, under censoring, for one and two-sided tests, one needs to consult the tables for $P\{\sup_{0<u\leq T} W^0(u) \leq a\}$ $(=A^+(a, T)$, say) and $P\{\sup_{0<u\leq T}|W^0(t)| \leq a\}$ $(=A(a, T)$, say), for various a (≥ 0) and T: $0 < T \leq 1$, where W^0 is a standard Brownian Bridge. By using the basic results in Anderson (1960), Schey (1977) considered these and has tabulated some of the entries. For example, for $T = 0.1, 0.5$ and 0.9, the one-sided critical values for $\alpha = 0.05$ are 0.5985, 1.133, and 1.224, respectively while the two-sided ones are 0.6825, 1.400 and 1.480, respectively. With more adaptations from Koziol and Byar (1975), other critical values for $\alpha = 0.10$ for the one-sided case corresponds to the entries for $\alpha = 0.05$ for the two-sided case, stated above. For $\alpha = 0.01$, the one-sided critical values for $T = 0.1, 0.5$ and 0.9 are 0.916, 1.658 and 1.731, respectively, while the two-sided values are 0.851, 1.552 and 1.628, respectively.

References

[1] Birnbaum, Z. W. (1952). Numerical tabulation of the distribution of Kolmogorov's statistics for finite sample size. *Jour. Amer. Statist. Assoc.* **47**, 425–441.

[2] Birnbaum, Z. W. and Hall, R. A. (1960). Small sample distribution for multi-sample statistics of the Smirnov type. *Ann Math. Statist.* **31**, 710–720.

[3] DeLong, D. (1980). Some asymptotic properties of a progressively censored nonparametric test for multiple regression. *Jour. Multivar. Anal.* **10**, 360–370.

[4] DeLong, D. (1981). Crossing probabilities for a square root boundary by a Bessel process. *Commun. Statist. Theor. Meth.* **A10**, 2197–2213.

[5] Hollander, M. and Wolfe, D. A. (1973). *Nonparametric Statistical Methods.* Wiley, New York.

[6] Iman, R. L., Quade, D. and Alexander, D. A. (1975). Exact probability levels for the Kruskal–Wallis test. In: H. L. Harter and D. B. Owen, eds., *Selected Tables in Mathematical Statistics*, Amer. Math. Soc., Providence, RI, pp. 329–384.

[7] Kendall, M. G. and Smith, B. (1939). The problem of m-rankings. *Ann. Math. Statist.* **10**, 275–287.

[8] Kiefer, J. (1959). K-sample analogues of the Kolmogorov–Smirnov and Cramer–V. Mises' tests. *Ann. Math. Statist.* **30**, 420–447.

[9] Koziol, J. A. and Byar, D. P. (1975). Percentage points of the asymptotic distributions of one- and two-sample K–S statistics for truncated or censored data. *Technometrics* **17**, 507–510.

[10] Michaelis, J. (1971). Schwellenwerte des Friedman-tests. *Biom. Zeit.* **13**, 118–129.

[11] Millar. L. H. (1956). Tables of percentage points of Kolmogorov statistics. *Jour. Amer. Statist. Assoc.* **51**, 111–121.

[12] Odeh, R. E. (1977). Extended tables of the distribution of Friedman's S-statistic in the two-way layout. *Commun. Statist. Ser. B* **6**, 29–48.

[13] Otten, A. (1973). The null distribution of Spearman's S when $n = 13(1)16$. *Statist. Neerland.* **27**, 19–20.

[14] Owen, D. B. (1962). *Handbook of Statistical Tables.* Addison-Wesley, Reading, MA.

[15] Quade, D. (1972). Average internal rank correlation. Tech. Report. Math. Centrum, Amsterdam.

[16] Schey, H. (1977). The asymptotic distribution of the one-sided Kolmogorov–Smirnov statistic for truncated data. *Commun. Statist. Theor. Meth.* **A6**, 1361–1366.

[17] Wilcoxon, F., Katti, S. K. and Wilcox, R. A. (1968). Critical values and probability levels for the Wilcoxon rank sum and the Wilcoxon signed rank test. Amer. Cynamid Co., Pearl River, New York.

Subject Index

Handbook of Statistics
Contents of Previous Volumes

Volume 1. Analysis of Variance
Edited by P. R. Krishnaiah
1980 xviii + 1002 pp.

Volume 2. Classification, Pattern Recognition and Reduction of Dimensionality
Edited by P. R. Krishnaiah and L. N. Kanal
1982 xxii + 903 pp.

Volume 3. Time Series in the Frequency Domain
Edited by D. R. Brillinger and P. R. Krishnaiah
1983 xiv + 485 pp.